Audi A6

Electrical Wiring Manual

A6 Sedan 1998, 1999, 2000
A6 Avant 1999, 2000
allroad quattro 2000

Audi platform code: C5

Bentley Publishers
Cambridge, Massachusetts

BENTLEY PUBLISHERS | Automotive Books & Manuals

Bentley Publishers, a division of Robert Bentley, Inc.
1734 Massachusetts Avenue
Cambridge, MA 02138 USA
800-423-4595 / 617-547-4170
www.
BentleyPublishers
.com

Information that makes the difference®

Technical Contact Information

We value your feedback. Technical comments and suggestions are helpful to us. Please send your comments and thoughts to Bentley Publishers e-mail:
tech@BentleyPublishers.com

From time to time, updates may be made to this manual. A listing of updates can be found on the web at:
www.BentleyPublishers.com/updates

WARNING—Important Safety Notice

Do not use this manual unless you are familiar with basic automotive repair procedures and safe workshop practices. This manual illustrates the workshop procedures required for most service work; it is not a substitute for full and up-to-date information from the vehicle manufacturer or for proper training as an automotive technician. Note that it is not possible for us to anticipate all of the ways or conditions under which vehicles may be serviced or to provide cautions as to all of the possible hazards that may result.

The vehicle manufacturer will continue to issue service information updates and parts retrofits after the editorial closing of this manual. Some of these updates and retrofits will apply to procedures and specifications in this manual. We regret that we cannot supply updates to purchasers of this manual.

We have endeavored to ensure the accuracy of the information in this manual. Please note, however, that considering the vast quantity and the complexity of the service information involved, we cannot warrant the accuracy or completeness of the information contained in this manual.

FOR THESE REASONS, NEITHER THE PUBLISHER NOR THE AUTHOR MAKES ANY WARRANTIES, EXPRESS OR IMPLIED, THAT THE INFORMATION IN THIS BOOK IS FREE OF ERRORS OR OMISSIONS AND WE EXPRESSLY DISCLAIM THE IMPLIED WARRANTIES OF MERCHANTABILITY AND OF FITNESS FOR A PARTICULAR PURPOSE, EVEN IF THE PUBLISHER OR AUTHOR HAVE BEEN ADVISED OF A PARTICULAR PURPOSE, AND EVEN IF A PARTICULAR PURPOSE IS INDICATED IN THE MANUAL. THE PUBLISHER AND AUTHOR ALSO DISCLAIM ALL LIABILITY FOR DIRECT, INDIRECT, INCIDENTAL OR CONSEQUENTIAL DAMAGES THAT RESULT FROM ANY USE OF THE EXAMPLES, INSTRUCTIONS OR OTHER INFORMATION IN THIS BOOK. IN NO EVENT SHALL OUR LIABILITY WHETHER IN TORT, CONTRACT OR OTHERWISE EXCEED THE COST OF THIS MANUAL.

Your common sense and good judgment are crucial to safe and successful service work. Read procedures through before starting them. Think about whether the condition of your car, your level of mechanical skill, or your level of reading comprehension might result in or contribute in some way to an occurrence which might cause you injury, damage your car, or result in an unsafe repair. If you have doubts for these or other reasons about your ability to perform safe repair work on your car, have the work done at an authorized Audi dealer or other qualified shop.

Part numbers listed in this manual are for identification purposes only, not for ordering. Always check with your authorized Audi dealer to verify part numbers and availability before beginning service work that may require new parts.

Before attempting any work on your Audi, read the warnings and cautions on page v and vi, and any warning or caution that accompanies a procedure in the manual. Review the warnings and cautions on page v and vi each time you prepare to work on your Audi.

Special tools required to perform certain service operations are identified in the manual and are recommended for use. Use of tools other than those recommended in this service manual may be detrimental to the car's safe operation as well as the safety of the person servicing the car.

Copies of this manual may be purchased from selected booksellers and automotive accessories and parts dealers, or directly from the publisher.

The publisher encourages comments from the reader of this manual. These communications have been and will be considered in the preparation of this and other manuals. Please write to Robert Bentley, Inc., Publishers at the address listed below.

This manual was prepared, published and distributed by Robert Bentley, Inc., 1734 Massachusetts Avenue, Cambridge, MA 02138. All information in this Manual is based on the latest product information available from Audi at the time of the editorial closing date. Audi has not reviewed and does not vouch for the accuracy or completeness of the technical specifications and work procedures described and given.

Library of Congress Cataloging-in-Publication Data

Audi A6 : electrical wiring manual : A6 sedan 1998, 1999, 2000 : A6 avant 1999, 2000:
allroad quattro 2000.
p. cm.
"Audi platform code: C5."
Includes index.
ISBN 0-8376-0166-5 (alk. paper)
1. Audi A6 automobile--Electric wiring--Charts, diagrams, etc. 2. Audi A6 automobile--Electric equipment--Maintenance and repair--Handbooks, manuals, etc. I. Robert Bentley, inc.

TL215.A8 A8947 2002
629.25'48--dc21 2002038229

Audi of America Literature Number: EWM A6 982000
Bentley Stock No. AW61

Editorial closing 2002.09

04 03 02 4 3 2 1

The paper used in this publication is acid free and meets the requirements of the National Standard for Information Sciences-Permanence of Paper for Printed Library Materials. ∞

Foreword

Service to Audi owners is a top priority of the Audi organization and has always included the continuing development and introduction of new and expanded services. In line with this purpose, Robert Bentley, Inc., in cooperation with Audi of America, Inc., has introduced this Audi A6 Electrical Wiring Manual. This manual covers Audi A6 models for the model years 1998 through 2000. This manual was created specifically to cover only those models built for sale in the United States and Canada.

For the Audi owner with basic mechanical skills and for independent auto service professionals, this manual includes the wiring diagrams that were available in an authorized Audi dealer service department as this manual went to press.

Some of the information in this manual applies only to cars of a particular model year or range of years. For example, "1999 m.y." refers to the 1999 model year. The model year does not necessarily match the calendar year in which the car was manufactured or sold. To be sure of the model year of a particular car, check the Vehicle Identification Number (VIN) on the car and use the table below to identify the particular model year.

The VIN is a unique sequence of 17 characters assigned by Audi to identify each individual car. WAULH54B6YN123456 is an example. Each of the 17 letters and numbers indicates certain facts about the car and its manufacture. Throughout this manual an abbreviated VIN (last 12 digits) are used to identify various running production changes. For example:
(➤ VIN 8D-YA-200000) means "up to" vehicles with a sequential chassis number of 200000.
(VIN 4B-YA-200001 ➤) means "from" vehicles with a sequential chassis number of 200001.
The "-" in the examples above represents an undefined character. This also can be represented with a blank space (i.e. VIN 4B YA 20000).

Your Audi's VIN can be found on a plate mounted on the top of the instrument panel, on the driver's side where the number can be seen through the windshield. The 10th character is the model year code. The letters "I", "O", "Q" and "U" are not used for model year designation. This manual covers Audi A6 models for the model years 1998 through 2000. The table below explains some of the various codes in the VIN numbers of models covered by this manual.

W A U	L	H	5	4 B	6	Y	N	1 2 3 4 5 6
Manufacturer Information	Body	Engine Type	Restraint Type	Body Type	Check Digit	Model Year	Factory	Sequential Chassis No.
WVW-German Audi passenger car	Depends on model year		Depends on type of air-bags and seatbelts	4B-A6	0-9 or X	W-1998 X-1999 Y-2000	A- EINgolstadt N- Neckarsulm	000001-Begin production 999999-End production

We have endeavored to insure the accuracy of the information in this manual. When the vast array of data presented in the manual is taken into account, however, no claim to infallibility can be made. We therefore cannot be responsible for the result of any errors that may have crept into the text. The publisher encourages comments from readers of this manual with regard to errors, and also, suggestions for improvement in the presentation of technical material. These communications have been and will be carefully considered in the preparation of this and other manuals. Please write or e-mail to Robert Bentley, Inc., at the address at the beginning of this manual.

Audi offers extensive warranties, especially on components of the fuel delivery and emission control systems. Therefore, before deciding to repair an Audi that may be covered wholly or in part by any warranties issued by Audi of America, Inc., consult your authorized Audi dealer. You may find that the dealer can make the repair either free or at minimum cost. If at any time a repair is needed that you feel is too difficult to do yourself, a trained Audi technician is ready to do the job for you.

Selected Books and Electronic Editions From Bentley Publishers

Audi

Audi: A History of Progress Chronicle of Audi AG
Audi AG ISBN 0-8376-0384-6

Audi A6 Electrical Wiring Manual: 1998-2000 *Bentley Publishers* ISBN 0-8376-0369-2
Audi of America Lit. No. EWM A6 982000

Audi A4 Service Manual: 1996-2001, 1.8L Turbo, 2.8L, including Avant and quattro
Bentley Publishers ISBN 0-8376-0371-4
Audi Part No. W42AUA4962001RM

Audi A4/S4: 1996-2001 Official Factory Repair Manual on CD-ROM
Audi of America ISBN 0-8376-0833-3
Audi of America Lit. No.: W42 AEB5 V025 CD

Audi TT Coupe, Roadster: 2000-2003 Official Factory Repair Manual on CD-ROM
Audi of America ISBN 0-8376-0839-2
Audi of America Lit. No.: W42 AETT V025 CD

Audi 90: 1993-1995 Official Factory Repair Manual on CD-ROM
Audi of America ISBN 0-8376-0843-0
Audi of America Lit. No.: W42 AE90 V025 CD

Audi 80, 90, Coupe Quattro Official Factory Repair Manual: 1988-1992, including 80 Quattro, 90 Quattro, Wagon, 20-valve
Audi of America ISBN 0-8376-0368-4
Audi Part No. LPV 800 604

Audi 100, 200 Official Factory Repair Manual: 1989-1991, including 100 Quattro, 200 Quattro, Wagon, Turbo and 20-valve
Audi of America ISBN 0-8376-0372-2
Audi Part No. LPV 800 701

Audi4000S, 4000CS, and Coupe GT Official Factory Repair Manual: 1984-1987 including Quattro and Quattro Turbo
Audi of America ISBN 0-8376-0373-0
Audi Part No. LPV 800 424

Audi 5000S, 5000CS Official Factory Repair Manual: 1984-1988, Gasoline, Turbo, and Turbo Diesel, including Wagon and Quattro
Audi of America ISBN 0-8376-0370-6
Audi Part No. LPV 800 445

Engineering

Bosch Fuel Injection and Engine Management *Charles O. Probst, SAE*
ISBN 0-8376-0300-5

Maximum Boost: Designing, Testing, and Installing Turbocharger Systems
Corky Bell ISBN 0-8376-0160-6

Race Car Aerodynamics *Joseph Katz*
ISBN 0-8376-0142-8

The Scientific Design of Exhaust and Intake Systems
Philip H. Smith and John C. Morrison
ISBN 0-8376-0309-9

Battle for the Beetle: The Story of the Battle for the Giant VW Factory and the Car that Became an Icon Around the Globe
Karl Ludvigsen ISBN 08376-0071-5

Volkswagen

Volkswagen Model Documentation
Joachim Kuch ISBN 0-8376-0078-2

Volkswagen Sport Tuning for Street and Competition
Per Schroeder ISBN 0-8376-0161-4

Jetta, Golf, GTI Service Manual: 1999-2002 2.0L Gasoline, 1.9L TDI Diesel, 2.8L VR6, 1.8L Turbo
Bentley Publishers ISBN 0-8376-0388-9
Volkswagen Part No. LPV 800 118

Golf, GTI, Jetta: 1999-2003, Jetta Wagon 2001-2003 Official Factory Repair Manual on CD-ROM
Volkswagen of America ISBN 0-8376-0841-4
VWoA Lit No.: W42 VEA4 V025 CD

Golf, GTI, Jetta: 1993-1999, Cabrio: 1995-2002 Official Factory Repair Manual on CD-ROM
Volkswagen of America ISBN 0-8376-0834-1
VWoA Lit No.: W42 VEA3 V025 CD

Passat: 1998-2003 Official Factory Repair Manual on CD-ROM
Volkswagen of America ISBN 0-8376-0837-6
VWoA Lit No.: W42 VEB5 V025 CD

Passat: 1998-2002 Service Manual
Volkswagen of America ISBN 0-8376-0379-X
Volkswagen Part No. LPV 800 208

New Beetle: 1998-2003, New Beetle Convertible 2003 Official Factory Repair Manual on CD-ROM
Volkswagen of America ISBN 0-8376-0838-4
VWoA Lit No.: W42 VENB V025 CD

New Beetle Service Manual: 1998-2002 1.8L turbo, 1.9L TDI diesel, 2.0L gasoline
Bentley Publishers ISBN 0-8376-0385-4
Volkswagen Part No. LPV 800 402

Jetta, Golf, GTI, Cabrio Service Manual: 1993-1999, including Jetta$_{III}$ and Golf$_{III}$
Bentley Publishers ISBN 0-8376-0366-8
Volkswagen Part No. LPV 800 116

Eurovan: 1992-2003 Official Factory Repair Manual on CD-ROM
Volkswagen of America ISBN 0-8376-0835-X
VWoA Lit No.: W42 VET4 V025 CD

Eurovan Official Factory Repair Manual: 1992-1999
Volkswagen of America ISBN 0-8376-0335-8
Volkswagen Part No. LPV 800 149

Passat Official Factory Repair Manual 1995-1997
Volkswagen of America ISBN 0-8376-0380-3
Volkswagen Part No. LPV 800 207

Passat: 1995-1997 Official Factory Repair Manual on CD-ROM
Bentley Publishers ISBN 0-8376-0842-2
VWoA Lit No.: W42 VEB4 V025 CD

Fox Service Manual 1987-1993
Bentley Publishers ISBN 0-8376-0363-3
Volkswagen Part No. LPV 800 504

GTI/Golf/Jetta Service Manual: 1985-1992 Gasoline, Diesel, and Turbo Diesel, including 16V
Bentley Publishers ISBN 0-8376-0342-0
Volkswagen Part No. LPV 800 112

Corrado Official Factory Repair Manual: 1990-1994 *Volkswagen of America*
ISBN 0-8376-0387-0
Volkswagen Part No. LPV 800 300

Passat Service Manual: 1990-1993 4-cylinder gasoline models, including GL and Wagon
Bentley Publishers ISBN 0-8376-0378-1
Volkswagen Part No. LPV 800 205

Cabriolet and Scirocco Service Manual: 1985-1993, including 16V
Bentley Publishers ISBN 0-8376-0362-5
Volkswagen Part No. LPV 800 113

Vanagon Official Factory Repair Manual: 1980-1991 including Diesel, Syncro, and Camper
Volkswagen of America ISBN 0-8376-0336-6
Volkswagen Part No. LPV 800 148

Rabbit, Scirocco, Jetta Service Manual: 1980-1984 Gasoline Models, Including Pickup Truck, Convertible, and GTI
Bentley Publishers ISBN 0-8376-0183-5
Volkswagen Part No. LPV 800 104

Rabbit, Jetta Service Manual: 1977-1984 Diesel Models, Including Pickup Truck and Turbo Diesel
Bentley Publishers ISBN 0-8376-0184-3
Volkswagen Part No. LPV 800 122

Super Beetle, Beetle and Karmann Ghia Official Service Manual Type 1: 1970-1979
Volkswagen of America ISBN 0-8376-0096-0
Volkswagen Part No. LPV 997 109

Transporter Workshop Manual: 1963-1967, Type 2, all models
Volkswagen of America ISBN 0-8376-0391-9
Volkswagen Part No. LPV 800 135

Transporter Workshop Manual: 1950-1962, Type 2, all models
Volkswagen of America ISBN 0-8376-0382-X
Volkswagen Part No. LPV 800 138

1200 Workshop Manual: 1961-1965, Type 11, 14 and 15, Beetle, Beetle Convertible, Karmann Ghia Coupe, and Karmann Ghia Convertible
Volkswagen of America ISBN 0-8376-0390-0
Volkswagen Part No. LPV 800 121

Workshop Manual: Types 11, 14, and 15 (Beetle & Karmann Ghia): 1958-1960
Volkswagen of America ISBN 0-8376-0392-7
Volkswagen Part No. LPV 800 137

Workshop Manual: Types 11, 14, and 15 (Beetle & Karmann Ghia): 1952-1957
Volkswagen of America ISBN 0-8376-0389-7
Volkswagen Part No. LPV 800 136

Please read these warnings and cautions before proceeding with maintenance and repair work.

WARNING—

- Some repairs may be beyond your capability. If you lack the skills, tools and equipment, or a suitable workplace for any procedure described in this manual, we suggest you leave such repairs to an authorized Audi dealer service department, or other qualified shop.

- Audi is constantly improving its cars. Sometimes these changes, both in parts and specifications, are made applicable to earlier models. Therefore, before starting any major jobs or repairs to components on which passenger safety may depend, consult your authorized Audi dealer about Technical Bulletins that may have been issued since the editorial closing of this manual.

- Do not re-use any fasteners that are worn or deformed in normal use. Many fasteners are designed to be used only once and become unreliable and may fail when used a second time. This includes, but is not limited to, nuts, bolts, washers, self-locking nuts or bolts, circlips and cotter pins. Always replace these fasteners with new parts.

- Never work under a lifted car unless it is solidly supported on stands designed for the purpose. Do not support a car on cinder blocks, hollow tiles or other props that may crumble under continuous load. Never work under a car that is supported solely by a jack. Never work under the car while the engine is running.

- If you are going to work under a car on the ground, make sure that the ground is level. Block the wheels to keep the car from rolling. Disconnect the battery negative (–) terminal (Ground strap) to prevent others from starting the car while you are under it.

- Never run the engine unless the work area is well ventilated. Carbon monoxide kills.

- Finger rings, bracelets and other jewelry should be removed so that they cannot cause electrical shorts, get caught in running machinery, or be crushed by heavy parts.

- Tie long hair behind your head. Do not wear a necktie, a scarf, loose clothing, or a necklace when you work near machine tools or running engines. If your hair, clothing, or jewelry were to get caught in the machinery, severe injury could result.

- Do not attempt to work on your car if you do not feel well. You increase the danger of injury to yourself and others if you are tired, upset or have taken medication or any other substance that may keep you from being fully alert.

- Illuminate your work area adequately but safely. Use a portable safety light for working inside or under the car. Make sure the bulb is enclosed by a wire cage. The hot filament of an accidentally broken bulb can ignite spilled fuel or oil.

- Catch draining fuel, oil, or brake fluid in suitable containers. Do not use food or beverage containers that might mislead someone into drinking from them. Store flammable fluids away from fire hazards. Wipe up spills at once, but do not store the oily rags, which can ignite and burn spontaneously.

- Always observe good workshop practices. Wear goggles when you operate machine tools or work with battery acid. Gloves or other protective clothing should be worn whenever the job requires working with harmful substances.

- Friction materials such as brake or clutch discs may contain asbestos fibers. Do not create dust by grinding, sanding, or by cleaning with compressed air. Avoid breathing asbestos fibers and asbestos dust. Breathing asbestos can cause serious diseases such as asbestosis or cancer, and may result in death.

- Disconnect the battery negative (–) terminal (ground strap) whenever you work on the fuel system or the electrical system. Do not smoke or work near heaters or other fire hazards. Keep an approved fire extinguisher handy.

- Interruption of voltage supply may erase adaptive settings in the engine management and automatic transmission control modules. Consult Repair Manual for additional engine specific information as well as procedure following interruption of voltage supply.

- Batteries give off explosive hydrogen gas during charging. Keep sparks, lighted matches and open flame away from the top of the battery. If hydrogen gas escaping from the cap vents is ignited, it will ignite gas trapped in the cells and cause the battery to explode.

- Connect and disconnect battery cables, jumper cables or a battery charger only with the ignition switched off, to prevent sparks. Do not disconnect the battery while the engine is running.

- Do not quick-charge the battery (for boost starting) for longer than one minute. Wait at least one minute before boosting the battery a second time.

- Do not allow battery charging voltage to exceed 16.5 volts. If the battery begins producing gas or boiling violently, reduce the charging rate. Boosting a sulfated battery at a high charging rate can cause an explosion.

- The air-conditioning system is filled with chemical refrigerant, which is hazardous. The A/C system should be serviced only by trained technicians using approved refrigerant recovery/recycling equipment, trained in related safety precautions, and familiar with regulations governing the discharging and disposal of automotive chemical refrigerants.

- Do not expose any part of the A/C system to high temperatures such as open flame. Excessive heat will increase system pressure and may cause the system to burst.

- Some aerosol tire inflators are highly flammable. Be extremely cautious when repairing a tire that may have been inflated using an aerosol tire inflator. Keep sparks, open flame or other sources of ignition away from the tire repair area. Inflate and deflate the tire at least four times before breaking the bead from the rim. Completely remove the tire from the rim before attempting any repair.

- Most cars covered by this manual are equipped with a supplemental restraint system (SRS), that automatically deploys an airbag in the event of a frontal impact. The airbag is inflated by an explosive device. Handled improperly or without adequate safeguards, it can be accidently activated and cause serious injury.

- To prevent personal injury or airbag system failure, **only factory trained Audi service technicians** should test, disassemble or service the airbag system.

continued on next page

Please read these warnings and cautions before proceeding with maintenance and repair work.

WARNING (continued) —

● Disconnect the power supply before working on the airbag system, or when doing repairs that require removing airbag system components. Disconnect the battery negative (–) terminal and cover the battery.

● Installation of aftermarket radios or original equipment radios not produced for use in model year 1998 and newer can affect various systems connected to the CAN-BUS. Incorrect radios can cause severe damage to VAG 1551/1552, VAS 5051/5052, or aftermarket diagnostic scan tools or computers used to run scan tool programs. If the radio is not correct for the vehicle, or if there is any doubt, disconnect radio wiring harness before connecting scan tool.

● On airbag-equipped cars, never apply stickers or any other type of covering on the steering wheel. Do not let chemical cleaners, oil or grease come into contact with vinyl covering of the airbag unit.

● Never open or otherwise attempt to repair airbag system parts. Always use new parts. Never leave airbag parts or the partially disassembled airbag system unattended.

● Never use a test light to conduct electrical tests on the airbag system. The system must only be tested by trained Audi Service technicians using the VAG 1551/1552 or 5051 Scan Tool (ST) or an approved equivalent. The airbag unit must never be electrically tested while it is not installed in the car.

● Do not expose the airbag unit to temperatures above 194°F (90°C), even for brief periods. Keep clear of heat sources such as hot plates, soldering irons, heat lamps and welding equipment.

● When driving or riding in an airbag-equipped vehicle, never hold test equipment in your hands or lap while the vehicle is in motion. Objects between you and the airbag can increase the risk of injury in an accident.

● Greases, lubricants and other automotive chemicals contain toxic substances, many of which are absorbed directly through the skin. Read manufacturer's instructions and warnings carefully. Use hand and eye protection. Avoid direct skin contact.

CAUTION—

● If you lack the skills, tools and equipment, or a suitable workshop for any procedure described in this manual, we suggest you leave such repairs to an authorized Audi dealer or other qualified shop. We especially urge you to consult an authorized Audi dealer before beginning repairs on any car that may still be covered wholly or in part by any of the extensive warranties issued by Audi of America.

● Audi offers extensive warranties, especially on components of fuel delivery and emission control systems. Therefore, before deciding to repair an Audi that may still be covered wholly or in part by any warranties issued by Audi of America, consult your authorized Audi dealer. You may find that he can make the repair for free, or at minimal cost.

● Audi part numbers listed in this manual are for identification purposes only, not for ordering. Always check with your authorized Audi dealer to verify part numbers and availability before beginning service work that may require new parts.

● Before starting a job, make certain that you have all the necessary tools and parts on hand. Read all the instructions thoroughly, do not attempt shortcuts. Use tools appropriate to the work and use only replacement parts meeting Audi specifications. Makeshift tools, parts and procedures will not make good repairs.

● Use pneumatic and electric tools only to loosen threaded parts and fasteners. Never use these tools to tighten fasteners, especially on light alloy parts. Always use a torque wrench to tighten fasteners to the tightening torque specification listed.

● Be mindful of the environment and ecology. Before you drain the crankcase, find out the proper way to dispose of the oil. Do not pour oil onto the ground, down a drain, or into a stream, pond or lake. Consult local ordinances that govern the disposal of wastes.

● On cars equipped with anti-theft radios, make sure you know the correct radio activation code before disconnecting the battery or removing the radio. If the wrong code is entered into the radio when power is restored, that radio may lock up and be rendered inoperable, even if the correct code is then entered.

● Connect and disconnect a battery charger only with the battery charger switched off.

● Do not quick-charge the battery (for boost starting) for longer than one minute. Wait at least one minute before boosting the battery a second time.

● Sealed or "maintenance free" batteries should be slow-charged only, at an amperage rate that is approximately 10% of the battery's ampere-hour (Ah) rating.

● Do not allow battery charging voltage to exceed 16.5 volts. If the battery begins producing gas or boiling violently, reduce the charging rate. Boosting a sulfated battery at a high charging rate can cause an explosion.

Wiring Diagrams

Section
WD

NOTE —

The index for this section is organized using a two-tiered system:

- *Pages WD 1 and WD 2 contain the broader overview index of the various wiring diagrams.*
- *Pages WD 3 through WD 17 contain more detailed indexes with specfic components listed chronologically.*
- *To find a particular component or system, begin with the model year and locate the specific system or use a standard equipment diagram that is most likely to contain the particular component or schematic of interest.*

Pages WD 18 and WD 19 contain a list of Warnings and Cautions that should be reviewed before working on any vehicle.

General Information

General Information	Page
- Warnings and Cautions	WD 18
- How to use wiring diagrams	WD 20
- Wiring diagram layout	WD 21
- How to read wiring diagrams	WD 22
- Symbols used in wiring diagrams	WD 24

1998 A6, Wiring Diagram System Index

1998 A6, Wiring Diagram System Index	Page
- 2.8 Liter 6-Cylinder Engine (engine code: AHA), and Standard Equipment	WD 3
- ABS	WD 3
- Airbag	WD 3
- Automatic Climate Control	WD 4
- Automatic Dimmer Inside and Outside Mirrors	WD 4
- Automatic Transmission 01V	WD 4
- Central Locking System, Anti-Theft System, Ultra-Sound Interior Monitor	WD 4
- Cruise Control	WD 4
- Daytime Running Lights with Fog Lights and Rear Fog Light (Canada)	WD 4
- Driver's Information System	WD 5
- Headlight Washer	WD 5
- Headlight with HID Gas Lamps and Automatic Headlight Beam Adjusting	WD 5
- Heated Seats	WD 5
- Heated Steering Wheel	WD 5
- Power Rear Window Shade	WD 5
- Power Seat and Mirror Adjustment, Mirror with/without fold-away Function, Heated Mirror	WD 5
- Power Seat and Mirror Adjustment with Memory, Mirror with/without fold-away Function	WD 5
- Power Sunroof with Automatic Preselection	WD 5
- Power Windows	WD 6
- Radio Concert, Bose, with/without CD-Changer	WD 6
- Radio Concert Stereo	WD 6
- Rear Wiring, Wagon	WD 6
- Servotronic	WD 6
- Telephone	WD 6

1999 A6, Wiring Diagram System Index

1999 A6, Wiring Diagram System Index	Page
- 2.8 Liter 6-Cylinder Engine (engine code: AHA), and Standard Equipment	WD 7
- 4.2 Liter 8-Cylinder Engine (engine code: ART)	WD 7
- ABS	WD 8
- Airbag	WD 8
- Automatic Climate Control	WD 8
- Automatic Dimmer Inside and Outside Mirrors	WD 8
- Automatic Transmission 01V	WD 8

1999 A6, Wiring Diagram System Index (continued)

	Page
- Central Locking System, Anti-Theft System, Ultra-Sound Interior Monitor	WD 8
- Cruise Control with Mechanical Throttle Actuation	WD 9
- Daytime Running Lights with Fog Lights and Rear Fog Light (Canada)	WD 9
- Driver's Information System	WD 9
- Garage Door Opener	WD 9
- Headlight Washer	WD 9
- Headlight with HID Gas Lamps and Automatic Headlight Beam Adjusting	WD 9
- Heated Seats	WD 9
- Heated Steering Wheel	WD 9
- Power Rear Window Shade	WD 9
- Power Seat and Mirror Adjustment, Mirror with/without fold-away Function, Heated Mirror	WD 9
- Power Seat and Mirror Adjustment with Memory, Mirror with/without fold-away Function	WD 10
- Power Sunroof with Automatic Preselection	WD 10
- Power Windows	WD 10
- Radio Concert, Bose, with/without CD-Changer	WD 10
- Radio Concert Stereo	WD 10
- Rear Wiring, Wagon	WD 10
- Servotronic	WD 10
- Telephone	WD 10

2000 A6, Wiring Diagram System Index

	Page
- 2.7 Liter 6-Cylinder Engine (engine code: APB)	WD 11
- 2.8 Liter 6-Cylinder Engine (engine code: ATQ)	WD 11
- 4.2 Liter 8-Cylinder Engine (engine code: ART)	WD 12
- ABS, EDS, ASR	WD 12
- ABS, EDS, ASR, ESP	WD 12
- Airbag	WD 12
- Automatic Climate Control	WD 12
- Automatic Dimmer Inside and Outside Mirrors	WD 13
- Automatic Transmission 01V, 01L	WD 13
- CAN-BUS	WD 13
- Central Locking System, Anti-Theft System, Ultra-Sound Interior Monitor	WD 13
- Coolant Fan	WD 13
- Cruise Control with Mechanical Throttle Actuation	WD 13
- Daytime Running Lights with Fog Lights and Rear Fog Light (Canada)	WD 13
- Driver's Information System	WD 14
- Garage Door Opener	WD 14
- Headlight Washer	WD 14
- Headlight with HID Gas Lamps and Automatic Headlight Beam Adjusting	WD 14
- Heated Seats	WD 14
- Heated Steering Wheel	WD 14
- Level Control System	WD 14
- Multi-Function Steering Wheel	WD 14
- Navigation	WD 14
- Parking Aid	WD 14
- Power Rear Window Shade	WD 15
- Power Seat and Mirror Adjustment with Memory, Mirror with/without fold-away Function	WD 15
- Power Seat and Mirror Adjustment, Mirror with/without fold-away Function, Heated Mirror	WD 15
- Power Sport Seats	WD 15
- Power Steering Column Adjustment with/without Memory	WD 15
- Power Sunroof with Automatic Preselection	WD 15
- Power Windows	WD 15
- Radio Concert Stereo, with/without CD-Changer	WD 15
- Radio Symphony, Bose, with/without CD-Changer	WD 16
- Rear Wiring, Wagon	WD 16
- Servotronic	WD 16
- Standard Equipment	WD 17
- Steering Wheel with Tiptronic Switch	WD 17
- Telephone	WD 17

1998 A6, Individual Wiring Diagram Index

Diagram #

- **2.8 Liter 6-Cylinder Engine (engine code: AHA), and Standard Equipment** 1
  - Fuse Panel, Micro Central Electric Panel, 13-Fold Relay Panel 1
  - 8-Fold Relay Panel, 3-Fold Relay Panel in E-Box plenum Chamber 1/1
  - Ground Connections 1/2
  - Battery, Starter, Generator, Starting Interlock Relay 1/3
  - Fuel Injectors, Motronic Engine Control Module, Power Output Stage, Leak Detection Pump 1/4
  - Motronic Engine Control Module, Throttle Valve Control Module, Engine Coolant Temperature Sensor, Knock Sensors 1/5
  - Motronic Engine Control Module, Camshaft Position Sensors 1/6
  - Motronic Engine Control Module, Heated Oxygen Sensors, Fuel Pump Relay 1/7
  - Motronic Engine Control Module, Mass Air Flow Sensor, Oxygen Sensor Behind Three Way Catalytic Converter 1/8
  - Coolant Fan, Secondary Air Injection Pump Motor 1/9
  - Instrument Cluster, Parking Brake Warning Light Switch, Brake Fluid Level Warning Switch 1/10
  - Instrument Cluster, Speedometer, Tachometer, Fuel Gauge, Fuel Pump 1/11
  - Instrument Cluster, Shear Elements 1/12
  - Instrument Cluster, Data Link Connector 1/13
  - Ignition/Starter Switch, Load Reduction Relay 1/14
  - Light Switch, Turn Signal Switch, Headlight Dimmer/Flasher Switch 1/15
  - Instrument Panel Light Dimmer Switch, Fog Light Relay, Front Fog Light 1/16
  - Emergency Flasher Switch 1/17
  - Windshield Wiper/Washer, Washer Nozzle Heater 1/18
  - Fuses, Fresh Air Blower, License Plate Light, Glove Compartment Light 1/19
  - Parking Light, Signal Light, Headlight, Lamp Control Module 1/20
  - Fuses, Brake Light Switch 1/21
  - Fuse, Horn, Cigarette Lighter 1/22
  - Auto Check System, Tail Light, Brake Light, Rear Fog Light 1/23
  - Connector Radio, Wire connection (Switch Illumination), Wire connection, (Vehicle Speed Signal) 1/24
  - Outside Air Temperature Display, Heated Rear Window 1/25

- **ABS** 8
  - Fuse Panel, 8-Fold Relay Panel 8
  - ABS Control Module (w/EDL), Front Wheel Speed Sensors 8/1
  - ABS Control Module (w/EDL), Solenoid Valve, Rear Wheel Speed Sensor 8/2
  - ABS Control Module (w/EDL), ABS Warning Light, Traction Control Indicator Light 8/3
  - ABS Control Module (w/EDL), Brake Light Switch, Anti Slip Control 8/4

- **Airbag** 6
  - Fuse Panel, Micro Central Electric Panel 6
  - Airbag, Left and Right Seat Belt Tensioner Igniter 6/1
  - Side Airbag Driver's and Passenger's Side, Belt Tensioner Rear 6/2
  - Side Airbag Rear Driver's and Passenger's Side 6/3

1998 A6, Individual Wiring Diagram Index (continued) — Diagram

- **Automatic Climate Control** 4
  - Fuse Panel, Micro Central Electric Panel, 13-Fold Relay Panel 4
  - 8-Fold Relay Panel 4/1
  - Climate Control 4/2
  - Climate Control, Control Module for Solar Operation, Solar Cells, Control Module for Fresh Air Blower, Fresh Air Blower 4/3
  - Climate Control, Outside Air Temperature Sensor, Sunlight Photo Sensor, A/C Clutch Relay 4/4
  - Climate Control, Fresh Air Temperature Sensor, Vent Temperature Sensor, Back Pressure and Right Temperature Regulator Flap Motor 4/5
  - Climate Control, Left Temperature Regulator Flap Motor, Defroster and Central Flap Motor, Outside Air Temperature Display 4/6
  - Climate Control, Instrument Panel Vent Illumination, Lighting for Rear Seat Outlet 4/7
  - Climate Control, Coolant Fan 4/8

- **Automatic Dimmer Inside and Outside Mirrors** 20
  - Fuse Panel, Micro Central Electric Panel 20
  - Automatic Dimmer Inside and Outside Mirror 20/1

- **Automatic Transmission 01V** 2
  - Fuse Panel, 13-Fold Relay Panel 2
  - Transmission Control Module, Solenoid Valves, Kickdown Switch 2/1
  - Transmission Control Module, Multi-Function Transmission Range Switch 2/2
  - Transmission Control Module, Tiptronic Switch, Shift Lock Solenoid 2/3
  - Starting Interlock Relay, Lever Light Relay 2/4

- **Central Locking System, Anti-Theft System, Ultra-Sound Interior Monitor** 3
  - Fuse Panel, 13-Fold Relay Panel 3
  - Alarm Horn, Alarm Switch, Antenna Wire for Central Locking and Anti-Theft Warning System 3/1
  - Switch for Interior Lock, Driver Side, Lock Unit for Central Locking, Driver Side 3/2
  - Lock Unit for Central Locking, Passenger Side, Right Door Warning Light 3/3
  - Left Door Warning Light 3/4
  - Lock Unit for Central Locking, Left Rear 3/5
  - Lock Unit for Central Locking. Right Rear, Starting Interlock Relay 3/6
  - Switch for Central Locking, Trunk Lid 3/7
  - Motor for Fuel Tank Lid Unlock 3/8
  - Control Module for Ultra-Sound Sensors 3/9
  - Door Warning Lights Front, Entry Lights Front 3/10
  - Door Warning Lights Rear, Entry Lights Rear 3/11
  - Footwell Lights 3/12
  - Interior Lights, Reading Lights 3/13
  - Make-Up Mirror 3/14

- **Cruise Control** 19
  - Fuse Panel, Micro Central Electric Panel 19
  - Cruise Control 19/1
  - Cruise Control 19/2

- **Daytime Running Lights with Rear Fog Light (Canada)** 11
  - Fuse Panel, Micro Central Electric Panel, 13-Fold Relay Panel 11
  - Light Switch, Front and Rear Fog Light Switch, Headlight Dimmer/Flasher Switch, Windshield Washer Fluid Level Warning Switch 11/1
  - Ignition/Starter Switch, Fog Light, License Plate Light 11/2
  - Headlight Left and Right 11/3

- **Driver's Information System** 5
  - Fuse Panel, 13-Fold Relay Panel 5
  - Auto Check System, Board Computer Function Selector Switch, ... 5/1
  - Auto Check System, Lamp Control Module,. 5/2
  - Lamp Control Module, Parking Light, Low Beam Headlight 5/3
  - Lamp Control Module, Tail Light, Brake Light 5/4

- **Headlight Washer** 21
  - Fuse Panel, Micro Central Electric Panel 21
  - Headlight Washer 21

- **Headlight with HID Gas Lamps and Automatic Headlight Beam Adjusting** 22
  - Fuse Panel, Micro Central Electric Panel, 13-Fold Relay Panel 22
  - Ignition/Starter Switch, Load Reduction Relay, ... 22/1
  - Light Switch, Turn Signal Switch, Headlight Dimmer/Flasher Switch 22/2
  - High Intensity Gas Discharge Lamp,
    Control Module for High-Intensity Gas Discharge Lamp 22/3
  - Control Unit for Beam Adjusting, Left Rear Level Control System Sensor 22/4
  - Headlight Beam Adjusting Motors 22/5

- **Heated Seats** 7
  - Fuse Panel, Micro Central Electric Panel 7
  - Heated Driver's Seat 7/1
  - Heated Passenger's Seat 7/2
  - Rear Heated Seat 7/3

- **Heated Steering Wheel** 24
  - Fuse Panel, Micro Central Electric Panel 24
  - Heated Steering Wheel 24/1

- **Power Rear Window Shade** 14
  - Fuse Panel, Micro Central Electric Panel 14
  - Power Rear Window Shade 14/1

- **Power Seat and Mirror Adjustment, Mirror with/without Fold-away Function, Heated Mirror** 10
  - Fuse Panel, Micro 13-Fold Relay Panel, 8-Fold Relay Panel 10
  - Mirror Adjustment Driver's Side 10/1
  - Mirror Adjustment Passenger's Side, Power Seat Circuit Breaker 10/2

- **Power Seat and Mirror Adjustment with Memory, Mirror with/without Fold-away Function** 9
  - Fuse Panel, Micro 13-Fold Relay Panel, 8-Fold Relay Panel 9
  - Memory Program Switch, Memory Seat Control Module 9/1
  - Power Seat Circuit Breaker 1,
    Driver's Seat Lumbar Support Adjustment Switch 9/2
  - Memory Seat Control Module, Driver's Seat Adjusting Switch 9/3
  - Power Seat Circuit Breaker 2,
    Front Passenger's Seat Lumbar Support Adjustment Switch 9/4
  - Mirror Adjustment Switch with and without Fold-Away Function 9/5
  - Mirror Adjustment Driver's Side, Mirror Memory Control Module 9/6
  - Mirror Adjustment Passenger's Side, Mirror Memory Control Module 9/7
  - Crash Sensor for Side Airbag, Ignitor for Side AIrbag 9/8

- **Power Sunroof with Automatic Preselection** 18
  - Fuse Panel, Micro Central Electric Panel 18
  - Power Sunroof with Automatic Preselection 18/1

1998 A6, Individual Wiring Diagram Index (continued) — Diagram

- **Power Windows** 12
  - Fuse Panel, Micro Central Electric Panel, 8-Fold Relay Panel 12
  - Left Front Power Window, Power Window Circuit Breaker 12/1
  - Right Front Power Window 12/2
  - Left Rear Power Window 12/3
  - Right Rear Power Window 12/4
  - Window Lockout Switch 12/5

- **Radio Concert, Bose, with/without CD-Changer** 16
  - Fuse Panel, Micro Central Electric Panel 16
  - Radio Concert, Bose, with and without CD Changer 16/1
  - Radio Concert, Bose, with and without CD Changer 16/2
  - Radio Concert, Bose, with and without CD Changer 16/3
  - Radio Concert, Bose, with and without CD Changer 16/4

- **Radio Concert Stereo** 17
  - Fuse Panel, Micro Central Electric Panel 17
  - Radio Concert, Power Antenna, Rear Window Defogger with Window Antenna 17/1
  - Radio Concert, CD Changer Unit, Front Speakers 17/2
  - Radio Concert, Right Rear Woofer/Amplifier, Rear Speakers 17/3

- **Rear Wiring, Wagon** 23
  - Fuse Panel, Micro Central Electric Panel, 13-Fold Relay Panel 23
  - License Plate Light, Luggage Compartment Light 23/1
  - Trunk Lock Alarm, Sensors for Broken Window Glass, Rear 23/2
  - Heated Rear Window, Rear Fog Light, High-Mounted Brake Light 23/3
  - Lamp Control Module, Tail Light, Rear Turn Signal 23/4
  - Windshield Washer Pump, Rear Window Wiper Motor, Windshield Wiper/Washer Switch 23/5

- **Servotronic** 13
  - Fuse Panel, 13-Fold Relay Panel 13
  - Servotronic 13/1

- **Telephone** 15
  - Fuse Panel, Micro Central Electric Panel 15
  - Telephone 15/1
  - Telephone 15/2

1999 A6, Individual Wiring Diagram Index — Diagram

- **2.8 Liter 6-Cylinder Engine (engine code: AHA), and Standard Equipment** 25
  - Fuse Panel, Micro Central Electric Panel, 13-Fold Relay Panel 25
  - 8-Fold Relay Panel, 3-Fold Relay Panel in E-Box plenum Chamber 25/1
  - Ground Connections 25/2
  - Battery, Starter, Generator, Starting Interlock Relay 25/3
  - Fuel Injectors, Motronic Engine Control Module, Power Output Stage, ... 25/4
  - Motronic Engine Control Module, Throttle Valve Control Module, Engine Coolant Temperature Sensor, Knock Sensors 25/5
  - Motronic Engine Control Module, Camshaft Position Sensors 25/6
  - Motronic Engine Control Module, Heated Oxygen Sensors, Fuel Pump Relay 25/7
  - Motronic Engine Control Module, Mass Air Flow Sensor, Oxygen Sensor Behind Three Way Catalytic Converter 25/8
  - Coolant Fan, Secondary Air Injection Pump Motor 25/9
  - Instrument Cluster, Parking Brake Warning Light Switch, Brake Fluid Level Warning Switch 25/10
  - Instrument Cluster, Speedometer, Tachometer, Fuel Gauge, Fuel Pump 25/11
  - Instrument Cluster, Shear Elements 25/12
  - Instrument Cluster, Data Link Connector 25/13
  - Ignition/Starter Switch, Load Reduction Relay 25/14
  - Light Switch, Turn Signal Switch, Headlight Dimmer/Flasher Switch 25/15
  - Instrument Panel Light Dimmer Switch, Fog Light Relay, Front Fog Light 25/16
  - Emergency Flasher Switch 25/17
  - Windshield Wiper/Washer, Washer Nozzle Heater 25/18
  - Fuses, Fresh Air Blower, License Plate Light, 12 Volt Socket 25/19
  - Parking Light, Signal Light, Headlight, Lamp Control Module 25/20
  - Fuses, Brake Light Switch 25/21
  - Fuse, Horn, Cigarette Lighter 25/22
  - Auto Check System, Tail Light, Brake Light, Rear Fog Light 25/23
  - Connector Radio, Wire Connection (Switch Illumination), Wire Connection, (Vehicle Speed Signal) 25/24
  - Outside Air Temperature Display, Heated Rear Window 25/25

- **4.2 Liter 8-Cylinder Engine (engine code: ART)** 48
  - Fuse Panel, Micro Central Electric Panel, 13-Fold Relay Panel 48
  - 8-Fold Relay Panel 48/1
  - Battery, Starter, Generator 48/2
  - Starting Interlock Relay, Motronic Engine Control Module 48/3
  - Fuel Injectors, Ignition Coils Cylinder 1 - 4 . . . . 48/4
  - Fuel Injectors, Ignition Coils Cylinder 5 - 8 . . . . 48/5
  - Throttle Valve Control Module, Throttle Position Sensor, Cruise Control Switch 48/6
  - Secondary Air Injection Pump, Evaporative Emission Canister Purge Regulator Valve 48/7
  - Heated Oxygen Sensors before Catalytic Converter 48/8
  - Mass Air Flow Sensor, Oxygen Sensors behind Three Way Catalytic Converter 48/9
  - Motronic Engine Control Module, Camshaft Position Sensor 48/10
  - Knock Sensors, Engine Speed Sensor 48/11
  - Brake Light Switch, Brake Pedal Switch 48/12
  - Oil Level Thermal Sensor, Leak Detection Pump 48/13
  - Engine Cooling 48/14
  - Instrument Cluster, Fuel Pump 48/15
  - Instrument Cluster, Oil Pressure Warning Light 48/16

1999 A6, Individual Wiring Diagram Index (continued) — Diagram

- **ABS** 32
  - Fuse Panel, 8-Fold Relay Panel 32
  - ABS Control Module (w/EDL), Front Wheel Speed Sensors. . . . 32/1
  - ABS Control Module (w/EDL), Solenoid Valve, Rear Wheel Speed Sensor 32/2
  - ABS Control Module (w/EDL), ABS Warning Light, Traction Control Indicator Light 32/3
  - ABS Control Module (w/EDL), Brake Light Switch, Anti Slip Control 32/4

- **Airbag** 31
  - Fuse Panel, Micro Central Electric Panel 31
  - Airbag, Left and Right Seat Belt Tensioner Igniter 31/1
  - Side Airbag Driver's and Passenger's Side, Belt Tensioner Rear 31/2
  - Side Airbag Rear Driver's and Passenger's Side, Head Airbag 31/3

- **Automatic Climate Control** 29
  - Fuse Panel, Micro Central Electric Panel, 13-Fold Relay Panel 29
  - 8-Fold Relay Panel 29/1
  - Climate Control 29/2
  - Climate Control, Control Module for Solar Operation, Solar Cells, Control Module for Fresh Air Blower, Fresh Air Blower 29/3
  - Climate Control, Outside Air Temperature Sensor, Sunlight Photo Sensor, A/C Clutch Relay 29/4
  - Climate Control, Fresh Air Temperature Sensor, Vent Temperature Sensor, Back Pressure and Right Temperature Regulator Flap Motor 29/5
  - Climate Control, Left Temperature Regulator Flap Motor, Defroster and Central Flap Motor, Outside Air Temperature Display 29/6
  - Climate Control, Instrument Panel Vent Illumination, Lighting for Rear Seat Outlet 29/7
  - Climate Control, Coolant Fan 29/8

- **Automatic Dimmer Inside and Outside Mirrors** 43
  - Fuse Panel, Micro Central Electric Panel 43
  - Automatic Dimmer Inside and Outside Mirror 43/1

- **Automatic Transmission 01V, 01L** 27
  - Fuse Panel, 13-Fold Relay Panel 27
  - Transmission Control Module, Solenoid Valves, Kickdown Switch 27/1
  - Transmission Control Module, Multi-Function Transmission Range Switch 27/2
  - Transmission Control Module, Tiptronic Switch, ... 27/3
  - Transmission Control Module, Shift Lock Solenoid, Tiptronic Switch 27/4
  - Starting Interlock Relay 27/5

- **Central Locking System, Anti-Theft System, Ultra-Sound Interior Monitor** 28
  - Fuse Panel, 13-Fold Relay Panel 28
  - Alarm Horn, Alarm Switch, Antenna Wire for Central Locking and Anti-Theft Warning System 28/1
  - Switch for Interior Lock, Driver Side, Lock Unit for Central Locking, Driver Side 28/2
  - Lock Unit for Central Locking, Passenger Side, Right Door Warning Light 28/3
  - Left Door Warning Light 28/4
  - Lock Unit for Central Locking, Left Rear 28/5
  - Lock Unit for Central Locking. Right Rear, Starting Interlock Relay 28/6
  - Switch for Central Locking, Trunk Lid 28/7
  - Motor for Fuel Tank Lid Unlock 28/8
  - Control Module for Ultra-Sound Sensors 28/9
  - Door Warning Lights Front, Entry Lights Front 28/10
  - Door Warning Lights Rear, Entry Lights Rear 28/11
  - Footwell Lights 28/12
  - Interior Lights, Reading Lights 28/13
  - Make-Up Mirror 28/14

1999 A6, Individual Wiring Diagram Index (continued) — Diagram

- **Cruise Control with Mechanical Throttle Actuation** 42
  - Fuse Panel, Micro Central Electric Panel 42
  - Cruise Control 42/1
  - Cruise Control 42/2

- **Daytime Running Lights with Rear Fog Light (Canada)** 35
  - Fuse Panel, Micro Central Electric Panel, 13-Fold Relay Panel 35
  - Light Switch, Front and Rear Fog Light Switch, Headlight Dimmer/Flasher Switch 35/1
  - Ignition/Starter Switch, Fog Light, License Plate Light 35/2
  - Headlight Left and Right 35/3

- **Driver's Information System** 30
  - Fuse Panel, 13-Fold Relay Panel 30
  - Auto Check System, Board Computer Function Selector Switch, Windshield Washer Fluid Level Warning Switch 30/1
  - Auto Check System, Lamp Control Module, ... 30/2
  - Lamp Control Module, Parking Light, Low Beam Headlight 30/3
  - Lamp Control Module, Tail Light, Brake Light 30/4

- **Garage Door Opener** 47
  - Fuse Panel, Micro Central Electric Panel 47
  - Garage Door Opener 47/1

- **Headlight Washer** 44
  - Fuse Panel, Micro Central Electric Panel 44
  - Headlight Washer 44/1

- **Headlight with HID Gas Lamps and Automatic Headlight Beam Adjusting** 45
  - Fuse Panel, Micro Central Electric Panel, 13-Fold Relay Panel 45
  - Ignition/Starter Switch, Load Reduction Relay, ... 45/1
  - Light Switch, Turn Signal Switch, Headlight Dimmer/Flasher Switch 45/2
  - High Intensity Gas Discharge Lamp, Control Module for High-Intensity Gas Discharge Lamp 45/3
  - Control Unit for Beam Adjusting, Left Rear Level Control System Sensor 45/4
  - Headlight Beam Adjusting Motors 45/5

- **Heated Seats** 7
  - Fuse Panel, Micro Central Electric Panel 7
  - Heated Driver's Seat 7/1
  - Heated Passenger's Seat 7/2
  - Rear Heated Seat 7/3

- **Heated Steering Wheel** 26
  - Fuse Panel, Micro Central Electric Panel 26
  - Heated Steering Wheel 26/1

- **Power Rear Window Shade** 14
  - Fuse Panel, Micro Central Electric Panel 14
  - Power Rear Window Shade 14/1

- **Power Seat and Mirror Adjustment, Mirror with/without Fold-away Function, Heated Mirror** 34
  - Fuse Panel, Micro 13-Fold Relay Panel, 8-Fold Relay Panel 34
  - Mirror Adjustment Driver's Side 34/1
  - Mirror Adjustment Passenger's Side, Power Seat Circuit Breaker 34/2

- **Power Seat and Mirror Adjustment with Memory, Mirror with/without Fold-away Function** . . . 33
  - Fuse Panel, Micro 13-Fold Relay Panel, 8-Fold Relay Panel . . . 33
  - Memory Program Switch, Memory Seat Control Module . . . 33/1
  - Power Seat Circuit Breaker 1, Driver's Seat Lumbar Support Adjustment Switch . . . 33/2
  - Memory Seat Control Module, Driver's Seat Adjusting Switch . . . 33/3
  - Power Seat Circuit Breaker 2, Front Passenger's Seat Lumbar Support Adjustment Switch . . . 33/4
  - Mirror Adjustment Switch with and without Fold-Away Function . . . 33/5
  - Mirror Adjustment Driver's Side, Mirror Memory Control Module . . . 33/6
  - Mirror Adjustment Passenger's Side, Mirror Memory Control Module . . . 33/7
  - Crash Sensor for Side Airbag, Ignitor for Side Airbag . . . 33/8

- **Power Sunroof with Automatic Preselection** . . . 41
  - Fuse Panel, Micro Central Electric Panel . . . 41
  - Power Sunroof with Automatic Preselection . . . 41/1

- **Power Windows** . . . 36
  - Fuse Panel, Micro Central Electric Panel, 8-Fold Relay Panel . . . 36/1
  - Left Front Power Window Control Module, Left Front Window Switch . . . 36/2
  - Right Front Power Window Control Module, Right Front Window Switch . . . 36/3
  - Left Rear Window Switch, Left Rear Power Window Control Module . . . 36/4
  - Right Rear Window Switch, Right Rear Power Window Control Module . . . 36/5
  - Window Lockout Switch . . . 36/6
  - Power Windows . . . 36/7

- **Radio Concert, Bose, with/without CD-Changer** . . . 39
  - Fuse Panel, Micro Central Electric Panel . . . 39
  - Radio Concert, Bose, with and without CD Changer . . . 39/1
  - Radio Concert, Bose, with and without CD Changer . . . 39/2
  - Radio Concert, Bose, with and without CD Changer . . . 39/3
  - Radio Concert, Bose, with and without CD Changer . . . 39/4

- **Radio Concert Stereo** . . . 40
  - Fuse Panel, Micro Central Electric Panel . . . 40
  - Radio Concert, Power Antenna, Rear Window Defogger with Window Antenna . . . 40/1
  - Radio Concert, CD Changer Unit, Front Speakers . . . 40/2
  - Radio Concert, Right Rear Woofer/Amplifier, Rear Speakers . . . 40/3

- **Rear Wiring, Wagon** . . . 46
  - Fuse Panel, 13-Fold Relay Panel, 8-Fold Relay Panel . . . 46
  - License Plate Light, Luggage Compartment Light . . . 46/1
  - Trunk Lock Alarm, Antenna for Radio/Telephone . . . 46/2
  - Heated Rear Window, Rear Fog Light, High-Mounted Brake Light, Sensors for Broken Window Glass, Rear . . . 46/3
  - Lamp Control Module, Tail Light, Rear Turn Signal . . . 46/4
  - Windshield Washer Pump, Rear Window Wiper Motor, Windshield Wiper/Washer Switch . . . 46/5

- **Servotronic** . . . 37
  - Fuse Panel, 13-Fold Relay Panel . . . 37
  - Servotronic . . . 37/1

- **Telephone** . . . 38
  - Fuse Panel, Micro Central Electric Panel . . . 38
  - Telephone . . . 38/1
  - Telephone . . . 38/2

2000 A6, Individual Wiring Diagram Index — Diagram

- **2.7 Liter 6-Cylinder Engine (engine code: APB)** 51/1
  - Fuse Panel, Micro Central Electric Panel, 13-Fold Relay Panel 51/1
  - 8-Fold Relay Panel, 3-Fold Relay Panel in E-Box plenum chamber 51/2
  - Battery, Starter, Generator, Ignition Starter Switch 51/3
  - Motronic Engine Control Module, Starting Interlock Relay, Clutch Pedal Position Switch 51/4
  - Motronic Engine Control Module, Power Output Stage, Ignition Coils 1 - 3 . . . . . 51/5
  - Motronic Engine Control Module, Power Output Stage, Ignition Coils 4 - 6 . . . . . 51/6
  - Motronic Engine Control Module, Knock Sensors, Engine Coolant Temperature Sensor 51/7
  - Motronic Engine Control Module, Camshaft Position Sensors 51/8
  - Motronic Engine Control Module, Fuel Pump Relay, Oxygen Sensors Heater 51/9
  - Motronic Engine Control Module, Mass Air Flow Sensor, Heated Oxygen Sensors 51/10
  - Motronic Engine Control Module, Valves, Secondary Air Injection Pump Motor 51/11
  - Motronic Engine Control Module, Brake Light Switch, Throttle Valve Control Module, Vacuum Vent Valve Brake 51/12
  - Motronic Engine Control Module, Cruise Control Switch, Leak Detection Pump 51/13
  - Coolant FC Control Module, Coolant Fan, Coolant Fan Control Thermal Switch 51/14
  - Instrument Cluster, Fuel pump 51/15
  - Instrument Cluster, Oil Level Thermal Sensor, Oil Pressure Warning Light 51/16

- **2.8 Liter 6-Cylinder Engine (engine code: ATQ)** 50/1
  - Fuse Panel, Micro Central Electric Panel, 13-Fold Relay Panel 50/1
  - 8-Fold Relay Panel, 3-Fold Relay Panel in E-Box Plenum Chamber 50/2
  - Battery, Starter, Generator, Starting Interlock Relay 50/3
  - Motronic Engine Control Module, Fuel Injectors, Ignition Coils, Spark Plugs 50/4
  - Motronic Engine Control Module, Knock Sensors, Throttle Valve Control Module 50/5
  - Motronic Engine Control Module, Transmission Control Module, Camshaft Position Sensor 50/6
  - Motronic Engine Control Module, Fuel Pump Relay, Heated Oxygen Sensors 50/7
  - Mass Air Flow Sensor, Heated Oxygen Sensors Behind Three Way Catalytic Converter 50/8
  - Motronic Engine Control Module, Valves for Camshaft Adjustment, Clutch Pedal Position Switch 50/9
  - Motronic Engine Control Module, Cruise Control 50/10
  - Motronic Engine Control Module, Secondary Air Pump, Throttle Position Sensor 50/11
  - Engine Cooling 50/12
  - Instrument Cluster, Fuel Pump 50/13
  - Instrument Cluster, Speedometer, Oil Pressure Warning Light, Engine Coolant Level/Temperature Warning Light 50/14
  - Brake Light Switch, Vacuum Vent Valve Brake 50/15

2000 A6, Individual Wiring Diagram Index (continued) — Diagram

- **4.2 Liter 8-Cylinder Engine (engine code: ART)** 52/1
  - Fuse Panel, Micro Central Electric Panel, 13-Fold Relay Panel 52/1
  - 8-Fold Relay Panel, 3-Fold Relay Panel in E-Box Plenum Chamber 52/2
  - Battery, Starter, Generator 52/3
  - Motronic Engine Control Module, Starting Interlock Relay 52/4
  - Motronic Engine Control Module, Fuel Injectors and Ignition Coils Cylinder 1 - 4 .... 52/5
  - Motronic Engine Control Module, Fuel Injectors and Ignition Coils Cylinder 5 - 8 .... 52/6
  - Motronic Engine Control Module, Throttle Valve Control Module, Throttle Position Sensor 52/7
  - Motronic Engine Control Module, Knock Sensor 1, Engine Speed Sensor, Engine Coolant Temperature Sensor 52/8
  - Motronic Engine Control Module, Camshaft Position Sensor, Brake Light Switch 52/9
  - Motronic Engine Control Module, Secondary Air Injection Pump Relay 52/10
  - Motronic Engine Control Module, Heated Oxygen Sensors, Mass Air Flow Sensor 52/11
  - Motronic Engine Control Module, Fuel Pump Relay 52/12
  - Motronic Engine Control Module, CAN-Bus 52/13
  - Heated Oxygen Sensors Behind Three Way Catalytic Converter 52/14
  - Engine Cooling with Coolant FC Control Module 52/15
  - Instrument Cluster, Fuel Pump, Fuel Gauge 52/16
  - Instrument Cluster 52/17

- **Airbag** 57/1
  - Fuse Panel, Micro Central Electric Panel 57/1
  - Airbag Control Module, Driver's Side Airbag Igniter, Passenger's Side Airbag Igniter 1 .... 57/2
  - Airbag Control Module, Side Airbag's 57/3
  - Airbag Control Module, Side Airbag's Rear, Head Airbag's 57/4

- **Anti-Lock Brake with Electronic Differential Lock (EDL/EDS) and Anti-Slip Regulation (ASR)** 56/1
  - Fuse Panel, 8-Fold Relay Panel 56/1
  - ABS Control Module, Speed Sensor 56/2
  - ABS Control Module, Inlet and Outlet Valves 56/3
  - ABS Control Module 56/4
  - ABS Control Module, Anti-Slip Control Switch 56/5
  - Brake Light Switch, Vacuum Vent Valve Brake 56/6

- **Anti-Lock Brakes with Electronic Differential Lock (EDL/EDS), Anti-Slip Regulation (ASR) and Electronic Stability Program (ESP, Traction Control)** 55/1
  - Fuse Panel, 8-Fold Relay Panel 55/1
  - ABS Control Module, ABS Wheel Speed Sensor 55/2
  - ABS Control Module, Brake Light Switch, Parking Brake Warning Light Switch 55/3
  - ABS Control Module, ABS Return Flow Pump Relay, ABS Solenoid Valve Relay 55/4
  - ABS Control Module, Steering Angle Sensor, Yaw Sensor 55/5
  - ABS Control Module, Sensor for Brake Pressure 55/6

2000 A6, Individual Wiring Diagram Index (continued) — Diagram

- **Automatic Climate Control** 62/1
  - Fuse Panel, Micro Central Electric Panel, 13-Fold Relay Panel 62/1
  - 8-Fold Relay Panel 62/2
  - A/C Control Head 62/3
  - A/C Control Head, Solar Cells, Fresh Air Blower 62/4
  - A/C Control Head, A/C Clutch Relay, Outside Air temperature Sensor 62/5
  - A/C Control Head, Temperature Sensor, Sensor for Air Quality 62/6
  - A/C Control Head, Flap Motor 62/7
  - A/C Control Head 62/8
  - Instrument Panel Vent Illumination 62/9
  - Coolant Fan 62/10
  - Coolant Fan, Coolant Fan Control Relay 62/11

- **Automatic Dimmer Inside and Outside Mirror** 82/1
  - Fuse Panel, Micro Central Electric Panel 82/1
  - Automatic Day/Night Interior Mirror, Multi-function Transmission Range Switch 82/2
  - Automatic Day/Night Interior Mirror 82/3

- **Automatic Transmission 01V, 01L** 53/1
  - Fuse Panel, 13-Fold Relay Panel 53/1
  - Transmission Control Module, Kick Down Switch 53/2
  - Transmission Control Module, Multi-Function Transmission Range Switch 53/3
  - Transmission Control Module 53/4
  - Transmission Control Module, Tiptronic Switch 53/5
  - Transmission Control Module, Shift Lock Solenoid, Transmission Vehicle Speed Sensor 53/6
  - Starting Interlock Relay 53/7

- **CAN-BUS Wiring** 74/1
  - Fuse Panel, Micro Central Electric Panel 74/1
  - CAN-Bus Wiring 74/2
  - CAN-Bus Wiring 74/3

- **Central Locking System, Anti Theft Alarm System, Ultra-Sound Interior Monitor** 54/1
  - Fuse Panel, 13-Fold Relay Panel 54/1
  - Control Module for Central Locking, Alarm Horn, Hood Alarm Switch 54/2
  - Control Module for Central Locking, Switch for Interior Lock Driver Side 54/3
  - Control Module for Central Locking, Door Contact Switch Driver's Side 54/4
  - Control Module for Central Locking, Door Contact Switch Passenger's Side 54/5
  - Control Module for Central Locking, Left and Rear Door Warning Light 54/6
  - Control Module for Central Locking, Left and Right Rear Door Contact Switch 54/7
  - Control Module for Central Locking, Window Motor's 54/8
  - Control Module for Central Locking, Truck Lock Alarm/Central Locking Switch 54/9
  - Control Module for Central Locking, Luggage Compartment Release Switch, Motor for Remote Unlock Rear Lid 54/10
  - Control Module for Central Locking, Window Antenna for Central Locking and Anti-Theft Warning System 54/11
  - Control Module for Interior Monitoring Sensors, Switch for Passenger Compartment Monitoring, Interior Monitoring Sensor 54/12
  - Control Module for Interior Monitoring Sensors, Interior Monitoring Sensor 54/13
  - Door Warning Light, Entry Light Front 54/14
  - Entry Light Rear, Door Warning Light Rear, Door Opener Lighting Rear 54/15
  - Footwell Lighting Rear, Footwell Light 54/16
  - Front Interior Light, Rear Reading Light, Rear Courtesy Lamp 54/17
  - Make-Up Mirror Light, Front Reading Light 54/18
  - Plus Connection 58s 54/19
  - Luggage Compartment Light, Rear Courtesy Lamp 54/20

2000 A6, Individual Wiring Diagram Index (continued) — Diagram

- **Coolant Fan** 73/1
  - Fuse Panel, 8-Fold Panel Relay 73/1
  - Coolant FC Control Module, Coolant Fan Control Thermal Switch 73/2
  - Coolant FC Control Module, After-Run Coolant Pump 73/3

- **Cruise Control with Mechanical Throttle Actuation** 78/1
  - Fuse Panel, Micro Central Electric Panel 78/1
  - Cruise Control, Control Module, Speedometer, Vehicle Speed Sensor, Cruise Control Vacuum Pump 78/2
  - Cruise Control Switch, Brake Light Switch 78/3

- **Daytime Running Lights with Fog Light and Rear Fog Light (Canada)** 65/1
  - Fuse Panel, Micro Central Electric Panel, 13-Fold Relay Panel 65/1
  - Ignition/Starter Switch, License Plate Light, Load Reduction Relay 65/2
  - Light Switch, Headlight Dimmer/Flasher Switch 65/3
  - Lamp Control Module, Front Fog Light, Left Beam Headlight 65/4
  - Lamp Control Module, Right Beam Headlight 65/5

- **Driver's Information System** 64/1
  - Fuse Panel, 13-Fold Relay Panel 64/1
  - Board Computer Function Selector Switch, Windshield Washer Fluid Level Warning Switch 64/2
  - Lamp Control Module, Outside Air Temperature Display 64/3
  - Lamp Control Module, Parking Light 64/4
  - Lamp Control Module, Brake Light, Tail Light 64/5
  - Outside Air Temperature Sensor 64/6

- **Garage Door Opener** 83/1
  - Fuse Panel, Micro Central Electric Panel 83/1
  - Garage Door Opener 83/2

- **Headlight Washer** 67/1
  - Fuse Panel, Micro Central Electric Panel 67/1
  - Headlight Washer 67/2

- **Headlight with High Intensity Gas Discharge Lamps and Automatic Headlight Beam Adjusting** 66/1
  - Fuse Panel, Micro Central Electric Panel, 13-Fold Relay Panel 66/1
  - Ignition/Starter Switch 66/2
  - Light Switch, Headlight Dimmer/Flasher Switch 66/3
  - Lamp Control Module, High Intensity Gas Discharge Lamp Right 66/4
  - Lamp Control Module, High Intensity Gas Discharge Lamp Left 66/5
  - Control Module for Headlight Beam Adjusting, Level Control System Sensor 66/6
  - Left Headlight Beam Adjusting Motor, Right Headlight Beam Adjusting Motor 66/7

- **Heated Seats** 71/1
  - Fuse Panel, Micro Central Electric Panel 71/1
  - Adjuster for Heated Driver's Seat, Driver's Backrest Heating Element 71/2
  - Adjuster for Heated Passenger's Seat, Passenger's Backrest Heating Element 71/3
  - Heated Seat Regulating Switch, Heated Seat Temperature Sensor 71/4

- **Heated Steering Wheel** 79/1
  - Fuse Panel, Micro Central Electric Panel 79/1
  - Heated Steering Wheel 79/2

2000 A6, Individual Wiring Diagram Index (continued) — Diagram

- **Level Control System** 77/1
  - Fuse Panel, Micro Central Electric Panel, 8 - Fold Relay Panel 77/1
  - Level Control System 77/2
  - Level Control System Control Module, Level Control System Sensor 77/3
  - Control Module for Headlight Beam Adjusting, Level Control System Sensor 77/4
  - Headlight Beam Adjusting Motor 77/5

- **Multi-Function Steering Wheel** 84/1
  - Fuse Panel, Micro Central Electric Panel, 13-Fold Relay Panel 84/1
  - Control Unit in Steering Wheel, Airbag Control Module 84/2
  - Control Module for Multi-function Steering Wheel, Radio 84/3
  - Control Module for Multi-function Steering Wheel 84/4
  - Control Module for Multi-function Steering Wheel, High and Low Tone Horn 84/5

- **Navigation** 61/1
  - Fuse Panel, Micro Central Electric Panel 61/1
  - Control Unit for Navigation with CD Drive 61/2
  - Control Unit for Navigation with CD Drive, Navigation Antenna 61/3
  - Control Unit for Navigation with CD Drive, Radio 61/4
  - Board Computer Function Selector Switch II 61/5

- **Parking Aid** 76/1
  - Fuse Panel, Micro Central Electric Panel 76/1
  - Control Module for Parking Aid, Warning Buzzer for Parking 76/2
  - Control Module for Parking Aid, Parking Sensors 76/3

- **Power Rear Window Shade** 85/1
  - Fuse Panel, Micro Central Electric Panel 85/1
  - Power Rear Window Shade 85/2

- **Power Seat and Mirror Adjustment, Mirror with/without Fold-away Function** 68/1
  - Fuse Panel, 13-Fold Relay Panel, 8-Fold Relay Panel 68/1
  - Memory Program Switch, Memory Operating Unit Light 68/2
  - Sensor for Slope Adjustment Driver Seat, Driver's Seat Front Height Adjusting Motor 68/3
  - Driver's Seat Lumbar Support Adjustment Switch, Driver's Seat Lumbar Support Height and Curvature Adjustment Motor 68/4
  - Memory Seat Control Module, Driver's Seat Adjusting Switch 68/5
  - Passenger's Seat Adjusting Switch 68/6
  - Power Seat Circuit Breaker 2, Front Passenger Seat Lumbar Support Adjustment Switch 68/7
  - Mirror Adjustment Switch with and without fold-away Function 68/8
  - Mirror Adjustment Driver's Side, Mirror Memory Control Module 68/9
  - Mirror Adjustment Passenger's Side, Mirror Memory Control Module 68/10

- **Power Seat and Mirror Adjustment, Mirror with/without Fold-away Function, Heated Mirror** 69/1
  - Fuse Panel, 13 - Fold Relay Panel, 8 - Fold Relay Panel 69/1
  - Driver's Backrest Adjustment Switch, Driver's Backrest Adjusting Motor 69/2
  - Driver's and Passenger's Seat Lumbar Support Adjustment Switch, Lumbar Support Height Adjustment Motor 69/3
  - Passenger's Seat Height Adjusting Switch, Passenger's Backrest Adjusting Motor 69/4
  - Airbag Control Module, Crash Sensor for Airbag, Igniter for Side Airbag 69/5

- **Power Sport Seats** .70/1
  - Fuse Panel, 8-Fold Relay Panel .70/1
  - Driver's Backrest Adjustment Switch, Driver's Seat Adjusting Motor .70/2
  - Passenger's Backrest Adjustment Switch, Passenger's Seat Adjusting Motor .70/3
  - Driver's and Passenger's Seat Lumbar Support Adjustment Switch, Lumbar Support Height Adjustment Motor .70/4

- **Power Steering Column Adjustment with/without Memory** .81/1
  - Fuse Panel, Micro Central Electric Panel .81/1
  - Control Module for Steering column-belt Height Adjustment, Motor for Steering Column Adjustment .81/2
  - Control Module for Steering column-belt Height Adjustment .81/3

- **Power Sunroof with Automatic Preselection** .86/1
  - Fuse Panel, Micro Central Electric Panel .86/1
  - Power Sunroof with Automatic Preselection .86/2

- **Power Windows** .72/1
  - Fuse Panel. Micro Central Electric Panel, 8 - Fold Relay Panel .72/1
  - Left Front Power Window Control Module, Left Front Window Switch .72/2
  - Right Front Power Window Control Module, Right Front Window Switch .72/3
  - Left Rear Window Switch, Left Rear Power Window Control Module .72/4
  - Right Rear Window Switch, Right Rear Power Window Control Module .72/5
  - Window Lockout Switch .72/6
  - Power Windows .72/7

- **Radio, Concert Stereo, with/without CD Changer** .58/1
  - Fuse Panel. Micro Central Electric Panel .58/1
  - Radio, Rear Window Defogger with Window Antenna .58/2
  - Radio, CD Changer Unit .58/3
  - Radio, Right Rear Woofer/Amplifier .58/4
  - Radio, Front Speaker .58/5
  - Radio, Rear Speaker .58/6

- **Radio, Symphony, Bose, with/without CD Changer** .59/1
  - Fuse Panel. Micro Central Electric Panel .59/1
  - Radio, Data Link Connector .59/2
  - Radio, CD Changer Unit, Speaker .59/3
  - Radio, Antenna Amplifier, Power Antenna .59/4
  - Radio, Right Rear Woofer/Amplifier .59/5
  - Right Rear Woofer/Amplifier .59/6
  - Control Unit for Aerial Selection, TV Antenna, Rear Window Defogger with Window Antenna .59/7
  - Control Unit for Aerial Selection, TV Antenna .59/8

- **Rear Wiring, Wagon** .63/1
  - Fuse Panel, 13-Fold Relay Panel, 8-Fold Relay Panel .63/1
  - Control Module for Central Locking, 12 V Socket, Luggage Compartment Light .63/2
  - Control Module for Central Locking, Trunk Lid Closed Position Switch, Motor for Remote Unlock Rear Lid .63/3
  - Sensors for Broken Window Glass Rear, Rear Window Defogger/Heat Element .63/4
  - Windshield Wiper/Washer Switch, Rear Window Wiper Motor .63/5
  - Lamp Control Module, Tail Light, Brake Light Left .63/6
  - Lamp Control Module, Left Rear Fog Light, Tail Light, Brake Light Right .63/7
  - Back-Up Light Switch, Multi-Function Transmission Range Switch .63/8
  - Roof Antenna .63/9
  - Antenna .63/10

2000 A6, Individual Wiring Diagram Index (continued) — Diagram

- **Servotronic** 75/1
  - Fuse Panel, 13-Fold Panel Relay 75/1
  - Instrument Cluster, Servotronic Control Module 75/2

- **Standard Equipment** 49/1
  - Fuse Panel, Micro Central Electric Panel, 13-Fold Relay Panel 49/1
  - 8-Fold Relay Panel 49/2
  - Ground Connections 49/3
  - Ignition/Starter Switch, Load Reduction Relay 49/4
  - Battery, Starter, Generator 49/5
  - Light Switch, Headlight Dimmer/Flasher Switch, Park Light Switch, Fog Light Switch 49/6
  - Instrument Panel Light Dimmer Switch, Headlight Adjuster 49/7
  - Emergency Flasher Switch 49/8
  - Windshield Wiper Motor, Windshield Washer Pump 49/9
  - Fuse Holder, Washer Jets, Fog Light 49/10
  - Fuse Holder, License Plate Light, Glove Compartment Light 49/11
  - Fuse Holder, Heated Mirror, Brake Light Switch 49/12
  - Fuse Holder, Dual Horn Relay 49/13
  - Fuse Holder, Cigarette Lighter, High Tone Horn, Low Tone Horn 49/14
  - Fuse Holder, Right Headlight 49/15
  - Fuse Holder, Left Headlight 49/16
  - Left Tail Light, Back-Up Light, High-Mount Brake Light 49/17
  - Right Tail Light, Fuel Pump, Rear Window Defogger/Heat Element 49/18
  - Instrument Cluster, Indicator Light 49/19
  - Instrument Cluster, Fuel Gauge, Engine Coolant Temperature Gauge, Tachometer, Speedometer 49/20
  - Instrument Cluster, Analog Clock, Brake Pad Wear Indicator 49/21
  - Instrument Cluster, Oil Pressure Switch, Engine Coolant Level Warning Switch, Brake Fluid Level Warning Switch 49/22
  - Instrument Cluster, Oil Level Thermal Sensor 49/23
  - Instrument Cluster, Data Link Connector Rapid Data Transfer 49/24
  - Connector Radio, Illumination for Storage Tray, Illumination for Cup Holder 49/25

- **Steering Wheel with Tiptronic Switch** 80/1
  - Fuse Panel, Micro Central Electric Panel 80/1
  - Control Unit in Steering Wheel, Horn Button 80/2
  - Control Unit in Steering Wheel, Airbag Control Module 80/3
  - Dual Horn Relay, High and Low Tone Horn 80/4

- **Telephone** 60/1
  - Fuse Panel, Micro Central Electric Panel 60/1
  - Telephone Transceiver, Telephone Handset, Telephone Microphone 60/2
  - Telephone Transceiver, Telephone Handset, Radio/Telephone Antenna 60/3
  - Telephone Transceiver, Speaker (Telephone, Navigation) 60/4

Please read these WARNINGS and CAUTIONS before proceeding with electrical system work.

WARNING—

• Some repairs may be beyond your capability. If you lack the skills, tools and equipment, or a suitable workplace for any procedure described in this manual, we suggest you leave such repairs to an authorized Audi dealer service department, or other qualified shop.

• Audi is constantly improving its cars. Sometimes these changes, both in parts and specifications, are made applicable to earlier models. Therefore, before starting any major jobs or repairs to components on which passenger safety may depend, consult your authorized Audi dealer about Technical Bulletins that may have been issued since the editorial closing of this manual.

• Do not re-use any fasteners that are worn or deformed in normal use. Some fasteners are designed to be used only once, and are unreliable and may fail if used a second time. This includes, but is not limited to, nuts, bolts, washers, circlips and cotter pins. Always replace these fasteners with new parts.

• Never work under a lifted car unless it is solidly supported on stands designed for the purpose. Do not support a car on cinder blocks, hollow tiles or other props that may crumble under continuous load. Never work under a car that is supported solely by a jack. Never work under the car while the engine is running.

• If you are going to work under a car on the ground, make sure that the ground is level. Block the wheels to keep the car from rolling. Disconnect the battery negative terminal (ground strap) to prevent others from starting the car while you are under it.

• Never run the engine unless the work area is well ventilated. Carbon monoxide (CO) kills.

• Do not attempt to work on your vehicle if you do not feel well. You increase the danger of injury to yourself and others if you are tired, upset or have taken medication or any other substance that may impair you or keep you from being fully alert.

• Disconnect the battery negative (–) terminal (ground strap) whenever you work on the fuel system or the electrical system. Do not smoke or work near heaters or other fire hazards. Keep an approved fire extinguisher handy.

• Batteries produce explosive hydrogen gas. Keep sparks, lighted matches and open flame away from top of battery. If hydrogen gas escaping from cap vents is ignited, it will ignite gas trapped in the cells and cause the battery to explode.

• Connect and disconnect battery cables, jumper cables or a battery charger only with the ignition switched off, to prevent sparks. Do not quick-charge the battery (for boost starting) for longer than one minute and do not allow charging voltage to exceed 16.5 volts. Wait at least one minute before boosting the battery a second time.

• Illuminate your work area adequately but safely. Use a portable safety light for working inside or under the car. Make sure the bulb is enclosed by a wire cage. The hot filament of an accidentally broken bulb can ignite spilled fuel or oil.

• Tie long hair behind your head. Do not wear a necktie, a scarf, loose clothing, or a necklace when you work near machine tools or running engines. If your hair, clothing, or jewelry were to get caught in the machinery, severe injury could result.

• Finger rings, bracelets and other jewelry should be removed so that they cannot cause electrical shorts, get caught in running machinery, or be crushed by heavy parts.

• Catch draining fuel, oil, or brake fluid in suitable containers. Do not use food or beverage containers that might mislead someone into drinking from them. Store flammable fluids away from fire hazards. Wipe up spills at once, but do not store the oily rags, which can ignite and burn spontaneously.

• Always observe good workshop practices. Wear goggles when you operate machine tools or work with battery acid. Gloves or other protective clothing should be worn whenever the job requires working with harmful substances.

• The Air-Conditioning (A/C) system is filled with chemical refrigerant, which is hazardous. The A/C system should be serviced only by trained technicians using approved refrigerant recovery/recycling equipment, trained in related safety precautions, and familiar with regulations governing the discharging and disposal of automotive chemical refrigerants.

• Do not expose any part of the A/C system to high temperatures such as an open flame. Excessive heat will increase system pressure and may cause the system to burst.

• Some aerosol tire inflators are highly flammable. Be extremely cautious when repairing a tire that may have been inflated using an aerosol tire inflator. Keep sparks, open flame or other sources of ignition away from the tire repair area. Inflate and deflate the tire at least four times before breaking the bead from the rim. Completely remove the tire from the rim before attempting any repair.

• Friction materials such as brake or clutch discs may contain asbestos fibers. Do not create dust by grinding, sanding, or by cleaning with compressed air. Avoid breathing asbestos fibers and asbestos dust. Breathing asbestos can cause serious diseases such as asbestosis or cancer, and may result in death.

• Greases, lubricants and other automotive chemicals contain toxic substances, many of which are absorbed directly through the skin. Read manufacturer's instructions and warnings carefully. Use hand and eye protection. Avoid direct skin contact.

• An airbag is an explosive device. Handled improperly or without adequate safeguards, it can be accidently activated and cause serious personal injury. To prevent personal injury or airbag system failure, **only trained Audi dealer service technicians** should test, disassemble or service the airbag system.

WARNINGS and CAUTIONS continued on next page ▶

Please read these WARNINGS and CAUTIONS before proceeding with electrical system work.

WARNING (continued) —

• ALWAYS disconnect the power supply before working on the airbag system, or when doing repairs that require removing airbag system components. Disconnect the battery negative (–) terminal and cover the battery.

• The airbag unit should not be removed, nor should any work be performed on the airbag system, without first reviewing the warnings and cautions that accompany the airbag system repair information in this manual.

• Never apply stickers or any other coverings on the steering wheel. Do not let chemical cleaners, oil or grease come into contact with vinyl covering of the airbag unit.

• Never open or otherwise attempt to repair airbag system parts. Always use new parts. Never leave airbag parts or the partially disassembled airbag system unattended.

• Always replace airbag system components that have been mechanically damaged (bubbles, cracks, etc.).

• Do not expose an airbag unit to temperature above 194°F (90°C), even for a brief period, while handling it during the repair process. Keep the unit clear of heat sources such as hot plates, soldering irons, heat lamps, welding equipment, or the like.

• Never use a test light or volt meter to conduct electrical tests on the airbag system. The airbag system must only be tested by trained Audi Service technicians using special airbag tester equipment. The airbag system must only be tested while it is installed in the car.

• Always set down a removed airbag unit so that the horn pad is facing upward.

• When driving or riding in an airbag-equipped vehicle, never hold test equipment in your hands or lap while the vehicle is in motion. Objects between you and the airbag can increase the risk of injury in an accident.

• The storage, transportation and disposal of airbag units are subject to the laws for flammable solids.

CAUTION—

• If you lack the skills, tools and equipment, or a suitable workshop for any procedure described in this manual, we suggest you leave such repairs to an authorized Audi dealer or other qualified shop. We especially urge you to consult an authorized Audi dealer before beginning repairs on any car that may still be covered wholly or in part by any of the extensive warranties issued by Audi of America, Inc.

• Audi is constantly improving its cars and sometimes these changes, both in parts and specifications, are made applicable to earlier models. Therefore part numbers in this manual are used for reference only. Always check with your authorized Audi dealer parts department for the latest information.

• Before starting a job, make certain that you have all the necessary tools and parts on hand. Read all the instructions thoroughly, do not attempt shortcuts. Use tools that are appropriate to the work and use only replacement parts meeting Audi specifications. Makeshift tools, parts and procedures will not make good repairs.

• Use pneumatic and electric tools only to loosen threaded parts and fasteners. Never use these tools to tighten fasteners, especially on light alloy parts. Always use a torque wrench to tighten fasteners to the tightening torque specification listed.

• Be mindful of the environment and ecology. Before you drain the crankcase, find out the proper way to dispose of the oil. Do not pour oil onto the ground, down a drain, or into a stream, pond or lake. Consult local ordinances that govern the disposal of wastes.

• When connecting cables or ground straps, all connections must be cleaned to bare metal. Be sure mounting bolts are tight.

• Do not replace fuse links with fuse links of other gauge or with standard sire. Always replace the entire length of the fuse link.

• Before doing any electrical welding on vehicles equipped with Anti-lock Brakes (ABS), disconnect the battery negative terminal (ground strap) and the ABS Control Module connector.

• On cars equipped with ABS, the ABS control unit cannot withstand temperatures from a paint-drying booth or a heat lamp in excess of 203°F (95°C) and should not be subjected to temperatures in excess of 185°F (85°C) for more than two hours.

• On cars equipped with anti-theft radios, make sure you know the correct radio activation code before disconnecting the battery or removing the radio. If the wrong code is entered into the radio when power is restored, that radio may lock up and be rendered inoperable, even if the correct code is then entered.

• When boost-charging the battery, first remove the fuses for the Engine Control Module (ECM), the Transmission Control Module (TCM), the ABS Control Module, and the On-Board computer. In cases where one or more of these components is not separately fused, disconnect the Control Module connector(s).

• Do not quick-charge the battery (for boost starting) for longer than one minute, and do not exceed 16.5 volts at the battery with the boosting cables attached. Wait at least one minute before boosting the battery a second time.

◀ **WARNINGS and CAUTIONS** continued from previous page

How to use wiring diagrams

This section presents several wiring diagram pages with detailed notes to explain diagram identification, layout, and how to read wiring diagrams. Following these pages is a listing of commonly used diagram symbols.

Because of the large number of components and the wide range of optional equipment the electrical systems are separated into individual wiring diagrams.

All cars make use of several diagrams including Standard Equipment, ABS, Airbag, and Central Locking, but beyond that, installed equipment will vary considerably and the individual system diagrams must be consulted.

NOTE —
Some components that are considered standard equipment in the USA and Canada have their own separate wiring diagrams. This is because this equipment is not standard in other parts of the world and therefore, not included in the standard wiring diagram.

Wiring diagrams are also known as current flow diagrams because they show the path of current, which is usually from the top of the page to the bottom. For this reason, the terms wiring diagram and current flow diagram are used interchangeably.

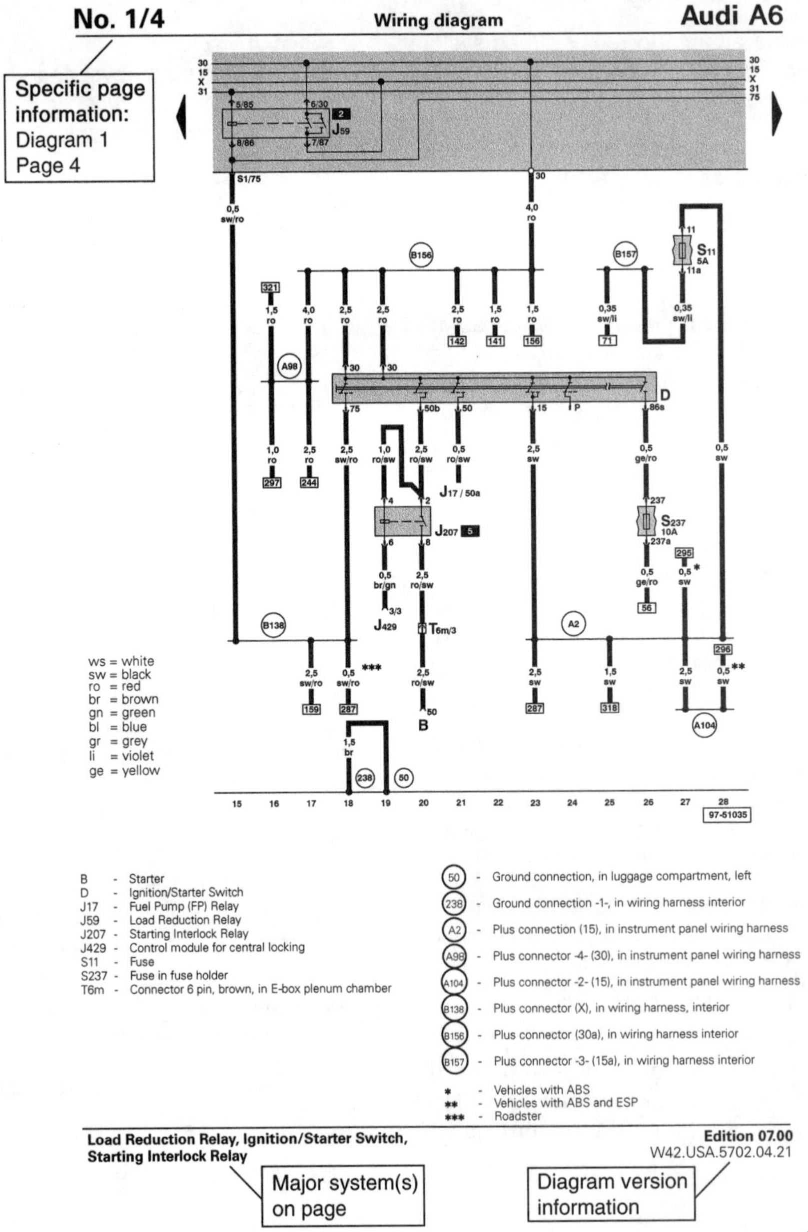

0024461

Wiring diagram layout

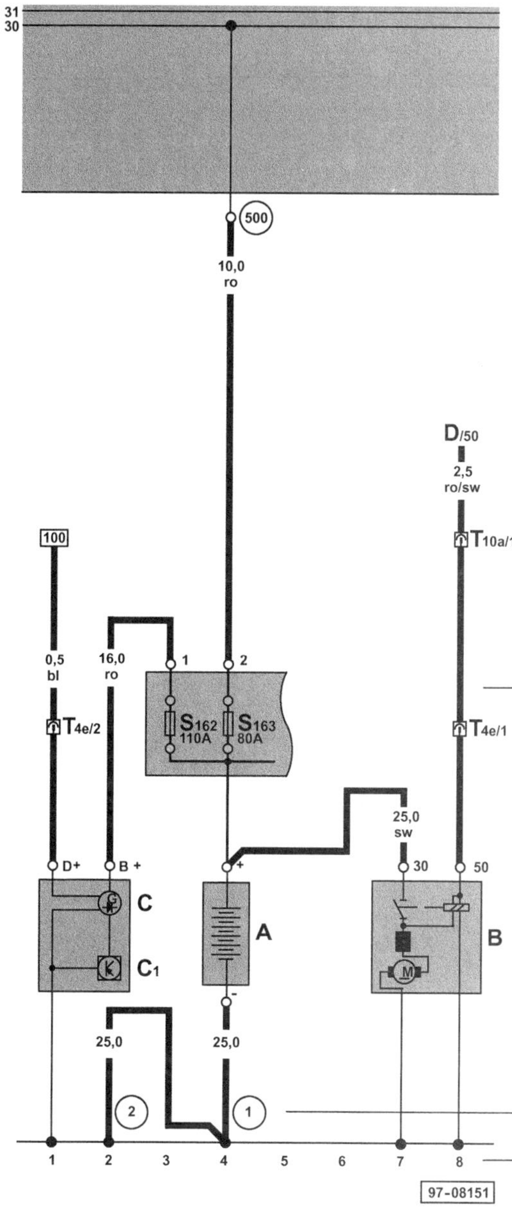

Relay panel
Indicated by grey area.

WIRING COLOR CODE		
ws	=	white
sw	=	black
ro	=	red
br	=	brown
gn	=	green
bl	=	blue
gr	=	grey
li	=	lilac
ge	=	yellow

Consumer circuit with wire routing
All switches and contacs are shown in the "off" position.

Vehicle ground
Numbers in circle indicate location on vehicle (see legend).

Current track number
Makes it easier to find the connections.

A - Battery
B - Starter
C - Generator (GEN)
C1 - Voltage Regulator (VR)
D - Ignition/Starter Switch
S162 - Fuse -1- (30) in fuse bracket / battery
S163 - Fuse -2- (30) in fuse bracket / battery
T4e - 4-Pin Connector, on transmission
T10a - 10-Pin Connector, on protective housing for control module, in engine compartment, left
(1) - Ground strap, battery to body
(2) - Ground strap, transmission to body
(500) - Screw connection -1- (30), on relay panel

Legend
In all wiring diagrams the same component designation (code) is used for a particular component; for example, always A for battery.

How to read wiring diagrams

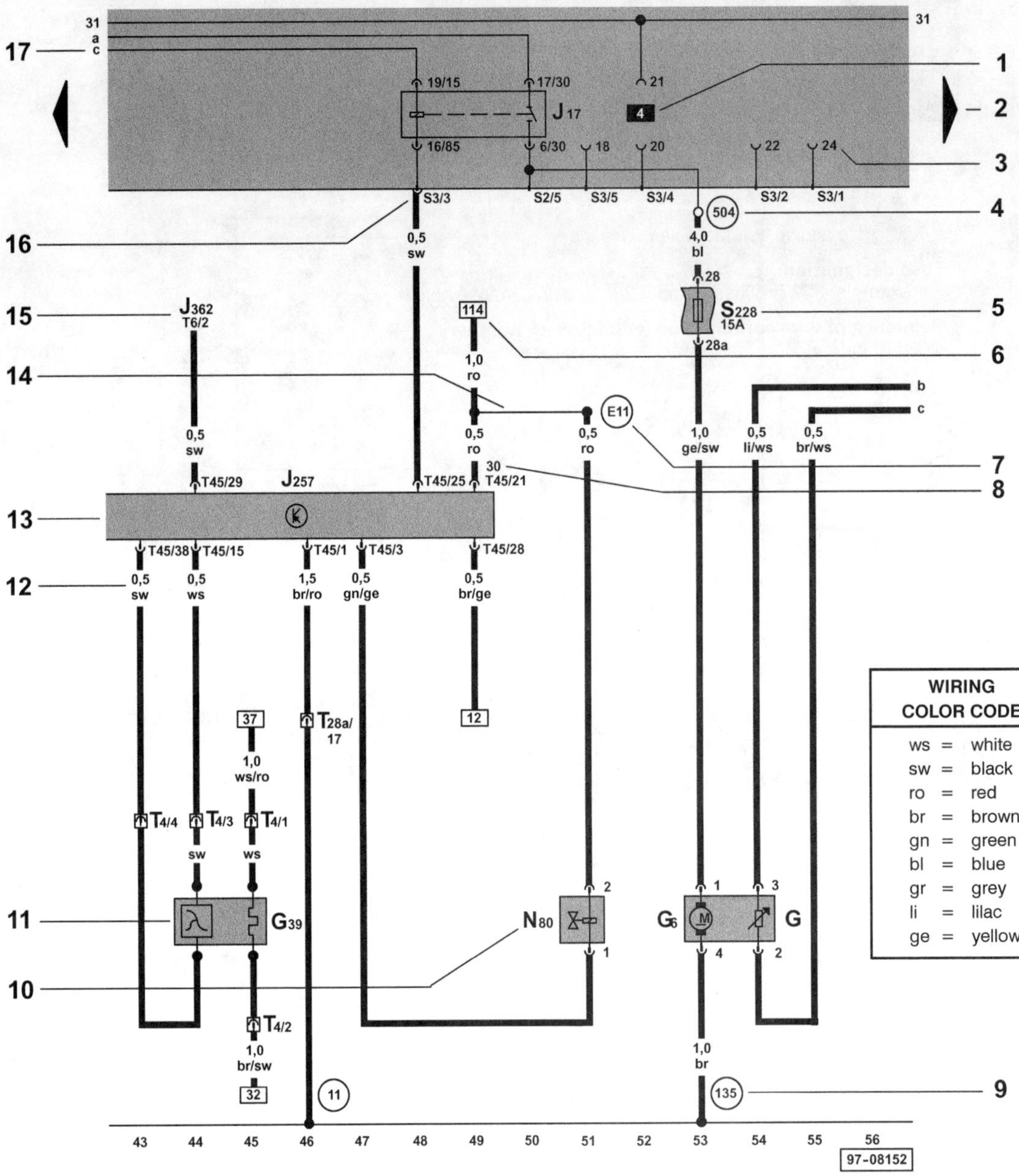

WIRING COLOR CODE		
ws	=	white
sw	=	black
ro	=	red
br	=	brown
gn	=	green
bl	=	blue
gr	=	grey
li	=	lilac
ge	=	yellow

1 - **Relay location number**
Indicates location on relay panel.

2 - **Arrow**
Indicates wiring circuit is continued on the previous and/or next page.

3 - **Connection designation - relay control module on relay panel**
Shows the individual terminals in a multi-point connector.
For example: contact 24 on terminal **4** on relay panel.

4 - **Screw connection on relay panel**
White circle shows a detachable connection.

5 - **Fuse designation**
For example: S228 = Fuse number 28, 15 amps, in fuse holder

6 - **Reference of wire continuation (current track number)**
Number in frame indicates current track where wire is continued.

7 - **Wire connection designation in wiring harness**
Location of wire connections are indicated in the legend.

8 - **Terminal designation**
Designation which appears on actual component and/or terminal number of a multi-point connector.

9 - **Ground connection designation in wire harness**
Locations of ground connections are indicated in legend.

10 - **Component designation**
Use legend at bottom of page to identify the component code.

11 - **Component symbols**

12 - **Wire cross-section size (in mm^2) and wire colors**
Abbreviations are explained in color chart beside the wiring diagram.

13 - **Component symbol with open drawing side**
Indicated component is continued on another wiring diagram. The number of corresponding wiring diagram can taken from list of contents.

14 - **Internal connections (thin lines)**
These connections are **not** wires. Internal connections are current carrying and are listed to allow tracing of current flow inside components and wiring harness.

15 - **Reference of continuation of wire to component**
For example: Control module for anti-theft immobilizer J362 on 6-Pin Connector, terminal 2

16 - **Relay panel connectors**
Shows wiring of multi-point or single connectors on relay panel
For example: S3/3 - Multi-point connector S3, terminal 3

17 - **Reference of internal connection continuation**
Letters indicate where connection continues on the previous and/or next page.

Symbols used in wiring diagrams

Fuse

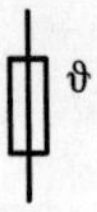
Thermo-fuse
(Circuit Breaker)

Battery

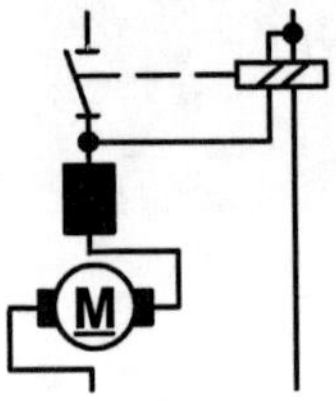
Starter

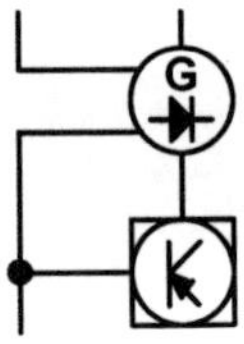
Generator (GEN)

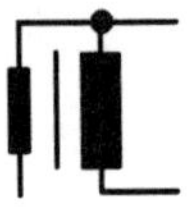
Ignition Coil

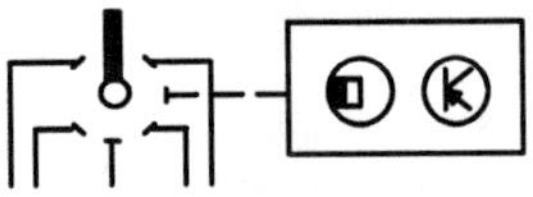
Distributor (electronic)

Spark plug connector
and plug

Glow plug
Heater element

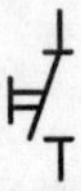
Switch
(manually operated)

Switch
(thermally operated)

Push putton switch
(manually operated)

Switch
(mechanically operated)

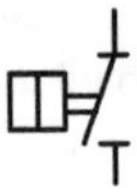
Switch
(pressure operated)

Multiple switch
(manually operated)

Resistance

Variable resistor
(Rheostat)

Resistor
temperature dependent

Heater element
temperature dependent

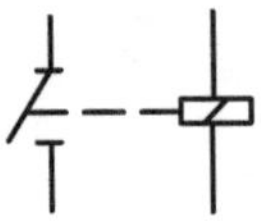
Relay

Symbol	Description	Symbol	Description
	Diode		Solenoid valve
	Zener diode	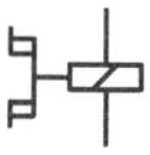	Magnetic clutch
	Diode light sensitive		Wire connector
	Light bulb		Pin connector
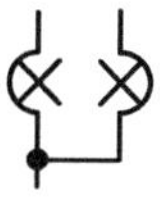	Light bulb (dual filament)		Multi-point connector at component
	LED		Internal connections in component
	Interior light		Wire connection detachable
	Instrument (Gauge)		Wire connection fixed
	Electronic control module		Wire connection in wiring harness
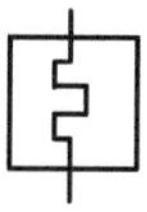	Rear window defogger heat element		Resistance wire
	Cigarette lighter	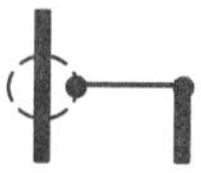	Shield wire

Symbols used in wiring diagrams

Control motor,
headlight range adjustment

Motor

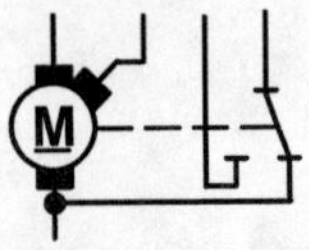
Wiper motor
2-speed

Crankshaft position sensor
(CKP)

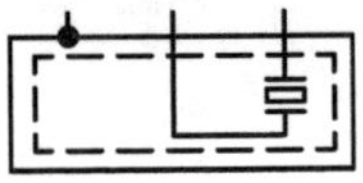
Knock sensor (KS)

Analog clock

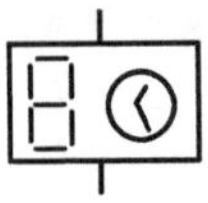
Digital clock

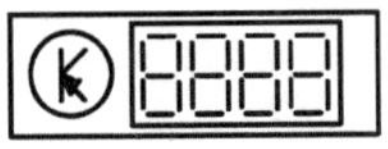
Multi-function
indicator

Airbag spiral spring

Speed sensor

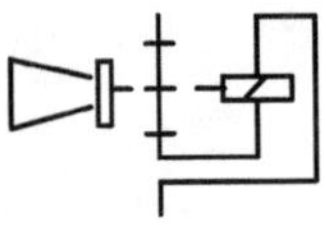
Horn

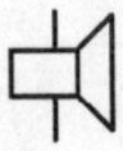
Speeker

Antenna with electronic
antenna amplifier

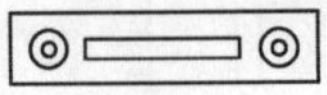
Radio

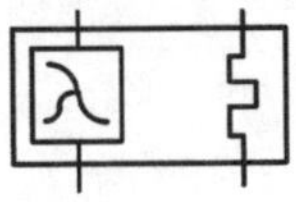
Heated oxygen sensor

(2,8 l - Injection Engine, 6-Cylinder), Code AHA

1998 m. y.

Fuse Panel

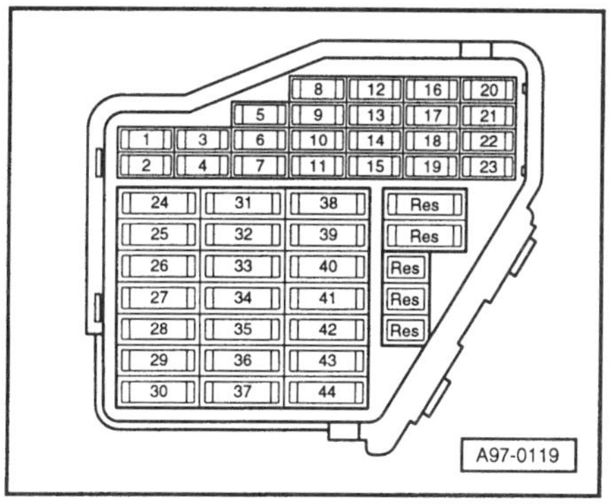

Fuse Colors:

30 A - Green
25 A - White
20 A - Yellow
15 A - Blue
10 A - Red
5 A - Beige

Starting with fuse position 23, fuses in the fuse holder are identified with 223 in the wiring diagram.

Micro Central Electric Panel

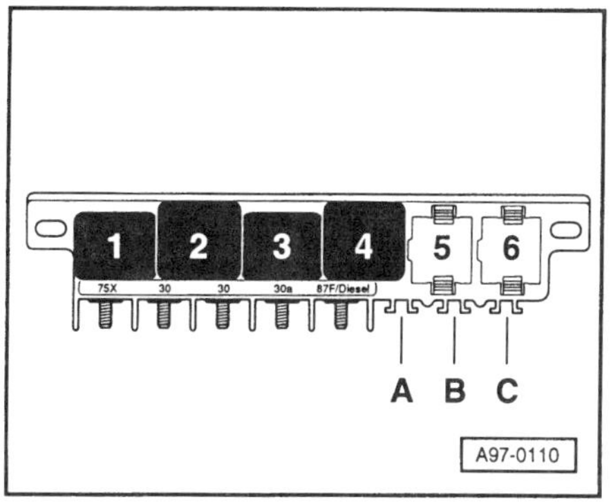

Relay Location:

1 - Dual Horn Relay, J4

2 - Load Reduction Relay, J59

4 - Fuel Pump (FP) Relay, J17

5 - Wiper/Washer Intermittent Relay, J31

6 - Wiper/Washer Intermittent Relay, J31

13 - Fold Relay Panel

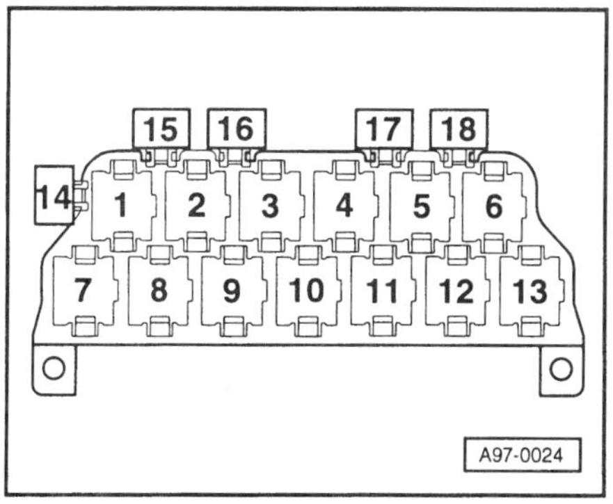

Relay Location:

4 - Starting Interlock Relay, J207

6 - Fog Light Relay, J5

17 - Fuse for secondary air pump, S130

8 - Fold Relay Panel

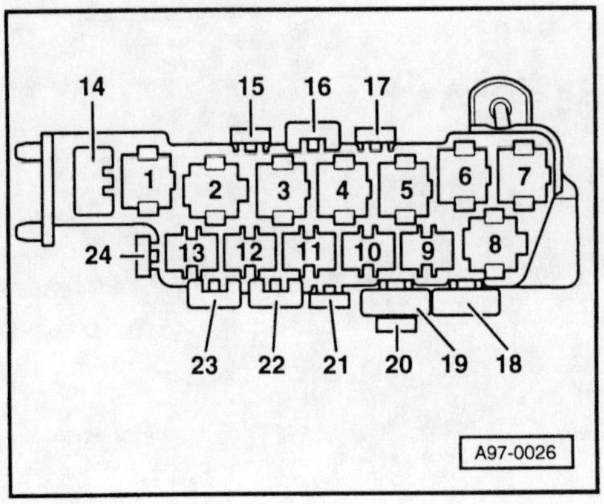

Relay Location:

- **2** - Second Speed Coolant Fan Control (FC) Relay, J101
- **3** - Coolant Fan control (FC) Relay, J26
- **20** - Coolant Fan Fuse, S42
- **21** - Control module fuse for coolant fan, S142

3 - Fold Relay Panel in E-Box plenum chamber

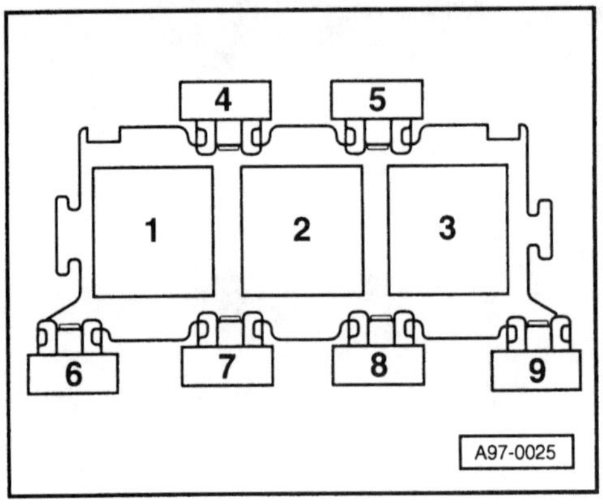

Relay Location:

- **2** - Secondary Air Injection (AIR) Pump Relay, J299

ws = white
sw = black
ro = red
gn = green
br = brown
bl = blue
gr = gray
li = violet
ge = yellow

(3) - Ground strap, engine to body

(12) - Ground connection, in engine compartment, left

(13) - Ground connection, in engine compartment, right

(43) - Ground connection, on right A-pillar, lower part

(44) - Ground connection, on left A-pillar, lower part

(81) - Ground connection -1-, in instrument panel wiring harness

(85) - Ground connection -1-, in engine compartment wiring harness

(86) - Ground connection -1-, in rear wiring harness

(135) - Ground connection -2-, in instrument panel wiring harness

(176) - Ground connection, in right headlight wiring harness

(179) - Ground connection, in left headlight wiring harness

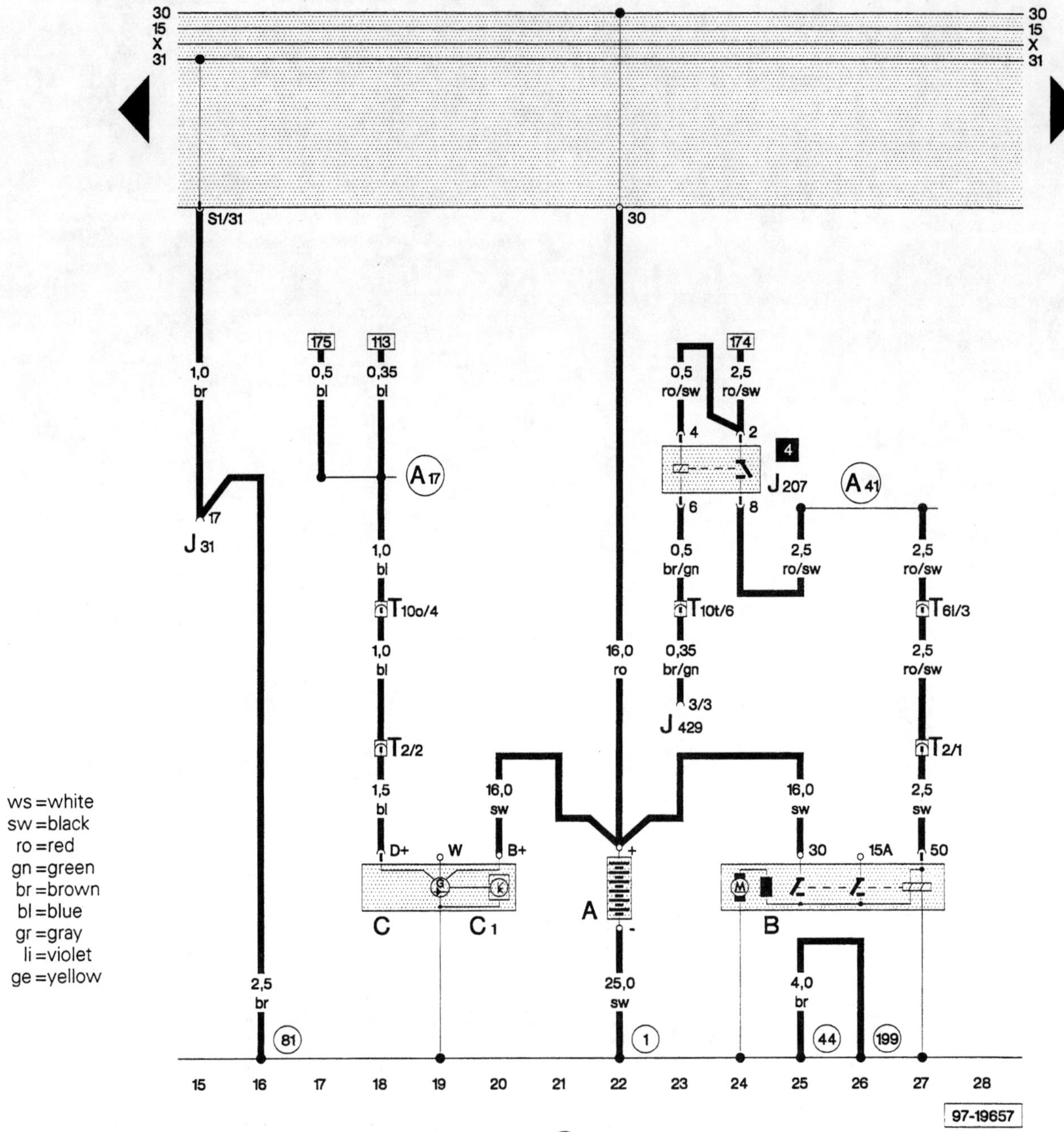

A - Battery
B - Starter
C - Generator (GEN)
C 1 - Voltage Regulator (VR)
J 31 - Wiper/Washer Intermittent Relay
J 207 - Starting Interlock Relay
J 429 - Control module for central locking
T 2 - Connector 2 pin, grey, in engine compartment, right
T 6 l - Connector 6 pin, red, connector station electronic box plenum chamber
T 10 o - Connector 10 pin, brown, connector station electronic box plenum chamber
T 10 t - Connector 10 pin, orange, connector station A pillar, right

(1) - Ground strap, battery to body
(44) - Ground connection, on left A-pillar, lower part
(81) - Ground connection -1-, in instrument panel wiring harness
(199) - Ground connection -3-, in instrument panel wiring harness
(A17) - Wire connection (61), in instrument panel wiring harness
(A41) - Plus connection (50), in instrument panel wiring harness

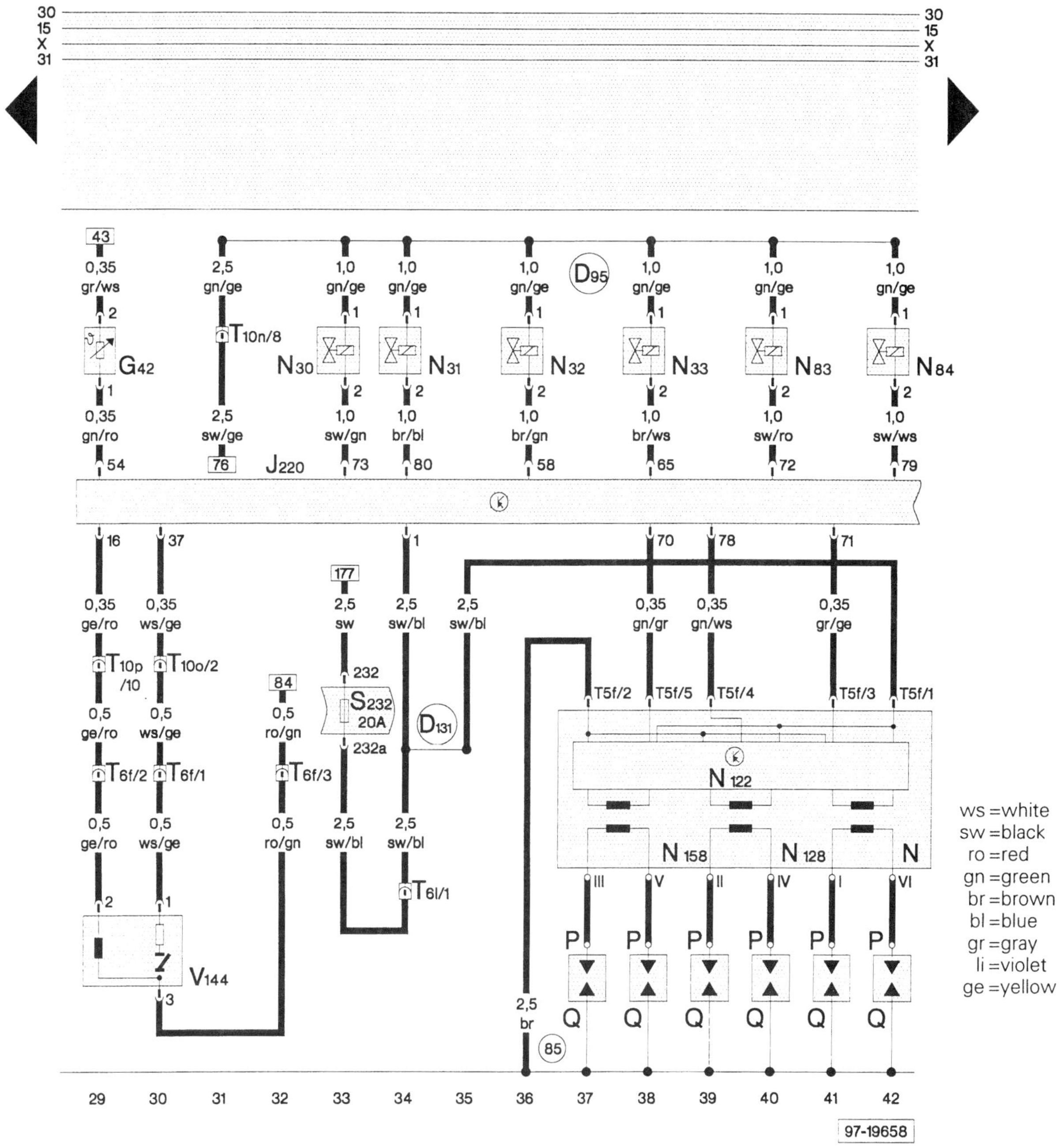

G 42 - Intake Air Temperature (IAT) Sensor
J 220 - Motronic Engine Control Module (ECM)
N - Ignition Coil
N 30 - Cylinder 1 Fuel Injector
N 31 - Cylinder 2 Fuel Injector
N 32 - Cylinder 3 Fuel Injector
N 33 - Cylinder 4 Fuel Injector
N 83 - Cylinder 5 Fuel Injector
N 84 - Cylinder 6 Fuel Injector
N 122 - Power Output Stage
N 128 - Ignition Coil 2
N 158 - Ignition Coil 3
P - Spark Plug Connectors
Q - Spark Plugs
S 232 - Fuse
T 5 f - Connector 5 pin, ignition coil unit
T 6 f - Connector 6 pin, darkbrown, connector station A pillar, right
T 6 l - Connector 6 pin, red, connector station electronic box plenum chamber
T 10 n - Connector 10 pin, yellow, connector station electronic box plenum chamber
T 10 o - Connector 10 pin, brown, connector station electronic box plenum chamber
T 10 p - Connector 10 pin, black, connector station electronic box plenum chamber
V 144 - Leak Detection Pump (LDP)

(85) - Ground connection -1-, in engine compartment wiring harness
(D95) - Wire connection (injectors), in engine compartment wiring harness
(D131) - Wire connection (15), above circuit Breaker, in engine compartment wiring harness

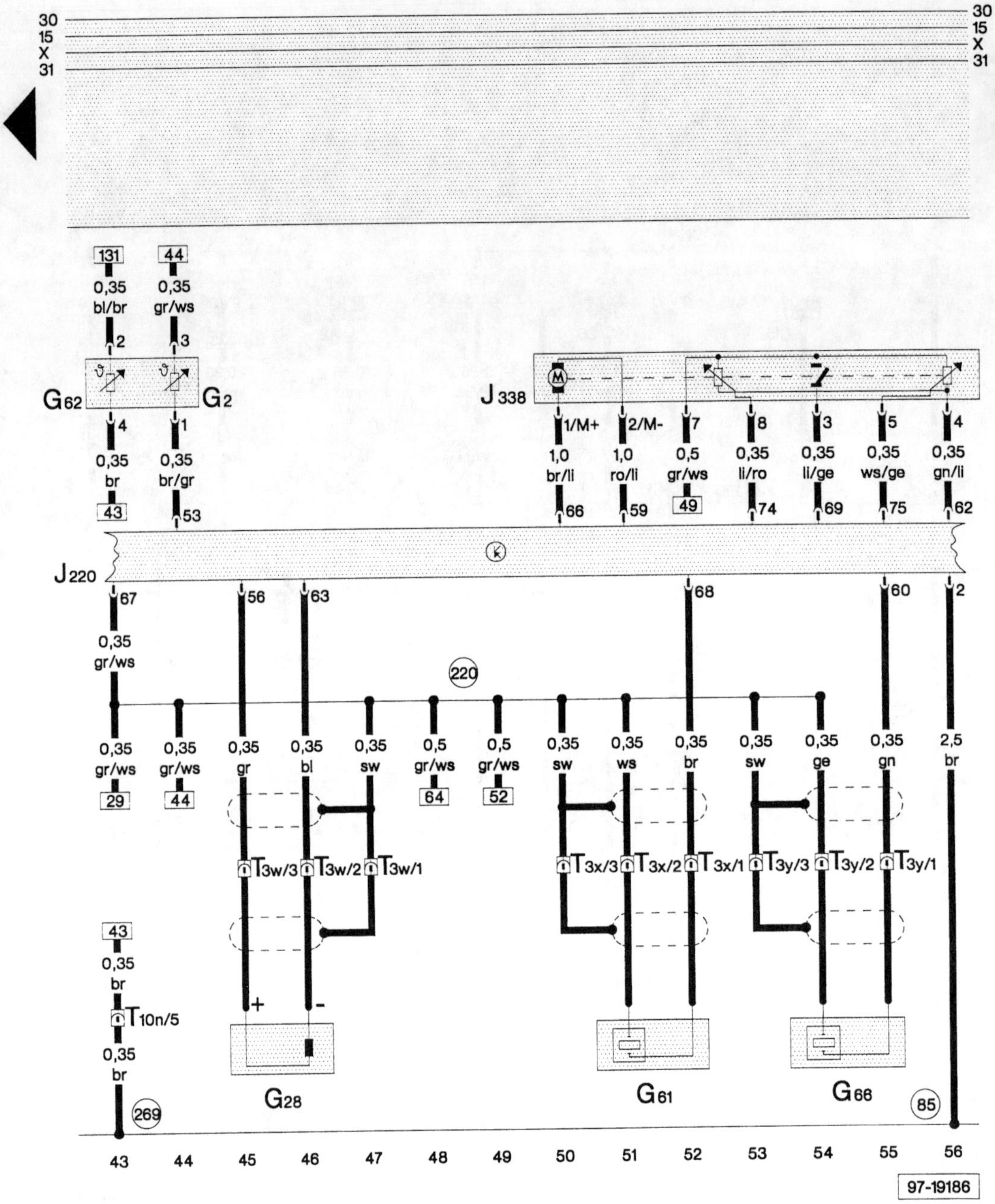

ws = white
sw = black
ro = red
gn = green
br = brown
bl = blue
gr = gray
li = violet
ge = yellow

G 2 - Engine Coolant Temperature (ECT) Sensor
G 28 - Engine Speed (RPM) Sensor
G 61 - Knock Sensor (KS) 1
G 62 - Engine Coolant Temperature (ECT) Sensor
G 66 - Knock Sensor (KS) 2
J 220 - Motronic Engine Control Module (ECM)
J 338 - Throttle Valve Control Module
T 3 w - Connector 3 pin, grey, on engine speed sensor
T 3 x - Connector 3 pin, blue, on knock sensor 1
T 3 y - Connector 3 pin, blue, on knock sensor 2
T 10 n - Connector 10 pin, yellow, connector station electronic box plenum chamber

(85) - Ground connection -1-, in engine compartment wiring harness
(220) - Ground connection (sensor ground), in engine compartment wiring harness
(269) - Ground connector (sensor ground) -1-, in instrument panel wiring harness

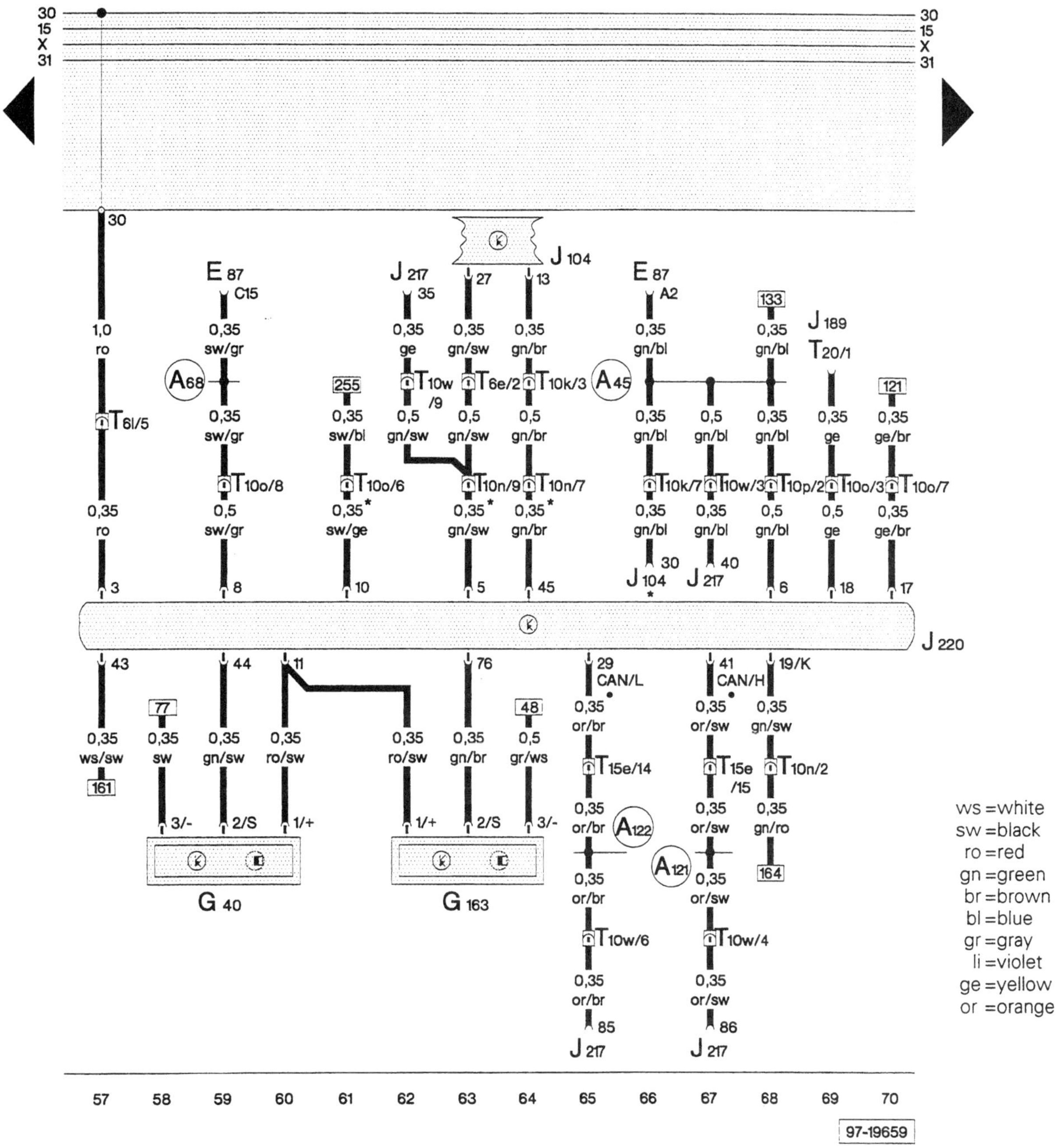

E 87 - A/C Control Head
G 40 - Camshaft Position (CMP) Sensor
G 163 - Camshaft Position (CMP) Sensor 2
J 104 - ABS Control Module (w/EDL)
J 189 - Auto Check System
J 217 - Transmission Control Module (TCM)
J 220 - Motronic Engine Control Module (ECM)
T 6 e - Connector 6 pin, grey, connector station A pillar, right
T 6 l - Connector 6 pin, red, connector station electronic box plenum chamber
T 10 k - Connector 10 pin, orange, connector station A pillar, left
T 10 n - Connector 10 pin, yellow, connector station electronic box plenum chamber
T 10 o - Connector 10 pin, brown, connector station electronic box plenum chamber
T 10 p - Connector 10 pin, black, connector station electronic box plenum chamber
T 10 w - Connector 10 pin, blue, connector station A pillar, right
T 15 e - Connector 15 pin, white, connector station electronic box plenum chamber
T 20 - Connector 20 pin, red/white, on instrument cluster

(A45) - Wire connection (RPM-signal), in instrument panel wiring harness
(A68) - Connector (C15, A/C), in instrument panel wiring harness
(A121) - Connection (high bus) in instrument panel wiring harness
(A122) - Connection (low bus) in instrument panel wiring harness

• - CAN-Bus
* - Vehicles with front wheel drive

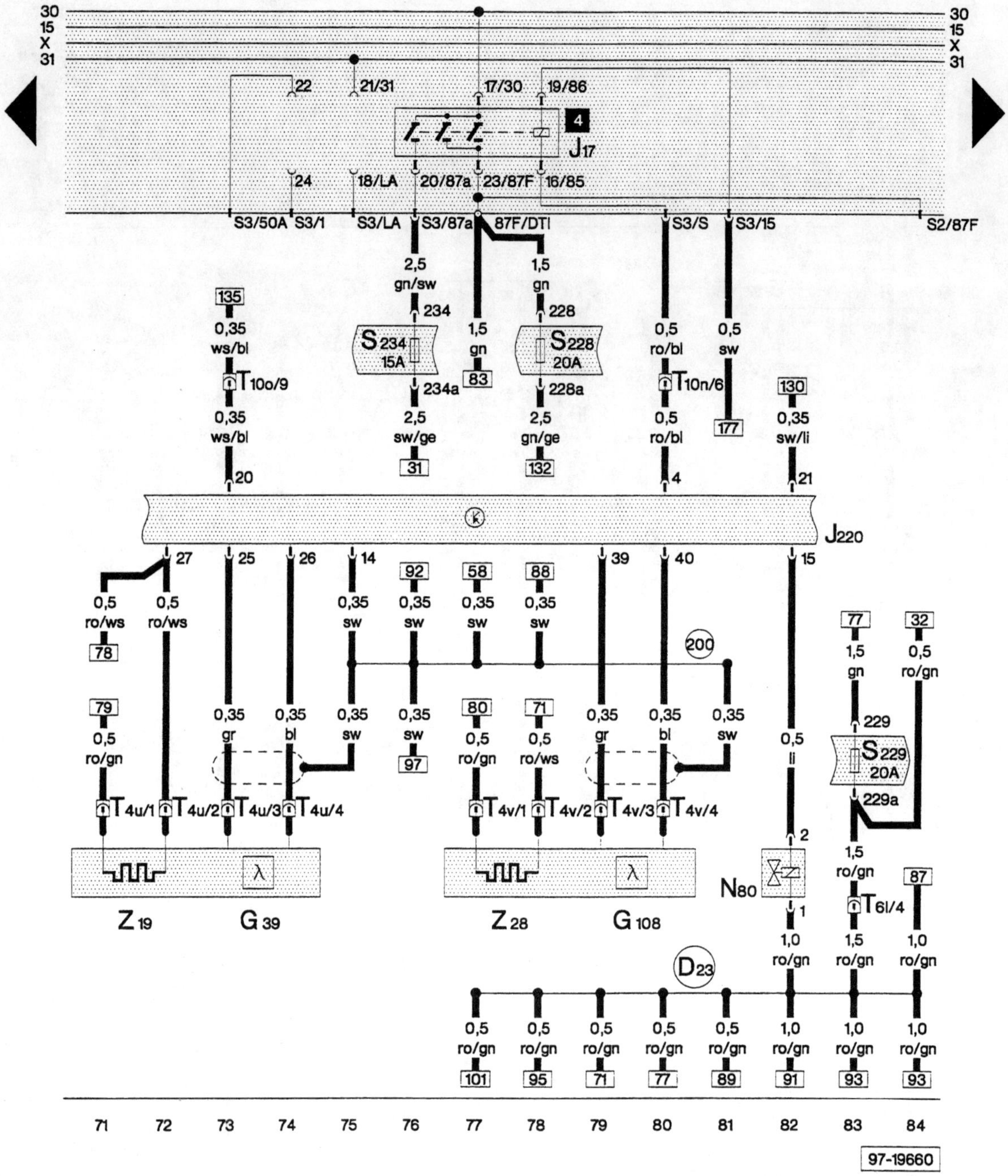

ws = white
sw = black
ro = red
gn = green
br = brown
bl = blue
gr = gray
li = violet
ge = yellow

G 39 - Heated Oxygen Sensor (HO2S)
G 108 - Heated Oxygen Sensor (HO2S) 2
J 17 - Fuel Pump (FP) Relay
J 220 - Motronic Engine Control Module (ECM)
N 80 - Evaporative Emission (EVAP) Canister Purge Regulator Valve
S 228 - Fuse
S 229 - Fuse
S 234 - Fuse
T 4 u - Connector 4 pin, black, Heated Oxygen Sensor
T 4 v - Connector 4 pin, black, Heated Oxygen Sensor 2
T 6 l - Connector 6 pin, red, connector station electronic box plenum chamber
T 10 n - Connector 10 pin, yellow, connector station electronic box plenum chamber
T 10 o - Connector 10 pin, brown, connector station electronic box plenum chamber
Z 19 - Oxygen Sensor (O2S) Heater
Z 28 - Oxygen Sensor (O2S) 2 Heater

(200) - Ground connection (shielding), in engine compartment wiring harness
(D23) - Connector (over fuse 229), in wiring harness front, right

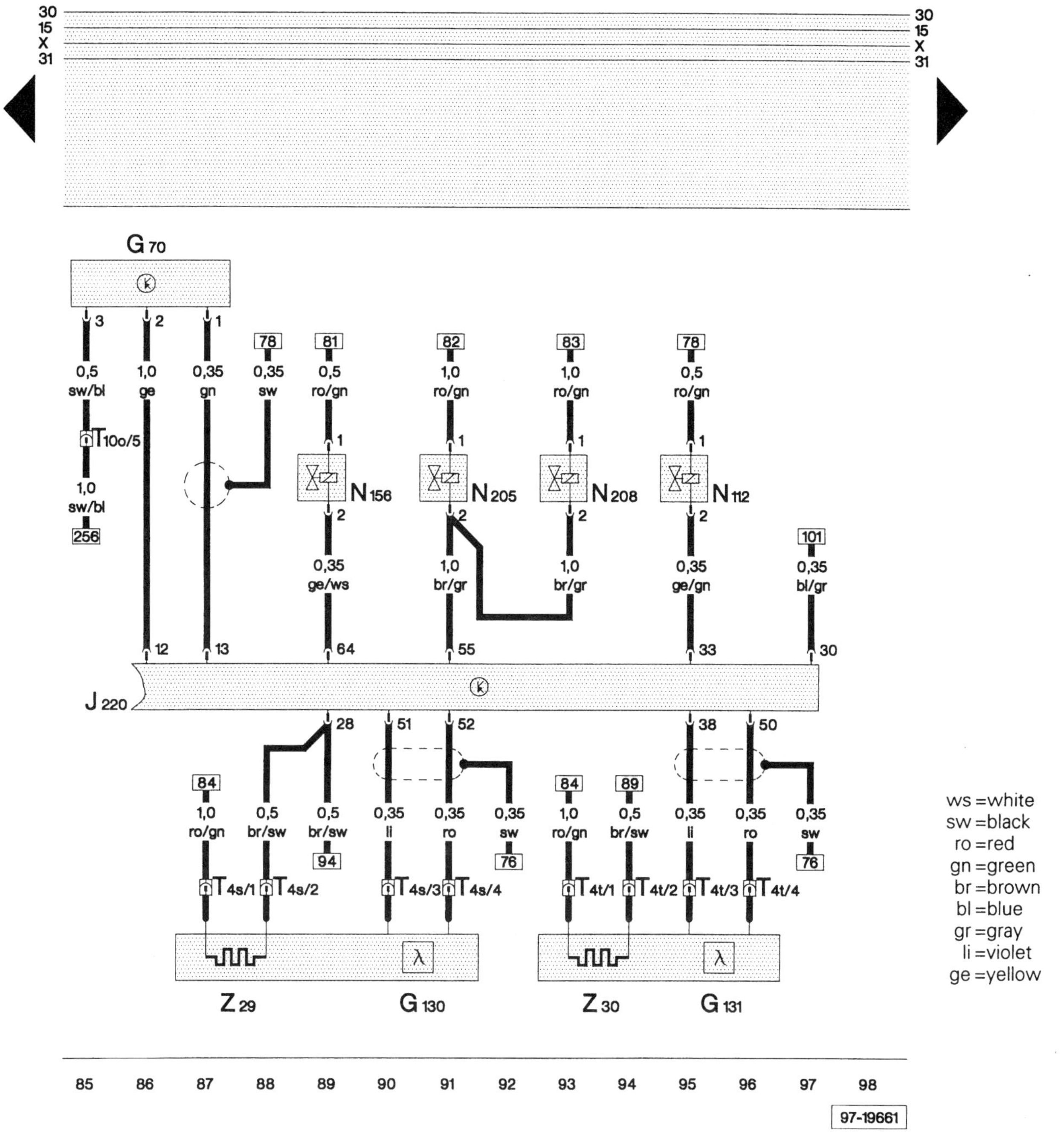

G 70 - Mass Air Flow (MAF) Sensor
G 130 - Oxygen Sensor (O2S) Behind Three Way Catalytic Converter (TWC)
G 131 - Oxygen Sensor (O2S) 2 Behind Three Way Catalytic Converter (TWC)
J 220 - Motronic Engine Control Module (ECM)
N 112 - Secondary Air Injection (AIR) Solenoid Valve
N 156 - Intake Manifold Change-Over Valve
N 205 - Valve -1- for camshaft adjustment
N 208 - Valve -2- for camshaft adjustment
T 4 s - Connector 4 pin, green, near oxygen sensor, behind Three Way Catalytic Converter (TWC)
T 4 t - Connector 4 pin, brown, near oxygen sensor 2, behind Three Way Catalytic Converter (TWC)
T 10 o - Connector 10 pin, brown, connector station electronic box plenum chamber
Z 29 - Heater for Lambda-probe 1
Z 30 - Heater for Lambda-probe 2

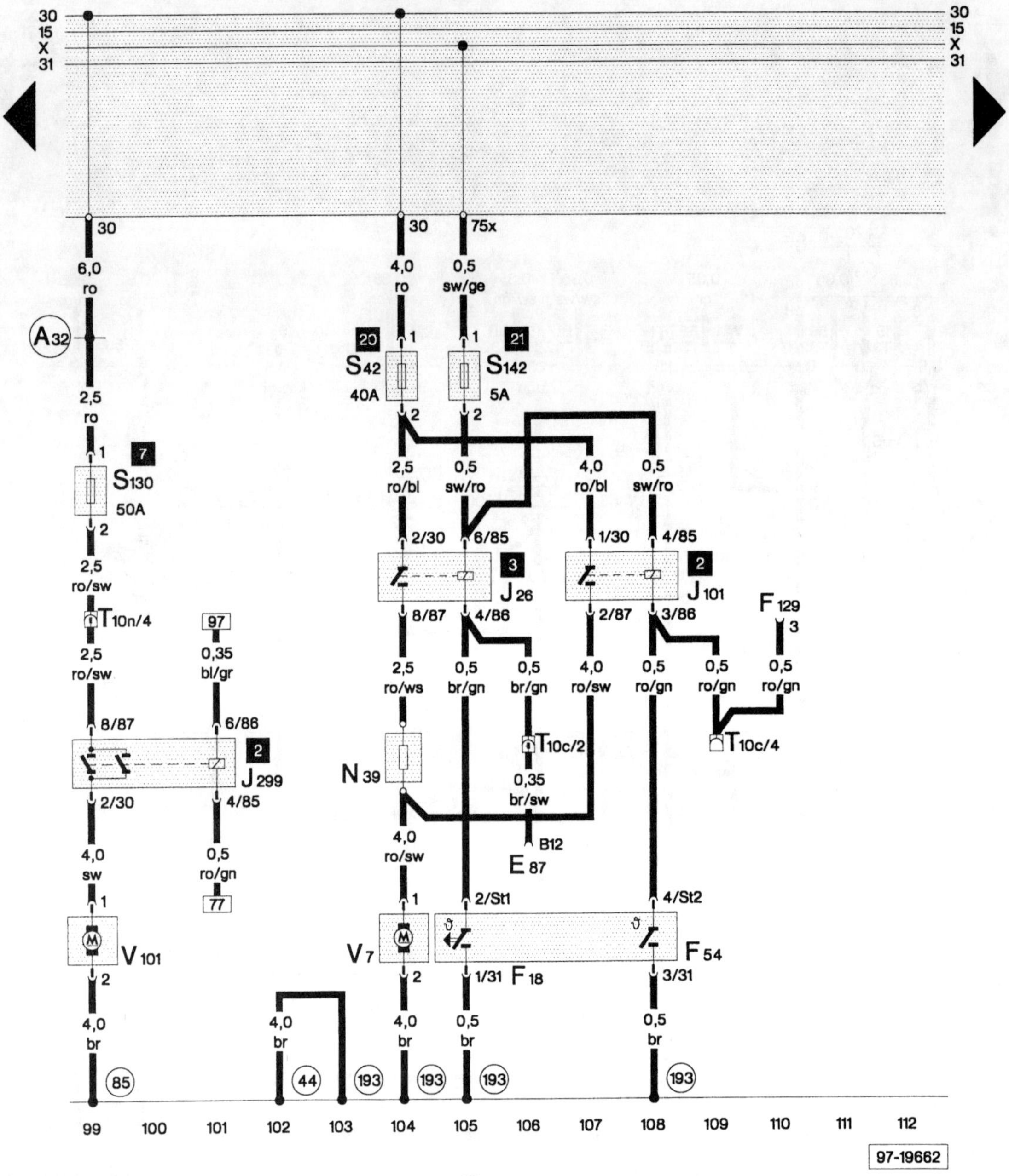

ws =white
sw =black
ro =red
gn =green
br =brown
bl =blue
gr =gray
li =violet
ge =yellow

E 87 - A/C Control Head
F 18 - Coolant Fan Control (FC) Thermal Switch
F 54 - Coolant Fan Control (FC) Thermal Switch
F 129 - A/C Pressure Switch
J 26 - Coolant Fan control (FC) Relay
J 101 - Second Speed Coolant Fan Control (FC) Relay
J 299 - Secondary Air Injection (AIR) Pump Relay
N 39 - Coolant Fan Control (FC) Series Resistance
S 42 - Coolant Fan Fuse
S 130 - Fuse for secondary air pump
S 142 - Control module fuse for coolant fan
T 6 a - Connector 6 pin, blue, connector station A pillar, left
T 10 c - Connector 10 pin, grey, connector station A pillar, left
T 10 n - Connector 10 pin, yellow, connector station electronic box plenum chamber
V 7 - Coolant Fan
V 101 - Secondary Air Injection (AIR) Pump Motor

(44) - Ground connection, on left A-pillar, lower part
(85) - Ground connection -1-, in engine compartment wiring harness
(193) - Ground connection -1-, in coolant fan wiring harness
(A32) - Plus connection (30), in instrument panel wiring harness

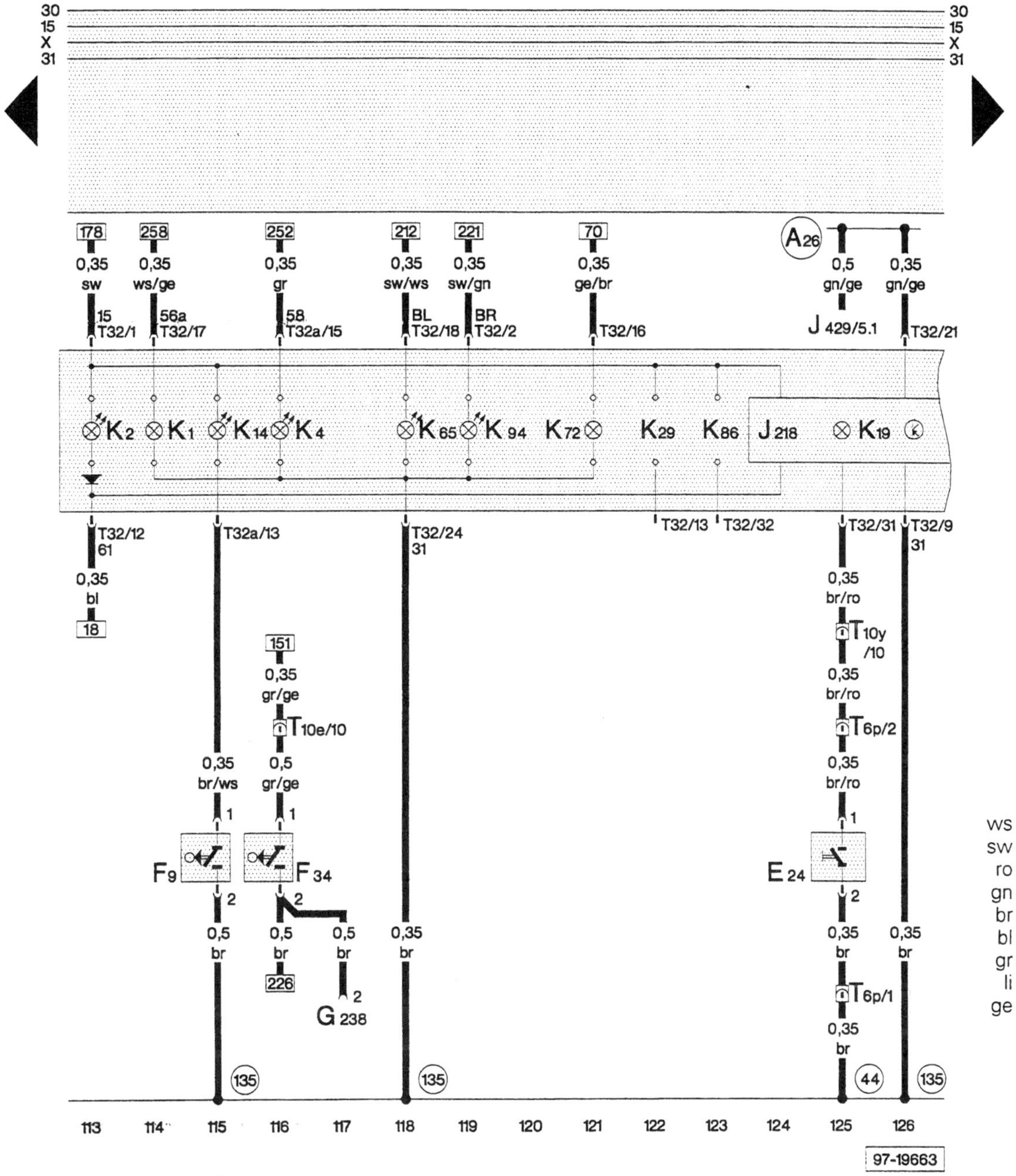

ws = white
sw = black
ro = red
gn = green
br = brown
bl = blue
gr = gray
li = violet
ge = yellow

E 24 - Left Seat Belt Switch
F 9 - Parking Brake Warning Light Switch
F 34 - Brake Fluid Level Warning Switch
G 238 - Air Quality Sensor
J 218 - Instrument Cluster Combination Processor
J 429 - Control module for central locking
K 1 - Headlight High Beam Indicator Light
K 2 - Generator (GEN) Warning Light
K 4 - Park Light Indicator Light
K 14 - Parking Brake Indicator Light
K 19 - Seat Belt Warning Light
K 29 - Glow Plug Indicator Light
K 65 - Left Turn Signal Indicator Light
K 72 - Warning light for catalyst monitoring
K 86 - Traction Control Indicator Light
K 94 - Right Turn Signal Indicator Light
T 6 p - Connector 6 pin, red, below driver's seat
T 10 e - Connector 10 pin, violet, connector station A pillar, left
T 10 y - Connector 10 pin, green, connector station A pillar, right
T 32 - Connector 32 pin, blue, on instrument cluster
T 32 a - Connector 32 pin, green, on instrument cluster

(44) - Ground connection, on left A-pillar, lower part

(135) - Ground connection -2-, in instrument panel wiring harness

(A26) - Wire connection (driver's door contact switch), in instrument panel wiring harness

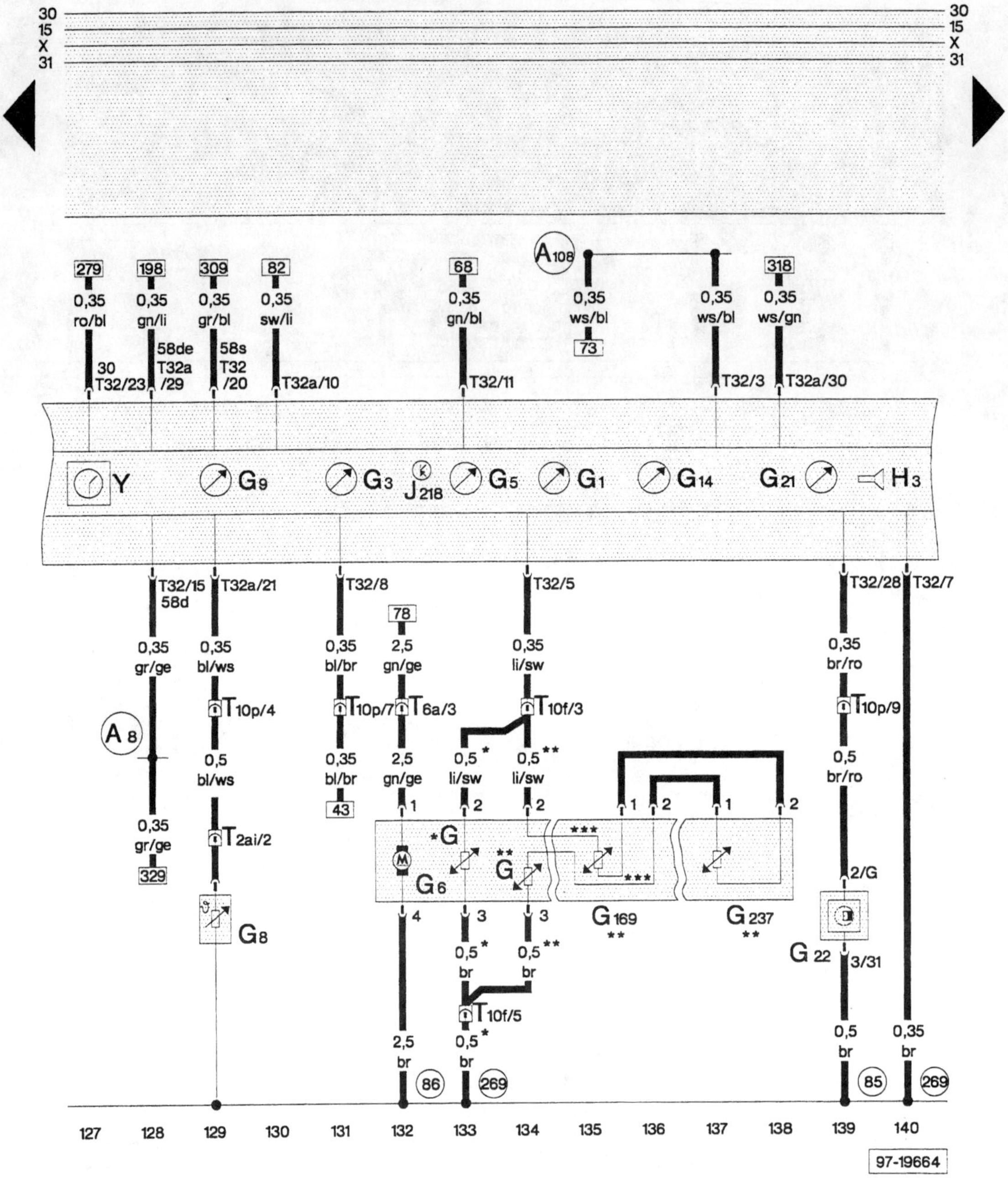

ws = white
sw = black
ro = red
gn = green
br = brown
bl = blue
gr = gray
li = violet
ge = yellow

G - Fuel Level Sensor
G 1 - Fuel Gauge
G 3 - Engine Coolant Temperature (ECT) Gauge
G 5 - Tachometer
G 6 - Fuel Pump (FP)
G 8 - Engine Oil Temperature Sensor
G 9 - Engine Oil Temperature Gauge
G 14 - Voltmeter
G 21 - Speedometer
G 22 - Speedometer Vehicle Speed Sensor (VSS)
G 169 - Fuel Level Sensor 2
G 237 - Fuel Level Sensor 3
H 3 - Warning Buzzer
J 218 - Instrument Cluster Combination Processor
T 2 ai - Connector 2 pin, black, oil sensor
T 6 a - Connector 6 pin, blue, connector station A pillar, left
T 10 f - Connector 10 pin, brown, connector station A pillar, left
T 10 p - Connector 10 pin, black, connector station electronic box plenum chamber
T 32 - Connector 32 pin, blue, on instrument cluster
T 32 a - Connector 32 pin, green, on instrument cluster
Y - Analog Clock

(85) - Ground connection -1-, in engine compartment wiring harness
(86) - Ground connection -1-, in rear wiring harness
(269) - Ground connector (sensor ground) -1-, in instrument panel wiring harness
(A8) - Plus connection (58d), in instrument panel wiring harness (Display illumination)
(A108) - Wire connection (vehicle speed signal), in instrument panel wiring harness

* - Vehicles with front-wheel drive
** - Vehicles with all-wheel drive

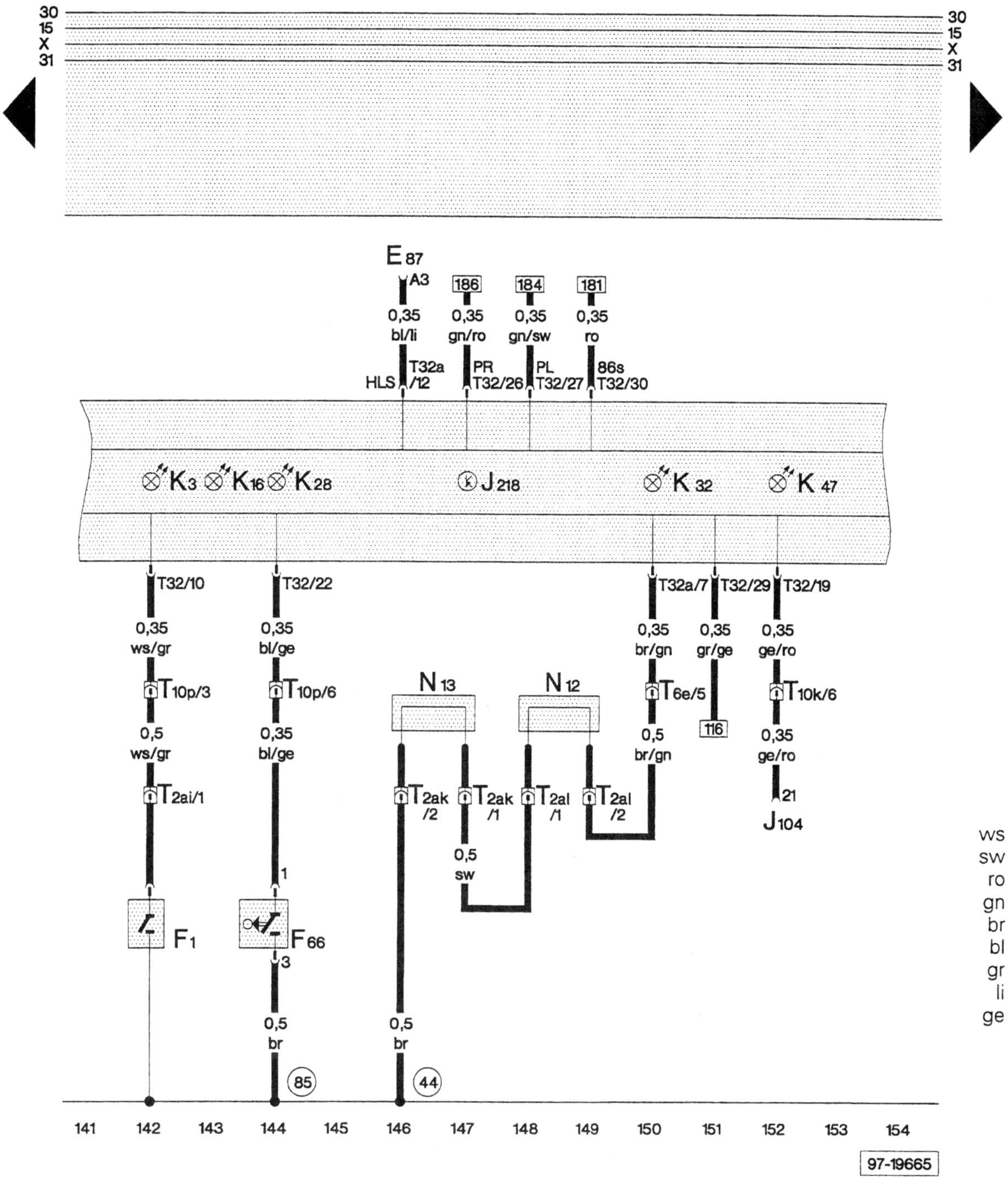

ws = white
sw = black
ro = red
gn = green
br = brown
bl = blue
gr = gray
li = violet
ge = yellow

E 87 - A/C Control Head
F 1 - Oil Pressure Switch
F 66 - Engine Coolant Level (ECL) Warning Switch
J 104 - ABS Control Module (w/EDL)
J 218 - Instrument Cluster Combination Processor
K 3 - Oil Pressure Warning Light
K 16 - Fuel Reserve Warning Light
K 28 - Engine Coolant Level/Temperature (ECL/ECT) Warning Light
K 32 - Brake Pad Wear Indicator Light
K 47 - ABS Warning Light
N 12 - Right Brake Pad Wear Indicator (Shear Element)
N 13 - Left Brake Pad Wear Indicator (Shear Element)
T 2 ai - Connector 2 pin, black, oil sensor
T 2 ak - Connector 2 pin, red, on firewall, left
T 2 al - Connector 2 pin, red, on firewall, right
T 6 e - Connector 6 pin, grey, connector station A pillar, right
T 10 k - Connector 10 pin, orange, connector station A pillar, left
T 10 p - Connector 10 pin, black, connector station electronic box plenum chamber
T 32 - Connector 32 pin, blue, on instrument cluster
T 32 a - Connector 32 pin, green, on instrument cluster

(44) - Ground connection, on left A-pillar, lower part

(85) - Ground connection -1-, in engine compartment wiring harness

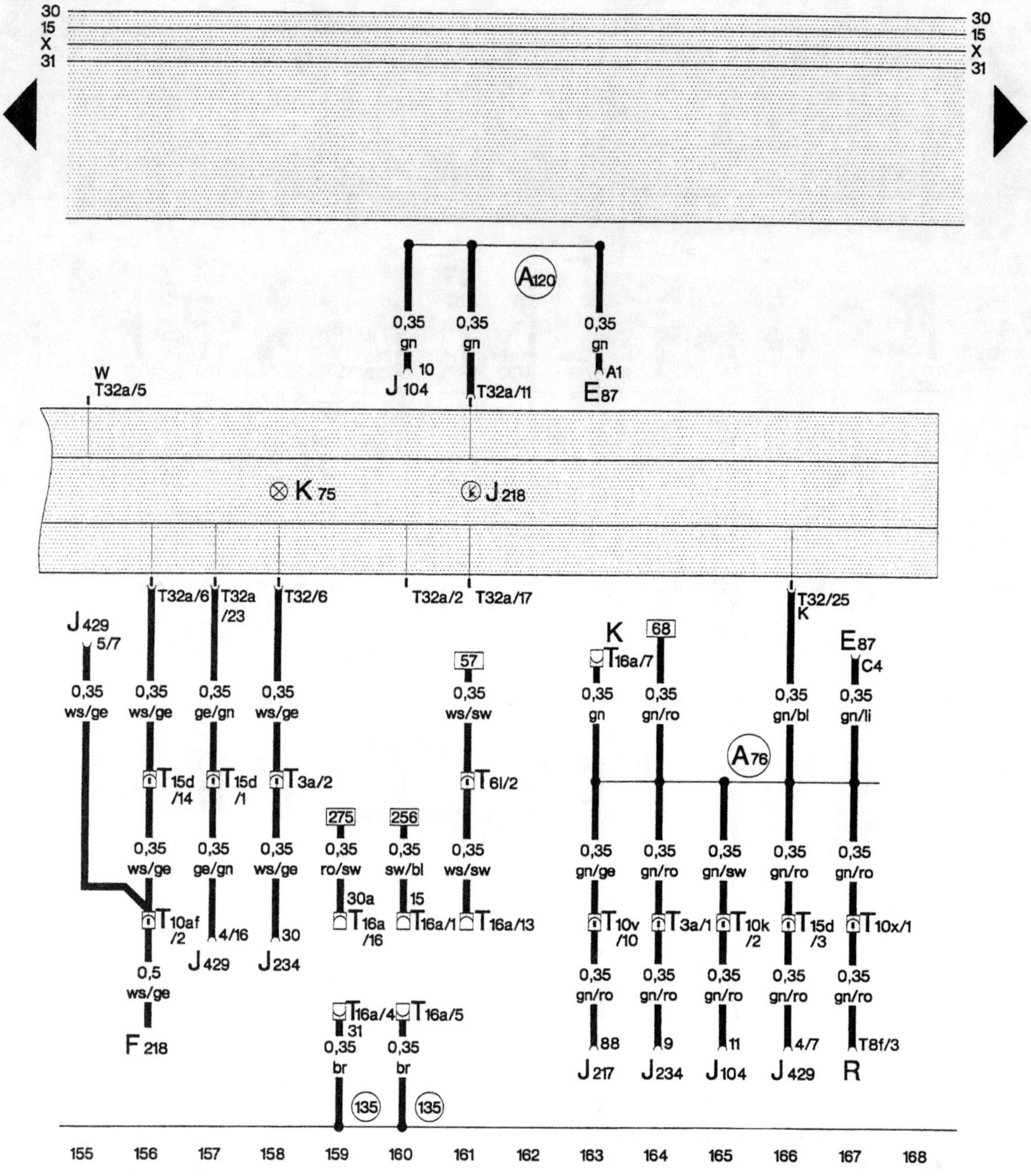

ws = white
sw = black
ro = red
gn = green
br = brown
bl = blue
gr = gray
li = violet
ge = yellow

E 87 - A/C Control Head
F 218 - Switch for central locking, trunk lid
J 104 - ABS Control Module (w/EDL)
J 217 - Transmission Control Module (TCM)
J 218 - Instrument Cluster Combination Processor
J 234 - Airbag Control Module
J 429 - Control module for central locking
K - DLC (Data Link Connector) Rapid Data Transfer
K 75 - Airbag Malfunction Indicator Lamp (MIL)
R - Radio
T 3 a - Connector 3 pin, black, connector station A pillar, right
T 6 l - Connector 6 pin, red, connector station electronic box plenum chamber
T 8 f - Connector 8 pin, black, connector radio III
T 10 k - Connector 10 pin, orange, connector station A pillar, left
T 10 v - Connector 10 pin, brown, connector station A pillar, right
T 10 x - Connector 10 pin, pink, connector station A pillar, right
T 15 d - Connector 15 pin, black, connector station A pillar, right
T 16 a - Connector 16 pin, data link connector
T 32 - Connector 32 pin, blue, on instrument cluster
T 32 a - Connector 32 pin, green, on instrument cluster

(135) - Ground connection -2-, in instrument panel wiring harness
(A76) - Connector (K-diagnosis wire), in instrument panel wiring harness
(A120) - Wire connection (Immobilization time), in instrument panel wiring harness

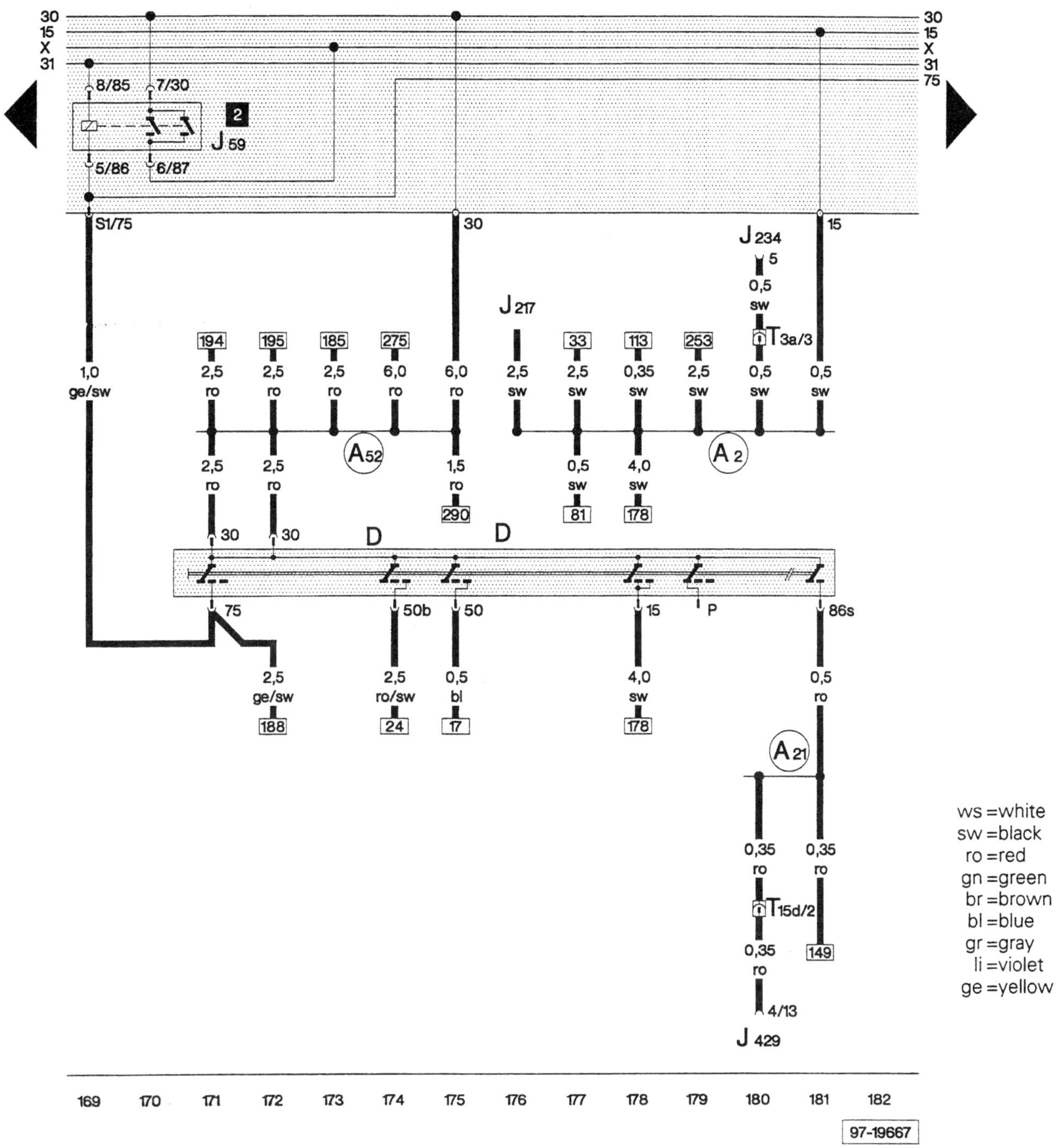

D - Ignition/Starter Switch
J 59 - Load Reduction Relay
J 217 - Transmission Control Module (TCM)
J 234 - Airbag Control Module
J 429 - Control module for central locking
T 3 a - Connector 3 pin, black, connector station A pillar, right
T 15 d - Connector 15 pin, black, connector station A pillar, right

(A2) - Plus connection (15), in instrument panel wiring harness
(A21) - Wire connection (86s), in instrument panel wiring harness
(A52) - Plus connection (30), in instrument panel wiring harness

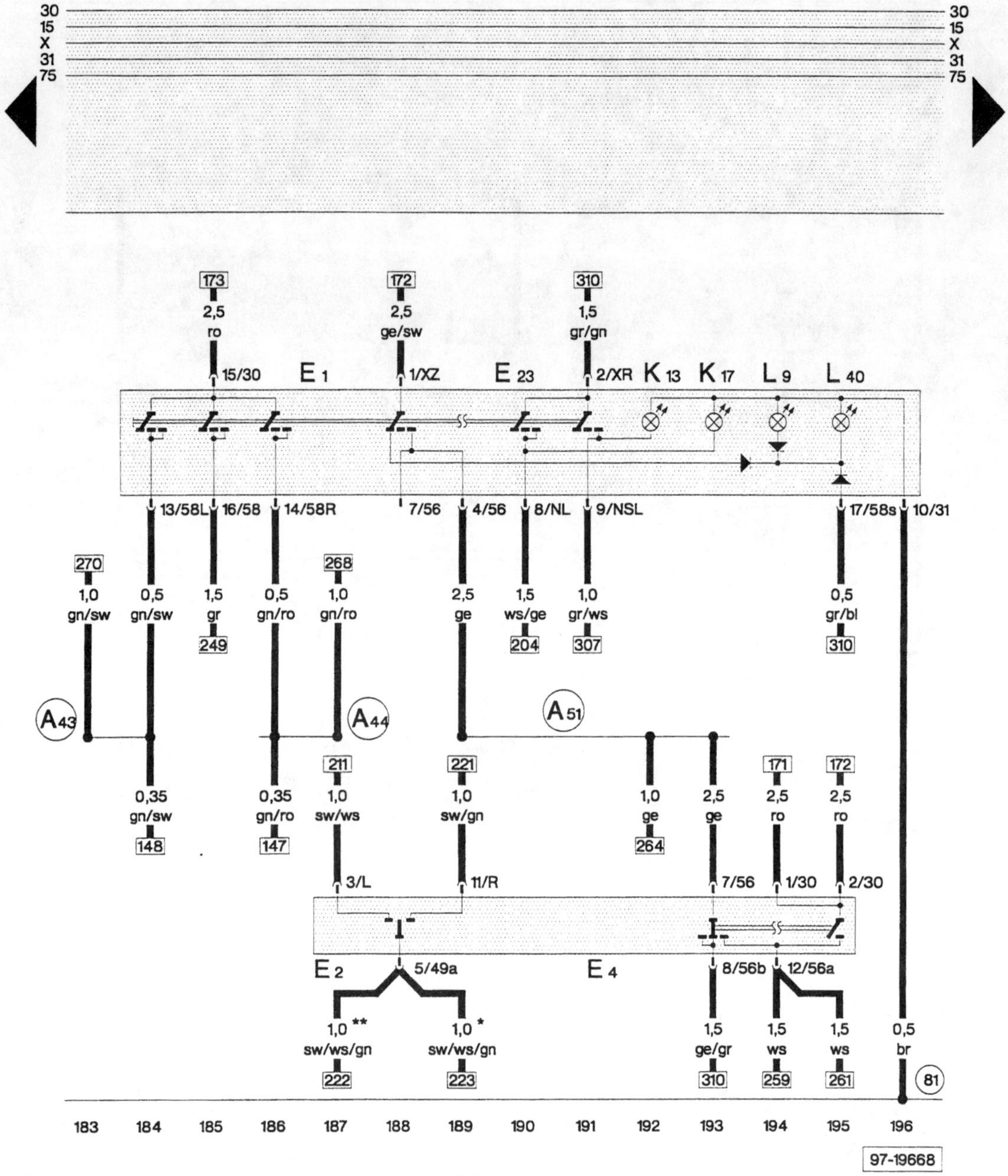

ws = white
sw = black
ro = red
gn = green
br = brown
bl = blue
gr = gray
li = violet
ge = yellow

E 1 - Light Switch
E 2 - Turn Signal Switch
E 4 - Headlight Dimmer/Flasher Switch
E 23 - Fog Light Switch
K 13 - Rear Fog Light Indicator Light
K 17 - Fog Light Indicator Light
L 9 - Headlight Switch Light
L 40 - Front And Rear Fog Light Switch Light

(81) - Ground connection -1-, in instrument panel wiring harness
(A43) - Wire connection (57L), in instrument panel wiring harness
(A44) - Wire connection (57R), in instrument panel wiring harness
(A51) - Wire connection (56), in instrument panel wiring harness

* - Chassis number upto 5000
** - Chassis number from 5001

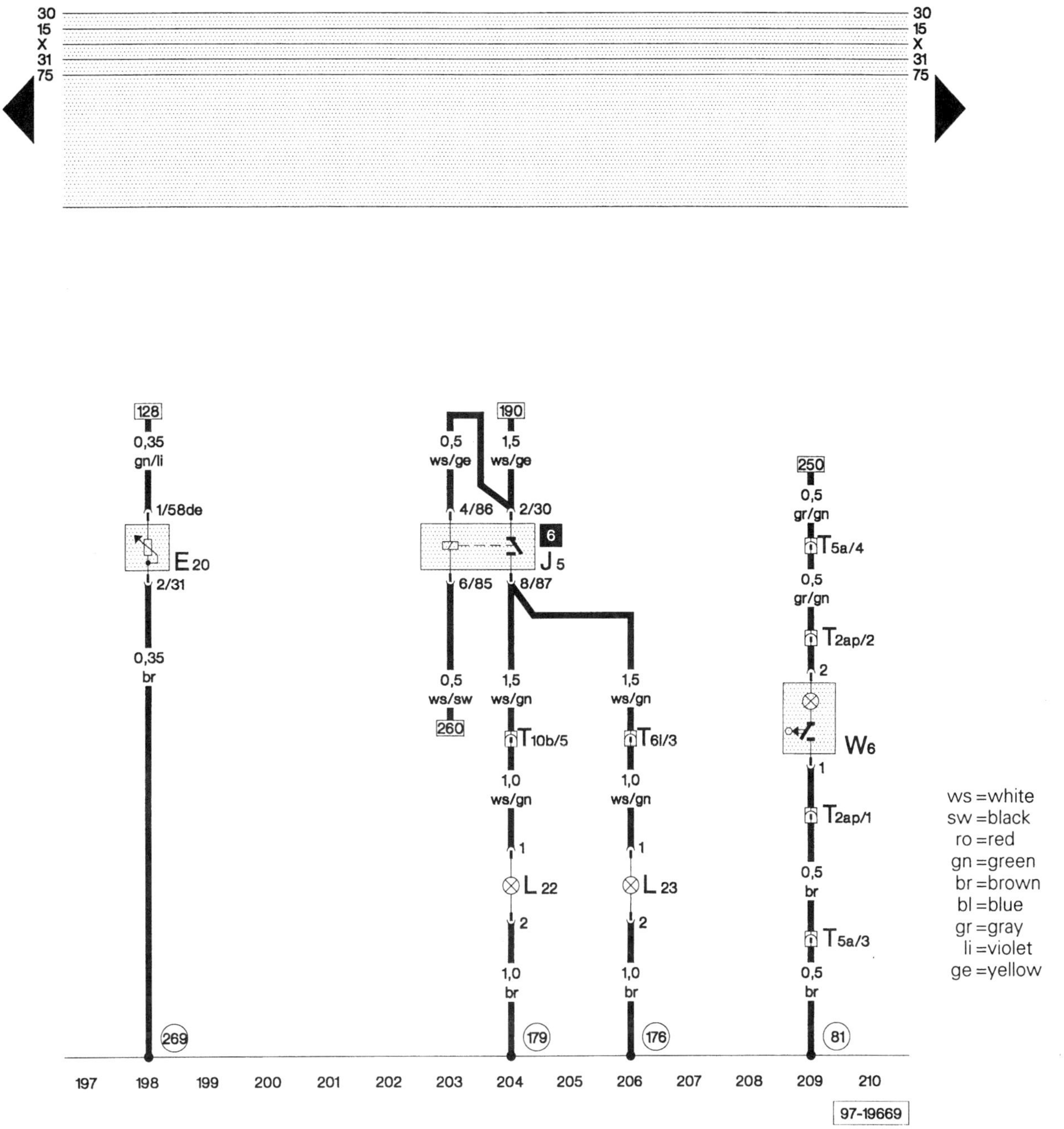

E 20 - Instrument Panel Light Dimmer Switch
J 5 - Fog Light Relay
L 22 - Left Front Fog Light
L 23 - Right Front Fog Light
T 2ap - Connector 2 pin, black, behind instrument panel, right
T 5 a - Connector 5 pin, black, behind instrument panel, right
T 6 i - Connector 6 pin, red, connector station A pillar, right
T 10 b - Connector 10 pin, red, connector station A pillar, left
W6 - Glove Compartment Light

(81) - Ground connection -1-, in instrument panel wiring harness
(176) - Ground connection, in right headlight wiring harness
(179) - Ground connection, in left headlight wiring harness
(269) - Ground connector (sensor ground) -1-, in instrument panel wiring harness

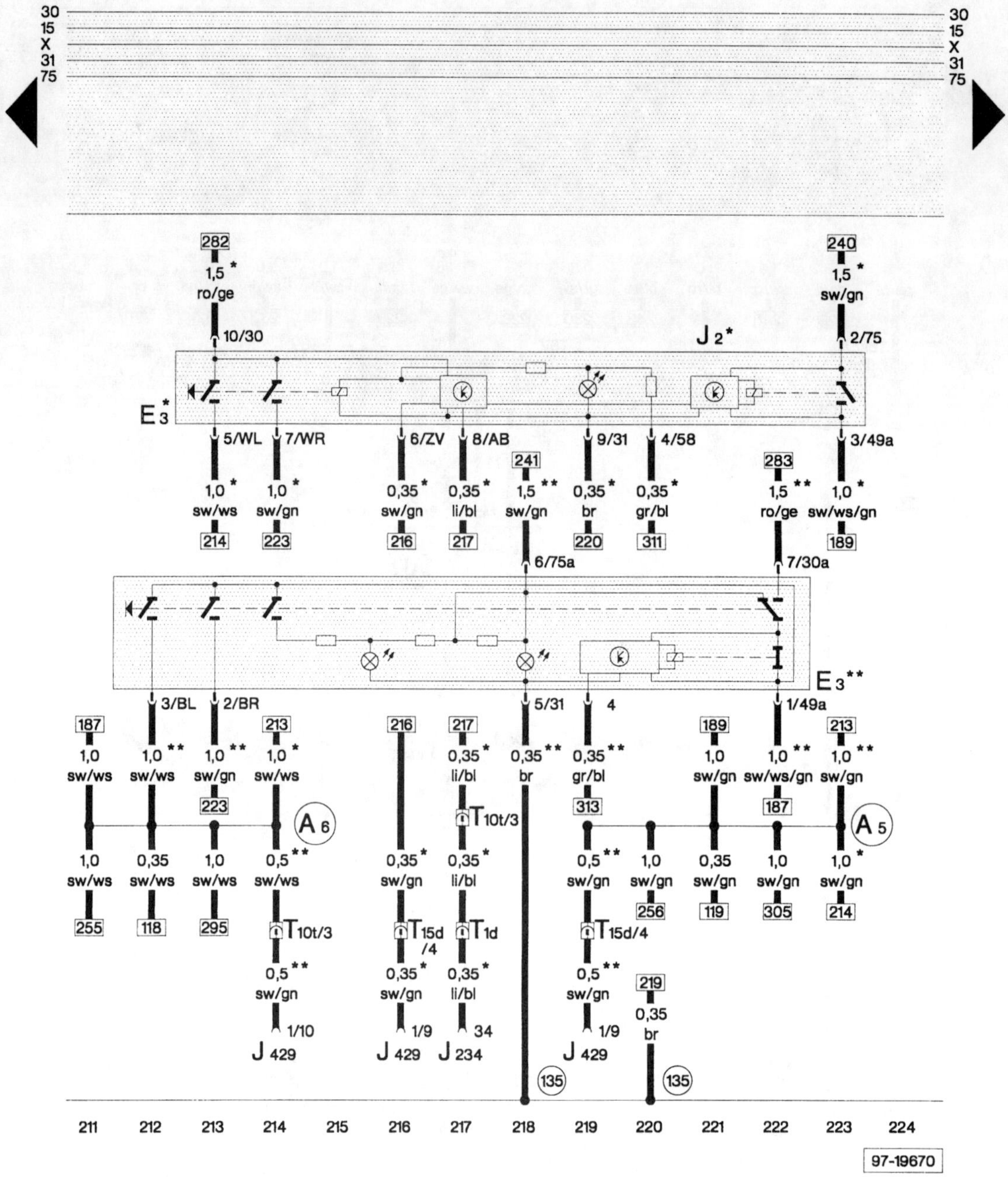

ws = white
sw = black
ro = red
gn = green
br = brown
bl = blue
gr = gray
li = violet
ge = yellow

E 3 - Emergency Flasher Switch
J 2 - Emergency Flasher Relay
J 234 - Airbag Control Module
J 429 - Control module for central locking
T 1 d - Single connector, brown, connector station A pillar right
T 10 t - Connector 10 pin, orange, connector station A pillar, right
T 15 d - Connector 15 pin, black, connector station A pillar, right

(135) - Ground connection -2-, in instrument panel wiring harness
(A5) - Plus connection (right turn signal), in instrument panel wiring harness
(A6) - Plus connection (left turn signal), in instrument panel wiring harness

* - Chassis number upto 5000
** - Chassis number from 5001

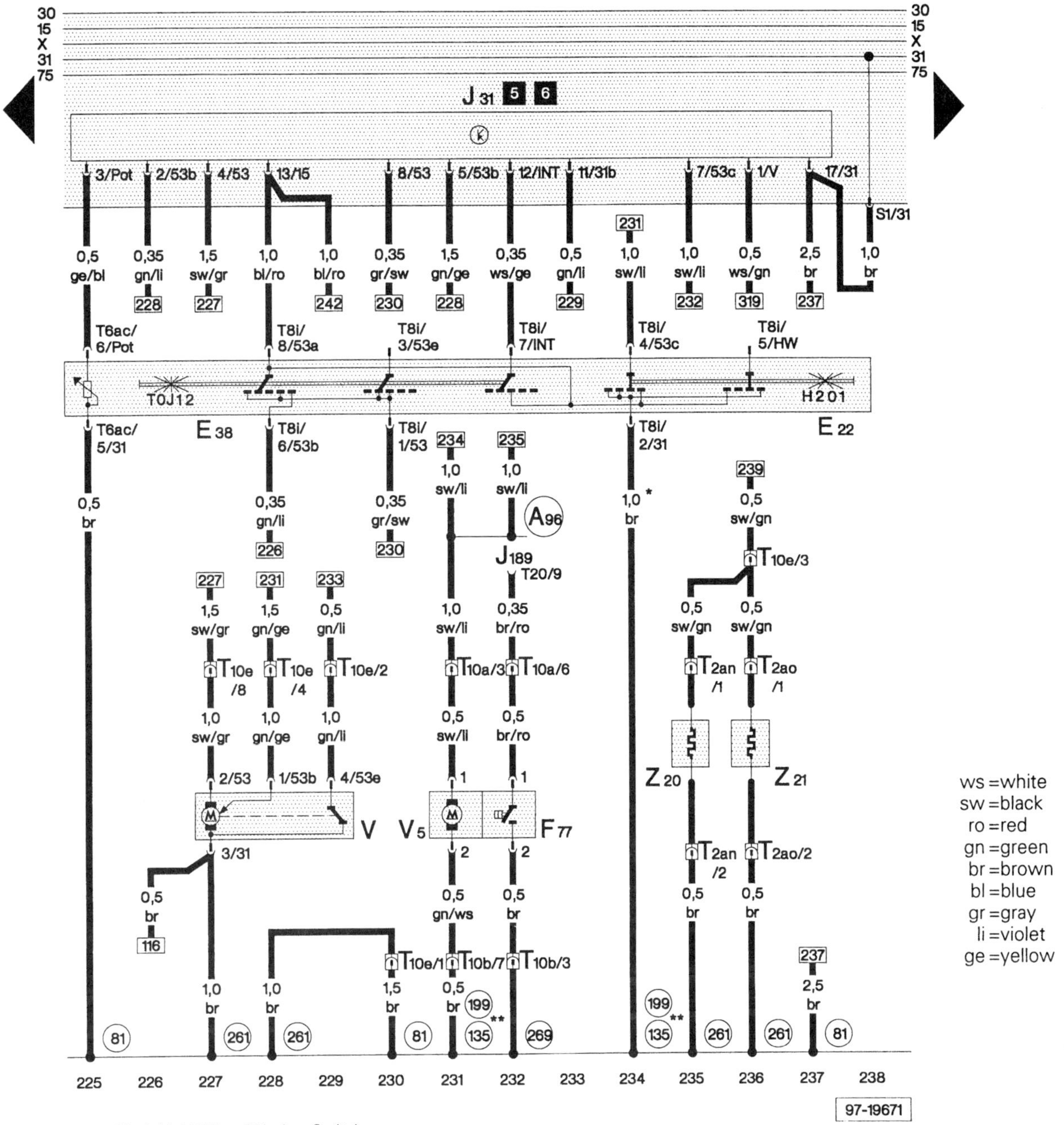

E 22 - Windshield Wiper/Washer Switch
E 38 - Windshield Wiper Intermittent Regulator
F 77 - Windshield Washer Fluid Level Warning Switch
J 31 - Wiper/Washer Intermittent Relay
J 189 - Auto Check System
T 2 an - Connector 2 pin, washer jet, left
T 2 ao - Connector 2 pin, washer jet, right
T 6 ac - Connector 6 pin, brown, on windshield wiper intermittent regulator
T 8 i - Connector 8 pin, on windshield wiper/washer switch
T 10 a - Connector 10 pin, pink, connector station A pillar, left
T 10 b - Connector 10 pin, red, connector station A pillar, left
T 10 e - Connector 10 pin, violet, connector station A pillar, left
T 20 - Connector 20 pin, red/white, on instrument cluster
V - Windshield Wiper Motor
V 5 - Windshield Washer Pump
Z 20 - Left Washer Nozzle Heater
Z 21 - Right Washer Nozzle Heater

(81) - Ground connection -1-, in instrument panel wiring harness
(135) - Ground connection -2-, in instrument panel wiring harness
(199) - Ground connection -3-, in instrument panel wiring harness
(261) - Ground connection, in wiring harness heated spray jet
(269) - Ground connector (sensor ground) -1-, in instrument panel wiring harness
(A96) - Connector (53a), in instrument panel wiring harness

* - only wagon
** - booth is possible

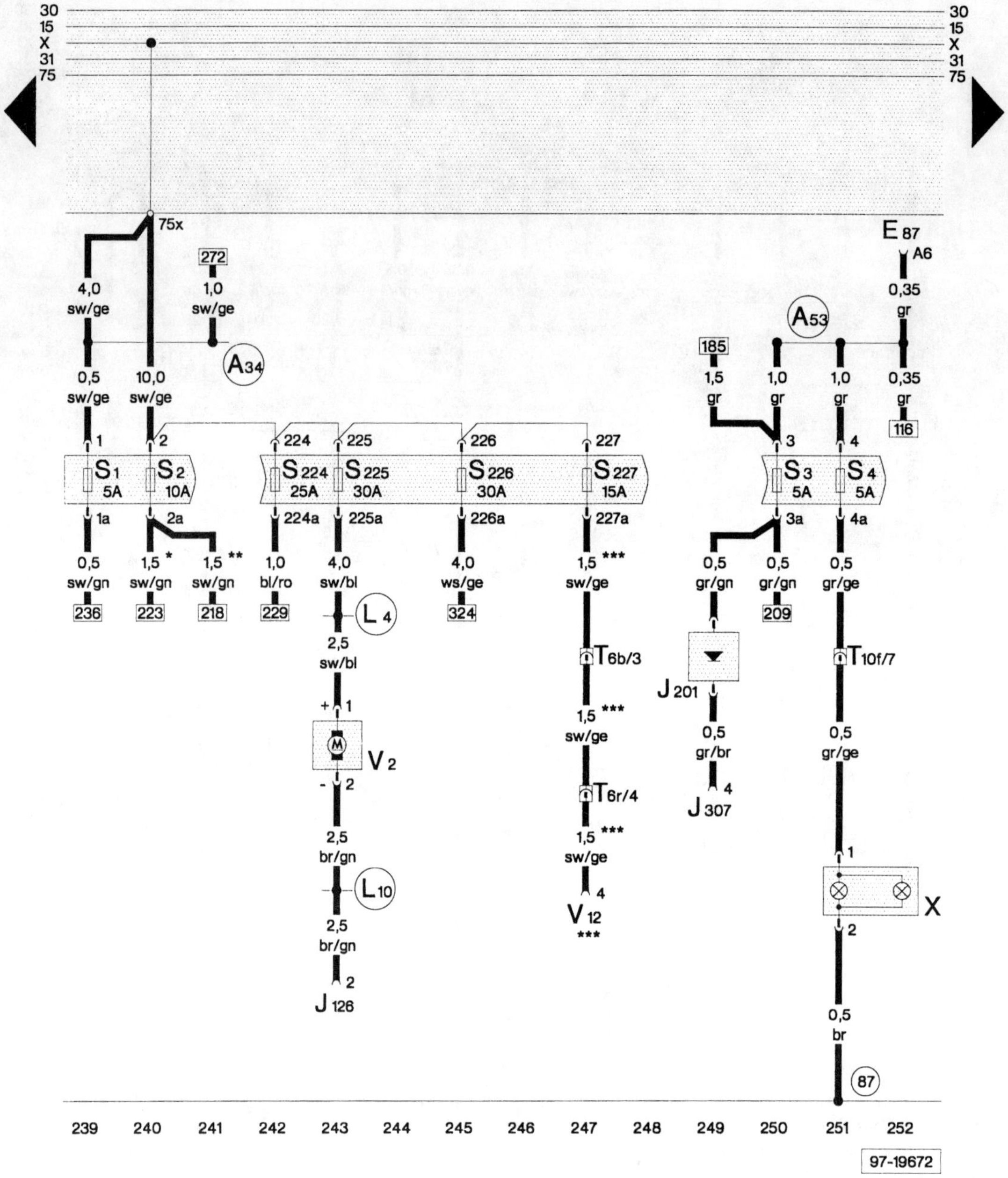

ws = white
sw = black
ro = red
gn = green
br = brown
bl = blue
gr = gray
li = violet
ge = yellow

E 87 - A/C Control Head
J 126 - Control module for fresh air blower
J 201 - Protection Diode
J 307 - Selector Lever Light Relay
S 1 - Fuse
S 2 - Fuse
S 3 - Fuse
S 4 - Fuse
S 224 - Fuse
S 225 - Fuse
S 226 - Fuse
S 227 - Fuse
T 6 b - Connector 6 pin, white, connector station A pillar, left
T 6 r - Connector 6 pin, brown, in luggage compartment
T 10 f - Connector 10 pin, brown, connector station A pillar, left
V 2 - Fresh Air Blower
V 12 - Rear Window Wiper Motor
X - License Plate Light

(87) - Ground connection -2-, in rear wiring harness
(A34) - Wire connection (75x), in instrument panel wiring harness
(A53) - Plus connection (58b), in instrument panel wiring harness
(L4) - Wire connection (75al), in A/C wiring harness
(L10) - Wire connection -2-, in A/C wiring harness

* - Chassis number upto 5000
** - Chassis number from 5001
*** - only wagon

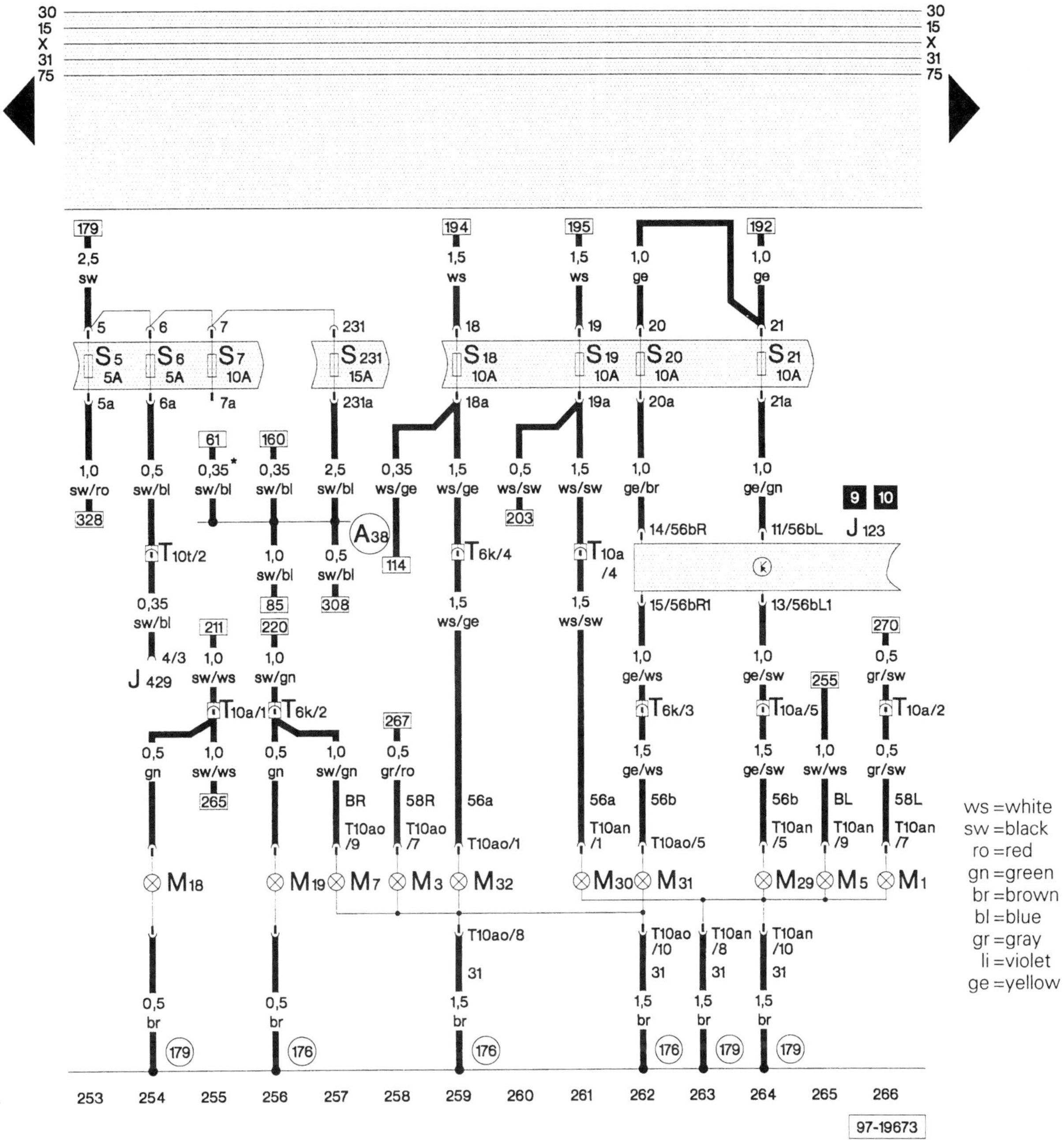

J 123 - Lamp Control Module
J 429 - Control module for central locking
M 1 - Left Parking Light
M 3 - Right Parking Light
M 5 - Left Front Turn Signal Light
M 7 - Right Front Turn Signal Light
M 18 - Left, Side Turn Signal Light
M 19 - Right, Side Turn Signal Light
M 29 - Left Low Beam Headlight
M 30 - Left High Beam Headlight
M 31 - Right Low Beam Headlight
M 32 - Right High Beam Headlight
S 5 - Fuse
S 6 - Fuse
S 7 - Fuse
S 18 - Fuse
S 19 - Fuse
S 20 - Fuse
S 21 - Fuse
S 231 - Fuse
T 6 k - Connector 6 pin, pink, connector station A pillar, right
T 10 a - Connector 10 pin, pink, connector station A pillar, left
T 10 t - Connector 10 pin, orange, connector station A pillar, right
T 10 an - Connector 10 pin, on headlight, left
T 10 ao - Connector 10 pin, on headlight, right

(176) - Ground connection, in right headlight wiring harness

(179) - Ground connection, in left headlight wiring harness

(A38) - Plus connection (15a II), in instrument panel wiring harness

* - Vehicles with front wheel drive

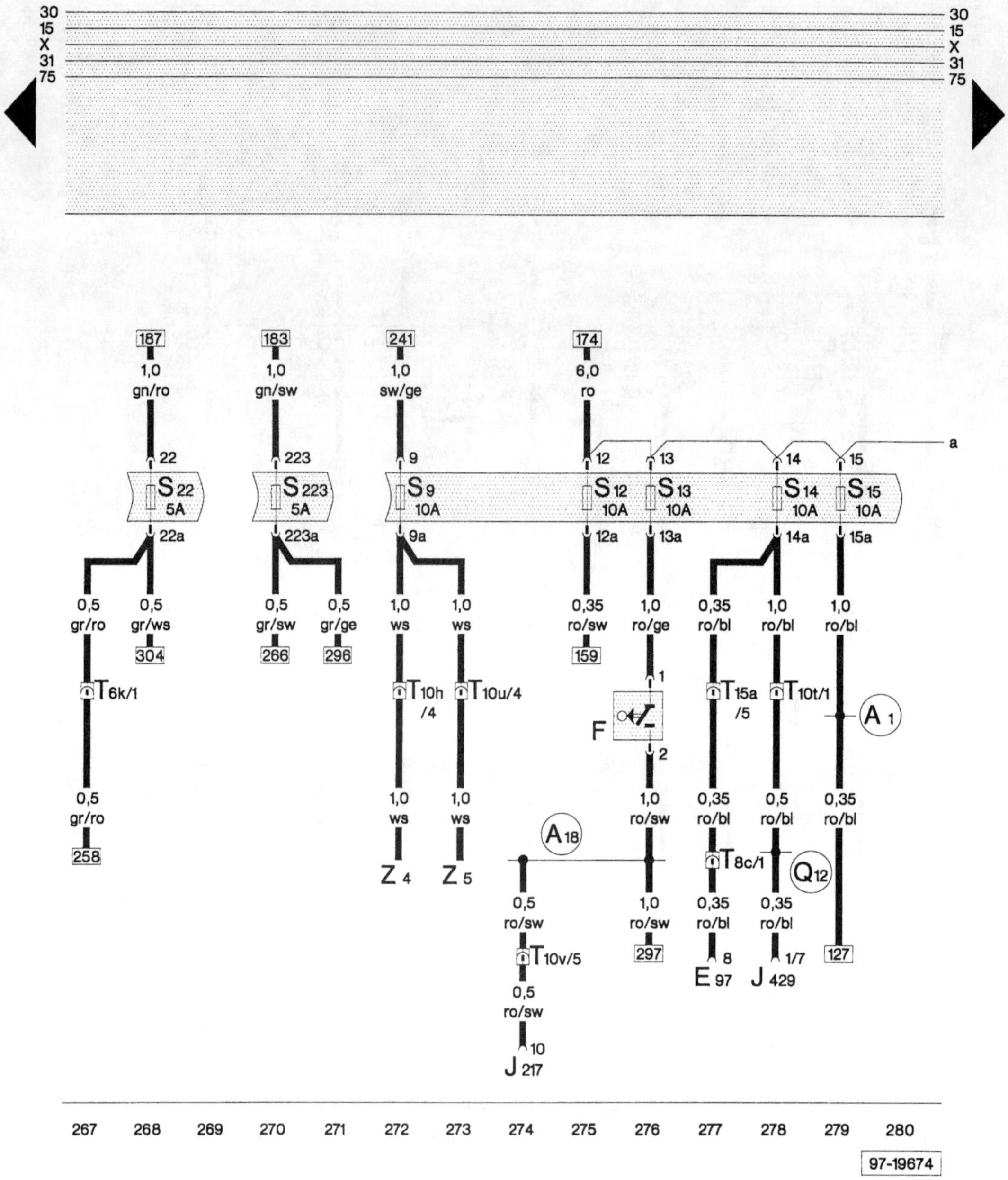

ws = white
sw = black
ro = red
gn = green
br = brown
bl = blue
gr = gray
li = violet
ge = yellow

E 97 - Memory Program Switch
F - Brake Light Switch
J 217 - Transmission Control Module (TCM)
J 429 - Control module for central locking
S 9 - Fuse
S 12 - Fuse
S 13 - Fuse
S 14 - Fuse
S 15 - Fuse
S 22 - Fuse
S 223 - Fuse
T 6 k - Connector 6 pin, pink, connector station A pillar, right
T 8 c - Connector 8 pin, black, in driver's door
T 10 h - Connector 10 pin, black, connector station A pillar, left
T 10 t - Connector 10 pin, orange, connector station A pillar, right
T 10 u - Connector 10 pin, black, connector station A pillar, right
T 10 v - Connector 10 pin, brown, connector station A pillar, right
T 15 a - Connector 15 pin, black, connector station A pillar, left

Z 4 - Heated outside mirror, driver side
Z 5 - Heated outside mirror, passenger side

(A1) - Plus connection (30a), in instrument panel wiring harness

(A18) - Wire connection (54), in instrument panel wiring harness

(Q12) - Plus connection (30a), in power window/central locking system and door contact switch wiring harness

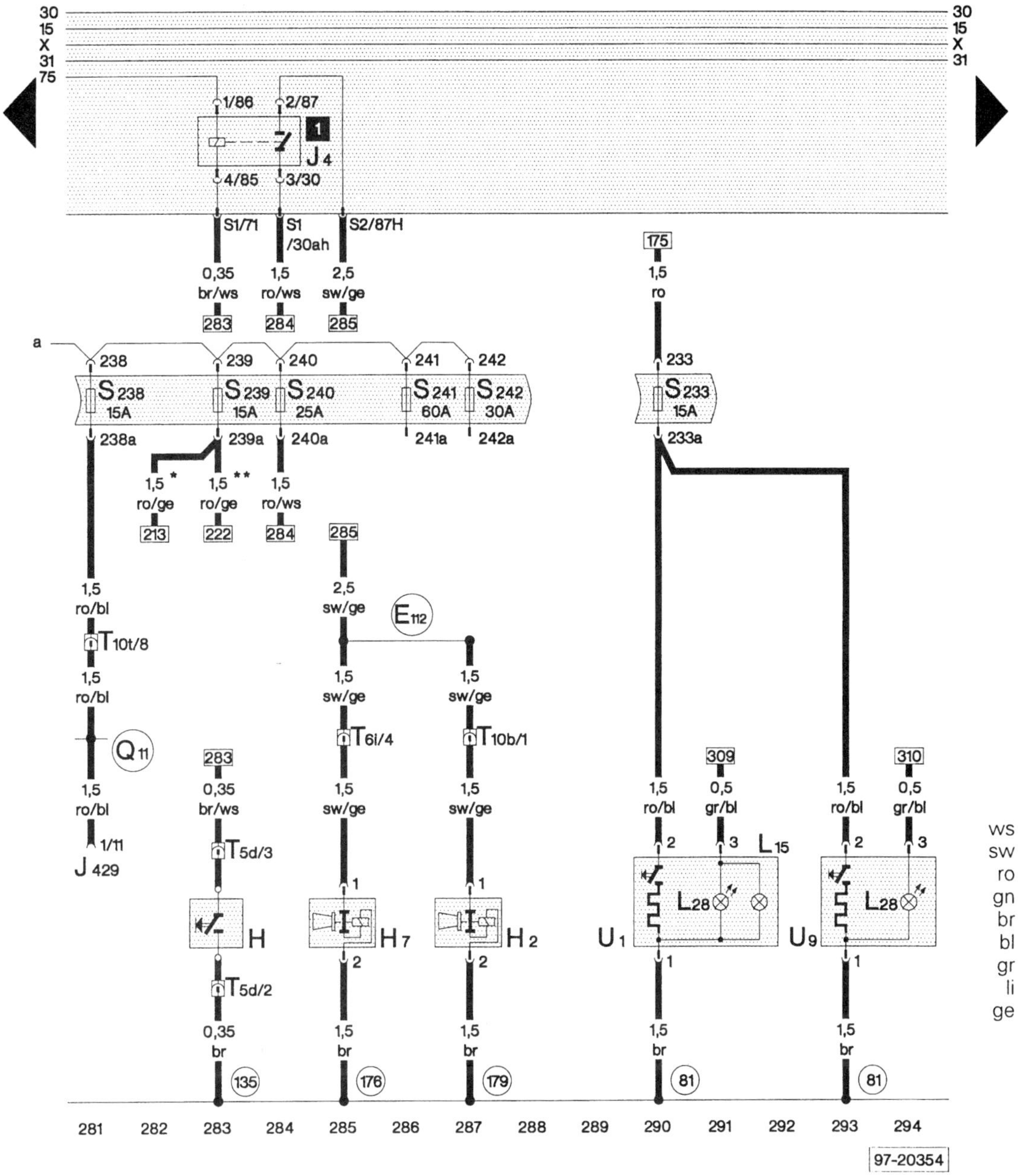

ws = white
sw = black
ro = red
gn = green
br = brown
bl = blue
gr = gray
li = violet
ge = yellow

H - Horn Button
H 2 - High Tone Horn
H 7 - Low Tone Horn
J 4 - Dual Horn Relay
J 429 - Control module for central locking
L 15 - Ashtray Light
L 28 - Cigarette Lighter Light
S 233 - Fuse
S 238 - Fuse
S 239 - Fuse
S 240 - Fuse
S 241 - Fuse
S 242 - Fuse
T 5 d - Connector 5 pin, behind steering column switch cover
T 6 i - Connector 6 pin, red, connector station A pillar, right
T 10 b - Connector 10 pin, red, connector station A pillar, left
T 10 t - Connector 10 pin, orange, connector station A pillar, right
U 1 - Cigarette Lighter
U 9 - Rear Cigarette Lighter

(81) - Ground connection -1-, in instrument panel wiring harness
(135) - Ground connection -2-, in instrument panel wiring harness
(176) - Ground connection, in right headlight wiring harness
(179) - Ground connection, in left headlight wiring harness
(E112) - Connector (Horn - 87h), in instrument panel wiring harness
(Q11) - Plus connection (30a), in power window/central locking system and door contact switch wiring harness

* - Chassis number upto 5000
** - Chassis number from 5001

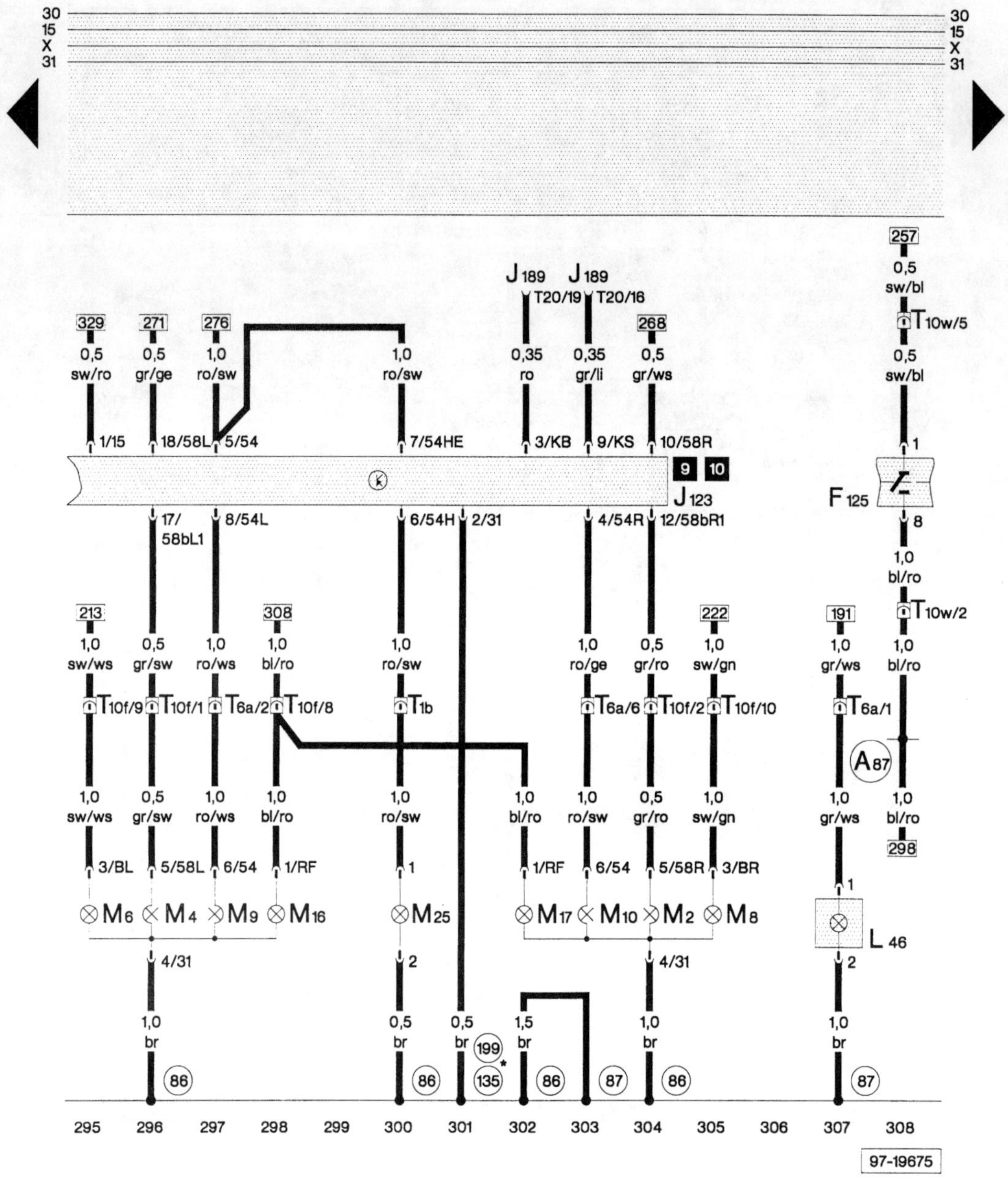

ws =white
sw =black
ro =red
gn =green
br =brown
bl =blue
gr =gray
li =violet
ge =yellow

F 125 - Multi-Function Transmission Range (TR) Switch
J 123 - Lamp Control Module
J 189 - Auto Check System
L 46 - Left Rear Fog Light
M 2 - Right Tail Light
M 4 - Left Tail Light
M 6 - Left Rear Turn Signal Light
M 8 - Right Rear Turn Signal Light
M 9 - Left Brake Light
M 10 - Right Brake Light
M 16 - Left Back-Up Light
M 17 - Right Back-Up Light
M 25 - High-mount Brake Light
T 1 b - Single connector, black, connector station A pillar right
T 6 a - Connector 6 pin, blue, connector station A pillar, left
T 10 f - Connector 10 pin, brown, connector station A pillar, left
T 10 w - Connector 10 pin, blue, connector station A pillar, right
T 20 - Connector 20 pin, red/white, on instrument cluster

(86) - Ground connection -1-, in rear wiring harness
(87) - Ground connection -2-, in rear wiring harness
(135) - Ground connection -2-, in instrument panel wiring harness
(199) - Ground connection -3-, in instrument panel wiring harness
(A87) - Connector (RF), in instrument panel wiring harness

* - booth is possible

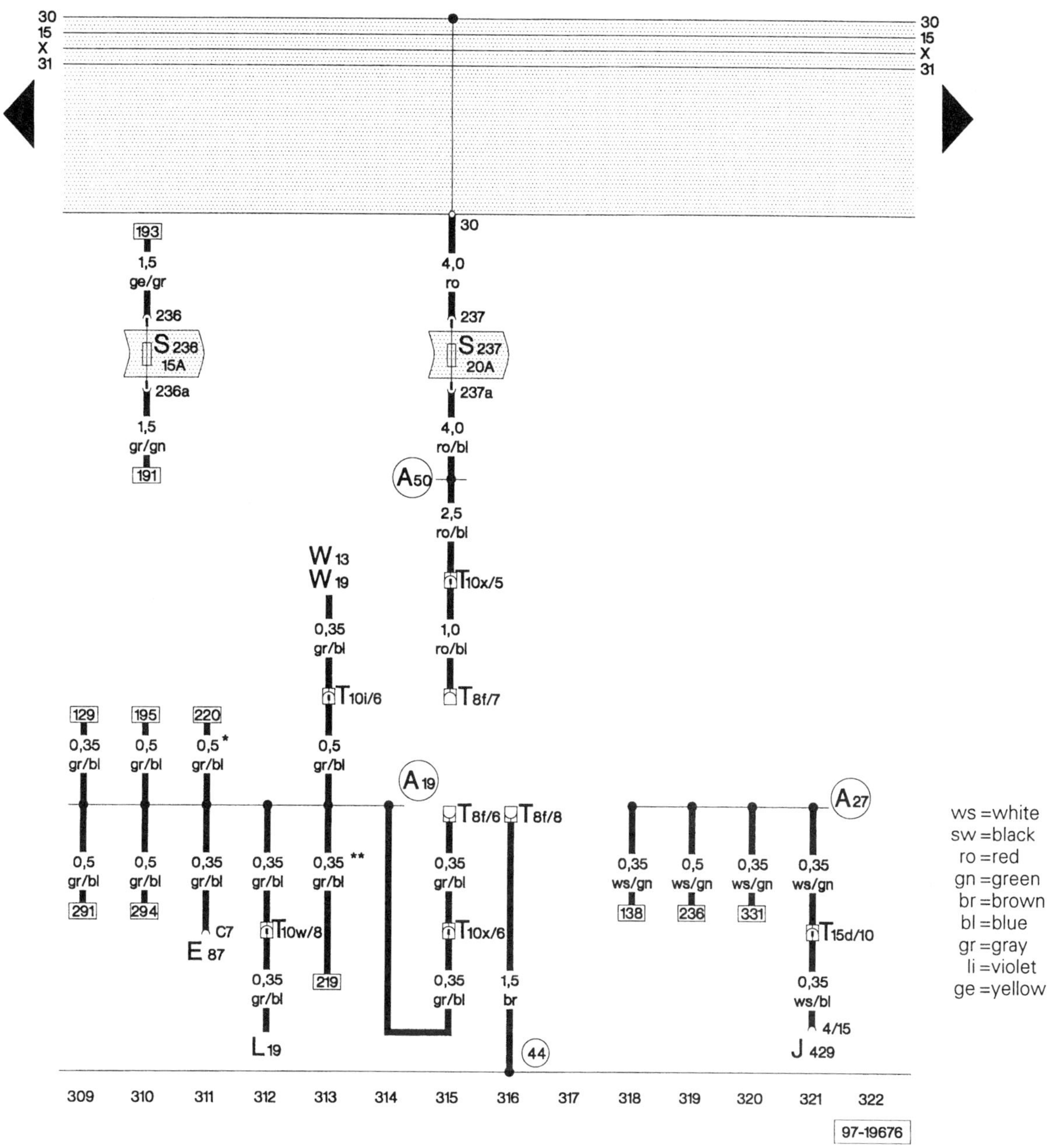

E 87 - A/C Control Head
J 429 - Control module for central locking
L 19 - Automatic Transmission Console Light
S 236 - Fuse
S 237 - Fuse
T 8 f - Connector 8 pin, black, connector radio III
T 10 i - Connector 10 pin, darkbrown, connector station A pillar, left
T 10 w - Connector 10 pin, blue, connector station A pillar, right
T 10 x - Connector 10 pin, pink, connector station A pillar, right
T 15 d - Connector 15 pin, black, connector station A pillar, right
W 13 - Right Front Map/Reading Light
W 19 - Left Front Reading Light

(44) - Ground connection, on left A-pillar, lower part

(A19) - Wire connection (58s), in instrument panel wiring harness (Switch illumination)

(A27) - Wire connection (vehicle speed signal), in instrument panel wiring harness

(A50) - Plus connection (30as), in instrument panel wiring harness

* - Chassis number upto 5000

** - Chassis number from 5001

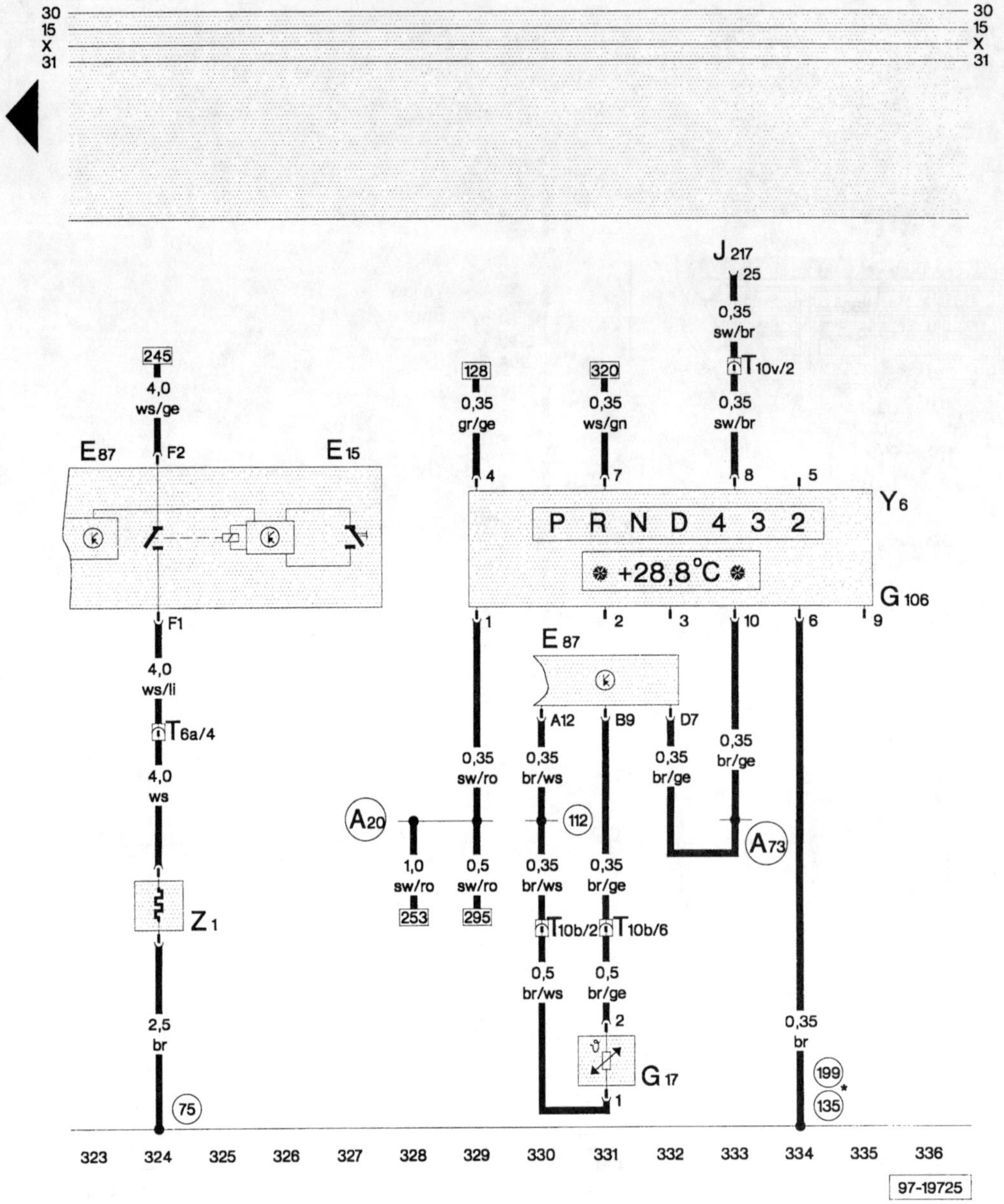

ws = white
sw = black
ro = red
gn = green
br = brown
bl = blue
gr = gray
li = violet
ge = yellow

E 15 - Rear Window Defogger Switch
E 87 - A/C Control Head
G 17 - Outside Air Temperature Sensor
G 106 - Outside Air Temperature Display
J 217 - Transmission Control Module (TCM)
T 6 a - Connector 6 pin, blue, connector station A pillar, left
T 10 b - Connector 10 pin, red, connector station A pillar, left
T 10 v - Connector 10 pin, brown, connector station A pillar, right
Y 6 - Transmission Range (TR) Selector Lever Display
Z 1 - Heated rear window

(75) - Ground connection, on right rear pillar
(112) - Ground connection -2-, in A/C wiring harness
(135) - Ground connection -2-, in instrument panel wiring harness
(199) - Ground connection -3-, in instrument panel wiring harness
(A20) - Wire connection (15a), in instrument panel wiring harness
(A73) - Connector (outside temperature indicator), in instrument panel wiring harness

* - booth is possible

Automatic Transmission (01V)

1998 m. y.

Fuse Panel

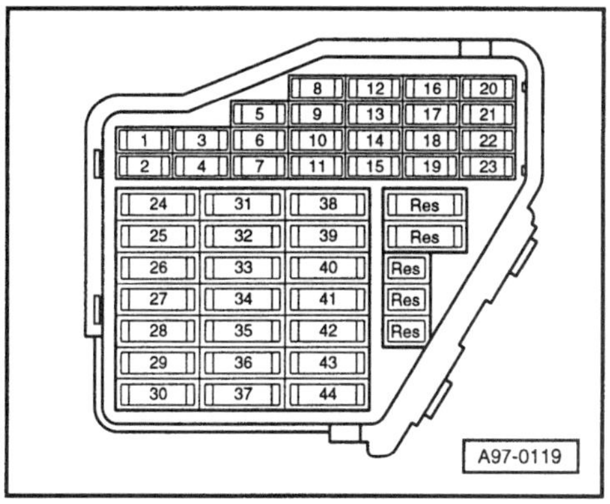

Fuse Colors:

30 A - Green
25 A - White
20 A - Yellow
15 A - Blue
10 A - Red
5 A - Beige

Starting with fuse position 23, fuses in the fuse holder are identified with 223 in the wiring diagram.

13 - Fold Relay Panel

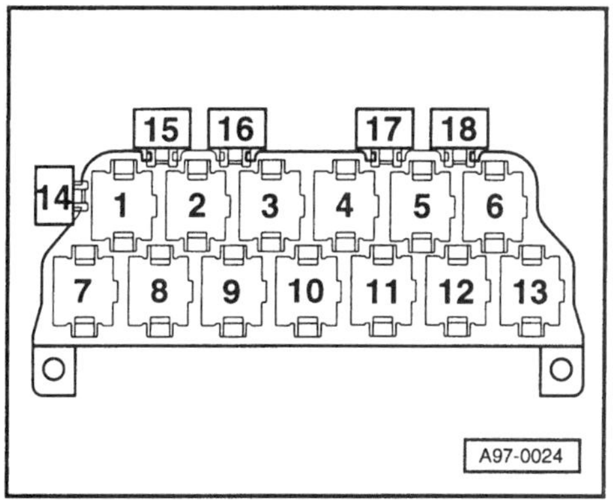

Relay Location:

1 - Selector Lever Light Relay, J307

4 - Starting Interlock Relay, J207

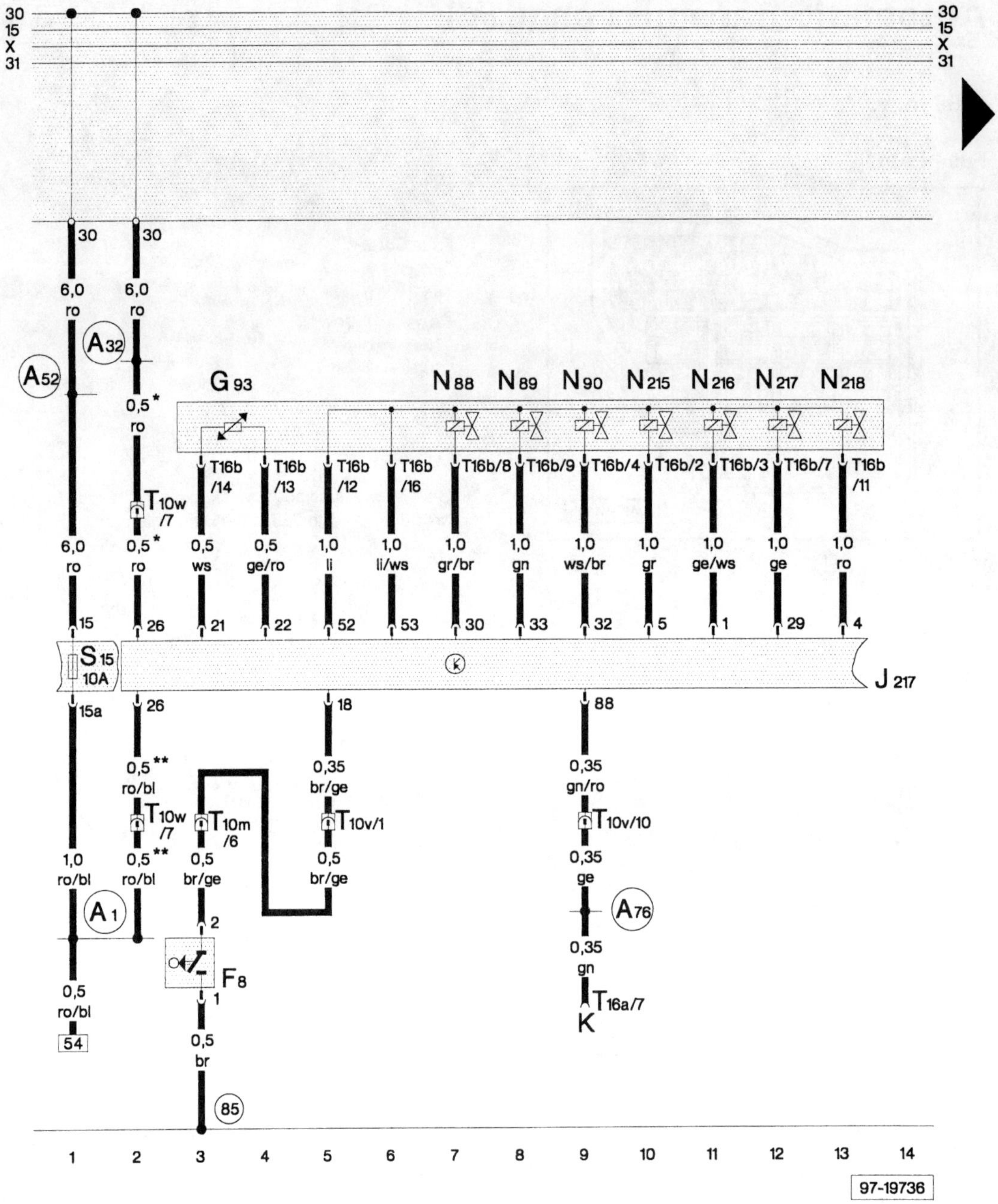

ws = white
sw = black
ro = red
gn = green
br = brown
bl = blue
gr = gray
li = violet
ge = yellow

F 8 - Kick Down Switch
G 93 - Transmission Fluid Temperature Sensor
J 217 - Transmission Control Module (TCM)
K - DLC (Data Link Connector) Rapid Data Transfer
N 88 - Solenoid Valve 1
N 89 - Solenoid Valve 2
N 90 - Solenoid Valve 3
N 215 - Pressure control valve -1- for auto. transmission
N 216 - Pressure control valve -2- for auto. transmission
N 217 - Pressure control valve -3- for auto. transmission
N 218 - Pressure control valve -4- for auto. transmission
S 15 - Fuse
T 10 m - Connector 10 pin, blue, connector station electronic box plenum chamber
T 10 v - Connector 10 pin, brown, connector station A pillar, right
T 10 w - Connector 10 pin, blue, connector station A pillar, right
T 16 a - Connector 16 pin, data link connector
T 16 b - Connector 16 pin, on transmission

(85) - Ground connection -1-, in engine compartment wiring harness
(A1) - Plus connection (30a), in instrument panel wiring harness
(A32) - Plus connection (30), in instrument panel wiring harness
(A52) - Plus connection (30), in instrument panel wiring harness
(A76) - Connector (K-diagnosis wire), in instrument panel wiring harness

* - for vehicles with power supply, wire connection 30, without fuse
** - for vehicles with power supply, wire connection 30, with fuse

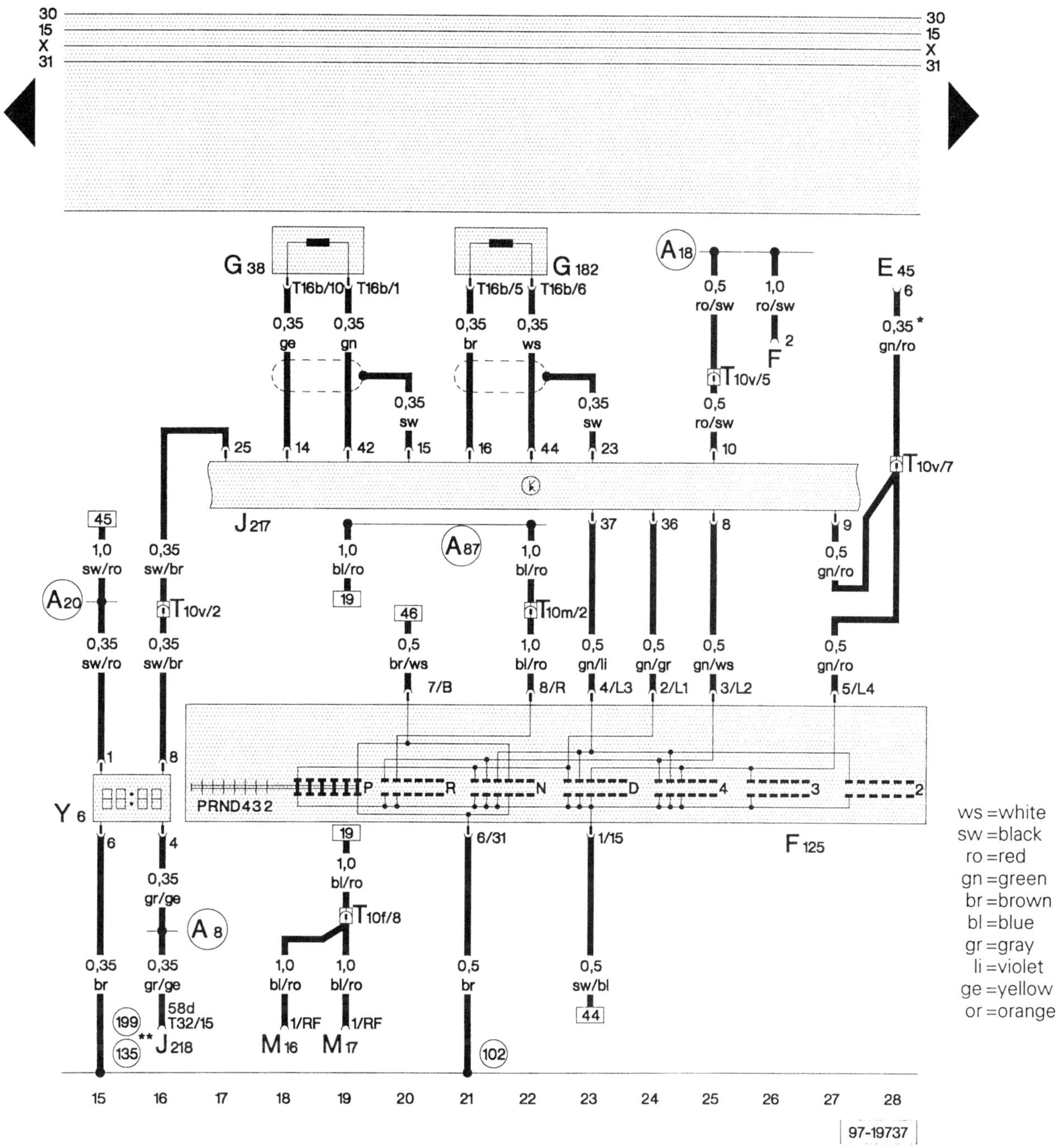

E 45 - Cruise Control Switch
F - Brake Light Switch
F 125 - Multi-Function Transmission Range (TR) Switch
G 38 - Transmission Vehicle Speed Sensor (VSS)
G 182 - Sensor for transmission RPM
J 217 - Transmission Control Module (TCM)
J 218 - Instrument Cluster Combination Processor
M 16 - Left Back-Up Light
M 17 - Right Back-Up Light
T 10 f - Connector 10 pin, brown, connector station A pillar, left
T 10 m - Connector 10 pin, blue, connector station electronic box plenum chamber
T 10 v - Connector 10 pin, brown, connector station A pillar, right
T 10 w - Connector 10 pin, blue, connector station A pillar, right
T 16 b - Connector 16 pin, on transmission
T 32 - Connector 32 pin, blue, on instrument cluster
Y 6 - Transmission Range (TR) Selector Lever Display

(102) - Ground connection, in transmission wiring harness
(135) - Ground connection -2-, in instrument panel wiring harness
(199) - Ground connection -3-, in instrument panel wiring harness
(A8) - Plus connection (58d), in instrument panel wiring harness (Display illumination)
(A18) - Wire connection (54), in instrument panel wiring harness
(A20) - Wire connection (15a), in instrument panel wiring harness
(A87) - Connector (RF), in instrument panel wiring harness
* - Vehicles with cruise control
** - booth is possible

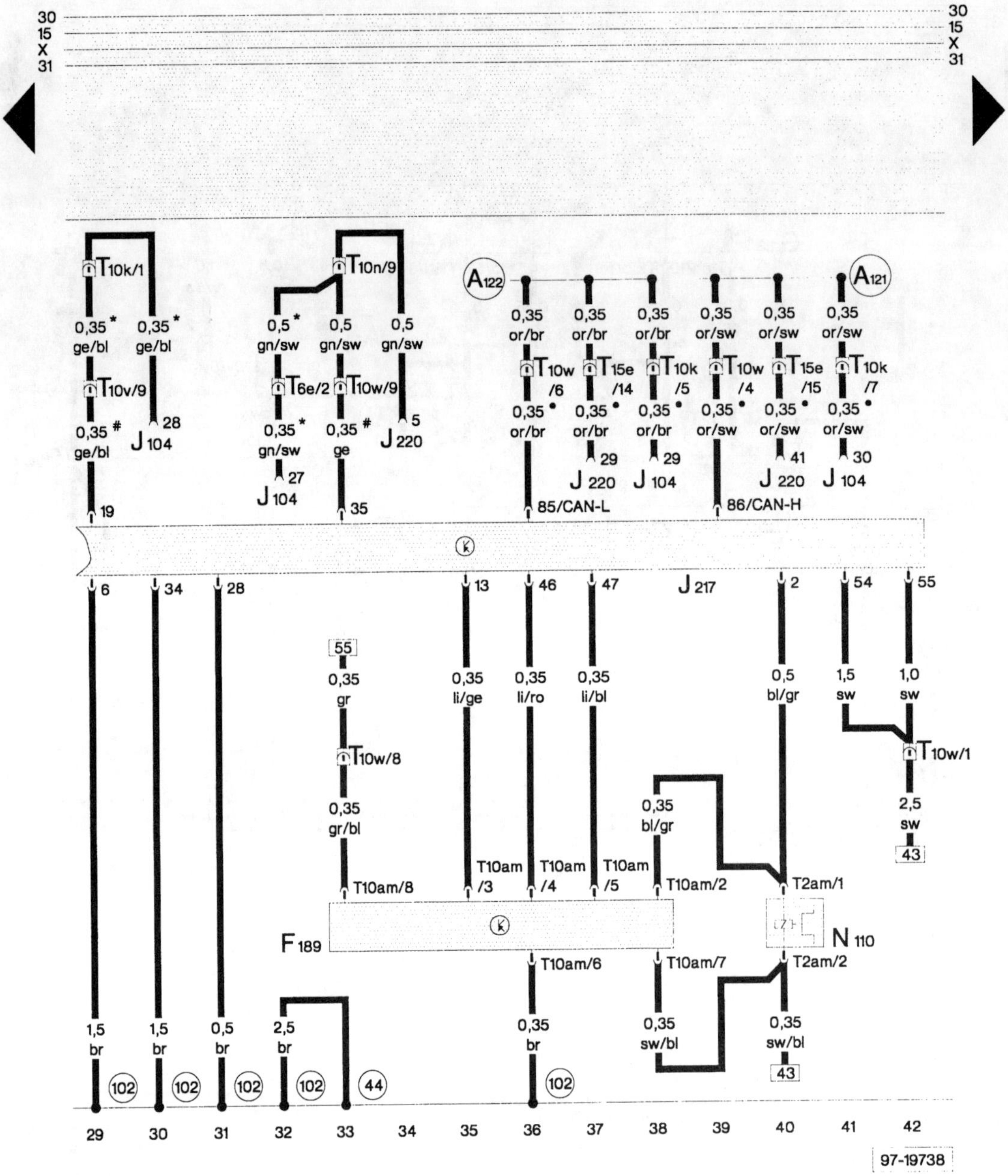

ws =white
sw =black
ro =red
gn =green
br =brown
bl =blue
gr =gray
li =violet
ge =yellow
or =orange

F 189 - Tiptronic Switch
J 104 - ABS Control Module (w/EDL)
J 217 - Transmission Control Module (TCM)
J 220 - Motronic Engine Control Module (ECM)
N 110 - Shift Lock Solenoid
T 2 am - Connector 2 pin, black, on selector lever
T 6 e - Connector 6 pin, grey, connector station A pillar, right
T 10 k - Connector 10 pin, orange, connector station A pillar, left
T 10 n - Connector 10 pin, yellow, connector station electronic box plenum chamber
T 10 v - Connector 10 pin, brown, connector station A pillar, right
T 10 w - Connector 10 pin, blue, connector station A pillar, right
T 10am - Connector 10 pin, on selector lever
T 15 e - Connector 15 pin, white, connector station electronic box plenum chamber

(44) - Ground connection, on left A-pillar, lower part
(102) - Ground connection, in transmission wiring harness
(A121) - Connection (high bus) in instrument panel wiring harness
(A122) - Connection (low bus) in instrument panel wiring harness

• - CAN-Bus
- Vehicles without CAN-Bus
* - Vehicles with front wheel drive

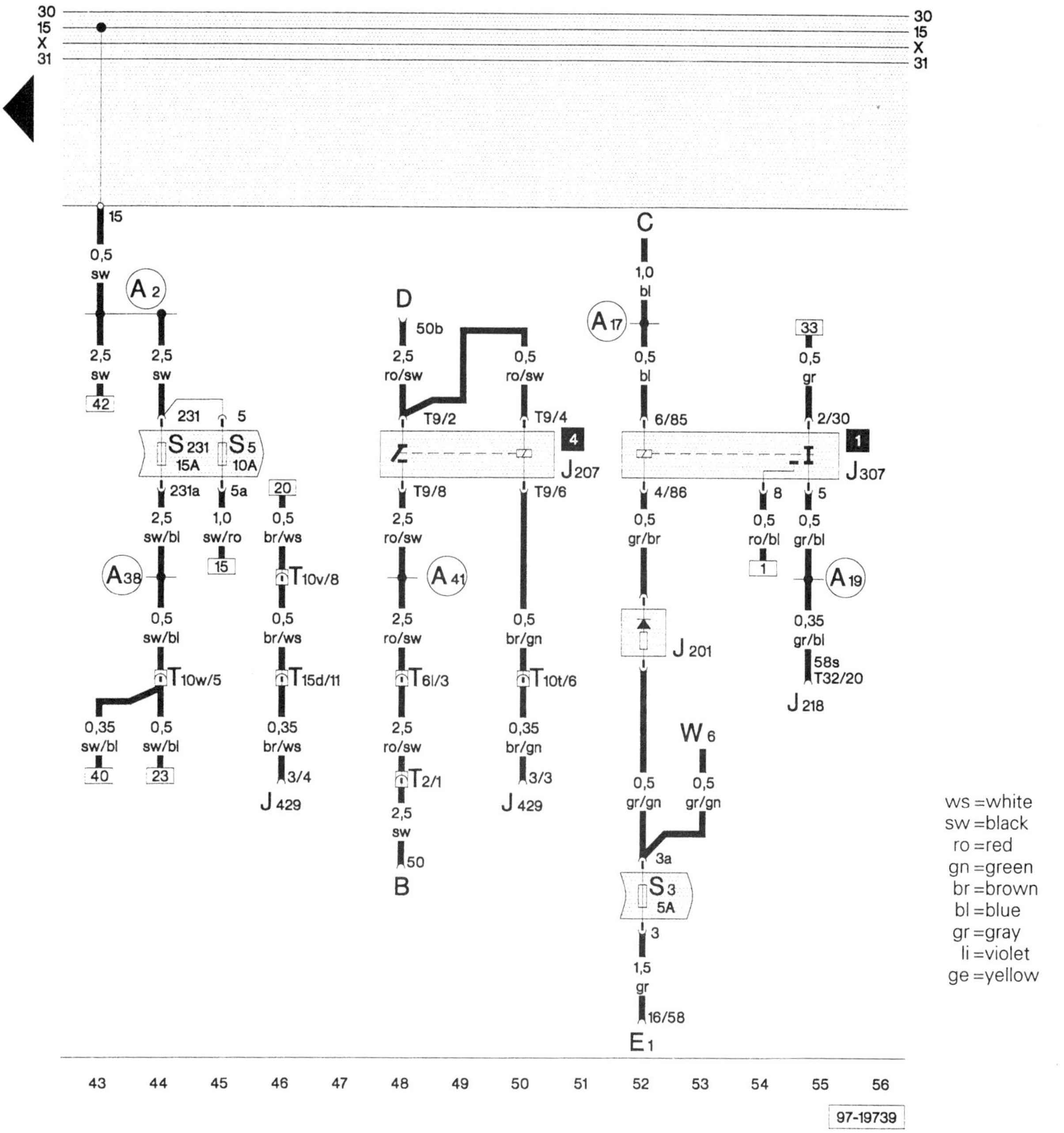

B - Starter
C - Generator (GEN)
D - Ignition/Starter Switch
E 1 - Light switch
J 201 - Protection Diode
J 207 - Starting Interlock Relay
J 218 - Instrument Cluster Combination Processor
J 307 - Selector Lever Light Relay
J 429 - Control module for central locking
S 3 - Fuse
S 5 - Fuse
S 231 - Fuse
T 2 - Connector 2 pin, grey, in engine compartment wiring loom, right side
T 6 l - Connector 6 pin, red, connector station electronic box plenum chamber
T 9 - Connector 9 pin, brown, on starting interlock relay
T 10 t - Connector 10 pin, orange, connector station A pillar, right
T 10 v - Connector 10 pin, brown, connector station A pillar, right
T 10 w - Connector 10 pin, blue, connector station A pillar, right
T 15 d - Connector 15 pin, black, connector station A pillar, right
T 32 - Connector 32 pin, blue, on instrument cluster
W 6 - Glove Compartment Light

(A2) - Plus connection (15), in instrument panel wiring harness
(A17) - Wire connection (61), in instrument panel wiring harness
(A19) - Wire connection (58s), in instrument panel wiring harness (Switch illumination)
(A38) - Plus connection (15a II), in instrument panel wiring harness
(A41) - Plus connection (50), in instrument panel wiring harness

Central Locking System/Anti Theft Alarm System/ Ultra-Sound Interior Monitor

1998 m. y.

Fuse Panel

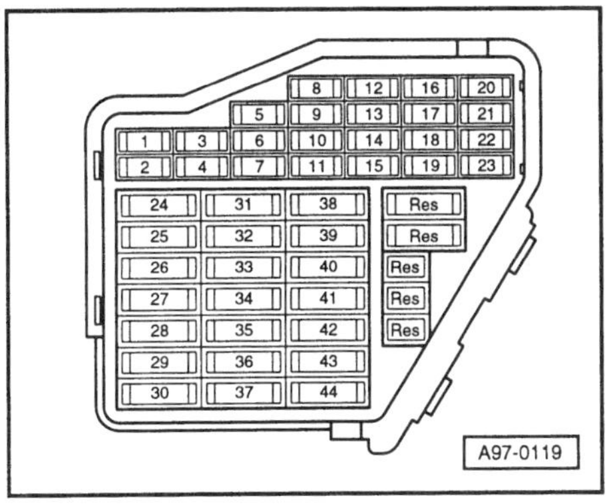

Fuse Colors:

30 A - Green
25 A - White
20 A - Yellow
15 A - Blue
10 A - Red
5 A - Beige

Starting with fuse position 23, fuses in the fuse holder are identified with 223 in the wiring diagram.

13 - Fold Relay Panel

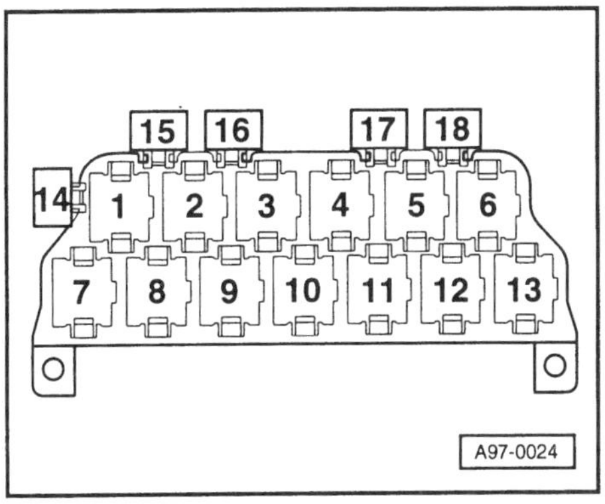

Relay Location:

4 - Starting Interlock Relay, J207

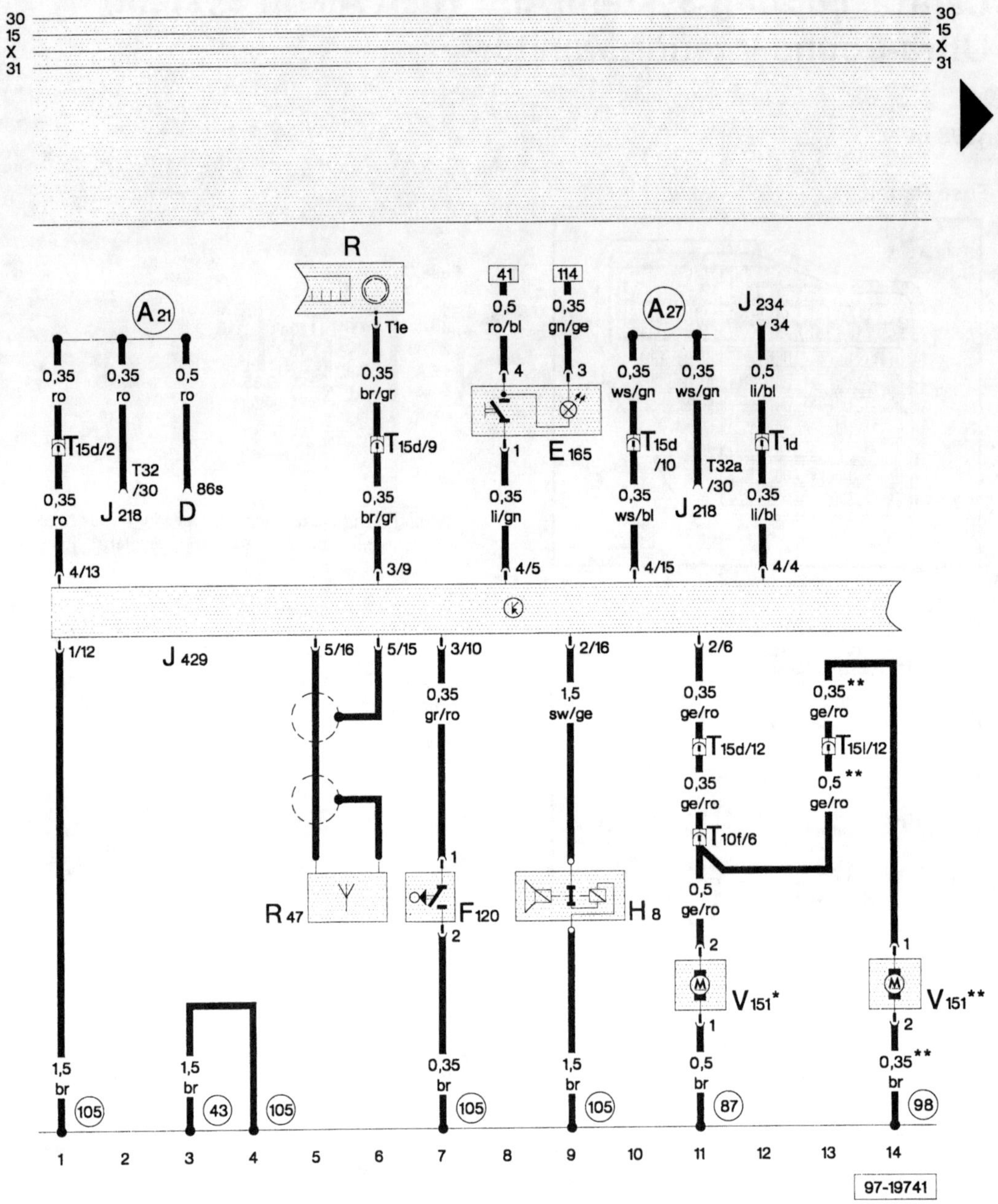

ws = white
sw = black
ro = red
gn = green
br = brown
bl = blue
gr = gray
li = violet
ge = yellow

D - Ignition/Starter Switch
E 165 - Trunk Lid Release Switch
F 120 - Hood Alarm Switch
H 8 - Alarm Horn
J 218 - Instrument Cluster Combination Processor
J 234 - Airbag Control Module
J 429 - Control module for central locking
R - Radio
R 47 - Antenna wire for central locking and anti-theft warning system
T 1 d - Single connector, brown, connector station A pillar right
T 1 e - Single connector, on Radio
T 10 f - Connector 10 pin, brown, connector station A pillar, left
T 15 d - Connector 15 pin, black, connector station A pillar, right
T 15 l - Connector 15 pin, black, in luggage compartment
T 32 - Connector 32 pin, blue, on instrument cluster
T 32 a - Connector 32 pin, green, on instrument cluster
V 151 - Motor for remote unlock rear lid

(43) - Ground connection, on right A-pillar, lower part

(87) - Ground connection -2-, in rear wiring harness

(98) - Ground connection, in rear lid wiring harness

(105) - Ground connection -1-, in central locking system wiring harness

(A21) - Wire connection (86s), in instrument panel wiring harness

(A27) - Wire connection (vehicle speed signal), in instrument panel wiring harness

* - only sedan
** - only wagon

Alarm Horn, Hood Alarm Switch, Antenna wire for Central Locking and Anti-Theft Warning System

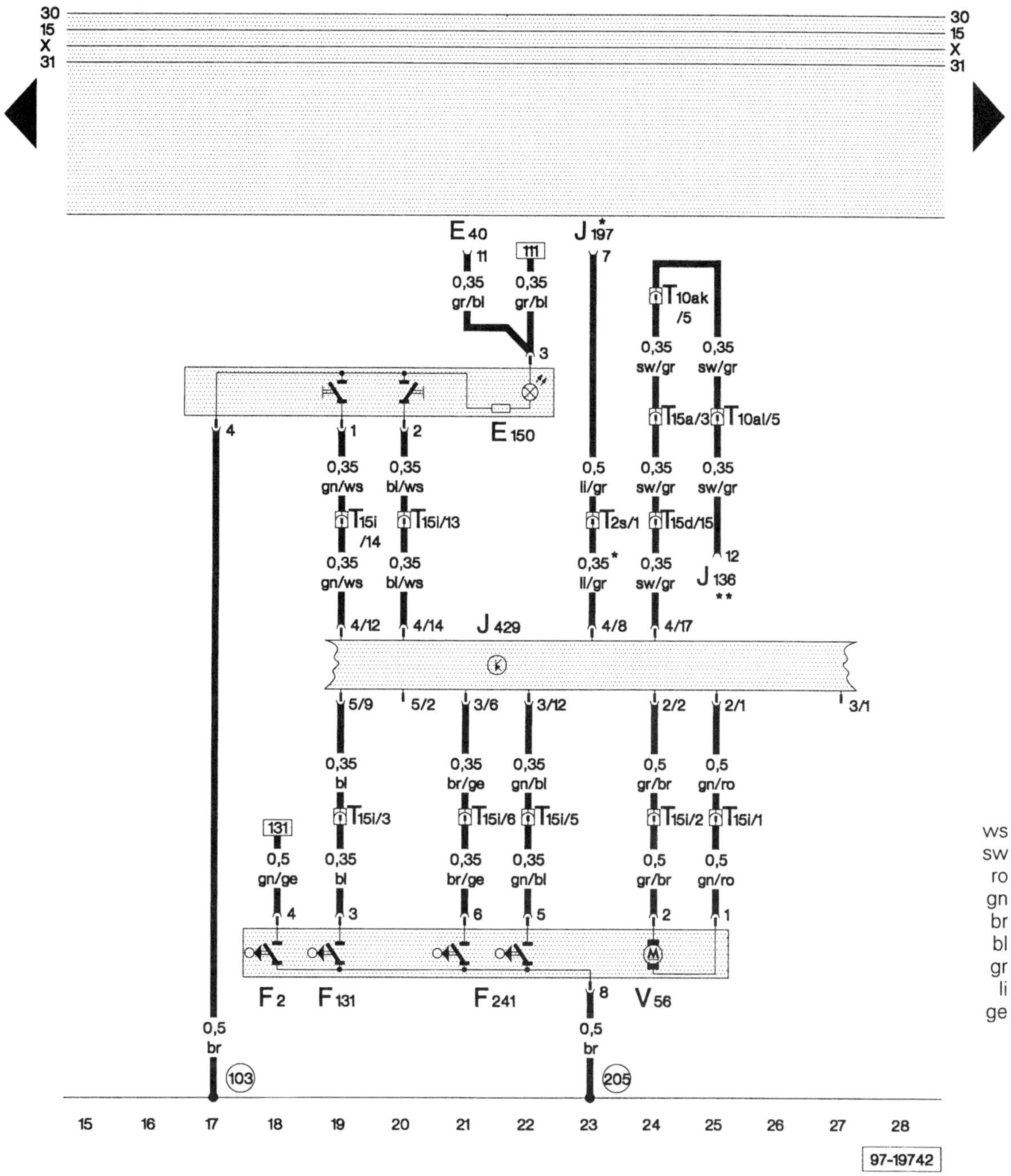

ws = white
sw = black
ro = red
gn = green
br = brown
bl = blue
gr = gray
li = violet
ge = yellow

E 40 - Left Front Window Switch
E 150 - Switch for interior lock, driver side
F 2 - Door Contact Switch - Driver's side
F 131 - Left Front Central Locking System Activator/Switch
F 241 - Switch in door lock, driver's side
J 136 - Memory Seat Control Module
J 197 - Level Control System Control Module
J 429 - Control module for central locking
T 2 s - Connector 2 pin, black, connector station A pillar, right
T 10 ak - Connector 10 pin, green, connector station A pillar, left
T 10 al - Connector 10 pin, brown, below driver's seat
T 15 a - Connector 15 pin, black, connector station A pillar, left
T 15 d - Connector 15 pin, black, connector station A pillar, right
T 15 i - Connector 15 pin, brown, connector station A pillar, left
V 56 - Driver's Door Central Locking System Motor

(103) - Ground connection -2-, in power window/central locking system/door contact switch wiring harness
(205) - Ground connection, in driver's door wiring harness

* - Vehicles with Level Control System
** - Vehicles with Memory Seat

Switch for Interior Lock, Driver Side, Lock Unit for Central Locking, Driver Side

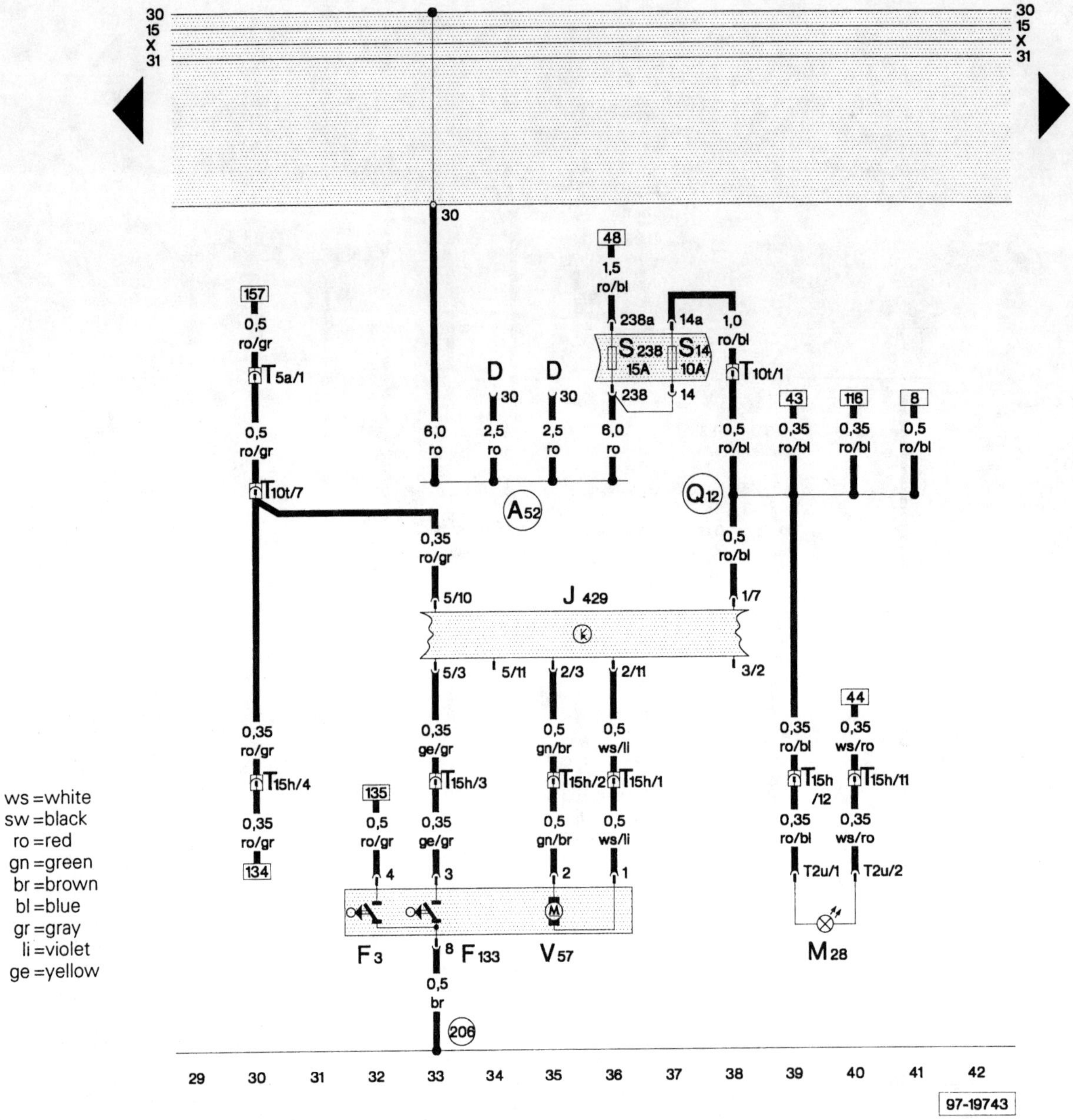

D - Ignition/Starter Switch
F 3 - Door Contact Switch - Passenger side
F 133 - Right Front Central Locking System Activator/Switch
J 429 - Control module for central locking
M 28 - Right Door Warning Light
S 14 - Fuse
S 238 - Fuse
T 2 u - Connector 2 pin, black, in front passenger's door)
T 5 a - Connector 5 pin, black, behind instrument panel, right
T 10 t - Connector 10 pin, orange, connector station A pillar, right
T 15 h - Connector 15 pin, brown, connector station A pillar, right
V 57 - Passenger's Door Central Locking System Motor

(206) - Ground connection, in passenger's door wiring harness
(A52) - Plus connection (30), in instrument panel wiring harness
(Q12) - Plus connection (30a), in power window/central locking system and door contact switch wiring harness

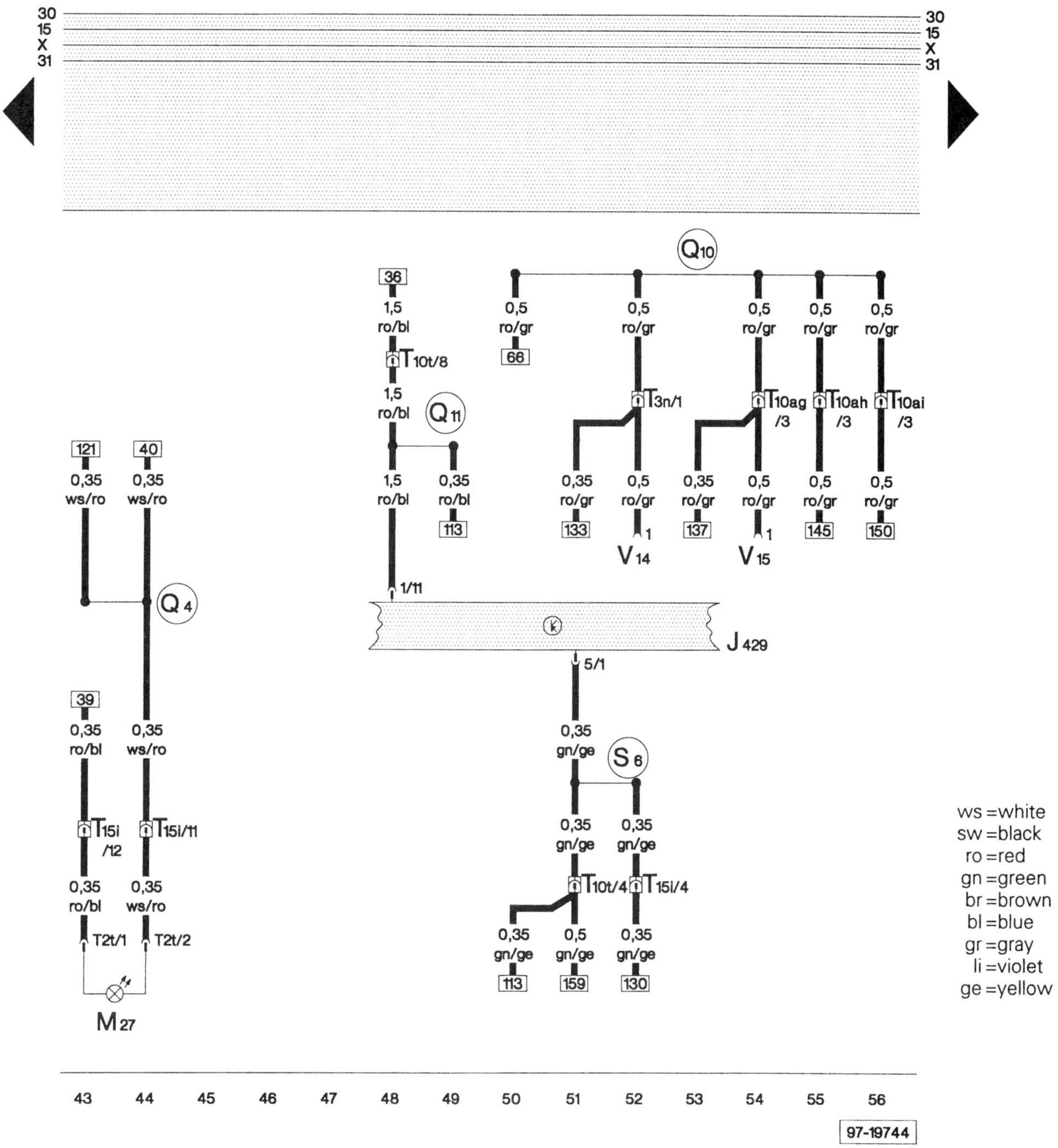

J 429 - Control module for central locking
M 27 - Left Door Warning Light
T 2 t - Connector 2 pin, black, in driver's door
T 3 n - Connector 3 pin, red, in driver's door
T 10 t - Connector 10 pin, orange, connector station A pillar, right
T 10 ag - Connector 10 pin, red, in passenger's door
T 10 ah - Connector 10 pin, red, in left rear door
T 10 ai - Connector 10 pin, red, in right rear door
T 15 i - Connector 15 pin, brown, connector station A pillar, left
V 14 - Left Window Motor
V 15 - Right Window Motor

Q4 - Plus connection (30), in power window/central locking system and door contact switch wiring harness
Q10 - Plus connection (87), in power window/central locking system and door contact switch wiring harness
Q11 - Plus connection (30a), in power window/entral locking system and door contact switch wiring harness
S6 - Wire connection (5/1), in central locking system wiring harness

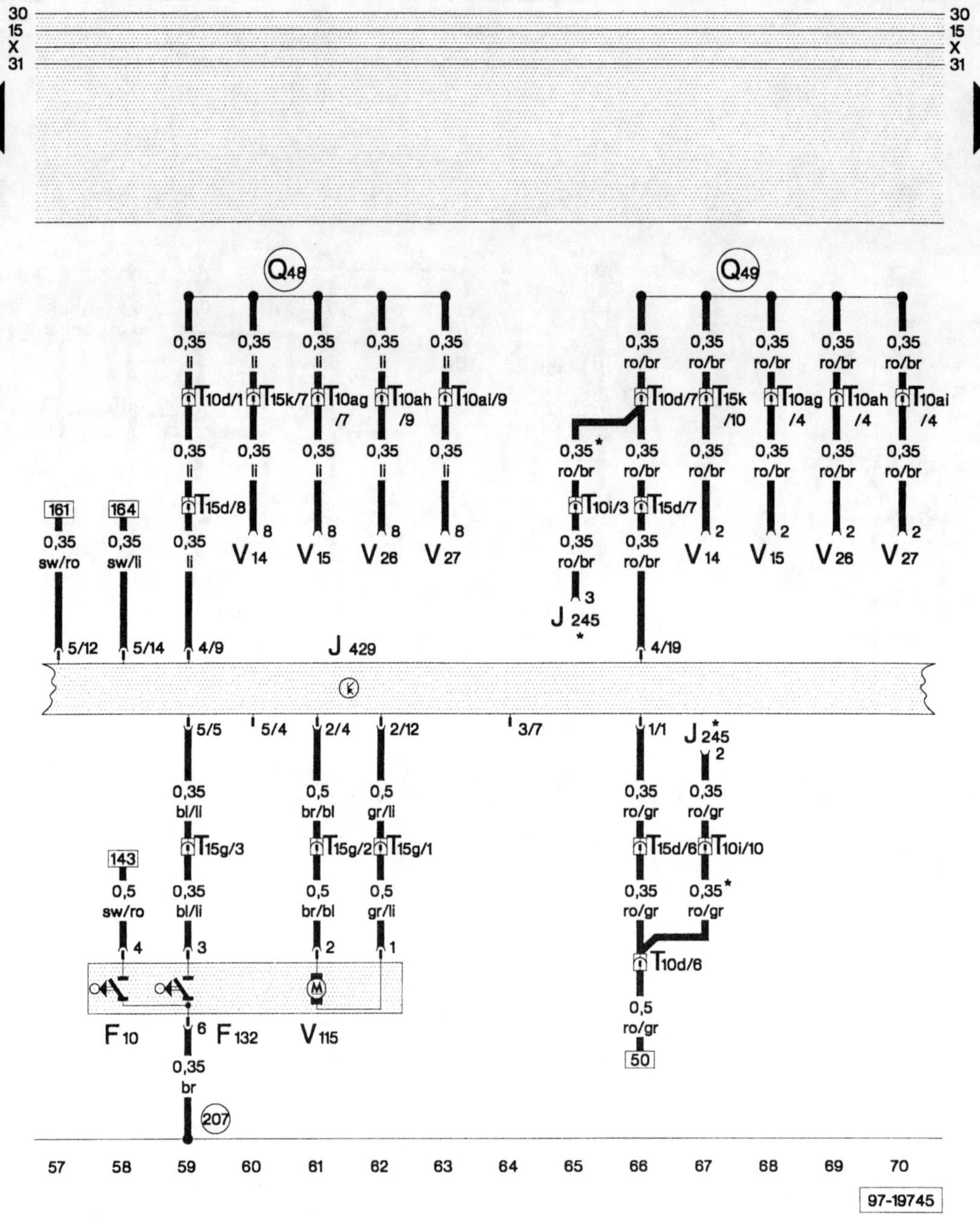

ws = white
sw = black
ro = red
gn = green
br = brown
bl = blue
gr = gray
li = violet
ge = yellow

F 10 - Left Rear Door Contact Switch
F 132 - Left Rear Central Locking System Activator/Switch
J 245 - Power Sunroof Control Module
J 429 - Control module for central locking
T 10 d - Connector 10 pin, green, connector station A pillar, left
T 10 i - Connector 10 pin, darkbrown, connector station A pillar, left
T 10 ag - Connector 10 pin, red, in passenger's door
T 10 ah - Connector 10 pin, red, in left rear door
T 10 ai - Connector 10 pin, red, in right rear door
T 15 d - Connector 15 pin, black, connector station A pillar, right
T 15 g - Connector 15 pin, blue, in connector station B pillar, left
T 15 k - Connector 15 pin, red, connector station A pillar, left
V 14 - Left Window Motor
V 15 - Right Window Motor
V 26 - Motor for Window Regulator, rear, left
V 27 - Motor for Window Regulator, rear, right
V 115 - Left Rear Central Locking System Motor

(207) - Ground connection, in left rear door wiring harness
(Q48) - Connection (open), in wiring harness interior
(Q49) - Connection (closed), in wiring harness interior

* - Vehicles with Power Sunroof

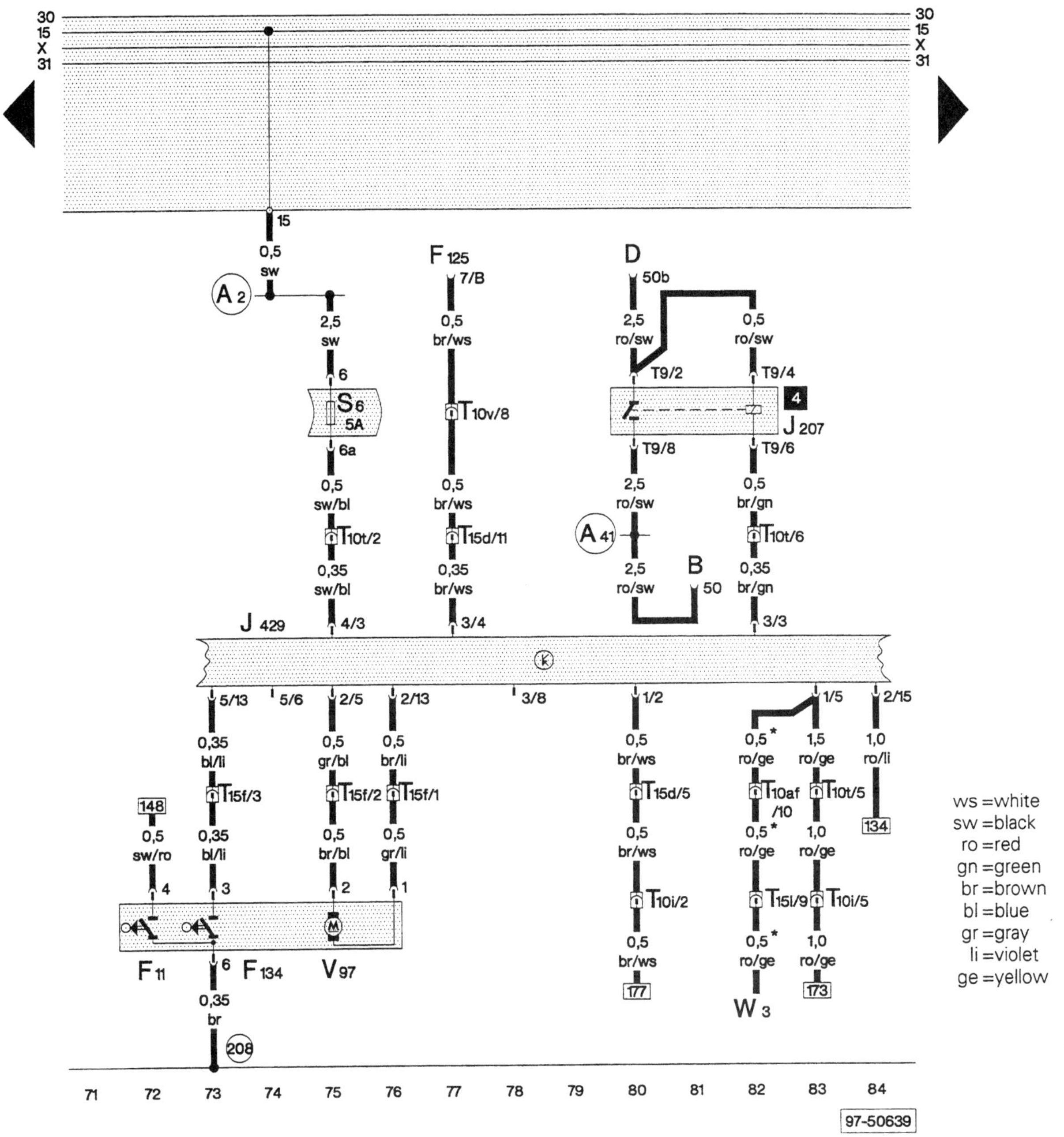

B - Starter
D - Ignition/Starter Switch
F 11 - Right Rear Door Contact Switch
F 125 - Multi-Function Transmission Range (TR) Switch
F 134 - Right Rear Central Locking System Activator/Switch
J 207 - Starting Interlock Relay
J 429 - Control module for central locking
S 6 - Fuse
T 9 - Connector 9 pin, brown, on starting interlock relay
T 10 i - Connector 10 pin, darkbrown, connector station A pillar, left
T 10 t - Connector 10 pin, orange, connector station A pillar, right
T 10 v - Connector 10 pin, brown, connector station A pillar, right
T 10 af - Connector 10 pin, violet, connector station A pillar, left
T 15 d - Connector 15 pin, black, connector station A pillar, right
T 15 f - Connector 15 pin, blue, in connector station B pillar, right
T 15 l - Connector 15 pin, black, in luggage compartment

V 97 - Double Cab Right Rear Door Central Locking System Motor
W 3 - Luggage Compartment Light

(208) - Ground connection, in right rear door wiring harness
(A2) - Plus connection (15), in instrument panel wiring harness
(A41) - Plus connection (50), in instrument panel wiring harness

* - only wagon

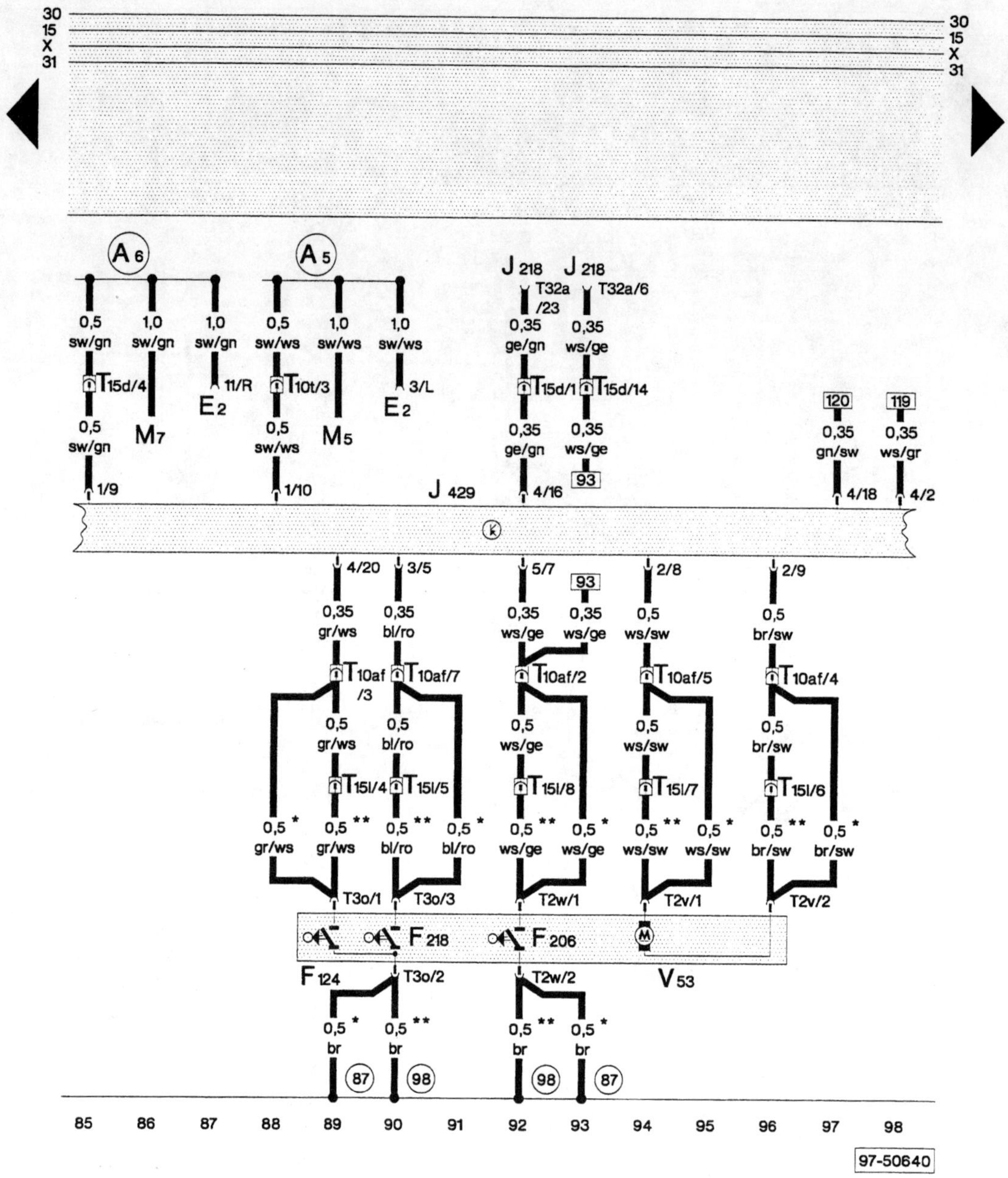

ws = white
sw = black
ro = red
gn = green
br = brown
bl = blue
gr = gray
li = violet
ge = yellow

E 2 - Turn Signal Switch
F 124 - Trunk Lock Alarm/Central Locking Switch
F 206 - Trunk Lid Closed Position Switch
F 218 - Switch for central locking, trunk lid
J 218 - Instrument Cluster Combination Processor
J 429 - Control module for central locking
M 5 - Left Front Turn Signal Light
M 7 - Right Front Turn Signal Light
T 2 v - Connector 2 pin, black, luggage compartment
T 2 w - Connector 2 pin, grey, luggage compartment
T 3 o - Connector 3 pin, black, in rear lid
T 10 t - Connector 10 pin, orange, connector station A pillar, right
T 10 af - Connector 10 pin, violet, connector station A pillar, left
T 15 d - Connector 15 pin, black, connector station A pillar, right
T 15 l - Connector 15 pin, black, in luggage compartment
T 32 a - Connector 32 pin, green, on instrument cluster
V 53 - Decklid Central Locking System Motor

(87) - Ground connection -2-, in rear wiring harness

(98) - Ground connection, in rear lid wiring harness

(A5) - Plus connection (right turn signal), in instrument panel wiring harness

(A6) - Plus connection (left turn signal), in instrument panel wiring harness

* - only sedan

** - only wagon

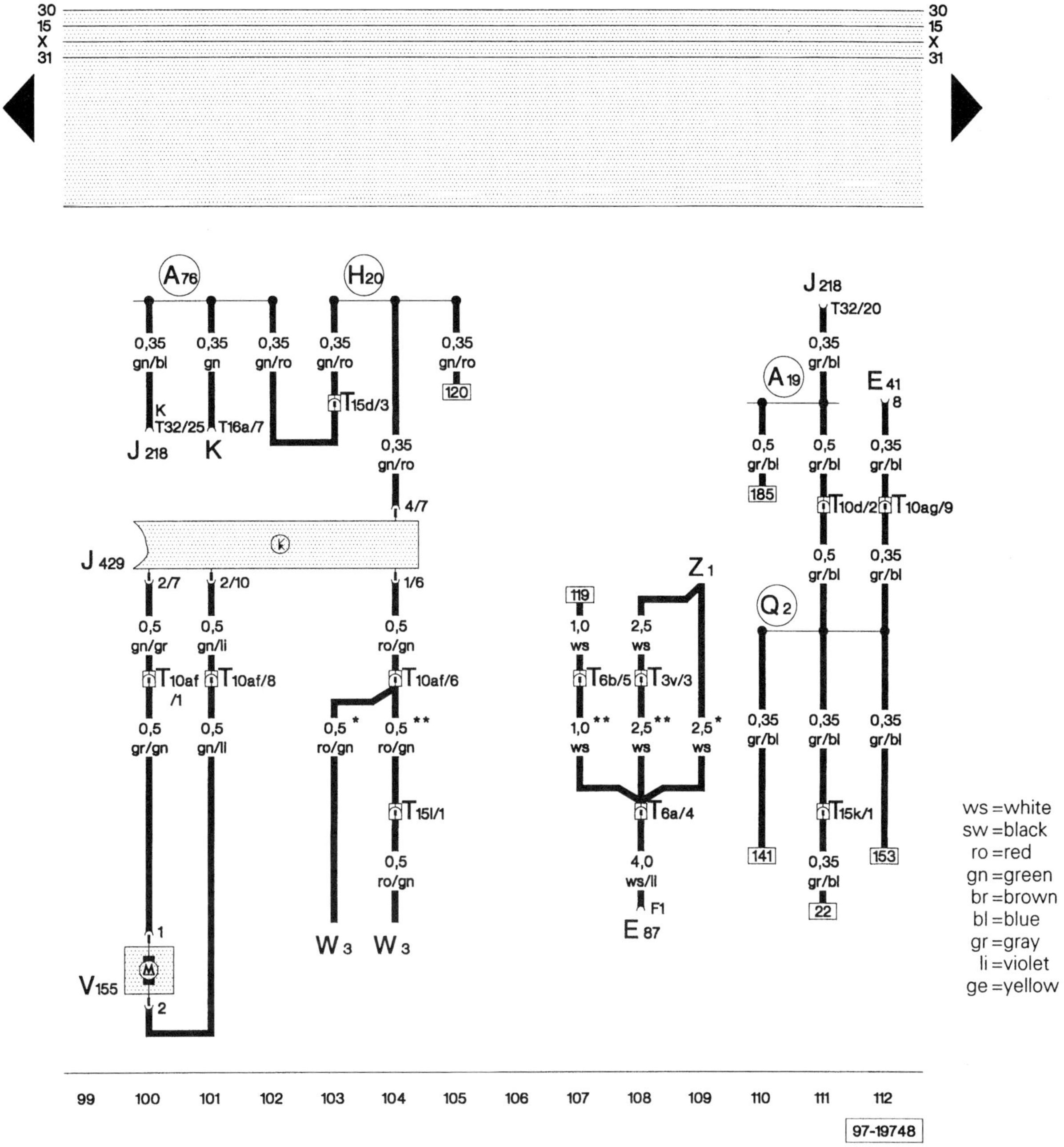

E 15 - Rear Window Defogger Switch
E 41 - Right Front Window Switch
E 87 - A/C Control Head
J 218 - Instrument Cluster Combination Processor
J 429 - Control module for central locking
K - DLC (Data Link Connector) Rapid Data Transfer
T 3 v - Connector 3 pin, black, luggage compartment
T 6 a - Connector 6 pin, blue, connector station A pillar, left
T 6 b - Connector 6 pin, white, connector station A pillar, left
T 10 d - Connector 10 pin, green, connector station A pillar, left
T 10 af - Connector 10 pin, violet, connector station A pillar, left
T 10 ag - Connector 10 pin, red, in passenger's door
T 15 d - Connector 15 pin, black, connector station A pillar, right
T 15 k - Connector 15 pin, red, connector station A pillar, left
T 15 l - Connector 15 pin, black, in luggage compartment
T 16 a - Connector 16 pin, data link connector
T 32 - Connector 32 pin, blue, on instrument cluster
V 155 - Motor for fuel tank lid unlock

W 3 - Luggage Compartment Light
Z 1 - Heated rear window

(A19) - Wire connection (58d), in instrument panel wiring harness

(A76) - Connector (K-diagnosis wire), in instrument panel wiring harness

(H20) - Wire connection -1-, in On-Board Diagnostic (OBD) wiring harness

(Q2) - Plus connection (58), in power window wiring harness

* - only sedan
** - only wagon

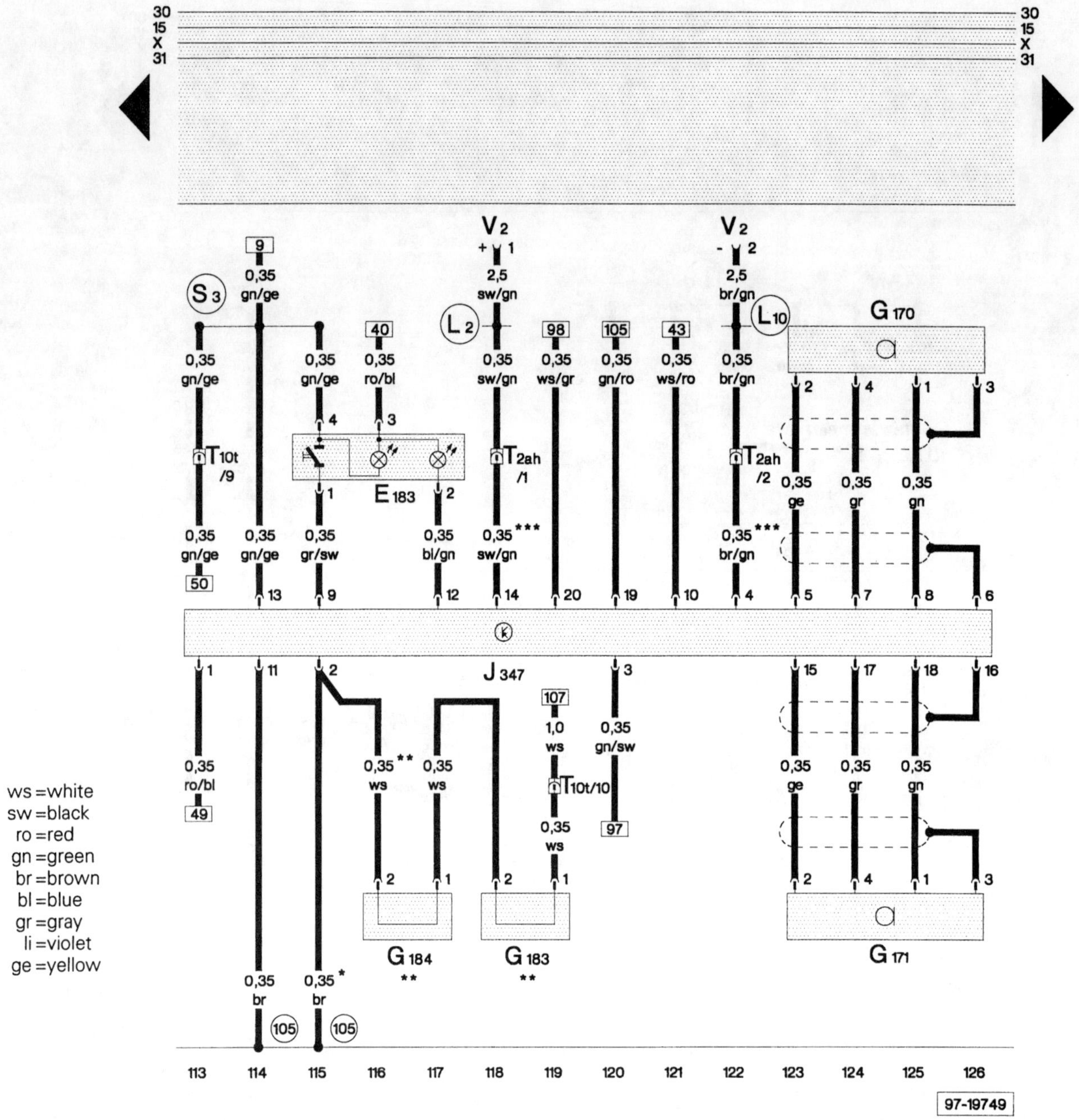

E 183 - Switch for passenger compartment monitoring
G 170 - Ultra - sound sensor, left for anti-theft warning system
G 171 - Ultra - sound sensor, right for anti-theft warning system
G 183 - Sensor for broken window glass rear left
G 184 - Sensor for broken window glass rear right
J 347 - Control Module for Ultra-Sound sensors
T 2 ah - Connector 2 pin, blue, connector station A pillar, right
T 10 t - Connector 10 pin, orange, connector station A pillar, right
V 2 - Fresh Air Blower

(105) - Ground connection -1-, in central locking system wiring harness
(L2) - Wire connection, in A/C wiring harness
(L10) - Wire connection -2-, in A/C wiring harness
(S3) - Wire connection -1-(switch/control module), central locking system wiring harness

* - only sedan
** - only wagon
*** - Vehicles with solar cells

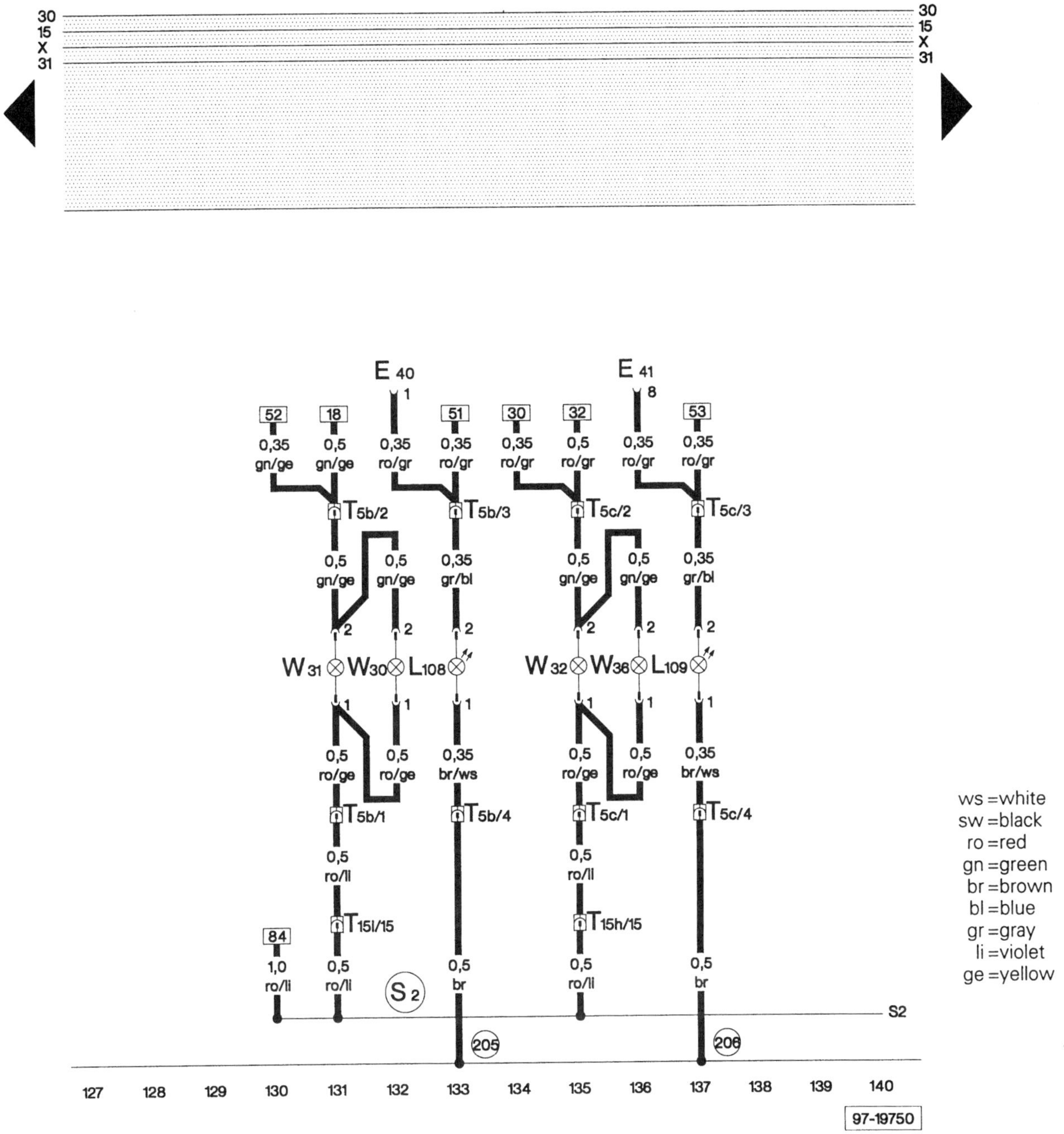

E 40 - Left Front Window Switch
E 41 - Right Front Window Switch
L 108 - Door Opener Illumination, Driver's Side
L 109 - Door Opener Illumination, Passenger's Side
T 5 b - Connector 5 pin, black, in driver's door
T 5 c - Connector 5 pin, black, in passenger's door
T 15 h - Connector 15 pin, brown, connector station A pillar, right
T 15 i - Connector 15 pin, brown, connector station A pillar, left
W 30 - Door warning light, driver side
W 31 - Entry light front, left
W 32 - Entry light front, right
W 36 - Door warning light, passenger side

(205) - Ground connection, in driver's door wiring harness
(206) - Ground connection, in passenger's door wiring harness
(S2) - Wire connection, in central locking system wiring harness

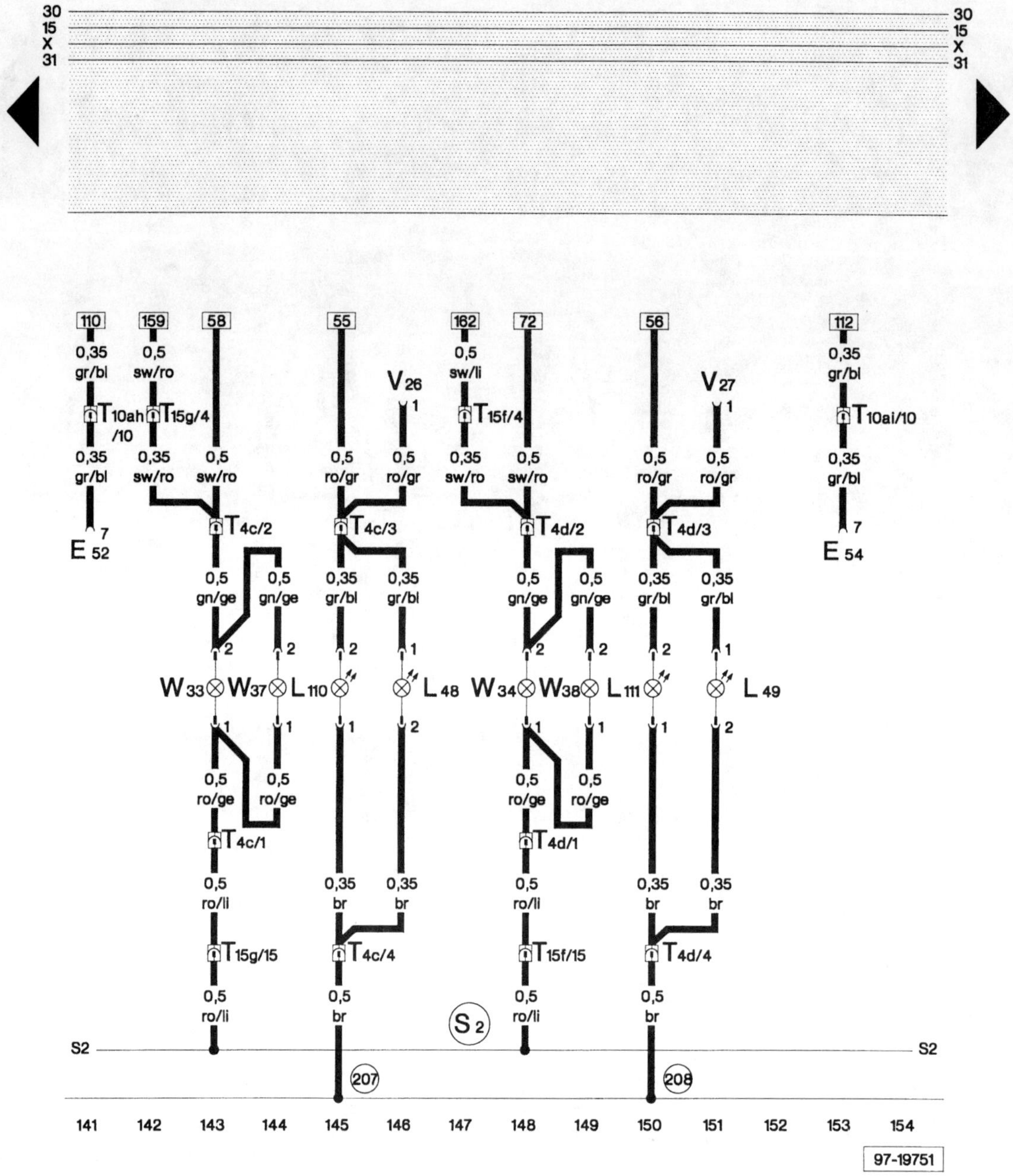

ws = white
sw = black
ro = red
gn = green
br = brown
bl = blue
gr = gray
li = violet
ge = yellow

E 52 - Left Rear Window Switch, (In LR Door)
E 54 - Right Rear Window Switch, (In RR Door)
L 48 - Left Rear Ashtray Light
L 49 - Right Rear Ashtray Light
L 110 - Door Opener Illumination, rear left
L 111 - Door Opener Illumination, rear right
T 4 c - Connector 4 pin, black, in left rear door
T 4 d - Connector 4 pin, black, in right rear door
T 10 ah - Connector 10 pin, red, in left rear door
T 10 ai - Connector 10 pin, red, in right rear door
T 15 f - Connector 15 pin, blue, in connector station B pillar, right
T 15 g - Connector 15 pin, blue, in connector station B pillar, left
V 26 - Motor for window regulator, rear, left
V 27 - Motor for window regulator, rear, right
W 33 - Entry light rear, left
W 34 - Entry light rear, right
W 37 - Door warning light rear, left
W 38 - Door warning light rear, right

(207) - Ground connection, in left rear door wiring harness
(208) - Ground connection, in right rear door wiring harness
(S2) - Wire connection, in central locking system wiring harness

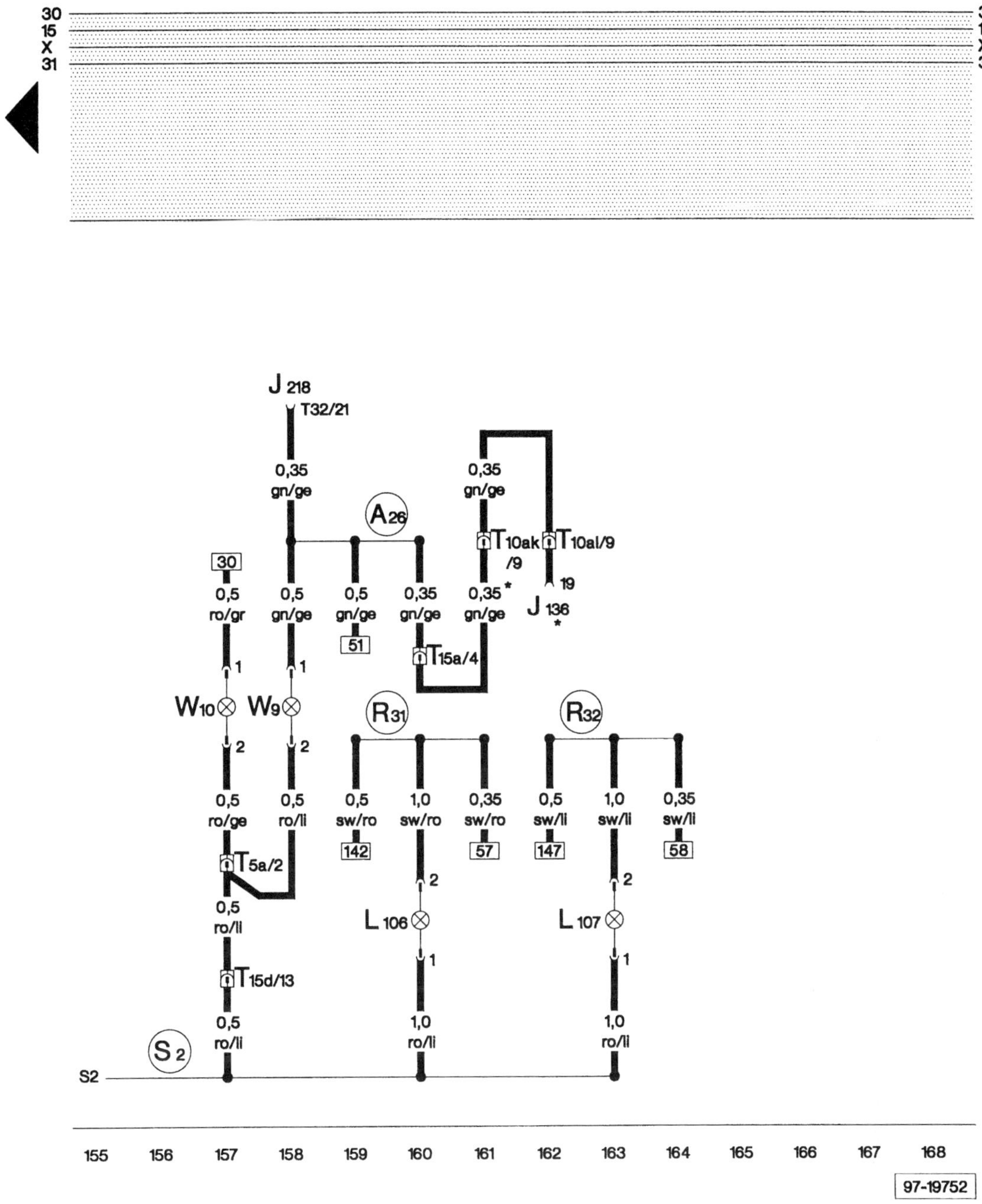

ws = white
sw = black
ro = red
gn = green
br = brown
bl = blue
gr = gray
li = violet
ge = yellow

J 136 - Memory Seat Control Module
J 218 - Instrument Cluster Combination Processor
L 106 - Lighting rear footwell, left
L 107 - Lighting rear footwell, right
T 5 a - Connector 5 pin, black, behind instrument panel, right
T 10 ak - Connector 10 pin, green, connector station A pillar, left
T 10 al - Connector 10 pin, brown, below driver's seat
T 15 a - Connector 15 pin, black, connector station A pillar, left
T 15 d - Connector 15 pin, black, connector station A pillar, right
T 32 - Connector 32 pin, blue, on instrument cluster
W 9 - Left Footwell Light
W 10 - Right Footwell Light

(A26) - Wire connection (driver's door contact switch), in instrument panel wiring harness
(R31) - Wire connection -1-, in left rear door wiring harness
(R32) - Wire connection -2-, in right rear door wiring harness
(S2) - Wire connection, in central locking system wiring harness

* - Vehicles with Memory Seat

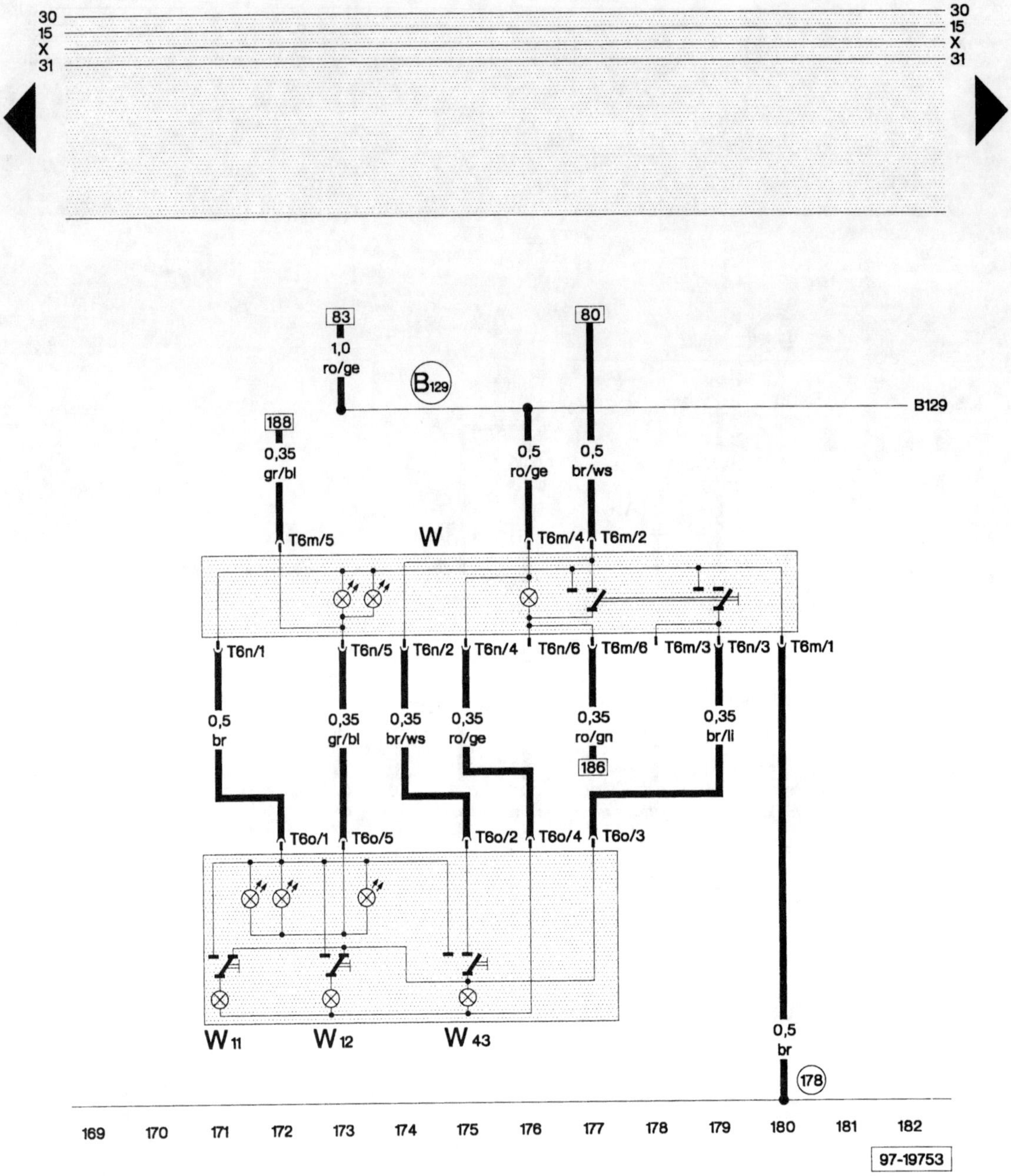

ws = white
sw = black
ro = red
gn = green
br = brown
bl = blue
gr = gray
li = violet
ge = yellow

T 6 m - Connector 6 pin, on interior light, front
T 6 n - Connector 6 pin, on interior light, front
T 6 o - Connector 6 pin, on interior light, rear
W - Front Interior Light
W 11 - Left Rear Reading Light
W 12 - Right Rear Reading Light
W 43 - Rear courtesy lamp

(178) - Ground connection, in sunroof wiring harness
(B129) - Connector (interior light 3/L), in wiring harness interior

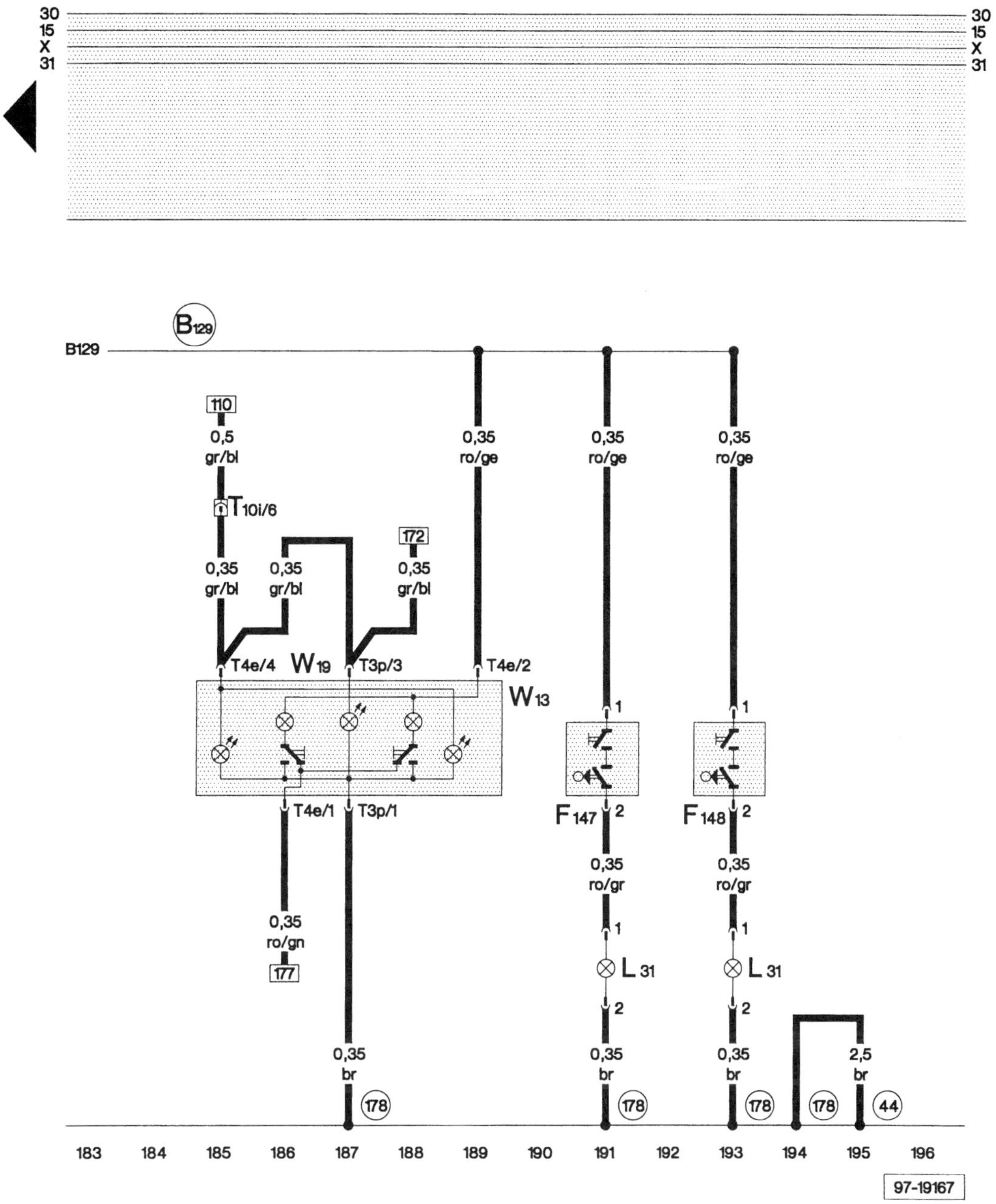

ws = white
sw = black
ro = red
gn = green
br = brown
bl = blue
gr = gray
li = violet
ge = yellow

F 147 - Left Make-Up Mirror Light Switch
F 148 - Right Make-Up Mirror Light Switch
L 31 - Make-up Mirror Light
T 3 p - Connector 3 pin, on reading lamp front right
T 4 e - Connector 4 pin, on reading lamp front left
T 10 i - Connector 10 pin, darkbrown, connector station A pillar, left
W 13 - Right Front Map/Reading Light
W 19 - Left Front Reading Light

(44) - Ground connection, on left A-pillar, lower part
(178) - Ground connection, in sunroof wiring harness
(B129) - Connector (interior light 3/L), in wiring harness interior

THIS PAGE INTENTIONALLY LEFT BLANK

Automatic Climate Control

1998 m. y.

Fuse Panel

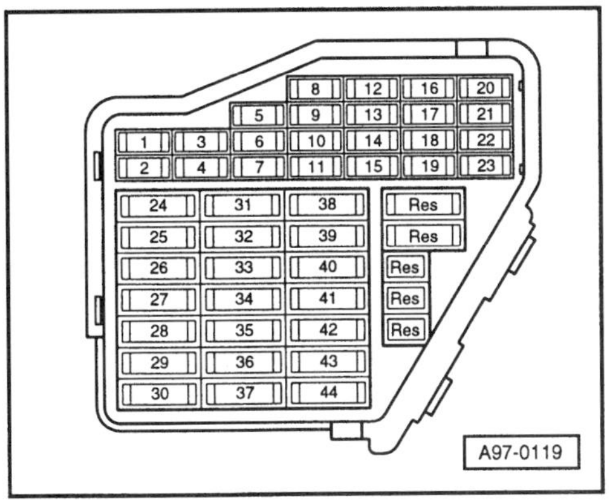

Fuse Colors:

30 A - Green
25 A - White
20 A - Yellow
15 A - Blue
10 A - Red
5 A - Beige

Starting with fuse position 23, fuses in the fuse holder are identified with 223 in the wiring diagram.

Micro Central Electric Panel

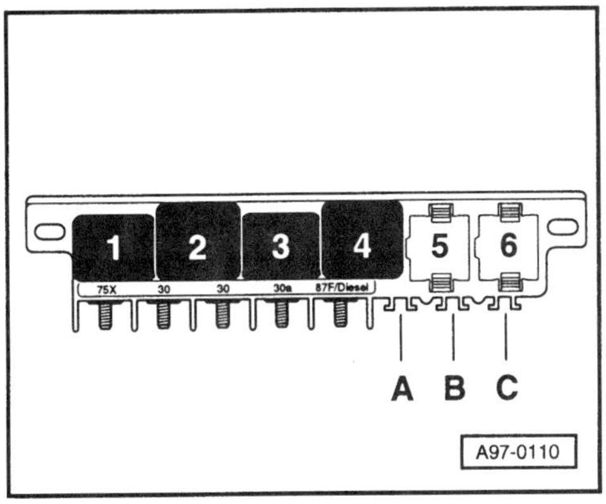

13 - Fold Relay Panel

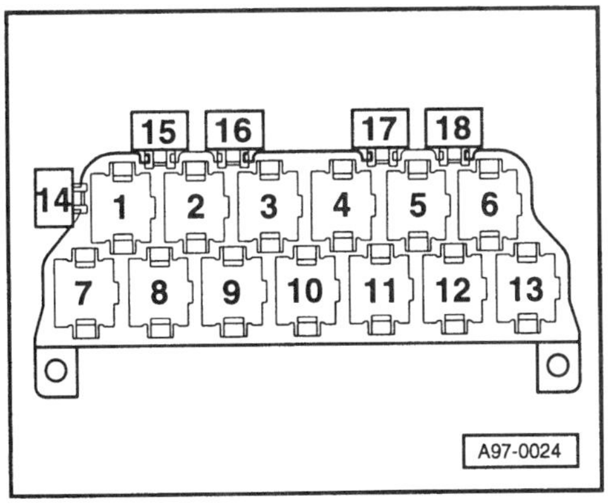

Relay Location:

3 - Solar Cell Separation Relay, J309

5 - A/C Clutch Relay, J44

8 - Fold Relay Panel

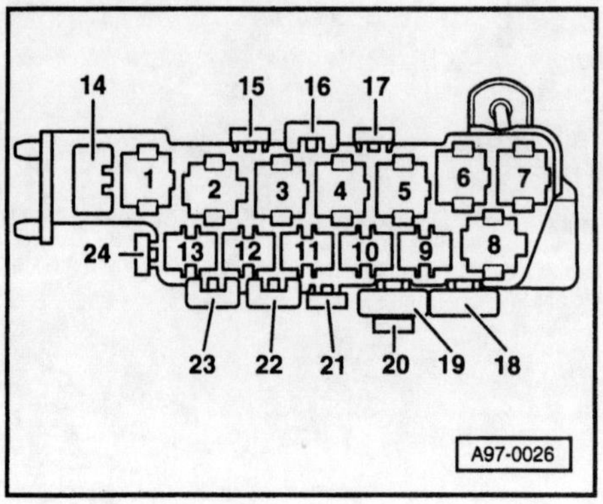

Relay Location:

- **2** - Second Speed Coolant Fan Control (FC) Relay, J101
- **3** - Coolant fan control (FC) Relay, J26
- **20** - Coolant fan fuse, S42
- **21** - Control module fuse for coolant fan, S142

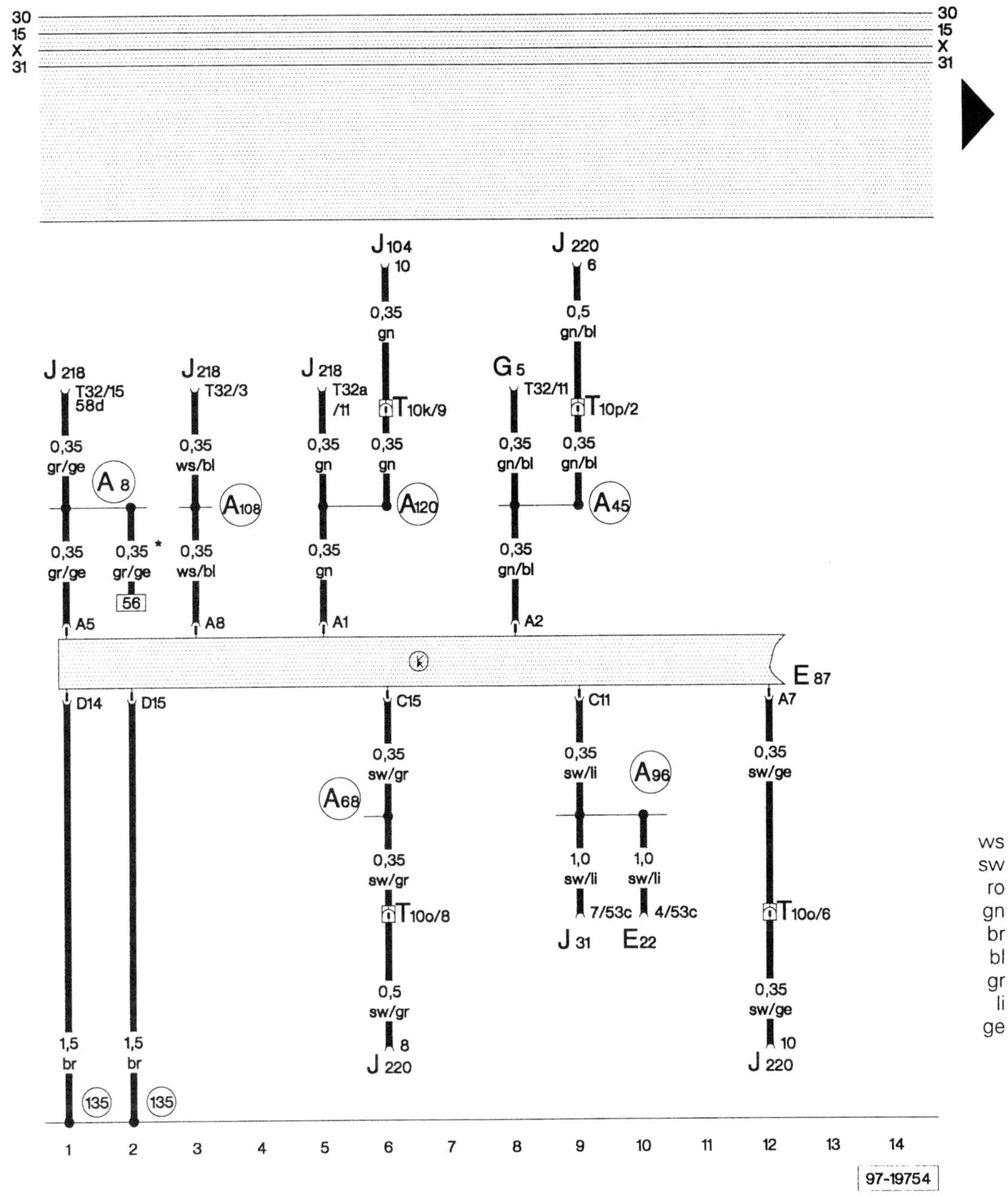

ws = white
sw = black
ro = red
gn = green
br = brown
bl = blue
gr = gray
li = violet
ge = yellow

E 22 - Windshield Wiper/Washer Switch
E 87 - A/C Control Head
G 5 - Tachometer
J 31 - Wiper/Washer Intermittent Relay
J 104 - ABS Control Module (w/EDL)
J 218 - Instrument Cluster Combination Processor
J 220 - Motronic Engine Control Module (ECM)
T 10 k - Connector 10 pin, orange, connector station A pillar, left
T 10 o - Connector 10 pin, brown, connector station electronic box plenum chamber
T 10 p - Connector 10 pin, black, connector station electronic box plenum chamber
T 32 - Connector 32 pin, blue, on instrument cluster
T 32 a - Connector 32 pin, green, on instrument cluster

(135) - Ground connection -2-, in instrument panel wiring harness
(A8) - Plus connection (58d), in instrument panel wiring harness (Display illumination)
(A45) - Wire connection (RPM-signal), in instrument panel wiring harness
(A68) - Connector (C15, A/C), in instrument panel wiring harness
(A96) - Connector (53a), in instrument panel wiring harness
(A108) - Wire connection (vehicle speed signal), in instrument panel wiring harness
(A120) - Wire connection (Immobilization time), in instrument panel wiring harness

* - Vehicles with Outside Air Temperature Display

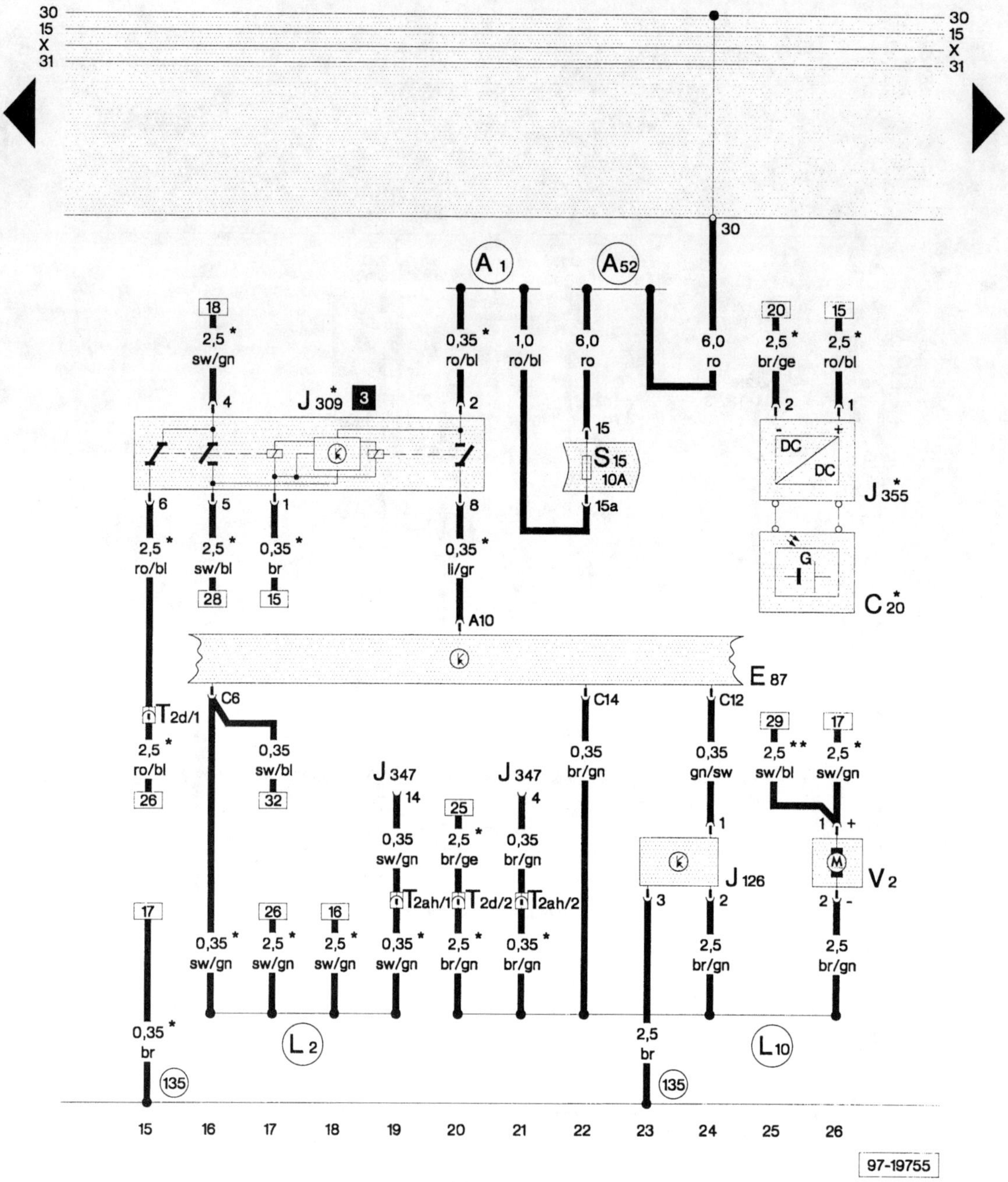

ws = white
sw = black
ro = red
gn = green
br = brown
bl = blue
gr = gray
li = violet
ge = yellow

C 20 - Solar Cells
E 87 - A/C Control Head
J 126 - Control module for fresh air blower
J 309 - Solar Cell Separation Relay
J 347 - Control Module for Ultra-Sound sensors
J 355 - Control Module for solar operation
S 15 - Fuse
T 2 d - Connector 2 pin, blue, connector station A pillar, left
T 2 ah - Connector 2 pin, blue, connector station A pillar, right
V 2 - Fresh Air Blower

(135) - Ground connection -2-, in instrument panel wiring harness
(A1) - Plus connection (30a), in instrument panel wiring harness
(A52) - Plus connection (30), in instrument panel wiring harness
(L2) - Wire connection, in A/C wiring harness
(L10) - Wire connection -2-, in A/C wiring harness

* - Vehicles with sunroof with Solar Cells
** - Vehicles without sunroof with Solar Cells

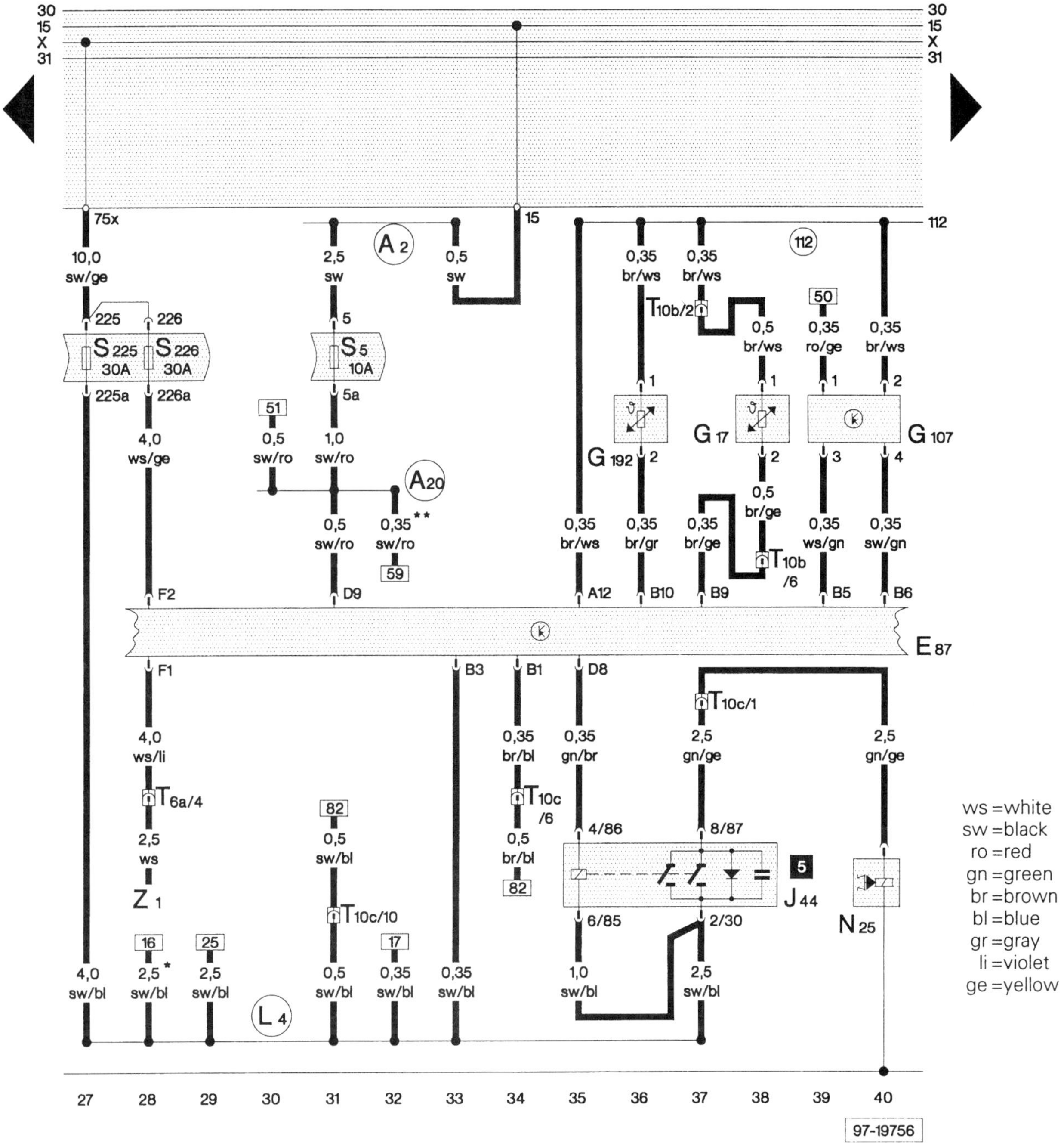

E 87 - A/C Control Head
G 17 - Outside Air Temperature Sensor
G 107 - Sunlight Photo Sensor
G 192 - Sender for outlet temperature, floor outlet
J 44 - A/C Clutch Relay
N 25 - A/C Clutch
S 5 - Fuse
S 225 - Fuse
S 226 - Fuse
T 6 a - Connector 6 pin, blue, connector station A pillar, left
T 10 b - Connector 10 pin, red, connector station A pillar, left
T 10 c - Connector 10 pin, grey, connector station A pillar, left
Z 1 - Heated rear window

(112) - Ground connection -2-, in A/C wiring harness
(A2) - Plus connection (15), in instrument panel wiring harness
(A20) - Wire connection (15a), in instrument panel wiring harness
(L4) - Wire connection (75al), in A/C wiring harness

* -Vehicles with sunroof with Solar Cells
** -Vehicles with Outside Air Temperature Display

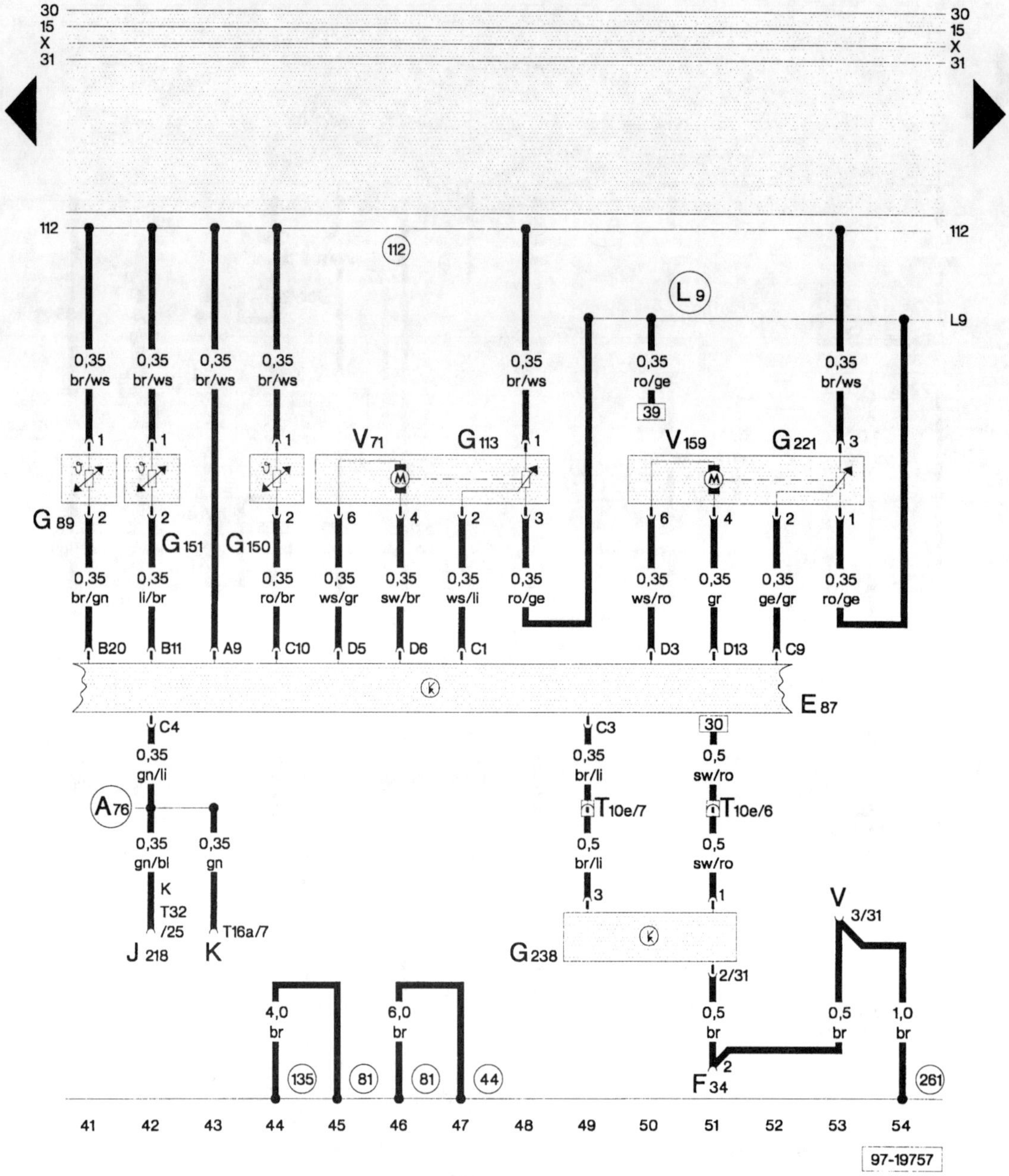

ws = white
sw = black
ro = red
gn = green
br = brown
bl = blue
gr = gray
li = violet
ge = yellow

E 87 - A/C Control Head
F 34 - Brake Fluid Level Warning Switch
G 89 - Fresh Air Intake Duct Temperature Sensor
G 113 - Back Pressure Flap Motor Position Sensor
G 150 - Left Vent Temperature Sensor
G 151 - Right Vent Temperature Sensor
G 221 - Right Temperature Regulator Flap Motor Position Sensor
G 238 - Sensor for Air Quality
J 218 - Instrument Cluster Combination Processor
K - DLC (Data Link Connector) Rapid Data Transfer
T 10 e - Connector 10 pin, violet, connector station A pillar, left
T 16 a - Connector 16 pin, data link connector
T 32 - Connector 32 pin, blue, on instrument cluster
V - Windshield Wiper Motor
V 71 - Air Flow Flap Motor
V 159 - Right Temperature Regulator Flap Motor

(44) - Ground connection, on left A-pillar, lower part
(81) - Ground connection -1-, in instrument panel wiring harness
(112) - Ground connection -2-, in A/C wiring harness
(135) - Ground connection -2-, in instrument panel wiring harness
(261) - Ground connection, in wiring harness heated spray jet
(A76) - Connector (K-diagnosis wire), in instrument panel wiring harness
(L9) - Wire connection -1-, in A/C wiring harness

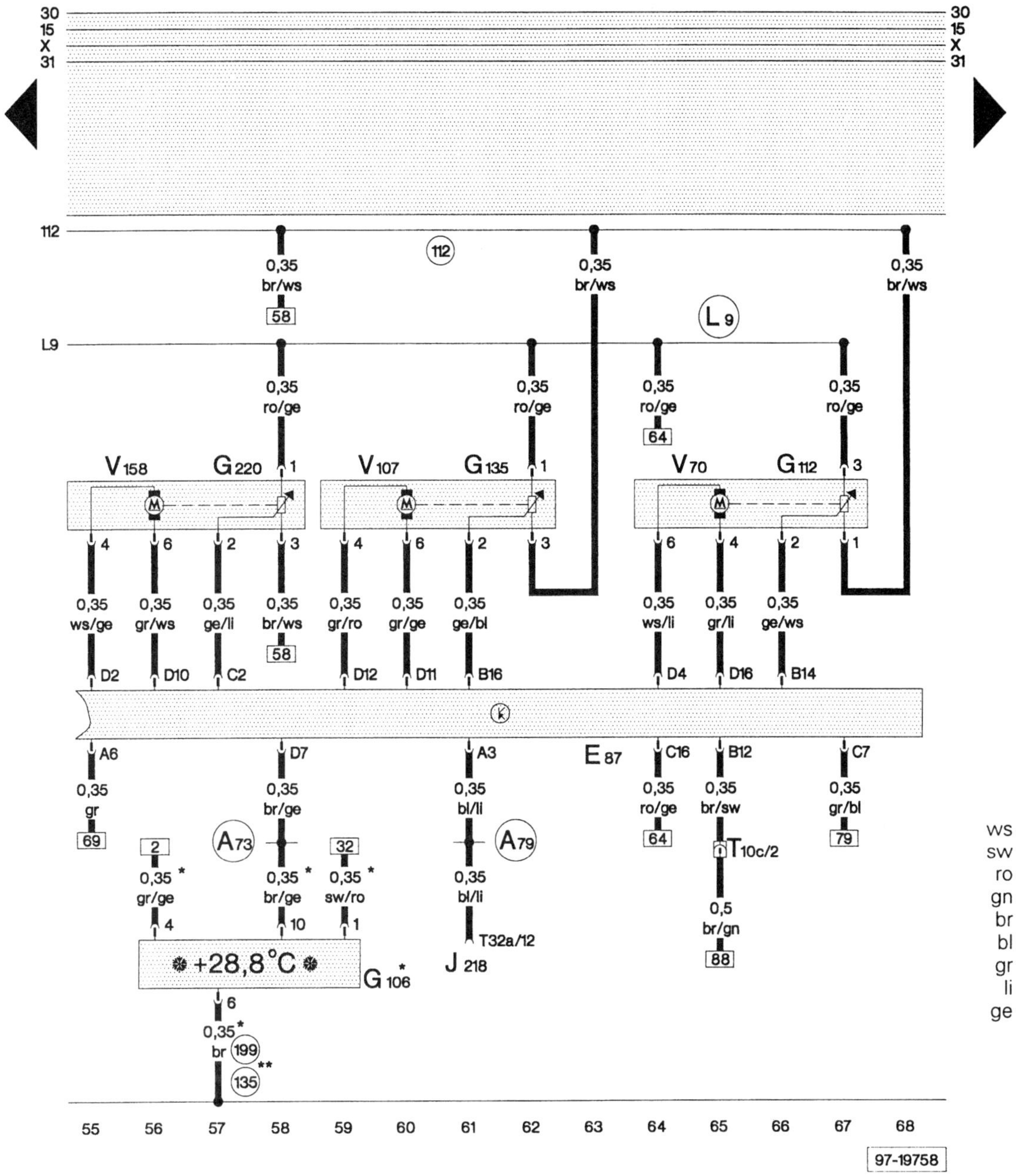

ws = white
sw = black
ro = red
gn = green
br = brown
bl = blue
gr = gray
li = violet
ge = yellow

E 87 - A/C Control Head
G 106 - Outside Air Temperature Display
G 112 - Central Flap Motor Position Sensor
G 135 - Defroster Flap Motor Position Sensor
G 220 - Left Temperature Regulator Flap Motor Position Sensor
J 218 - Instrument Cluster Combination Processor
T 10 c - Connector 10 pin, grey, connector station A pillar, left
T 32 a - Connector 32 pin, green, on instrument cluster
V 70 - Central Air Flap Motor
V 107 - Defroster Flap Motor
V 158 - Left Temperature Regulator Flap Motor

(112) - Ground connection -2-, in A/C wiring harness
(135) - Ground connection -2-, in instrument panel wiring harness
(199) - Ground connection -3-, in instrument panel wiring harness
(A73) - Connector (outside temperature indicator), in instrument panel wiring harness
(A79) - Connector (engine overheat warning light), in instrument panel wiring harness
(L9) - Wire connection -1-, in A/C wiring harness

* - Vehicles with Outside Air Temperature Display
** - booth is possible

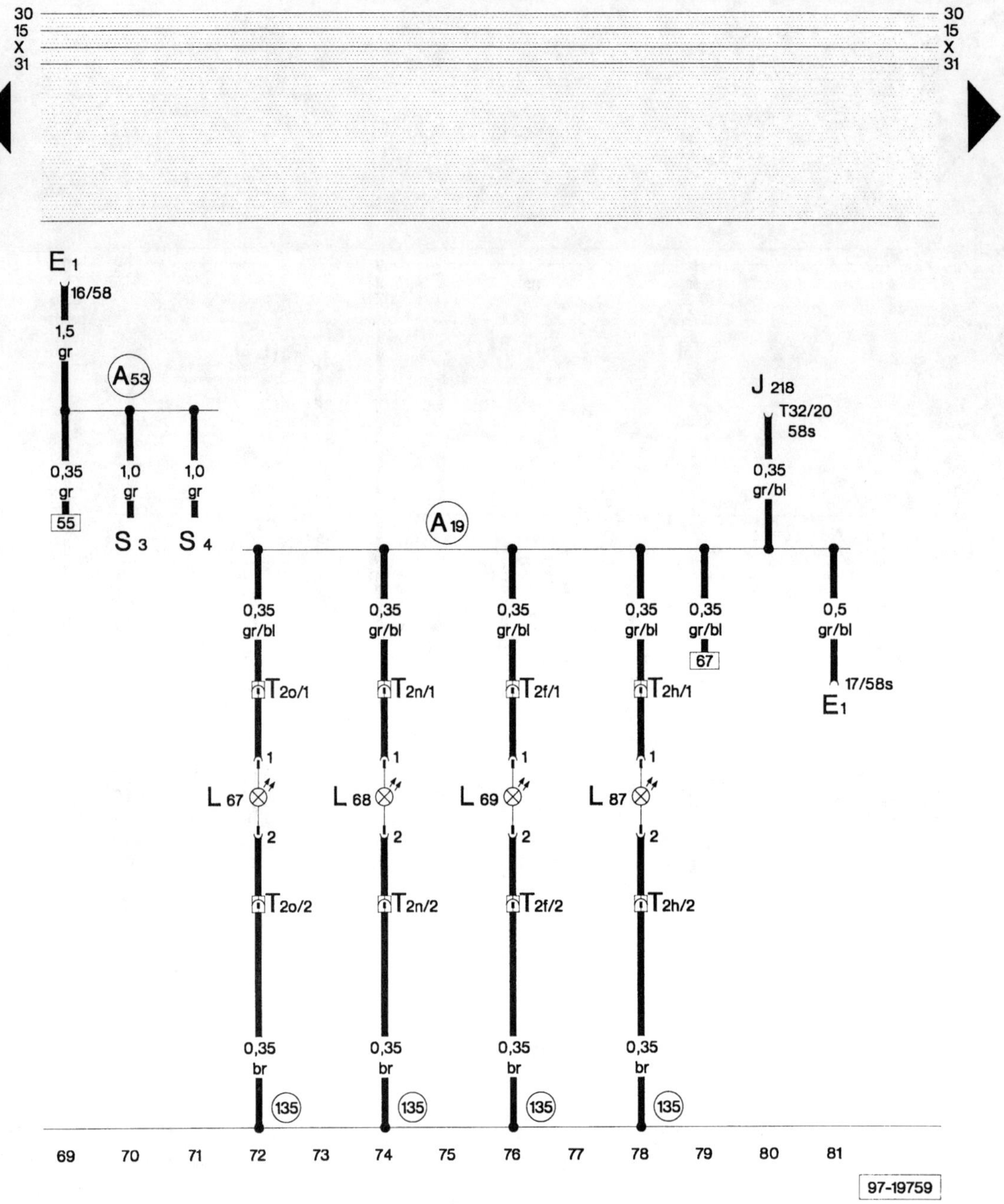

ws = white
sw = black
ro = red
gn = green
br = brown
bl = blue
gr = gray
li = violet
ge = yellow

E 1 - Light switch
J 218 - Instrument Cluster Combination Processor
L 67 - Left Instrument Panel Vent Illumination
L 68 - Center Instrument Panel Vent Illumination
L 69 - Right Instrument Panel Vent Illumination
L 87 - Lighting for rearseat outlet center
S 3 - Fuse
S 4 , - Fuse
T 2 f - Connector 2 pin, brown, behind instrument panel , right
T 2 h - Connector 2 pin, black, in console
T 2 n - Connector 2 pin, brown, behind instrument panel, center
T 2 o - Connector 2 pin, behind instrument panel, left
T 32 - Connector 32 pin, blue, on instrument cluster

(135) - Ground connection -2-, in instrument panel wiring harness
(A19) - Wire connection (58s), in instrument panel wiring harness (Switch illumination)
(A53) - Plus connection (58b), in instrument panel wiring harness

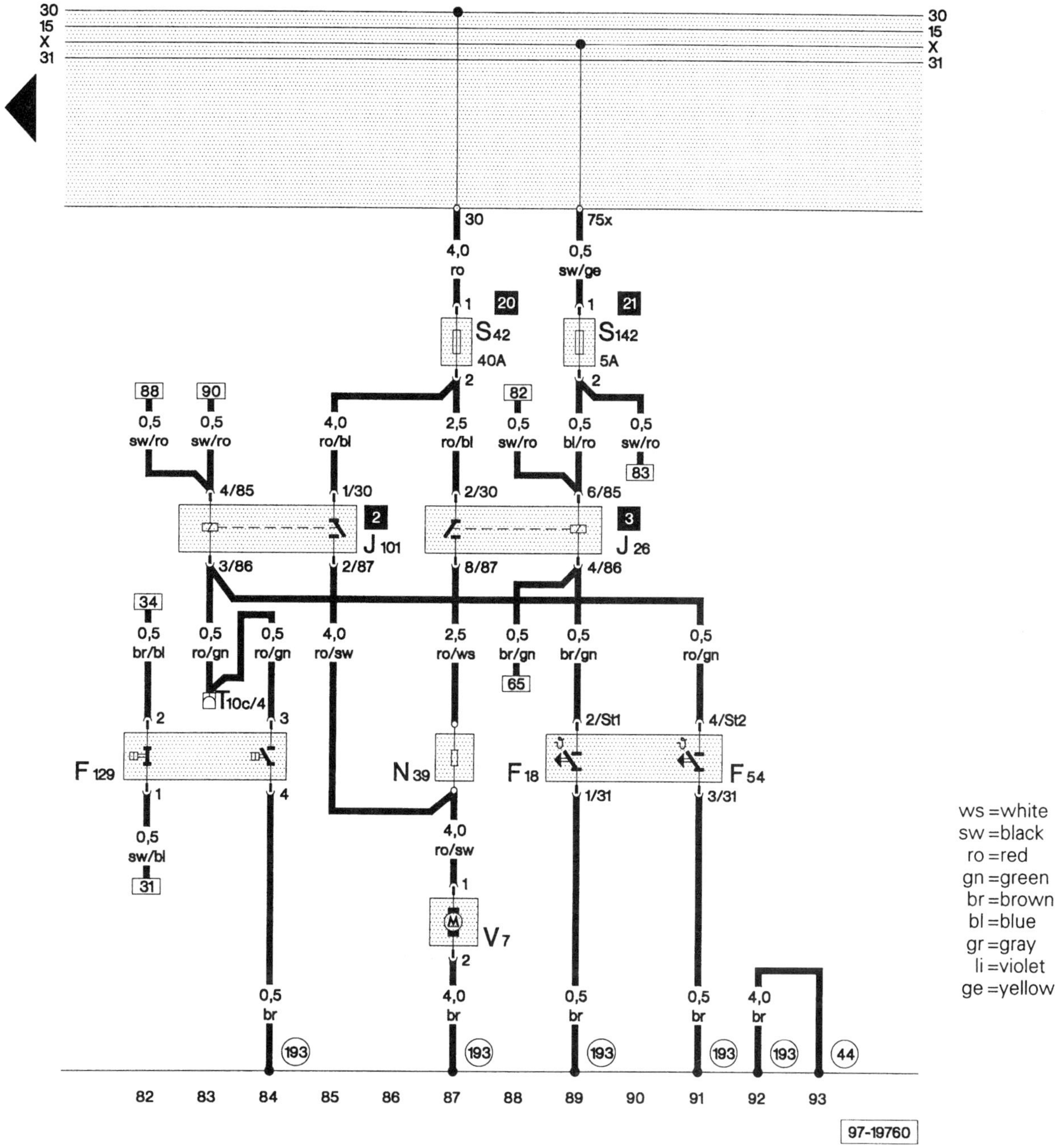

F 18 - Coolant Fan Control (FC) Thermal Switch
F 54 - Coolant Fan Control (FC) Thermal Switch
F 129 - A/C Pressure Switch
J 26 - Coolant Fan control (FC) Relay
J 101 - Second Speed Coolant Fan Control (FC) Relay
N 39 - Coolant Fan Control (FC) Series Resistance
S 42 - Coolant Fan Fuse
S 142 - Control module fuse for coolant fan
T 10 c - Connector 10 pin, grey, connector station A pillar, left
V 7 - Coolant Fan

(44) - Ground connection, on left A-pillar, lower part
(193) - Ground connection -1-, in coolant fan wiring harness

THIS PAGE INTENTIONALLY LEFT BLANK

Driver's Information System

1998 m. y.

Fuse Panel

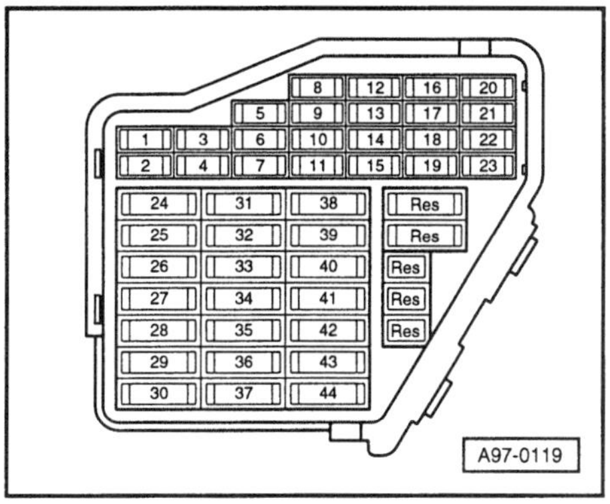

Fuse Colors:

30 A - Green
25 A - White
20 A - Yellow
15 A - Blue
10 A - Red
5 A - Beige

Starting with fuse position 23, fuses in the fuse holder are identified with 223 in the wiring diagram.

13 - Fold Relay Panel

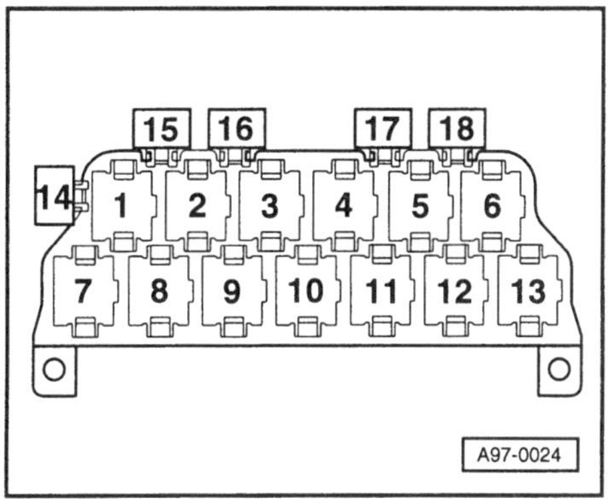

Relay Location:

9 - Lamp Control Module, J123

10 - Lamp Control Module, J123

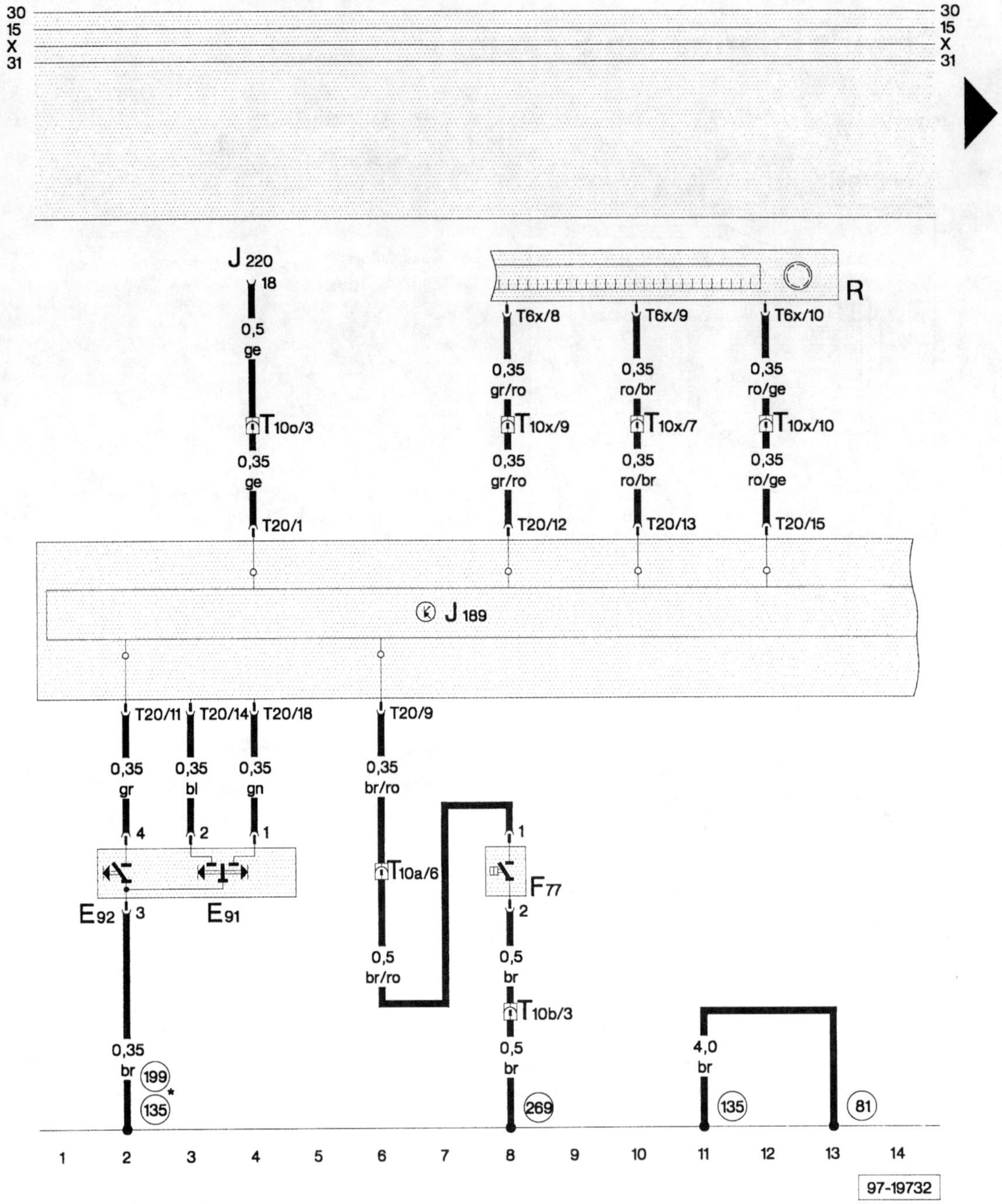

ws = white
sw = black
ro = red
gn = green
br = brown
bl = blue
gr = gray
li = violet
ge = yellow

E 91 - Board Computer Function Selector Switch
E 92 - Board Computer Reset Button
F 77 - Windshield Washer Fluid Level Warning Switch
J 189 - Auto Check System
J 220 - Motronic Engine Control Module (ECM)
R - Radio
T 6 x - Connector 6 pin, green, connector radio I - 2
T 10 a - Connector 10 pin, pink, connector station A pillar, left
T 10 b - Connector 10 pin, red, connector station A pillar, left
T 10 o - Connector 10 pin, brown, connector station electronic box plenum chamber
T 10 x - Connector 10 pin, pink, connector station A pillar, right
T 20 - Connector 20 pin, red/white, on instrument cluster

(81) - Ground connection -1-, in instrument panel wiring harness

(135) - Ground connection -2-, in instrument panel wiring harness

(199) - Ground connection -3-, in instrument panel wiring harness

(269) - Ground connector (sensor ground) -1-, in instrument panel wiring harness

* - booth is possible

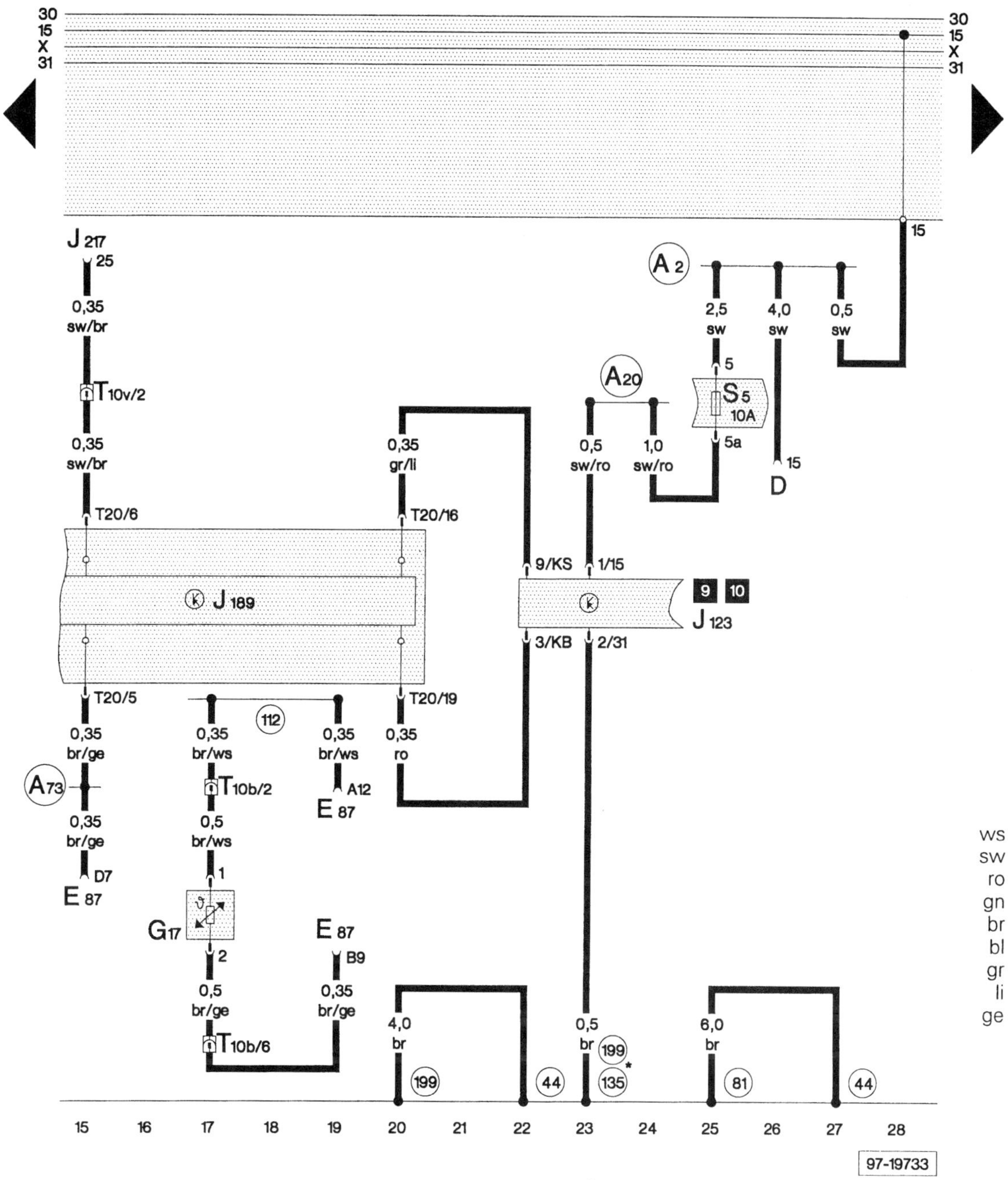

ws = white
sw = black
ro = red
gn = green
br = brown
bl = blue
gr = gray
li = violet
ge = yellow

D - Ignition/Starter Switch
E 87 - A/C Control Head
G 17 - Outside Air Temperature Sensor
J 123 - Lamp Control Module
J 189 - Auto Check System
J 217 - Transmission Control Module (TCM)
S 5 - Fuse
T 10 b - Connector 10 pin, red, connector station A pillar, left
T 10 v - Connector 10 pin, brown, connector station A pillar, right
T 20 - Connector 20 pin, red/white, on instrument cluster

(44) - Ground connection, on left A-pillar, lower part

(81) - Ground connection -1-, in instrument panel wiring harness

(112) - Ground connection -2-, in A/C wiring harness

(135) - Ground connection -2-, in instrument panel wiring harness

(199) - Ground connection -3-, in instrument panel wiring harness

(A2) - Plus connection (15), in instrument panel wiring harness

(A20) - Wire connection (15a), in instrument panel wiring harness

(A73) - Connector (outside temperature indicator), in instrument panel wiring harness

* - booth is possible

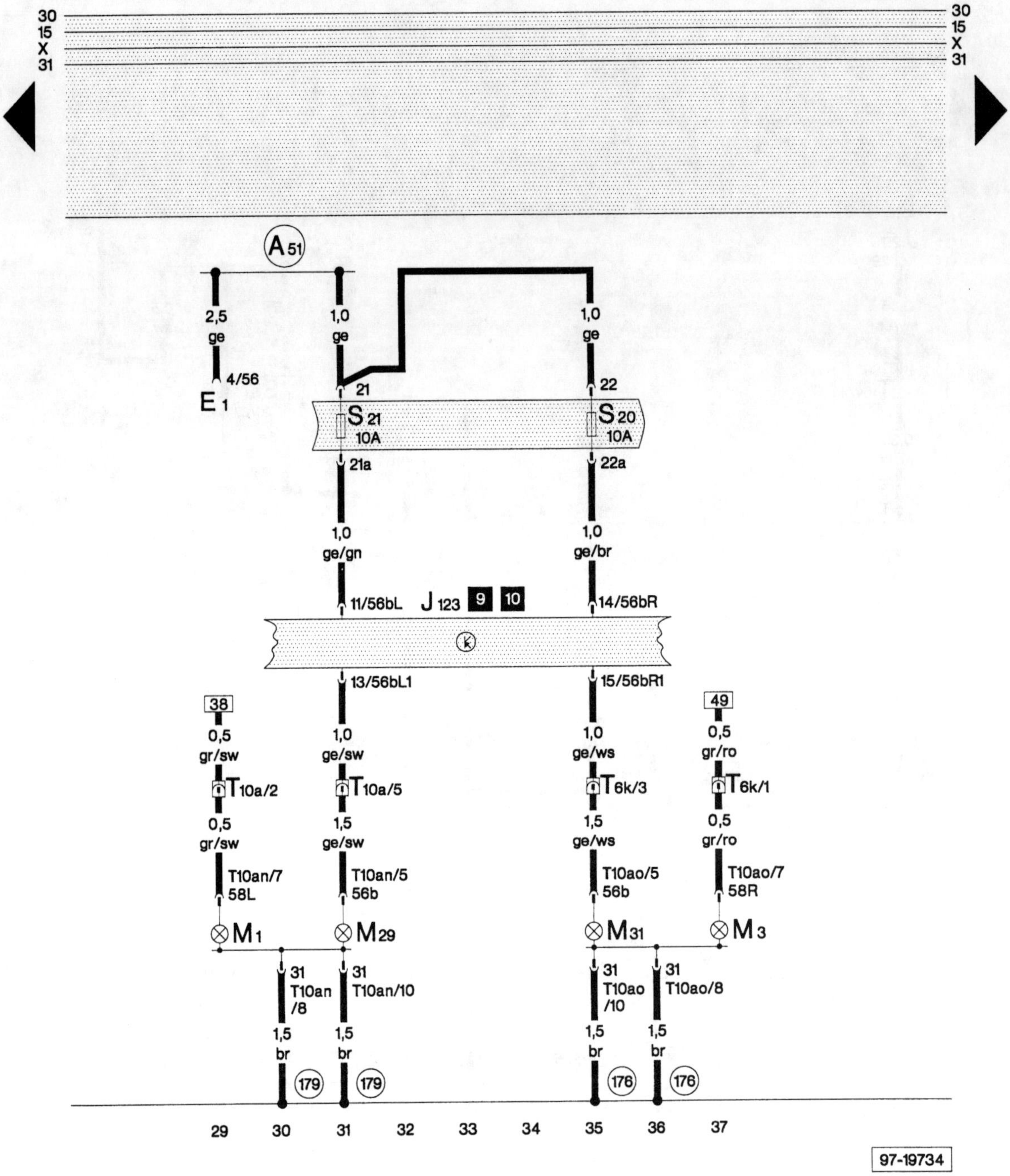

ws = white
sw = black
ro = red
gn = green
br = brown
bl = blue
gr = gray
li = violet
ge = yellow

E 1 - Light switch
J 123 - Lamp Control Module
M 1 - Left Parking Light
M 3 - Right Parking Light
M 29 - Left Low Beam Headlight
M 31 - Right Low Beam Headlight
S 20 - Fuse
S 21 - Fuse
T 6 k - Connector 6 pin, pink, connector station A pillar, right
T 10 a - Connector 10 pin, pink, connector station A pillar, left
T 10 an - Connector 10 pin, on headlight, left
T 10 ao - Connector 10 pin, on headlight, right

(176) - Ground connection, in right headlight wiring harness

(179) - Ground connection, in left headlight wiring harness

(A51) - Wire connection (56), in instrument panel wiring harness

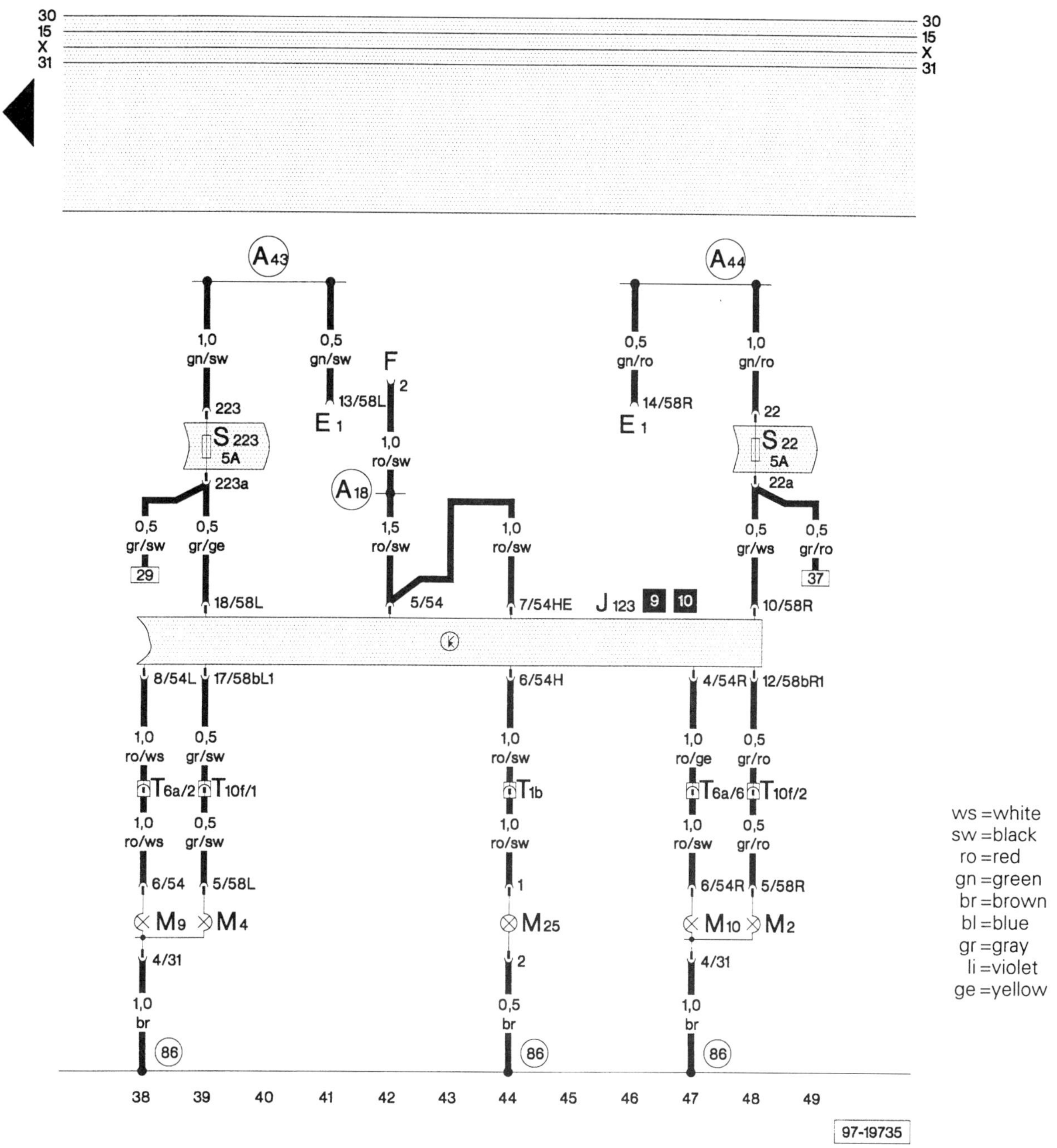

E 1 - Light switch
F - Brake Light Switch
J 123 - Lamp Control Module
M 2 - Right Tail Light
M 4 - Left Tail Light
M 9 - Left Brake Light
M 10 - Right Brake Light
M 25 - High-mount Brake Light
S 22 - Fuse
S 223 - Fuse
T 1 b - Single connector, black, connector station A pillar right
T 6 a - Connector 6 pin, blue, connector station A pillar, left
T 10 f - Connector 10 pin, brown, connector station A pillar, left

(86) - Ground connection -1-, in rear wiring harness
(A18) - Wire connection (54), in instrument panel wiring harness
(A43) - Wire connection (57L), in instrument panel wiring harness
(A44) - Wire connection (57R), in instrument panel wiring harness

Airbag

1998 m. y.

Fuse Panel

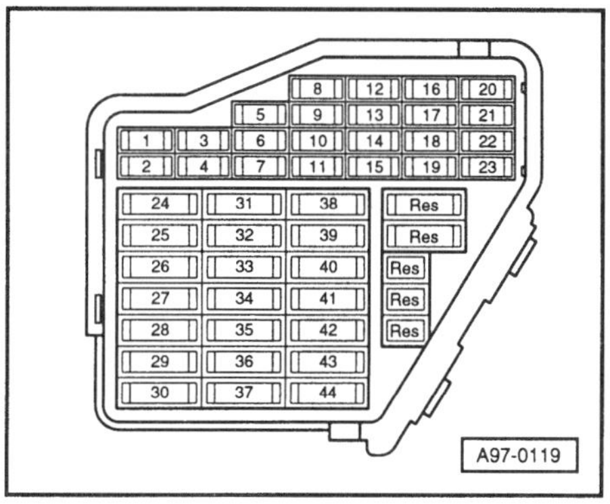

Fuse Colors:

30 A - Green
25 A - White
20 A - Yellow
15 A - Blue
10 A - Red
5 A - Beige

Starting with fuse position 23, fuses in the fuse holder are identified with 223 in the wiring diagram.

Micro Central Electric Panel

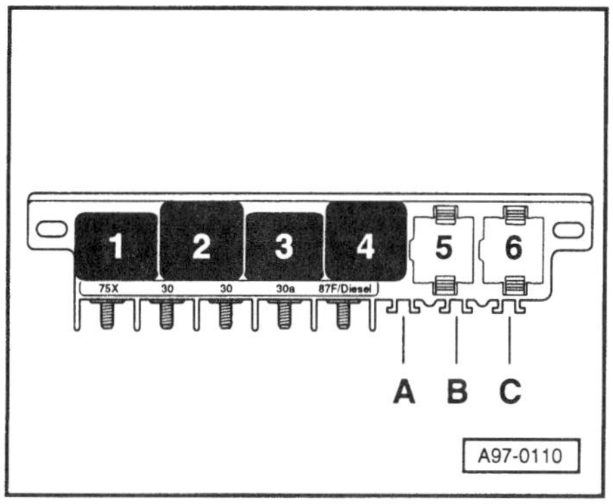

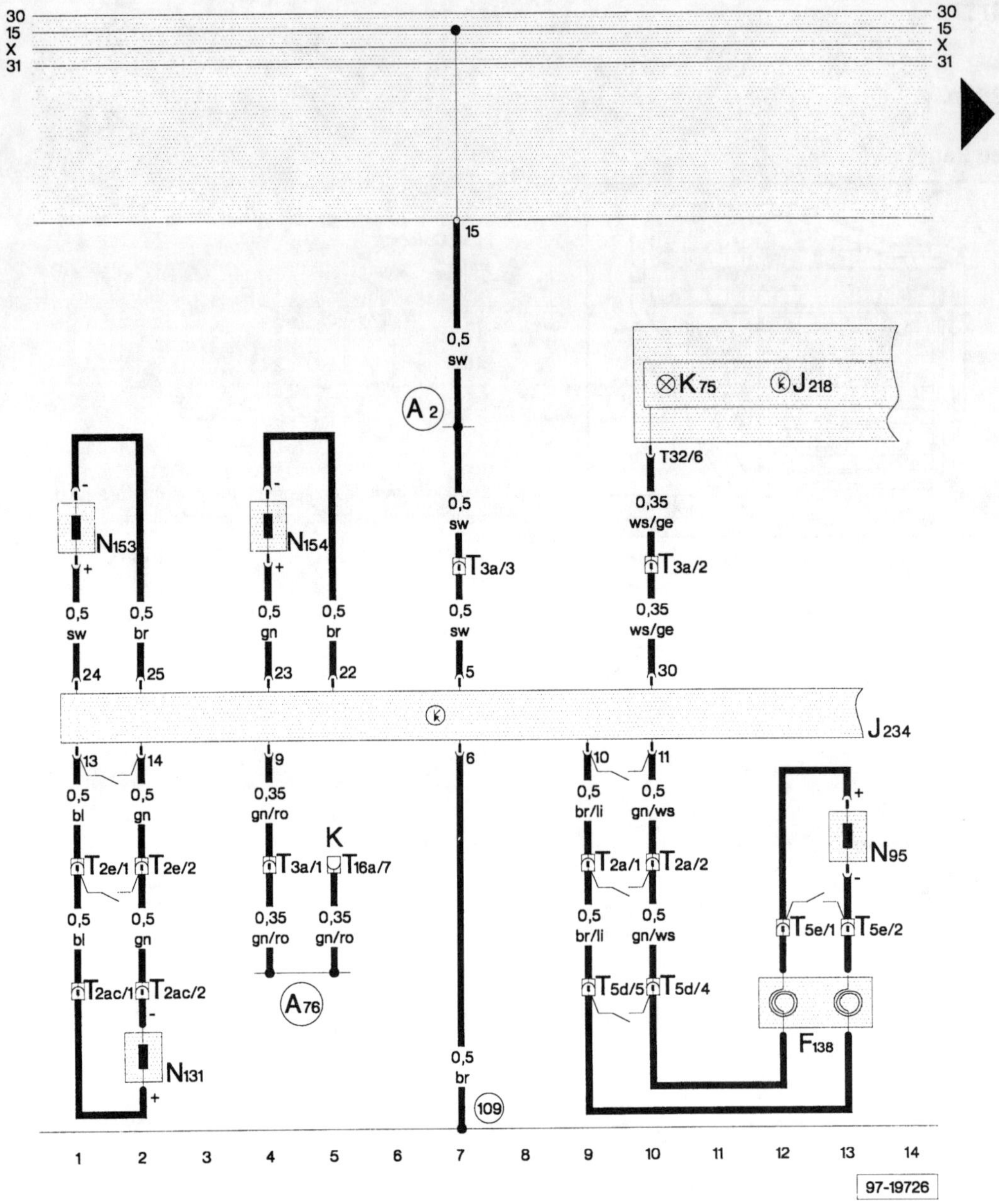

ws = white
sw = black
ro = red
gn = green
br = brown
bl = blue
gr = gray
li = violet
ge = yellow

F 138 - Airbag Spiral Spring/Return Spring With Slip Ring
J 218 - Instrument Cluster Combination Processor
J 234 - Airbag Control Module
K - DLC (Data Link Connector) Rapid Data Transfer
K 75 - Airbag Malfunction Indicator Lamp (MIL)
N 95 - Driver's Side Airbag Igniter
N 131 - Passenger's Side Airbag Igniter 1
N 153 - Left Seat Belt Tensioner Igniter
N 154 - Right Seat Belt Tensioner Igniter
T 2 a - Connector 2 pin, yellow, connector station A pillar, left
T 2 e - Connector 2 pin, yellow, connector station A pillar, right
T 2 ac - Connector 2 pin, yellow, behind instrument panel, right
T 3 a - Connector 3 pin, black, connector station A pillar, right
T 5 d - Connector 5 pin, behind steering column switch cover
T 5 e - Connector 5 pin, in steering wheel
T 16 a - Connector 16 pin, data link connector
T 32 - Connector 32 pin, blue, on instrument cluster

(109) - Ground connection, in airbag wiring harness
(A2) - Plus connection (15), in instrument panel wiring harness
(A76) - Connector (K-diagnosis wire), in instrument panel wiring harness

- Bridge closes when disconnecting connector (safety feature)

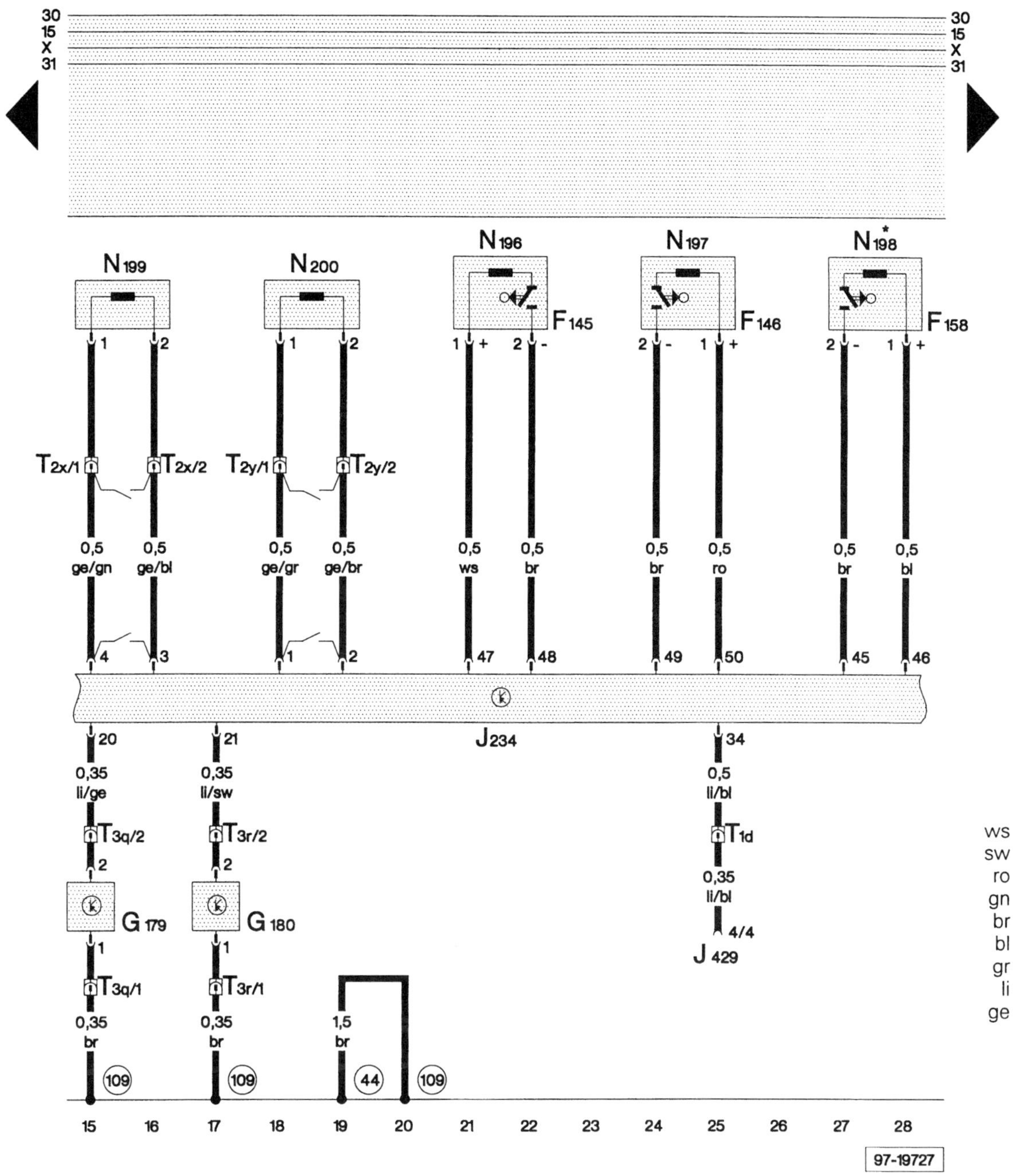

ws = white
sw = black
ro = red
gn = green
br = brown
bl = blue
gr = gray
li = violet
ge = yellow

F 145 - Left Rear B-pillar Microswitch
F 146 - Right Rear B-pillar Microswitch
F 158 - Seat Belt Tensioner Switch (1)
G 179 - Crash sensor for side airbag, driver's side
G 180 - Crash sensor for side airbag, passenger side
J 234 - Airbag Control Module
J 429 - Control module for central locking
N 196 - Igniter for belt tensioner rear, driver's side
N 197 - Igniter for belt tensioner rear, passenger side
N 198 - Igniter for belt tensioner, rear, center
N 199 - Igniter for side airbag, driver's side
N 200 - Igniter for side airbag, passenger side
T 1 d - Single connector, brown, connector station A pillar right
T 2 x - Connector 2 pin, yellow, below driver's seat
T 2 y - Connector 2 pin, yellow, below passenger's seat
T 3 q - Connector 3 pin, below driver's seat
T 3 r - Connector 3 pin, below passenger's seat

(44) - Ground connection, on left A-pillar, lower part

(109) - Ground connection, in airbag wiring harness

* - Only undivided seatback rear saet

- Bridge closes when disconnecting connector (safety feature)

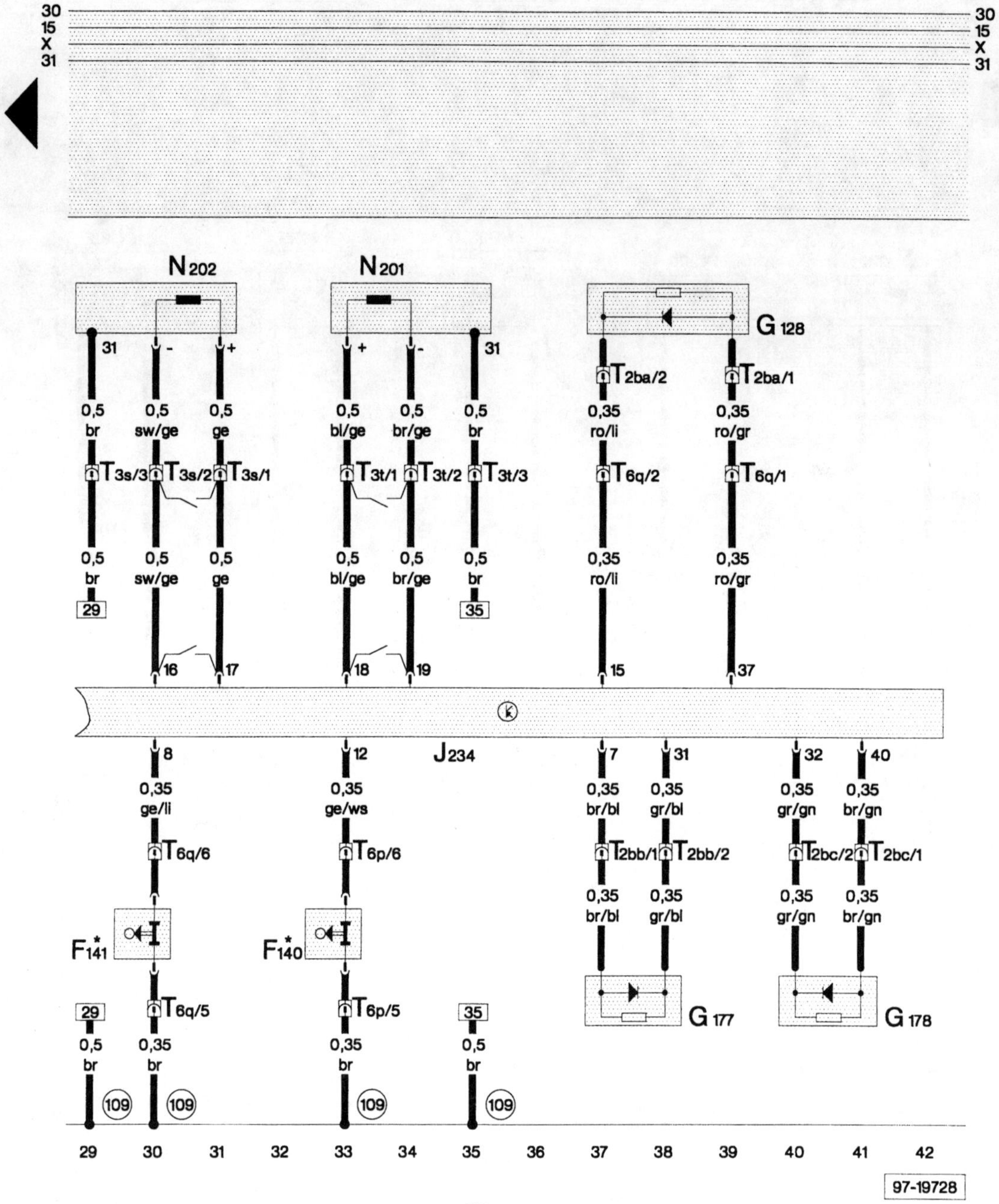

ws = white
sw = black
ro = red
gn = green
br = brown
bl = blue
gr = gray
li = violet
ge = yellow

F 140 - Left Front A-pillar Microswitch
F 141 - Right Front A-pillar Microswitch
G 128 - Seat occupiet sensor, passenger side
G 177 - Sensor for occupied sensor seat, rear, driver side
G 178 - Sensor for occupied sensor seat, rear, passenger side
J 234 - Airbag Control Module
N 201 - Igniter for side airbag rear, driver's side
N 202 - Igniter for side airbag rear, passenger side
T 2 ba - Connector 2 pin, in passenger's seat
T 2 bb - Connector 2 pin, in rear seat
T 2 bc - Connector 2 pin, in rear seat
T 3 s - Connector 3 pin, orange, in rear seat
T 3 t - Connector 3 pin, orange, in rear seat
T 6 p - Connector 6 pin, red, below driver's seat
T 6 q - Connector 6 pin, red, below passenger's seat

(109) - Ground connection, in airbag wiring harness

* - Fasten Seat Belt

Heated Seat

1998 m. y.

Fuse Panel

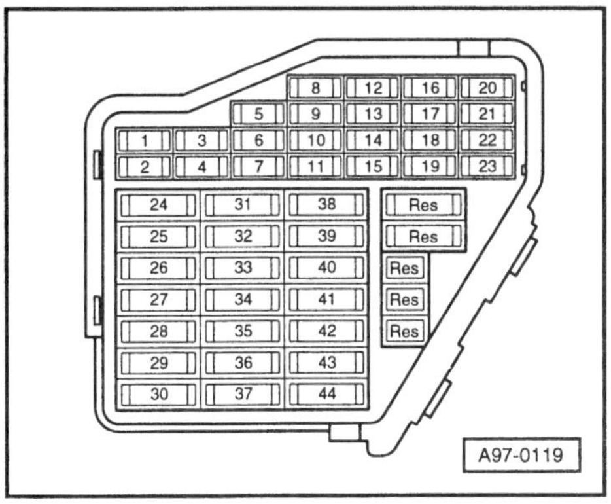

Fuse Colors:

30 A - Green
25 A - White
20 A - Yellow
15 A - Blue
10 A - Red
5 A - Beige

Starting with fuse position 23, fuses in the fuse holder are identified with 223 in the wiring diagram.

Micro Central Electric Panel

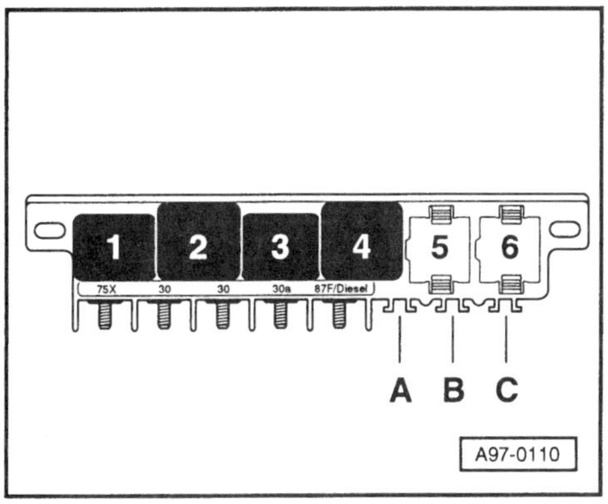

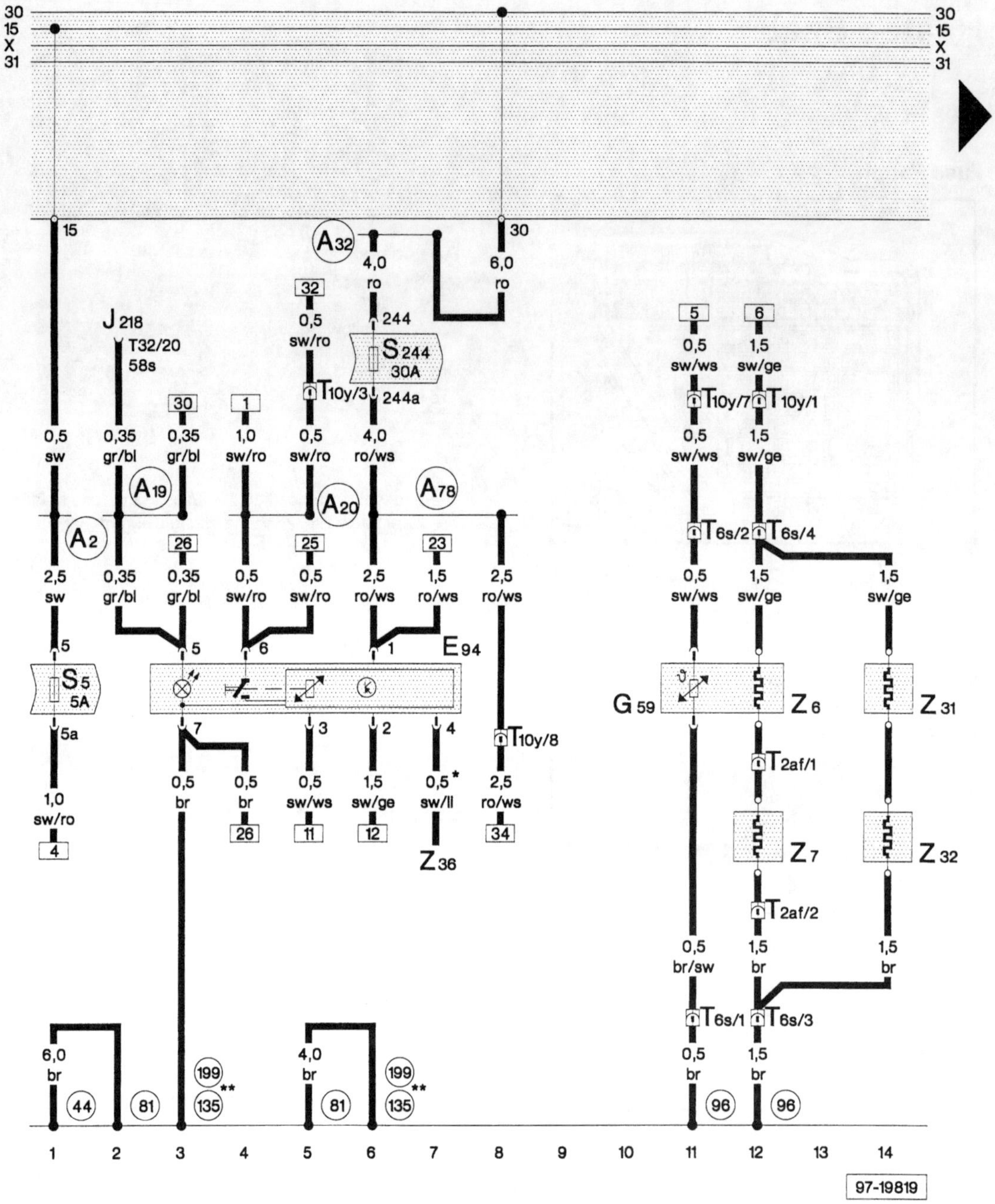

ws = white
sw = black
ro = red
gn = green
br = brown
bl = blue
gr = gray
li = violet
ge = yellow

E 94 - Adjuster for heated driver's seat
G 59 - Driver's Heated Seat Temperature Sensor
J 218 - Instrument Cluster Combination Processor
S 5 - Fuse
S 244 - Fuse
T 2 af - Connector 2 pin, black, below driver's seat
T 6 s - Connector 6 pin, green, below driver's seat
T 10 y - Connector 10 pin, green, connector station A pillar, right
T 32 - Connector 32 pin, blue, on instrument cluster
Z 6 - Driver's Seat Heating Element
Z 7 - Driver's Backrest Heating Element
Z 31 - Facial heater mounted to backrest, driver's side
Z 32 - Facial heater mounted to driver's seatback
Z 36 - Heated steering wheel

(44) - Ground connection, on left A-pillar, lower part

(81) - Ground connection -1-, in instrument panel wiring harness

(96) - Ground connection -1-, in heated seats wiring harness

(135) - Ground connection -2-, in instrument panel wiring harness

(199) - Ground connection -3-, in instrument panel wiring harness

(A2) - Plus connection (15), in instrument panel wiring harness

(A19) - Wire connection (58s), in instrument panel wiring harness (Switch illumination)

(A20) - Wire connection (15a), in instrument panel wiring harness

(A32) - Plus connection (30), in instrument panel wiring harness

(A78) - Connector (30a, S244), in instrument panel wiring harness

* - Vehicles with heated steering wheel
** - booth is possible

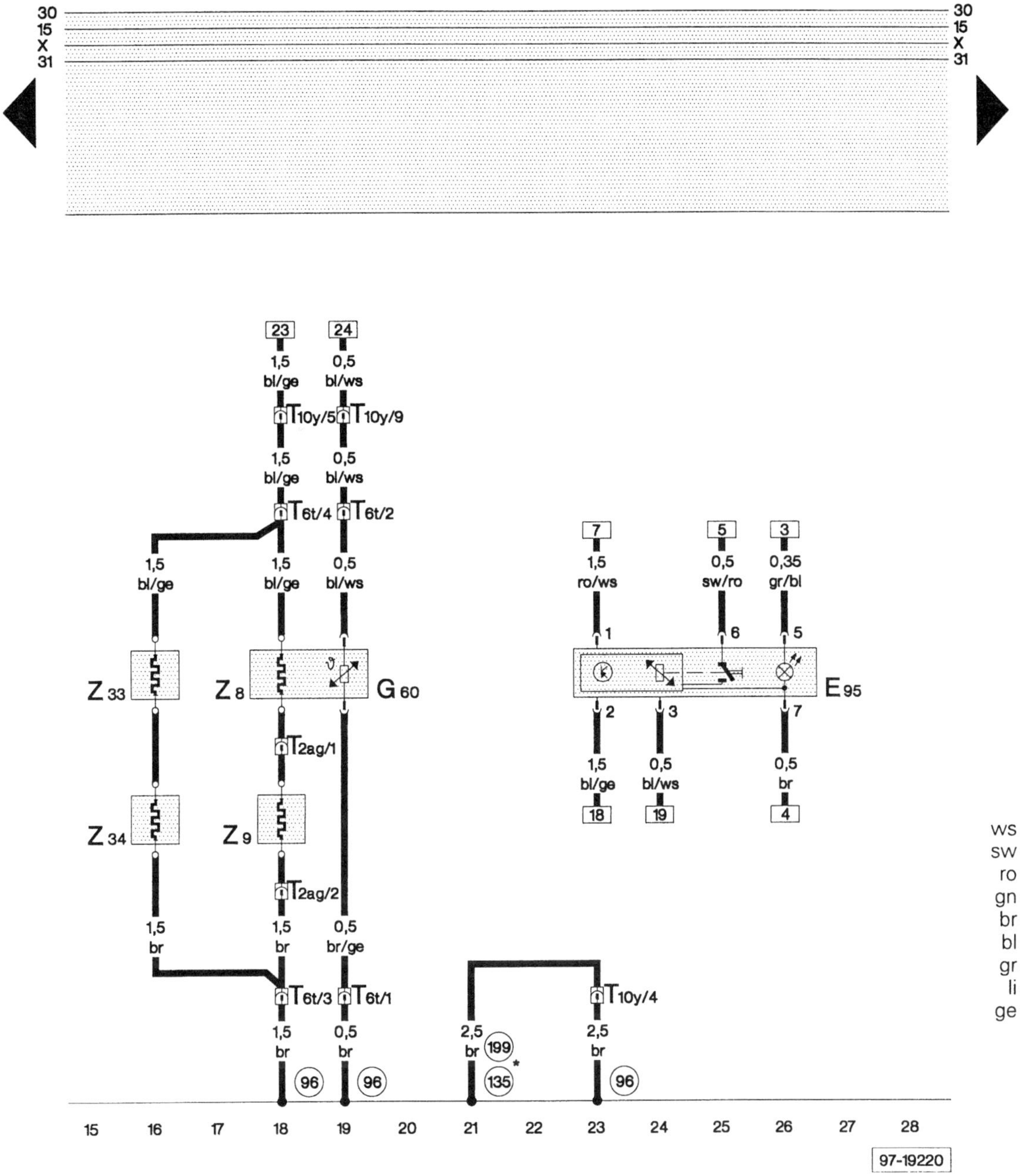

ws = white
sw = black
ro = red
gn = green
br = brown
bl = blue
gr = gray
li = violet
ge = yellow

E 95 - Adjuster for heated passenger's seat
G 60 - Passenger's Heated Seat Temperature Sensor
T 2 ag - Connector 2 pin, black, below passenger's seat
T 6 t - Connector 6 pin, green, below passenger's seat
T 10 y - Connector 10 pin, green, connector station A pillar, right
Z 8 - Passenger's Seat Heating Element
Z 9 - Passenger's Backrest Heating Element
Z 33 - Facial heater, passenger side
Z 34 - Facial heater, passenger seatback

(96) - Ground connection -1-, in heated seats wiring harness
(135) - Ground connection -2-, in instrument panel wiring harness
(199) - Ground connection -3-, in instrument panel wiring harness

* - booth ist possible

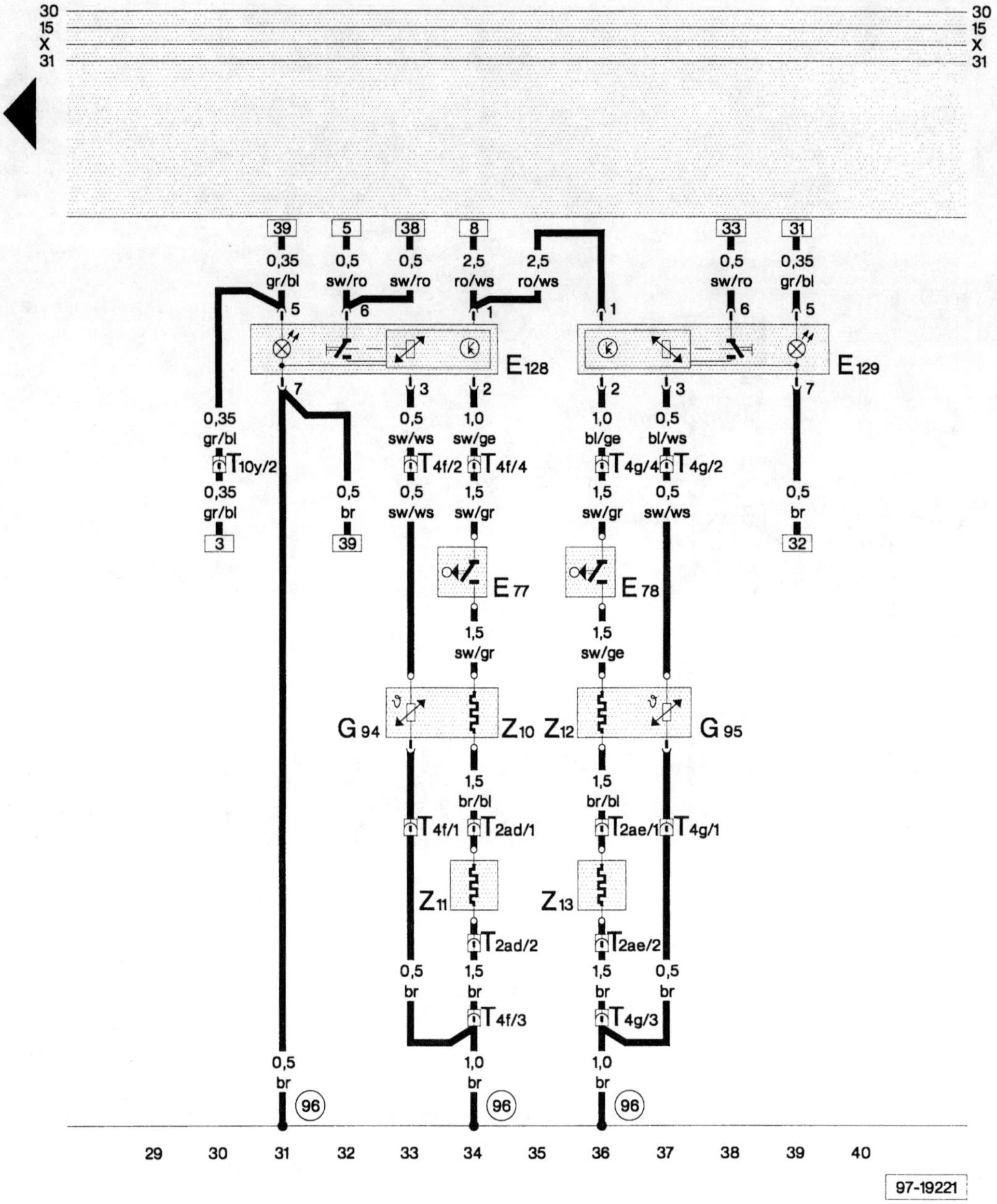

ws = white
sw = black
ro = red
gn = green
br = brown
bl = blue
gr = gray
li = violet
ge = yellow

E 77 - Left Rear Heated Seat Switch
E 78 - Right Rear Heated Seat Switch
E 128 - Left Rear Heated Seat Regulating Switch
E 129 - Right Rear Heated Seat Regulating Switch
G 94 - Left Rear Heated Seat Temperature Sensor
G 95 - Right Rear Heated Seat Temperature Sensor
T 2 ad - Connector 2 pin, black, below rear seat, left
T 2 ae - Connector 2 pin, black, below rear seat, right
T 4 f - Connector 4 pin, black, below rear seat, left
T 4 g - Connector 4 pin, black, below rear seat, right
T 10 y - Connector 10 pin, green, connector station A pillar, right
Z 10 - Left Rear Seat Heating Element
Z 11 - Left Rear Backrest Heating Element
Z 12 - Right Rear Seat Heating Element
Z 13 - Right Rear Backrest Heating Element

(96) - Ground connection -1-, in heated seats wiring harness

ABS

1998 m. y.

Fuse Panel

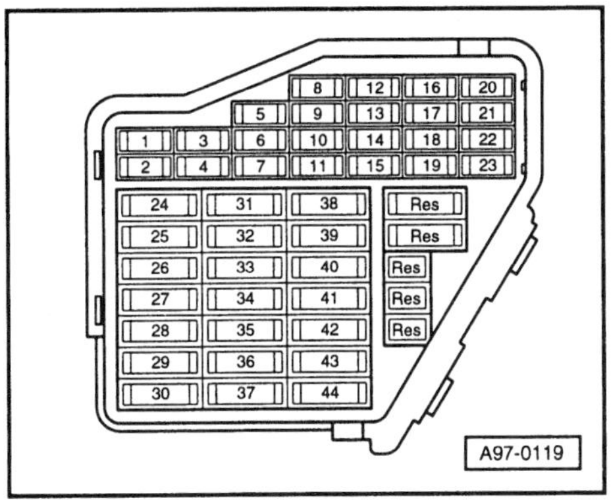

Fuse Colors:

30 A - Green
25 A - White
20 A - Yellow
15 A - Blue
10 A - Red
5 A - Beige

Starting with fuse position 23, fuses in the fuse holder are identified with 223 in the wiring diagram.

8 - Fold Relay Panel

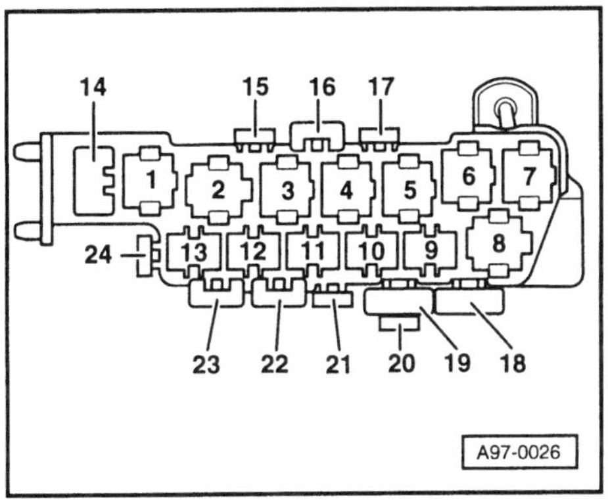

Relay Location:

18 - Fuse 1 for control module - ABS, S123

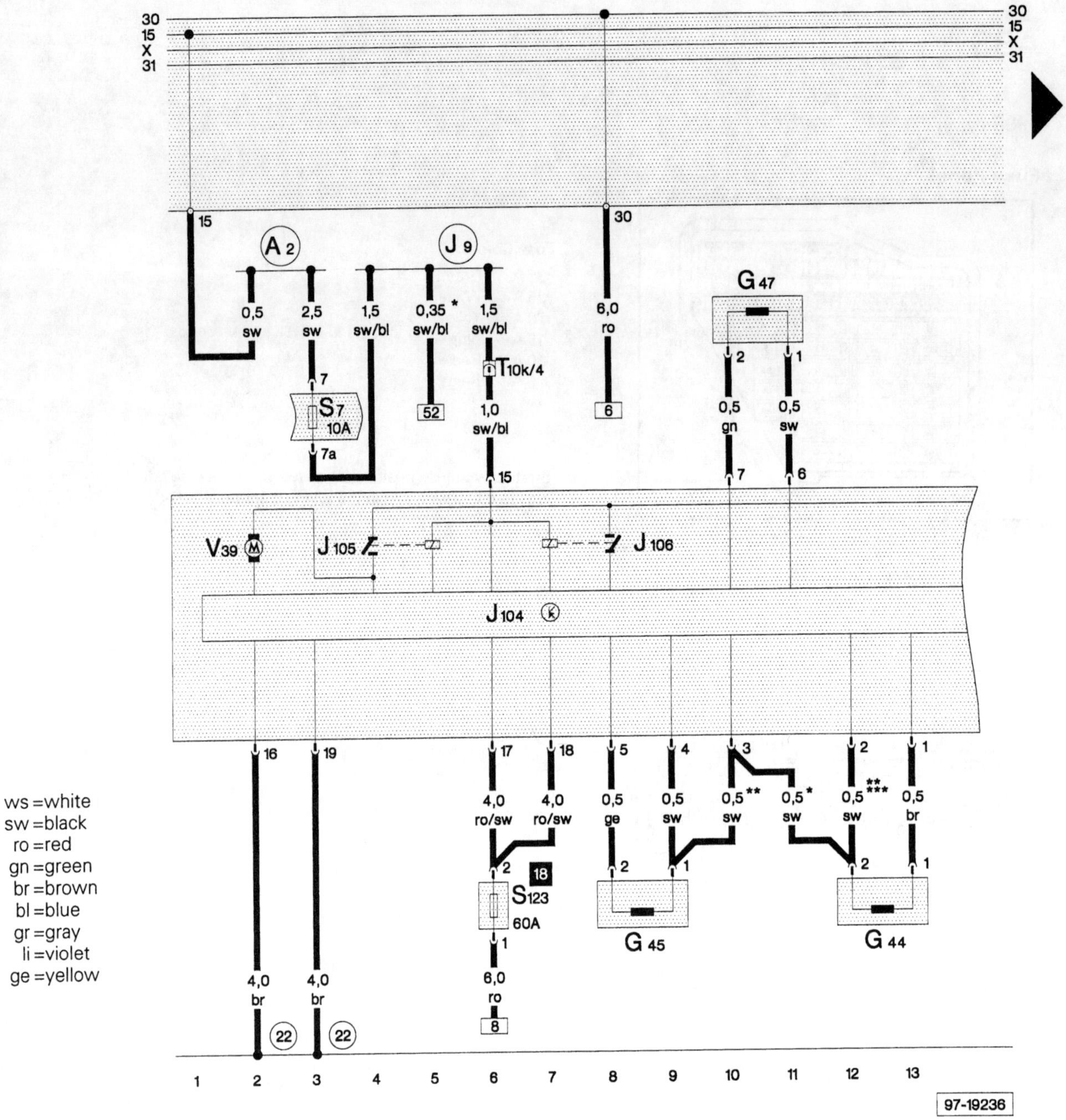

G 44 - Right Rear ABS Wheel Speed Sensor
G 45 - Right Front ABS Wheel Speed Sensor
G 47 - Left Front ABS Wheel Speed Sensor
J 104 - ABS Control Module (w/EDL)
J 105 - ABS Return Flow Pump Relay
J 106 - ABS Solenoid Valve Relay
S 7 - Fuse
S 123 - Fuse 1 for control module - ABS
T 10 k - Connector 10 pin, orange, connector station A pillar, left
V 39 - ABS Return Flow Pump

(22) - Ground connection, on hydraulic unit

(A2) - Plus connection (15), in instrument panel wiring harness

(J9) - Wire connection -1- (15a), in ABS wiring harness

* - Vehicles without Anti-Slip Control
** - Vehicles with Anti-Slip Control
*** - Vehicles with all-wheel drive

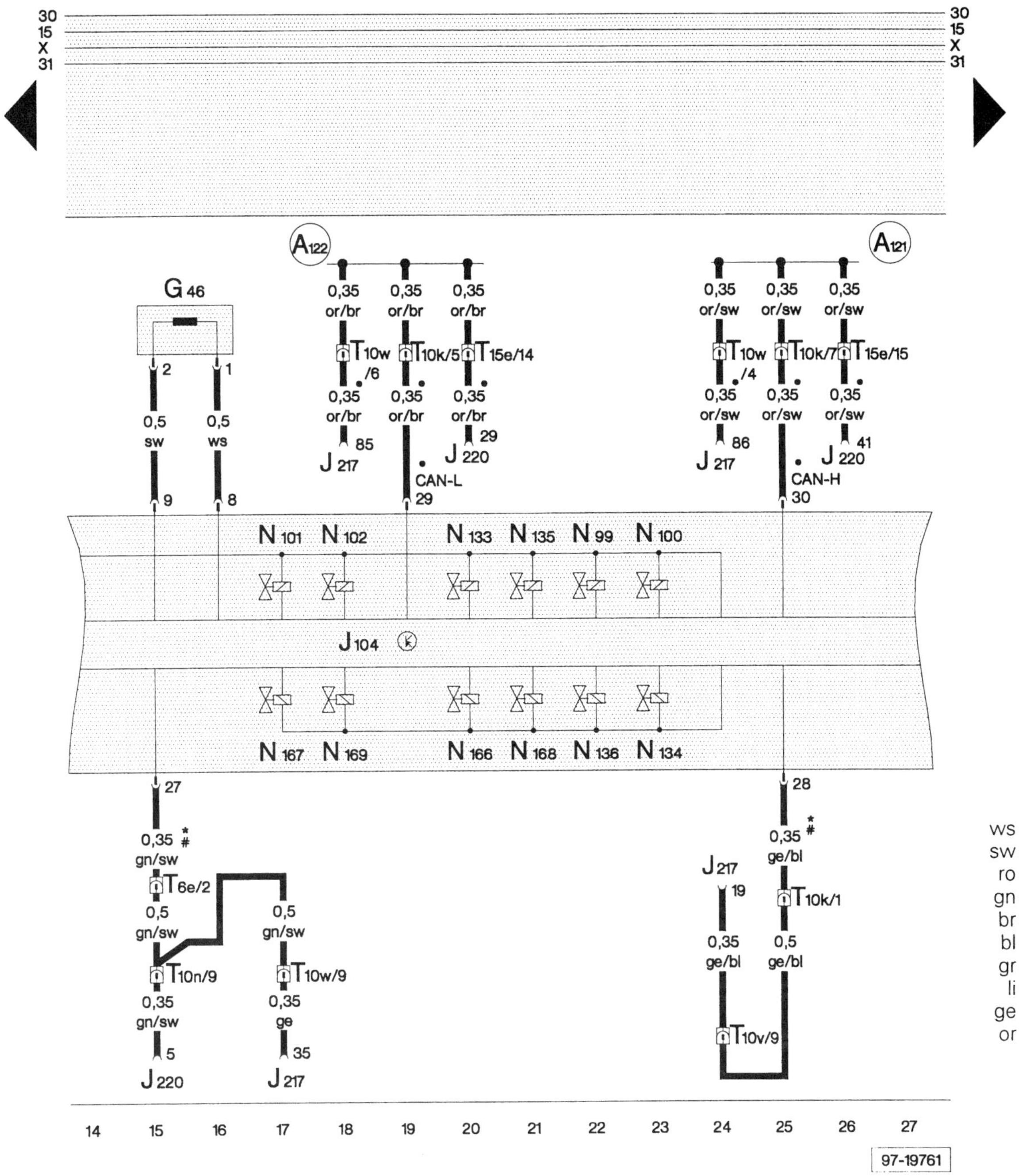

ws = white
sw = black
ro = red
gn = green
br = brown
bl = blue
gr = gray
li = violet
ge = yellow
or = orange

G 46 - Left Rear ABS Wheel Speed Sensor
J 104 - ABS Control Module (w/EDL)
J 217 - Transmission Control Module (TCM)
J 220 - Motronic Engine Control Module (ECM)
N 99 - Right Front ABS Inlet Valve
N 100 - Right Front ABS Outlet Valve
N 101 - Left Front ABS Inlet Valve
N 102 - Left Front ABS Outlet Valve
N 133 - Right Rear ABS Inlet Valve
N 134 - Left Rear ABS Inlet Valve
N 135 - Right Rear ABS Outlet Valve
N 136 - Left Rear ABS Outlet Valve
N 166 - Right Front Traction Control Switch-Over Valve
N 167 - Right Front Traction Control Outlet Valve
N 168 - Left Front Traction Control Switch-Over Valve
N 169 - Left Front Traction Control Outlet Valve
T 6 e - Connector 6 pin, grey, connector station A pillar, right
T 10 k - Connector 10 pin, orange, connector station A pillar, left
T 10 n - Connector 10 pin, yellow, connector station electronic box plenum chamber
T 10 v - Connector 10 pin, brown, connector station A pillar, right
T 10 w - Connector 10 pin, blue, connector station A pillar, right
T 15 e - Connector 15 pin, white, connector station electronic box plenum chamber

(A121) - Connector (High-Bus), in instrument panel wiring harness
(A122) - Connector (Low-Bus), in instrument panel wiring harness

• - CAN-Bus
* - Vehicles with Anti-Slip Control
# - Vehicles without CAN-Bus

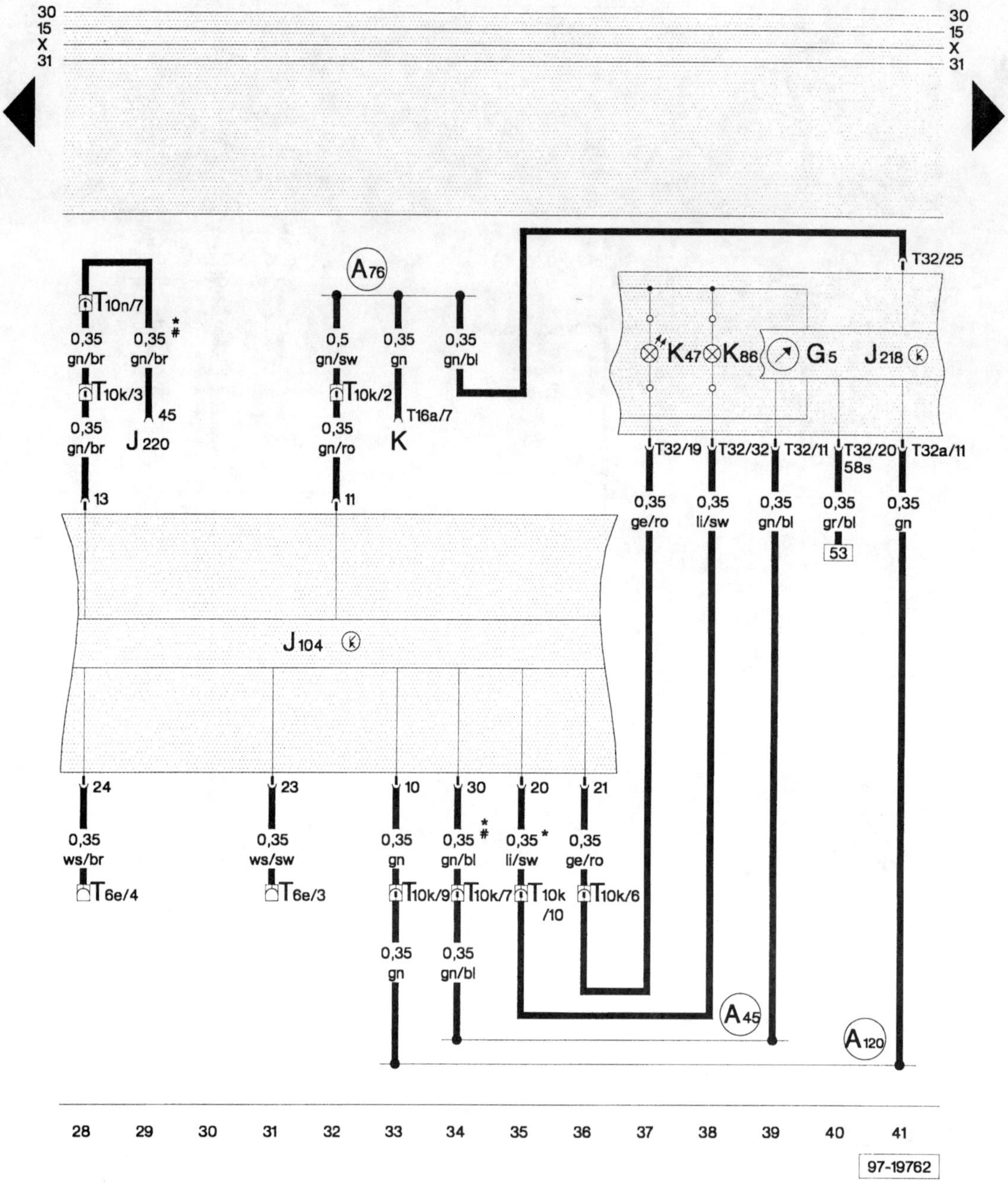

ws = white
sw = black
ro = red
gn = green
br = brown
bl = blue
gr = gray
li = violet
ge = yellow

G 5 - Tachometer
J 104 - ABS Control Module (w/EDL)
J 218 - Instrument Cluster Combination Processor
J 220 - Motronic Engine Control Module (ECM)
K - DLC (Data Link Connector) Rapid Data Transfer
K 47 - ABS Warning Light
K 86 - Traction Control Indicator Light
T 6 e - Connector 6 pin, grey, connector station A pillar, right
T 10 k - Connector 10 pin, orange, connector station A pillar, left
T 10 n - Connector 10 pin, yellow, connector station electronic box plenum chamber
T 16 a - Connector 16 pin, data link connector
T 32 - Connector 32 pin, blue, on instrument cluster
T 32 a - Connector 32 pin, green, on instrument cluster

(A45) - Wire connection (RPM-signal), in instrument panel wiring harness
(A76) - Connector (K-diagnosis wire), in instrument panel wiring harness
(A120) - Wire connection (Immobilization time), in instrument panel wiring harness

* - Vehicles with Anti-Slip Control
# - Vehicles without CAN-Bus

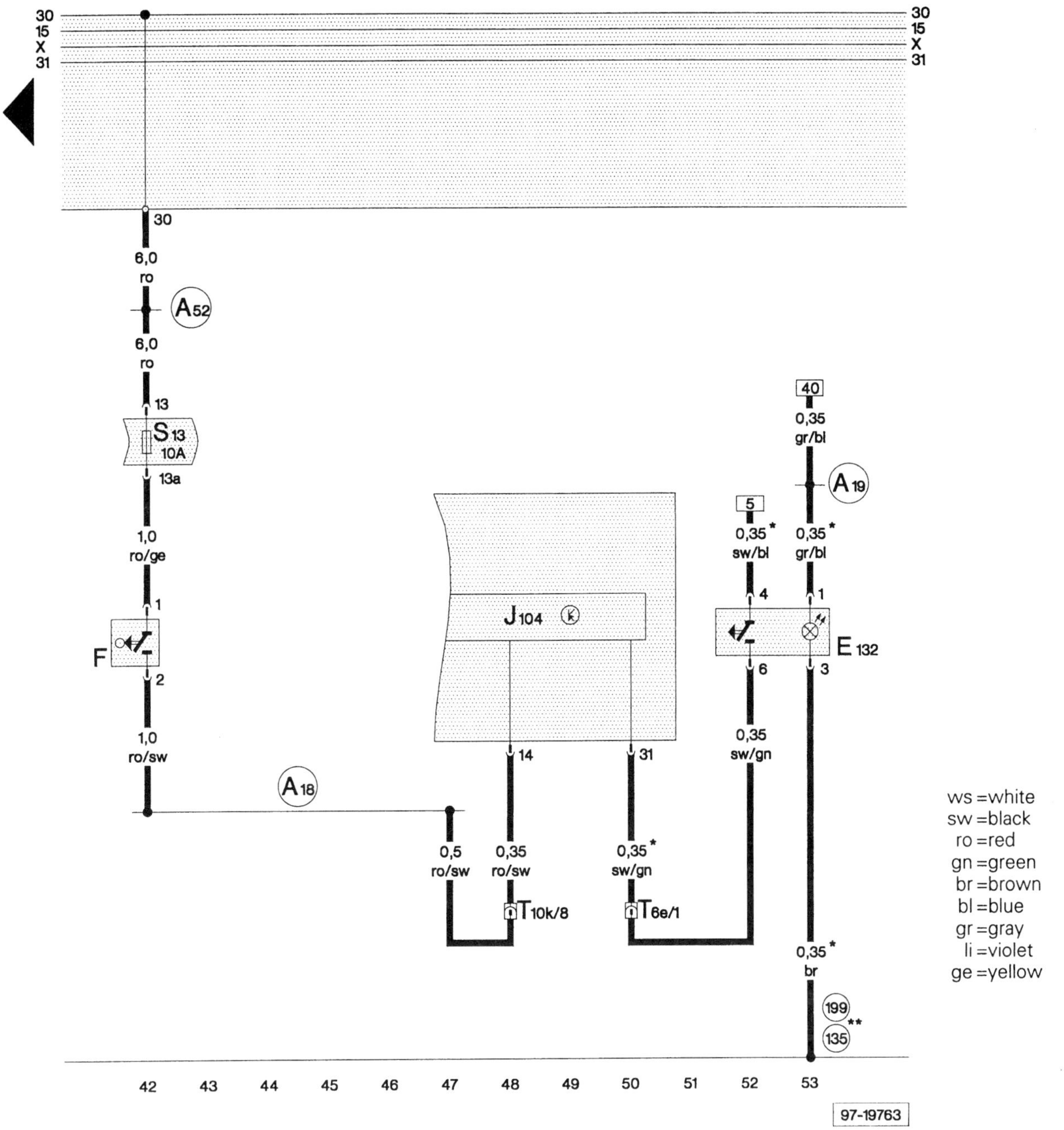

E 132 - Anti-Slip Control Switch
F - Brake Light Switch
J 104 - ABS Control Module (w/EDL)
S 13 - Fuse
T 6 e - Connector 6 pin, grey, connector station A pillar, right
T 10 k - Connector 10 pin, orange, connector station A pillar, left
T 10 l - Connector 10 pin, white, connector station electronic box plenum chamber
T 15 e - Connector 15 pin, white, connector station electronic box plenum chamber

(135) - Ground connection -2-, in instrument panel wiring harness
(199) - Ground connection -3-, in instrument panel wiring harness
(A18) - Wire connection (54), in instrument panel wiring harness
(A19) - Wire connection (58s), in instrument panel wiring harness (Switch illumination)
(A52) - Plus connection (30), in instrument panel wiring harness

* - Vehicles with Anti-Slip Control
** - booth is possible

THIS PAGE INTENTIONALLY LEFT
BLANK

Power Seat and Mirror Adjustment with Memory, Mirror with and without fold-away Function

1998 m. y.

Fuse Panel

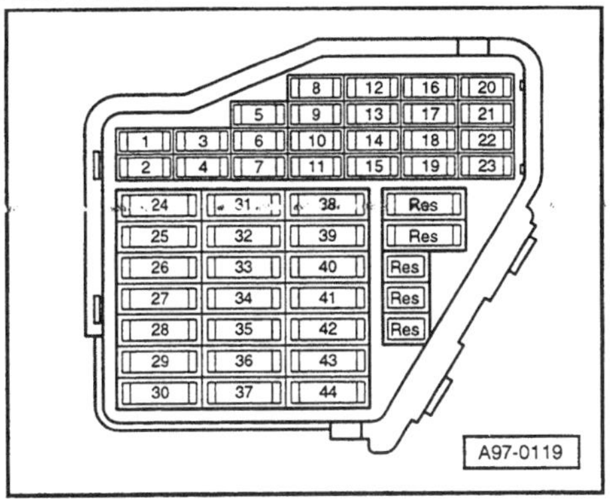

Fuse Colors:

30 A - Green
25 A - White
20 A - Yellow
15 A - Blue
10 A - Red
7,5 A - Brown
5 A - Beige

Starting with fuse position 23, fuses in the fuse holder are identified with 223 in the wiring diagram.

13 - Fold Relay Panel

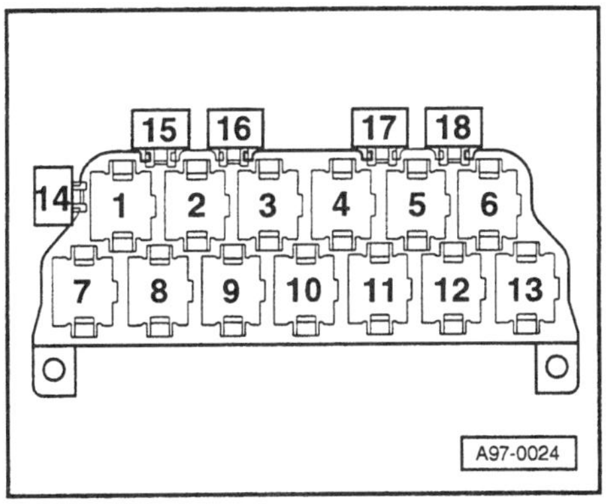

Relay Location:

11 - Control Module for Mirror fold-away Function, J351

12 - Control Module for Mirror fold-away Function, J351

8 - Fold Relay Panel

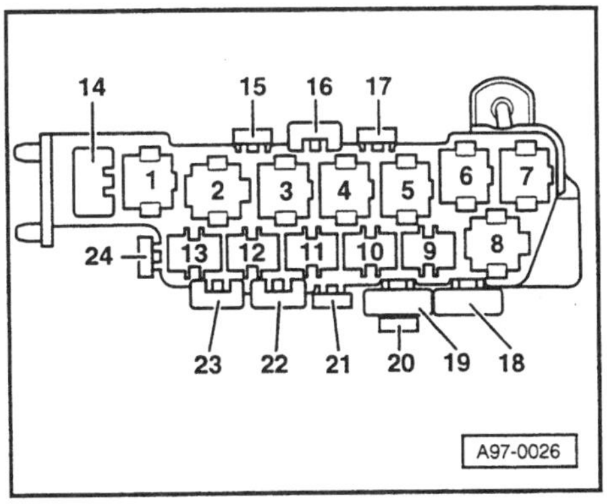

Fuse Location:

14 - Power Seat Circuit Breaker 1, S44

24 - Power Seat Circuit Breaker 2, S80

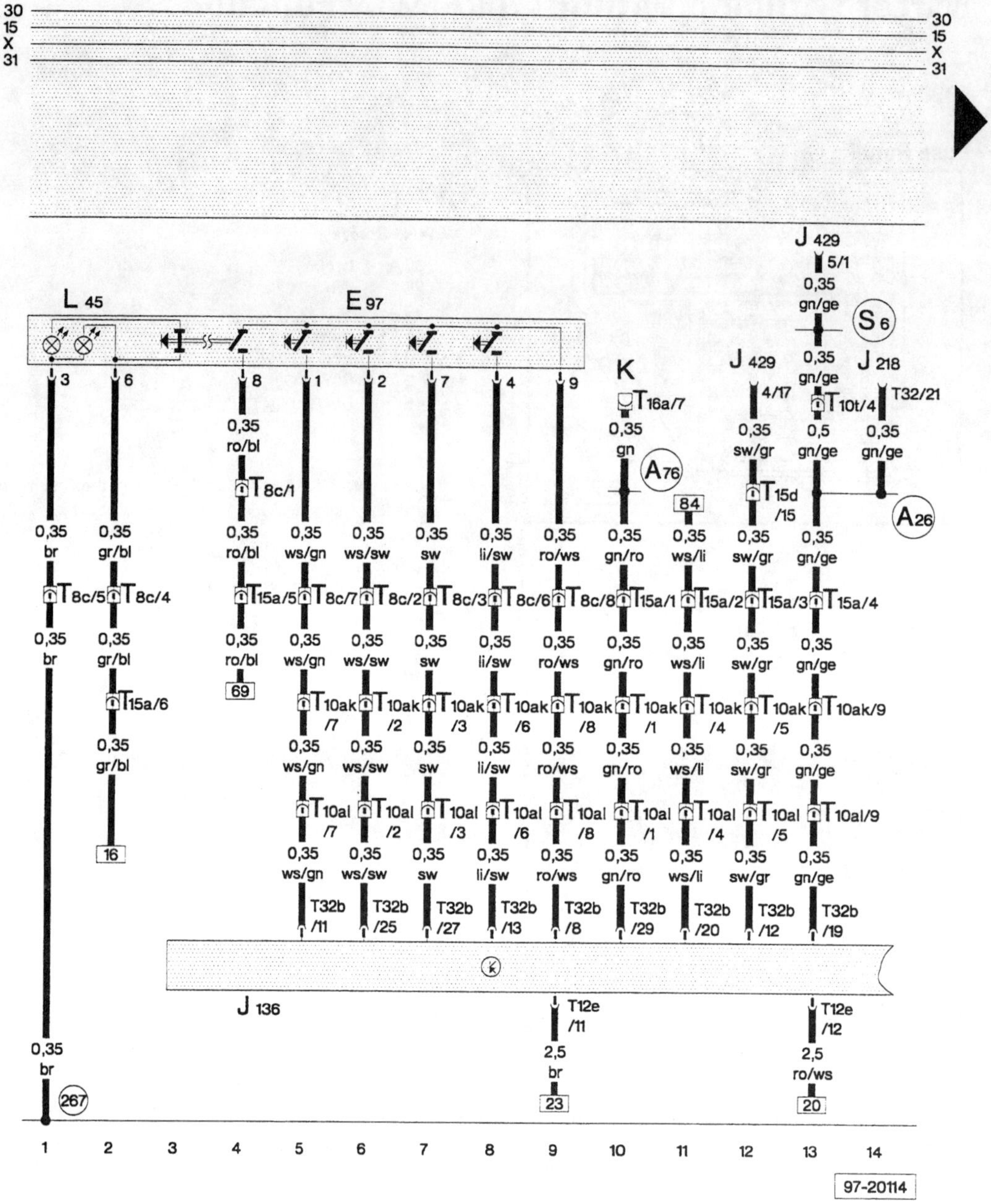

ws = white
sw = black
ro = red
gn = green
br = brown
bl = blue
gr = gray
li = violet
ge = yellow

E 97 - Memory Program Switch
J 136 - Memory Seat Control Module
J 218 - Instrument Cluster Combination Processor
J 429 - Control module for central locking
K - DLC (Data Link Connector) Rapid Data Transfer
L 45 - Memory Operating Unit Light
T 8 c - Connector 8 pin, black, in driver's door
T 10 t - Connector 10 pin, orange, connector station A pillar, right
T 10 ak - Connector 10 pin, green, connector station A pillar, left
T 10 al - Connector 10 pin, brown, below driver's seat
T 12 e - Connector 12 pin, black, in driver's seat
T 15 a - Connector 15 pin, black, connector station A pillar, left
T 15 d - Connector 15 pin, black, connector station A pillar, right
T 16 a - Connector 16 pin, data link connector
T 32 - Connector 32 pin, blue, on instrument cluster
T 32 b - Connector 32 pin, blue, in driver's seat

(267) - Ground connection -2-, in wiring harness Door cable - driver side
(A26) - Wire connection (driver's door contact switch), in instrument panel wiring harness
(A76) - Connector (K-diagnosis wire), in instrument panel wiring harness
(S6) - Wire connection (Pin 5/1), in central locking system wiring harness

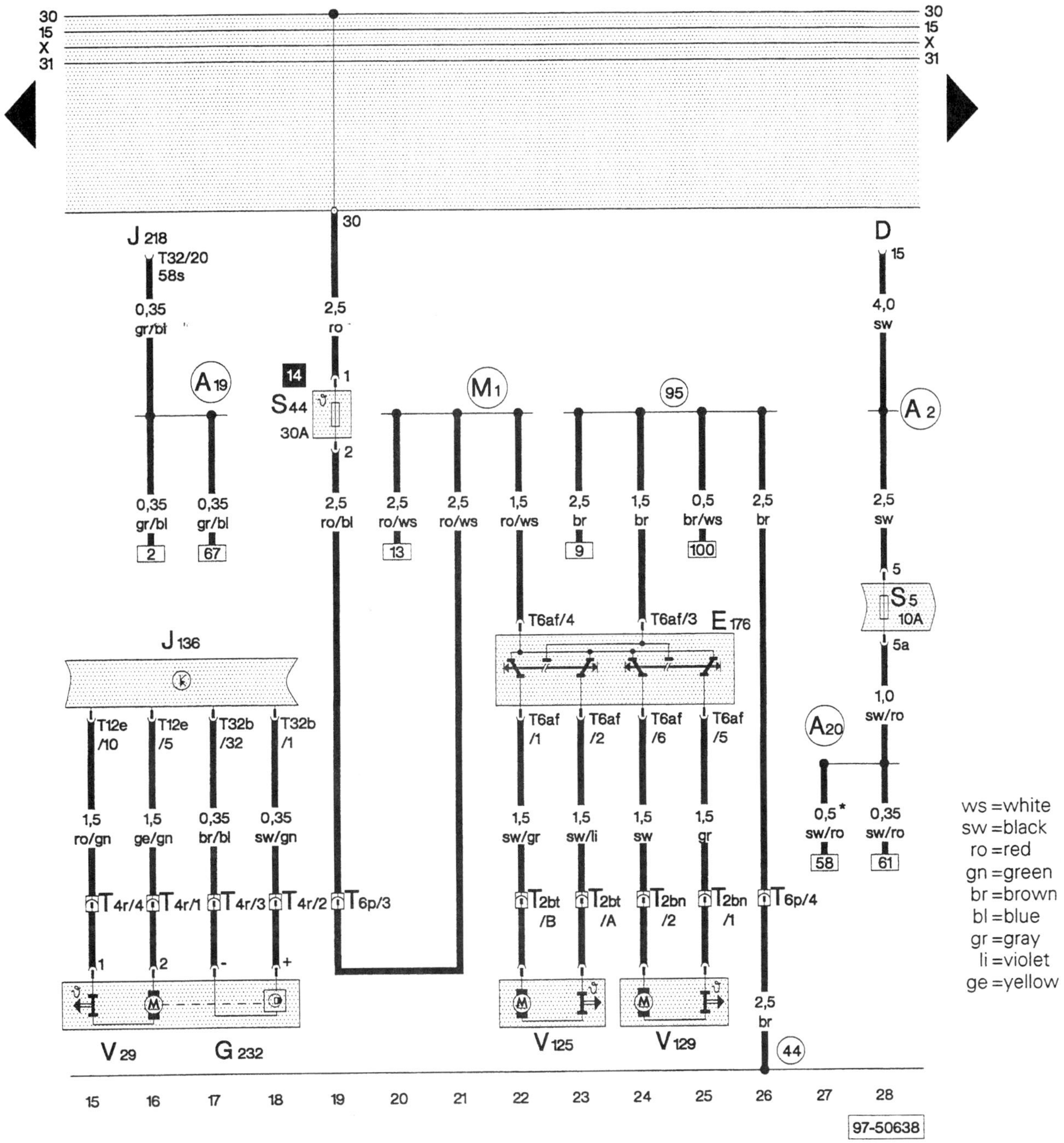

D - Ignition/Starter Switch
E 176 - Driver's Seat Lumbar Support Adjustment Switch
G 232 - Sensor for slope adjustment, driver seat
J 136 - Memory Seat Control Module
J 218 - Instrument Cluster Combination Processor
S 5 - Fuse
S 44 - Power Seat Circuit Breaker 1
T 2 bt - Connector 2 pin, black, in driver's seat
T 2 bu - Connector 2 pin, black, in driver's seat
T 4 r - Connector 4 pin, black, in driver's seat
T 6 p - Connector 6 pin, red, below driver's seat
T 6 af - Connector 6 pin, black, in driver's seat
T 12 e - Connector 12 pin, black, in driver's seat
T 32 - Connector 32 pin, blue, on instrument cluster
T 32 b - Connector 32 pin, blue, in driver's seat
V 29 - Driver's Seat Front Height Adjusting Motor
V 125 - Driver's Seat Lumbar Support Curvature Adjustment Motor
V 129 - Driver's Seat Lumbar Support Height Adjustment Motor

(44) - Ground connection, on left A-pillar, lower part

(95) - Ground connection -1-, in power seat wiring harness

(A2) - Plus connection (15), in instrument panel wiring harness

(A19) - Wire connection (58s), in instrument panel wiring harness (Switch illumination)

(A20) - Wire connection (15a), in instrument panel wiring harness

(M1) - Plus connection -1- (30), in power seats wiring harness

* - mirror with fold-away Function

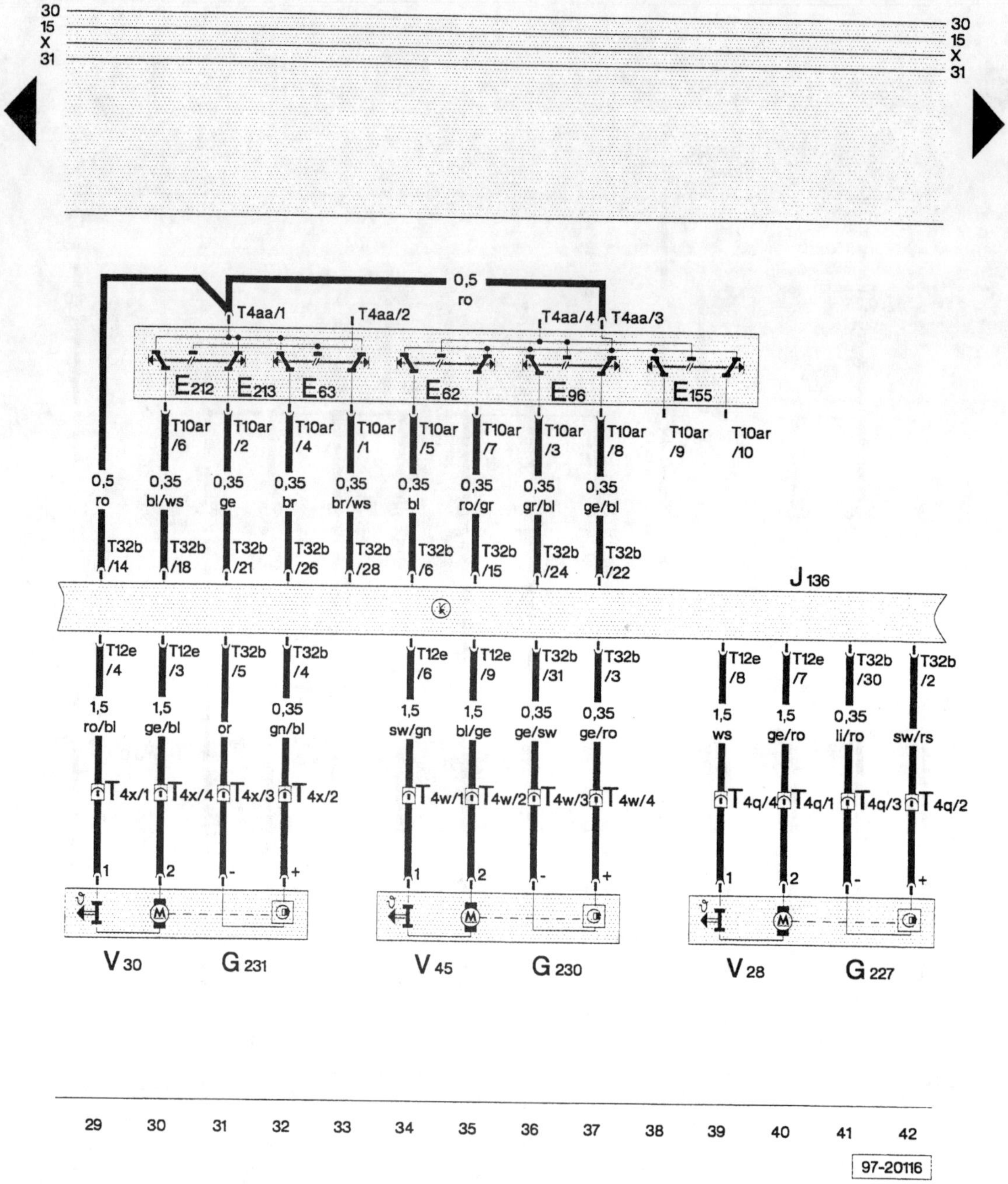

ws =white
sw =black
ro =red
gn =green
br =brown
bl =blue
gr =gray
li =violet
ge =yellow

E 62 - Driver's Seat Front Height Adjusting Switch
E 63 - Driver's Seat Rear Height Adjusting Switch
E 96 - Driver's Backrest Adjustment Switch
E 155 - Driver's Headrest Adjustment Switch
E 212 - Button for length adjustment, forward, driver seat
E 213 - Button for length adjustment, backward, driver seat
G 227 - Sensor for fore and aft adjustment, driver seat
G 230 - Sensor for seat back adjustment, driver seat
G 231 - Sensor for height adjustment, driver seat
J 136 - Memory Seat Control Module
T 4 q - Connector 4 pin, black, in driver's seat
T 4 w - Connector 4 pin, black, in driver's seat
T 4 x - Connector 4 pin, black, in driver's seat
T 4 aa - Connector 4 pin, black, in driver's seat
T 10 ar - Connector 10 pin, black, in driver's seat
T 12 e - Connector 12 pin, black, in driver's seat
T 32 b - Connector 32 pin, blue, in driver's seat
V 28 - Driver's Seat Fore/Aft Adjusting Motor
V 30 - Driver's Seat Rear Height Adjusting Motor
V 45 - Driver's Backrest Adjusting Motor

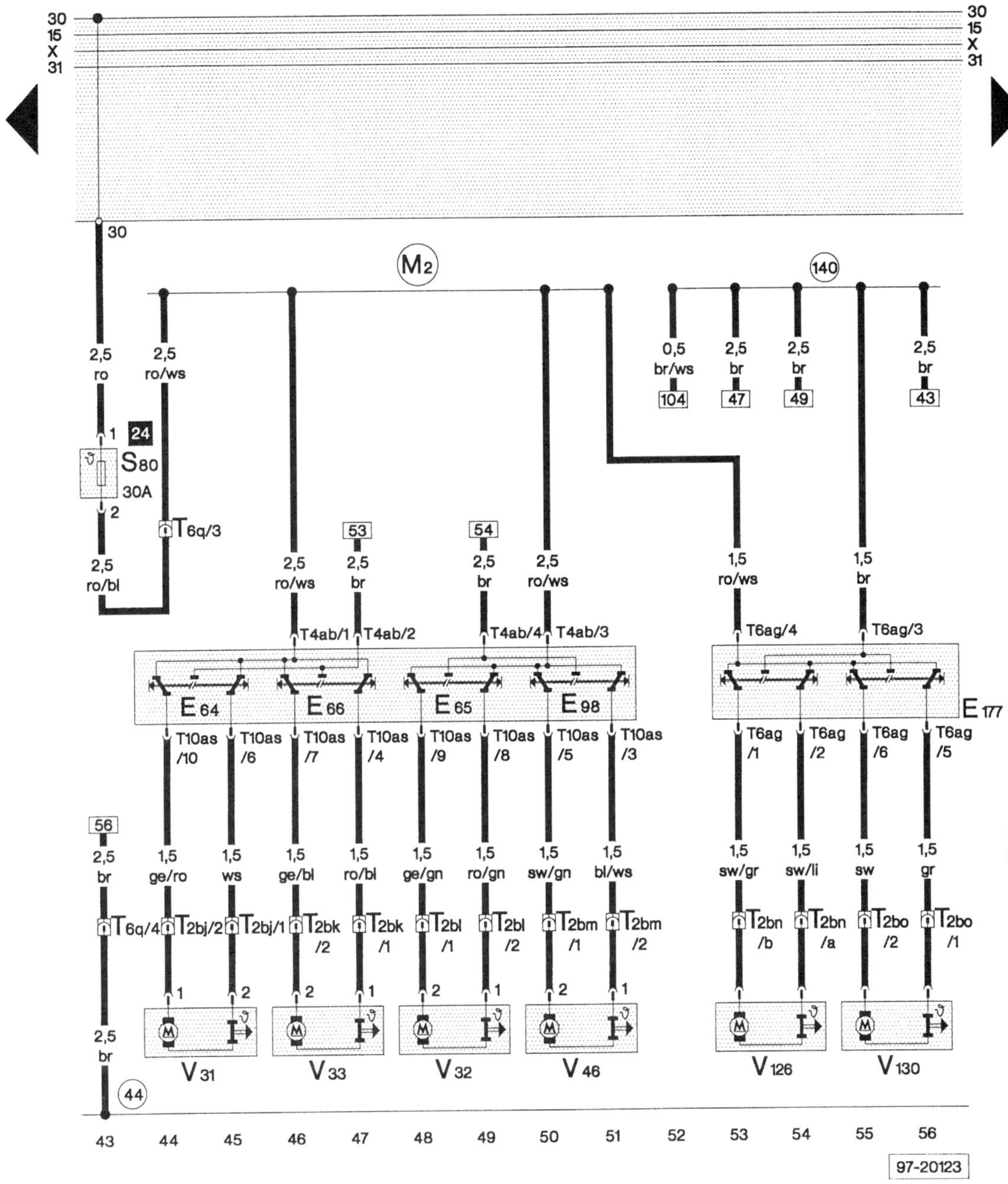

ws = white
sw = black
ro = red
gn = green
br = brown
bl = blue
gr = gray
li = violet
ge = yellow

E 64 - Passenger's Seat Fore/Aft Adjusting Switch
E 65 - Passenger's Seat Front Height Adjusting Switch
E 66 - Passenger's Seat Rear Height Adjusting Switch
E 98 - Passenger Backrest Adjustment Switch
E 177 - Front Passenger Seat Lumbar Support Adjustment Switch
S 80 - Power Seat Circuit Breaker 2
T 2 bj - Connector 2 pin, black, in passenger's seat
T 2 bk - Connector 2 pin, black, in passenger's seat
T 2 bl - Connector 2 pin, black, in passenger's seat
T 2 bm - Connector 2 pin, black, in passenger's seat
T 2 bn - Connector 2 pin, black, in passenger's seat
T 2 bo - Connector 2 pin, black, in passenger's seat
T 4 ab - Connector 4 pin, black, in passenger's seat
T 6 q - Connector 6 pin, red, below passenger's seat
T 6 ag - Connector 6 pin, black, in passenger's seat
T 10 as - Connector 10 pin, black, in passenger's seat
V 31 - Passenger's Seat Fore/Aft Adjusting Motor
V 32 - Passenger's Seat Front Height Adjusting Motor
V 33 - Passenger's Seat Rear Height Adjusting Motor
V 46 - Passenger's Backrest Adjusting Motor
V 126 - Front Passenger Seat Lumbar Support Curvature Adjustment Motor
V 130 - Front Passenger Seat Lumbar Support Height Adjustment Motor

(44) - Ground connection, on left A-pillar, lower part
(140) - Ground connection -2-, in power seat wiring harness
(M2) - Plus connection -2- (30), in power seats wiring harness

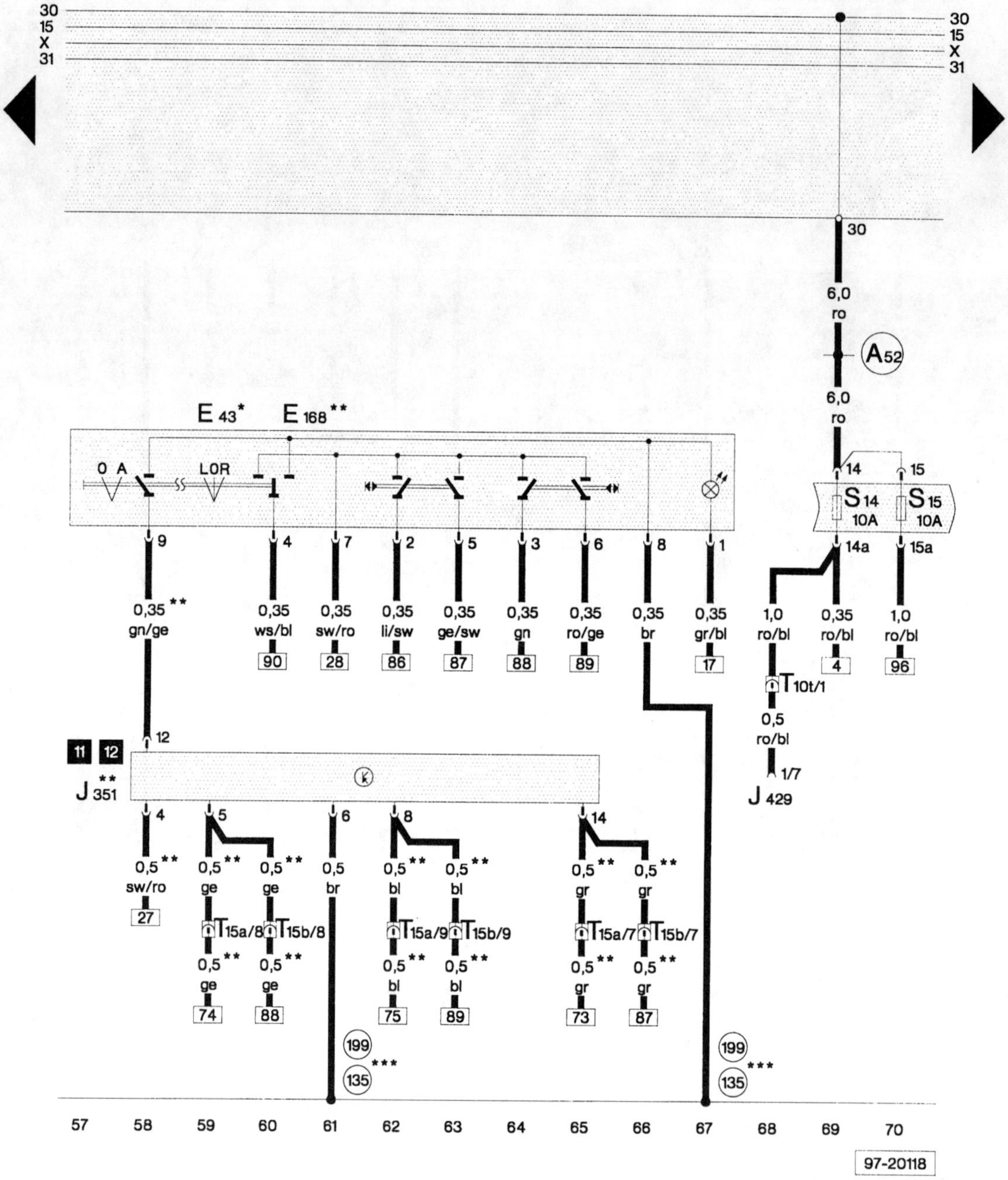

ws = white
sw = black
ro = red
gn = green
br = brown
bl = blue
gr = gray
li = violet
ge = yellow

E 43 - Mirror Adjustment Switch
E 168 - Switch for mirror adjustment with fold-away Function
J 351 - Control Module for Mirror fold-away Function
J 429 - Control module for central locking
S 14 - Fuse
S 15 - Fuse
T 10 t - Connector 10 pin, orange, connector station A pillar, right
T 15 a - Connector 15 pin, black, connector station A pillar, left
T 15 b - Connector 15 pin, black, connector station A pillar, right

(135) - Ground connection -2-, in instrument panel wiring harness
(199) - Ground connection -3-, in instrument panel wiring harness
(A52) - Plus connection (30), in instrument panel wiring harness

* - mirror without fold-away Function
** - mirror with fold-away Function
*** - booth is possible

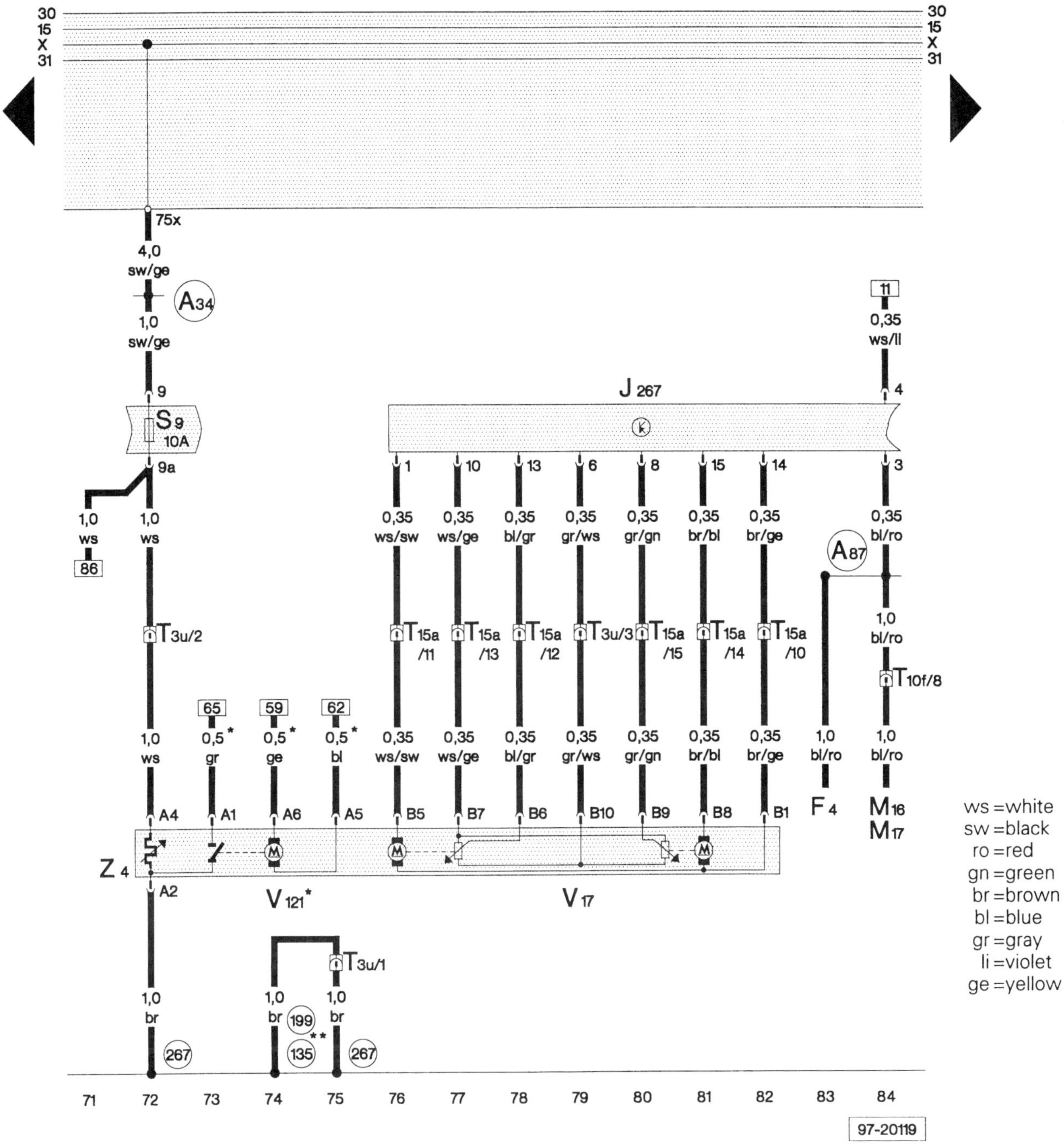

F 4 - Back-Up Light Switch
J 267 - Mirror Memory Control Module
M 16 - Left Back-Up Light
M 17 - Right Back-Up Light
S 9 - Fuse
T 3 u - Connector 3 pin, black, connector station A pillar, left
T 10 f - Connector 10 pin, brown, connector station A pillar, left
T 15 a - Connector 15 pin, black, connector station A pillar, left
V 17 - Driver's Side Mirror Adjustment Motor
V 121 - Motor for fold-in Mirror driver`s side
Z 4 - Heated outside mirror, driver side

(135) - Ground connection -2-, in instrument panel wiring harness
(199) - Ground connection -3-, in instrument panel wiring harness
(267) - Ground connection -2-, in wiring harness Door cable - driver side
(A34) - Wire connection (75x), in instrument panel wiring harness
(A87) - Connector (RF), in instrument panel wiring harness

* - mirror with fold-away Function
** - booth is possible

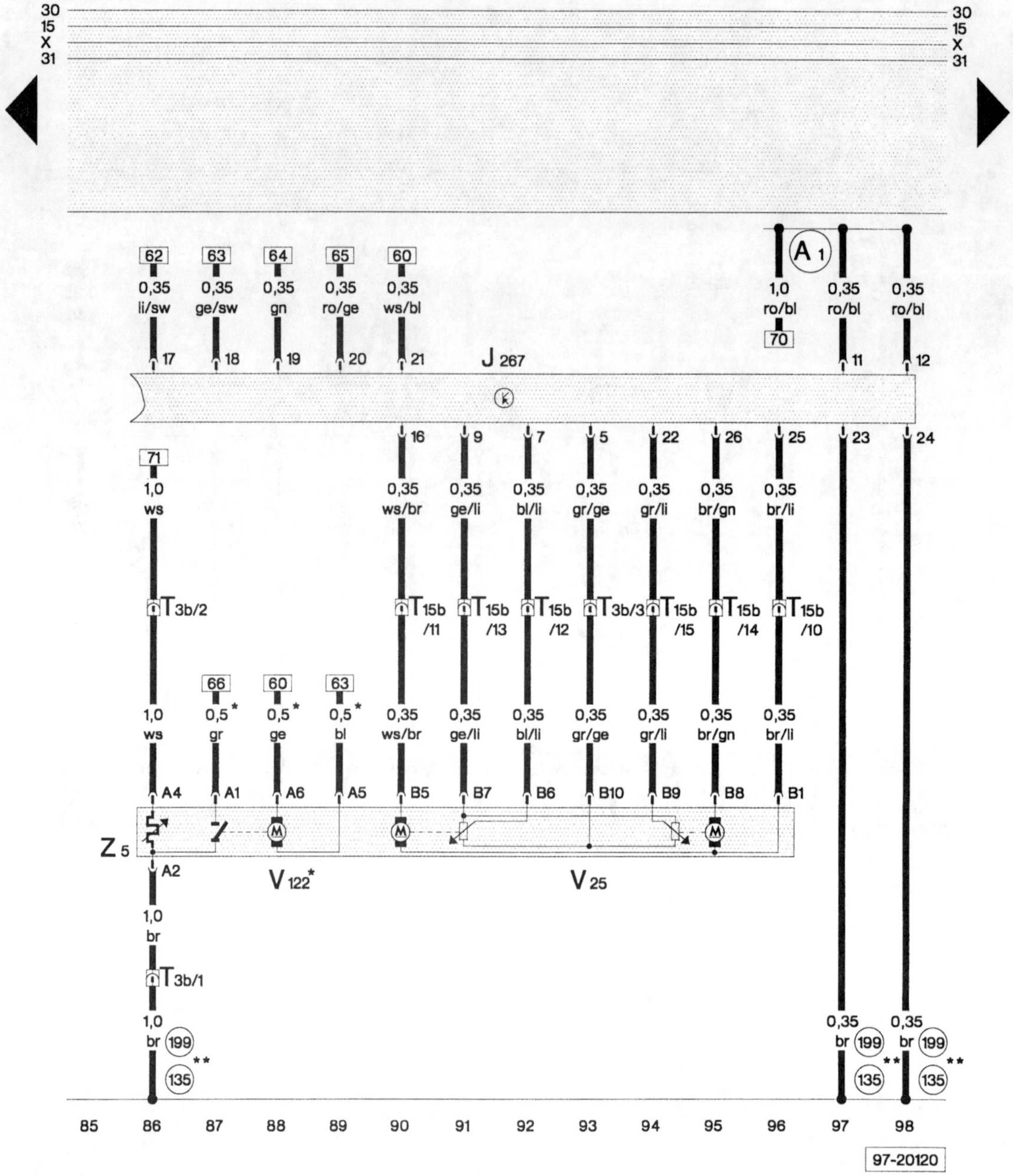

ws = white
sw = black
ro = red
gn = green
br = brown
bl = blue
gr = gray
li = violet
ge = yellow

J 267 - Mirror Memory Control Module
T 3 b - Connector 3 pin, black, connector station A pillar, right
T 15 b - Connector 15 pin, black, connector station A pillar, right
V 25 - Passenger's Side Mirror Adjustment Motor
V 122 - Motor for fold-in Mirror passenger side
Z 5 - Heated outside mirror, passenger side

(135) - Ground connection -2-, in instrument panel wiring harness
(199) - Ground connection -3-, in instrument panel wiring harness
(A1) - Plus connection (30a), in instrument panel wiring harness

* - mirror with fold-away Function
** - booth is possible

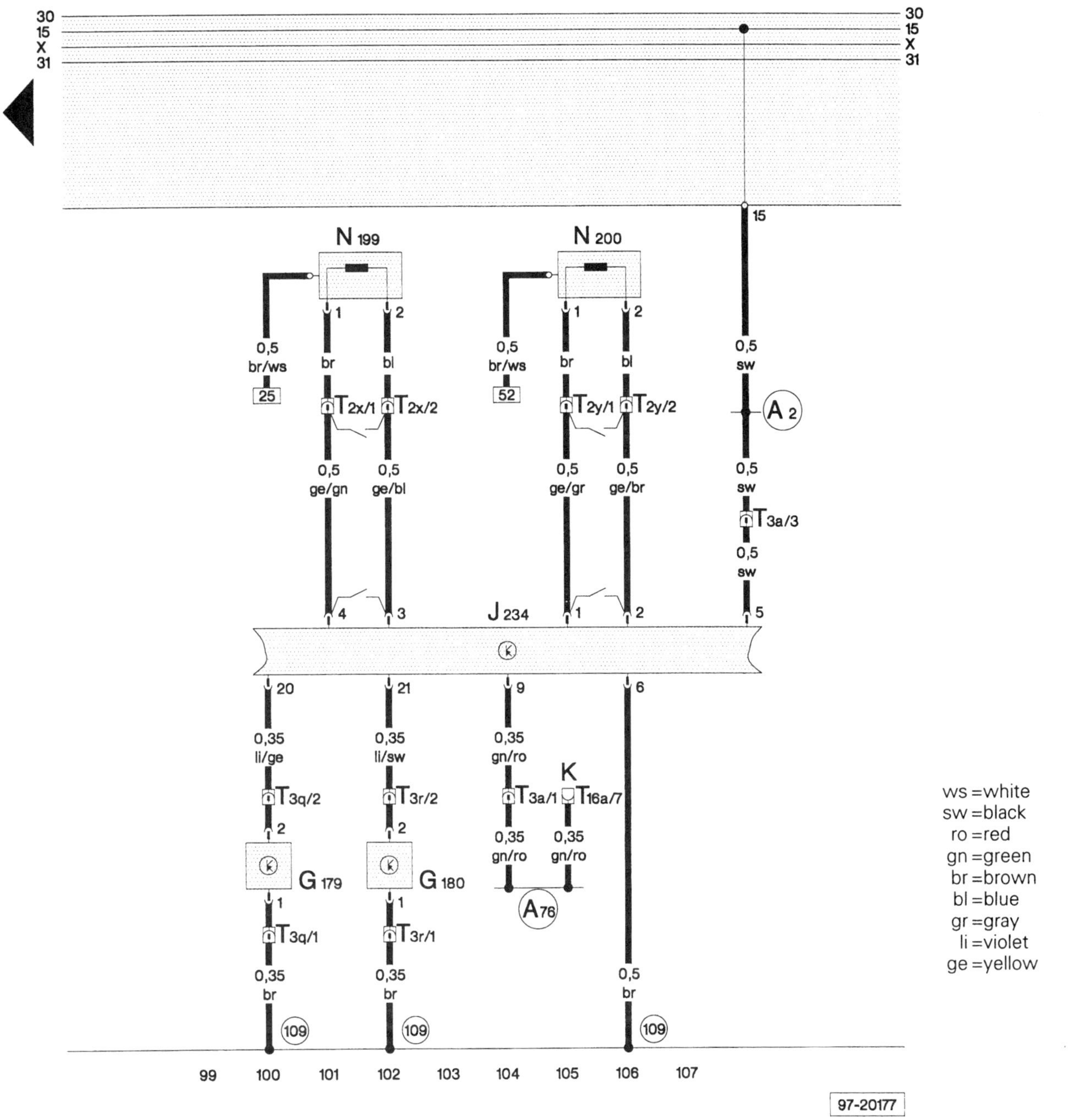

ws = white
sw = black
ro = red
gn = green
br = brown
bl = blue
gr = gray
li = violet
ge = yellow

G 179 - Crash sensor for side airbag, driver's side
G 180 - Crash sensor for side airbag, passenger side
J 234 - Airbag Control Module
K - DLC (Data Link Connector) Rapid Data Transfer
N 199 - Igniter for side airbag, driver's side
N 200 - Igniter for side airbag, passenger side
T 1 d - Single connector, brown, connector station A pillar right
T 2 x - Connector 2 pin, yellow, below driver's seat
T 2 y - Connector 2 pin, yellow, below passenger's seat
T 3 a - Connector 3 pin, black, connector station A pillar, right
T 3 q - Connector 3 pin, below driver's seat
T 3 r - Connector 3 pin, below passenger's seat
T 16 a - Connector 16 pin, data link connector

(109) - Ground connection, in airbag wiring harness

(A2) - Plus connection (15), in instrument panel wiring harness

(A76) - Connector (K-diagnosis wire), in instrument panel wiring harness

- Bridge closes when disconnecting connector (safety feature)

THIS PAGE INTENTIONALLY LEFT

BLANK

Power Seat and Mirror Adjustment, Mirror with and without fold-away Function, Heated Mirror

1998 m. y.

Fuse Panel

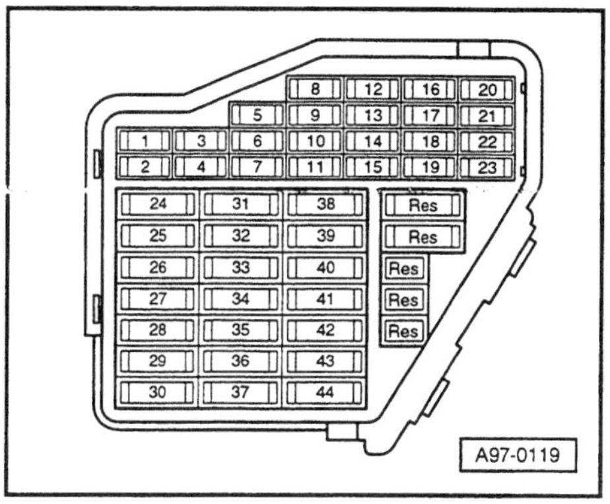

Fuse Colors:

30 A - Green
25 A - White
20 A - Yellow
15 A - Blue
10 A - Red
7,5 A - Brown
5 A - Beige

Starting with fuse position 23, fuses in the fuse holder are identified with 223 in the wiring diagram.

13 - Fold Relay Panel

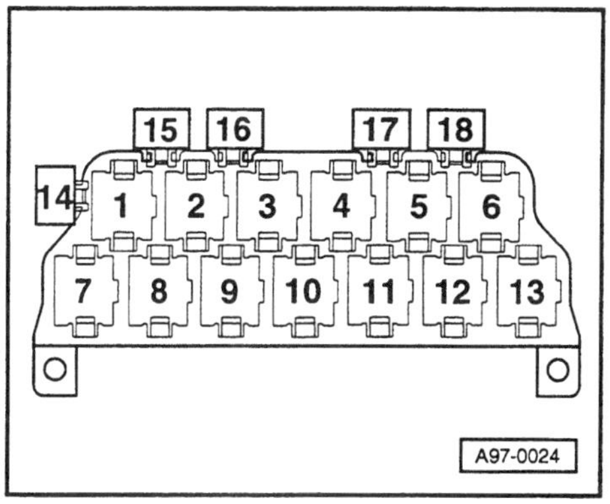

Relay Location:

11 - Control Module for Mirror fold-away Function, J351

12 - Control Module for Mirror fold-away Function, J351

8 - Fold Relay Panel

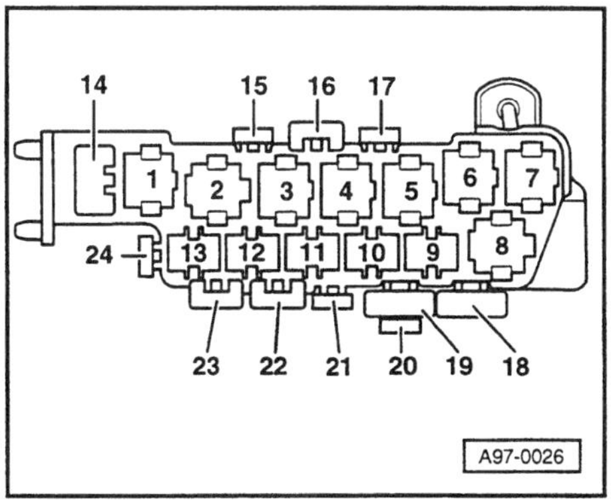

Fuse Location:

14 - Power Seat Circuit Breaker 1, S44

24 - Power Seat Circuit Breaker 2, S80

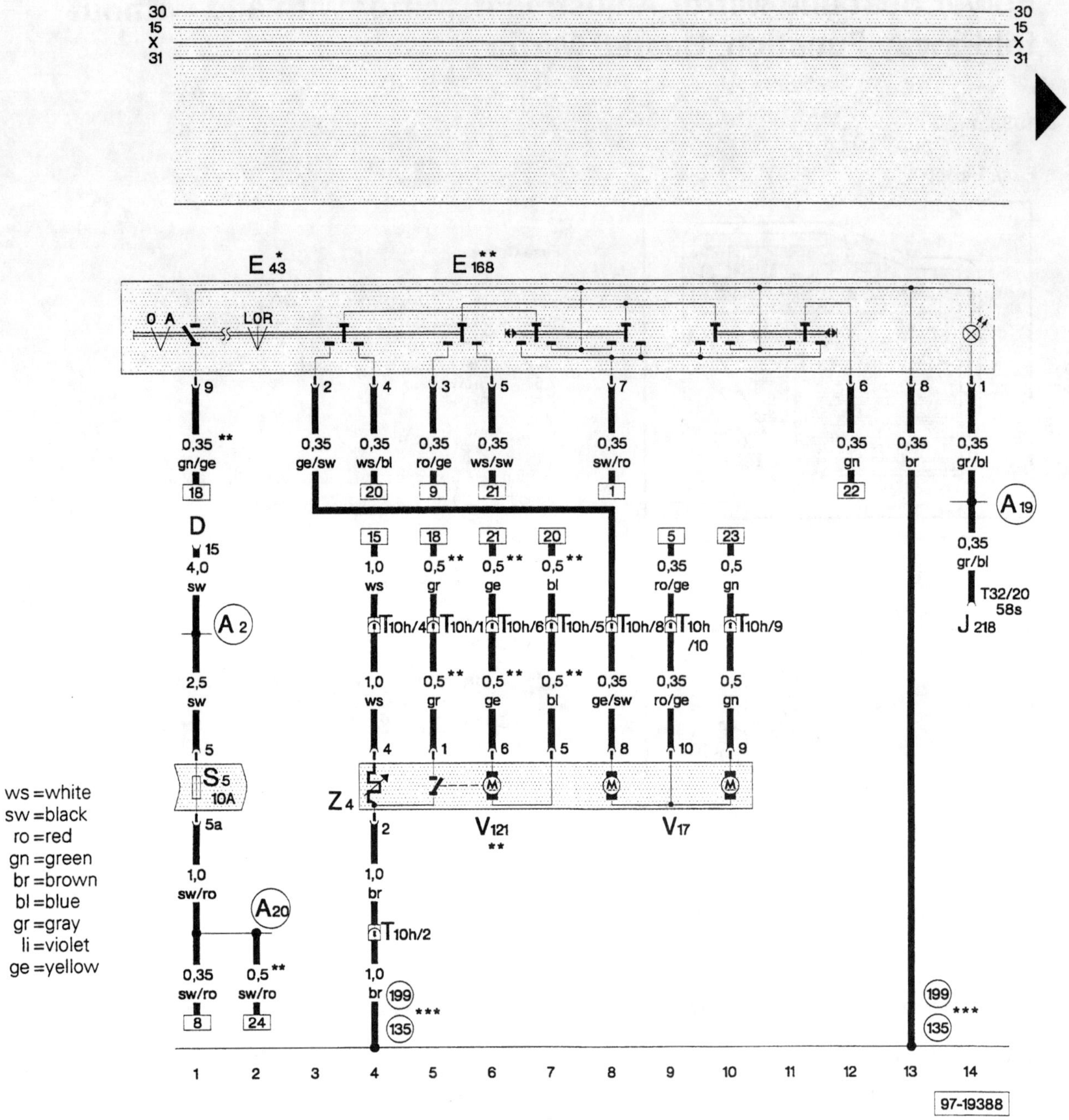

ws = white
sw = black
ro = red
gn = green
br = brown
bl = blue
gr = gray
li = violet
ge = yellow

D	- Ignition/Starter Switch
E 43	- Mirror Adjustment Switch
E 168	- Switch for mirror adjustment with fold-away Function
J 218	- Instrument Cluster Combination Processor
S 5	- Fuse
T 10 h	- Connector 10 pin, black, connector station A pillar, left
T 32	- Connector 32 pin, blue, on instrument cluster
V 17	- Driver's Side Mirror Adjustment Motor
V 121	- Motor for fold-in Mirror driver`s side
Z 4	- Heated outside mirror, driver side

(135)	- Ground connection -2-, in instrument panel wiring harness
(199)	- Ground connection -3-, in instrument panel wiring harness
(A2)	- Plus connection (15), in instrument panel wiring harness
(A19)	- Wire connection (58s), in instrument panel wiring harness (Switch illumination)
(A20)	- Wire connection (15a), in instrument panel wiring harness

* - mirror without fold-away Function
** - mirror with fold-away Function
*** - booth is possible

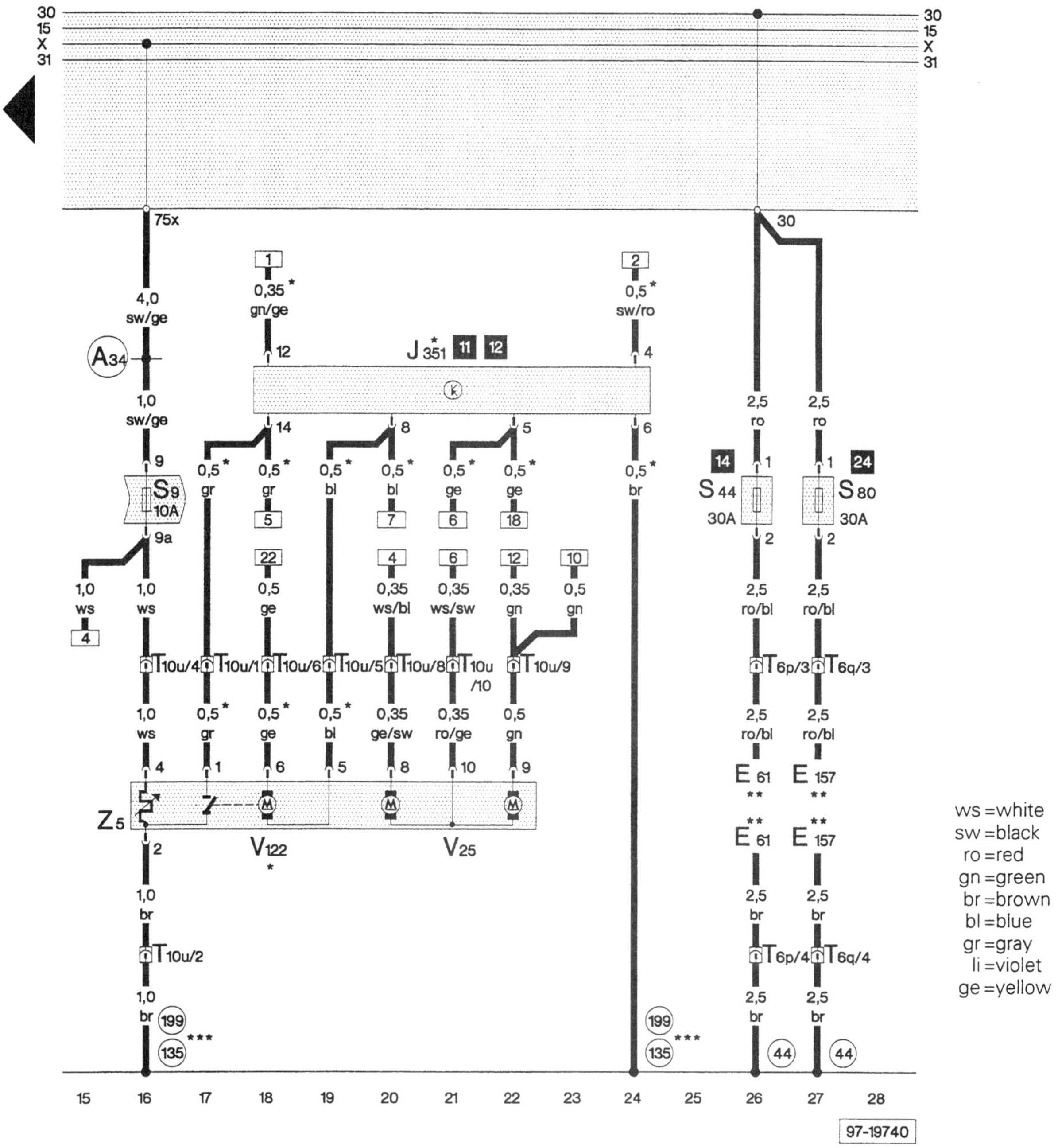

E 61 - Driver's Seat Fore/Aft and Height Adjusting Switch
E 157 - Passenger's Seat Adjusting Switch
J 351 - Control Module for Mirror fold-away Function
S 9 - Fuse
S 44 - Power Seat Circuit Breaker 1
S 80 - Power Seat Circuit Breaker 2
T 6 p - Connector 6 pin, red, below driver's seat
T 6 q - Connector 6 pin, red, below passenger's seat
T 10 u - Connector 10 pin, black, connector station A pillar, right
V 25 - Passenger's Side Mirror Adjustment Motor
V 122 - Motor for fold-in Mirror passenger side
Z 5 - Heated outside mirror, passenger side

(44) - Ground connection, on left A-pillar, lower part
(199) - Ground connection -3-, in instrument panel wiring harness
(135) - Ground connection -2-, in instrument panel wiring harness
(A34) - Wire connection (75x), in instrument panel wiring harness

* - Mirror with fold-away Function
** - see Wiring Diagram power seat
*** - booth is possible

Daytime Running Lights with Fog Light and Rear Fog Light (Canada)

1998 m. y.

Fuse Panel

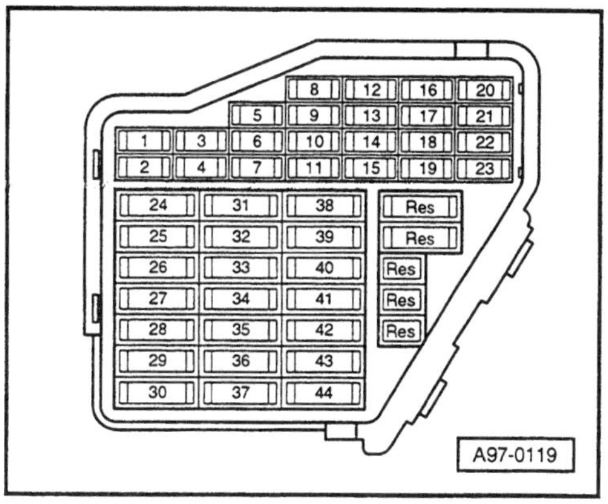

Fuse Colors:

30 A - Green
25 A - White
20 A - Yellow
15 A - Blue
10 A - Red
7,5 A - Brown
5 A - Beige

Starting with fuse position 23, fuses in the fuse holder are identified with 223 in the wiring diagram.

Micro Central Electric Panel

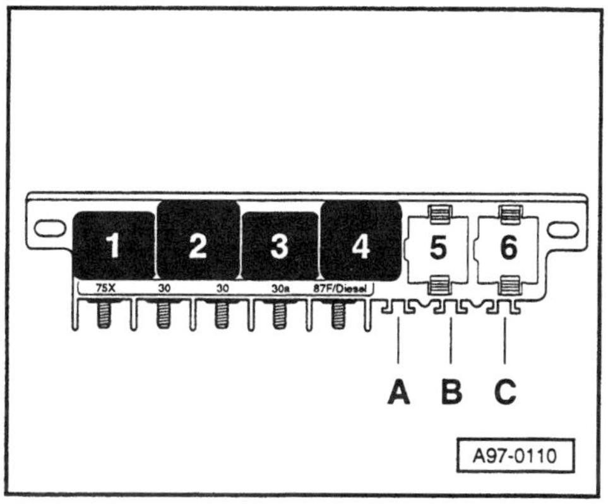

Relay Location:

2 - Load Reduction Relay, J59

13 - Fold Relay Panel

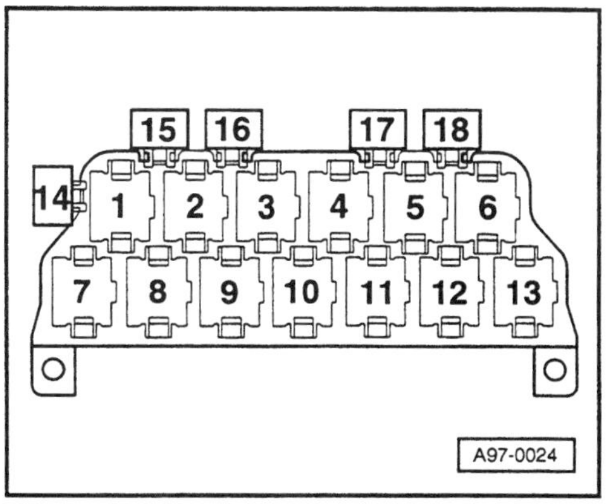

Relay Location:

6 - Fog Light Relay, J5

9 - Lamp Control Module, J123

10 - Lamp Control Module, J123

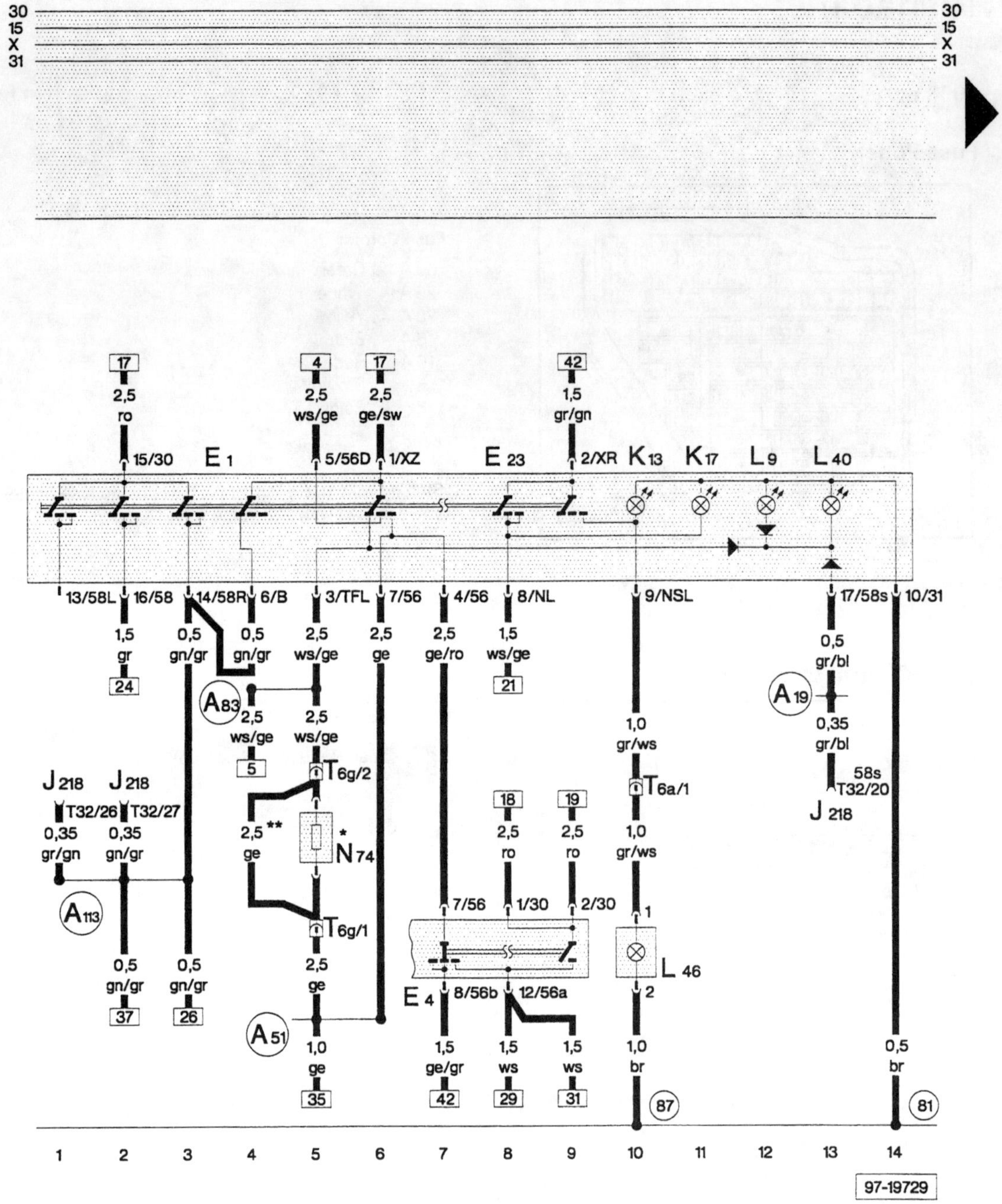

ws = white
sw = black
ro = red
gn = green
br = brown
bl = blue
gr = gray
li = violet
ge = yellow

E 1	- Light switch
E 4	- Headlight Dimmer/Flasher Switch
E 23	- Fog Light Switch
J 218	- Instrument Cluster Combination Processor
K 13	- Rear Fog Light Indicator Light
K 17	- Fog Light Indicator Light
L 9	- Headlight Switch Light
L 40	- Front And Rear Fog Light Switch Light
L 46	- Left Rear Fog Light
N 74	- Series Resistance Wiring
T 6 a	- Connector 6 pin, blue, connector station A pillar, left
T 6 g	- Connector 6 pin, violet, connector station A pillar, right
T 32	- Connector 32 pin, blue, on instrument cluster

81	- Ground connection -1-, in instrument panel wiring harness
87	- Ground connection -2-, in rear wiring harness
A19	- Wire connection (58s), in instrument panel wiring harness (Switch illumination)
A51	- Wire connection (56), in instrument panel wiring harness
A83	- Connector (dayttime running lights), in instrument panel wiring harness
A113	- Connector -2- (daytime running lights) , in instrument panel wiring harness

* - without High intensity gas discharge lamps
** - with High intensity gas discharge lamps

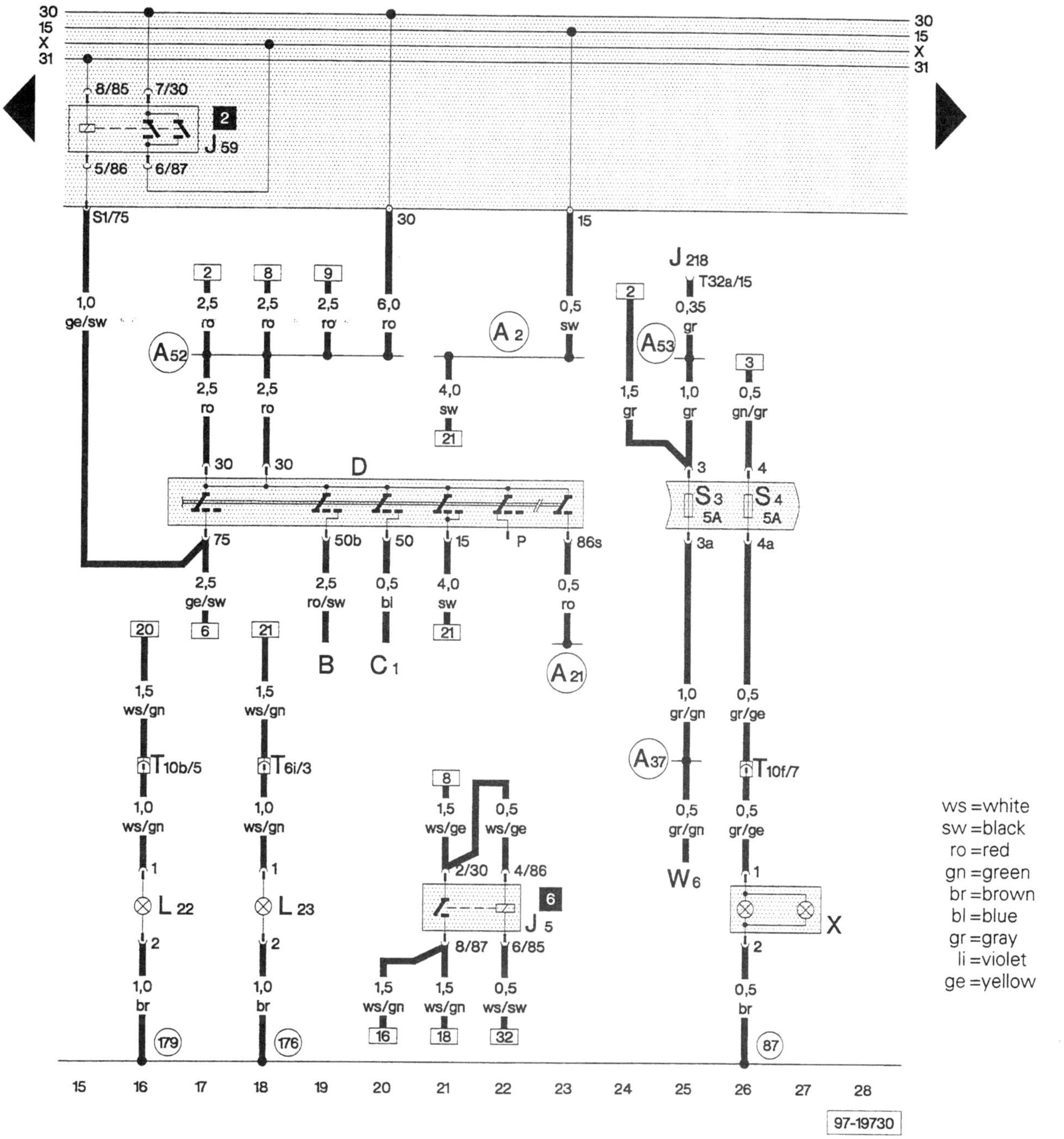

B - Starter
C 1 - Voltage Regulator (VR)
D - Ignition/Starter Switch
J 5 - Fog Light Relay
J 59 - Load Reduction Relay
J 218 - Instrument Cluster Combination Processor
L 22 - Left Front Fog Light
L 23 - Right Front Fog Light
S 3 - Fuse
S 4 - Fuse
T 6 i - Connector 6 pin, red, connector station A pillar, right
T 10 b - Connector 10 pin, red, connector station A pillar, left
T 10 f - Connector 10 pin, brown, connector station A pillar, left
T 32 a - Connector 32 pin, green, on instrument cluster
W 6 - Glove Compartment Light
X - License Plate Light

(87) - Ground connection -2-, in rear wiring harness
(176) - Ground connection, in right headlight wiring harness
(179) - Ground connection, in left headlight wiring harness
(A2) - Plus connection (15), in instrument panel wiring harness
(A21) - Wire connection (86s), in instrument panel wiring harness
(A37) - Wire connection (58a), in instrument panel wiring harness
(A52) - Plus connection (30), in instrument panel wiring harness
(A53) - Plus connection (58b), in instrument panel wiring harness

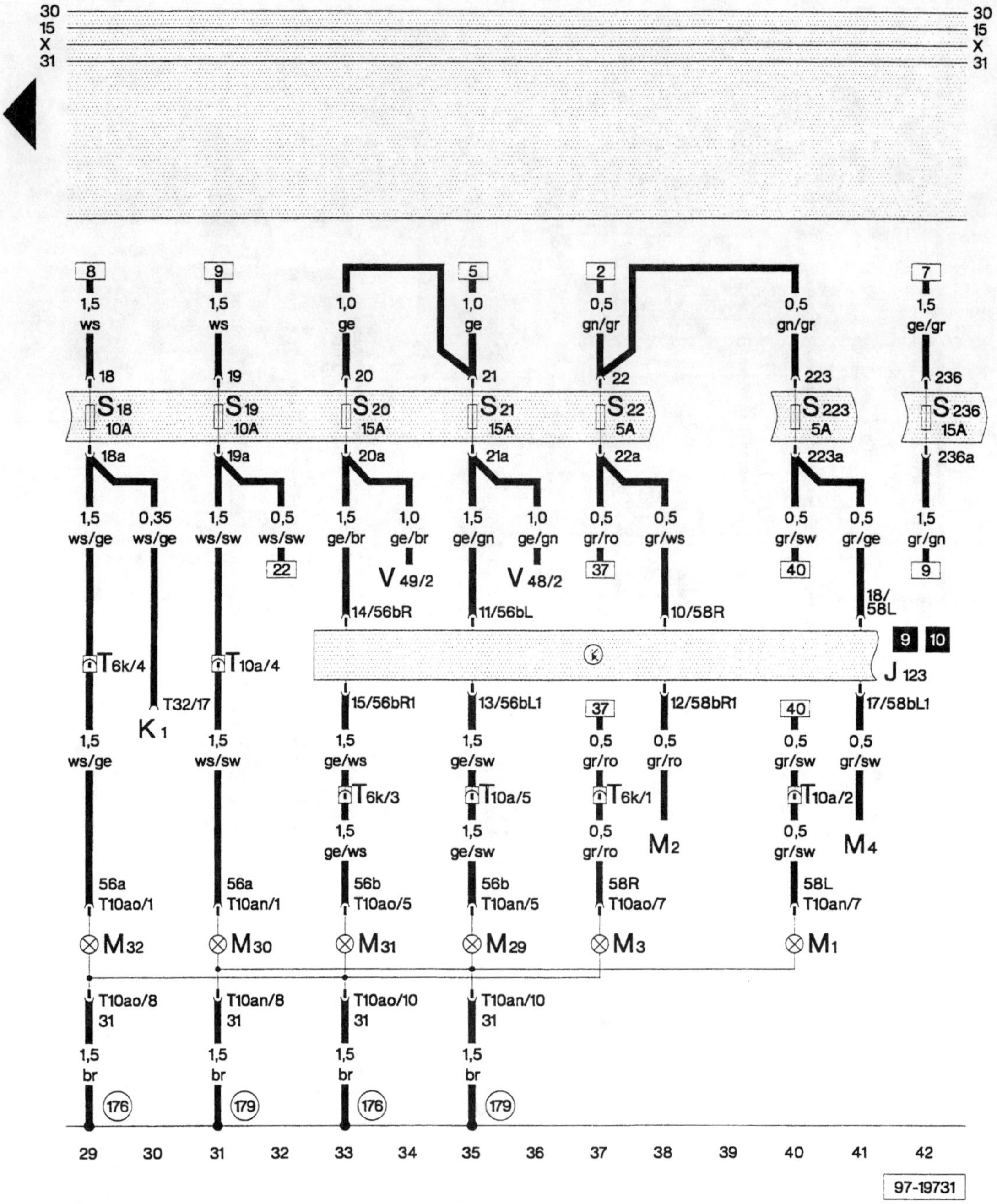

ws = white
sw = black
ro = red
gn = green
br = brown
bl = blue
gr = gray
li = violet
ge = yellow

J 123 - Lamp Control Module
K 1 - Headlight High Beam Indicator Light
M 1 - Left Parking Light
M 2 - Right Tail Light
M 3 - Right Parking Light
M 4 - Left Tail Light
M 29 - Left Low Beam Headlight
M 30 - Left High Beam Headlight
M 31 - Right Low Beam Headlight
M 32 - Right High Beam Headlight
S 18 - Fuse
S 19 - Fuse
S 20 - Fuse
S 21 - Fuse
S 22 - Fuse
S 223 - Fuse
S 236 - Fuse
T 6 k - Connector 6 pin, pink, connector station A pillar, right
T 10 a - Connector 10 pin, pink, connector station A pillar, left
T 10 an - Connector 10 pin, on headlight, left
T 10 ao - Connector 10 pin, on headlight, right
T 32 - Connector 32 pin, blue, on instrument cluster
V 48 - Left Headlight Beam Adjusting Motor
V 49 - Right Headlight Beam Adjusting Motor

(176) - Ground connection, in right headlight wiring harness

(179) - Ground connection, in left headlight wiring harness

Power Window

1998 m. y.

Fuse Panel

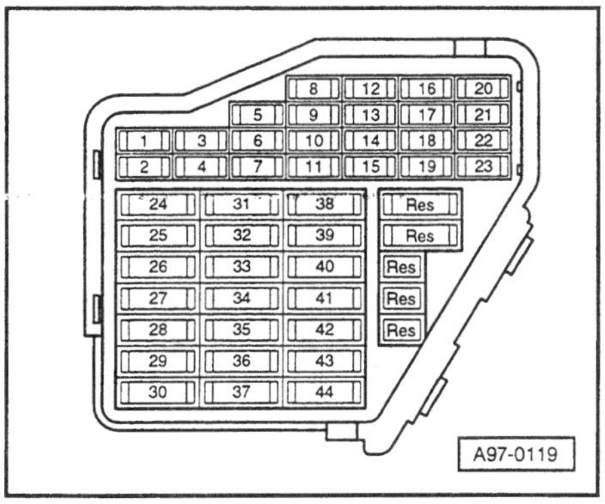

Fuse Colors:

30 A - Green
25 A - White
20 A - Yellow
15 A - Blue
10 A - Red
7,5 A - Brown
5 A - Beige

Starting with fuse position 23, fuses in the fuse holder are identified with 223 in the wiring diagram.

Micro Central Electric Panel

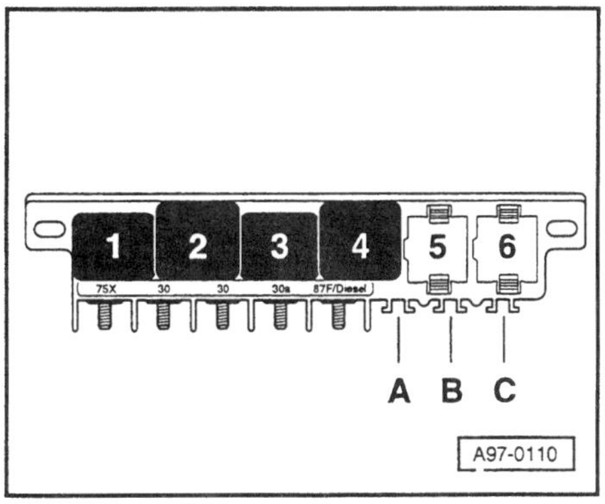

8 - Fold Relay Panel

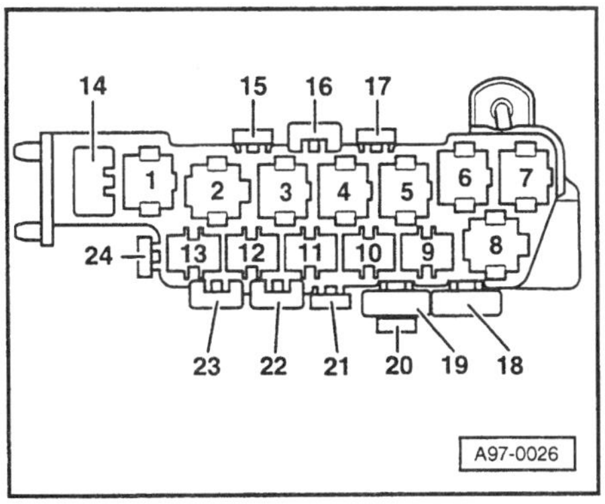

Fuse Location:

16 - Power Window Circuit Breaker Front, S37

17 - Power Window Circuit Breaker Rear, S43

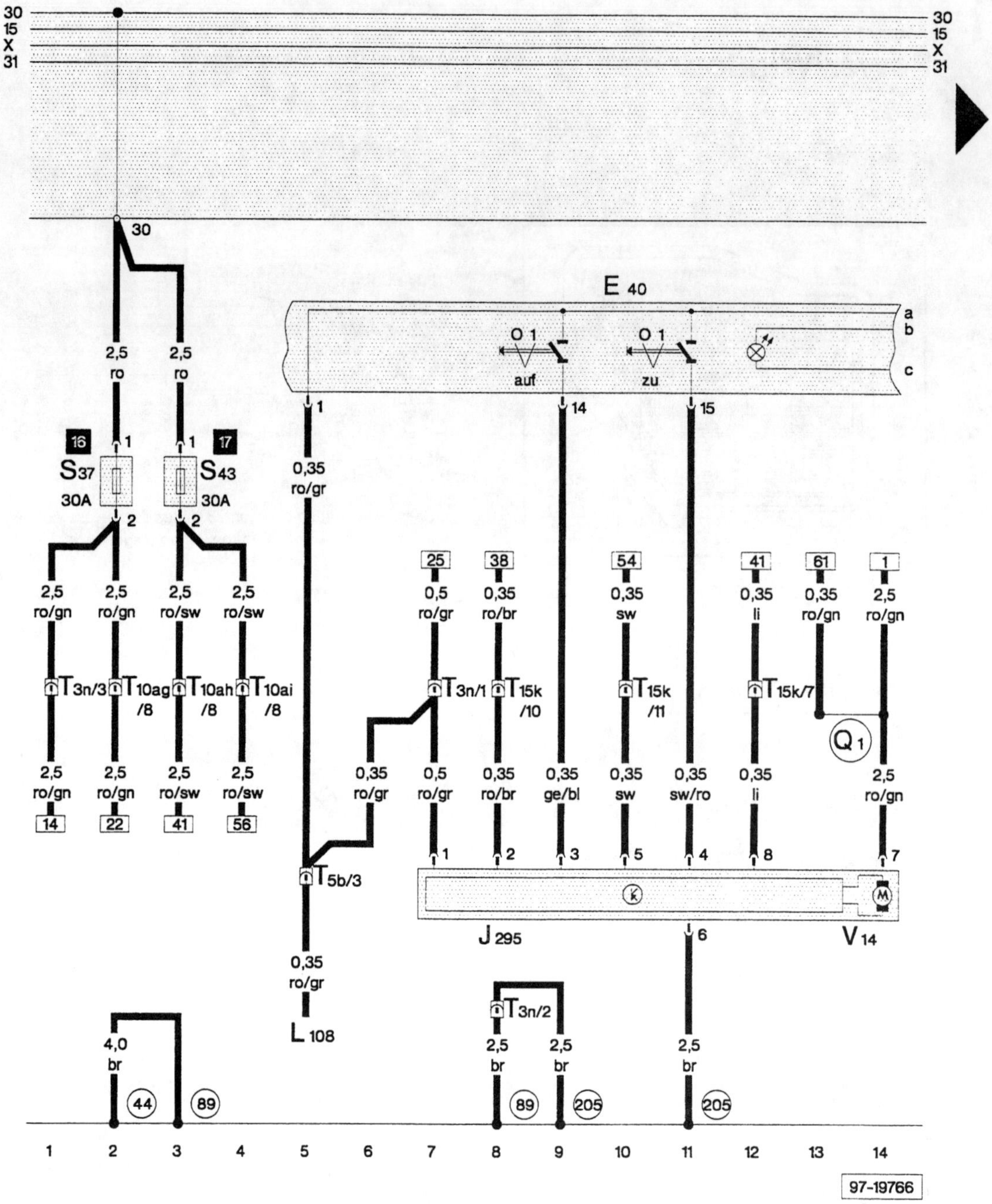

ws = white
sw = black
ro = red
gn = green
br = brown
bl = blue
gr = gray
li = violet
ge = yellow

E 40 - Left Front Window Switch
J 295 - Left Front Power Window Control Module
L 108 - Door Opener Illumination, Driver's Side
S 37 - Power Window Circuit Breaker Front
S 43 - Power Window Circuit Breaker Rear
T 3 n - Connector 3 pin, red, connector station A pillar, left
T 5 b - Connector 5 pin, black, in driver's door
T 10 ag - Connector 10 pin, red, connector station A pillar, right
T 10 ah - Connector 10 pin, red, in B pillar, left
T 10 ai - Connector 10 pin, red, in B pillar, right
T 15 k - Connector 15 pin, red, connector station A pillar, left
V 14 - Left Window Motor

(44) - Ground connection, on left A-pillar, lower part
(89) - Ground connection -1-, in power window wiring harness
(205) - Ground connection, in driver's door wiring harness
(Q1) - Plus connection (30), in power window wiring harness

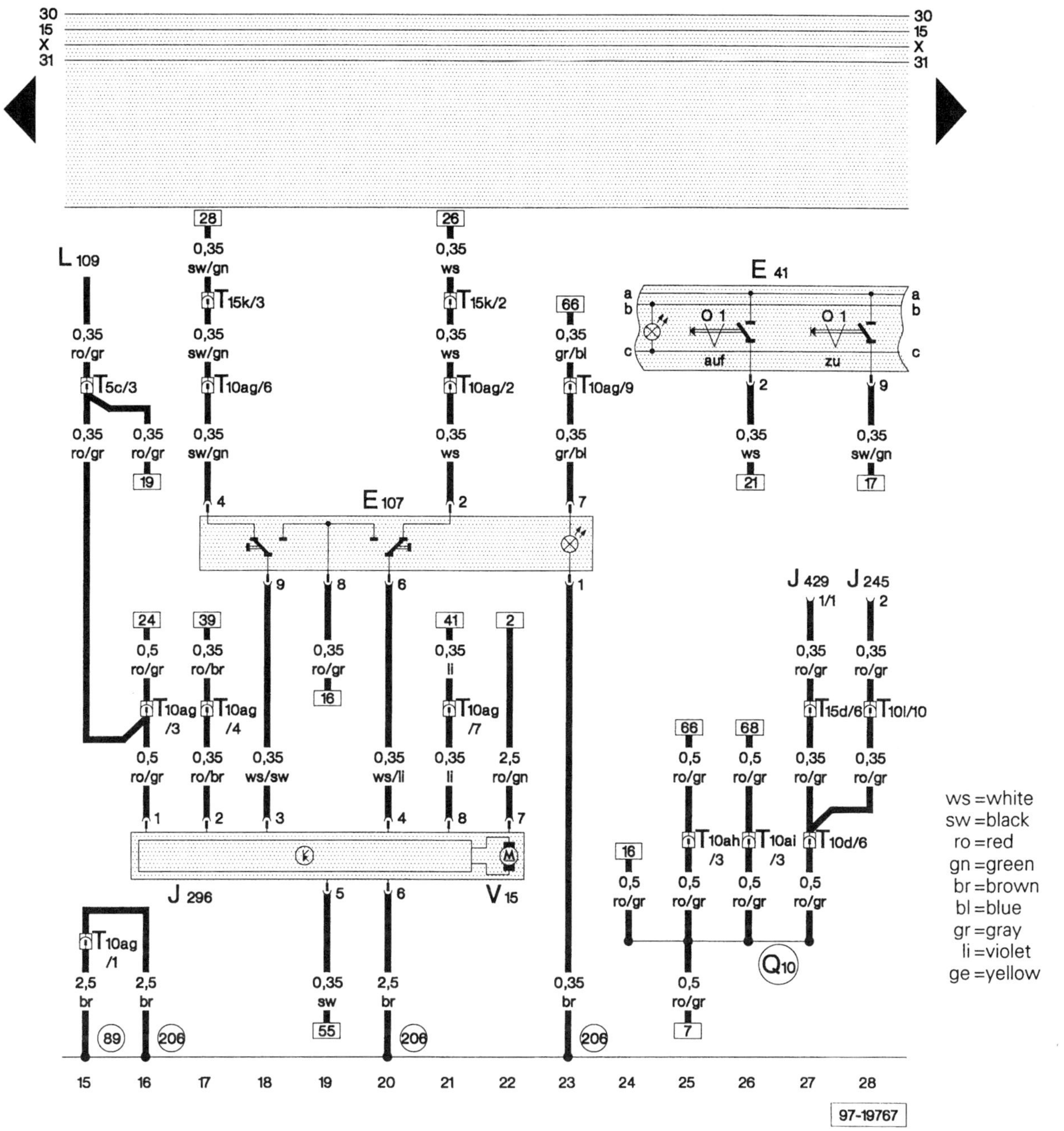

E 41 - Right Front Window Switch
E 107 - Switch for window regulator, in passenger door
J 245 - Power Sunroof Control Module
J 296 - Right Front Power Window Control Module
J 429 - Control module for central locking
L 109 - Door Opener Illumination, Passenger's Side
T 5 c - Connector 5 pin, black, in passenger's door
T 10 d - Connector 10 pin, green, connector station A pillar, left
T 10 i - Connector 10 pin, darkbrown, connector station A pillar, left
T 10 ag - Connector 10 pin, red, in passenger's door
T 10 ah - Connector 10 pin, red, in left rear door
T 10 ai - Connector 10 pin, red, in right rear door
T 15 d - Connector 15 pin, black, connector station A pillar, right
T 15 k - Connector 15 pin, red, connector station A pillar, left
V 15 - Right Window Motor

(89) - Ground connection -1-, in power window wiring harness
(206) - Ground connection, in passenger's door wiring harness
(Q10) - Plus connection (87), in power window/central locking system and door contact switch wiring harness

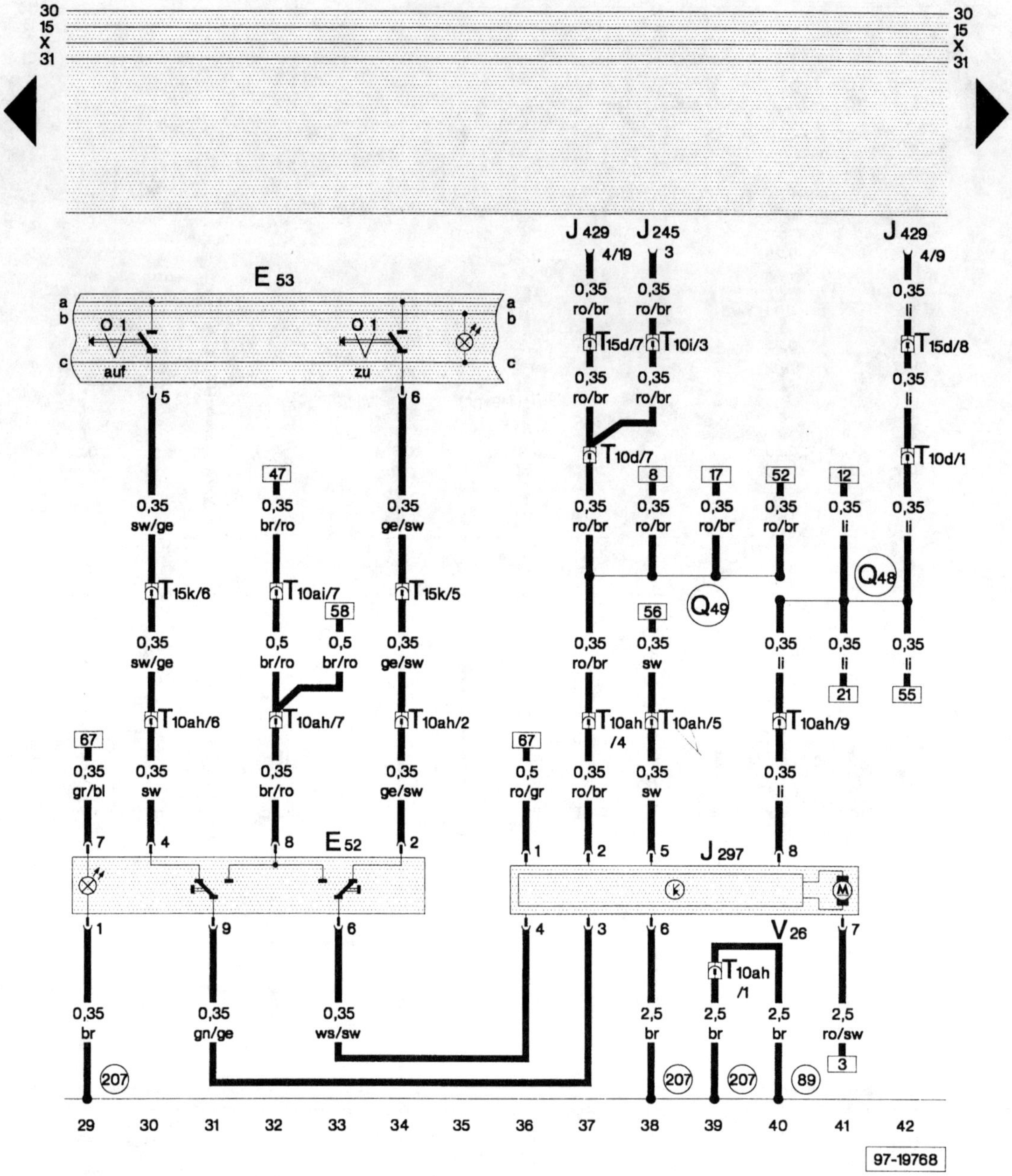

ws = white
sw = black
ro = red
gn = green
br = brown
bl = blue
gr = gray
li = violet
ge = yellow

E 52 - Left Rear Window Switch, (In LR Door)
E 53 - Left Rear Window Switch, (In Console)
J 245 - Power Sunroof Control Module
J 297 - Left Rear Power Window Control Module
J 429 - Control module for central locking
T 10 d - Connector 10 pin, green, connector station A pillar, left
T 10 i - Connector 10 pin, darkbrown, connector station A pillar, left
T 10 ah - Connector 10 pin, red, in left rear door
T 10 ai - Connector 10 pin, red, in right rear door
T 15 d - Connector 15 pin, black, connector station A pillar, right
T 15 k - Connector 15 pin, red, connector station A pillar, left
V 26 - Motor for window regulator, rear, left

(89) - Ground connection -1-, in power window wiring harness
(207) - Ground connection, in left rear door wiring harness
(Q48) - Connector (open), in wiring harness interior
(Q49) - Connector (closed), in wiring harness interior

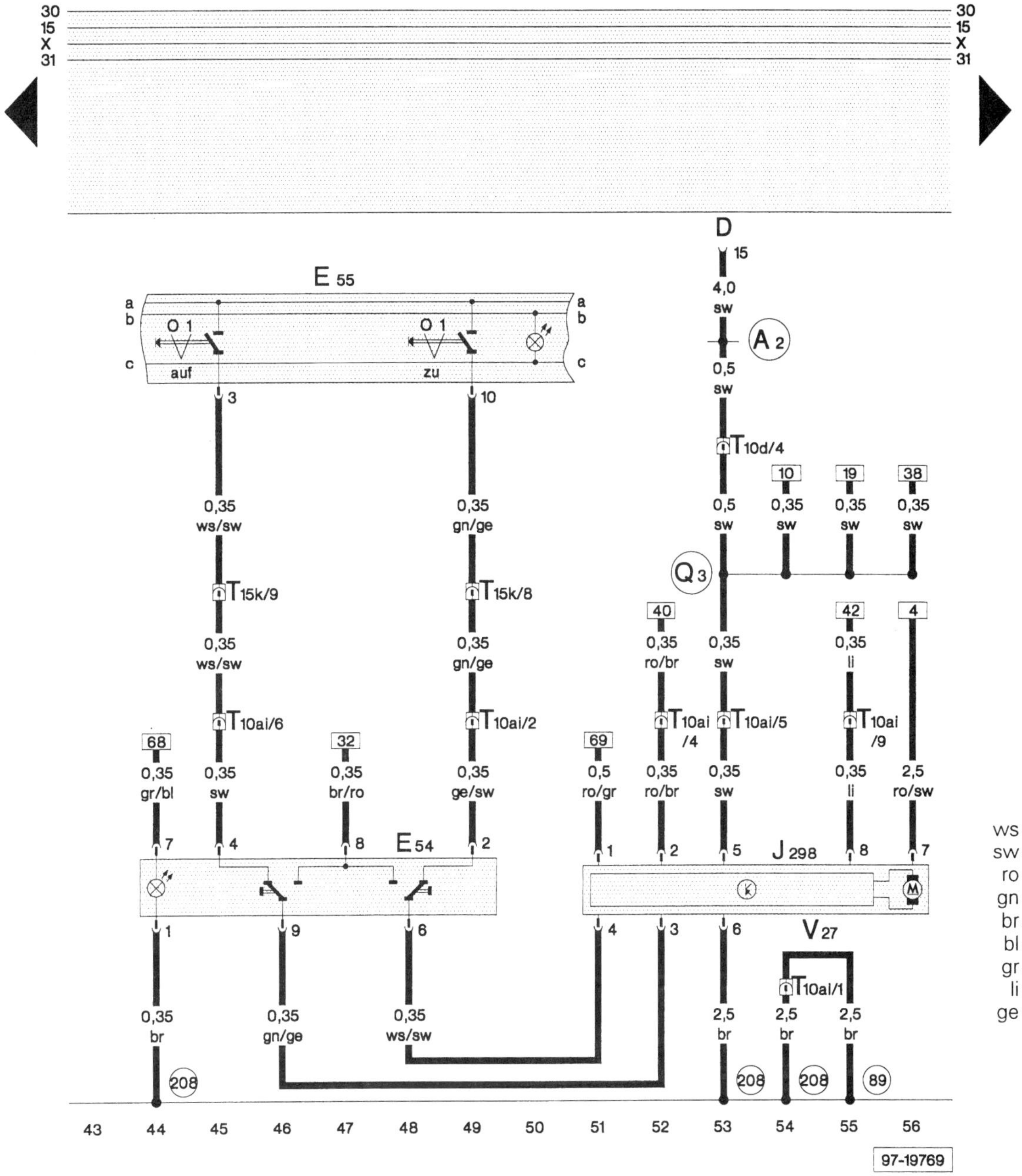

ws = white
sw = black
ro = red
gn = green
br = brown
bl = blue
gr = gray
li = violet
ge = yellow

D - Ignition/Starter Switch
E 54 - Right Rear Window Switch, (In RR Door)
E 55 - Right Rear Window Switch, (In Console)
J 298 - Right Rear Power Window Control Module
T 10 d - Connector 10 pin, green, connector station A pillar, left
T 10 ai - Connector 10 pin, red, in B pillar, right
T 15 k - Connector 15 pin, red, connector station A pillar, left
V 27 - Motor for window regulator, rear, right

(89) - Ground connection -1-, in power window wiring harness
(208) - Ground connection, in right rear door wiring harness
(A2) - Plus connection (15), in instrument panel wiring harness
(Q3) - Plus connection (15), in power window wiring harness

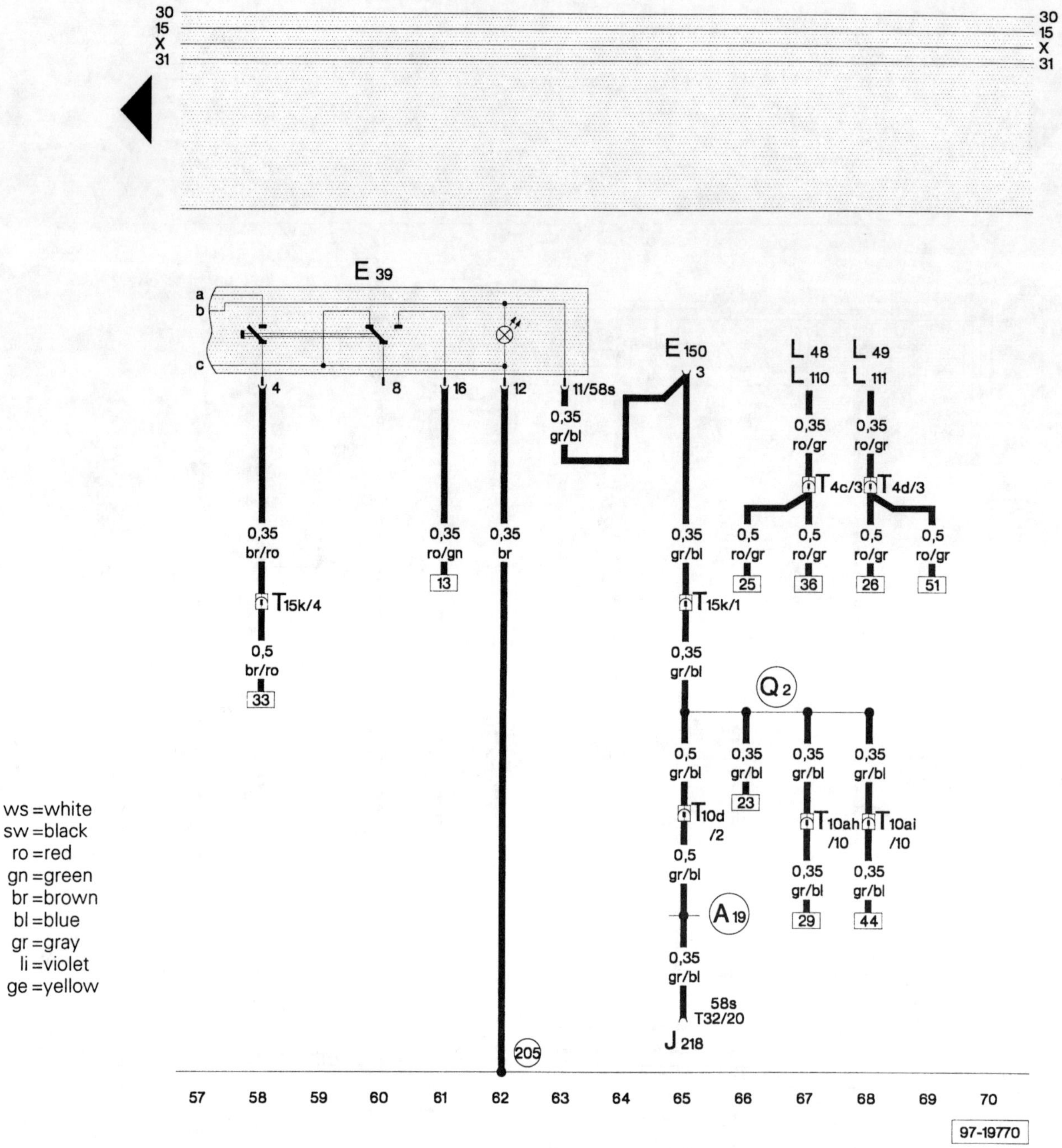

E 39 - Window Lockout Switch
E 150 - Switch for interior lock, driver side
J 218 - Instrument Cluster Combination Processor
L 48 - Left Rear Ashtray Light
L 49 - Right Rear Ashtray Light
L 110 - Door Opener Illumination, rear left
L 111 - Door Opener Illumination, rear right
T 4 c - Connector 4 pin, black, in left rear door
T 4 d - Connector 4 pin, black, in right rear door
T 10 d - Connector 10 pin, green, connector station A pillar, left
T 10 ah - Connector 10 pin, red, in B pillar, left
T 10 ai - Connector 10 pin, red, in B pillar, right
T 15 k - Connector 15 pin, red, connector station A pillar, left
T 32 - Connector 32 pin, blue, on instrument cluster

(205) - Ground connection, in driver's door wiring harness
(A19) - Wire connection (58s), in instrument panel wiring harness (Switch illumination)
(Q2) - Plus connection (58s), in power window wiring harness

Servotronic

1998 m. y.

Fuse Panel

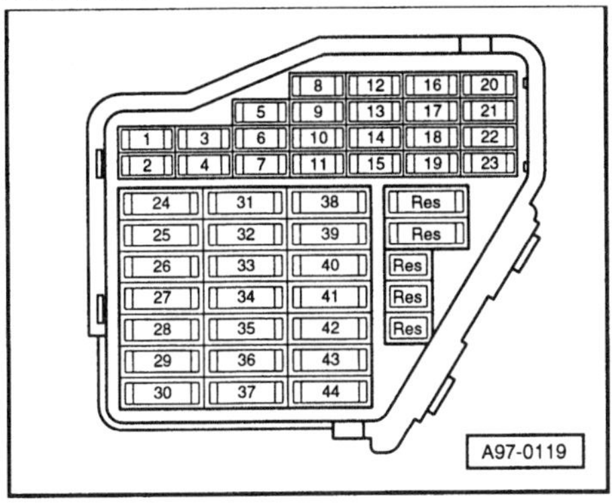

Fuse Colors:

30 A - Green
25 A - White
20 A - Yellow
15 A - Blue
10 A - Red
5 A - Beige

Starting with fuse position 23, fuses in the fuse holder are identified with 223 in the wiring diagram.

13 - Fold Relay Panel

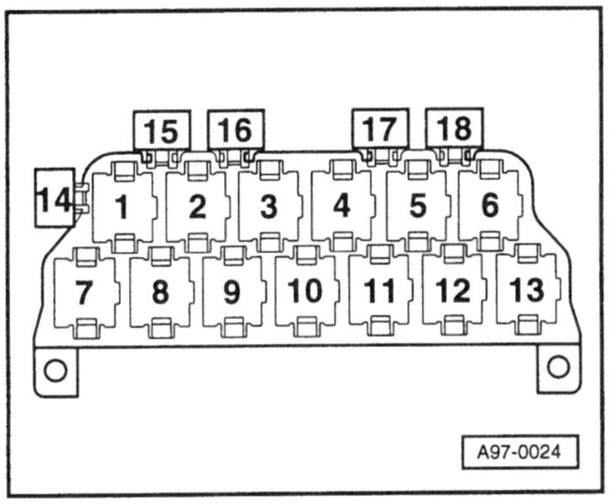

Relay Location:

13 - Servotronic Control Module, J236

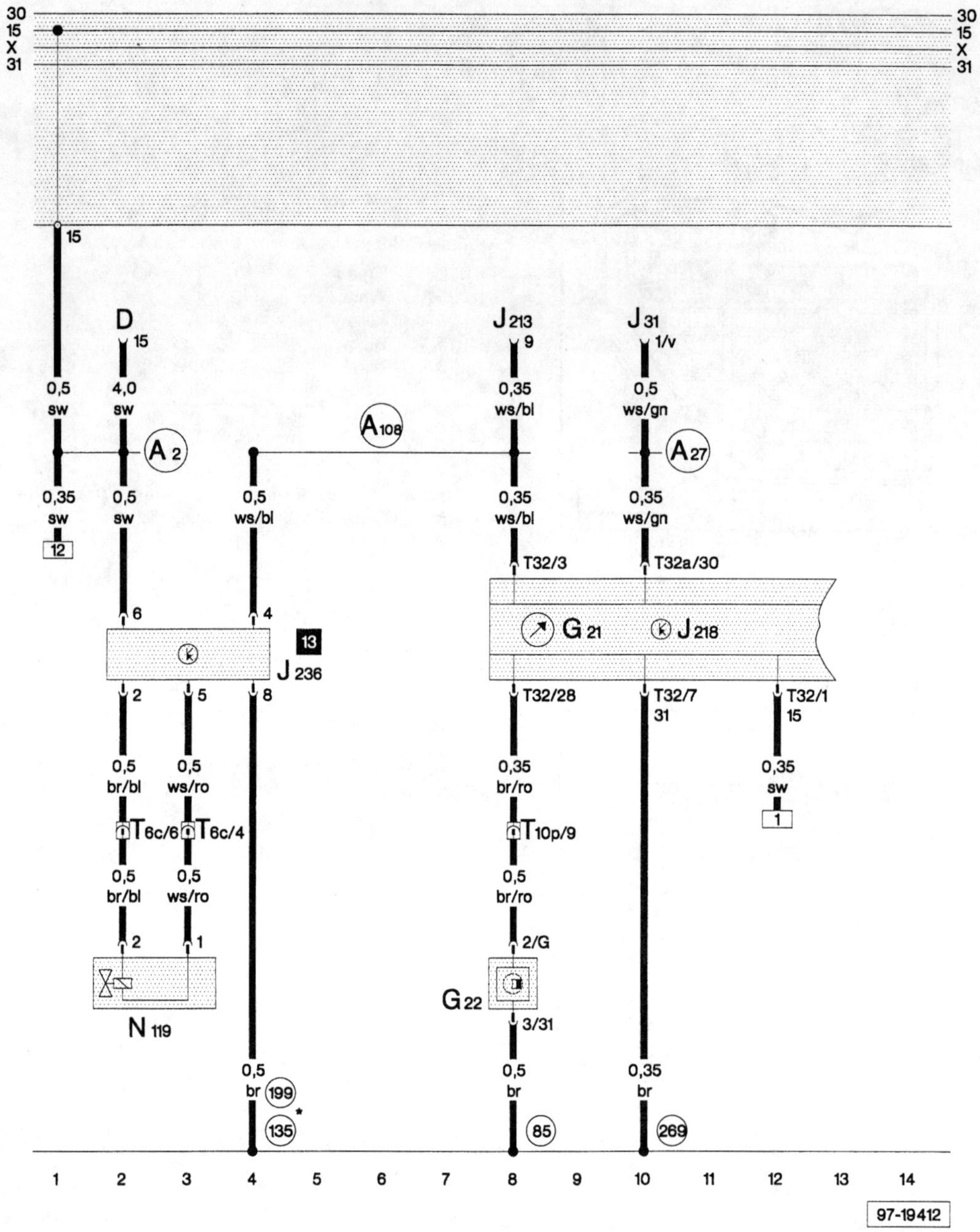

ws = white
sw = black
ro = red
gn = green
br = brown
bl = blue
gr = gray
li = violet
ge = yellow

D - Ignition/Starter Switch
G 21 - Speedometer
G 22 - Speedometer Vehicle Speed Sensor (VSS)
J 31 - Wiper/Washer Intermittent Relay
J 213 - Cruise Control, Control Module
J 218 - Instrument Cluster Combination Processor
J 236 - Servotronic Control Module
N '119 - Servotronic Solenoid Valve
T 6 c - Connector 6 pin, black, connector station A pillar, left
T 10 p - Connector 10 pin, black, connector station electronic box plenum chamber
T 32 - Connector 32 pin, blue, on instrument cluster
T 32 a - Connector 32 pin, green, on instrument cluster

(85) - Ground connection -1-, in engine compartment wiring harness
(135) - Ground connection -2-, in instrument panel wiring harness
(199) - Ground connection -3-, in instrument panel wiring harness
(269) - Ground connector (sensor ground) -1-, in instrument panel wiring harness
(A2) - Plus connection (15), in instrument panel wiring harness
(A27) - Wire connection (vehicle speed signal), in instrument panel wiring harness
(A108) - Wire connection (vehicle speed signal), in instrument panel wiring harness

* - booth is possible

Power Rear Window Shade

1998 m. y.

Fuse Panel

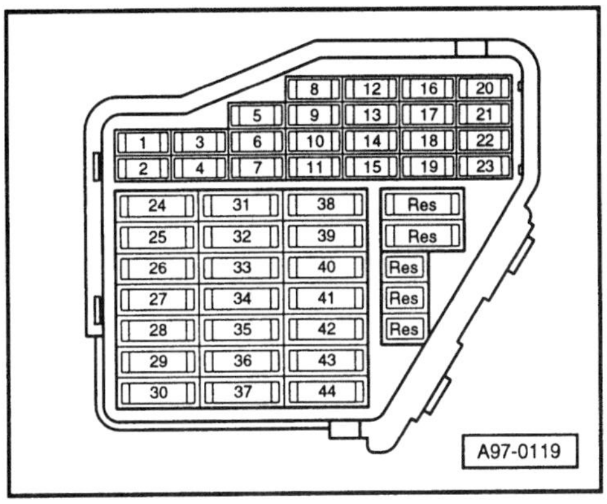

Fuse Colors:

30 A - Green
25 A - White
20 A - Yellow
15 A - Blue
10 A - Red
5 A - Beige

Starting with fuse position 23, fuses in the fuse holder are identified with 223 in the wiring diagram.

Micro Central Electric Panel

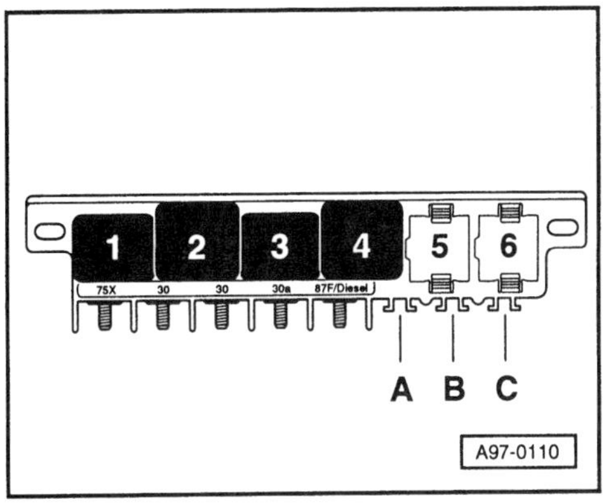

Fuse Location:

B - Rear Shade Fuse, S100

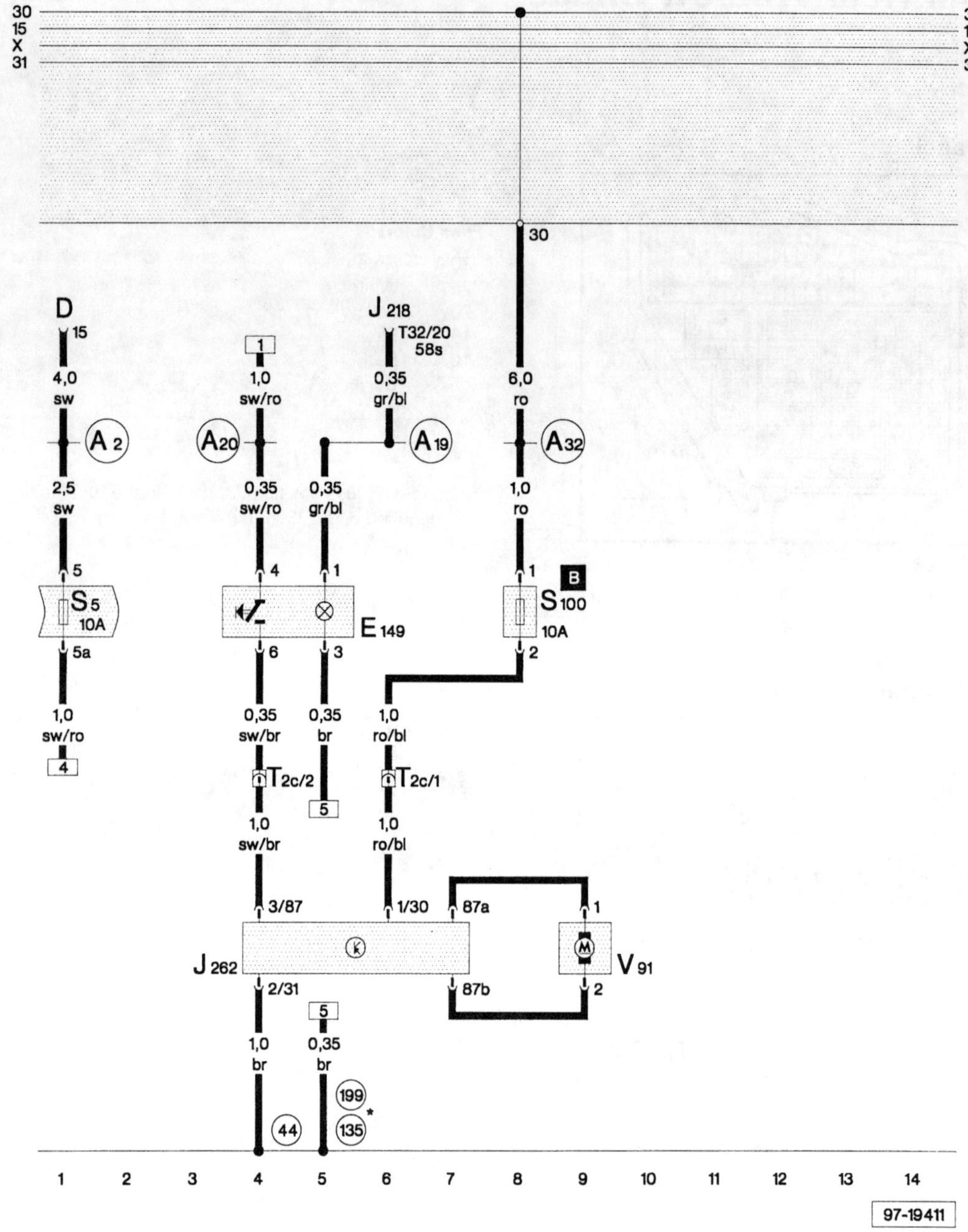

ws = white
sw = black
ro = red
gn = green
br = brown
bl = blue
gr = gray
li = violet
ge = yellow

- D - Ignition/Starter Switch
- E 149 - Rear Window Shade Switch
- J 218 - Instrument Cluster Combination Processor
- J 262 - Rear Window Shade Control Module
- S 5 - Fuse
- S 100 - Rear Shade Fuse
- T 2 c - Connector 2 pin, brown, connector station A pillar, left
- T 32 - Connector 32 pin, blue, on instrument cluster
- V 91 - Rear Window Shade Motor

- (44) - Ground connection, on left A-pillar, lower part
- (135) - Ground connection -2-, in instrument panel wiring harness
- (199) - Ground connection -3-, in instrument panel wiring harness
- (A2) - Plus connection (15), in instrument panel wiring harness
- (A19) - Wire connection (58s), in instrument panel wiring harness (Switch illumination)
- (A20) - Wire connection (15a), in instrument panel wiring harness
- (A32) - Plus connection (30), in instrument panel wiring harness

* - booth is possible

Telephone

1998 m. y.

Fuse Panel

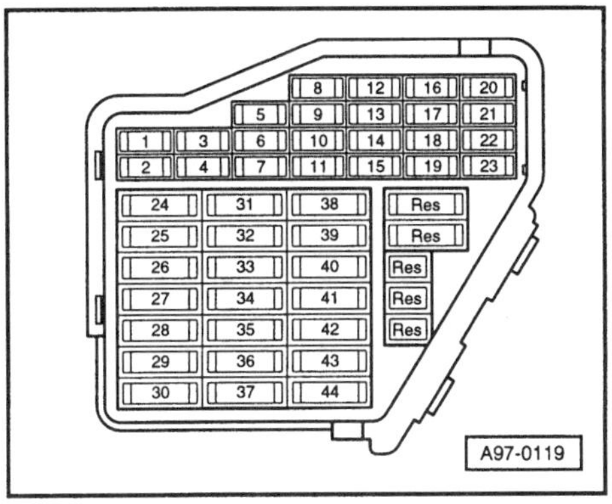

Fuse Colors:

30 A - Green
25 A - White
20 A - Yellow
15 A - Blue
10 A - Red
5 A - Beige

Starting with fuse position 23, fuses in the fuse holder are identified with 223 in the wiring diagram.

Micro Central Electric Panel

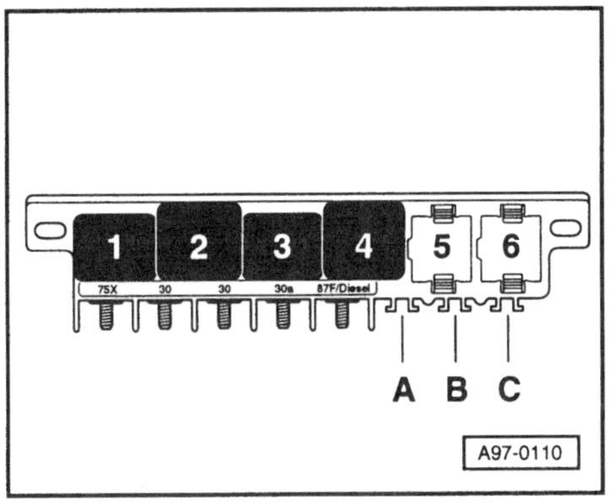

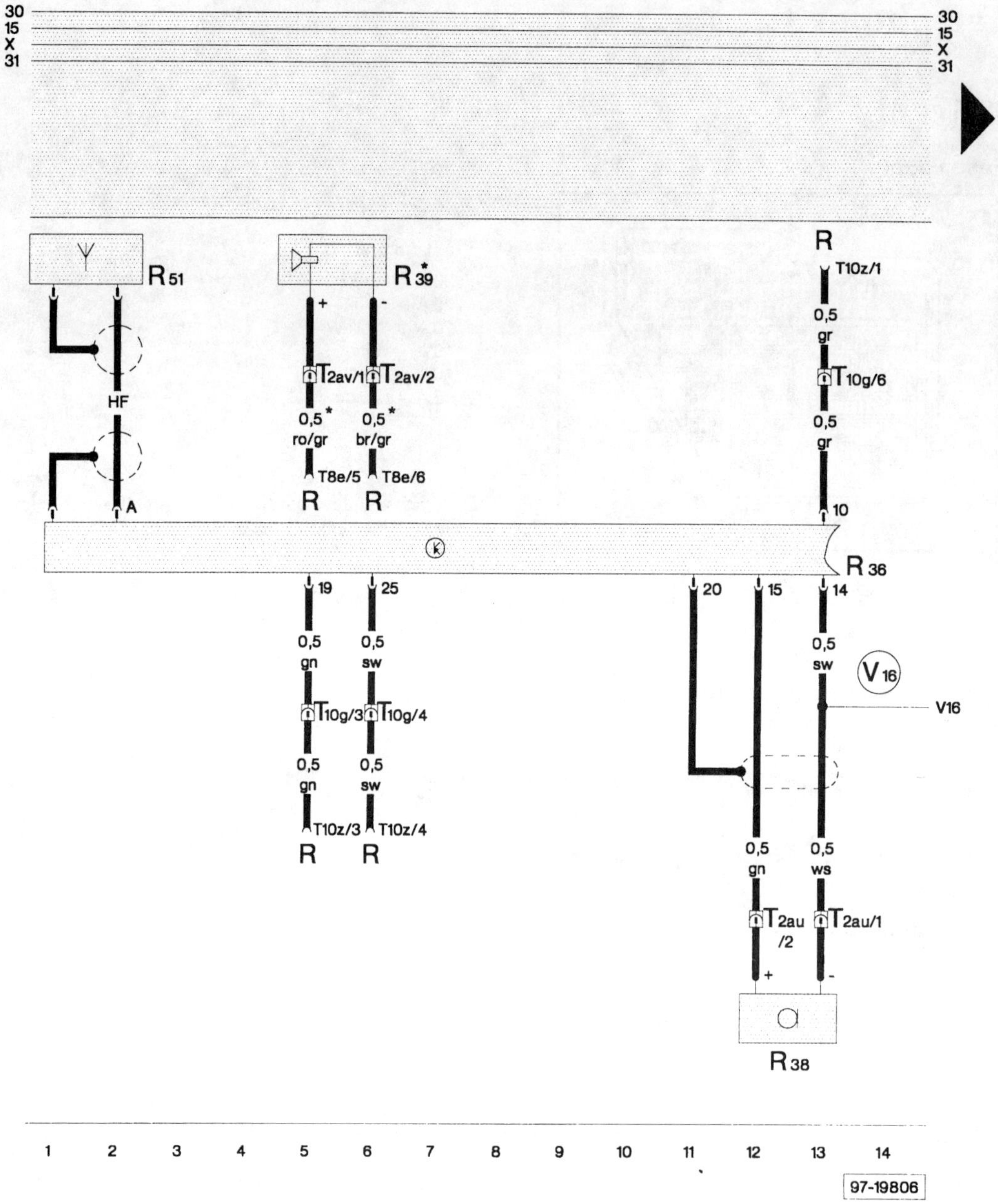

ws = white
sw = black
ro = red
gn = green
br = brown
bl = blue
gr = gray
li = violet
ge = yellow

R - Radio
R 36 - Telephone Transceiver
R 38 - Telephone Microphone
R 39 - Telephone Speaker
R 51 - Antenna for Radio/Telephone
T 2 au - Connector 2 pin, for Microphone, A pillar, left
T 2 av - Connector 2 pin, black, for loudspeaker, B pillar, left
T 8 e - Connector 8 pin, brown, connector radio II
T 10 g - Connector 10 pin, white, connector station A pillar, left
T 10 z - Connector 10 pin, red, connector radio IV

(V16) - Wire connection (shielding), in telephone wiring harness

* - Vehicles with Radio-Bose

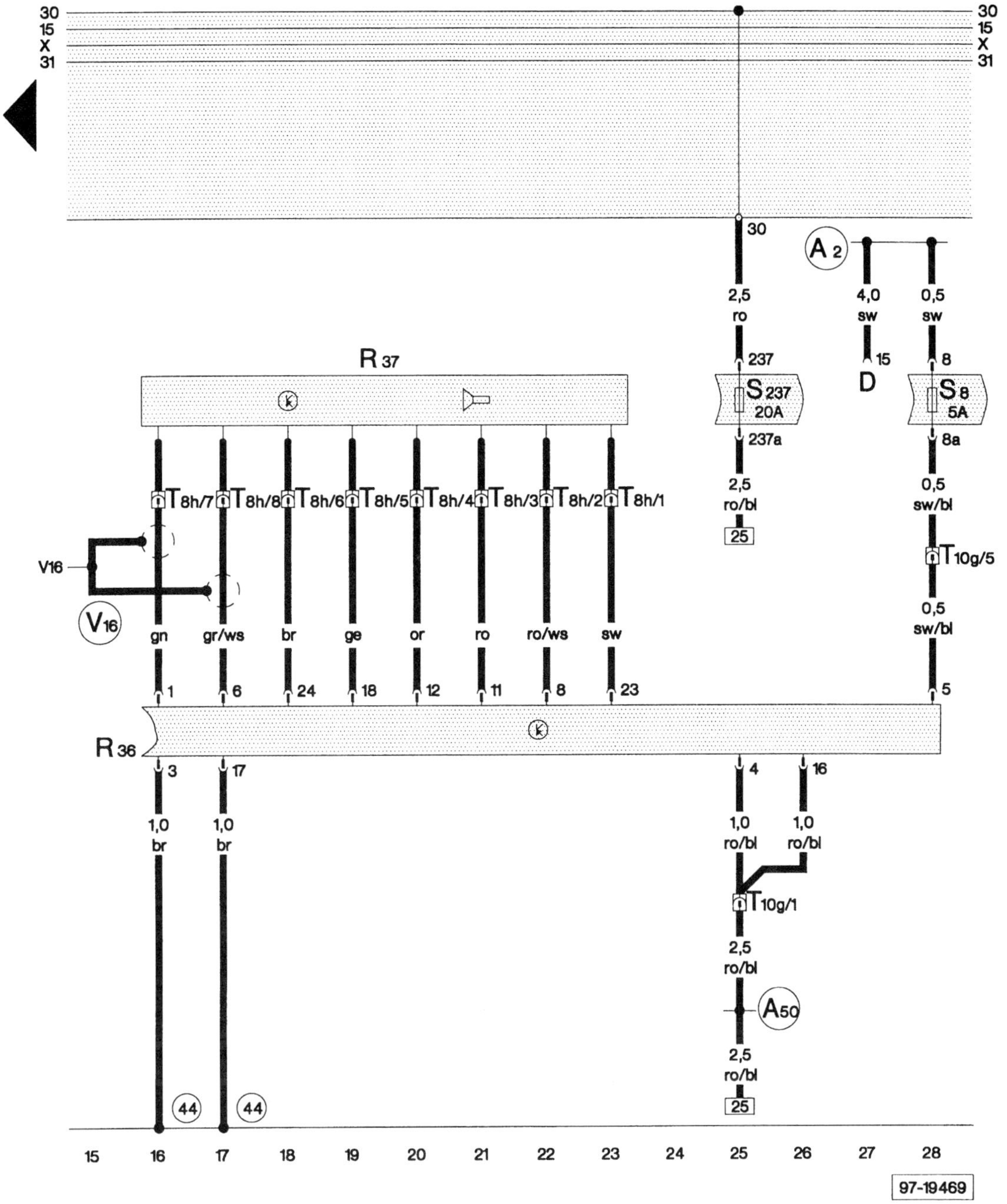

ws = white
sw = black
ro = red
gn = green
br = brown
bl = blue
gr = gray
li = violet
ge = yellow

D - Ignition/Starter Switch
R 36 - Telephone Transceiver
R 37 - Telephone Handset
S 8 - Fuse
S 237 - Fuse
T 8 h - Connector 8 pin, Telephone Handset, in console
T 10 g - Connector 10 pin, white, connector station A pillar, left

(44) - Ground connection, on left A-pillar, lower part
(A2) - Plus connection (15), in instrument panel wiring harness
(A50) - Plus connection (30as), in instrument panel wiring harness
(V16) - Wire connection (shielding), in telephone wiring harness

THIS PAGE INTENTIONALLY LEFT BLANK

Radio Concert, Bose with and without CD Changer

1998 m. y.

Fuse Panel

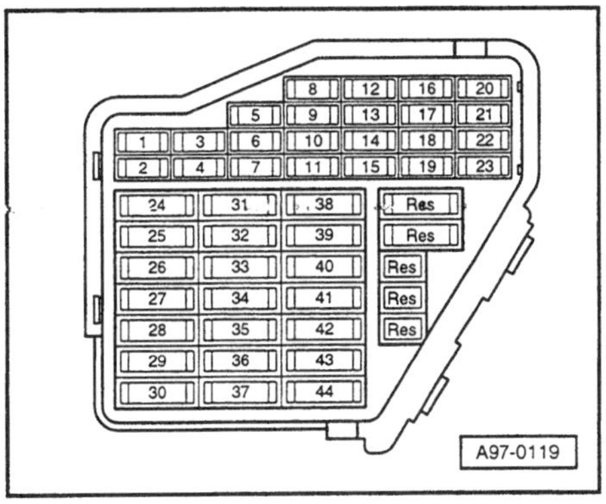

Fuse Colors:

30 A - Green
25 A - White
20 A - Yellow
15 A - Blue
10 A - Red
7,5 A - Brown
5 A - Beige

Starting with fuse position 23, fuses in the fuse holder are identified with 223 in the wiring diagram.

Micro Central Electric Panel

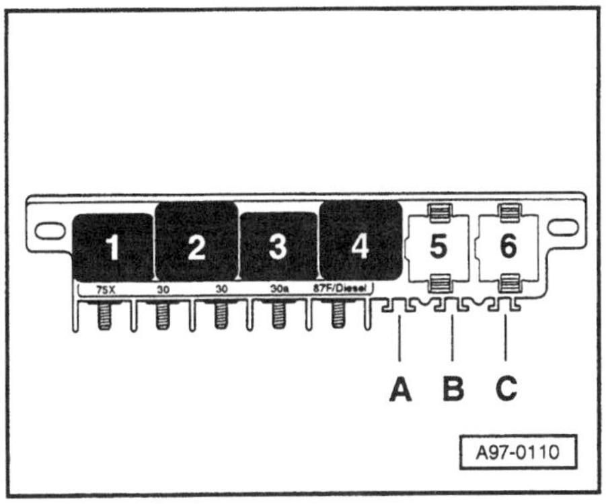

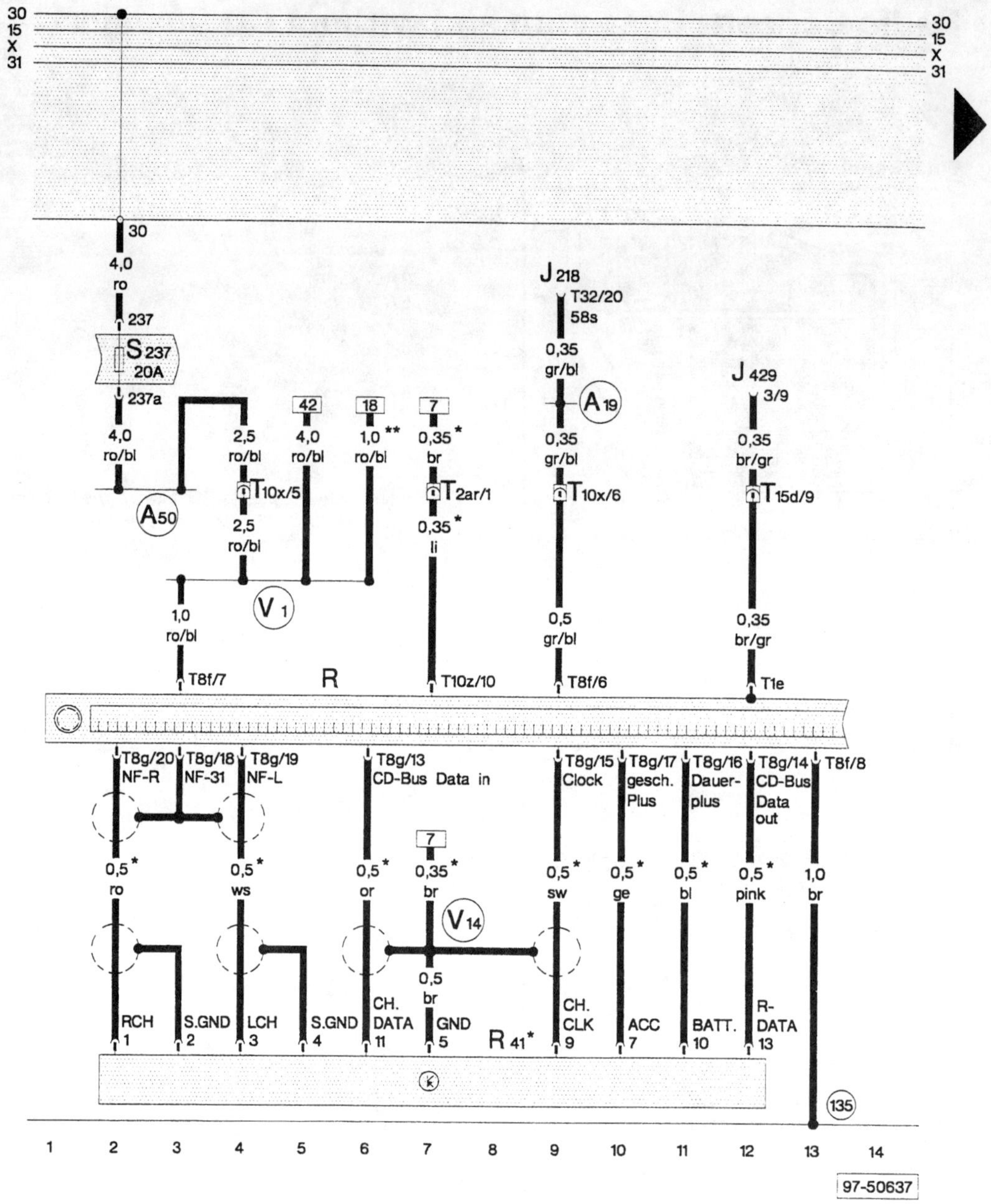

ws = white
sw = black
ro = red
gn = green
br = brown
bl = blue
gr = gray
li = violet
ge = yellow
or = orange

J 218 - Instrument Cluster Combination Processor
J 429 - Control module for central locking
R - Radio
R 41 - CD Changer Unit
T 1 e - Single connector, on Radio
T 2 ar - Connector 2 pin, black, behind instrument panel, center
T 8 f - Connector 8 pin, black, connector radio III
T 8 g - Connector 8 pin, blue, connector radio I - 3
T 10 x - Connector 10 pin, pink, connector station A pillar, right
T 10 z - Connector 10 pin, red, connector radio IV
T 15 d - Connector 15 pin, black, connector station A pillar, right
T 32 - Connector 32 pin, blue, on instrument cluster

(133) - Ground connection, in speaker wiring harness
(A19) - Wire connection (58s), in instrument panel wiring harness (Switch illumination)
(A50) - Plus connection (30as), in instrument panel wiring harness
(V1) - Plus connection (30), in rear speaker wiring harness
(V14) - Wire connection (shielding), in CD-changer wiring harness

* - Vehicles with CD Changer
** - Automatic Antenna

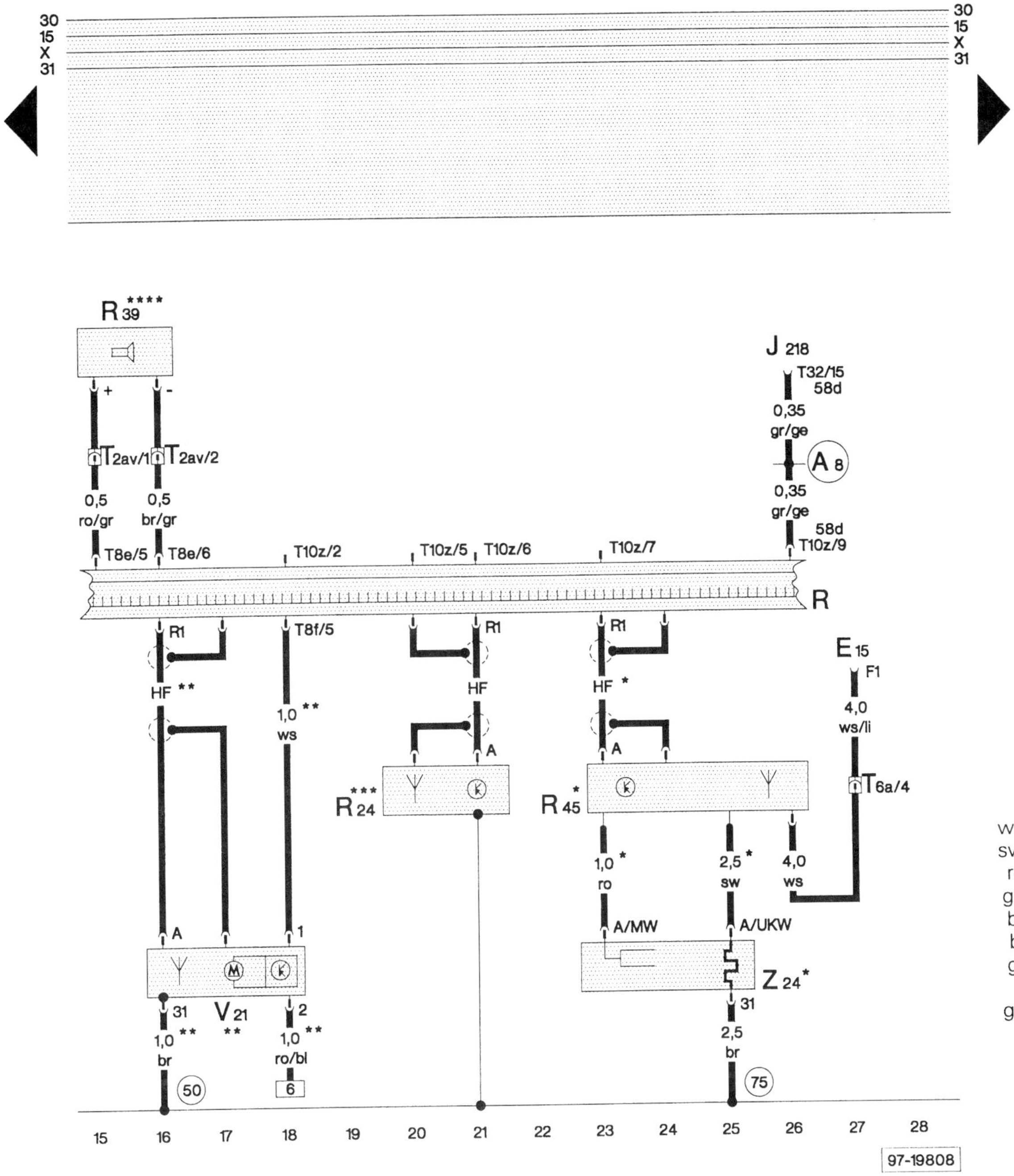

ws = white
sw = black
ro = red
gn = green
br = brown
bl = blue
gr = gray
li = violet
ge = yellow

E 15 - Rear Window Defogger Switch
J 218 - Instrument Cluster Combination Processor
R - Radio
R 1 - Antenna Connection
R 24 - Antenna Amplifier
R 39 - Telephone Speaker
R 45 - Telephone/Radio Antenna Amplifier
T 2 av - Connector 2 pin, black, for loudspeaker, B pillar, left
T 6 a - Connector 6 pin, blue, connector station A pillar, left
T 8 e - Connector 8 pin, brown, connector radio II
T 8 f - Connector 8 pin, black, connector radio III
T 10 z - Connector 10 pin, red, connector radio IV
T 32 - Connector 32 pin, blue, on instrument cluster
V 21 - Power Antenna
Z 24 - Rear Window Defogger With Window Antenna

(50) - Ground connection, in luggage compartment, left

(75) - Ground connection, on right rear pillar

(A8) - Plus connection (58d), in instrument panel wiring harness (Display illumination)

* - Rear Window Antenna
** - Automatic Antenna
*** - Wagon
**** - Vehicles with Telephone

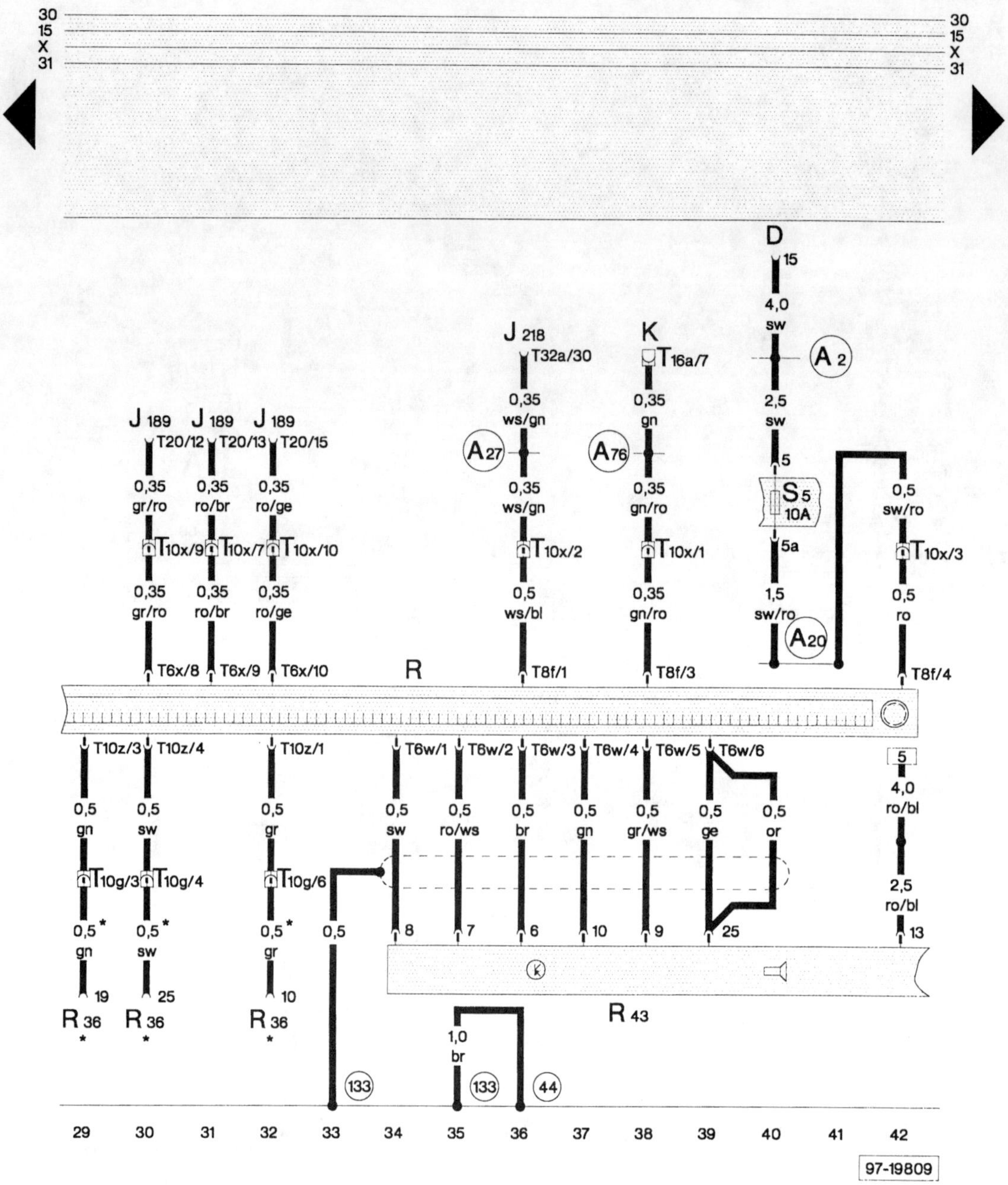

ws = white
sw = black
ro = red
gn = green
br = brown
bl = blue
gr = gray
li = violet
ge = yellow

D - Ignition/Starter Switch
J 189 - Auto Check System
J 218 - Instrument Cluster Combination Processor
K - DLC (Data Link Connector) Rapid Data Transfer
R - Radio
R 36 - Telephone Transceiver
R 43 - Right Rear Woofer/Amplifier
S 5 - Fuse
T 6 w - Connector 6 pin, yellow, connector radio I - 1
T 6 x - Connector 6 pin, green, connector radio I - 2
T 8 f - Connector 8 pin, black, connector radio III
T 10 g - Connector 10 pin, white, connector station A pillar, left
T 10 x - Connector 10 pin, pink, connector station A pillar, right
T 10 z - Connector 10 pin, red, connector radio IV
T 16 a - Connector 16 pin, data link connector
T 20 - Connector 20 pin, red/white, on instrument cluster
T 32 a - Connector 32 pin, green, on instrument cluster

(44) - Ground connection, on left A-pillar, lower part
(133) - Ground connection, in speaker wiring harness
(A2) - Plus connection (15), in instrument panel wiring harness
(A20) - Wire connection (15a), in instrument panel wiring harness
(A27) - Wire connection (vehicle speed signal), in instrument panel wiring harness
(A76) - Connector (K-diagnosis wire), in instrument panel wiring harness

* - Vehicles with Telephone

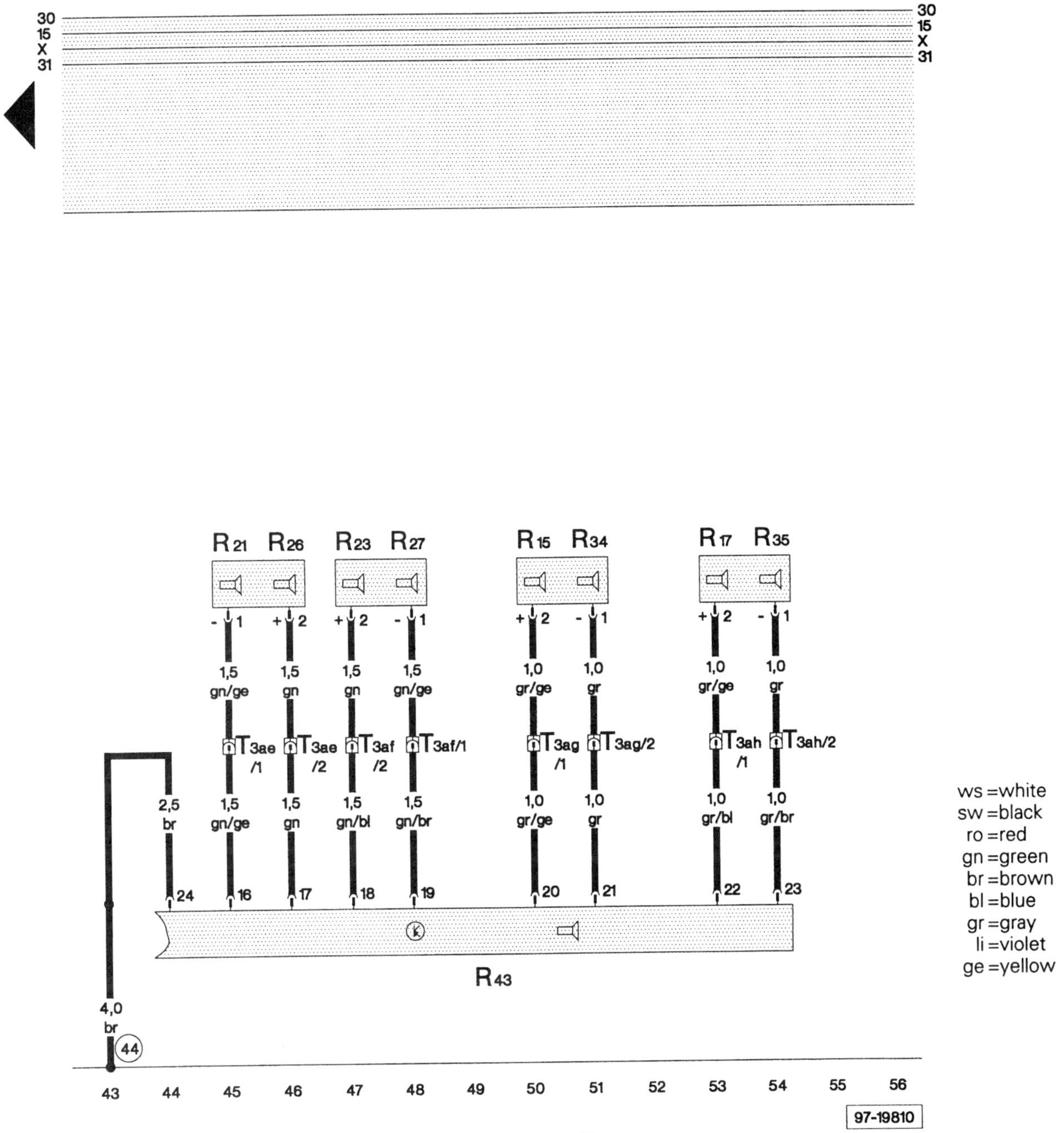

R 15 - Left Rear Woofer
R 17 - Right Rear Woofer
R 21 - Left Front Woofer
R 23 - Right Front Woofer
R 26 - Left Front Midrange Speaker
R 27 - Right Front Midrange Speaker
R 34 - Left Rear Midrange Speaker
R 35 - Right Rear Midrange Speaker
R 43 - Right Rear Woofer/Amplifier
T 3 ae - Connector 3 pin, brown, in driver's door
T 3 af - Connector 3 pin, brown, in front passenger door
T 3 ag - Connector 3 pin, blue, in rear left door
T 3 ah - Connector 3 pin, blue, in rear right door

(44) - Ground connection, on left A-pillar, lower part

Radio-Stereo, Concert

1998 m. y.

Fuse Panel

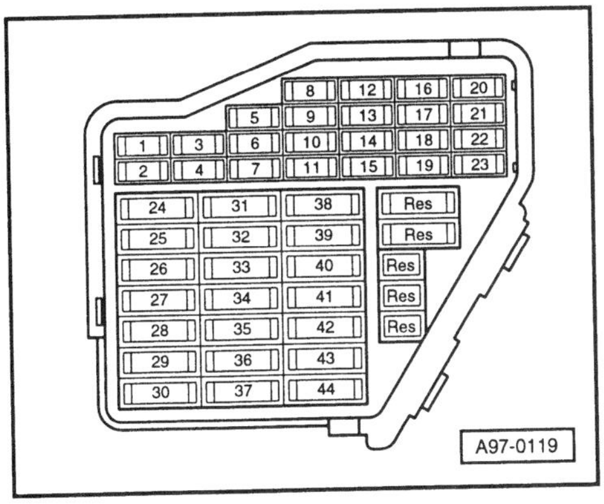

Fuse Colors:

30 A - Green
25 A - White
20 A - Yellow
15 A - Blue
10 A - Red
5 A - Beige

Starting with fuse position 23, fuses in the fuse holder are identified with 223 in the wiring diagram.

Micro Central Electric Panel

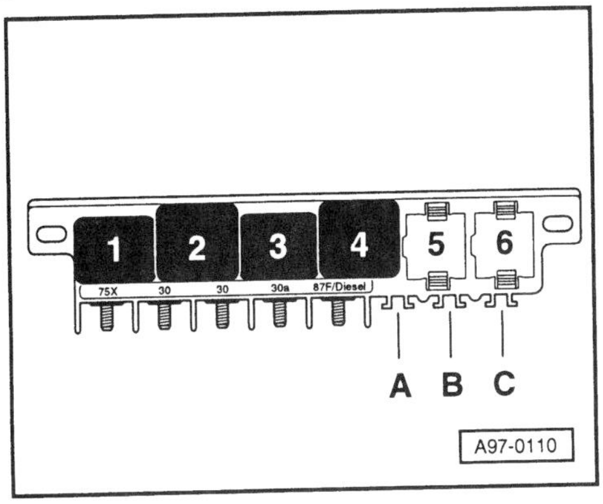

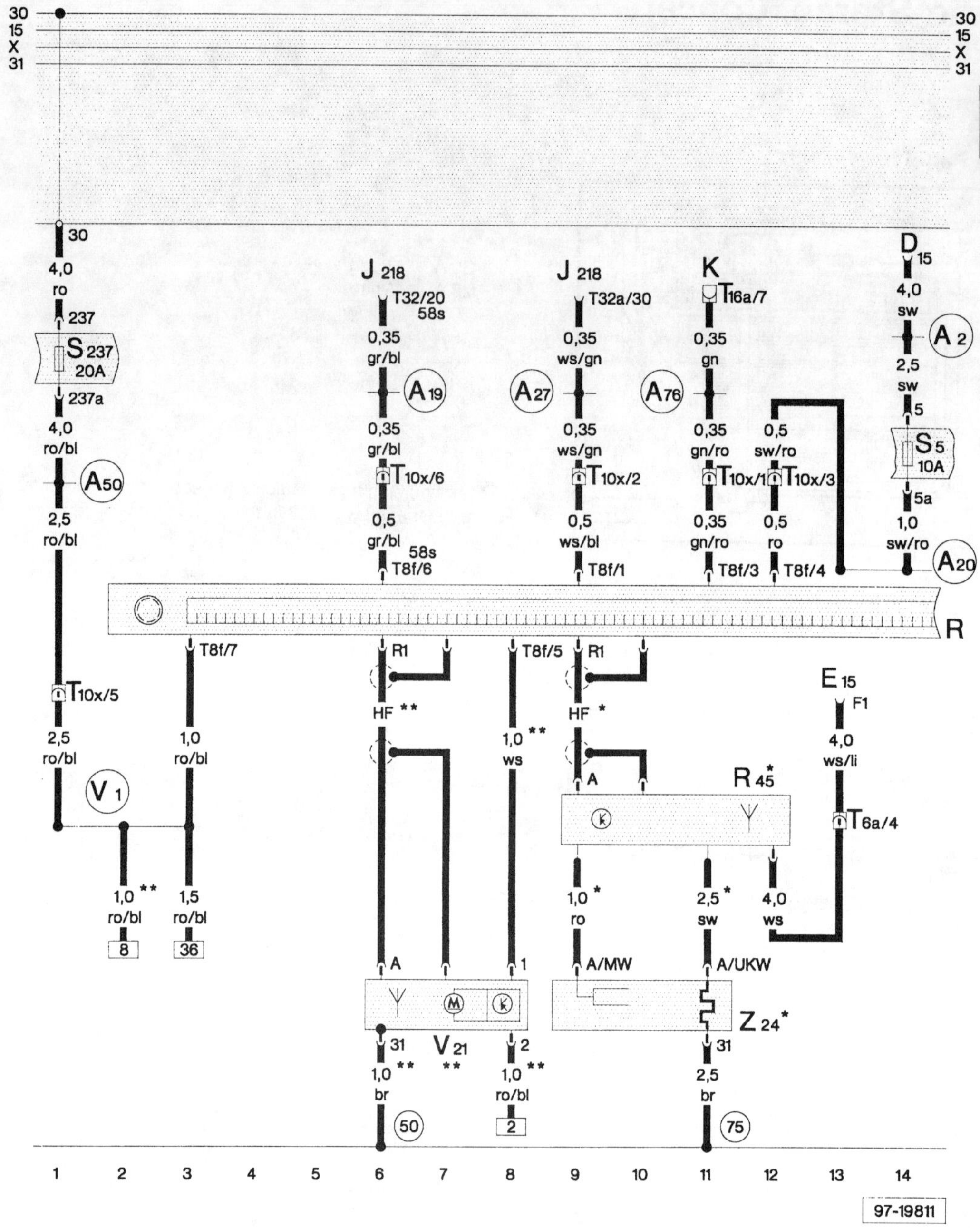

ws = white
sw = black
ro = red
gn = green
br = brown
bl = blue
gr = gray
li = violet
ge = yellow

- D - Ignition/Starter Switch
- E 15 - Rear Window Defogger Switch
- J 218 - Instrument Cluster Combination Processor
- K - DLC (Data Link Connector) Rapid Data Transfer
- R - Radio
- R 1 - Antenna Connection
- R 45 - Telephone/Radio Antenna Amplifier
- S 5 - Fuse
- S 237 - Fuse
- T 6 a - Connector 6 pin, blue, connector station A pillar, left
- T 8 f - Connector 8 pin, black, connector radio III
- T 10 x - Connector 10 pin, pink, connector station A pillar, right
- T 16 a - Connector 16 pin, data link connector
- T 32 - Connector 32 pin, blue, on instrument cluster
- T 32 a - Connector 32 pin, green, on instrument cluster
- V 21 - Power Antenna
- Z 24 - Rear Window Defogger With Window Antenna
- (50) - Ground connection, in luggage compartment, left
- (75) - Ground connection, on right rear pillar
- (A2) - Plus connection (15), in instrument panel wiring harness
- (A19) - Wire connection (58s), in instrument panel wiring harness (Switch illumination)
- (A20) - Wire connection (15a), in instrument panel wiring harness
- (A27) - Wire connection (vehicle speed signal), in instrument panel wiring harness
- (A50) - Plus connection (30as), in instrument panel wiring harness
- (A76) - Connector (K-diagnosis wire), in instrument panel wiring harness
- (V1) - Plus connection (30), in rear speaker wiring harness
- * - Rear Window Antenna
- ** - Automatic Antenna

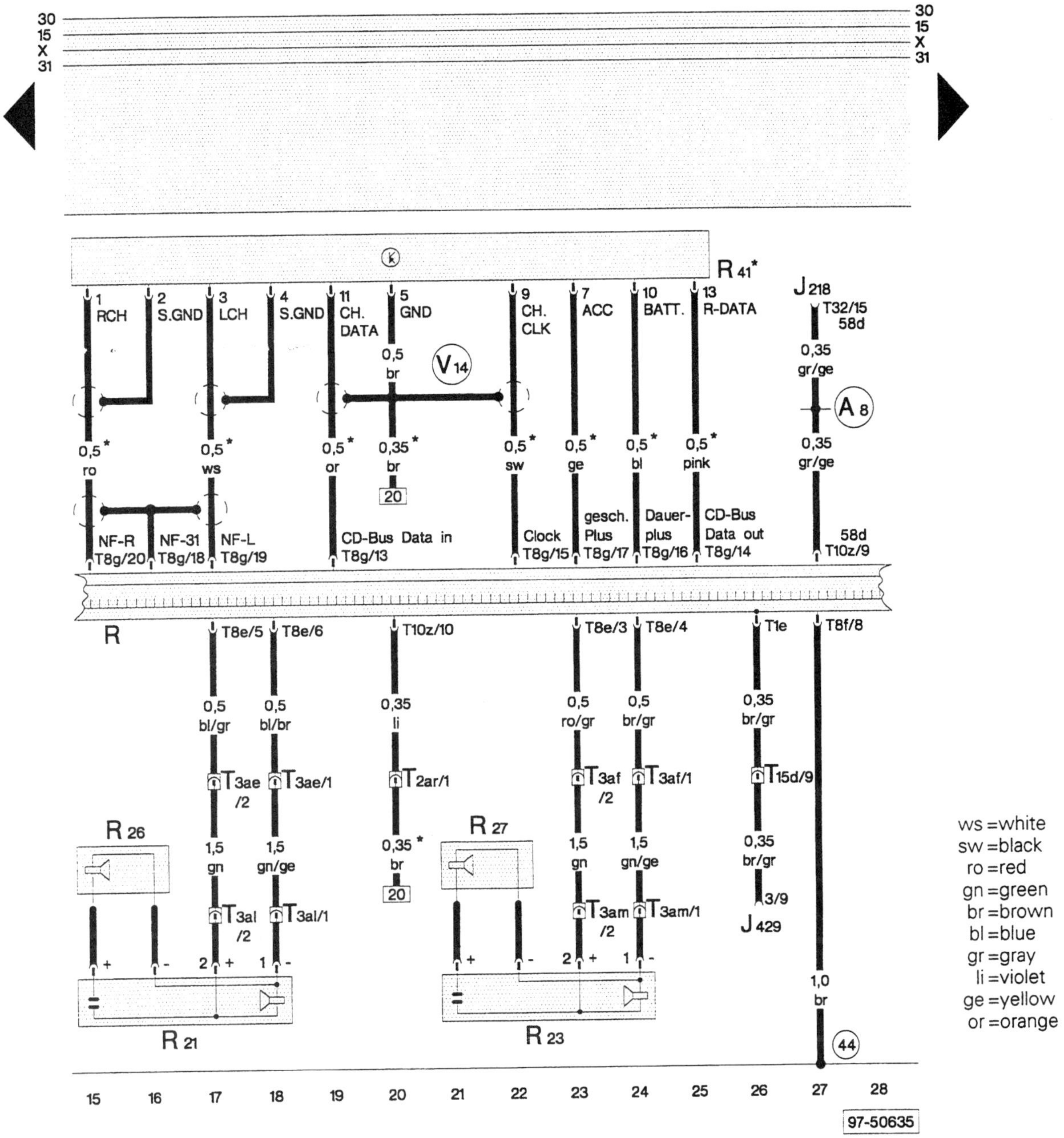

J 218 - Instrument Cluster Combination Processor
J 429 - Control module for central locking
R - Radio
R 21 - Left Front Woofer
R 23 - Right Front Woofer
R 26 - Left Front Midrange Speaker
R 27 - Right Front Midrange Speaker
R 41 - CD Changer Unit
T 1 e - Single connector, on Radio
T 2 ar - Connector 2 pin, black, behind instrument panel, center
T 3 ae - Connector 3 pin, brown, connector station A pillar, left
T 3 af - Connector 3 pin, brown, connector station A pillar, right
T 3 al - Connector 3 pin, black, in driver's door
T 3 am - Connector 3 pin, black, in passenger's door
T 8 e - Connector 8 pin, brown, connector radio II
T 8 f - Connector 8 pin, black, connector radio III
T 8 g - Connector 8 pin, blue, connector radio I - 3
T 10 z - Connector 10 pin, red, connector radio IV
T 15 d - Connector 15 pin, black, connector station A pillar, right
T 32 - Connector 32 pin, blue, on instrument cluster

(44) - Ground connection, on left A-pillar, lower part

(A8) - Plus connection (58d), in instrument panel wiring harness (Display illumination)

(V14) - Wire connection (shielding), in CD Changer wiring harness

* - Vehicles with CD Changer

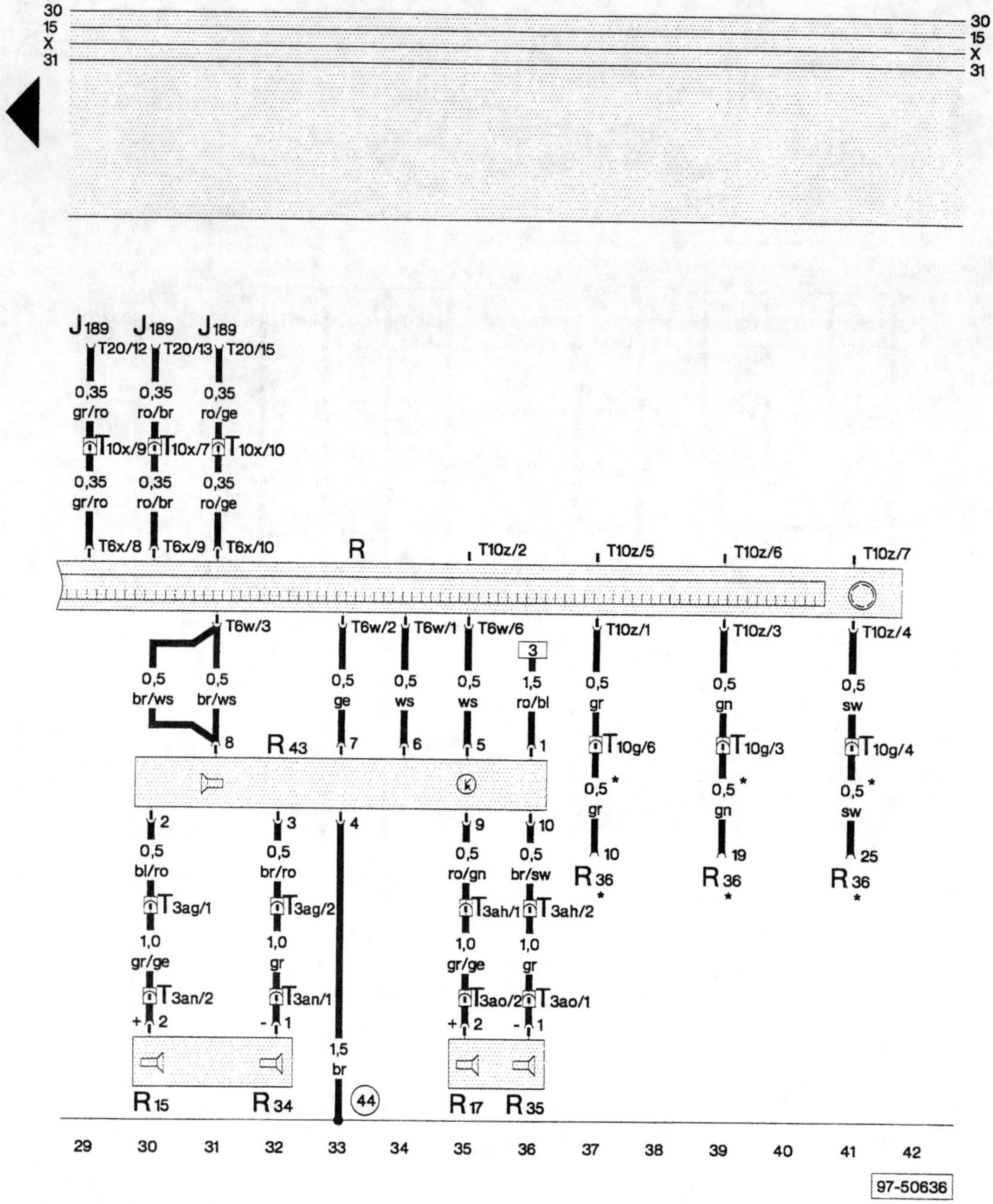

ws =white
sw =black
ro =red
gn =green
br =brown
bl =blue
gr =gray
li =violet
ge =yellow

J 189 - Auto Check System
R - Radio
R 15 - Left Rear Woofer
R 17 - Right Rear Woofer
R 34 - Left Rear Midrange Speaker
R 35 - Right Rear Midrange Speaker
R 36 - Telephone Transceiver
R 43 - Right Rear Woofer/Amplifier
T 3 ag - Connector 3 pin, blue, in B pillar, left
T 3 ah - Connector 3 pin, blue, in B pillar, right
T 3 an - Connector 3 pin, black, in left rear door
T 3 ao - Connector 3 pin, black, in right rear door
T 6 w - Connector 6 pin, yellow, connector radio I - 1
T 6 x - Connector 6 pin, green, connector radio I - 2
T 10 g - Connector 10 pin, white, connector station A pillar, left
T 10 x - Connector 10 pin, pink, connector station A pillar, right
T 10 z - Connector 10 pin, red, connector radio IV
T 20 - Connector 20 pin, red/white, on instrument cluster

(44) - Ground connection, on left A-pillar, lower part

* - Vehicles with Telephone

Power Sunroof with Automatic Preselection

1998 m. y.

Fuse Panel

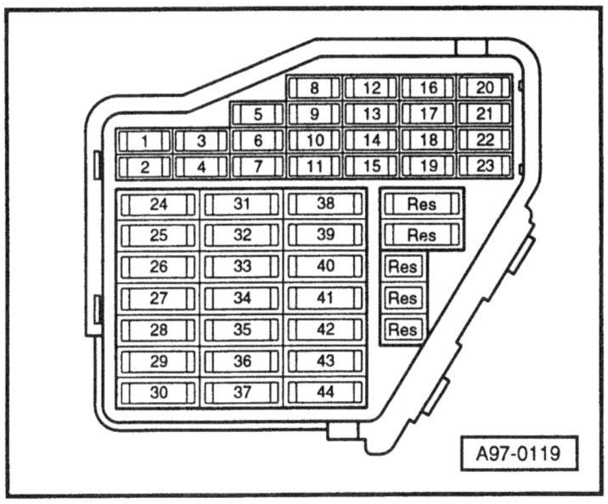

Fuse Colors:

30 A	-	Green
25 A	-	White
20 A	-	Yellow
15 A	-	Blue
10 A	-	Red
5 A	-	Beige

Starting with fuse position 23, fuses in the fuse holder are identified with 223 in the wiring diagram.

Micro Central Electric Panel

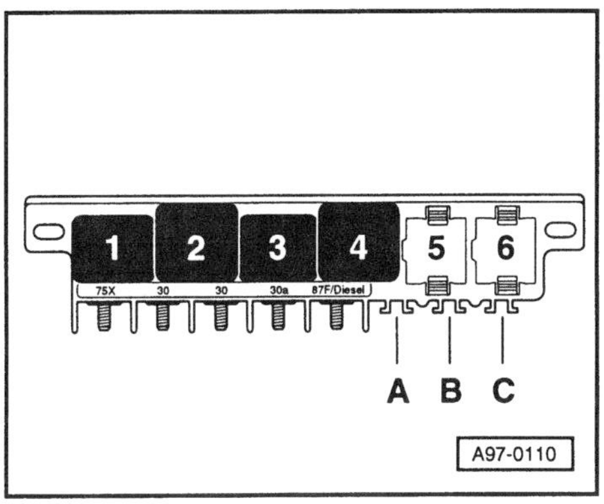

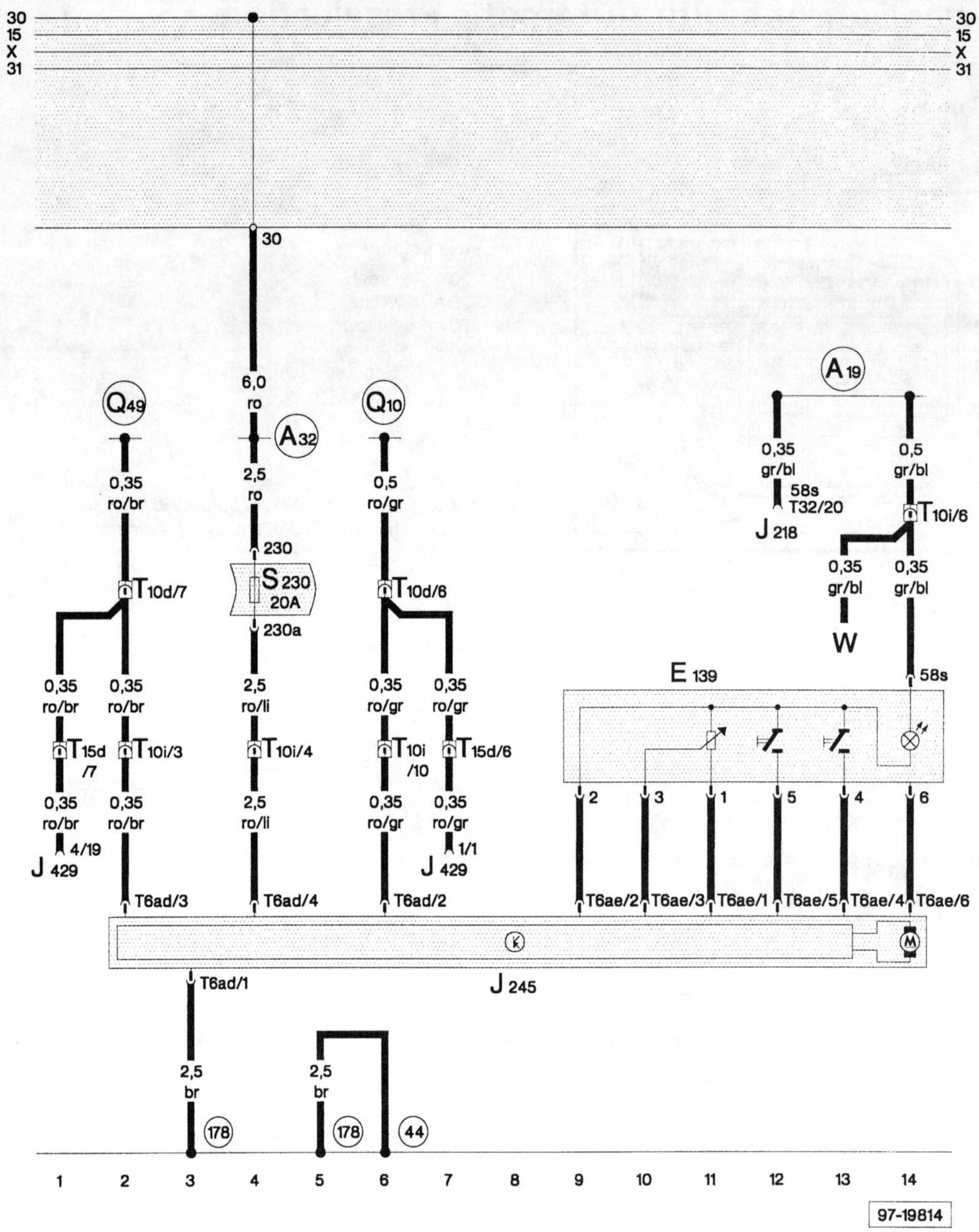

ws = white
sw = black
ro = red
gn = green
br = brown
bl = blue
gr = gray
li = violet
ge = yellow

E 139 - Sunroof Regulator
J 218 - Instrument Cluster Combination Processor
J 245 - Power Sunroof Control Module
J 429 - Control module for central locking
S 230 - Fuse
T 6 ad - Connector 6 pin, blue, on power sunroof control module
T' 6 ae - Connector 6 pin, on power sunroof control module
T 10 d - Connector 10 pin, green, connector station A pillar, left
T 10 i - Connector 10 pin, darkbrown, connector station A pillar, left
T 15 d - Connector 15 pin, black, connector station A pillar, right
T 32 - Connector 32 pin, blue, on instrument cluster
W - Front Interior Light

(44) - Ground connection, on left A-pillar, lower part
(178) - Ground connection, in sunroof wiring harness
(A19) - Wire connection (58s), in instrument panel wiring harness (Switch illumination)
(A32) - Plus connection (30), in instrument panel wiring harness
(Q10) - Plus connection (87), in power window/central locking system and door contact switch wiring harness
(Q49) - Connection (closed), in wiring harness interior

Cruise Control

1998 m. y.

Fuse Panel

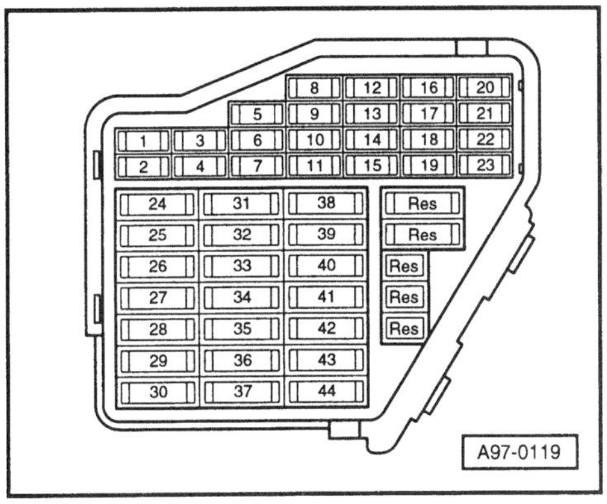

Fuse Colors:

30 A - Green
25 A - White
20 A - Yellow
15 A - Blue
10 A - Red
5 A - Beige

Starting with fuse position 23, fuses in the fuse holder are identified with 223 in the wiring diagram.

Micro Central Electric Panel

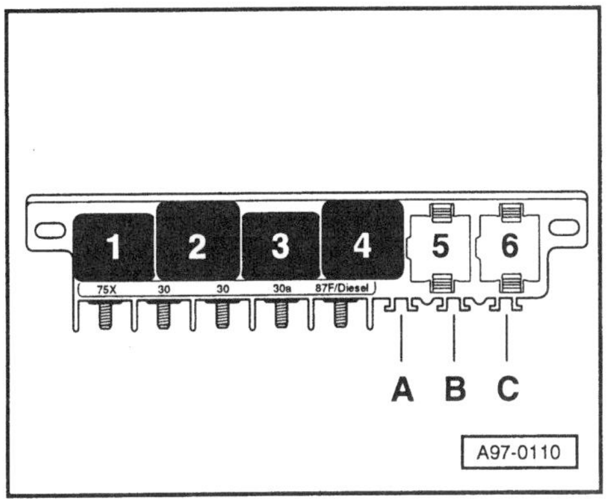

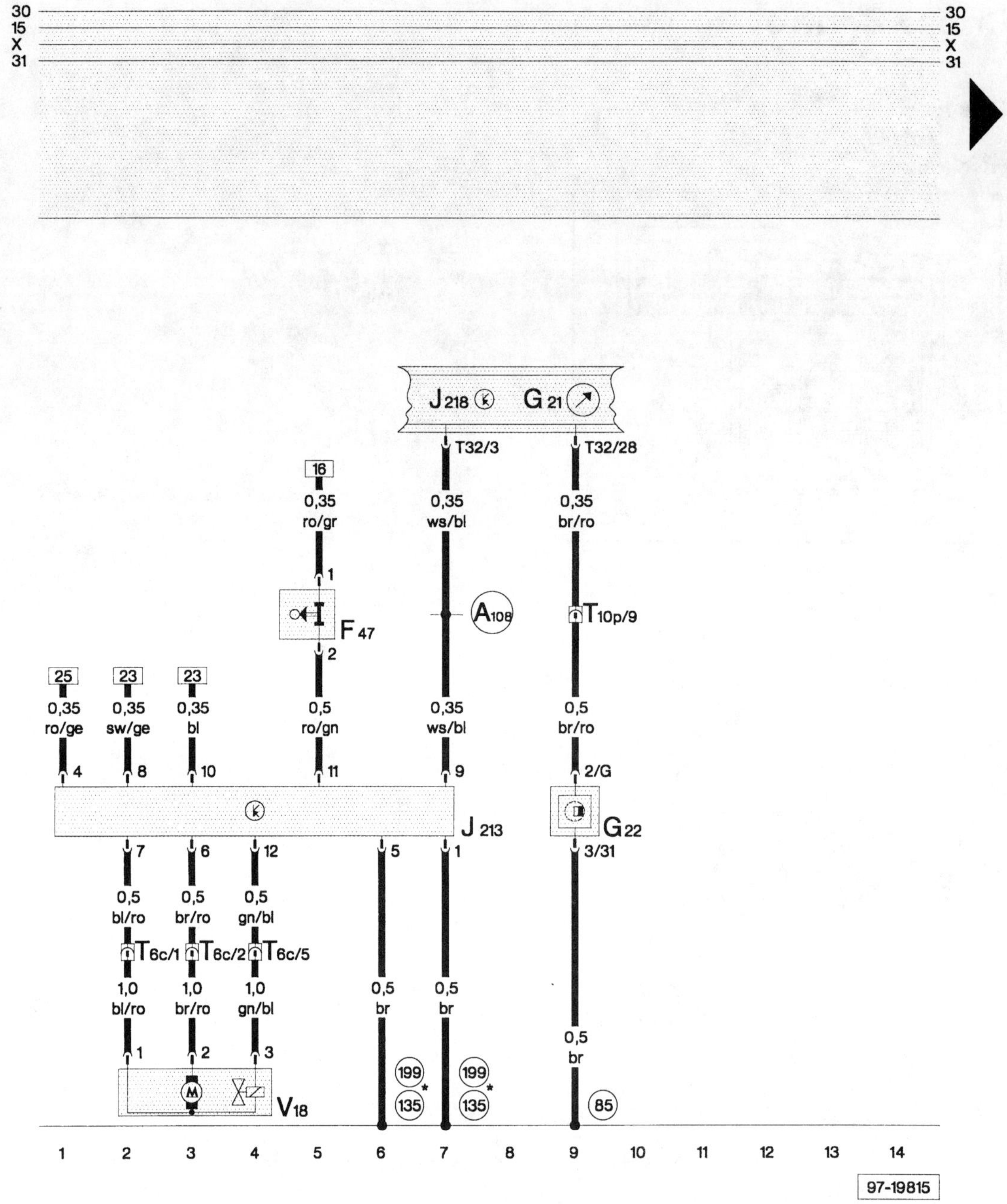

ws = white
sw = black
ro = red
gn = green
br = brown
bl = blue
gr = gray
li = violet
ge = yellow

F 47 - Vacuum Vent Valve, Brake
G 21 - Speedometer
G 22 - Speedometer Vehicle Speed Sensor (VSS)
J 213 - Cruise Control, Control Module
J 218 - Instrument Cluster Combination Processor
T 6 c - Connector 6 pin, black, connector station A pillar, left
T 10 p - Connector 10 pin, black, connector station electronic box plenum chamber
T 32 - Connector 32 pin, blue, on instrument cluster
V 18 - Cruise Control Vacuum Pump

(85) - Ground connection -1-, in engine compartment wiring harness
(135) - Ground connection -2-, in instrument panel wiring harness
(199) - Ground connection -3-, in instrument panel wiring harness
(A108) - Wire connection (vehicle speed signal), in instrument panel wiring harness

* - booth is possible

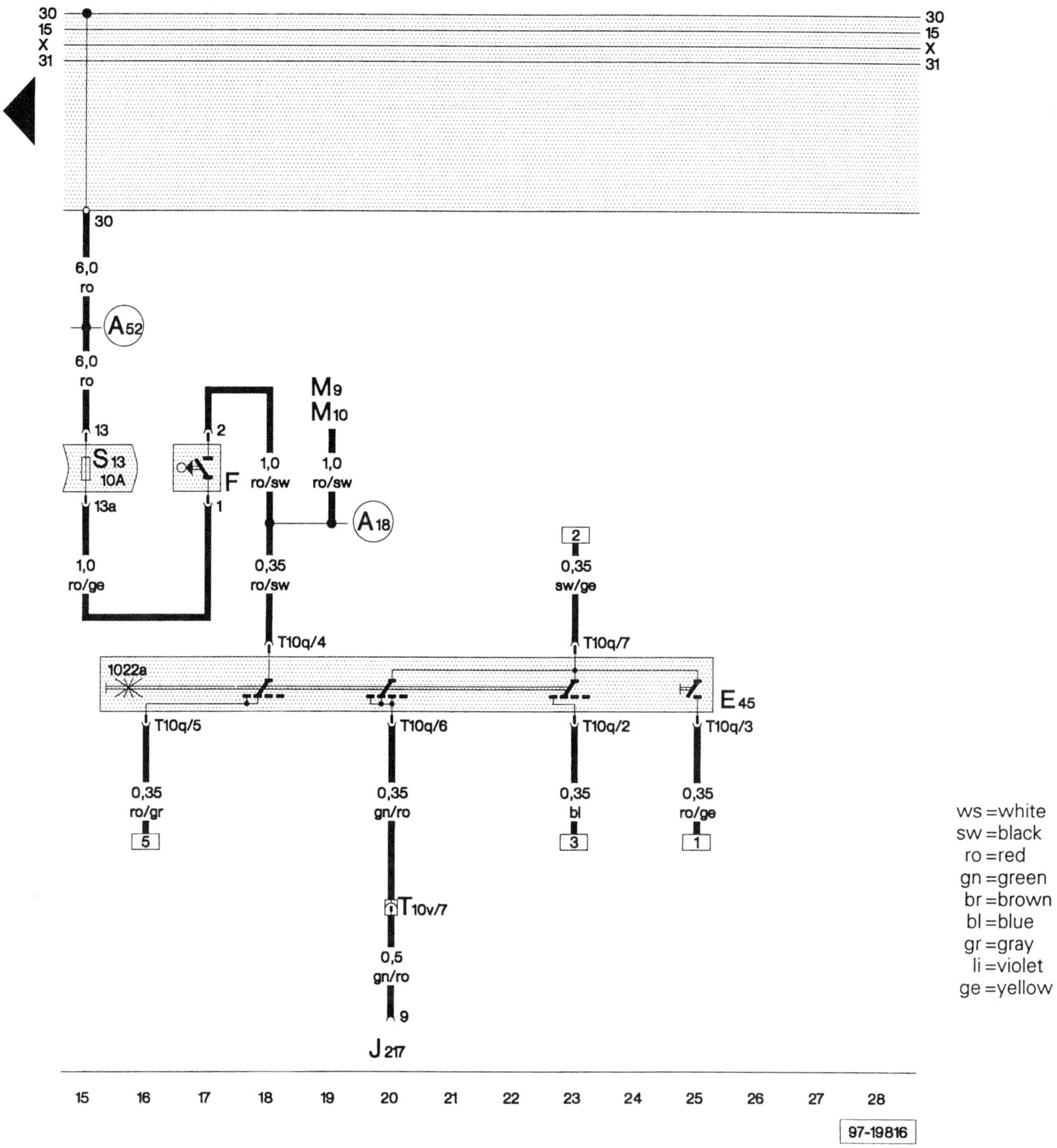

E 45 - Cruise Control Switch
F - Brake Light Switch
J 217 - Transmission Control Module (TCM)
M 9 - Left Brake Light
M 10 - Right Brake Light
S 13 - Fuse
T 10 q - Connector 10 pin, black, on cruise control switch
T 10 v - Connector 10 pin, brown, connector station A pillar, right

(A18) - Wire connection (54), in instrument panel wiring harness

(A52) - Plus connection (30), in instrument panel wiring harness

THIS PAGE INTENTIONALLY LEFT BLANK

Automatic Dimmer Inside and Outside Mirror

1998 m. y.

Fuse Panel

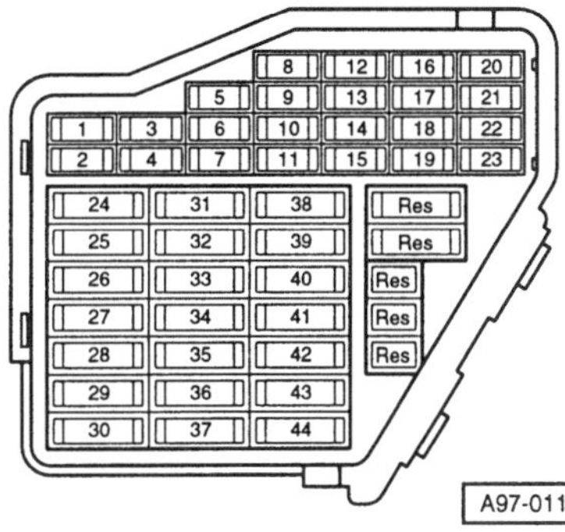

A97-0119

Fuse Colors:

30 A - Green
25 A - White
20 A - Yellow
15 A - Blue
10 A - Red
5 A - Beige

Starting with fuse position 23, fuses in the fuse holder are identified with 223 in the wiring diagram.

Micro Central Electric Panel

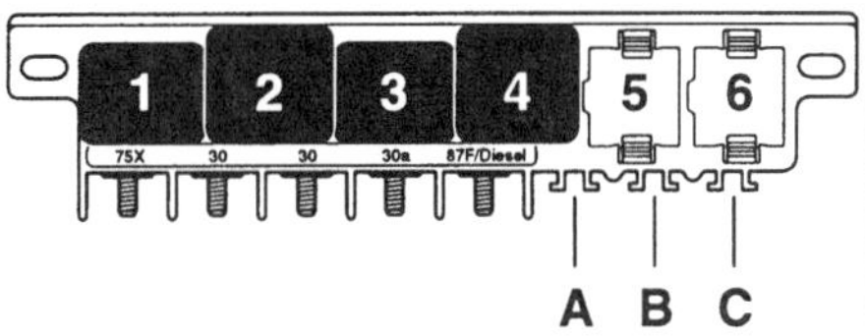

A97-0110

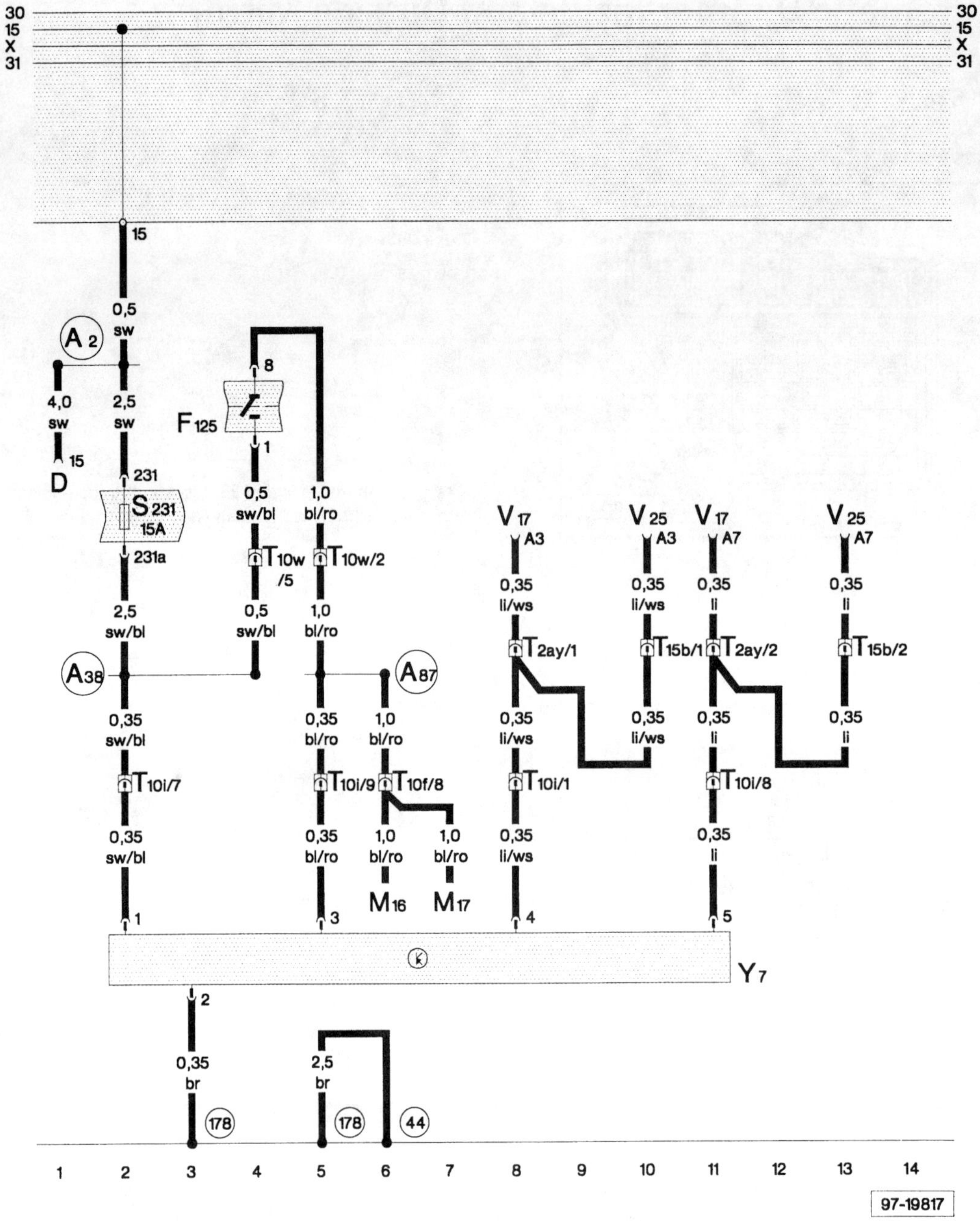

ws = white
sw = black
ro = red
gn = green
br = brown
bl = blue
gr = gray
li = violet
ge = yellow

D - Ignition/Starter Switch
F 125 - Multi-Function Transmission Range (TR) Switch
M 16 - Left Back-Up Light
M 17 - Right Back-Up Light
S 231 - Fuse
T 2 ay - Connector 2 pin, green, connector station A pillar, left
T 10 f - Connector 10 pin, brown, connector station A pillar, left
T 10 i - Connector 10 pin, darkbrown, connector station A pillar, left
T 10 w - Connector 10 pin, blue, connector station A pillar, right
T 15 b - Connector 15 pin, black, connector station A pillar, right
V 17 - Driver's Side Mirror Adjustment Motor
V 25 - Passenger's Side Mirror Adjustment Motor
Y 7 - Automatic day/night interior mirror

(44) - Ground connection, on left A-pillar, lower part
(178) - Ground connection, in sunroof wiring harness
(A2) - Plus connection (15), in instrument panel wiring harness
(A38) - Plus connection (15a II), in instrument panel wiring harness
(A87) - Connector (reverse lamp), in instrument panel wiring harness

Headlight Washer

1998 m. y.

Fuse Panel

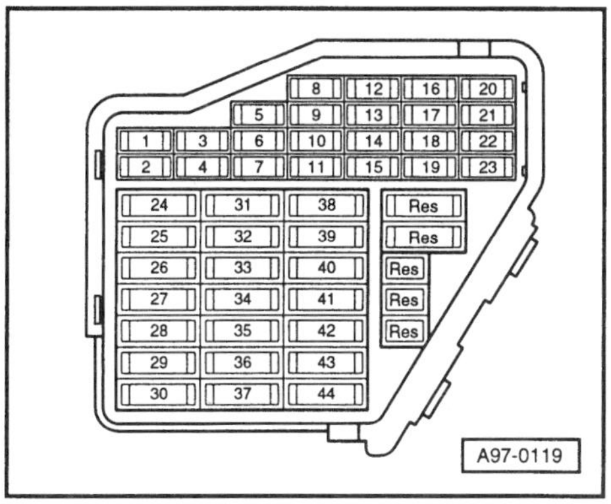

Fuse Colors:

30 A - Green
25 A - White
20 A - Yellow
15 A - Blue
10 A - Red
5 A - Beige

Starting with fuse position 23, fuses in the fuse holder are identified with 223 in the wiring diagram.

Micro Central Electric Panel

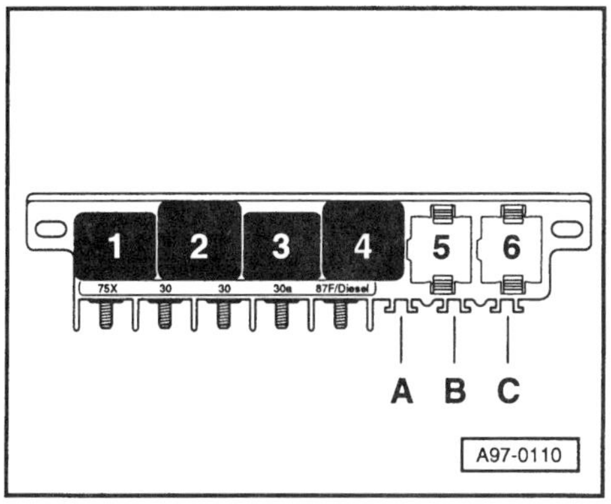

Relay Location:

5 - Wiper/Washer Intermittent Relay, J31

6 - Wiper/Washer Intermittent Relay, J31

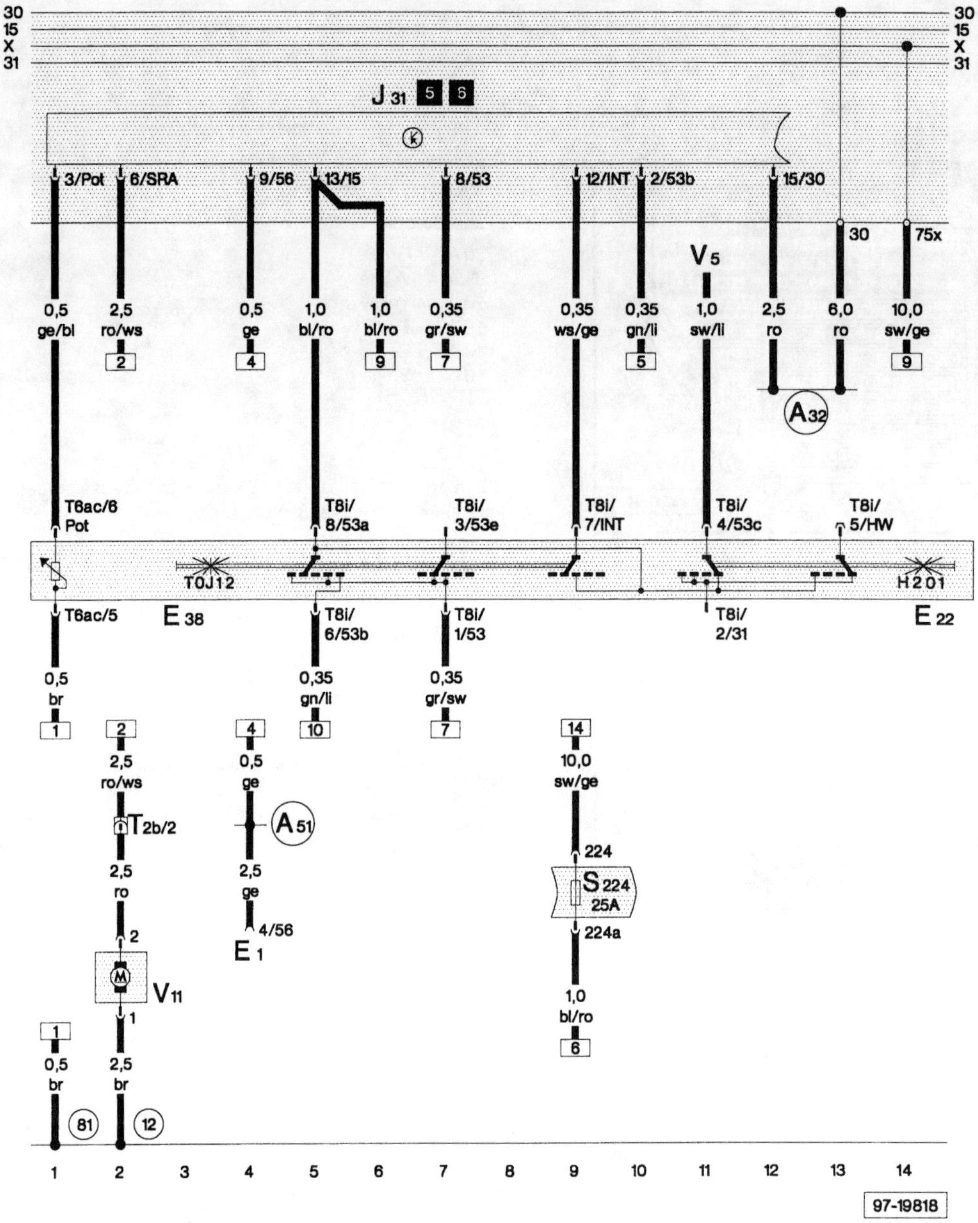

ws = white
sw = black
ro = red
gn = green
br = brown
bl = blue
gr = gray
li = violet
ge = yellow

E 1 - Light switch
E 22 - Windshield Wiper/Washer Switch
E 38 - Windshield Wiper Intermittent Regulator
J 31 - Wiper/Washer Intermittent Relay
S 224 - Fuse
T 2 b - Connector 2 pin, black, connector station A pillar, left
T 6 ac - Connector 6 pin, brown, on windshield wiper intermittent regulator
T 8 i - Connector 8 pin, on windshield wiper/washer switch
V 5 - Windshield Washer Pump
V 11 - Headlight Washer Pump

(12) - Ground connection, in engine compartment, left
(81) - Ground connection -1-, in instrument panel wiring harness
(A32) - Plus connection (30), in instrument panel wiring harness
(A51) - Wire connection (56), in instrument panel wiring harness

Headlight with Hight Intensity Gas discharge Lamps and automatic Headlight Beam Adjusting

1998 m. y.

Fuse Panel

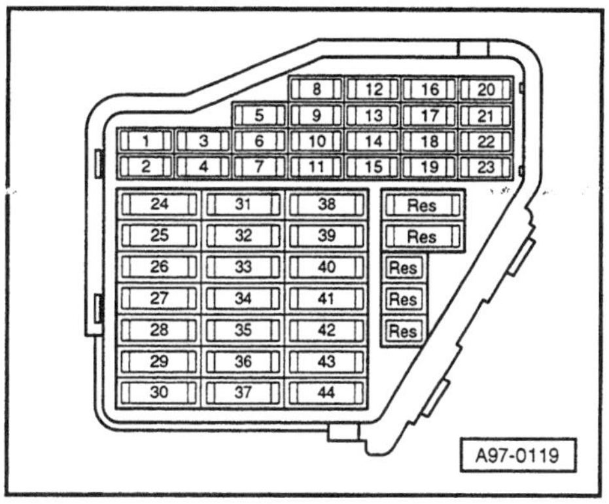

Fuse Colors:

30 A - Green
25 A - White
20 A - Yellow
15 A - Blue
10 A - Red
7,5 A - Brown
5 A - Beige

Starting with fuse position 23, fuses in the fuse holder are identified with 223 in the wiring diagram.

Micro Central Electric Panel

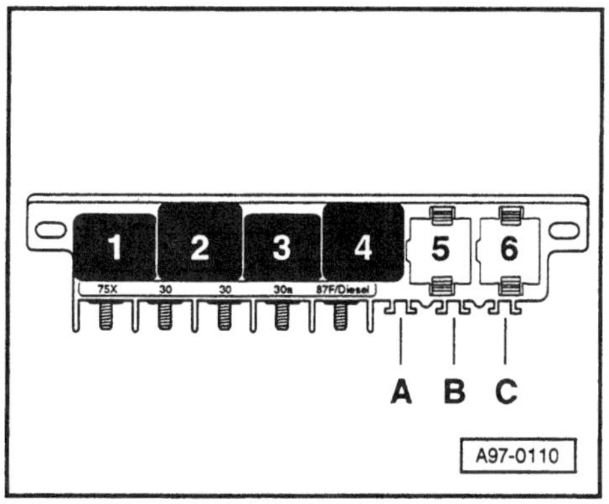

Relay Location:

2 - Load Reduction Relay, J59

13 - Fold Relay Panel

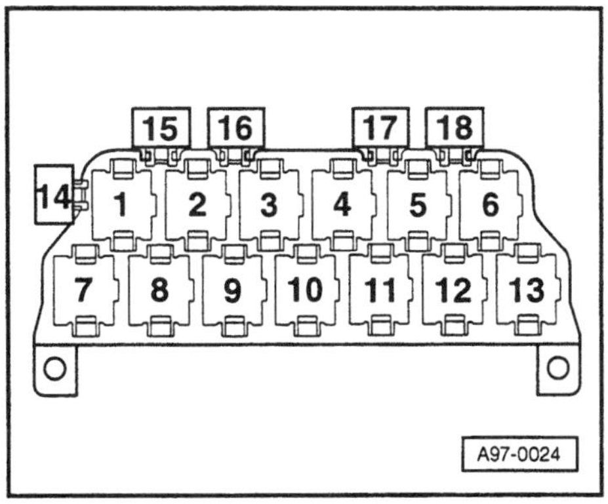

Relay Location:

9 - Lamp Control Module, J123

10 - Lamp Control Module, J123

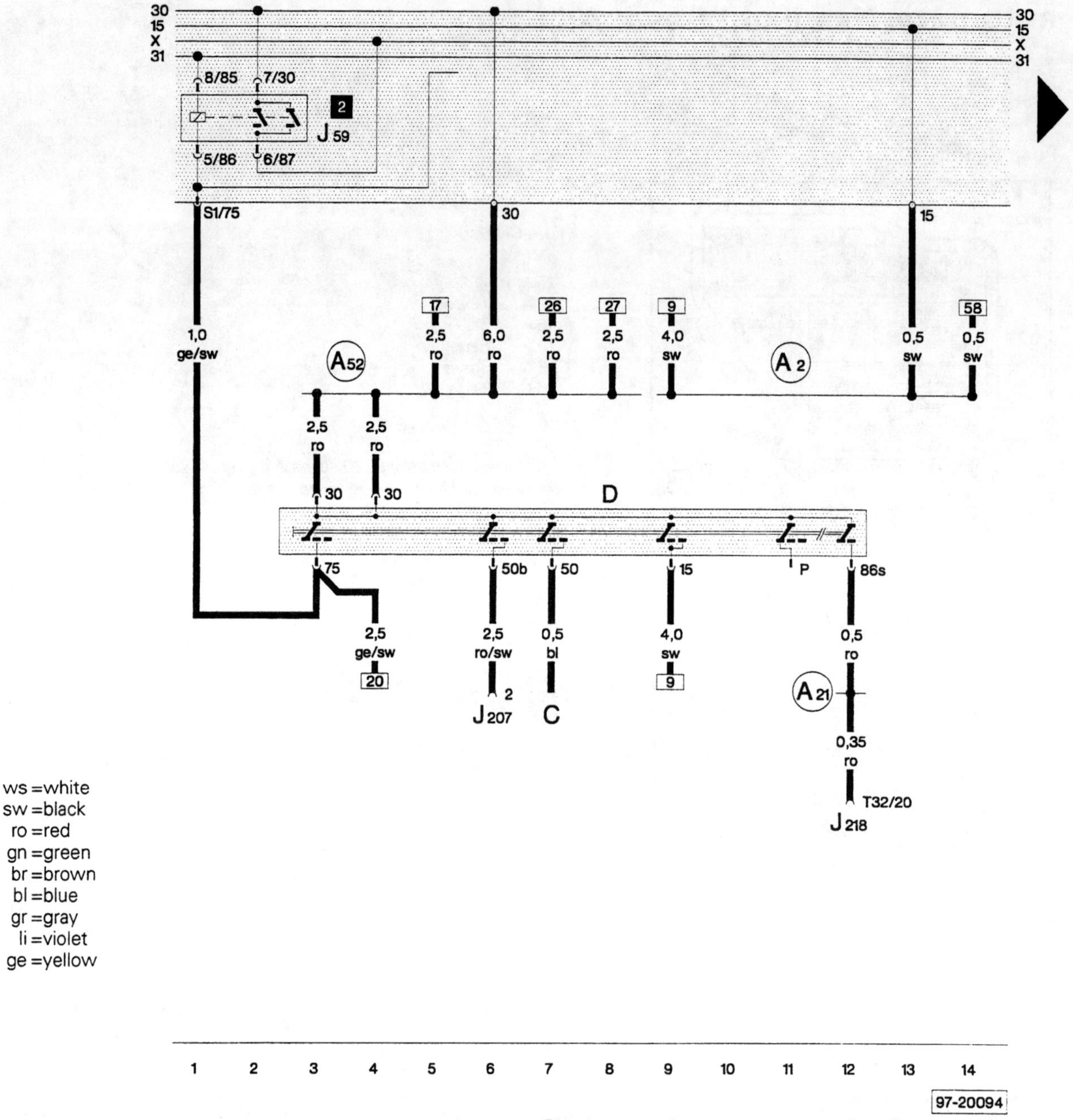

C	- Generator (GEN)
D	- Ignition/Starter Switch
J 59	- Load Reduction Relay
J 207	- Starting Interlock Relay
J 218	- Instrument Cluster Combination Processor
T 32	- Connector 32 pin, blue, on instrument cluster

A2	- Plus connection (15), in instrument panel wiring harness
A21	- Wire connection (86s), in instrument panel wiring harness
A52	- Plus connection (30), in instrument panel wiring harness

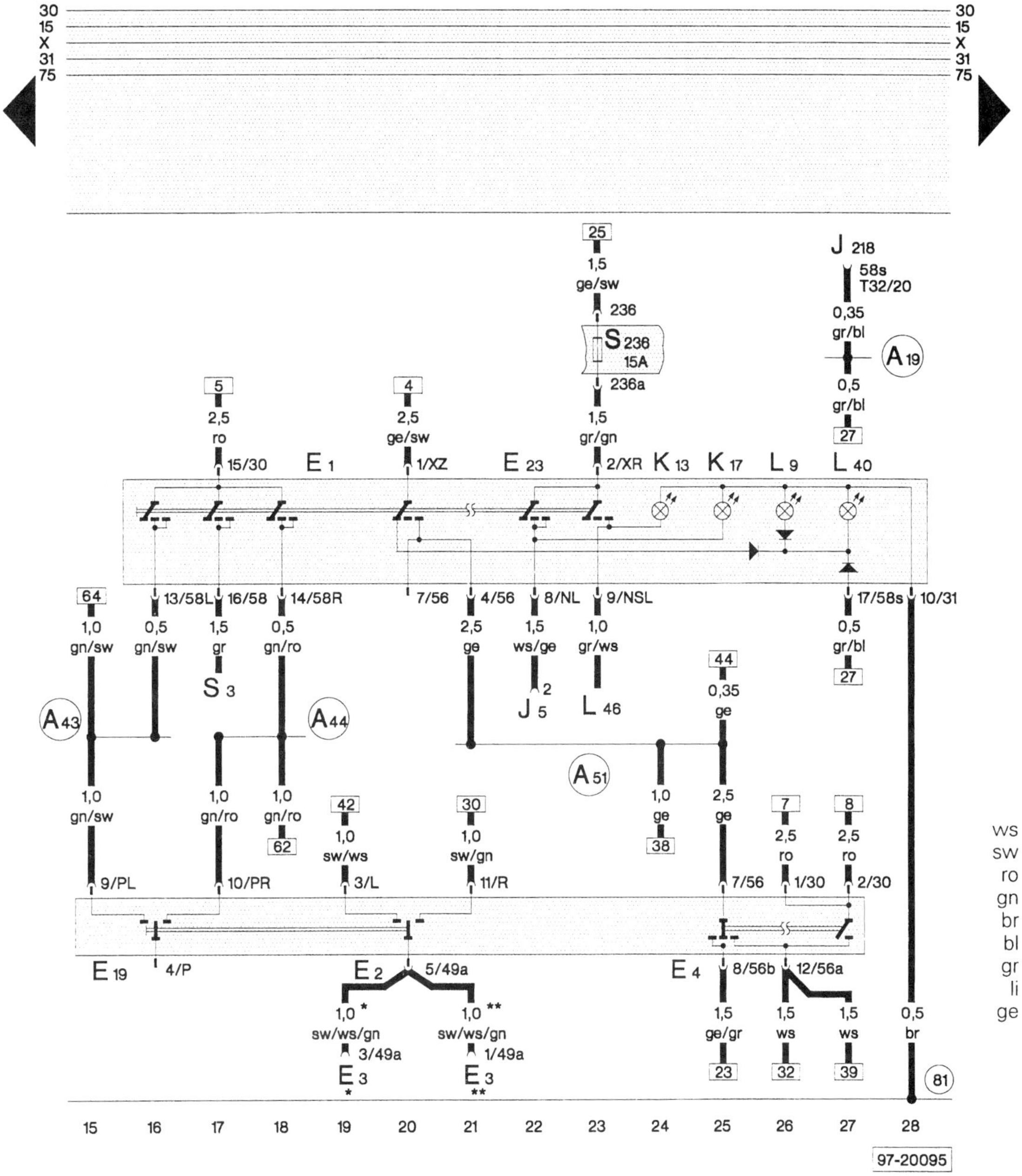

ws = white
sw = black
ro = red
gn = green
br = brown
bl = blue
gr = gray
li = violet
ge = yellow

E 1 - Light Switch
E 2 - Turn Signal Switch
E 3 - Emergency Flasher Switch
E 4 - Headlight Dimmer/Flasher Switch
E 19 - Park Light Switch
E 23 - Fog Light Switch
J 5 - Fog Light Relay
J 218 - Instrument Cluster Combination Processor
K 13 - Rear Fog Light Indicator Light
K 17 - Fog Light Indicator Light
L 9 - Headlight Switch Light
L 40 - Front And Rear Fog Light Switch Light
L 46 - Left Rear Fog Light
S 3 - Fuse
S 236 - Fuse
T 32 - Connector 32 pin, blue, on instrument cluster

(81) - Ground connection -1-, in instrument panel wiring harness
(A19) - Wire connection (58s), in instrument panel wiring harness (Switch illumination)
(A43) - Wire connection (57l), in instrument panel wiring harness
(A44) - Wire connection (57r), in instrument panel wiring harness
(A51) - Wire connection (56), in instrument panel wiring harness

* - Chassis number upto 5000
** - Chassis number from 5001

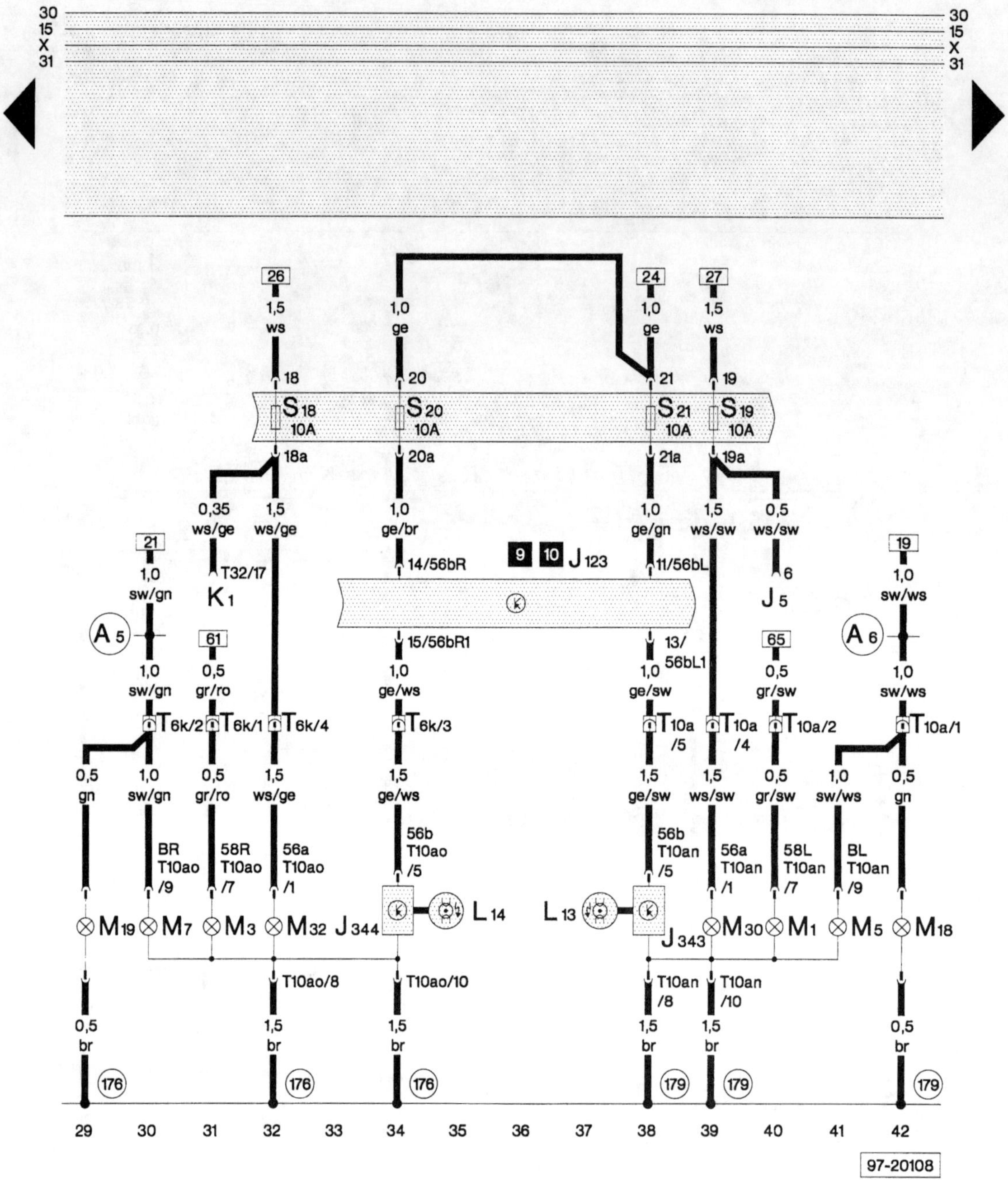

ws = white
sw = black
ro = red
gn = green
br = brown
bl = blue
gr = gray
li = violet
ge = yellow

J 5 - Fog Light Relay
J 123 - Lamp Control Module
J 343 - Control Module for High-intensity gas discharge lamp
J 344 - Control Module for High-intensity gas discharge lamp
K 1 - Headlight High Beam Indicator Light
L 13 - High intensity gas discharge lamp left
L 14 - High intensity gas discharge lamp right
M 1 - Left Parking Light
M 3 - Right Parking Light
M 5 - Left Front Turn Signal Light
M 7 - Right Front Turn Signal Light
M 18 - Left, Side Turn Signal Light
M 19 - Right, Side Turn Signal Light
M 30 - Left High Beam Headlight
M 32 - Right High Beam Headlight
S 18 - Fuse
S 19 - Fuse
S 20 - Fuse
S 21 - Fuse
T 6 k - Connector 6 pin, pink, connector station A pillar, right
T 10 a - Connector 10 pin, pink, connector station A pillar, left
T 10 an - Connector 10 pin, on headlight, left
T 10 ao - Connector 10 pin, on headlight, right
T 32 - Connector 32 pin, blue, on instrument cluster

(176) - Ground connection, in right headlight wiring harness
(179) - Ground connection, in left headlight wiring harness
(A5) - Plus connection (right turn signal), in instrument panel wiring harness
(A6) - Plus connection (left turn signal), in instrument panel wiring harness

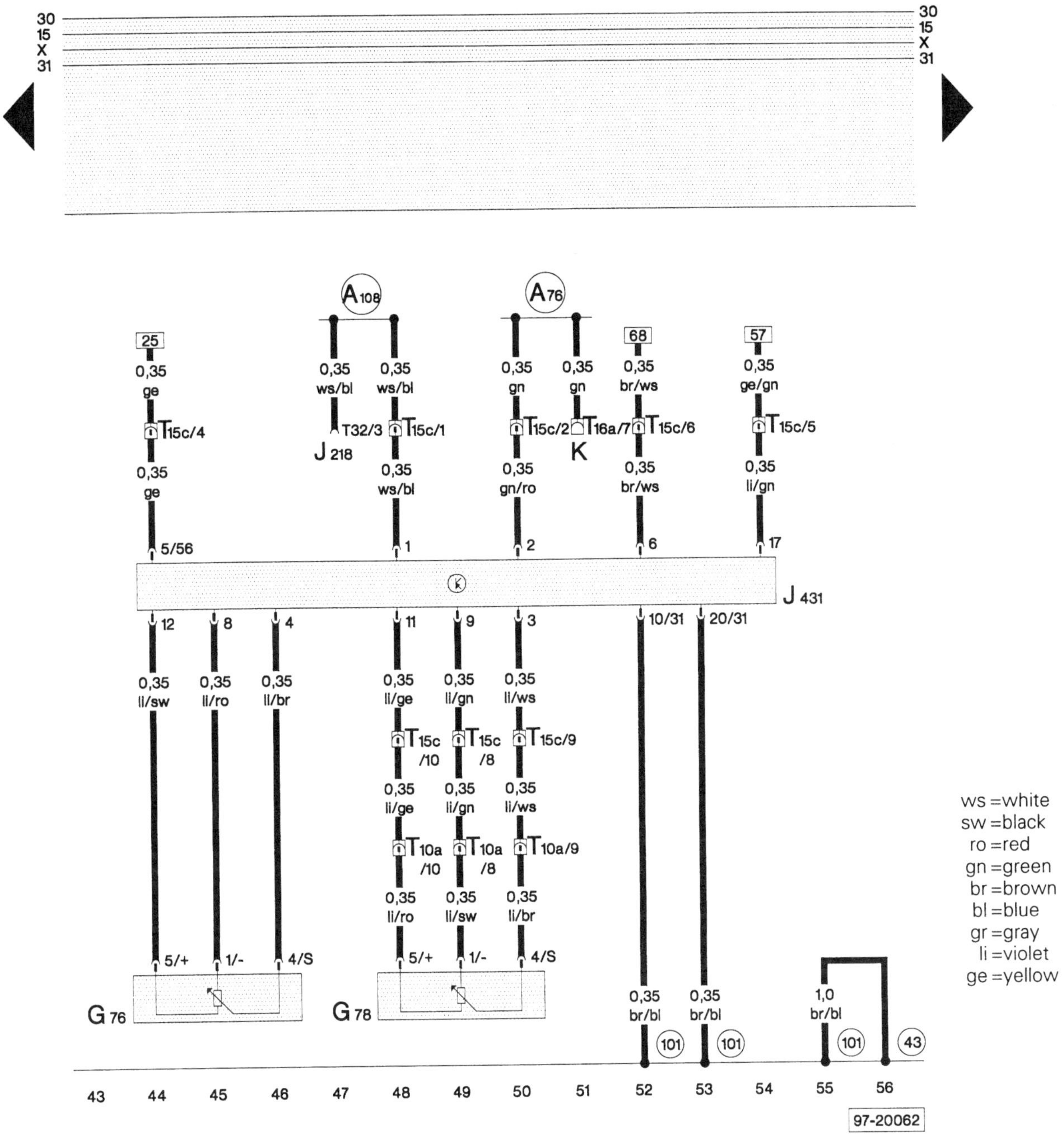

G 76 - Left Rear Level Control System Sensor
G 78 - Left Front Level Control System Sensor
J 218 - Instrument Cluster Combination Processor
J 431 - Control Unit for Beam Adjusting
K - DLC (Data Link Connector) Rapid Data Transfer
T 10 a - Connector 10 pin, pink, connector station A pillar, left
T 15 c - Connector 15 pin, white, connector station A pillar, right
T 16 a - Connector 16 pin, data link connector
T 32 - Connector 32 pin, blue, on instrument cluster

(43) - Ground connection, on right A-pillar, lower part
(101) - Ground connection, in headlight range adjustment wiring harness
(A76) - Connector (K-diagnosis wire), in instrument panel wiring harness
(A108) - Wire connection (vehicle speed signal), in instrument panel wiring harness

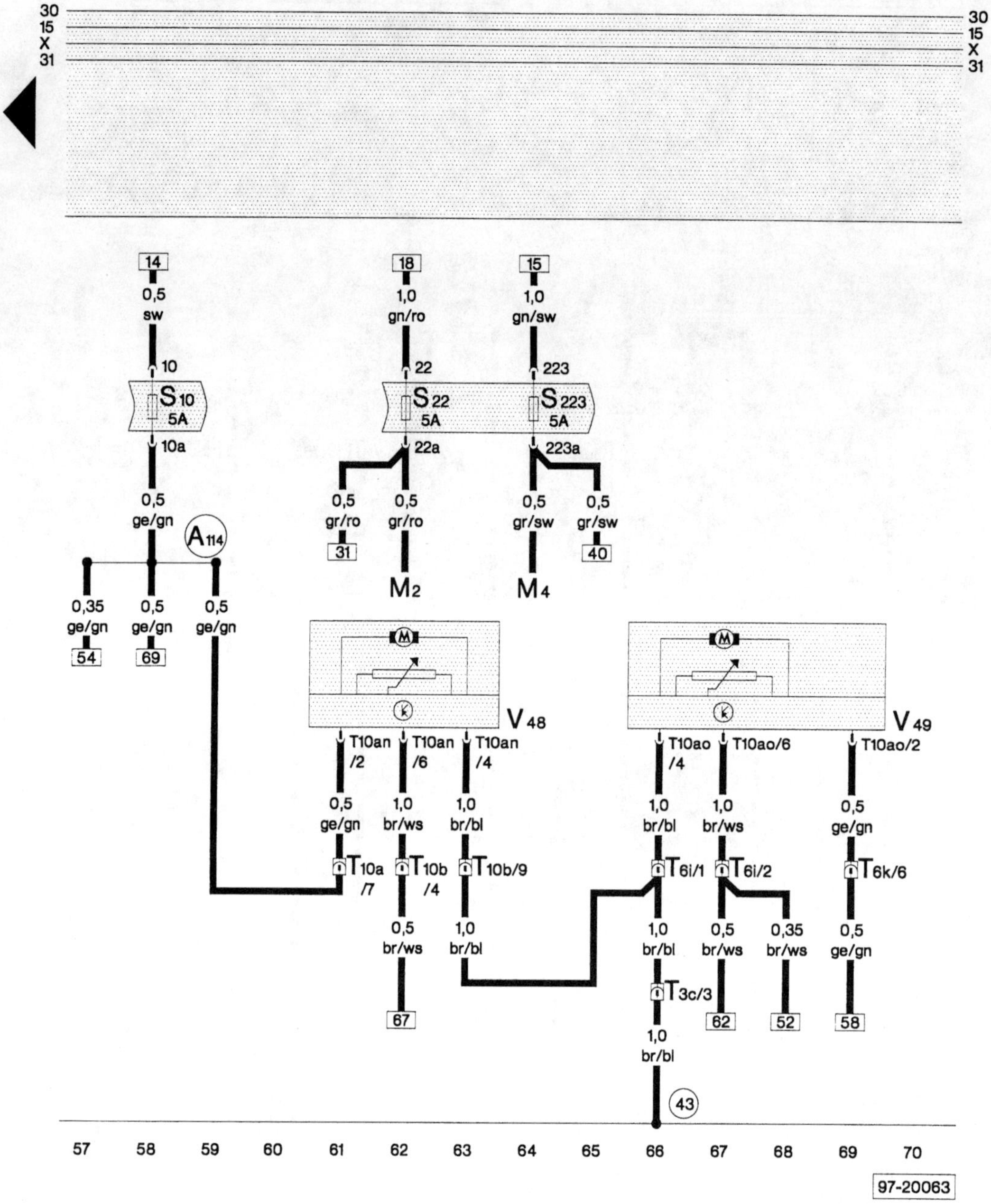

ws = white
sw = black
ro = red
gn = green
br = brown
bl = blue
gr = gray
li = violet
ge = yellow

M 2 - Right Tail Light
M 4 - Left Tail Light
S 10 - Fuse
S 22 - Fuse
S 223 - Fuse
T 3 c - Connector 3 pin, white, connector station A pillar, right
T 6 i - Connector 6 pin, red, connector station A pillar, right
T 6 k - Connector 6 pin, pink, connector station A pillar, right
T 10 a - Connector 10 pin, pink, connector station A pillar, left
T 10 b - Connector 10 pin, red, connector station A pillar, left
T 10 an - Connector 10 pin, on headlight, left
T 10 ao - Connector 10 pin, on headlight, right
V 48 - Left Headlight Beam Adjusting Motor
V 49 - Right Headlight Beam Adjusting Motor

(43) - Ground connection, on right A-pillar, lower part

(A114) - Connection (Beam Adjusting), in instrument panel wiring harness

Rear Wiring, Wagon

1998 m. y.

Fuse Panel

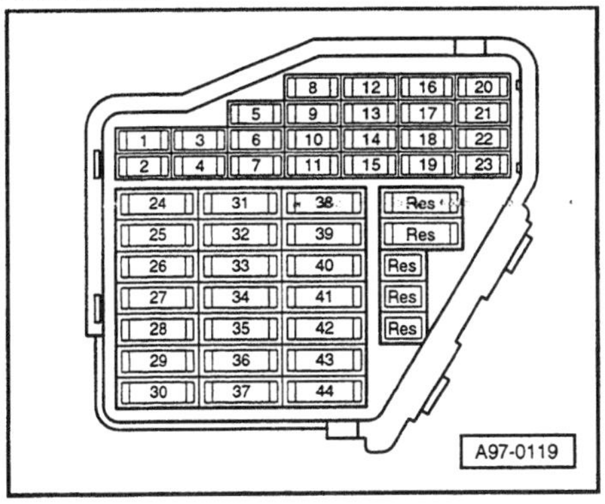

Fuse Colors:

30 A - Green
25 A - White
20 A - Yellow
15 A - Blue
10 A - Red
7,5 A - Brown
5 A - Beige

Starting with fuse position 23, fuses in the fuse holder are identified with 223 in the wiring diagram.

13 - Fold Relay Panel

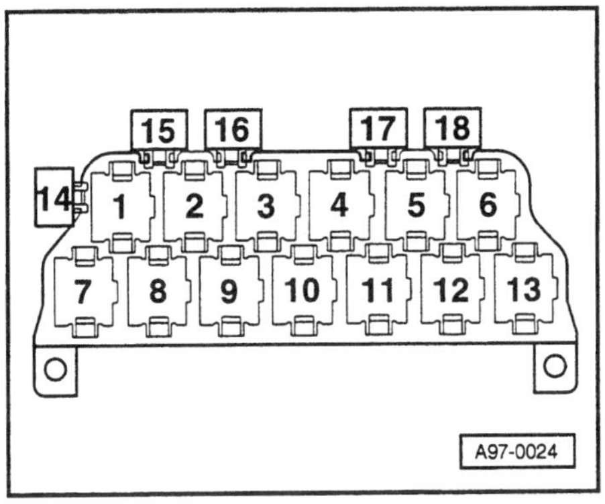

Relay Location:

9 - Lamp Control Module, J123

10 - Lamp Control Module, J123

8 - Fold Relay Panel

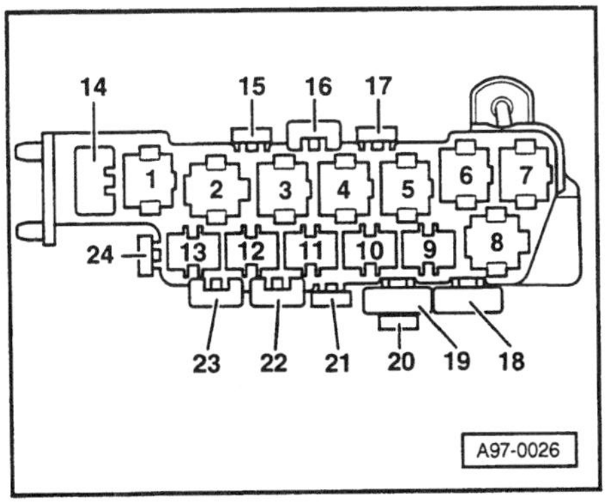

Fuse Location:

15 - Fuse for Luggage Compartment Socket, S184

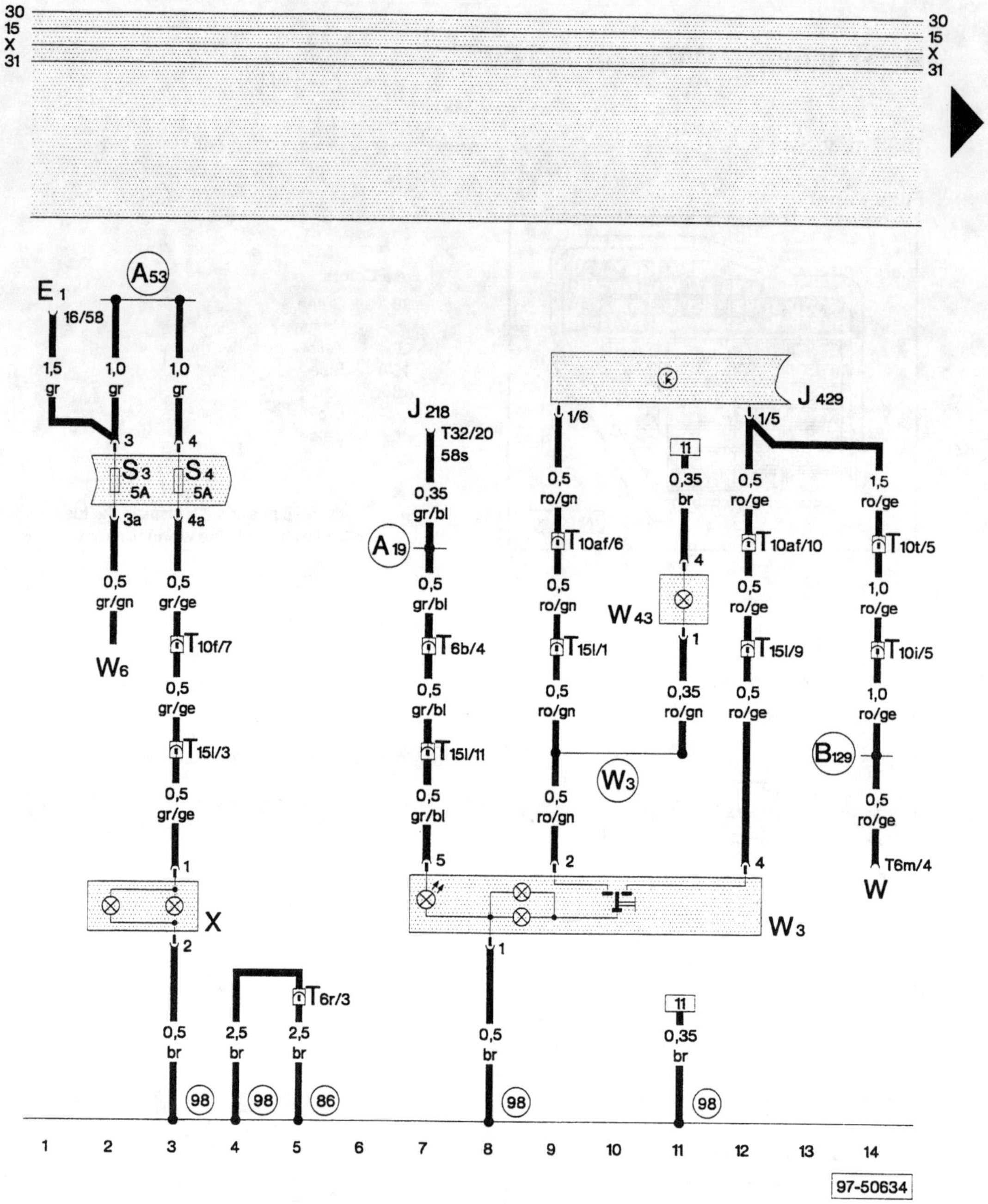

ws = white
sw = black
ro = red
gn = green
br = brown
bl = blue
gr = gray
li = violet
ge = yellow

E 1 - Light Switch
J 218 - Instrument Cluster Combination Processor
J 429 - Control module for central locking
S 3 - Fuse
S 4 - Fuse
T 6 b - Connector 6 pin, white, connector station A pillar, left
T 6 m - Connector 6 pin, on interior light, front
T 6 r - Connector 6 pin, brown, in luggage compartment
T 10 f - Connector 10 pin, brown, connector station A pillar, left
T 10 i - Connector 10 pin, darkbrown, connector station A pillar, left
T 10 t - Connector 10 pin, orange, connector station A pillar, right
T 10 af - Connector 10 pin, violet, connector station A pillar, left
T 15 l - Connector 15 pin, black, in luggage compartment
T 32 - Connector 32 pin, blue, on instrument cluster
W - Front Interior Light
W 3 - Luggage Compartment Light
W 6 - Glove Compartment Light
W 43 - Rear courtesy lamp, in rear lid
X - License Plate Light

(86) - Ground connection -1-, in rear wiring harness
(98) - Ground connection, in rear lid wiring harness
(A19) - Wire connection (58s), in instrument panel wiring harness (Switch illumination)
(A53) - Plus connection (58b), in instrument panel wiring harness
(B129) - Connector (interior light 3/L), in wiring harness interior
(W3) - Wire connection, in rear wiring harness

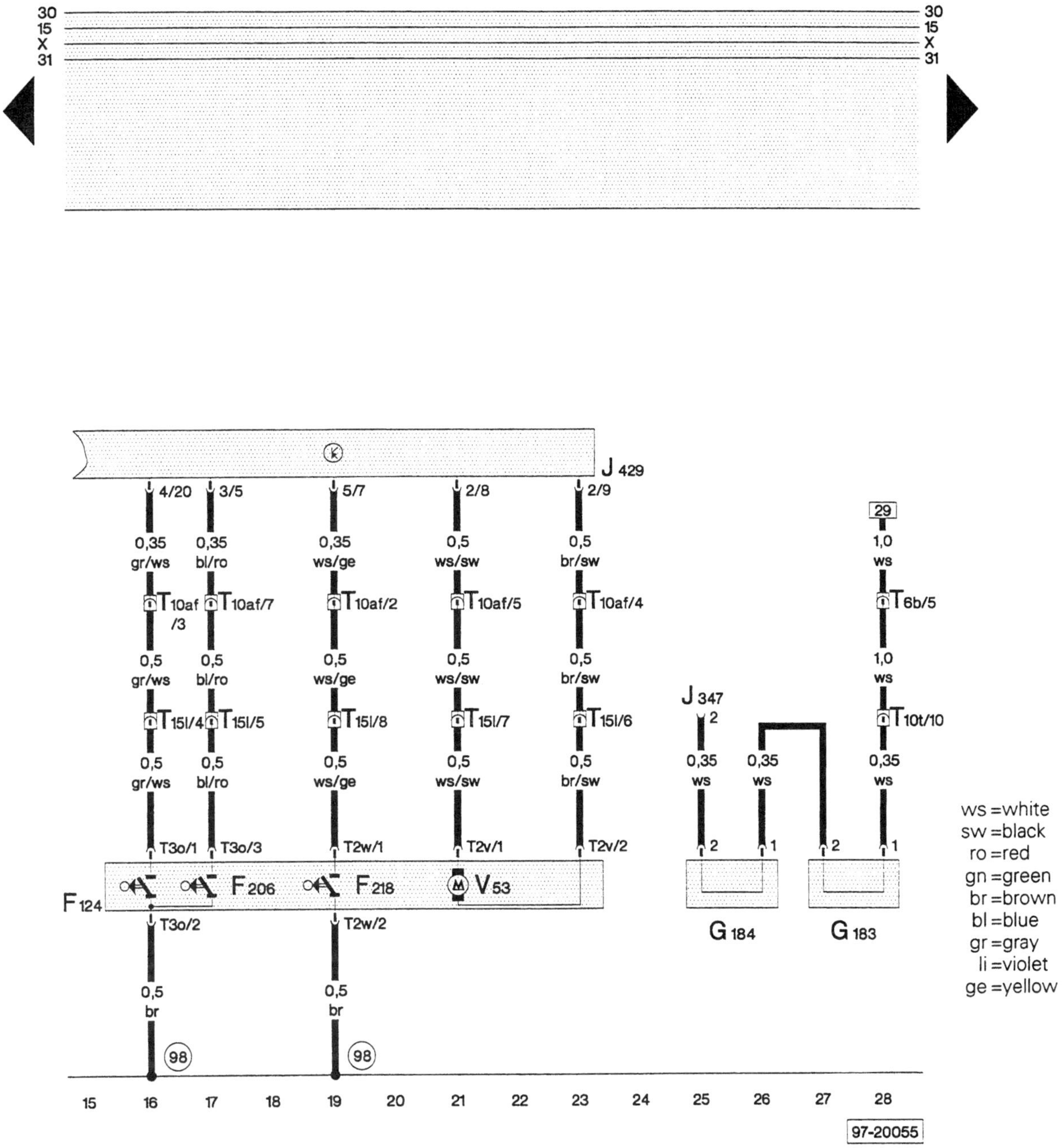

F 124 - Trunk Lock Alarm/Central Locking Switch
F 206 - Trunk Lid Closed Position Switch
F 218 - Switch for central locking, trunk lid
G 183 - Sensor for broken window glass rear left
G 184 - Sensor for broken window glass rear right
J 347 - Control Module for Ultra-Sound sensors
J 429 - Control module for central locking
T 2 v - Connector 2 pin, black, luggage compartment
T 2 w - Connector 2 pin, grey, luggage compartment
T 3 o - Connector 3 pin, black, in rear lid
T 6 b - Connector 6 pin, white, connector station A pillar, left
T 10 t - Connector 10 pin, orange, connector station A pillar, right
T 10 af - Connector 10 pin, violet, connector station A pillar, left
T 15 l - Connector 15 pin, black, in luggage compartment
V 53 - Decklid Central Locking System Motor

(98) - Ground connection, in rear lid wiring harness

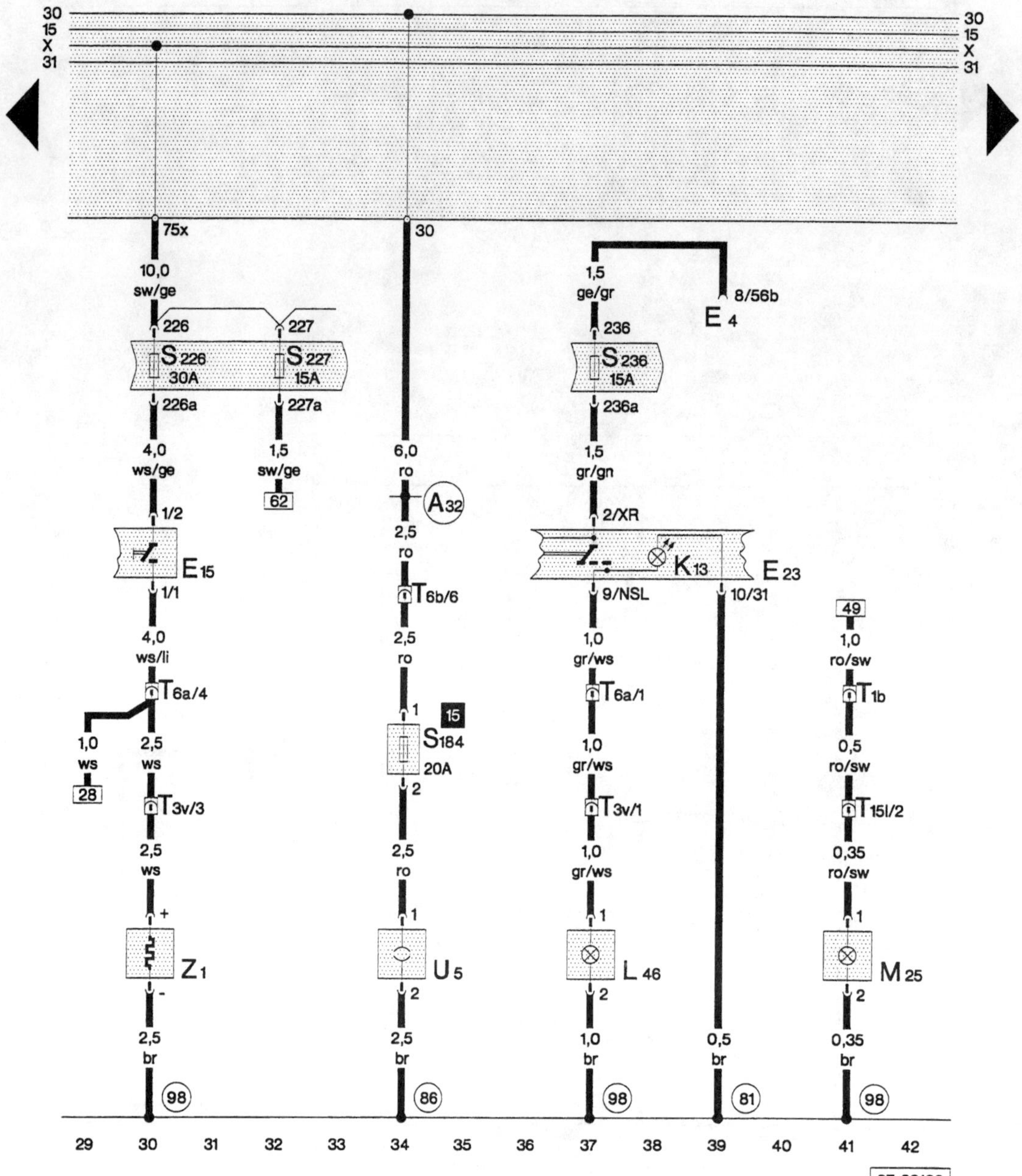

E 4 - Headlight Dimmer/Flasher Switch
E 15 - Rear Window Defogger Switch
E 23 - Fog Light Switch
K 13 - Rear Fog Light Indicator Light
L 46 - Left Rear Fog Light
M 25 - High-mount Brake Light
S 184 - Fuse for Luggage Compartment Socket
S 226 - Fuse
S 227 - Fuse
S 236 - Fuse
T 1 b - Single connector, black, connector station A pillar right
T 3 v - Connector 3 pin, black, in luggage compartment
T 6 a - Connector 6 pin, blue, connector station A pillar, left
T 6 b - Connector 6 pin, white, connector station A pillar, left
T 15 l - Connector 15 pin, black, in luggage compartment
U 5 - 12 V Socket, in Luggage Compartment
Z 1 - Heated rear window

(81) - Ground connection -1-, in instrument panel wiring harness
(86) - Ground connection -1-, in rear wiring harness
(98) - Ground connection, in rear lid wiring harness
(A32) - Plus connection (30), in instrument panel wiring harness

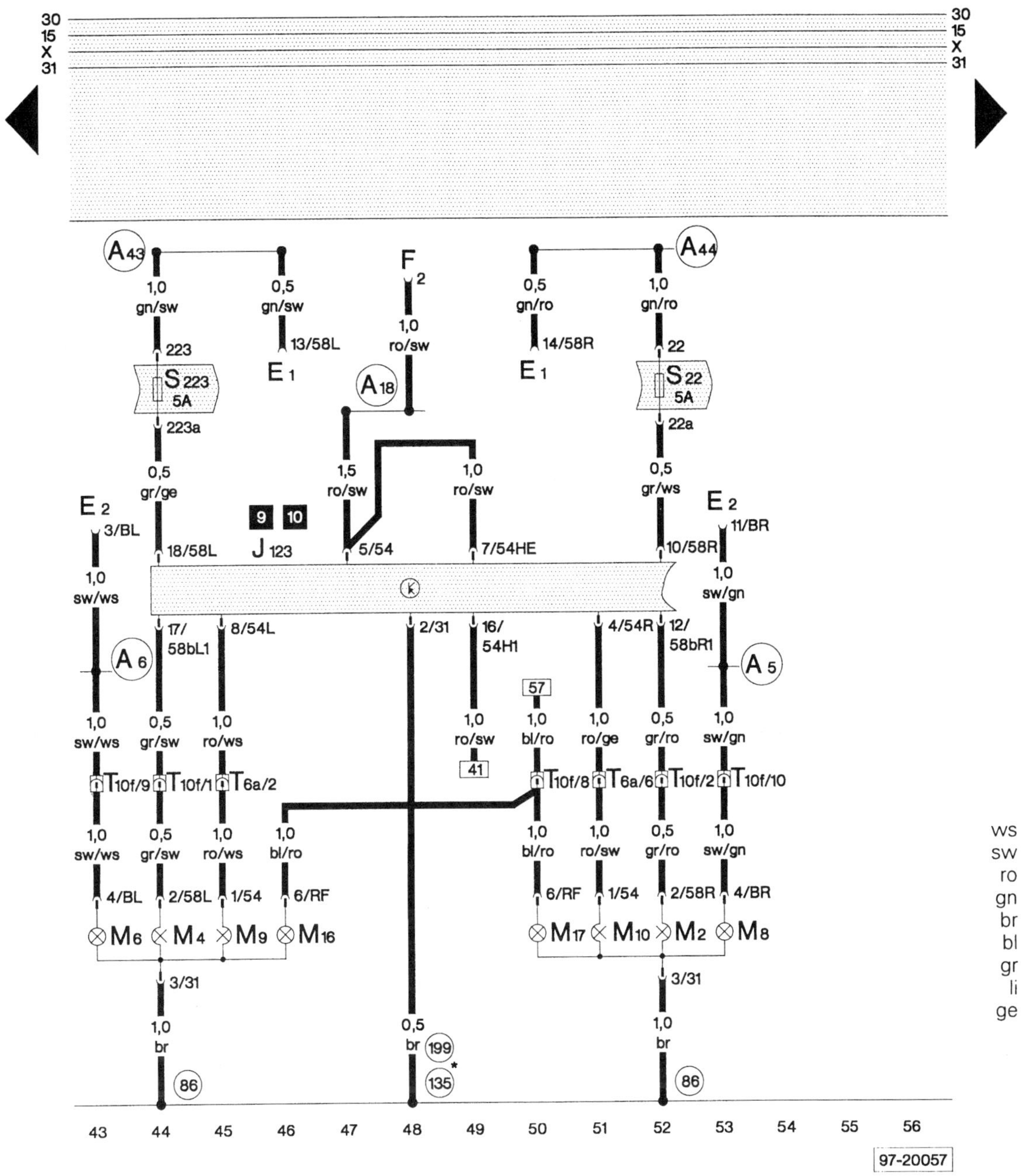

ws = white
sw = black
ro = red
gn = green
br = brown
bl = blue
gr = gray
li = violet
ge = yellow

E 1 - Light Switch
E 2 - Turn Signal Switch
F - Brake Light Switch
J 123 - Lamp Control Module
M 2 - Right Tail Light
M 4 - Left Tail Light
M 6 - Left Rear Turn Signal Light
M 8 - Right Rear Turn Signal Light
M 9 - Left Brake Light
M 10 - Right Brake Light
M 16 - Left Back-Up Light
M 17 - Right Back-Up Light
S 22 - Fuse
S 223 - Fuse
T 6 a - Connector 6 pin, blue, connector station A pillar, left
T 10 f - Connector 10 pin, brown, connector station A pillar, left

(86) - Ground connection -1-, in rear wiring harness
(135) - Ground connection -2-, in instrument panel wiring harness
(199) - Ground connection -3-, in instrument panel wiring harness
(A5) - Plus connection (right turn signal), in instrument panel wiring harness
(A6) - Plus connection (left turn signal), in instrument panel wiring harness
(A18) - Wire connection (54), in instrument panel wiring harness
(A43) - Wire connection (57l), in instrument panel wiring harness
(A44) - Wire connection (57r), in instrument panel wiring harness

* - booth is possible

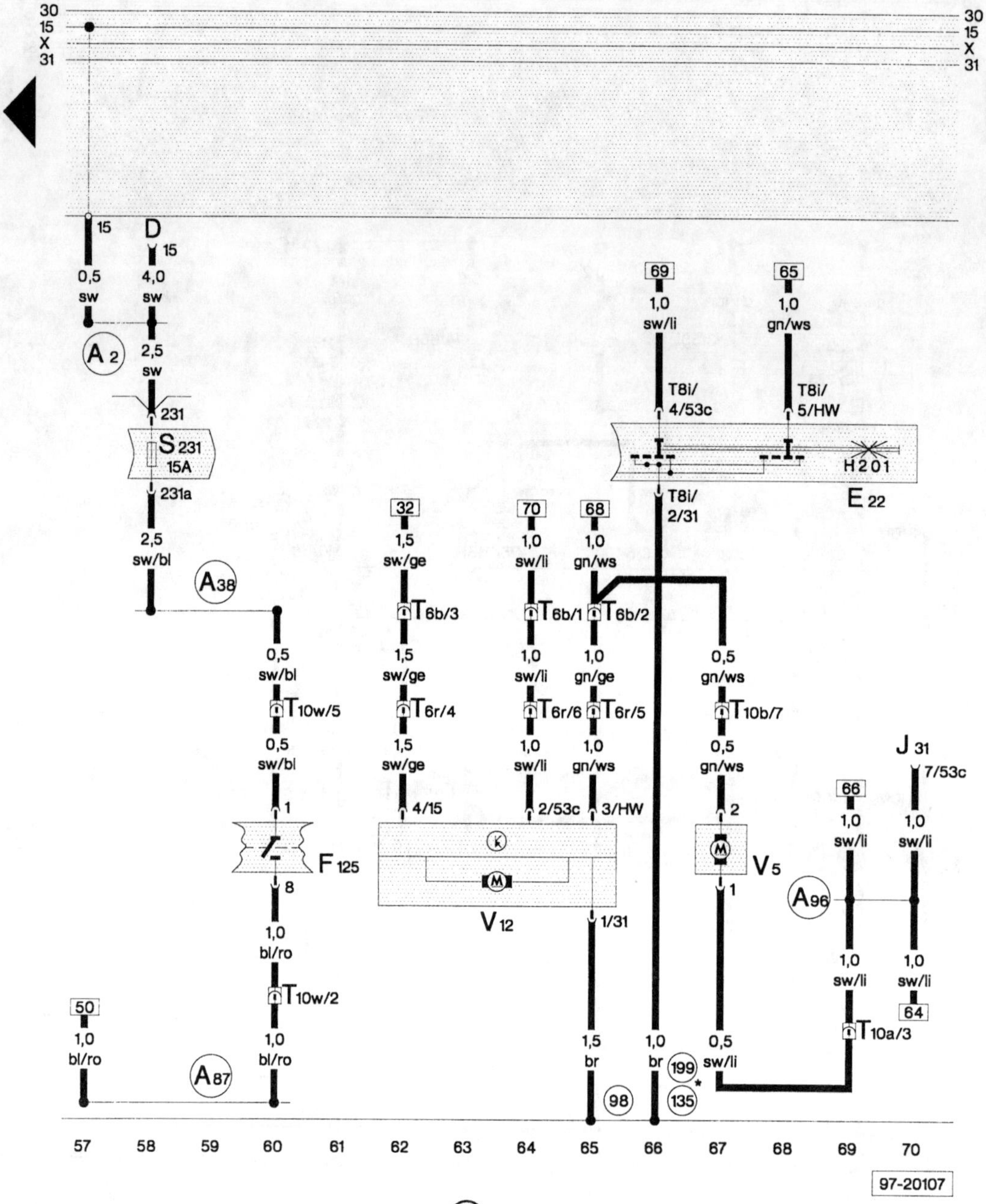

ws = white
sw = black
ro = red
gn = green
br = brown
bl = blue
gr = gray
li = violet
ge = yellow

D - Ignition/Starter Switch
E 22 - Windshield Wiper/Washer Switch
F 125 - Multi-Function Transmission Range (TR) Switch
J 31 - Wiper/Washer Intermittent Relay
S 231 - Fuse
T 6 b - Connector 6 pin, white, connector station A pillar, left
T 6 r - Connector 6 pin, brown, in luggage compartment
T 8 i - Connector 8 pin, on windshield wiper/washer switch
T 10 a - Connector 10 pin, pink, connector station A pillar, left
T 10 b - Connector 10 pin, red, connector station A pillar, left
T 10 w - Connector 10 pin, blue, connector station A pillar, right
V 5 - Windshield Washer Pump
V 12 - Rear Window Wiper Motor

(98) - Ground connection, in rear lid wiring harness
(135) - Ground connection -2-, in instrument panel wiring harness
(199) - Ground connection -3-, in instrument panel wiring harness
(A2) - Plus connection (15), in instrument panel wiring harness
(A38) - Plus connection (15a II), in instrument panel wiring harness
(A87) - Connector (RF), in instrument panel wiring harness
(A96) - Connector (53a), in instrument panel wiring harness

* - booth is possible

Heated Steering Wheel

1998 m. y.

Fuse Panel

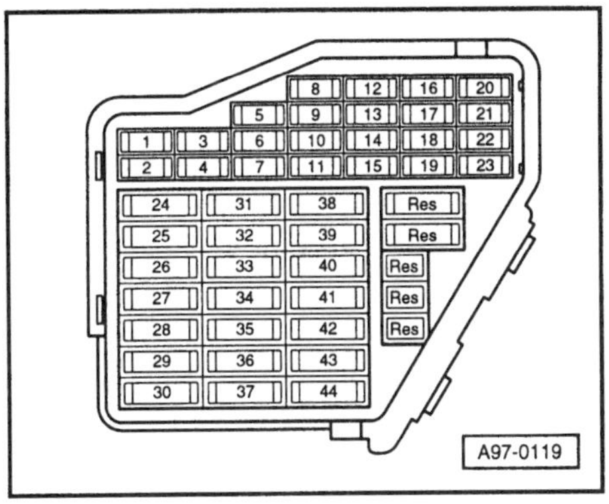

Fuse Colors:

30 A - Green
25 A - White
20 A - Yellow
15 A - Blue
10 A - Red
5 A - Beige

Starting with fuse position 23, fuses in the fuse holder are identified with 223 in the wiring diagram.

Micro Central Electric Panel

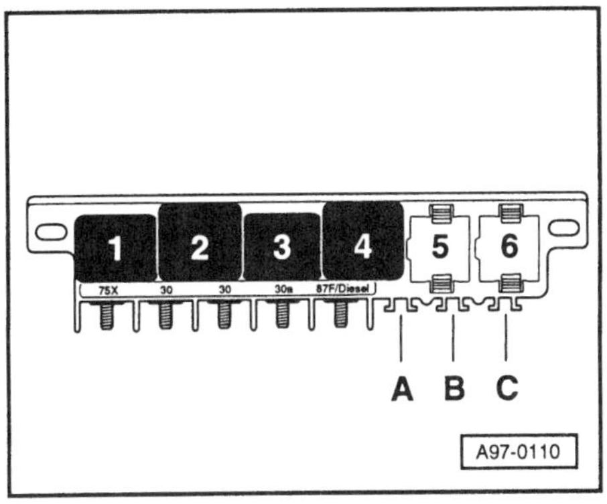

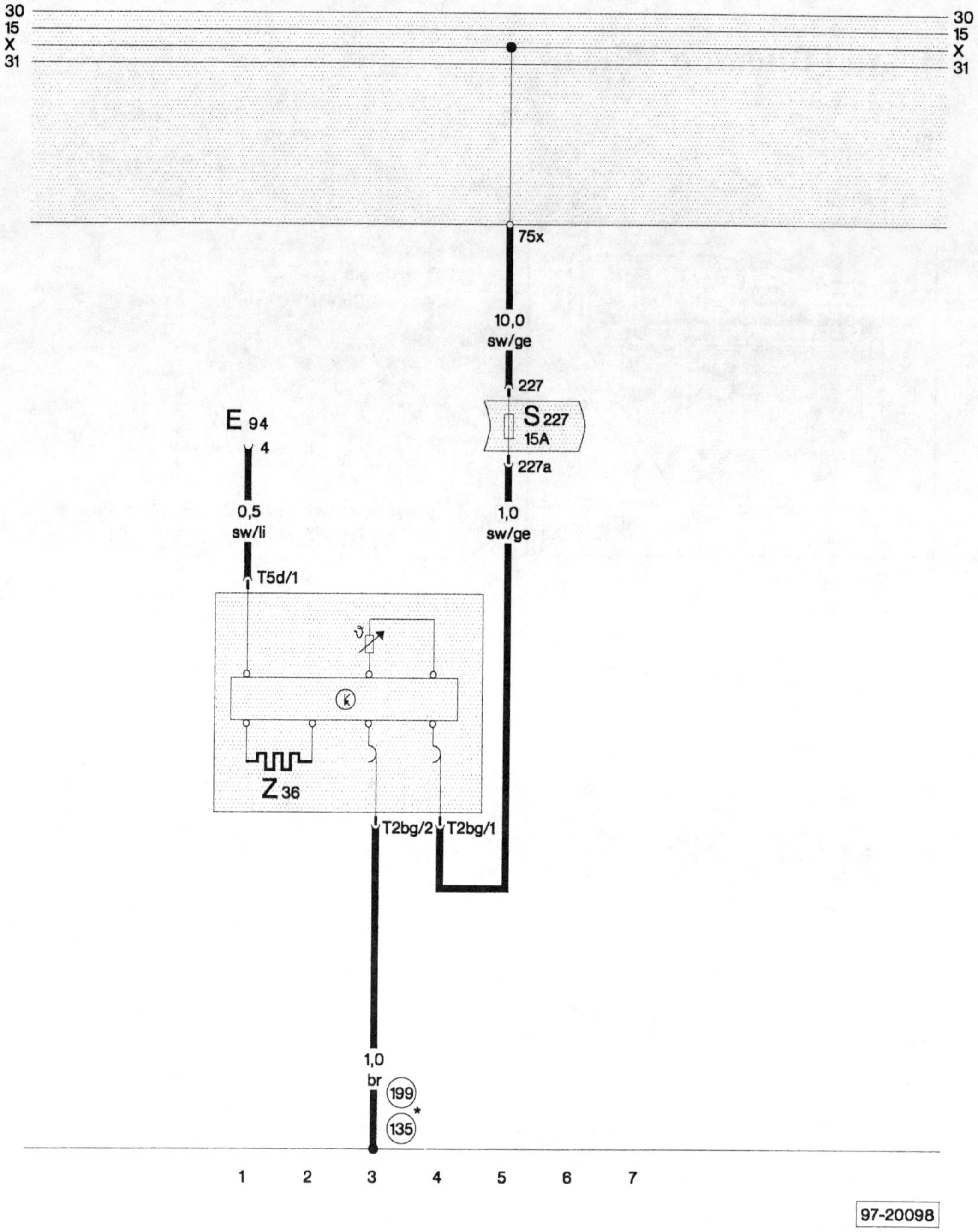

ws =white
sw =black
ro =red
gn =green
br =brown
bl =blue
gr =gray
li =violet
ge =yellow

E 94 - Adjuster for heated driver's seat
S 227 - Fuse
T 2 bg - Connector 2 pin, behind steering column switch cover
T 5 d - Connector 5 pin, behind steering column switch cover
Z 36 - Heated steering wheel

(135) - Ground connection -2-, in instrument panel wiring harness

(199) - Ground connection -3-, in instrument panel wiring harness

* - booth is possible

(2,8 l - Injection Engine, 6-Cylinder), Code AHA

1999 m. y.

Fuse Panel

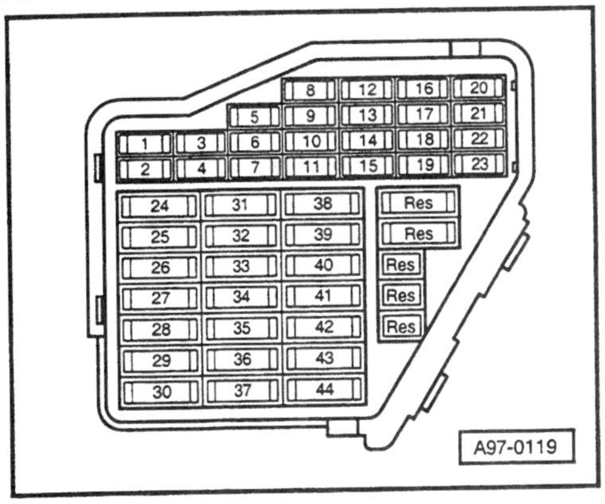

Fuse Colors:

30 A - Green
25 A - White
20 A - Yellow
15 A - Blue
10 A - Red
7,5 A - Brown
5 A - Beige

Starting with fuse position 23, fuses in the fuse holder are identified with 223 in the wiring diagram.

Micro Central Electric Panel

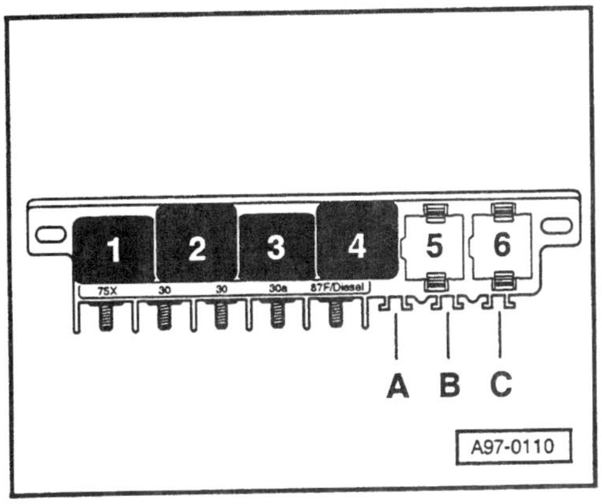

Relay Location:

- **1** - Dual Horn Relay, J4
- **2** - Load Reduction Relay, J59
- **4** - Fuel Pump (FP) Relay, J17
- **5** - Wiper/Washer Intermittent Relay, J31
- **6** - Wiper/Washer Intermittent Relay, J31

13 - Fold Relay Panel

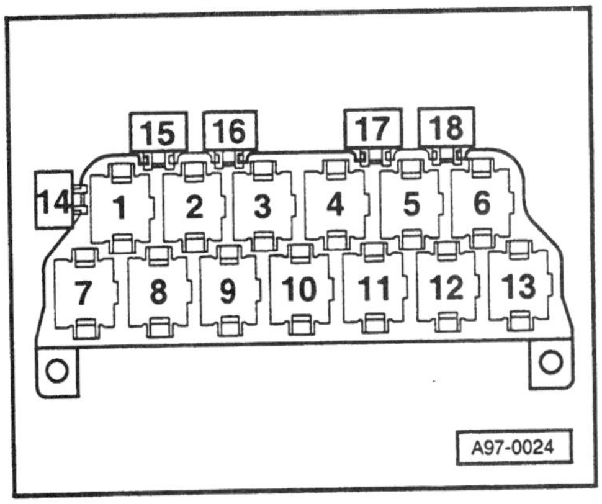

Relay Location:

- **4** - Starting Interlock Relay, J207
- **6** - Fog Light Relay, J5
- **17** - Fuse for secondary air pump, S130 upto october 1998

8 - Fold Relay Panel

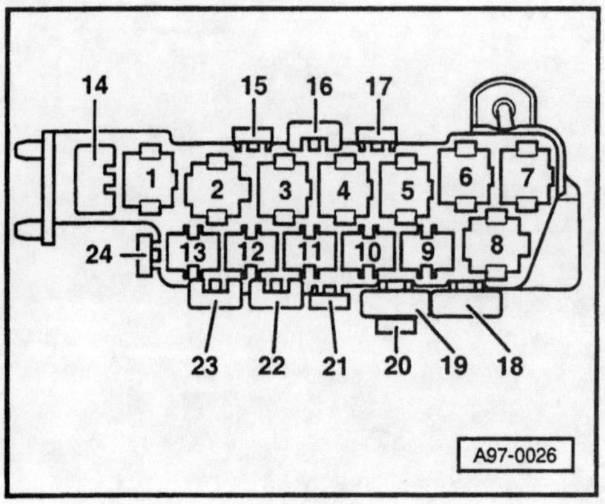

Relay Location:

- **15** - Fuse for Luggage Compartment Socket, S184
- **20** - Coolant Fan Fuse, S42
- **21** - Control module fuse for coolant fan, S142

3 - Fold Relay Panel in E-Box plenum chamber

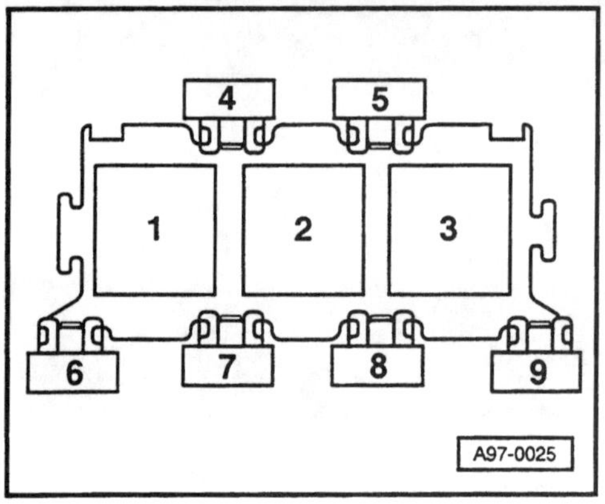

Relay Location:

- **2** - Secondary Air Injection (AIR) Pump Relay, J299
- **7** - Fuse for secondary air pump, S130
  from november 1998

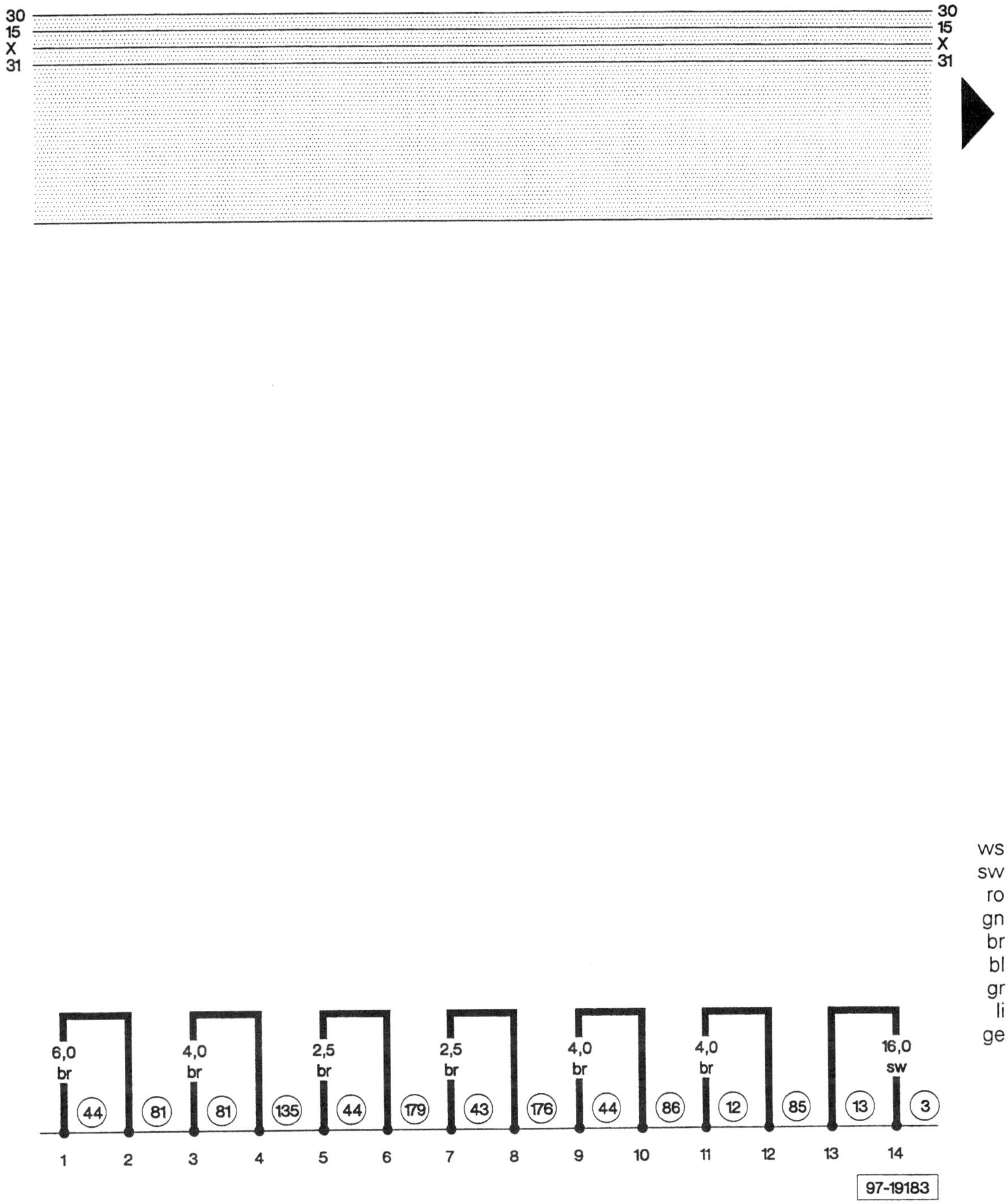

ws = white
sw = black
ro = red
gn = green
br = brown
bl = blue
gr = gray
li = violet
ge = yellow

(3) - Ground strap, engine to body

(12) - Ground connection, in engine compartment, left

(13) - Ground connection, in engine compartment, right

(43) - Ground connection, on right A-pillar, lower part

(44) - Ground connection, on left A-pillar, lower part

(81) - Ground connection -1-, in instrument panel wiring harness

(85) - Ground connection -1-, in engine compartment wiring harness

(86) - Ground connection -1-, in rear wiring harness

(135) - Ground connection -2-, in instrument panel wiring harness

(176) - Ground connection, in right headlight wiring harness

(179) - Ground connection, in left headlight wiring harness

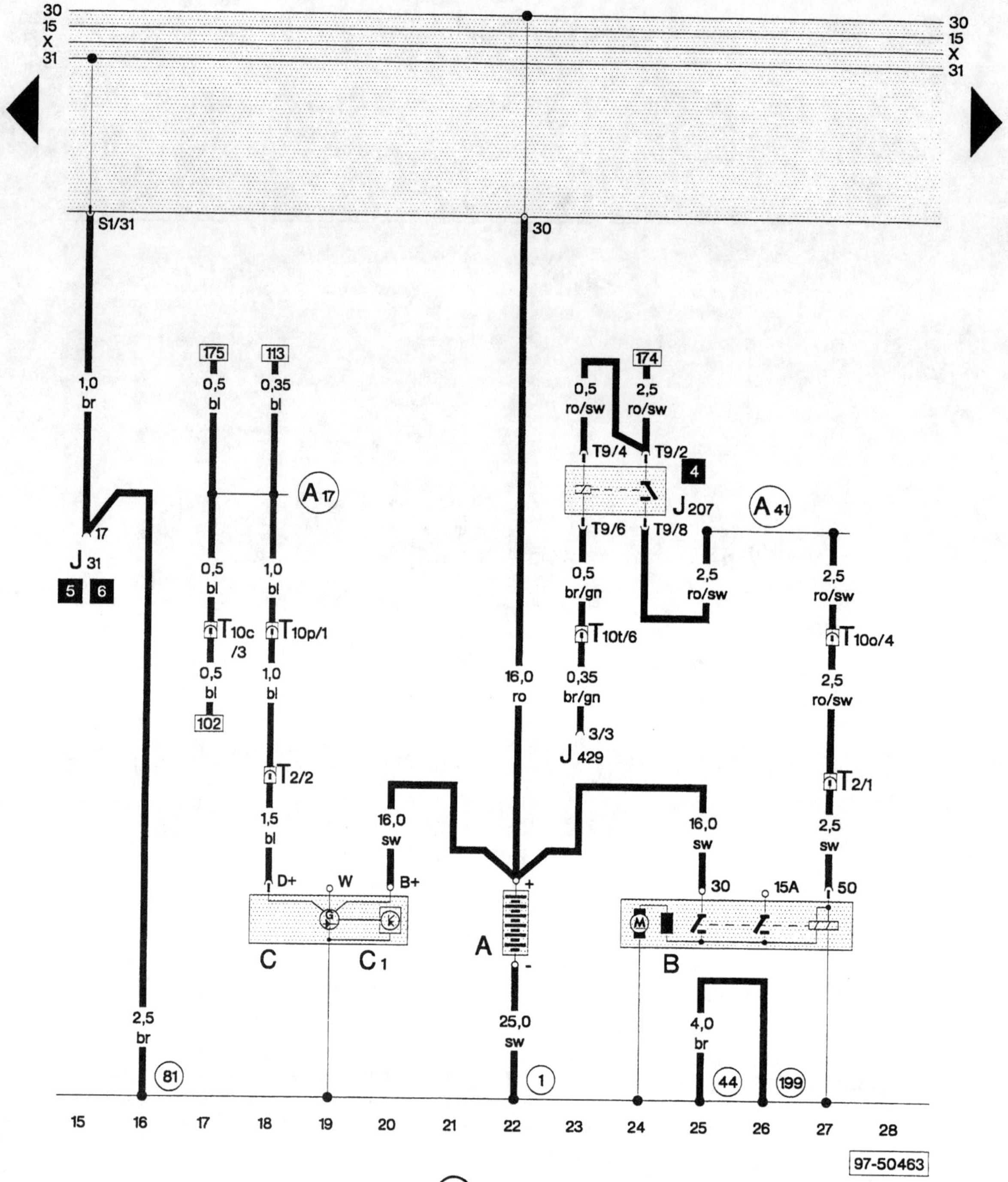

ws = white
sw = black
ro = red
gn = green
br = brown
bl = blue
gr = gray
li = violet
ge = yellow

A - Battery
B - Starter
C - Generator (GEN)
C 1 - Voltage Regulator (VR)
J 31 - Wiper/Washer Intermittent Relay
J 207 - Starting Interlock Relay
J 429 - Control module for central locking
T 2 - Connector 2 pin, grey, in engine compartment, right
T 9 - Connector 9 pin, brown, on starting interlock relay
T 10 c - Connector 10 pin, grey, connector station A pillar, left
T 10 o - Connector 10 pin, brown, connector station electronic box plenum chamber
T 10 p - Connector 10 pin, black, connector station electronic box plenum chamber
T 10 t - Connector 10 pin, orange, connector station A pillar, right

(1) - Ground strap, battery to body
(44) - Ground connection, on left A-pillar, lower part
(81) - Ground connection -1-, in instrument panel wiring harness
(199) - Ground connection -3-, in instrument panel wiring harness
(A17) - Wire connection (61), in instrument panel wiring harness
(A41) - Plus connection (50), in instrument panel wiring harness

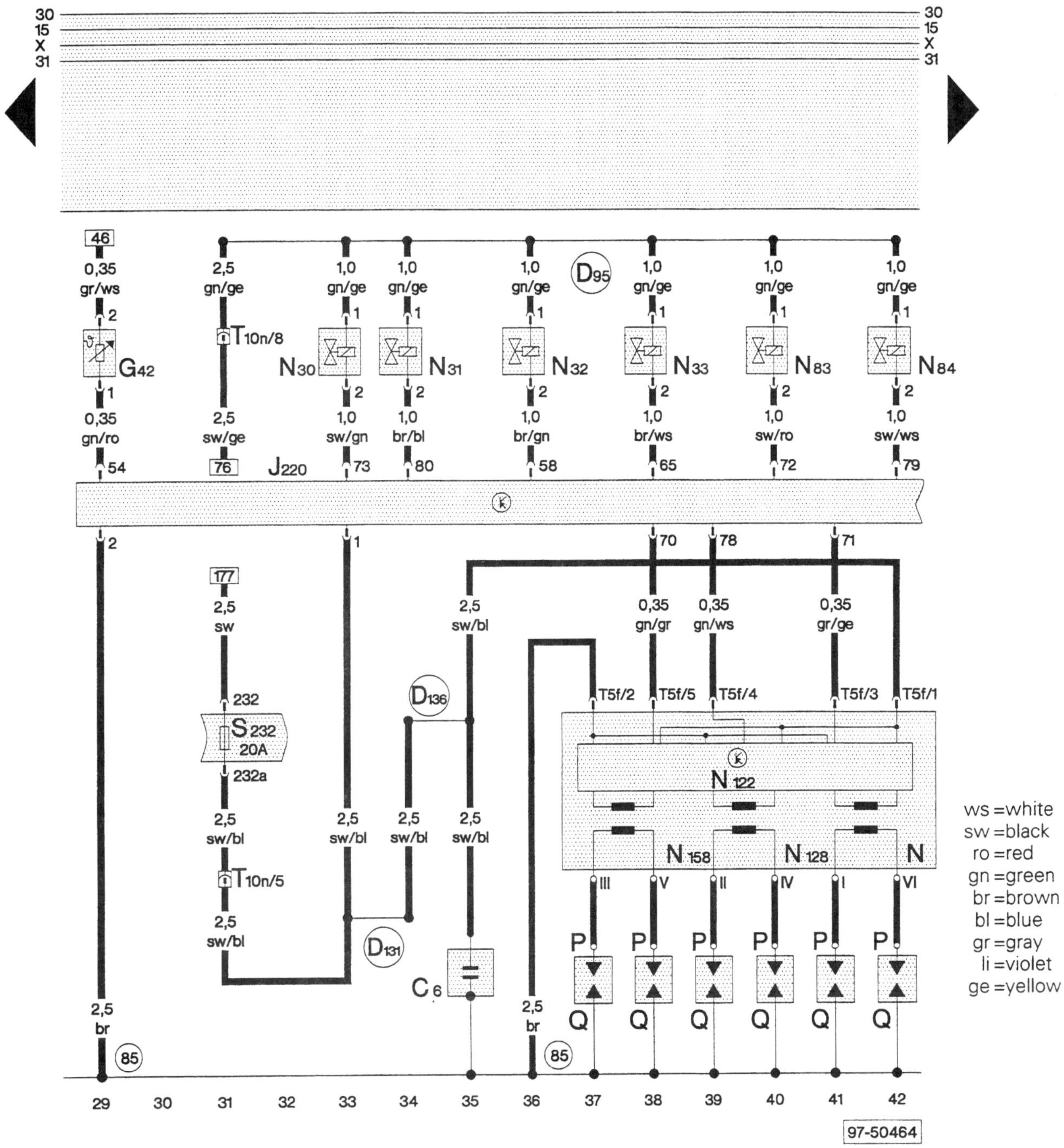

C 6 - Distributor Ignition (DI) Capacitor
G 42 - Intake Air Temperature (IAT) Sensor
J 220 - Motronic Engine Control Module (ECM)
N - Ignition Coil
N 30 - Cylinder 1 Fuel Injector
N 31 - Cylinder 2 Fuel Injector
N 32 - Cylinder 3 Fuel Injector
N 33 - Cylinder 4 Fuel Injector
N 83 - Cylinder 5 Fuel Injector
N 84 - Cylinder 6 Fuel Injector
N 122 - Power Output Stage
N 128 - Ignition Coil 2
N 158 - Ignition Coil 3
P - Spark Plug Connectors
Q - Spark Plugs
S 232 - Fuse
T 5 f - Connector 5 pin, ignition coil unit

T 10 n - Connector 10 pin, orange, connector station electronic box plenum chamber

(85) - Ground connection -1-, in engine compartment wiring harness
(D95) - Wire connection (injectors), in engine compartment wiring harness
(D131) - Wire connection (15), above fuse, in engine compartment wiring harness
(D136) - Plus connection -2- (15a), in wiring harness engine compartment

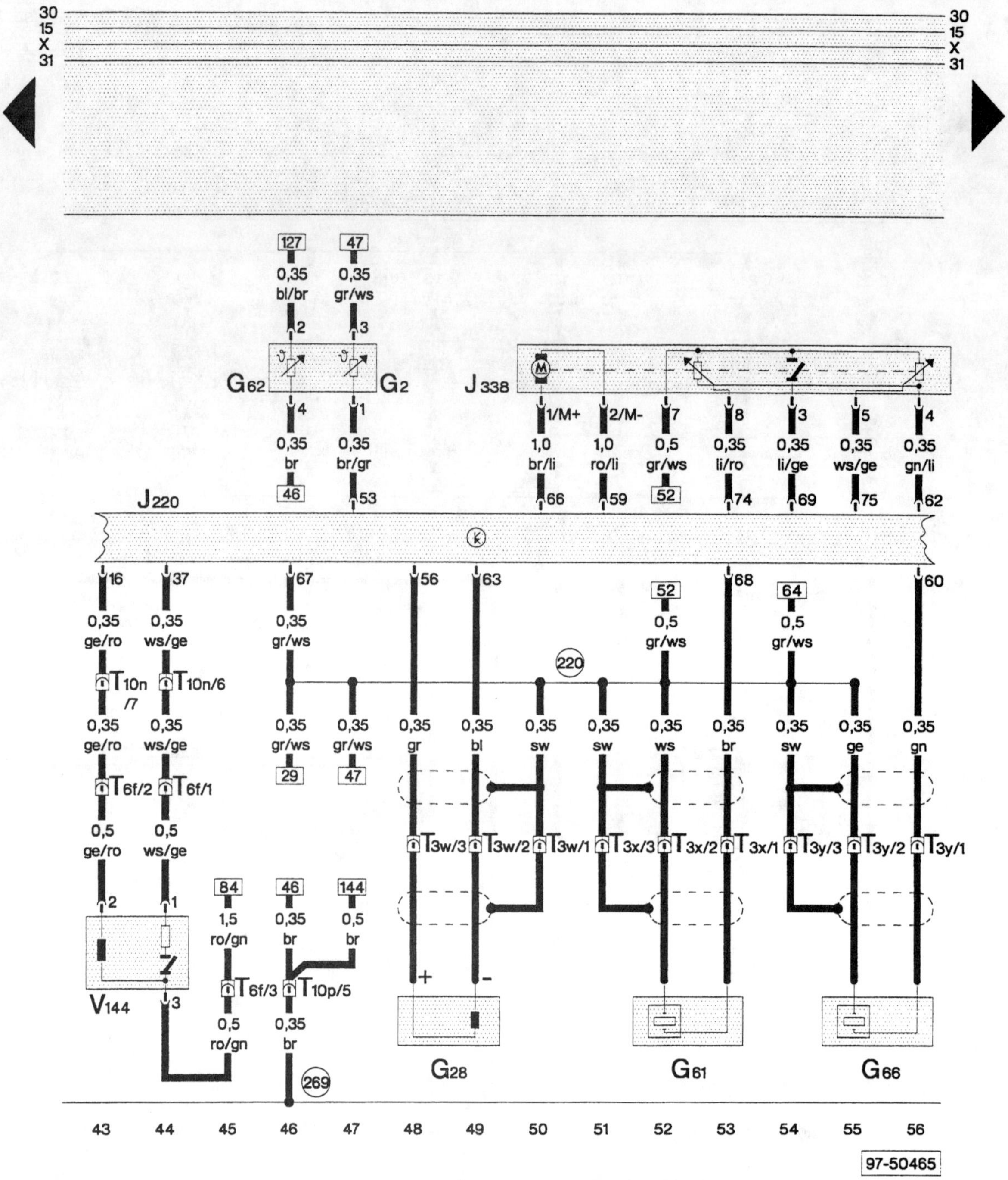

ws = white
sw = black
ro = red
gn = green
br = brown
bl = blue
gr = gray
li = violet
ge = yellow

G 2 - Engine Coolant Temperature (ECT) Sensor
G 28 - Engine Speed (RPM) Sensor
G 61 - Knock Sensor (KS) 1
G 62 - Engine Coolant Temperature (ECT) Sensor
G 66 - Knock Sensor (KS) 2
J 220 - Motronic Engine Control Module (ECM)
J 338 - Throttle Valve Control Module
T 3 w - Connector 3 pin, grey, on engine speed sensor
T 3 x - Connector 3 pin, blue, on knock sensor 1
T 3 y - Connector 3 pin, blue, on knock sensor 2
T 6 f - Connector 6 pin, darkbrown, connector station A pillar, right
T 10 n - Connector 10 pin, orange, connector station electronic box plenum chamber
T 10 p - Connector 10 pin, black, connector station electronic box plenum chamber
V 144 - Leak Detection Pump (LDP)

(220) - Ground connection (sensor ground), in engine compartment wiring harness
(269) - Ground connector (sensor ground) -1-, in instrument panel wiring harness

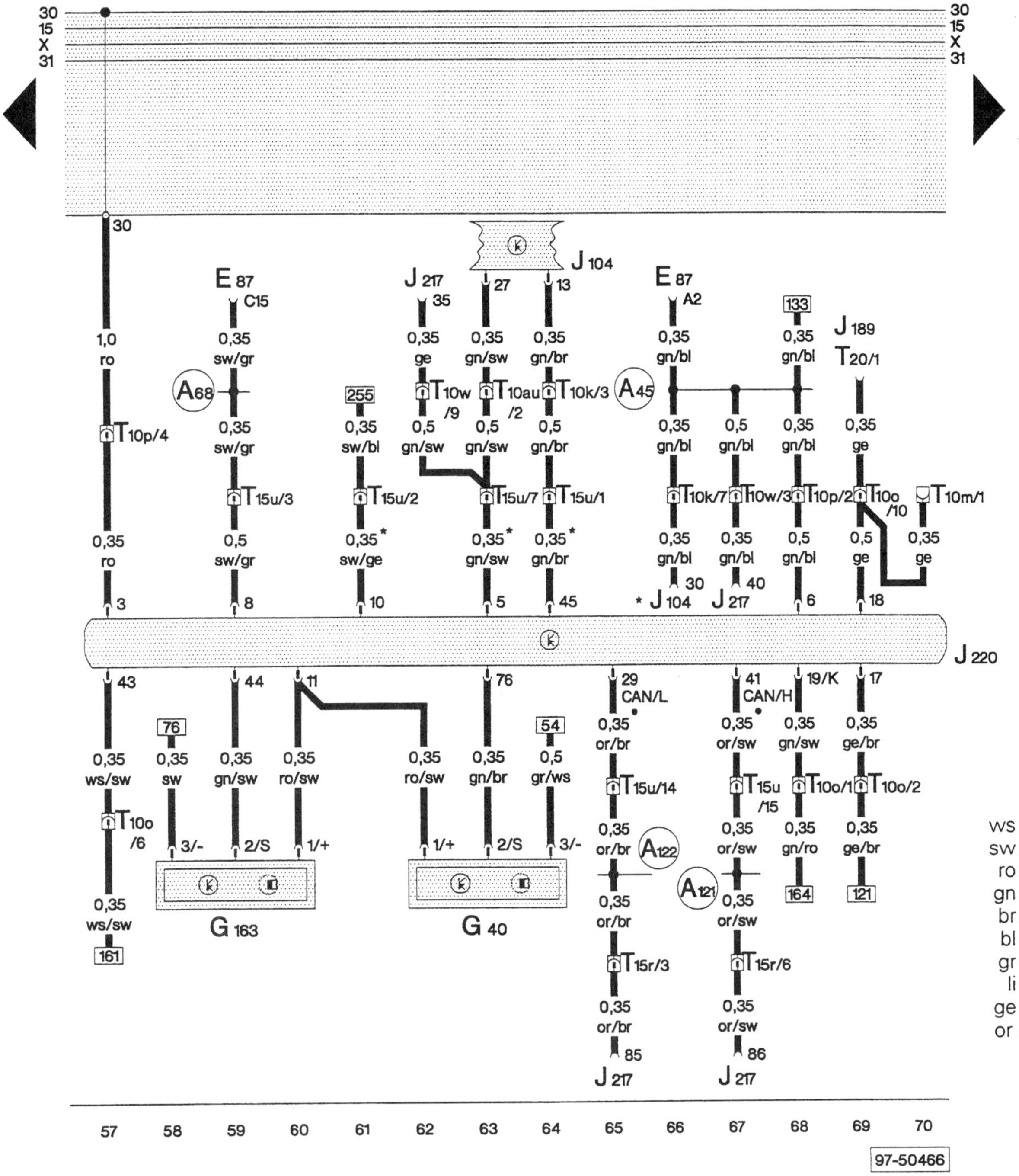

ws = white
sw = black
ro = red
gn = green
br = brown
bl = blue
gr = gray
li = violet
ge = yellow
or = orange

E 87 - A/C Control Head
G 40 - Camshaft Position (CMP) Sensor
G 163 - Camshaft Position (CMP) Sensor 2
J 104 - ABS Control Module (w/EDL)
J 189 - Auto Check System
J 217 - Transmission Control Module (TCM)
J 220 - Motronic Engine Control Module (ECM)
T 10 k - Connector 10 pin, orange, connector station A pillar, left
T 10 m - Connector 10 pin, blue, connector station electronic box plenum chamber
T 10 o - Connector 10 pin, brown, connector station electronic box plenum chamber
T 10 p - Connector 10 pin, black, connector station electronic box plenum chamber
T 10 w - Connector 10 pin, blue, connector station A pillar, right
T 10 au - Connector 10 pin, grey, connector station A pillar, right
T 15 r - Connector 15 pin, brown, connector station A pillar, right
T 15 u - Connector 15 pin, red, connector station electronic box plenum chamber
T 20 - Connector 20 pin, red/white, on instrument cluster

(A45) - Wire connection (RPM-signal), in instrument panel wiring harness
(A68) - Connector (C15, A/C), in instrument panel wiring harness
(A121) - Connection (high bus) in instrument panel wiring harness
(A122) - Connection (low bus) in instrument panel wiring harness

• - CAN-Bus
* - Vehicles with front wheel drive

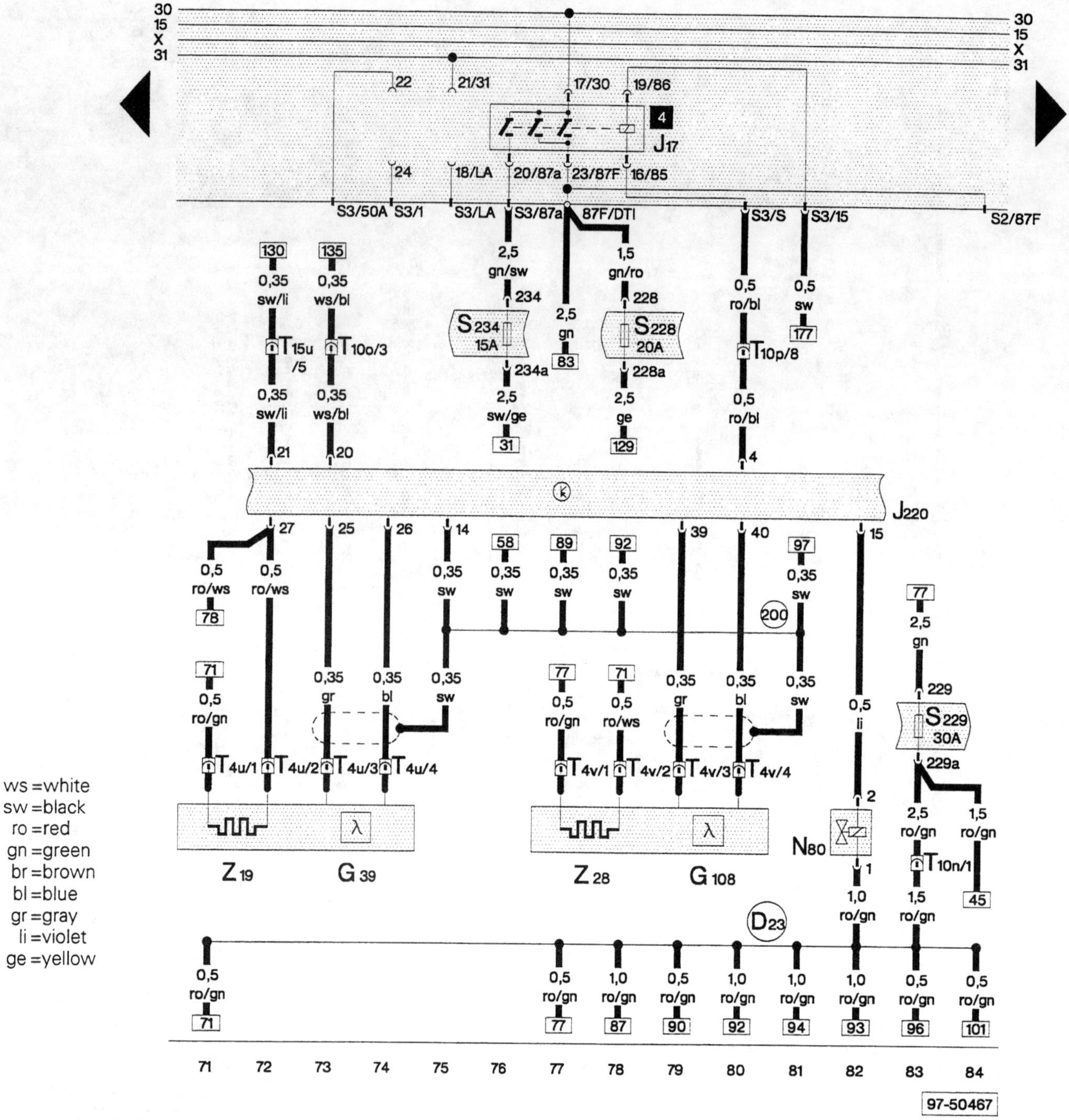

G 39 - Heated Oxygen Sensor (HO2S)
G 108 - Heated Oxygen Sensor (HO2S) 2
J 17 - Fuel Pump (FP) Relay
J 220 - Motronic Engine Control Module (ECM)
N 80 - Evaporative Emission (EVAP) Canister Purge Regulator Valve
S 228 - Fuse
S 229 - Fuse
S 234 - Fuse
T 4 u - Connector 4 pin, black, Heated Oxygen Sensor
T 4 v - Connector 4 pin, black, Heated Oxygen Sensor 2
T 10 n - Connector 10 pin, orange, connector station electronic box plenum chamber
T 10 o - Connector 10 pin, brown, connector station electronic box plenum chamber
T 10 p - Connector 10 pin, black, connector station electronic box plenum chamber
T 15 u - Connector 15 pin, red, connector station electronic box plenum chamber
Z 19 - Oxygen Sensor (O2S) Heater
Z 28 - Oxygen Sensor (O2S) 2 Heater

(200) - Ground connection (shielding), in engine compartment wiring harness
(D23) - Connector (over fuse 229), in wiring harness front, right

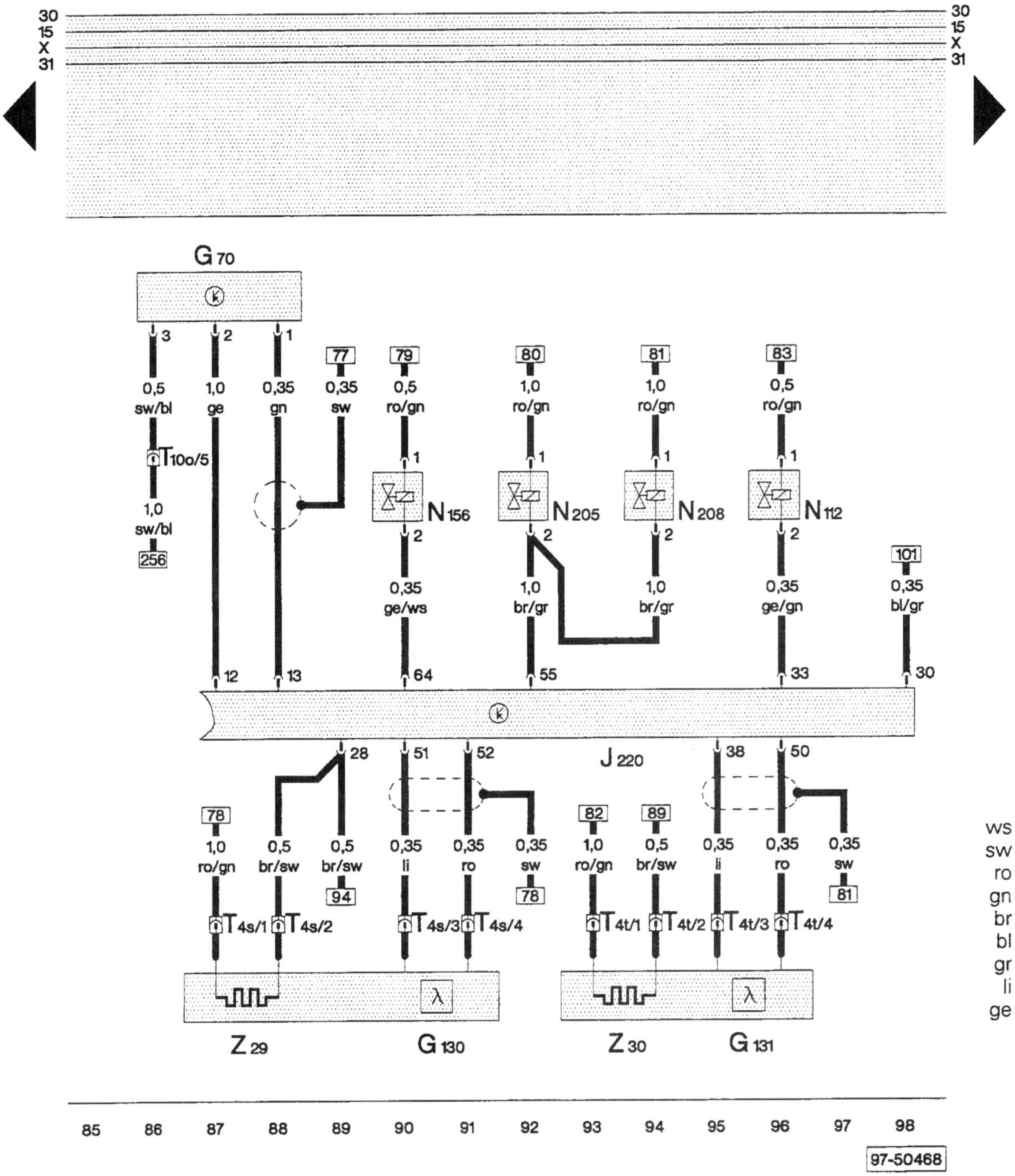

ws = white
sw = black
ro = red
gn = green
br = brown
bl = blue
gr = gray
li = violet
ge = yellow

G 70 - Mass Air Flow (MAF) Sensor
G 130 - Oxygen Sensor (O2S) Behind Three Way Catalytic Converter (TWC)
G 131 - Oxygen Sensor (O2S) 2 Behind Three Way Catalytic Converter (TWC)
J 220 - Motronic Engine Control Module (ECM)
N 112 - Secondary Air Injection (AIR) Solenoid Valve
N 156 - Intake Manifold Change-Over Valve
N 205 - Valve -1- for camshaft adjustment
N 208 - Valve -2- for camshaft adjustment
T 4 s - Connector 4 pin, green, near oxygen sensor, behind Three Way Catalytic Converter (TWC)
T 4 t - Connector 4 pin, brown, near oxygen sensor 2, behind Three Way Catalytic Converter (TWC)
T 10 o - Connector 10 pin, brown, connector station electronic box plenum chamber
Z 29 - Heater for Lambda-probe 1
Z 30 - Heater for Lambda-probe 2

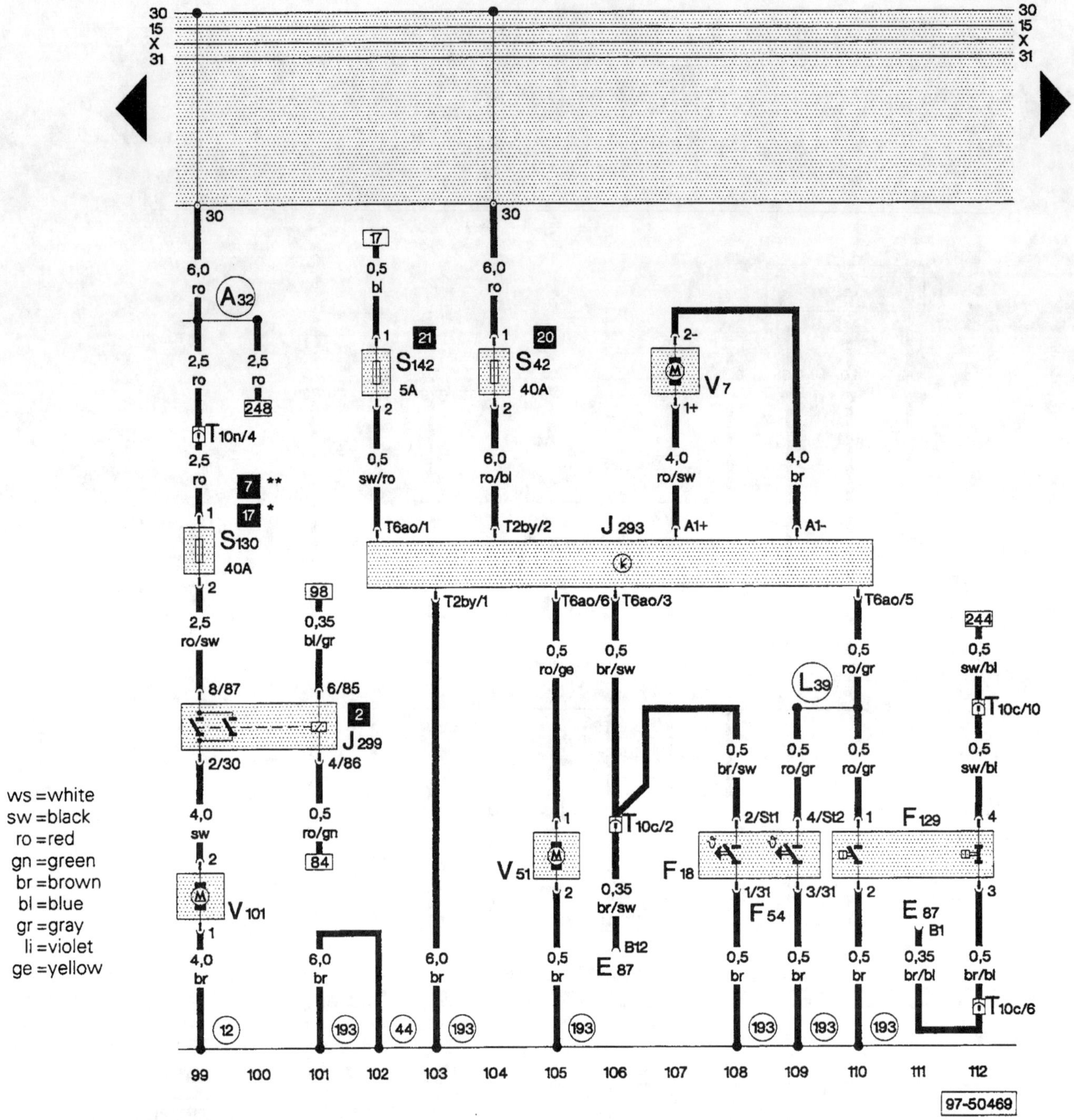

E 87 - A/C Control Head
F 18 - Coolant Fan Control (FC) Thermal Switch
F 54 - Coolant Fan Control (FC) Thermal Switch
F 129 - A/C Pressure Switch
J 293 - Coolant FC (Fan Control) Control Module
J 299 - Secondary Air Injection (AIR) Pump Relay
S 42 - Coolant Fan Fuse
S 130 - Fuse for secondary air pump
S 142 - Control module fuse for coolant fan
T 2 by - Connector 2 pin, on Coolant FC (Fan Control) Control Module
T 6 ao - Connector 6 pin, on Coolant FC (Fan Control) Control Module
T 10 c - Connector 10 pin, grey, connector station A pillar, left
T 10 n - Connector 10 pin, orange, connector station electronic box plenum chamber
V 7 - Coolant Fan
V 51 - After-Run Coolant Pump

V 101 - Secondary Air Injection (AIR) Pump Motor

(12) - Ground connection, in engine compartment, left

(44) - Ground connection, on left A-pillar, lower part

(193) - Ground connection -1-, in coolant fan wiring harness

(A32) - Plus connection (30), in instrument panel wiring harness

(L39) - Wire connection -1-(A/C Pressure Switch), in Climatronic wiring harness

* - upto october 1998
** - from november 1998

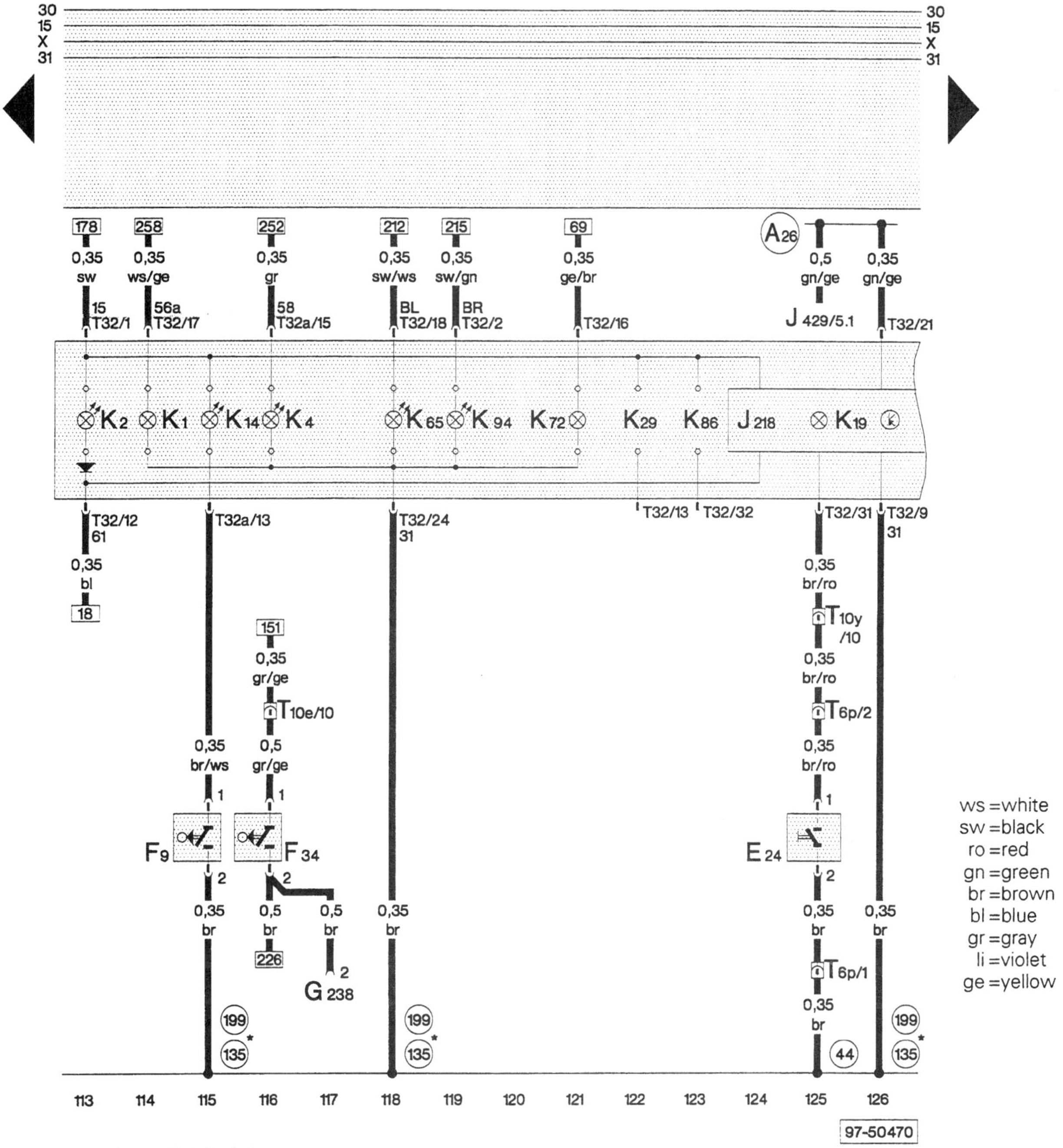

E 24 - Left Seat Belt Switch
F 9 - Parking Brake Warning Light Switch
F 34 - Brake Fluid Level Warning Switch
G 238 - Sensor for air quality
J 218 - Instrument Cluster Combination Processor
J 429 - Control module for central locking
K 1 - Headlight High Beam Indicator Light
K 2 - Generator (GEN) Warning Light
K 4 - Park Light Indicator Light
K 14 - Parking Brake Indicator Light
K 19 - Seat Belt Warning Light
K 29 - Glow Plug Indicator Light
K 65 - Left Turn Signal Indicator Light
K 72 - Warning light for catalyst monitoring
K 86 - Traction Control Indicator Light
K 94 - Right Turn Signal Indicator Light
T 6 p - Connector 6 pin, red, below driver's seat
T 10 e - Connector 10 pin, violet, connector station A pillar, left
T 10 y - Connector 10 pin, green, connector station A pillar, right
T 32 - Connector 32 pin, blue, on instrument cluster
T 32 a - Connector 32 pin, green, on instrument cluster

(44) - Ground connection, on left A-pillar, lower part

(135) - Ground connection -2-, in instrument panel wiring harness

(199) - Ground connection -3-, in instrument panel wiring harness

(A26) - Wire connection (driver's door contact switch), in instrument panel wiring harness

* - booth is possible

Instrument Cluster, Parking Brake Warning Light Switch, Brake Fluid Level Warning Switch

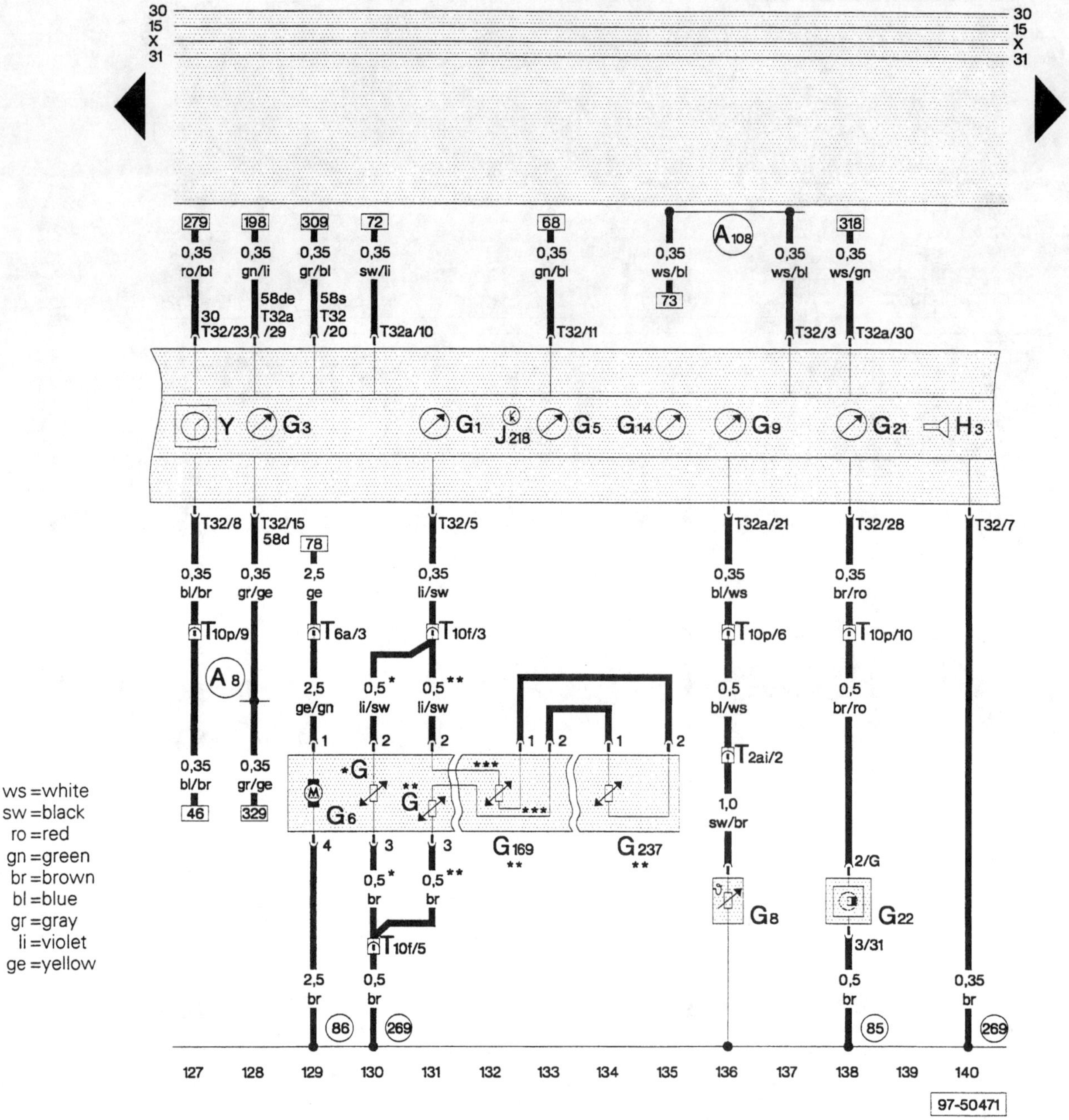

G - Fuel Level Sensor
G 1 - Fuel Gauge
G 3 - Engine Coolant Temperature (ECT) Gauge
G 5 - Tachometer
G 6 - Fuel Pump (FP)
G 8 - Engine Oil Temperature Sensor
G 9 - Engine Oil Temperature Gauge
G 14 - Voltmeter
G 21 - Speedometer
G 22 - Speedometer Vehicle Speed Sensor (VSS)
G 169 - Fuel Level Sensor 2
G 237 - Fuel Level Sensor 3
H 3 - Warning Buzzer
J 218 - Instrument Cluster Combination Processor
T 2 ai - Connector 2 pin, black, oil sensor
T 6 a - Connector 6 pin, blue, connector station A pillar, left
T 10 f - Connector 10 pin, brown, connector station A pillar, left
T 10 p - Connector 10 pin, black, connector station electronic box plenum chamber
T 32 - Connector 32 pin, blue, on instrument cluster
T 32 a - Connector 32 pin, green, on instrument cluster
Y - Analog Clock

(85) - Ground connection -1-, in engine compartment wiring harness
(86) - Ground connection -1-, in rear wiring harness
(269) - Ground connector (sensor ground) -1-, in instrument panel wiring harness
(A8) - Plus connection (58d), in instrument panel wiring harness (Display illumination)
(A108) - Wire connection (vehicle speed signal), in instrument panel wiring harness

* - Vehicles with front-wheel drive
** - Vehicles with all-wheel drive
*** - Wiring inside fuel tank

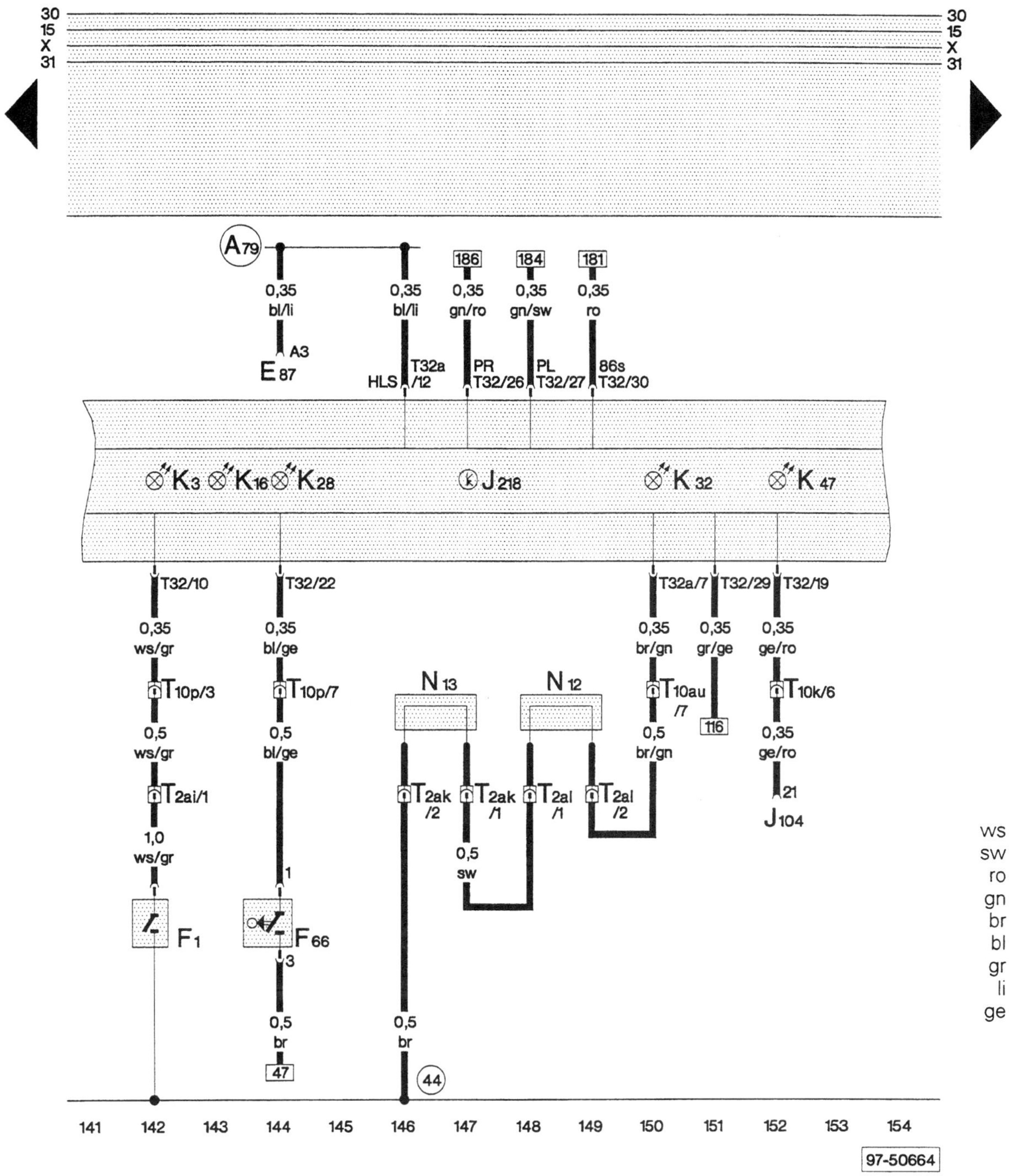

ws = white
sw = black
ro = red
gn = green
br = brown
bl = blue
gr = gray
li = violet
ge = yellow

E 87 - A/C Control Head
F 1 - Oil Pressure Switch
F 66 - Engine Coolant Level (ECL) Warning Switch
J 104 - ABS Control Module (w/EDL)
J 218 - Instrument Cluster Combination Processor
K 3 - Oil Pressure Warning Light
K 16 - Fuel Reserve Warning Light
K 28 - Engine Coolant Level/Temperature (ECL/ECT) Warning Light
K 32 - Brake Pad Wear Indicator Light
K 47 - ABS Warning Light
N 12 - Right Brake Pad Wear Indicator (Shear Element)
N 13 - Left Brake Pad Wear Indicator (Shear Element)
T 2 ai - Connector 2 pin, black, oil sensor
T 2 ak - Connector 2 pin, red, on firewall, left
T 2 al - Connector 2 pin, red, on firewall, right
T 10 k - Connector 10 pin, orange, connector station A pillar, left
T 10 p - Connector 10 pin, black, connector station electronic box plenum chamber
T 10 au - Connector 10 pin, grey, connector station A pillar, right
T 32 - Connector 32 pin, blue, on instrument cluster
T 32 a - Connector 32 pin, green, on instrument cluster

(44) - Ground connection, on left A-pillar, lower part

(A79) - Connector (engine overheat warning light), in instrument panel wiring harness

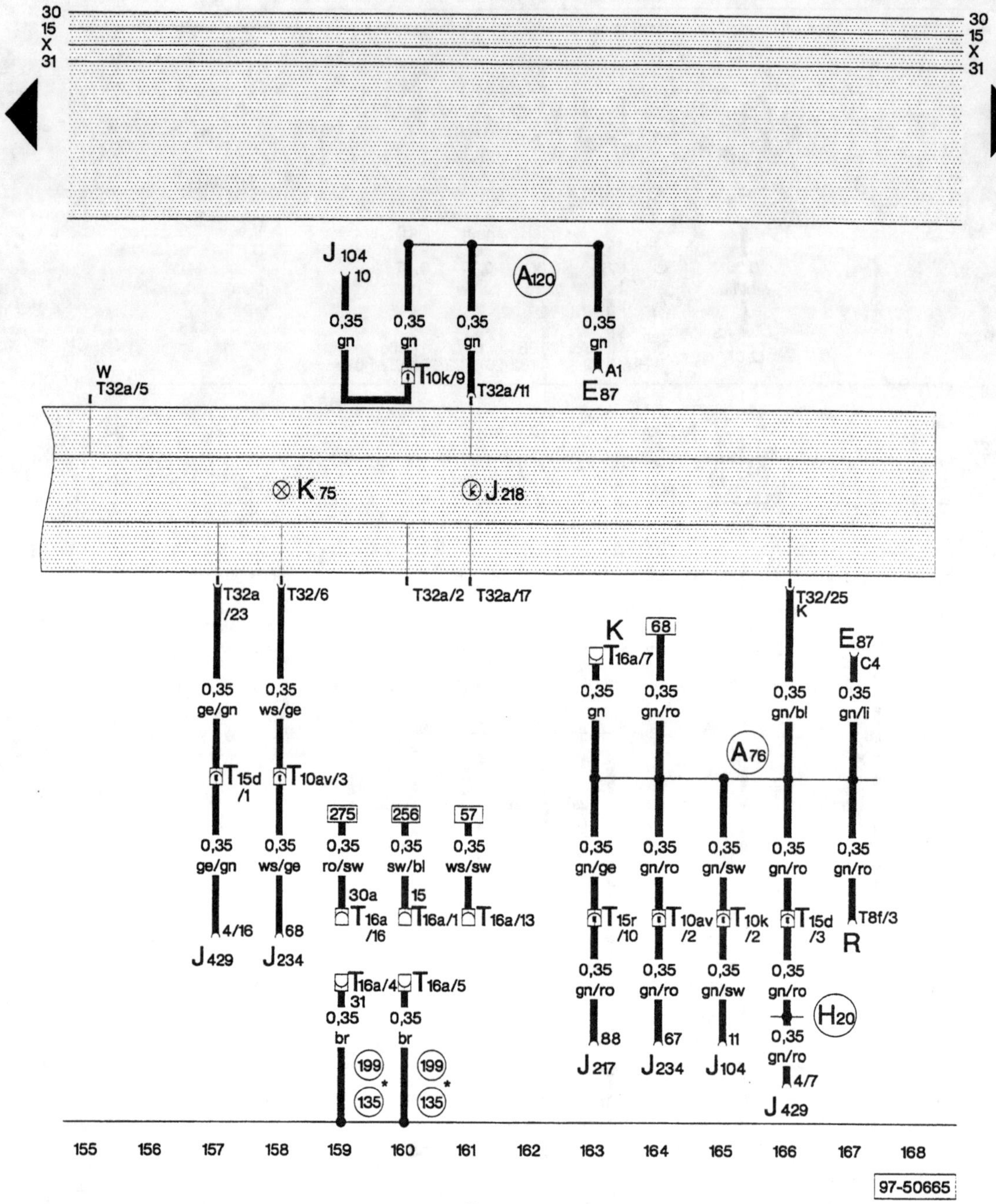

ws = white
sw = black
ro = red
gn = green
br = brown
bl = blue
gr = gray
li = violet
ge = yellow

E 87 - A/C Control Head
J 104 - ABS Control Module (w/EDL)
J 217 - Transmission Control Module (TCM)
J 218 - Instrument Cluster Combination Processor
J 234 - Airbag Control Module
J 429 - Control module for central locking
K - DLC (Data Link Connector) Rapid Data Transfer
K 75 - Airbag Malfunction Indicator Lamp (MIL)
R - Radio
T 8 f - Connector 8 pin, black, connector radio III
T 10 k - Connector 10 pin, orange, connector station A pillar, left
T 10 av - Connector 10 pin, yellow, connector station A pillar, right
T 15 d - Connector 15 pin, black, connector station A pillar, right
T 15 r - Connector 15 pin, brown, connector station A pillar, right
T 16 a - Connector 16 pin, data link connector
T 32 - Connector 32 pin, blue, on instrument cluster
T 32 a - Connector 32 pin, green, on instrument cluster

(135) - Ground connection -2-, in instrument panel wiring harness

(199) - Ground connection -3-, in instrument panel wiring harness

(A76) - Connector (K-diagnosis wire), in instrument panel wiring harness

(A120) - Connection (standing time) in instrument panel wiring harness

(H20) - Wire connection -1-, in On-Board Diagnostic (OBD) wiring harness

* - booth is possible

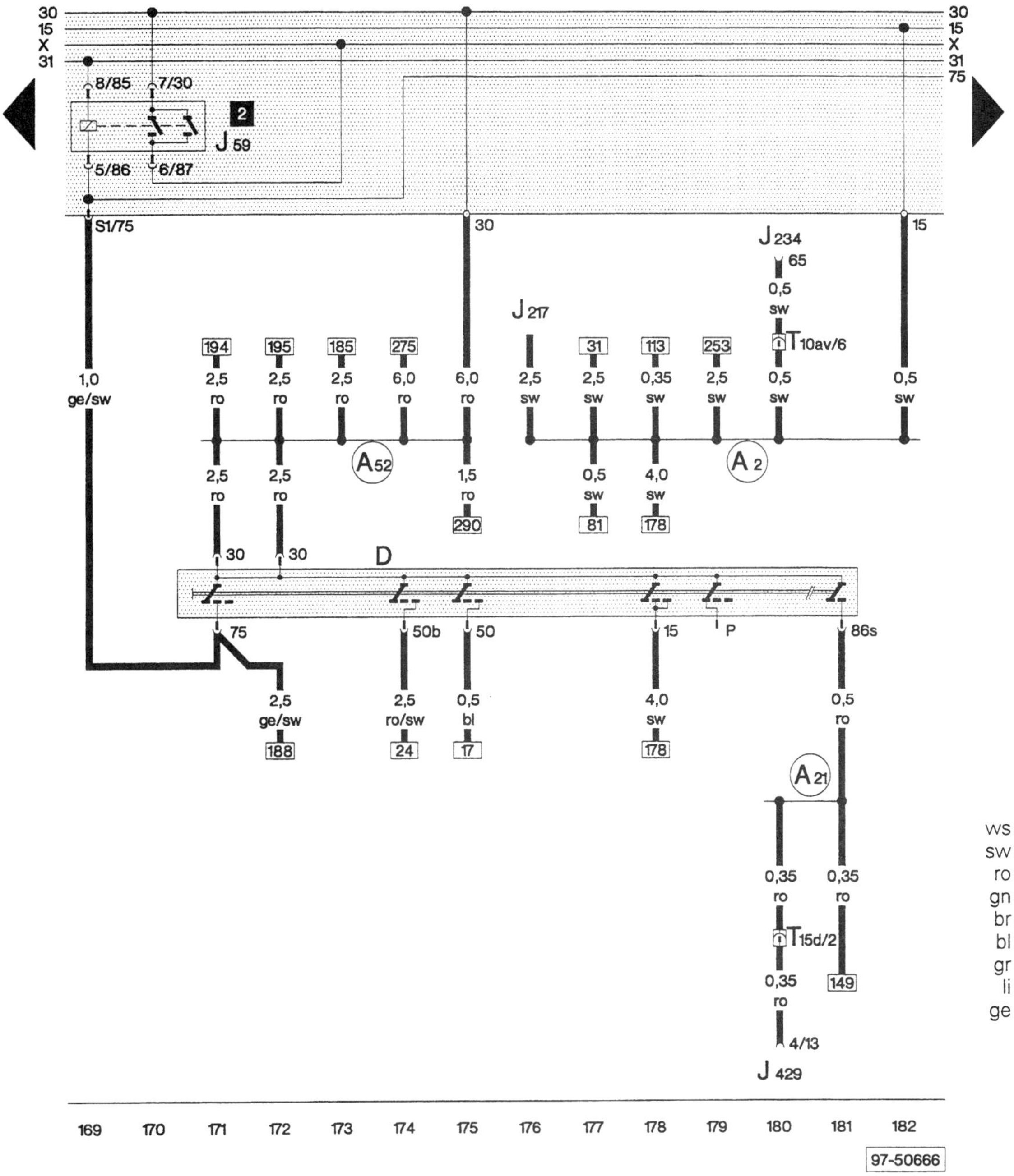

ws = white
sw = black
ro = red
gn = green
br = brown
bl = blue
gr = gray
li = violet
ge = yellow

D - Ignition/Starter Switch
J 59 - Load Reduction Relay
J 217 - Transmission Control Module (TCM)
J 234 - Airbag Control Module
J 429 - Control module for central locking
T 10 av - Connector 10 pin, yellow, connector station A pillar, right
T 15 d - Connector 15 pin, black, connector station A pillar, right

(A2) - Plus connection (15), in instrument panel wiring harness
(A21) - Wire connection (86s), in instrument panel wiring harness
(A52) - Plus connection (30), in instrument panel wiring harness

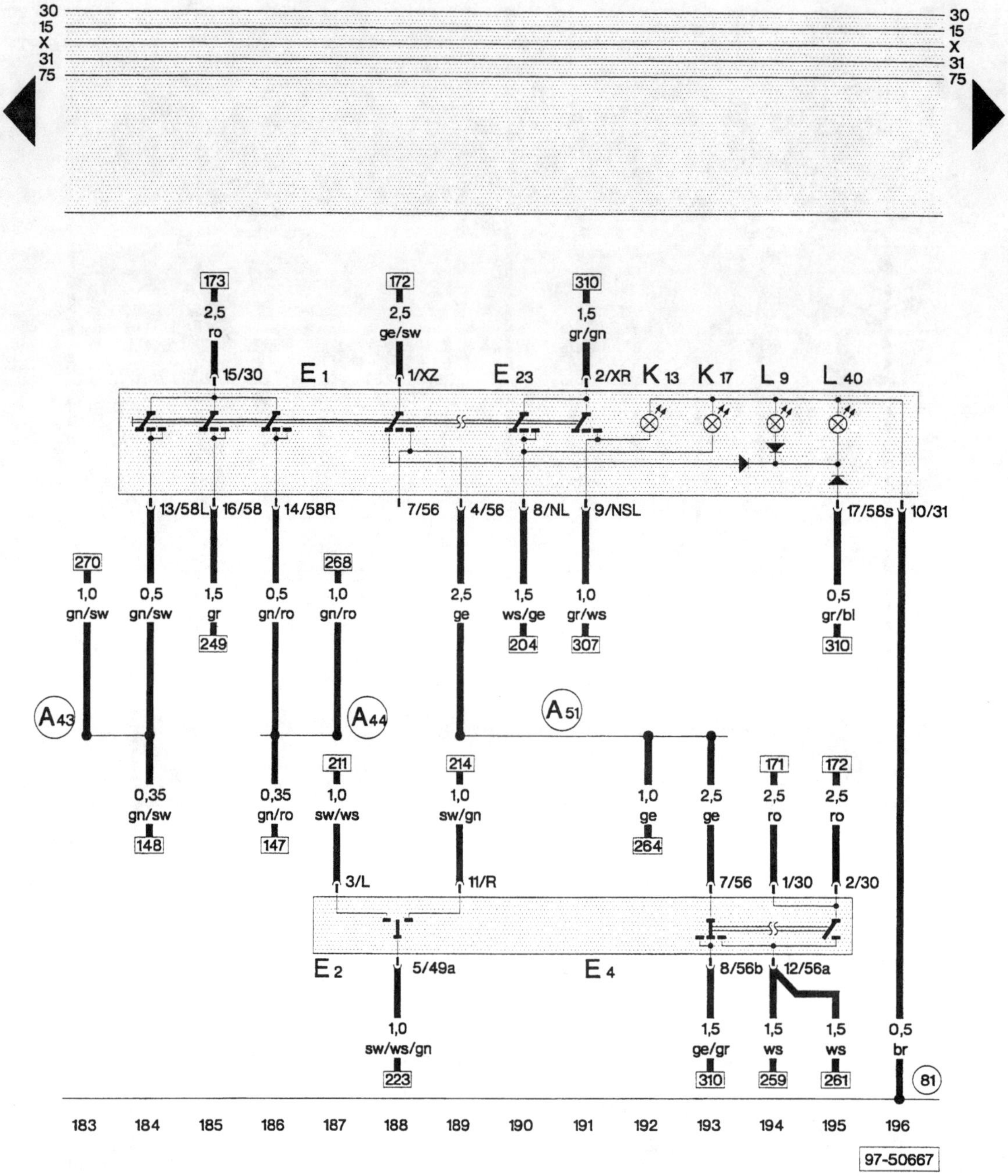

ws = white
sw = black
ro = red
gn = green
br = brown
bl = blue
gr = gray
li = violet
ge = yellow

E 1 - Light Switch
E 2 - Turn Signal Switch
E 4 - Headlight Dimmer/Flasher Switch
E 23 - Fog Light Switch
K 13 - Rear Fog Light Indicator Light
K 17 - Fog Light Indicator Light
L 9 - Headlight Switch Light
L 40 - Front And Rear Fog Light Switch Light

(81) - Ground connection -1-, in instrument panel wiring harness
(A43) - Wire connection (57L), in instrument panel wiring harness
(A44) - Wire connection (57R), in instrument panel wiring harness
(A51) - Wire connection (56), in instrument panel wiring harness

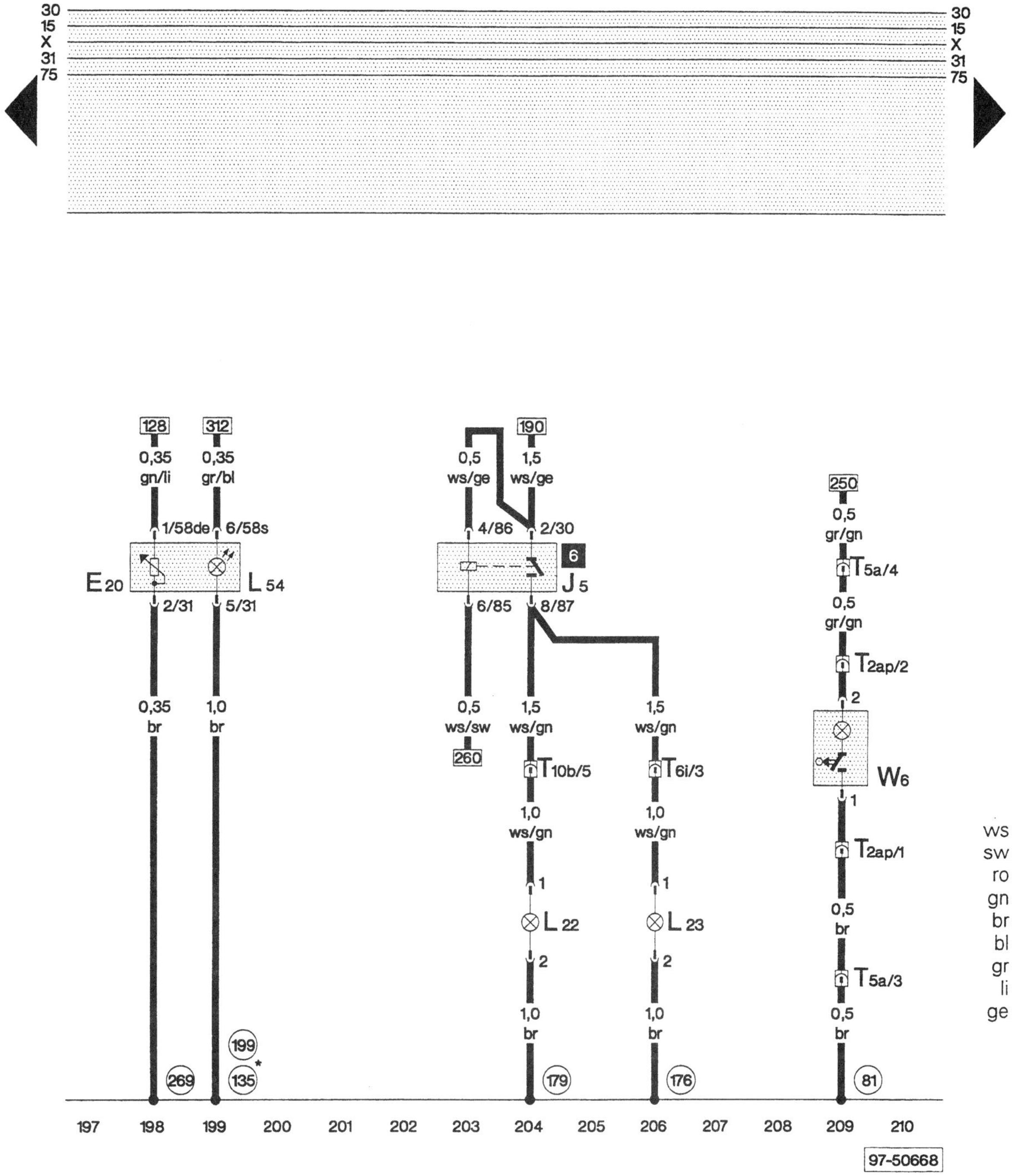

ws = white
sw = black
ro = red
gn = green
br = brown
bl = blue
gr = gray
li = violet
ge = yellow

E 20 - Instrument Panel Light Dimmer Switch
J 5 - Fog Light Relay
L 22 - Left Front Fog Light
L 23 - Right Front Fog Light
L 54 - Headlight Adjuster Control Light
T 2ap - Connector 2 pin, black, behind instrument panel, right
T 5 a - Connector 5 pin, black, behind instrument panel, right
T 6 i - Connector 6 pin, red, connector station A pillar, right
T 10 b - Connector 10 pin, red, connector station A pillar, left
W6 - Glove Compartment Light

(81) - Ground connection -1-, in instrument panel wiring harness
(135) - Ground connection -2-, in instrument panel wiring harness
(176) - Ground connection, in right headlight wiring harness
(179) - Ground connection, in left headlight wiring harness
(199) - Ground connection -3-, in instrument panel wiring harness
(269) - Ground connector (sensor ground) -1-, in instrument panel wiring harness

* - booth is possible

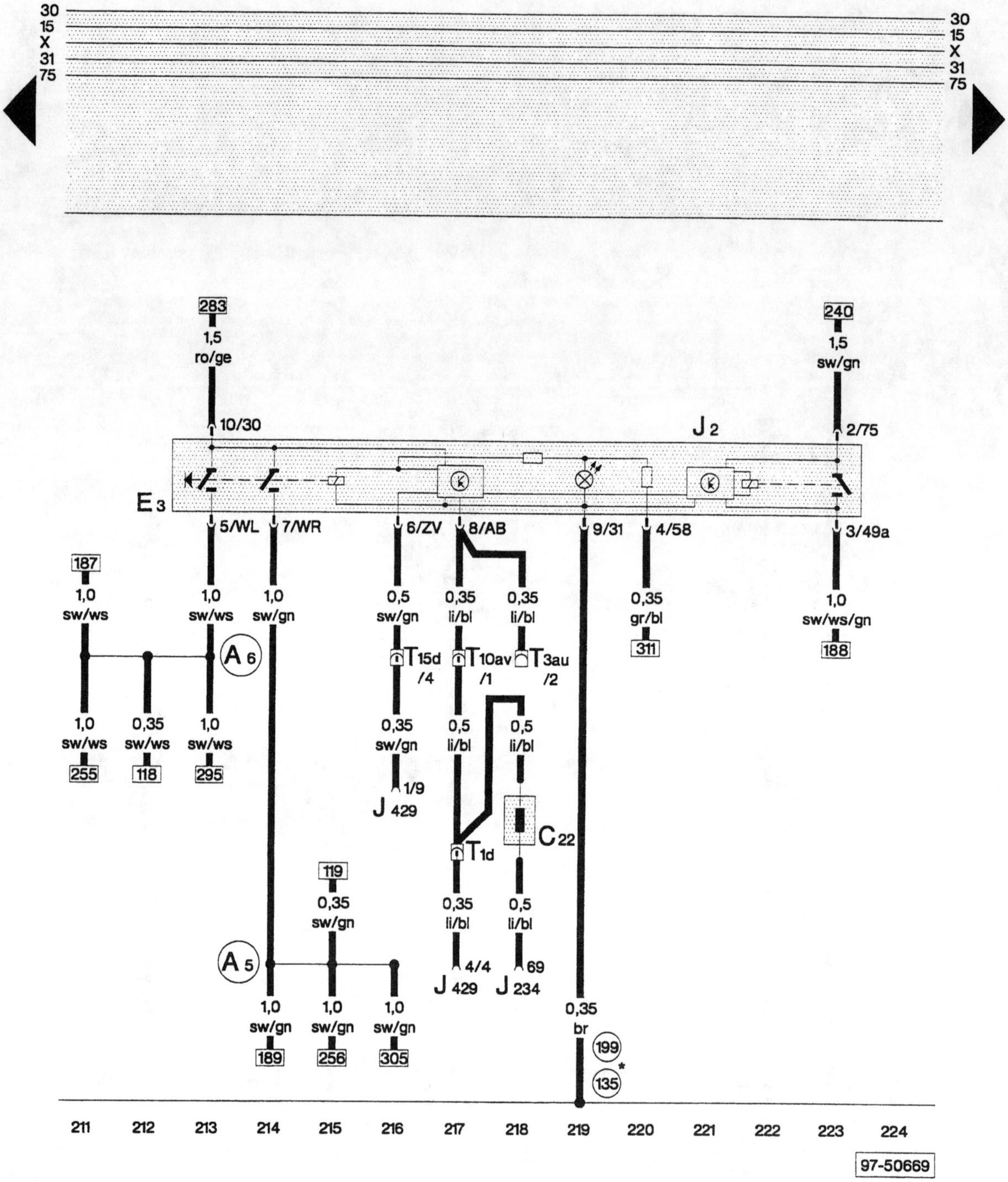

ws = white
sw = black
ro = red
gn = green
br = brown
bl = blue
gr = gray
li = violet
ge = yellow

C 22 - Interference filter
E 3 - Emergency Flasher Switch
J 2 - Emergency Flasher Relay
J 234 - Airbag Control Module
J 429 - Control module for central locking
T 1 d - Single connector, brown, connector station A pillar right
T 3 au - Connector 3 pin, red, connector station electronic box plenum chamber
T 10 av - Connector 10 pin, yellow, connector station A pillar, right
T 15 d - Connector 15 pin, black, connector station A pillar, right

(135) - Ground connection -2-, in instrument panel wiring harness
(199) - Ground connection -3-, in instrument panel wiring harness
(A5) - Plus connection (right turn signal), in instrument panel wiring harness
(A6) - Plus connection (left turn signal), in instrument panel wiring harness

* - booth is possible

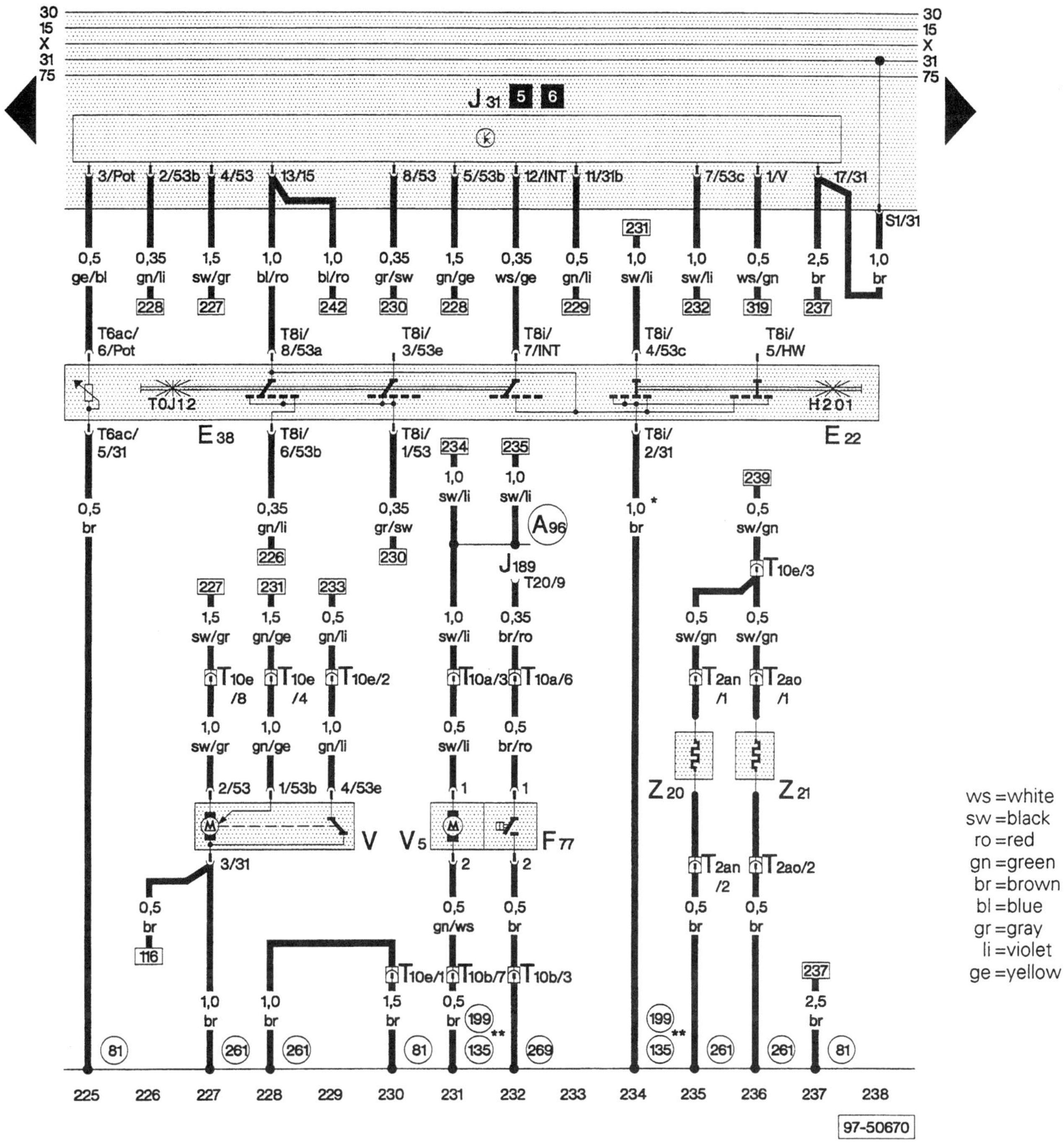

E 22 - Windshield Wiper/Washer Switch
E 38 - Windshield Wiper Intermittent Regulator
F 77 - Windshield Washer Fluid Level Warning Switch
J 31 - Wiper/Washer Intermittent Relay
J 189 - Auto Check System
T 2 an - Connector 2 pin, washer jet, left
T 2 ao - Connector 2 pin, washer jet, right
T 6 ac - Connector 6 pin, brown, on windshield wiper intermittent regulator
T 8 i - Connector 8 pin, on windshield wiper/washer switch
T 10 a - Connector 10 pin, pink, connector station A pillar, left
T 10 b - Connector 10 pin, red, connector station A pillar, left
T 10 e - Connector 10 pin, violet, connector station A pillar, left
T 20 - Connector 20 pin, red/white, on instrument cluster
V - Windshield Wiper Motor
V 5 - Windshield Washer Pump
Z 20 - Left Washer Nozzle Heater
Z 21 - Right Washer Nozzle Heater

(81) - Ground connection -1-, in instrument panel wiring harness
(135) - Ground connection -2-, in instrument panel wiring harness
(199) - Ground connection -3-, in instrument panel wiring harness
(261) - Ground connection, in wiring harness heated spray jet
(269) - Ground connector (sensor ground) -1-, in instrument panel wiring harness
(A96) - Connector (53a), in instrument panel wiring harness

* - only wagon
** - booth is possible

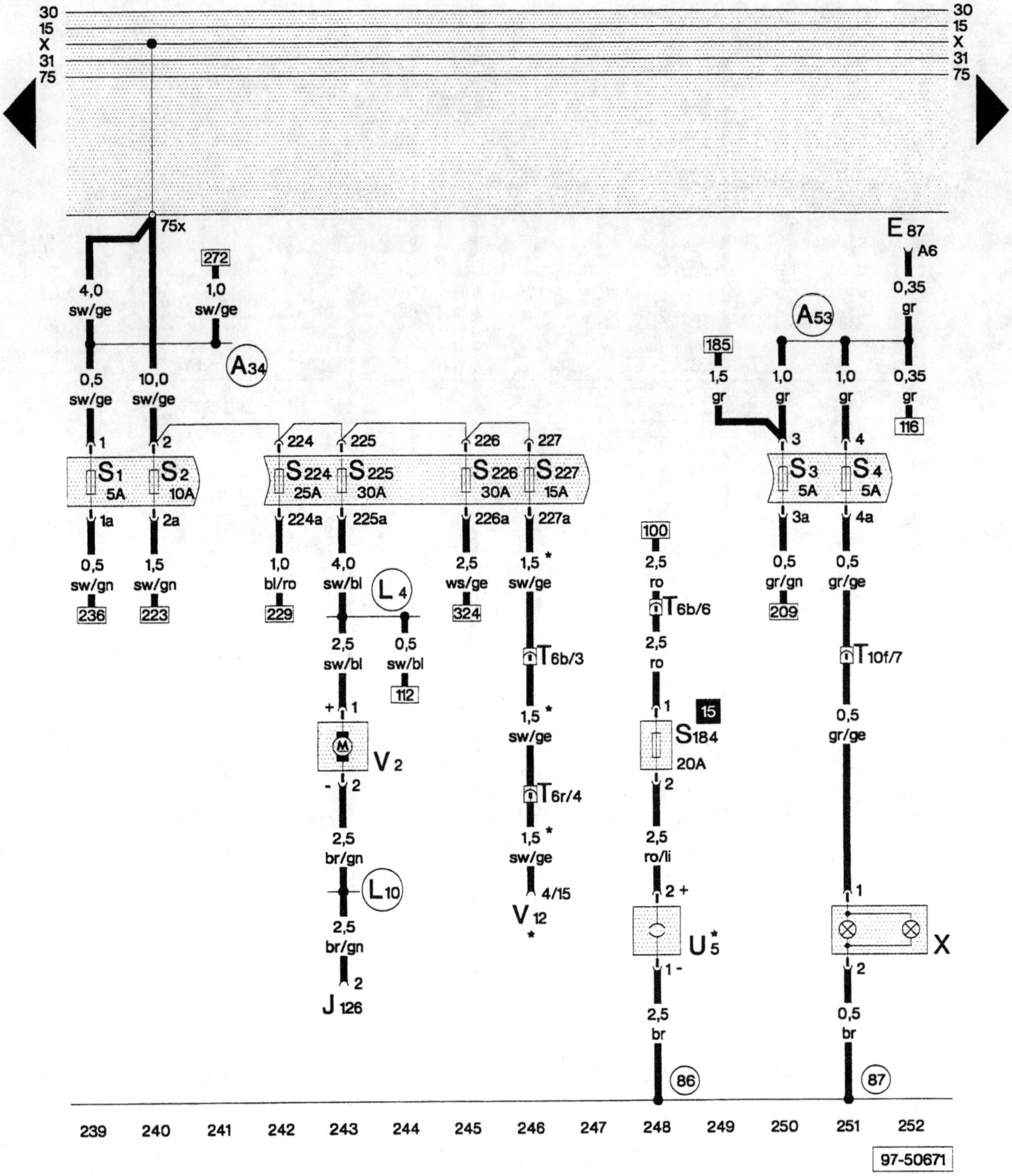

ws = white
sw = black
ro = red
gn = green
br = brown
bl = blue
gr = gray
li = violet
ge = yellow

E 87	- A/C Control Head
J 126	- Control module for fresh air blower
S 1	- Fuse
S 2	- Fuse
S 3	- Fuse
S 4	- Fuse
S 184	- Fuse for Luggage Compartment Socket
S 224	- Fuse
S 225	- Fuse
S 226	- Fuse
S 227	- Fuse
T 6 b	- Connector 6 pin, white, connector station A pillar, left
T 6 r	- Connector 6 pin, brown, in luggage compartment
T 10 f	- Connector 10 pin, brown, connector station A pillar, left
U 5	- 12 V Socket, in Luggage Compartment
V 2	- Fresh Air Blower
V 12	- Rear Window Wiper Motor
X	- License Plate Light

(86)	- Ground connection -1-, in rear wiring harness
(87)	- Ground connection -2-, in rear wiring harness
(A34)	- Wire connection (75x), in instrument panel wiring harness
(A53)	- Plus connection (58b), in instrument panel wiring harness
(L4)	- Wire connection (75al), in A/C wiring harness
(L10)	- Wire connection -2-, in A/C wiring harness

* - only wagon

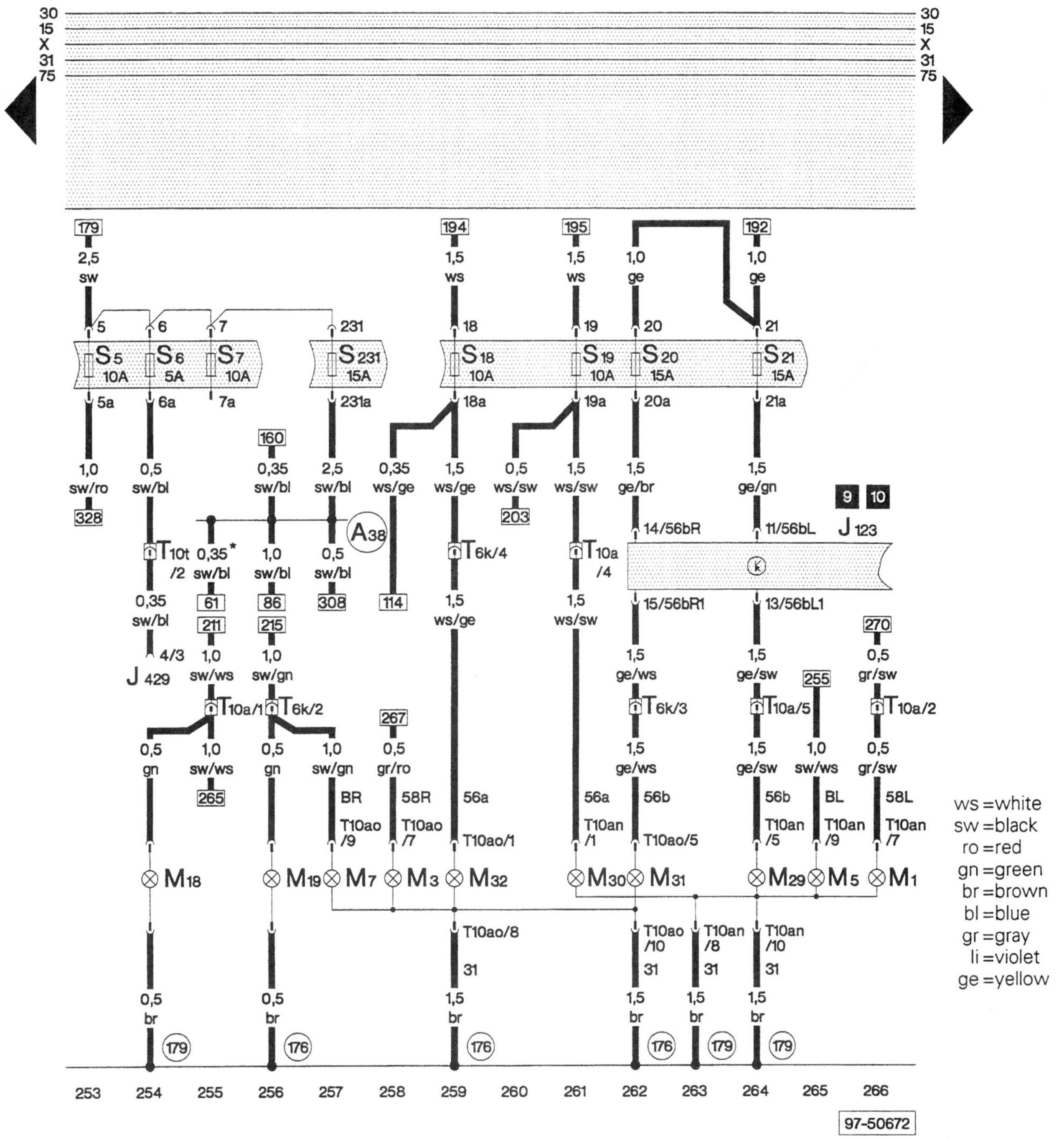

J 123 - Lamp Control Module
J 429 - Control module for central locking
M 1 - Left Parking Light
M 3 - Right Parking Light
M 5 - Left Front Turn Signal Light
M 7 - Right Front Turn Signal Light
M 18 - Left, Side Turn Signal Light
M 19 - Right, Side Turn Signal Light
M 29 - Left Low Beam Headlight
M 30 - Left High Beam Headlight
M 31 - Right Low Beam Headlight
M 32 - Right High Beam Headlight
S 5 - Fuse
S 6 - Fuse
S 7 - Fuse
S 18 - Fuse
S 19 - Fuse
S 20 - Fuse
S 21 - Fuse
S 231 - Fuse
T 6 k - Connector 6 pin, pink, connector station A pillar, right
T 10 a - Connector 10 pin, pink, connector station A pillar, left
T 10 t - Connector 10 pin, orange, connector station A pillar, right
T 10 an - Connector 10 pin, on headlight, left
T 10 ao - Connector 10 pin, on headlight, right

(176) - Ground connection, in right headlight wiring harness

(179) - Ground connection, in left headlight wiring harness

(A38) - Plus connection (15a II), in instrument panel wiring harness

* - Vehicles with front wheel drive

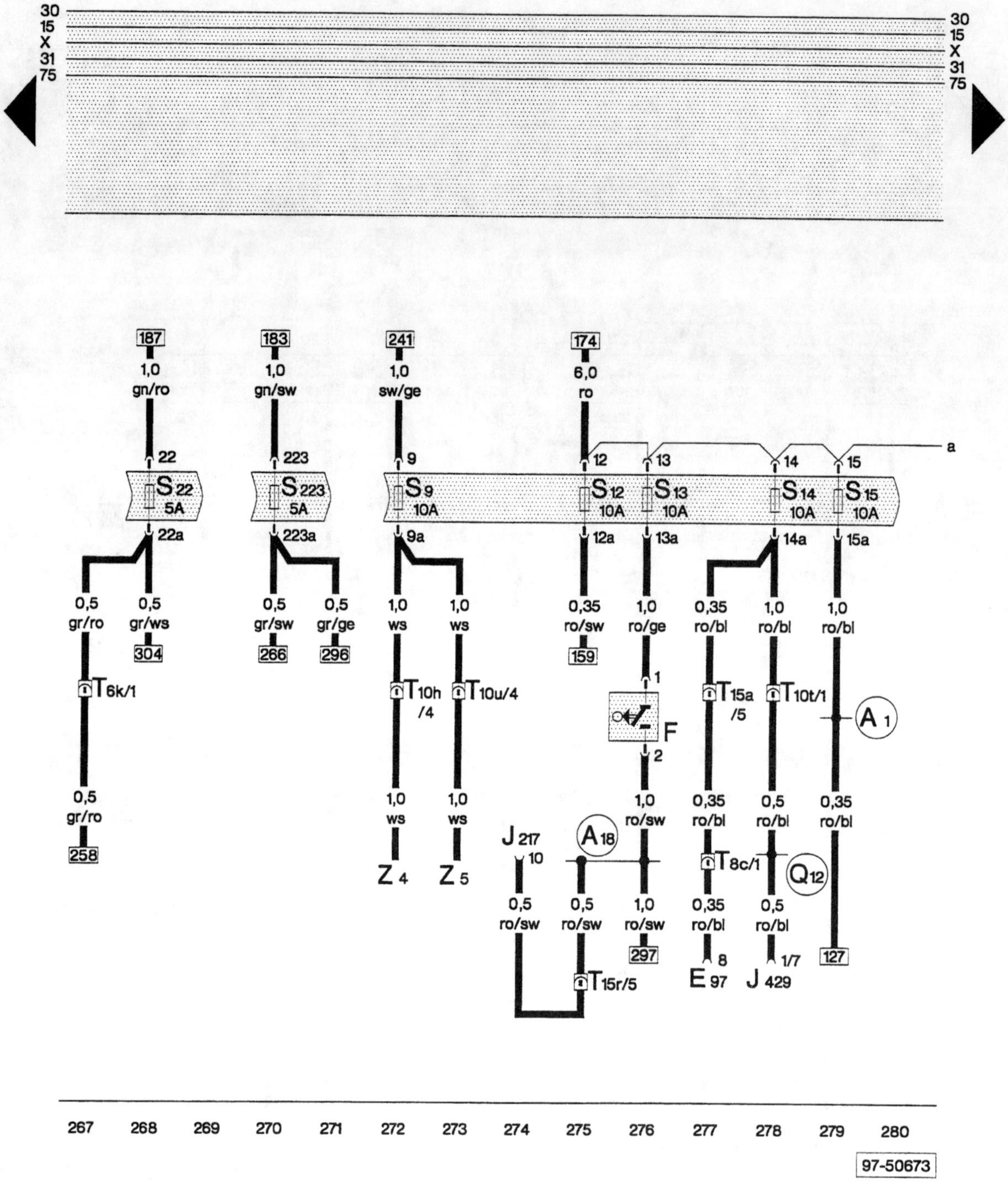

ws = white
sw = black
ro = red
gn = green
br = brown
bl = blue
gr = gray
li = violet
ge = yellow

E 97 - Memory Program Switch
F - Brake Light Switch
J 217 - Transmission Control Module (TCM)
J 429 - Control module for central locking
S 9 - Fuse
S 12 - Fuse
S 13 - Fuse
S 14 - Fuse
S 15 - Fuse
S 22 - Fuse
S 223 - Fuse
T 6 k - Connector 6 pin, pink, connector station A pillar, right
T 8 c - Connector 8 pin, black, in driver's door
T 10 h - Connector 10 pin, black, connector station A pillar, left
T 10 t - Connector 10 pin, orange, connector station A pillar, right
T 10 u - Connector 10 pin, black, connector station A pillar, right
T 15 a - Connector 15 pin, black, connector station A pillar, left
T 15 r - Connector 15 pin, brown, connector station A pillar, right

Z 4 - Heated outside mirror, driver side
Z 5 - Heated outside mirror, passenger side

(A1) - Plus connection (30a), in instrument panel wiring harness
(A18) - Wire connection (54), in instrument panel wiring harness
(Q12) - Plus connection (30a), in power window/central locking system and door contact switch wiring harness

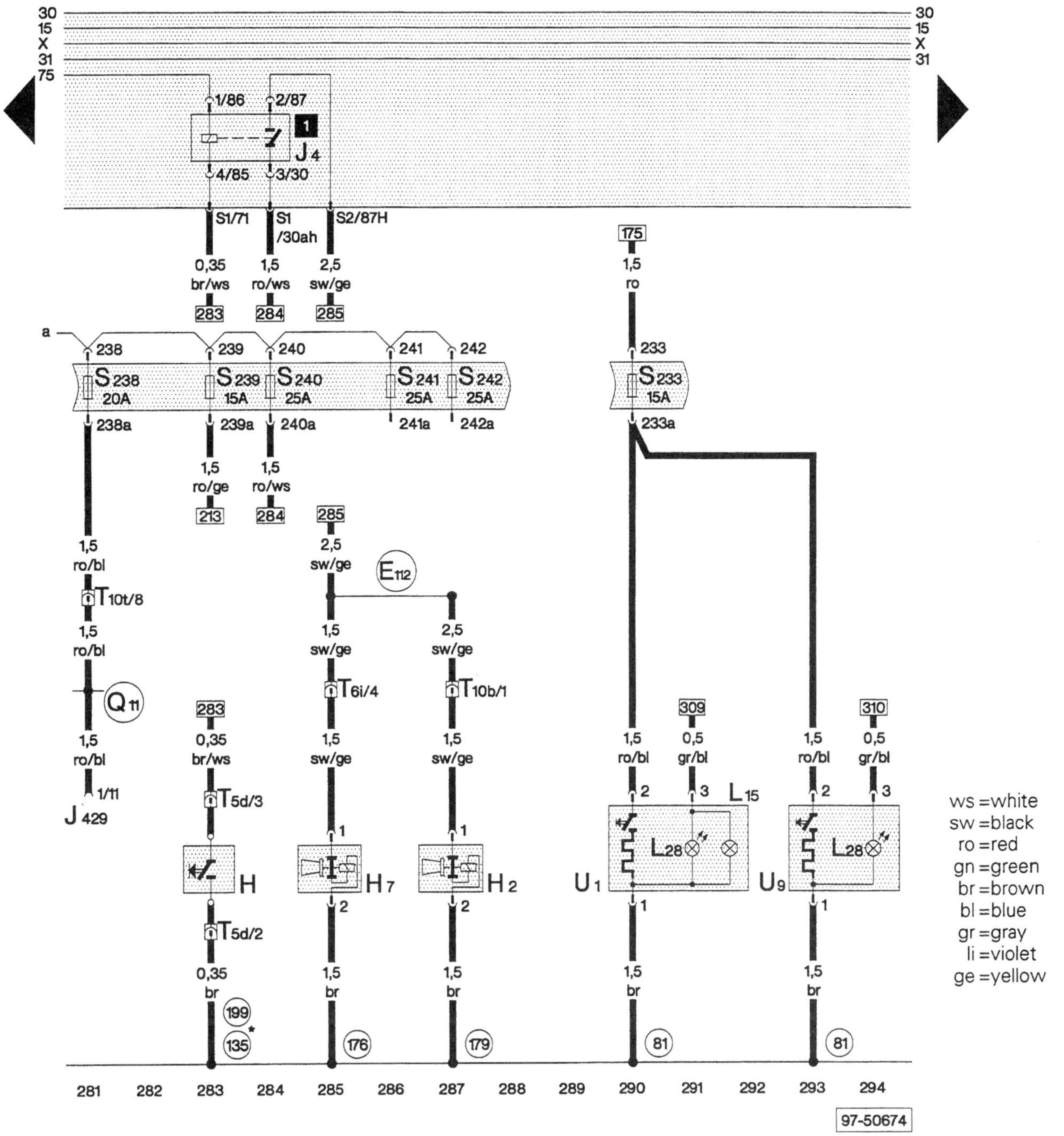

H - Horn Button
H 2 - High Tone Horn
H 7 - Low Tone Horn
J 4 - Dual Horn Relay
J 429 - Control module for central locking
L 15 - Ashtray Light
L 28 - Cigarette Lighter Light
S 233 - Fuse
S 238 - Fuse
S 239 - Fuse
S 240 - Fuse
S 241 - Fuse
S 242 - Fuse
T 5 d - Connector 5 pin, behind steering column switch cover
T 6 i - Connector 6 pin, red, connector station A pillar, right
T 10 b - Connector 10 pin, red, connector station A pillar, left
T 10 t - Connector 10 pin, orange, connector station A pillar, right
U 1 - Cigarette Lighter
U 9 - Rear Cigarette Lighter

(81) - Ground connection -1-, in instrument panel wiring harness
(135) - Ground connection -2-, in instrument panel wiring harness
(176) - Ground connection, in right headlight wiring harness
(179) - Ground connection, in left headlight wiring harness
(199) - Ground connection -3-, in instrument panel wiring harness
(E112) - Connector (Horn - 87h), in instrument panel wiring harness
(Q11) - Plus connection (30a), in power window/central locking system and door contact switch wiring harness

* - booth is possible

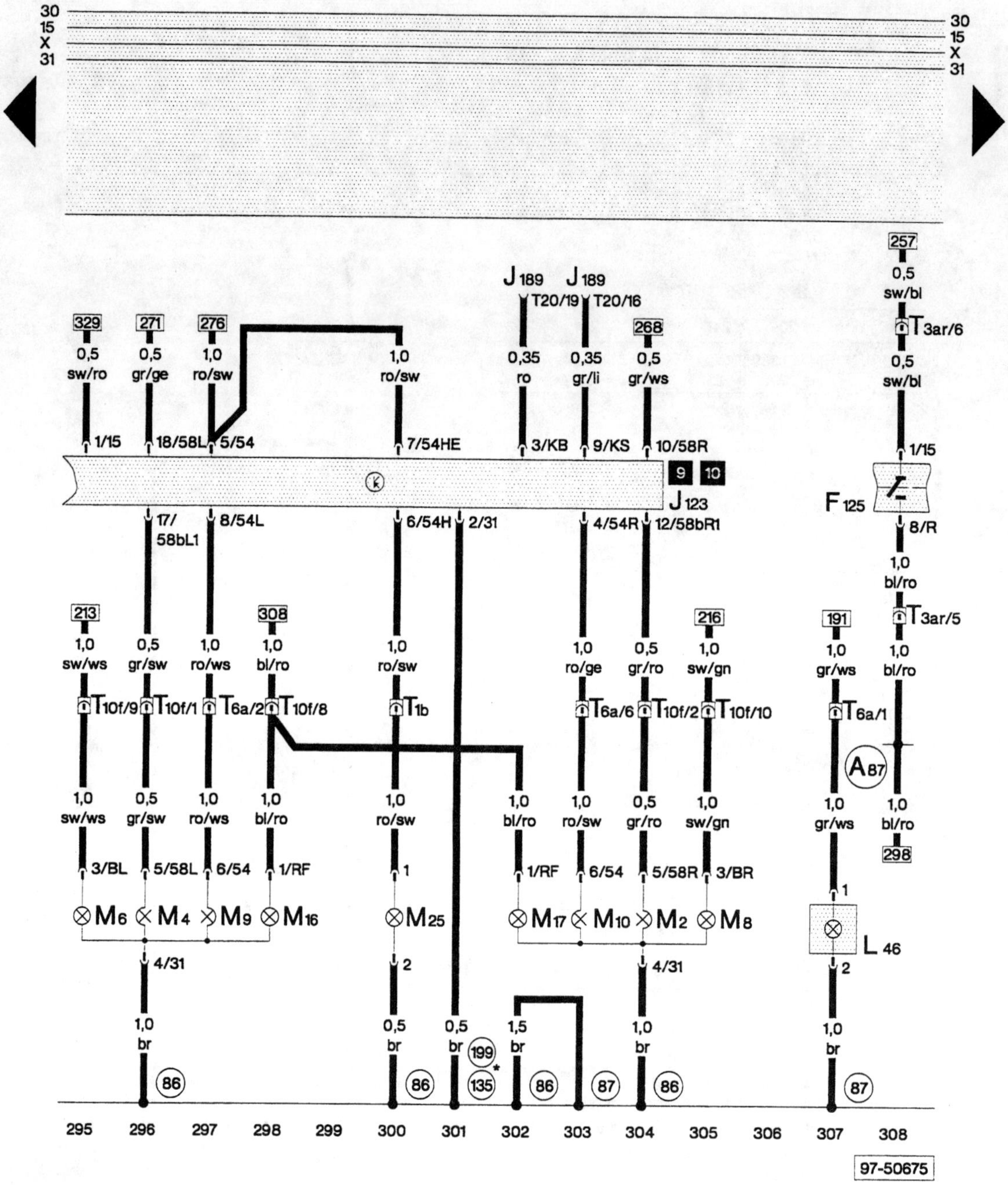

ws = white
sw = black
ro = red
gn = green
br = brown
bl = blue
gr = gray
li = violet
ge = yellow

F 125 - Multi-Function Transmission Range (TR) Switch
J 123 - Lamp Control Module
J 189 - Auto Check System
L 46 - Left Rear Fog Light
M 2 - Right Tail Light
M 4 - Left Tail Light
M 6 - Left Rear Turn Signal Light
M 8 - Right Rear Turn Signal Light
M 9 - Left Brake Light
M 10 - Right Brake Light
M 16 - Left Back-Up Light
M 17 - Right Back-Up Light
M 25 - High-mount Brake Light
T 1 b - Single connector, black, connector station A pillar right
T 3 ar - Connector 3 pin, brown, connector station A pillar, right
T 6 a - Connector 6 pin, blue, connector station A pillar, left
T 10 f - Connector 10 pin, brown, connector station A pillar, left
T 20 - Connector 20 pin, red/white, on instrument cluster

(86) - Ground connection -1-, in rear wiring harness
(87) - Ground connection -2-, in rear wiring harness
(135) - Ground connection -2-, in instrument panel wiring harness
(199) - Ground connection -3-, in instrument panel wiring harness
(A87) - Connector (RF), in instrument panel wiring harness

* - booth is possible

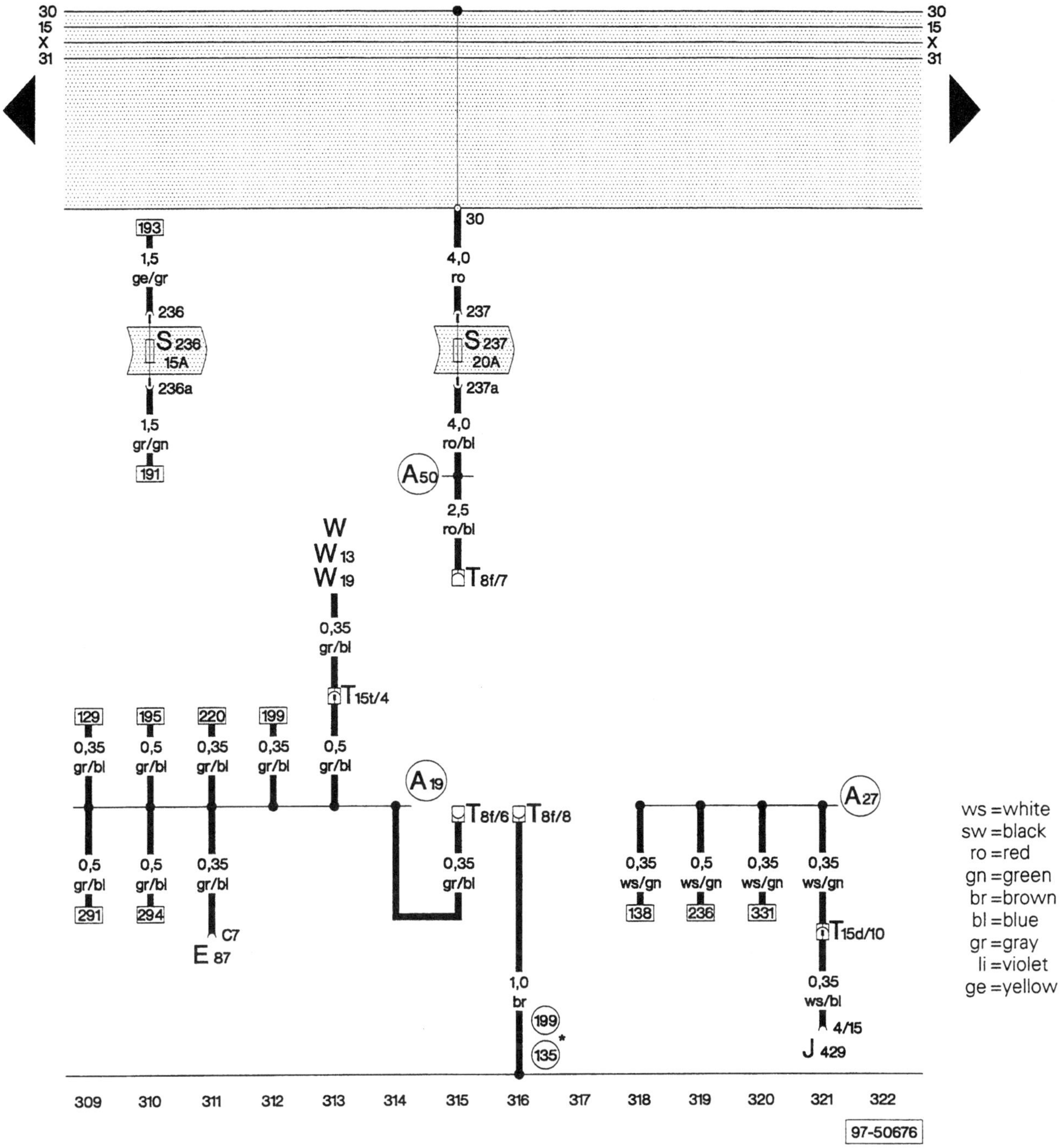

E 87 - A/C Control Head
J 429 - Control module for central locking
S 236 - Fuse
S 237 - Fuse
T 8 f - Connector 8 pin, black, connector radio III
T 15 d - Connector 15 pin, black, connector station A pillar, right
T 15 t - Connector 15 pin, grey, connector station A pillar, left
W - Front Interior Light
W 13 - Right Front Map/Reading Light
W 19 - Left Front Reading Light

(135) - Ground connection -2-, in instrument panel wiring harness
(199) - Ground connection -3-, in instrument panel wiring harness
(A19) - Wire connection (58s), in instrument panel wiring harness
(A27) - Wire connection (vehicle speed signal), in instrument panel wiring harness
(A50) - Plus connection (30as), in instrument panel wiring harness

* - booth is possible

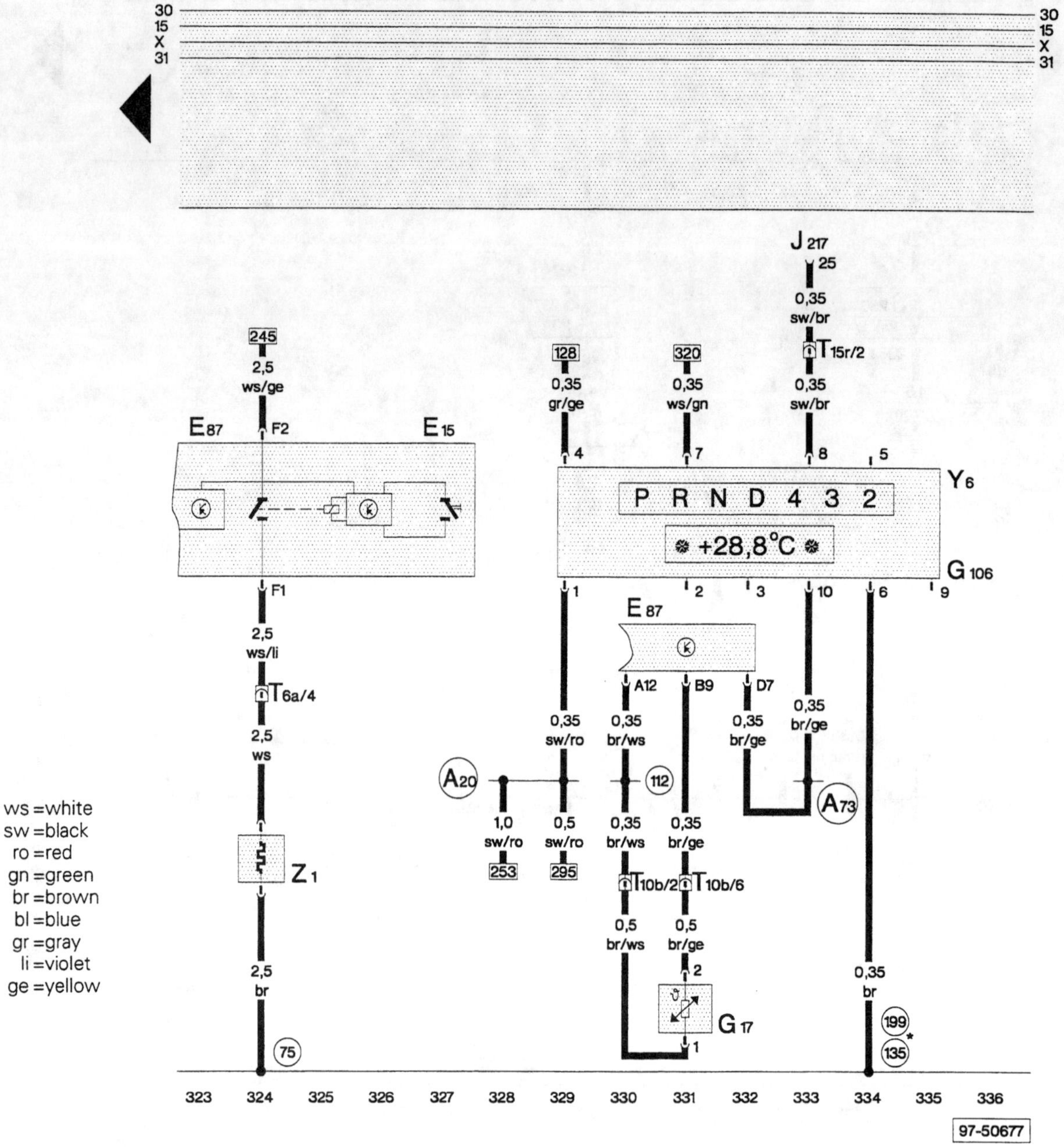

E 15 - Rear Window Defogger Switch
E 87 - A/C Control Head
G 17 - Outside Air Temperature Sensor
G 106 - Outside Air Temperature Display
J 217 - Transmission Control Module (TCM)
T 6 a - Connector 6 pin, blue, connector station A pillar, left
T 10 b - Connector 10 pin, red, connector station A pillar, left
T 15 r - Connector 15 pin, brown, connector station A pillar, right
Y 6 - Transmission Range (TR) Selector Lever Display
Z 1 - Heated rear window

(75) - Ground connection, on right rear pillar
(112) - Ground connection -2-, in A/C wiring harness
(135) - Ground connection -2-, in instrument panel wiring harness
(199) - Ground connection -3-, in instrument panel wiring harness
(A20) - Wire connection (15a), in instrument panel wiring harness
(A73) - Connector (outside temperature indicator), in instrument panel wiring harness

* - booth is possible

Heated Steering Wheel

1999 m. y.

Fuse Panel

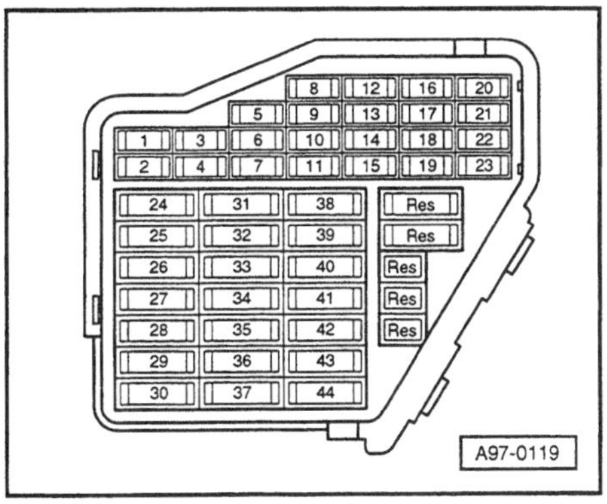

Fuse Colors:

30 A - Green
25 A - White
20 A - Yellow
15 A - Blue
10 A - Red
7,5 A - Brown
5 A - Beige

Starting with fuse position 23, fuses in the fuse holder are identified with 223 in the wiring diagram.

Micro Central Electric Panel

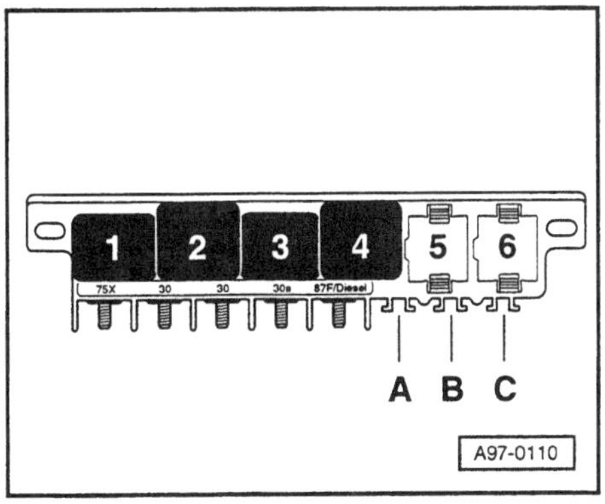

Relay Location:

1 - Dual Horn Relay, J4

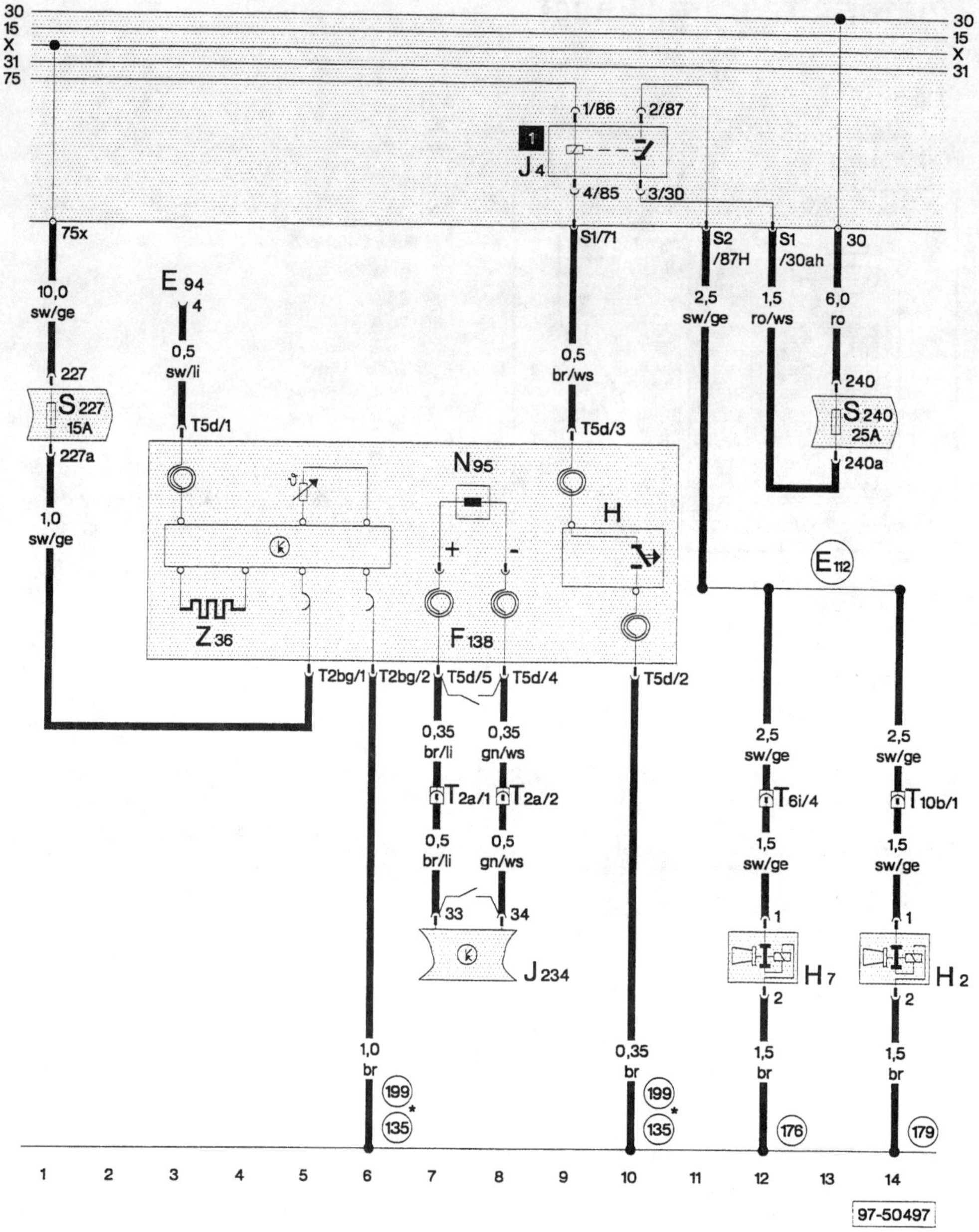

ws = white
sw = black
ro = red
gn = green
br = brown
bl = blue
gr = gray
li = violet
ge = yellow

E 94 - Adjuster for heated driver's seat
F 138 - Airbag Spiral Spring/Return Spring With Slip Ring
H - Horn Button
H 2 - High Tone Horn
H 7 - Low Tone Horn
J 4 - Dual Horn Relay
J 234 - Airbag Control Module
N 95 - Driver's Side Airbag Igniter
S 227 - Fuse
S 240 - Fuse
T 2 a - Connector 2 pin, yellow, connector station A pillar, left
T 2 bg - Connector 2 pin, behind steering column switch cover
T 5 d - Connector 5 pin, behind steering column switch cover
T 6 i - Connector 6 pin, red, connector station A pillar, right
T 10 b - Connector 10 pin, red, connector station A pillar, left
Z 36 - Heated steering wheel

(135) - Ground connection -2-, in instrument panel wiring harness
(176) - Ground connection, in right headlight wiring harness
(179) - Ground connection, in left headlight wiring harness
(199) - Ground connection -3-, in instrument panel wiring harness
(E112) - Connector (Horn - 87h), in instrument panel wiring harness

* - booth is possible

- Bridge closes when disconnecting connector (safety feature)

Automatic Transmission (01V), (01L)

1999 m. y.

Fuse Panel

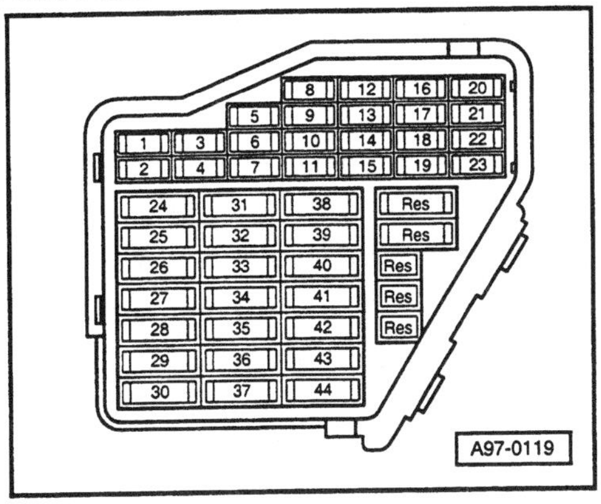

Fuse Colors:

30 A - Green
25 A - White
20 A - Yellow
15 A - Blue
10 A - Red
7,5 A - Brown
5 A - Beige

Starting with fuse position 23, fuses in the fuse holder are identified with 223 in the wiring diagram.

13 - Fold Relay Panel

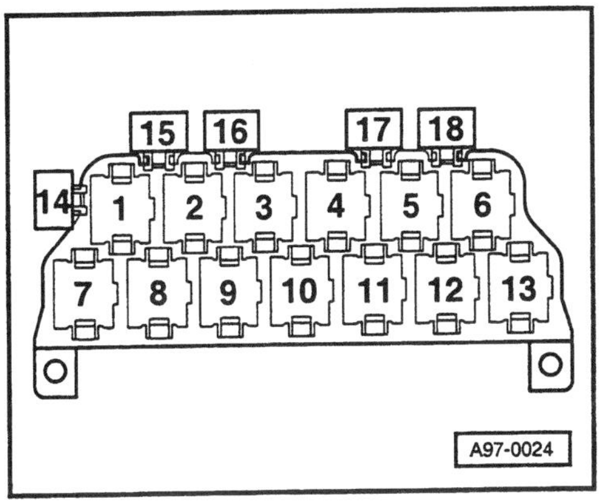

Relay Location:

4 - Starting Interlock Relay, J207

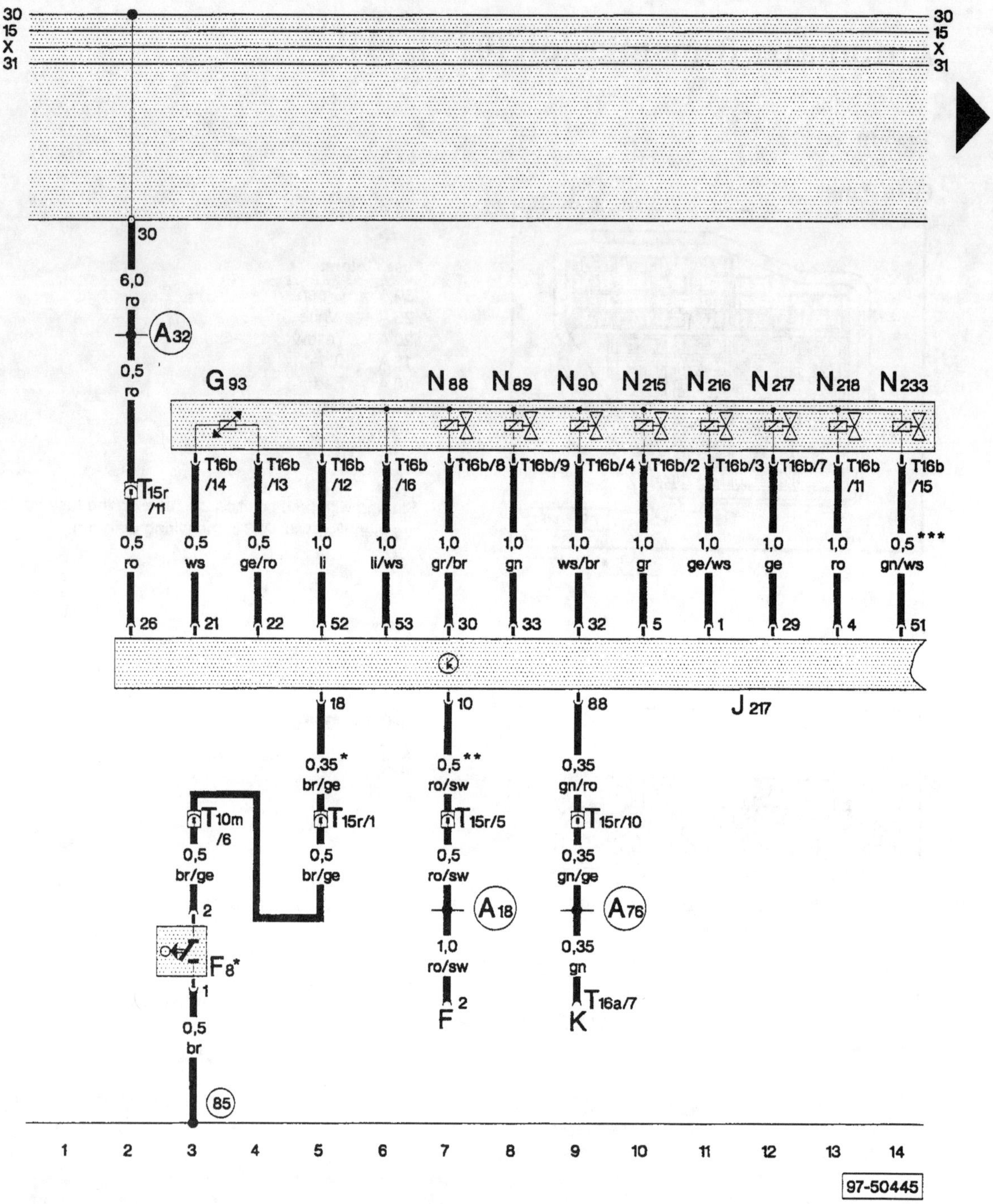

ws = white
sw = black
ro = red
gn = green
br = brown
bl = blue
gr = gray
li = violet
ge = yellow

F - Brake Light Switch
F 8 - Kick Down Switch
G 93 - Transmission Fluid Temperature Sensor
J 217 - Transmission Control Module (TCM)
K - DLC (Data Link Connector) Rapid Data Transfer
N 88 - Solenoid Valve 1
N 89 - Solenoid Valve 2
N 90 - Solenoid Valve 3
N 215 - Pressure control valve -1- for auto. transmission
N 216 - Pressure control valve -2- for auto. transmission
N 217 - Pressure control valve -3- for auto. transmission
N 218 - Pressure control valve -4- for auto. transmission
N 233 - Pressure control valve -5- for auto. transmission
T 10 m - Connector 10 pin, blue, connector station electronic box plenum chamber
T 15 r - Connector 15 pin, brown, connector station A pillar, right
T 16 a - Connector 16 pin, data link connector
T 16 b - Connector 16 pin, on transmission

(85) - Ground connection -1-, in engine compartment wiring harness
(A18) - Wire connection (54), in instrument panel wiring harness
(A32) - Plus connection (30), in instrument panel wiring harness
(A76) - Connector (K-diagnosis wire), in instrument panel wiring harness

* - Vehicles with mechanical throttle actuation
** - only automatic transmission 01V
*** - only automatic transmission 01L

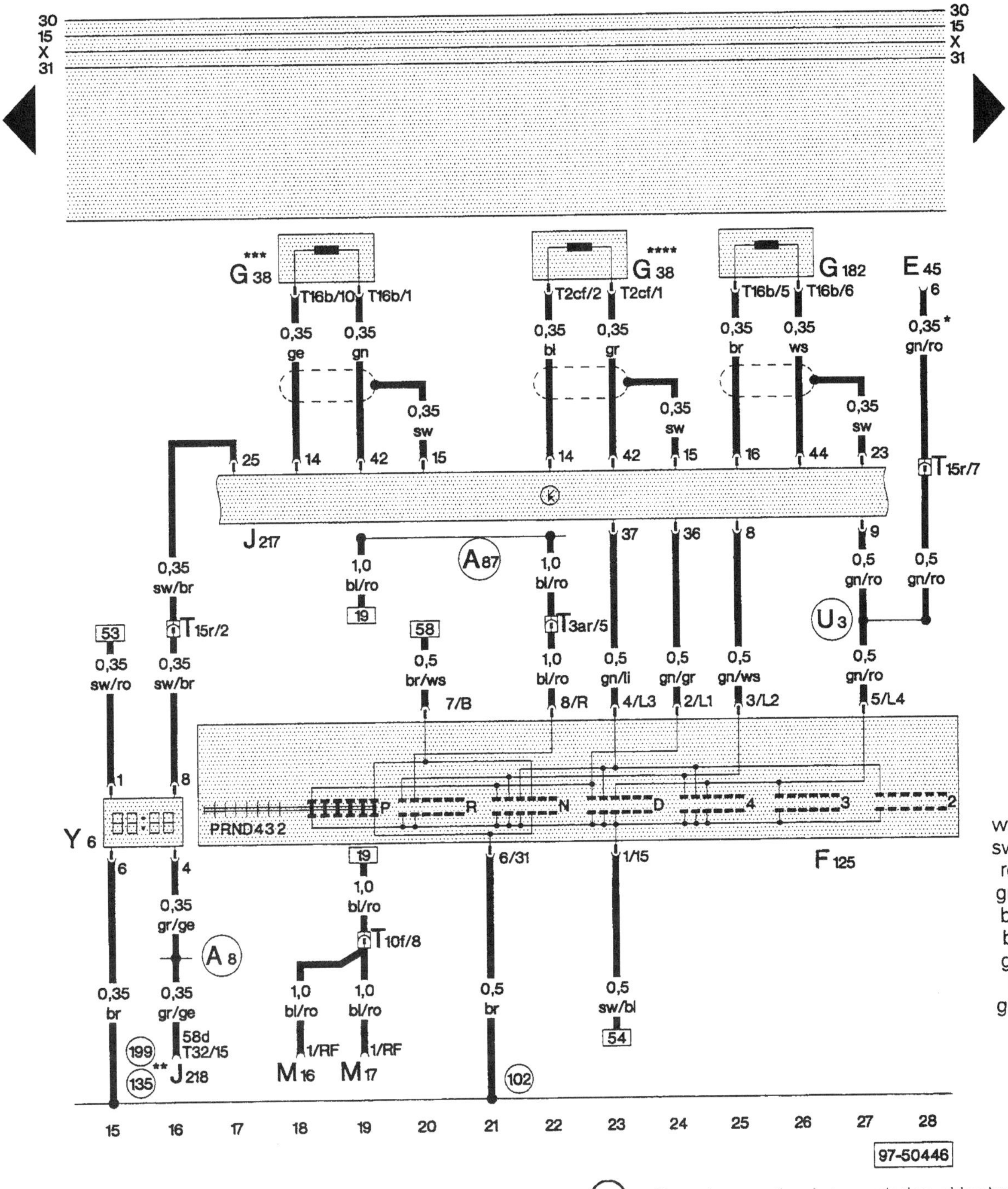

ws = white
sw = black
ro = red
gn = green
br = brown
bl = blue
gr = gray
li = violet
ge = yellow

E 45 - Cruise Control Switch
F 125 - Multi-Function Transmission Range (TR) Switch
G 38 - Transmission Vehicle Speed Sensor (VSS)
G 182 - Sensor for transmission RPM
J 217 - Transmission Control Module (TCM)
J 218 - Instrument Cluster Combination Processor
M 16 - Left Back-Up Light
M 17 - Right Back-Up Light
T 2 cf - Connector 2 pin, black, on transmission
T 3 ar - Connector 3 pin, brown, connector station A pillar, right
T 10 f - Connector 10 pin, brown, connector station A pillar, left
T 15 r - Connector 15 pin, brown, connector station A pillar, right
T 16 b - Connector 16 pin, on transmission
T 32 - Connector 32 pin, blue, on instrument cluster
Y 6 - Transmission Range (TR) Selector Lever Display

(102) - Ground connection, in transmission wiring harness
(135) - Ground connection -2-, in instrument panel wiring harness
(199) - Ground connection -3-, in instrument panel wiring harness
(A8) - Plus connection (58d), in instrument panel wiring harness (Display illumination)
(A87) - Connector (RF), in instrument panel wiring harness
(U3) - Wire connection -1-, in automatic transmission wiring harness

* - Vehicles with cruise control
** - booth is possible
*** - only automatic transmission 01V
**** - only automatic transmission 01L

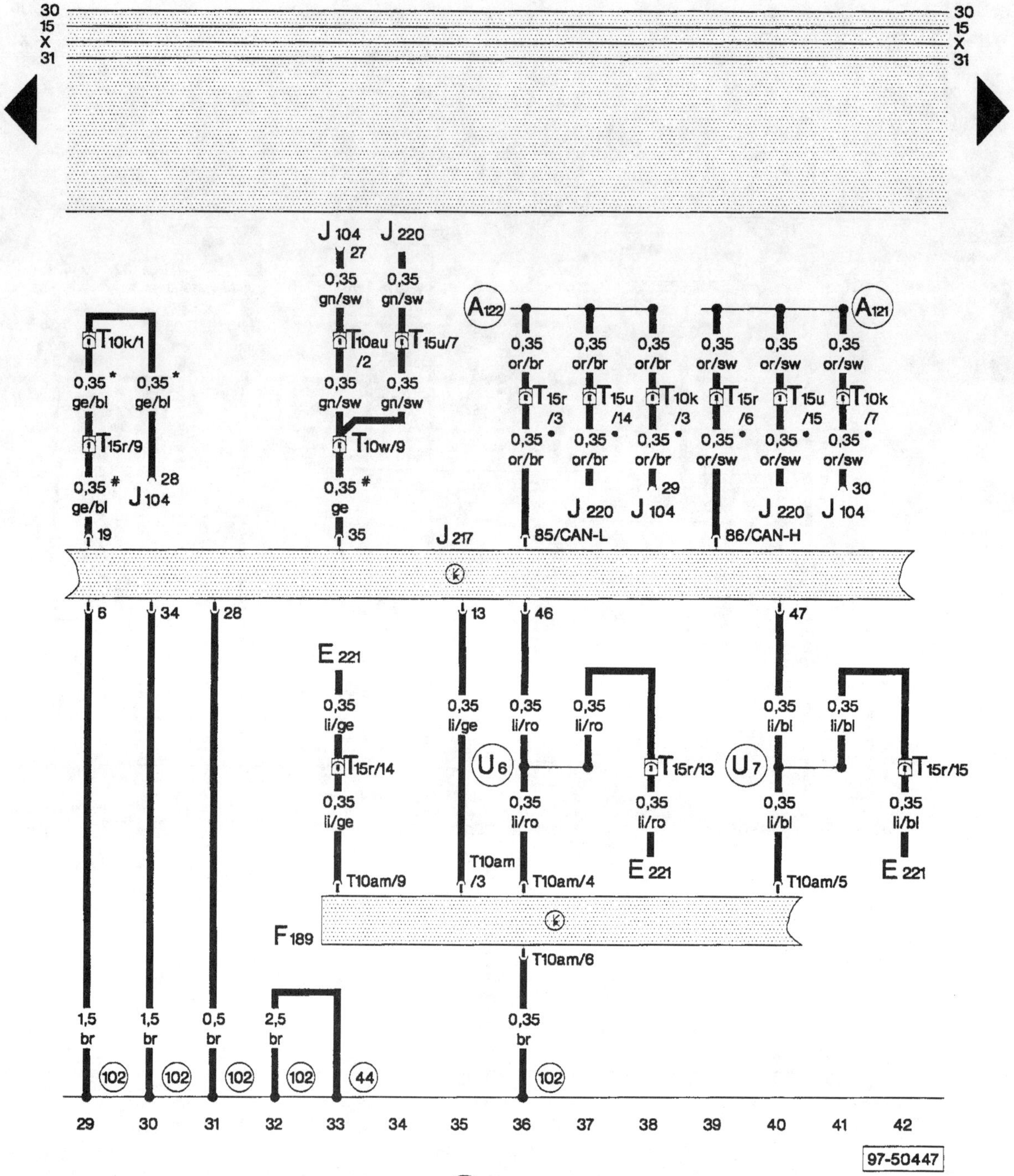

ws = white
sw = black
ro = red
gn = green
br = brown
bl = blue
gr = gray
li = violet
ge = yellow
or = orange

E 221 - Control unit in steering wheel
F 189 - Tiptronic Switch
J 104 - ABS Control Module (w/EDL)
J 217 - Transmission Control Module (TCM)
J 220 - Motronic Engine Control Module (ECM)
T 10 k - Connector 10 pin, orange, connector station A pillar, left
T 10 w - Connector 10 pin, blue, connector station A pillar, right
T 10am - Connector 10 pin, on selector lever
T 10au - Connector 10 pin, grey, connector station A pillar, right
T 15 r - Connector 15 pin, brown, connector station A pillar, right
T 15 u - Connector 15 pin, red, connector station electronic box plenum chamber

(44) - Ground connection, on left A-pillar, lower part
(102) - Ground connection, in transmission wiring harness
(A121) - Connection (high bus) in instrument panel wiring harness
(A122) - Connection (low bus) in instrument panel wiring harness
(U6) - Wire connection -2-, in automatic transmission wiring harness
(U7) - Wire connection -3-, in automatic transmission wiring harness

• - CAN-Bus
- Vehicles without CAN-Bus
* - Vehicles with front wheel drive

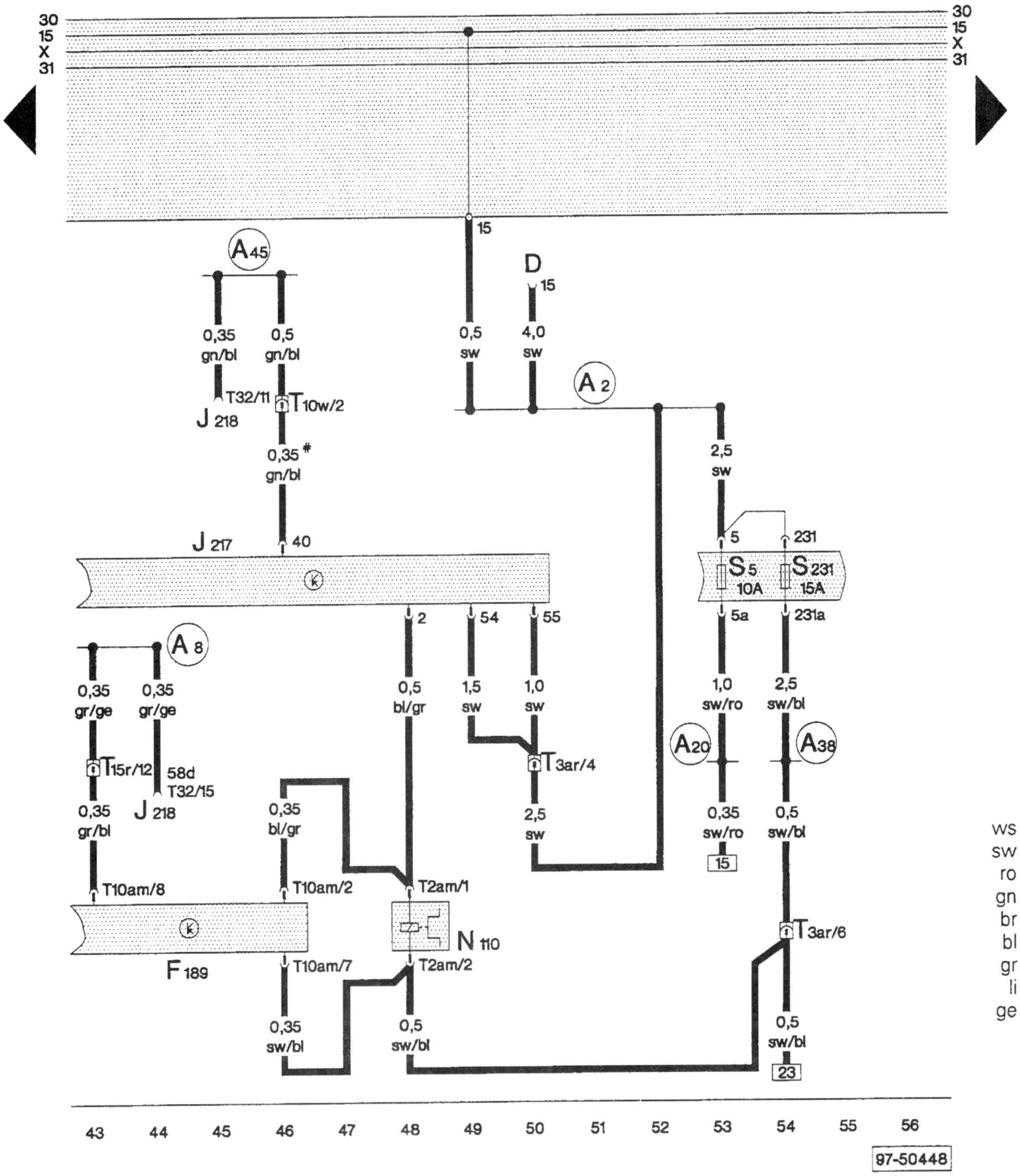

ws = white
sw = black
ro = red
gn = green
br = brown
bl = blue
gr = gray
li = violet
ge = yellow

D - Ignition/Starter Switch
F 189 - Tiptronic Switch
J 217 - Transmission Control Module (TCM)
J 218 - Instrument Cluster Combination Processor
N 110 - Shift Lock Solenoid
S 5 - Fuse
S 231 - Fuse
T 2 am - Connector 2 pin, black, on selector lever
T 3 ar - Connector 3 pin, brown, connector station A pillar, right
T 10 w - Connector 10 pin, blue, connector station A pillar, right
T 10am - Connector 10 pin, on selector lever
T 15 r - Connector 15 pin, brown, connector station A pillar, right
T 32 - Connector 32 pin, blue, on instrument cluster

(A2) - Plus connection (15), in instrument panel wiring harness
(A8) - Plus connection (58d), in instrument panel wiring harness
(A20) - Wire connection (15a), in instrument panel wiring harness
(A38) - Plus connection (15a (l)), in instrument panel wiring harness
(A45) - Wire connection (RPM-signal), in instrument panel wiring harness

- Vehicles without CAN-Bus

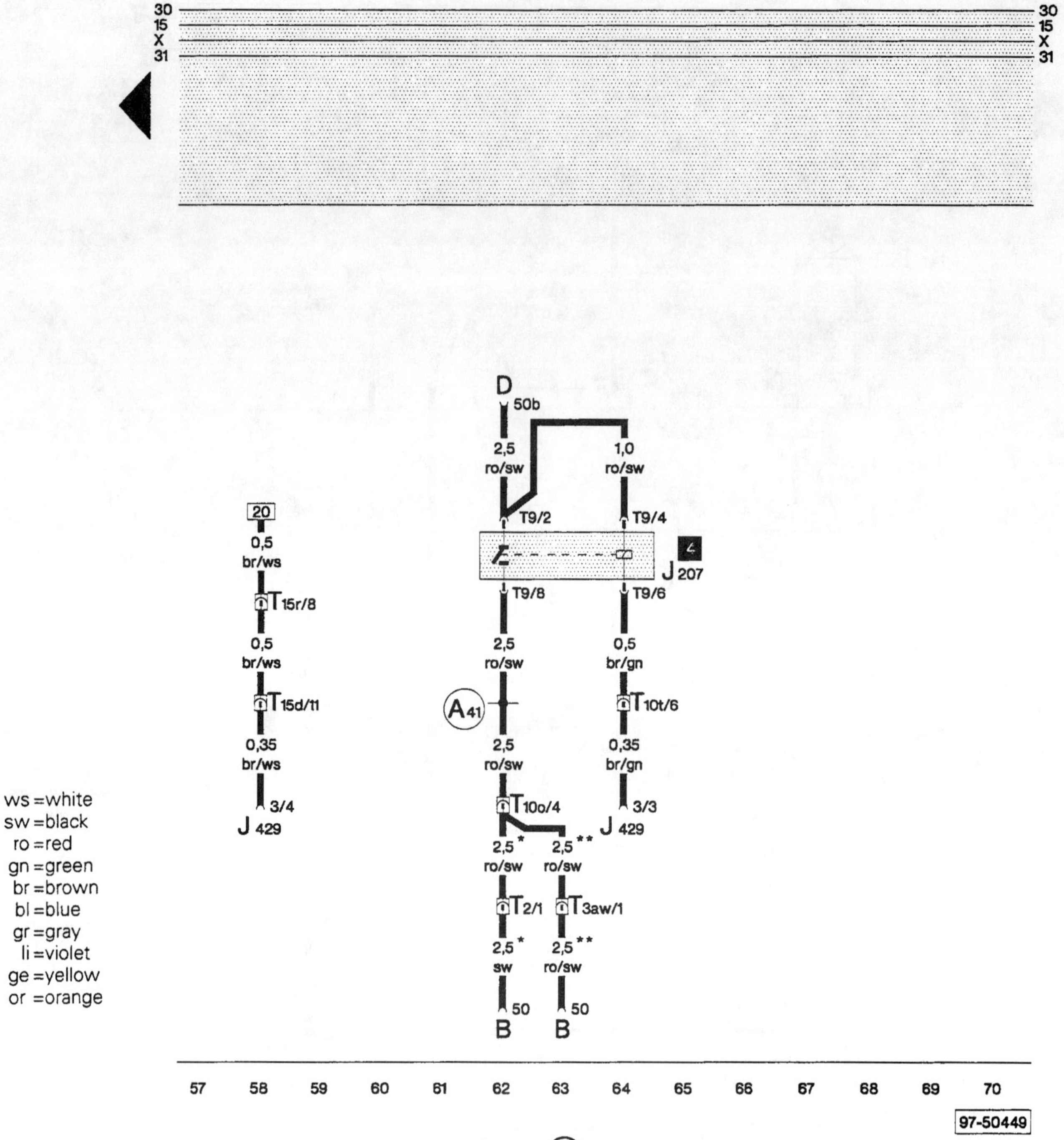

B - Starter
D - Ignition/Starter Switch
J 207 - Starting Interlock Relay
J 429 - Control module for central locking
T 2 - Connector 2 pin, grey, in engine compartment, right
T 3 aw - Connector 3 pin, black, in engine compartment, right
T 9 - Connector 9 pin, brown, on starting interlock relay
T 10 o - Connector 10 pin, brown, connector station electronic box plenum chamber
T 10 t - Connector 10 pin, orange, connector station A pillar, right
T 15 d - Connector 15 pin, black, connector station A pillar, right
T 15 r - Connector 15 pin, brown, connector station A pillar, right

(A41) - Plus connection (50), in instrument panel wiring harness

* - Vehicles with 6 cylinder engine
** - Vehicles with 8 cylinder engine

Central Locking System/Anti Theft Alarm System/ Ultra-Sound Interior Monitor

1999 m. y.

Fuse Panel

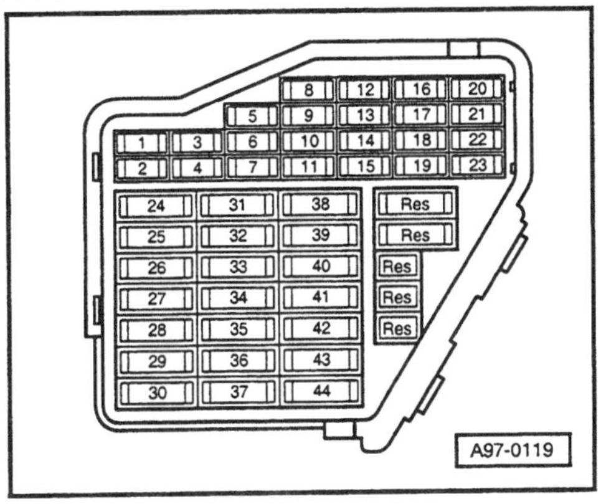

Fuse Colors:

30 A - Green
25 A - White
20 A - Yellow
15 A - Blue
10 A - Red
7,5 A - Brown
5 A - Beige

Starting with fuse position 23, fuses in the fuse holder are identified with 223 in the wiring diagram.

13 - Fold Relay Panel

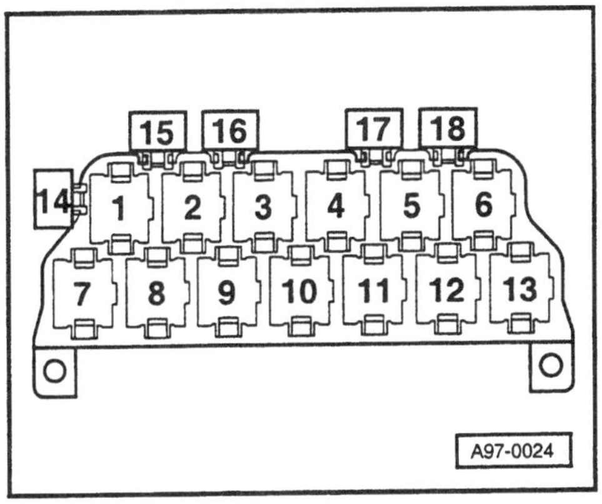

Relay Location:

4 - Starting Interlock Relay, J207

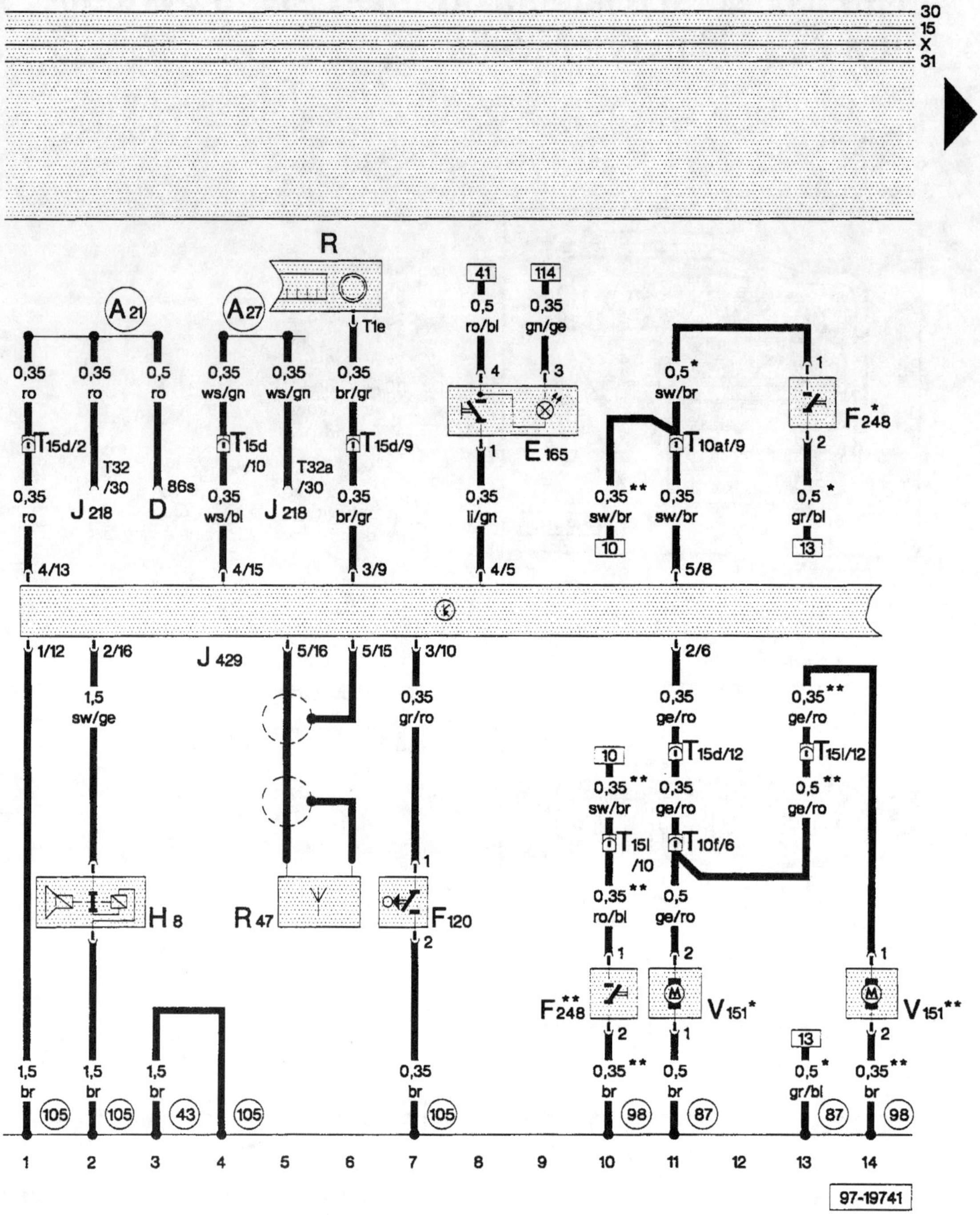

ws = white
sw = black
ro = red
gn = green
br = brown
bl = blue
gr = gray
li = violet
ge = yellow

- D - Ignition/Starter Switch
- E 165 - Trunk Lid Release Switch
- F 120 - Hood Alarm Switch
- F 248 - Switch for unlocking, rear lid lock cylinder
- H 8 - Alarm Horn
- J 218 - Instrument Cluster Combination Processor
- J 429 - Control module for central locking
- R - Radio
- R 47 - Antenna wire for central locking and anti-theft warning system
- T 1 e - Single connector, on Radio
- T 10 f - Connector 10 pin, brown, connector station A pillar, left
- T 15 d - Connector 15 pin, black, connector station A pillar, right
- T 15 l - Connector 15 pin, black, in luggage compartment
- T 32 - Connector 32 pin, blue, on instrument cluster
- T 32 a - Connector 32 pin, green, on instrument cluster
- V 151 - Motor for remote unlock rear lid

- (43) - Ground connection, on right A-pillar, lower part
- (87) - Ground connection -2-, in rear wiring harness
- (98) - Ground connection, in rear lid wiring harness
- (105) - Ground connection -1-, in central locking system wiring harness
- (A21) - Wire connection (86s), in instrument panel wiring harness
- (A27) - Wire connection (vehicle speed signal), in instrument panel wiring harness

- * - only sedan
- ** - only wagon

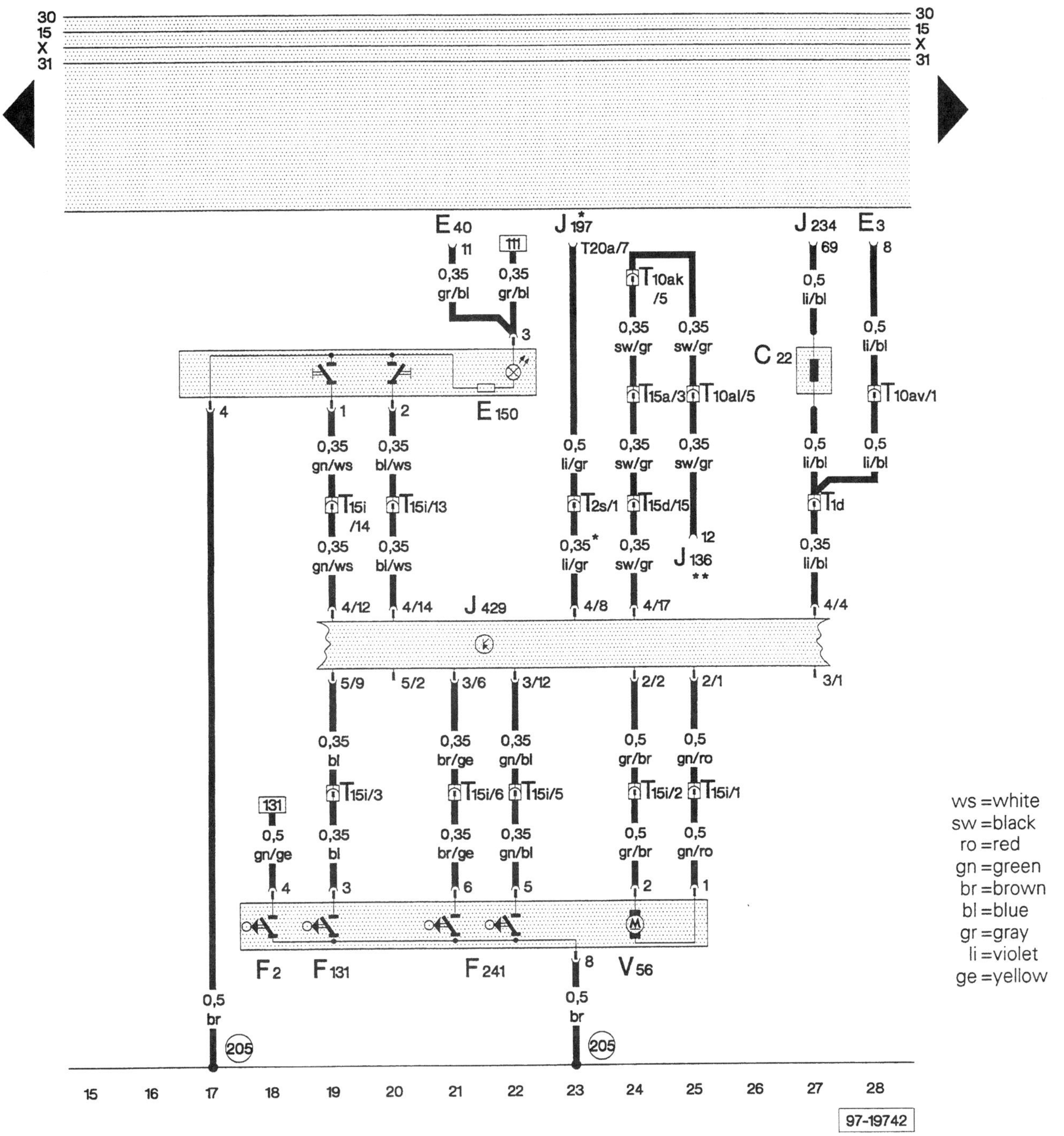

C 22 - Interference filter
E 3 - Emergency Flasher Switch
E 40 - Left Front Window Switch
E 150 - Switch for interior lock, driver side
F 2 - Door Contact Switch - Driver's side
F 131 - Left Front Central Locking System Activator/Switch
F 241 - Switch in door lock, driver's side
J 136 - Memory Seat Control Module
J 197 - Level Control System Control Module
J 234 - Airbag Control Module
J 429 - Control module for central locking
T 1 d - Single connector, brown, connector station A pillar right
T 2 s - Connector 2 pin, black, connector station A pillar, right
T 10 ak - Connector 10 pin, green, connector station A pillar, left
T 10 al - Connector 10 pin, brown, below driver's seat
T 10 av - Connector 10 pin, yellow, connector station A pillar, right
T 15 a - Connector 15 pin, black, connector station A pillar, left
T 15 d - Connector 15 pin, black, connector station A pillar, right
T 15 i - Connector 15 pin, brown, connector station A pillar, left
T 20 a - Connector 20 pin, red, on Control Module for Level Control
V 56 - Driver's Door Central Locking System Motor

(205) - Ground connection, in driver's door wiring harness

* - Vehicles with Level Control System
** - Vehicles with Memory Seat

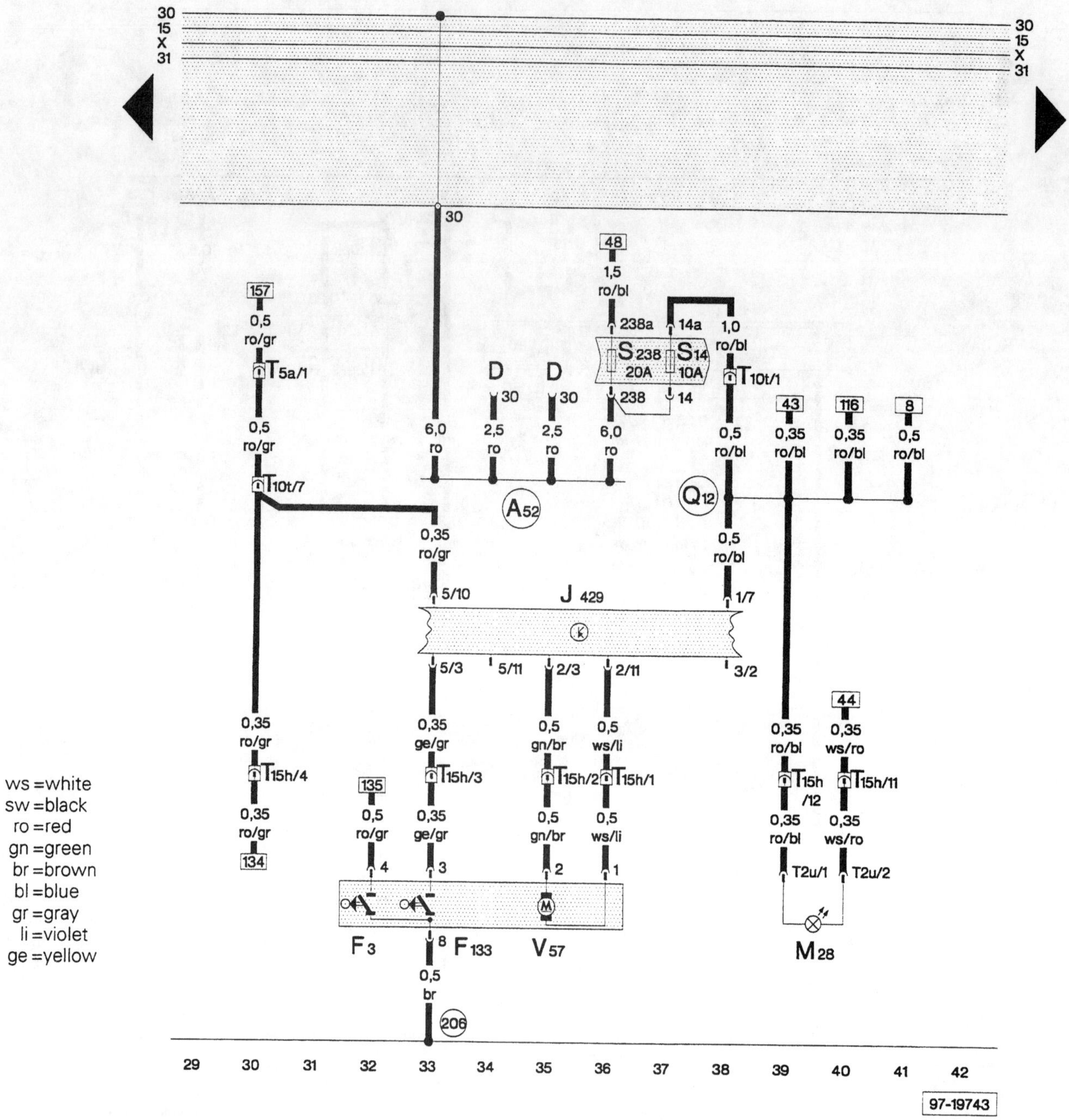

D - Ignition/Starter Switch
F 3 - Door Contact Switch - Passenger side
F 133 - Right Front Central Locking System Activator/Switch
J 429 - Control module for central locking
M 28 - Right Door Warning Light
S 14 - Fuse
S 238 - Fuse
T 2 u - Connector 2 pin, black, in front passenger's door)
T 5 a - Connector 5 pin, black, behind instrument panel, right
T 10 t - Connector 10 pin, orange, connector station A pillar, right
T 15 h - Connector 15 pin, brown, connector station A pillar, right
V 57 - Passenger's Door Central Locking System Motor

(206) - Ground connection, in passenger's door wiring harness
(A52) - Plus connection (30), in instrument panel wiring harness
(Q12) - Plus connection (30a), in power window/central locking system and door contact switch wiring harness

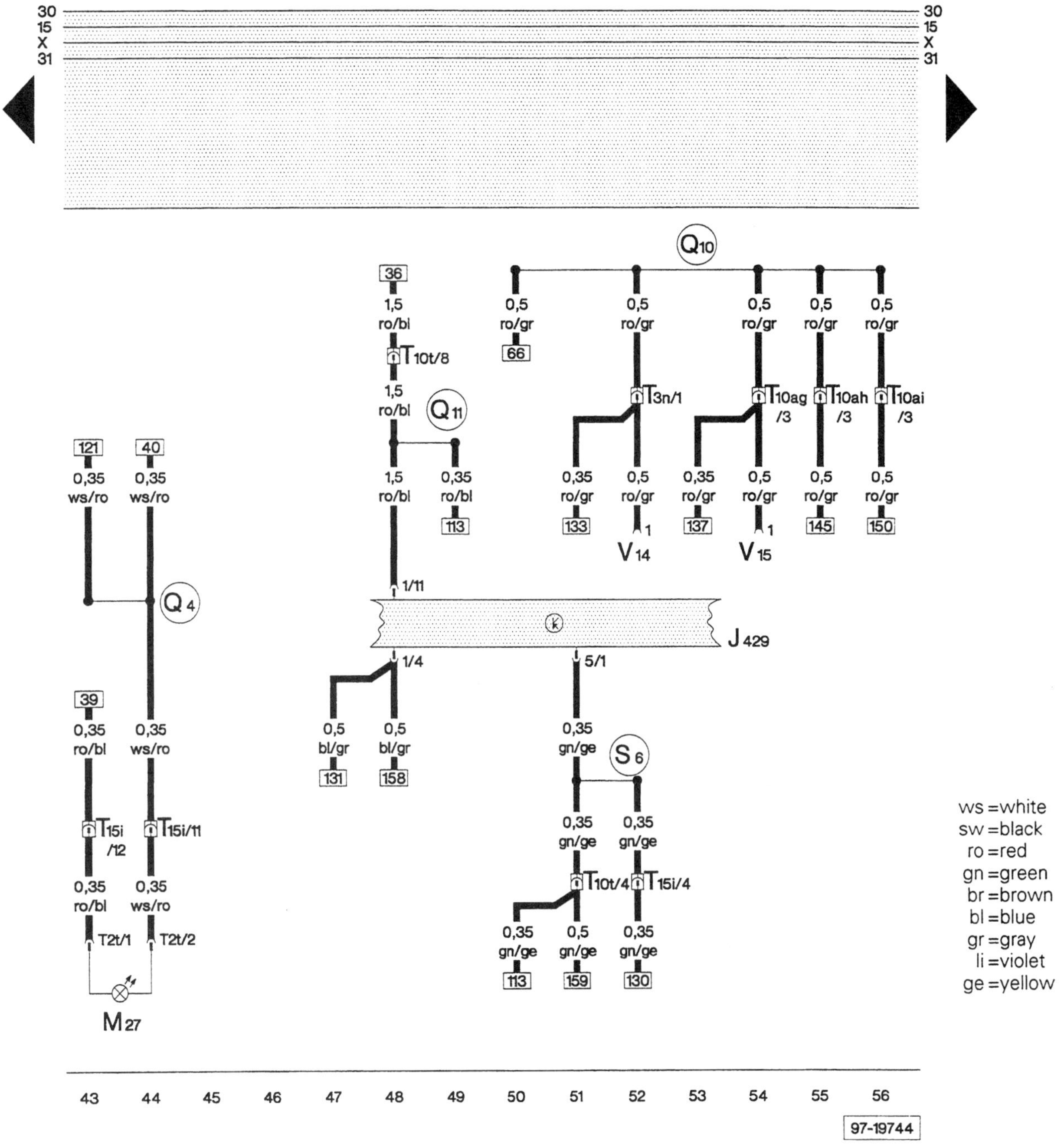

J 429 - Control module for central locking
M 27 - Left Door Warning Light
T 2 t - Connector 2 pin, black, in driver's door
T 3 n - Connector 3 pin, red, connector station A pillar, left
T 10 t - Connector 10 pin, orange, connector station A pillar, right
T 10 ag - Connector 10 pin, red, connector station A pillar, right
T 10 ah - Connector 10 pin, red, in B pillar, left
T 10 ai - Connector 10 pin, red, in B pillar, right
T 15 i - Connector 15 pin, brown, connector station A pillar, left
V 14 - Left Window Motor
V 15 - Right Window Motor

Q4 - Plus connection (30), in power window/central locking system and door contact switch wiring harness
Q10 - Plus connection (87), in power window/central locking system and door contact switch wiring harness
Q11 - Plus connection (30a), in power window/central locking system and door contact switch wiring harness
S6 - Wire connection (Pin 5/1), in central locking system wiring harness

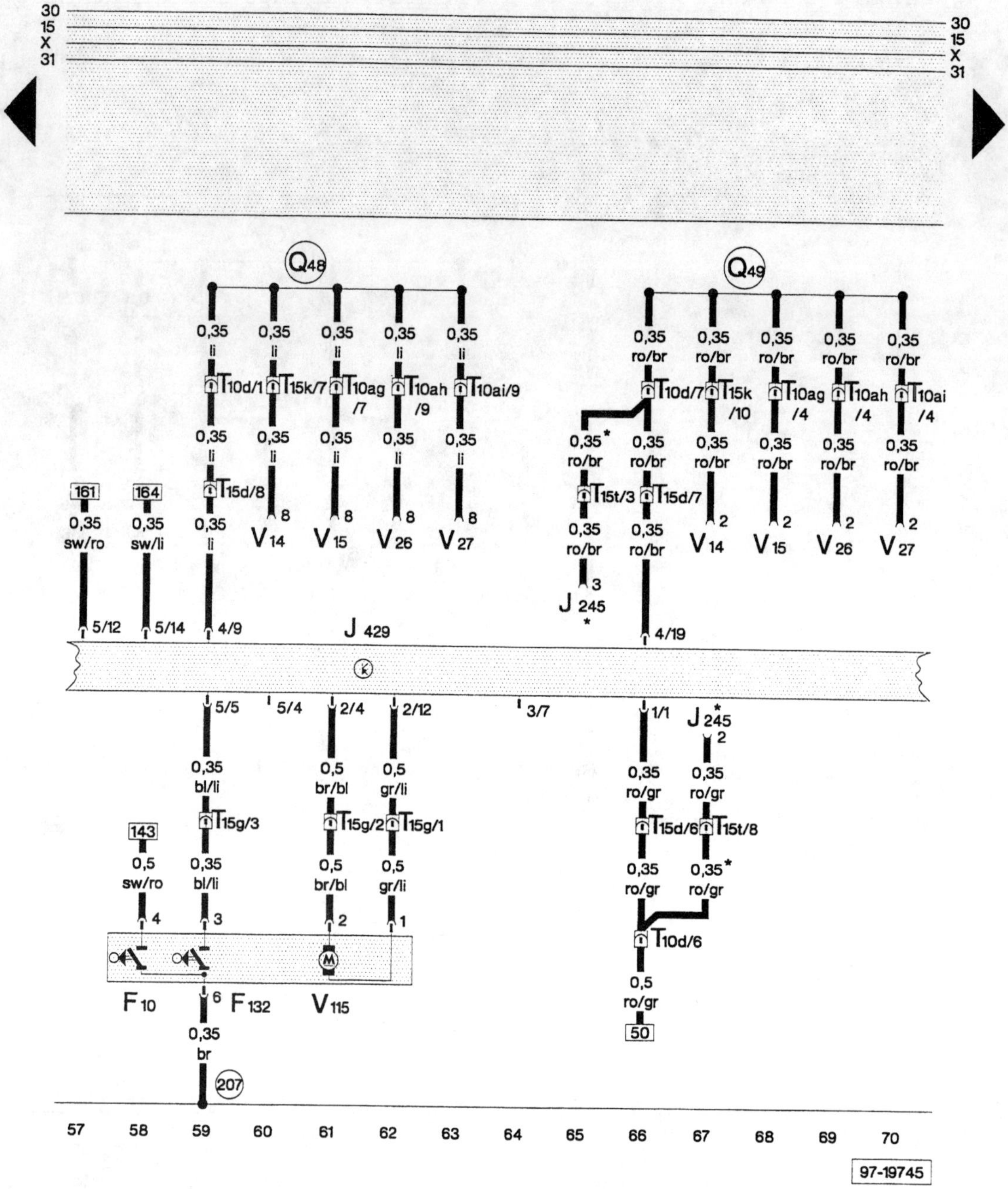

ws = white
sw = black
ro = red
gn = green
br = brown
bl = blue
gr = gray
li = violet
ge = yellow

F 10 - Left Rear Door Contact Switch
F 132 - Left Rear Central Locking System Activator/Switch
J 245 - Power Sunroof Control Module
J 429 - Control module for central locking
T 10 d - Connector 10 pin, green, connector station A pillar, left
T 10 ag - Connector 10 pin, red, connector station A pillar, right
T 10 ah - Connector 10 pin, red, in B pillar, left
T 10 ai - Connector 10 pin, red, in B pillar, right
T 15 d - Connector 15 pin, black, connector station A pillar, right
T 15 g - Connector 15 pin, blue, in connector station B pillar, left
T 15 k - Connector 15 pin, red, connector station A pillar, left
T 15 t - Connector 15 pin, grey, connector station A pillar, left
V 14 - Left Window Motor
V 15 - Right Window Motor
V 26 - Motor for Window Regulator, rear, left
V 27 - Motor for Window Regulator, rear, right
V 115 - Left Rear Central Locking System Motor

(207) - Ground connection, in left rear door wiring harness
(Q48) - Connection (open), in wiring harness interior
(Q49) - Connection (closed), in wiring harness interior

* - Vehicles with Power Sunroof

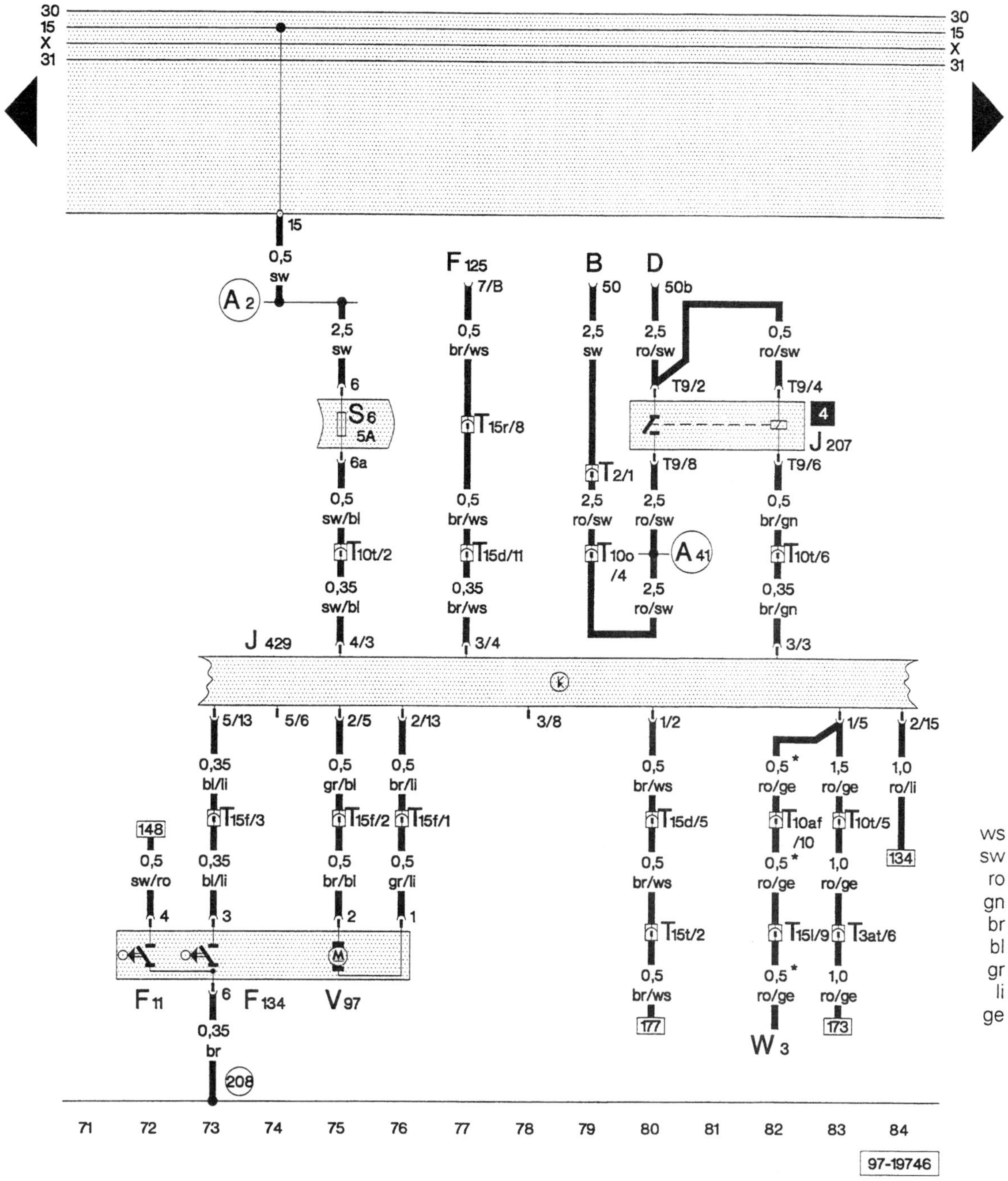

ws = white
sw = black
ro = red
gn = green
br = brown
bl = blue
gr = gray
li = violet
ge = yellow

- B - Starter
- D - Ignition/Starter Switch
- F 11 - Right Rear Door Contact Switch
- F 125 - Multi-Function Transmission Range (TR) Switch
- F 134 - Right Rear Central Locking System Activator/Switch
- J 207 - Starting Interlock Relay
- J 429 - Control module for central locking
- S 6 - Fuse
- T 2 - Connector 2 pin, grey, in engine compartment, right
- T 3 at - Connector 3 pin, grey, connector station A pillar, left
- T 9 - Connector 9 pin, brown, on starting interlock relay
- T 10 o - Connector 10 pin, brown, connector station electronic box plenum chamber
- T 10 t - Connector 10 pin, orange, connector station A pillar, right
- T 10 af - Connector 10 pin, violet, connector station A pillar, left
- T 15 d - Connector 15 pin, black, connector station A pillar, right
- T 15 f - Connector 15 pin, blue, in connector station B pillar, right
- T 15 l - Connector 15 pin, black, in luggage compartment
- T 15 r - Connector 15 pin, brown, connector station A pillar, right
- V 97 - Double Cab Right Rear Door Central Locking System Motor
- W 3 - Luggage Compartment Light

- (208) - Ground connection, in right rear door wiring harness
- (A2) - Plus connection (15), in instrument panel wiring harness
- (A41) - Plus connection (50), in instrument panel wiring harness

- * - only wagon

Lock Unit for Central Locking, rear, right, Starting Interlock Relay

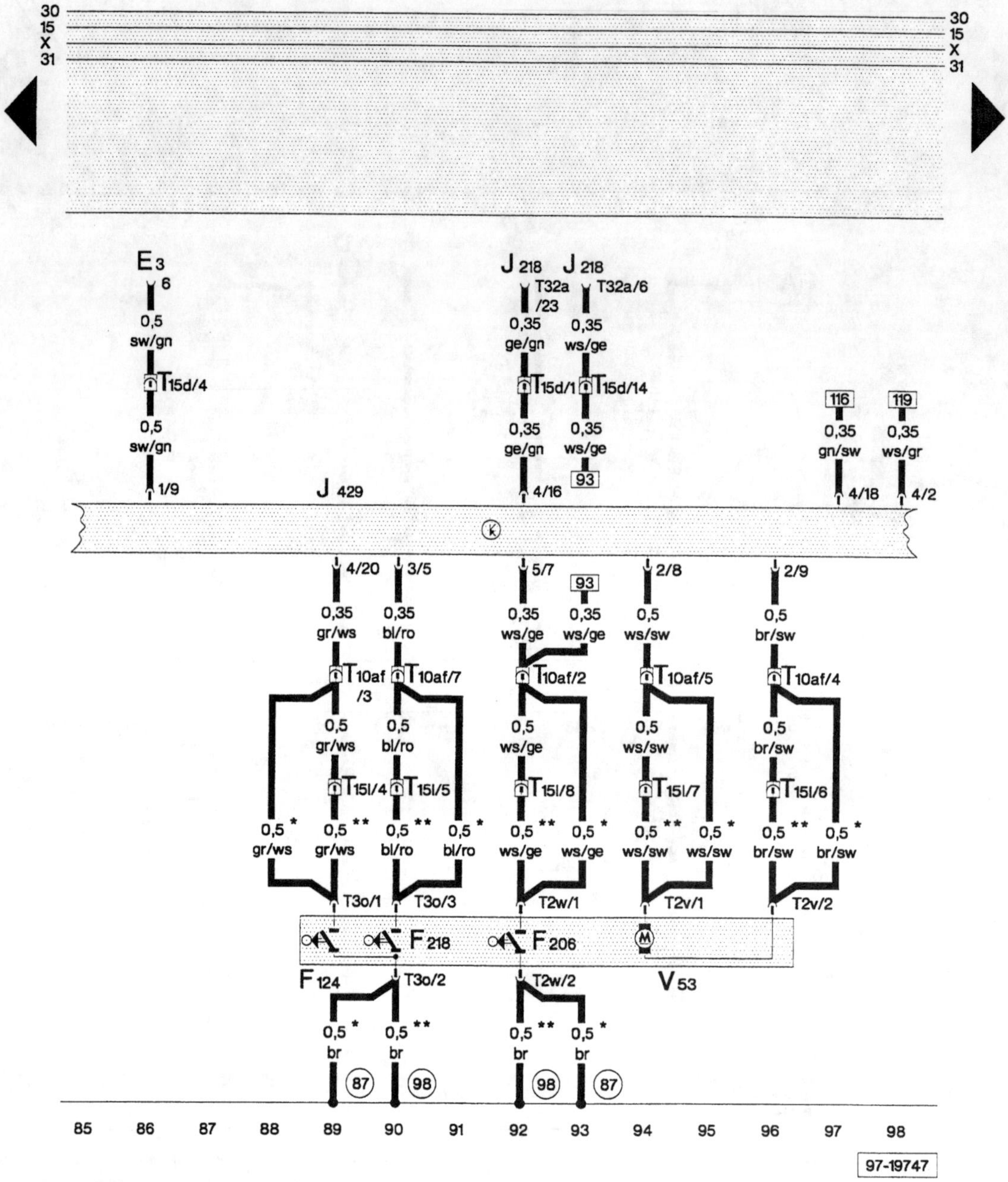

ws = white
sw = black
ro = red
gn = green
br = brown
bl = blue
gr = gray
li = violet
ge = yellow

E 3 - Emergency Flasher Switch
F 124 - Trunk Lock Alarm/Central Locking Switch
F 206 - Trunk Lid Closed Position Switch
F 218 - Switch for central locking, trunk lid
J 218 - Instrument Cluster Combination Processor
J 429 - Control module for central locking
T 2 v - Connector 2 pin, black, luggage compartment
T 2 w - Connector 2 pin, grey, luggage compartment
T 3 o - Connector 3 pin, black, in rear lid
T 10 af - Connector 10 pin, violet, connector station A pillar, left
T 15 d - Connector 15 pin, black, connector station A pillar, right
T 15 l - Connector 15 pin, black, in luggage compartment
T 32 a - Connector 32 pin, green, on instrument cluster
V 53 - Decklid Central Locking System Motor

(87) - Ground connection -2-, in rear wiring harness
(98) - Ground connection, in rear lid wiring harness

* - only sedan
** - only wagon

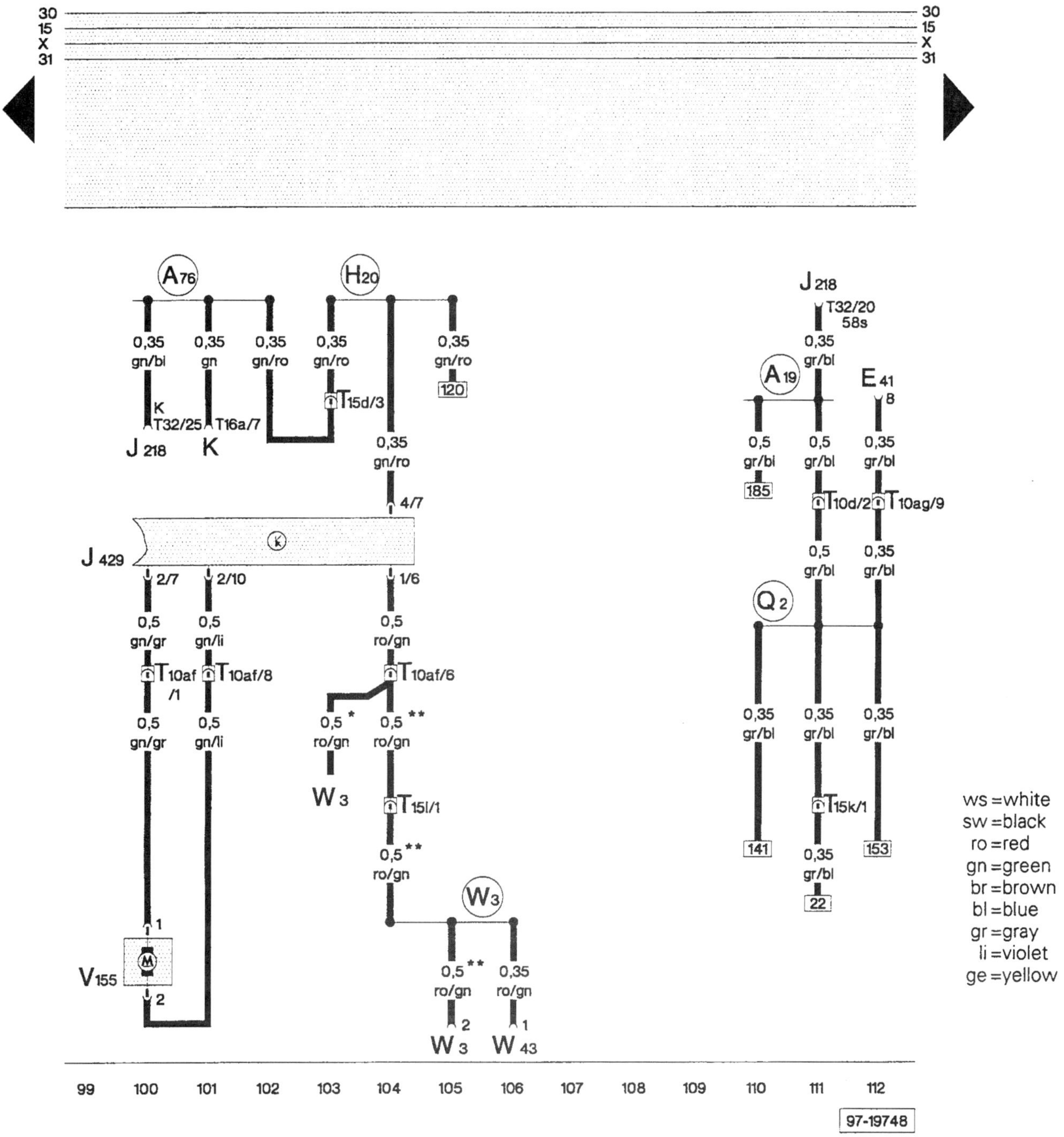

- E 41 - Right Front Window Switch
- J 218 - Instrument Cluster Combination Processor
- J 429 - Control module for central locking
- K - DLC (Data Link Connector) Rapid Data Transfer
- T 10 d - Connector 10 pin, green, connector station A pillar, left
- T 10 af - Connector 10 pin, violet, connector station A pillar, left
- T 10 ag - Connector 10 pin, red, connector station A pillar, right
- T 15 d - Connector 15 pin, black, connector station A pillar, right
- T 15 k - Connector 15 pin, red, connector station A pillar, left
- T 15 l - Connector 15 pin, black, in luggage compartment
- T 16 a - Connector 16 pin, data link connector
- T 32 - Connector 32 pin, blue, on instrument cluster
- V 155 - Motor for fuel tank lid unlock
- W 3 - Luggage Compartment Light
- W 43 - Rear courtesy lamp, in rear lid

- (A19) - Wire connection (58s), in instrument panel wiring harness
- (A76) - Connector (K-diagnosis wire), in instrument panel wiring harness
- (H20) - Wire connection -1-, in On-Board Diagnostic (OBD) wiring harness
- (Q2) - Plus connection (58), in power window wiring harness
- (W3) - Wire connection, in rear wiring harness

* - only sedan
** - only wagon

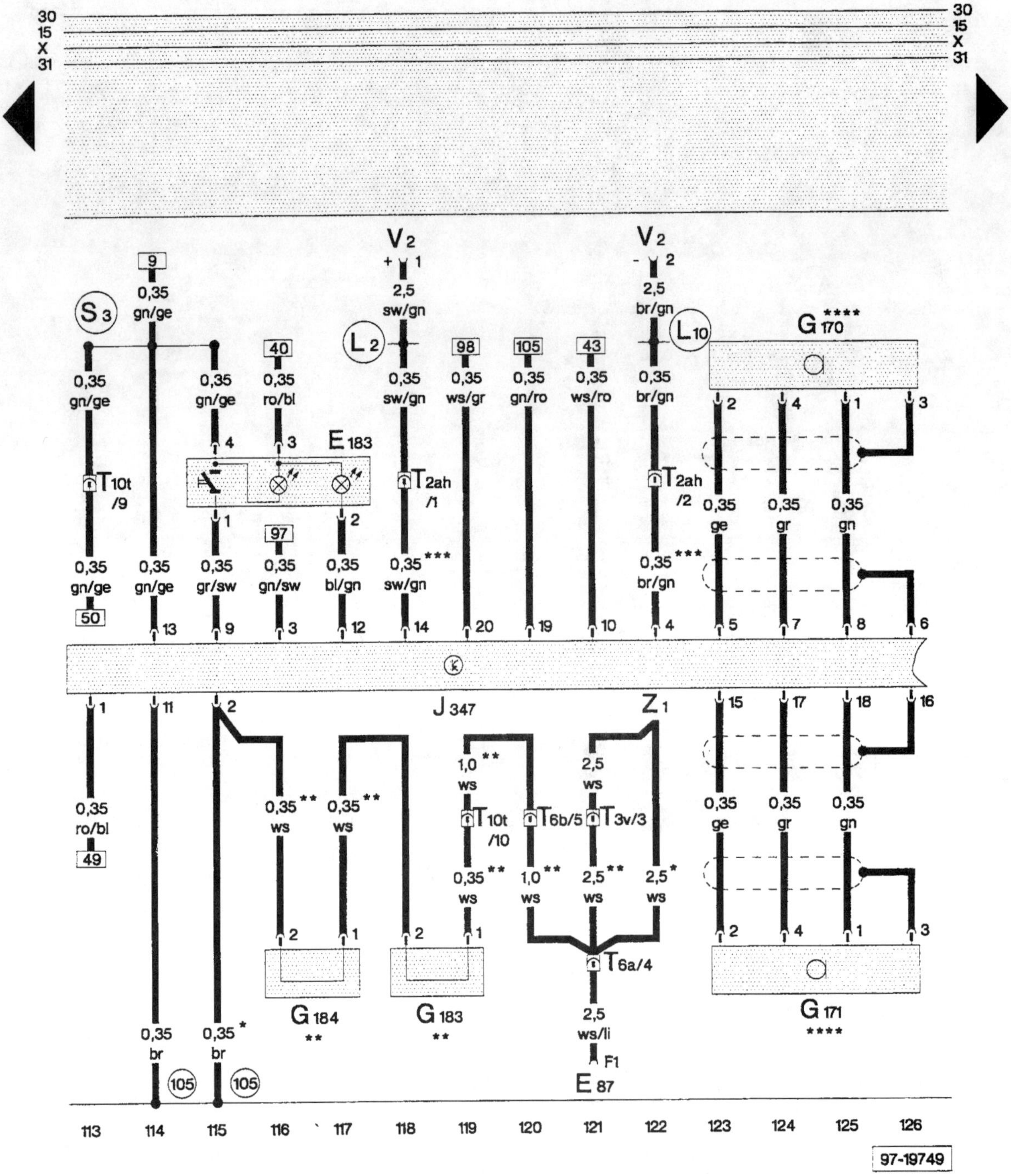

ws = white
sw = black
ro = red
gn = green
br = brown
bl = blue
gr = gray
li = violet
ge = yellow

E 87 - A/C Control Head
E 183 - Switch for passenger compartment monitoring
G 170 - Ultra - sound sensor, left for anti-theft warning system
G 171 - Ultra - sound sensor, right for anti-theft warning system
G 183 - Sensor for broken window glass rear left
G 184 - Sensor for broken window glass rear right
J 347 - Control Module for Ultra-Sound sensors
T 2 ah - Connector 2 pin, blue, connector station A pillar, right
T 3 v - Connector 3 pin, black, in luggage compartment
T 6 a - Connector 6 pin, blue, connector station A pillar, left
T 6 b - Connector 6 pin, white, connector station A pillar, left
T 10 t - Connector 10 pin, orange, connector station A pillar, right
V 2 - Fresh Air Blower
Z 1 - Heated rear window

(105) - Ground connection -1-, in central locking system wiring harness
(L2) - Wire connection, in A/C wiring harness
(L10) - Wire connection -2-, in A/C wiring harness
(S3) - Wire connection -1-(switch/control module), central locking system wiring harness

* - only sedan
** - only wagon
*** - Vehicles with solar cells
**** - Vehicles without head airbag

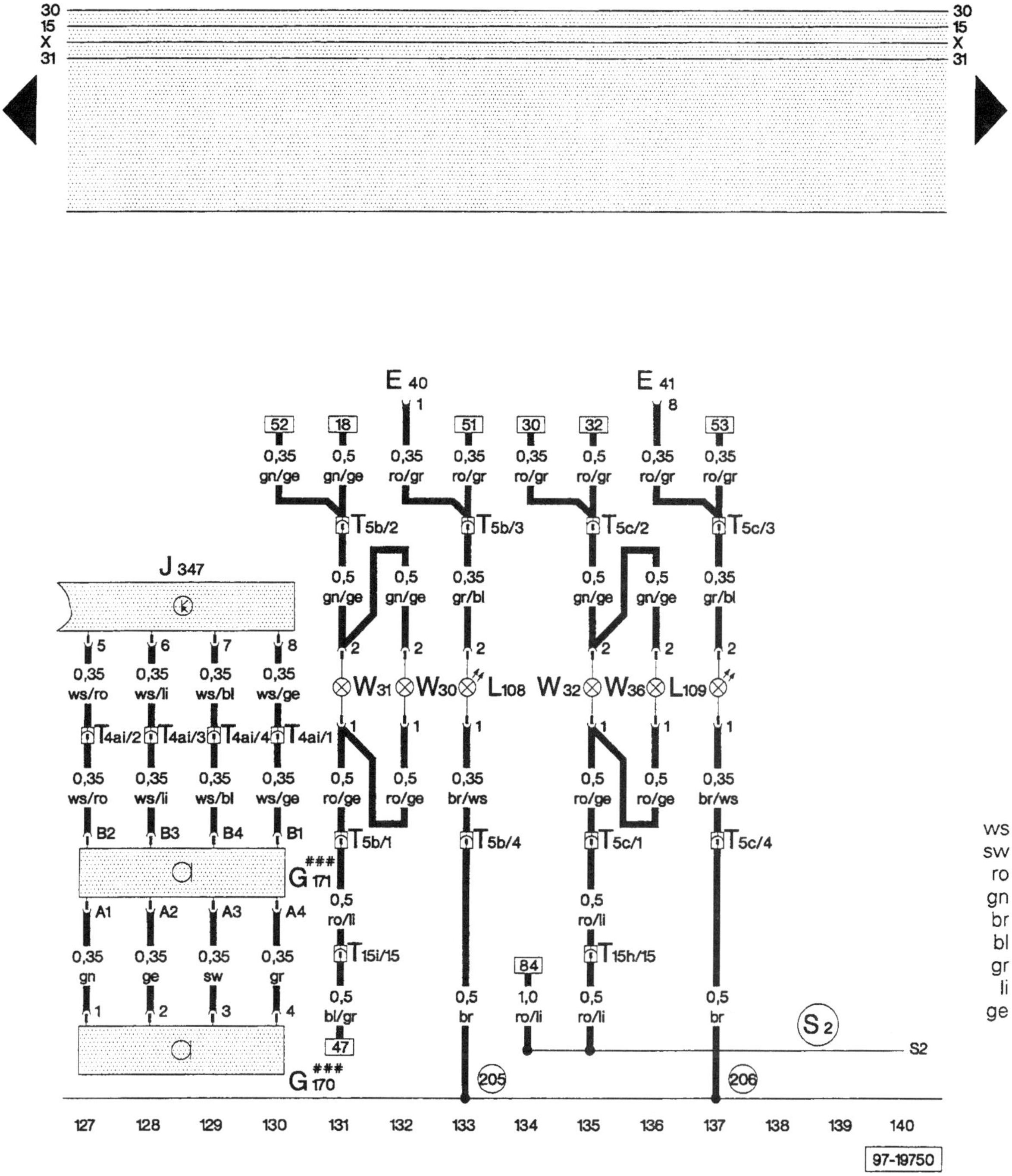

ws = white
sw = black
ro = red
gn = green
br = brown
bl = blue
gr = gray
li = violet
ge = yellow

E 40 - Left Front Window Switch
E 41 - Right Front Window Switch
G 170 - Ultra-sound sensor, left for anti-theft warning system
G 171 - Ultra-sound sensor, right for anti-theft warning system
J 347 - Control Module for Ultra-Sound sensors
L 108 - Door Opener Illumination, Driver's Side
L 109 - Door Opener Illumination, Passenger's Side
T 4 ai - Connector 4 pin, on C pillar, right
T 5 b - Connector 5 pin, black, in driver's door
T 5 c - Connector 5 pin, black, in passenger's door
T 15 h - Connector 15 pin, brown, connector station A pillar, right
T 15 i - Connector 15 pin, brown, connector station A pillar, left
W 30 - Door warning light, driver side
W 31 - Entry light front, left
W 32 - Entry light front, right
W 36 - Door warning light, passenger side

(205) - Ground connection, in driver's door wiring harness
(206) - Ground connection, in passenger's door wiring harness
(S2) - Wire connection, in central locking system wiring harness

* - Vehicles with head airbag

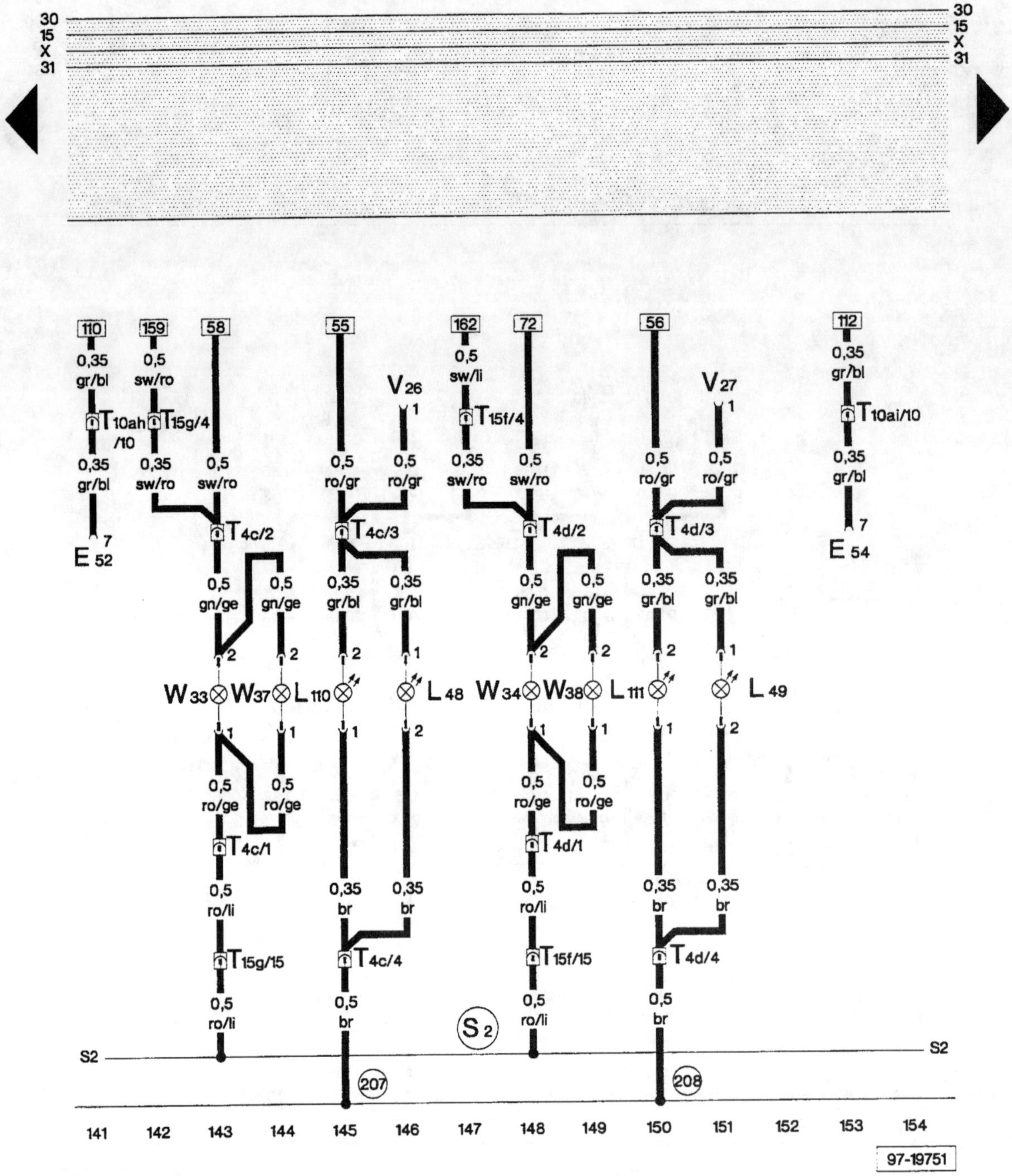

ws = white
sw = black
ro = red
gn = green
br = brown
bl = blue
gr = gray
li = violet
ge = yellow

E 52	- Left Rear Window Switch, (In LR Door)
E 54	- Right Rear Window Switch, (In RR Door)
L 48	- Left Rear Ashtray Light
L 49	- Right Rear Ashtray Light
L 110	- Door Opener Illumination, rear left
L 111	- Door Opener Illumination, rear right
T 4 c	- Connector 4 pin, black, in left rear door
T 4 d	- Connector 4 pin, black, in right rear door
T 10 ah	- Connector 10 pin, red, in B pillar, left
T 10 ai	- Connector 10 pin, red, in B pillar, right
T 15 f	- Connector 15 pin, blue, in connector station B pillar, right
T 15 g	- Connector 15 pin, blue, in connector station B pillar, left
V 26	- Motor for window regulator, rear, left
V 27	- Motor for window regulator, rear, right
W 33	- Entry light rear, left
W 34	- Entry light rear, right
W 37	- Door warning light rear, left
W 38	- Door warning light rear, right
(207)	- Ground connection, in left rear door wiring harness
(208)	- Ground connection, in right rear door wiring harness
(S2)	- Wire connection, in central locking system wiring harness

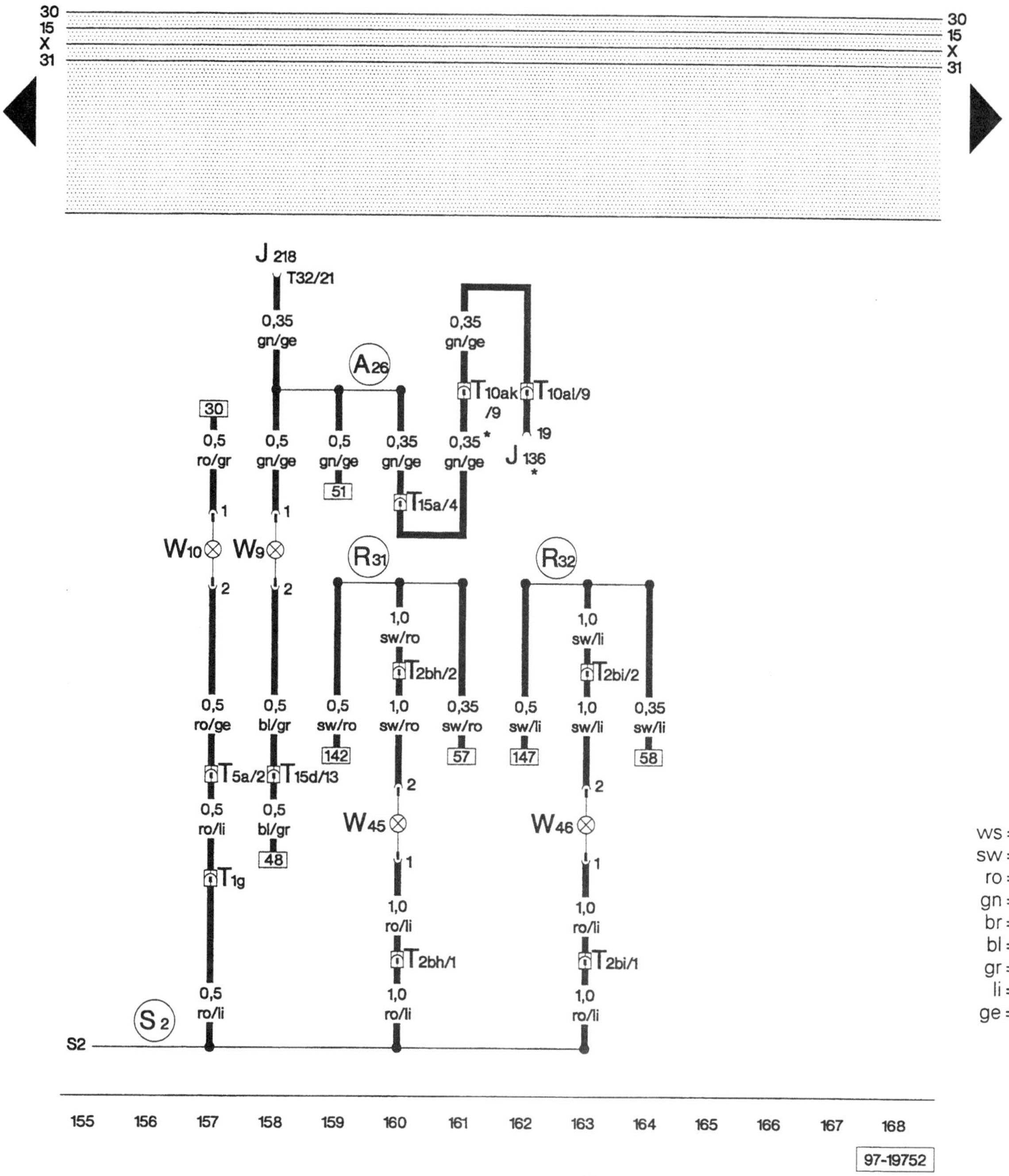

ws = white
sw = black
ro = red
gn = green
br = brown
bl = blue
gr = gray
li = violet
ge = yellow

J 136 - Memory Seat Control Module
J 218 - Instrument Cluster Combination Processor
T 1 g - Single connector, black, connector station A pillar, right
T 5 a - Connector 5 pin, black, behind instrument panel, right
T 2 bh - Connector 2 pin, black, below driver's seat
T 2 bi - Connector 2 pin, black, below passenger's seat
T 10 ak - Connector 10 pin, green, connector station A pillar, left
T 10 al - Connector 10 pin, brown, below driver's seat
T 15 a - Connector 15 pin, black, connector station A pillar, left
T 15 d - Connector 15 pin, black, connector station A pillar, right
T 32 - Connector 32 pin, blue, on instrument cluster
W 9 - Left Footwell Light
W 10 - Right Footwell Light
W 45 - Left Footwell Light, rear
W 46 - Right Footwell Light, rear

(A26) - Wire connection (driver's door contact switch), in instrument panel wiring harness
(R31) - Wire connection -1-, in left rear door wiring harness
(R32) - Wire connection -2-, in right rear door wiring harness
(S2) - Wire connection, in central locking system wiring harness

* - Vehicles with Memory Seat

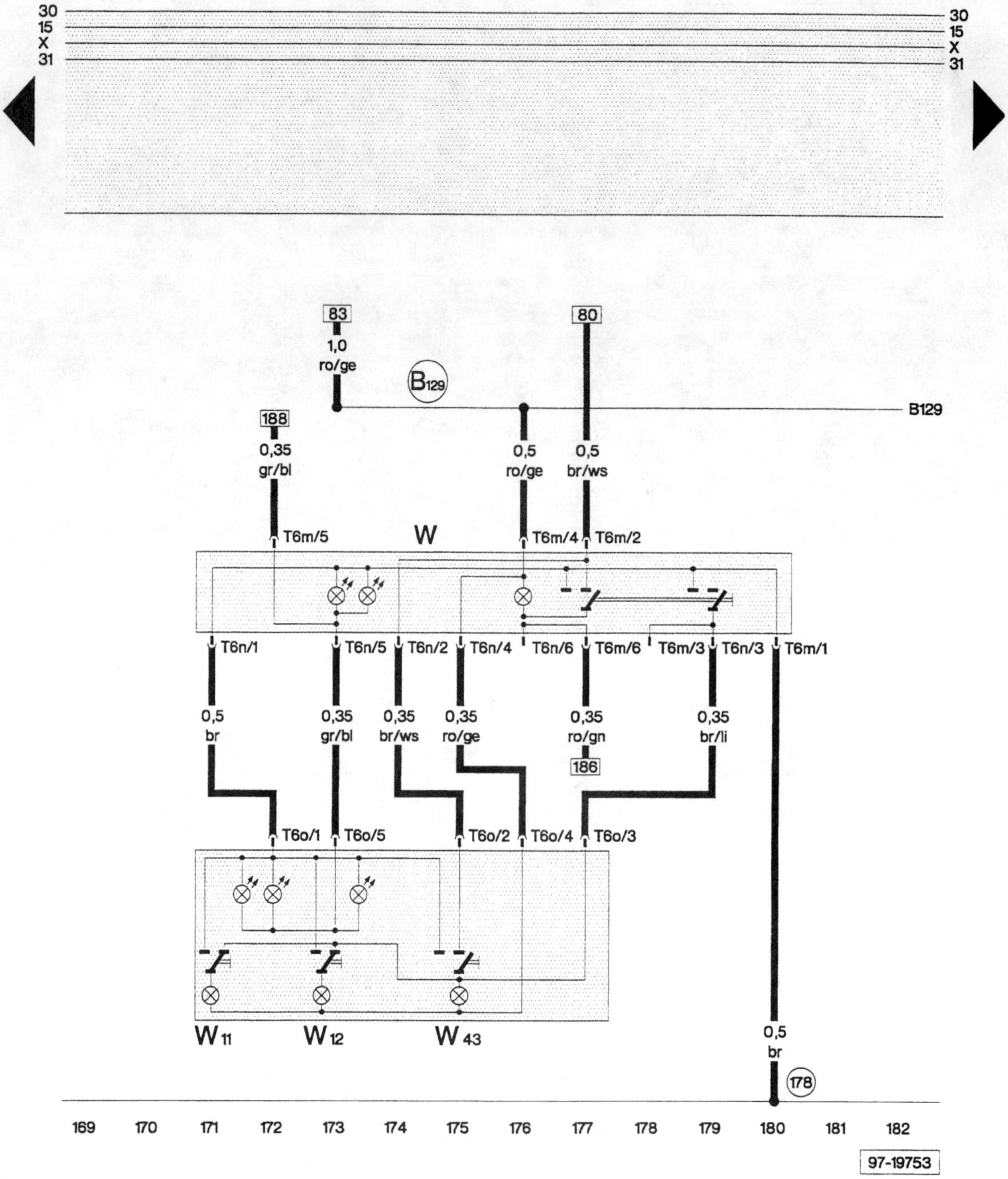

ws = white
sw = black
ro = red
gn = green
br = brown
bl = blue
gr = gray
li = violet
ge = yellow

T 6 m - Connector 6 pin, on interior light, front
T 6 n - Connector 6 pin, on interior light, front
T 6 o - Connector 6 pin, on interior light, rear
W - Front Interior Light
W 11 - Left Rear Reading Light
W 12 - Right Rear Reading Light
W 43 - Rear courtesy lamp

(178) - Ground connection, in sunroof wiring harness
(B129) - Connector (interior light 3/L), in wiring harness interior

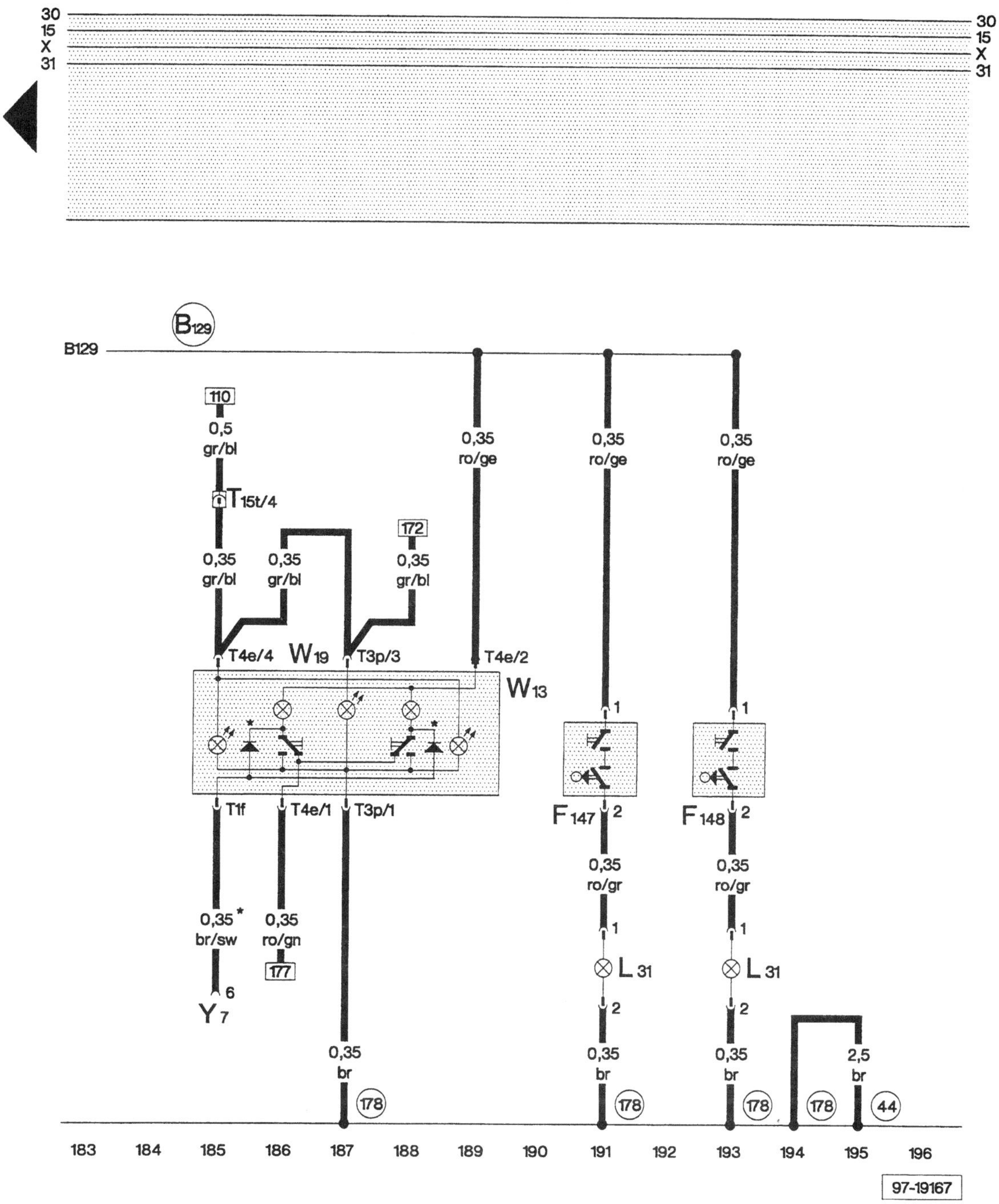

ws = white
sw = black
ro = red
gn = green
br = brown
bl = blue
gr = gray
li = violet
ge = yellow

F 147 - Left Make-Up Mirror Light Switch
F 148 - Right Make-Up Mirror Light Switch
L 31 - Make-up Mirror Light
T 1 f - Single connector, black, on front interior light
T 3 p - Connector 3 pin, on reading lamp front right
T 4 e - Connector 4 pin, on reading lamp front left
T 15 t - Connector 15 pin, grey, connector station A pillar, left
W 13 - Right Front Map/Reading Light
W 19 - Left Front Reading Light
Y 7 - Automatic day/night interior mirror

(44) - Ground connection, on left A-pillar, lower part
(178) - Ground connection, in sunroof wiring harness
(B129) - Connector (interior light 3/L), in wiring harness interior

* - Vehicles with Automatic Dimmer Inside Mirror

THIS PAGE INTENTIONALLY LEFT

BLANK

Automatic Climate Control

1999 m. y.

Fuse Panel

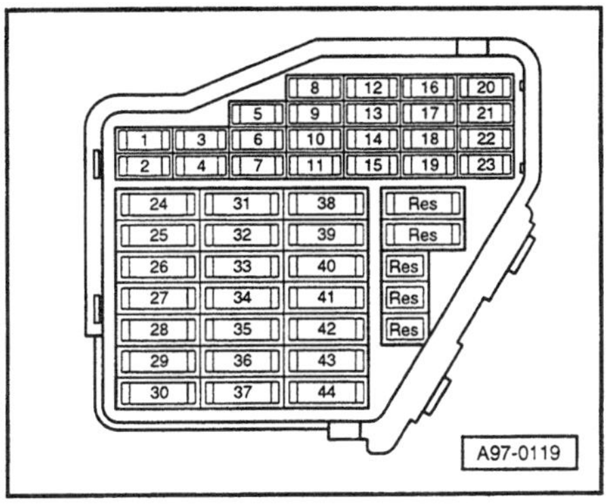

Fuse Colors:

30 A - Green
25 A - White
20 A - Yellow
15 A - Blue
10 A - Red
7,5 A - Brown
5 A - Beige

Starting with fuse position 23, fuses in the fuse holder are identified with 223 in the wiring diagram.

Micro Central Electric Panel

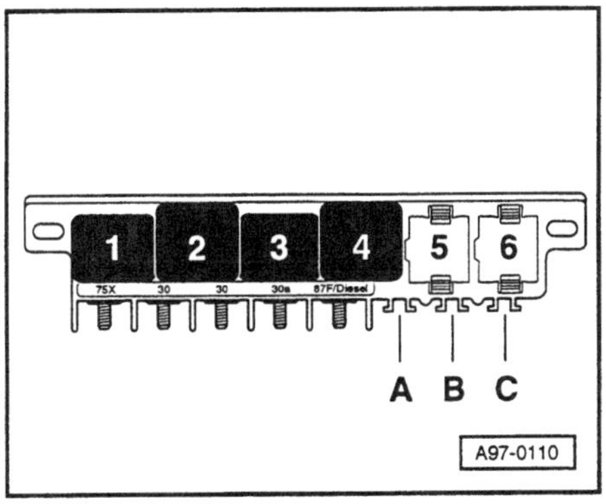

13 - Fold Relay Panel

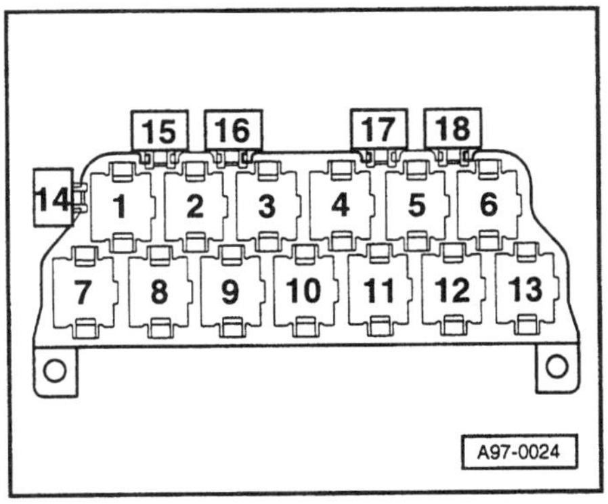

Relay Location:

3 - Solar Cell Separation Relay, J309

5 - A/C Clutch Relay, J44

8 - Fold Relay Panel

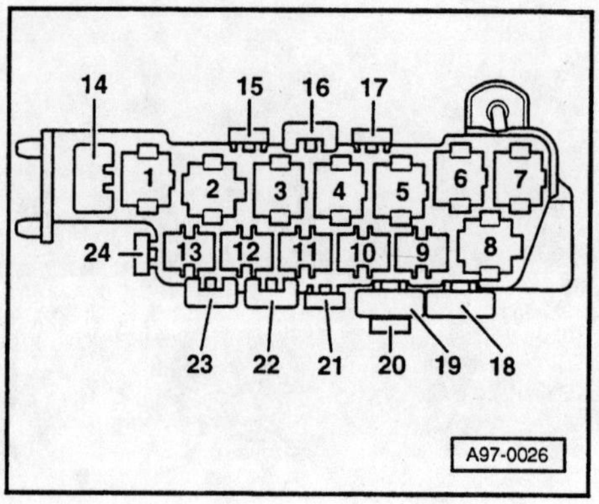

Relay Location:

20 - Coolant fan fuse, S42

21 - Control module fuse for coolant fan, S142

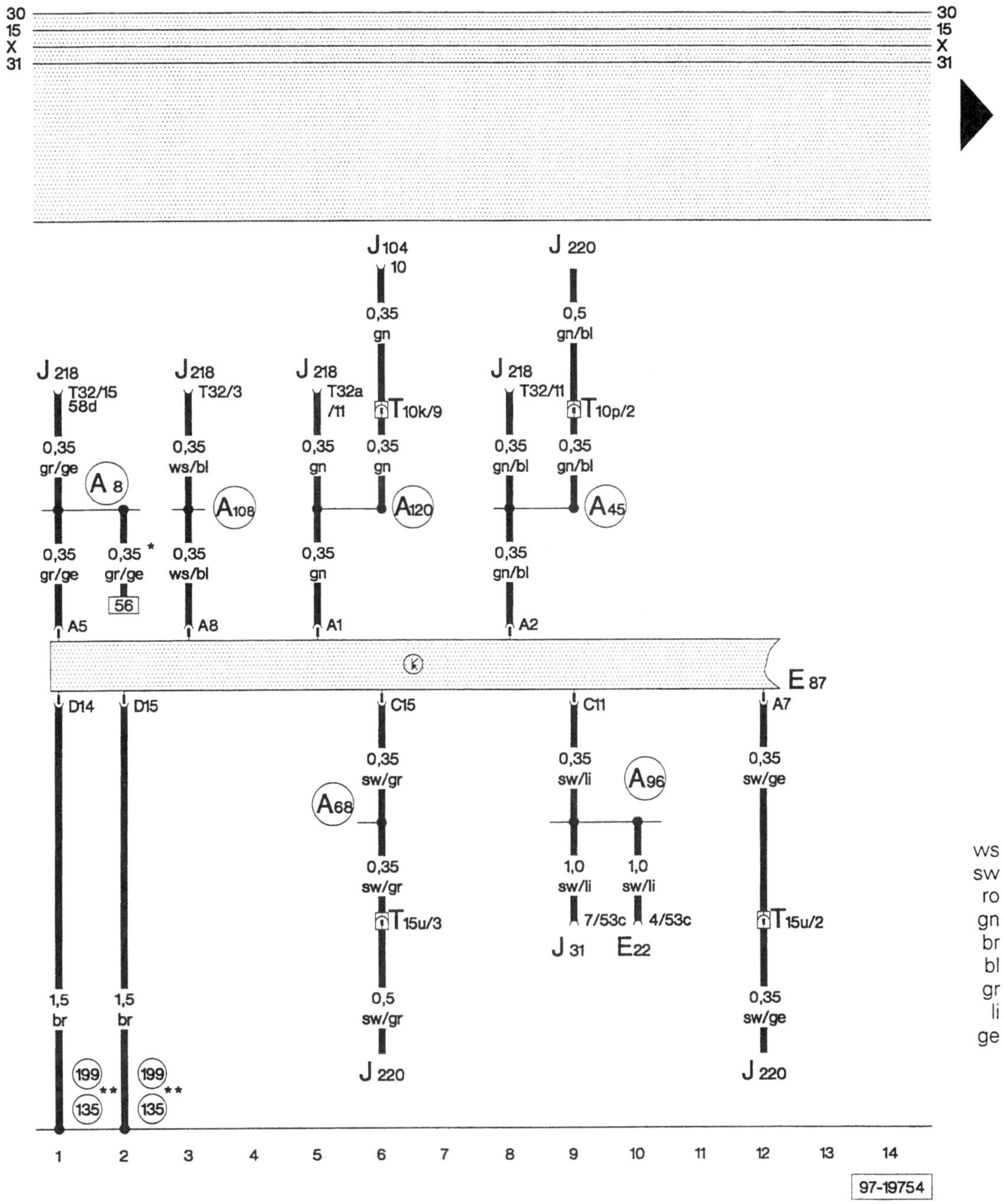

ws = white
sw = black
ro = red
gn = green
br = brown
bl = blue
gr = gray
li = violet
ge = yellow

E 22 - Windshield Wiper/Washer Switch
E 87 - A/C Control Head
J 31 - Wiper/Washer Intermittent Relay
J 104 - ABS Control Module (w/EDL)
J 218 - Instrument Cluster Combination Processor
J 220 - Motronic Engine Control Module (ECM)
T 10 k - Connector 10 pin, orange, connector station A pillar, left
T 10 p - Connector 10 pin, black, connector station electronic box plenum chamber
T 15 u - Connector 15 pin, red, connector station electronic box plenum chamber
T 32 - Connector 32 pin, blue, on instrument cluster
T 32 a - Connector 32 pin, green, on instrument cluster

(135) - Ground connection -2-, in instrument panel wiring harness

(199) - Ground connection -3-, in instrument panel wiring harness

(A8) - Plus connection (58d), in instrument panel wiring harness (Display illumination)

(A45) - Wire connection (RPM-signal), in instrument panel wiring harness

(A68) - Connector (C15, A/C), in instrument panel wiring harness

(A96) - Connector (53a), in instrument panel wiring harness

(A108) - Wire connection (vehicle speed signal), in instrument panel wiring harness

(A120) - Wire connection (Immobilization time), in instrument panel wiring harness

* - Vehicles with Outside Air Temperature Display
** - booth is possible

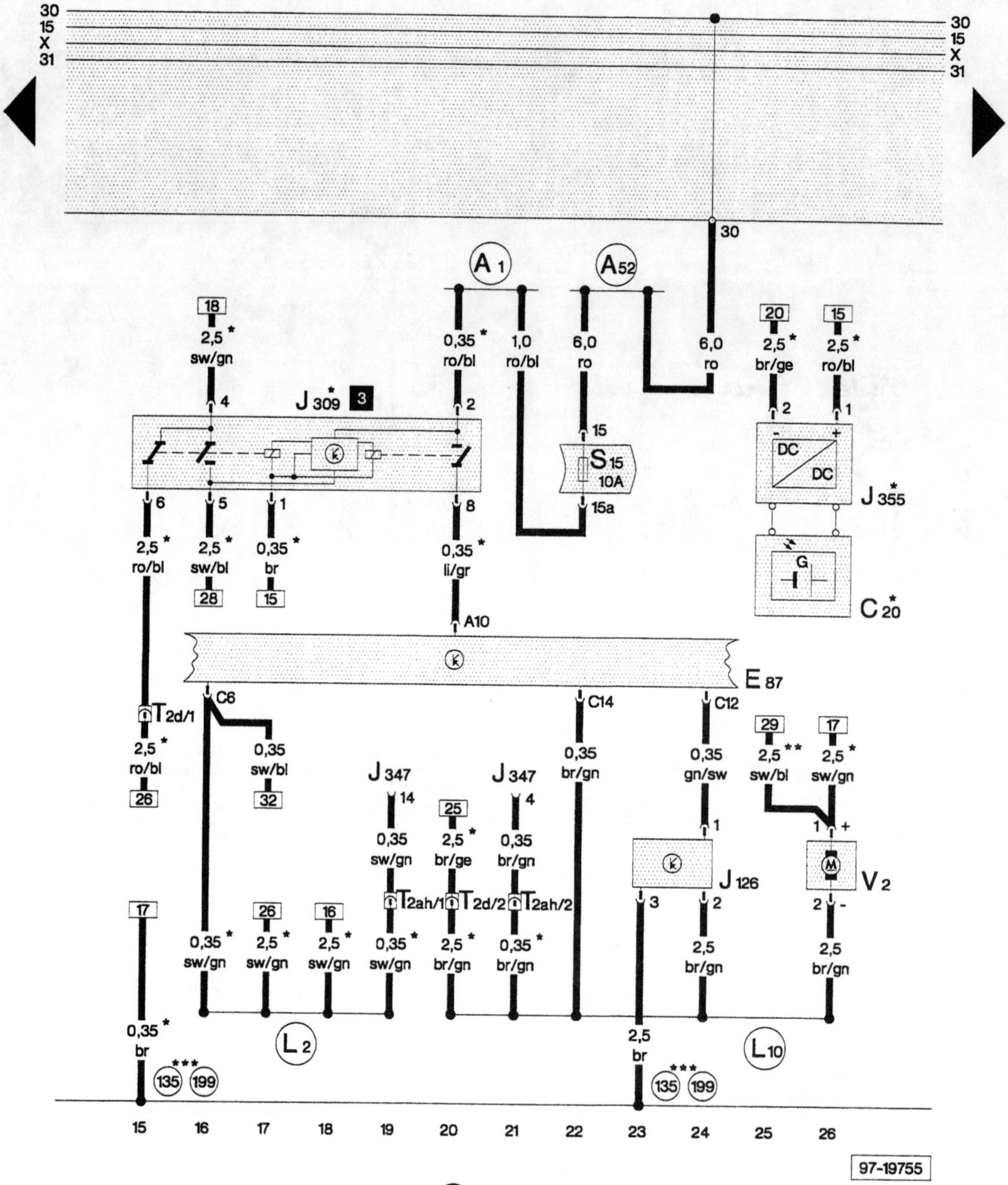

ws = white
sw = black
ro = red
gn = green
br = brown
bl = blue
gr = gray
li = violet
ge = yellow

C 20 - Solar Cells
E 87 - A/C Control Head
J 126 - Control module for fresh air blower
J 309 - Solar Cell Separation Relay
J 347 - Control Module for Ultra-Sound sensors
J 355 - Control Module for solar operation
S 15 - Fuse
T 2 d - Connector 2 pin, blue, connector station A pillar, left
T 2 ah - Connector 2 pin, blue, connector station A pillar, right
V 2 - Fresh Air Blower

(135) - Ground connection -2-, in instrument panel wiring harness
(199) - Ground connection -3-, in instrument panel wiring harness
(A1) - Plus connection (30a), in instrument panel wiring harness
(A52) - Plus connection (30), in instrument panel wiring harness
(L2) - Wire connection, in A/C wiring harness
(L10) - Wire connection -2-, in A/C wiring harness

* - Vehicles with sunroof with Solar Cells
** - Vehicles without sunroof with Solar Cells
*** - booth is possible

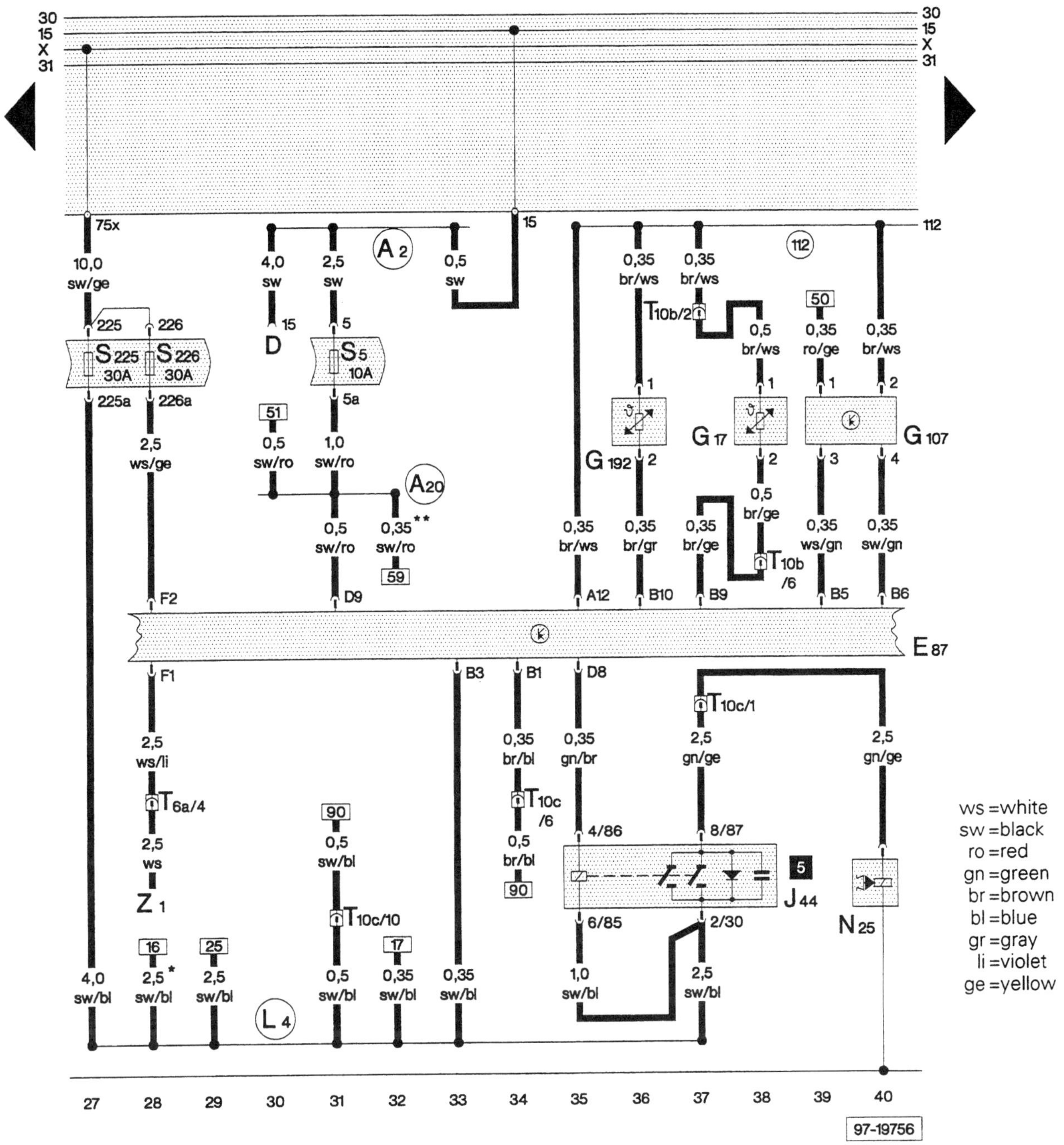

ws = white
sw = black
ro = red
gn = green
br = brown
bl = blue
gr = gray
li = violet
ge = yellow

D - Ignition/Starter Switch
E 87 - A/C Control Head
G 17 - Outside Air Temperature Sensor
G 107 - Sunlight Photo Sensor
G 192 - Sender for outlet temperature, floor outlet
J 44 - A/C Clutch Relay
N 25 - A/C Clutch
S 5 - Fuse
S 225 - Fuse
S 226 - Fuse
T 6 a - Connector 6 pin, blue, connector station A pillar, left
T 10 b - Connector 10 pin, red, connector station A pillar, left
T 10 c - Connector 10 pin, grey, connector station A pillar, left
Z 1 - Heated rear window

(112) - Ground connection -2-, in A/C wiring harness
(A2) - Plus connection (15), in instrument panel wiring harness
(A20) - Wire connection (15a), in instrument panel wiring harness
(L4) - Wire connection (75al), in A/C wiring harness

* - Vehicles with sunroof with Solar Cells
** - Vehicles with Outside Air Temperature Display

ws = white
sw = black
ro = red
gn = green
br = brown
bl = blue
gr = gray
li = violet
ge = yellow

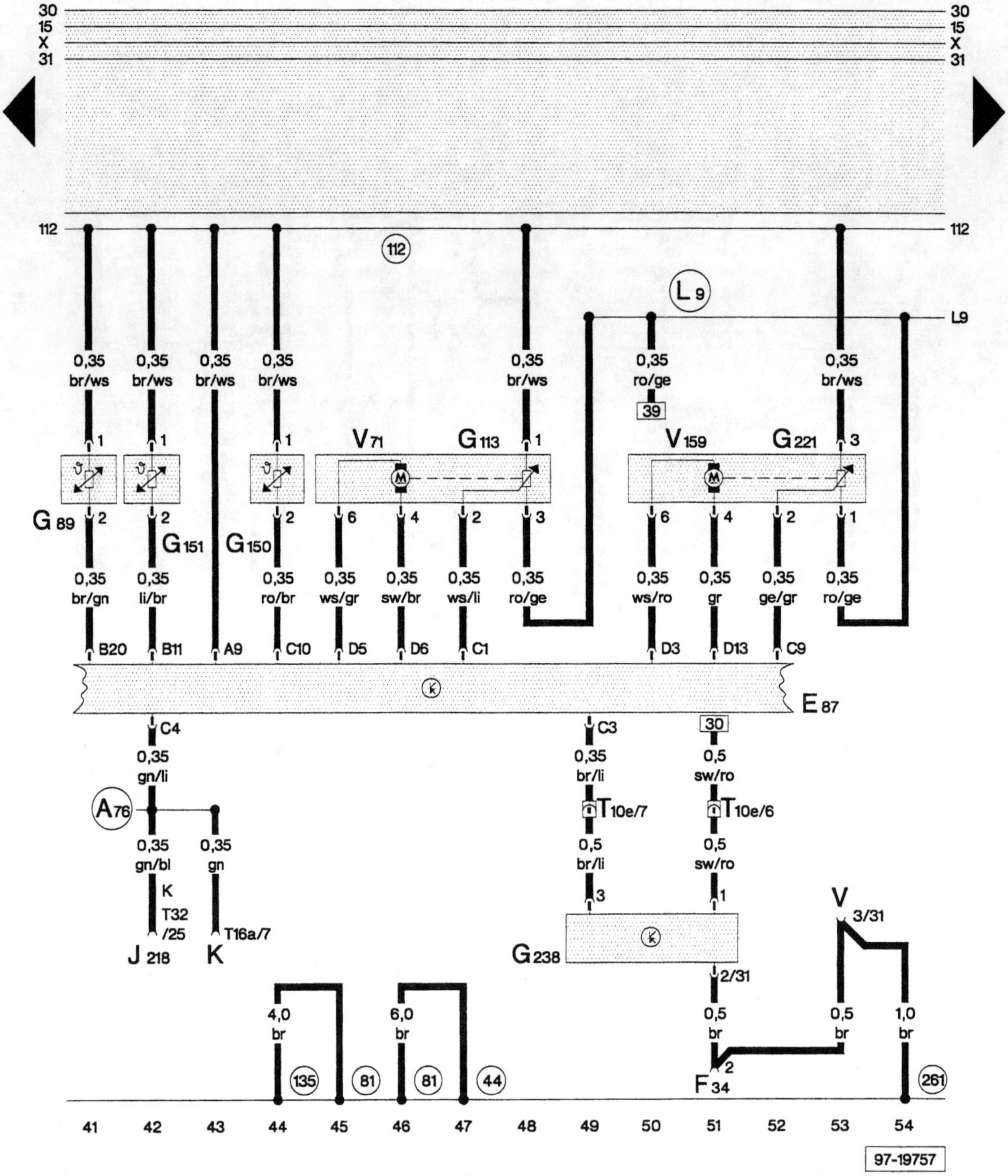

E 87 - A/C Control Head
F 34 - Brake Fluid Level Warning Switch
G 89 - Fresh Air Intake Duct Temperature Sensor
G 113 - Back Pressure Flap Motor Position Sensor
G 150 - Left Vent Temperature Sensor
G 151 - Right Vent Temperature Sensor
G 221 - Right Temperature Regulator Flap Motor Position Sensor
G 238 - Sensor for Air Quality
J 218 - Instrument Cluster Combination Processor
K - DLC (Data Link Connector) Rapid Data Transfer
T 10 e - Connector 10 pin, violet, connector station A pillar, left
T 16 a - Connector 16 pin, data link connector
T 32 - Connector 32 pin, blue, on instrument cluster
V - Windshield Wiper Motor
V 71 - Air Flow Flap Motor
V 159 - Right Temperature Regulator Flap Motor

(44) - Ground connection, on left A-pillar, lower part
(81) - Ground connection -1-, in instrument panel wiring harness
(112) - Ground connection -2-, in A/C wiring harness
(135) - Ground connection -2-, in instrument panel wiring harness
(261) - Ground connection, in wiring harness heated spray jet
(A76) - Connector (K-diagnosis wire), in instrument panel wiring harness
(L9) - Wire connection -1-, in A/C wiring harness

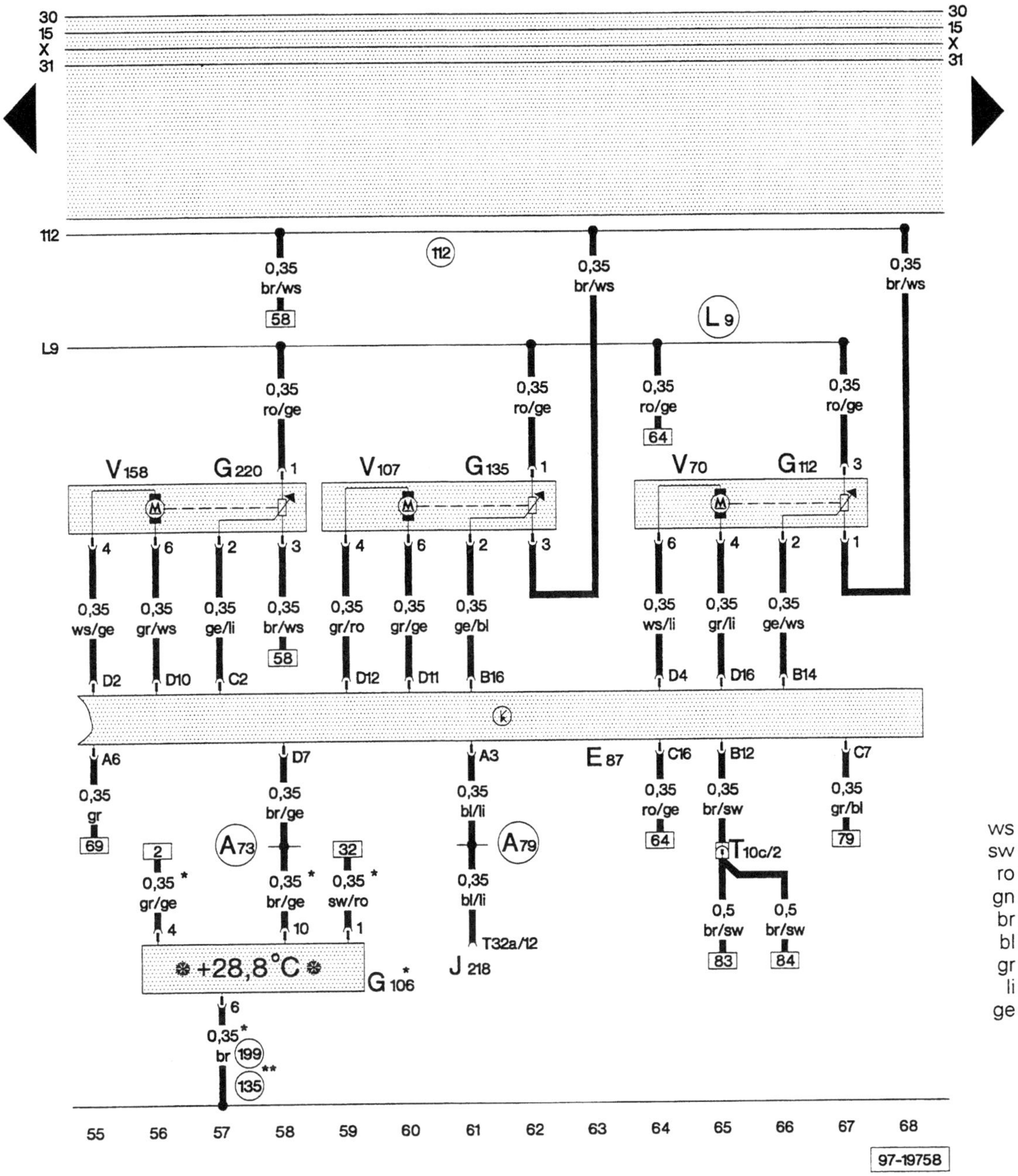

ws = white
sw = black
ro = red
gn = green
br = brown
bl = blue
gr = gray
li = violet
ge = yellow

- E 87 - A/C Control Head
- G 106 - Outside Air Temperature Display
- G 112 - Central Flap Motor Position Sensor
- G 135 - Defroster Flap Motor Position Sensor
- G 220 - Left Temperature Regulator Flap Motor Position Sensor
- J 218 - Instrument Cluster Combination Processor
- T 10 c - Connector 10 pin, grey, connector station A pillar, left
- T 32 a - Connector 32 pin, green, on instrument cluster
- V 70 - Central Air Flap Motor
- V 107 - Defroster Flap Motor
- V 158 - Left Temperature Regulator Flap Motor

- (112) - Ground connection -2-, in A/C wiring harness
- (135) - Ground connection -2-, in instrument panel wiring harness
- (199) - Ground connection -3-, in instrument panel wiring harness
- (A73) - Connector (outside temperature indicator), in instrument panel wiring harness
- (A79) - Connector (engine overheat warning light), in instrument panel wiring harness
- (L9) - Wire connection -1-, in A/C wiring harness

- * - Vehicles with Outside Air Temperature Display
- ** - booth is possible

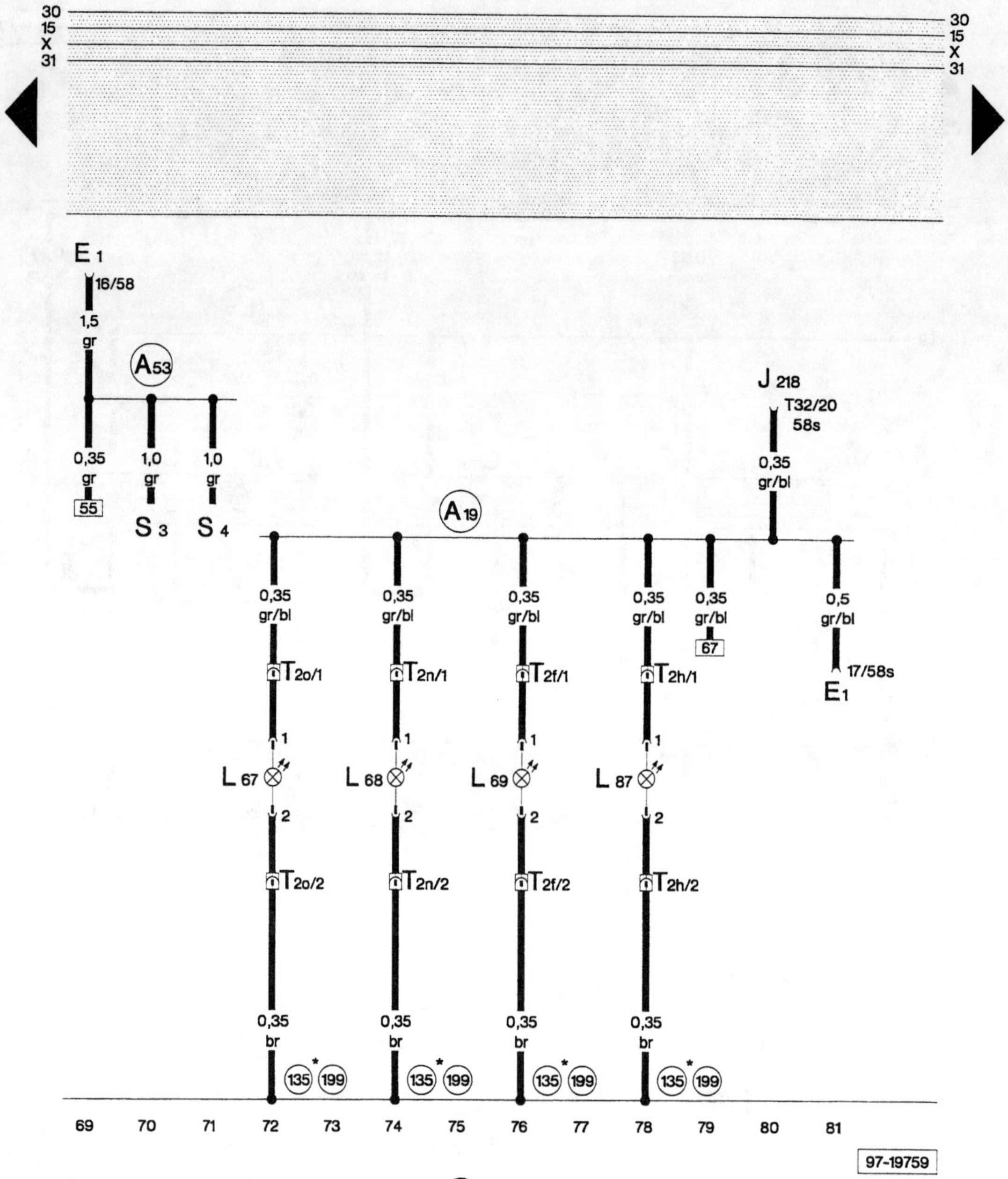

ws = white
sw = black
ro = red
gn = green
br = brown
bl = blue
gr = gray
li = violet
ge = yellow

E 1 - Light switch
J 218 - Instrument Cluster Combination Processor
L 67 - Left Instrument Panel Vent Illumination
L 68 - Center Instrument Panel Vent Illumination
L 69 - Right Instrument Panel Vent Illumination
L 87 - Lighting for rearseat outlet center
S 3 - Fuse
S 4 - Fuse
T 2 f - Connector 2 pin, brown, behind instrument panel, right
T 2 h - Connector 2 pin, black, in console
T 2 n - Connector 2 pin, brown, behind instrument panel, center
T 2 o - Connector 2 pin, behind instrument panel, left
T 32 - Connector 32 pin, blue, on instrument cluster

(135) - Ground connection -2-, in instrument panel wiring harness
(199) - Ground connection -3-, in instrument panel wiring harness
(A19) - Wire connection (58s), in instrument panel wiring harness (Switch illumination)
(A53) - Plus connection (58b), in instrument panel wiring harness

* - booth is possible

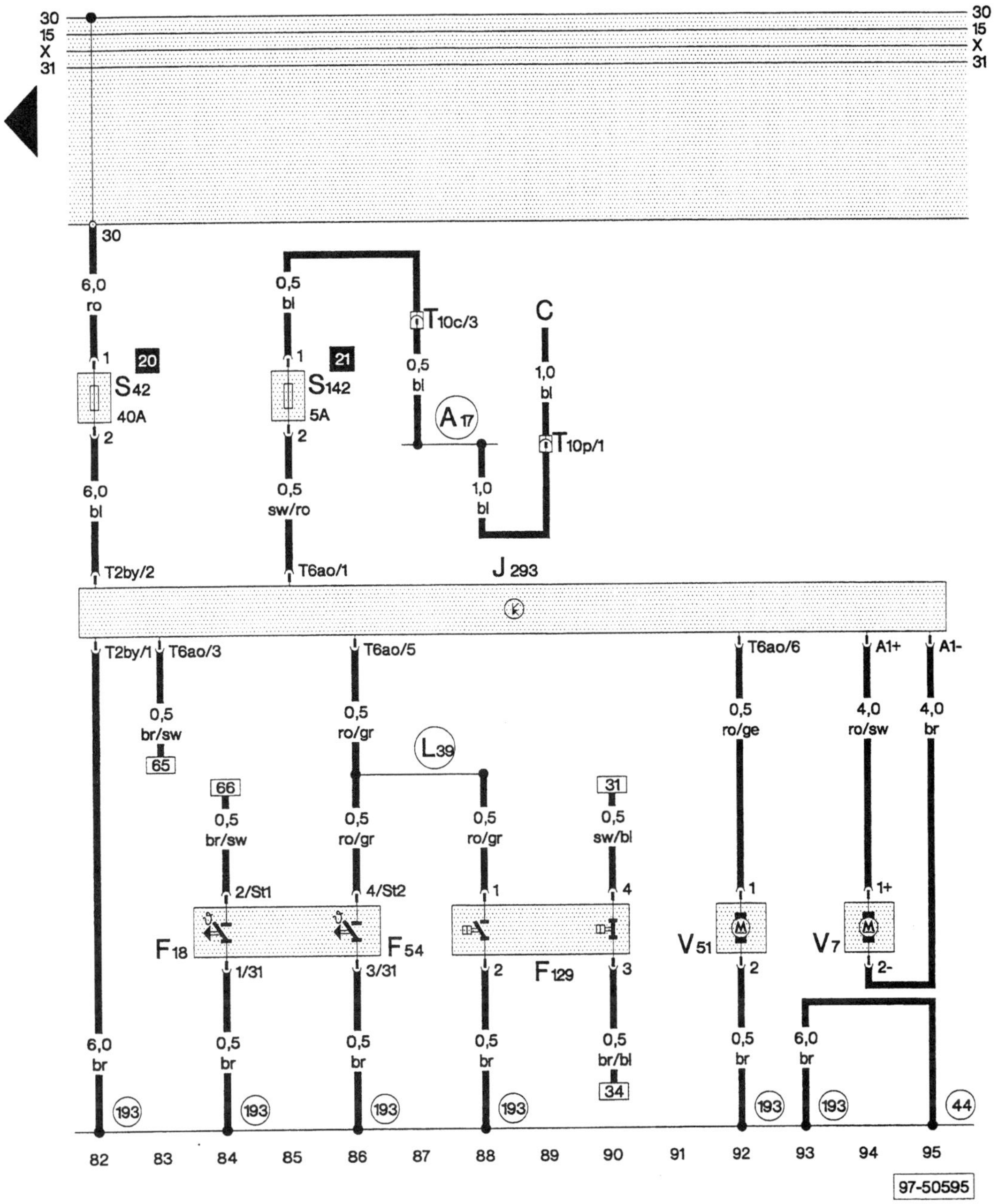

ws =white
sw =black
ro =red
gn =green
br =brown
bl =blue
gr =gray
li =violet
ge =yellow

C - Generator (GEN)
F 18 - Coolant Fan Control (FC) Thermal Switch
F 54 - Coolant Fan Control (FC) Thermal Switch
F 129 - A/C Pressure Switch
J 293 - Coolant FC (Fan Control) Control Module
S 42 - Coolant Fan Fuse
S 142 - Control module fuse for coolant fan
T 2 by - Connector 2 pin, on Coolant FC (Fan Control) Control Module
T 6 ao - Connector 6 pin, on Coolant FC (Fan Control) Control Module
T 10 c - Connector 10 pin, grey, connector station A pillar, left
T 10 p - Connector 10 pin, black, connector station electronic box plenum chamber
V 7 - Coolant Fan
V 51 - After-Run Coolant Pump

(44) - Ground connection, on left A-pillar, lower part
(193) - Ground connection -1-, in coolant fan wiring harness
(A17) - Wire connection (61), in instrument panel wiring harness
(L39) - Wire connection -1-(A/C Pressure Switch), in Climatronic wiring harness

Driver's Information System

1999 m. y.

Fuse Panel

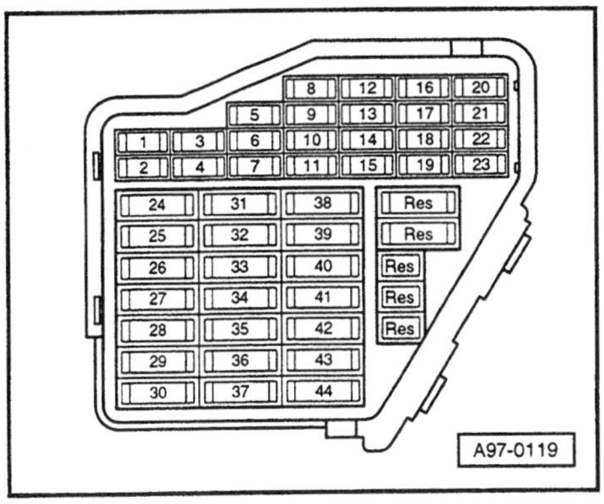

Fuse Colors:

30 A - Green
25 A - White
20 A - Yellow
15 A - Blue
10 A - Red
7,5 A - Brown
5 A - Beige

Starting with fuse position 23, fuses in the fuse holder are identified with 223 in the wiring diagram.

13 - Fold Relay Panel

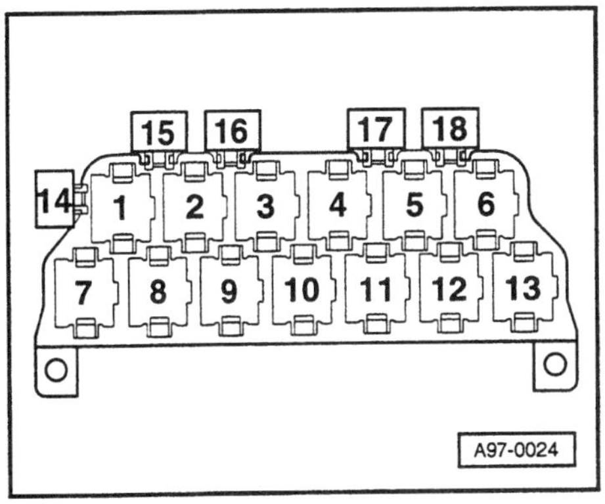

Relay Location:

9 - Lamp Control Module, J123

10 - Lamp Control Module, J123

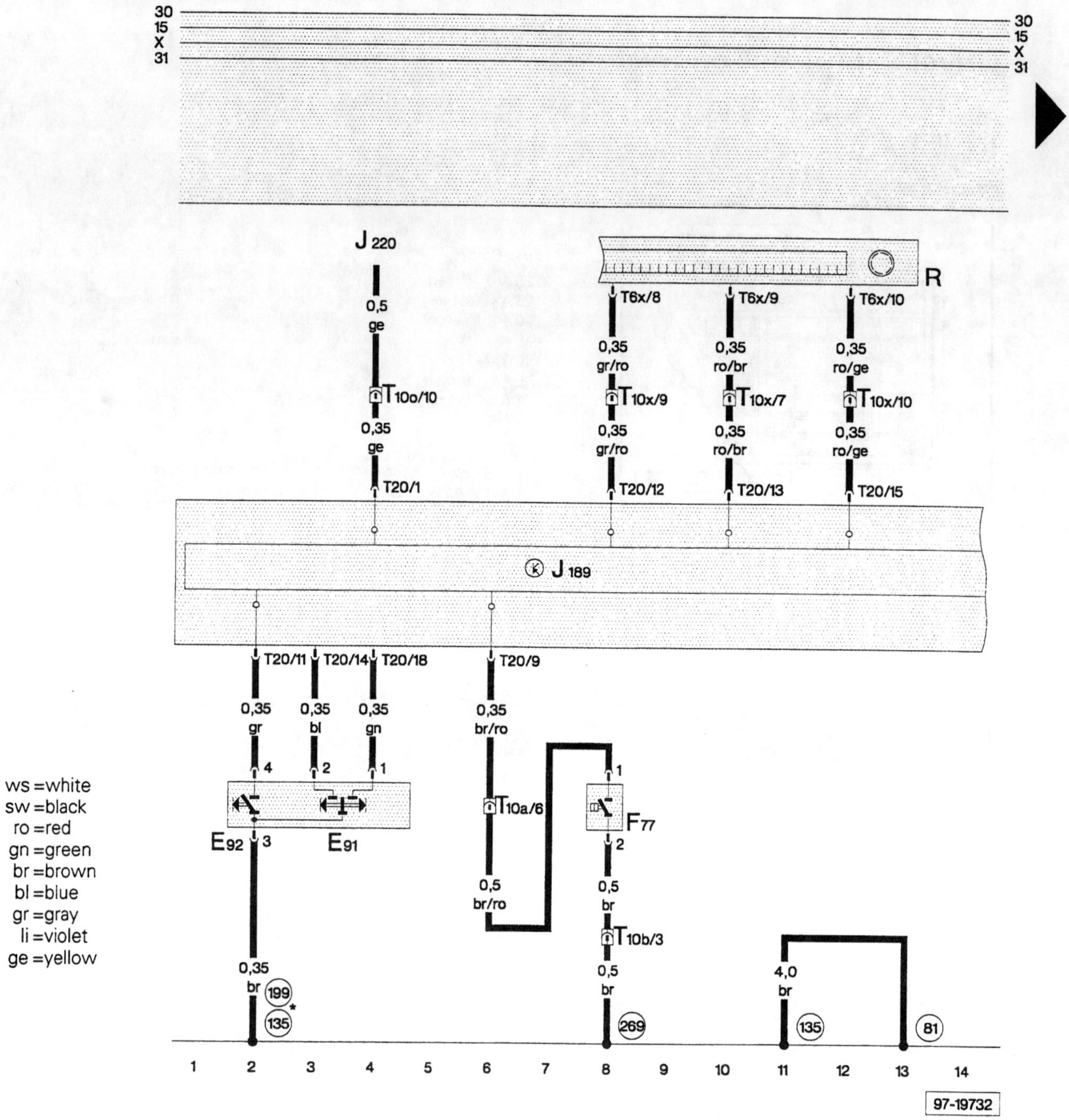

- E 91 - Board Computer Function Selector Switch
- E 92 - Board Computer Reset Button
- F 77 - Windshield Washer Fluid Level Warning Switch
- J 189 - Auto Check System
- J 220 - Motronic Engine Control Module (ECM)
- R - Radio
- T 6 x - Connector 6 pin, green, connector radio I - 2
- T 10 a - Connector 10 pin, pink, connector station A pillar, left
- T 10 b - Connector 10 pin, red, connector station A pillar, left
- T 10 o - Connector 10 pin, brown, connector station electronic box plenum chamber
- T 10 x - Connector 10 pin, pink, connector station A pillar, right
- T 20 - Connector 20 pin, red/white, on instrument cluster

- (81) - Ground connection -1-, in instrument panel wiring harness
- (135) - Ground connection -2-, in instrument panel wiring harness
- (199) - Ground connection -3-, in instrument panel wiring harness
- (269) - Ground connector (sensor ground) -1-, in instrument panel wiring harness

* - booth is possible

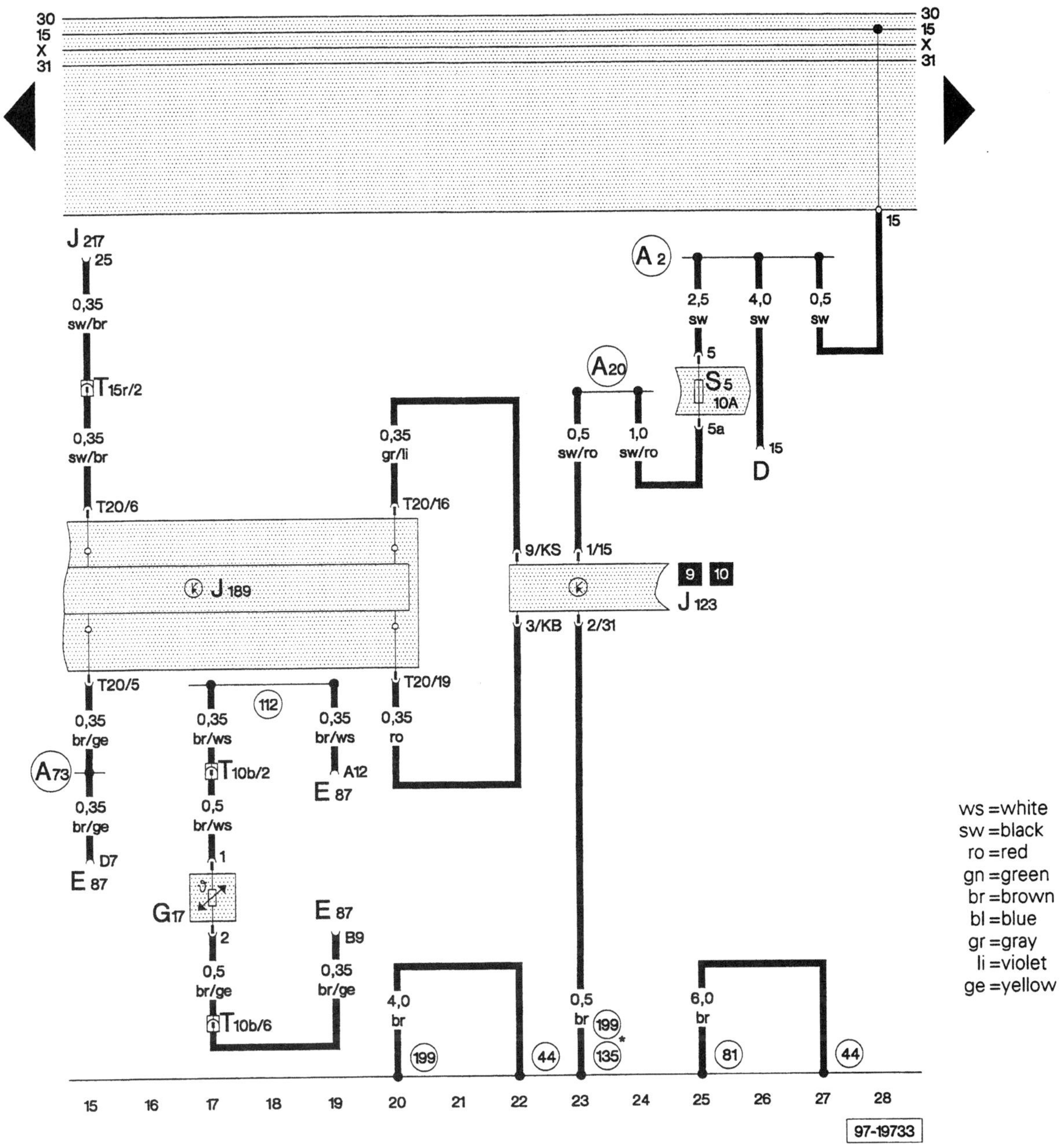

D - Ignition/Starter Switch
E 87 - A/C Control Head
G 17 - Outside Air Temperature Sensor
J 123 - Lamp Control Module
J 189 - Auto Check System
J 217 - Transmission Control Module (TCM)
S 5 - Fuse
T 10 b - Connector 10 pin, red, connector station A pillar, left
T 15 r - Connector 15 pin, brown, connector station A pillar, right
T 20 - Connector 20 pin, red/white, on instrument cluster

(44) - Ground connection, on left A-pillar, lower part

(81) - Ground connection -1-, in instrument panel wiring harness

(112) - Ground connection -2-, in A/C wiring harness

(135) - Ground connection -2-, in instrument panel wiring harness

(199) - Ground connection -3-, in instrument panel wiring harness

(A2) - Plus connection (15), in instrument panel wiring harness

(A20) - Wire connection (15a), in instrument panel wiring harness

(A73) - Connector (outside temperature indicator), in instrument panel wiring harness

* - booth is possible

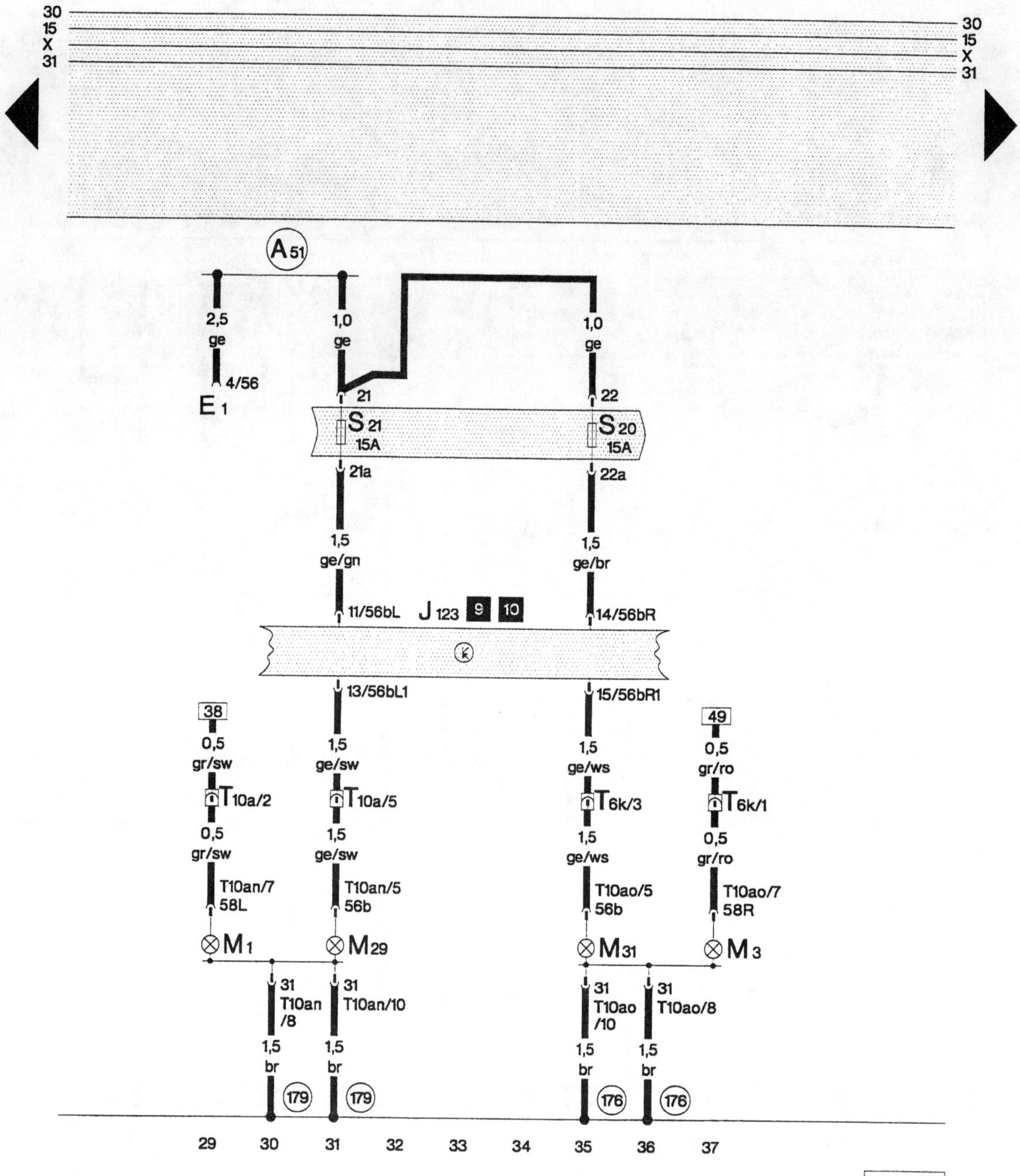

ws = white
sw = black
ro = red
gn = green
br = brown
bl = blue
gr = gray
li = violet
ge = yellow

E 1 - Light switch
J 123 - Lamp Control Module
M 1 - Left Parking Light
M 3 - Right Parking Light
M 29 - Left Low Beam Headlight
M 31 - Right Low Beam Headlight
S 20 - Fuse
S 21 - Fuse
T 6 k - Connector 6 pin, pink, connector station A pillar, right
T 10 a - Connector 10 pin, pink, connector station A pillar, left
T 10 an - Connector 10 pin, on headlight, left
T 10 ao - Connector 10 pin, on headlight, right

(176) - Ground connection, in right headlight wiring harness
(179) - Ground connection, in left headlight wiring harness
(A51) - Wire connection (56), in instrument panel wiring harness

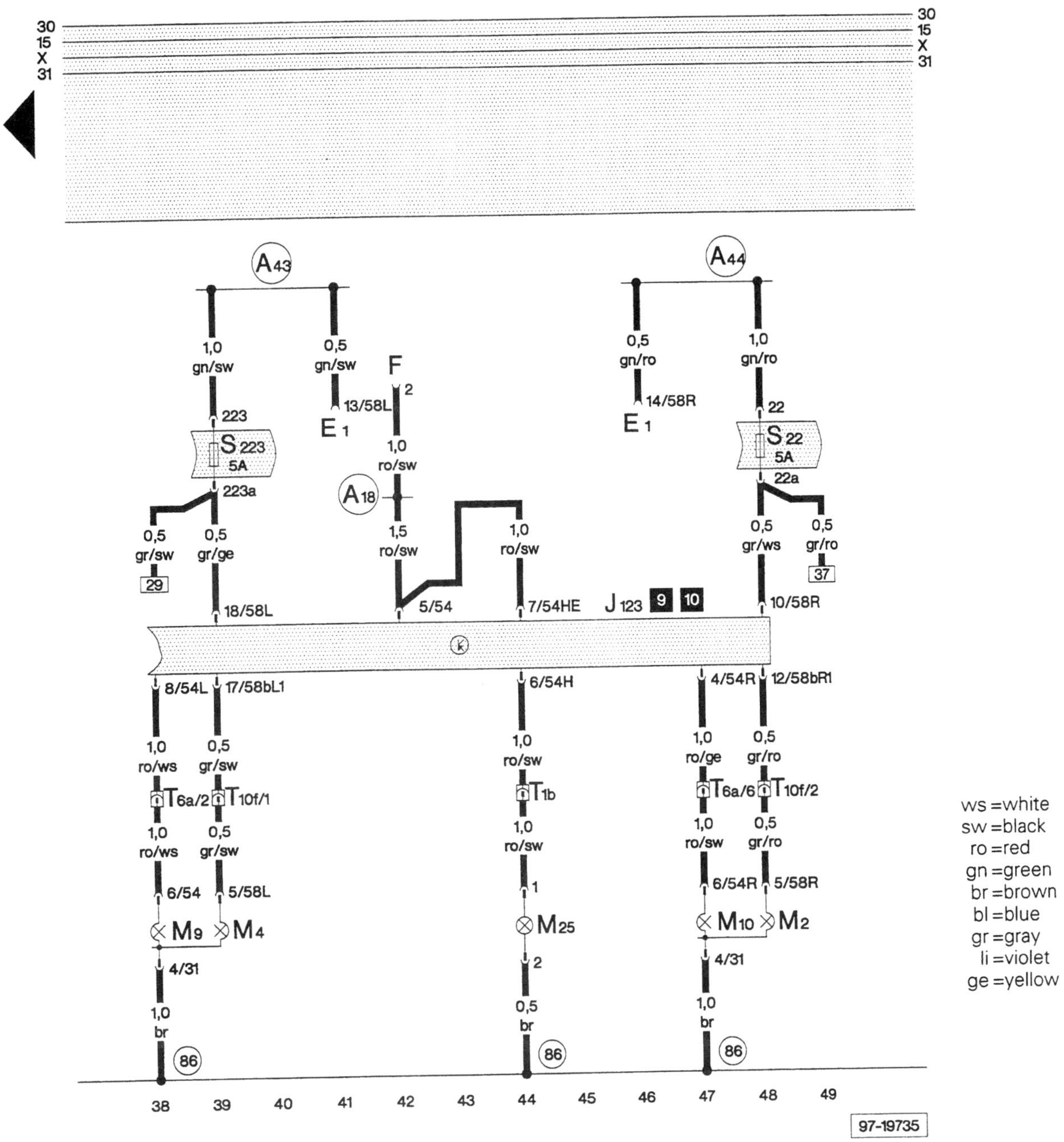

- E 1 - Light switch
- F - Brake Light Switch
- J 123 - Lamp Control Module
- M 2 - Right Tail Light
- M 4 - Left Tail Light
- M 9 - Left Brake Light
- M 10 - Right Brake Light
- M 25 - High-mount Brake Light
- S 22 - Fuse
- S 223 - Fuse
- T 1 b - Single connector, black, connector station A pillar right
- T 6 a - Connector 6 pin, blue, connector station A pillar, left
- T 10 f - Connector 10 pin, brown, connector station A pillar, left

- (86) - Ground connection -1-, in rear wiring harness
- (A18) - Wire connection (54), in instrument panel wiring harness
- (A43) - Wire connection (57L), in instrument panel wiring harness
- (A44) - Wire connection (57R), in instrument panel wiring harness

Airbag

1999 m. y.

Fuse Panel

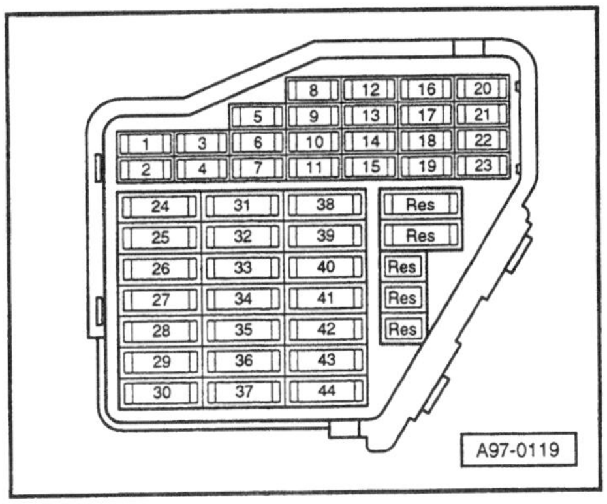

Fuse Colors:

30 A - Green
25 A - White
20 A - Yellow
15 A - Blue
10 A - Red
7,5 A - Brown
5 A - Beige

Starting with fuse position 23, fuses in the fuse holder are identified with 223 in the wiring diagram.

Micro Central Electric Panel

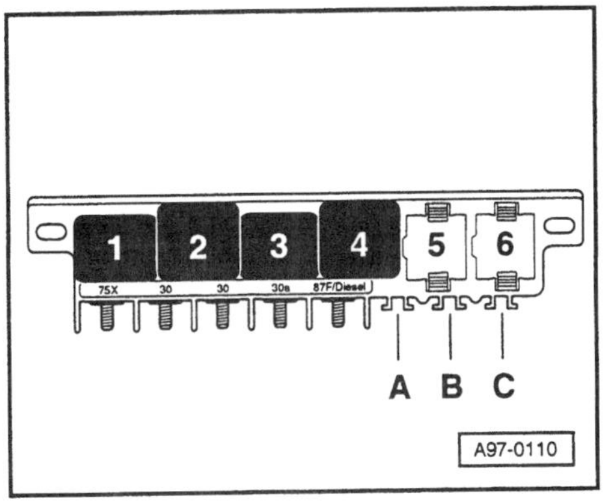

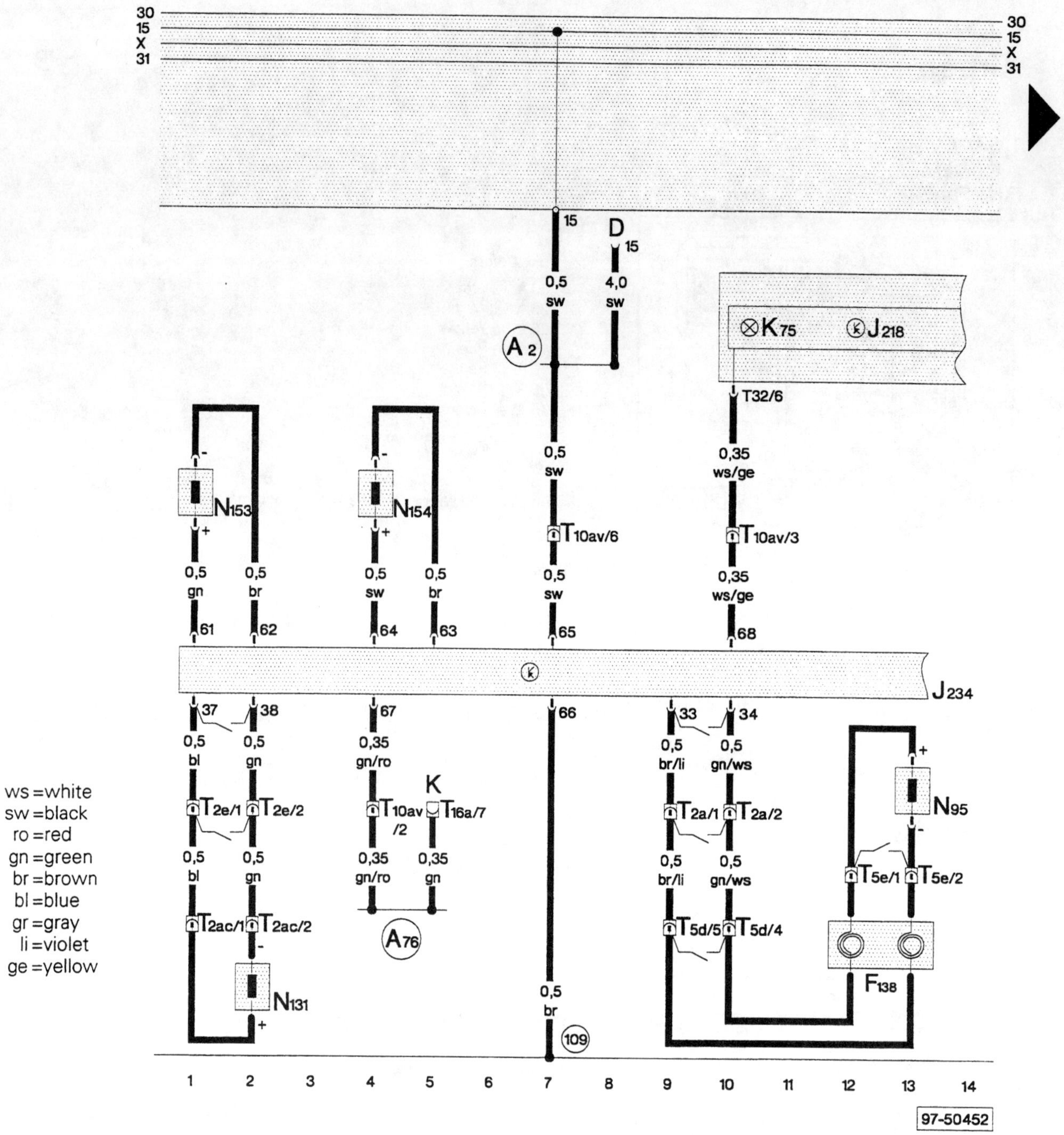

D - Ignition/Starter Switch
F 138 - Airbag Spiral Spring/Return Spring With Slip Ring
J 218 - Instrument Cluster Combination Processor
J 234 - Airbag Control Module
K - DLC (Data Link Connector) Rapid Data Transfer
K 75 - Airbag Malfunction Indicator Lamp (MIL)
N 95 - Driver's Side Airbag Igniter
N 131 - Passenger's Side Airbag Igniter 1
N 153 - Left Seat Belt Tensioner Igniter
N 154 - Right Seat Belt Tensioner Igniter
T 2 a - Connector 2 pin, yellow, connector station A pillar, left
T 2 e - Connector 2 pin, yellow, connector station A pillar, right
T 2 ac - Connector 2 pin, yellow, behind instrument panel, right
T 5 d - Connector 5 pin, behind steering column switch cover
T 5 e - Connector 5 pin, in steering wheel
T 10 av - Connector 10 pin, yellow, connector station A pillar, right
T 16 a - Connector 16 pin, data link connector
T 32 - Connector 32 pin, blue, on instrument cluster

(109) - Ground connection, in airbag wiring harness

(A2) - Plus connection (15), in instrument panel wiring harness

(A76) - Connector (K-diagnosis wire), in instrument panel wiring harness

- Bridge closes when disconnecting connector (safety feature)

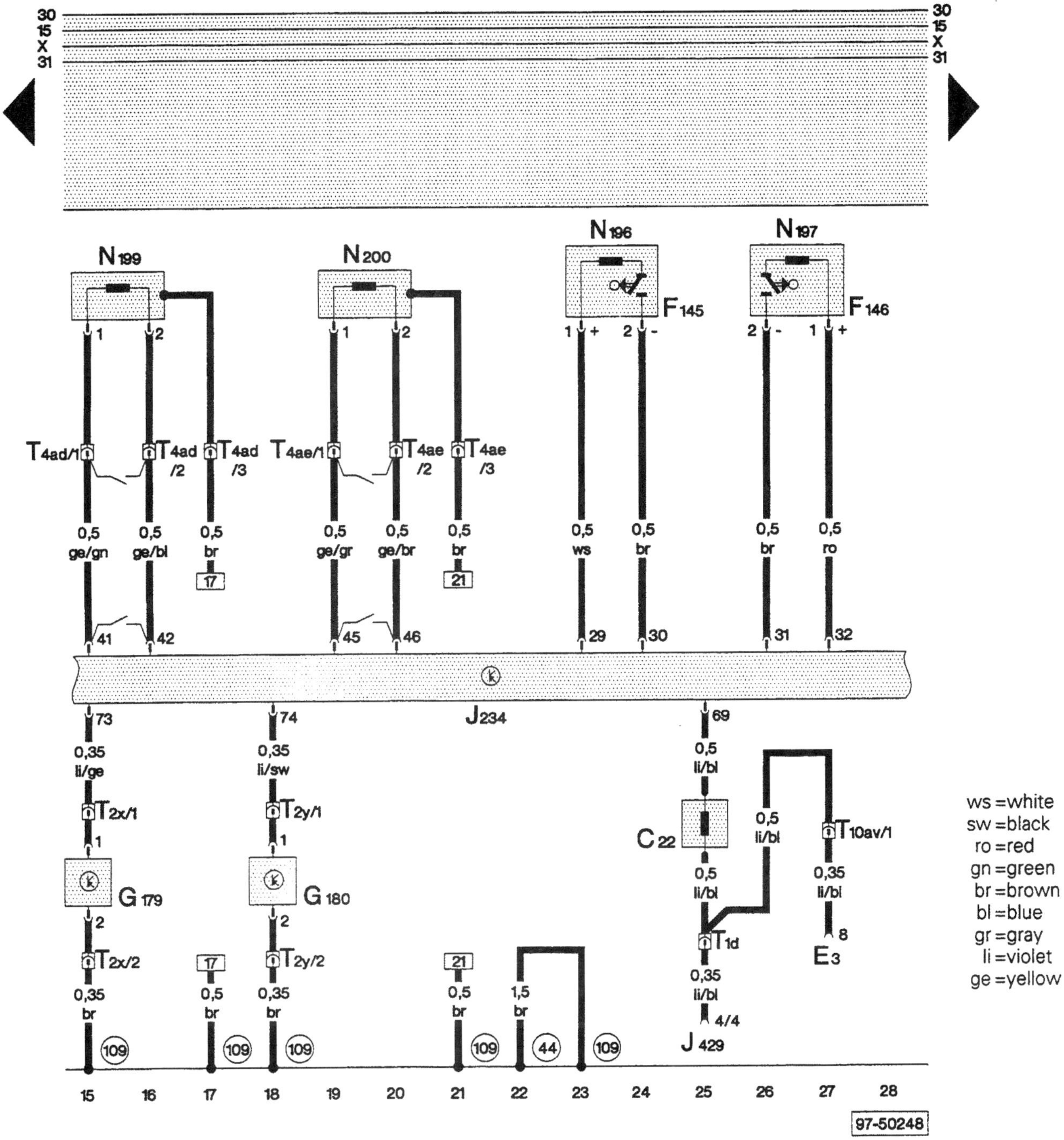

C 22 - Interference filter
E 3 - Emergency Flasher Switch
F 145 - Left Rear B-pillar Microswitch
F 146 - Right Rear B-pillar Microswitch
G 179 - Crash sensor for side airbag, driver's side
G 180 - Crash sensor for side airbag, passenger side
J 234 - Airbag Control Module
J 429 - Control module for central locking
N 196 - Igniter for belt tensioner rear, driver's side
N 197 - Igniter for belt tensioner rear, passenger side
N 199 - Igniter for side airbag, driver's side
N 200 - Igniter for side airbag, passenger side
T 1 d - Single connector, brown, connector station A pillar right
T 2 x - Connector 2 pin, yellow, below driver's seat
T 2 y - Connector 2 pin, yellow, below passenger's seat
T 4 ad - Connector 4 pin, yellow, below driver's seat
T 4 ae - Connector 4 pin, yellow, below passenger's seat
T 10av - Connector 10 pin, yellow, connector station A pillar, right

(44) - Ground connection, on left A-pillar, lower part
(109) - Ground connection, in airbag wiring harness

- Bridge closes when disconnecting connector (safety feature)

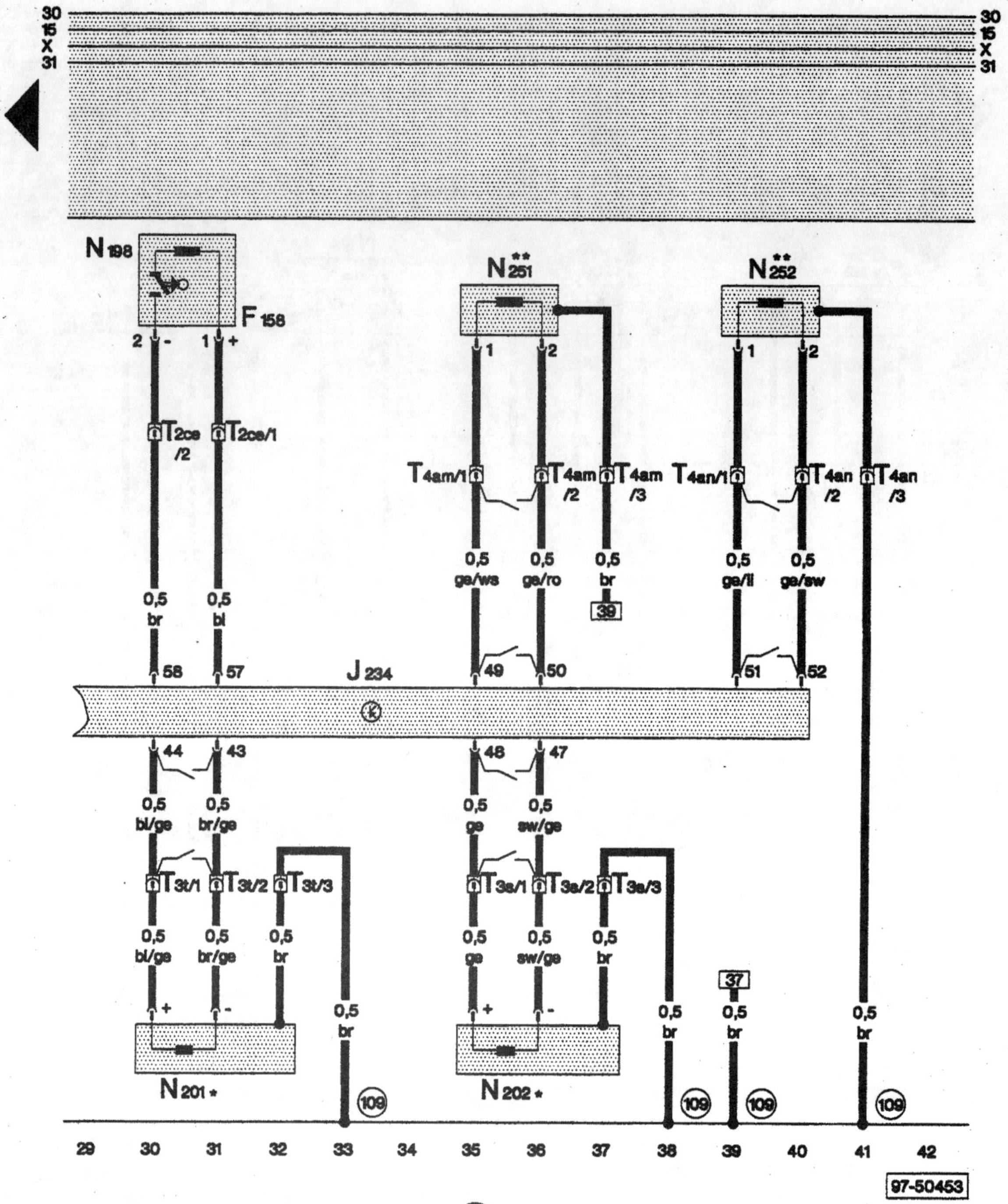

ws = white
sw = black
ro = red
gn = green
br = brown
bl = blue
gr = gray
li = violet
ge = yellow

F 158 - Seat Belt Tensioner Switch (1)
J 234 - Airbag Control Module
N 198 - Igniter for belt tensioner, rear, center
N 201 - Igniter for side airbag rear, driver's side
N 202 - Igniter for side airbag rear, passenger side
N 251 - Igniter for head airbag, driver side
N 252 - Igniter for head airbag, passenger side
T 2 ce - Connector 2 pin, yellow, in rear seat
T 3 s - Connector 3 pin, orange, in rear seat
T 3 t - Connector 3 pin, orange, in rear seat
T 4 am - Connector 4 pin, yellow, in C pillar, left
T 4 an - Connector 4 pin, yellow, in C pillar, right

(109) - Ground connection, in airbag wiring harness

- Bridge closes when disconnecting connector (safety feature)
* - Vehicles with sideairbag rear
** - Vehicles with head airbag

ABS

1999 m. y.

Fuse Panel

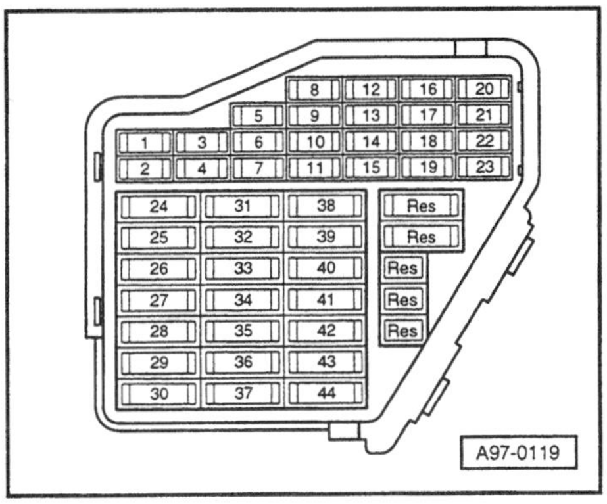

Fuse Colors:

30 A - Green
25 A - White
20 A - Yellow
15 A - Blue
10 A - Red
7,5 A - Brown
5 A - Beige

Starting with fuse position 23, fuses in the fuse holder are identified with 223 in the wiring diagram.

8 - Fold Relay Panel

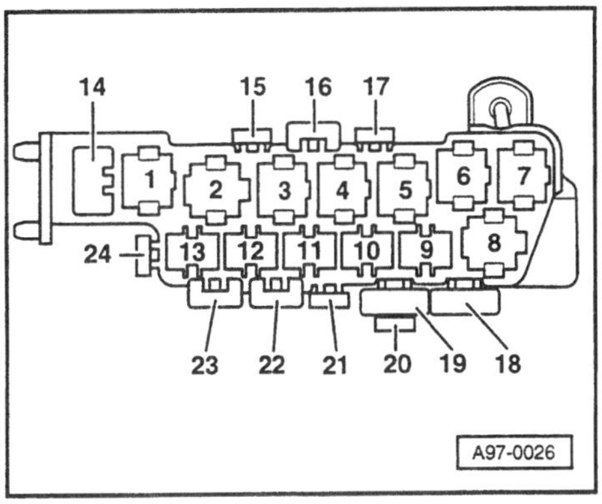

Relay Location:

18 - Fuse 1 for control module - ABS, S123

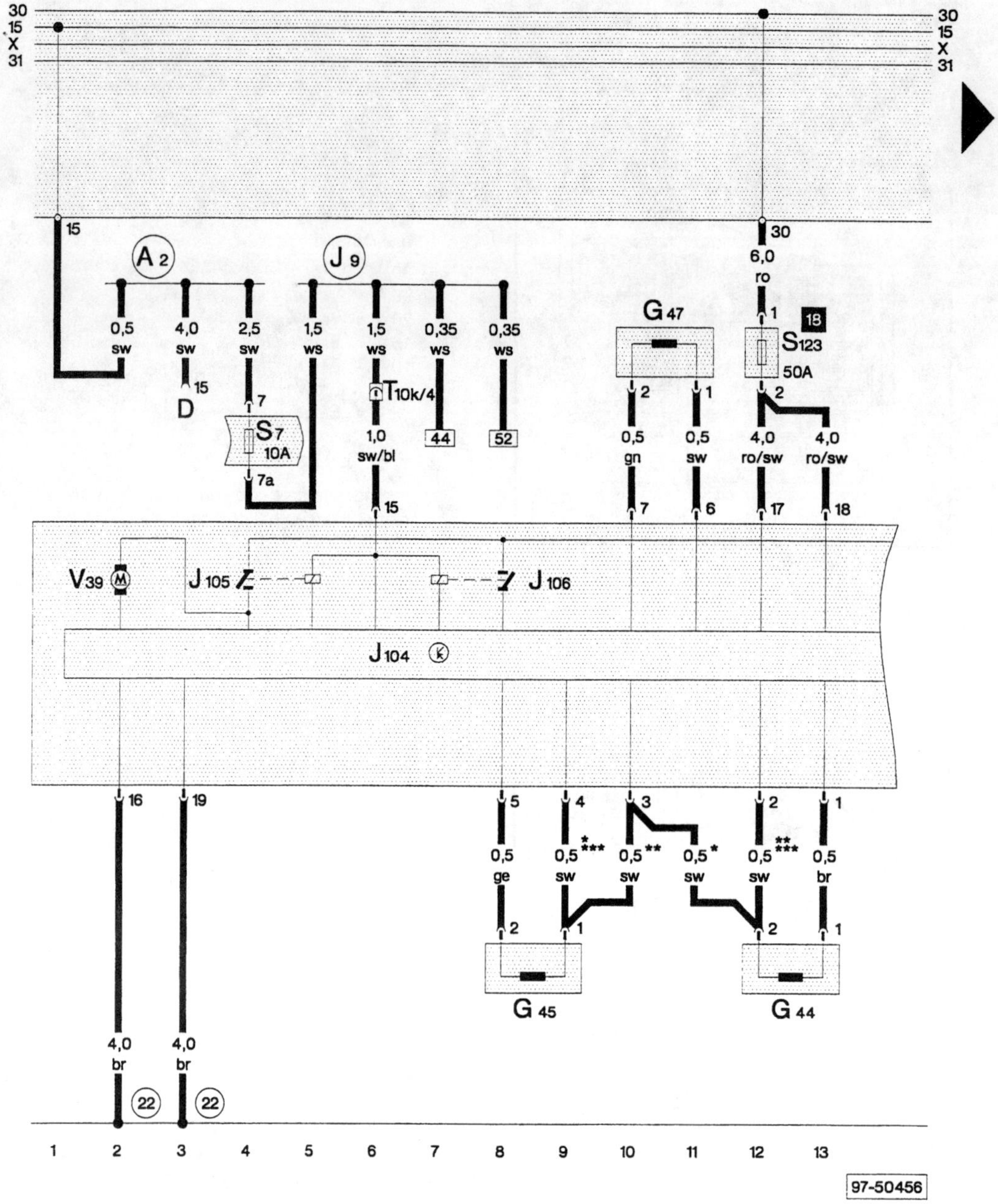

D - Ignition/Starter Switch
G 44 - Right Rear ABS Wheel Speed Sensor
G 45 - Right Front ABS Wheel Speed Sensor
G 47 - Left Front ABS Wheel Speed Sensor
J 104 - ABS Control Module (w/EDL)
J 105 - ABS Return Flow Pump Relay
J 106 - ABS Solenoid Valve Relay
S 7 - Fuse
S 123 - Fuse 1 for control module - ABS
T 10 k - Connector 10 pin, orange, connector station A pillar, left
V 39 - ABS Return Flow Pump

(22) - Ground connection, on hydraulic unit
(A2) - Plus connection (15), in instrument panel wiring harness
(J9) - Wire connection -1- (15a), in ABS wiring harness

* - Vehicles without Anti-Slip Control
** - Vehicles with Anti-Slip Control
*** - Vehicles with all-wheel drive

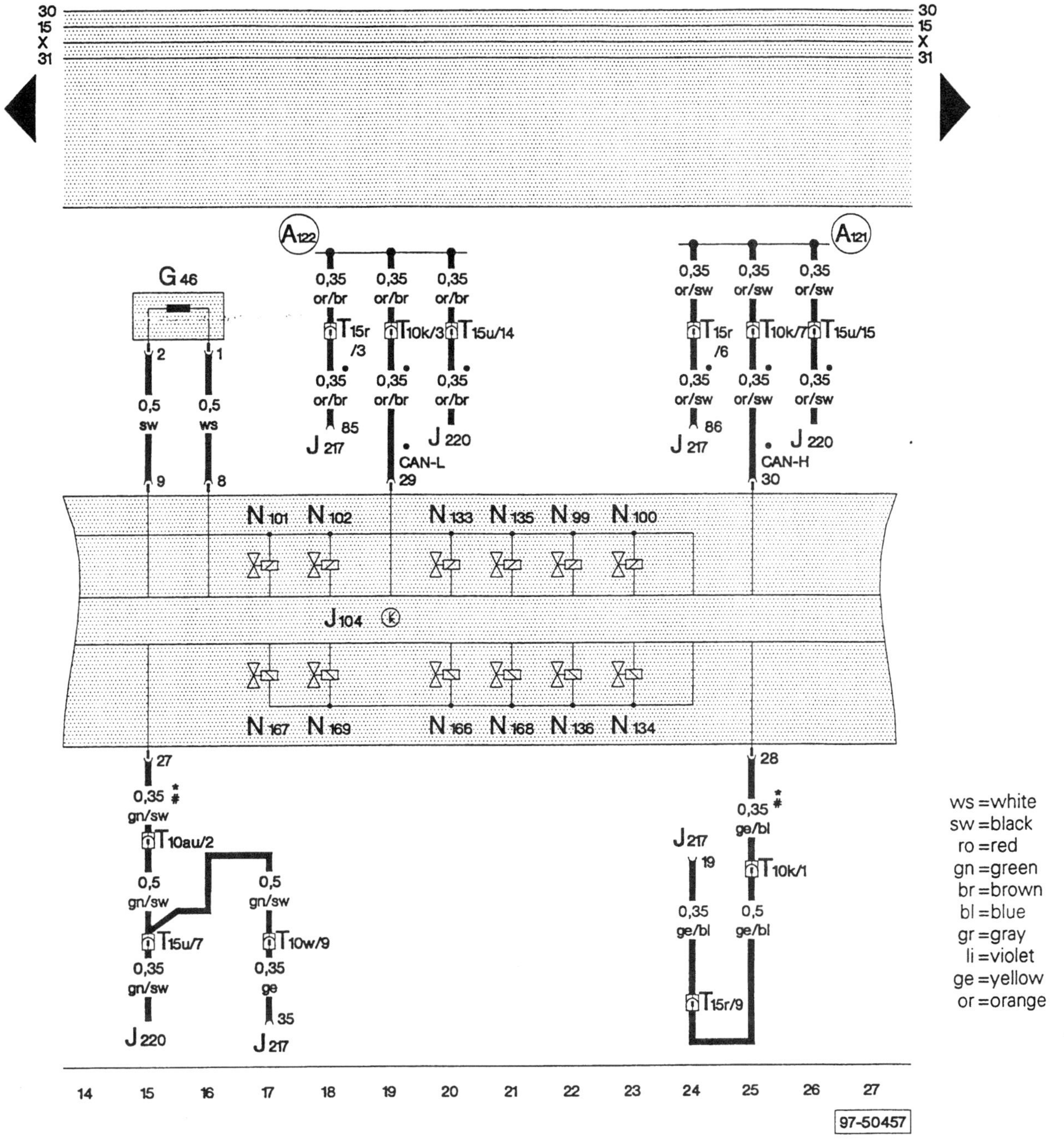

G 46 - Left Rear ABS Wheel Speed Sensor
J 104 - ABS Control Module (w/EDL)
J 217 - Transmission Control Module (TCM)
J 220 - Motronic Engine Control Module (ECM)
N 99 - Right Front ABS Inlet Valve
N 100 - Right Front ABS Outlet Valve
N 101 - Left Front ABS Inlet Valve
N 102 - Left Front ABS Outlet Valve
N 133 - Right Rear ABS Inlet Valve
N 134 - Left Rear ABS Inlet Valve
N 135 - Right Rear ABS Outlet Valve
N 136 - Left Rear ABS Outlet Valve
N 166 - Right Front Traction Control Switch-Over Valve
N 167 - Right Front Traction Control Outlet Valve
N 168 - Left Front Traction Control Switch-Over Valve
N 169 - Left Front Traction Control Outlet Valve
T 10 k - Connector 10 pin, orange, connector station A pillar, left
T 10 w - Connector 10 pin, blue, connector station A pillar, right

T 10 au - Connector 10 pin, grey, connector station A pillar, right
T 15 r - Connector 15 pin, brown, connector station A pillar, right
T 15 u - Connector 15 pin, red, connector station electronic box plenum chamber

(A121) - Connector (High-Bus), in instrument panel wiring harness
(A122) - Connector (Low-Bus), in instrument panel wiring harness

• - CAN-Bus
* - Vehicles with Anti-Slip Control
# - Vehicles without CAN-Bus

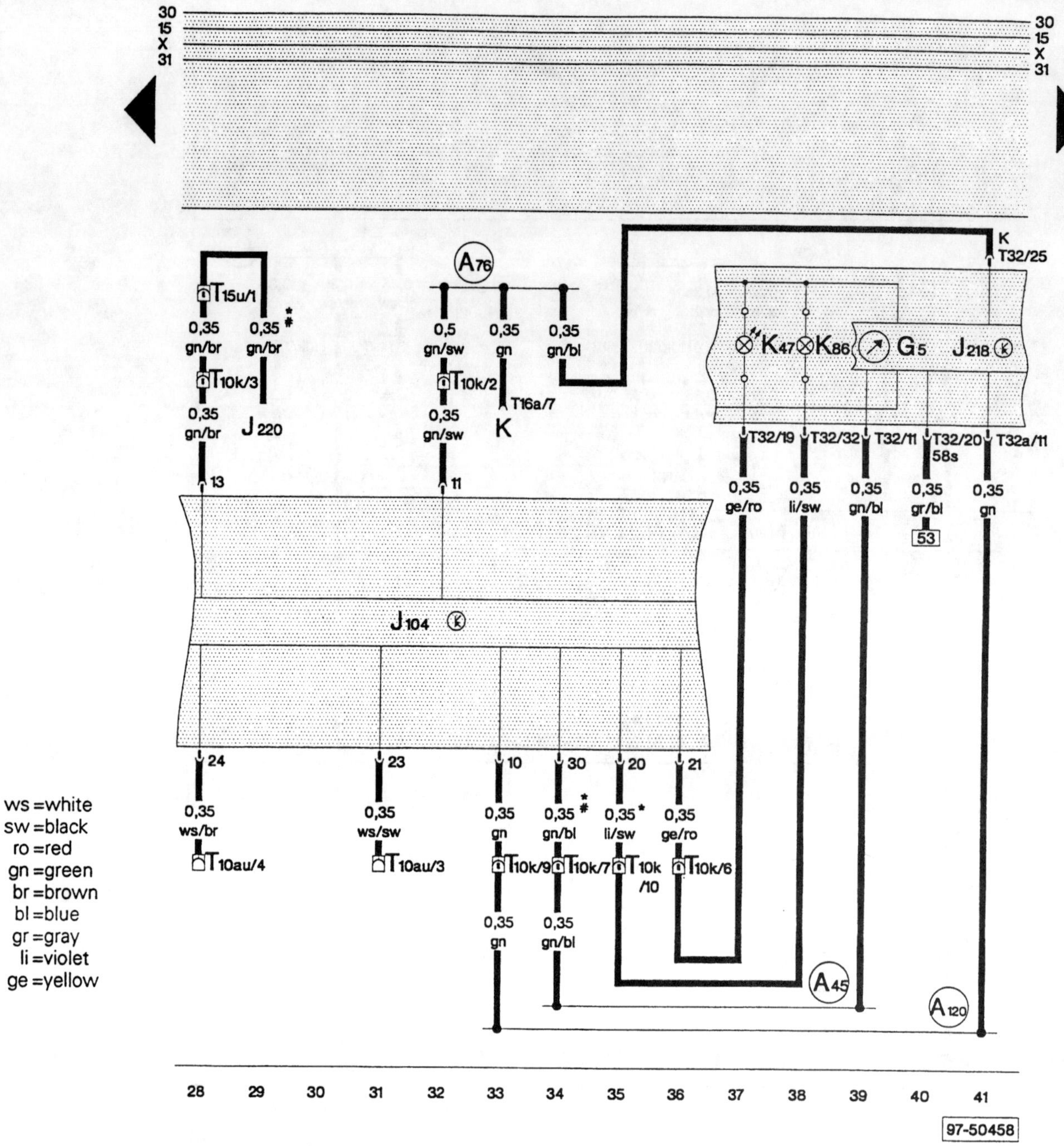

G 5 - Tachometer
J 104 - ABS Control Module (w/EDL)
J 218 - Instrument Cluster Combination Processor
J 220 - Motronic Engine Control Module (ECM)
K - DLC (Data Link Connector) Rapid Data Transfer
K 47 - ABS Warning Light
K 86 - Traction Control Indicator Light
T 10 k - Connector 10 pin, orange, connector station A pillar, left
T 10 au - Connector 10 pin, grey, connector station A pillar, right
T 15 u - Connector 15 pin, red, connector station electronic box plenum chamber
T 16 a - Connector 16 pin, data link connector
T 32 - Connector 32 pin, blue, on instrument cluster
T 32 a - Connector 32 pin, green, on instrument cluster

(A45) - Wire connection (RPM-signal), in instrument panel wiring harness
(A76) - Connector (K-diagnosis wire), in instrument panel wiring harness
(A120) - Wire connection (Immobilization time), in instrument panel wiring harness

* - Vehicles with Anti-Slip Control
# - Vehicles without CAN-Bus

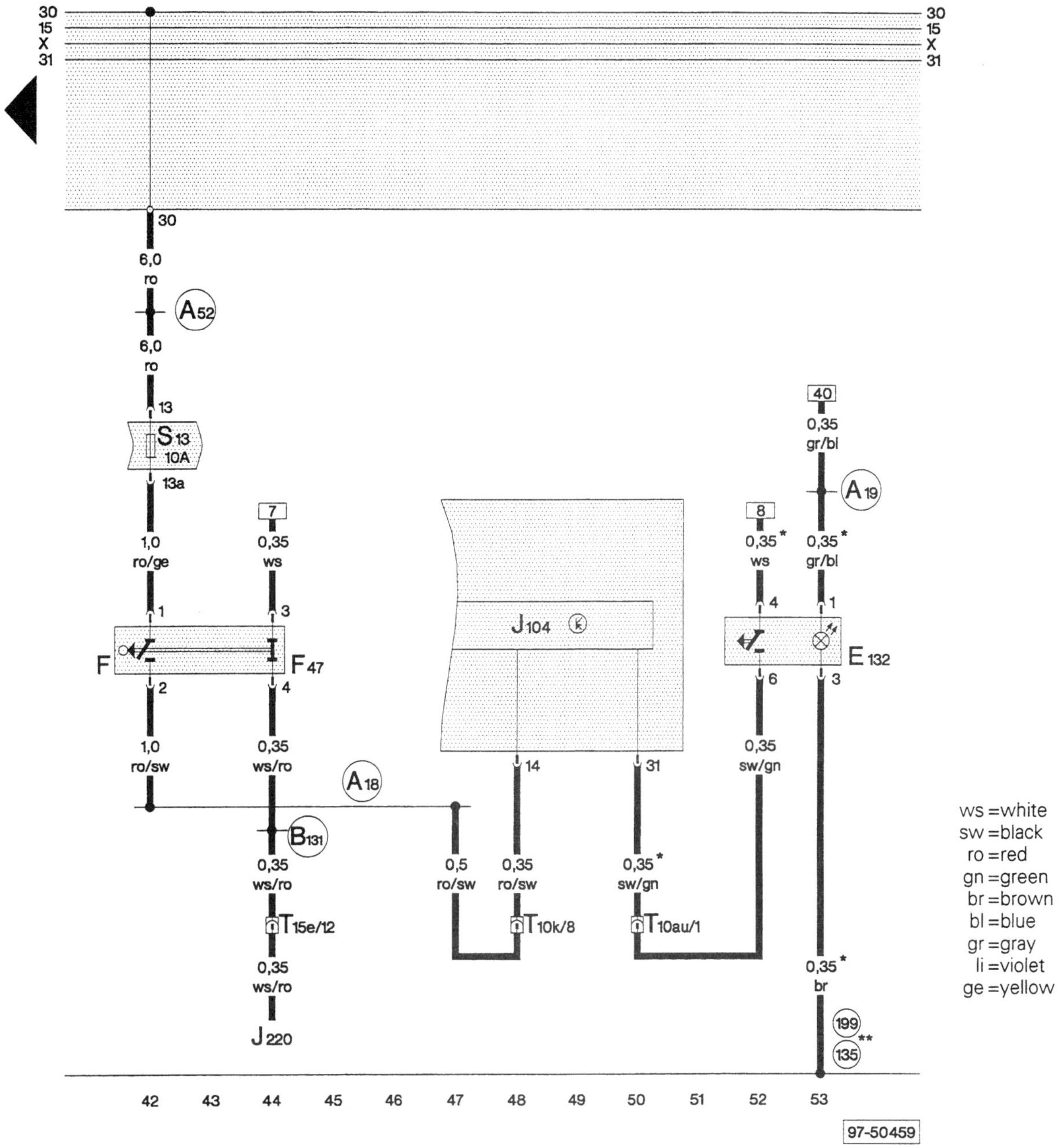

- E 132 - Anti-Slip Control Switch
- F - Brake Light Switch
- F 47 - Brake Pedal Switch
- J 104 - ABS Control Module (w/EDL)
- J 220 - Motronic Engine Control Module (ECM)
- S 13 - Fuse
- T 10 k - Connector 10 pin, orange, connector station A pillar, left
- T 10 au - Connector 10 pin, grey, connector station A pillar, right
- T 15 e - Connector 15 pin, white, connector station electronic box plenum chamber

- (135) - Ground connection -2-, in instrument panel wiring harness
- (199) - Ground connection -3-, in instrument panel wiring harness
- (A18) - Wire connection (54), in instrument panel wiring harness
- (A19) - Wire connection (58s), in instrument panel wiring harness (Switch illumination)
- (A52) - Plus connection (30), in instrument panel wiring harness
- (B131) - Connector (54), in wiring harness interior

* -Vehicles with Anti-Slip Control
** -booth is possible

Power Seat and Mirror Adjustment with Memory, Mirror with and without fold-away Function

1999 m. y.

Fuse Panel

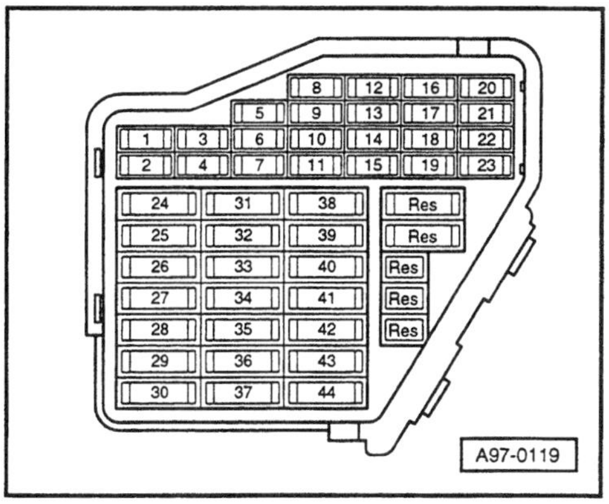

Fuse Colors:

30 A - Green
25 A - White
20 A - Yellow
15 A - Blue
10 A - Red
7,5 A - Brown
5 A - Beige

Starting with fuse position 23, fuses in the fuse holder are identified with 223 in the wiring diagram.

13 - Fold Relay Panel

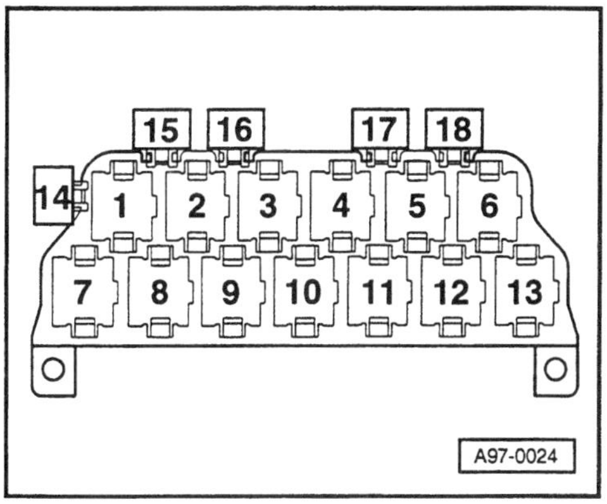

Relay Location:

11 - Control Module for Mirror fold-away Function, J351

12 - Control Module for Mirror fold-away Function, J351

8 - Fold Relay Panel

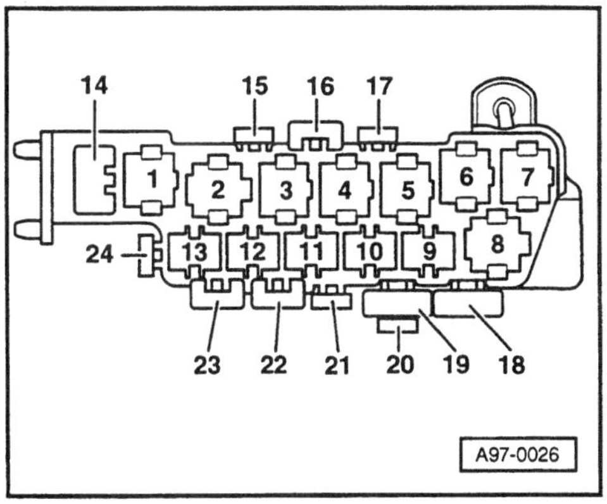

Fuse Location:

14 - Power Seat Circuit Breaker 1, S44

24 - Power Seat Circuit Breaker 2, S80

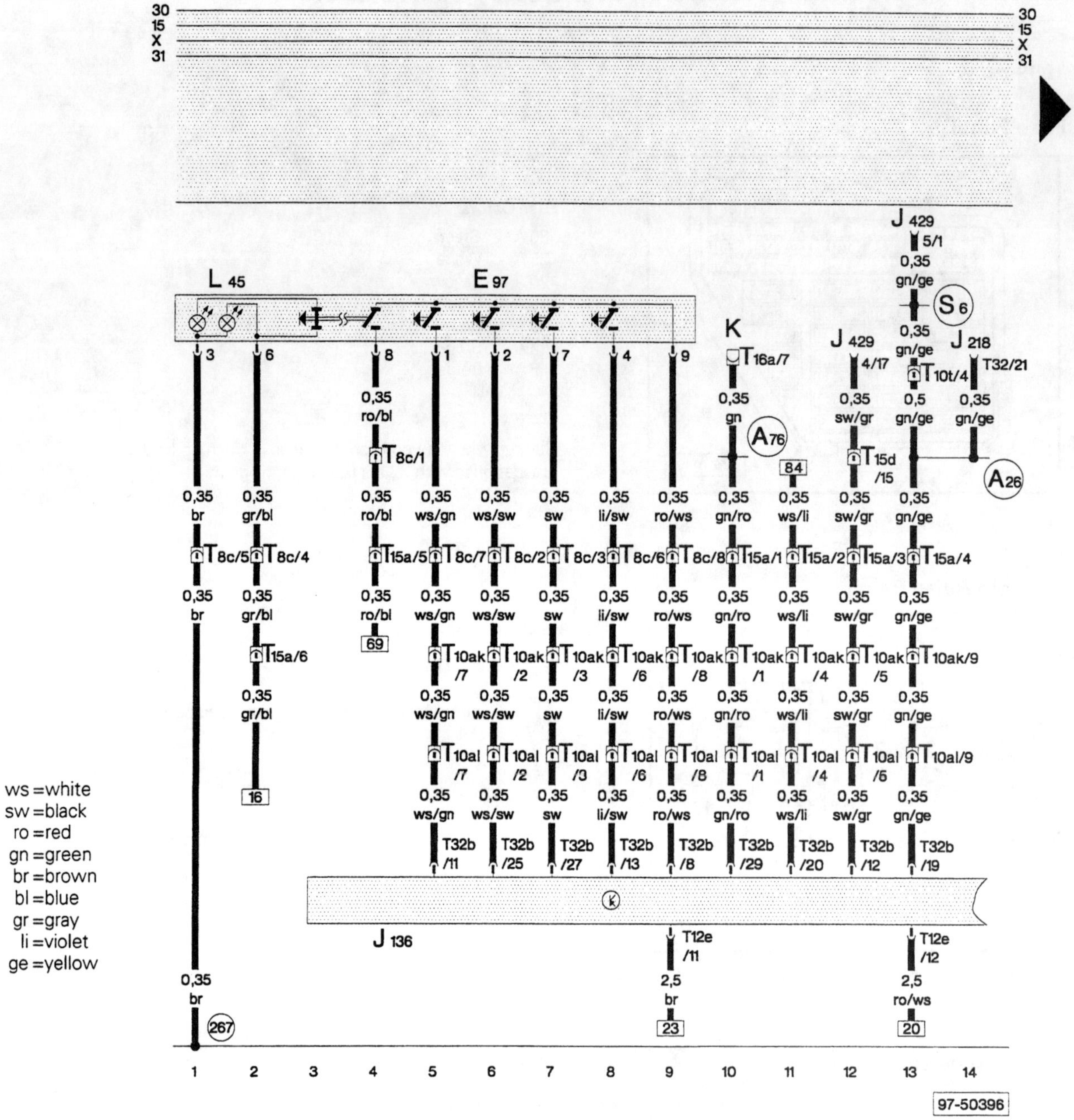

E 97 - Memory Program Switch
J 136 - Memory Seat Control Module
J 218 - Instrument Cluster Combination Processor
J 429 - Control module for central locking
K - DLC (Data Link Connector) Rapid Data Transfer
L 45 - Memory Operating Unit Light
T 8 c - Connector 8 pin, black, in driver's door
T 10 t - Connector 10 pin, orange, connector station A pillar, right
T 10 ak - Connector 10 pin, green, connector station A pillar, left
T 10 al - Connector 10 pin, brown, below driver's seat
T 12 e - Connector 12 pin, black, in driver's seat
T 15 a - Connector 15 pin, black, connector station A pillar, left
T 15 d - Connector 15 pin, black, connector station A pillar, right
T 16 a - Connector 16 pin, data link connector
T 32 - Connector 32 pin, blue, on instrument cluster
T 32 b - Connector 32 pin, blue, in driver's seat

(267) - Ground connection -2-, in wiring harness Door cable - driver side
(A26) - Wire connection (driver's door contact switch), in instrument panel wiring harness
(A76) - Connector (K-diagnosis wire), in instrument panel wiring harness
(S6) - Wire connection (Pin 5/1), in central locking system wiring harness

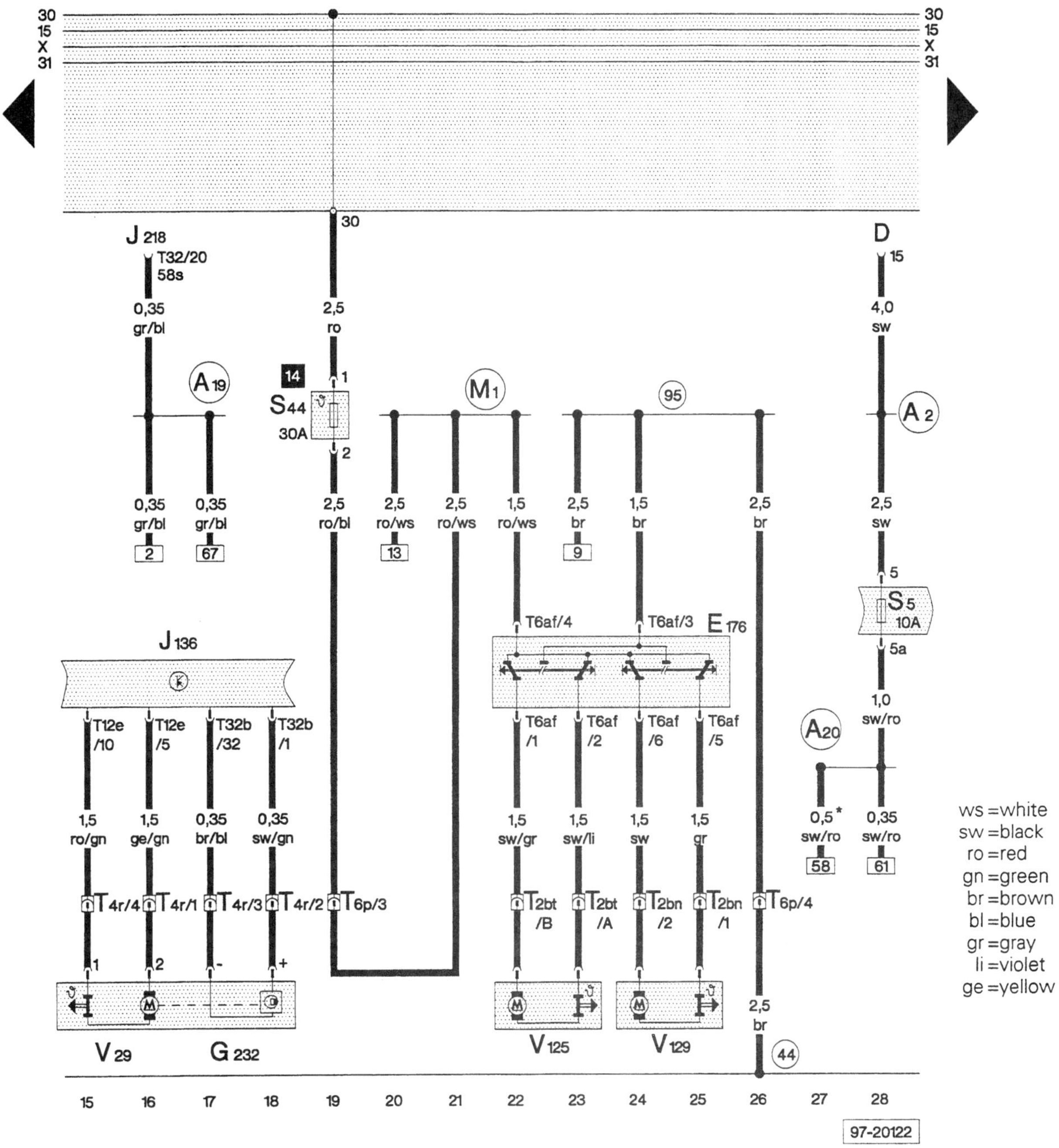

D - Ignition/Starter Switch
E 176 - Driver's Seat Lumbar Support Adjustment Switch
G 232 - Sensor for slope adjustment, driver seat
J 136 - Memory Seat Control Module
J 218 - Instrument Cluster Combination Processor
S 5 - Fuse
S 44 - Power Seat Circuit Breaker 1
T 2 bt - Connector 2 pin, black, in driver's seat
T 2 bu - Connector 2 pin, black, in driver's seat
T 4 r - Connector 4 pin, black, in driver's seat
T 6 p - Connector 6 pin, red, below driver's seat
T 6 af - Connector 6 pin, black, in driver's seat
T 12 e - Connector 12 pin, black, in driver's seat
T 32 - Connector 32 pin, blue, on instrument cluster
T 32 b - Connector 32 pin, blue, in driver's seat
V 29 - Driver's Seat Front Height Adjusting Motor
V 125 - Driver's Seat Lumbar Support Curvature Adjustment Motor
V 129 - Driver's Seat Lumbar Support Height Adjustment Motor

(44) - Ground connection, on left A-pillar, lower part
(95) - Ground connection -1-, in power seat wiring harness
(A2) - Plus connection (15), in instrument panel wiring harness
(A19) - Wire connection (58s), in instrument panel wiring harness (Switch illumination)
(A20) - Wire connection (15a), in instrument panel wiring harness
(M1) - Plus connection -1- (30), in power seats wiring harness

* - mirror with fold-away Function

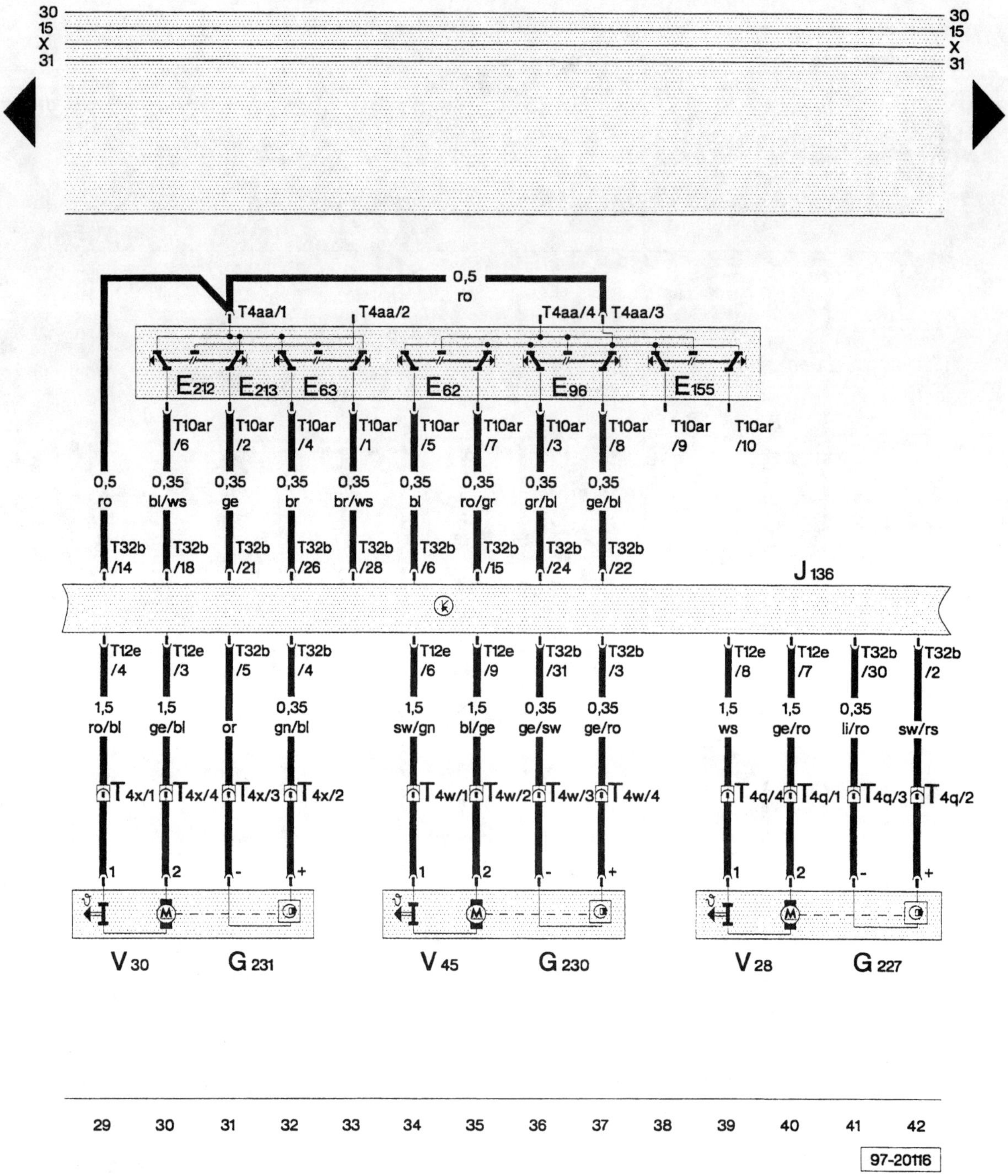

ws = white
sw = black
ro = red
gn = green
br = brown
bl = blue
gr = gray
li = violet
ge = yellow

- E 62 - Driver's Seat Front Height Adjusting Switch
- E 63 - Driver's Seat Rear Height Adjusting Switch
- E 96 - Driver's Backrest Adjustment Switch
- E 155 - Driver's Headrest Adjustment Switch
- E 212 - Button for length adjustment, forward, driver seat
- E 213 - Button for length adjustment, backward, driver seat
- G 227 - Sensor for fore and aft adjustment, driver seat
- G 230 - Sensor for seat back adjustment, driver seat
- G 231 - Sensor for height adjustment, driver seat
- J 136 - Memory Seat Control Module
- T 4 q - Connector 4 pin, black, in driver's seat
- T 4 w - Connector 4 pin, black, in driver's seat
- T 4 x - Connector 4 pin, black, in driver's seat
- T 4 aa - Connector 4 pin, black, in driver's seat
- T 10 ar - Connector 10 pin, black, in driver's seat
- T 12 e - Connector 12 pin, black, in driver's seat
- T 32 b - Connector 32 pin, blue, in driver's seat
- V 28 - Driver's Seat Fore/Aft Adjusting Motor
- V 30 - Driver's Seat Rear Height Adjusting Motor
- V 45 - Driver's Backrest Adjusting Motor

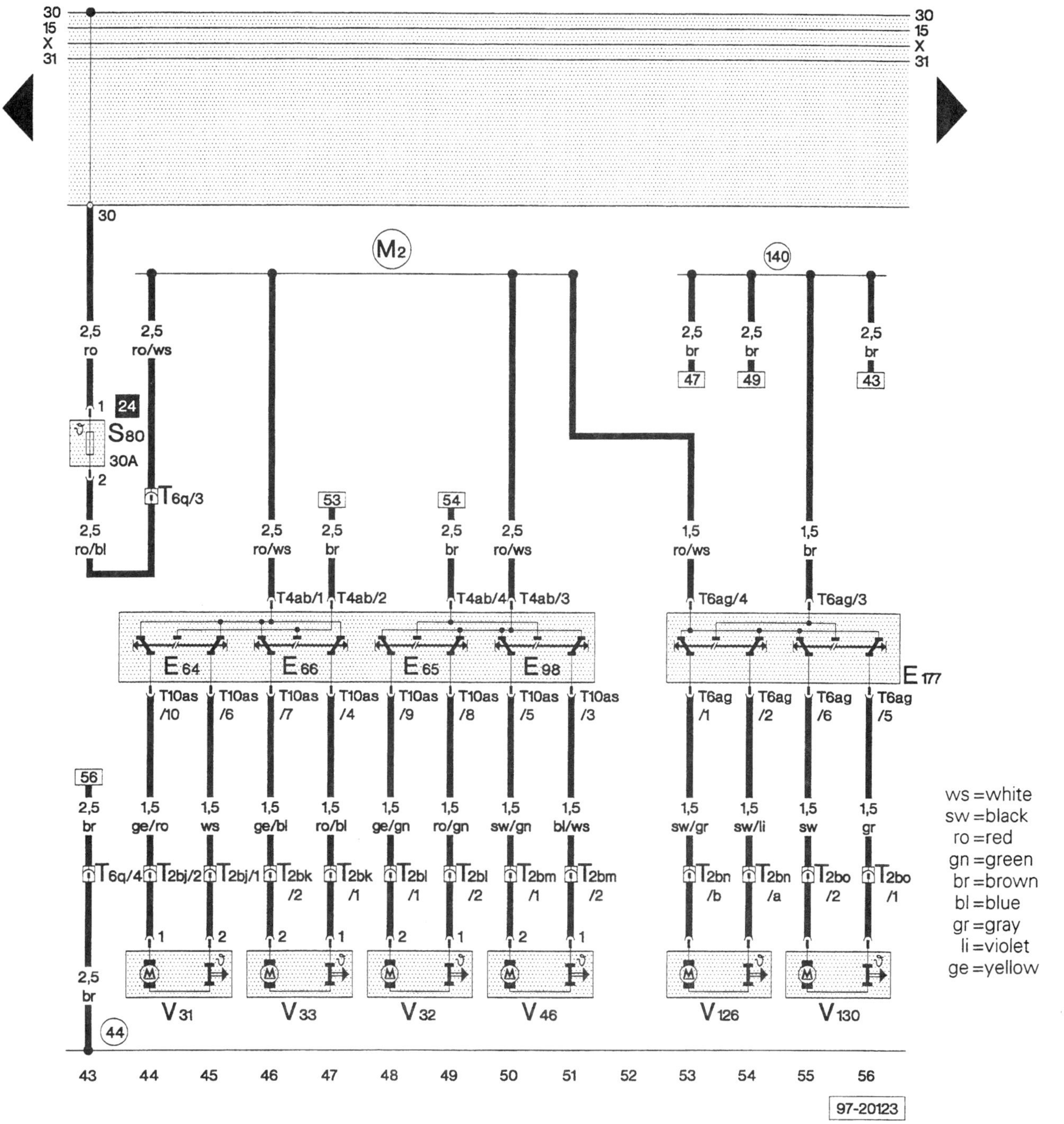

E 64 - Passenger's Seat Fore/Aft Adjusting Switch
E 65 - Passenger's Seat Front Height Adjusting Switch
E 66 - Passenger's Seat Rear Height Adjusting Switch
E 98 - Passenger Backrest Adjustment Switch
E 177 - Front Passenger Seat Lumbar Support Adjustment Switch
S 80 - Power Seat Circuit Breaker 2
T 2 bj - Connector 2 pin, black, in passenger's seat
T 2 bk - Connector 2 pin, black, in passenger's seat
T 2 bl - Connector 2 pin, black, in passenger's seat
T 2 bm - Connector 2 pin, black, in passenger's seat
T 2 bn - Connector 2 pin, black, in passenger's seat
T 2 bo - Connector 2 pin, black, in passenger's seat
T 4 ab - Connector 4 pin, black, in passenger's seat
T 6 q - Connector 6 pin, red, below passenger's seat
T 6 ag - Connector 6 pin, black, in passenger's seat
T 10 as - Connector 10 pin, black, in passenger's seat
V 31 - Passenger's Seat Fore/Aft Adjusting Motor
V 32 - Passenger's Seat Front Height Adjusting Motor
V 33 - Passenger's Seat Rear Height Adjusting Motor
V 46 - Passenger's Backrest Adjusting Motor
V 126 - Front Passenger Seat Lumbar Support Curvature Adjustment Motor
V 130 - Front Passenger Seat Lumbar Support Height Adjustment Motor

(44) - Ground connection, on left A-pillar, lower part
(140) - Ground connection -2-, in power seat wiring harness
(M2) - Plus connection -2- (30), in power seats wiring harness

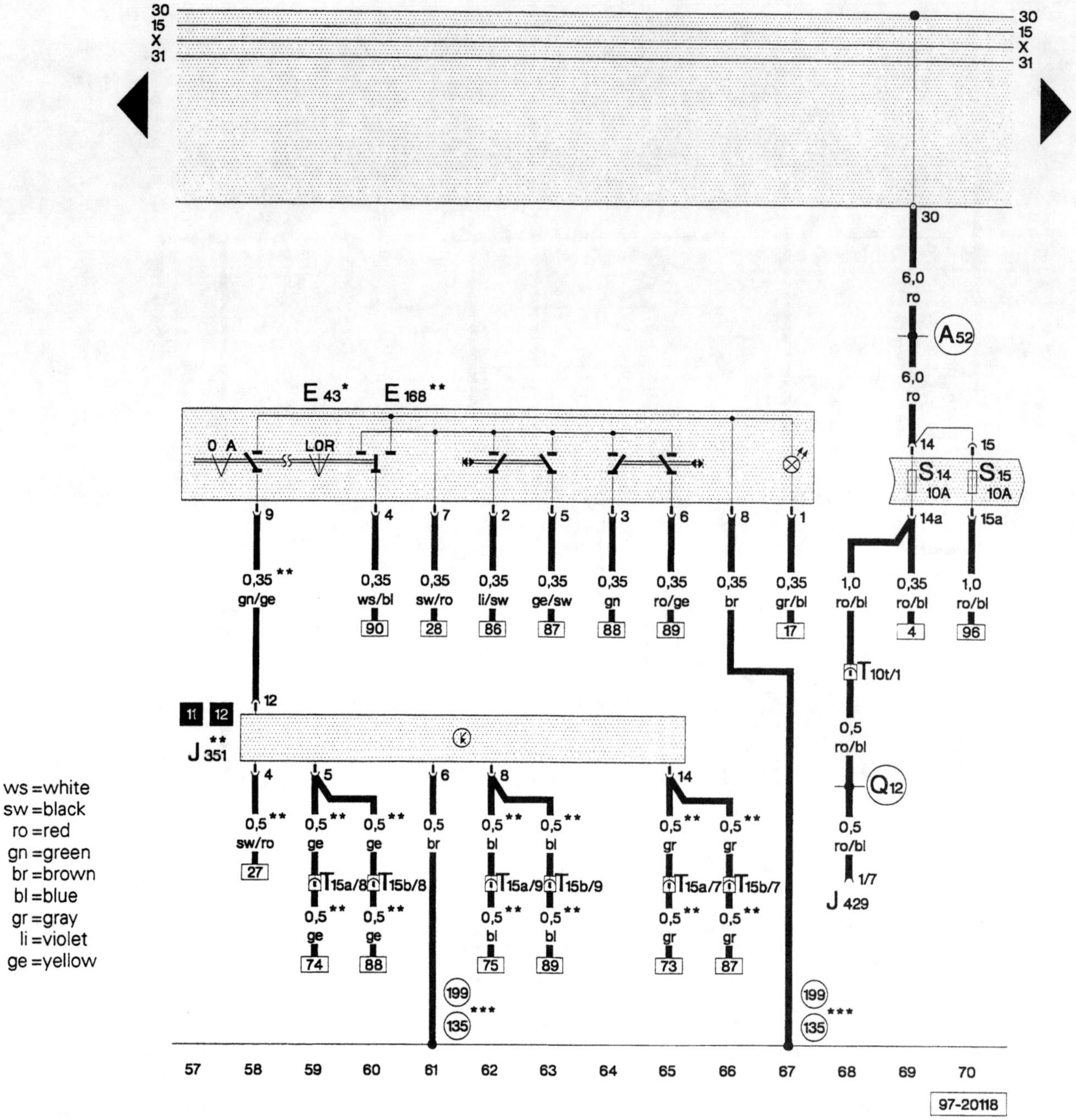

E 43 - Mirror Adjustment Switch
E 168 - Switch for mirror adjustment with fold-away Function
J 351 - Control Module for Mirror fold-away Function
J 429 - Control module for central locking
S 14 - Fuse
S 15 - Fuse
T 10 t - Connector 10 pin, orange, connector station A pillar, right
T 15 a - Connector 15 pin, black, connector station A pillar, left
T 15 b - Connector 15 pin, black, connector station A pillar, right

(135) - Ground connection -2-, in instrument panel wiring harness
(199) - Ground connection -3-, in instrument panel wiring harness
(A52) - Plus connection (30), in instrument panel wiring harness
(Q12) - Plus connection (30a), in power window/central locking system and door contact switch wiring harness

* - mirror without fold-away Function
** - mirror with fold-away Function
*** - booth is possible

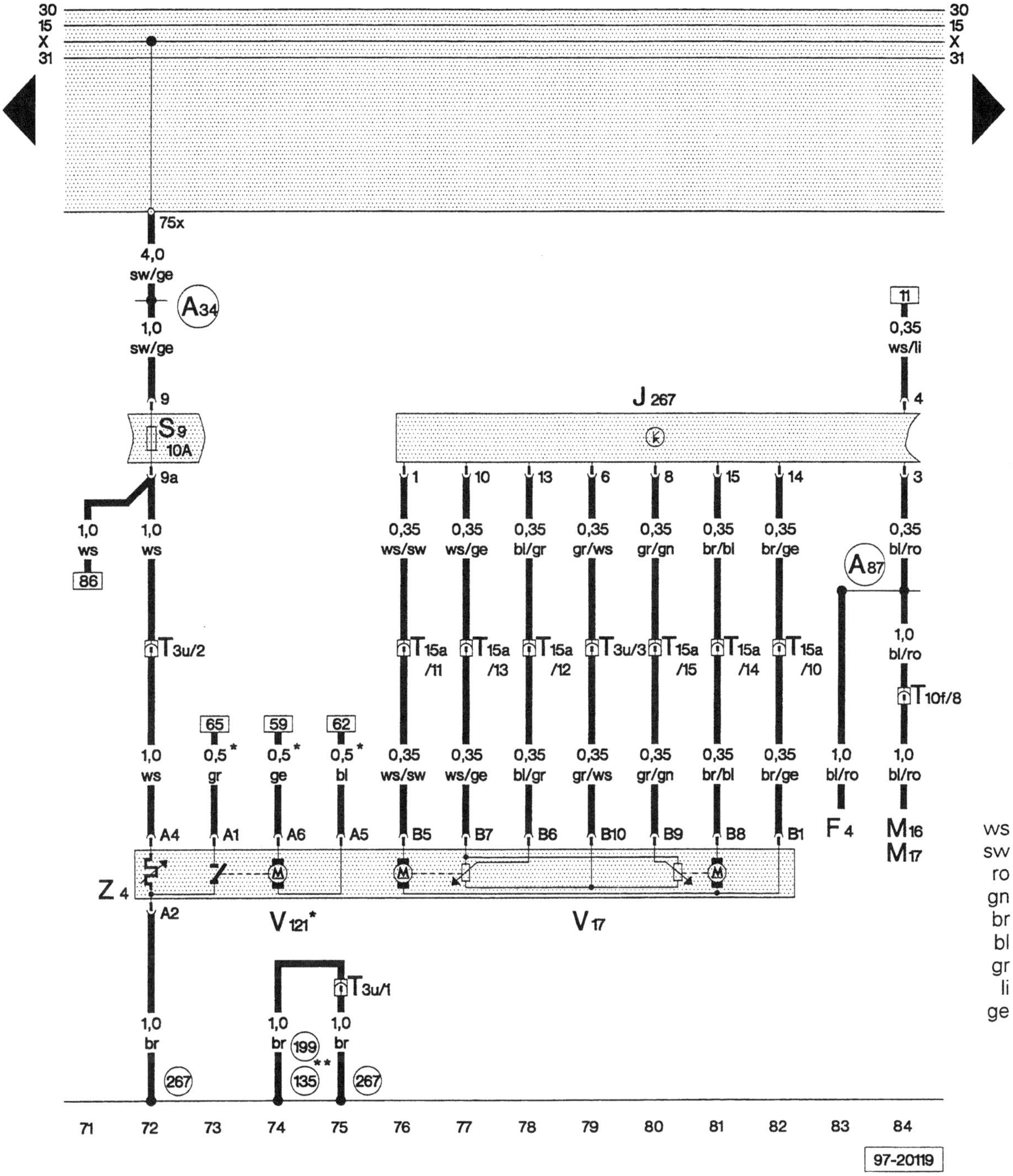

ws = white
sw = black
ro = red
gn = green
br = brown
bl = blue
gr = gray
li = violet
ge = yellow

F 4 - Back-Up Light Switch
J 267 - Mirror Memory Control Module
M 16 - Left Back-Up Light
M 17 - Right Back-Up Light
S 9 - Fuse
T 3 u - Connector 3 pin, black, connector station A pillar, left
T 10 f - Connector 10 pin, brown, connector station A pillar, left
T 15 a - Connector 15 pin, black, connector station A pillar, left
V 17 - Driver's Side Mirror Adjustment Motor
V 121 - Motor for fold-in Mirror driver's side
Z 4 - Heated outside mirror, driver side

(135) - Ground connection -2-, in instrument panel wiring harness
(199) - Ground connection -3-, in instrument panel wiring harness
(267) - Ground connection -2-, in wiring harness Door cable - driver side
(A34) - Wire connection (75x), in instrument panel wiring harness
(A87) - Connector (RF), in instrument panel wiring harness

* - mirror with fold-away Function
** - booth is possible

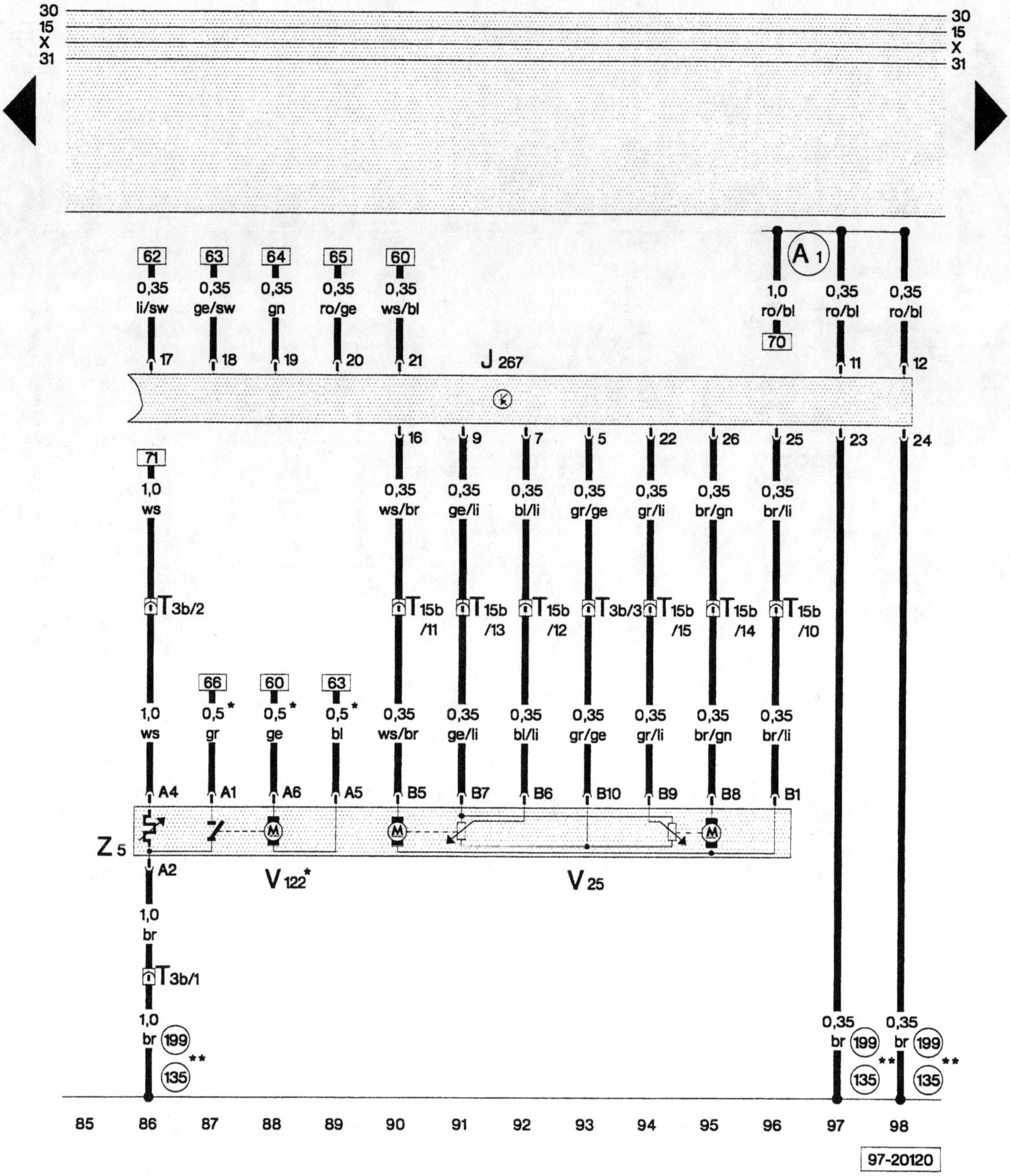

ws = white
sw = black
ro = red
gn = green
br = brown
bl = blue
gr = gray
li = violet
ge = yellow

J 267 - Mirror Memory Control Module
T 3 b - Connector 3 pin, black, connector station A pillar, right
T 15 b - Connector 15 pin, black, connector station A pillar, right
V 25 - Passenger's Side Mirror Adjustment Motor
V 122 - Motor for fold-in Mirror passenger side
Z 5 - Heated outside mirror, passenger side

(135) - Ground connection -2-, in instrument panel wiring harness
(199) - Ground connection -3-, in instrument panel wiring harness
(A1) - Plus connection (30a), in instrument panel wiring harness

* - mirror with fold-away Function
** - booth is possible

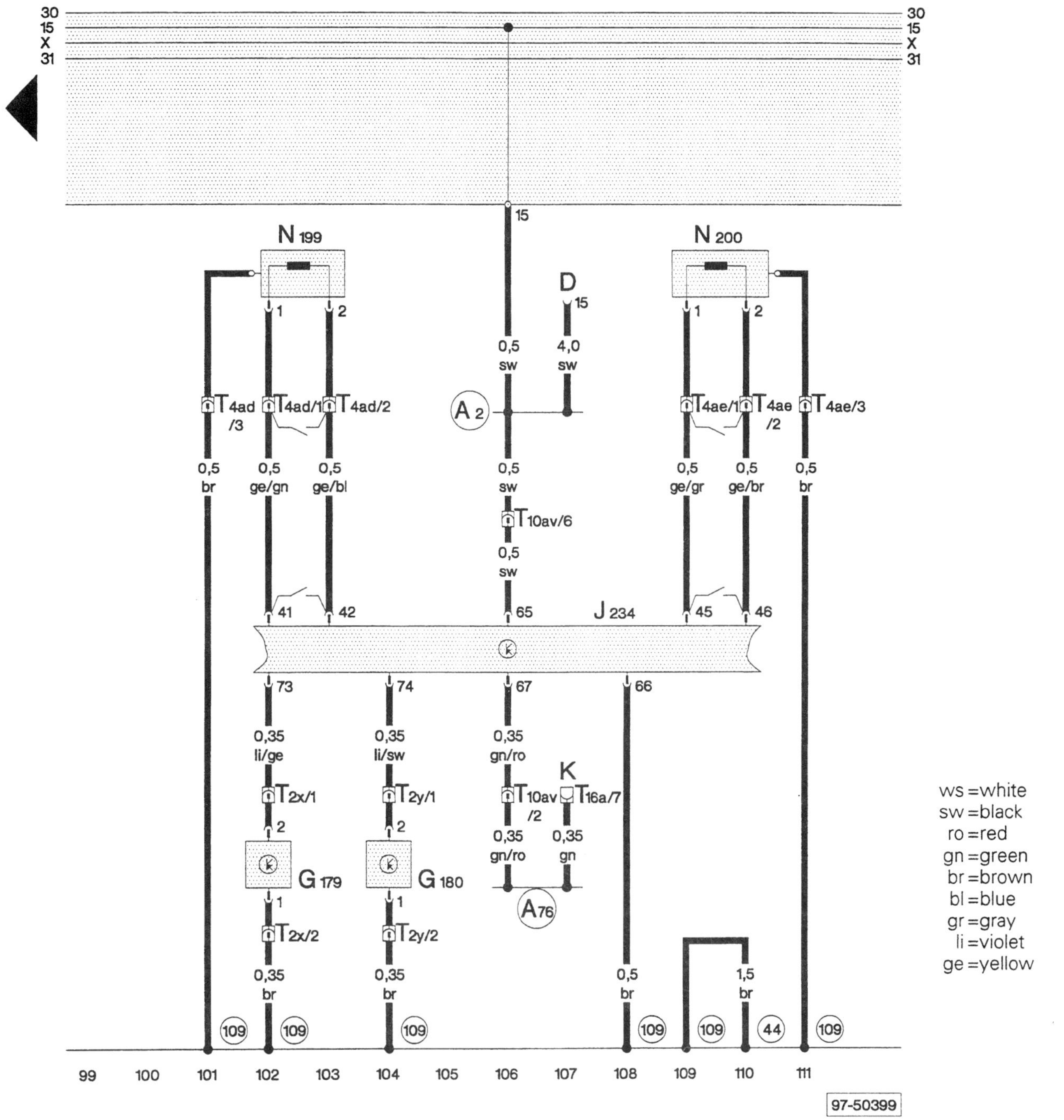

D - Ignition/Starter Switch
G 179 - Crash sensor for side airbag, driver's side
G 180 - Crash sensor for side airbag, passenger side
J 234 - Airbag Control Module
K - DLC (Data Link Connector) Rapid Data Transfer
N 199 - Igniter for side airbag, driver's side
N 200 - Igniter for side airbag, passenger side
T 2 x - Connector 2 pin, yellow, below driver's seat
T 2 y - Connector 2 pin, yellow, below passenger's seat
T 4 ad - Connector 4 pin, yellow, below driver's seat
T 4 ae - Connector 4 pin, yellow, below passenger's seat
T 10 av - Connector 10 pin, yellow, connector station A pillar, right
T 16 a - Connector 16 pin, data link connector

(44) - Ground connection, on left A-pillar, lower part
(109) - Ground connection, in airbag wiring harness
(A2) - Plus connection (15), in instrument panel wiring harness
(A76) - Connector (K-diagnosis wire), in instrument panel wiring harness

- Bridge closes when disconnecting connector (safety feature)

THIS PAGE INTENTIONALLY LEFT BLANK

Power Seat and Mirror Adjustment, Mirror with and without fold-away Function, Heated Mirror

1999 m. y.

Fuse Panel

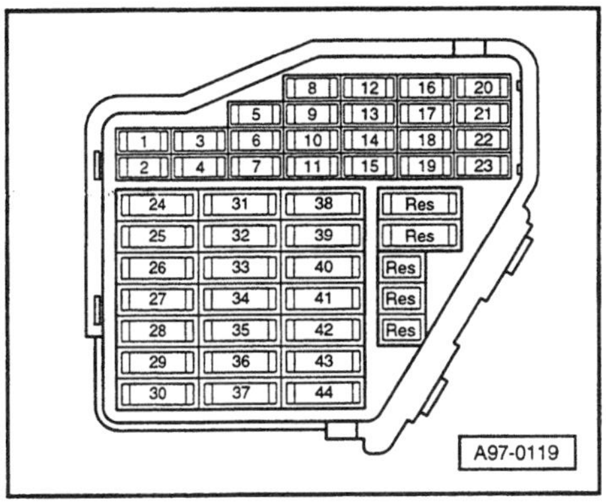

Fuse Colors:

30 A - Green
25 A - White
20 A - Yellow
15 A - Blue
10 A - Red
7,5 A - Brown
5 A - Beige

Starting with fuse position 23, fuses in the fuse holder are identified with 223 in the wiring diagram.

13 - Fold Relay Panel

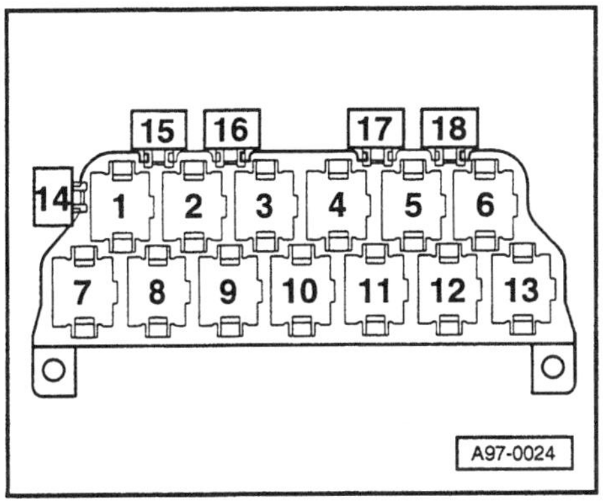

Relay Location:

11 - Control Module for Mirror fold-away Function, J351

12 - Control Module for Mirror fold-away Function, J351

8 - Fold Relay Panel

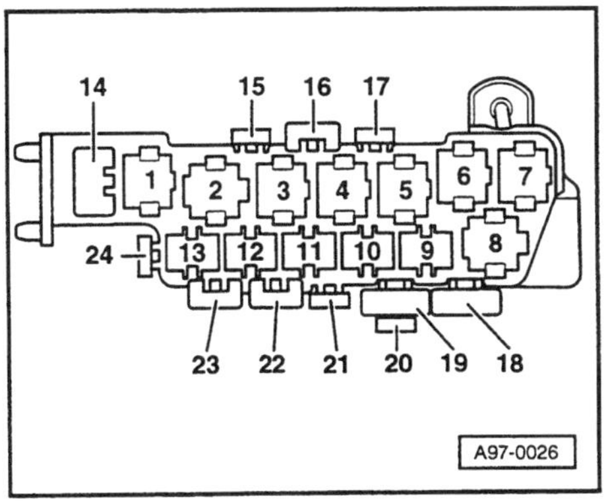

Fuse Location:

14 - Power Seat Circuit Breaker 1, S44

24 - Power Seat Circuit Breaker 2, S80

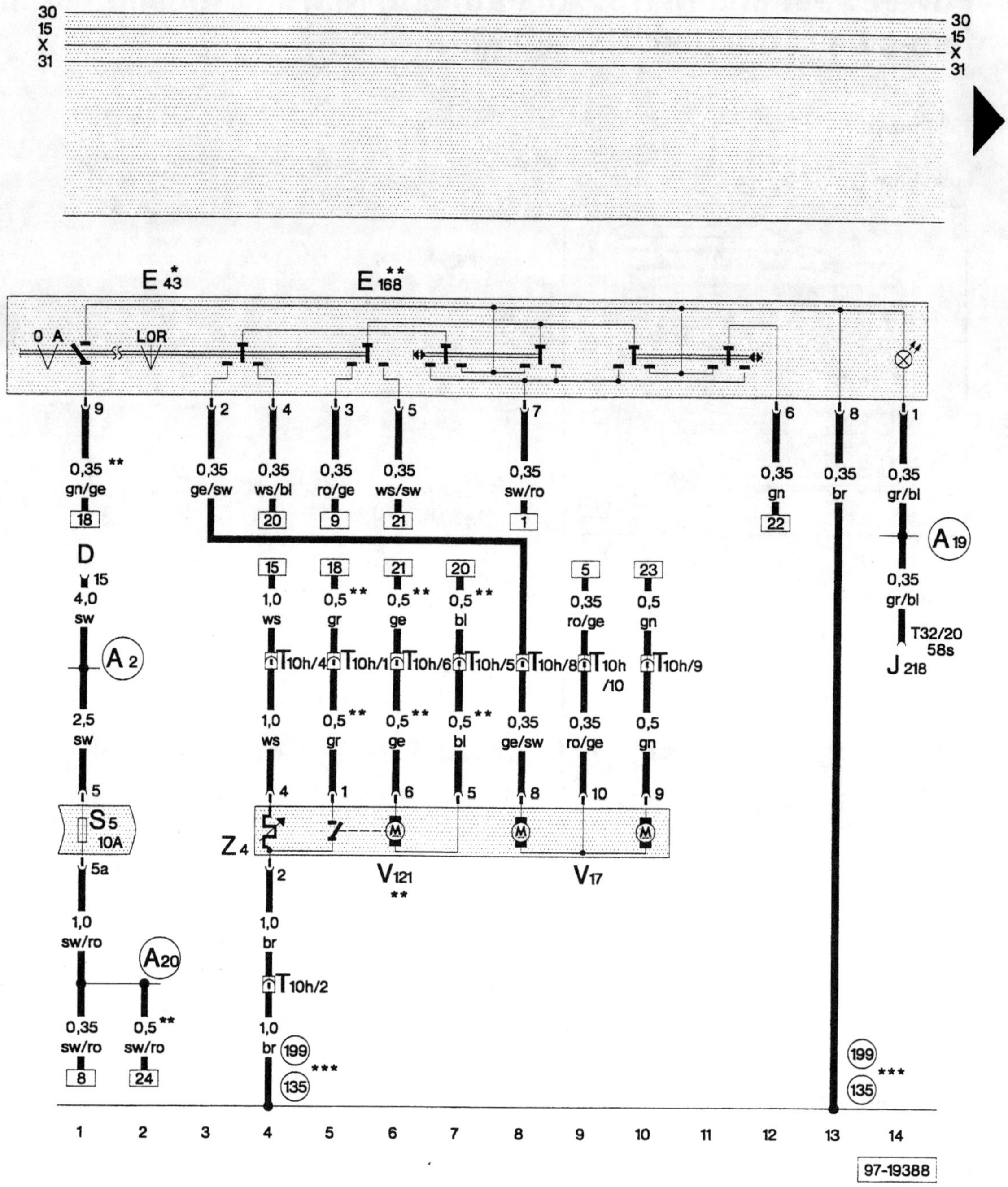

ws = white
sw = black
ro = red
gn = green
br = brown
bl = blue
gr = gray
li = violet
ge = yellow

D - Ignition/Starter Switch
E 43 - Mirror Adjustment Switch
E 168 - Switch for mirror adjustment with fold-away Function
J 218 - Instrument Cluster Combination Processor
S 5 - Fuse
T 10 h - Connector 10 pin, black, connector station A pillar, left
T 32 - Connector 32 pin, blue, on instrument cluster
V 17 - Driver's Side Mirror Adjustment Motor
V 121 - Motor for fold-in Mirror driver`s side
Z 4 - Heated outside mirror, driver side

(135) - Ground connection -2-, in instrument panel wiring harness
(199) - Ground connection -3-, in instrument panel wiring harness
(A2) - Plus connection (15), in instrument panel wiring harness
(A19) - Wire connection (58s), in instrument panel wiring harness (Switch illumination)
(A20) - Wire connection (15a), in instrument panel wiring harness

* - mirror without fold-away Function
** - mirror with fold-away Function
*** - booth is possible

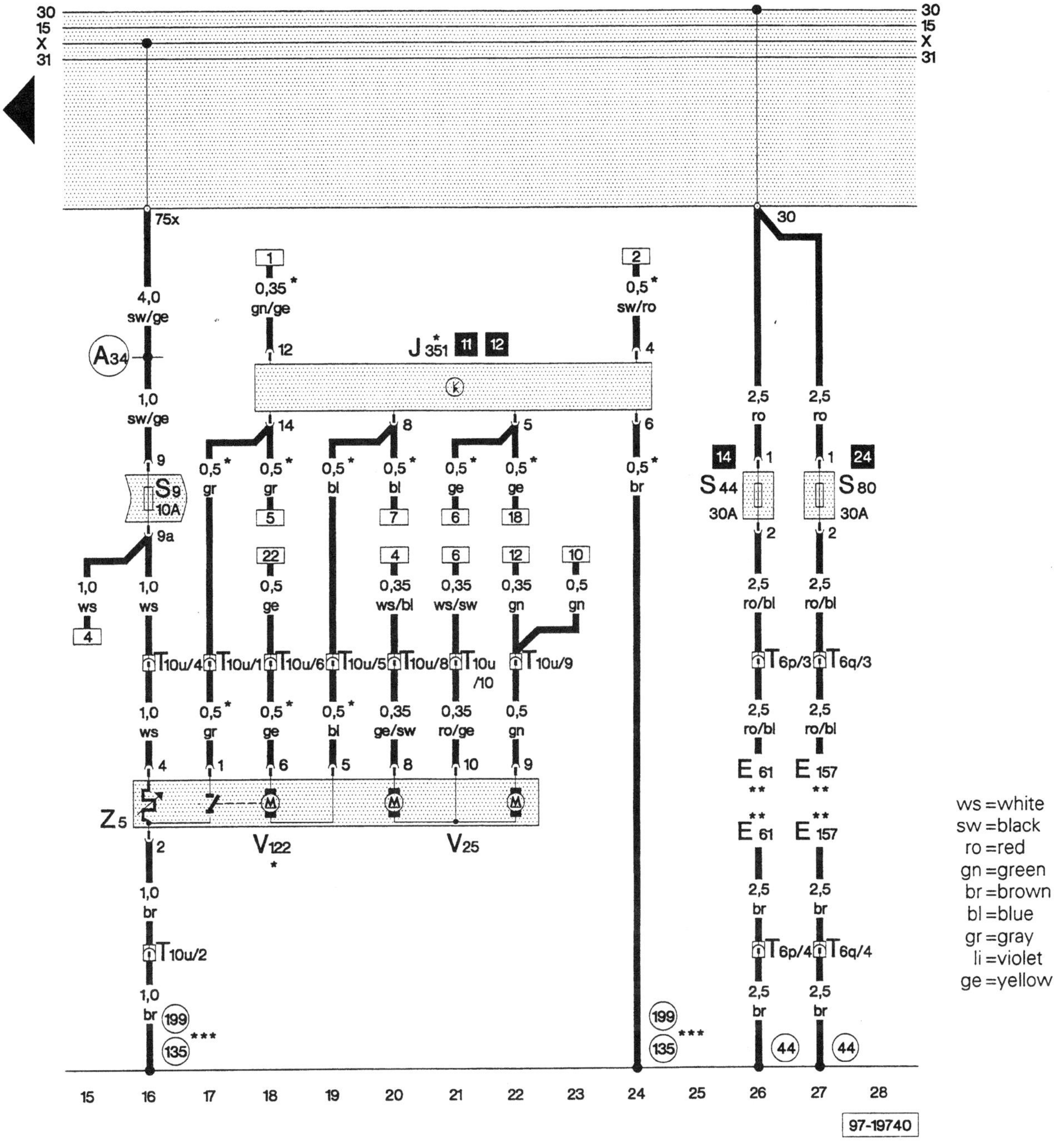

E 61 - Driver's Seat Fore/Aft and Height Adjusting Switch
E 157 - Passenger's Seat Adjusting Switch
J 351 - Control Module for Mirror fold-away Function
S 9 - Fuse
S 44 - Power Seat Circuit Breaker 1
S 80 - Power Seat Circuit Breaker 2
T 6 p - Connector 6 pin, red, below driver's seat
T 6 q - Connector 6 pin, red, below passenger's seat
T 10 u - Connector 10 pin, black, connector station A pillar, right
V 25 - Passenger's Side Mirror Adjustment Motor
V 122 - Motor for fold-in Mirror passenger side
Z 5 - Heated outside mirror, passenger side

(44) - Ground connection, on left A-pillar, lower part
(199) - Ground connection -3-, in instrument panel wiring harness
(135) - Ground connection -2-, in instrument panel wiring harness
(A34) - Wire connection (75x), in instrument panel wiring harness

* - Mirror with fold-away Function
** - see Wiring Diagram power seat
*** - booth is possible

THIS PAGE INTENTIONALLY LEFT BLANK

Daytime Running Lights with Fog Light and Rear Fog Light (Canada)

1999 m. y.

Fuse Panel

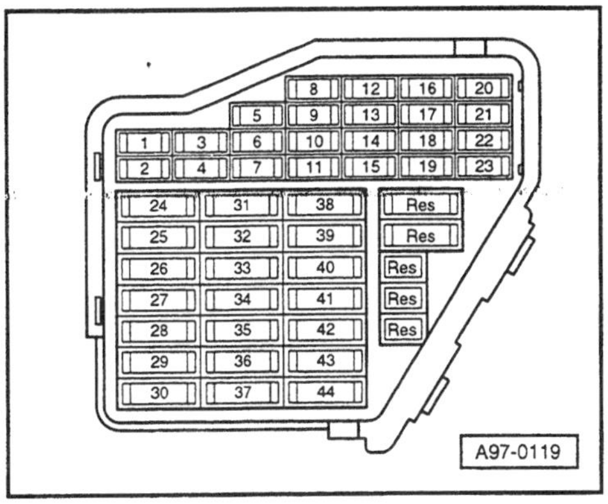

Fuse Colors:

30 A - Green
25 A - White
20 A - Yellow
15 A - Blue
10 A - Red
7,5 A - Brown
5 A - Beige

Starting with fuse position 23, fuses in the fuse holder are identified with 223 in the wiring diagram.

Micro Central Electric Panel

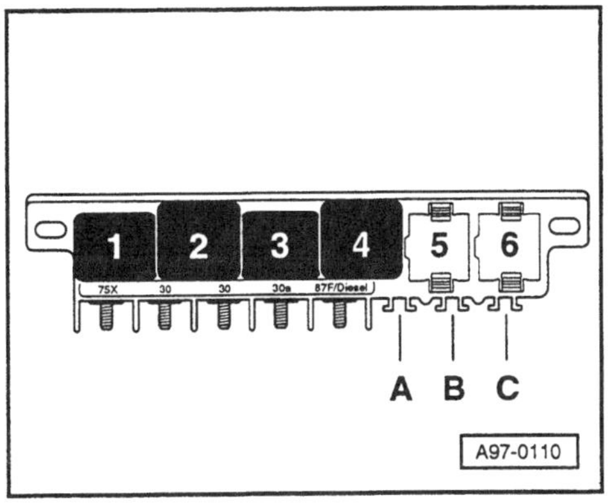

Relay Location:

2 - Load Reduction Relay, J59

13 - Fold Relay Panel

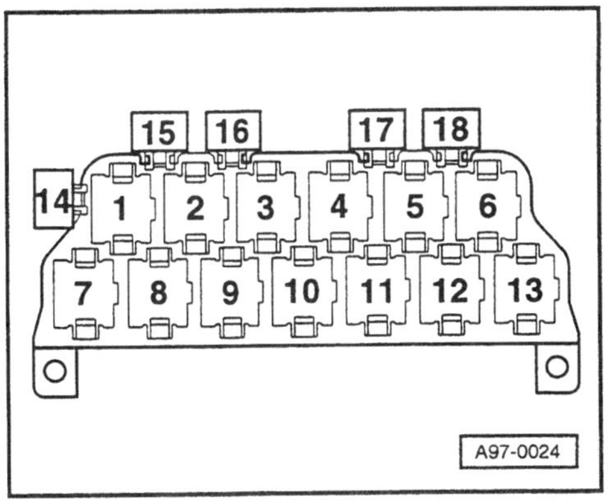

Relay Location:

6 - Fog Light Relay, J5

9 - Lamp Control Module, J123

10 - Lamp Control Module, J123

16 - Fog Light Circuit Breaker, S28

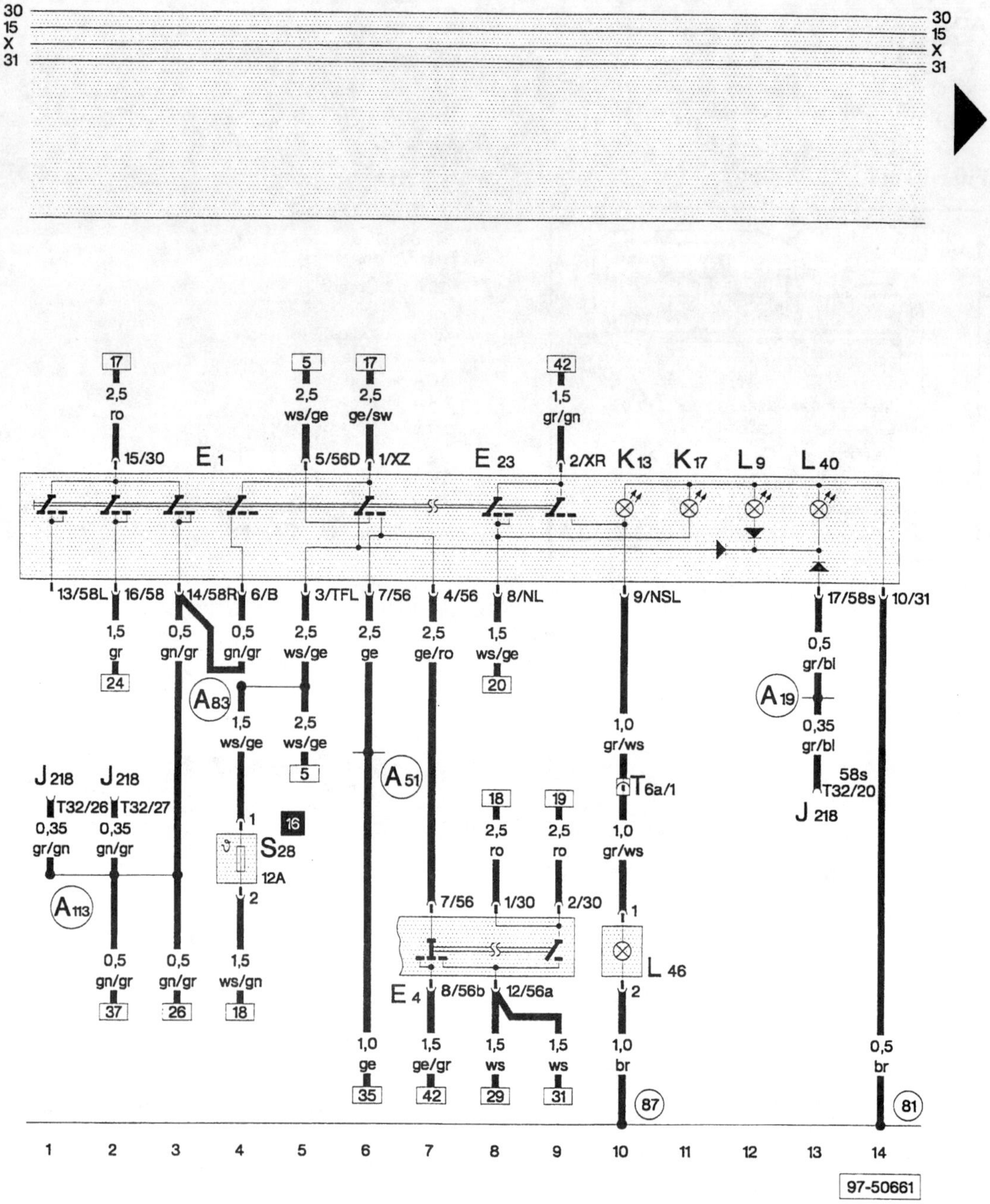

ws = white
sw = black
ro = red
gn = green
br = brown
bl = blue
gr = gray
li = violet
ge = yellow

E 1 - Light switch
E 4 - Headlight Dimmer/Flasher Switch
E 23 - Fog Light Switch
J 218 - Instrument Cluster Combination Processor
K 13 - Rear Fog Light Indicator Light
K 17 - Fog Light Indicator Light
L 9 - Headlight Switch Light
L 40 - Front And Rear Fog Light Switch Light
L 46 - Left Rear Fog Light
S 28 - Fog Light Circuit Breaker
T 6 a - Connector 6 pin, blue, connector station A pillar, left
T 32 - Connector 32 pin, blue, on instrument cluster

(81) - Ground connection -1-, in instrument panel wiring harness
(87) - Ground connection -2-, in rear wiring harness
(A19) - Wire connection (58s), in instrument panel wiring harness (Switch illumination)
(A51) - Wire connection (56), in instrument panel wiring harness
(A83) - Connector (dayttime running lights), in instrument panel wiring harness
(A113) - Connector -2- (daytime running lights) , in instrument panel wiring harness

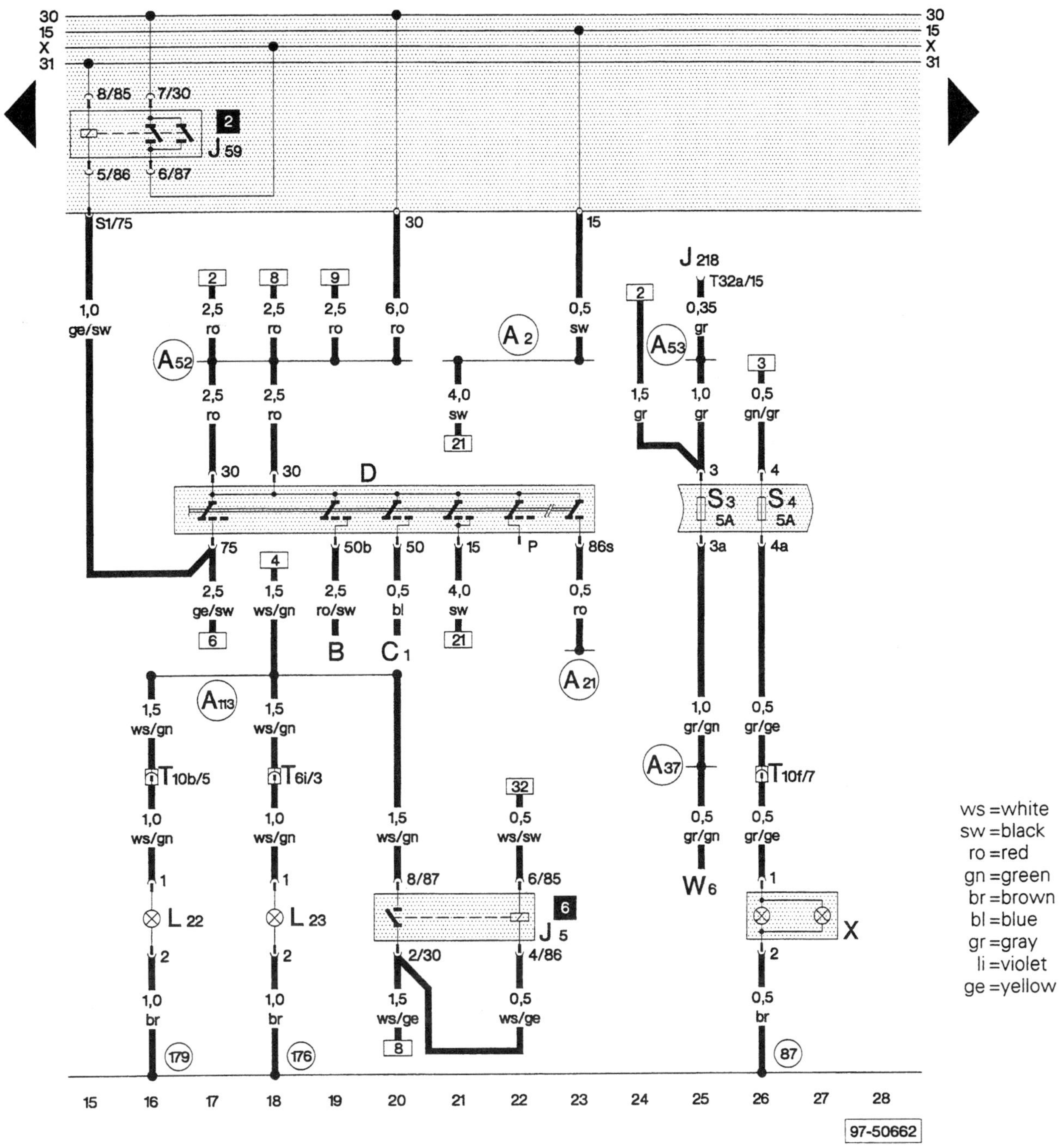

B - Starter
C 1 - Voltage Regulator (VR)
D - Ignition/Starter Switch
J 5 - Fog Light Relay
J 59 - Load Reduction Relay
J 218 - Instrument Cluster Combination Processor
L 22 - Left Front Fog Light
L 23 - Right Front Fog Light
S 3 - Fuse
S 4 - Fuse
T 6 i - Connector 6 pin, red, connector station A pillar, right
T 10 b - Connector 10 pin, red, connector station A pillar, left
T 10 f - Connector 10 pin, brown, connector station A pillar, left
T 32 a - Connector 32 pin, green, on instrument cluster
W 6 - Glove Compartment Light
X - License Plate Light

(87) - Ground connection -2-, in rear wiring harness
(176) - Ground connection, in right headlight wiring harness
(179) - Ground connection, in left headlight wiring harness
(A2) - Plus connection (15), in instrument panel wiring harness
(A21) - Wire connection (86s), in instrument panel wiring harness
(A37) - Wire connection (58a), in instrument panel wiring harness
(A52) - Plus connection (30), in instrument panel wiring harness
(A53) - Plus connection (58b), in instrument panel wiring harness
(A113) - Connection 2 (daytime running lights), in instrument panel wiring harness

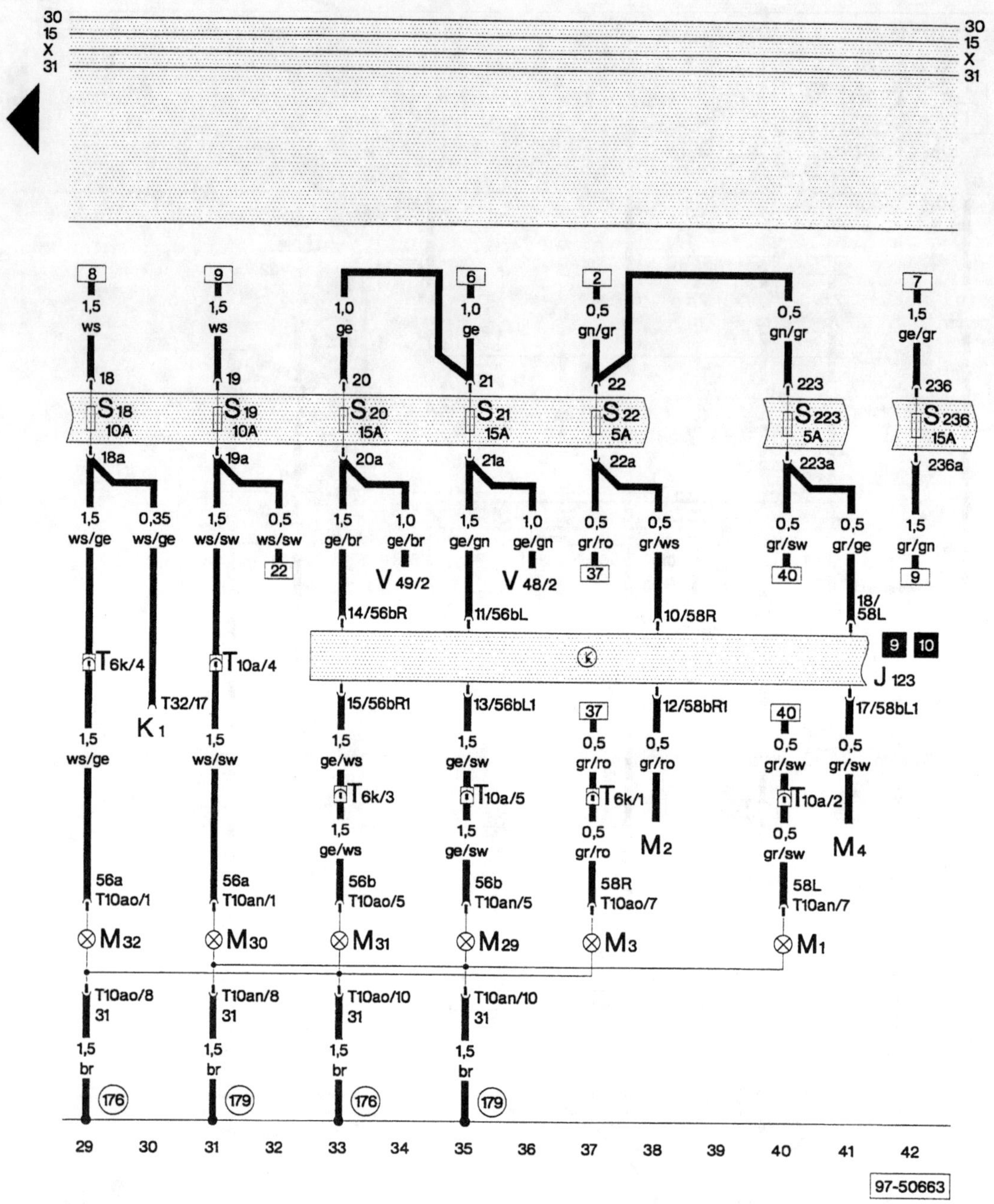

ws =white
sw =black
ro =red
gn =green
br =brown
bl =blue
gr =gray
li =violet
ge =yellow

J 123 - Lamp Control Module
K 1 - Headlight High Beam Indicator Light
M 1 - Left Parking Light
M 2 - Right Tail Light
M 3 - Right Parking Light
M 4 - Left Tail Light
M 29 - Left Low Beam Headlight
M 30 - Left High Beam Headlight
M 31 - Right Low Beam Headlight
M 32 - Right High Beam Headlight
S 18 - Fuse
S 19 - Fuse
S 20 - Fuse
S 21 - Fuse
S 22 - Fuse
S 223 - Fuse
S 236 - Fuse
T 6 k - Connector 6 pin, pink, connector station A pillar, right
T 10 a - Connector 10 pin, pink, connector station A pillar, left
T 10 an - Connector 10 pin, on headlight, left
T 10 ao - Connector 10 pin, on headlight, right
T 32 - Connector 32 pin, blue, on instrument cluster
V 48 - Left Headlight Beam Adjusting Motor
V 49 - Right Headlight Beam Adjusting Motor

(176) - Ground connection, in right headlight wiring harness

(179) - Ground connection, in left headlight wiring harness

Power Windows

1999 m. y.

Fuse Panel

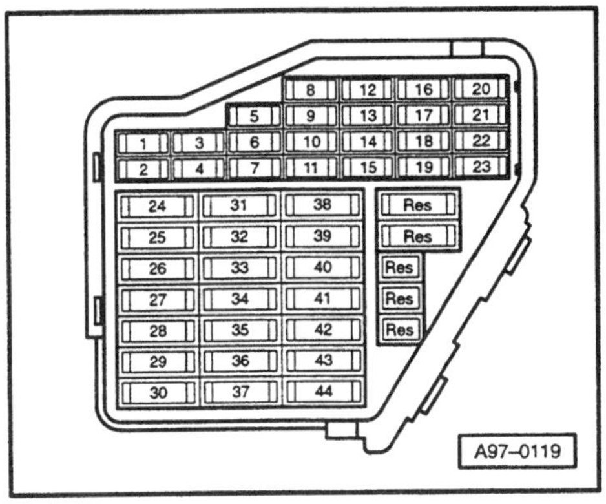

Fuse Colors:

30 A - Green
25 A - White
20 A - Yellow
15 A - Blue
10 A - Red
7,5 A - Brown
5 A - Beige

Starting with fuse position 23, fuses in the fuse holder are identified with 223 in the wiring diagram.

Micro Central Electric Panel

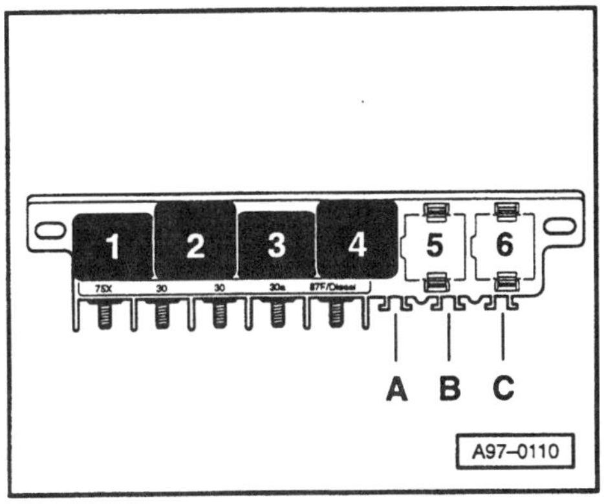

8 - Fold Relay Panel

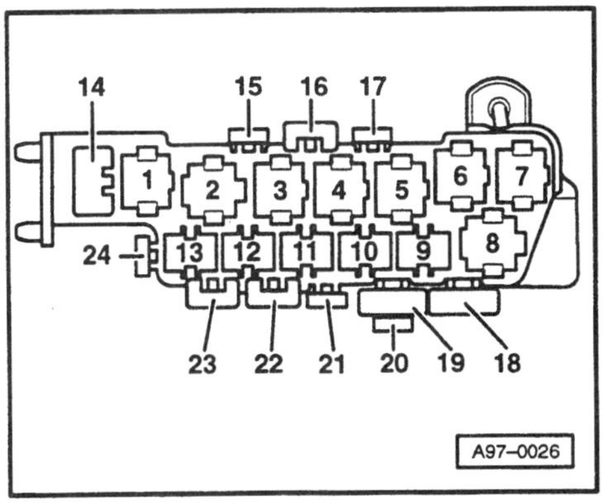

Fuse Location:

16 –Power Window Circuit Breaker Front, S37

17 –Power Window Circuit Breaker Rear, S43

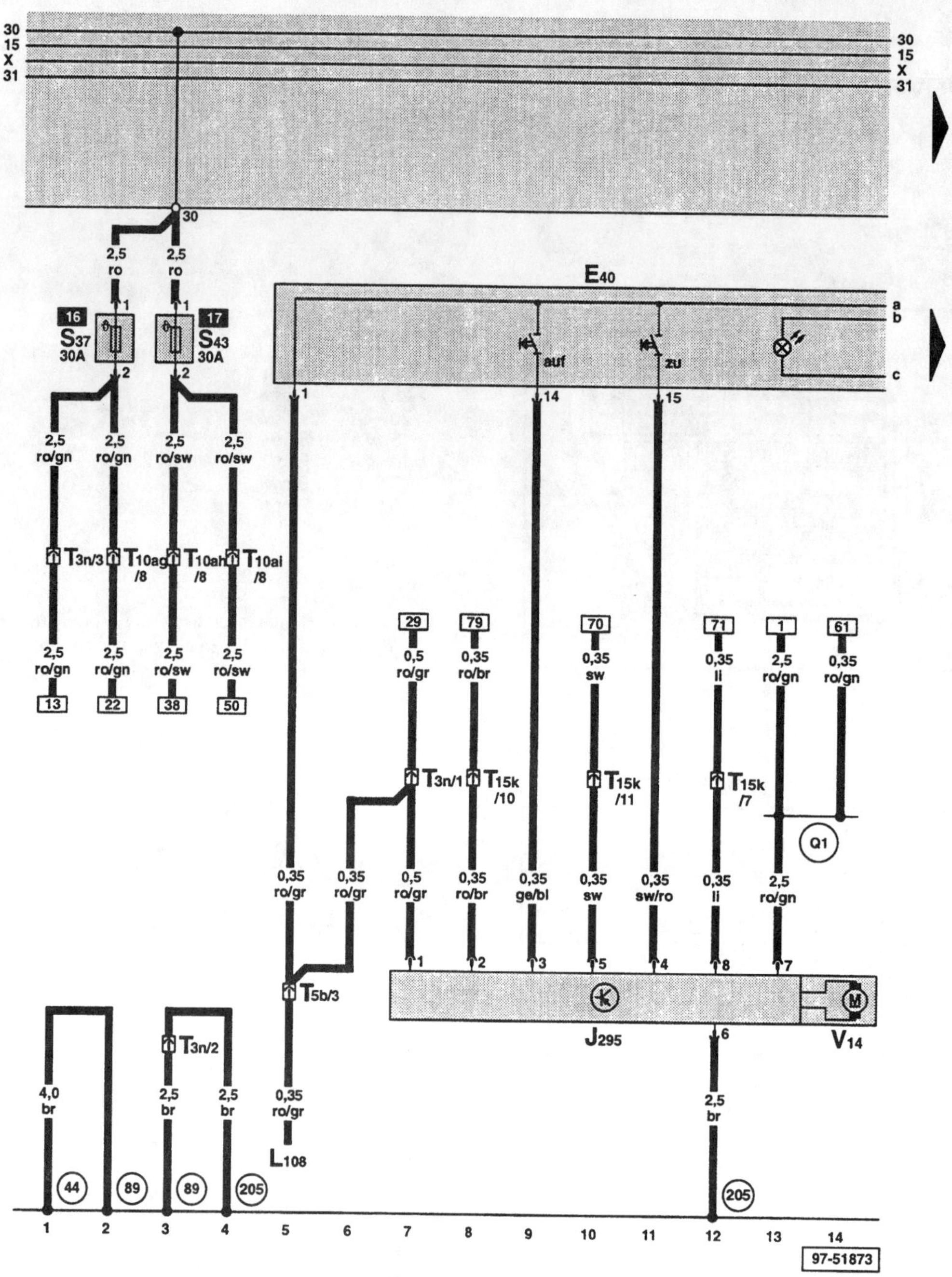

ws = white
sw = black
ro = red
br = brown
gn = green
bl = blue
gr = grey
li = violet
ge = yellow

E40 – Left Front Window Switch
J295 – Left Front Power Window Control Module
L108 – Door opener lighting, driver side
S37 – Power Window Circuit Breaker Front
S43 – Power Window Circuit Breaker Rear
T3n – Connector 3 pin, red, connector station A pillar, left
T5b – Connector 5 pin, black, in driver's door
T10ag – Connector 10 pin, red, connector station A pillar, right
T10ah – Connector 10 pin, red, in B pillar, left
T10ai – Connector 10 pin, red, in B pillar, right
T15k – Connector 15 pin, red, connector station A pillar, left
V14 – Left Window Motor

(44) – Ground connection, on left A-pillar, lower part

(89) – Ground connection -1-, in power window wiring harness

(205) – Ground connection, in driver's door wiring harness

(Q1) – Plus connection (30), in power window wiring harness

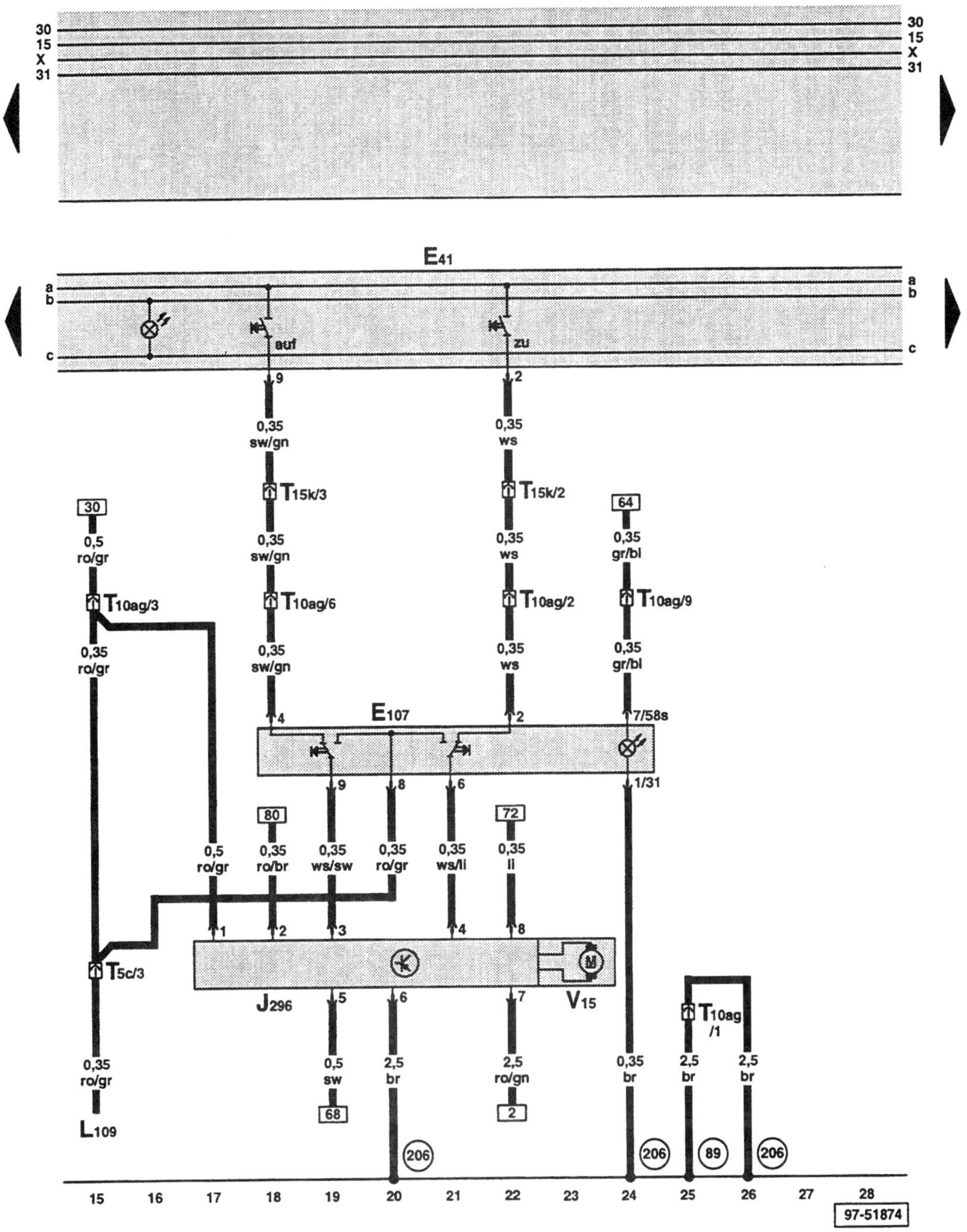

ws = white
sw = black
ro = red
br = brown
gn = green
bl = blue
gr = grey
li = violet
ge = yellow

E41 – Right Front Window Switch
E107 – Switch for window regulator, in passenger door
J296 – Right Front Power Window Control Module
L109 – Door opener lighting, passenger side
T5c – Connector 5 pin, black, in passenger's door
T10ag – Connector 10 pin, red, connector station A pillar, right
T15k – Connector 15 pin, red, connector station A pillar, left
V15 – Right Window Motor

(89) – Ground connection -1-, in power window wiring harness

(206) – Ground connection, in passenger's door wiring harness

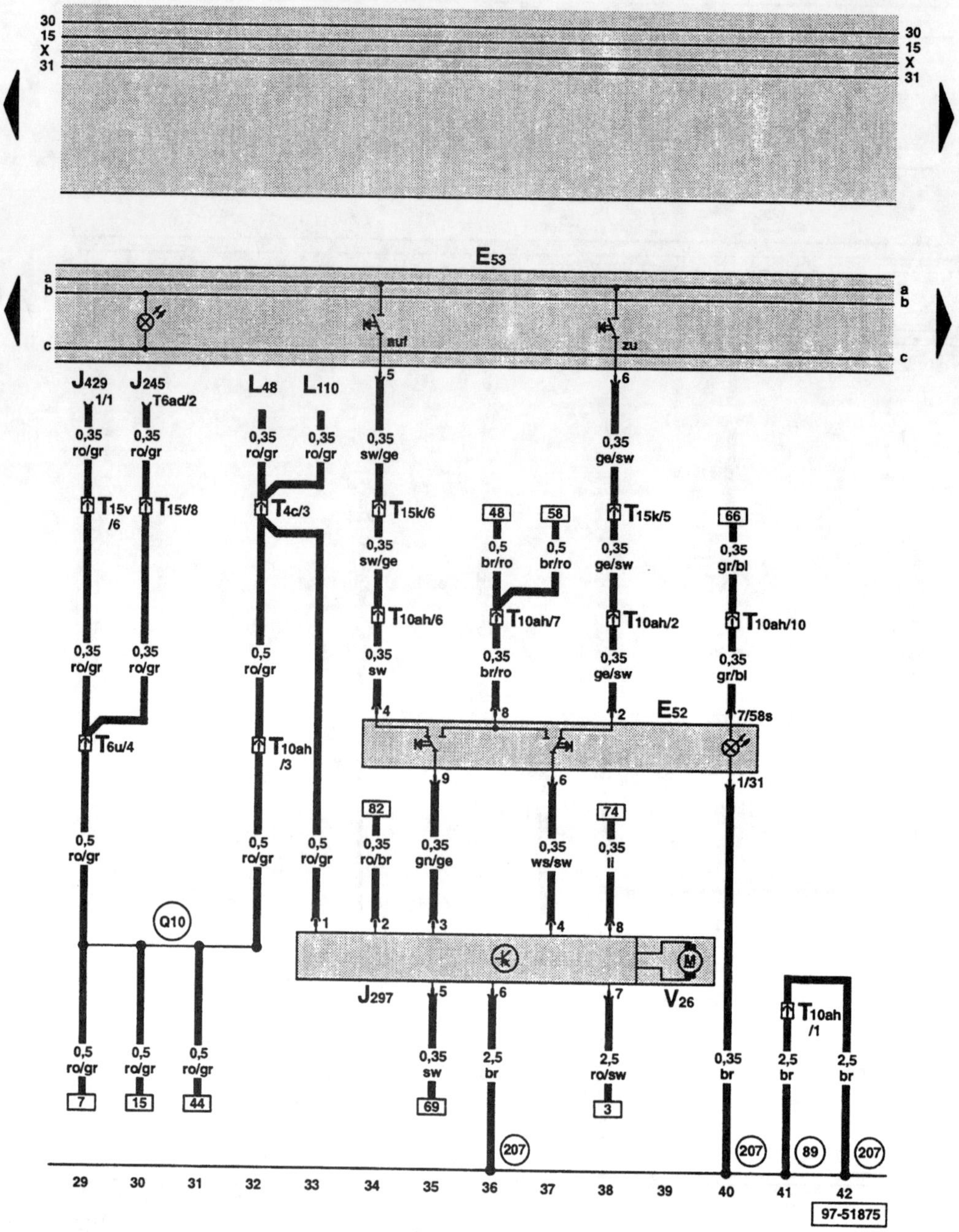

ws = white
sw = black
ro = red
br = brown
gn = green
bl = blue
gr = grey
li = violet
ge = yellow

E52 – Left Rear Window Switch, (In LR Door)
E53 – Left Rear Window Switch, (In Console)
J245 – Power Sunroof Control Module
J297 – Left Rear Power Window Control Module
J429 – Control module for central locking
L48 – Left Rear Ashtray Light
L110 – Door opener lighting, left rear
T4c – Connector 4 pin, black, in left rear door
T6u – Connector 6 pin, grey, connector station A pillar, left
T6ad – Connector 6 pin, blue, on power sunroof control module
T10ah – Connector 10 pin, red, in B pillar, left
T15k – Connector 15 pin, red, connector station A pillar, left
T15t – Connector 15 pin, grey, connector station A pillar, left
T15v – Connector 15 pin, orange, connector station A pillar, right
V26 – Left Rear Door Window Motor

(89) – Ground connection -1-, in power window wiring harness

(207) – Ground connection, in left rear door wiring harness

(Q10) – Plus connection (87), in power window/central locking system and door contact switch wiring harness

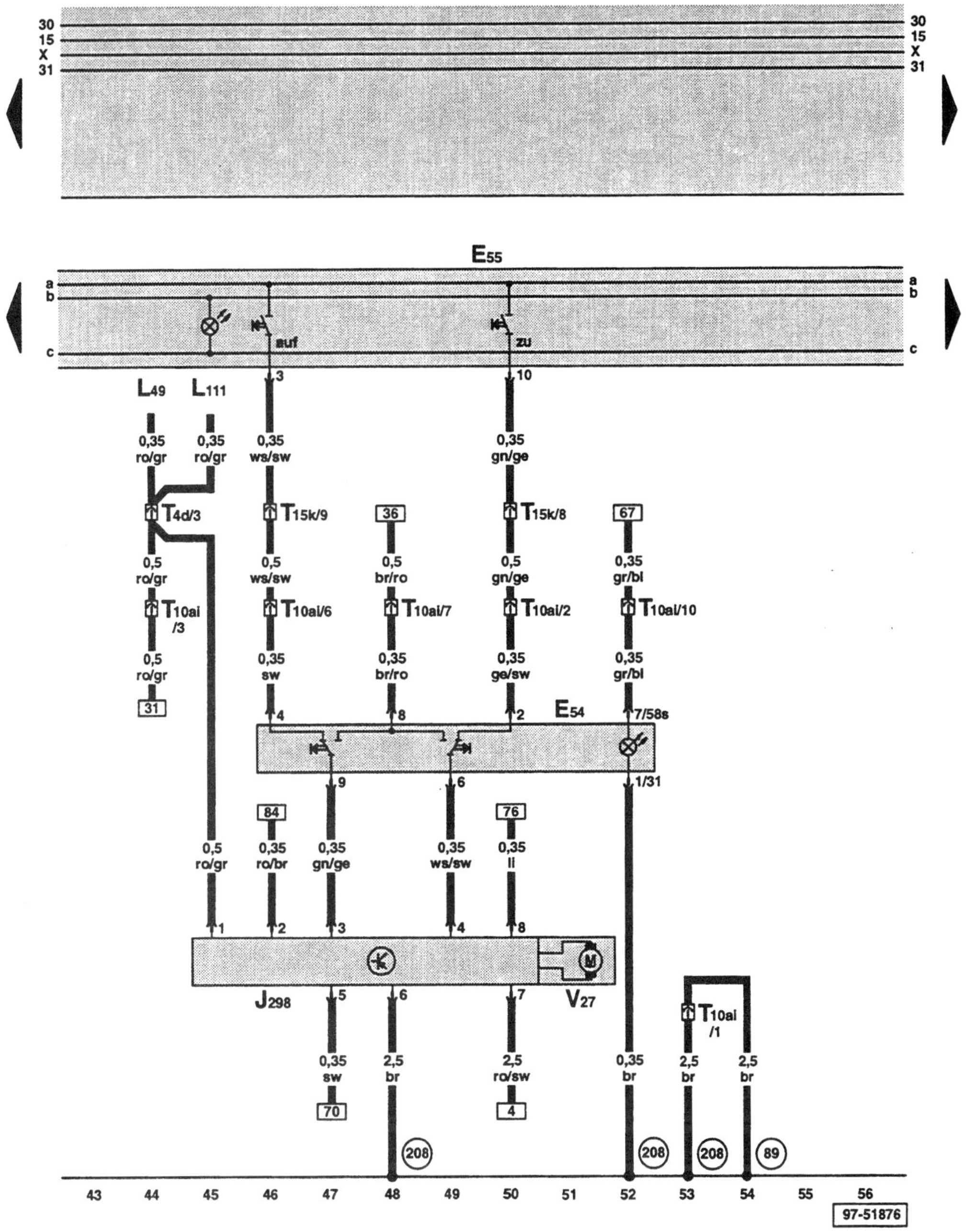

ws = white
sw = black
ro = red
br = brown
gn = green
bl = blue
gr = grey
li = violet
ge = yellow

E54 – Right Rear Window Switch, (In RR Door)
E55 – Right Rear Window Switch, (In Console)
J298 – Right Rear Power Window Control Module
L49 – Right Rear Ashtray Light
L111 – Door opener lighting, right rear
T4d – Connector 4 pin, black, in right rear door
T10ai – Connector 10 pin, red, in B pillar, right
T15k – Connector 15 pin, red, connector station A pillar, left
V27 – Right Rear Door Window Motor

(89) – Ground connection -1-, in power window wiring harness

(208) – Ground connection, in right rear door wiring harness

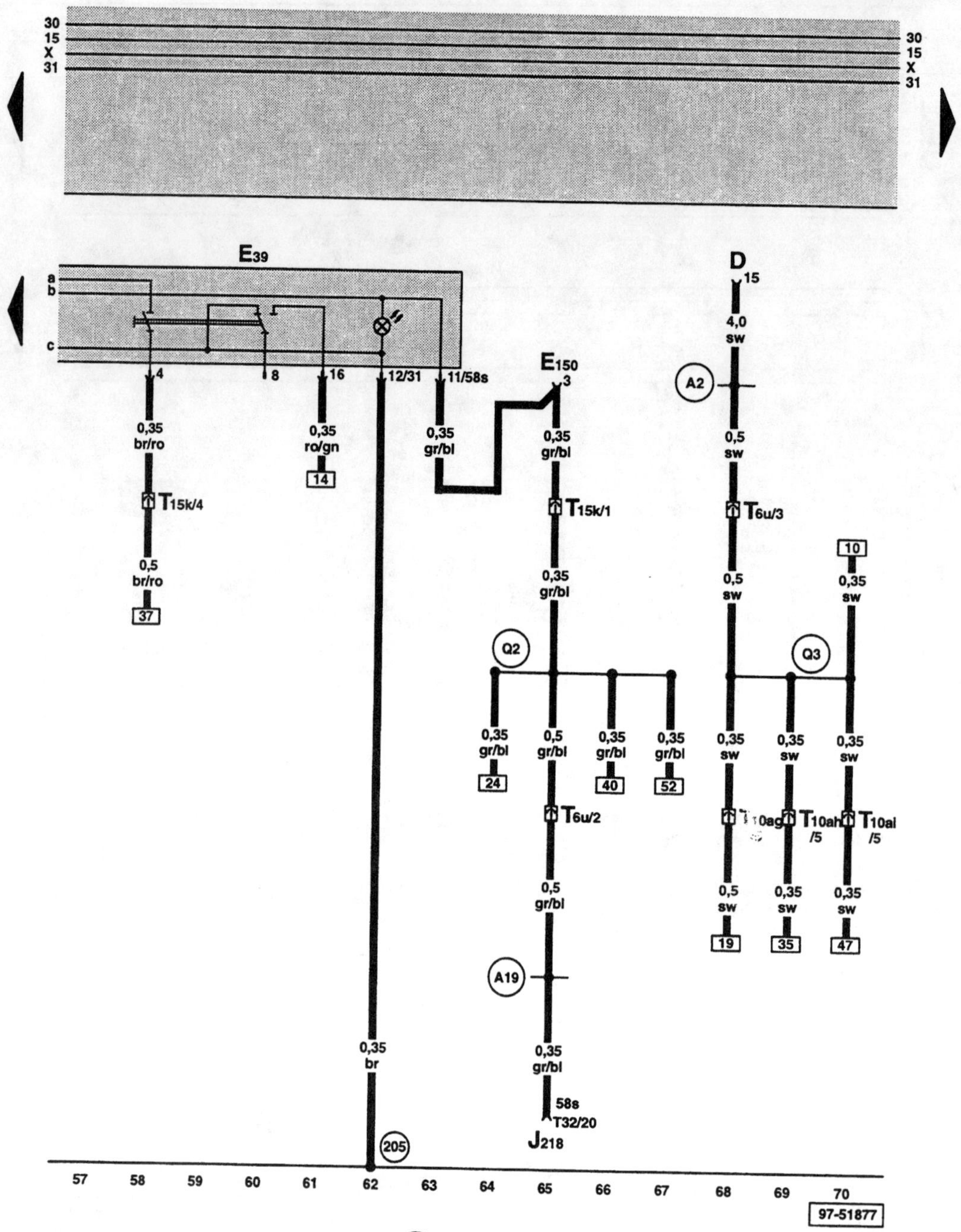

ws = white
sw = black
ro = red
br = brown
gn = green
bl = blue
gr = grey
li = violet
ge = yellow

D – Ignition/Starter Switch
E39 – Window Lockout Switch
E150 – Switch for interior lock, driver side
J218 – Instrument Cluster Combination Processor
T6u – Connector 6 pin, grey, connector station A pillar, left
T10ag – Connector 10 pin, red, connector station A pillar, right
T10ah – Connector 10 pin, red, in B pillar, left
T10ai – Connector 10 pin, red, in B pillar, right
T15k – Connector 15 pin, red, connector station A pillar, left
T32 – Connector 32 pin, blue, on instrument cluster

(205) – Ground connection, in driver's door wiring harness
(A2) – Plus connection (15), in instrument panel wiring harness
(A19) – Wire connection (58s), in instrument panel wiring harness
(Q2) – Plus connection (58), in power window wiring harness
(Q3) – Plus connection (75s), in power window wiring harness

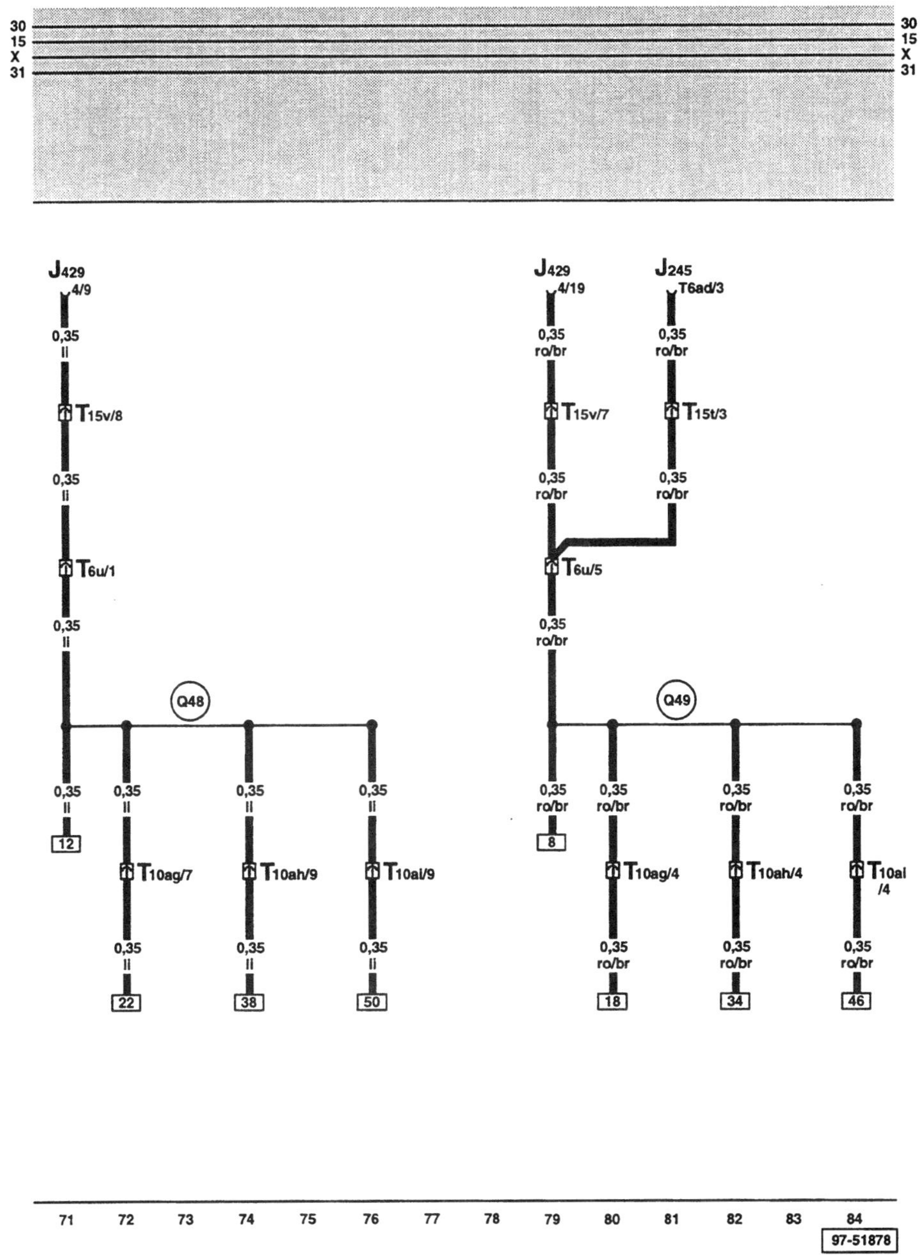

ws = white
sw = black
ro = red
br = brown
gn = green
bl = blue
gr = grey
li = violet
ge = yellow

J245 – Power Sunroof Control Module
J429 – Control module for central locking
T6u – Connector 6 pin, grey, connector station A pillar, left
T6ad – Connector 6 pin, blue, on power sunroof control module
T10ag – Connector 10 pin, red, connector station A pillar, right
T10ah – Connector 10 pin, red, in B pillar, left
T10ai – Connector 10 pin, red, in B pillar, right
T15t – Connector 15 pin, grey, connector station A pillar, left
T15v – Connector 15 pin, orange, connector station A pillar, right

(Q48) – Connection (open), in wiring harness interior

(Q49) – Connection (closed), in wiring harness interior

THIS PAGE INTENTIONALLY LEFT BLANK

Servotronic

1999 m. y.

Fuse Panel

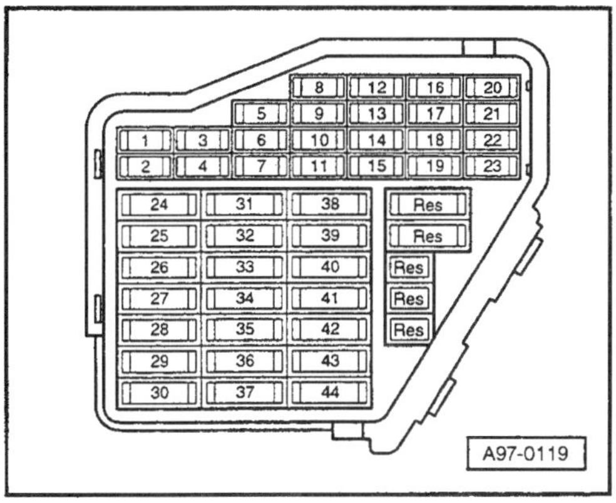

Fuse Colors:

30 A - Green
25 A - White
20 A - Yellow
15 A - Blue
10 A - Red
7,5 A - Brown
5 A - Beige

Starting with fuse position 23, fuses in the fuse holder are identified with 223 in the wiring diagram.

13 - Fold Relay Panel

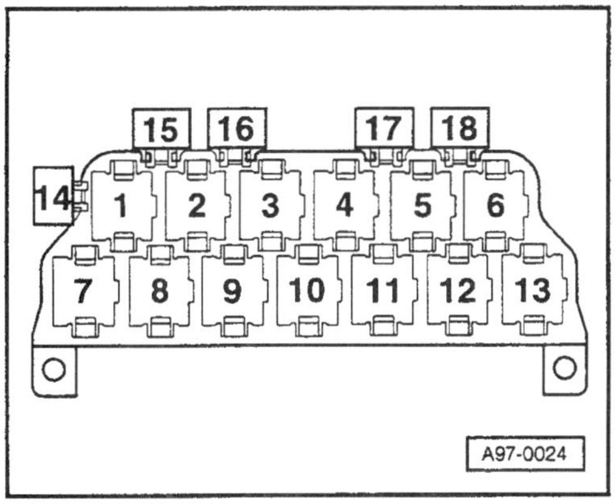

Relay Location:

13 - Servotronic Control Module, J236

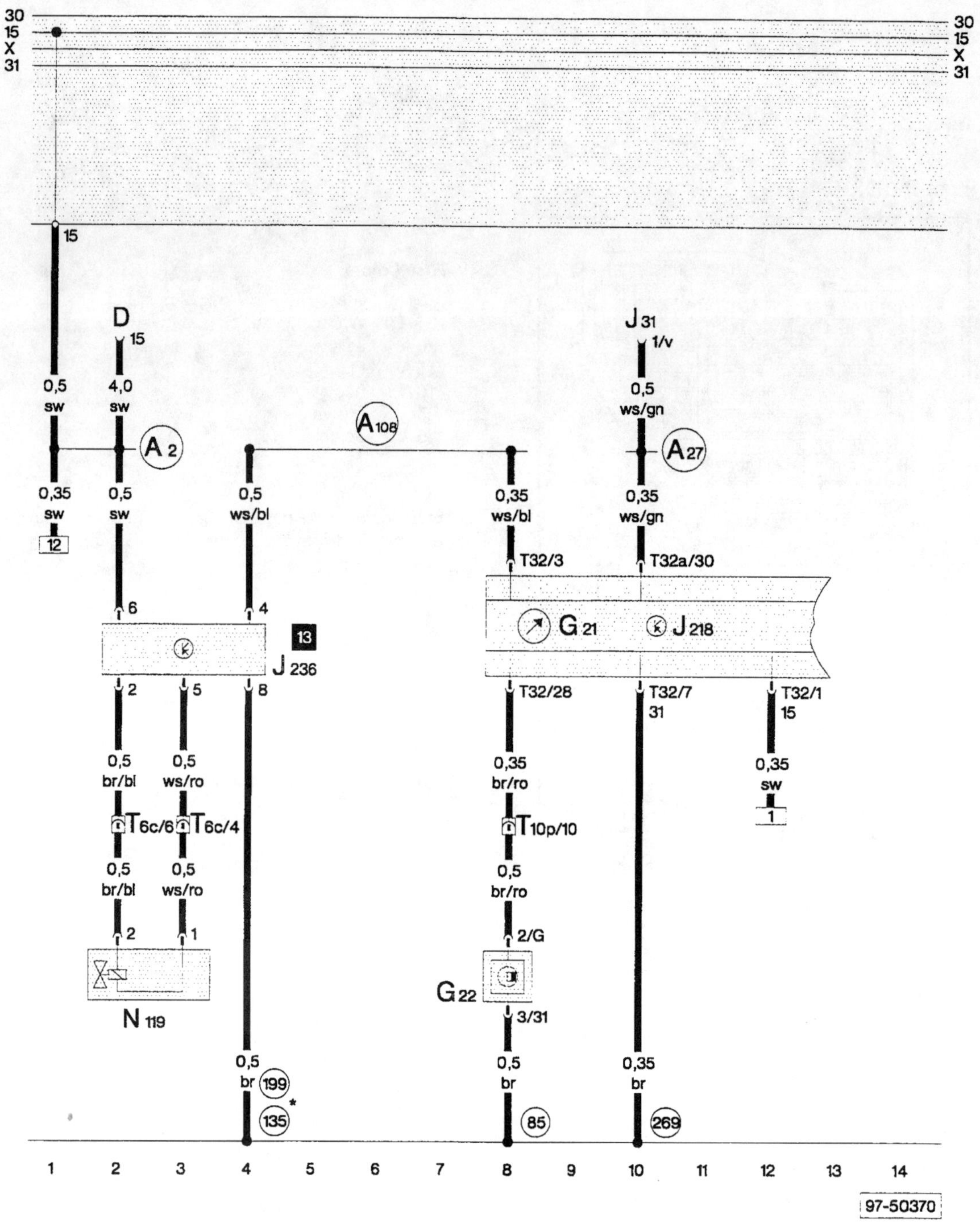

ws = white
sw = black
ro = red
gn = green
br = brown
bl = blue
gr = gray
li = violet
ge = yellow

D - Ignition/Starter Switch
G 21 - Speedometer
G 22 - Speedometer Vehicle Speed Sensor (VSS)
J 31 - Wiper/Washer Intermittent Relay
J 218 - Instrument Cluster Combination Processor
J 236 - Servotronic Control Module
N 119 - Servotronic Solenoid Valve
T 6 c - Connector 6 pin, black, connector station A pillar, left
T 10 p - Connector 10 pin, black, connector station electronic box plenum chamber
T 32 - Connector 32 pin, blue, on instrument cluster
T 32 a - Connector 32 pin, green, on instrument cluster

(85) - Ground connection -1-, in engine compartment wiring harness
(135) - Ground connection -2-, in instrument panel wiring harness
(199) - Ground connection -3-, in instrument panel wiring harness
(269) - Ground connector (sensor ground) -1-, in instrument panel wiring harness
(A2) - Plus connection (15), in instrument panel wiring harness
(A27) - Wire connection (vehicle speed signal), in instrument panel wiring harness
(A108) - Wire connection (vehicle speed signal), in instrument panel wiring harness

* - booth is possible

Telephone

1999 m. y.

Fuse Panel

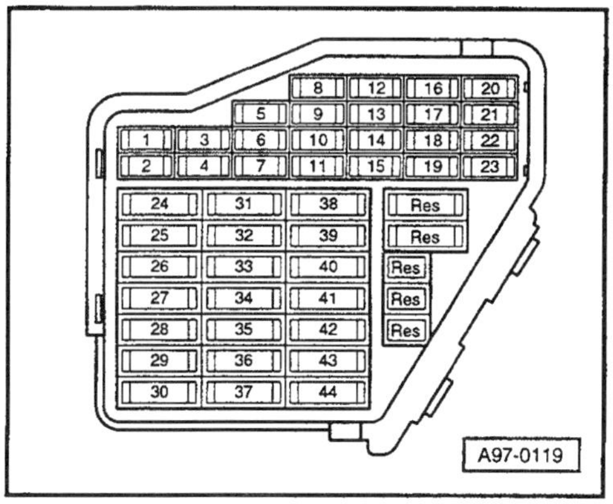

Fuse Colors:

30 A - Green
25 A - White
20 A - Yellow
15 A - Blue
10 A - Red
7,5 A - Brown
5 A - Beige

Starting with fuse position 23, fuses in the fuse holder are identified with 223 in the wiring diagram.

Micro Central Electric Panel

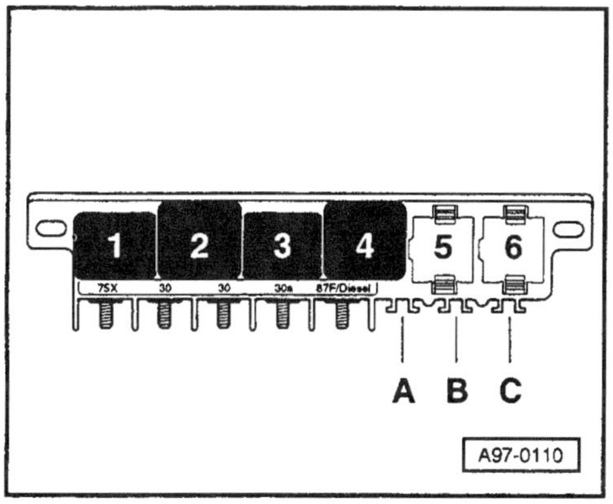

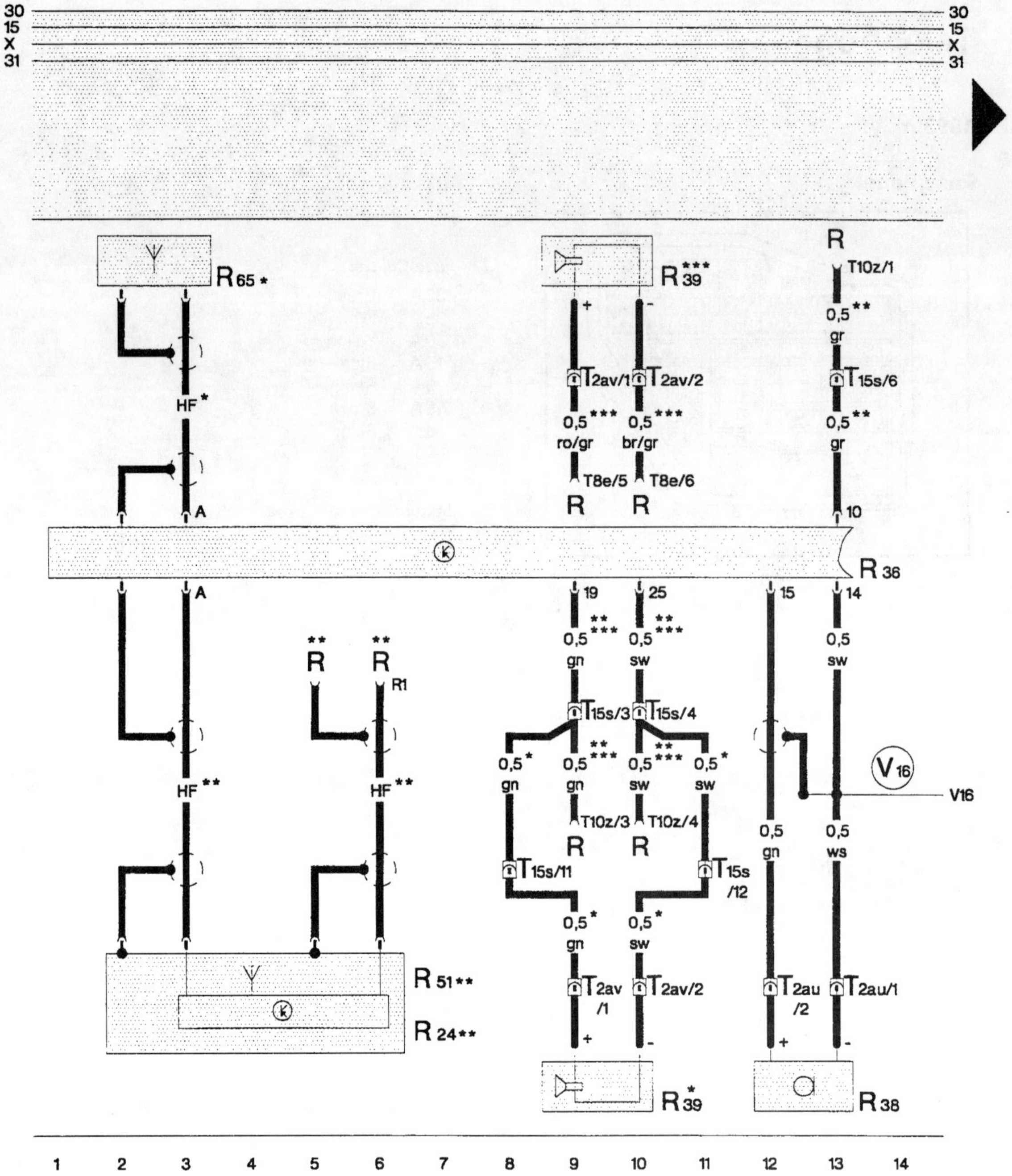

ws = white
sw = black
ro = red
gn = green
br = brown
bl = blue
gr = gray
li = violet
ge = yellow

97-19806

R - Radio
R 1 - Antenna Connection
R 24 - Antenna Amplifier
R 36 - Telephone Transceiver
R 38 - Telephone Microphone
R 39 - Telephone Speaker
R 51 - Antenna for Radio/Telephone
R 65 - Antenna for telephone
T 2 au - Connector 2 pin, for Microphone, A pillar, left
T 2 av - Connector 2 pin, black, for loudspeaker, B pillar, left
T 8 e - Connector 8 pin, brown, connector radio II
T 10 z - Connector 10 pin, red, connector radio IV
T 15 s - Connector 15 pin, brown, connector station A pillar, left

(V16) - Wire connection (shielding), in telephone wiring harness

* - Vehicles without Radio
** - Vehicles with Radio
*** - Vehicles with Radio-Bose

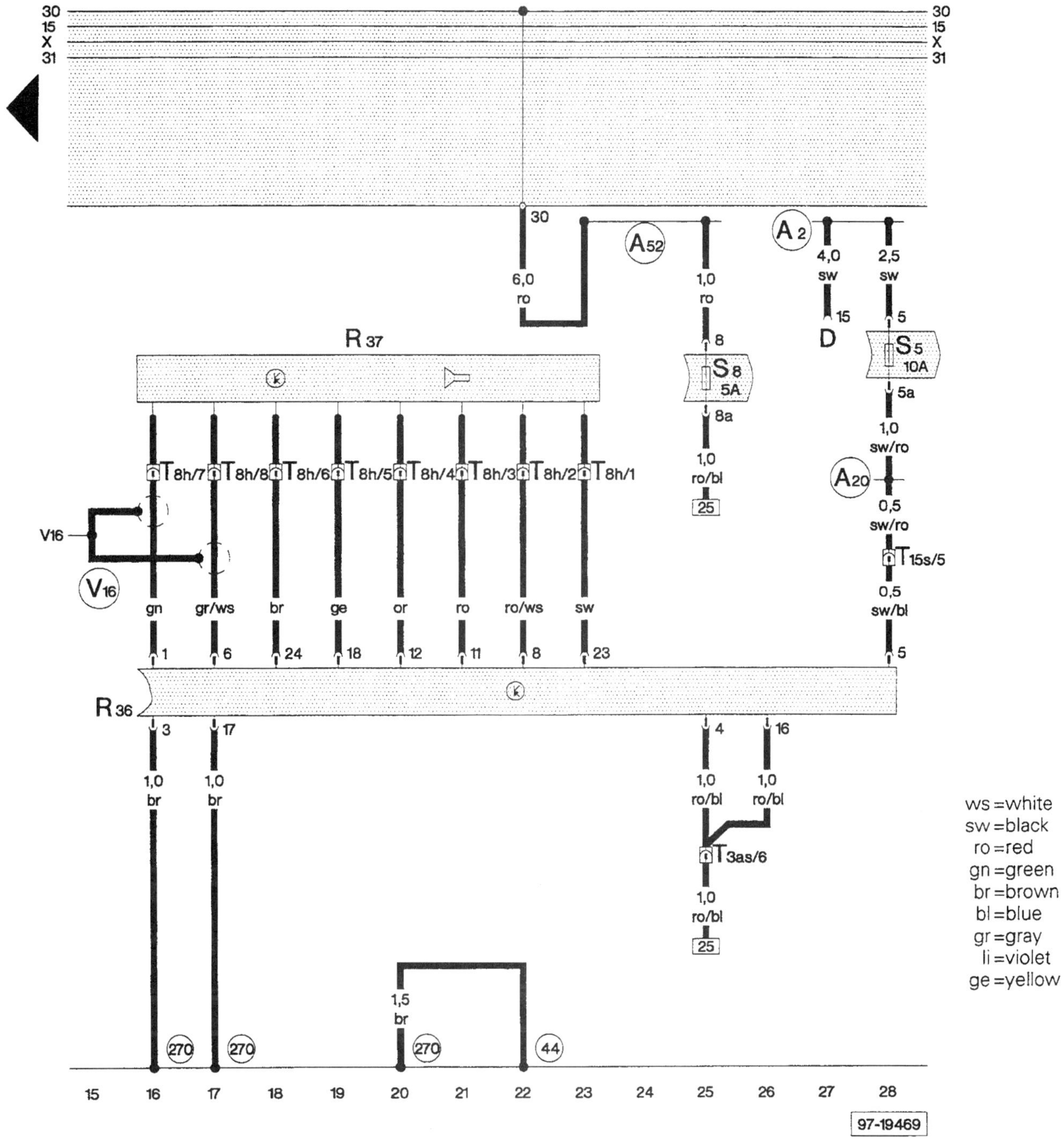

D - Ignition/Starter Switch
R 36 - Telephone Transceiver
R 37 - Telephone Handset
S 5 - Fuse
S 8 - Fuse
T 3 as - Connector 3 pin, brown, connector station A pillar, left
T 8 h - Connector 8 pin, Telephone Handset, in console
T 15 s - Connector 15 pin, brown, connector station A pillar, left

(44) - Ground connection, on left A-pillar, lower part
(270) - Ground connector, in wiring harness telephone
(A2) - Plus connection (15), in instrument panel wiring harness
(A20) - Wire connection (15a), in instrument panel wiring harness
(A52) - Plus connection (30), in instrument panel wiring harness
(V16) - Wire connection (shielding), in telephone wiring harness

Radio Concert, Bose with and without CD Changer

1999 m. y.

Fuse Panel

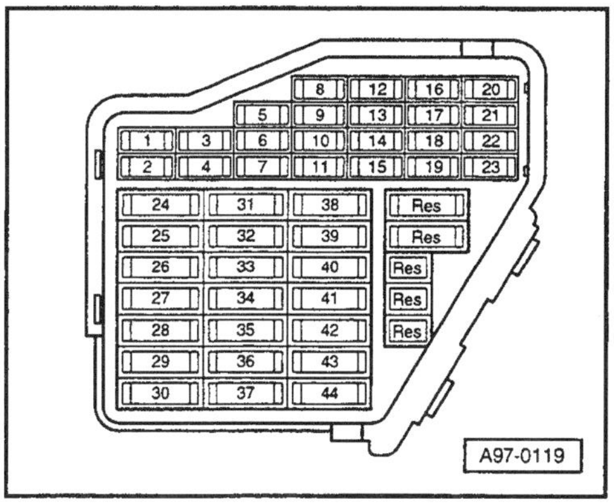

Fuse Colors:

30 A - Green
25 A - White
20 A - Yellow
15 A - Blue
10 A - Red
7,5 A - Brown
5 A - Beige

Starting with fuse position 23, fuses in the fuse holder are identified with 223 in the wiring diagram.

Micro Central Electric Panel

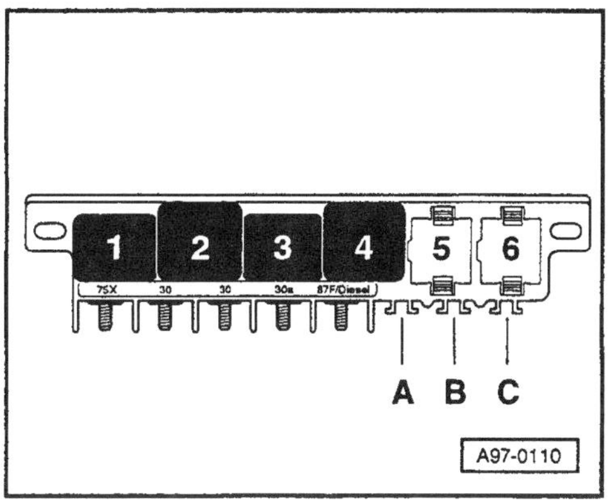

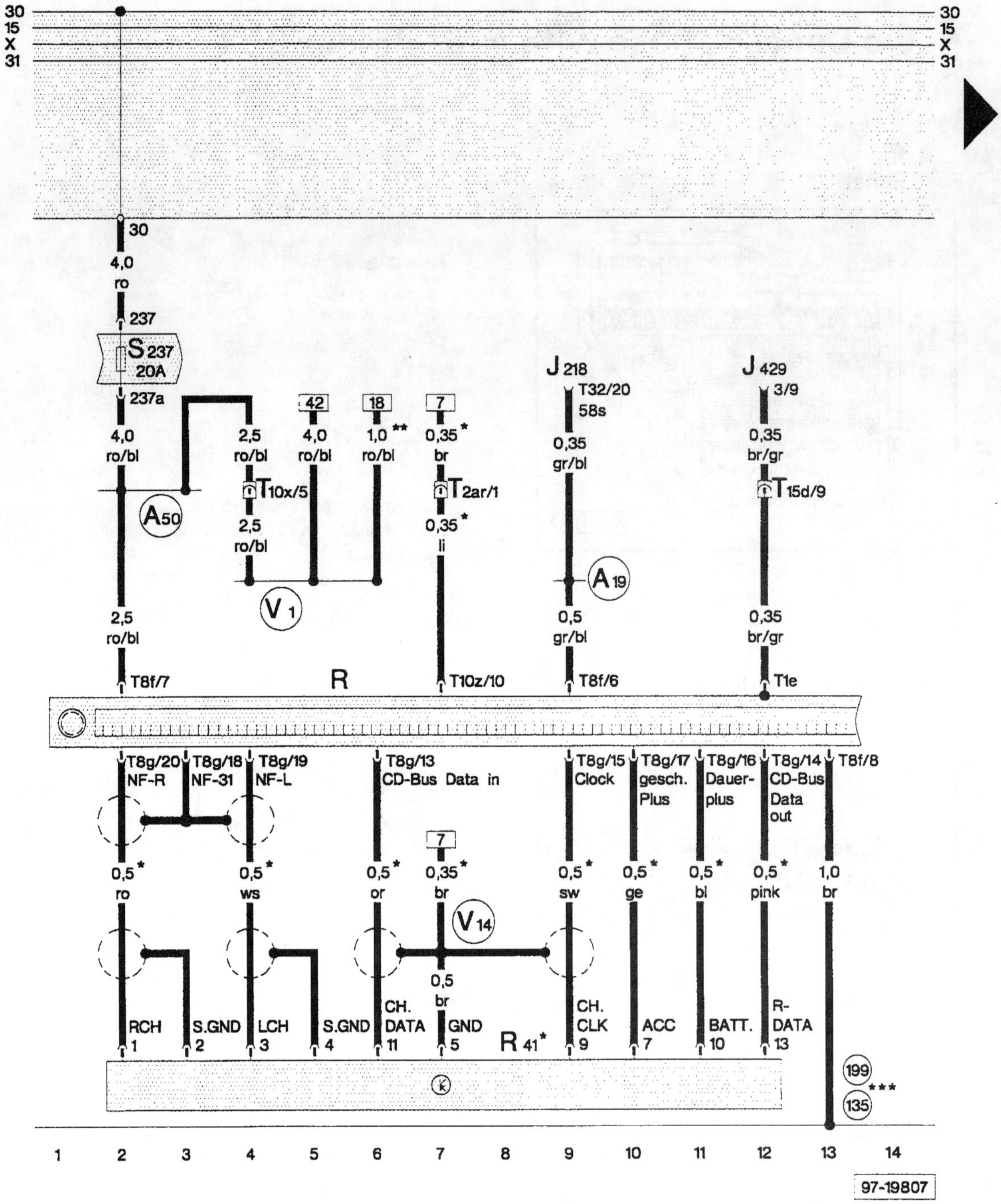

ws = white
sw = black
ro = red
gn = green
br = brown
bl = blue
gr = gray
li = violet
ge = yellow
or = orange

J 218 - Instrument Cluster Combination Processor
J 429 - Control module for central locking
R - Radio
R 41 - CD Changer Unit
T 1 e - Single connector, on Radio
T 2 ar - Connector 2 pin, black, behind instrument panel, center
T 8 f - Connector 8 pin, black, connector radio III
T 8 g - Connector 8 pin, blue, connector radio I - 3
T 10 x - Connector 10 pin, pink, connector station A pillar, right
T 10 z - Connector 10 pin, red, connector radio IV
T 15 d - Connector 15 pin, black, connector station A pillar, right
T 32 - Connector 32 pin, blue, on instrument cluster

(135) - Ground connection -2-, in instrument panel wiring harness
(199) - Ground connection -3-, in instrument panel wiring harness
(A19) - Wire connection (58s), in instrument panel wiring harness (Switch illumination)
(A50) - Plus connection (30as), in instrument panel wiring harness
(V1) - Plus connection (30), in rear speaker wiring harness
(V14) - Wire connection (shielding), in CD-changer wiring harness

* - Vehicles with CD Changer
** - Automatic Antenna
*** - booth is possible

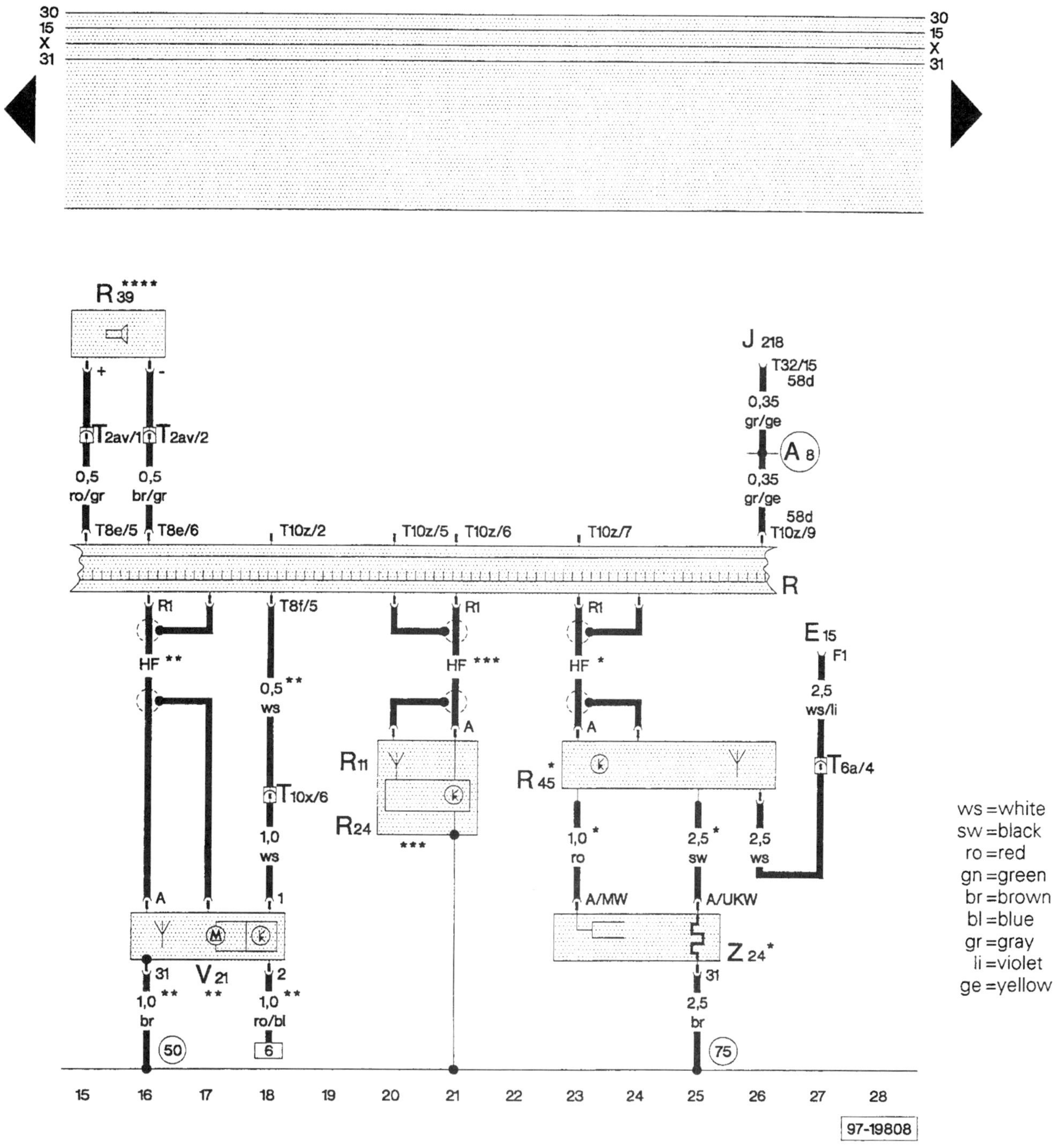

E 15		- Rear Window Defogger Switch
J 218		- Instrument Cluster Combination Processor
R		- Radio
R 1		- Antenna Connection
R 11		- Antenna
R 24		- Antenna Amplifier
R 39		- Telephone Speaker
R 45		- Telephone/Radio Antenna Amplifier
T 2	av	- Connector 2 pin, black, for loudspeaker, B pillar, left
T 6	a	- Connector 6 pin, blue, connector station A pillar, left
T 8	e	- Connector 8 pin, brown, connector radio II
T 8	f	- Connector 8 pin, black, connector radio III
T 10	x	- Connector 10 pin, pink, connector station A pillar, right
T 10	z	- Connector 10 pin, red, connector radio IV
T 32		- Connector 32 pin, blue, on instrument cluster
V 21		- Power Antenna
Z 24		- Rear Window Defogger With Window Antenna

(50) - Ground connection, in luggage compartment, left

(75) - Ground connection, on right rear pillar

(A8) - Plus connection (58d), in instrument panel wiring harness (Display illumination)

* - Rear Window Antenna
** - Automatic Antenna
*** - Wagon
**** - Vehicles with Telephone

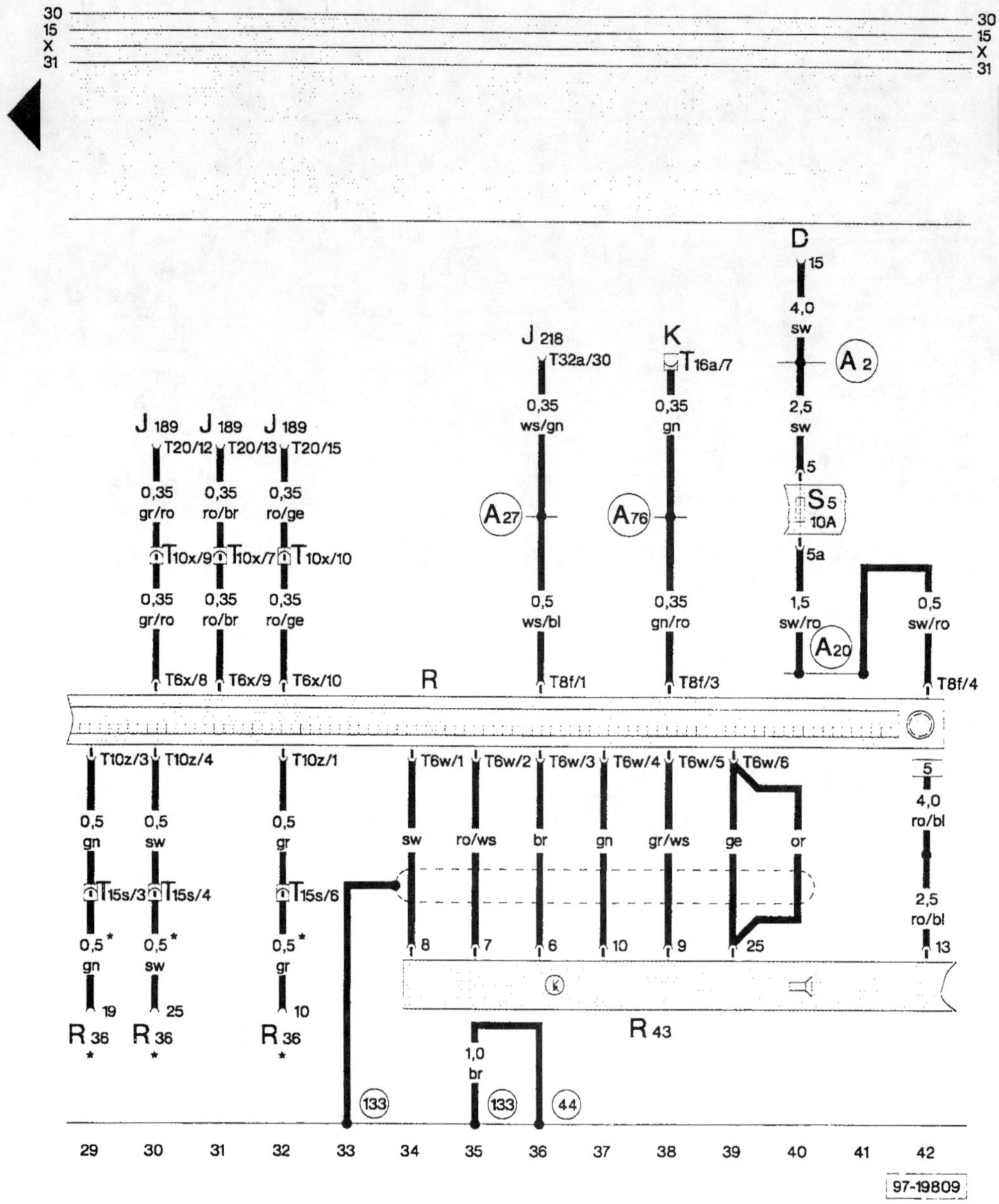

ws = white
sw = black
ro = red
gn = green
br = brown
bl = blue
gr = gray
li = violet
ge = yellow
or = orange

D - Ignition/Starter Switch
J 189 - Auto Check System
J 218 - Instrument Cluster Combination Processor
K - DLC (Data Link Connector) Rapid Data Transfer
R - Radio
R 36 - Telephone Transceiver
R 43 - Right Rear Woofer/Amplifier
S 5 - Fuse
T 6 w - Connector 6 pin, yellow, connector radio I - 1
T 6 x - Connector 6 pin, green, connector radio I - 2
T 8 f - Connector 8 pin, black, connector radio III
T 10 x - Connector 10 pin, pink, connector station A pillar, right
T 10 z - Connector 10 pin, red, connector radio IV
T 15 s - Connector 15 pin, brown, connector station A pillar, left
T 16 a - Connector 16 pin, data link connector
T 20 - Connector 20 pin, red/white, on instrument cluster
T 32 a - Connector 32 pin, green, on instrument cluster

(44) - Ground connection, on left A-pillar, lower part
(133) - Ground connection, in speaker wiring harness
(A2) - Plus connection (15), in instrument panel wiring harness
(A20) - Wire connection (15a), in instrument panel wiring harness
(A27) - Wire connection (vehicle speed signal), in instrument panel wiring harness
(A76) - Connector (K-diagnosis wire), in instrument panel wiring harness

* - Vehicles with Telephone

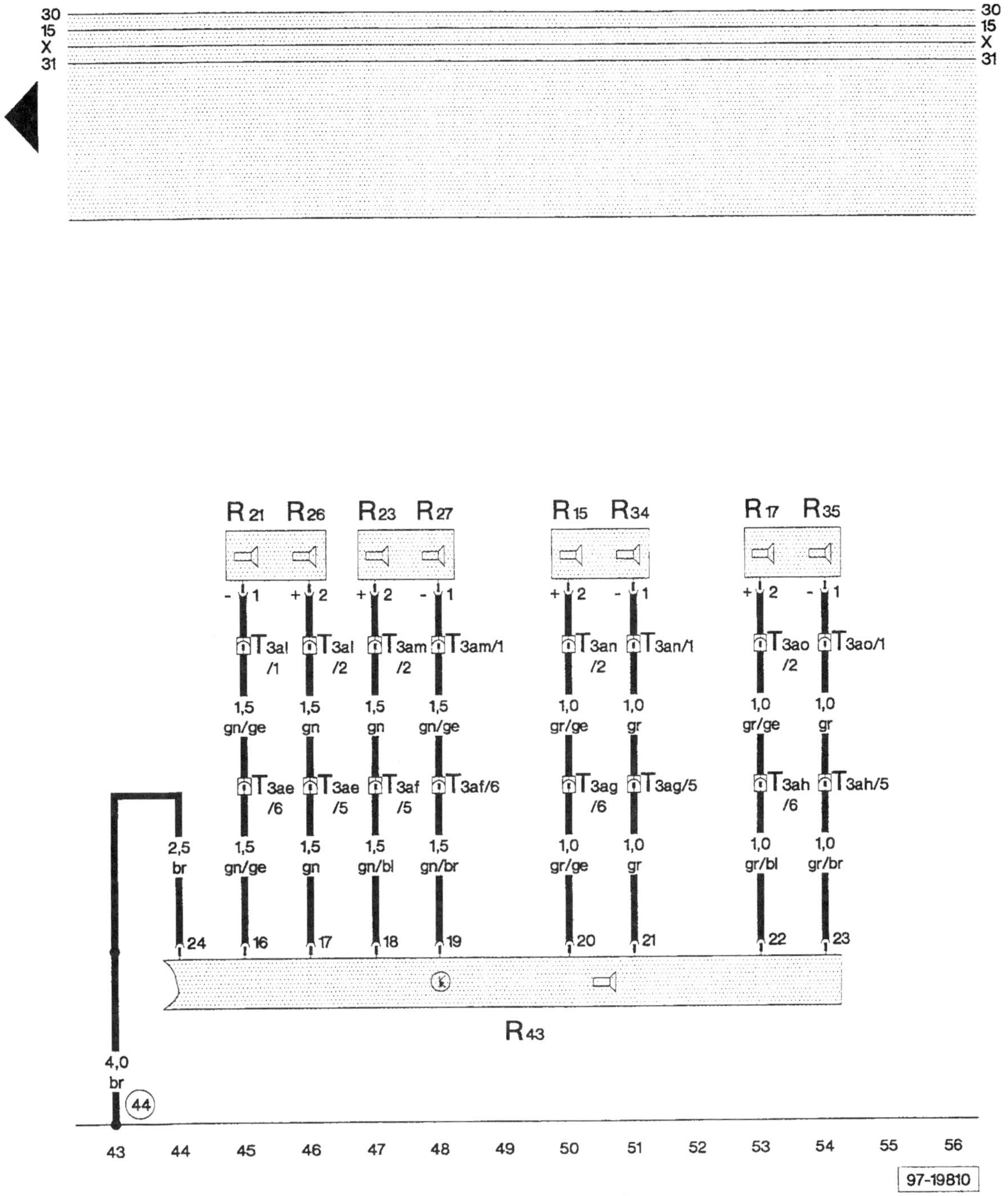

ws = white
sw = black
ro = red
gn = green
br = brown
bl = blue
gr = gray
li = violet
ge = yellow

44 - Ground connection, on left A-pillar, lower part

R 15 - Left Rear Woofer
R 17 - Right Rear Woofer
R 21 - Left Front Woofer
R 23 - Right Front Woofer
R 26 - Left Front Midrange Speaker
R 27 - Right Front Midrange Speaker
R 34 - Left Rear Midrange Speaker
R 35 - Right Rear Midrange Speaker
R 43 - Right Rear Woofer/Amplifier
T 3 ae - Connector 3 pin, brown, connector station A pillar, left
T 3 af - Connector 3 pin, brown, connector station A pillar, right
T 3 ag - Connector 3 pin, blue, in B pillar left
T 3 ah - Connector 3 pin, blue, in B pillar right
T 3 al - Connector 3 pin, brown, in driver's door
T 3 am - Connector 3 pin, brown, in front passenger door
T 3 an - Connector 3 pin, blue, in rear left door
T 3 ao - Connector 3 pin, blue, in rear right door

Radio-Stereo, Concert

1999 m. y.

Fuse Panel

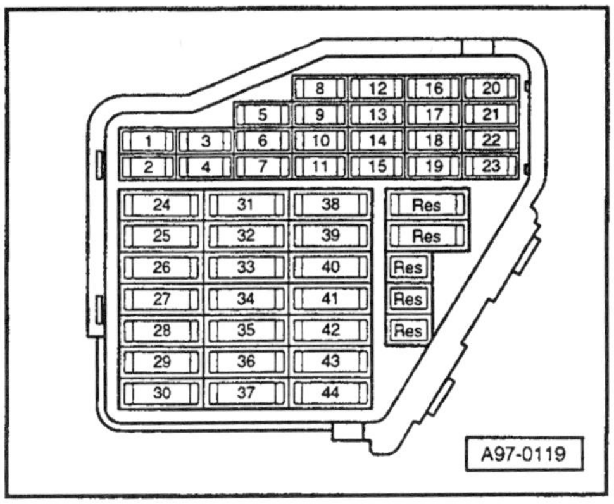

Fuse Colors:

30 A - Green
25 A - White
20 A - Yellow
15 A - Blue
10 A - Red
7,5 A - Brown
5 A - Beige

Starting with fuse position 23, fuses in the fuse holder are identified with 223 in the wiring diagram.

Micro Central Electric Panel

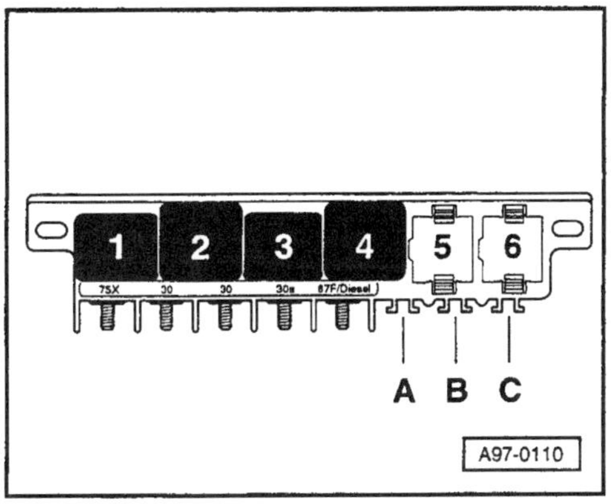

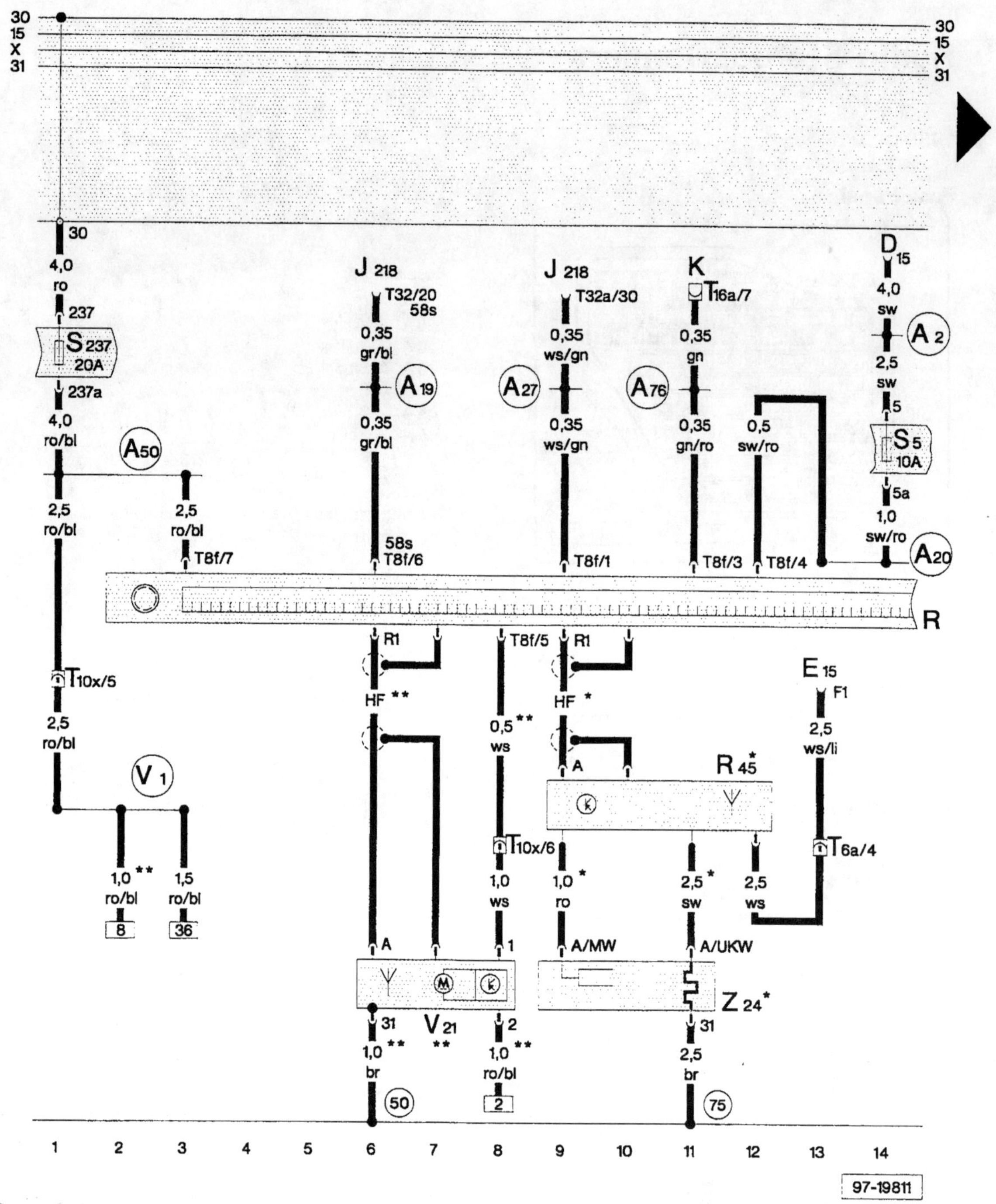

ws = white
sw = black
ro = red
gn = green
br = brown
bl = blue
gr = gray
li = violet
ge = yellow

D - Ignition/Starter Switch
E 15 - Rear Window Defogger Switch
J 218 - Instrument Cluster Combination Processor
K - DLC (Data Link Connector) Rapid Data Transfer
R - Radio
R 1 - Antenna Connection
R 45 - Telephone/Radio Antenna Amplifier
S 5 - Fuse
S 237 - Fuse
T 6 a - Connector 6 pin, blue, connector station A pillar, left
T 8 f - Connector 8 pin, black, connector radio III
T 10 x - Connector 10 pin, pink, connector station A pillar, right
T 16 a - Connector 16 pin, data link connector
T 32 - Connector 32 pin, blue, on instrument cluster
T 32 a - Connector 32 pin, green, on instrument cluster
V 21 - Power Antenna
Z 24 - Rear Window Defogger With Window Antenna

(50) - Ground connection, in luggage compartment, left

(75) - Ground connection, on right rear pillar

(A2) - Plus connection (15), in instrument panel wiring harness

(A19) - Wire connection (58s), in instrument panel wiring harness (Switch illumination)

(A20) - Wire connection (15a), in instrument panel wiring harness

(A27) - Wire connection (vehicle speed signal), in instrument panel wiring harness

(A50) - Plus connection (30as), in instrument panel wiring harness

(A76) - Connector (K-diagnosis wire), in instrument panel wiring harness

(V1) - Plus connection (30), in rear speaker wiring harness

* - Rear Window Antenna
** - Automatic Antenna

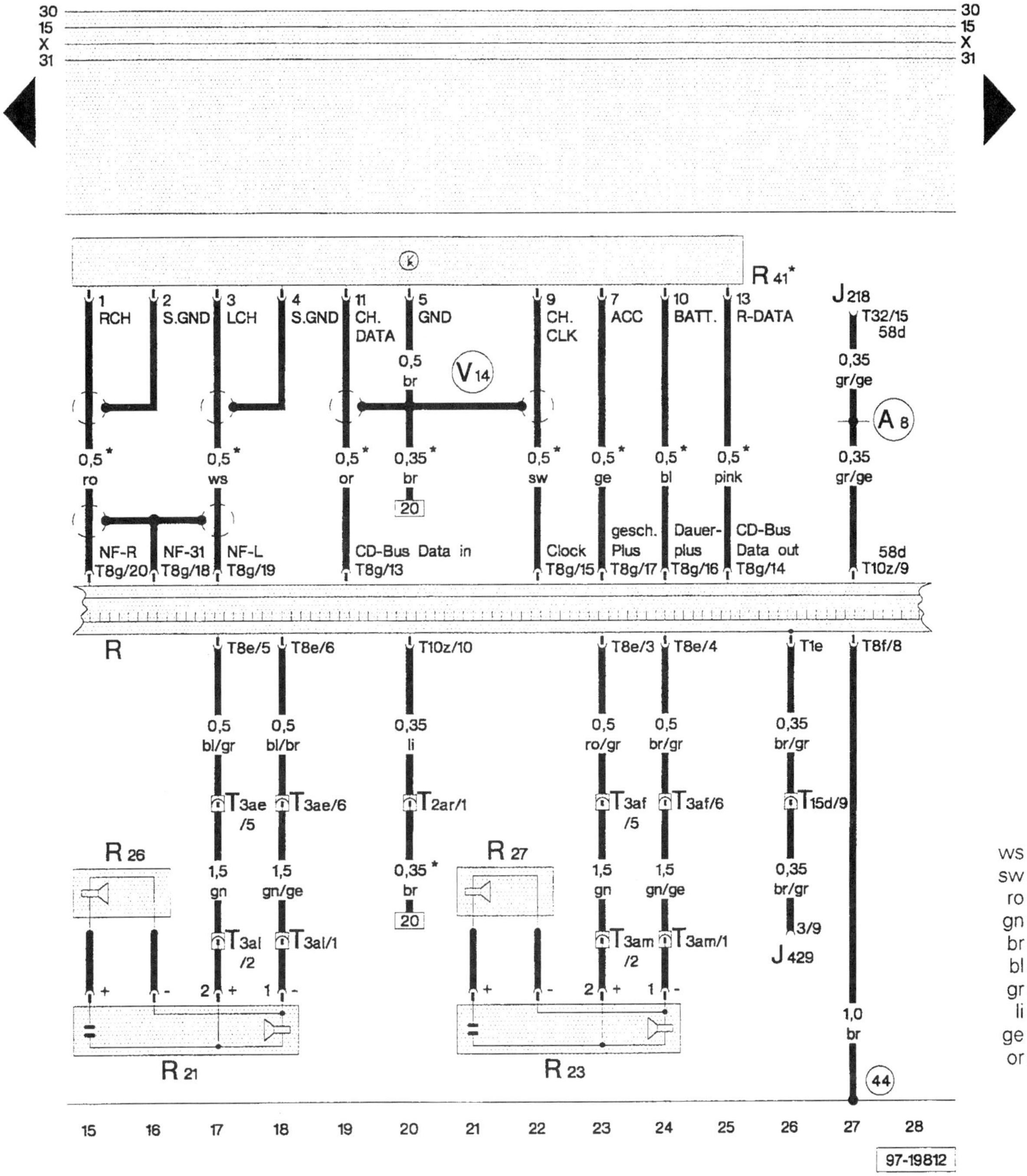

ws = white
sw = black
ro = red
gn = green
br = brown
bl = blue
gr = gray
li = violet
ge = yellow
or = orange

J 218 - Instrument Cluster Combination Processor
J 429 - Control module for central locking
R - Radio
R 21 - Left Front Woofer
R 23 - Right Front Woofer
R 26 - Left Front Midrange Speaker
R 27 - Right Front Midrange Speaker
R 41 - CD Changer Unit
T 1 e - Single connector, on Radio
T 2 ar - Connector 2 pin, black, behind instrument panel, center
T 3 ae - Connector 3 pin, brown, connector station A pillar, left
T 3 af - Connector 3 pin, brown, connector station A pillar, right
T 3 al - Connector 3 pin, black, in driver's door
T 3 am - Connector 3 pin, black, in passenger's door
T 8 e - Connector 8 pin, brown, connector radio II
T 8 f - Connector 8 pin, black, connector radio III
T 8 g - Connector 8 pin, blue, connector radio I - 3
T 10 z - Connector 10 pin, red, connector radio IV
T 15 d - Connector 15 pin, black, connector station A pillar, right
T 32 - Connector 32 pin, blue, on instrument cluster

(44) - Ground connection, on left A-pillar, lower part

(A8) - Plus connection (58d), in instrument panel wiring harness (Display illumination)

(V14) - Wire connection (shielding), in CD Changer wiring harness

* - Vehicles with CD Changer

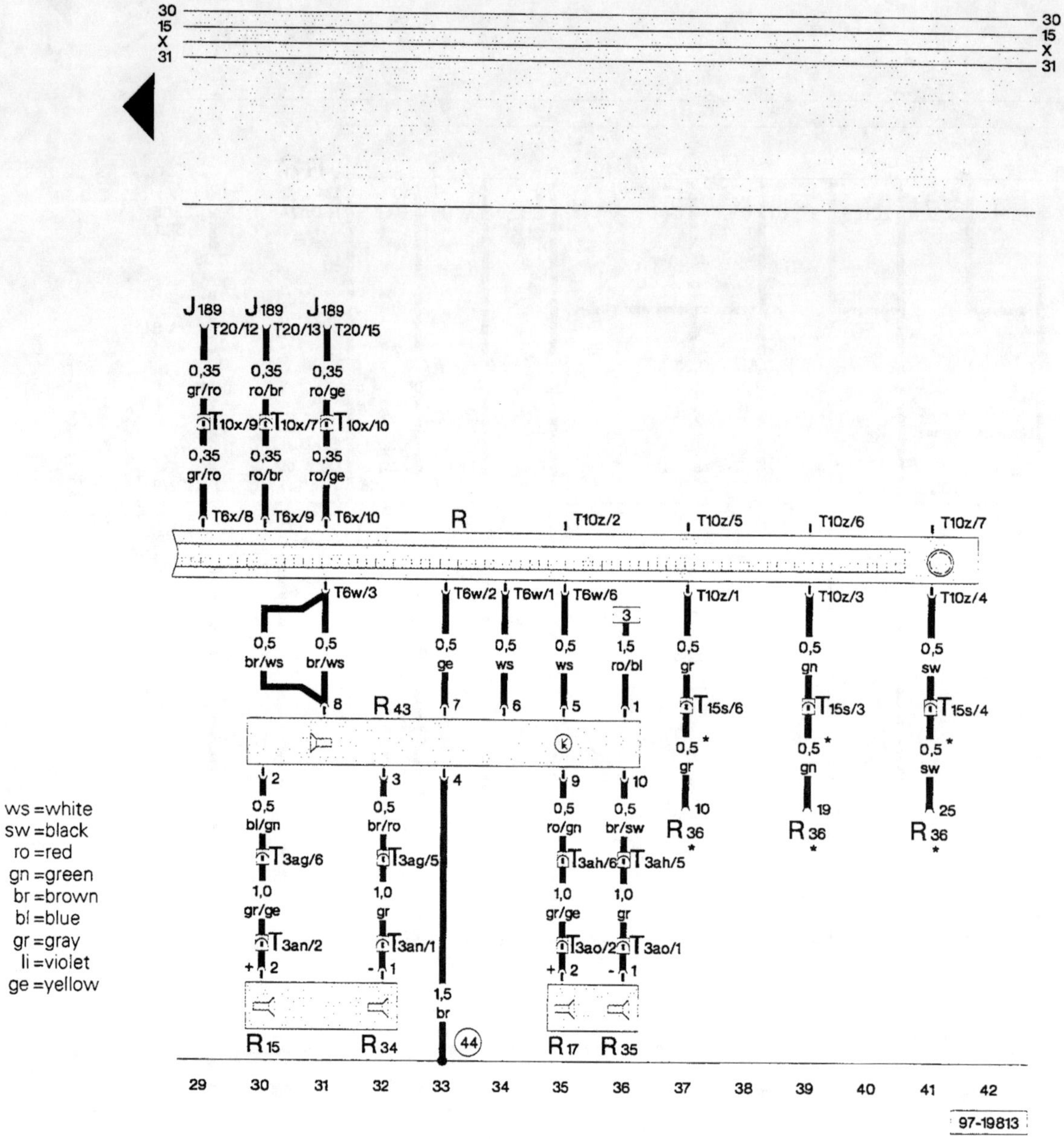

J 189 - Auto Check System
R - Radio
R 15 - Left Rear Woofer
R 17 - Right Rear Woofer
R 34 - Left Rear Midrange Speaker
R 35 - Right Rear Midrange Speaker
R 36 - Telephone Transceiver
R 43 - Right Rear Woofer/Amplifier
T 3 ag - Connector 3 pin, blue, in B pillar, left
T 3 ah - Connector 3 pin, blue, in B pillar, right
T 3 an - Connector 3 pin, black, in left rear door
T 3 ao - Connector 3 pin, black, in right rear door
T 6 w - Connector 6 pin, yellow, connector radio I - 1
T 6 x - Connector 6 pin, green, connector radio I - 2
T 10 x - Connector 10 pin, pink, connector station A pillar, right
T 10 z - Connector 10 pin, red, connector radio IV
T 15 s - Connector 15 pin, brown, connector station A pillar, left
T 20 - Connector 20 pin, red/white, on instrument cluster

(44) - Ground connection, on left A-pillar, lower part

* - Vehicles with Telephone

Power Sunroof with Automatic Preselection

1999 m. y.

Fuse Panel

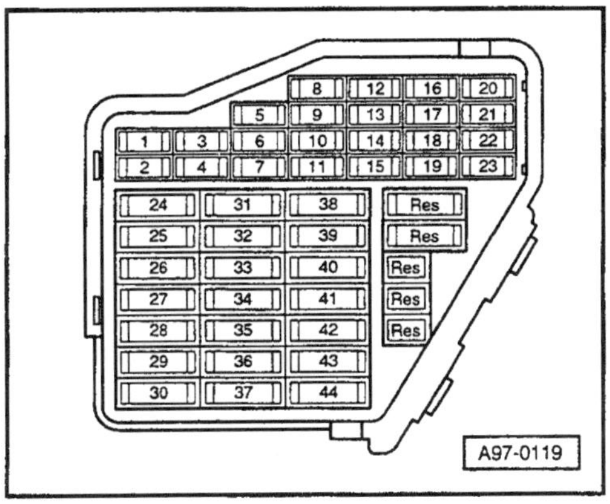

Fuse Colors:

30 A - Green
25 A - White
20 A - Yellow
15 A - Blue
10 A - Red
7,5 A - Brown
5 A - Beige

Starting with fuse position 23, fuses in the fuse holder are identified with 223 in the wiring diagram.

Micro Central Electric Panel

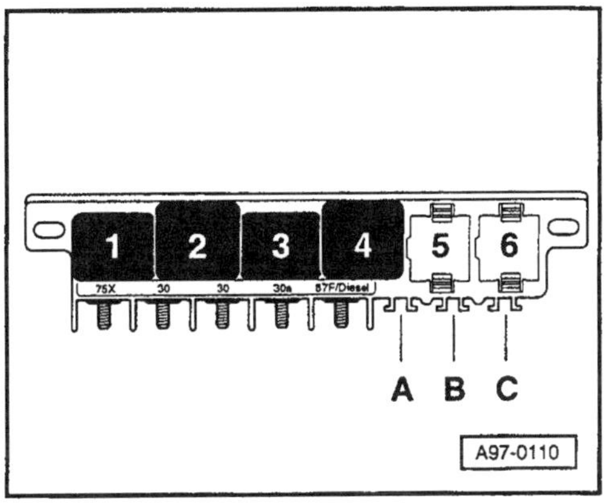

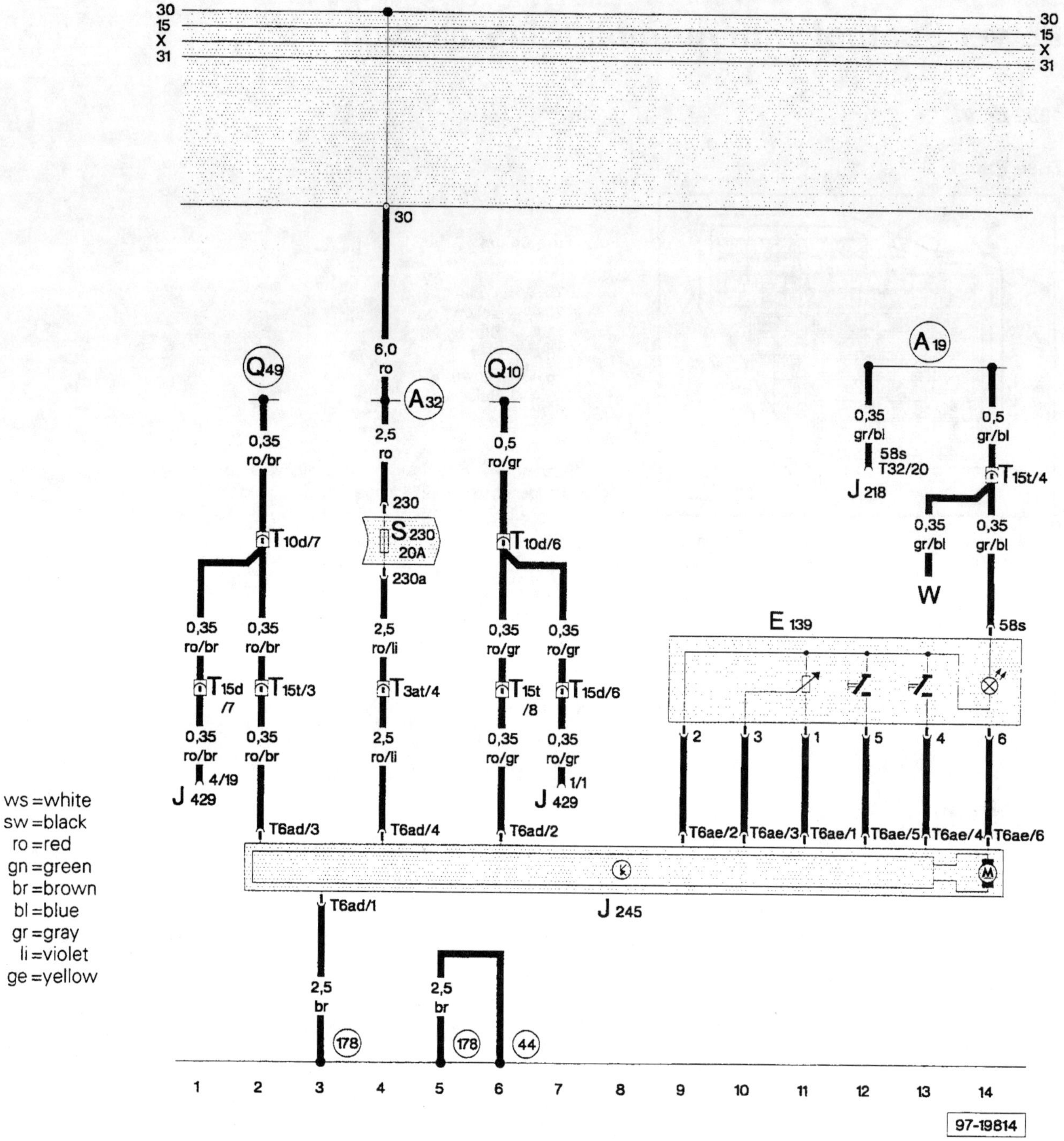

E 139 - Sunroof Regulator
J 218 - Instrument Cluster Combination Processor
J 245 - Power Sunroof Control Module
J 429 - Control module for central locking
S 230 - Fuse
T 3 at - Connector 3 pin, grey, connector station A pillar, left
T 6 ad - Connector 6 pin, blue, on power sunroof control module
T 6 ae - Connector 6 pin, on power sunroof control module
T 10 d - Connector 10 pin, green, connector station A pillar, left
T 15 d - Connector 15 pin, black, connector station A pillar, right
T 15 t - Connector 15 pin, grey, connector station A pillar, left
T 32 - Connector 32 pin, blue, on instrument cluster
W - Front Interior Light

(44) - Ground connection, on left A-pillar, lower part
(178) - Ground connection, in sunroof wiring harness
(A19) - Wire connection (58s), in instrument panel wiring harness (Switch illumination)
(A32) - Plus connection (30), in instrument panel wiring harness
(Q10) - Plus connection (87), in power window/central locking system and door contact switch wiring harness
(Q49) - Connection (closed), in wiring harness interior

Cruise Control with Mechanical Throttle Actuation

1999 m. y.

Fuse Panel

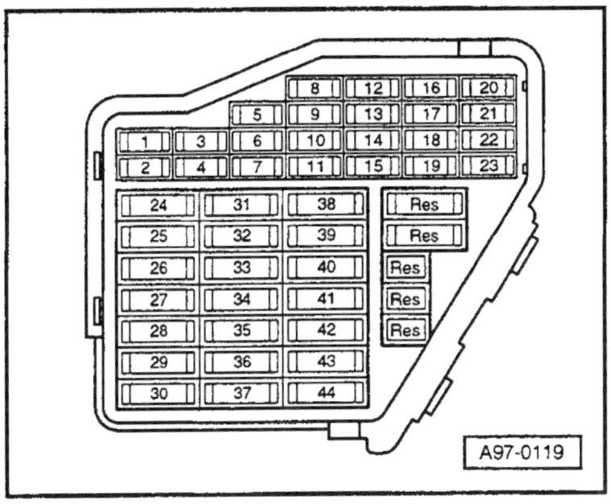

Fuse Colors:

30 A - Green
25 A - White
20 A - Yellow
15 A - Blue
10 A - Red
7,5 A - Brown
5 A - Beige

Starting with fuse position 23, fuses in the fuse holder are identified with 223 in the wiring diagram.

Micro Central Electric Panel

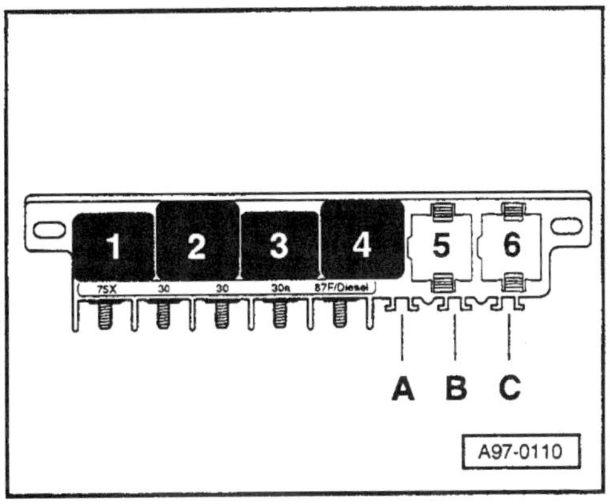

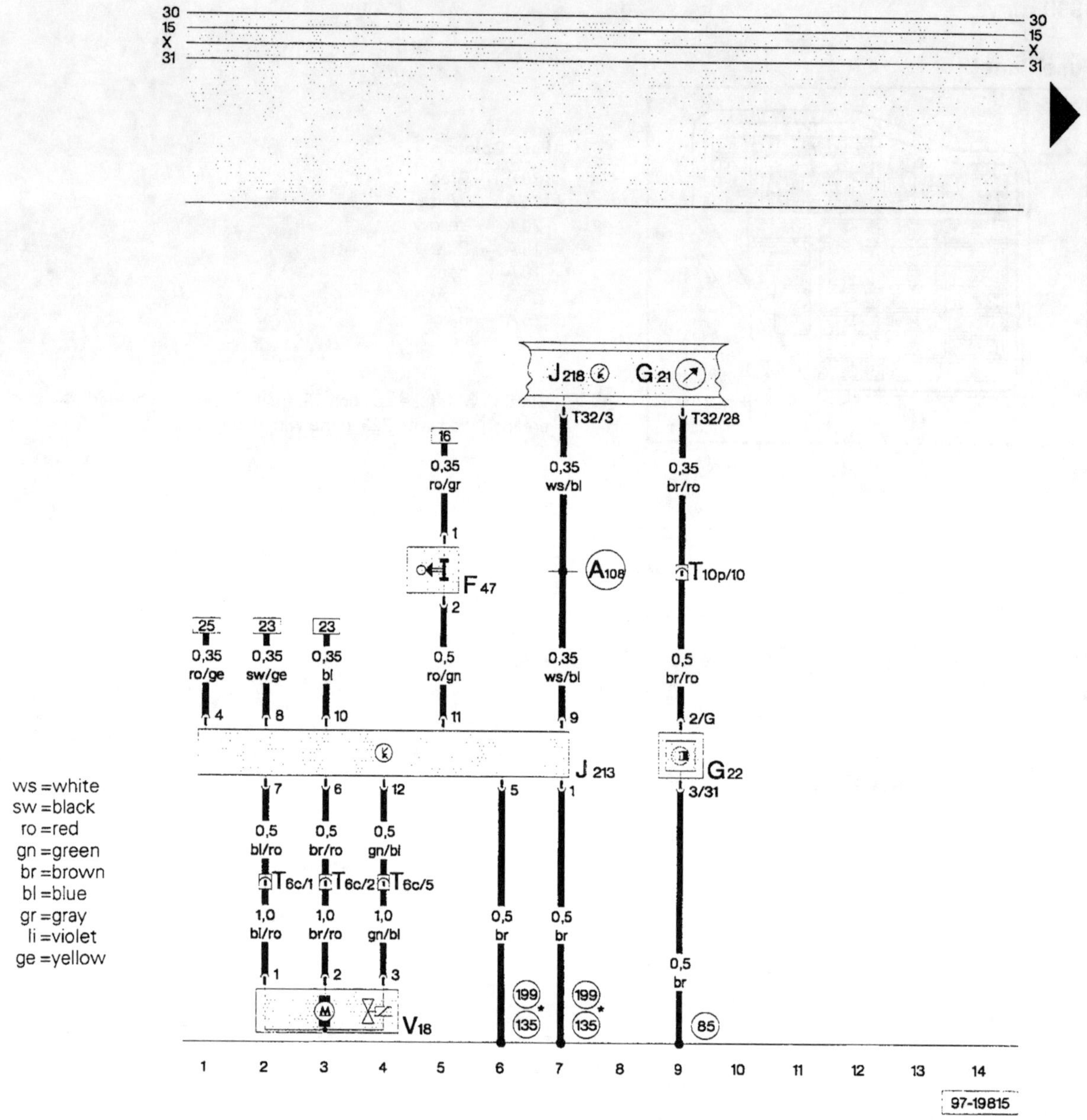

F 47 - Brake Pedal Switch
G 21 - Speedometer
G 22 - Speedometer Vehicle Speed Sensor (VSS)
J 213 - Cruise Control, Control Module
J 218 - Instrument Cluster Combination Processor
T 6 c - Connector 6 pin, black, connector station A pillar, left
T 10 p - Connector 10 pin, black, connector station electronic box plenum chamber
T 32 - Connector 32 pin, blue, on instrument cluster
V 18 - Cruise Control Vacuum Pump

(85) - Ground connection -1-, in engine compartment wiring harness
(135) - Ground connection -2-, in instrument panel wiring harness
(199) - Ground connection -3-, in instrument panel wiring harness
(A108) - Wire connection (vehicle speed signal), in instrument panel wiring harness

* - booth is possible

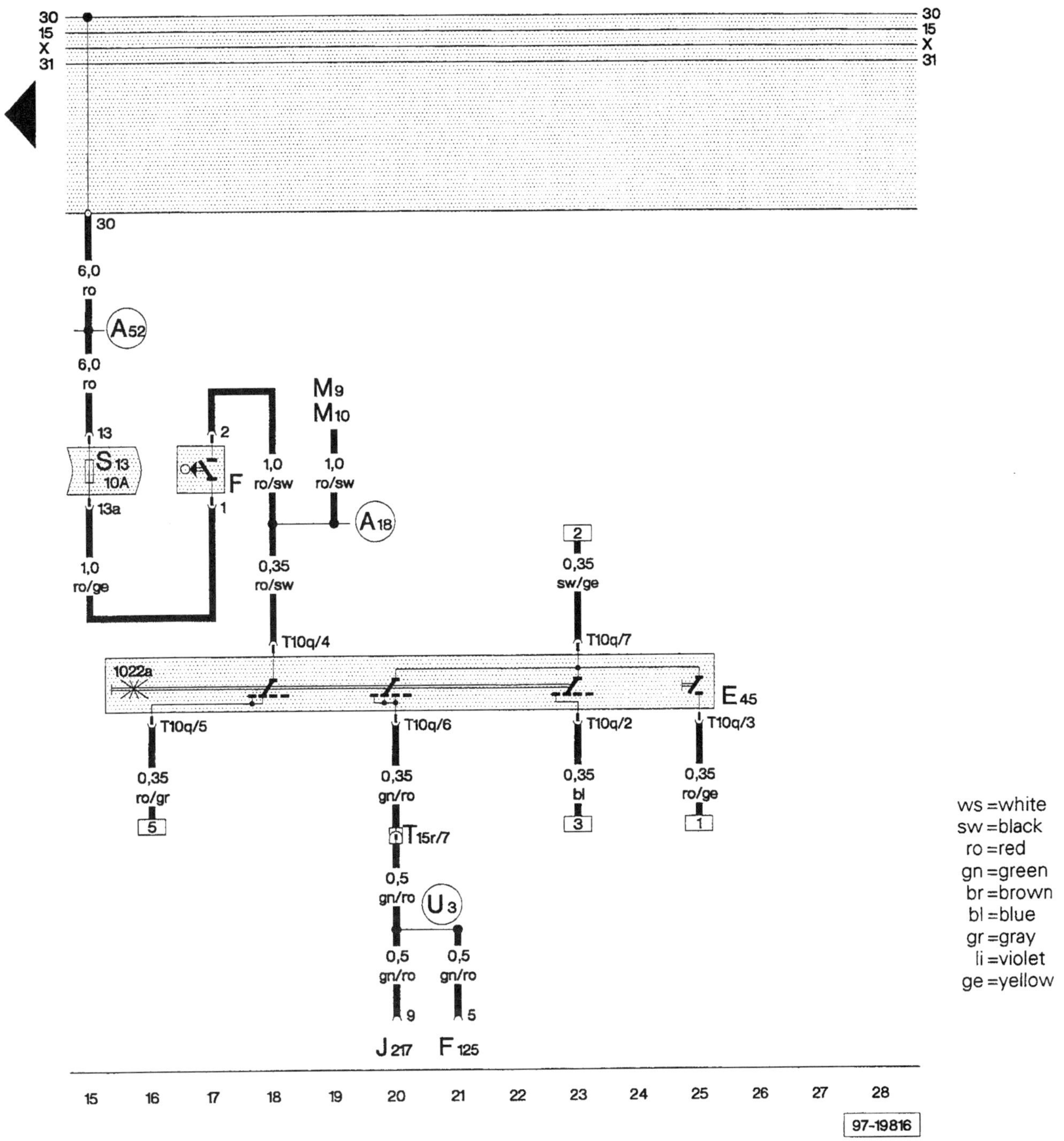

E 45 - Cruise Control Switch
F - Brake Light Switch
F 125 - Multi-Function Transmission Range (TR) Switch
J 217 - Transmission Control Module (TCM)
M 9 - Left Brake Light
M 10 - Right Brake Light
S 13 - Fuse
T 10 q - Connector 10 pin, black, on cruise control switch
T 15 r - Connector 15 pin, brown, connector station A pillar, right

(A18) - Wire connection (54), in instrument panel wiring harness

(A52) - Plus connection (30), in instrument panel wiring harness

(U3) - Wire connection -1-, in automatic transmission wiring harness

THIS PAGE INTENTIONALLY LEFT

BLANK

Automatic Dimmer Inside and Outside Mirror

1999 m. y.

Fuse Panel

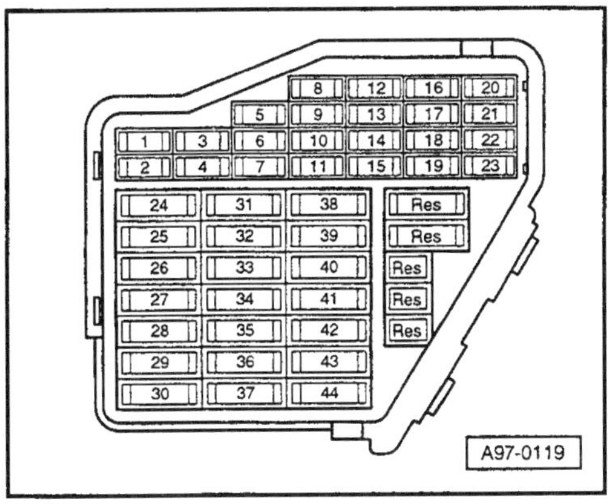

Fuse Colors:

30 A - Green
25 A - White
20 A - Yellow
15 A - Blue
10 A - Red
7,5 A - Brown
5 A - Beige

Starting with fuse position 23, fuses in the fuse holder are identified with 223 in the wiring diagram.

Micro Central Electric Panel

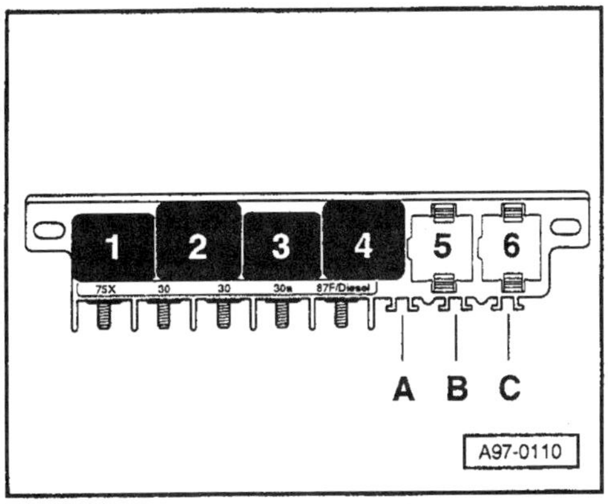

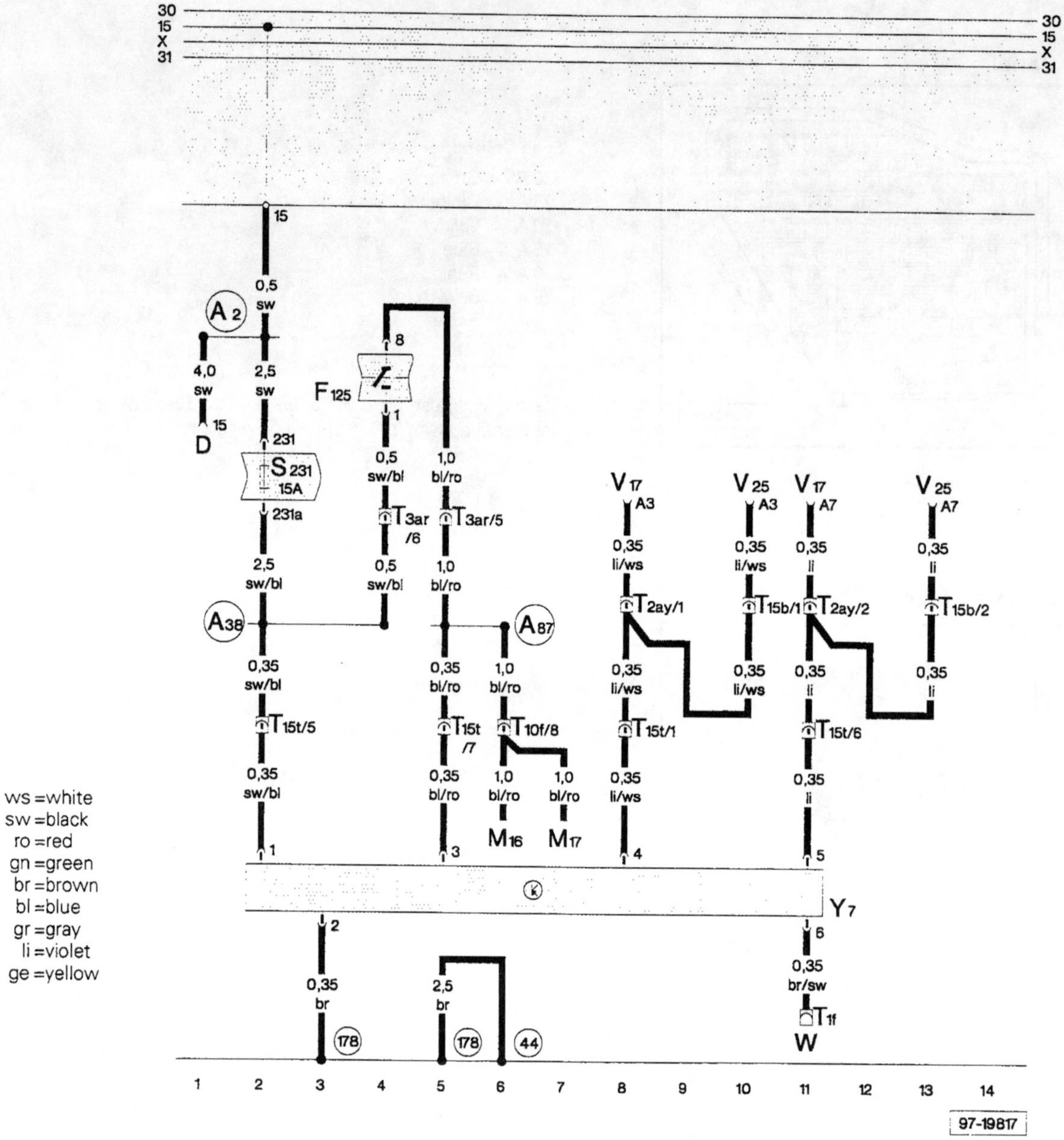

D - Ignition/Starter Switch
F 125 - Multi-Function Transmission Range (TR) Switch
M 16 - Left Back-Up Light
M 17 - Right Back-Up Light
S 231 - Fuse
T 1 f - Single connector, black, on front interior light
T 2 ay - Connector 2 pin, green, connector station A pillar, left
T 3 ar - Connector 3 pin, brown, connector station A pillar, right
T 10 f - Connector 10 pin, brown, connector station A pillar, left
T 10 w - Connector 10 pin, blue, connector station A pillar, right
T 15 b - Connector 15 pin, black, connector station A pillar, right
T 15 t - Connector 15 pin, grey, connector station A pillar, left
V 17 - Driver's Side Mirror Adjustment Motor
V 25 - Passenger's Side Mirror Adjustment Motor
W - Front Interior Light
Y 7 - Automatic day/night interior mirror

(44) - Ground connection, on left A-pillar, lower part
(178) - Ground connection, in sunroof wiring harness
(A2) - Plus connection (15), in instrument panel wiring harness
(A38) - Plus connection (15a II), in instrument panel wiring harness
(A87) - Connector (reverse lamp), in instrument panel wiring harness

Headlight Washer

1999 m. y.

Fuse Panel

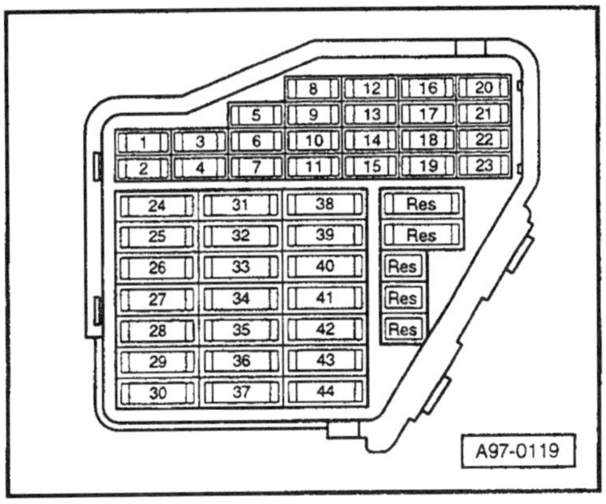

Fuse Colors:

30 A - Green
25 A - White
20 A - Yellow
15 A - Blue
10 A - Red
7,5 A - Brown
5 A - Beige

Starting with fuse position 23, fuses in the fuse holder are identified with 223 in the wiring diagram.

Micro Central Electric Panel

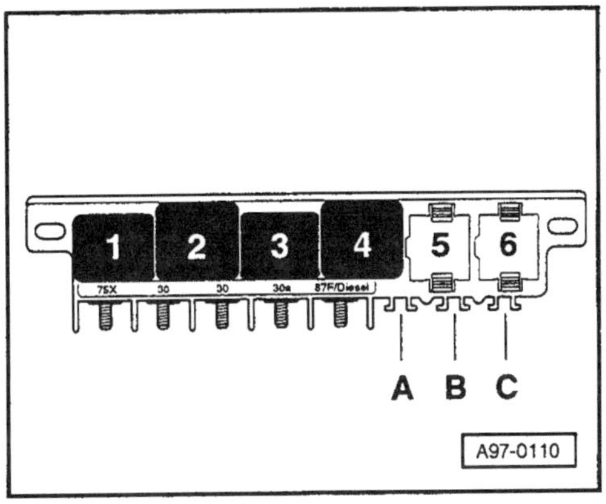

Relay Location:

5 - Wiper/Washer Intermittent Relay, J31

6 - Wiper/Washer Intermittent Relay, J31

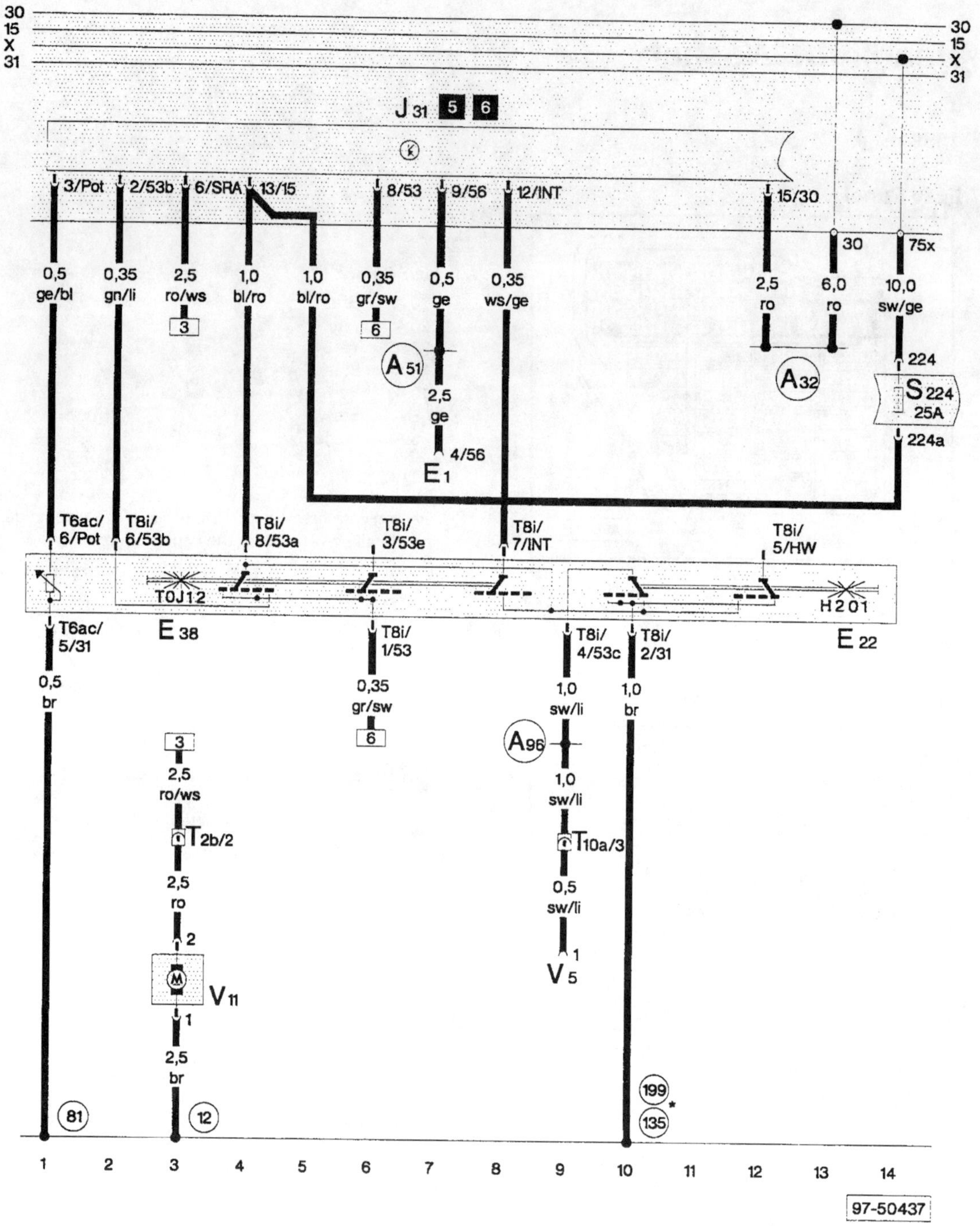

ws = white
sw = black
ro = red
gn = green
br = brown
bl = blue
gr = gray
li = violet
ge = yellow

E 1 - Light Switch
E 22 - Windshield Wiper/Washer Switch
E 38 - Windshield Wiper Intermittent Regulator
J 31 - Wiper/Washer Intermittent Relay
S 224 - Fuse
T 2 b - Connector 2 pin, black, connector station A pillar, left
T 6 ac - Connector 6 pin, brown, on windshield wiper intermittent regulator
T 8 i - Connector 8 pin, on windshield wiper/washer switch
T 10 a - Connector 10 pin, pink, connector station A pillar, left
V 5 - Windshield Washer Pump
V 11 - Headlight Washer Pump

(12) - Ground connection, in engine compartment, left
(81) - Ground connection -1-, in instrument panel wiring harness
(135) - Ground connection -2-, in instrument panel wiring harness
(199) - Ground connection -3-, in instrument panel wiring harness
(A32) - Plus connection (30), in instrument panel wiring harness
(A51) - Wire connection (56), in instrument panel wiring harness
(A96) - Connector (53a), in instrument panel wiring harness

* - booth is possible

Headlight with Hight Intensity Gas discharge Lamps and automatic Headlight Beam Adjusting

1999 m. y.

Fuse Panel

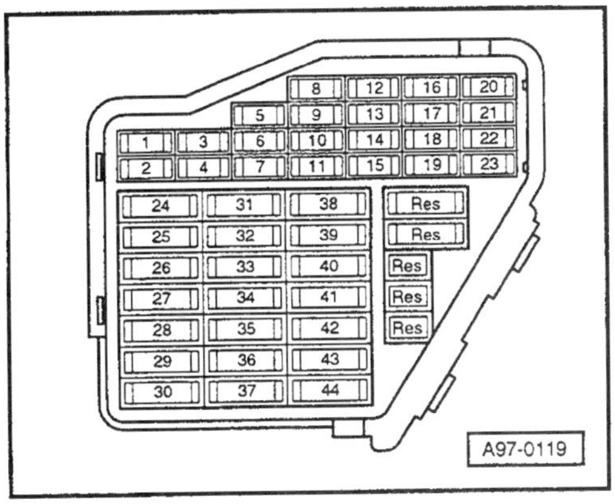

Fuse Colors:

30 A - Green
25 A - White
20 A - Yellow
15 A - Blue
10 A - Red
7,5 A - Brown
5 A - Beige

Starting with fuse position 23, fuses in the fuse holder are identified with 223 in the wiring diagram.

Micro Central Electric Panel

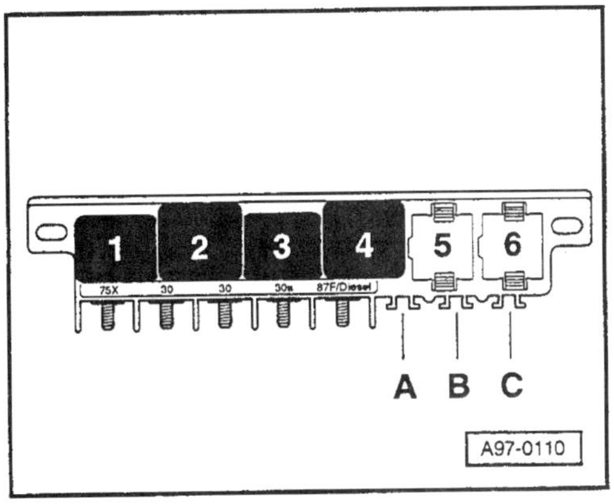

Relay Location:

2 - Load Reduction Relay, J59

13 - Fold Relay Panel

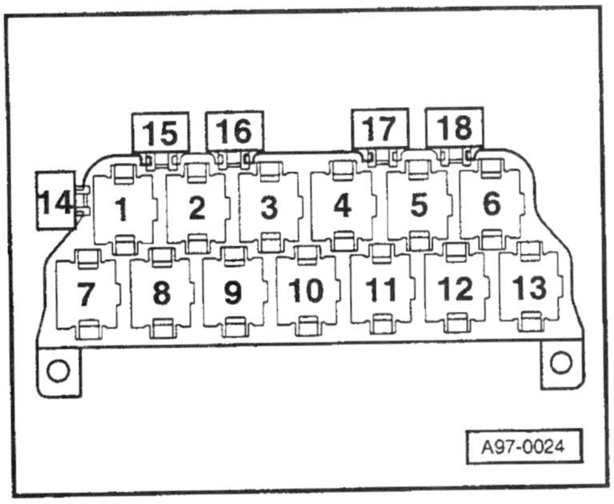

Relay Location:

9 - Lamp Control Module, J123

10 - Lamp Control Module, J123

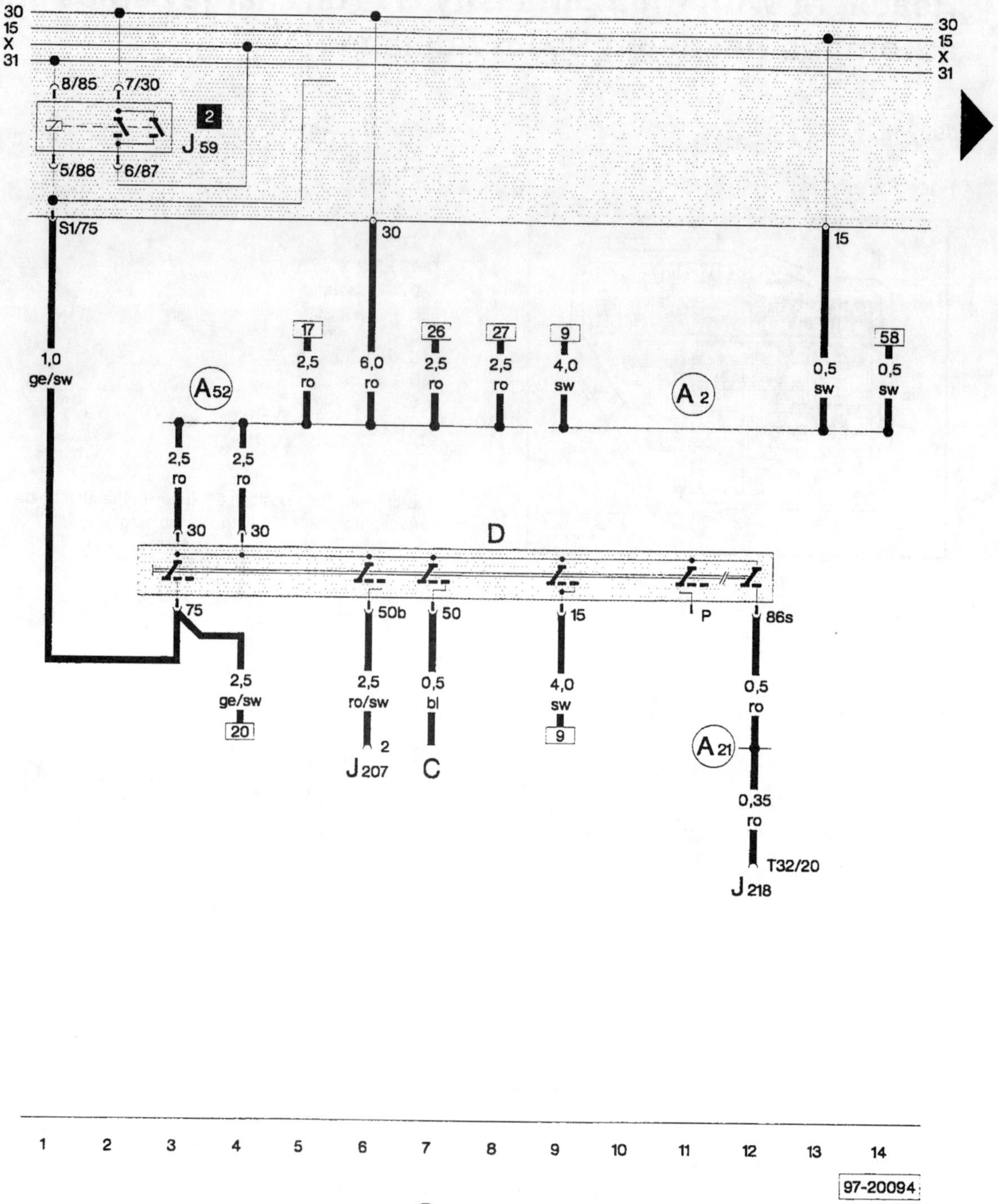

ws = white
sw = black
ro = red
gn = green
br = brown
bl = blue
gr = gray
li = violet
ge = yellow

C - Generator (GEN)
D - Ignition/Starter Switch
J 59 - Load Reduction Relay
J 207 - Starting Interlock Relay
J 218 - Instrument Cluster Combination Processor
T 32 - Connector 32 pin, blue, on instrument cluster

A2 - Plus connection (15), in instrument panel wiring harness
A21 - Wire connection (86s), in instrument panel wiring harness
A52 - Plus connection (30), in instrument panel wiring harness

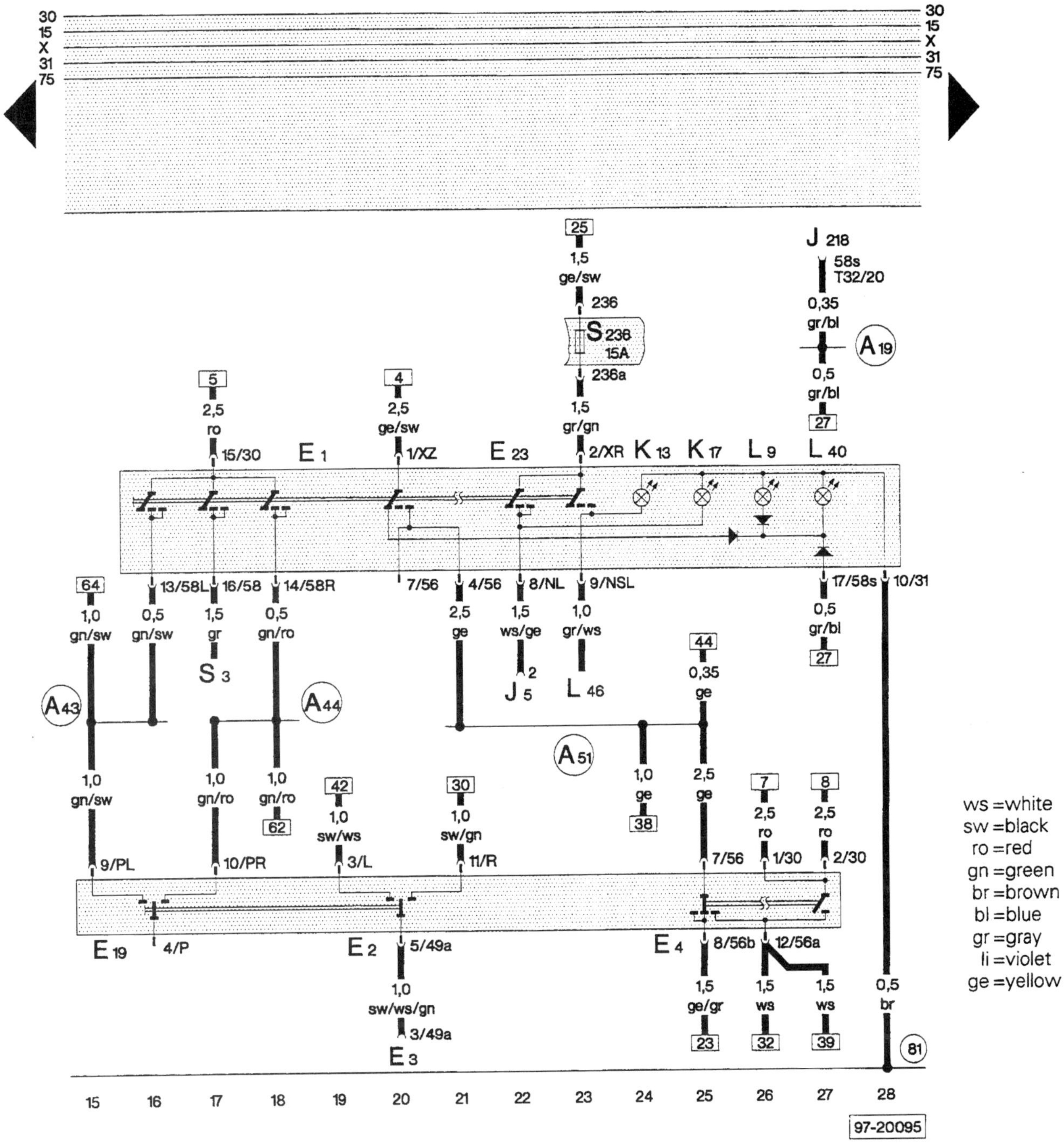

E 1 - Light Switch
E 2 - Turn Signal Switch
E 3 - Emergency Flasher Switch
E 4 - Headlight Dimmer/Flasher Switch
E 19 - Park Light Switch
E 23 - Fog Light Switch
J 5 - Fog Light Relay
J 218 - Instrument Cluster Combination Processor
K 13 - Rear Fog Light Indicator Light
K 17 - Fog Light Indicator Light
L 9 - Headlight Switch Light
L 40 - Front And Rear Fog Light Switch Light
L 46 - Left Rear Fog Light
S 3 - Fuse
S 236 - Fuse
T 32 - Connector 32 pin, blue, on instrument cluster

(81) - Ground connection -1-, in instrument panel wiring harness

(A19) - Wire connection (58s), in instrument panel wiring harness (Switch illumination)

(A43) - Wire connection (57l), in instrument panel wiring harness

(A44) - Wire connection (57r), in instrument panel wiring harness

(A51) - Wire connection (56), in instrument panel wiring harness

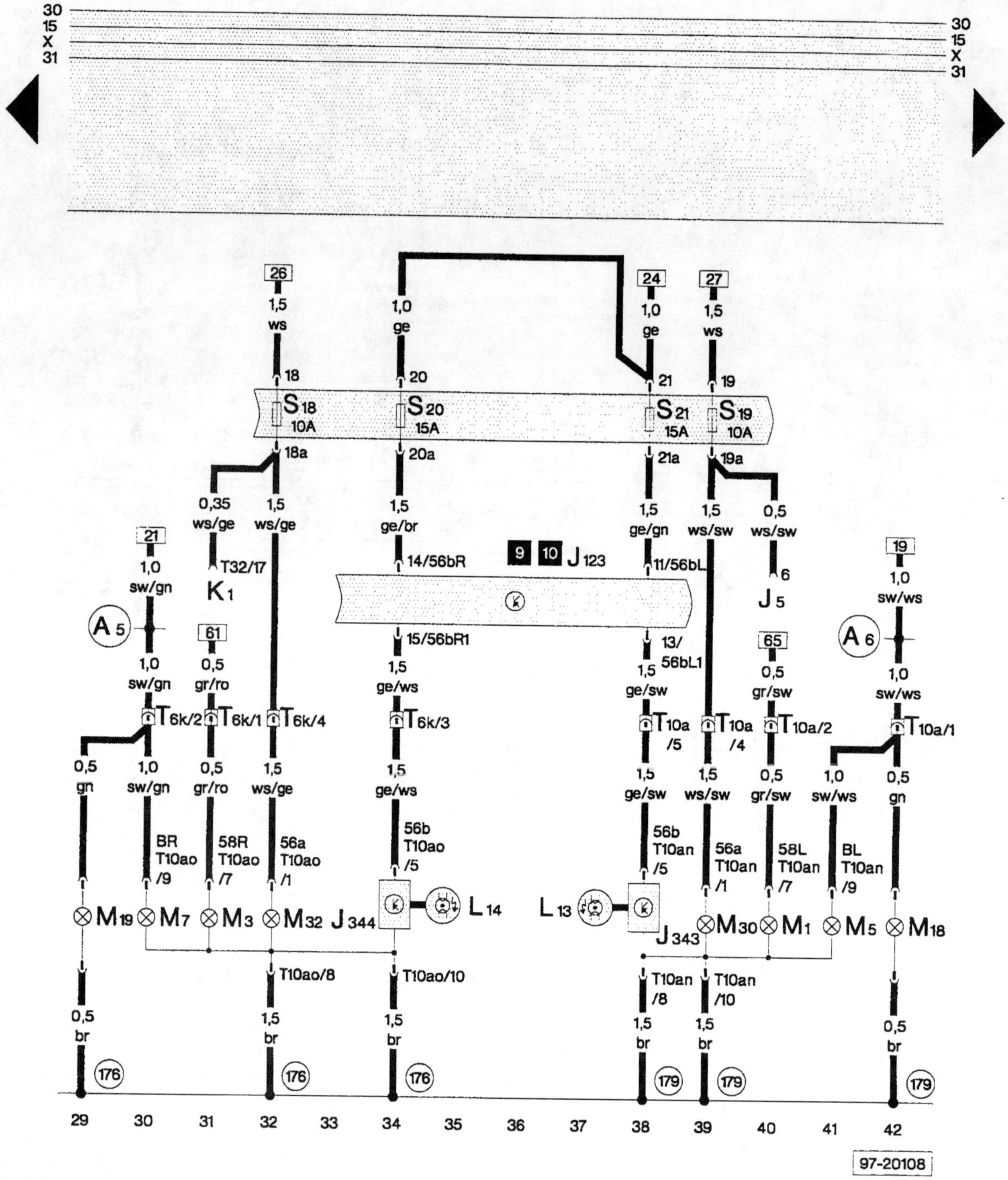

ws = white
sw = black
ro = red
gn = green
br = brown
bl = blue
gr = gray
li = violet
ge = yellow

J 5 - Fog Light Relay
J 123 - Lamp Control Module
J 343 - Control Module for High-intensity gas discharge lamp
J 344 - Control Module for High-intensity gas discharge lamp
K 1 - Headlight High Beam Indicator Light
L 13 - High intensity gas discharge lamp left
L 14 - High intensity gas discharge lamp right
M 1 - Left Parking Light
M 3 - Right Parking Light
M 5 - Left Front Turn Signal Light
M 7 - Right Front Turn Signal Light
M 18 - Left, Side Turn Signal Light
M 19 - Right, Side Turn Signal Light
M 30 - Left High Beam Headlight
M 32 - Right High Beam Headlight
S 18 - Fuse
S 19 - Fuse
S 20 - Fuse
S 21 - Fuse
T 6 k - Connector 6 pin, pink, connector station A pillar, right
T 10 a - Connector 10 pin, pink, connector station A pillar, left
T 10 an - Connector 10 pin, on headlight, left
T 10 ao - Connector 10 pin, on headlight, right
T 32 - Connector 32 pin, blue, on instrument cluster

(176) - Ground connection, in right headlight wiring harness
(179) - Ground connection, in left headlight wiring harness
(A5) - Plus connection (right turn signal), in instrument panel wiring harness
(A6) - Plus connection (left turn signal), in instrument panel wiring harness

High intensitiy gas discharge lamp,
Control Module for High intensity gas discharge lamp

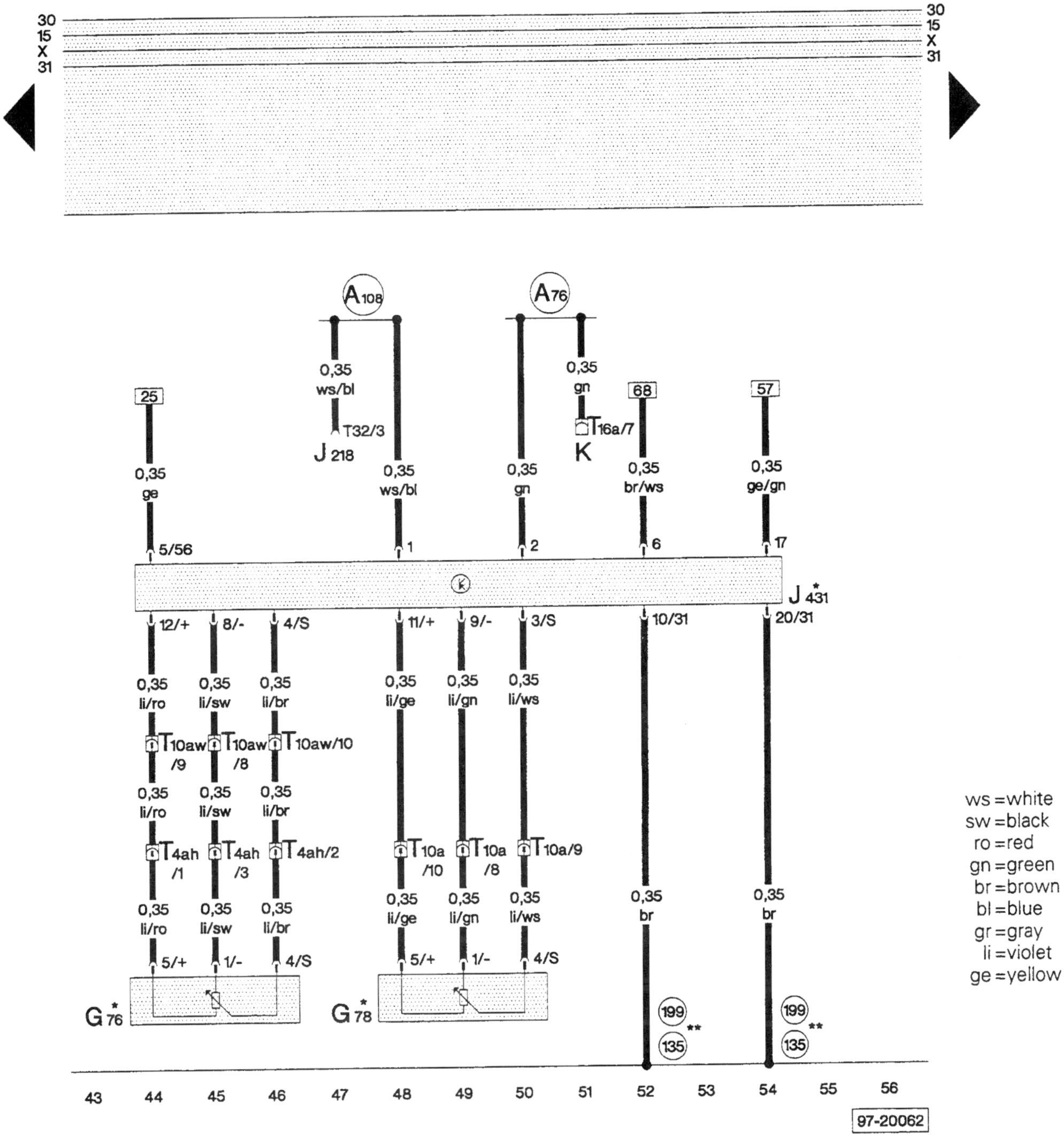

G 76 - Left Rear Level Control System Sensor
G 78 - Left Front Level Control System Sensor
J 218 - Instrument Cluster Combination Processor
J 431 - Control Unit for Beam Adjusting
K - DLC (Data Link Connector) Rapid Data Transfer
T 4 ah - Connector 4 pin, black, below rear seat, left
T 10 a - Connector 10 pin, pink, connector station A pillar, left
T 10 aw - Connector 10 pin, black, connector station A pillar, right
T 16 a - Connector 16 pin, data link connector
T 32 - Connector 32 pin, blue, on instrument cluster

(135) - Ground connection -2-, in instrument panel wiring harness
(199) - Ground connection -3-, in instrument panel wiring harness
(A76) - Connector (K-diagnosis wire), in instrument panel wiring harness
(A108) - Wire connection (vehicle speed signal), in instrument panel wiring harness

* - Vehicles with automatic beam adjusting
** - booth is possible

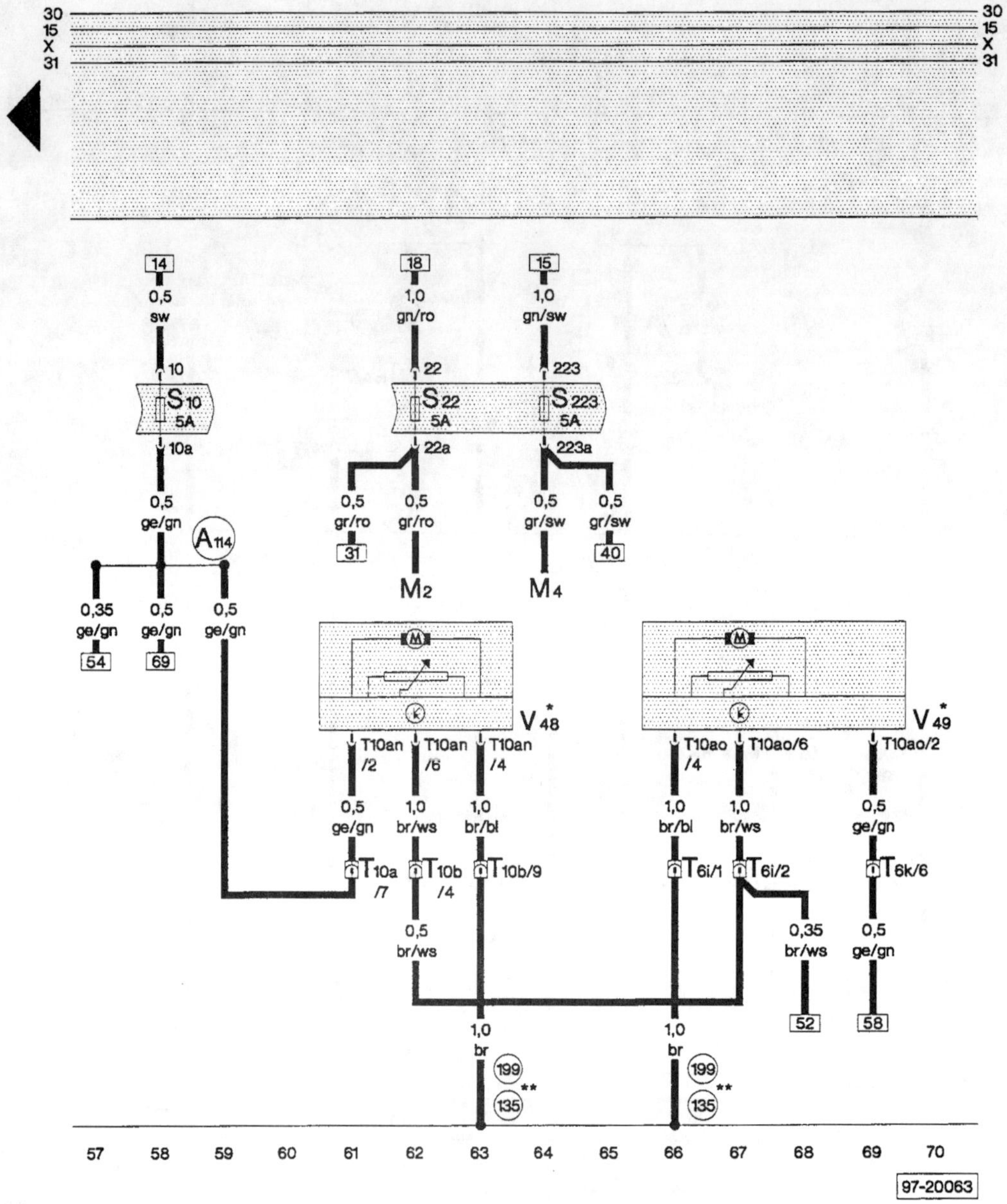

ws = white
sw = black
ro = red
gn = green
br = brown
bl = blue
gr = gray
li = violet
ge = yellow

M 2 - Right Tail Light
M 4 - Left Tail Light
S 10 - Fuse
S 22 - Fuse
S 223 - Fuse
T 6 i - Connector 6 pin, red, connector station A pillar, right
T 6 k - Connector 6 pin, pink, connector station A pillar, right
T 10 a - Connector 10 pin, pink, connector station A pillar, left
T 10 b - Connector 10 pin, red, connector station A pillar, left
T 10 an - Connector 10 pin, on headlight, left
T 10 ao - Connector 10 pin, on headlight, right
V 48 - Left Headlight Beam Adjusting Motor
V 49 - Right Headlight Beam Adjusting Motor

(135) - Ground connection -2-, in instrument panel wiring harness

(199) - Ground connection -3-, in instrument panel wiring harness

(A114) - Connection (Beam Adjusting), in instrument panel wiring harness

* - Vehicles with automatic beam adjusting
** - booth is possible

Rear Wiring, Wagon

1999 m. y.

Fuse Panel

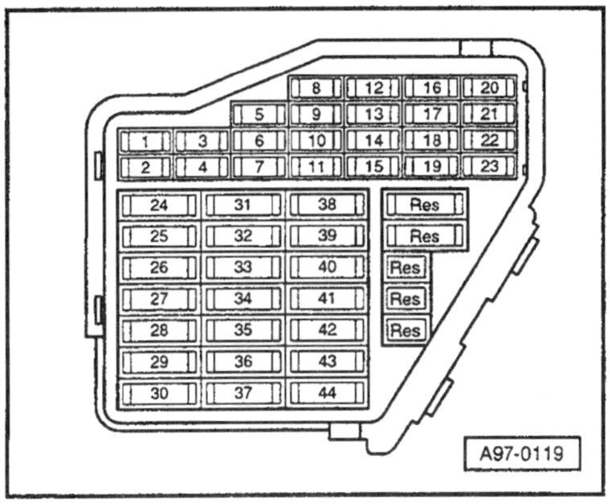

Fuse Colors:

30 A - Green
25 A - White
20 A - Yellow
15 A - Blue
10 A - Red
7,5 A - Brown
5 A - Beige

Starting with fuse position 23, fuses in the fuse holder are identified with 223 in the wiring diagram.

13 - Fold Relay Panel

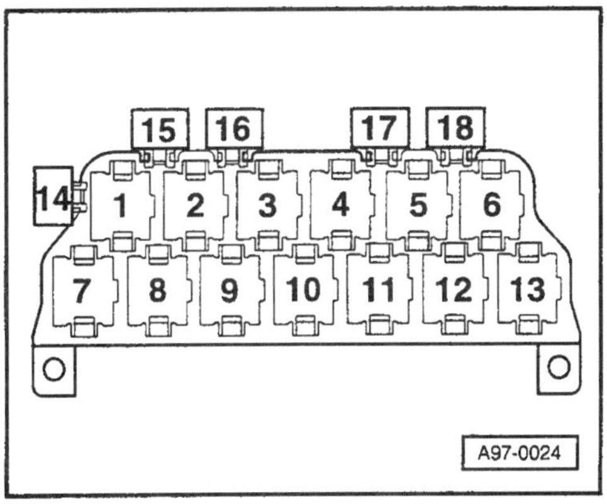

Relay Location:

9 - Lamp Control Module, J123

10 - Lamp Control Module, J123

8 - Fold Relay Panel

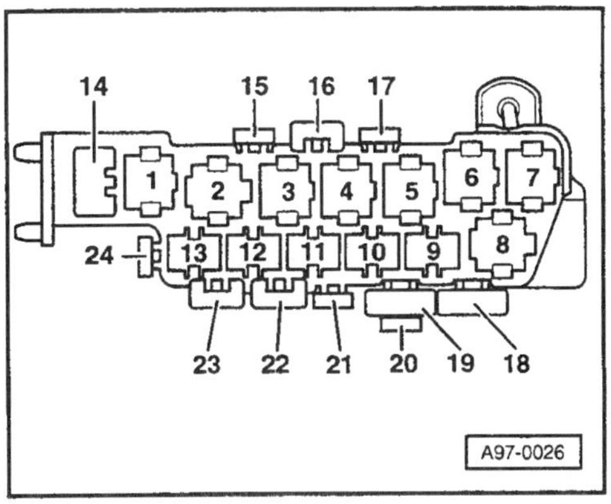

Fuse Location:

15 - Fuse for Luggage Compartment Socket, S184

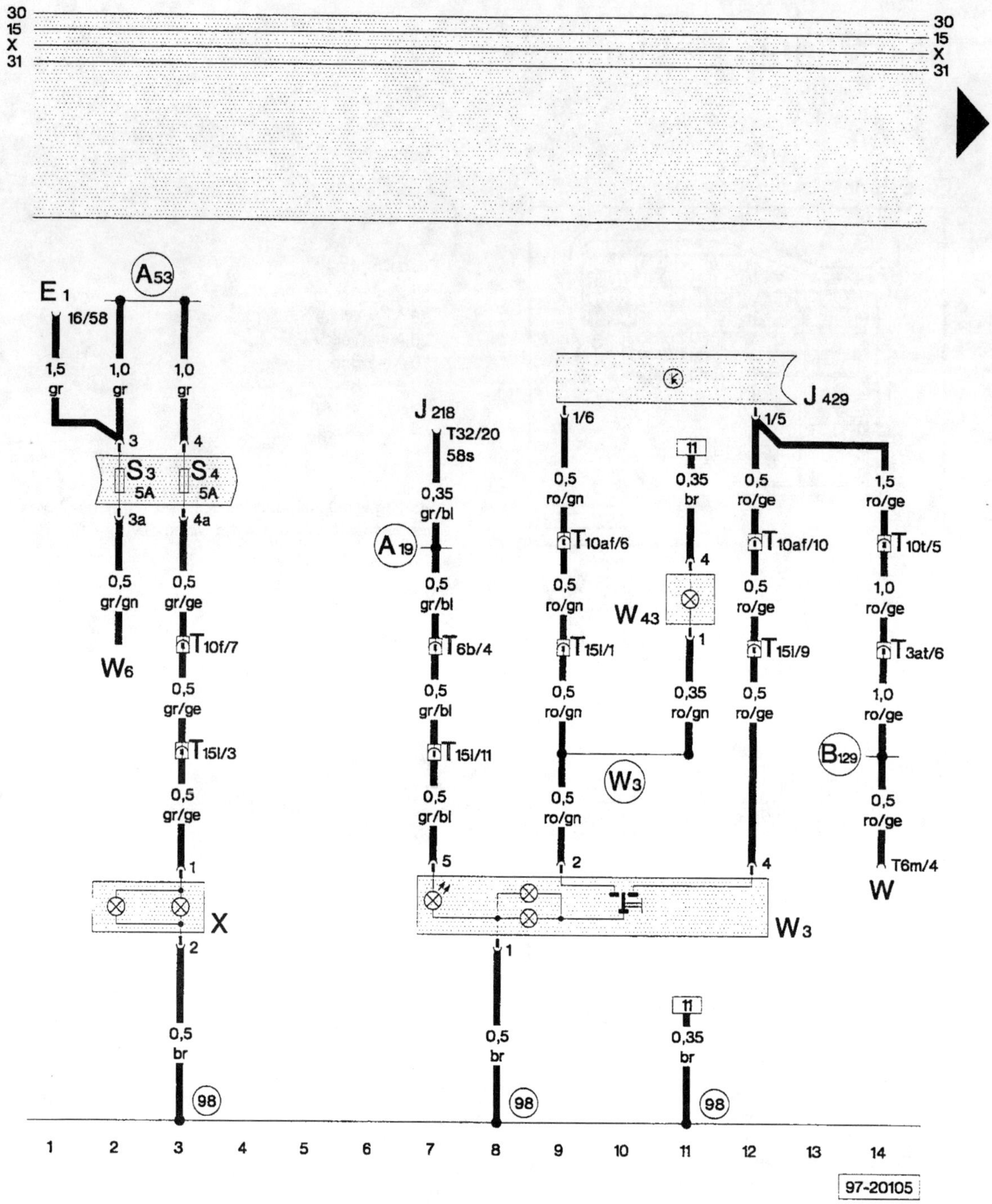

ws = white
sw = black
ro = red
gn = green
br = brown
bl = blue
gr = gray
li = violet
ge = yellow

E 1 - Light Switch
J 218 - Instrument Cluster Combination Processor
J 429 - Control module for central locking
S 3 - Fuse
S 4 - Fuse
T 3 at - Connector 3 pin, grey, connector station A pillar, left
T 6 b - Connector 6 pin, white, connector station A pillar, left
T 6 m - Connector 6 pin, on interior light, front
T 10 f - Connector 10 pin, brown, connector station A pillar, left
T 10 t - Connector 10 pin, orange, connector station A pillar, right
T 10 af - Connector 10 pin, violet, connector station A pillar, left
T 15 l - Connector 15 pin, black, in luggage compartment
T 32 - Connector 32 pin, blue, on instrument cluster
W - Front Interior Light
W 3 - Luggage Compartment Light
W 6 - Glove Compartment Light
W 43 - Rear courtesy lamp, in rear lid
X - License Plate Light

(98) - Ground connection, in rear lid wiring harness
(A19) - Wire connection (58s), in instrument panel wiring harness (Switch illumination)
(A53) - Plus connection (58b), in instrument panel wiring harness
(B129) - Connector (interior light 3/L), in wiring harness interior
(W3) - Wire connection, in rear wiring harness

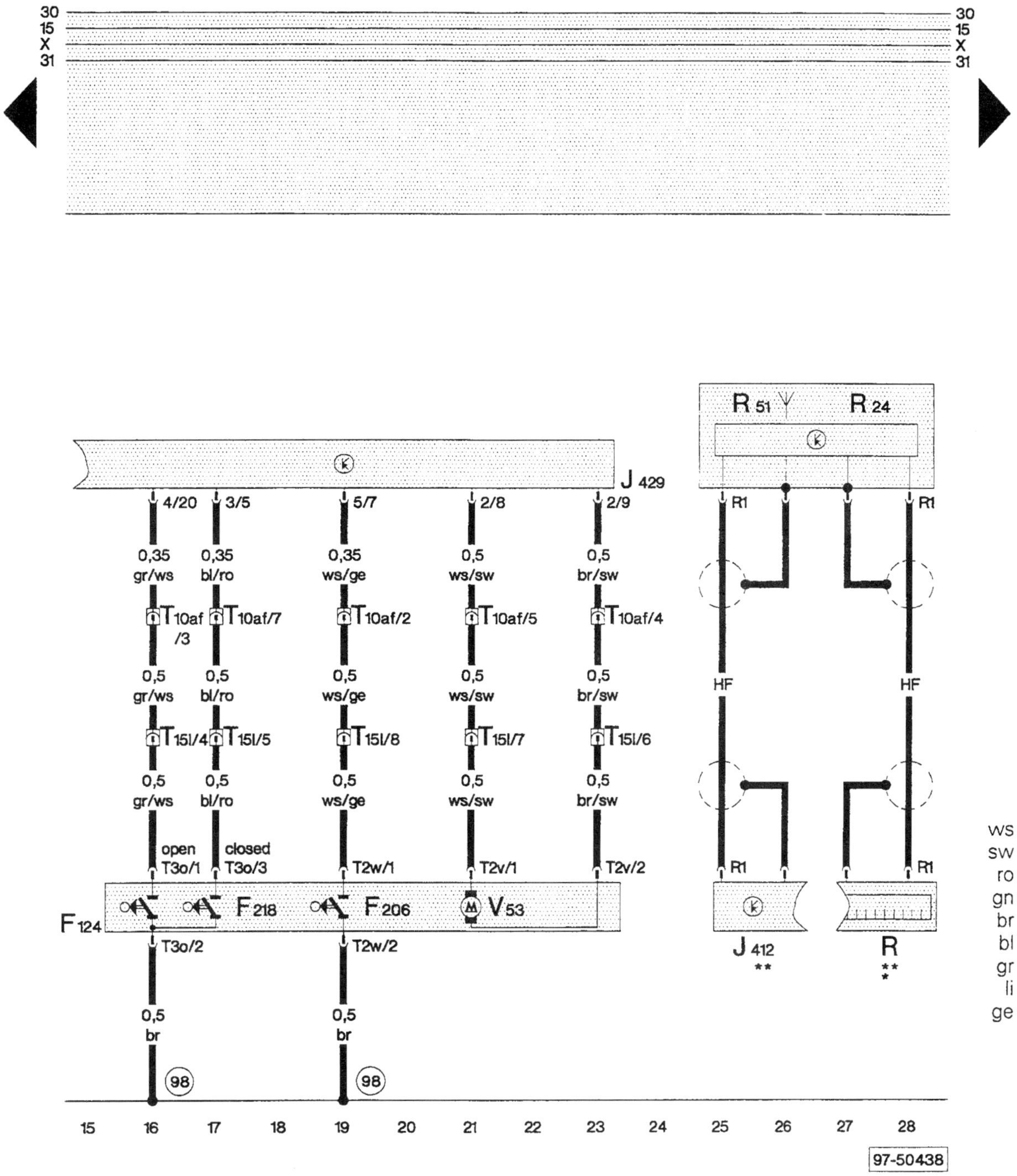

ws =white
sw =black
ro =red
gn =green
br =brown
bl =blue
gr =gray
li =violet
ge =yellow

F 124 - Trunk Lock Alarm/Central Locking Switch
F 206 - Trunk Lid Closed Position Switch
F 218 - Switch for central locking, trunk lid
J 412 - Control module for operating electronics, telephone
J 429 - Control module for central locking
R - Radio
R 1 - Antenna Connection
R 24 - Antenna Amplifier
R 51 - Antenna for radio/telephone
T 2 v - Connector 2 pin, black, luggage compartment
T 2 w - Connector 2 pin, grey, luggage compartment
T 3 o - Connector 3 pin, black, in rear lid
T 10 af - Connector 10 pin, violet, connector station A pillar, left
T 15 l - Connector 15 pin, black, in luggage compartment
V 53 - Decklid Central Locking System Motor

(98) - Ground connection, in rear lid wiring harness

* - Vehicles with Radio
** - Vehicles with Radio and Telephone

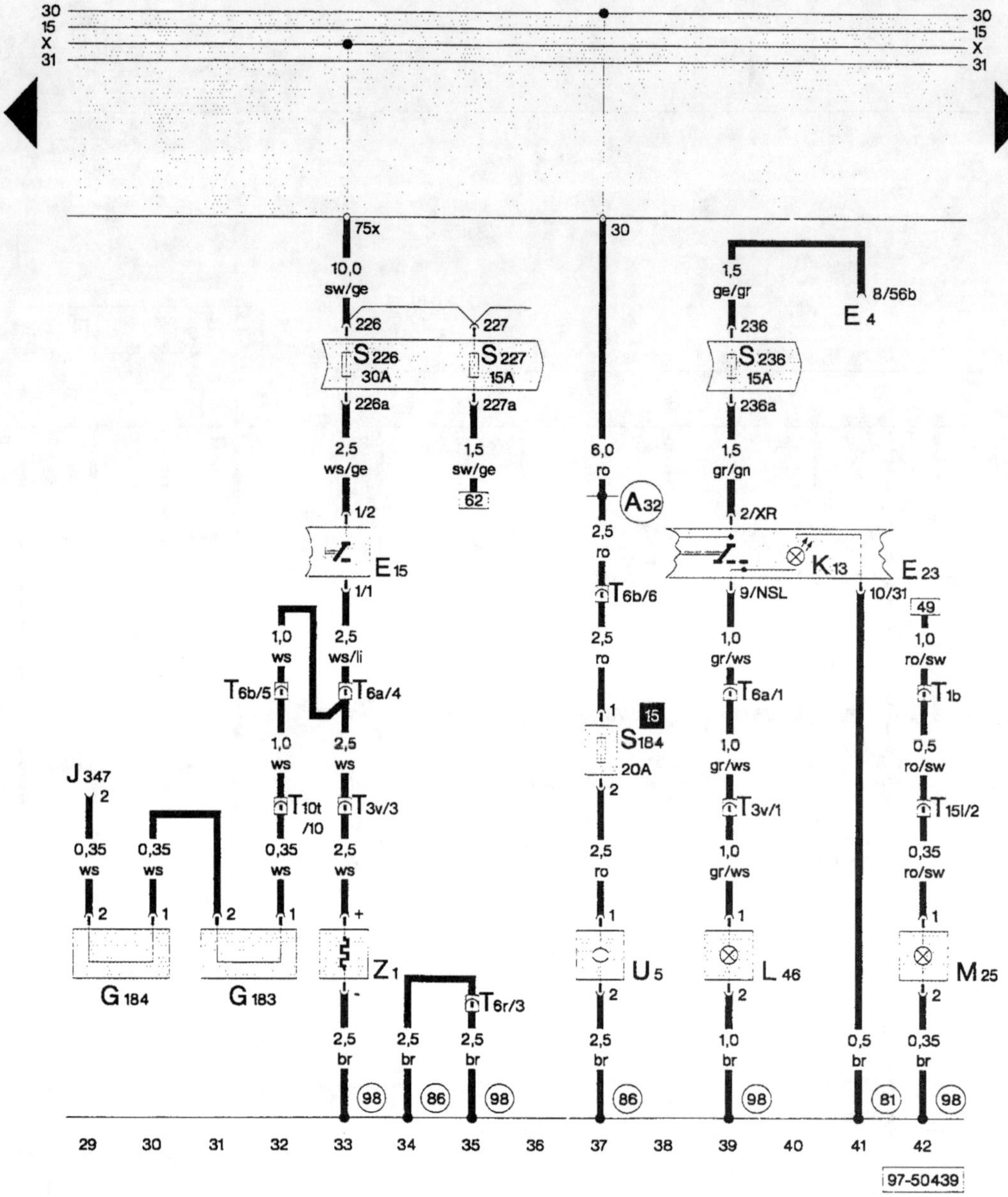

ws = white
sw = black
ro = red
gn = green
br = brown
bl = blue
gr = gray
li = violet
ge = yellow

E 4 - Headlight Dimmer/Flasher Switch
E 15 - Rear Window Defogger Switch
E 23 - Fog Light Switch
G 183 - Sensor for broken window glass rear left
G 184 - Sensor for broken window glass rear right
J 347 - Control Module for Ultra-Sound sensors
K 13 - Rear Fog Light Indicator Light
L 46 - Left Rear Fog Light
M 25 - High-mount Brake Light
S 184 - Fuse for Luggage Compartment Socket
S 226 - Fuse
S 227 - Fuse
S 236 - Fuse
T 1 b - Single connector, black, connector station A pillar right
T 3 v - Connector 3 pin, black, in luggage compartment
T 6 a - Connector 6 pin, blue, connector station A pillar, left
T 6 b - Connector 6 pin, white, connector station A pillar, left
T 6 r - Connector 6 pin, brown, in luggage compartment
T 10 t - Connector 10 pin, orange, connector station A pillar, right
T 15 l - Connector 15 pin, black, in luggage compartment
U 5 - 12 V Socket, in Luggage Compartment
Z 1 - Heated rear window

(81) - Ground connection -1-, in instrument panel wiring harness
(86) - Ground connection -1-, in rear wiring harness
(98) - Ground connection, in rear lid wiring harness
(A32) - Plus connection (30), in instrument panel wiring harness

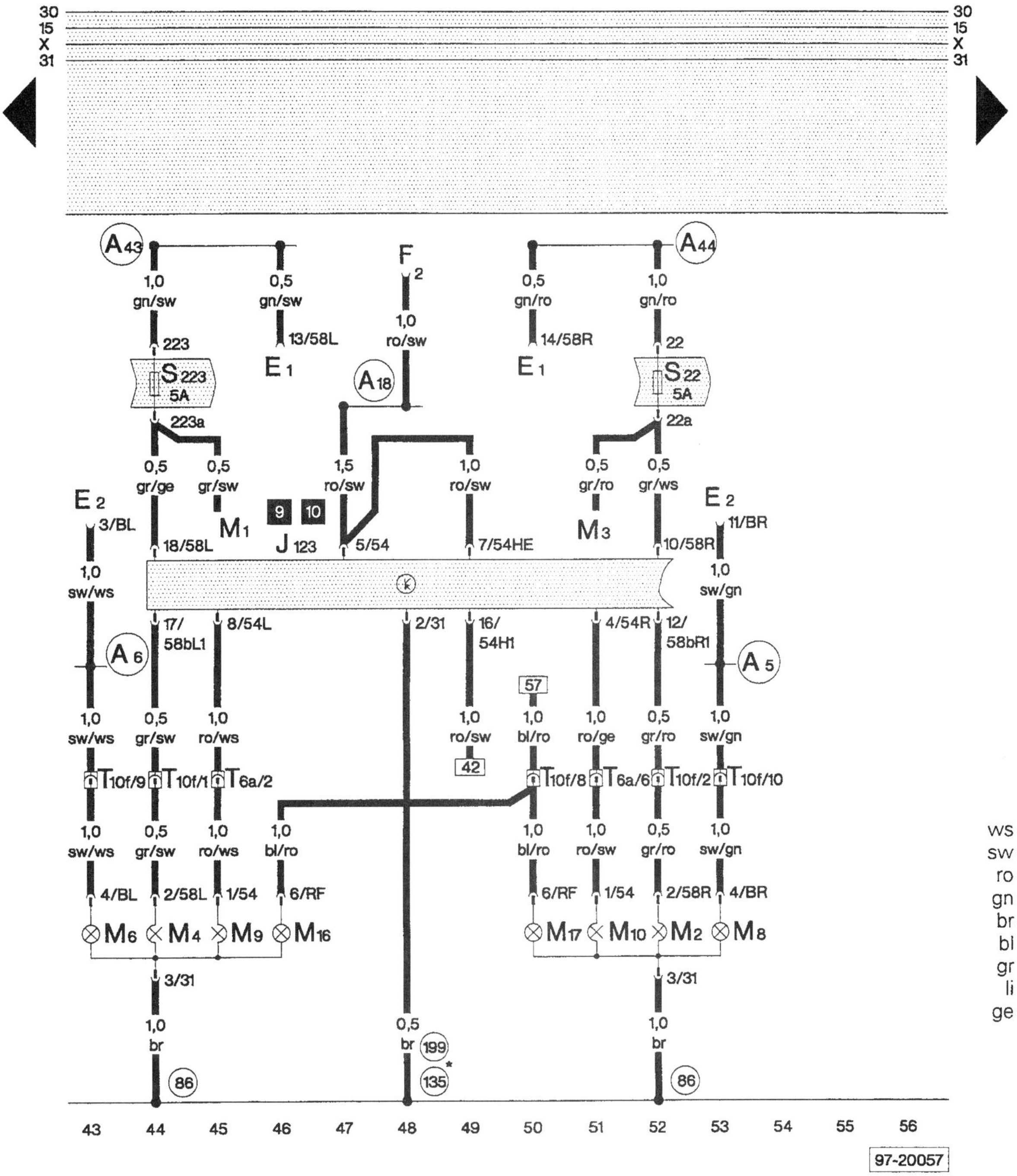

ws = white
sw = black
ro = red
gn = green
br = brown
bl = blue
gr = gray
li = violet
ge = yellow

E 1 - Light Switch
E 2 - Turn Signal Switch
F - Brake Light Switch
J 123 - Lamp Control Module
M 1 - Left Parking Light
M 2 - Right Tail Light
M 3 - Right Parking Light
M 4 - Left Tail Light
M 6 - Left Rear Turn Signal Light
M 8 - Right Rear Turn Signal Light
M 9 - Left Brake Light
M 10 - Right Brake Light
M 16 - Left Back-Up Light
M 17 - Right Back-Up Light
S 22 - Fuse
S 223 - Fuse
T 6 a - Connector 6 pin, blue, connector station A pillar, left
T 10 f - Connector 10 pin, brown, connector station A pillar, left

(86) - Ground connection -1-, in rear wiring harness
(135) - Ground connection -2-, in instrument panel wiring harness
(199) - Ground connection -3-, in instrument panel wiring harness
(A5) - Plus connection (right turn signal), in instrument panel wiring harness
(A6) - Plus connection (left turn signal), in instrument panel wiring harness
(A18) - Wire connection (54), in instrument panel wiring harness
(A43) - Wire connection (57l), in instrument panel wiring harness
(A44) - Wire connection (57r), in instrument panel wiring harness
* - booth is possible

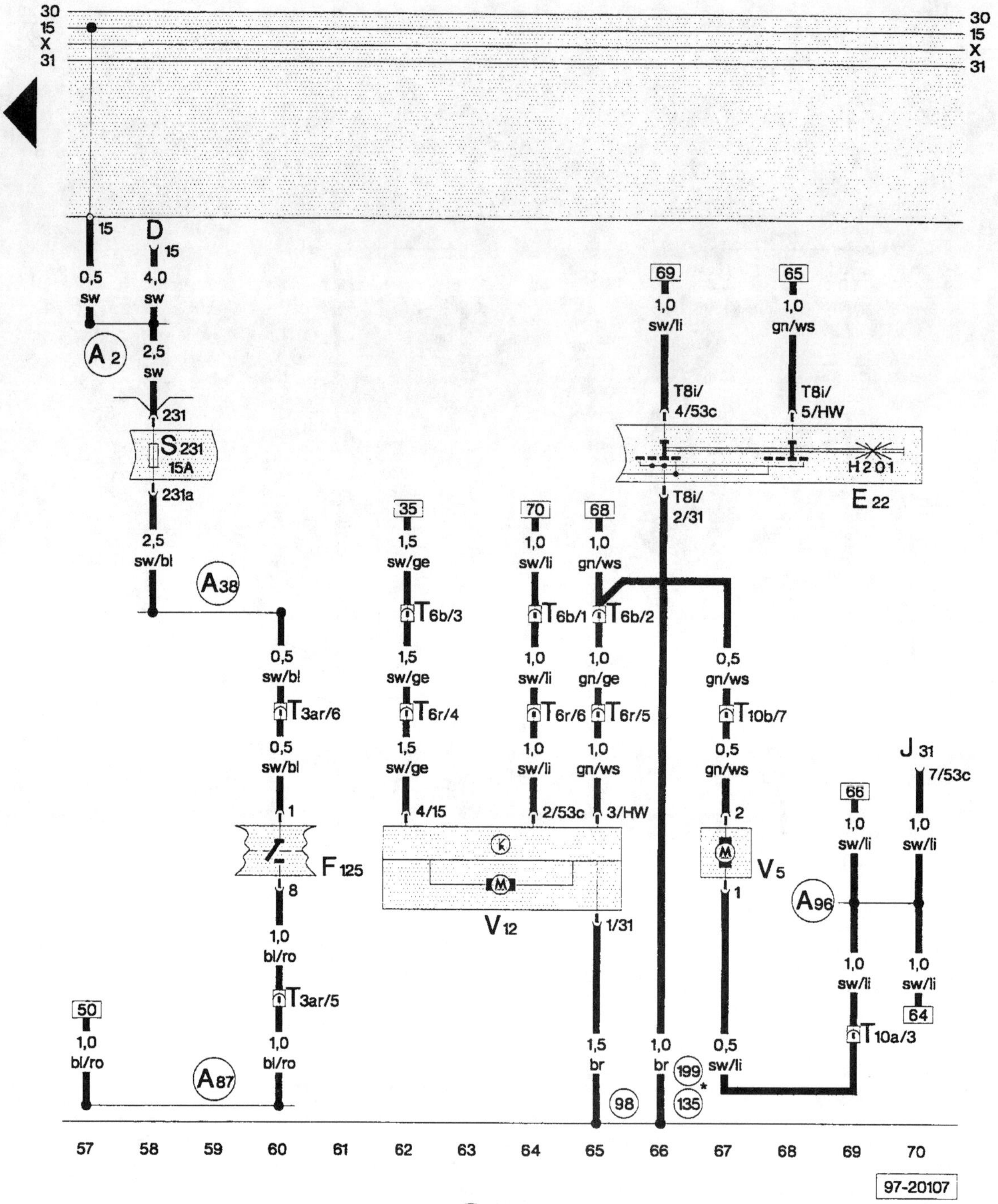

ws = white
sw = black
ro = red
gn = green
br = brown
bl = blue
gr = gray
li = violet
ge = yellow

D		- Ignition/Starter Switch
E 22		- Windshield Wiper/Washer Switch
F 125		- Multi-Function Transmission Range (TR) Switch
J 31		- Wiper/Washer Intermittent Relay
S 231		- Fuse
T 3	ar	- Connector 3 pin, brown, connector station A pillar, right
T 6	b	- Connector 6 pin, white, connector station A pillar, left
T 6	r	- Connector 6 pin, brown, in luggage compartment
T 8	i	- Connector 8 pin, on windshield wiper/washer switch
T 10	a	- Connector 10 pin, pink, connector station A pillar, left
T 10	b	- Connector 10 pin, red, connector station A pillar, left
V 5		- Windshield Washer Pump
V 12		- Rear Window Wiper Motor

(98)	- Ground connection, in rear lid wiring harness
(135)	- Ground connection -2-, in instrument panel wiring harness
(199)	- Ground connection -3-, in instrument panel wiring harness
(A2)	- Plus connection (15), in instrument panel wiring harness
(A38)	- Plus connection (15a II), in instrument panel wiring harness
(A87)	- Connector (RF), in instrument panel wiring harness
(A96)	- Connector (53a), in instrument panel wiring harness

* - booth is possible

Garage Door Opener

1999 m. y.

Fuse Panel

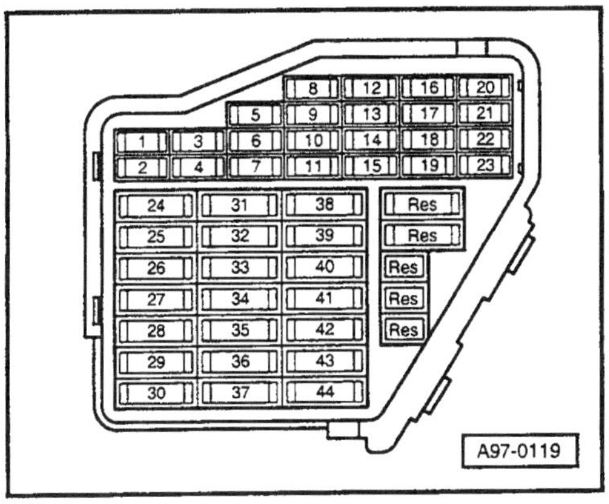

Fuse Colors:

30 A - Green
25 A - White
20 A - Yellow
15 A - Blue
10 A - Red
7,5 A - Brown
5 A - Beige

Starting with fuse position 23, fuses in the fuse holder are identified with 223 in the wiring diagram.

Micro Central Electric Panel

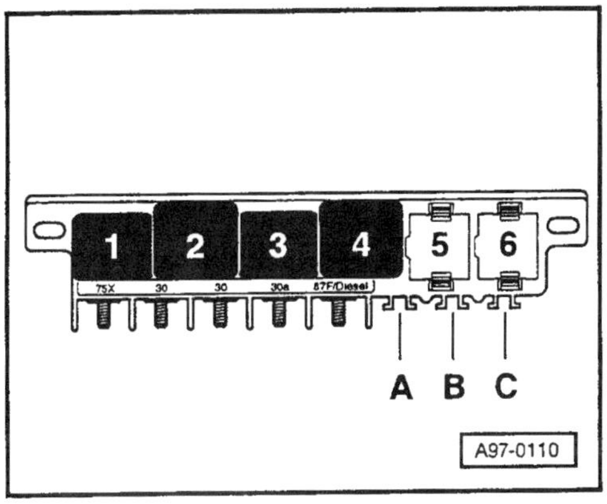

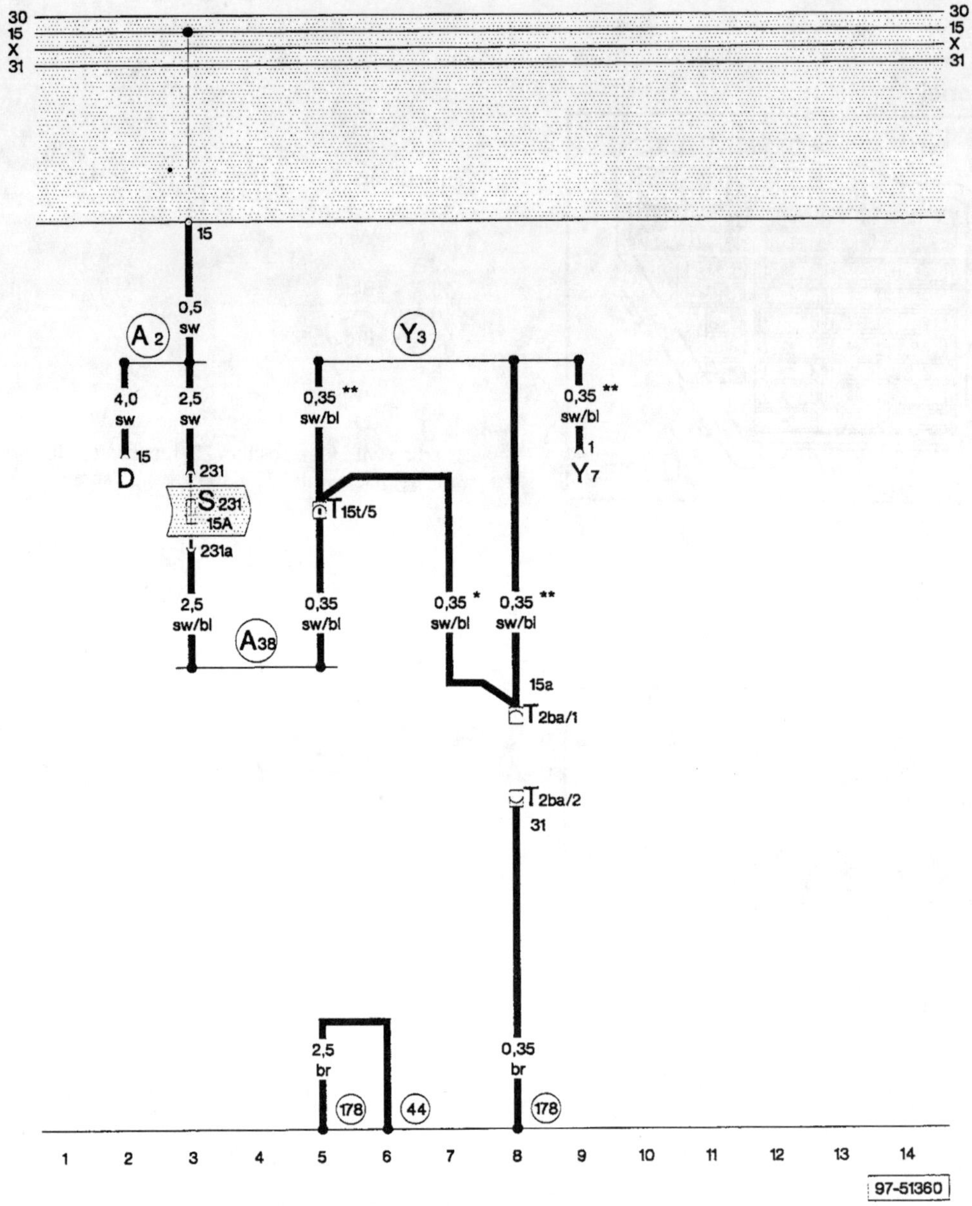

ws = white
sw = black
ro = red
gn = green
br = brown
bl = blue
gr = gray
li = violet
ge = yellow

D - Ignition/Starter Switch
S 231 - Fuse
T 2 ba - Connector 2 pin, brouwn, for garage door opener, near left make-up mirror
T 15 t - Connector 15 pin, grey, connector station A pillar, left
Y 7 - Automatic day/night interior mirror

(44) - Ground connection, on left A-pillar, lower part
(178) - Ground connection, in sunroof wiring harness
(A2) - Plus connection (15), in instrument panel wiring harness
(A38) - Plus connection (15a II), in instrument panel wiring harness
(Y3) - Plus connection (15), in interior light/sunroof wiring harness

* - Vehicles without Automatic day/night interior mirror
** - Vehicles with Automatic day/night interior mirror

(4,2 l - Injection Engine, 8-Cylinder), Code ART

1999 m. y.

Fuse Panel

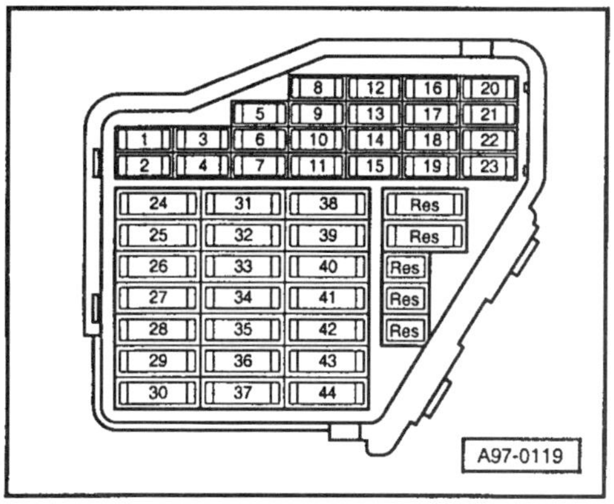

Fuse Colors:

30 A - Green
25 A - White
20 A - Yellow
15 A - Blue
10 A - Red
7,5 A - Brown
5 A - Beige

Starting with fuse position 23, fuses in the fuse holder are identified with 223 in the wiring diagram.

Micro Central Electric Panel

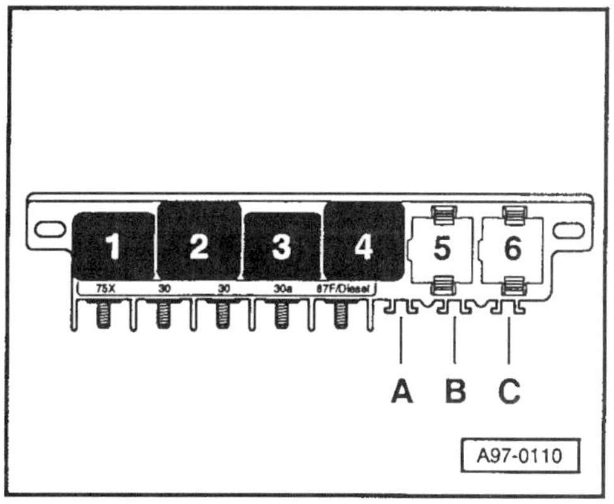

Relay Location:

4 - Fuel Pump (FP) Relay, J17

5 - Wiper/Washer Intermittent Relay, J31

6 - Wiper/Washer Intermittent Relay, J31

13 - Fold Relay Panel

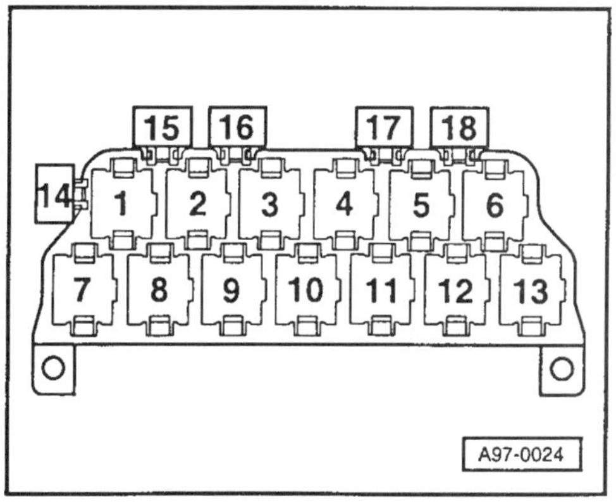

Relay Location:

4 - Starting Interlock Relay, J207

8 - Fold Relay Panel

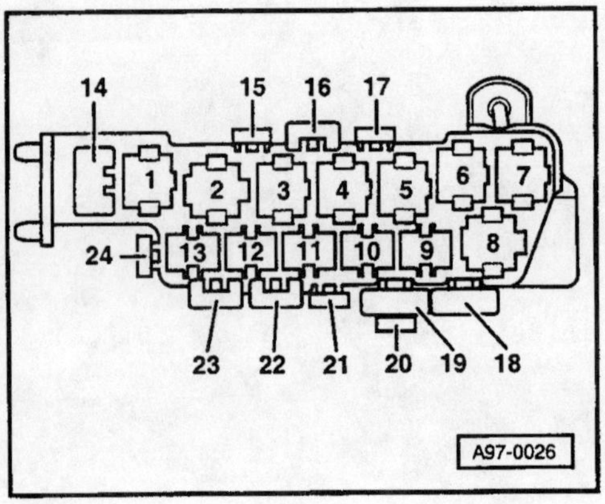

Relay Location:

- **20** - Coolant Fan Fuse, S42
- **21** - Control module fuse for coolant fan, S142

3 - Fold Relay Panel in E-Box plenum chamber

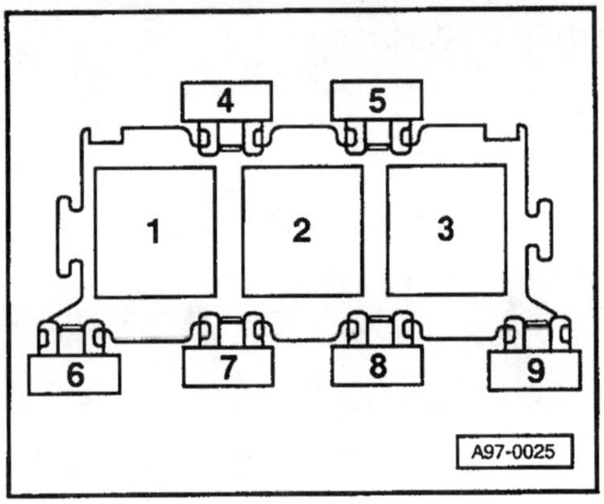

Relay Location:

- **2** - Secondary Air Injection (AIR) Pump Relay, J299
- **7** - Fuse for secondary air pump, S130

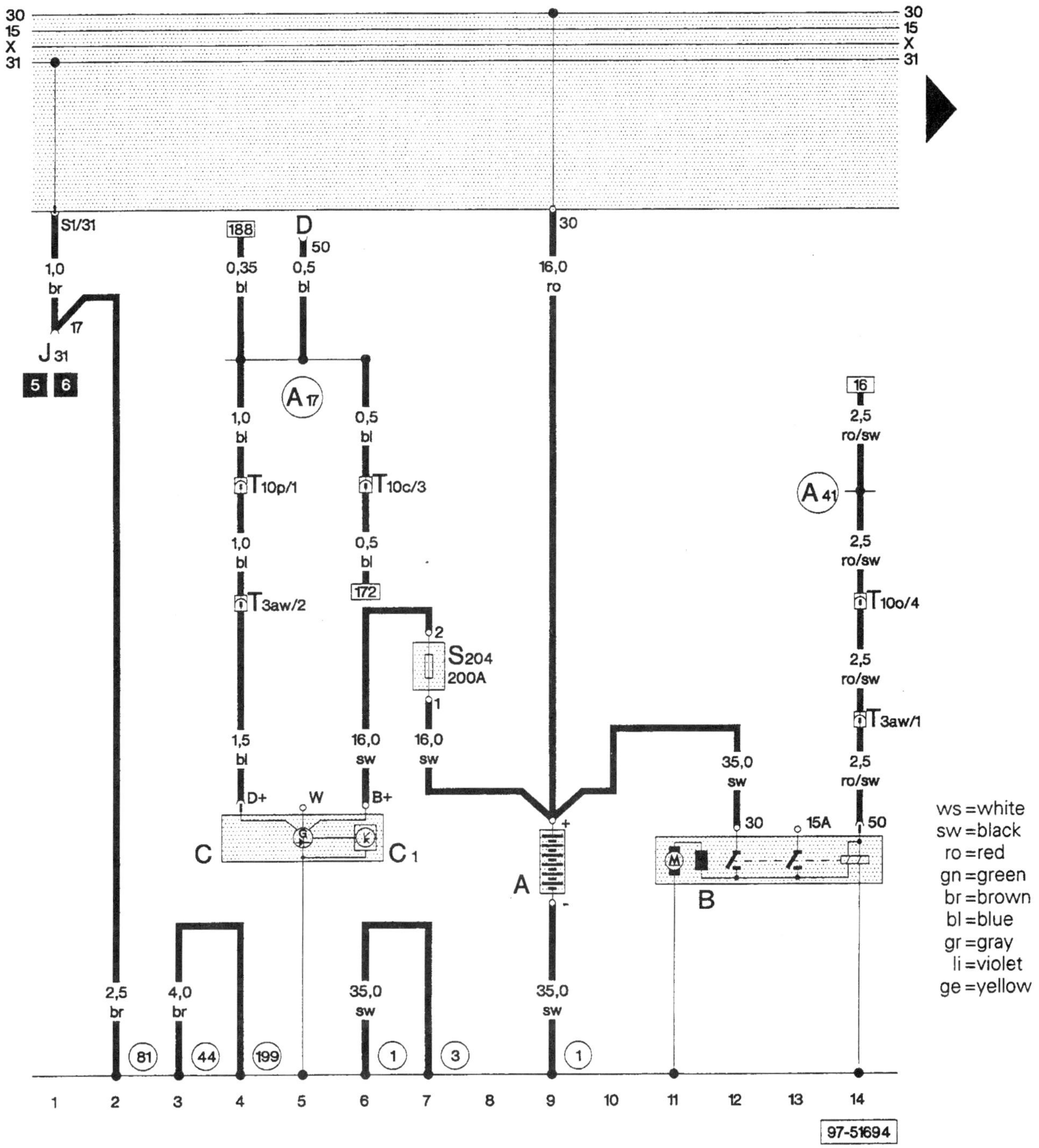

A - Battery
B - Starter
C - Generator (GEN)
C 1 - Voltage Regulator (VR)
D - Ignition/Starter Switch
J 31 - Wiper/Washer Intermittent Relay
S 204 - Fuse 1 (30)
T 3 aw - Connector 3 pin, in engine compartment, right
T 10 c - Connector 10 pin, grey, connector station A pillar, left
T 10 o - Connector 10 pin, brown, connector station electronic box plenum chamber
T 10 p - Connector 10 pin, black, connector station electronic box plenum chamber

(1) - Ground strap, battery to body
(3) - Ground strap, engine to body
(44) - Ground connection, on left A-pillar, lower part
(81) - Ground connection -1-, in instrument panel wiring harness
(199) - Ground connection -3-, in instrument panel wiring harness
(A17) - Wire connection (61), in instrument panel wiring harness
(A41) - Plus connection (50), in instrument panel wiring harness

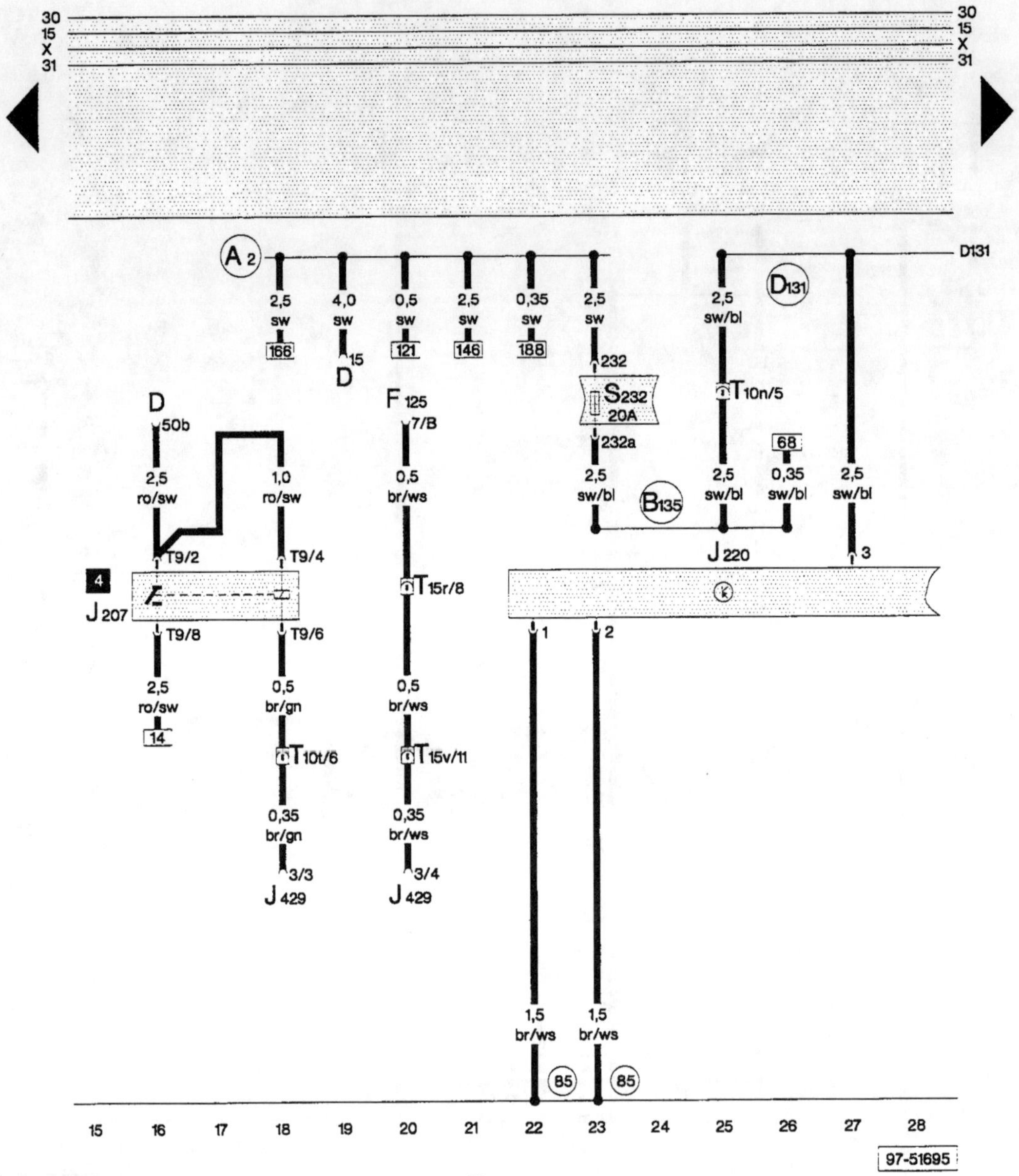

ws =white
sw =black
ro =red
gn =green
br =brown
bl =blue
gr =gray
li =violet
ge =yellow

D - Ignition/Starter Switch
F 125 - Multi-Function Transmission Range (TR) Switch
J 207 - Starting Interlock Relay
J 220 - Motronic Engine Control Module (ECM)
J 429 - Control module for central locking
S 232 - Fuse
T 9 - Connector 9 pin, brown, on starting interlock relay
T 10 n - Connector 10 pin, orange, connector station electronic box plenum chamber
T 10 t - Connector 10 pin, orange, connector station A pillar, right
T 15 v - Connector 15 pin, orange, connector station A pillar, right

(85) - Ground connection -1-, in engine compartment wiring harness
(A2) - Plus connection (15), in instrument panel wiring harness
(B135) - Connector (15a), in wiring harness interior
(D131) - Wire connection (15), above fuse, in engine compartment wiring harness

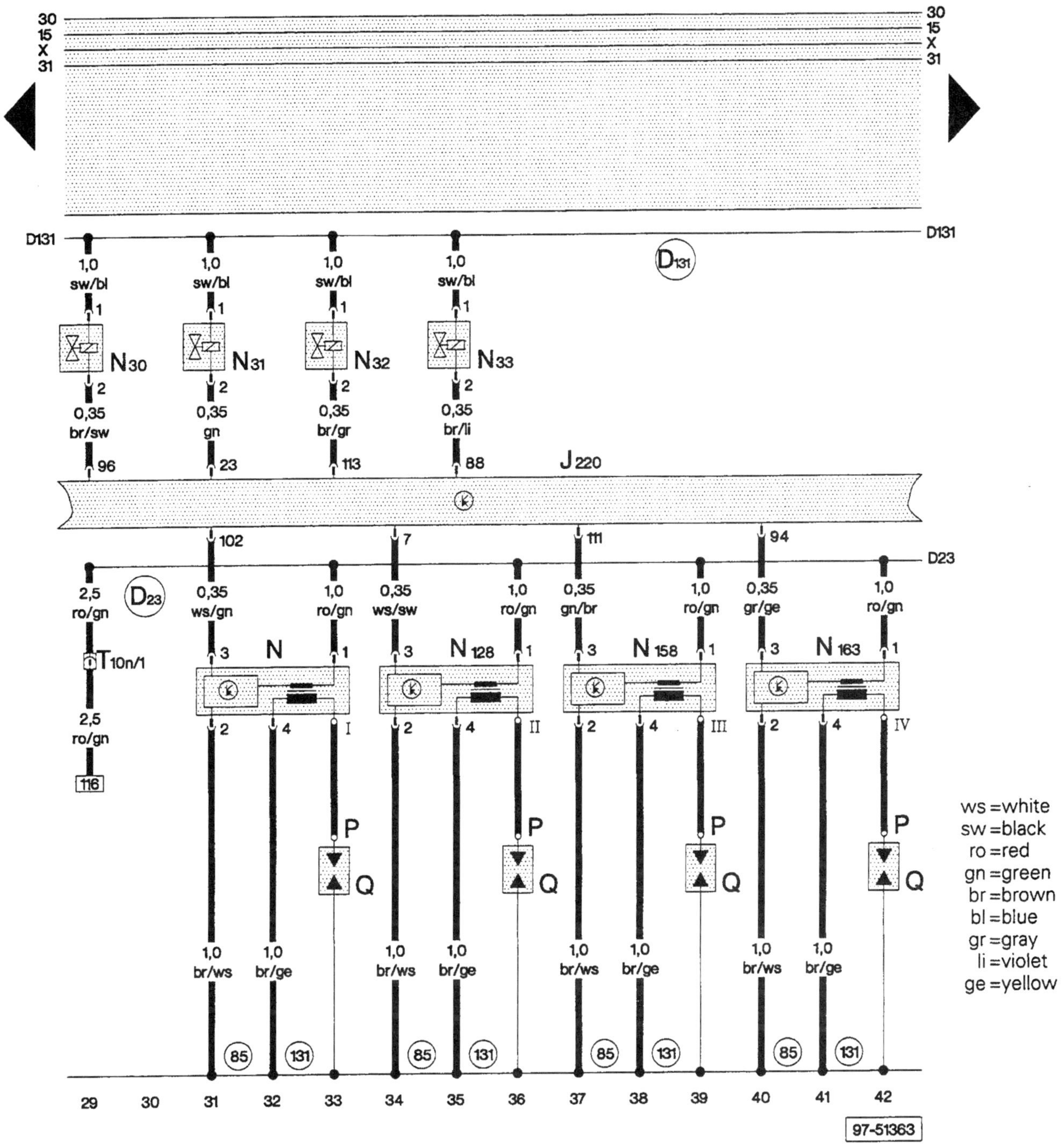

J 220 - Motronic Engine Control Module (ECM)
N - Ignition Coil
N 30 - Cylinder 1 Fuel Injector
N 31 - Cylinder 2 Fuel Injector
N 32 - Cylinder 3 Fuel Injector
N 33 - Cylinder 4 Fuel Injector
N 128 - Ignition Coil 2
N 158 - Ignition Coil 3
N 163 - Ignition Coil 4
P - Spark Plug Connectors
Q - Spark Plugs
T 10 n - Connector 10 pin, orange, connector station electronic box plenum chamber

(85) - Ground connection -1-, in engine compartment wiring harness
(131) - Ground connection -2-, in engine compartment wiring harness
(D23) - Connector (over fuse 229), in wiring harness front, right
(D131) - Wire connection (15), above fuse, in engine compartment wiring harness

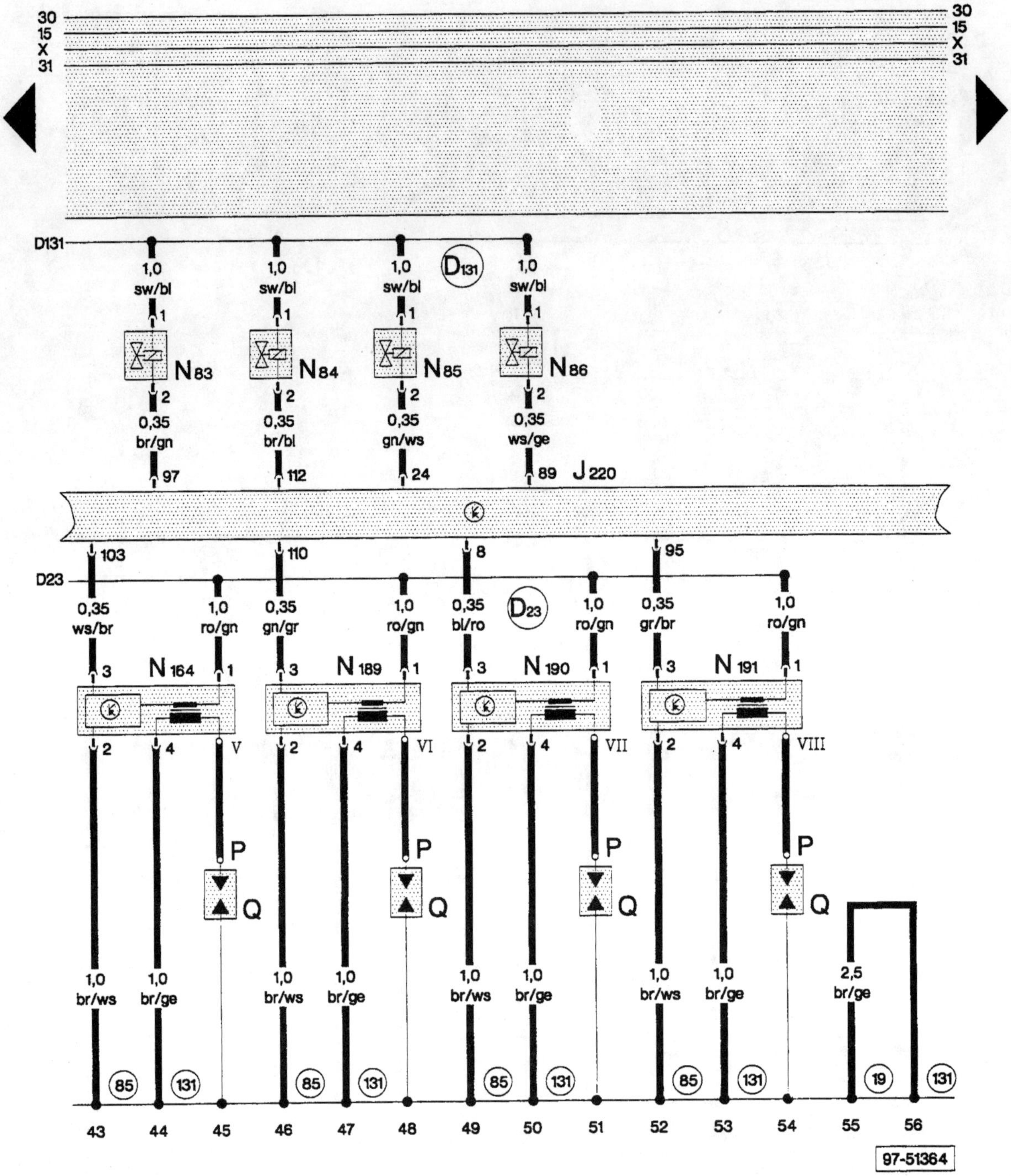

ws = white
sw = black
ro = red
gn = green
br = brown
bl = blue
gr = gray
li = violet
ge = yellow

J 220 - Motronic Engine Control Module (ECM)
N 83 - Cylinder 5 Fuel Injector
N 84 - Cylinder 6 Fuel Injector
N 85 - Cylinder 7 Fuel Injector
N 86 - Cylinder 8 Fuel Injector
N 164 - Ignition Coil 5
N 189 - Ignition Coil 6
N 190 - Ignition Coil 7
N 191 - Ignition Coil 8
P - Spark Plug Connectors
Q - Spark Plugs

(19) - Ground connection, near ignition coil

(85) - Ground connection -1-, in engine compartment wiring harness

(131) - Ground connection -2-, in engine compartment wiring harness

(D23) - Connector (over fuse 229), in wiring harness front, right

(D131) - Wire connection (15), above fuse, in engine compartment wiring harness

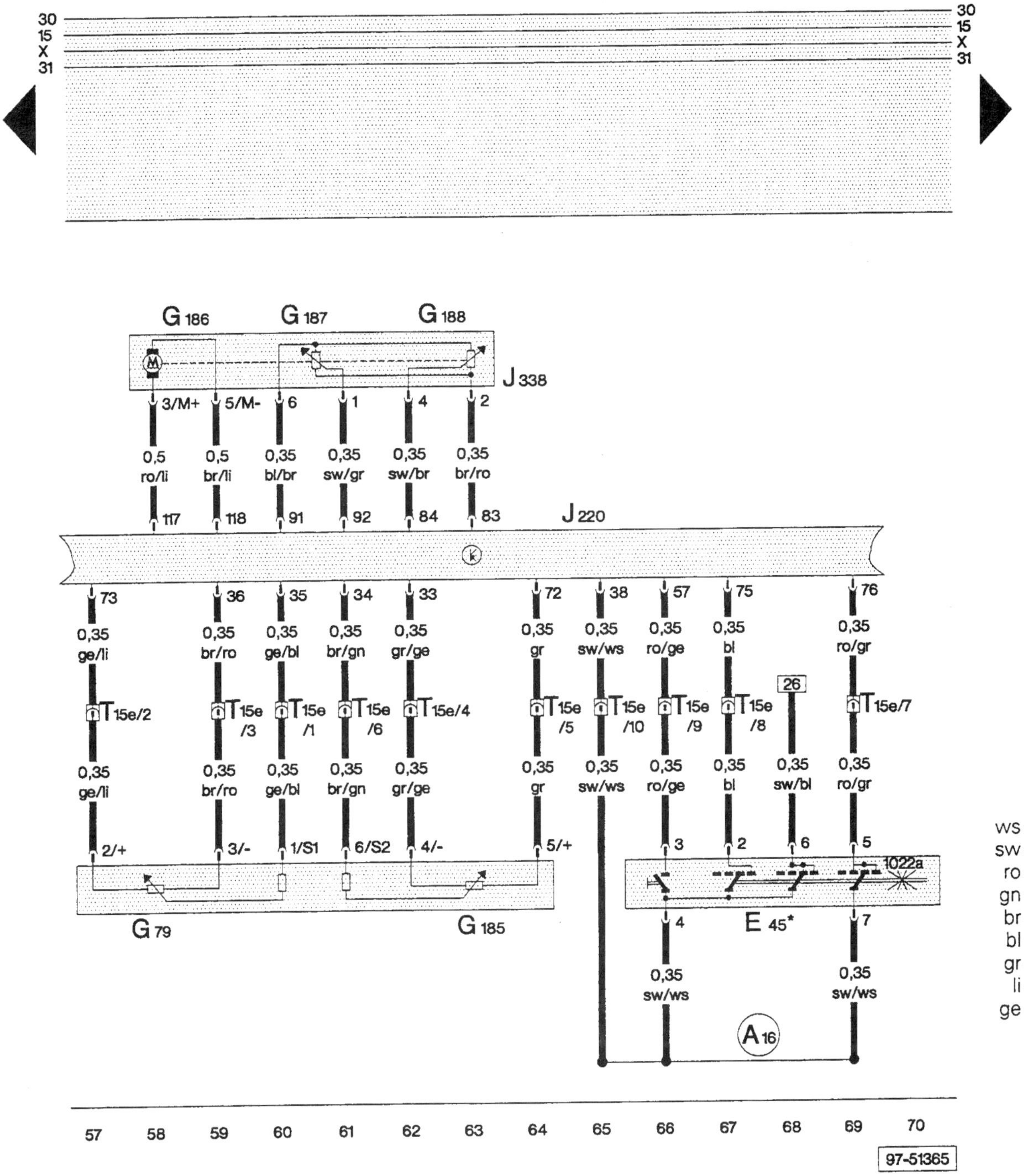

ws = white
sw = black
ro = red
gn = green
br = brown
bl = blue
gr = gray
li = violet
ge = yellow

E 45 - Cruise Control Switch
G 79 - Throttle Position (TP) Sensor
G 185 - Sender -2- for accelerator pedal position
G 186 - Throttle drive (power accelerator actuation)
G 187 - Angle sensor -1- for throttle drive (power accelerator actuation)
G 188 - Angle sensor -2- for throttle drive (power accelerator actuation)
J 220 - Motronic Engine Control Module (ECM)
J 338 - Throttle Valve Control Module
T 15 e - Connector 15 pin, white, connector station electronic box plenum chamber

(A16) - Wire connection (cruise control), in instrument panel wiring harness

* - Vehicles with cruise control

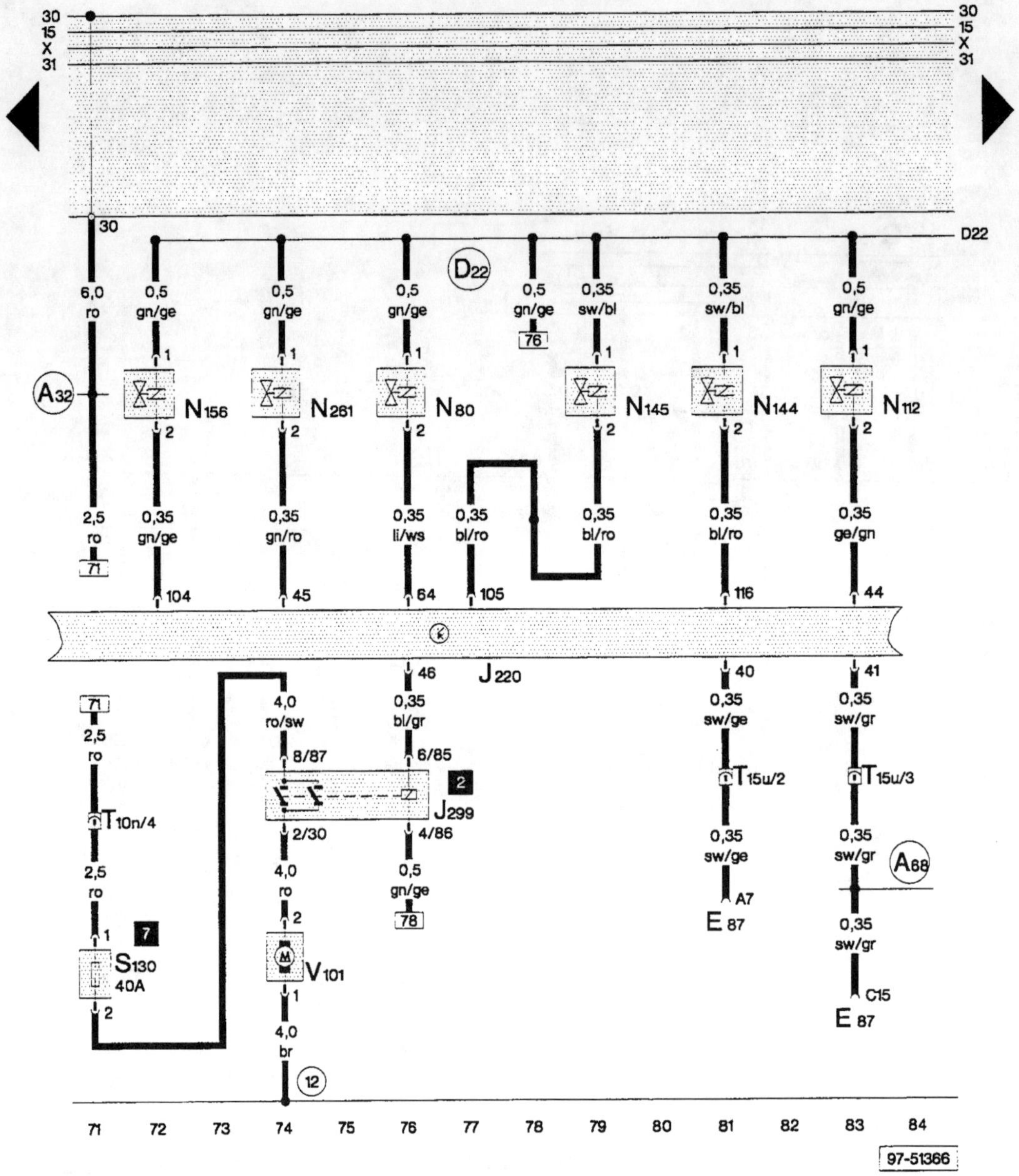

ws = white
sw = black
ro = red
gn = green
br = brown
bl = blue
gr = gray
li = violet
ge = yellow

E 87 - A/C Control Head
J 220 - Motronic Engine Control Module (ECM)
J 299 - Secondary Air Injection (AIR) Pump Relay
N 80 - Evaporative Emission (EVAP) Canister Purge Regulator Valve
N 112 - Secondary Air Injection (AIR) Solenoid Valve
N 144 - Left Electro-Hydraulic Engine Mount Solenoid Valve
N 145 - Right Electro-Hydraulic Engine Mount Solenoid Valve
N 156 - Intake Manifold Change-Over Valve
N 261 - Intake Manifold Tuning (IMT) Valve -2-
S 130 - Fuse for secondary air pump
T 10 n - Connector 10 pin, orange, connector station electronic box plenum chamber
T 15 u - Connector 15 pin, red, connector station electronic box plenum chamber
V 101 - Secondary Air Injection (AIR) Pump Motor

(12) - Ground connection, in engine compartment, left
(A32) - Plus connection (30), in instrument panel wiring harness
(A68) - Connector (C15, A/C), in instrument panel wiring harness
(D22) - Connector (over fuse 234), in wiring harness front, right

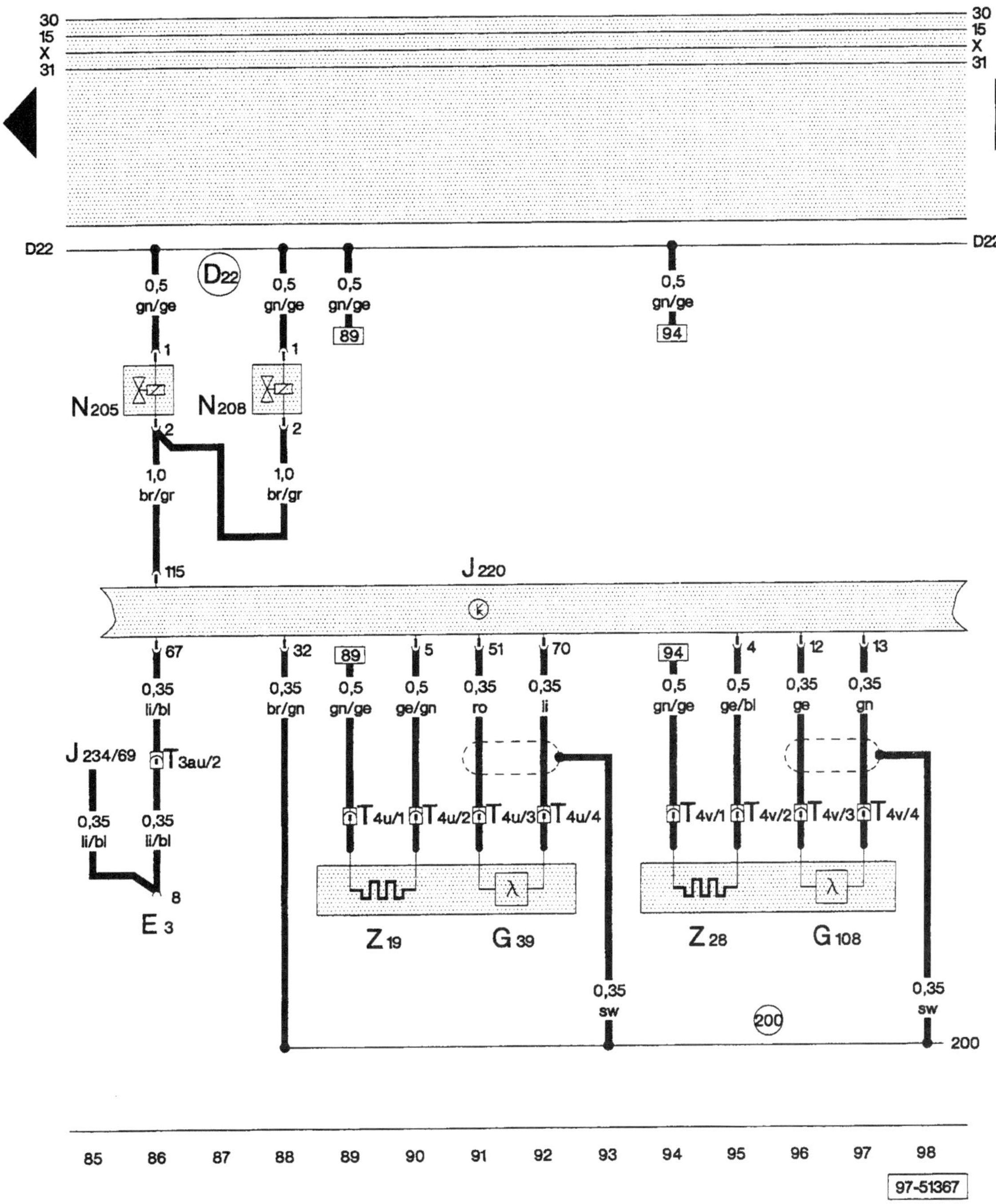

ws = white
sw = black
ro = red
gn = green
br = brown
bl = blue
gr = gray
li = violet
ge = yellow

E 3 - Emergency Flasher Switch
G 39 - Heated Oxygen Sensor (HO2S)
G 108 - Heated Oxygen Sensor (HO2S) 2
J 220 - Motronic Engine Control Module (ECM)
J 234 - Airbag Control Module
N 205 - Valve -1- for camshaft adjustment
N 208 - Valve -2- for camshaft adjustment
T 3 au - Connector 3 pin, red, connector station electronic box plenum chamber
T 4 u - Connector 4 pin, black, heated oxygen sensor
T 4 v - Connector 4 pin, black, heated oxygen sensor 2
Z 19 - Oxygen Sensor (O2S) Heater
Z 28 - Oxygen Sensor (O2S) 2 Heater

(200) - Ground connection (shielding), in engine compartment wiring harness
(D22) - Connector (over fuse 234), in wiring harness front, right

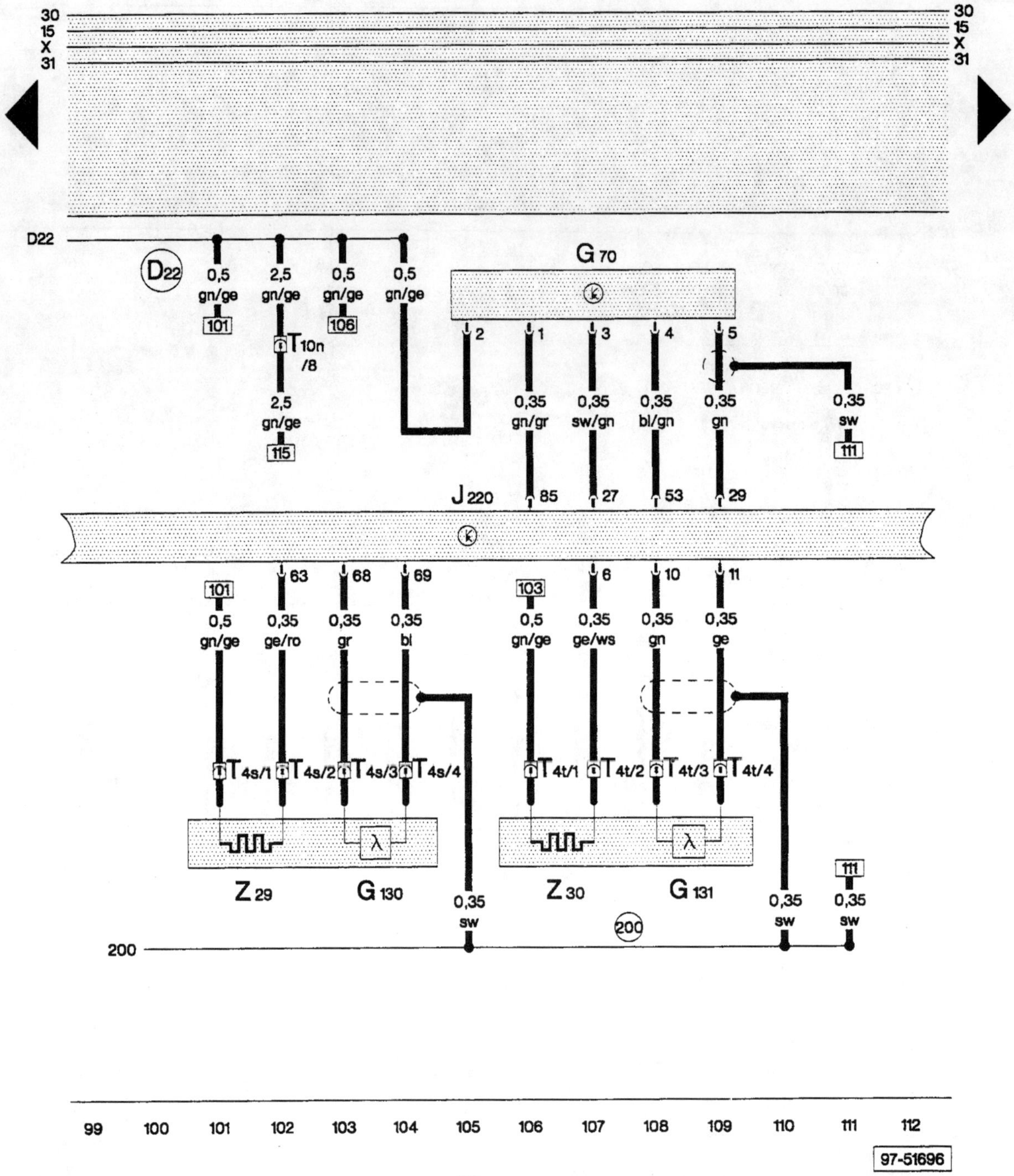

ws = white
sw = black
ro = red
gn = green
br = brown
bl = blue
gr = gray
li = violet
ge = yellow

G 70 - Mass Air Flow (MAF) Sensor
G 130 - Oxygen Sensor (O2S) Behind Three Way Catalytic Converter (TWC)
G 131 - Oxygen Sensor (O2S) 2 Behind Three Way Catalytic Converter (TWC)
J 220 - Motronic Engine Control Module (ECM)
T 4 s - Connector 4 pin, green, near oxygen sensor, behind Three Way Catalytic Converter (TWC)
T 4 t - Connector 4 pin, brown, near oxygen sensor 2, behind Three Way Catalytic Converter (TWC)
T 10 n - Connector 10 pin, orange, connector station electronic box plenum chamber
Z 29 - Heater for Lambda-probe 1
Z 30 - Heater for Lambda-probe 2

(200) - Ground connection (shielding), in engine compartment wiring harness
(D22) - Connector (over fuse 234), in wiring harness front, right

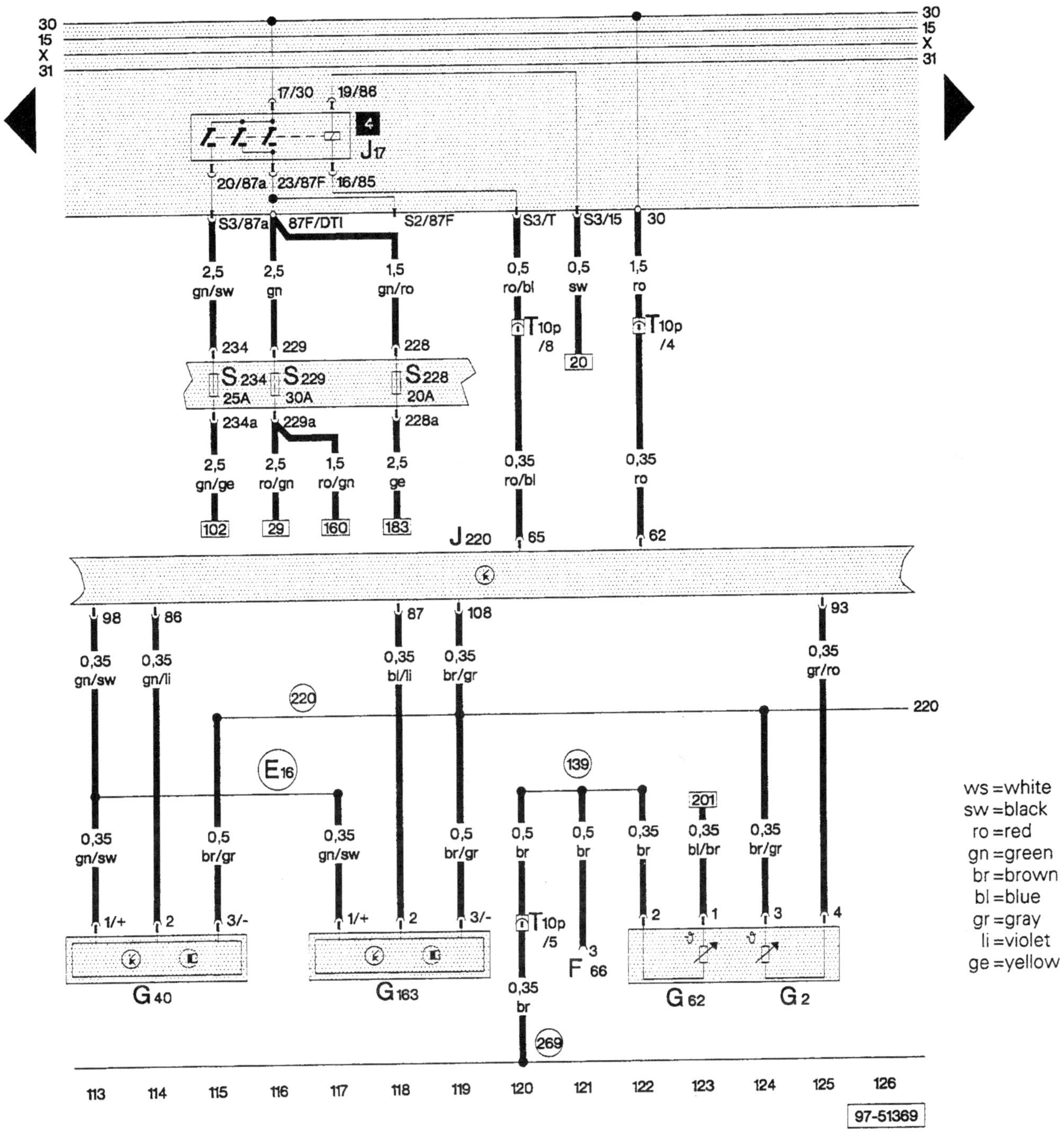

F 66 - Engine Coolant Level (ECL) Warning Switch
G 2 - Engine Coolant Temperature (ECT) Sensor
G 40 - Camshaft Position (CMP) Sensor
G 62 - Engine Coolant Temperature (ECT) Sensor
G 163 - Camshaft Position (CMP) Sensor 2
J 17 - Fuel Pump (FP) Relay
J 220 - Motronic Engine Control Module (ECM)
S 228 - Fuse
S 229 - Fuse
S 234 - Fuse
T 10 p - Connector 10 pin, black, connector station electronic box plenum chamber

(139) - Ground connection (sensor Ground), in Motronic Multiport Fuel Injection (MFI) wiring harness
(220) - Ground connection (sensor ground), in engine compartment wiring harness
(269) - Ground connector (sensor ground) -1-, in instrument panel wiring harness
(E16) - Plus connection (5 Volts), in Motronic Multiport Fuel Injection (MFI) wiring harness

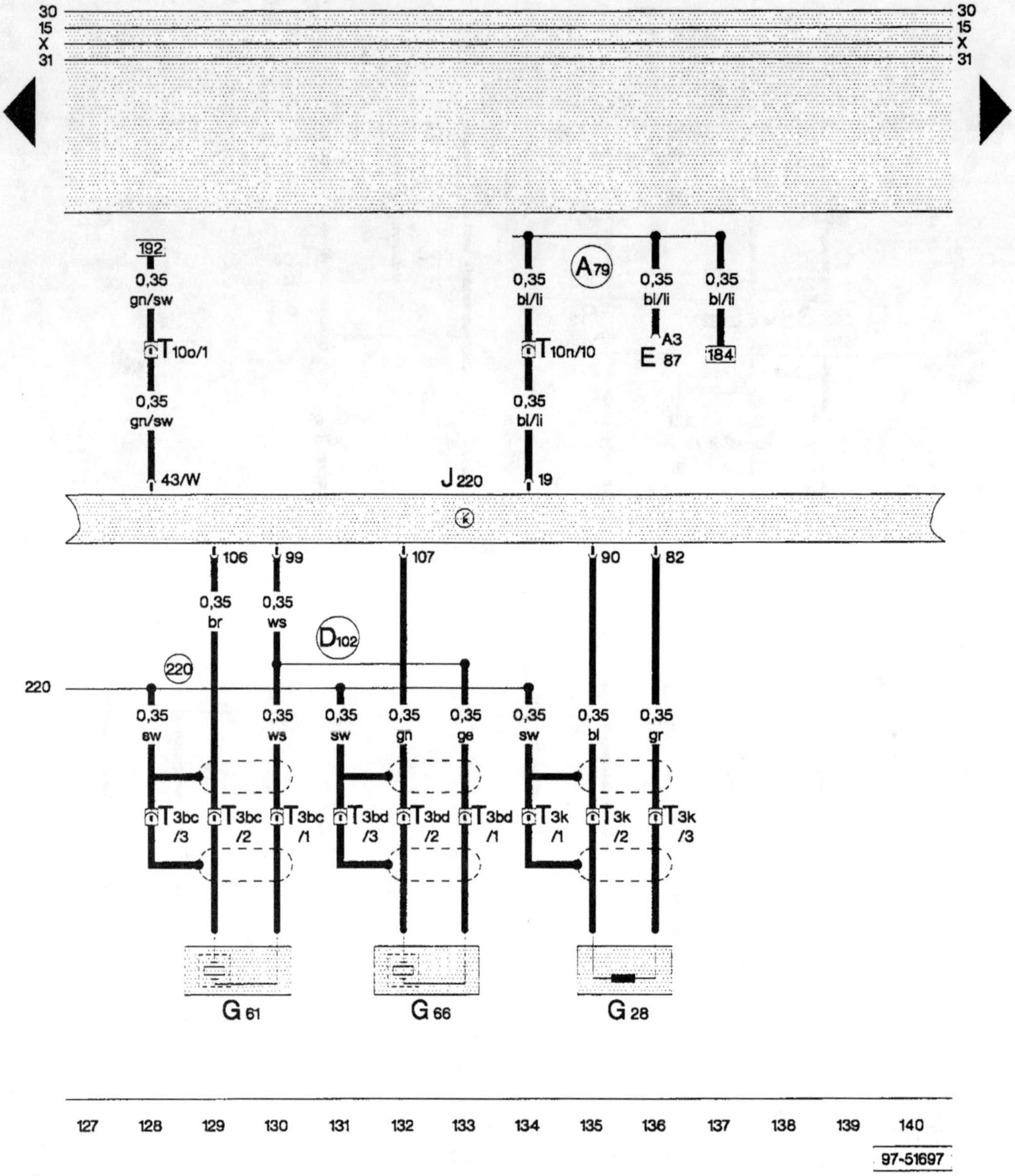

ws = white
sw = black
ro = red
gn = green
br = brown
bl = blue
gr = gray
li = violet
ge = yellow

E 87 - A/C Control Head
G 28 - Engine Speed (RPM) Sensor
G 61 - Knock Sensor (KS) 1
G 66 - Knock Sensor (KS) 2
J 220 - Motronic Engine Control Module (ECM)
T 3 k - Connector 3 pin, grey, on engine speed sensor
T 3 bc - Connector 3 pin, green, on knock sensor 1
T 3 bd - Connector 3 pin, green, on knock sensor 2
T 10 n - Connector 10 pin, orange, connector station electronic box plenum chamber
T 10 o - Connector 10 pin, brown, connector station electronic box plenum chamber

(220) - Ground connection (sensor ground), in engine compartment wiring harness
(A79) - Connector (engine overheat warning light), in instrument panel wiring harness
(D102) - Wire connection -2-, in engine compartment wiring harness

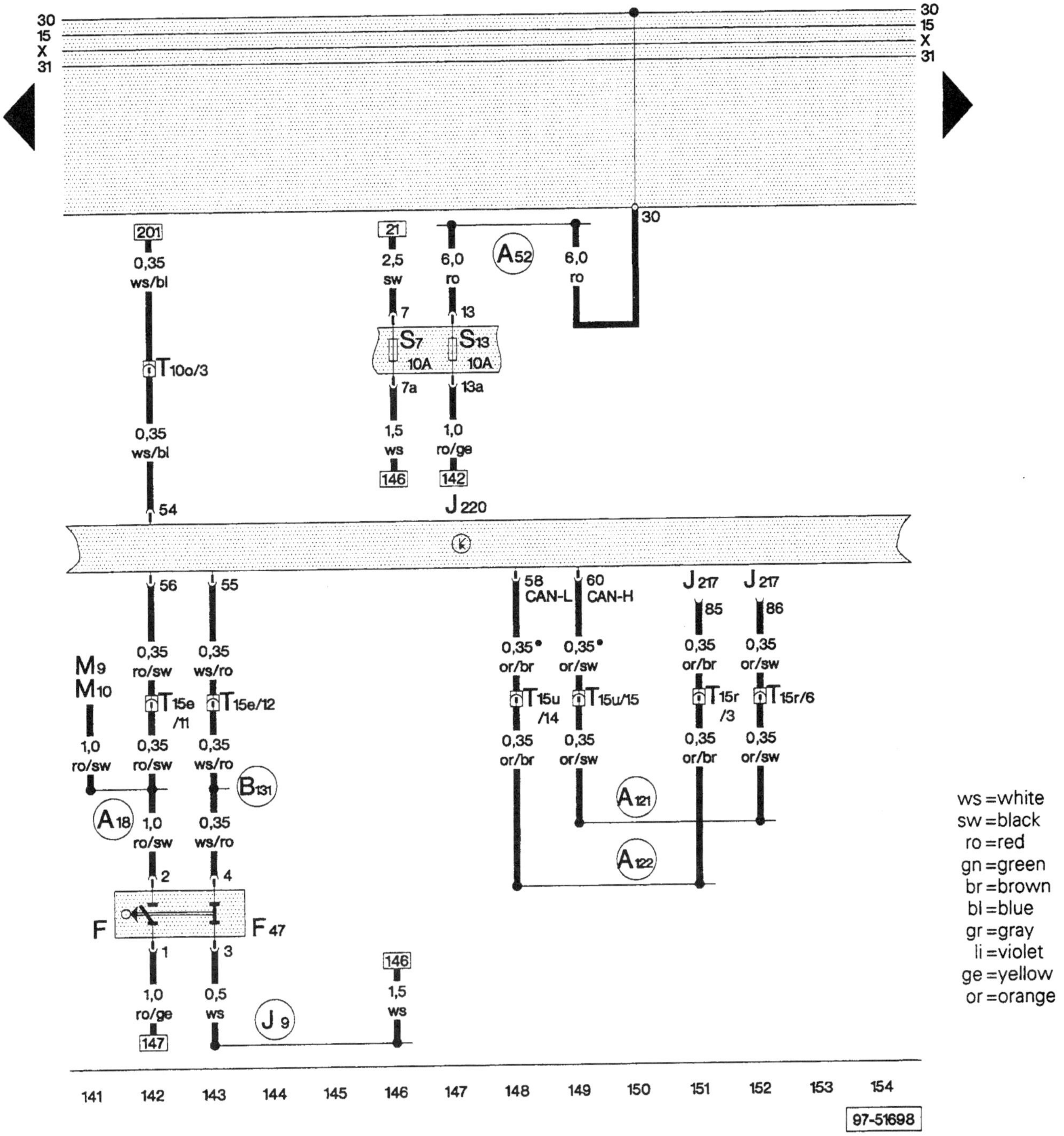

ws = white
sw = black
ro = red
gn = green
br = brown
bl = blue
gr = gray
li = violet
ge = yellow
or = orange

F - Brake Light Switch
F 47 - Brake Pedal Switch
J 217 - Transmission Control Module (TCM)
J 220 - Motronic Engine Control Module (ECM)
M 9 - Left Brake Light
M 10 - Right Brake Light
S 7 - Fuse
S 13 - Fuse
T 10 o - Connector 10 pin, brown, connector station electronic box plenum chamber
T 15 e - Connector 15 pin, white, connector station electronic box plenum chamber
T 15 r - Connector 15 pin, brown, connector station A pillar, right
T 15 u - Connector 15 pin, red, connector station electronic box plenum chamber

(A18) - Wire connection (54), in instrument panel wiring harness
(A52) - Plus connection (30), in instrument panel wiring harness
(A121) - Connection (high bus) in instrument panel wiring harness
(A122) - Connection (low bus) in instrument panel wiring harness
(B131) - Connector (54), in wiring harness interior
(J9) - Wire connection -1- (15a), in ABS wiring harness

• - CAN-Bus (Data-Bus)

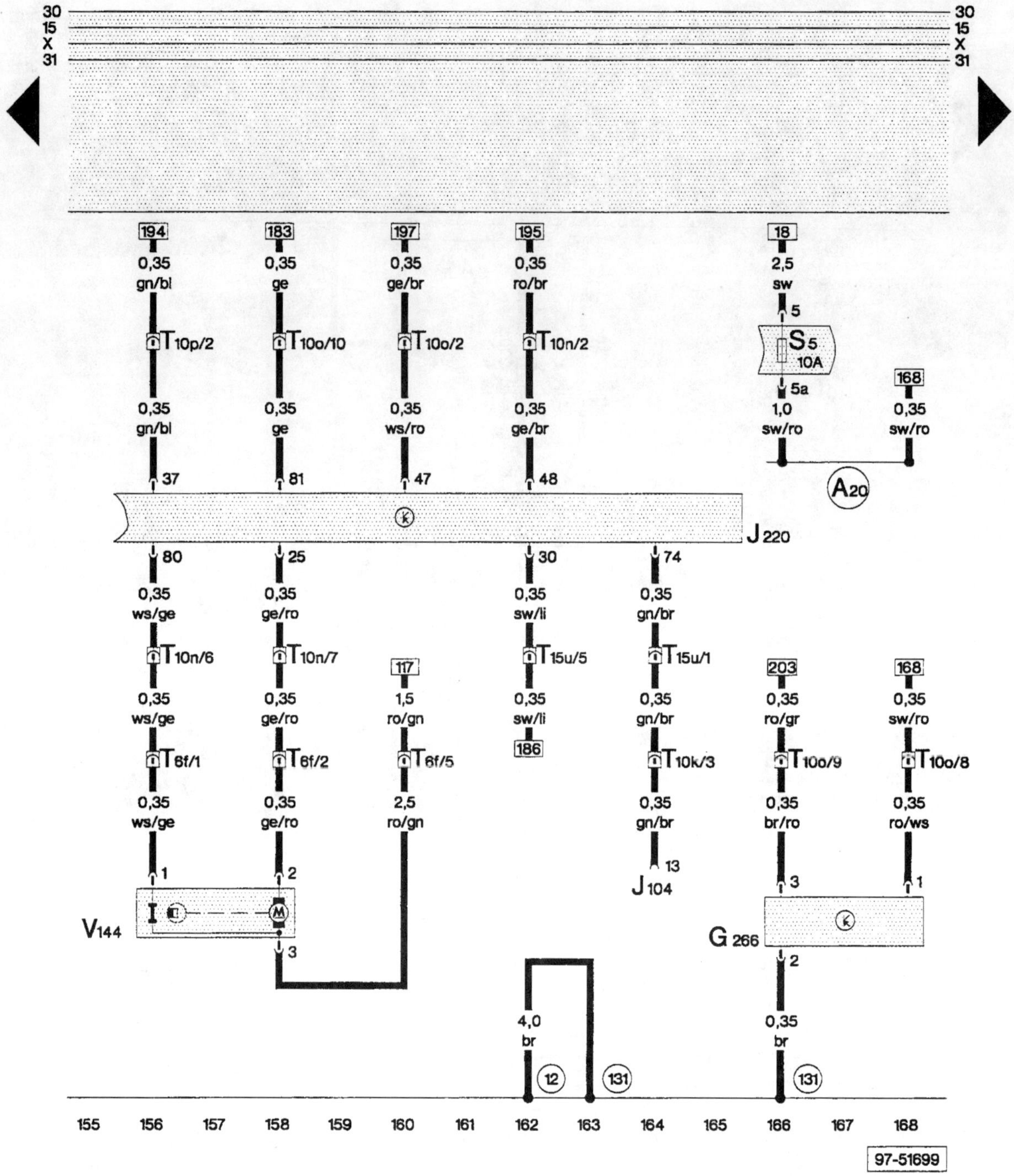

ws = white
sw = black
ro = red
gn = green
br = brown
bl = blue
gr = gray
li = violet
ge = yellow

G 266 - Oil Level Thermal Sensor (SRI)
J 104 - ABS Control Module (w/EDL)
J 220 - Motronic Engine Control Module (ECM)
S 5 - Fuse
T 6 f - Connector 6 pin, darkbrown, connector station A pillar, right
T 10 k - Connector 10 pin, orange, connector station A pillar, left
T 10 n - Connector 10 pin, orange, connector station electronic box plenum chamber
T 10 o - Connector 10 pin, brown, connector station electronic box plenum chamber
T 10 p - Connector 10 pin, black, connector station electronic box plenum chamber
T 15 u - Connector 15 pin, red, connector station electronic box plenum chamber
V 144 - Leak Detection Pump (LDP)

(12) - Ground connection, in engine compartment, left
(131) - Ground connection -2-, in engine compartment wiring harness
(A20) - Wire connection (15a), in instrument panel wiring harness

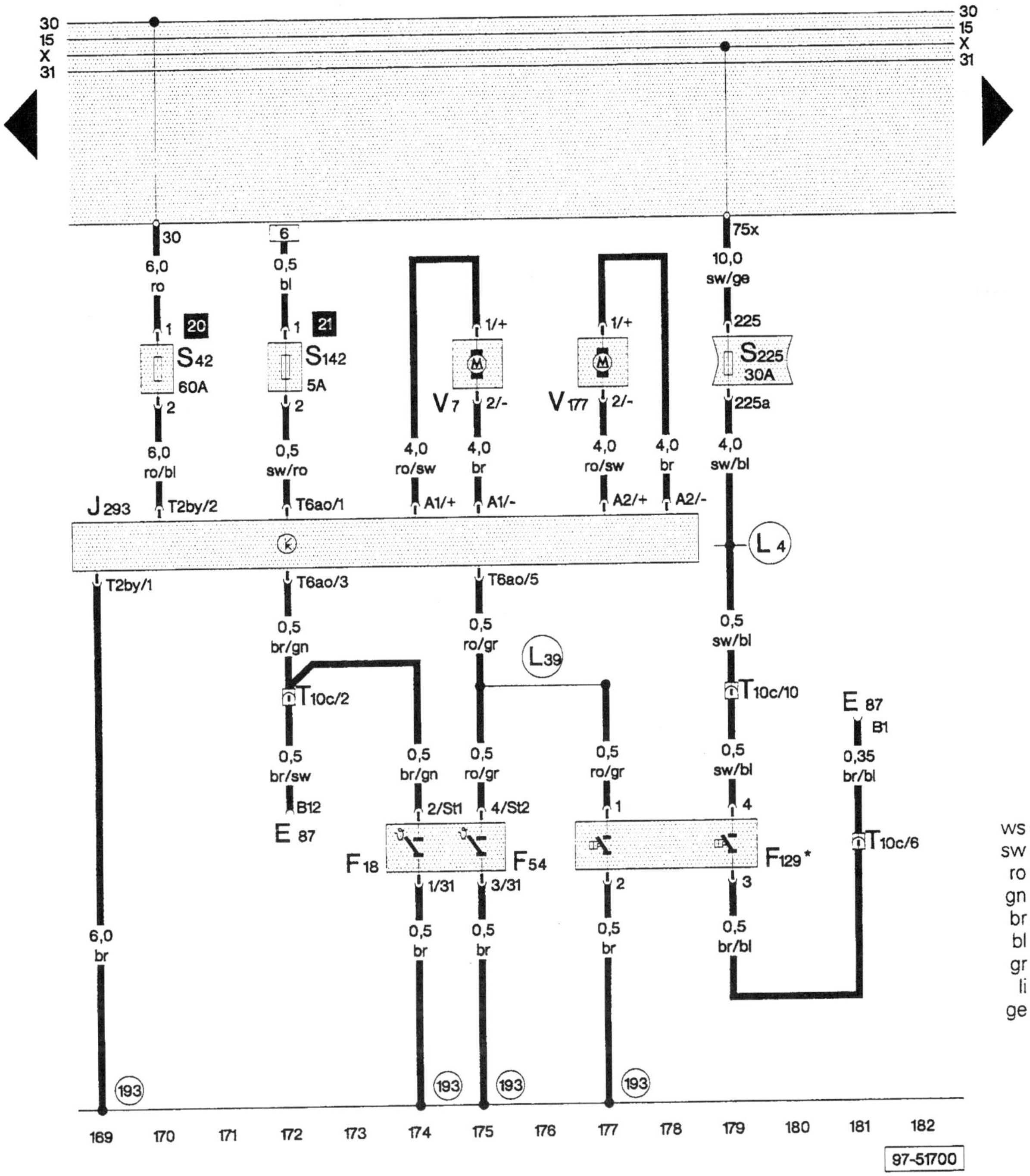

ws = white
sw = black
ro = red
gn = green
br = brown
bl = blue
gr = gray
li = violet
ge = yellow

E 87 - A/C Control Head
F 18 - Coolant Fan Control (FC) Thermal Switch
F 54 - Coolant Fan Control (FC) Thermal Switch
F 129 - A/C Pressure Switch
J 293 - Coolant FC (Fan Control) Control Module
S 42 - Coolant Fan Fuse
S 142 - Control module fuse for coolant fan
S 225 - Fuse
T 2 by - Connector 2 pin, on Coolant FC (Fan Control) Control Module
T 6 ao - Connector 6 pin, on Coolant FC (Fan Control) Control Module
T 10 c - Connector 10 pin, grey, connector station A pillar, left
V 7 - Coolant Fan
V 177 - Coolant Fan -2-

(193) - Ground connection -1-, in coolant fan wiring harness

(L4) - Wire connection (75al), in A/C wiring harness

(L39) - Wire connection -1-(A/C Pressure Switch), in Climatronic wiring harness

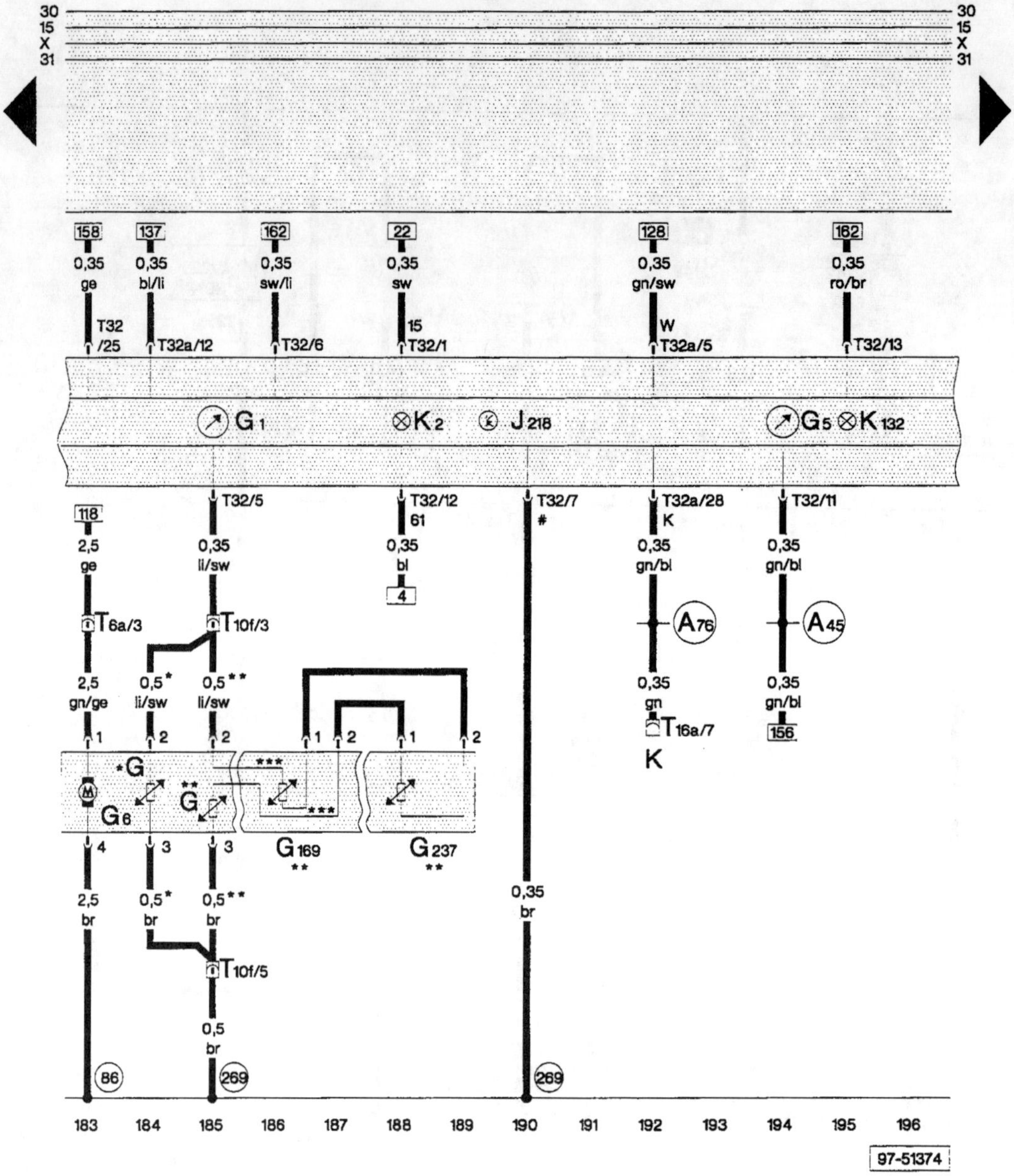

ws = white
sw = black
ro = red
gn = green
br = brown
bl = blue
gr = gray
li = violet
ge = yellow

G	- Fuel Level Sensor
G 1	- Fuel Gauge
G 5	- Tachometer
G 6	- Fuel Pump (FP)
G 169	- Sender 2 for fuel gauge
G 237	- Sensor -3- for fuel supply
J 218	- Instrument Cluster Combination Processor
K	- DLC (Data Link Connector) Rapid Data Transfer
K 2	- Generator (GEN) Warning Light
K 132	- Fault light for power accelerator activation
T 6 a	- Connector 6 pin, blue, connector station A pillar, left
T 10 f	- Connector 10 pin, brown, connector station A pillar, left
T 16 a	- Connector 16 pin, data link connector
T 32	- Connector 32 pin, blue, on instrument cluster
T 32 a	- Connector 32 pin, green, on instrument cluster

(86)	- Ground connection -1-, in rear wiring harness
(269)	- Ground connector (sensor ground) -1-, in instrument panel wiring harness
(A45)	- Wire connection (RPM-signal), in instrument panel wiring harness
(A76)	- Connector (K-diagnosis wire), in instrument panel wiring harness

*	- Vehicles with front-wheel drive
**	- Vehicles with all-wheel drive
***	- Wire in fuel tank
#	- Exit sensor ground

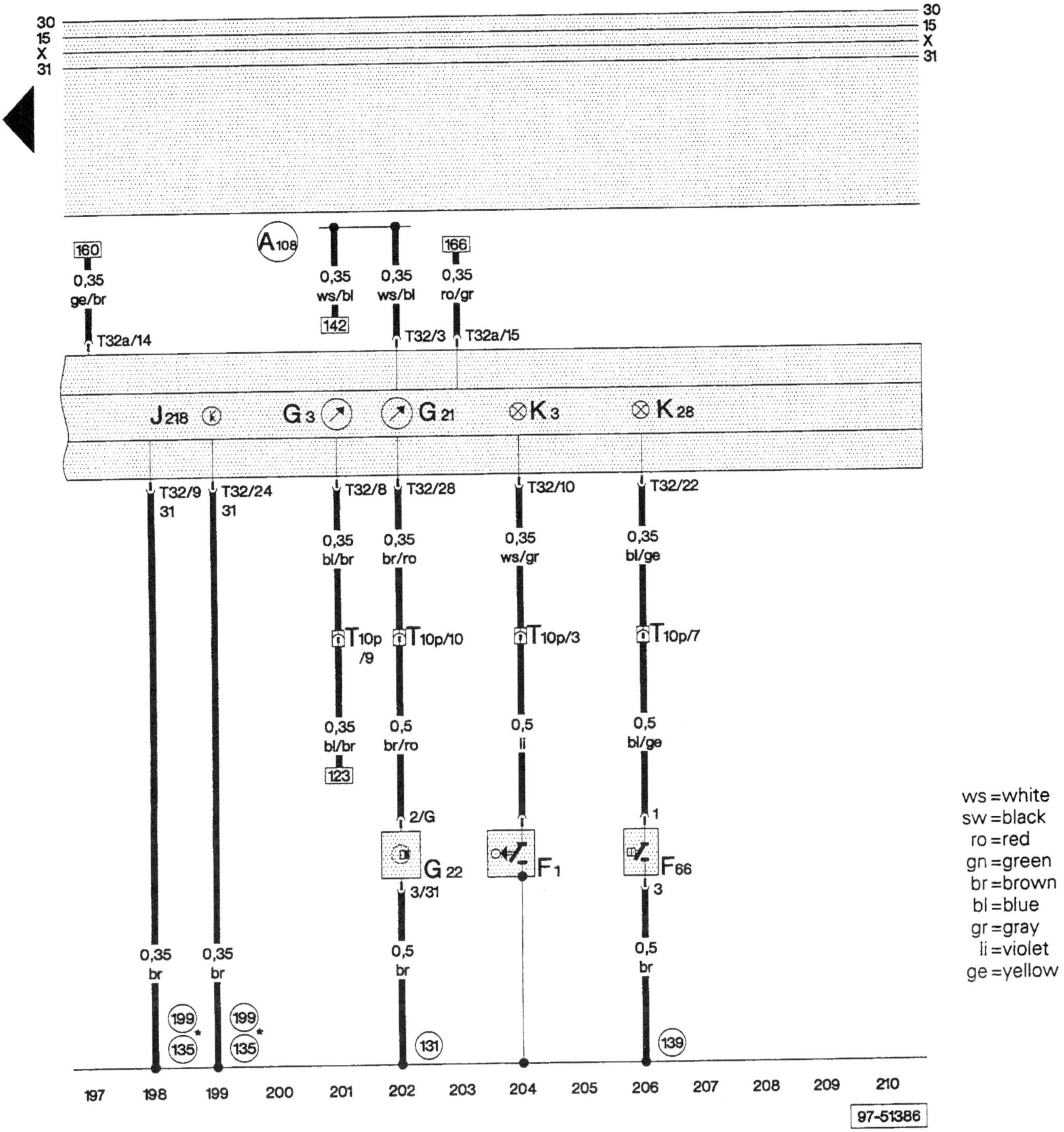

F 1 - Oil Pressure Switch
F 66 - Engine Coolant Level (ECL) Warning Switch
G 3 - Engine Coolant Temperature (ECT) Gauge
G 21 - Speedometer
G 22 - Speedometer Vehicle Speed Sensor (VSS)
J 218 - Instrument Cluster Combination Processor
K 3 - Oil Pressure Warning Light
K 28 - Engine Coolant Level/Temperature (ECL/ECT) Warning Light
T 10 p - Connector 10 pin, black, connector station electronic box plenum chamber
T 32 - Connector 32 pin, blue, on instrument cluster
T 32 a - Connector 32 pin, green, on instrument cluster

(131) - Ground connection -2-, in engine compartment wiring harness
(135) - Ground connection -2-, in instrument panel wiring harness
(139) - Ground connection (sensor Ground), in Motronic Multiport Fuel Injection (MFI) wiring harness
(199) - Ground connection -3-, in instrument panel wiring harness
(A108) - Connector (speed signal), in instrument panel wiring harness

* - booth is possible

THIS PAGE INTENTIONALLY LEFT

BLANK

Standard Equipment

2000 m. y.

Fuse Panel

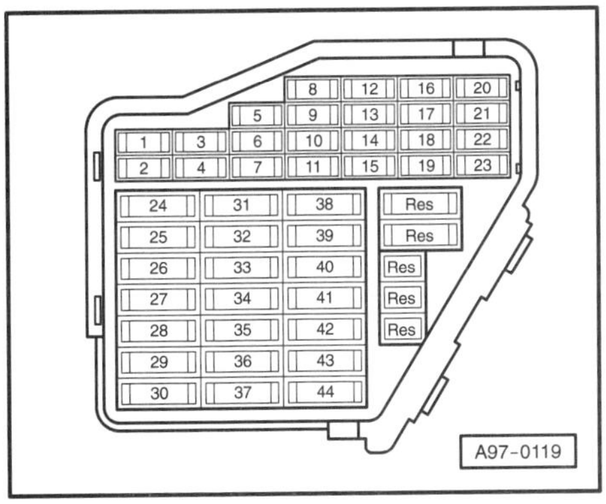

Fuse Colors:

30 A - Green
25 A - White
20 A - Yellow
15 A - Blue
10 A - Red
7,5 A - Brown
5 A - Beige

Starting with fuse position 23, fuses in the fuse holder are identified with 223 in the wiring diagram.

Micro Central Electric Panel

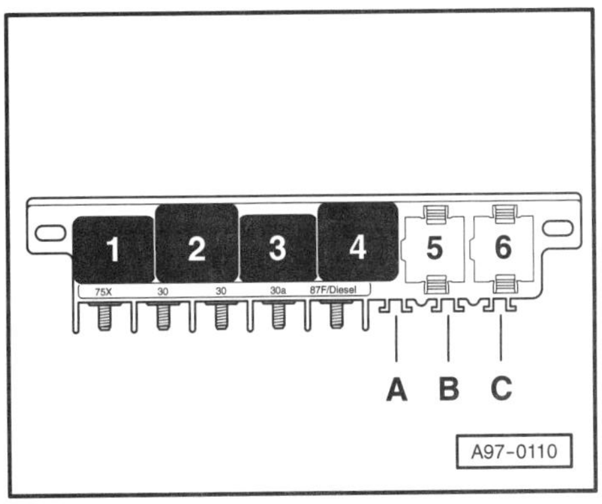

Relay Location:

- **1** - Dual Horn Relay, J4
- **2** - Load Reduction Relay, J59
- **4** - Fuel Pump (FP) Relay, J17
- **5** - Wiper/Washer Intermittent Relay, J31
- **6** - Wiper/Washer Intermittent Relay, J31

13 - Fold Relay Panel

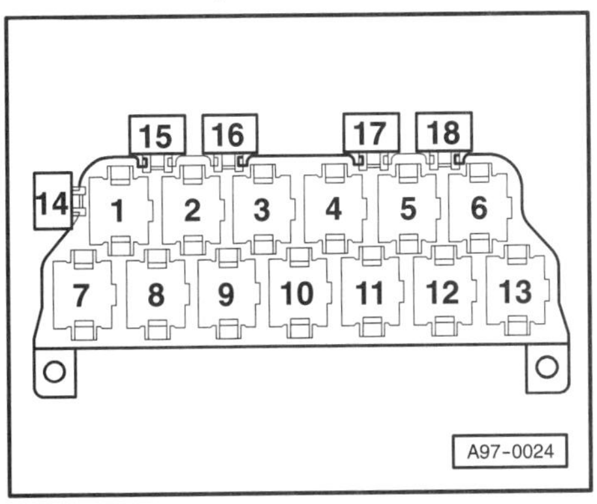

Relay Location:

- **4** - Starting Interlock Relay, J207
- **6** - Fog Light Relay, J5

8 - Fold Relay Panel

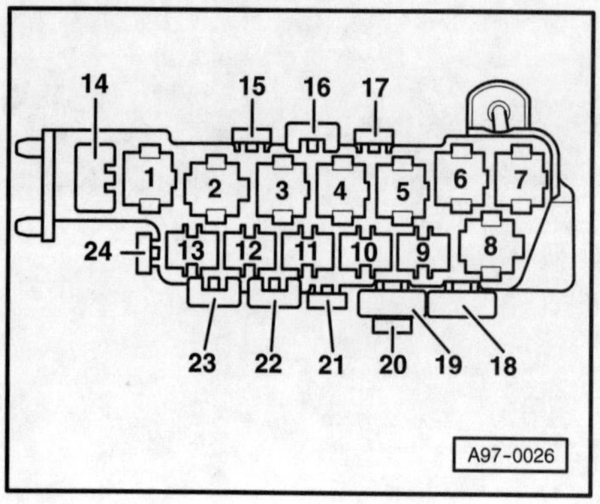

Fuse Location:

15 - Fuse for Luggage Compartment Socket, S184

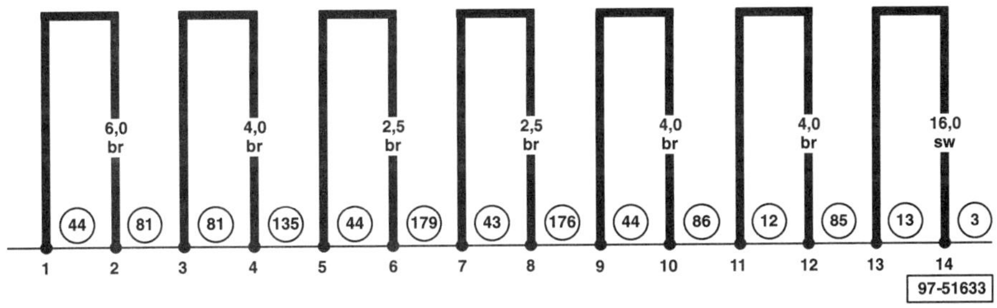

ws = white
sw = black
ro = red
br = brown
gn = green
bl = blue
gr = grey
li = violet
ge = yellow

(3) - Ground strap, engine to body

(12) - Ground connection, in engine compartment, left

(13) - Ground connection, in engine compartment, right

(43) - Ground connection, on right A-pillar, lower part

(44) - Ground connection, on left A-pillar, lower part

(81) - Ground connection -1-, in instrument panel wiring harness

(85) - Ground connection -1-, in engine compartment wiring harness

(86) - Ground connection -1-, in rear wiring harness

(135) - Ground connection -2-, in instrument panel wiring harness

(176) - Ground connection, in right headlight wiring harness

(179) - Ground connection, in left headlight wiring harness

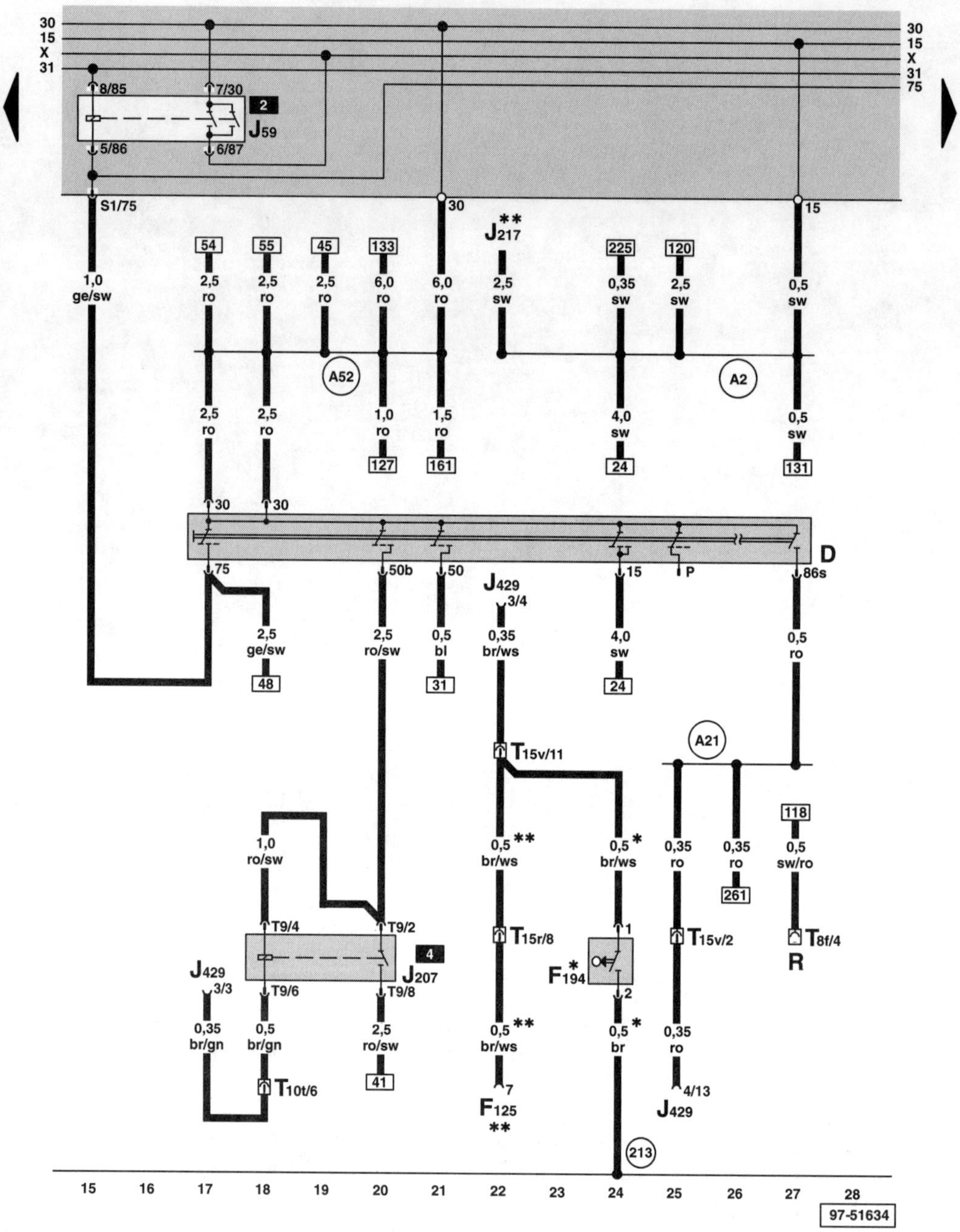

ws = white
sw = black
ro = red
br = brown
gn = green
bl = blue
gr = grey
li = violet
ge = yellow

D - Ignition/Starter Switch
F125 - Multi-Function Transmission Range (TR) Switch
F194 - Clutch Pedal Position (CPP) Switch
J59 - Load Reduction Relay
J207 - Starting Interlock Relay
J217 - Transmission Control Module (TCM)
J429 - Control module for central locking
R - Radio
T8f - Connector 8 pin, black, connector radio III
T9 - Connector 9 pin, brown, on starting interlock relay
T10t - Connector 10 pin, orange, connector station A pillar, right
T15r - Connector 15 pin, blue, connector station A pillar, right
T15v - Connector 15 pin, orange, connector station A pillar, right

(213) - Ground connection, in instrument panel wiring harness
(A2) - Plus connection (15), in instrument panel wiring harness
(A21) - Wire connection (86s), in instrument panel wiring harness
(A52) - Plus connection -2- (30), in instrument panel wiring harness

* - Vehicles with manually shifted transmission
** - Vehicles with automatic transmission

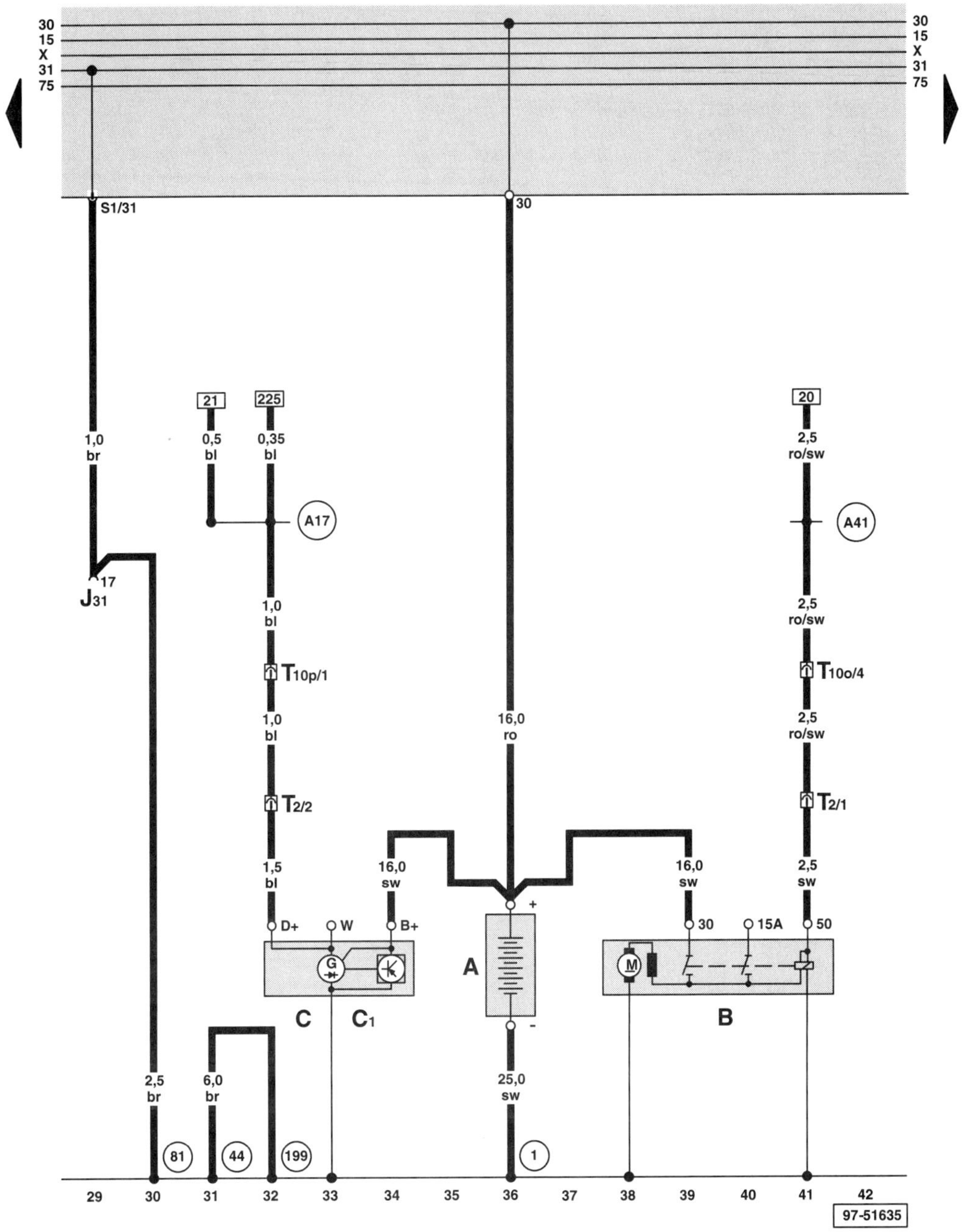

ws = white
sw = black
ro = red
br = brown
gn = green
bl = blue
gr = grey
li = violet
ge = yellow

A - Battery
B - Starter
C - Generator (GEN)
C1 - Voltage Regulator (VR)
J31 - Wiper/Washer Intermittent Relay
T2 - Connector 2 pin, grey, in engine compartment, right
T10o - Connector 10 pin, brown, connector station electronic box plenum chamber
T10p - Connector 10 pin, black, connector station electronic box plenum chamber

(1) - Ground strap, battery to body
(44) - Ground connection, on left A-pillar, lower part
(81) - Ground connection -1-, in instrument panel wiring harness
(199) - Ground connection -3-, in instrument panel wiring harness
(A17) - Wire connection (61), in instrument panel wiring harness
(A41) - Plus connection (50), in instrument panel wiring harness

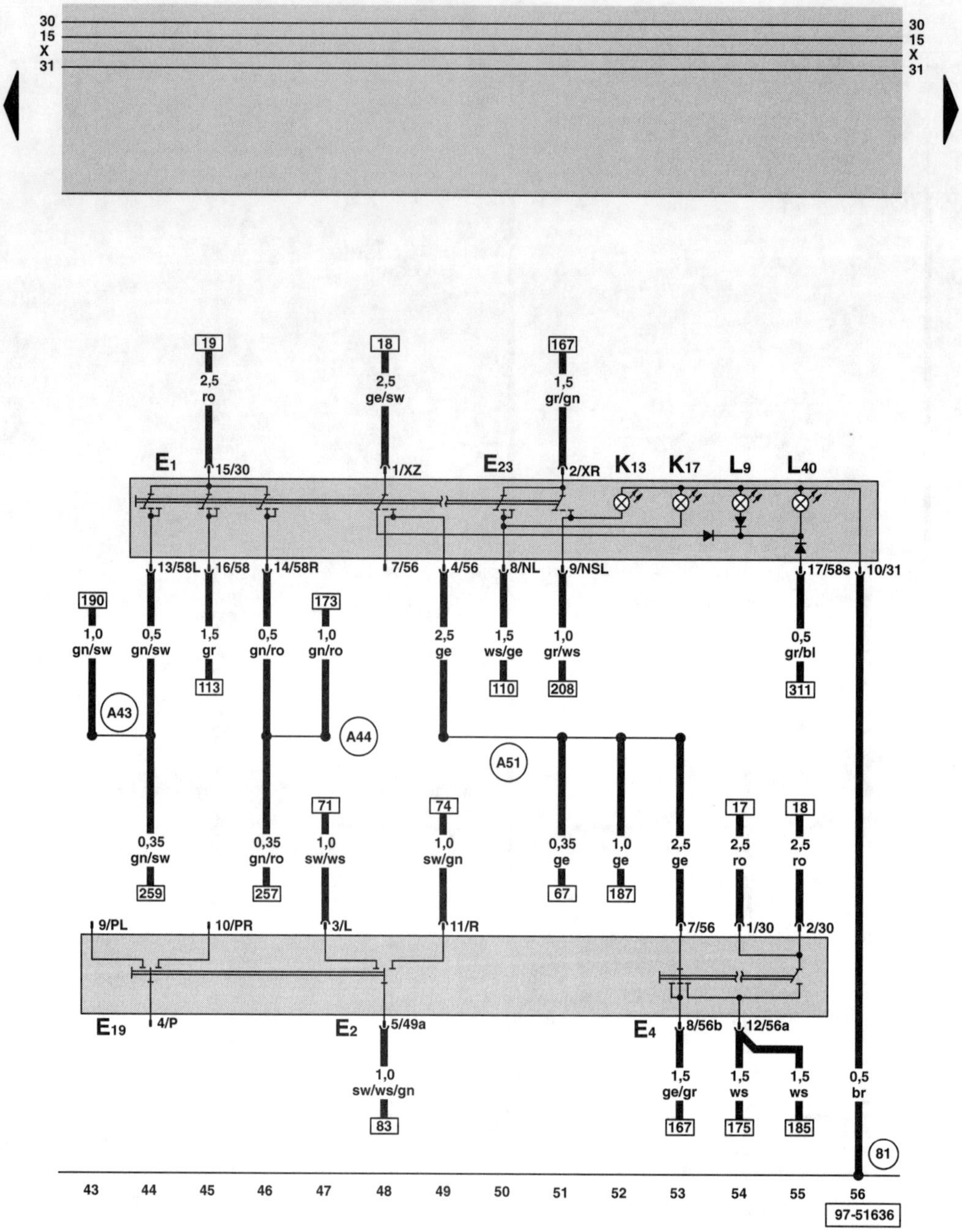

ws = white
sw = black
ro = red
br = brown
gn = green
bl = blue
gr = grey
li = violet
ge = yellow

E1 - Light switch
E2 - Turn Signal Switch
E4 - Headlight Dimmer/Flasher Switch
E19 - Park Light Switch
E23 - Fog Light Switch
K13 - Rear Fog Light Indicator Light
K17 - Fog Light Indicator Light
L9 - Headlight Switch Light
L40 - Front And Rear Fog Light Switch Light

(81) - Ground connection -1-, in instrument panel wiring harness
(A43) - Wire connection (57l), in instrument panel wiring harness
(A44) - Wire connection (57r), in instrument panel wiring harness
(A51) - Wire connection (56), in instrument panel wiring harness

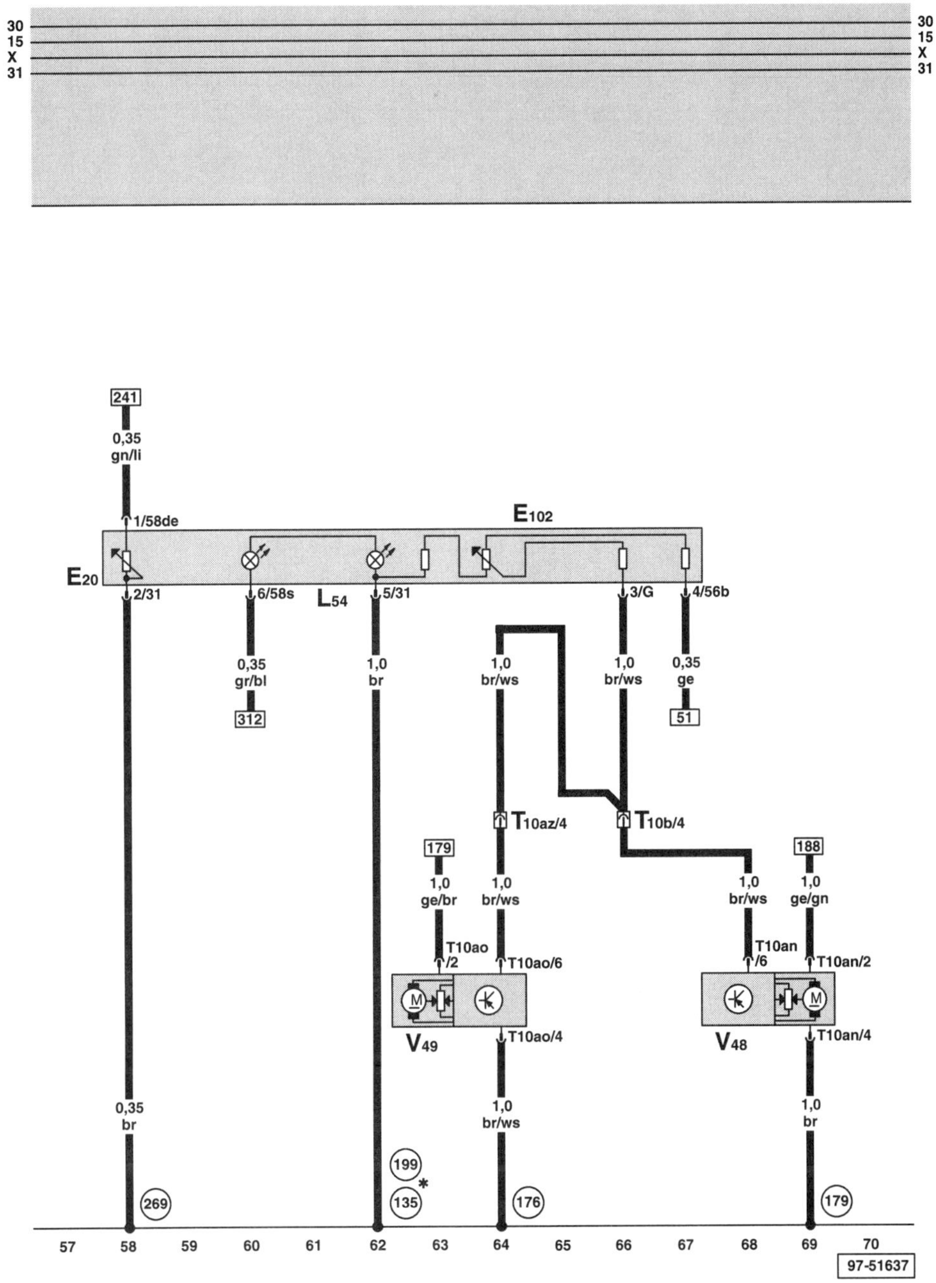

ws = white
sw = black
ro = red
br = brown
gn = green
bl = blue
gr = grey
li = violet
ge = yellow

E20 - Instrument Panel Light Dimmer Switch
E102 - Headlight Adjuster
L54 - Headlight Adjuster Control Light
T10b - Connector 10 pin, red, connector station A pillar, left
T10an - Connector 10 pin, on headlight, left
T10ao - Connector 10 pin, on headlight, right
T10az - Connector 10 pin, red, connector station A pillar, right
V48 - Left Headlight Beam Adjusting Motor
V49 - Right Headlight Beam Adjusting Motor

(135) - Ground connection -2-, in instrument panel wiring harness
(176) - Ground connection, in right headlight wiring harness
(179) - Ground connection, in left headlight wiring harness
(199) - Ground connection -3-, in instrument panel wiring harness
(269) - Ground connector (sensor ground) -1-, in instrument panel wiring harness

* - Both are possible

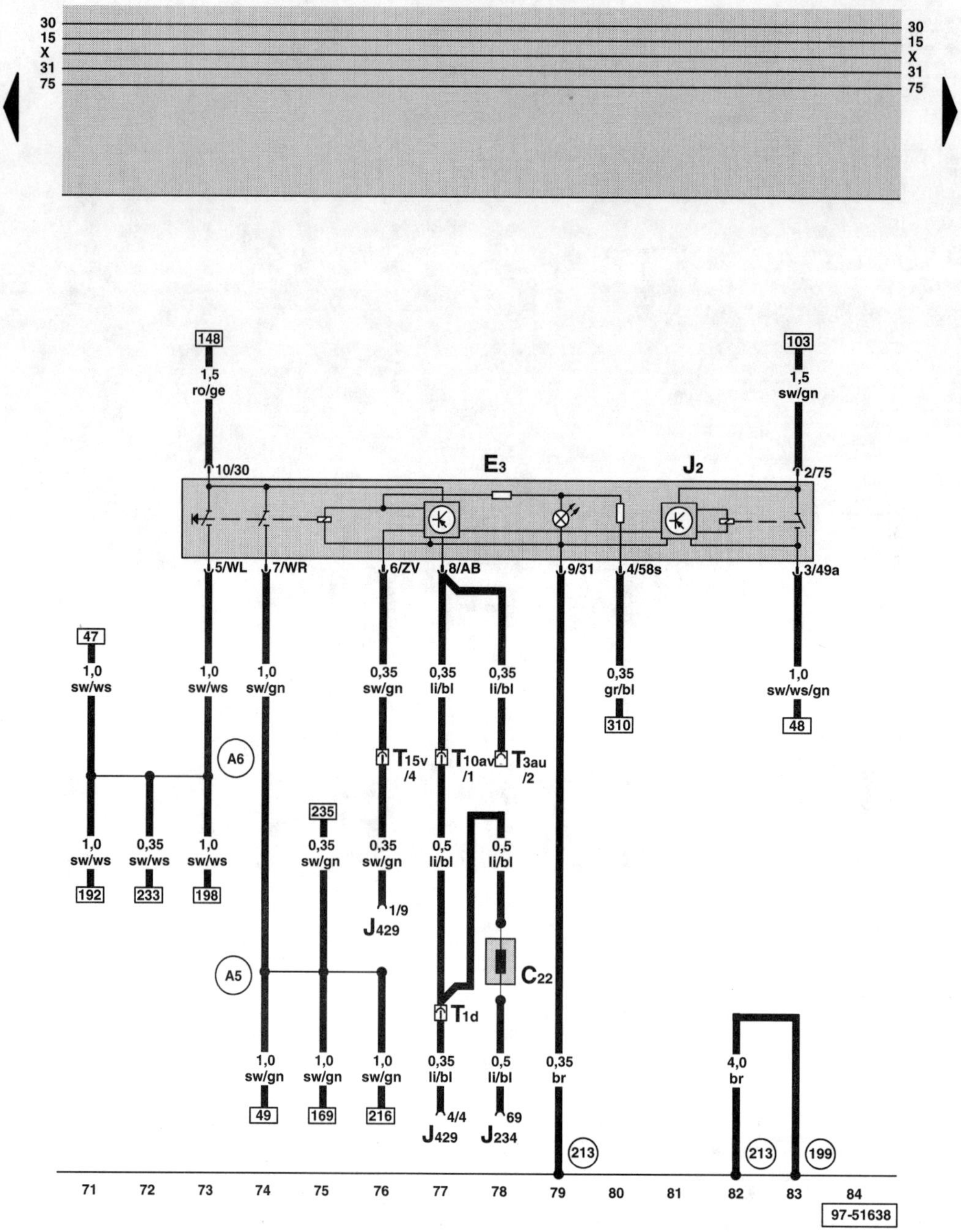

ws = white
sw = black
ro = red
br = brown
gn = green
bl = blue
gr = grey
li = violet
ge = yellow

C22 - Interference filter
E3 - Emergency Flasher Switch
J2 - Emergency Flasher Relay
J234 - Airbag Control Module
J429 - Control module for central locking
T1d - Single connector, brown, connector station A pillar right
T3au - Connector 3 pin, red, connector station electronic box plenum chamber
T10av - Connector 10 pin, yellow, connector station A pillar, right
T15v - Connector 15 pin, orange, connector station A pillar, right

(199) - Ground connection -3-, in instrument panel wiring harness
(213) - Ground connection, in instrument panel wiring harness
(A5) - Plus connection (right turn signal), in instrument panel wiring harness
(A6) - Plus connection (left turn signal), in instrument panel wiring harness

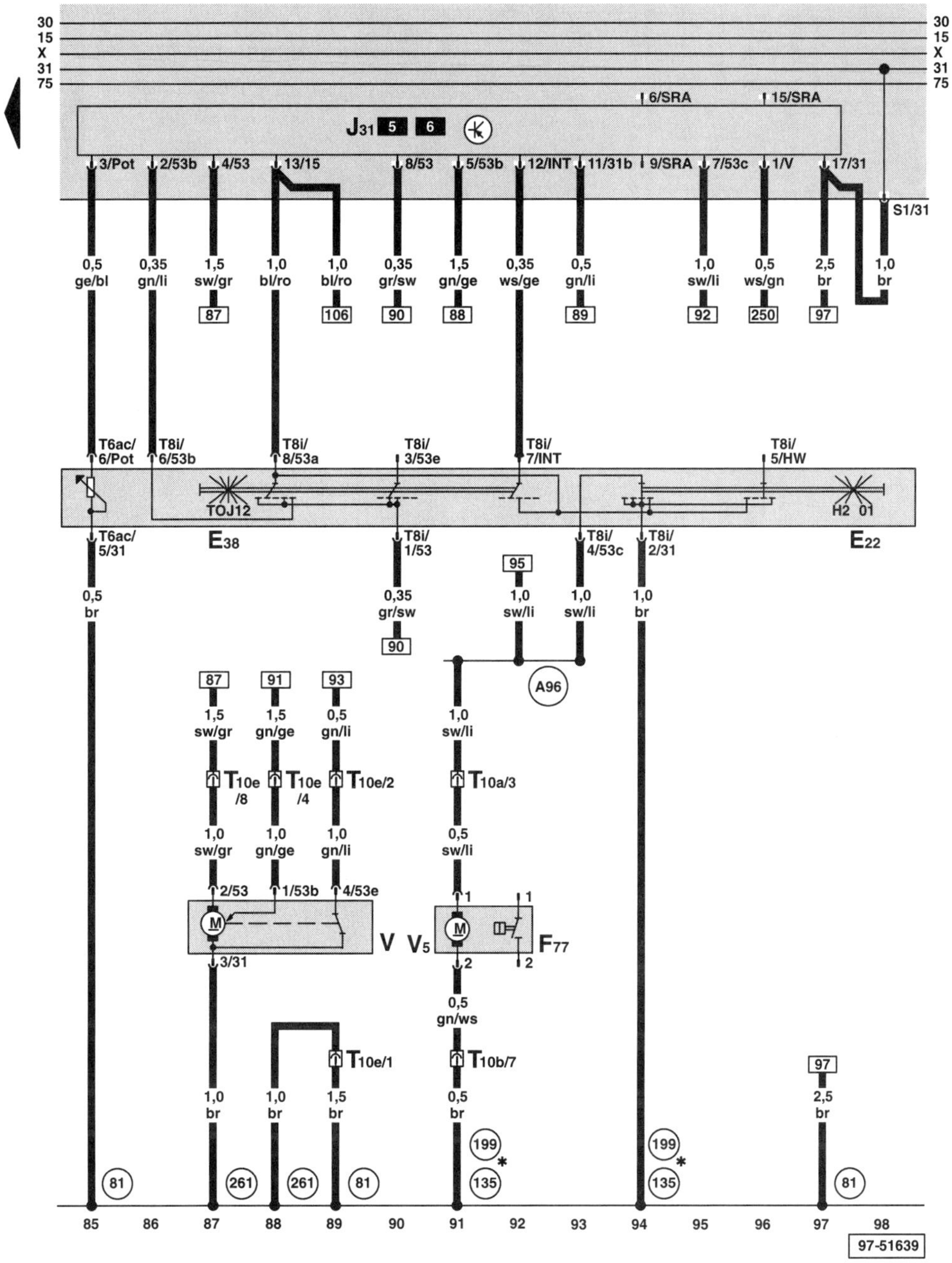

ws = white
sw = black
ro = red
br = brown
gn = green
bl = blue
gr = grey
li = violet
ge = yellow

E22 - Windshield Wiper/Washer Switch
E38 - Windshield Wiper Intermittent Regulator
F77 - Windshield Washer Fluid Level Warning Switch
J31 - Wiper/Washer Intermittent Relay
T6ac - Connector 6 pin, brown, on windshield wiper intermittent regulator
T8i - Connector 8 pin, on windshield wiper/washer switch
T10a - Connector 10 pin, pink, connector station A pillar, left
T10b - Connector 10 pin, red, connector station A pillar, left
T10e - Connector 10 pin, violet, connector station A pillar, left
V - Windshield Wiper Motor
V5 - Windshield Washer Pump

(81) - Ground connection -1-, in instrument panel wiring harness
(135) - Ground connection -2-, in instrument panel wiring harness
(199) - Ground connection -3-, in instrument panel wiring harness
(261) - Ground connection, in wiring harness heated spray jet
(A96) - Connector (53a), in instrument panel wiring harness
* - Both are possible

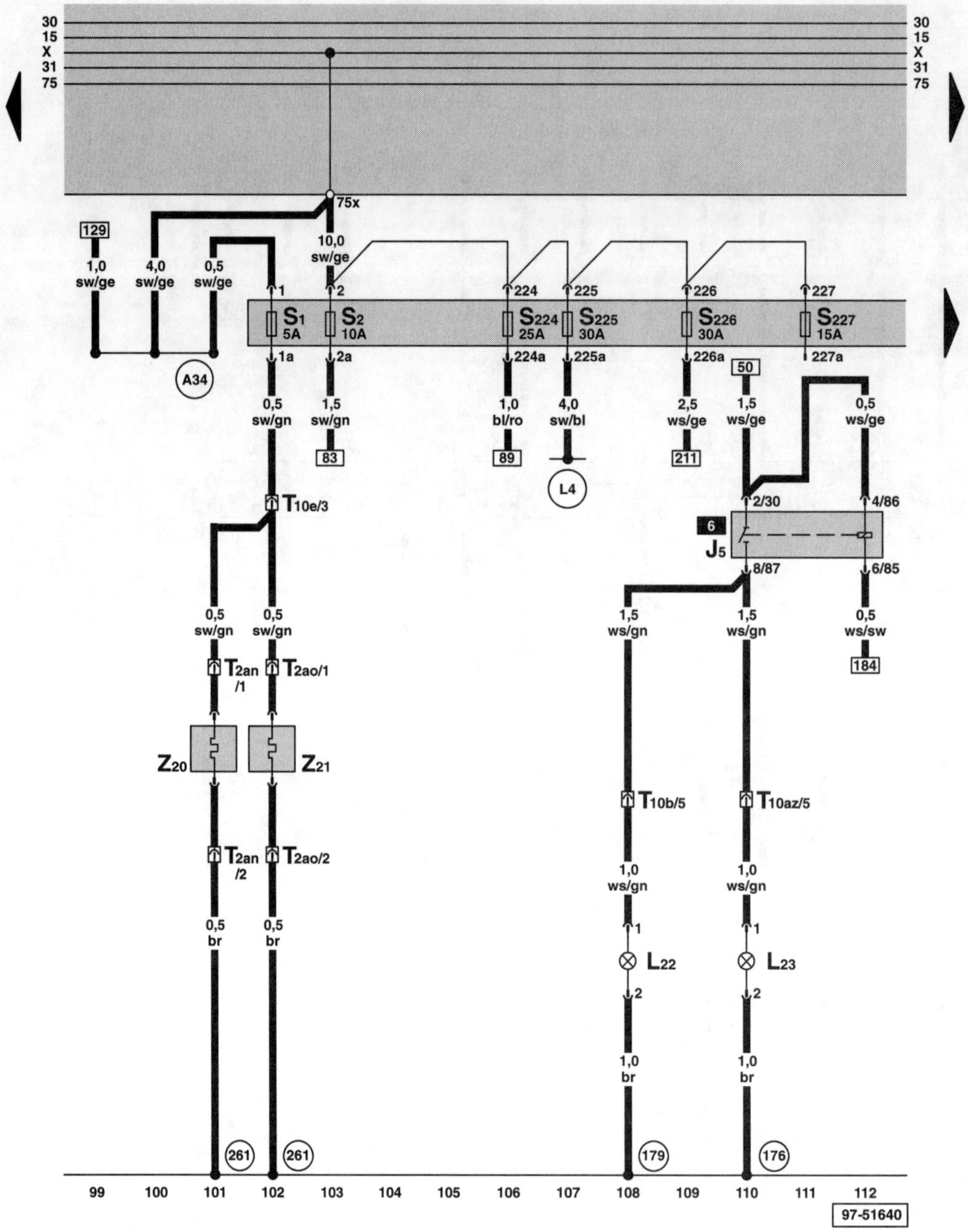

ws = white
sw = black
ro = red
br = brown
gn = green
bl = blue
gr = grey
li = violet
ge = yellow

J5 - Fog Light Relay
L22 - Left Front Fog Light
L23 - Right Front Fog Light
S1 - Fuse
S2 - Fuse
S224 - Fuse 24 in fuse holder
S225 - Fuse 25 in fuse holder
S226 - Fuse 26 in fuse holder
S227 - Fuse 27 in fuse holder
T2an - Connector 2 pin, washer jet, left
T2ao - Connector 2 pin, washer jet, right
T10b - Connector 10 pin, red, connector station A pillar, left
T10e - Connector 10 pin, violet, connector station A pillar, left
T10az - Connector 10 pin, red, connector station A pillar, right
Z20 - Left Washer Nozzle Heater
Z21 - Right Washer Nozzle Heater

(176) - Ground connection, in right headlight wiring harness
(179) - Ground connection, in left headlight wiring harness
(261) - Ground connection, in wiring harness heated spray jet
(A34) - Wire connection (75x), in instrument panel wiring harness
(L4) - Wire connection (75al), in A/C wiring harness

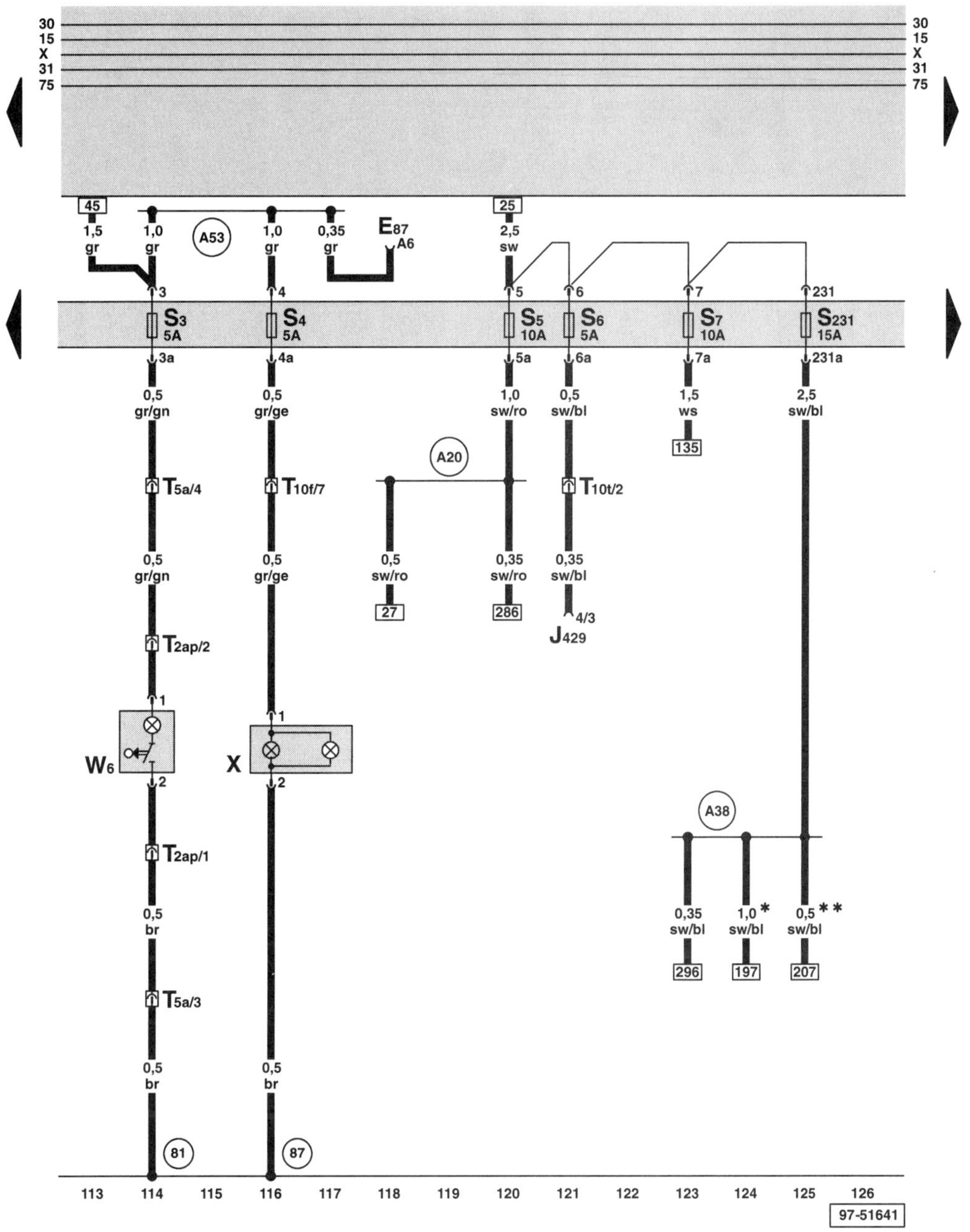

ws = white
sw = black
ro = red
br = brown
gn = green
bl = blue
gr = grey
li = violet
ge = yellow

E87 - A/C Control Head
J429 - Control module for central locking
S3 - Fuse
S4 - Fuse
S5 - Fuse
S6 - Fuse
S7 - Fuse
S231 - Fuse 31 in fuse holder
T2ap - Connector 2 pin, black, behind instrument panel, right
T5a - Connector 5 pin, black, behind instrument panel, right
T10f - Connector 10 pin, brown, connector station A pillar, left
T10t - Connector 10 pin, orange, connector station A pillar, right
W6 - Glove Compartment Light
X - License Plate Light

(81) - Ground connection -1-, in instrument panel wiring harness
(87) - Ground connection -2-, in rear wiring harness
(A20) - Wire connection (15a), in instrument panel wiring harness
(A38) - Plus connection (15a II), in instrument panel wiring harness
(A53) - Plus connection (58b), in instrument panel wiring harness

* - Vehicles with manually shifted transmission
** - Vehicles with automatic transmission

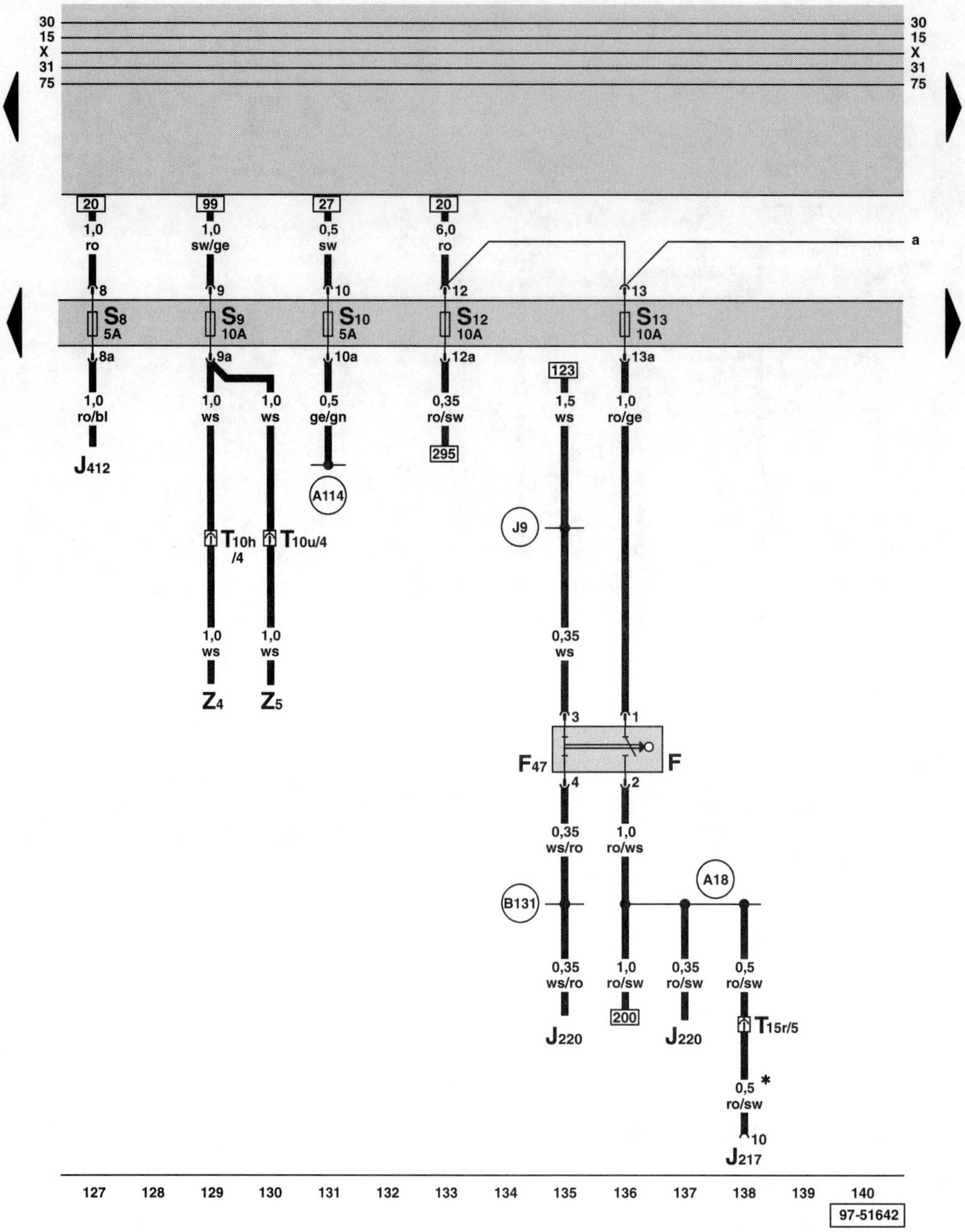

ws = white
sw = black
ro = red
br = brown
gn = green
bl = blue
gr = grey
li = violet
ge = yellow

F - Brake Light Switch
F47 - Vacuum Vent Valve, Brake
J217 - Transmission Control Module (TCM)
J220 - Motronic Engine Control Module (ECM)
J412 - Control module for operating electronics, telephone
S8 - Fuse
S9 - Fuse
S10 - Fuse
S12 - Fuse
S13 - Fuse
T10h - Connector 10 pin, black, connector station A pillar, left
T10u - Connector 10 pin, black, connector station A pillar, right
T15r - Connector 15 pin, blue, connector station A pillar, right
Z4 - Driver's Side Heated Mirror
Z5 - Passenger's Side Heated Mirror

(A18) - Wire connection (54), in instrument panel wiring harness
(A114) - Connection (headlight beam adjusting) in instrument panel wiring harness
(B131) - Connector (54), in wiring harness interior
(J9) - Wire connection -1- (15a), in ABS wiring harness
* - Vehicles with automatic transmission 01V

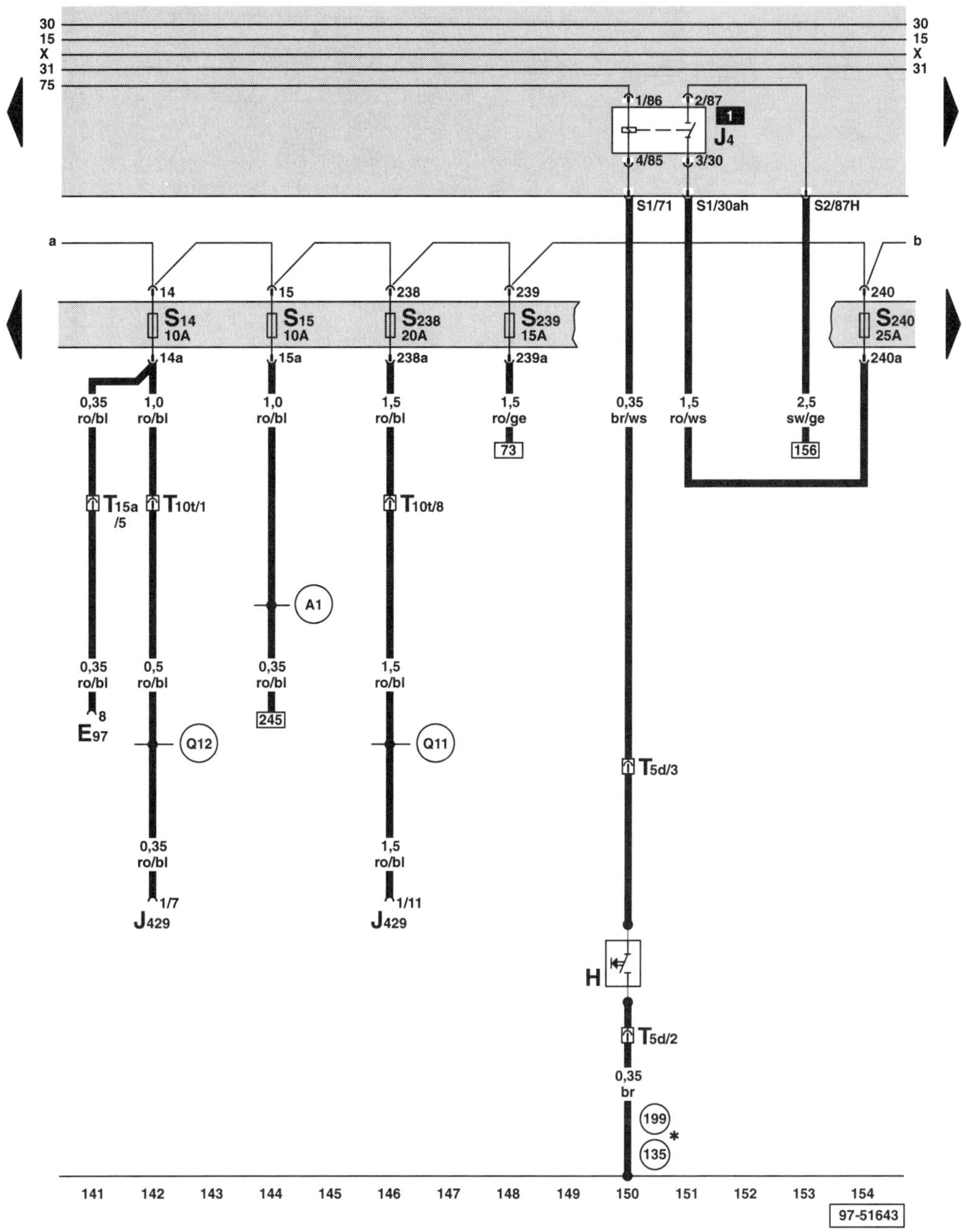

ws = white
sw = black
ro = red
br = brown
gn = green
bl = blue
gr = grey
li = violet
ge = yellow

E97 - Memory Program Switch
H - Horn Button
J4 - Dual Horn Relay
J429 - Control module for central locking
S14 - Fuse
S15 - Fuse
S238 - Fuse 38 in fuse holder
S239 - Fuse 39 in fuse holder
S240 - Fuse 40 in fuse holder
T5d - Connector 5 pin, behind steering column switch cover
T10t - Connector 10 pin, orange, connector station A pillar, right
T15a - Connector 15 pin, black, connector station A pillar, left

(135) - Ground connection -2-, in instrument panel wiring harness
(199) - Ground connection -3-, in instrument panel wiring harness
(A1) - Plus connection (30a), in instrument panel wiring harness
(Q11) - Plus connection (30a), in power window/central locking system and door contact switch wiring harness
(Q12) - Plus connection (30a), in power window/central locking system and door contact switch wiring harness
* - Both are possible

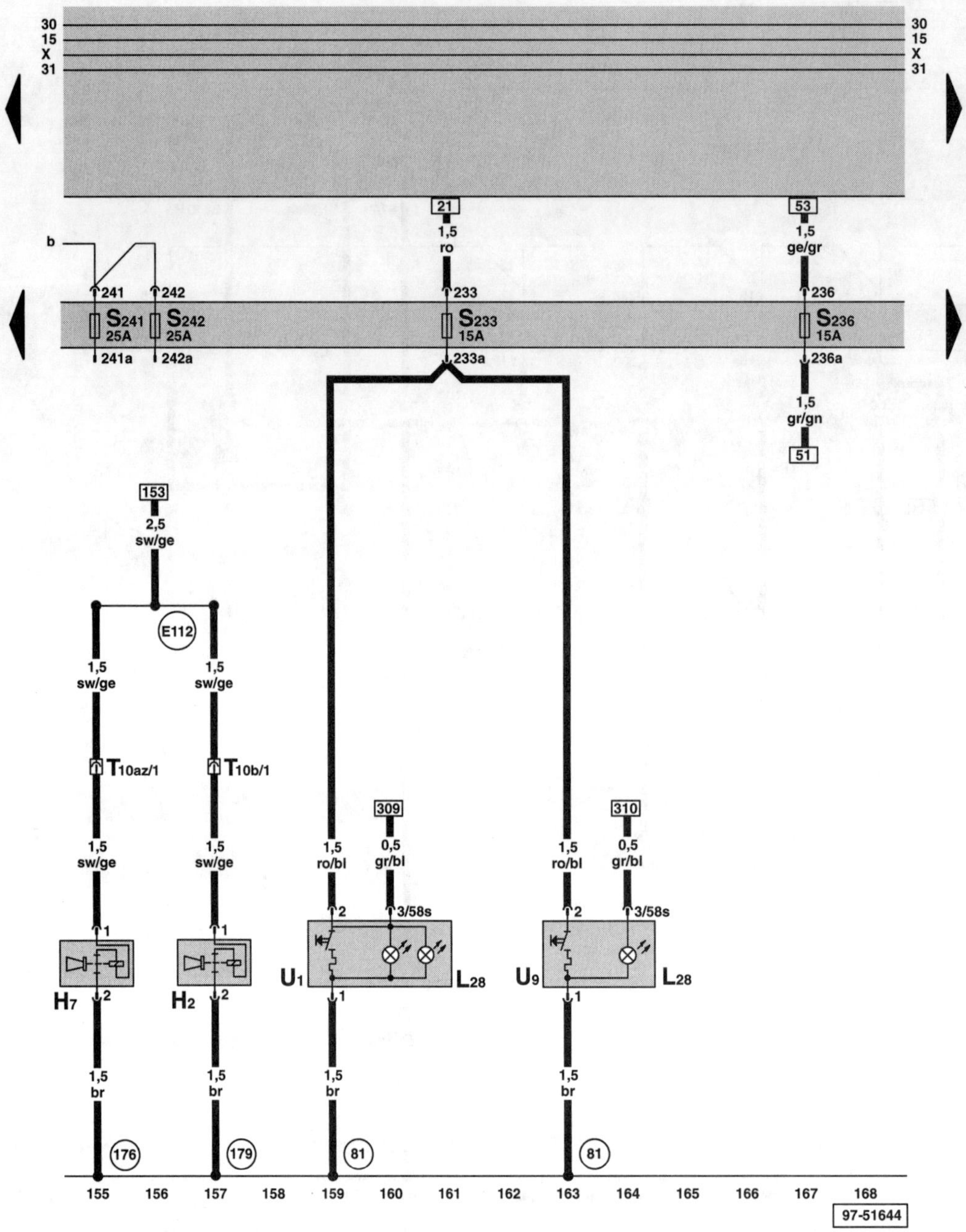

ws = white
sw = black
ro = red
br = brown
gn = green
bl = blue
gr = grey
li = violet
ge = yellow

- H2 - High Tone Horn
- H7 - Low Tone Horn
- L28 - Cigarette Lighter Light
- S233 - Fuse 33 in fuse holder
- S236 - Fuse 36 in fuse holder
- S241 - Fuse 41 in fuse holder
- S242 - Fuse 42 in fuse holder
- T10b - Connector 10 pin, red, connector station A pillar, left
- T10az - Connector 10 pin, red, connector station A pillar, right
- U1 - Cigarette Lighter
- U9 - Rear Cigarette Lighter

- (81) - Ground connection -1-, in instrument panel wiring harness
- (176) - Ground connection, in right headlight wiring harness
- (179) - Ground connection, in left headlight wiring harness
- (E112) - Connector (Horn - 87h), in instrument panel wiring harness

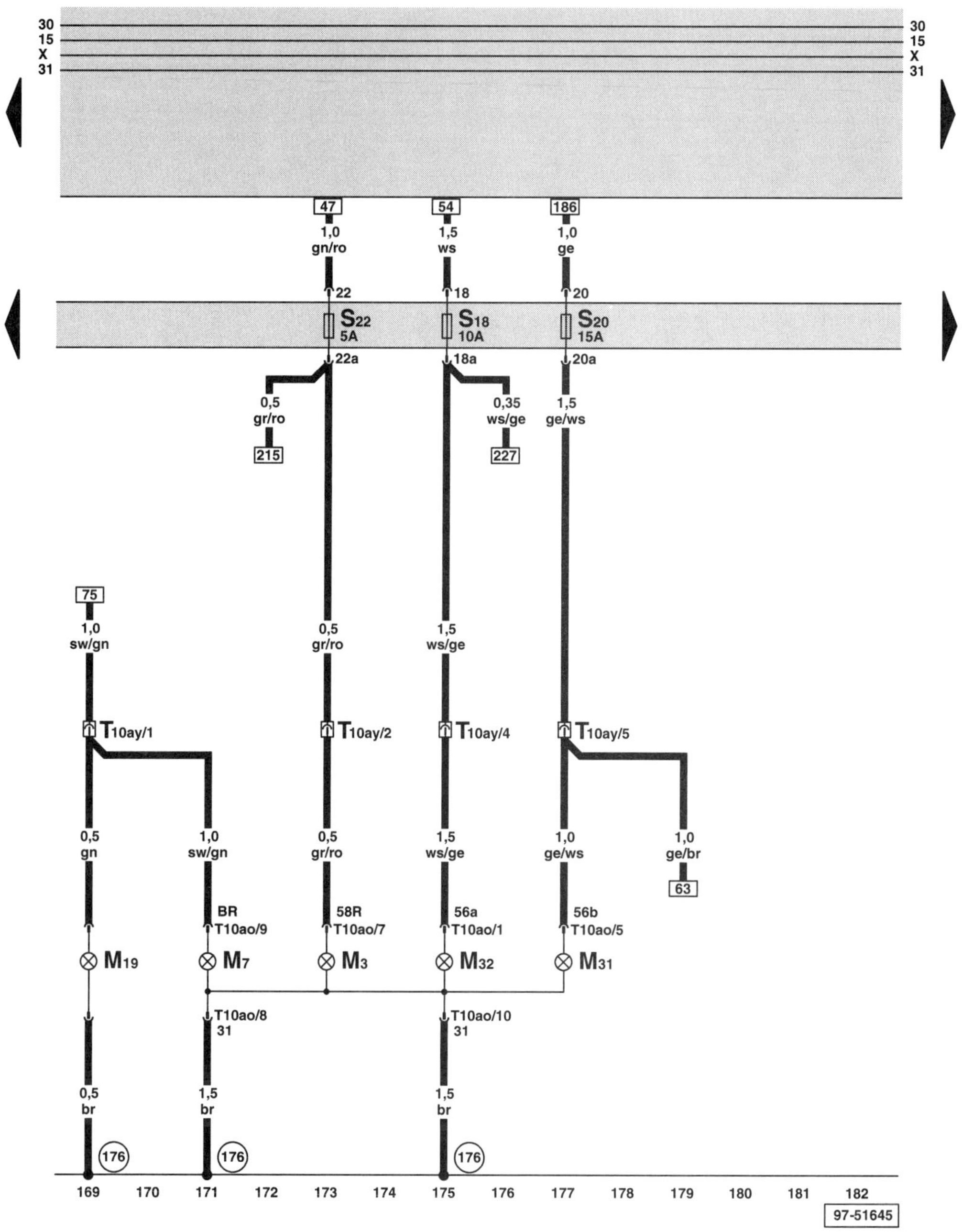

ws = white
sw = black
ro = red
br = brown
gn = green
bl = blue
gr = grey
li = violet
ge = yellow

M3 - Right Parking Light
M7 - Right Front Turn Signal Light
M19 - Right, Side Turn Signal Light
M31 - Right Low Beam Headlight
M32 - Right High Beam Headlight
S18 - Fuse
S20 - Fuse
S22 - Fuse
T10ao - Connector 10 pin, on headlight, right
T10ay - Connector 10 pin, pink, connector station A pillar, right

(176) - Ground connection, in right headlight wiring harness

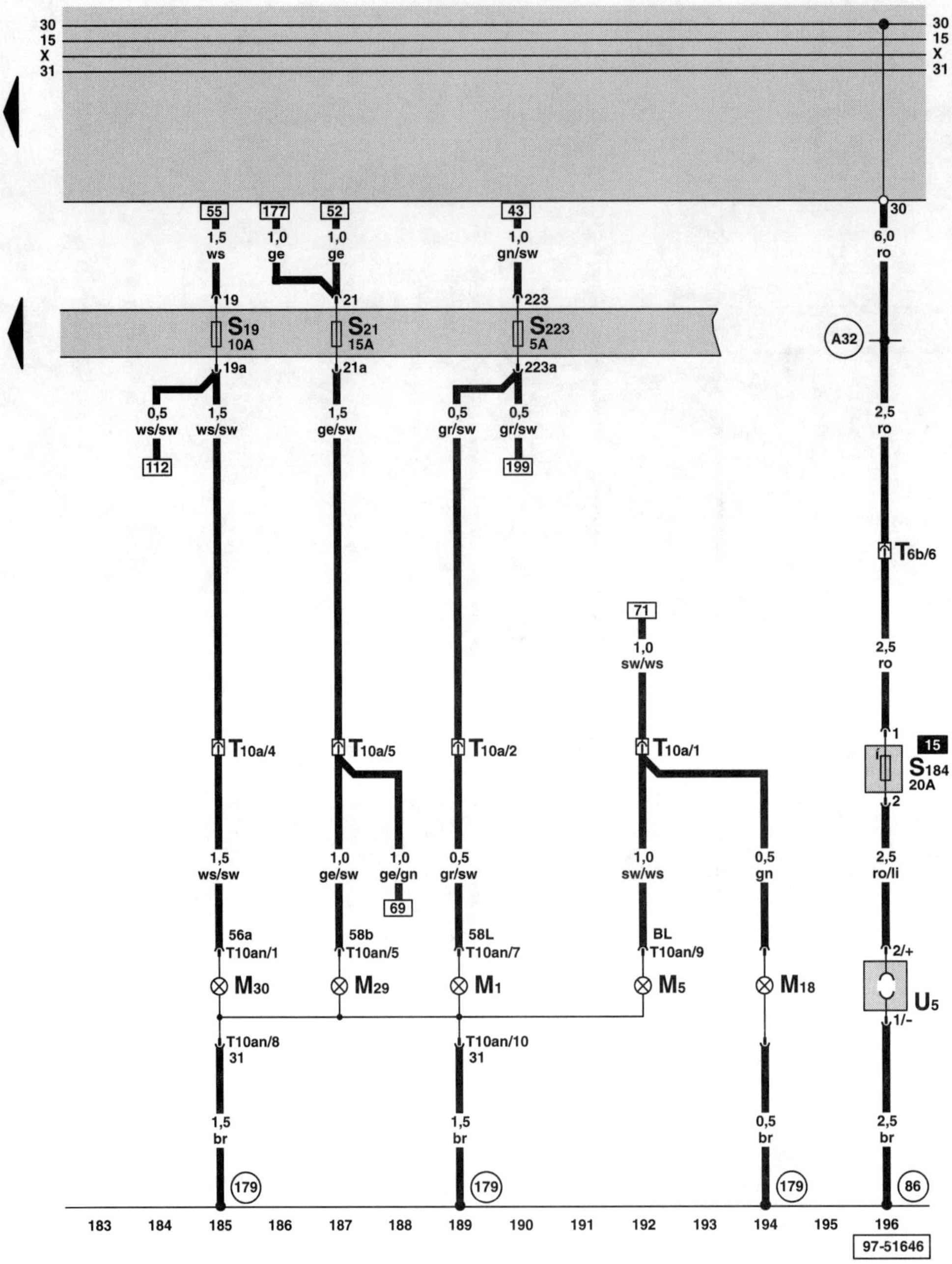

ws = white
sw = black
ro = red
br = brown
gn = green
bl = blue
gr = grey
li = violet
ge = yellow

M1 - Left Parking Light
M5 - Left Front Turn Signal Light
M18 - Left, Side Turn Signal Light
M29 - Left Low Beam Headlight
M30 - Left High Beam Headlight
S19 - Fuse
S21 - Fuse
S184 - Fuse for Luggage Compartment Socket
S223 - Fuse 23 in fuse holder
T6b - Connector 6 pin, white, connector station A pillar, left
T10a - Connector 10 pin, pink, connector station A pillar, left
T10an - Connector 10 pin, on headlight, left
U5 - 12 V Socket

(86) - Ground connection -1-, in rear wiring harness

(179) - Ground connection, in left headlight wiring harness

(A32) - Plus connection (30), in instrument panel wiring harness

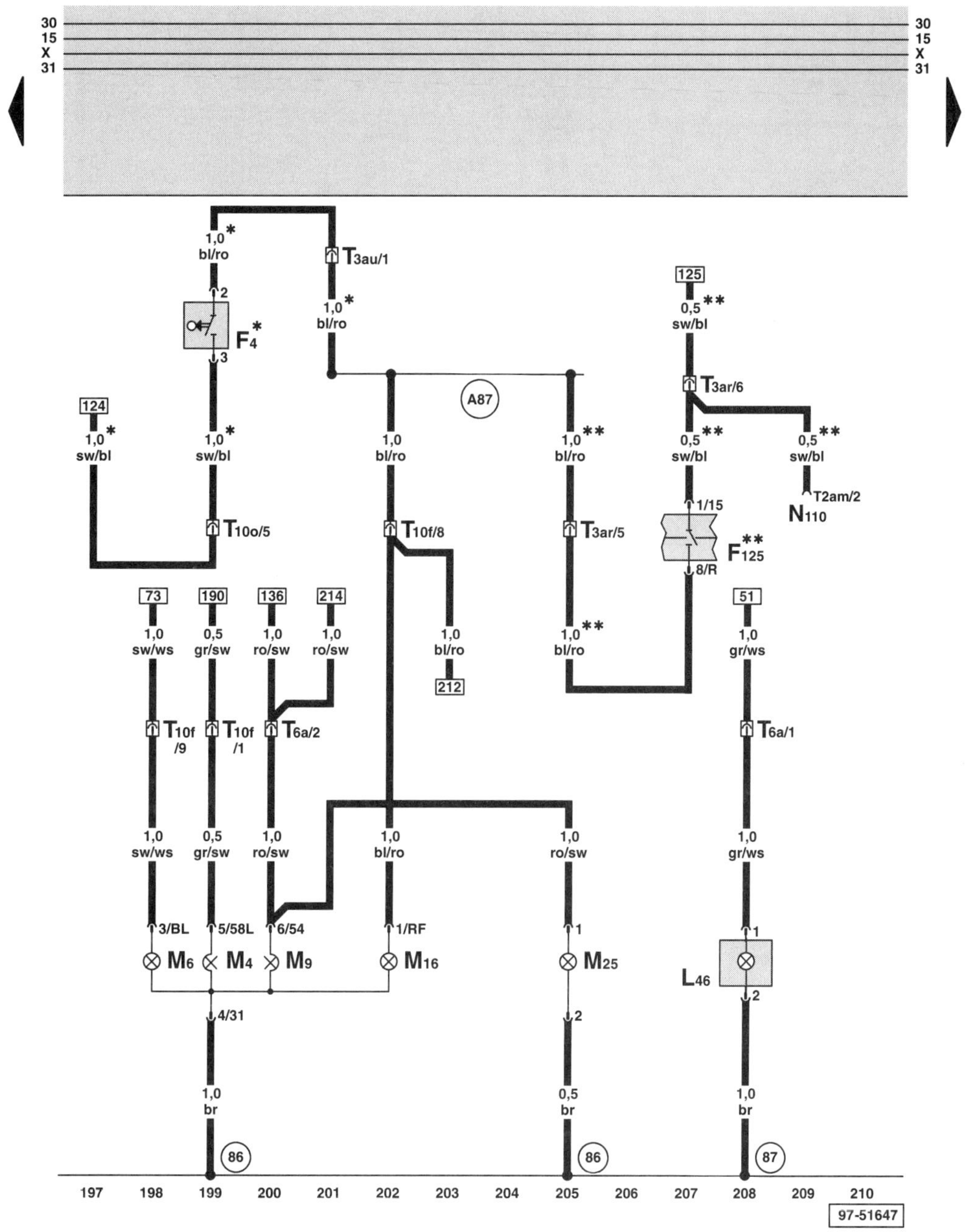

ws = white
sw = black
ro = red
br = brown
gn = green
bl = blue
gr = grey
li = violet
ge = yellow

F4 - Back-Up Light Switch
F125 - Multi-Function Transmission Range (TR) Switch
L46 - Left Rear Fog Light
M4 - Left Tail Light
M6 - Left Rear Turn Signal Light
M9 - Left Brake Light
M16 - Left Back-Up Light
M25 - High-mount Brake Light
N110 - Shift Lock Solenoid
T2am - Connector 2 pin, black, on selector lever
T3ar - Connector 3 pin, blue, connector station A pillar, right
T3au - Connector 3 pin, red, connector station electronic box plenum chamber
T6a - Connector 6 pin, blue, connector station A pillar, left
T10f - Connector 10 pin, brown, connector station A pillar, left
T10o - Connector 10 pin, brown, connector station electronic box plenum chamber

(86) - Ground connection -1-, in rear wiring harness
(87) - Ground connection -2-, in rear wiring harness
(A87) - Connector (RF), in instrument panel wiring harness
* - Vehicles with manually shifted transmission
** - Vehicles with automatic transmission

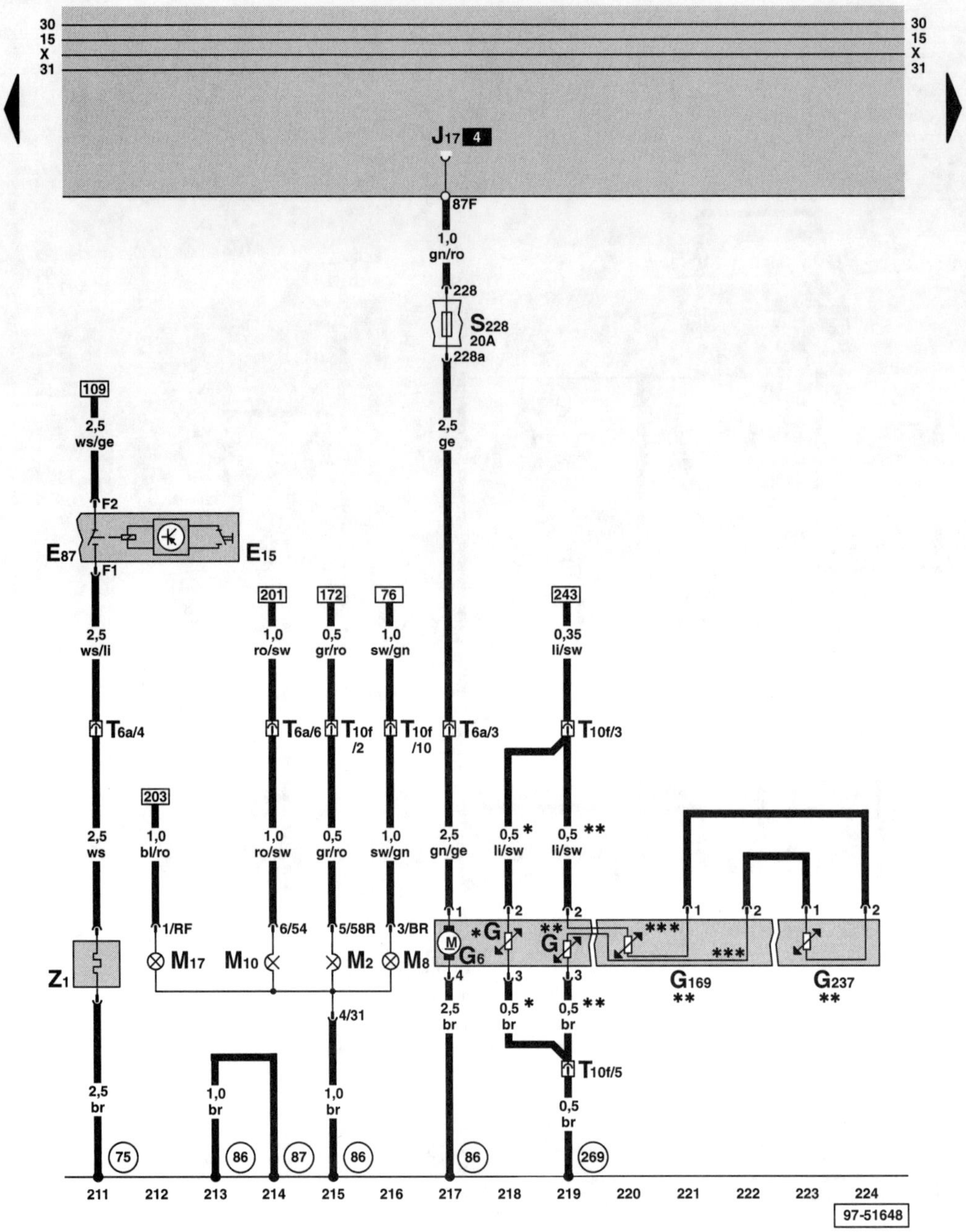

ws = white
sw = black
ro = red
br = brown
gn = green
bl = blue
gr = grey
li = violet
ge = yellow

E15 - Rear Window Defogger Switch
E87 - A/C Control Head
G - Sender for fuel gauge
G6 - Fuel Pump (FP)
G169 - Sender 2 for fuel gauge
G237 - Sensor -3- for fuel supply
J17 - Fuel Pump (FP) Relay
M2 - Right Tail Light
M8 - Right Rear Turn Signal Light
M10 - Right Brake Light
M17 - Right Back-Up Light
S228 - Fuse 28 in fuse holder
T6a - Connector 6 pin, blue, connector station A pillar, left
T10f - Connector 10 pin, brown, connector station A pillar, left
Z1 - Rear Window Defogger/Heat Element

(75) - Ground connection, on right rear pillar

(86) - Ground connection -1-, in rear wiring harness

(87) - Ground connection -2-, in rear wiring harness

(269) - Ground connector (sensor ground) -1-, in instrument panel wiring harness

* - Vehicles with front wheel drive

** - Vehicles with all wheel drive

*** - Wiring inside fuel tank

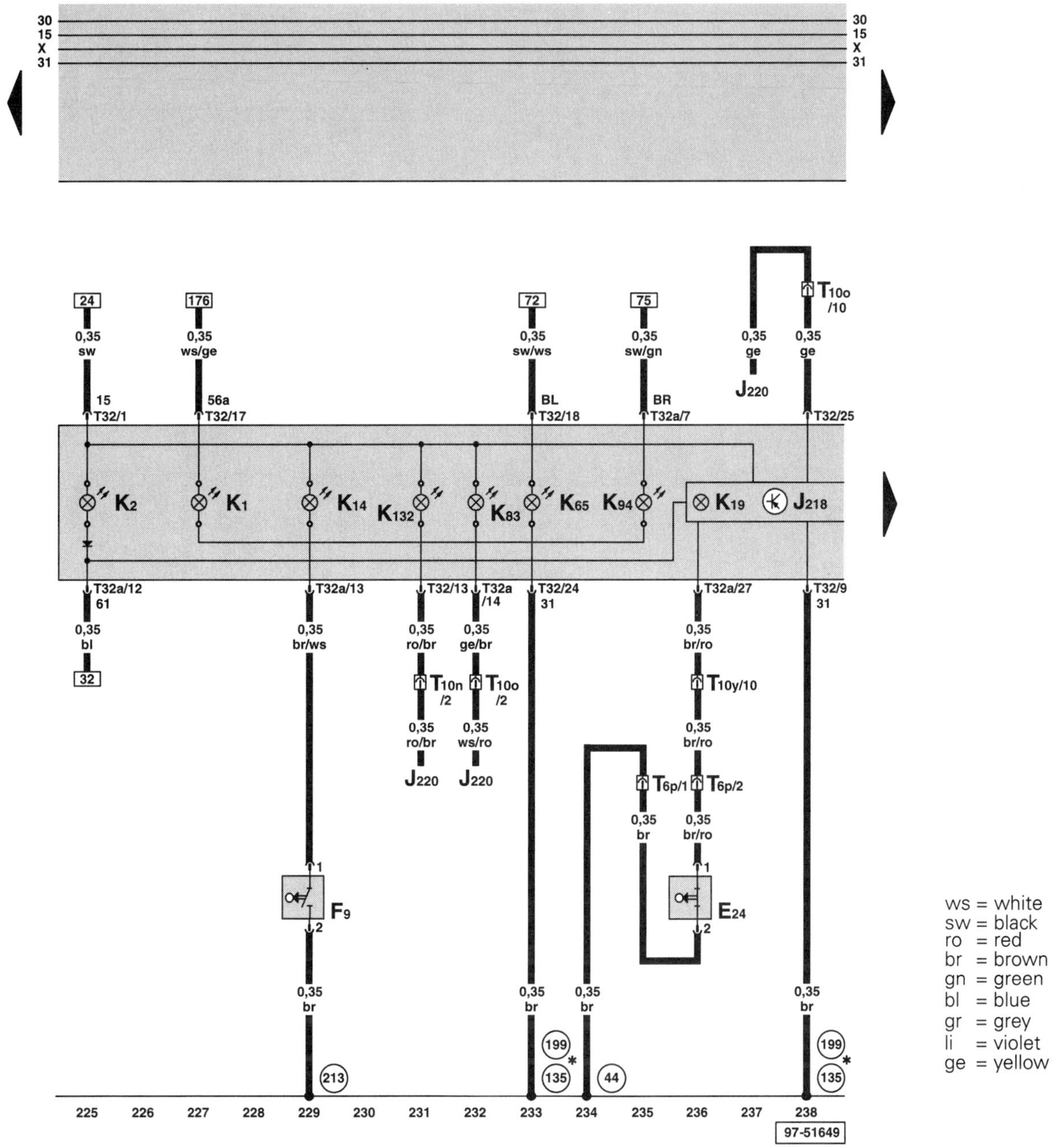

E24 - Left Seat Belt Switch
F9 - Parking Brake Warning Light Switch
J218 - Instrument Cluster Combination Processor
J220 - Motronic Engine Control Module (ECM)
K1 - Headlight High Beam Indicator Light
K2 - Generator (GEN) Warning Light
K14 - Parking Brake Indicator Light
K19 - Seat Belt Warning Light
K65 - Left Turn Signal Indicator Light
K83 - Malfunction Indicator Lamp (MIL)
K94 - Right Turn Signal Indicator Light
K132 - Fault light for power accelerator activation
T6p - Connector 6 pin, red, below driver's seat
T10n - Connector 10 pin, orange, connector station electronic box plenum chamber
T10o - Connector 10 pin, brown, connector station electronic box plenum chamber
T10y - Connector 10 pin, green, connector station A pillar, right
T32 - Connector 32 pin, blue, on instrument cluster
T32a - Connector 32 pin, green, on instrument cluster

(44) - Ground connection, on left A-pillar, lower part
(135) - Ground connection -2-, in instrument panel wiring harness
(199) - Ground connection -3-, in instrument panel wiring harness
(213) - Ground connection, in instrument panel wiring harness
* - Both are possible

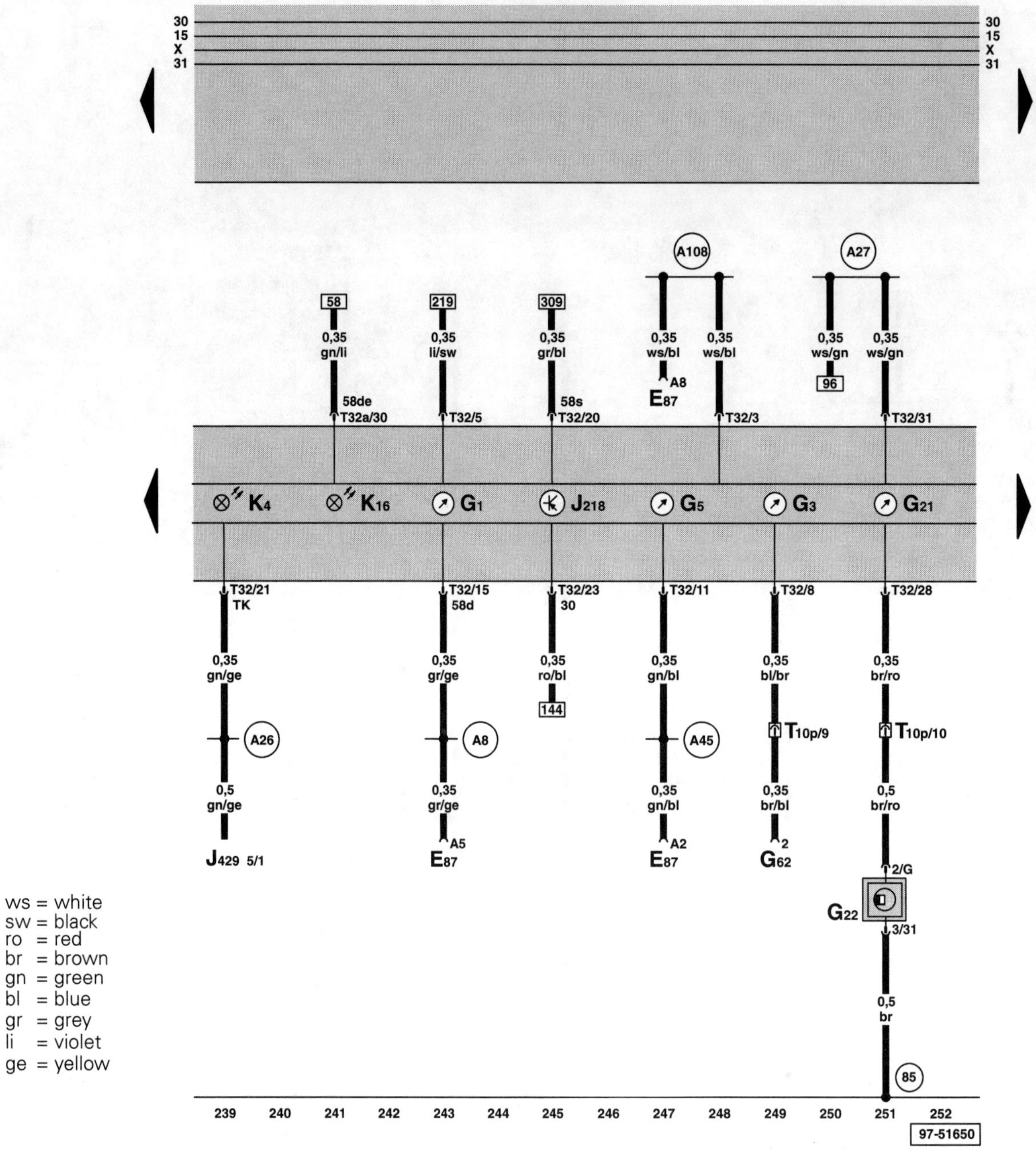

E87 - A/C Control Head
G1 - Fuel Gauge
G3 - Engine Coolant Temperature (ECT) Gauge
G5 - Tachometer
G21 - Speedometer
G22 - Speedometer Vehicle Speed Sensor (VSS)
G62 - Engine Coolant Temperature (ECT) Sensor
J218 - Instrument Cluster Combination Processor
J429 - Control module for central locking
K4 - Park Light Indicator Light
K16 - Fuel Reserve Warning Light
T10p - Connector 10 pin, black, connector station electronic box plenum chamber
T32 - Connector 32 pin, blue, on instrument cluster
T32a - Connector 32 pin, green, on instrument cluster

(85) - Ground connection -1-, in engine compartment wiring harness
(A8) - Plus connection (58d), in instrument panel wiring harness
(A26) - Wire connection (driver's door contact switch), in instrument panel wiring harness
(A27) - Wire connection (vehicle speed signal), in instrument panel wiring harness
(A45) - Wire connection (RPM-signal), in instrument panel wiring harness
(A108) - Connector (speed signal), in instrument panel wiring harness

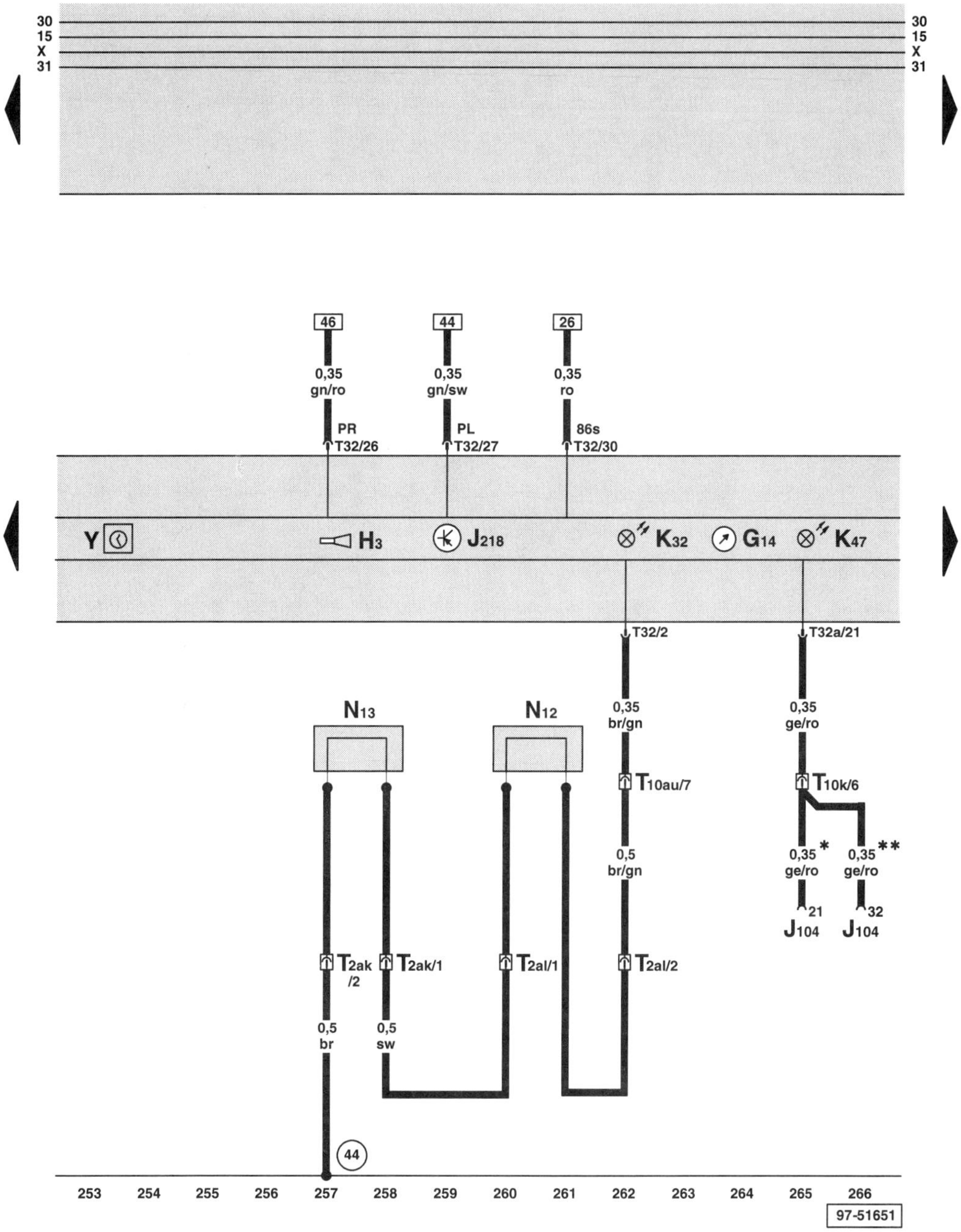

ws = white
sw = black
ro = red
br = brown
gn = green
bl = blue
gr = grey
li = violet
ge = yellow

G14 - Voltmeter
H3 - Warning Buzzer
J104 - ABS Control Module (w/EDL)
J218 - Instrument Cluster Combination Processor
K32 - Brake Pad Wear Indicator Light
K47 - ABS Warning Light
N12 - Right Brake Pad Wear Indicator (Shear Element)
N13 - Left Brake Pad Wear Indicator (Shear Element)
T2ak - Connector 2 pin, red, on firewall, left
T2al - Connector 2 pin, red, on firewall, right
T10k - Connector 10 pin, orange, connector station A pillar, left
T10au - Connector 10 pin, grey, connector station A pillar, right
T32 - Connector 32 pin, blue, on instrument cluster
T32a - Connector 32 pin, green, on instrument cluster
Y - Analog Clock

(44) - Ground connection, on left A-pillar, lower part

* - Vehicles with ABS 5.3
** - Vehicles with ABS and Electronic Stability Program (ESP)

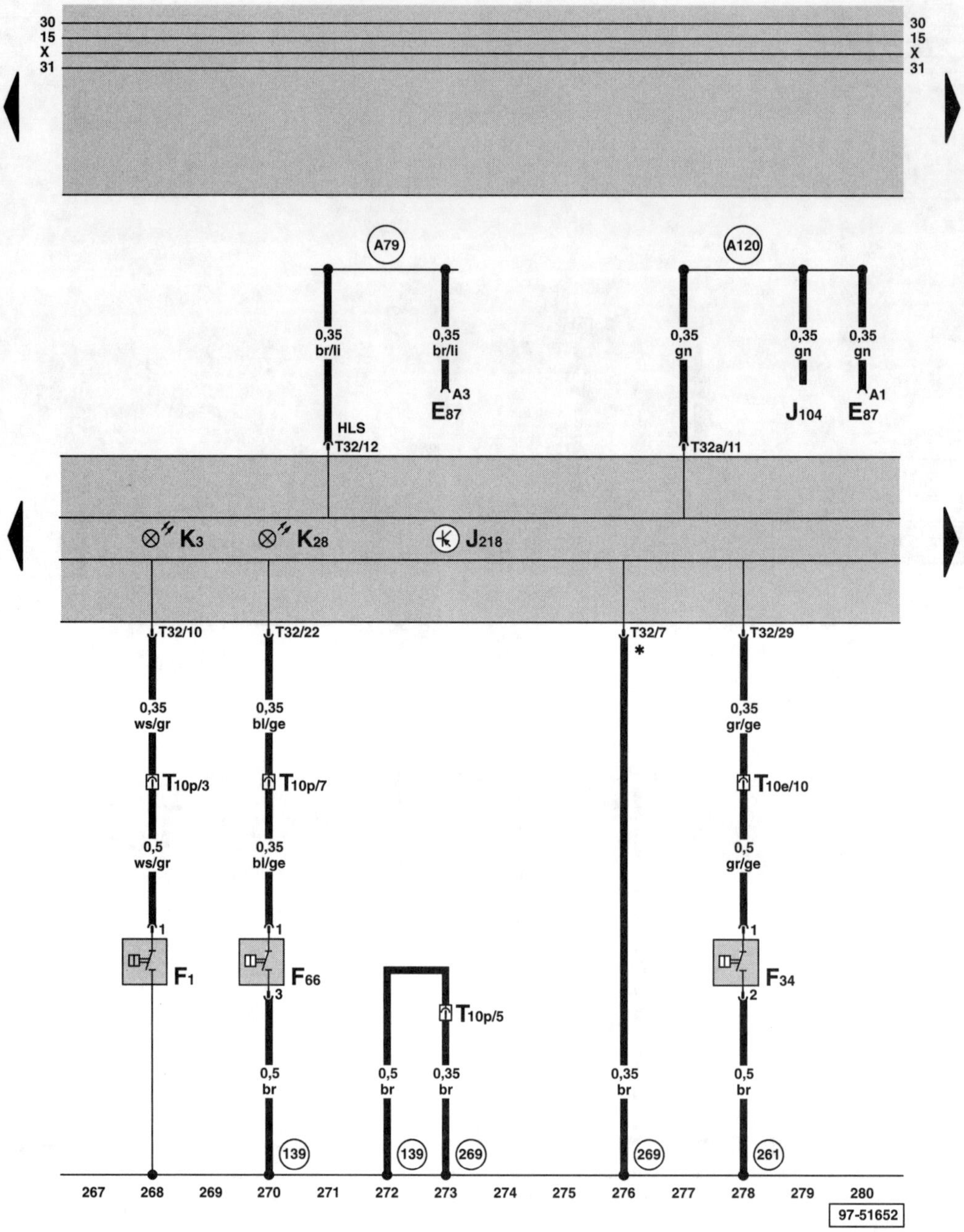

ws = white
sw = black
ro = red
br = brown
gn = green
bl = blue
gr = grey
li = violet
ge = yellow

E87 - A/C Control Head
F1 - Oil Pressure Switch
F34 - Brake Fluid Level Warning Switch
F66 - Engine Coolant Level (ECL) Warning Switch
J104 - ABS Control Module (w/EDL)
J218 - Instrument Cluster Combination Processor
K3 - Oil Pressure Warning Light
K28 - Engine Coolant Level/Temperature (ECL/ECT) Warning Light
T10e - Connector 10 pin, violet, connector station A pillar, left
T10p - Connector 10 pin, black, connector station electronic box plenum chamber
T32 - Connector 32 pin, blue, on instrument cluster
T32a - Connector 32 pin, green, on instrument cluster

(139) - Ground connection (sensor Ground), in Motronic Multiport Fuel Injection (MFI) wiring harness
(261) - Ground connection, in wiring harness heated spray jet
(269) - Ground connector (sensor ground) -1-, in instrument panel wiring harness
(A79) - Connector (engine overheat warning light), in instrument panel wiring harness
(A120) - Connection (standing time) in instrument panel wiring harness
* - Sensor ground, output

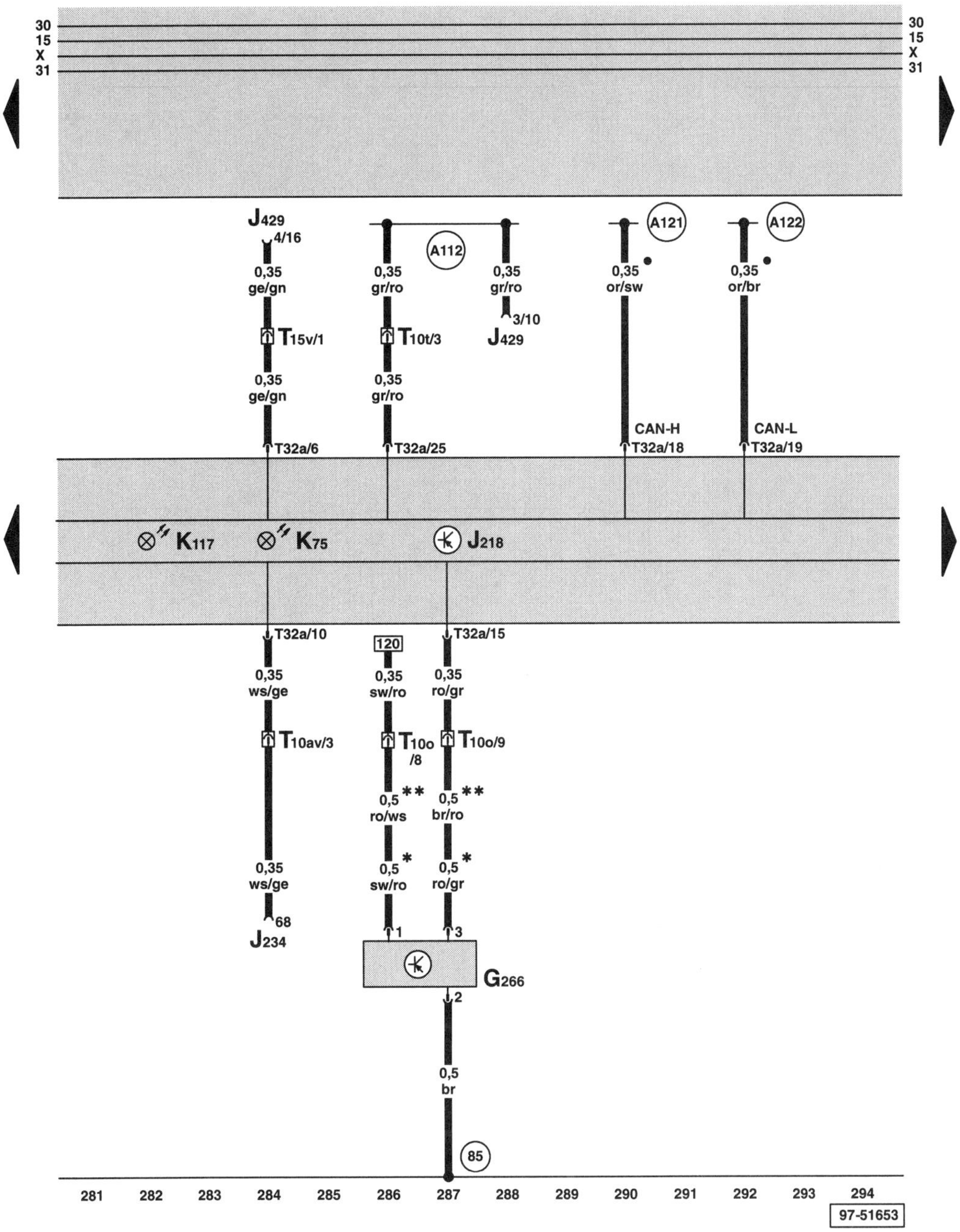

ws = white
sw = black
ro = red
br = brown
gn = green
bl = blue
gr = grey
li = violet
ge = yellow
or = orange

G266 - Oil Level Thermal Sensor (SRI)
J218 - Instrument Cluster Combination Processor
J234 - Airbag Control Module
J429 - Control module for central locking
K75 - Airbag Malfunction Indicator Lamp (MIL)
K117 - Warning light for anti-theft imobilizer
T10o - Connector 10 pin, brown, connector station electronic box plenum chamber
T10t - Connector 10 pin, orange, connector station A pillar, right
T10av - Connector 10 pin, yellow, connector station A pillar, right
T15v - Connector 15 pin, orange, connector station A pillar, right
T32 - Connector 32 pin, blue, on instrument cluster
T32a - Connector 32 pin, green, on instrument cluster

(85) - Ground connection -1-, in engine compartment wiring harness
(A112) - Connection (driver's cab lock), in instrument panel wiring harness
(A121) - Connection (high bus) in instrument panel wiring harness
(A122) - Connection (low bus) in instrument panel wiring harness
• - CAN-Bus (Data-Bus)
* - Vehicles with 6 cylinder engine
** - Vehicles with 8 cylinder engine

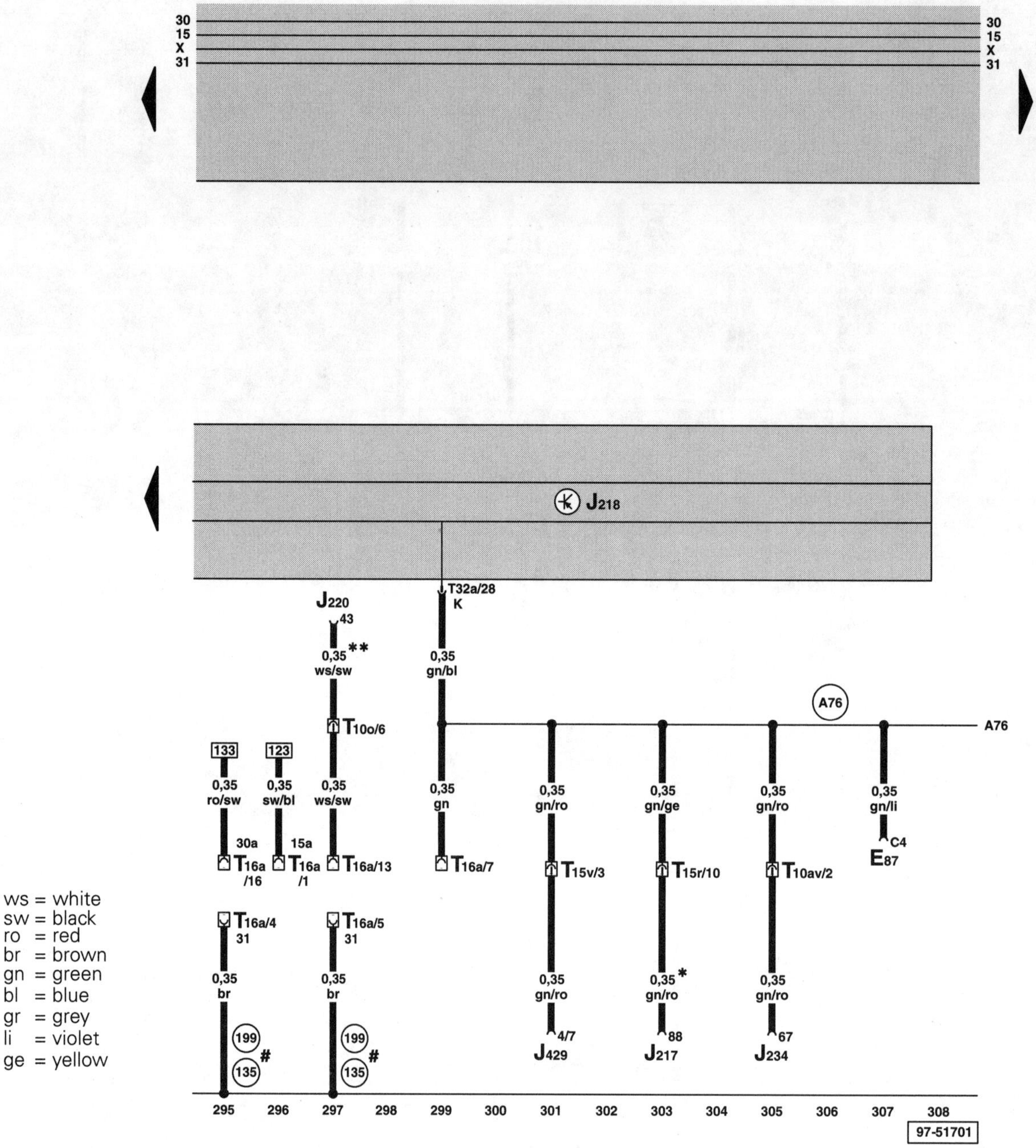

E87 - A/C Control Head
J217 - Transmission Control Module (TCM)
J218 - Instrument Cluster Combination Processor
J220 - Motronic Engine Control Module (ECM)
J234 - Airbag Control Module
J429 - Control module for central locking
T10o - Connector 10 pin, brown, connector station electronic box plenum chamber
T10av - Connector 10 pin, yellow, connector station A pillar, right
T15r - Connector 15 pin, blue, connector station A pillar, right
T15v - Connector 15 pin, orange, connector station A pillar, right
T16a - Connector 16 pin, Data Link Connector (DLC)
T32a - Connector 32 pin, green, on instrument cluster

(135) - Ground connection -2-, in instrument panel wiring harness
(199) - Ground connection -3-, in instrument panel wiring harness
(A76) - Connector (K-diagnosis wire), in instrument panel wiring harness

* - Vehicles with automatic transmission
** - Vehicles with 6 cylinder engine
# - Both are possible

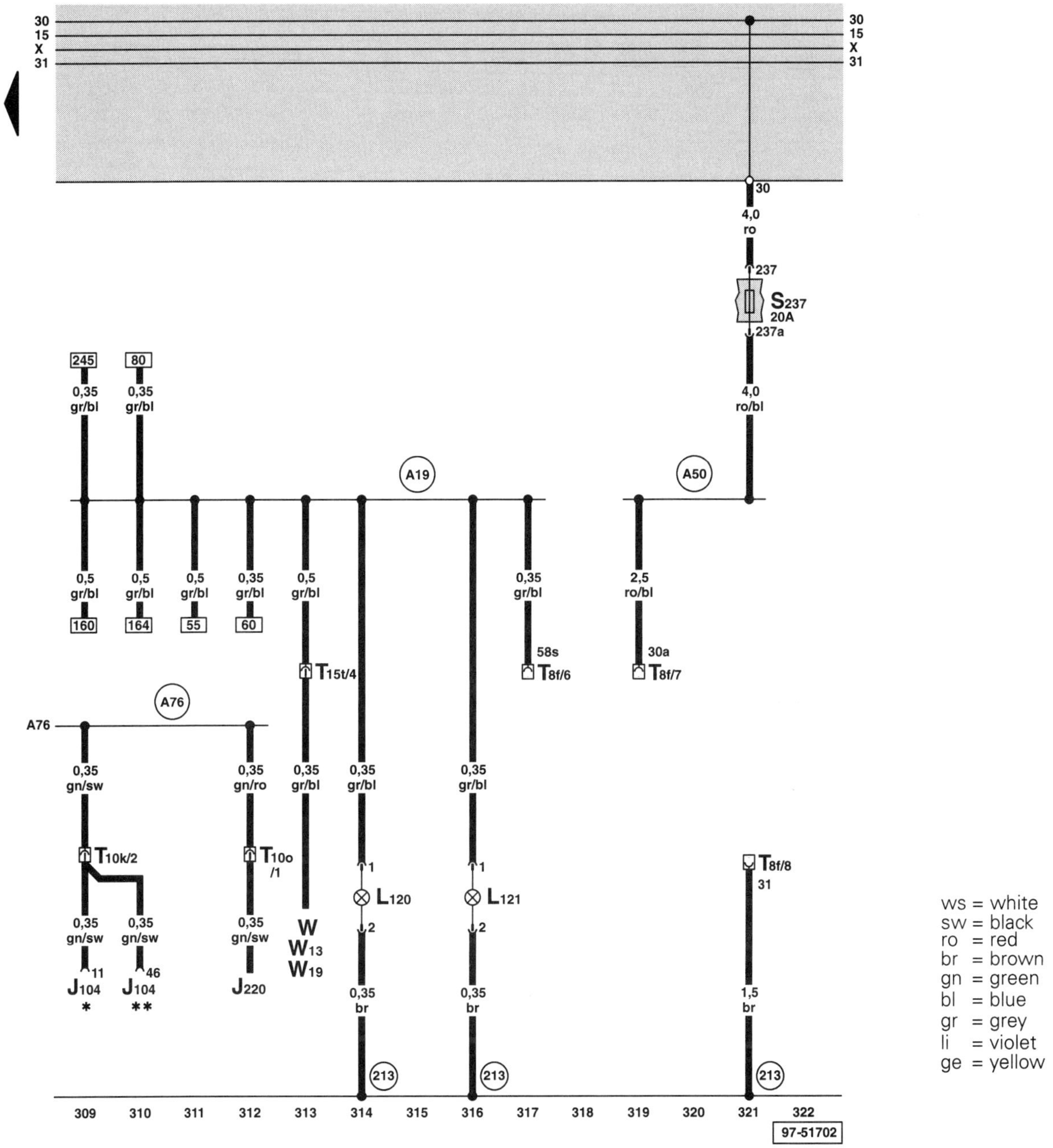

J104	-	ABS Control Module (w/EDL)
J220	-	Motronic Engine Control Module (ECM)
L120	-	Illumination for storage tray
L121	-	Illumination for cup holder
S237	-	Fuse 37 in fuse holder
T8f	-	Connector 8 pin, black, connector radio III
T10k	-	Connector 10 pin, orange, connector station A pillar, left
T10o	-	Connector 10 pin, brown, connector station electronic box plenum chamber
T15t	-	Connector 15 pin, grey, connector station A pillar, left
W	-	Front Interior Light
W13	-	Right Front Map/Reading Light
W19	-	Left Front Reading Light

(213)	-	Ground connection, in instrument panel wiring harness
(A19)	-	Wire connection (58s), in instrument panel wiring harness
(A50)	-	Plus connection (30as), in instrument panel wiring harness
(A76)	-	Connector (K-diagnosis wire), in instrument panel wiring harness
*	-	Vehicles with ABS 5.3
**	-	Vehicles with ABS and Electronic Stability Program (ESP)

THIS PAGE INTENTIONALLY LEFT
BLANK

(2,8 l - Injection Engine, 6 Cylinder), Code ATQ

2000 m. y.

Fuse Panel

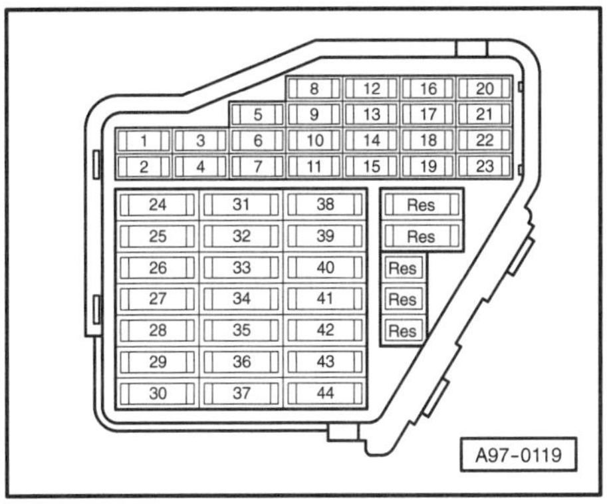

Fuse Colors:

30 A - Green
25 A - White
20 A - Yellow
15 A - Blue
10 A - Red
7,5 A - Brown
5 A - Beige

Starting with fuse position 23, fuses in the fuse holder are identified with 223 in the wiring diagram.

Micro Central Electric Panel

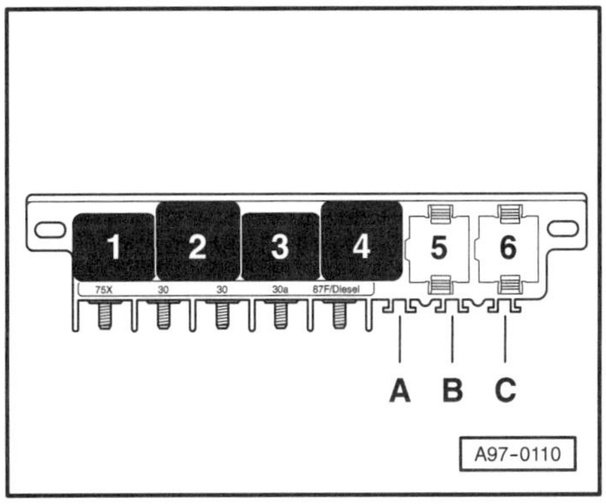

Relay Location:

4 -Fuel Pump (FP) Relay , J17

13 - Fold Relay Panel

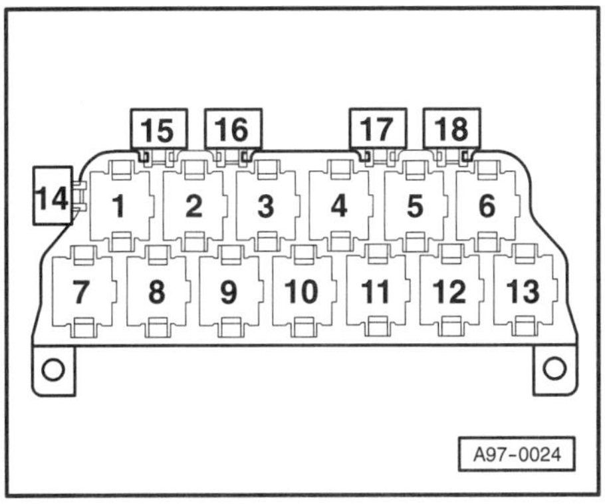

Relay Location:

4 -Starting Interlock Relay, J207

8 - Fold Relay Panel

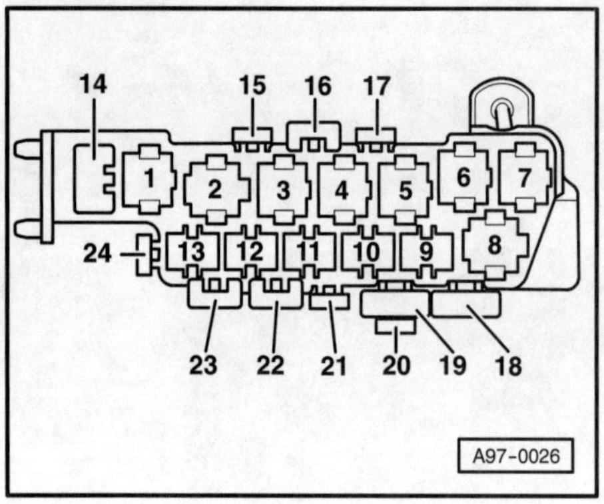

Fuse Location:

20 -Coolant Fan Fuse, S42

21 -Control module fuse for coolant fan, S142

3 - Fold Relay Panel in E-Box plenum chamber

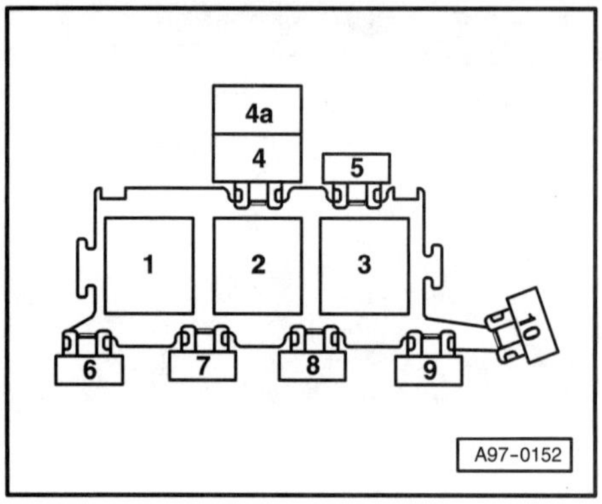

Relay Location:

2 -Secondary Air Injection (AIR) Pump Relay, J299

7 -Fuse for secondary air pump , S130

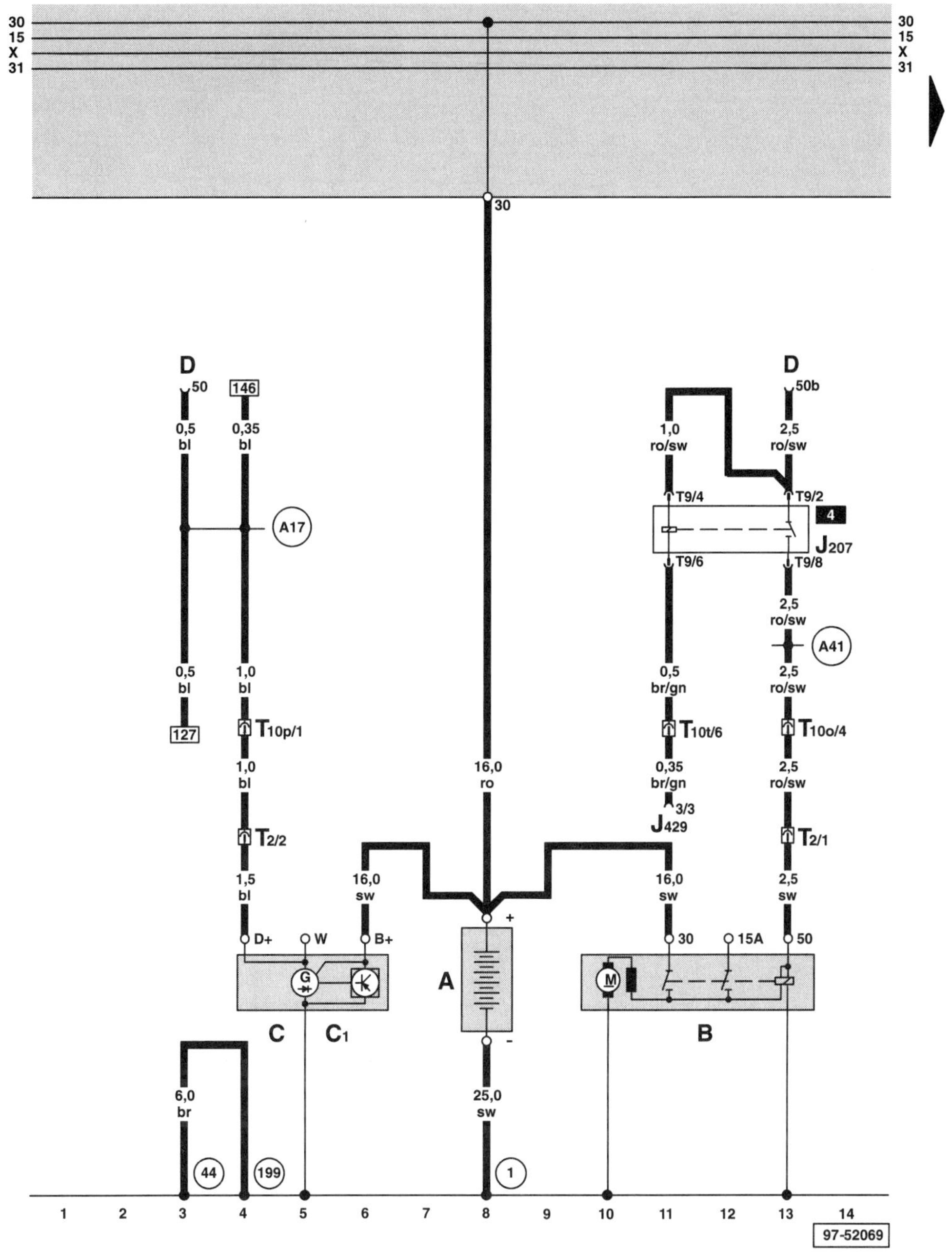

ws = white
sw = black
ro = red
br = brown
gn = green
bl = blue
gr = grey
li = violet
ge = yellow

A - Battery
B - Starter
C - Generator (GEN)
C1 - Voltage Regulator (VR)
D - Ignition/Starter Switch
J207 - Starting Interlock Relay
J429 - Control module for central locking
T2 - Connector 2 pin, grey, in engine compartment, right
T9 - Connector 9 pin, brown, on starting interlock relay
T10o - Connector 10 pin, brown, connector station electronic box plenum chamber
T10p - Connector 10 pin, black, connector station electronic box plenum chamber
T10t - Connector 10 pin, orange, connector station A pillar, right

(1) - Ground strap, battery to body
(44) - Ground connection, on left A-pillar, lower part
(199) - Ground connection -3-, in instrument panel wiring harness
(A17) - Wire connection (61), in instrument panel wiring harness
(A41) - Plus connection (50), in instrument panel wiring harness

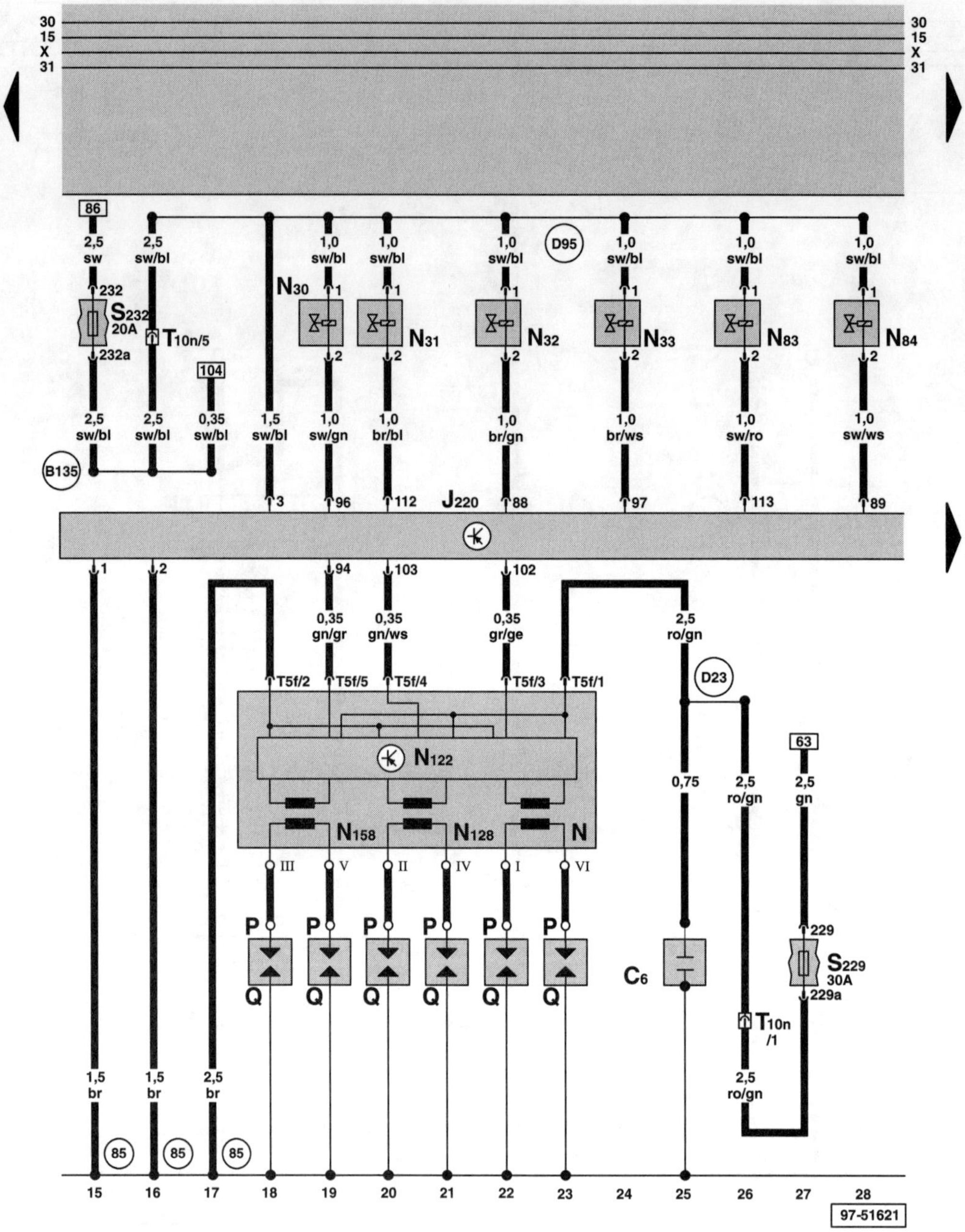

ws = white
sw = black
ro = red
br = brown
gn = green
bl = blue
gr = grey
li = violet
ge = yellow

C6	-	Distributor Ignition (DI) Capacitor
J220	-	Motronic Engine Control Module (ECM)
N	-	Ignition Coil
N30	-	Cylinder 1 Fuel Injector
N31	-	Cylinder 2 Fuel Injector
N32	-	Cylinder 3 Fuel Injector
N33	-	Cylinder 4 Fuel Injector
N83	-	Cylinder 5 Fuel Injector
N84	-	Cylinder 6 Fuel Injector
N122	-	Power Output Stage
N128	-	Ignition Coil 2
N158	-	Ignition Coil 3
P	-	Spark Plug Connectors
Q	-	Spark Plugs
S229	-	Fuse 29 in fuse holder
S232	-	Fuse 32 in fuse holder
T5f	-	Connector 5 pin, ignition coil unit
T10n	-	Connector 10 pin, orange, connector station electronic box plenum chamber

(85)	-	Ground connection -1-, in engine compartment wiring harness
(B135)	-	Connector (15a), in wiring harness interior
(D23)	-	Connector (over fuse 229), in wiring harness front, right
(D95)	-	Wire connection (injectors), in engine compartment wiring harness

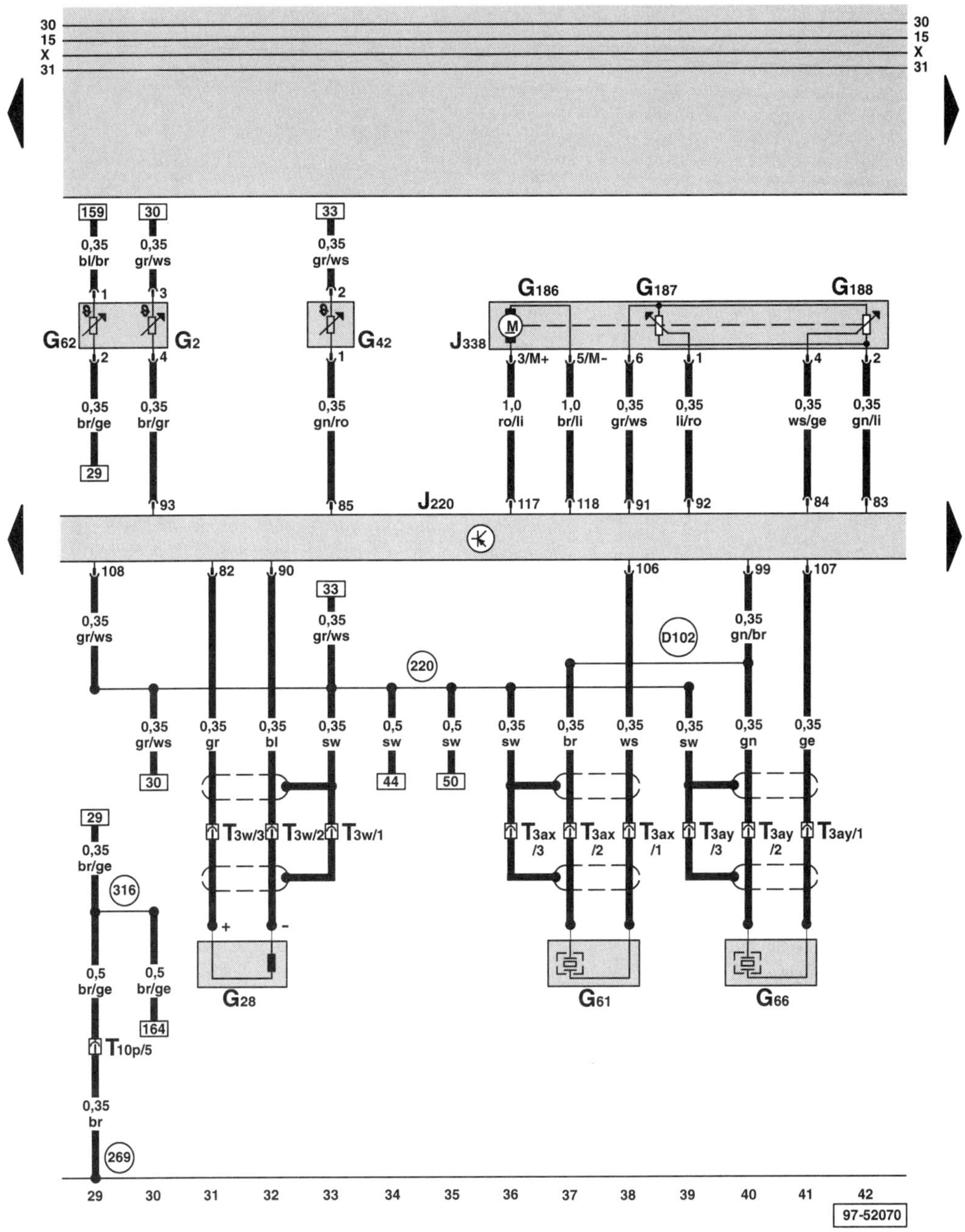

ws = white
sw = black
ro = red
br = brown
gn = green
bl = blue
gr = grey
li = violet
ge = yellow

- G2 - Engine Coolant Temperature (ECT) Sensor
- G28 - Engine Speed (RPM) Sensor
- G42 - Intake Air Temperature (IAT) Sensor
- G61 - Knock Sensor (KS) 1
- G62 - Engine Coolant Temperature (ECT) Sensor
- G66 - Knock Sensor (KS) 2
- G186 - Throttle drive (power accelerator actuation)
- G187 - Angel sensor -1- for throttle drive (power accelerator actuation)
- G188 - Angel sensor -2- for throttle drive (power accelerator actuation)
- J220 - Motronic Engine Control Module (ECM)
- J338 - Throttle Valve Control Module
- T3w - Connector 3 pin, grey, on engine speed sensor
- T3ax - Connector 3 pin, brown, on knock sensor 1
- T3ay - Connector 3 pin, brown, on knock sensor 2
- T10p - Connector 10 pin, black, connector station electronic box plenum chamber

- (220) - Ground connection (sensor ground), in engine compartment wiring harness
- (269) - Ground connector (sensor ground) -1-, in instrument panel wiring harness
- (316) - Sensor Ground Connection 2 (in engine wiring harness)
- (D102) - Wire connection -2-, in engine compartment wiring harness

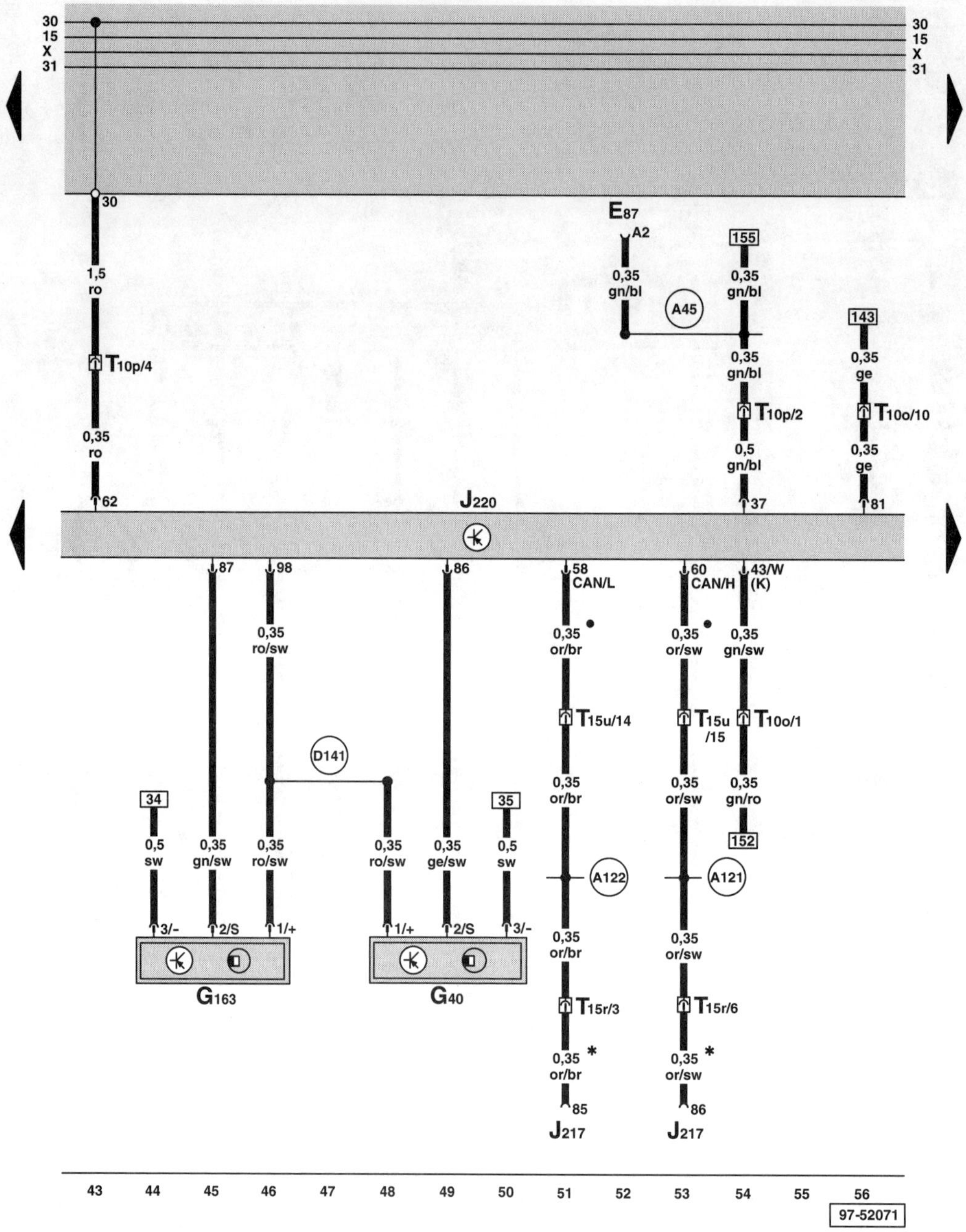

ws = white
sw = black
ro = red
br = brown
gn = green
bl = blue
gr = grey
li = violet
ge = yellow
or = orange

E87 - A/C Control Head
G40 - Camshaft Position (CMP) Sensor
G163 - Camshaft Position (CMP) Sensor 2
J217 - Transmission Control Module (TCM)
J220 - Motronic Engine Control Module (ECM)
T10o - Connector 10 pin, brown, connector station electronic box plenum chamber
T10p - Connector 10 pin, black, connector station electronic box plenum chamber
T15r - Connector 15 pin, blue, connector station A pillar, right
T15u - Connector 15 pin, red, connector station electronic box plenum chamber

(A45) - Wire connection (RPM-signal), in instrument panel wiring harness
(A121) - Connection (high bus) in instrument panel wiring harness
(A122) - Connection (low bus) in instrument panel wiring harness
(D141) - Connector (5V), in wiring harness, engine pre-wiring
• - CAN-Bus (Data-Bus)
* - Vehicles with automatic transmission

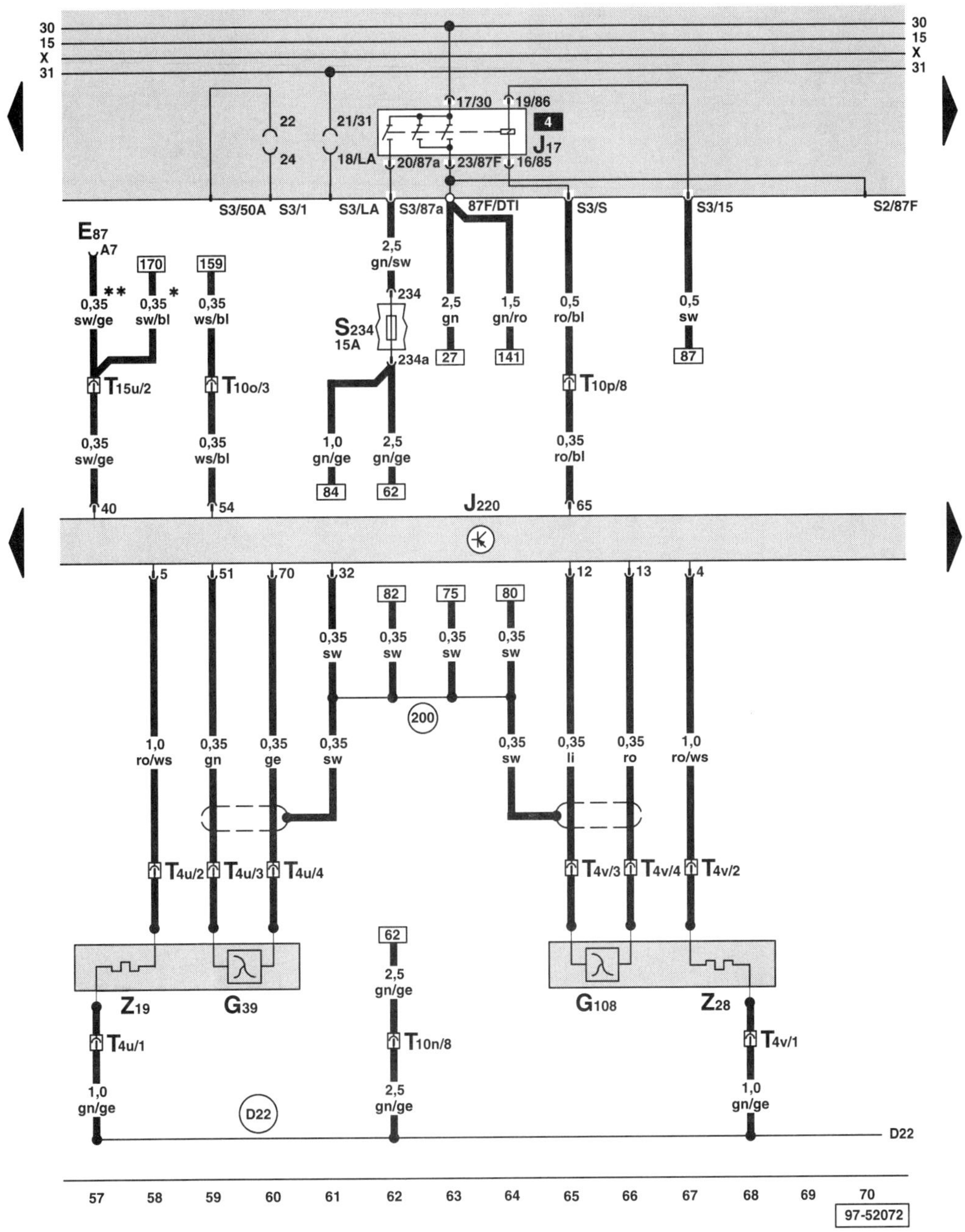

ws = white
sw = black
ro = red
br = brown
gn = green
bl = blue
gr = grey
li = violet
ge = yellow

E87 - A/C Control Head
G39 - Heated Oxygen Sensor (HO2S) 2
G108 - Heated Oxygen Sensor (HO2S) 2
J17 - Fuel Pump (FP) Relay
J220 - Motronic Engine Control Module (ECM)
S234 - Fuse 34 in fuse holder
T4u - Connector 4 pin, black, heated oxygen sensor
T4v - Connector 4 pin, black, heated oxygen sensor 2
T10n - Connector 10 pin, orange, connector station electronic box plenum chamber
T10o - Connector 10 pin, brown, connector station electronic box plenum chamber
T10p - Connector 10 pin, black, connector station electronic box plenum chamber
T15u - Connector 15 pin, red, connector station electronic box plenum chamber
Z19 - Oxygen Sensor (O2S) Heater
Z28 - Oxygen Sensor (O2S) 2 Heater

(200) - Ground connection (shielding), in engine compartment wiring harness
(D22) - Connector (over fuse 234), in wiring harness front, right

* - Vehicles with front wheel drive
** - Vehicles with all wheel drive and manually shifted transmission

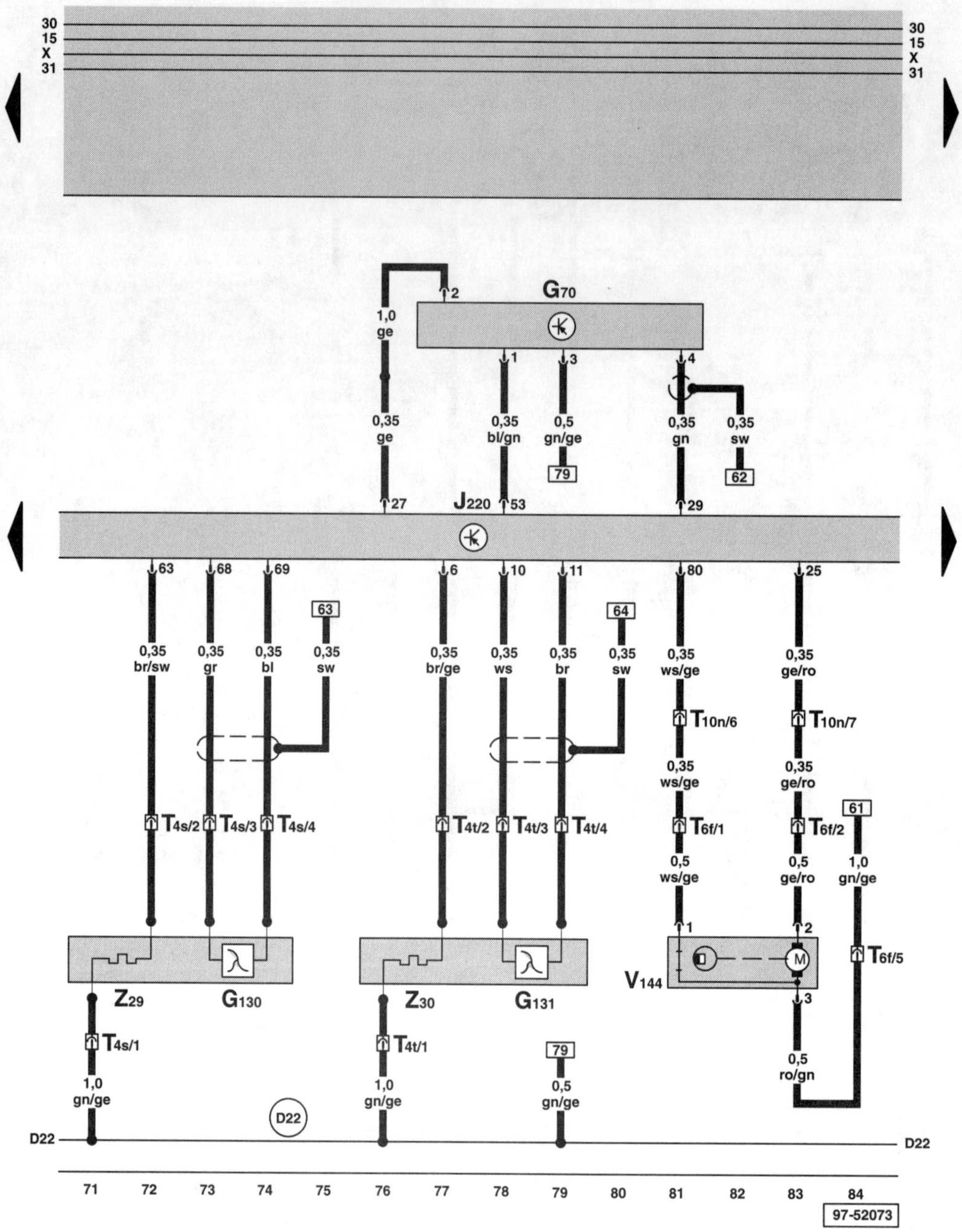

ws = white
sw = black
ro = red
br = brown
gn = green
bl = blue
gr = grey
li = violet
ge = yellow

G70 - Mass Air Flow (MAF) Sensor
G130 - Heated Oxygen Sensor (O2S) Behind Three Way Catalytic Converter (TWC)
G131 - Heated Oxygen Sensor (O2S) 2 Behind Three Way Catalytic Converter (TWC)
J220 - Motronic Engine Control Module (ECM)
T4s - Connector 4 pin, green, near oxygen sensor, behind Three Way Catalytic Converter (TWC)
T4t - Connector 4 pin, brown, near oxygen sensor 2, behind Three Way Catalytic Converter (TWC)
T6f - Connector 6 pin, darkbrown, connector station A pillar, right
T10n - Connector 10 pin, orange, connector station electronic box plenum chamber
V144 - Leak Detection Pump (LDP)
Z29 - Oxygen Sensor (O2S) Heater
Z30 - Oxygen Sensor (O2S) 2 Heater

(D22) - Connector (over fuse 234), in wiring harness front, right

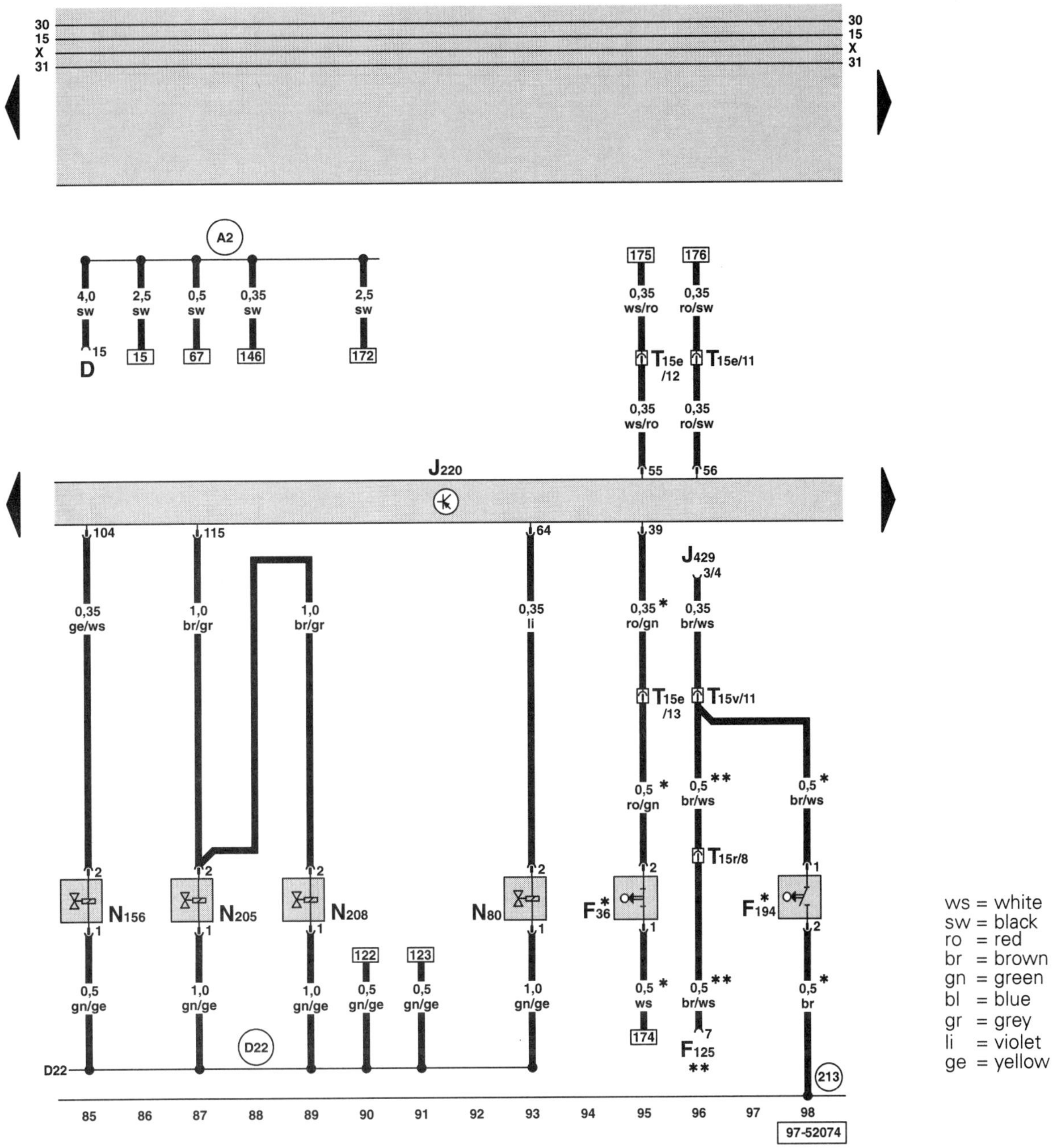

D - Ignition/Starter Switch
F36 - Clutch Vacuum Vent Valve Switch
F125 - Multi-Function Transmission Range (TR) Switch
F194 - Clutch Pedal Position (CPP) Switch
J220 - Motronic Engine Control Module (ECM)
J429 - Control module for central locking
N80 - Evaporative Emission (EVAP) Canister Purge Regulator Valve
N156 - Intake Manifold Change-Over Valve
N205 - Valve -1- for camshaft adjustment
N208 - Valve -2- for camshaft adjustment
T15e - Connector 15 pin, white, connector station electronic box plenum chamber
T15r - Connector 15 pin, blue, connector station A pillar, right
T15v - Connector 15 pin, orange, connector station A pillar, right

(213) - Ground connection, in instrument panel wiring harness
(A2) - Plus connection (15), in instrument panel wiring harness
(D22) - Connector (over fuse 234), in wiring harness front, right
* - Vehicles with manually shifted transmission
** - Vehicles with automatic transmission

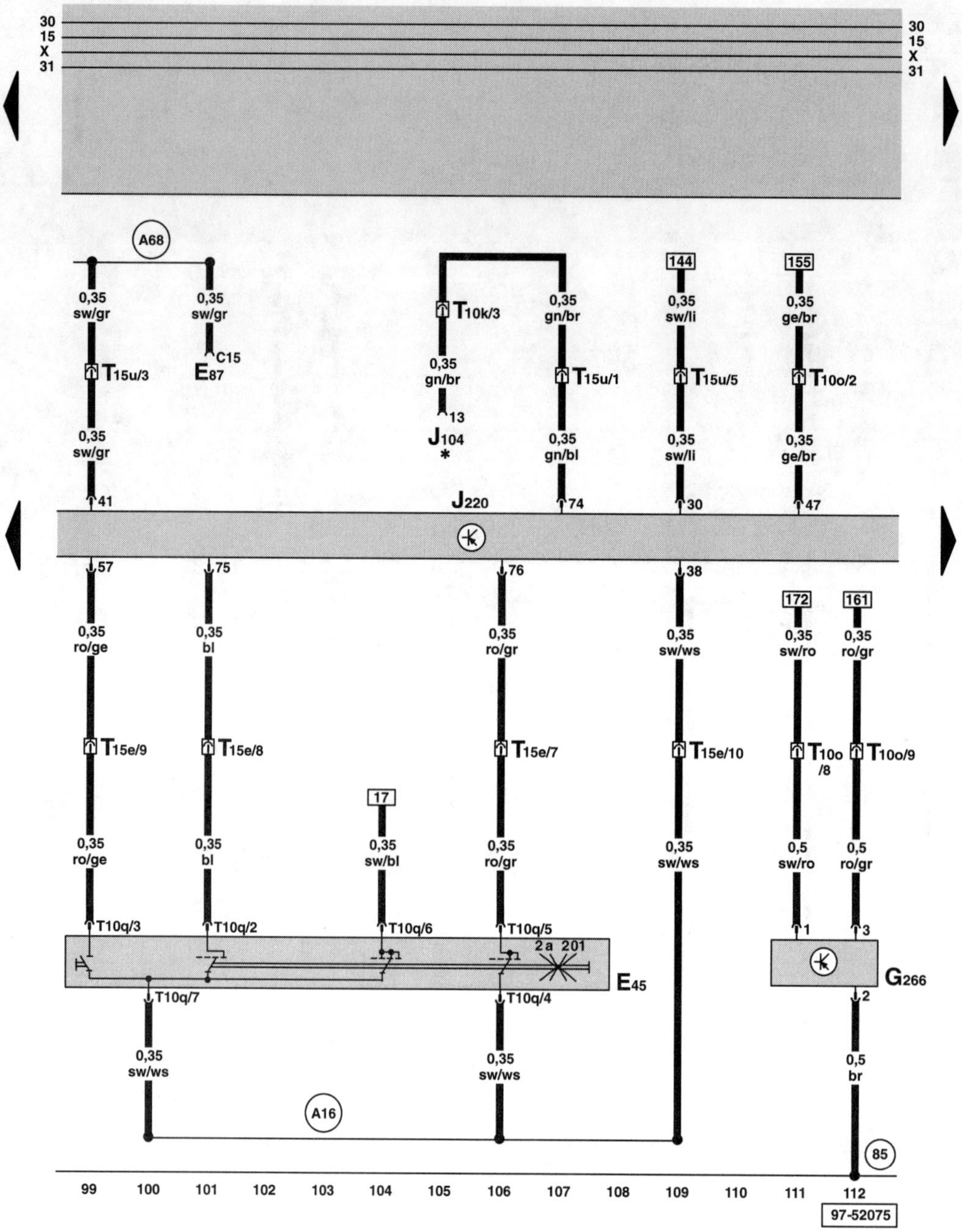

ws = white
sw = black
ro = red
br = brown
gn = green
bl = blue
gr = grey
li = violet
ge = yellow

E45 - Cruise Control Switch
E87 - A/C Control Head
G266 - Oil Level Thermal Sensor (SRI)
J104 - ABS Control Module (w/EDL)
J220 - Motronic Engine Control Module (ECM)
T10k - Connector 10 pin, orange, connector station A pillar, left
T10o - Connector 10 pin, brown, connector station electronic box plenum chamber
T10q - Connector 10 pin, black, on cruise control switch
T15e - Connector 15 pin, white, connector station electronic box plenum chamber
T15u - Connector 15 pin, red, connector station electronic box plenum chamber

(85) - Ground connection -1-, in engine compartment wiring harness
(A16) - Wire connection (cruise control), in instrument panel wiring harness
(A68) - Connector (C15, A/C), in instrument panel wiring harness
* - Vehicles with ABS 5.3

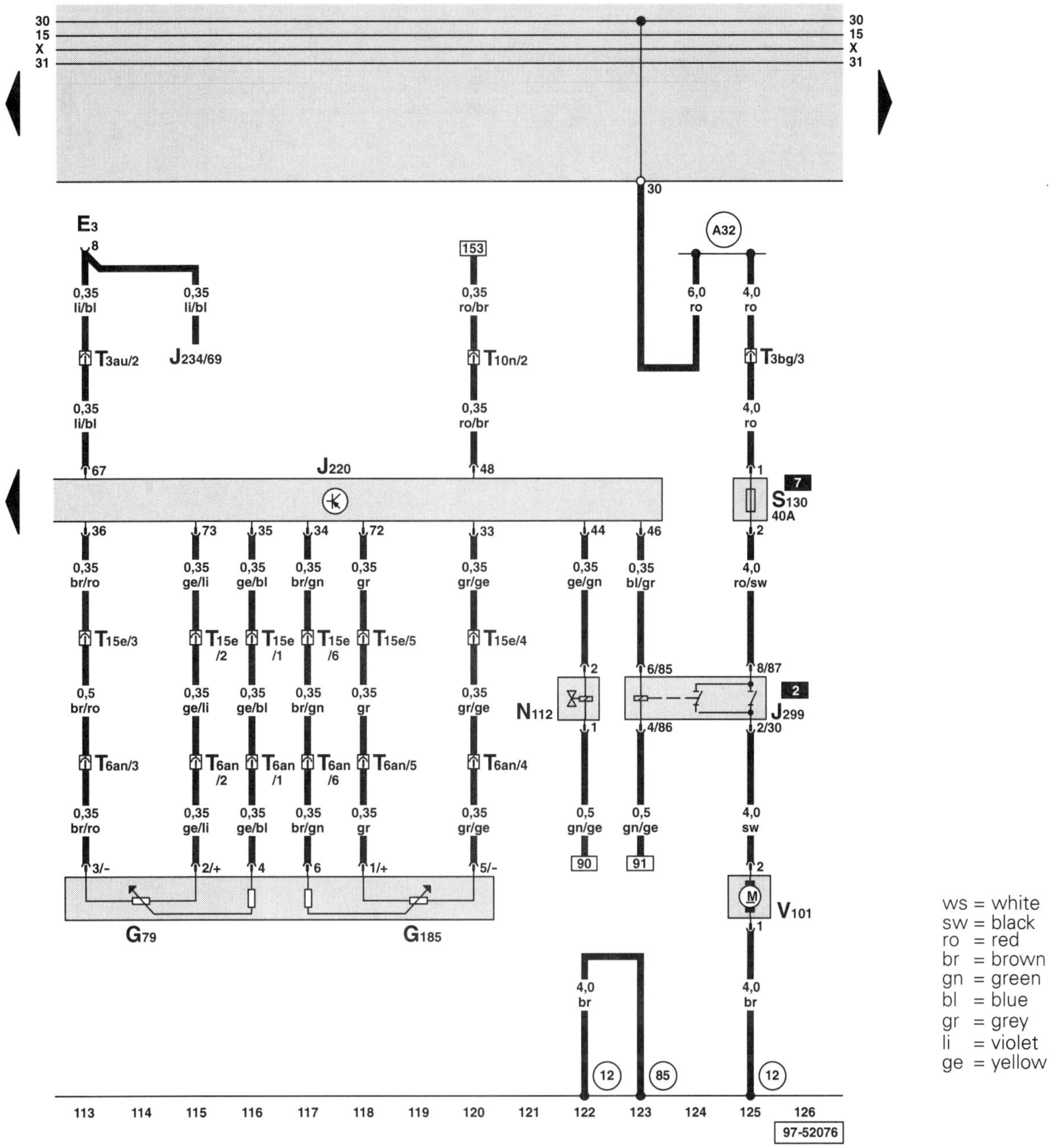

E3 - Emergency Flasher Switch
G79 - Throttle Position (TP) Sensor
G185 - Sensor -2- for accelerator pedal position
J220 - Motronic Engine Control Module (ECM)
J234 - Airbag Control Module
J299 - Secondary Air Injection (AIR) Pump Relay
N112 - Secondary Air Injection (AIR) Solenoid Valve
S130 - Fuse for secondary air pump
T3au - Connector 3 pin, red, connector station electronic box plenum chamber
T3bg - Connector 3 pin, white, connector station electronic box plenum chamber
T6an - Connector 6 pin, in front footwell, left
T10n - Connector 10 pin, orange, connector station electronic box plenum chamber
T15e - Connector 15 pin, white, connector station electronic box plenum chamber
V101 - Secondary Air Injection (AIR) Pump Motor

(12) - Ground connection, in engine compartment, left
(85) - Ground connection -1-, in engine compartment wiring harness
(A32) - Plus connection (30), in instrument panel wiring harness

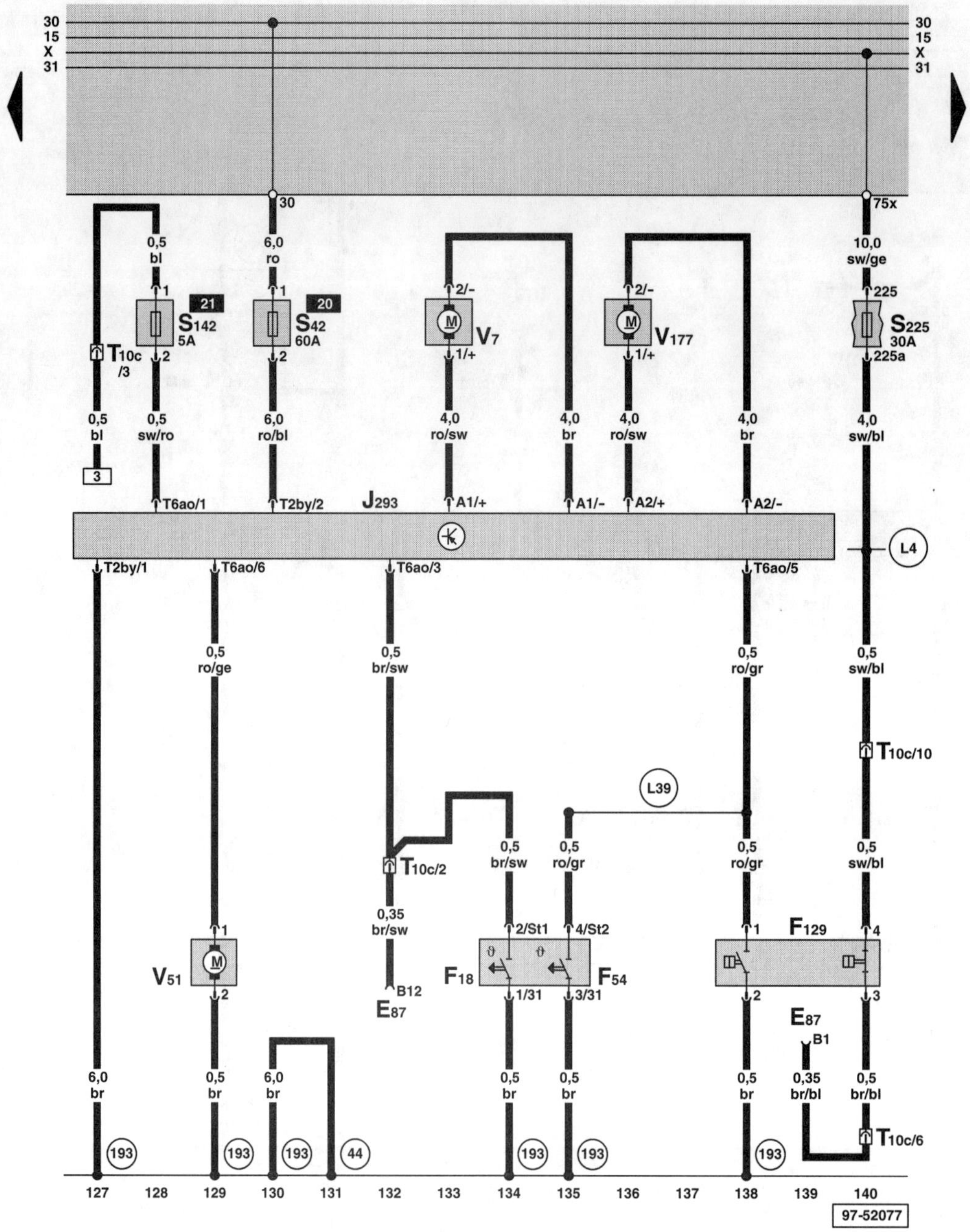

ws = white
sw = black
ro = red
br = brown
gn = green
bl = blue
gr = grey
li = violet
ge = yellow

E87 - A/C Control Head
F18 - Coolant Fan Control (FC) Thermal Switch
F54 - Coolant Fan Control (FC) Thermal Switch
F129 - A/C Pressure Switch
J293 - Coolant FC (Fan Control) Control Module
S42 - Coolant Fan Fuse
S142 - Control module fuse for coolant fan
S225 - Fuse 25 in fuse holder
T2by - Connector 2 pin, on Coolant FC (Fan Control) Control Module
T6ao - Connector 6 pin, on Coolant FC (Fan Control) Control Module
T10c - Connector 10 pin, grey, connector station A pillar, left
V7 - Coolant Fan
V51 - After-Run Coolant Pump
V177 - Coolant fan -2-

(44) - Ground connection, on left A-pillar, lower part
(193) - Ground connection -1-, in coolant fan wiring harness
(L4) - Wire connection (75al), in A/C wiring harness
(L39) - Wire connection -1- (A/C Pressure Switch), in Climatronic wiring harness

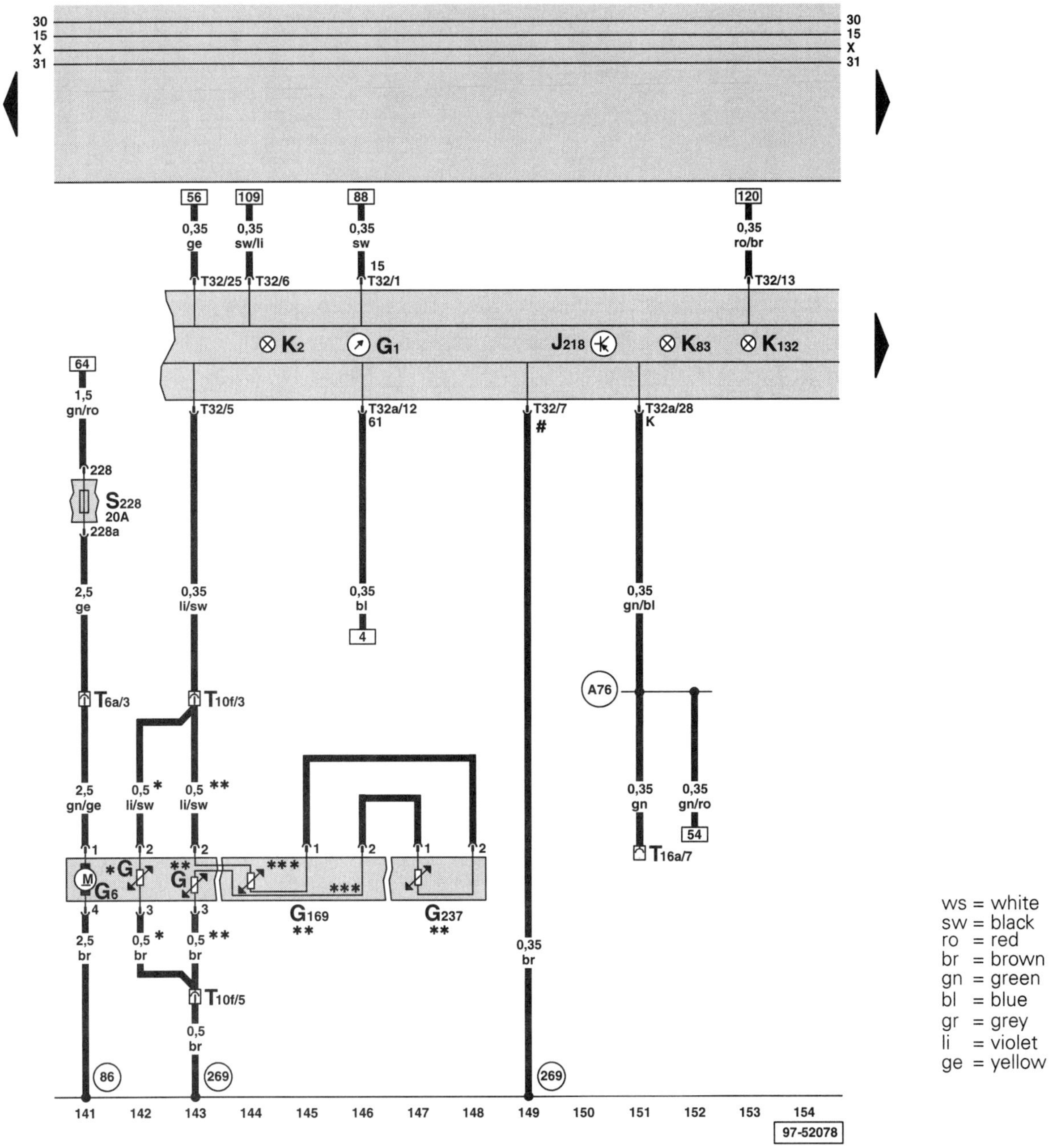

G - Sender for fuel gauge
G1 - Fuel Gauge
G6 - Fuel Pump (FP)
G169 - Sender 2 for fuel gauge
G237 - Sensor -3- for fuel supply
J218 - Instrument Cluster Combination Processor
K2 - Generator (GEN) Warning Light
K83 - Malfunction Indicator Lamp (MIL)
K132 - Fault light for power accelerator activation
S228 - Fuse 28 in fuse holder
T6a - Connector 6 pin, blue, connector station A pillar, left
T10f - Connector 10 pin, brown, connector station A pillar, left
T16a - Connector 16 pin, Data Link Connector
T32 - Connector 32 pin, blue, on instrument cluster
T32a - Connector 32 pin, green, on instrument cluster

(86) - Ground connection -1-, in rear wiring harness

(269) - Ground connector (sensor ground) -1-, in instrument panel wiring harness

(A76) - Connector (K-diagnosis wire), in instrument panel wiring harness

* - Vehicles with front wheel drive

** - Vehicles with all wheel drive

*** - Wiring inside fuel tank

- Sensor ground, output

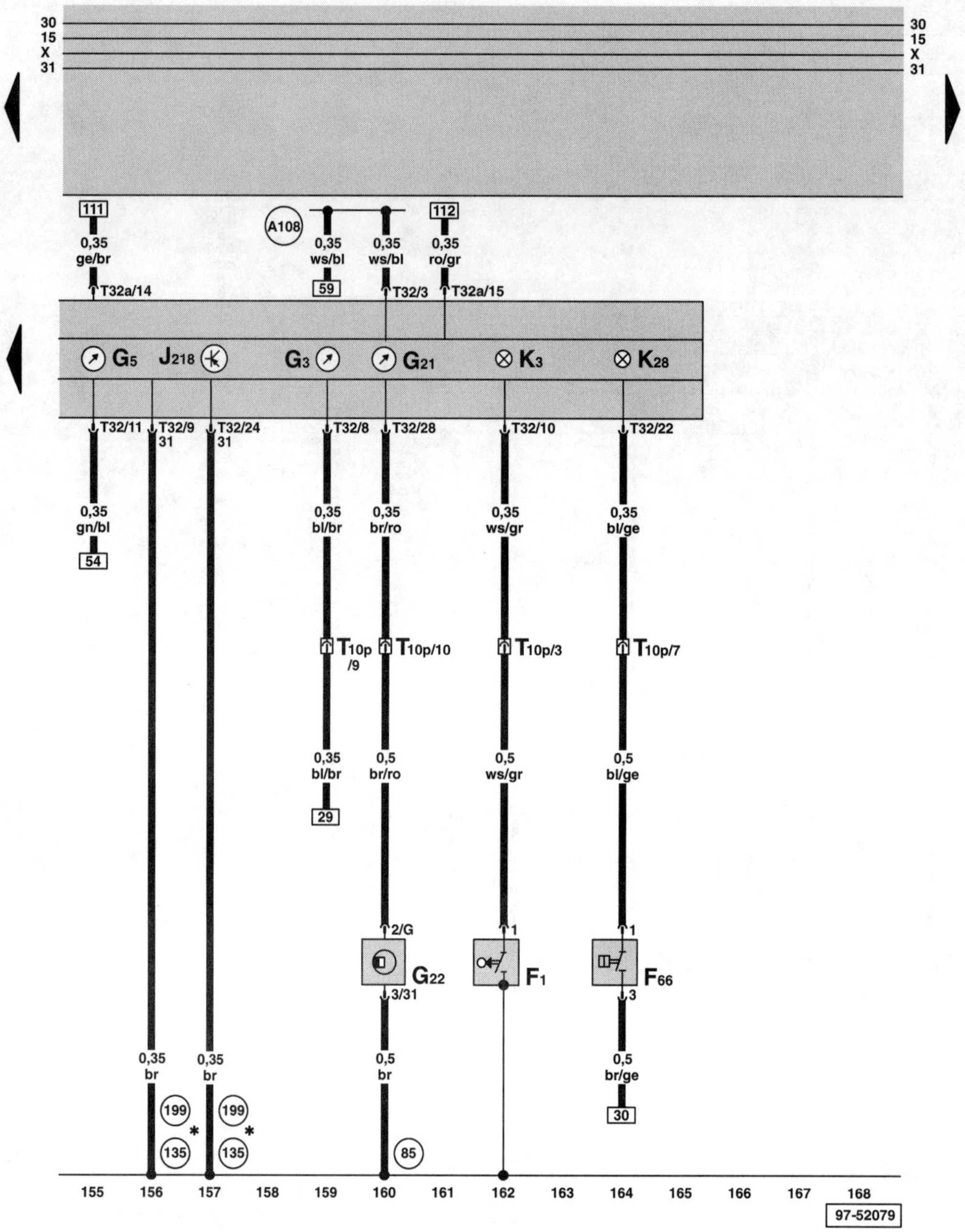

ws = white
sw = black
ro = red
br = brown
gn = green
bl = blue
gr = grey
li = violet
ge = yellow

F1 - Oil Pressure Switch
F66 - Engine Coolant Level (ECL) Warning Switch
G3 - Engine Coolant Temperature (ECT) Gauge
G5 - Tachometer
G21 - Speedometer
G22 - Speedometer Vehicle Speed Sensor (VSS)
J218 - Instrument Cluster Combination Processor
K3 - Oil Pressure Warning Light
K28 - Engine Coolant Level/Temperature (ECL/ECT) Warning Light
T10p - Connector 10 pin, black, connector station electronic box plenum chamber
T32 - Connector 32 pin, blue, on instrument cluster
T32a - Connector 32 pin, green, on instrument cluster

(85) - Ground connection -1-, in engine compartment wiring harness
(135) - Ground connection -2-, in instrument panel wiring harness
(199) - Ground connection -3-, in instrument panel wiring harness
(A108) - Connector (speed signal), in instrument panel wiring harness
* - Both are possible

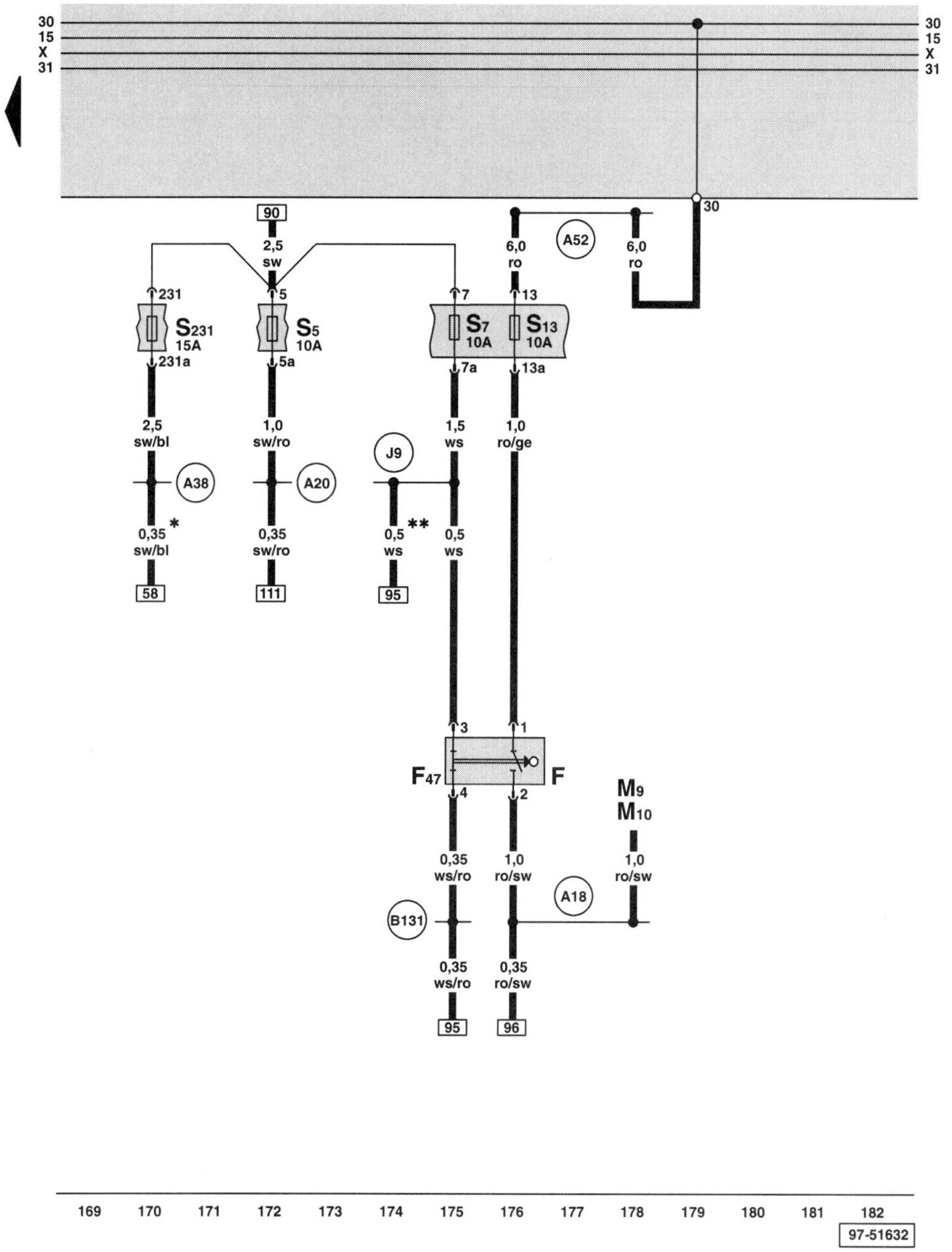

ws = white
sw = black
ro = red
br = brown
gn = green
bl = blue
gr = grey
li = violet
ge = yellow

F - Brake Light Switch
F47 - Vacuum Vent Valve, Brake
M9 - Left Brake Light
M10 - Right Brake Light
S5 - Fuse
S7 - Fuse
S13 - Fuse
S231 - Fuse 31 in fuse holder

(A18) - Wire connection (54), in instrument panel wiring harness
(A20) - Wire connection (15a), in instrument panel wiring harness
(A38) - Plus connection (15a II), in instrument panel wiring harness
(A52) - Plus connection -2- (30), in instrument panel wiring harness
(B131) - Connector (54), in wiring harness interior
(J9) - Wire connection -1- (15a), in ABS wiring harness

* - Vehicles with front wheel drive
** - Vehicles with manually shifted transmission

(2,7 l - Injection Engine, 6 Cylinder), Code APB

2000 m. y.

Fuse Panel

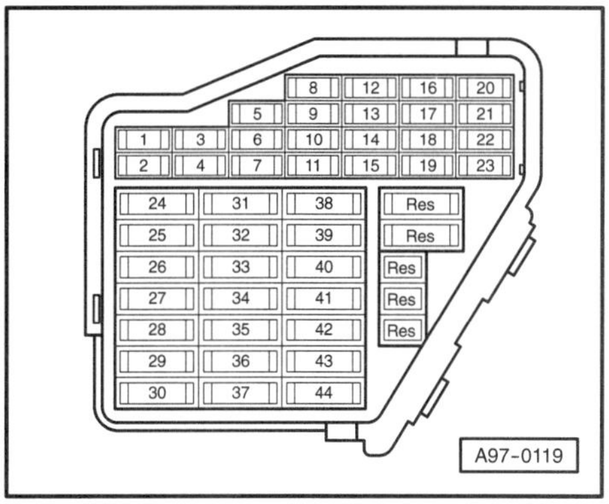

Fuse Colors:

30 A - Green
25 A - White
20 A - Yellow
15 A - Blue
10 A - Red
7,5 A - Brown
5 A - Beige

Starting with fuse position 23, fuses in the fuse holder are identified with 223 in the wiring diagram.

Micro Central Electric Panel

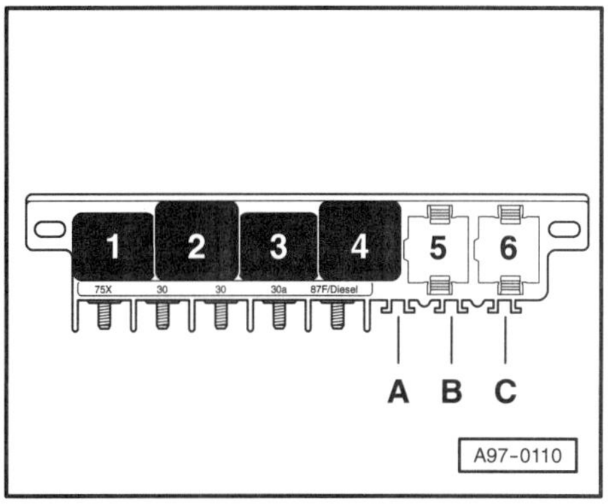

Relay Location:

4 -Fuel Pump (FP) Relay , J17

13 - Fold Relay Panel

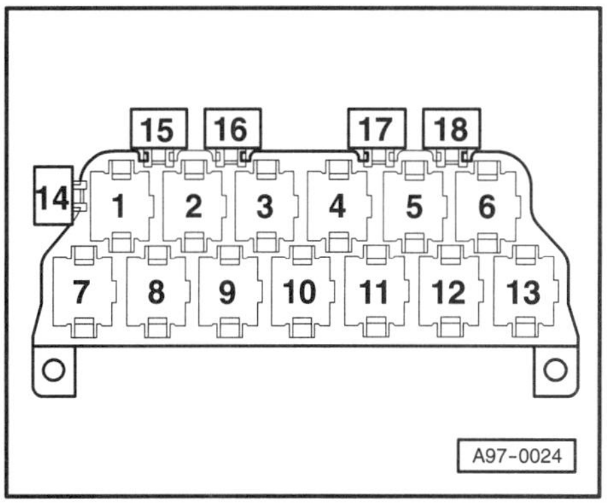

Relay Location:

4 -Starting Interlock Relay, J207

8 - Fold Relay Panel

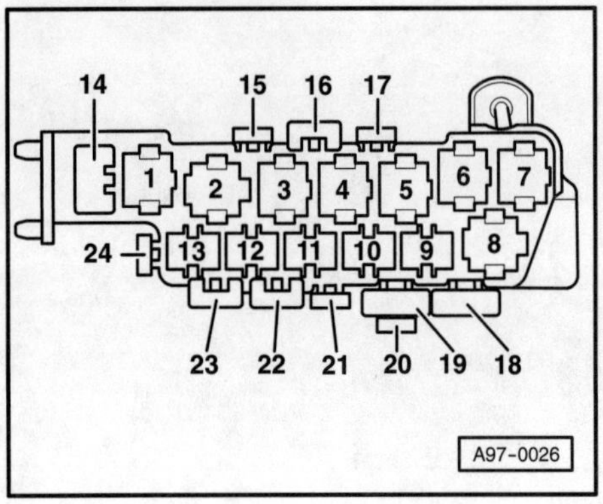

Fuse Location:

20 -Coolant Fan Fuse, S42

21 -Control module fuse for coolant fan, S142

3 - Fold Relay Panel in E-Box plenum chamber

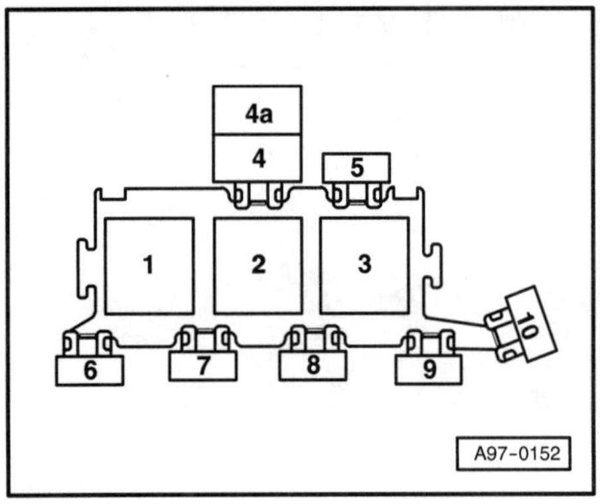

Relay Location:

2 -Secondary Air Injection (AIR) Pump Relay, J299

7 -Fuse for secondary air pump , S130

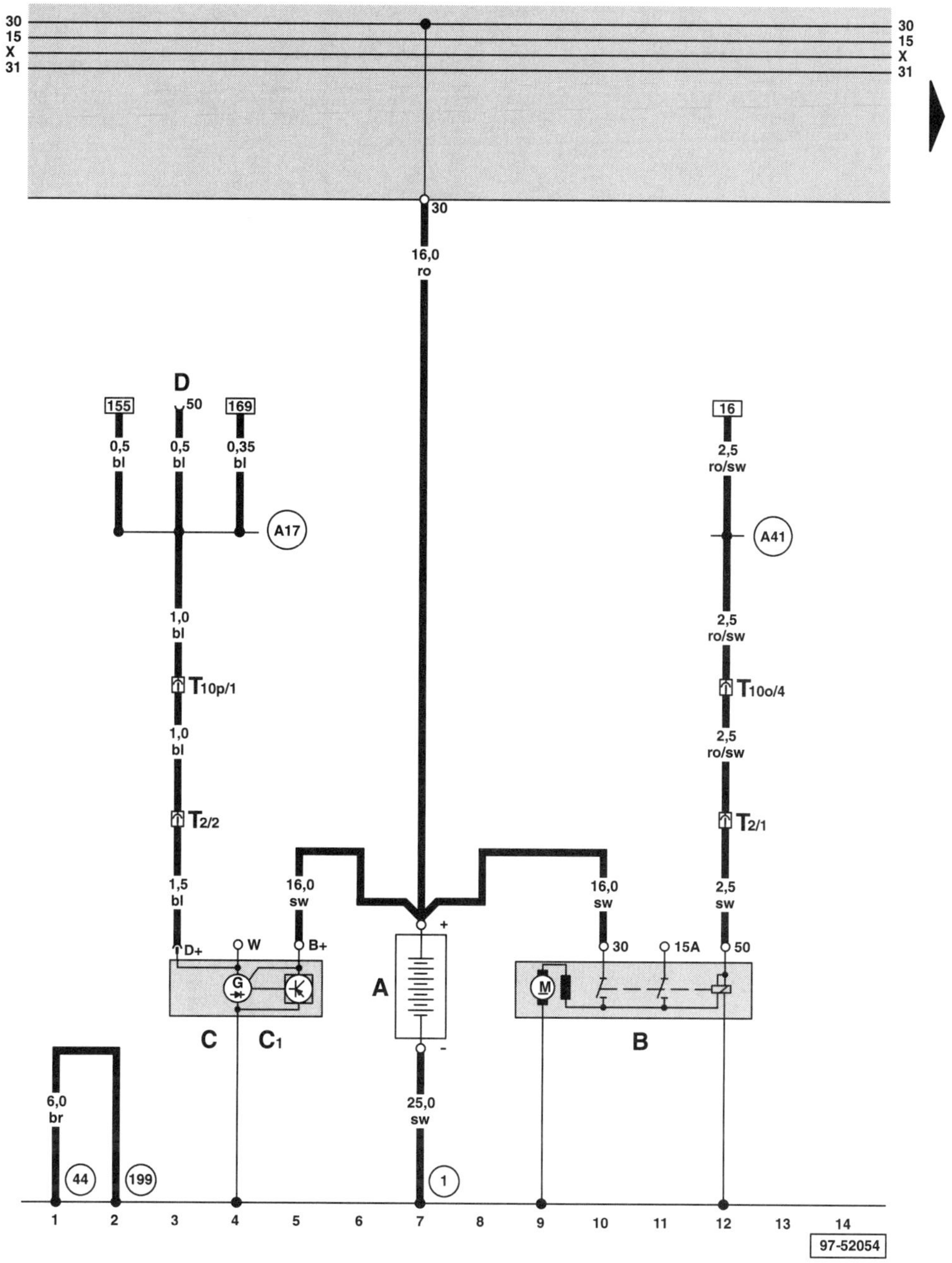

ws = white
sw = black
ro = red
br = brown
gn = green
bl = blue
gr = grey
li = violet
ge = yellow

A - Battery
B - Starter
C - Generator (GEN)
C1 - Voltage Regulator (VR)
D - Ignition/Starter Switch
T2 - Connector 2 pin, grey, in engine compartment, right
T10o - Connector 10 pin, brown, connector station electronic box plenum chamber
T10p - Connector 10 pin, black, connector station electronic box plenum chamber

(1) - Ground strap, battery to body
(44) - Ground connection, on left A-pillar, lower part
(199) - Ground connection -3-, in instrument panel wiring harness
(A17) - Wire connection (61), in instrument panel wiring harness
(A41) - Plus connection (50), in instrument panel wiring harness

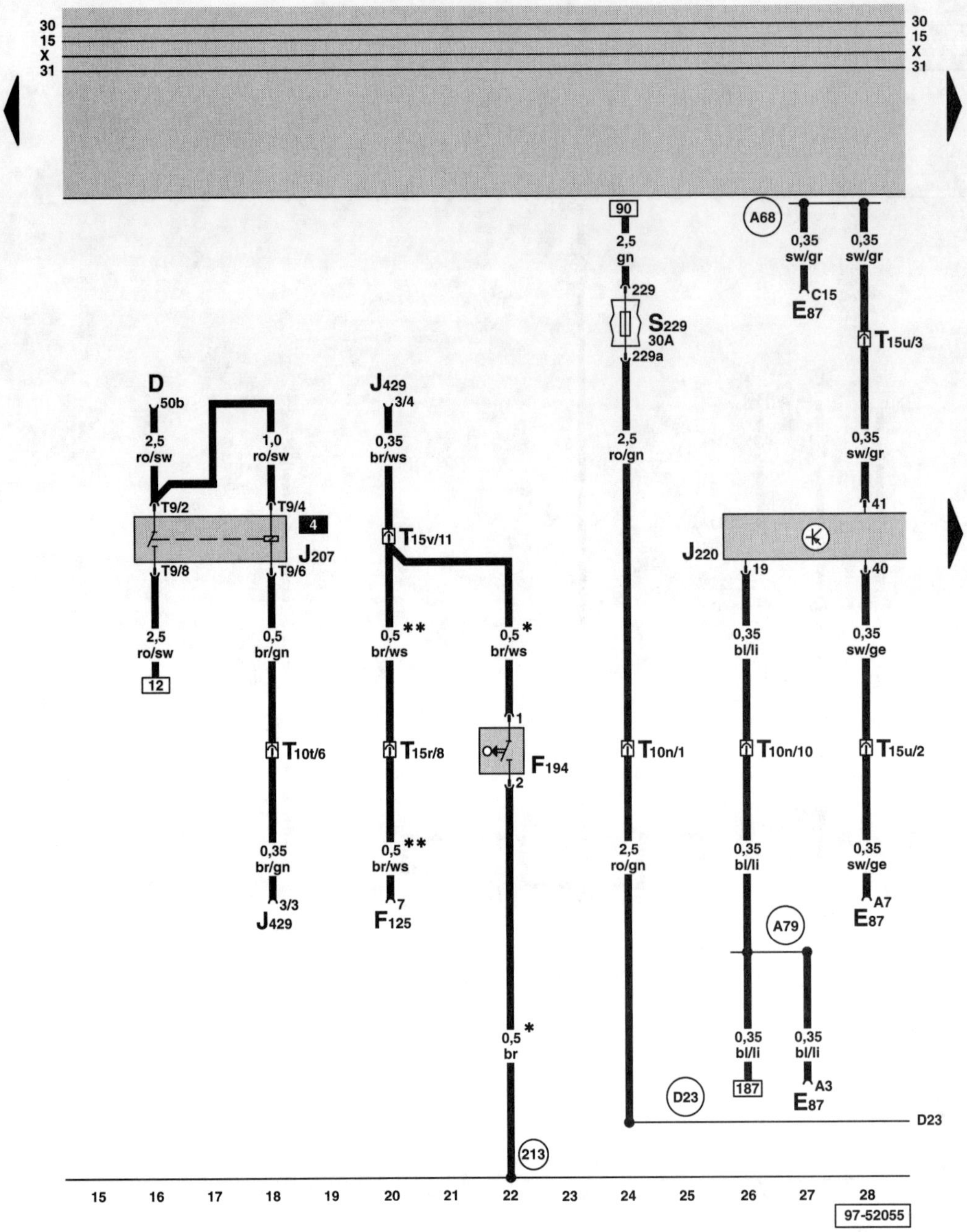

ws = white
sw = black
ro = red
br = brown
gn = green
bl = blue
gr = grey
li = violet
ge = yellow

- D - Ignition/Starter Switch
- E87 - A/C Control Head
- F125 - Multi-Function Transmission Range (TR) Switch
- F194 - Clutch Pedal Position (CPP) Switch
- J207 - Starting Interlock Relay
- J220 - Motronic Engine Control Module (ECM)
- J429 - Control module for central locking
- S229 - Fuse 29 in fuse holder
- T9 - Connector 9 pin, brown, on starting interlock relay
- T10n - Connector 10 pin, orange, connector station electronic box plenum chamber
- T10t - Connector 10 pin, orange, connector station A pillar, right
- T15r - Connector 15 pin, blue, connector station A pillar, right
- T15u - Connector 15 pin, red, connector station electronic box plenum chamber
- T15v - Connector 15 pin, orange, connector station A pillar, right

- (213) - Ground connection, in instrument panel wiring harness
- (A68) - Connector (C15, A/C), in instrument panel wiring harness
- (A79) - Connector (engine overheat warning light), in instrument panel wiring harness
- (D23) - Connector (over fuse 229), in wiring harness front, right
- * - Vehicles with manually shifted transmission
- ** - Vehicles with automatic transmission

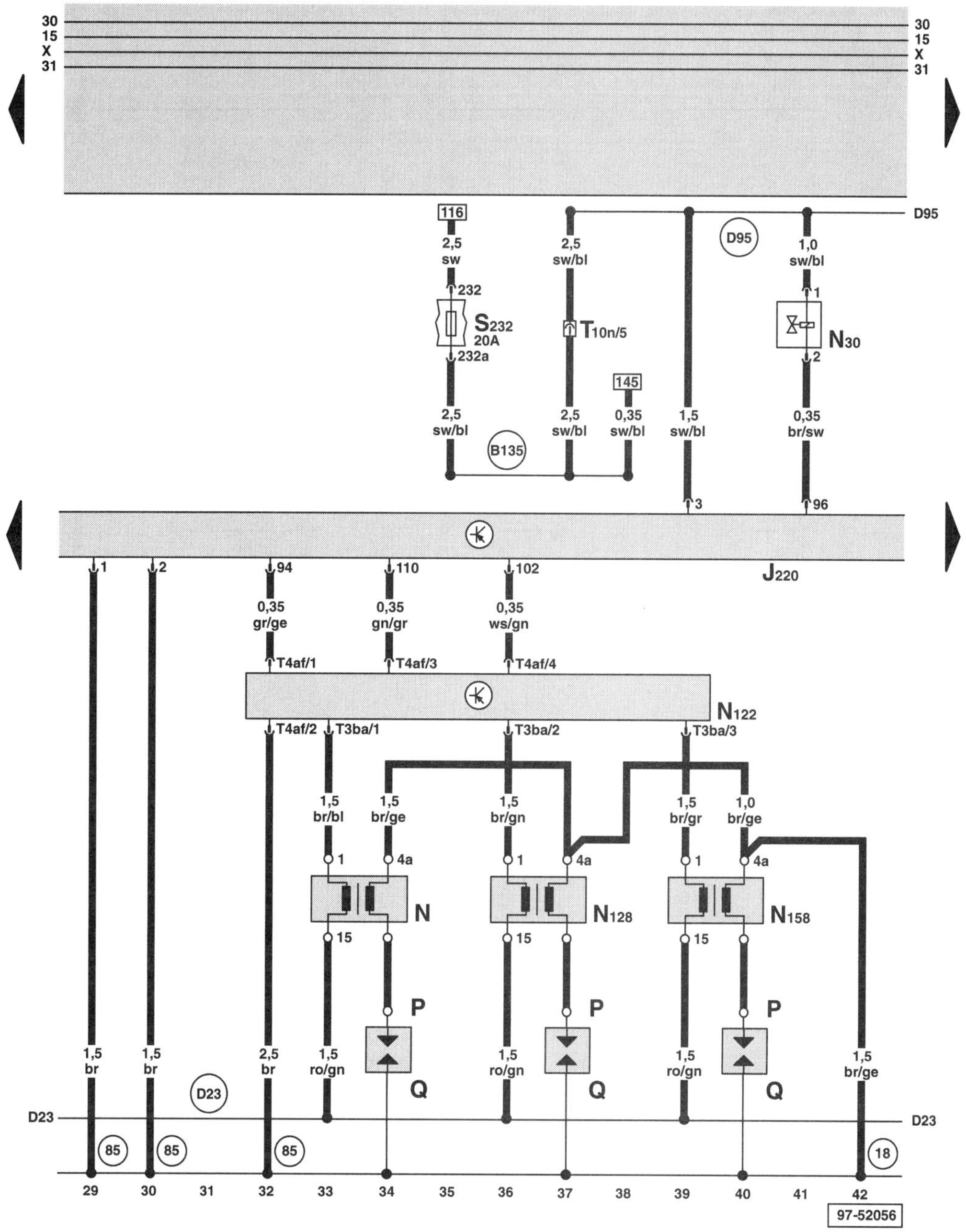

ws = white
sw = black
ro = red
br = brown
gn = green
bl = blue
gr = grey
li = violet
ge = yellow

J220 - Motronic Engine Control Module (ECM)
N - Ignition Coil
N30 - Cylinder 1 Fuel Injector
N122 - Power Output Stage
N128 - Ignition Coil 2
N158 - Ignition Coil 3
P - Spark Plug Connectors
Q - Spark Plugs
S232 - Fuse 32 in fuse holder
T3ba - Connector 3 pin, black, on Power Output Stage 1
T4af - Connector 4 pin, black, on Power Output Stage 1
T10n - Connector 10 pin, orange, connector station electronic box plenum chamber

(18) - Ground connection, on engine block
(85) - Ground connection -1-, in engine compartment wiring harness
(B135) - Connector (15a), in wiring harness interior
(D23) - Connector (over fuse 229), in wiring harness front, right
(D95) - Wire connection (injectors), in engine compartment wiring harness

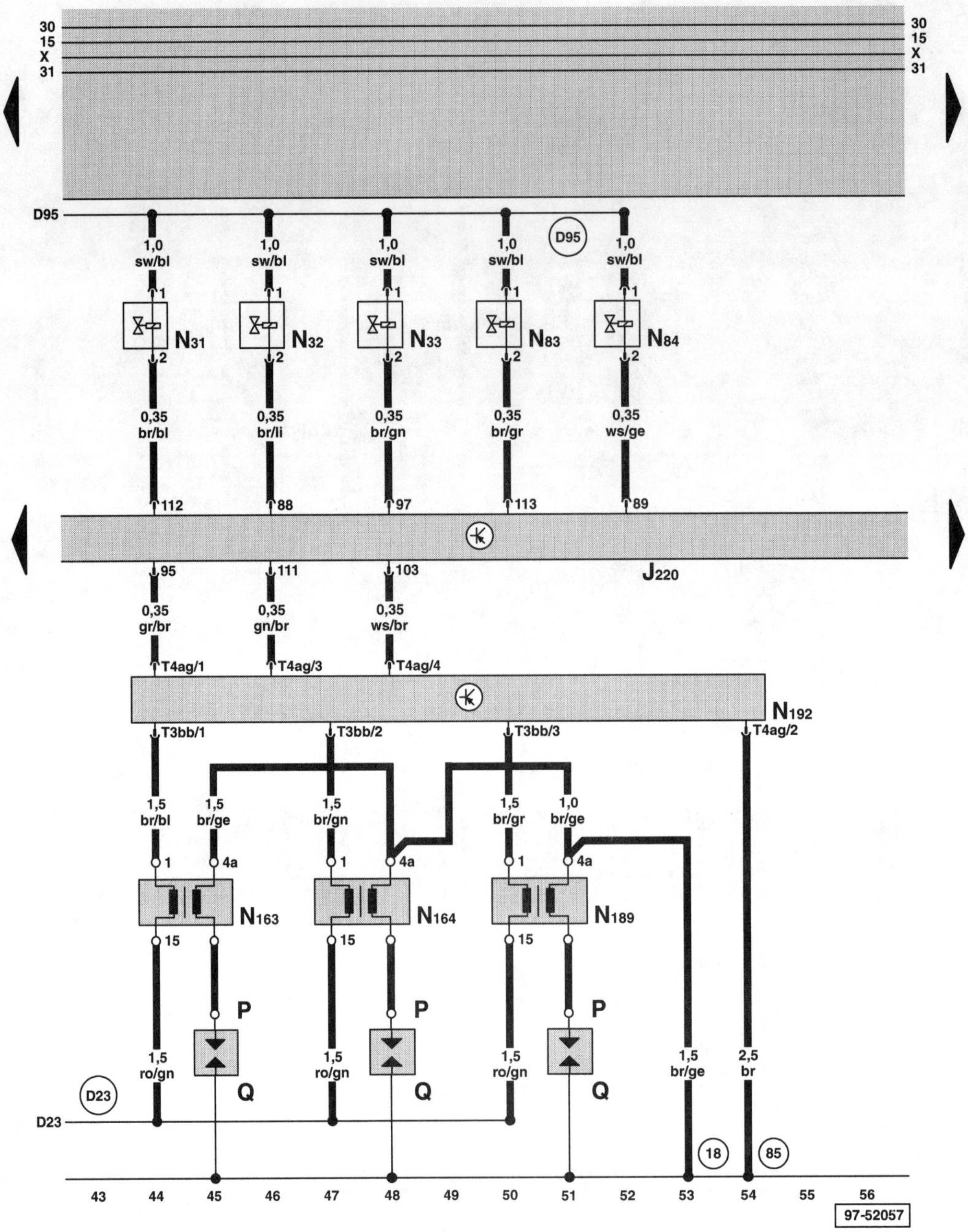

ws = white
sw = black
ro = red
br = brown
gn = green
bl = blue
gr = grey
li = violet
ge = yellow

J220 - Motronic Engine Control Module (ECM)
N31 - Cylinder 2 Fuel Injector
N32 - Cylinder 3 Fuel Injector
N33 - Cylinder 4 Fuel Injector
N83 - Cylinder 5 Fuel Injector
N84 - Cylinder 6 Fuel Injector
N163 - Ignition Coil 4
N164 - Ignition Coil 5
N189 - Ignition Coil 6
N192 - Power Output Stage 2
P - Spark Plug Connectors
Q - Spark Plugs
T3bb - Connector 3 pin, brown, on Power Output Stage 2
T4ag - Connector 4 pin, brown, on Power Output Stage 2

(18) - Ground connection, on engine block

(85) - Ground connection -1-, in engine compartment wiring harness

(D23) - Connector (over fuse 229), in wiring harness front, right

(D95) - Wire connection (injectors), in engine compartment wiring harness

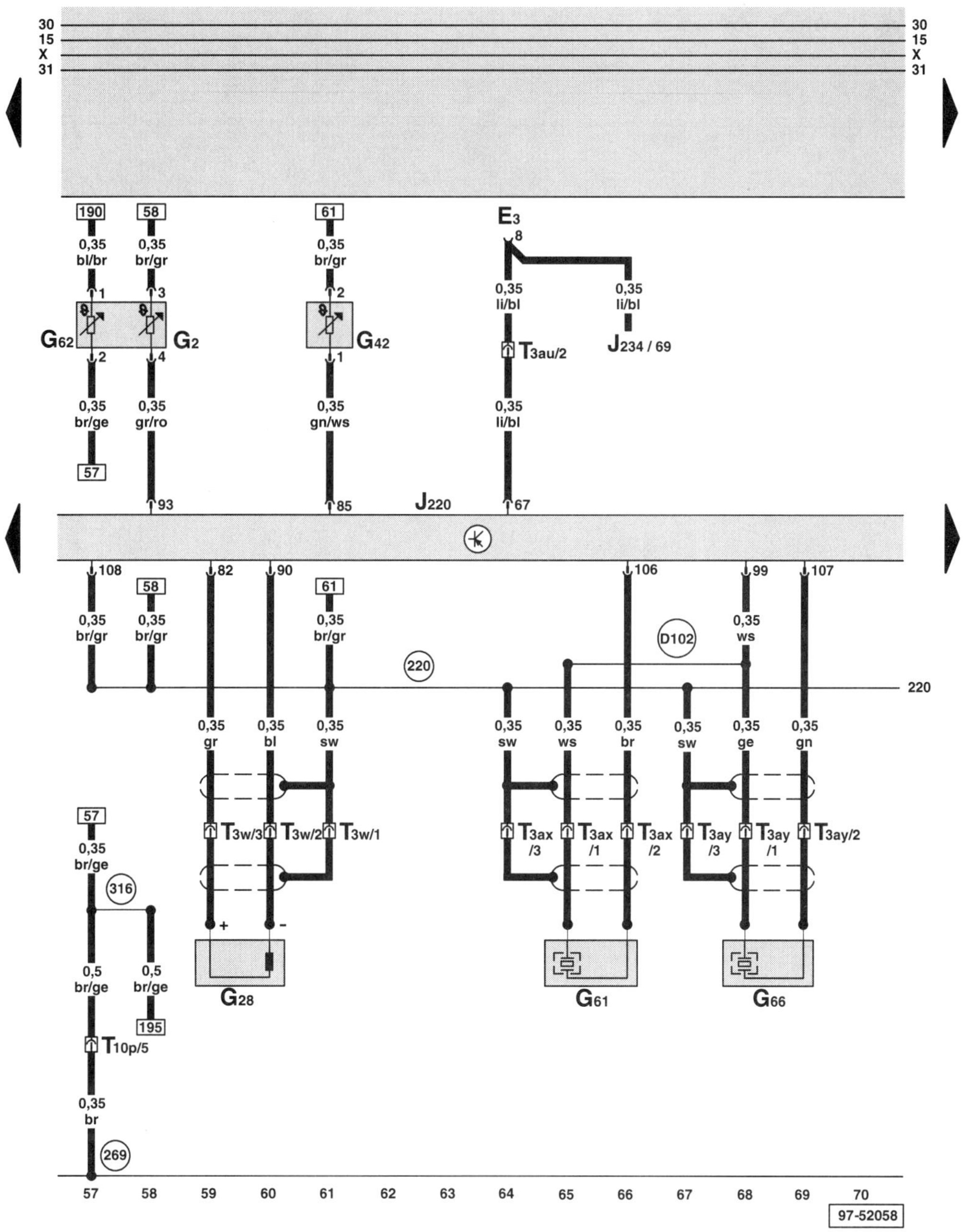

ws = white
sw = black
ro = red
br = brown
gn = green
bl = blue
gr = grey
li = violet
ge = yellow

E3 - Emergency Flasher Switch
G2 - Engine Coolant Temperature (ECT) Sensor
G28 - Engine Speed (RPM) Sensor
G42 - Intake Air Temperature (IAT) Sensor
G61 - Knock Sensor (KS) 1
G62 - Engine Coolant Temperature (ECT) Sensor
G66 - Knock Sensor (KS) 2
J220 - Motronic Engine Control Module (ECM)
J234 - Airbag Control Module
T3w - Connector 3 pin, grey, on engine speed sensor
T3au - Connector 3 pin, red, connector station electronic box plenum chamber
T3ax - Connector 3 pin, brown, on knock sensor 1
T3ay - Connector 3 pin, brown, on knock sensor 2
T10p - Connector 10 pin, black, connector station electronic box plenum chamber

(220) - Ground connection (sensor ground), in engine compartment wiring harness
(269) - Ground connector (sensor ground) -1-, in instrument panel wiring harness
(316) - Sensor Ground Connection 2 (in engine wiring harness)
(D102) - Wire connection -2-, in engine compartment wiring harness

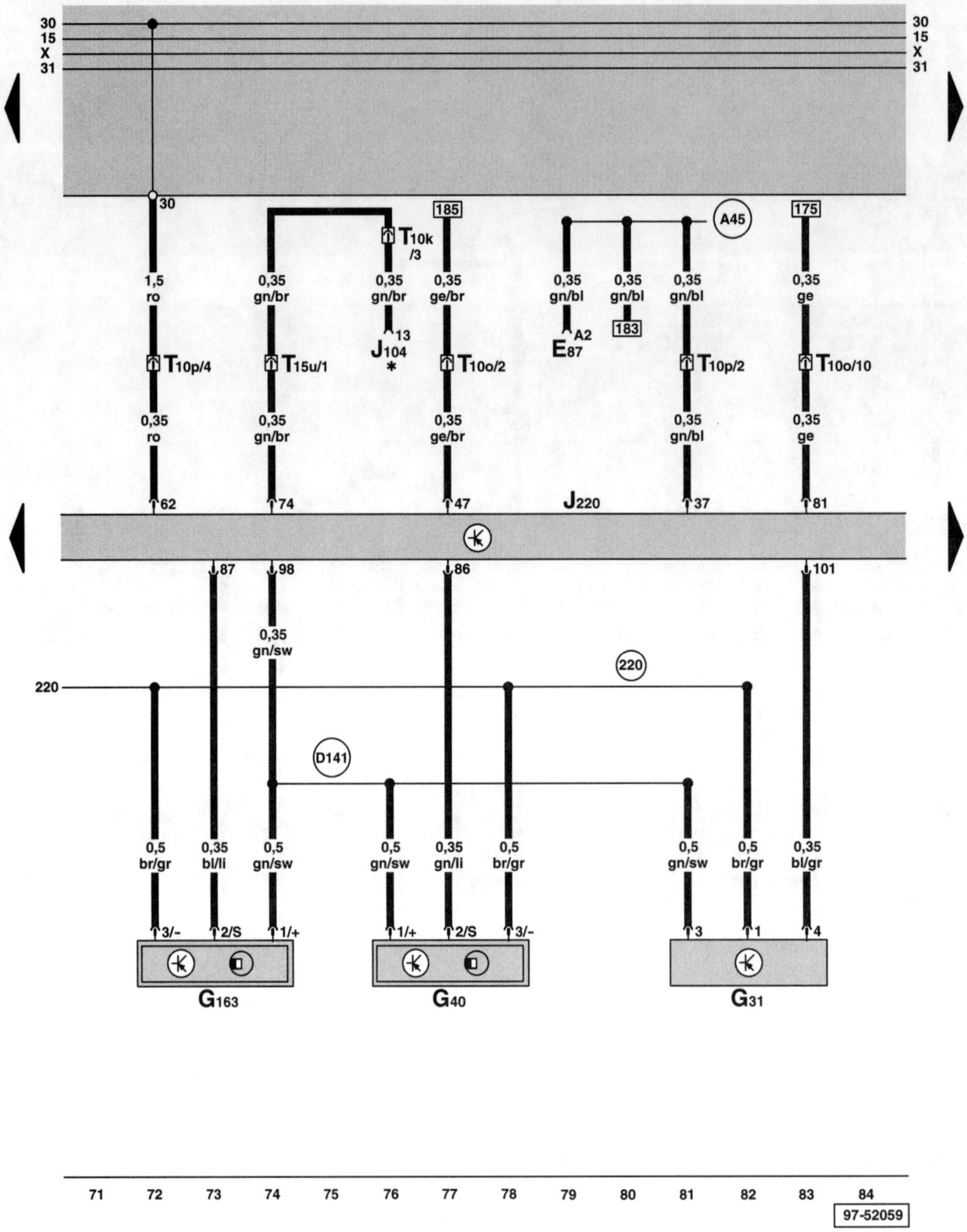

ws = white
sw = black
ro = red
br = brown
gn = green
bl = blue
gr = grey
li = violet
ge = yellow

E87 - A/C Control Head
G31 - Charge Air Pressure Sensor
G40 - Camshaft Position (CMP) Sensor
G163 - Camshaft Position (CMP) Sensor 2
J104 - ABS Control Module (w/EDL)
J220 - Motronic Engine Control Module (ECM)
T10k - Connector 10 pin, orange, connector station A pillar, left
T10o - Connector 10 pin, brown, connector station electronic box plenum chamber
T10p - Connector 10 pin, black, connector station electronic box plenum chamber
T15u - Connector 15 pin, red, connector station electronic box plenum chamber

(220) - Ground connection (sensor ground), in engine compartment wiring harness
(A45) - Wire connection (RPM-signal), in instrument panel wiring harness
(D141) - Connector (5V), in wiring harness, engine pre-wiring
* - Vehicles with ABS 5.3

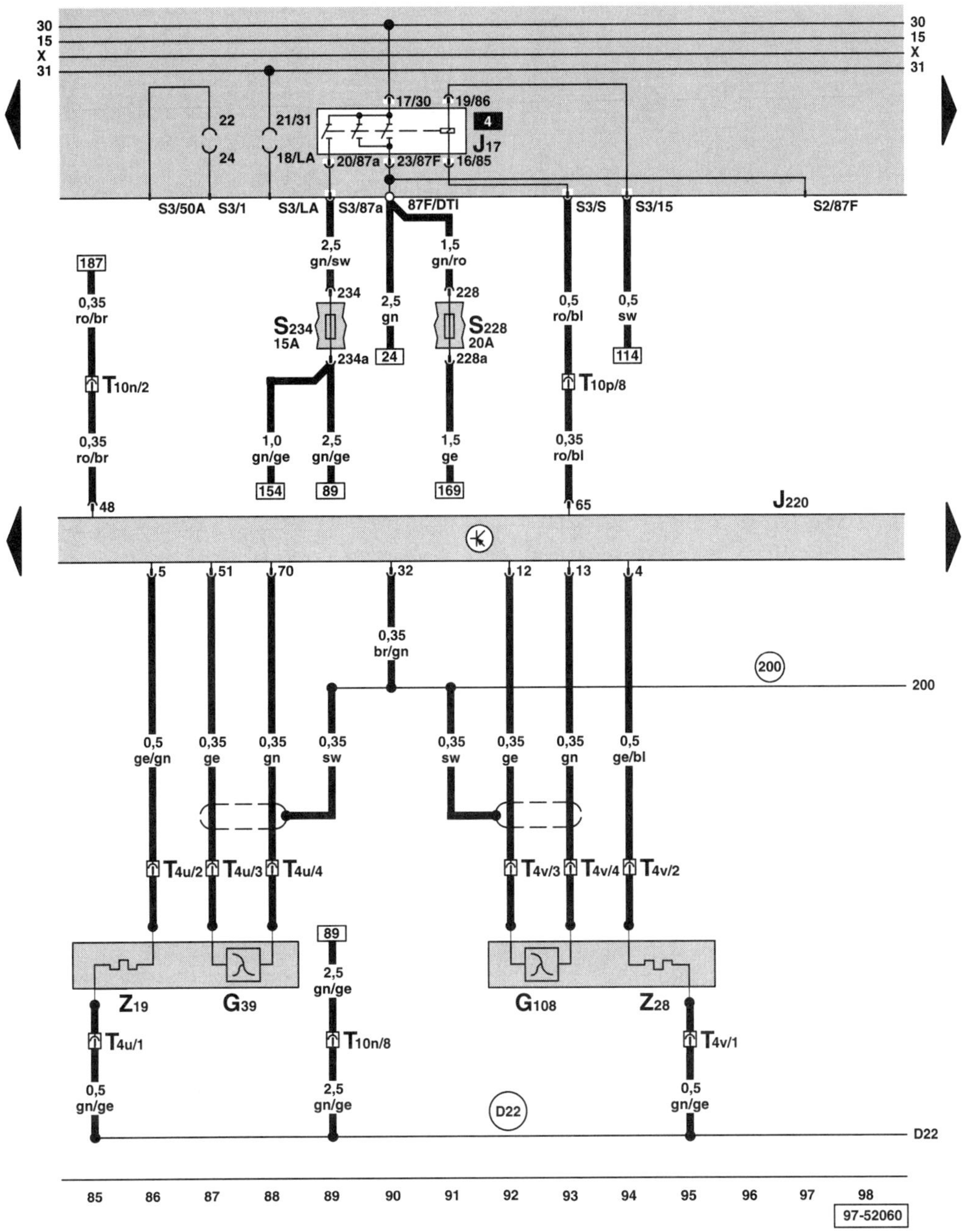

ws = white
sw = black
ro = red
br = brown
gn = green
bl = blue
gr = grey
li = violet
ge = yellow

G39 - Heated Oxygen Sensor (HO2S) 2
G108 - Heated Oxygen Sensor (HO2S) 2
J17 - Fuel Pump (FP) Relay
J220 - Motronic Engine Control Module (ECM)
S228 - Fuse 28 in fuse holder
S234 - Fuse 34 in fuse holder
T4u - Connector 4 pin, black, heated oxygen sensor
T4v - Connector 4 pin, black, heated oxygen sensor 2
T10n - Connector 10 pin, orange, connector station electronic box plenum chamber
T10p - Connector 10 pin, black, connector station electronic box plenum chamber
Z19 - Oxygen Sensor (O2S) Heater
Z28 - Oxygen Sensor (O2S) 2 Heater

(200) - Ground connection (shielding), in engine compartment wiring harness
(D22) - Connector (over fuse 234), in wiring harness front, right

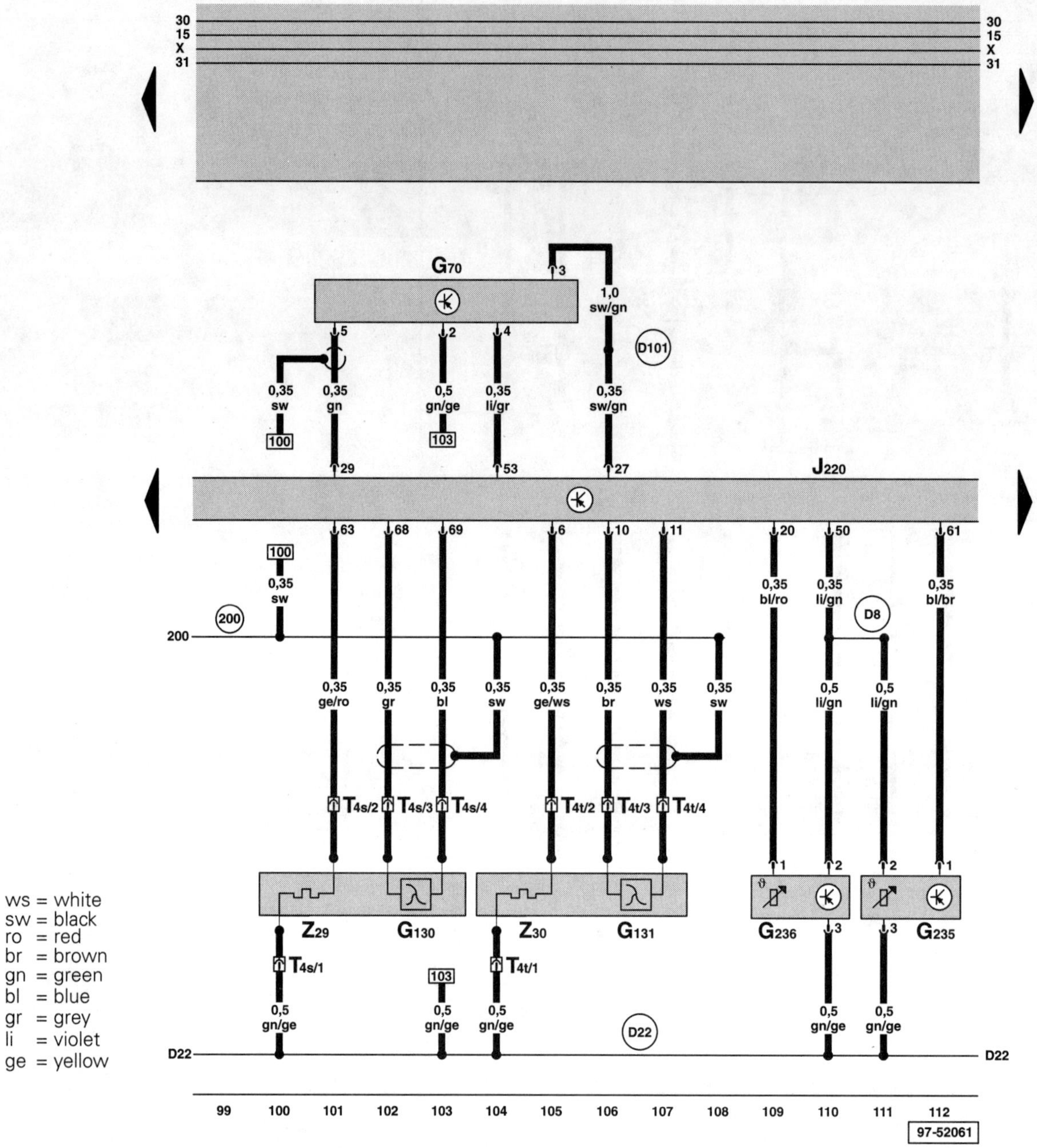

G70 - Mass Air Flow (MAF) Sensor
G130 - Heated Oxygen Sensor (O2S) Behind Three Way Catalytic Converter (TWC)
G131 - Heated Oxygen Sensor (O2S) 2 Behind Three Way Catalytic Converter (TWC)
G235 - Sensor -1- for exhaust temperature
G236 - Sensor -2- for exhaust temperature
J220 - Motronic Engine Control Module (ECM)
T4s - Connector 4 pin, green, near oxygen sensor, behind Three Way Catalytic Converter (TWC)
T4t - Connector 4 pin, brown, near oxygen sensor 2, behind Three Way Catalytic Converter (TWC)
Z29 - Oxygen Sensor (O2S) Heater
Z30 - Oxygen Sensor (O2S) 2 Heater

(200) - Ground connection (shielding), in engine compartment wiring harness
(D8) - Wire connection (temperature sensor/potentiometer), in right front wiring harness
(D22) - Connector (over fuse 234), in wiring harness front, right
(D101) - Wire connection -1-, in engine compartment wiring harness

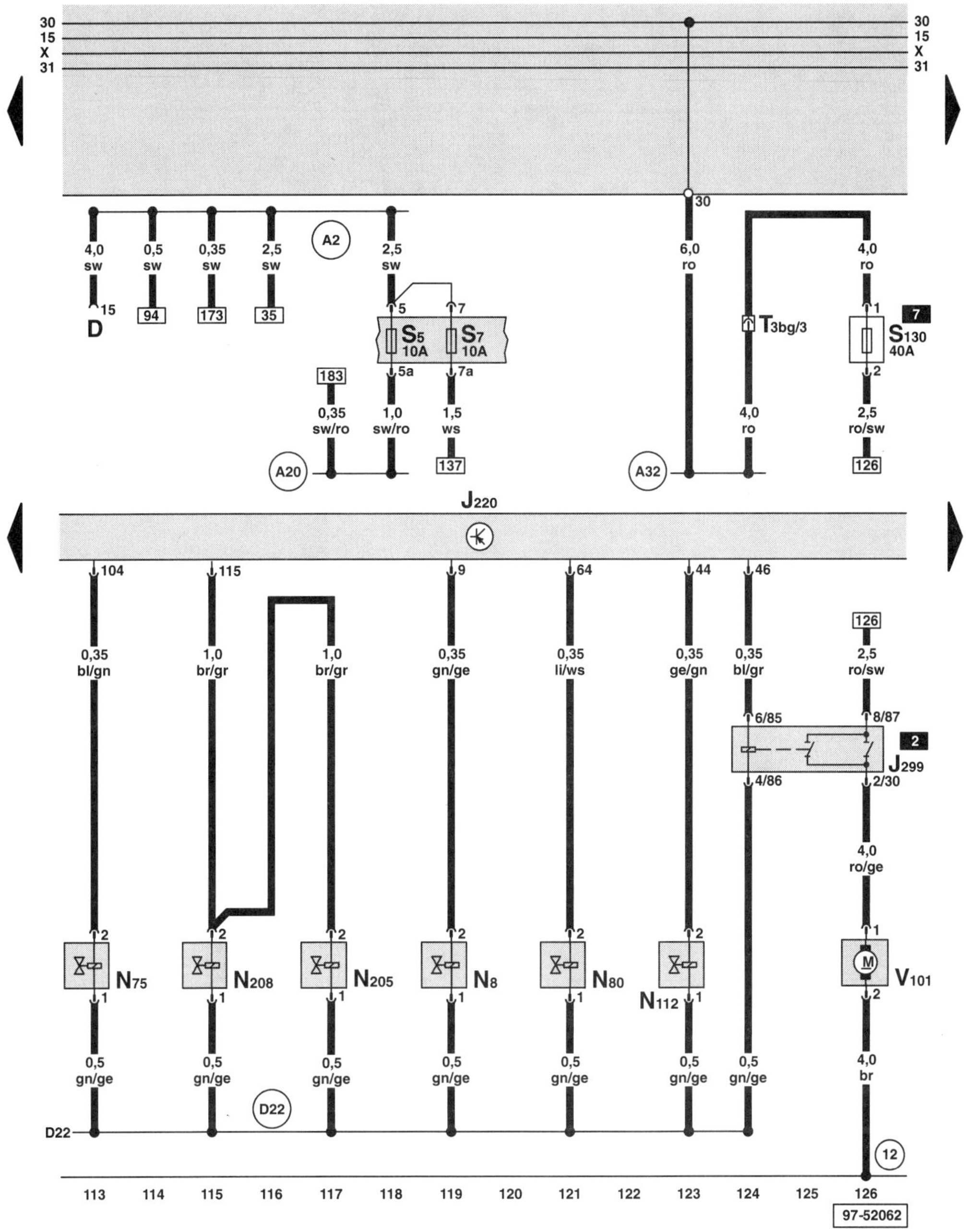

ws = white
sw = black
ro = red
br = brown
gn = green
bl = blue
gr = grey
li = violet
ge = yellow

D - Ignition/Starter Switch
J220 - Motronic Engine Control Module (ECM)
J299 - Secondary Air Injection (AIR) Pump Relay
N8 - Central Idle, Idle Air Control (IAC) Cut-Off Valve
N75 - Wastegate Bypass Regulator Valve
N80 - Evaporative Emission (EVAP) Canister Purge Regulator Valve
N112 - Secondary Air Injection (AIR) Solenoid Valve
N205 - Valve -1- for camshaft adjustment
N208 - Valve -2- for camshaft adjustment
S5 - Fuse
S7 - Fuse
S130 - Fuse for secondary air pump
T3bg - Connector 3 pin, white, connector station electronic box plenum chamber
V101 - Secondary Air Injection (AIR) Pump Motor

(12) - Ground connection, in engine compartment, left

(A2) - Plus connection (15), in instrument panel wiring harness

(A20) - Wire connection (15a), in instrument panel wiring harness

(A32) - Plus connection (30), in instrument panel wiring harness

(D22) - Connector (over fuse 234), in wiring harness front, right

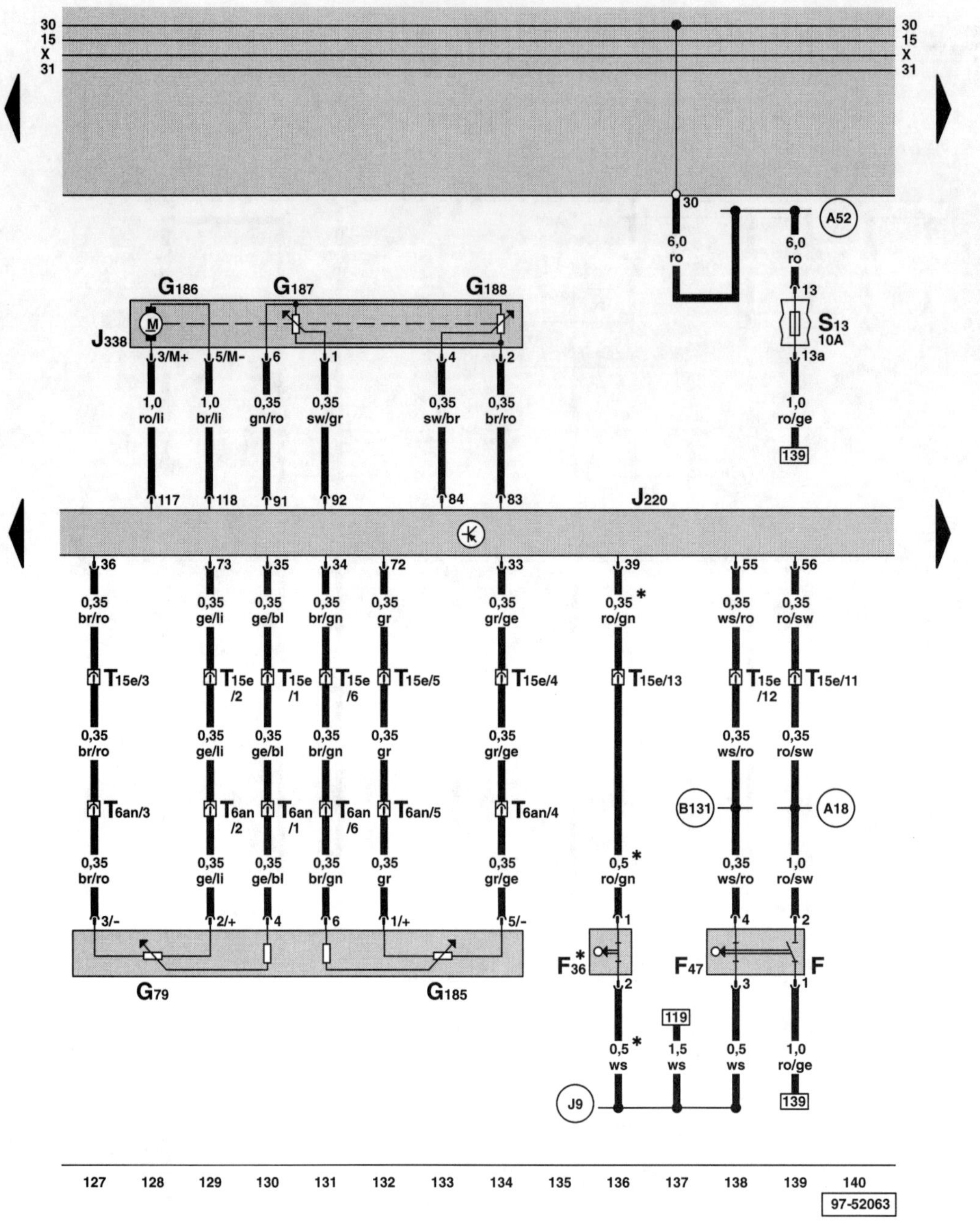

ws = white
sw = black
ro = red
br = brown
gn = green
bl = blue
gr = grey
li = violet
ge = yellow

- F - Brake Light Switch
- F36 - Clutch Vacuum Vent Valve Switch
- F47 - Vacuum Vent Valve, Brake
- G79 - Throttle Position (TP) Sensor
- G185 - Sensor -2- for accelerator pedal position
- G186 - Throttle drive (power accelerator actuation)
- G187 - Angel sensor -1- for throttle drive (power accelerator actuation)
- G188 - Angel sensor -2- for throttle drive (power accelerator actuation)
- J220 - Motronic Engine Control Module (ECM)
- J338 - Throttle Valve Control Module
- S13 - Fuse
- T6an - Connector 6 pin, in front footwell, left
- T15e - Connector 15 pin, white, connector station electronic box plenum chamber

- (A18) - Wire connection (54), in instrument panel wiring harness
- (A52) - Plus connection -2- (30), in instrument panel wiring harness
- (B131) - Connector (54), in wiring harness interior
- (J9) - Wire connection -1- (15a), in ABS wiring harness
- * - Vehicles with manually shifted transmission

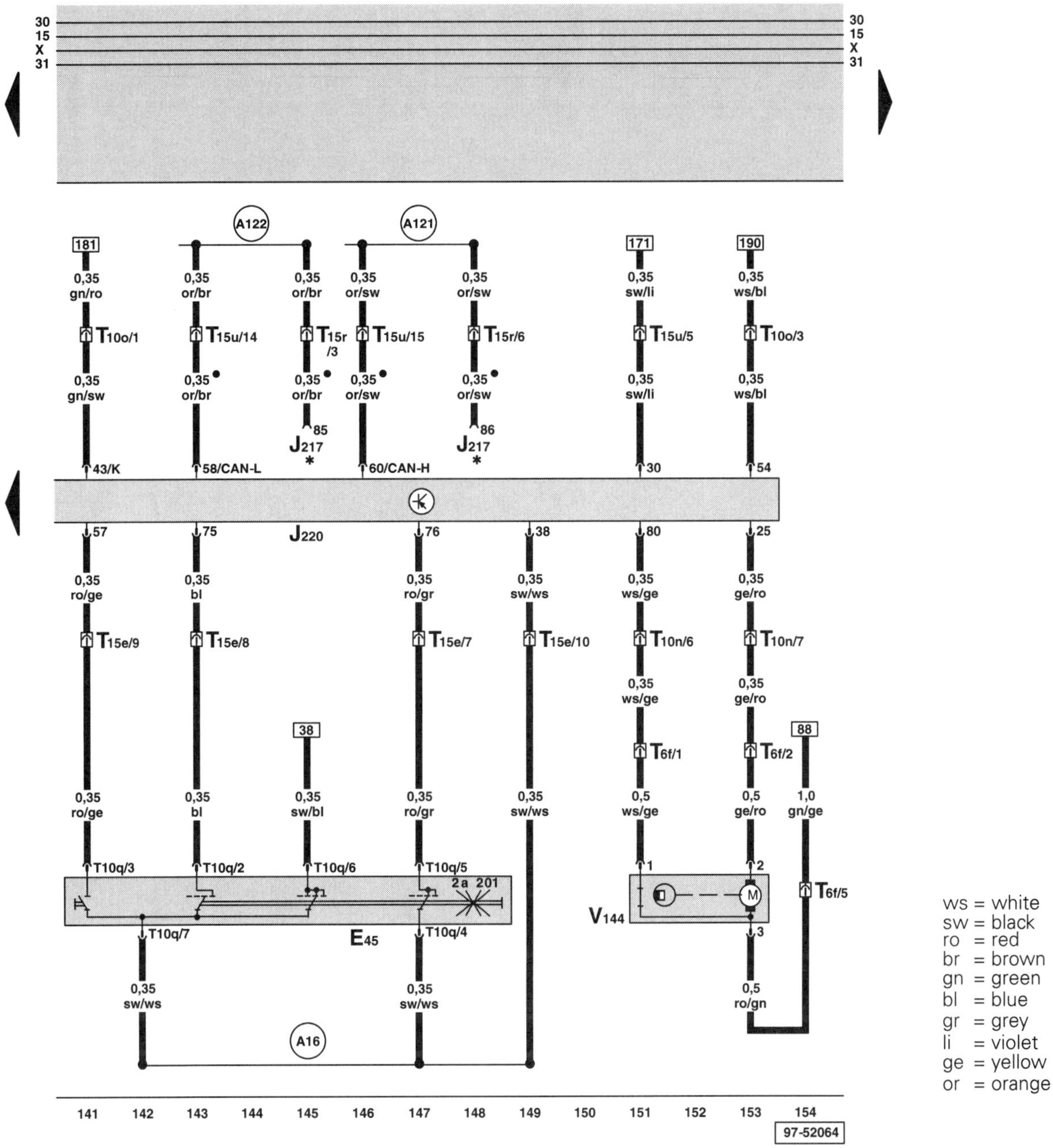

E45 - Cruise Control Switch
J217 - Transmission Control Module (TCM)
J220 - Motronic Engine Control Module (ECM)
T6f - Connector 6 pin, darkbrown, connector station A pillar, right
T10n - Connector 10 pin, orange, connector station electronic box plenum chamber
T10o - Connector 10 pin, brown, connector station electronic box plenum chamber
T10q - Connector 10 pin, black, on cruise control switch
T15e - Connector 15 pin, white, connector station electronic box plenum chamber
T15r - Connector 15 pin, blue, connector station A pillar, right
T15u - Connector 15 pin, red, connector station electronic box plenum chamber
V144 - Leak Detection Pump (LDP)

(A16) - Wire connection (cruise control), in instrument panel wiring harness
(A121) - Connection (high bus) in instrument panel wiring harness
(A122) - Connection (low bus) in instrument panel wiring harness
• - CAN-Bus (Data-Bus)
* - Vehicles with automatic transmission

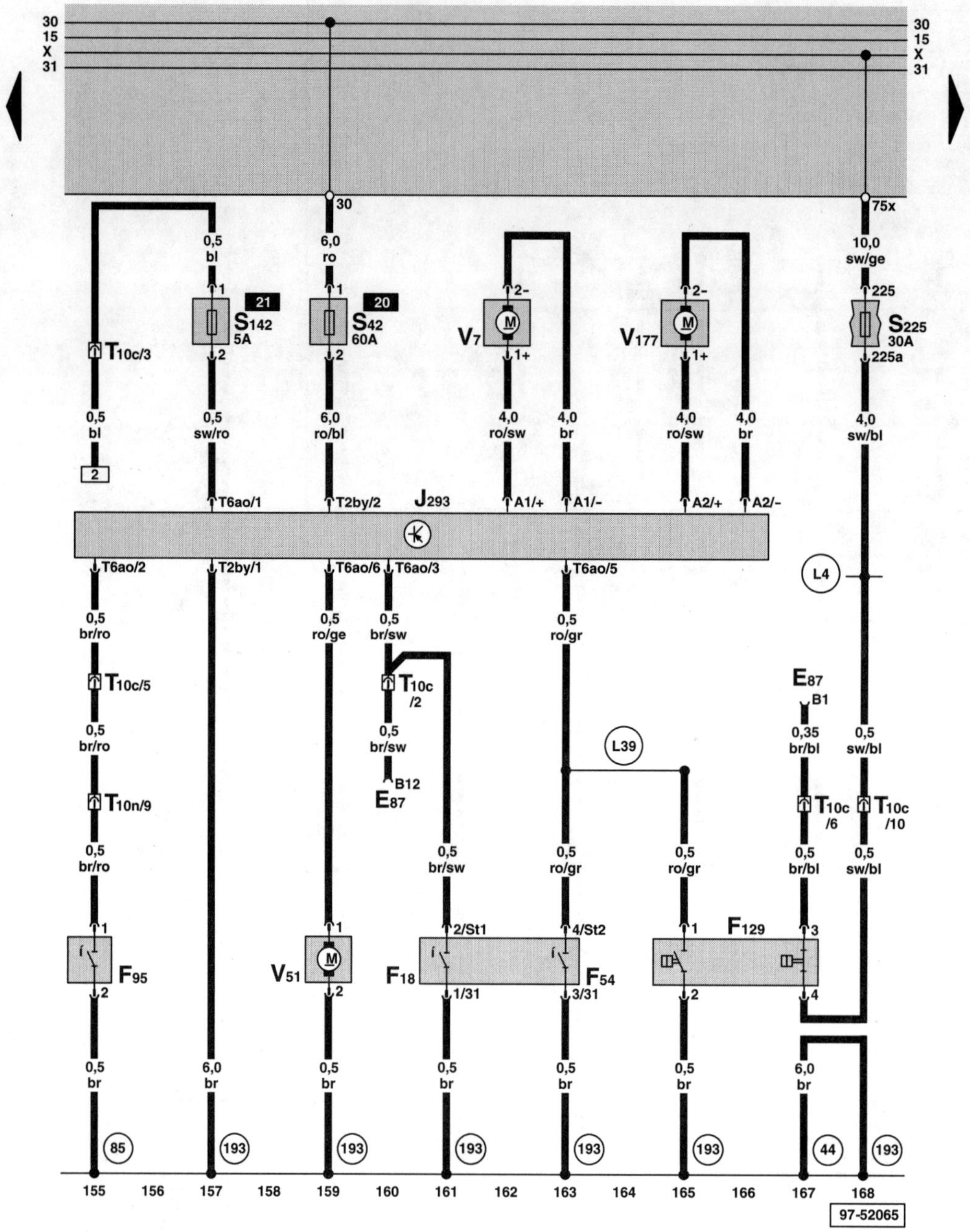

ws = white
sw = black
ro = red
br = brown
gn = green
bl = blue
gr = grey
li = violet
ge = yellow

E87 - A/C Control Head
F18 - Coolant Fan Control (FC) Thermal Switch
F54 - Coolant Fan Control (FC) Thermal Switch
F95 - After-Run Coolant Thermal Switch
F129 - A/C Pressure Switch
J293 - Coolant FC (Fan Control) Control Module
S42 - Coolant Fan Fuse
S142 - Control module fuse for coolant fan
S225 - Fuse 25 in fuse holder
T2by - Connector 2 pin, on Coolant FC (Fan Control) Control Module
T6ao - Connector 6 pin, on Coolant FC (Fan Control) Control Module
T10c - Connector 10 pin, grey, connector station A pillar, left
V7 - Coolant Fan
V51 - After-Run Coolant Pump
V177 - Coolant fan -2-

(44) - Ground connection, on left A-pillar, lower part
(85) - Ground connection -1-, in engine compartment wiring harness
(193) - Ground connection -1-, in coolant fan wiring harness
(L4) - Wire connection (75al), in A/C wiring harness
(L39) - Wire connection -1- (A/C Pressure Switch), in Climatronic wiring harness

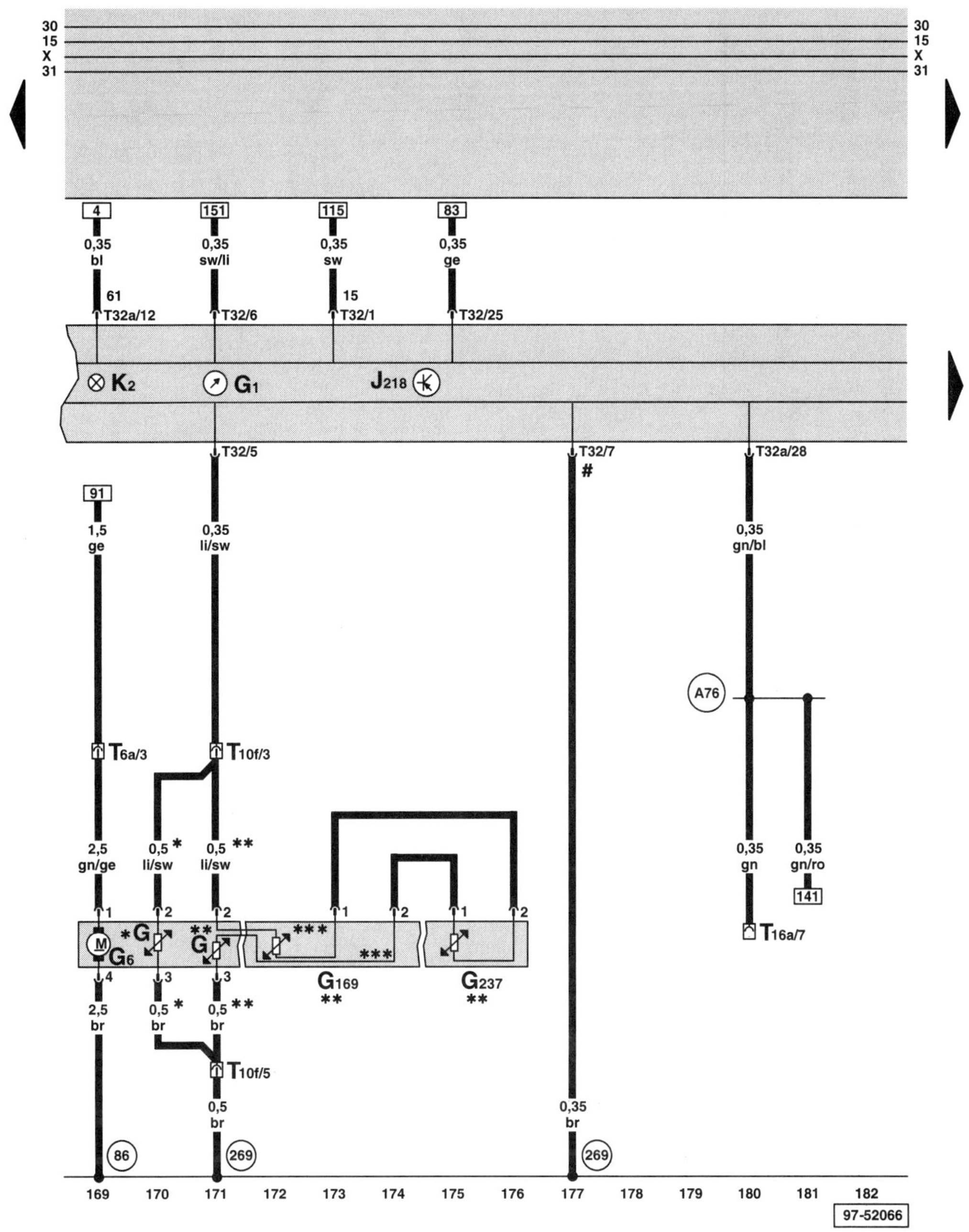

ws = white
sw = black
ro = red
br = brown
gn = green
bl = blue
gr = grey
li = violet
ge = yellow

G - Sender for fuel gauge
G1 - Fuel Gauge
G6 - Fuel Pump (FP)
G169 - Sender 2 for fuel gauge
G237 - Sensor -3- for fuel supply
J218 - Instrument Cluster Combination Processor
K2 - Generator (GEN) Warning Light
T6a - Connector 6 pin, blue, connector station A pillar, left
T10f - Connector 10 pin, brown, connector station A pillar, left
T16a - Connector 16 pin, Data Link Connector (DLC)
T32 - Connector 32 pin, blue, on instrument cluster
T32a - Connector 32 pin, green, on instrument cluster

(86) - Ground connection -1-, in rear wiring harness
(269) - Ground connector (sensor ground) -1-, in instrument panel wiring harness
(A76) - Connector (K-diagnosis wire), in instrument panel wiring harness

* - Vehicles with front wheel drive
** - Vehicles with all wheel drive
*** - Wiring inside fuel tank
# - Sensor ground, output

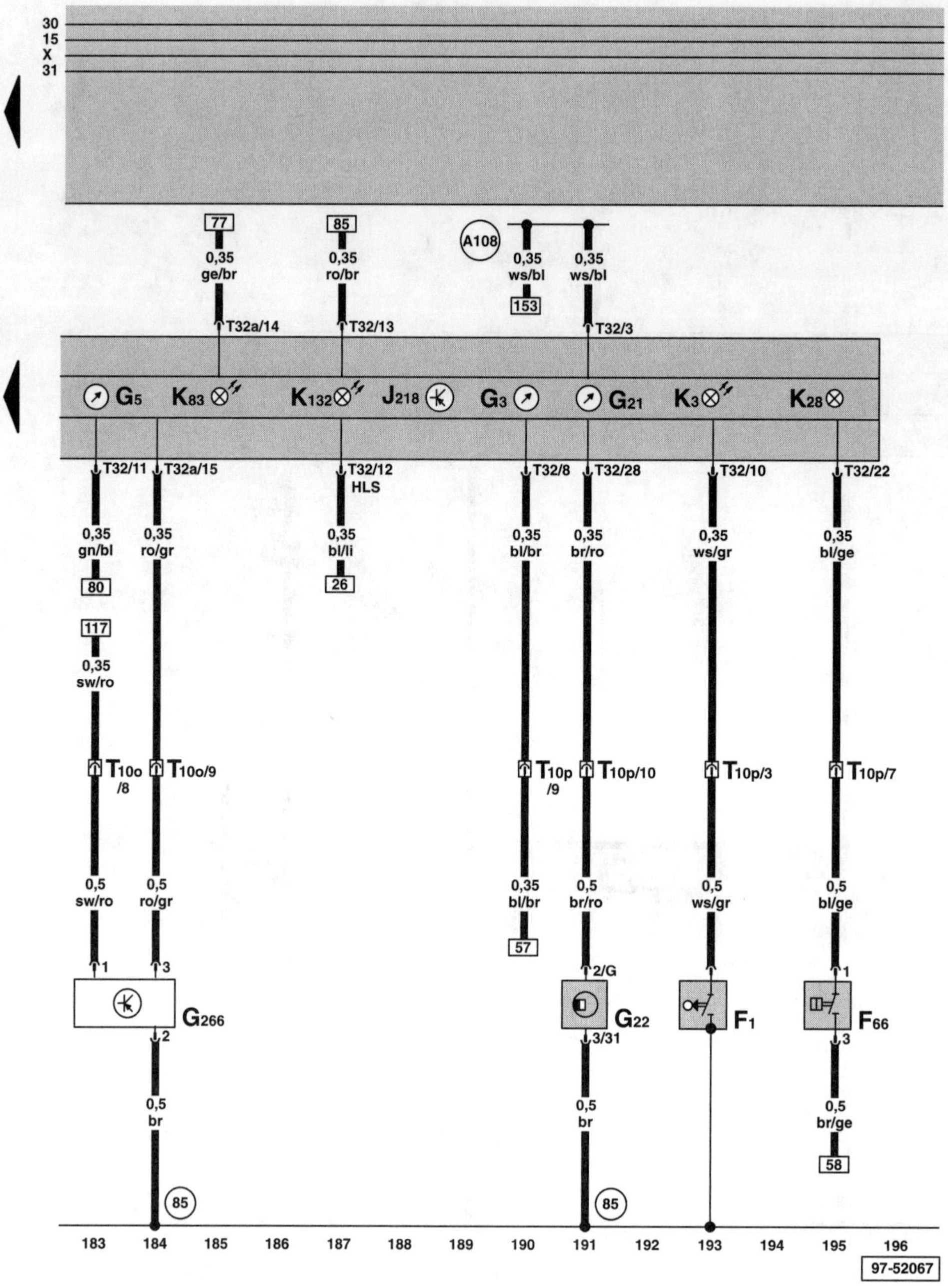

ws = white
sw = black
ro = red
br = brown
gn = green
bl = blue
gr = grey
li = violet
ge = yellow

- F1 - Oil Pressure Switch
- F66 - Engine Coolant Level (ECL) Warning Switch
- G3 - Engine Coolant Temperature (ECT) Gauge
- G5 - Tachometer
- G21 - Speedometer
- G22 - Speedometer Vehicle Speed Sensor (VSS)
- G266 - Oil Level Thermal Sensor (SRI)
- J218 - Instrument Cluster Combination Processor
- K3 - Oil Pressure Warning Light
- K28 - Engine Coolant Level/Temperature (ECL/ECT) Warning Light
- K83 - Malfunction Indicator Lamp (MIL)
- K132 - Fault light for power accelerator activation
- T10o - Connector 10 pin, brown, connector station electronic box plenum chamber
- T10p - Connector 10 pin, black, connector station electronic box plenum chamber
- T32 - Connector 32 pin, blue, on instrument cluster
- T32a - Connector 32 pin, green, on instrument cluster

- (85) - Ground connection -1-, in engine compartment wiring harness
- (A108) - Connector (speed signal), in instrument panel wiring harness

(4,2 l - Injection Engine, 8 Cylinder), Code ART

2000 m. y.

Fuse Panel

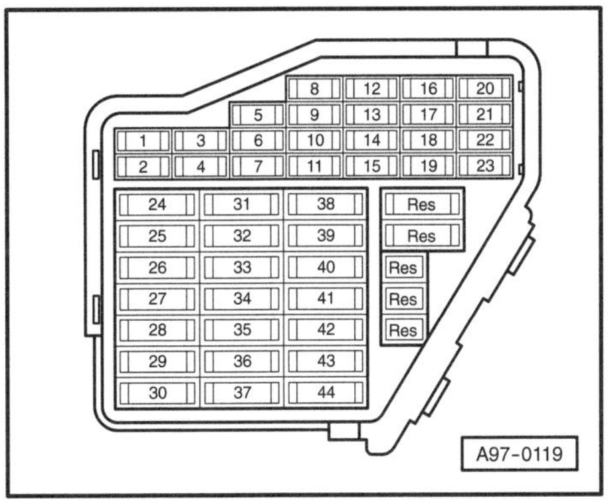

Fuse Colors:

30 A - Green
25 A - White
20 A - Yellow
15 A - Blue
10 A - Red
7,5 A - Brown
5 A - Beige

Starting with fuse position 23, fuses in the fuse holder are identified with 223 in the wiring diagram.

Micro Central Electric Panel

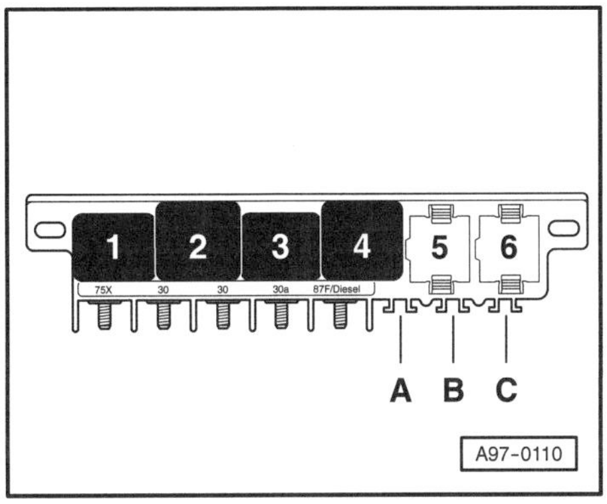

Relay Location:

4 -Fuel Pump (FP) Relay , J17

5 -Wiper/Washer Intermittent Relay, J31

6 -Wiper/Washer Intermittent Relay, J31

13 - Fold Relay Panel

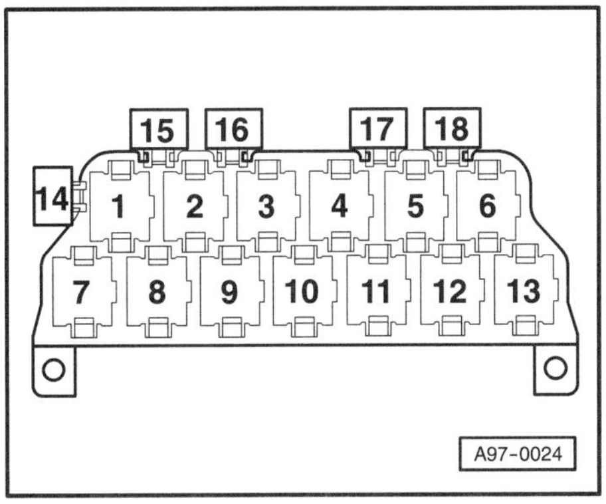

Relay Location:

4 -Starting Interlock Relay, J207

8 - Fold Relay Panel

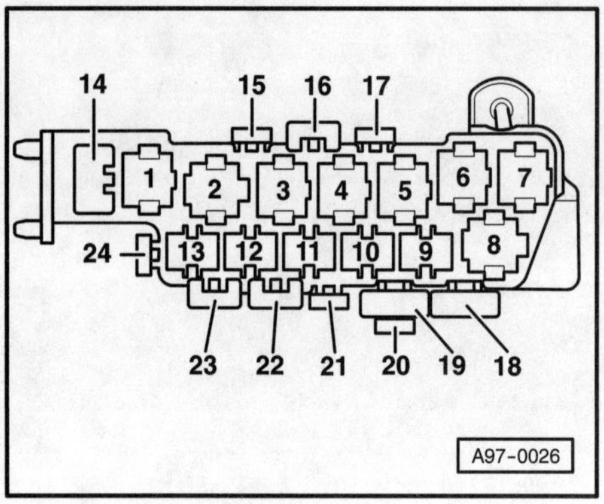

Fuse Location:

20 -Coolant Fan Fuse, S42

21 -Control module fuse for coolant fan, S142

3 - Fold Relay Panel in E-Box plenum chamber

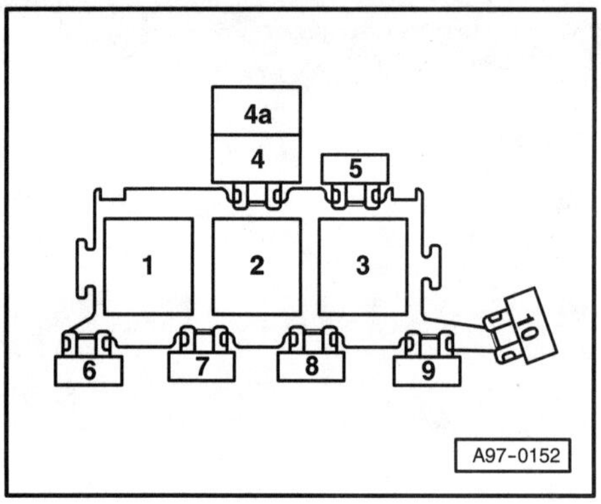

Relay Location:

2 -Secondary Air Injection (AIR) Pump Relay, J299

7 -Fuse for secondary air pump , S130

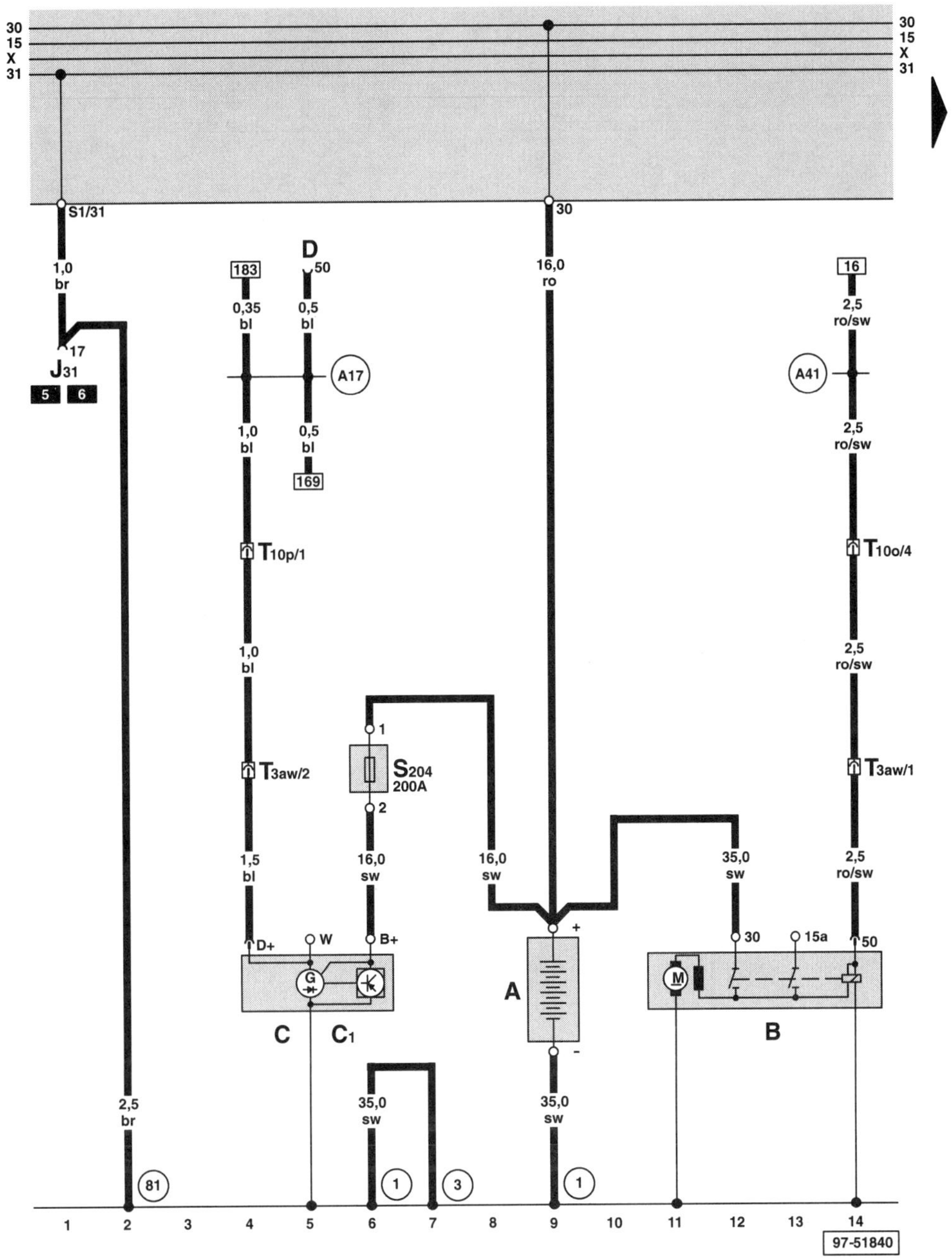

ws = white
sw = black
ro = red
br = brown
gn = green
bl = blue
gr = grey
li = violet
ge = yellow

A - Battery
B - Starter
C - Generator (GEN)
C1 - Voltage Regulator (VR)
D - Ignition/Starter Switch
J31 - Wiper/Washer Intermittent Relay
S204 - Fuse 1 (30)
T3aw - Connector 3 pin, black, in engine compartment, right
T10o - Connector 10 pin, brown, connector station electronic box plenum chamber
T10p - Connector 10 pin, black, connector station electronic box plenum chamber

(1) - Ground strap, battery to body
(3) - Ground strap, engine to body
(81) - Ground connection -1-, in instrument panel wiring harness
(A17) - Wire connection (61), in instrument panel wiring harness
(A41) - Plus connection (50), in instrument panel wiring harness

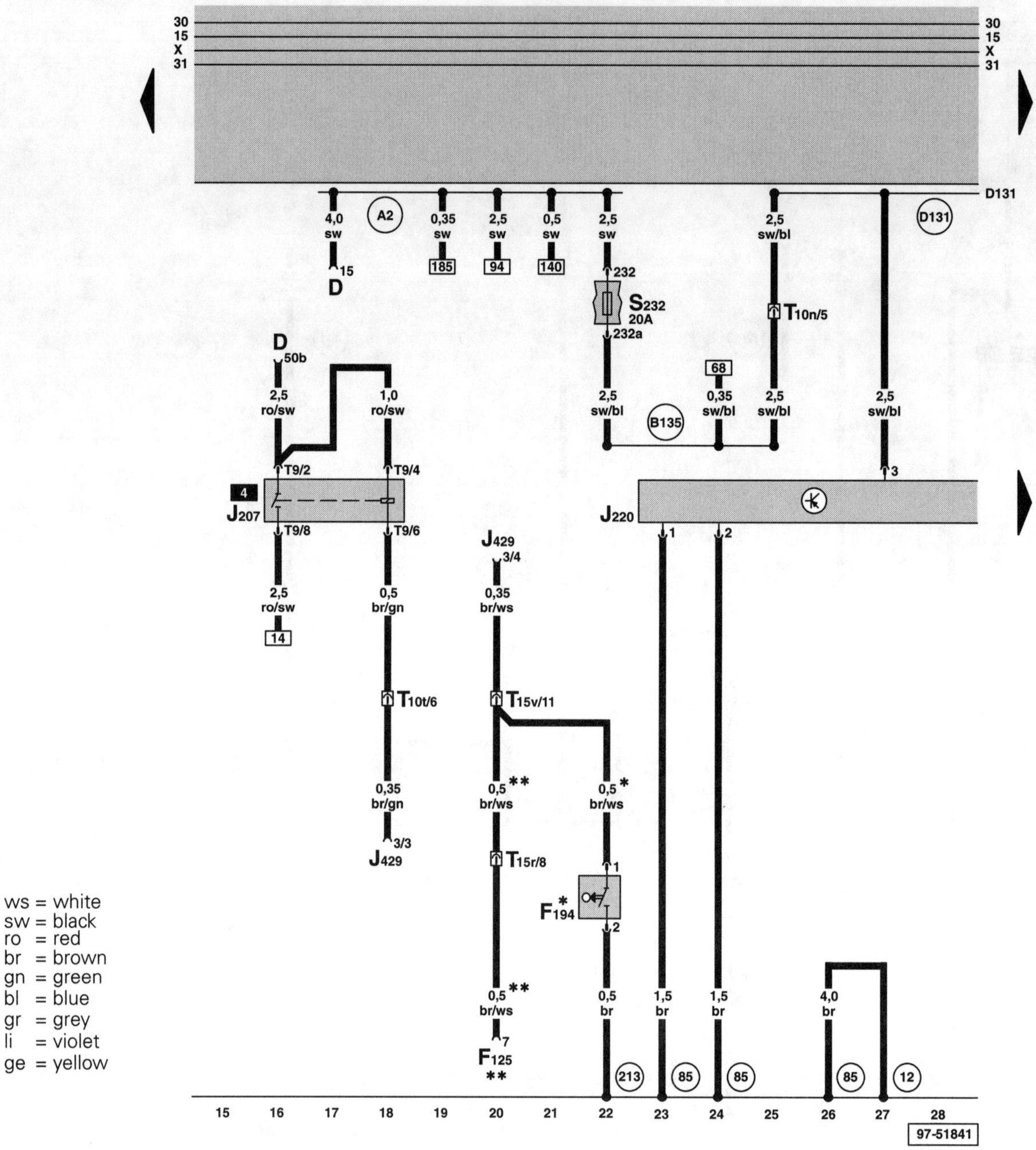

D	-	Ignition/Starter Switch
F125	-	Multi-Function Transmission Range (TR) Switch
F194	-	Clutch Pedal Position (CPP) Switch
J207	-	Starting Interlock Relay
J220	-	Motronic Engine Control Module (ECM)
J429	-	Control module for central locking
S232	-	Fuse 32 in fuse holder
T9	-	Connector 9 pin, brown, on starting interlock relay
T10n	-	Connector 10 pin, orange, connector station electronic box plenum chamber
T10t	-	Connector 10 pin, orange, connector station A pillar, right
T15r	-	Connector 15 pin, blue, connector station A pillar, right
T15v	-	Connector 15 pin, orange, connector station A pillar, right

12	-	Ground connection, in engine compartment, left
85	-	Ground connection -1-, in engine compartment wiring harness
213	-	Ground connection, in instrument panel wiring harness
A2	-	Plus connection (15), in instrument panel wiring harness
B135	-	Connector (15a), in wiring harness interior
D131	-	Wire connection (15), over fuse 232, in engine compartment wiring harness
*	-	Vehicles with manually shifted transmission
**	-	Vehicles with automatic transmission

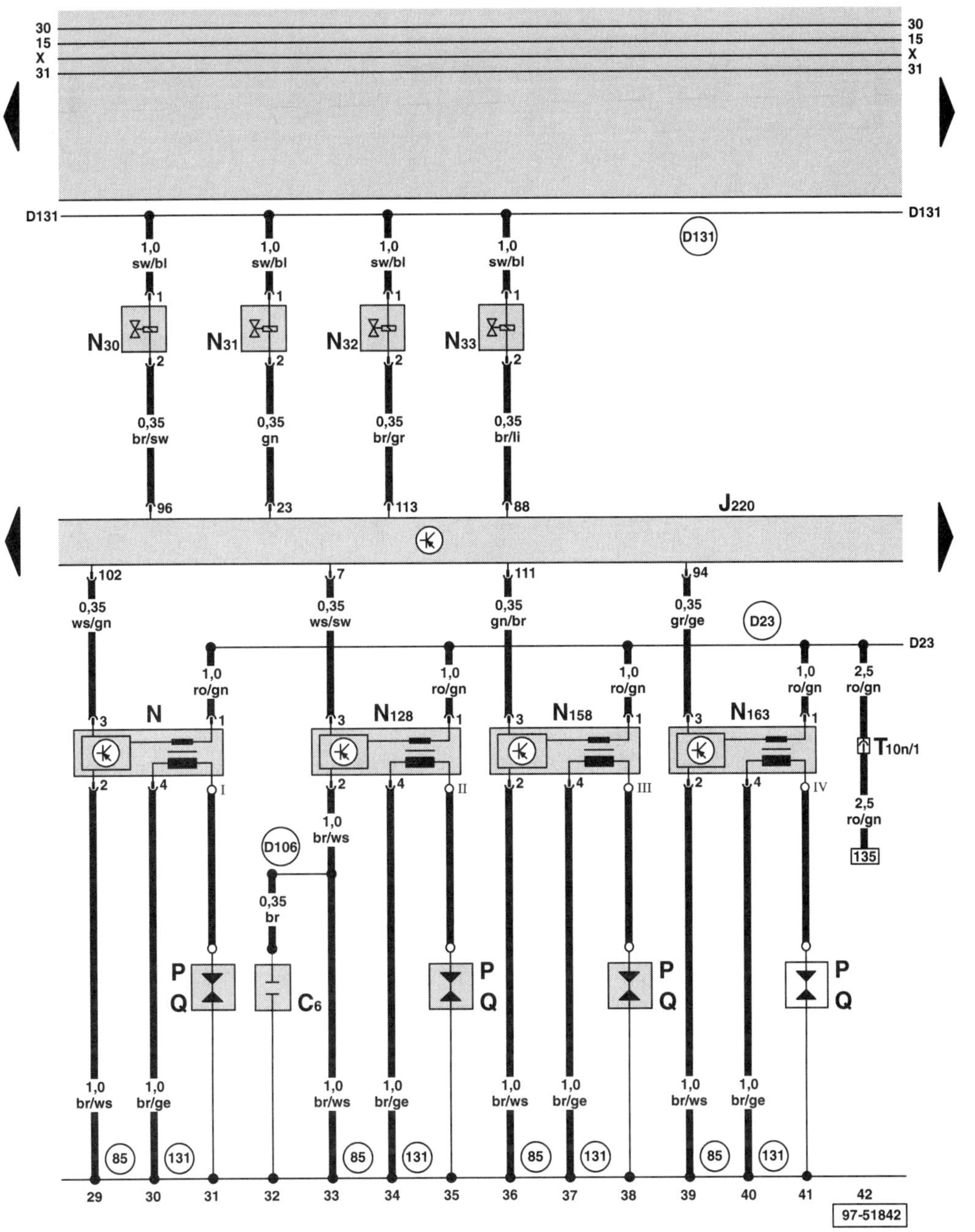

ws = white
sw = black
ro = red
br = brown
gn = green
bl = blue
gr = grey
li = violet
ge = yellow

C6 - Distributor Ignition (DI) Capacitor
J220 - Motronic Engine Control Module (ECM)
N - Ignition Coil
N30 - Cylinder 1 Fuel Injector
N31 - Cylinder 2 Fuel Injector
N32 - Cylinder 3 Fuel Injector
N33 - Cylinder 4 Fuel Injector
N128 - Ignition Coil 2
N158 - Ignition Coil 3
N163 - Ignition Coil 4
P - Spark Plug Connectors
Q - Spark Plugs
T10n - Connector 10 pin, orange, connector station electronic box plenum chamber

(85) - Ground connection -1-, in engine compartment wiring harness
(131) - Ground connection -2-, in engine compartment wiring harness
(D23) - Connector (over fuse 229), in wiring harness front, right
(D106) - Wire connection -4-, in engine compartment wiring harness
(D131) - Wire connection (15), over fuse 232, in engine compartment wiring harness

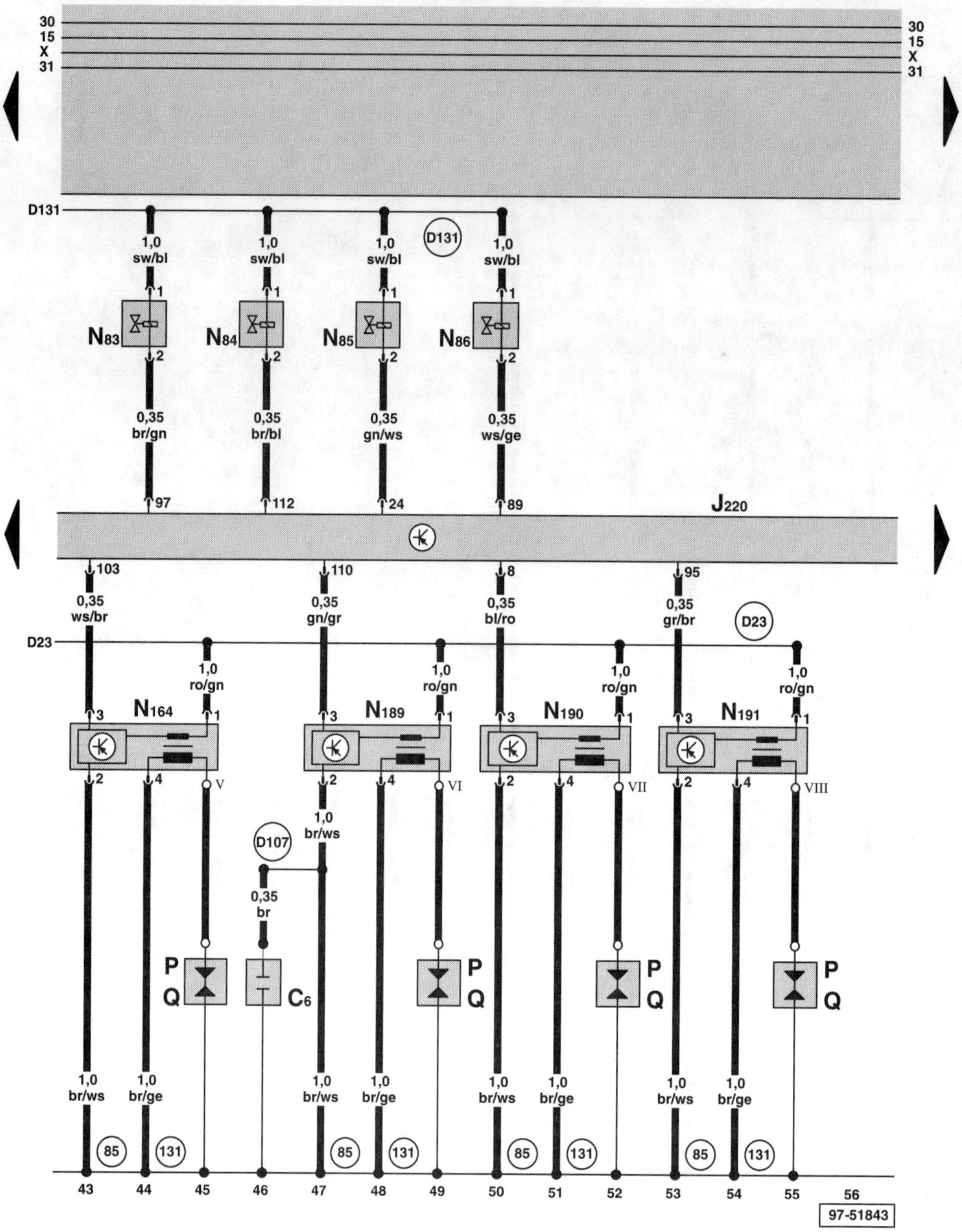

ws = white
sw = black
ro = red
br = brown
gn = green
bl = blue
gr = grey
li = violet
ge = yellow

C6 - Distributor Ignition (DI) Capacitor
J220 - Motronic Engine Control Module (ECM)
N83 - Cylinder 5 Fuel Injector
N84 - Cylinder 6 Fuel Injector
N85 - Cylinder 7 Fuel Injector
N86 - Cylinder 8 Fuel Injector
N164 - Ignition Coil 5
N189 - Ignition Coil 6
N190 - Ignition Coil 7
N191 - Ignition Coil 8
P - Spark Plug Connectors
Q - Spark Plugs

(85) - Ground connection -1-, in engine compartment wiring harness
(131) - Ground connection -2-, in engine compartment wiring harness
(D23) - Connector (over fuse 229), in wiring harness front, right
(D107) - Wire connection -5-, in engine compartment wiring harness
(D131) - Wire connection (15), over fuse 232, in engine compartment wiring harness

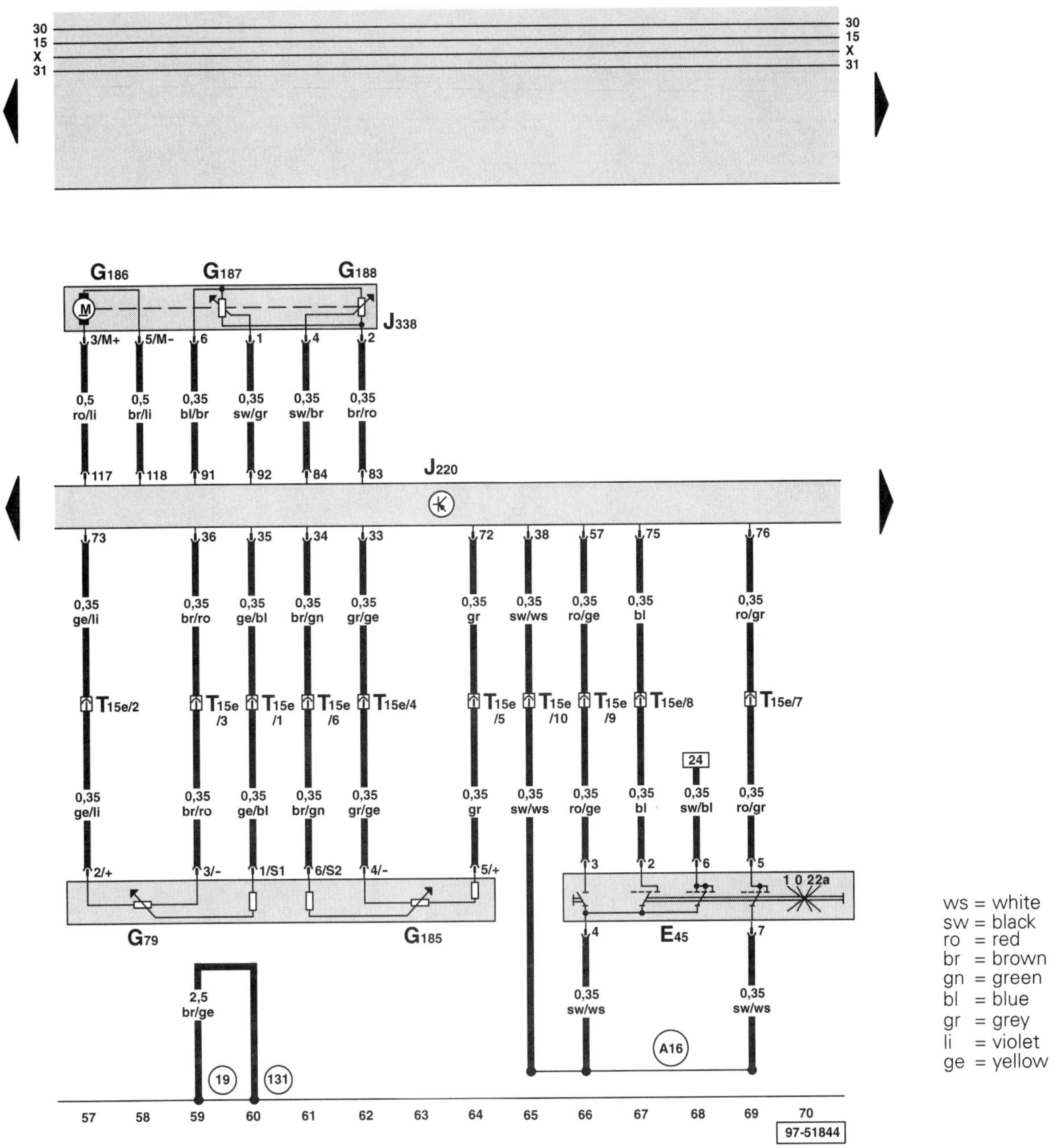

ws = white
sw = black
ro = red
br = brown
gn = green
bl = blue
gr = grey
li = violet
ge = yellow

E45 - Cruise Control Switch
G79 - Throttle Position (TP) Sensor
G185 - Sensor -2- for accelerator pedal position
G186 - Throttle drive (power accelerator actuation)
G187 - Angel sensor -1- for throttle drive (power accelerator actuation)
G188 - Angel sensor -2- for throttle drive (power accelerator actuation)
J220 - Motronic Engine Control Module (ECM)
J338 - Throttle Valve Control Module
T15e - Connector 15 pin, white, connector station electronic box plenum chamber

(19) - Ground connection, near ignition coil
(131) - Ground connection -2-, in engine compartment wiring harness
(A16) - Wire connection (cruise control), in instrument panel wiring harness

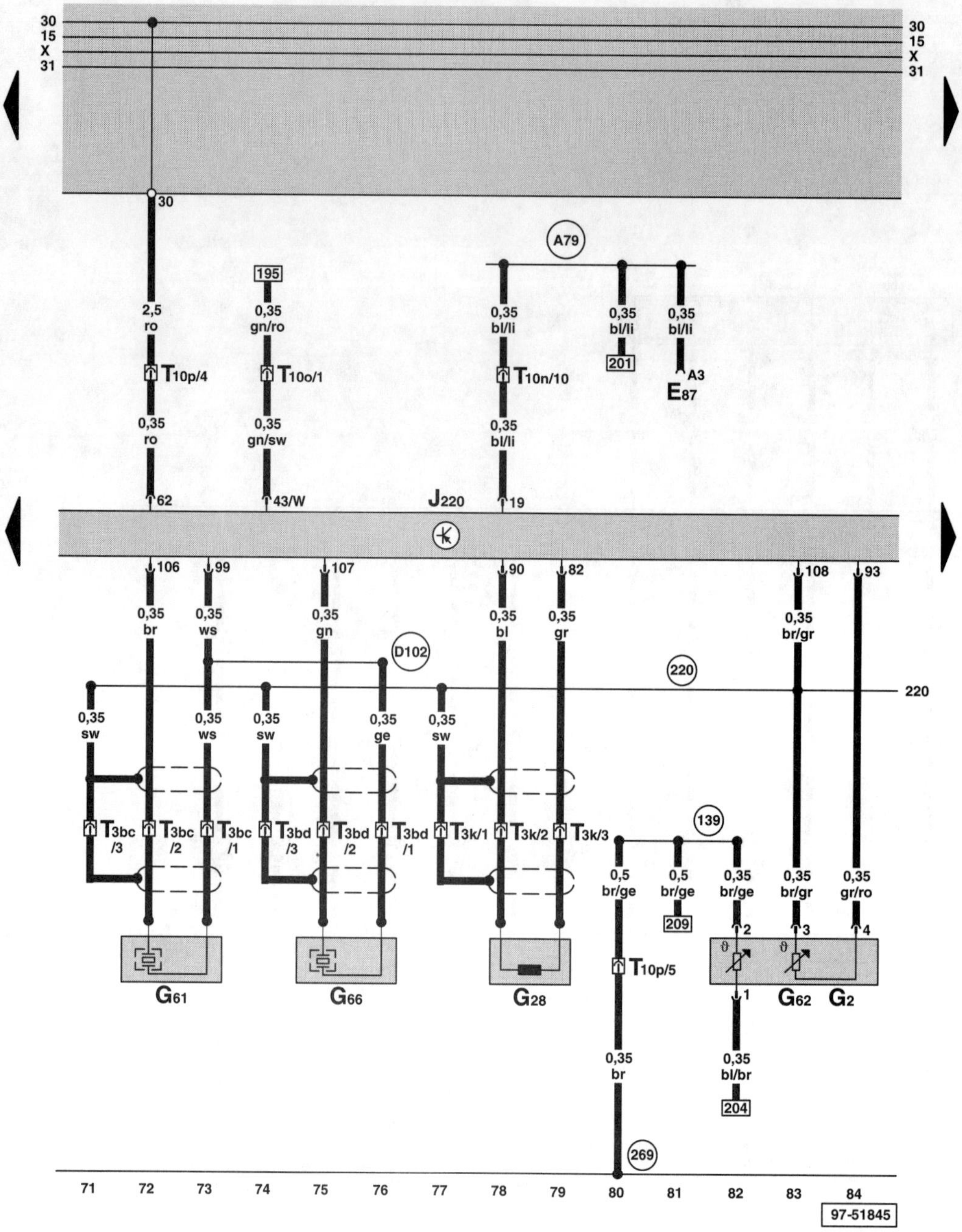

ws = white
sw = black
ro = red
br = brown
gn = green
bl = blue
gr = grey
li = violet
ge = yellow

E87 - A/C Control Head
G2 - Engine Coolant Temperature (ECT) Sensor
G28 - Engine Speed (RPM) Sensor
G61 - Knock Sensor (KS) 1
G62 - Engine Coolant Temperature (ECT) Sensor
G66 - Knock Sensor (KS) 2
J220 - Motronic Engine Control Module (ECM)
T3k - Connector 3 pin, grey, on Engine Speed (RPM) Sensor
T3bc - Connector 3 pin, green, on Knock Sensor 1
T3bd - Connector 3 pin, green, on Knock Sensor 2
T10n - Connector 10 pin, orange, connector station electronic box plenum chamber
T10o - Connector 10 pin, brown, connector station electronic box plenum chamber
T10p - Connector 10 pin, black, connector station electronic box plenum chamber

(139) - Ground connection (sensor Ground), in Motronic Multiport Fuel Injection (MFI) wiring harness
(220) - Ground connection (sensor ground), in engine compartment wiring harness
(269) - Ground connector (sensor ground) -1-, in instrument panel wiring harness
(A79) - Connector (engine overheat warning light), in instrument panel wiring harness
(D102) - Wire connection -2-, in engine compartment wiring harness

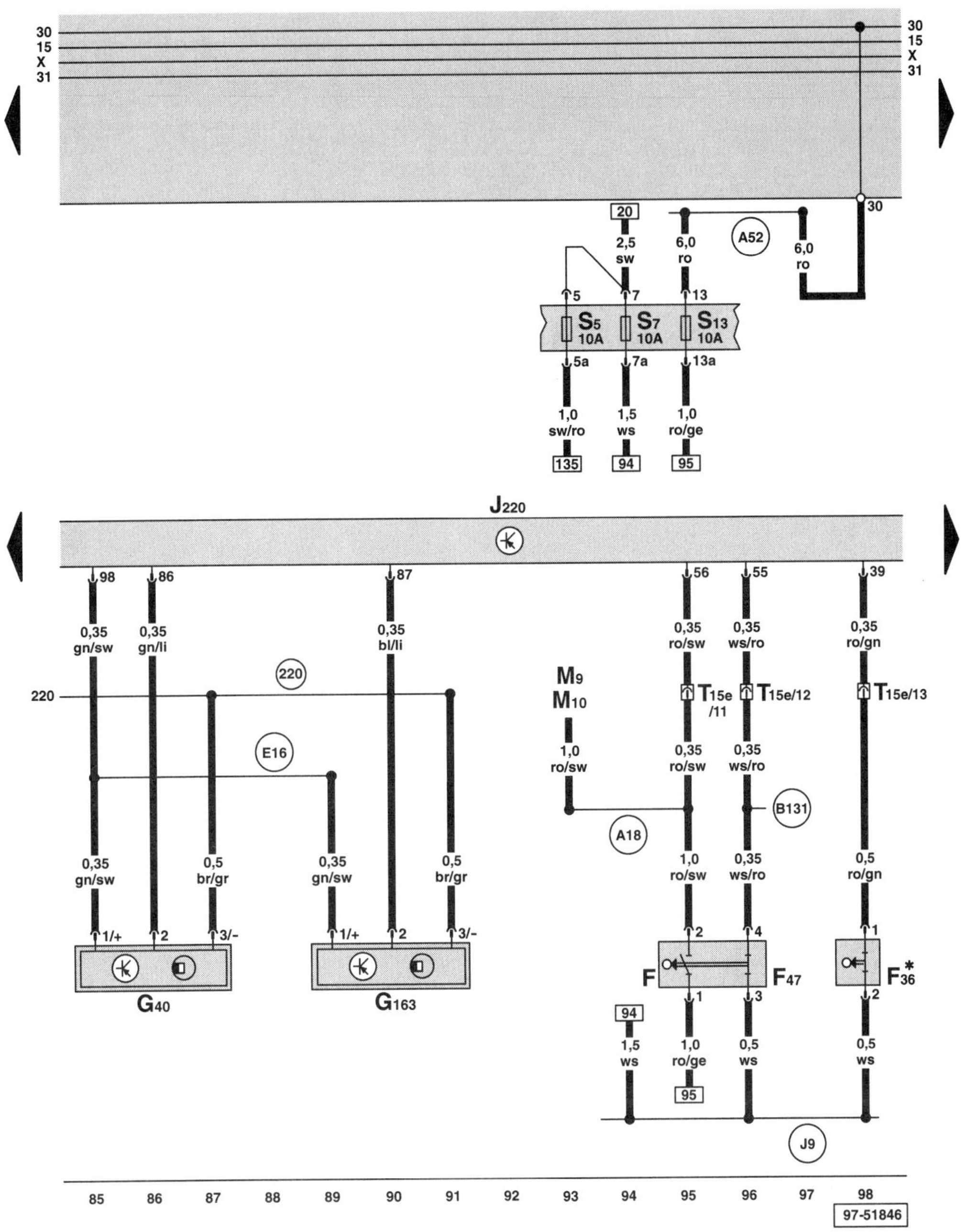

ws = white
sw = black
ro = red
br = brown
gn = green
bl = blue
gr = grey
li = violet
ge = yellow

F - Brake Light Switch
F36 - Clutch Vacuum Vent Valve Switch
F47 - Vacuum Vent Valve, Brake
G40 - Camshaft Position (CMP) Sensor
G163 - Camshaft Position (CMP) Sensor 2
J220 - Motronic Engine Control Module (ECM)
M9 - Left Brake Light
M10 - Right Brake Light
S5 - Fuse
S7 - Fuse
S13 - Fuse
T15e - Connector 15 pin, white, connector station electronic box plenum chamber

(220) - Ground connection (sensor ground), in engine compartment wiring harness
(A18) - Wire connection (54), in instrument panel wiring harness
(A52) - Plus connection -2- (30), in instrument panel wiring harness
(B131) - Connector (54), in wiring harness interior
(E16) - Plus connection (5 Volts), in Motronic Multiport Fuel Injection (MFI) wiring harness
(J9) - Wire connection -1- (15a), in ABS wiring harness
* - Vehicles with manually shifted transmission

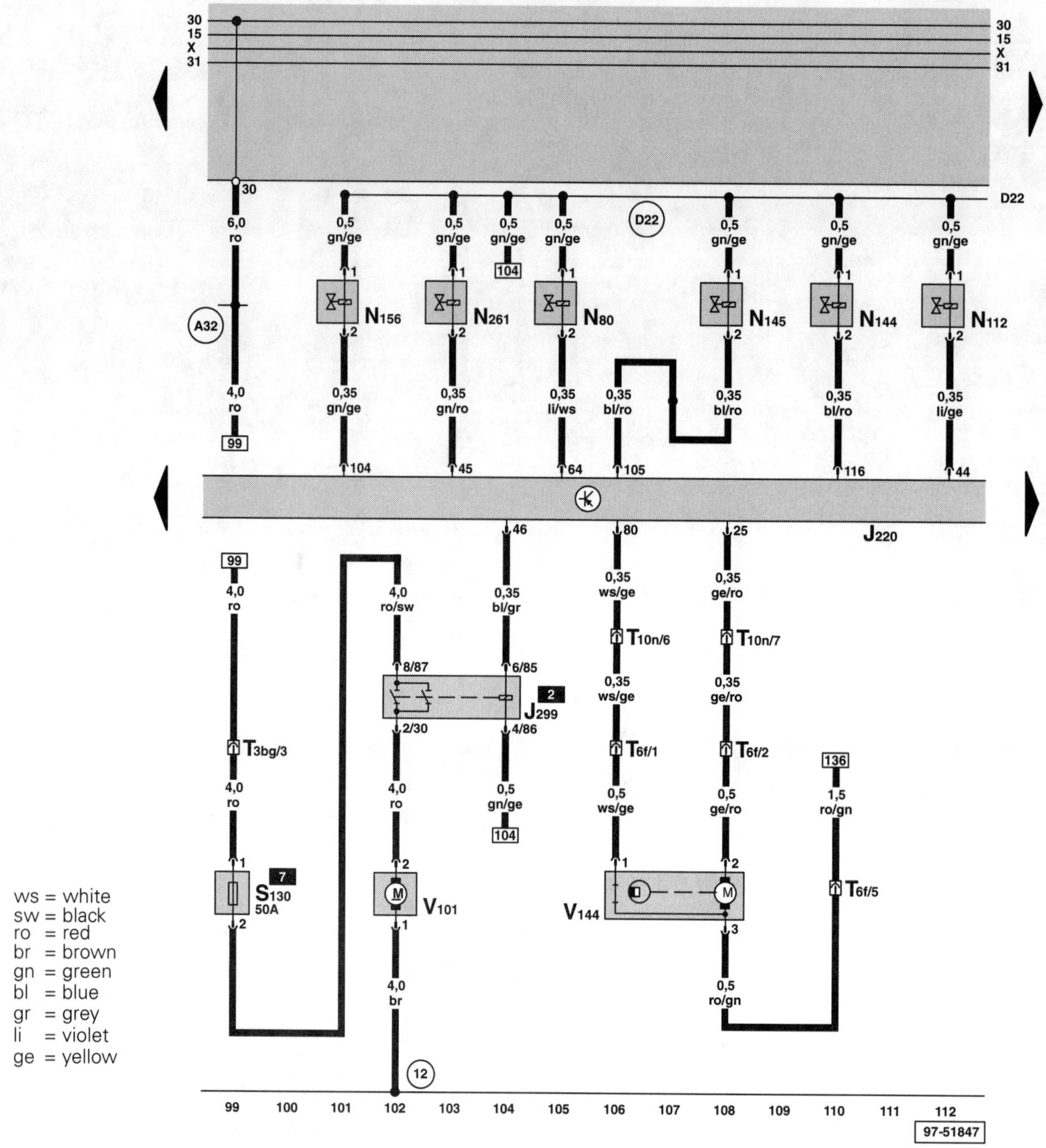

J220 - Motronic Engine Control Module (ECM)
J299 - Secondary Air Injection (AIR) Pump Relay
N80 - Evaporative Emission (EVAP) Canister Purge Regulator Valve
N112 - Secondary Air Injection (AIR) Solenoid Valve
N144 - Left Electro-Hydraulic Engine Mount Solenoid Valve
N145 - Right Electro-Hydraulic Engine Mount Solenoid Valve
N156 - Intake Change-Over Valve
N261 - Intake Manifold Tuning (IMT) Valve -2-
S130 - Fuse for secondary air pump
T3bg - Connector 3 pin, white, connector station electronic box plenum chamber
T6f - Connector 6 pin, darkbrown, connector station A pillar, right
T10n - Connector 10 pin, orange, connector station electronic box plenum chamber
V101 - Secondary Air Injection (AIR) Pump Motor
V144 - Leak Detection Pump (LDP)

(12) - Ground connection, in engine compartment, left

(A32) - Plus connection (30), in instrument panel wiring harness

(D22) - Connector (over fuse 234), in wiring harness front, right

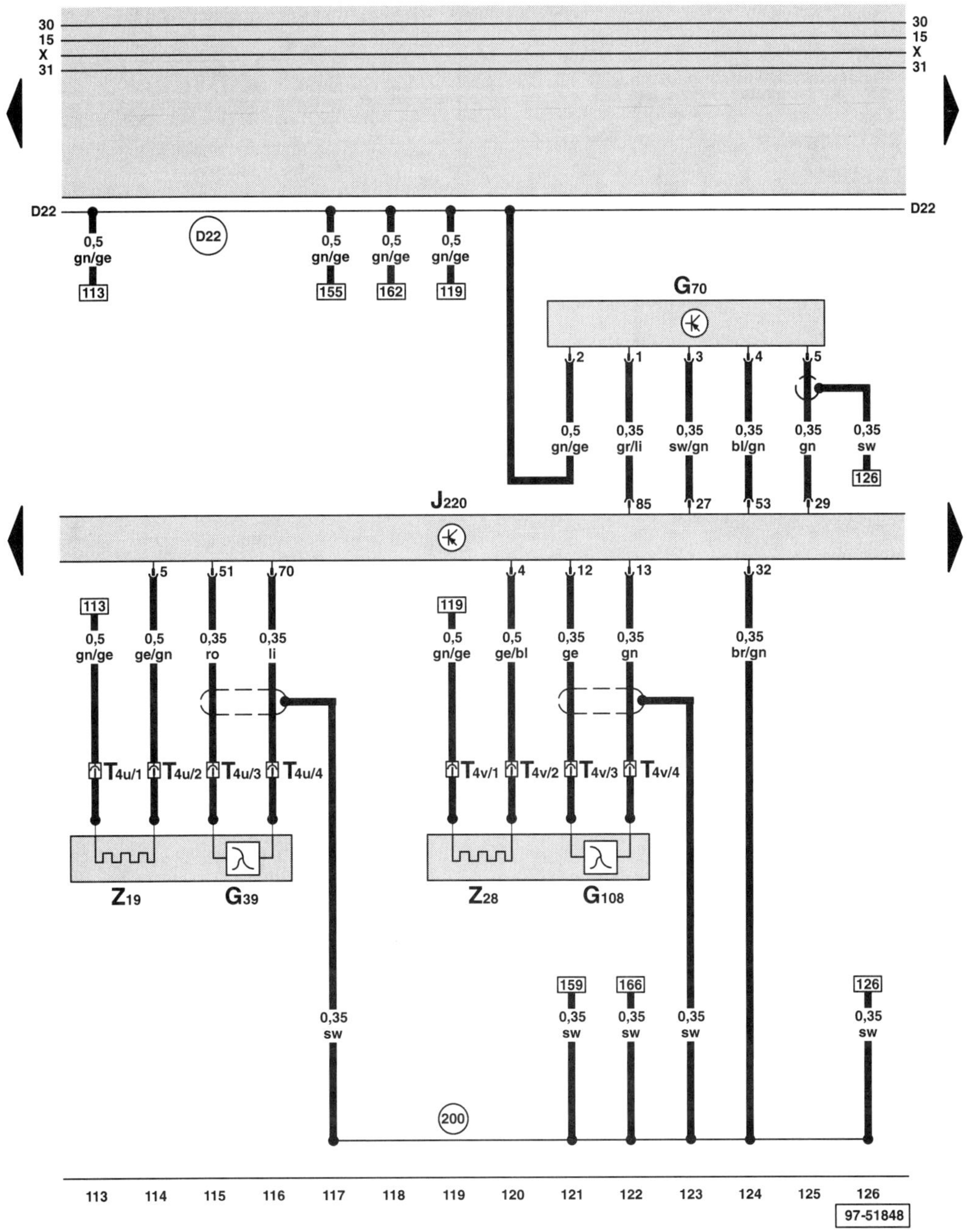

ws = white
sw = black
ro = red
br = brown
gn = green
bl = blue
gr = grey
li = violet
ge = yellow

G39 - Heated Oxygen Sensor (HO2S) 2
G70 - Mass Air Flow (MAF) Sensor
G108 - Heated Oxygen Sensor (HO2S) 2
J220 - Motronic Engine Control Module (ECM)
T4u - Connector 4 pin, black, heated oxygen sensor
T4v - Connector 4 pin, black, heated oxygen sensor 2
Z19 - Oxygen Sensor (O2S) Heater
Z28 - Oxygen Sensor (O2S) 2 Heater

(200) - Ground connection (shielding), in engine compartment wiring harness
(D22) - Connector (over fuse 234), in wiring harness front, right

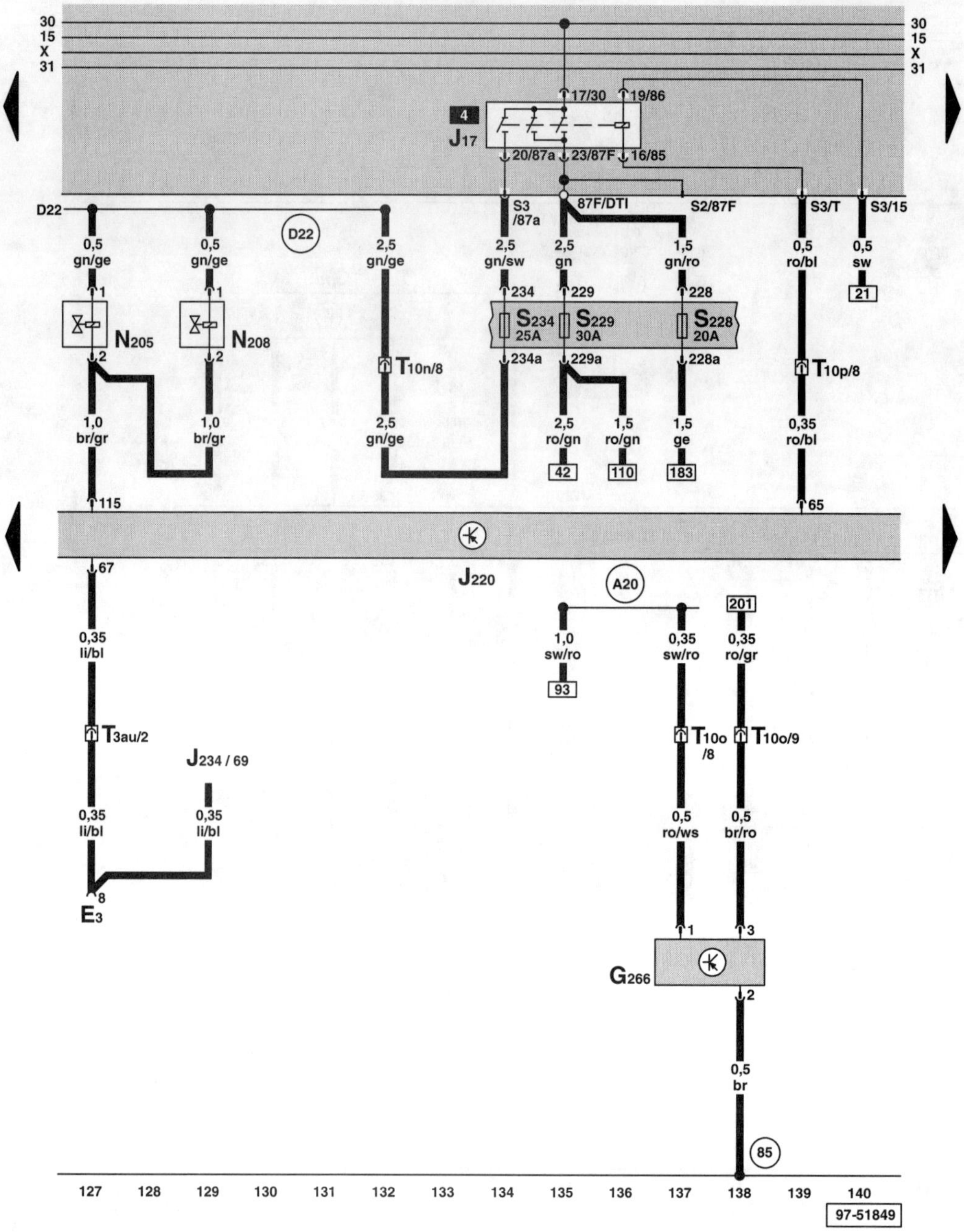

ws = white
sw = black
ro = red
br = brown
gn = green
bl = blue
gr = grey
li = violet
ge = yellow

- E3 - Emergency Flasher Switch
- G266 - Oil Level Thermal Sensor (SRI)
- J17 - Fuel Pump (FP) Relay
- J220 - Motronic Engine Control Module (ECM)
- J234 - Airbag Control Module
- N205 - Valve -1- for camshaft adjustment
- N208 - Valve -2- for camshaft adjustment
- S228 - Fuse 28 in fuse holder
- S229 - Fuse 29 in fuse holder
- S234 - Fuse 34 in fuse holder
- T3au - Connector 3 pin, red, connector station electronic box plenum chamber
- T10n - Connector 10 pin, orange, connector station electronic box plenum chamber
- T10o - Connector 10 pin, brown, connector station electronic box plenum chamber
- T10p - Connector 10 pin, black, connector station electronic box plenum chamber

- (85) - Ground connection -1-, in engine compartment wiring harness
- (A20) - Wire connection (15a), in instrument panel wiring harness
- (D22) - Connector (over fuse 234), in wiring harness front, right

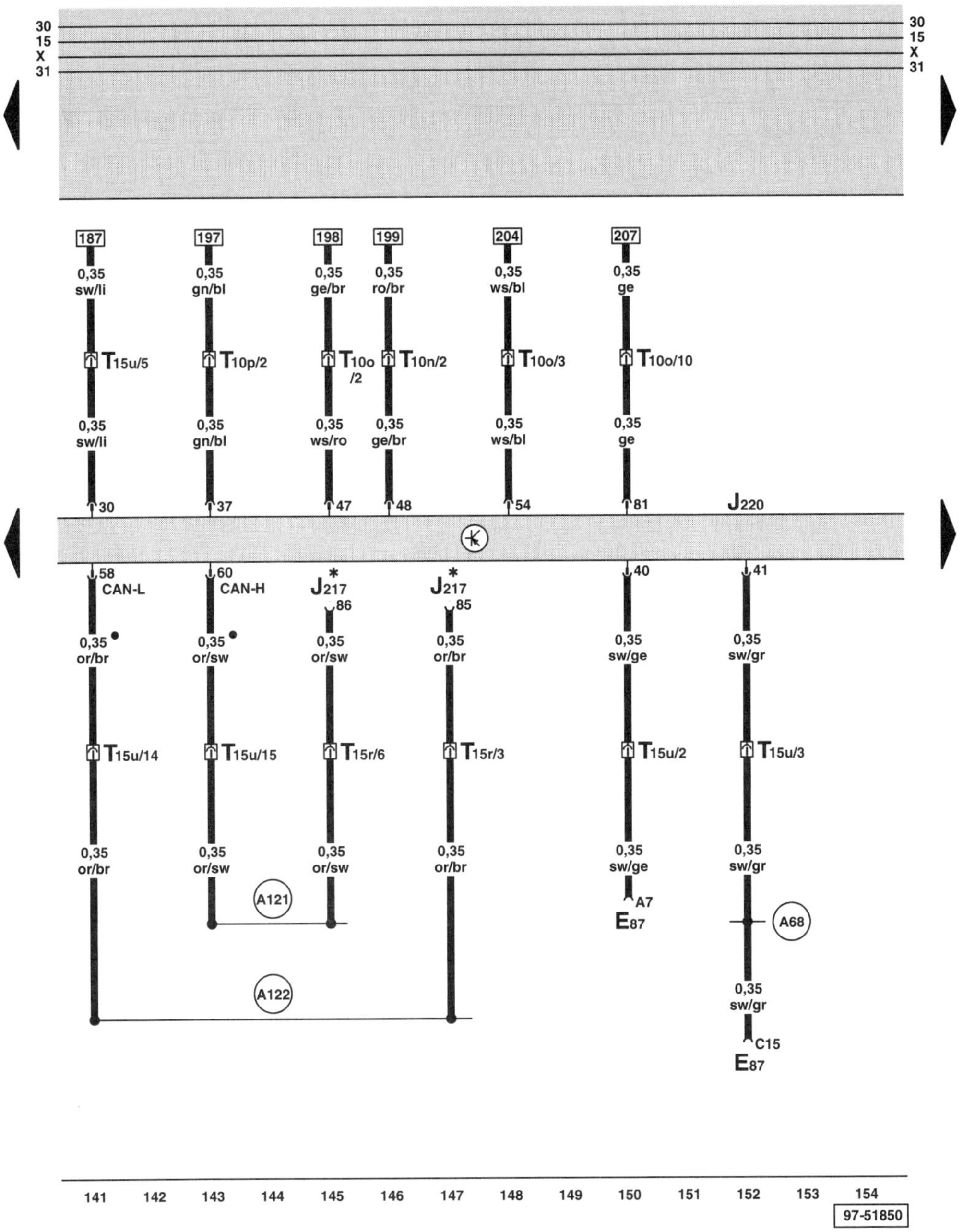

ws = white
sw = black
ro = red
br = brown
gn = green
bl = blue
gr = grey
li = violet
ge = yellow
or = orange

E87 - A/C Control Head
J217 - Transmission Control Module (TCM)
J220 - Motronic Engine Control Module (ECM)
T10n - Connector 10 pin, orange, connector station electronic box plenum chamber
T10o - Connector 10 pin, brown, connector station electronic box plenum chamber
T10p - Connector 10 pin, black, connector station electronic box plenum chamber
T15r - Connector 15 pin, blue, connector station A pillar, right
T15u - Connector 15 pin, red, connector station electronic box plenum chamber

(A68) - Connector (C15, A/C), in instrument panel wiring harness
(A121) - Connection (high bus) in instrument panel wiring harness
(A122) - Connection (low bus) in instrument panel wiring harness
• - CAN-Bus (Data-Bus)
* - Vehicles with automatic transmission

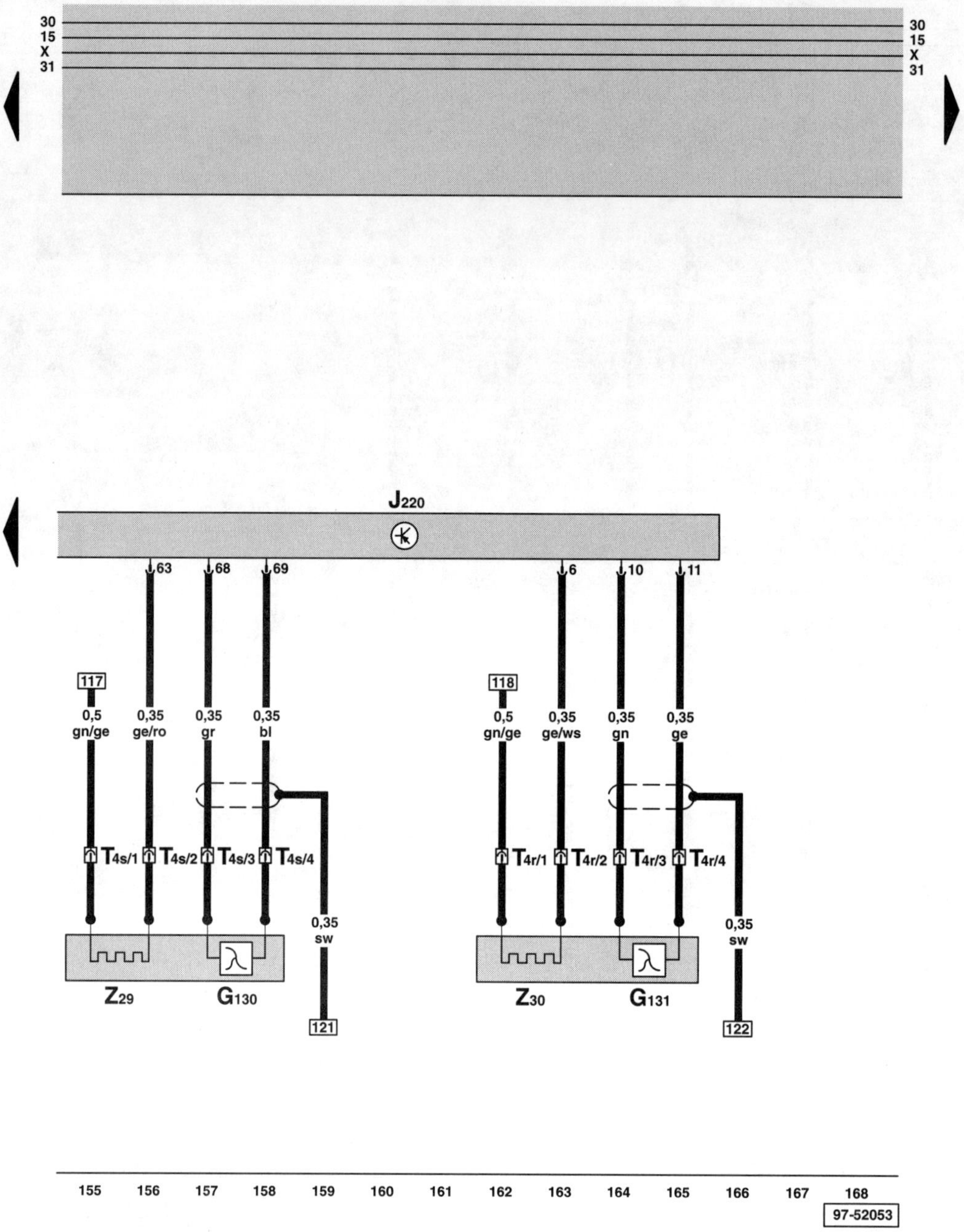

ws = white
sw = black
ro = red
br = brown
gn = green
bl = blue
gr = grey
li = violet
ge = yellow

G130 - Heated Oxygen Sensor (O2S) Behind Three Way Catalytic Converter (TWC)
G131 - Heated Oxygen Sensor (O2S) 2 Behind Three Way Catalytic Converter (TWC)
J220 - Motronic Engine Control Module (ECM)
T4r - Connector 4 pin, brown, in driver's seat
T4s - Connector 4 pin, green, near oxygen sensor, behind Three Way Catalytic Converter (TWC)
Z29 - Oxygen Sensor (O2S) Heater
Z30 - Oxygen Sensor (O2S) 2 Heater

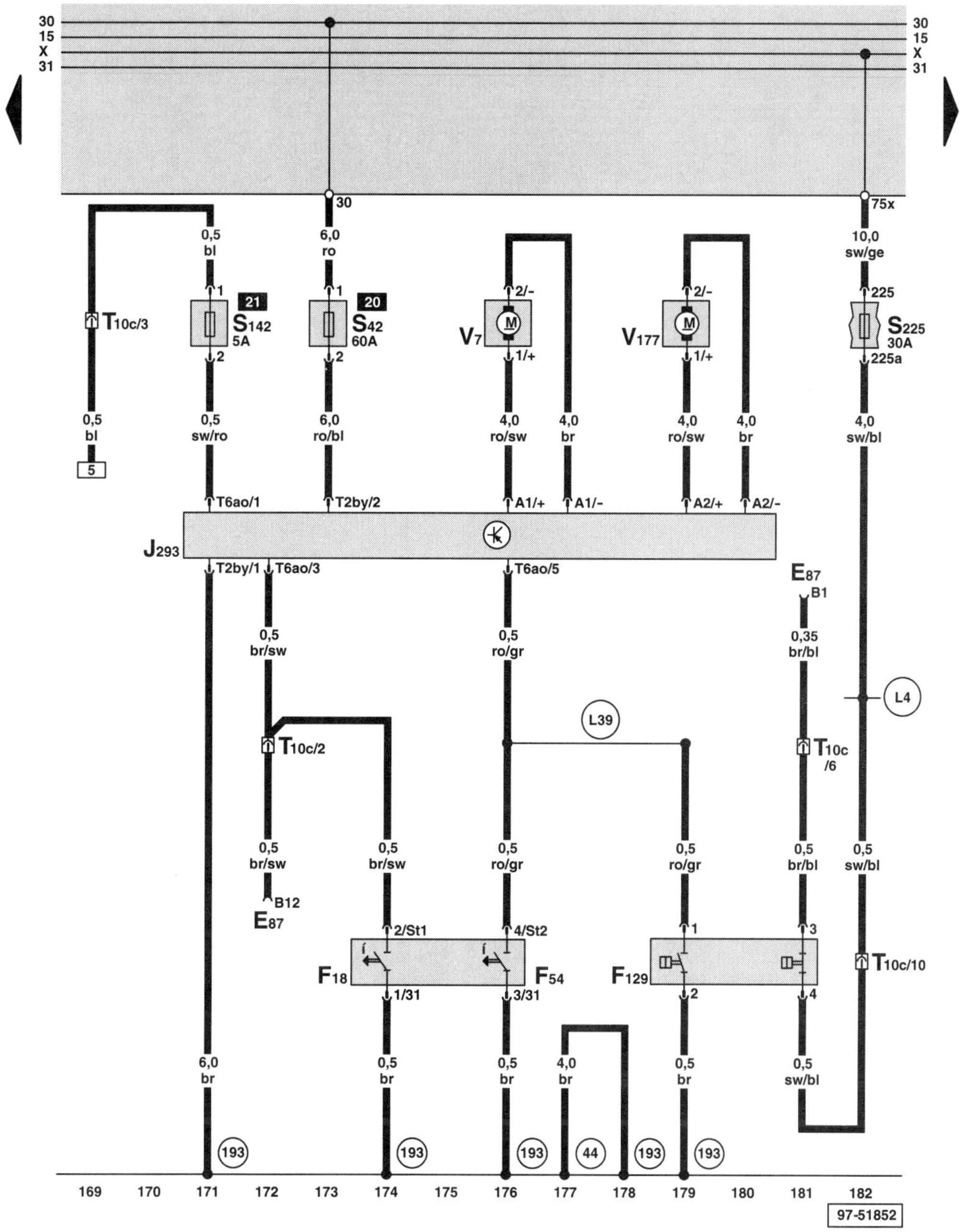

ws = white
sw = black
ro = red
br = brown
gn = green
bl = blue
gr = grey
li = violet
ge = yellow

E87 - A/C Control Head
F18 - Coolant Fan Control (FC) Thermal Switch
F54 - Coolant Fan Control (FC) Thermal Switch
F129 - A/C Pressure Switch
J293 - Coolant FC (Fan Control) Control Module
S42 - Coolant Fan Fuse
S142 - Control module fuse for coolant fan
S225 - Fuse 25 in fuse holder
T2by - Connector 2 pin, on Coolant FC (Fan Control) Control Module
T6ao - Connector 6 pin, on Coolant FC (Fan Control) Control Module
T10c - Connector 10 pin, grey, connector station A pillar, left
V7 - Coolant Fan
V177 - Coolant fan -2-

(44) - Ground connection, on left A-pillar, lower part

(193) - Ground connection -1-, in coolant fan wiring harness

(L4) - Wire connection (75al), in A/C wiring harness

(L39) - Wire connection -1- (A/C Pressure Switch), in Climatronic wiring harness

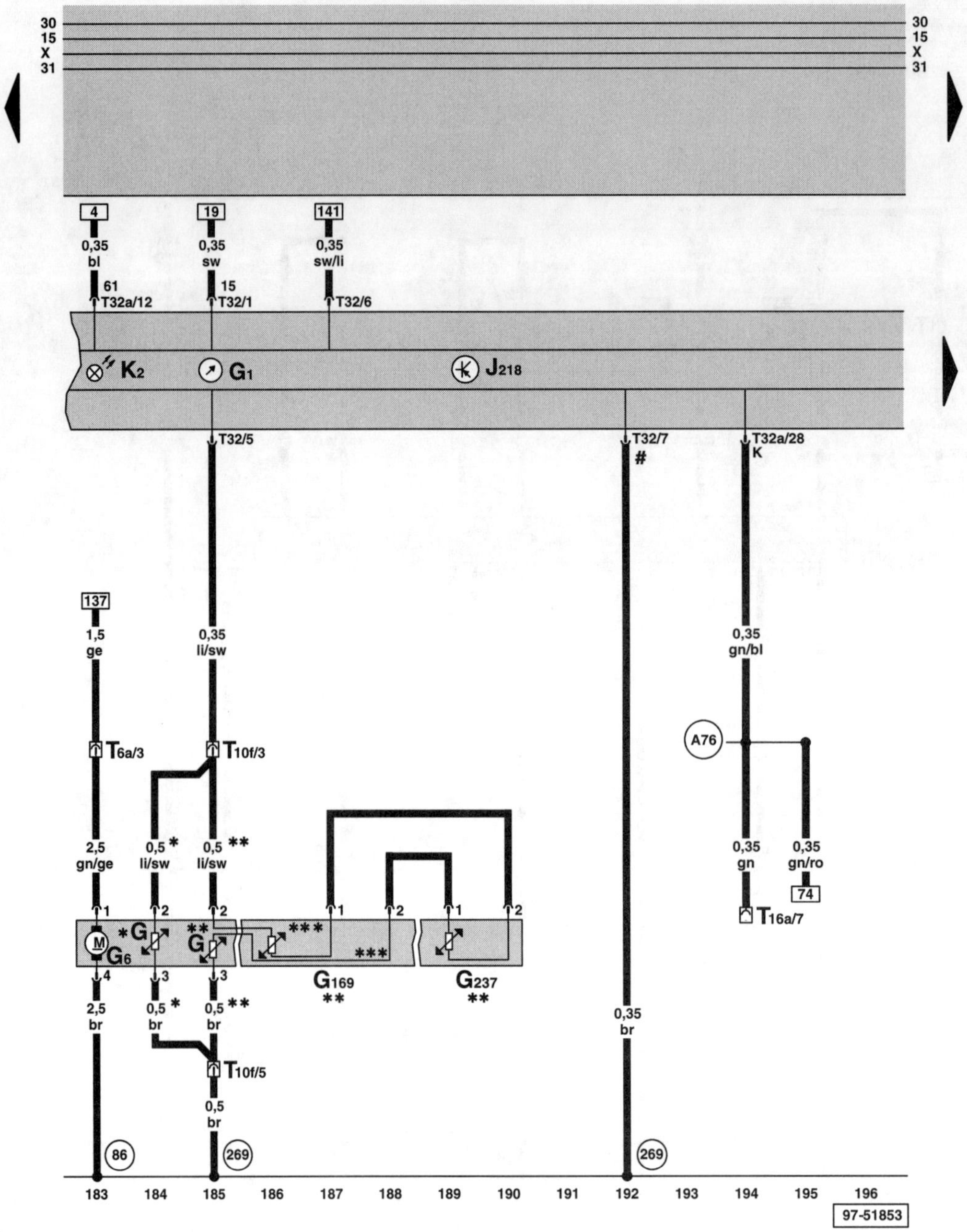

ws = white
sw = black
ro = red
br = brown
gn = green
bl = blue
gr = grey
li = violet
ge = yellow

G - Sender for fuel gauge
G1 - Fuel Gauge
G6 - Fuel Pump (FP)
G169 - Sender 2 for fuel gauge
G237 - Sensor -3- for fuel supply
J218 - Instrument Cluster Combination Processor
K2 - Generator (GEN) Warning Light
T6a - Connector 6 pin, blue, connector station A pillar, left
T10f - Connector 10 pin, brown, connector station A pillar, left
T16a - Connector 16 pin, Data Link Connector (DLC)
T32 - Connector 32 pin, blue, on instrument cluster
T32a - Connector 32 pin, green, on instrument cluster

(86) - Ground connection -1-, in rear wiring harness
(269) - Ground connector (sensor ground) -1-, in instrument panel wiring harness
(A76) - Connector (K-diagnosis wire), in instrument panel wiring harness

* - Vehicles with front wheel drive
** - Vehicles with all wheel drive
*** - Wiring inside fuel tank
# - Sensor ground, output

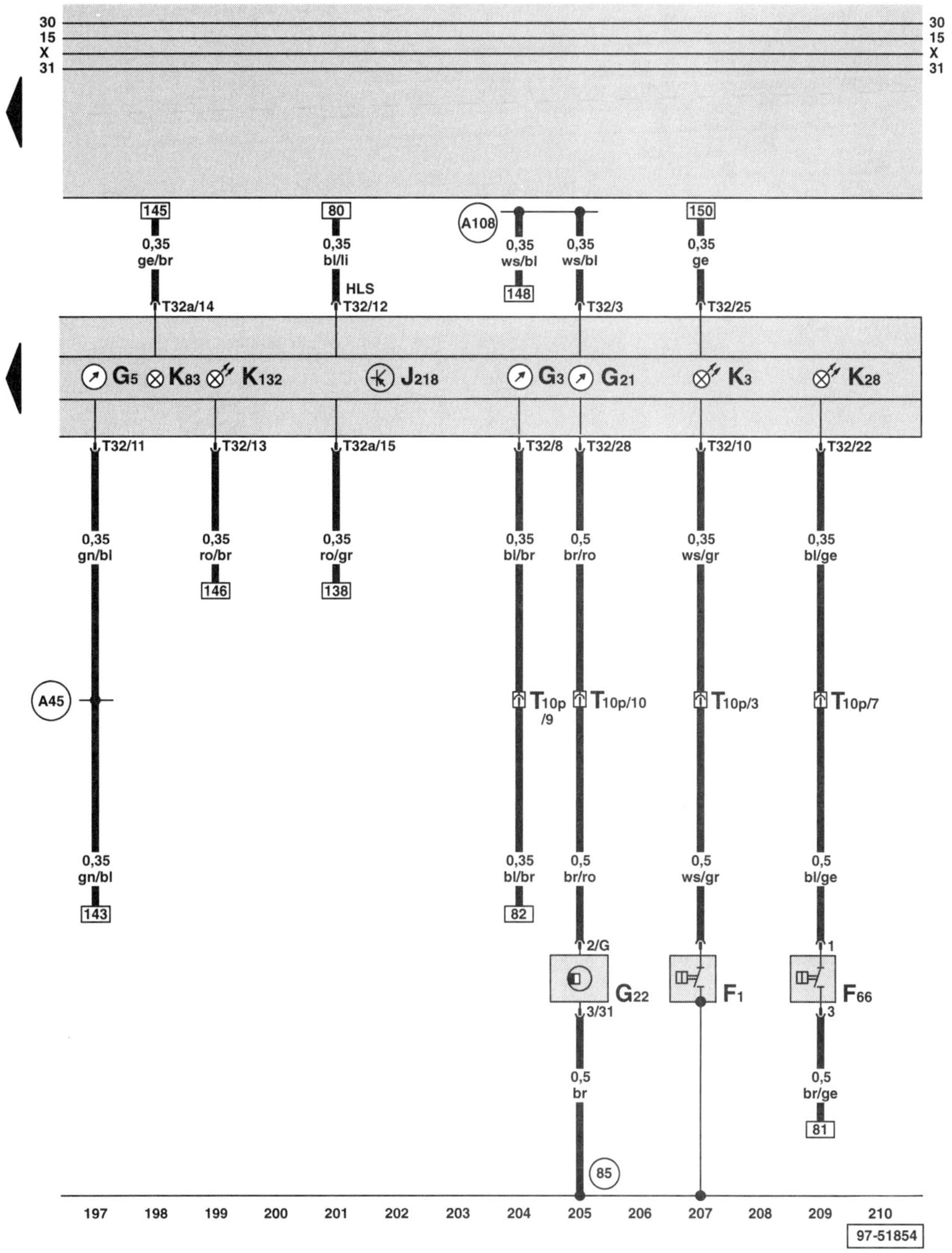

ws = white
sw = black
ro = red
br = brown
gn = green
bl = blue
gr = grey
li = violet
ge = yellow

- F1 - Oil Pressure Switch
- F66 - Engine Coolant Level (ECL) Warning Switch
- G3 - Engine Coolant Temperature (ECT) Gauge
- G5 - Tachometer
- G21 - Speedometer
- G22 - Speedometer Vehicle Speed Sensor (VSS)
- J218 - Instrument Cluster Combination Processor
- K3 - Oil Pressure Warning Light
- K28 - Engine Coolant Level/Temperature (ECL/ECT) Warning Light
- K83 - Malfunction Indicator Lamp (MIL)
- K132 - Fault light for power accelerator activation
- T10p - Connector 10 pin, black, connector station electronic box plenum chamber
- T32 - Connector 32 pin, blue, on instrument cluster
- T32a - Connector 32 pin, green, on instrument cluster

- (85) - Ground connection -1-, in engine compartment wiring harness
- (A45) - Wire connection (RPM-signal), in instrument panel wiring harness
- (A108) - Connector (speed signal), in instrument panel wiring harness

THIS PAGE INTENTIONALLY LEFT BLANK

Automatic Transmission (01V), (01L)

2000 m. y.

Fuse Panel

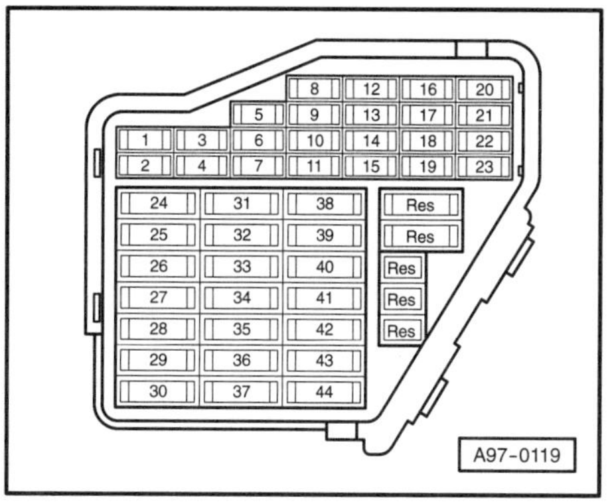

Fuse Colors:

30 A - Green
25 A - White
20 A - Yellow
15 A - Blue
10 A - Red
7,5 A - Brown
5 A - Beige

Starting with fuse position 23, fuses in the fuse holder are identified with 223 in the wiring diagram.

13 - Fold Relay Panel

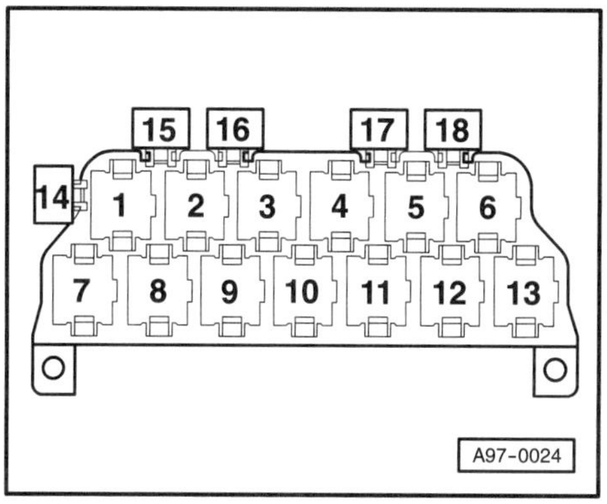

Relay Location:

4 -Starting Interlock Relay, J207

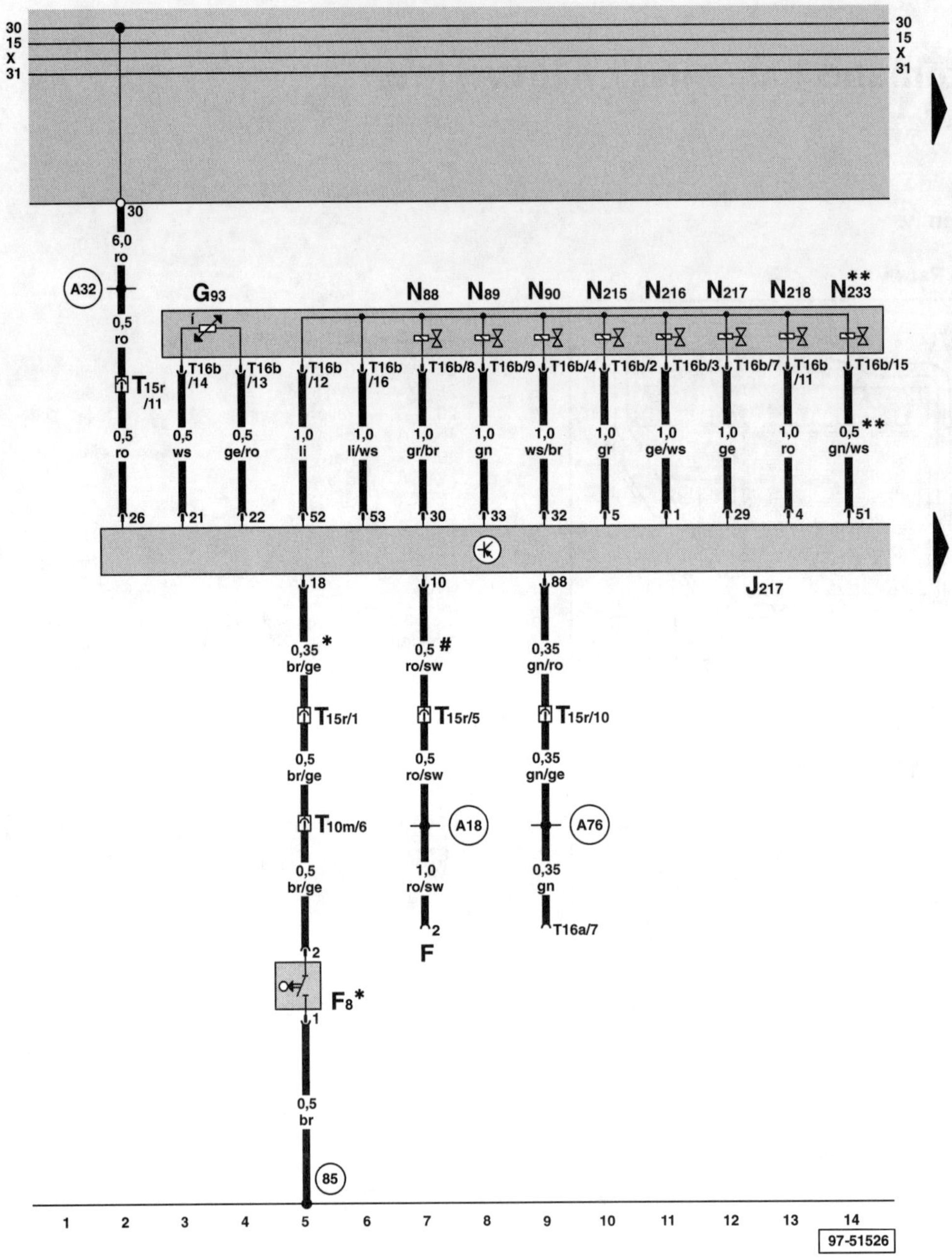

ws = white
sw = black
ro = red
br = brown
gn = green
bl = blue
gr = grey
li = violet
ge = yellow

F - Brake Light Switch
F8 - Kick Down Switch
G93 - Transmission Fluid Temperature Sensor
J217 - Transmission Control Module (TCM)
N88 - Solenoid Valve 1
N89 - Solenoid Valve 2
N90 - Solenoid Valve 3
N215 - Pressure control valve -1- for auto. transmission
N216 - Pressure control valve -2- for auto. transmission
N217 - Pressure control valve -3- for auto. transmission
N218 - Pressure control valve -4- for auto. transmission
N233 - Pressure control valve -5- for auto. transmission
T10m - Connector 10 pin, blue, connector station electronic box plenum chamber
T15r - Connector 15 pin, blue, connector station A pillar, right
T16a - Connector 16 pin, Data Link Connector (DLC)
T16b - Connector 16 pin, on transmission

(85) - Ground connection -1-, in engine compartment wiring harness
(A18) - Wire connection (54), in instrument panel wiring harness
(A32) - Plus connection (30), in instrument panel wiring harness
(A76) - Connector (K-diagnosis wire), in instrument panel wiring harness

* - Vehicles with mechanical throttle actuation
** - Vehicles with automatic transmission 01V
# - Vehicles with automatic transmission 01L

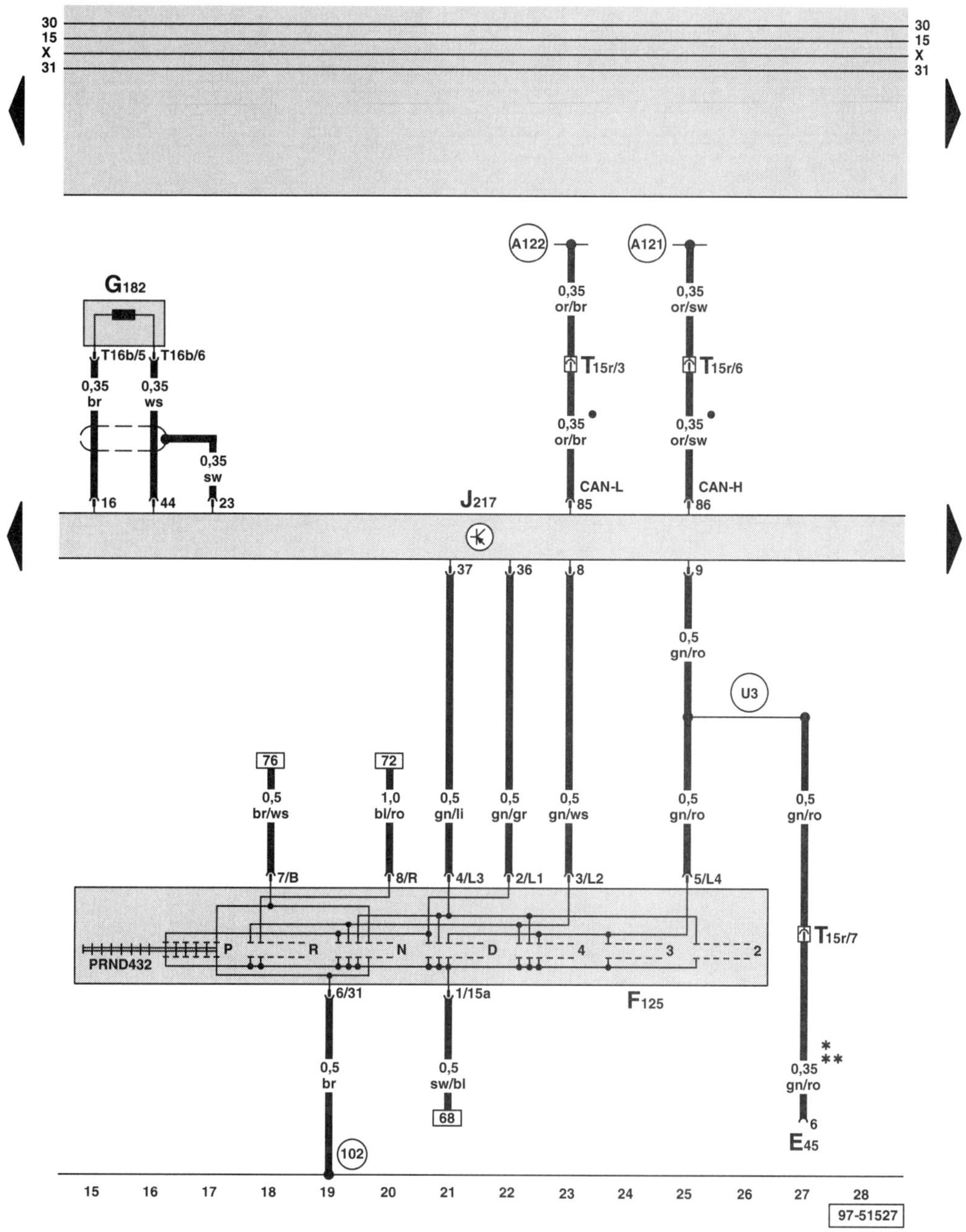

ws = white
sw = black
ro = red
br = brown
gn = green
bl = blue
gr = grey
li = violet
ge = yellow
or = orange

E45 - Cruise Control Switch
F125 - Multi-Function Transmission Range (TR) Switch
G182 - Sensor for transmission RPM
J217 - Transmission Control Module (TCM)
T15r - Connector 15 pin, blue, connector station A pillar, right
T16b - Connector 16 pin, on transmission

(102) - Ground connection, in transmission wiring harness
(A121) - Connection (high bus) in instrument panel wiring harness
(A122) - Connection (low bus) in instrument panel wiring harness
(U3) - Wire connection -1-, in automatic transmission wiring harness

● - CAN-Bus (Data-Bus)
* - Vehicles with cruise control
** - Vehicles with mechanical throttle actuation

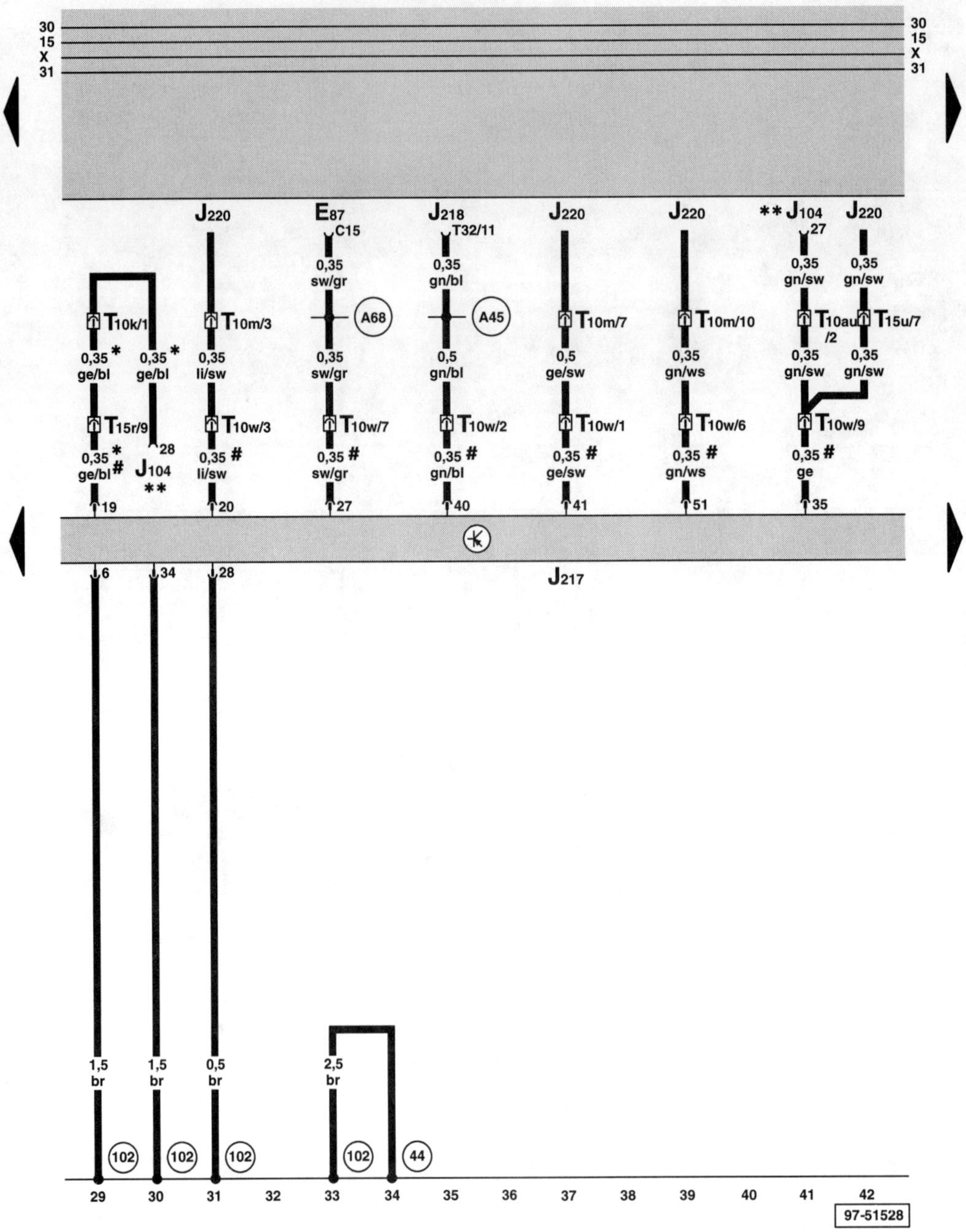

ws = white
sw = black
ro = red
br = brown
gn = green
bl = blue
gr = grey
li = violet
ge = yellow

E87 - A/C Control Head
J104 - ABS Control Module (w/EDL)
J217 - Transmission Control Module (TCM)
J218 - Instrument Cluster Combination Processor
J220 - Motronic Engine Control Module (ECM)
T10k - Connector 10 pin, orange, connector station A pillar, left
T10m - Connector 10 pin, blue, connector station electronic box plenum chamber
T10w - Connector 10 pin, blue, connector station A pillar, right
T10au - Connector 10 pin, grey, connector station A pillar, right
T15r - Connector 15 pin, blue, connector station A pillar, right
T15u - Connector 15 pin, red, connector station electronic box plenum chamber
T32 - Connector 32 pin, blue, on instrument cluster

(44) - Ground connection, on left A-pillar, lower part
(102) - Ground connection, in transmission wiring harness
(A45) - Wire connection (RPM-signal), in instrument panel wiring harness
(A68) - Connector (C8, A/C), in instrument panel wiring harness

* - Vehicles with Anti-Slip control
** - Vehicles with ABS 5.3
# - Vehicles only without CAN-Bus

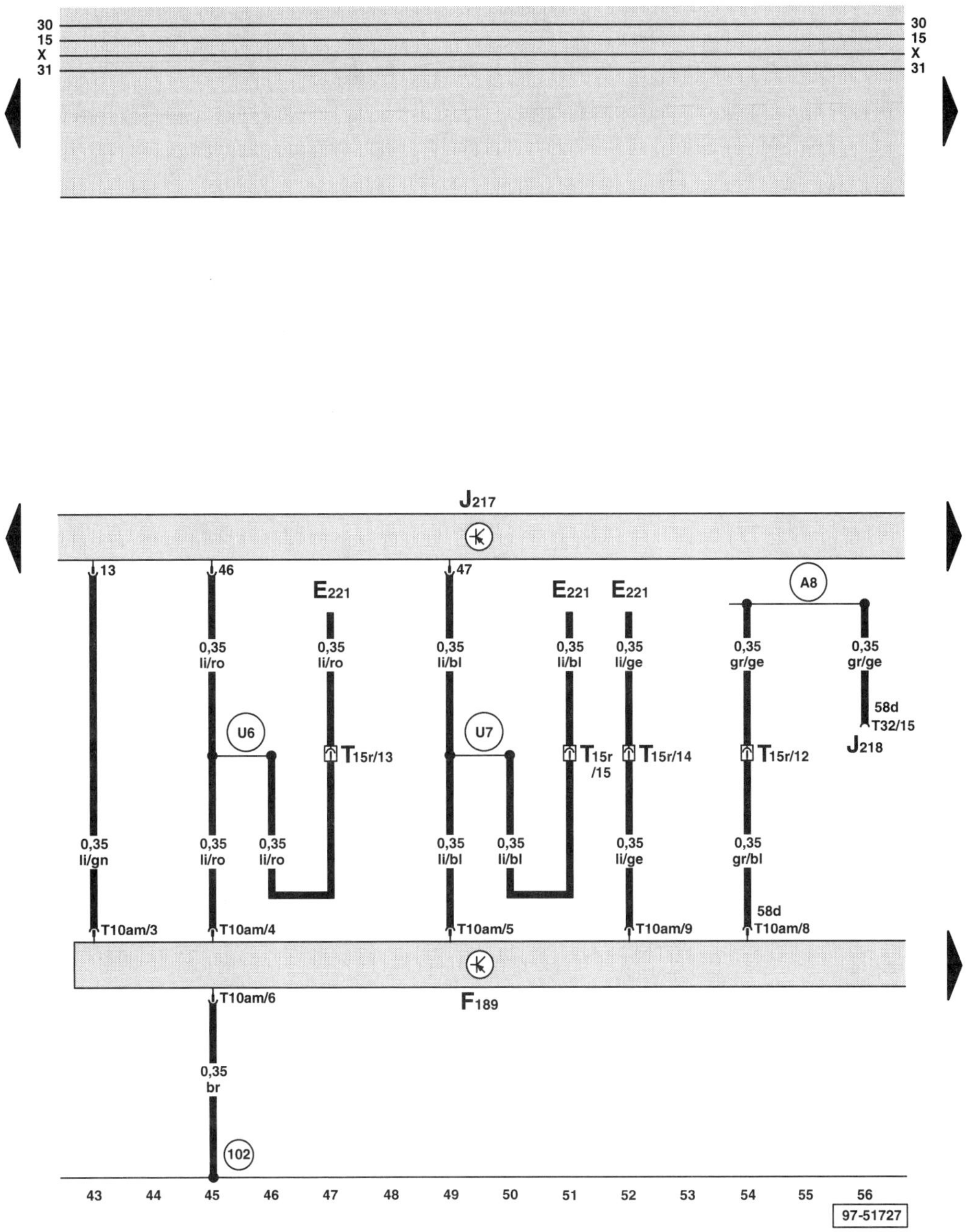

ws = white
sw = black
ro = red
br = brown
gn = green
bl = blue
gr = grey
li = violet
ge = yellow

E221 - Control unit in steering wheel
F189 - Tiptronic Switch
J217 - Transmission Control Module (TCM)
J218 - Instrument Cluster Combination Processor
T10am- Connector 10 pin, on selector lever
T15r - Connector 15 pin, blue, connector station A pillar, right
T32 - Connector 32 pin, blue, on instrument cluster

(102) - Ground connection, in transmission wiring harness
(A8) - Plus connection (58d), in instrument panel wiring harness
(U6) - Wire connection -2-, in automatic transmission wiring harness
(U7) - Wire connection -3-, in automatic transmission wiring harness

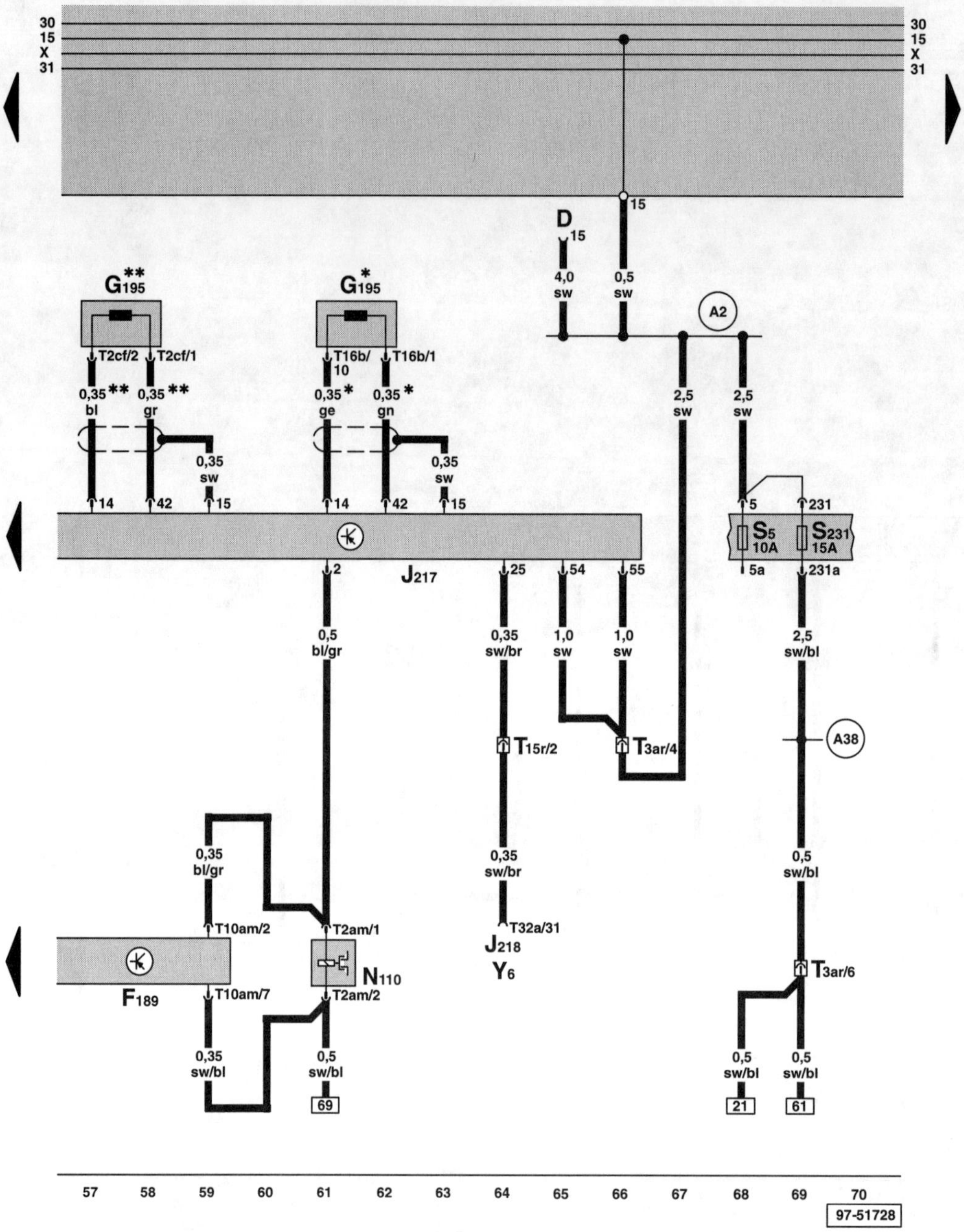

ws = white
sw = black
ro = red
br = brown
gn = green
bl = blue
gr = grey
li = violet
ge = yellow

D - Ignition/Starter Switch
F189 - Tiptronic Switch
G195 - Sender for transmission output RPM
J217 - Transmission Control Module (TCM)
J218 - Instrument Cluster Combination Processor
N110 - Shift Lock Solenoid
S5 - Fuse
S231 - Fuse 31 in fuse holder
T2am - Connector 2 pin, black, on selector lever
T2cf - Connector 2 pin, black, on transmission
T3ar - Connector 3 pin, blue, connector station A pillar, right
T10am - Connector 10 pin, on selector lever
T15r - Connector 15 pin, blue, connector station A pillar, right
T16b - Connector 16 pin, on transmission
T32a - Connector 32 pin, green, on instrument cluster
Y6 - Transmission Range (TR) Selector Lever Display

(A2) - Plus connection (15), in instrument panel wiring harness

(A38) - Plus connection (15a II), in instrument panel wiring harness

* - Vehicles with automatic transmission 01V

** - Vehicles with automatic transmission 01L

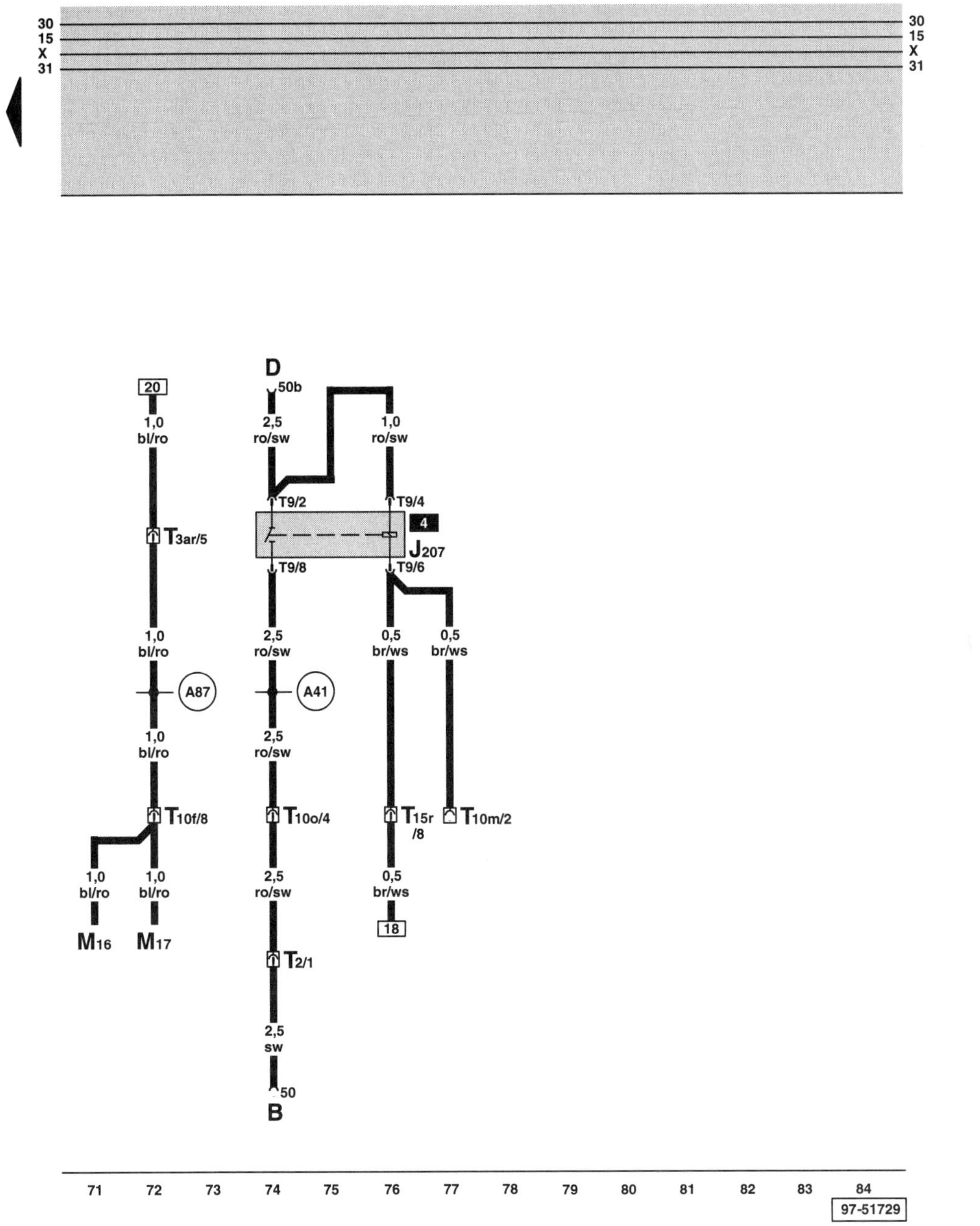

ws = white
sw = black
ro = red
br = brown
gn = green
bl = blue
gr = grey
li = violet
ge = yellow

B - Starter
D - Ignition/Starter Switch
J207 - Starting Interlock Relay
M16 - Left Back-Up Light
M17 - Right Back-Up Light
T2 - Connector 2 pin, grey, in engine compartment, right
T3ar - Connector 3 pin, blue, connector station A pillar, right
T9 - Connector 9 pin, brown, on starting interlock relay
T10f - Connector 10 pin, brown, connector station A pillar, left
T10m - Connector 10 pin, blue, connector station electronic box plenum chamber
T10o - Connector 10 pin, brown, connector station electronic box plenum chamber
T15r - Connector 15 pin, blue, connector station A pillar, right

(A41) - Plus connection (50), in instrument panel wiring harness

(A87) - Connector (RF), in instrument panel wiring harness

THIS PAGE INTENTIONALLY LEFT BLANK

Central Locking System / Anti Theft Alarm System / Ultra-Sound Interior Monitor

2000 m. y.

Fuse Panel

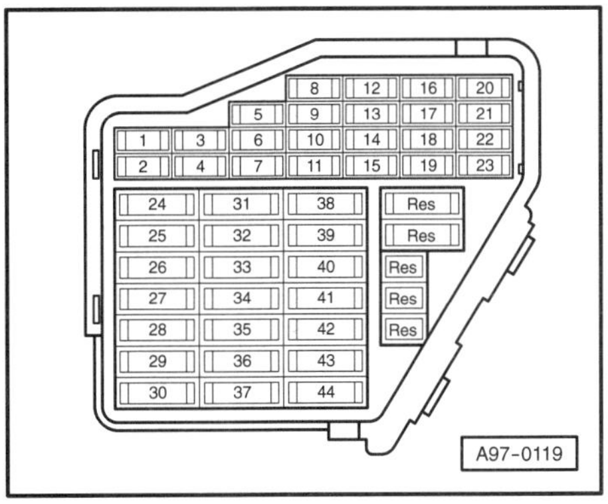

Fuse Colors:

30 A - Green
25 A - White
20 A - Yellow
15 A - Blue
10 A - Red
7,5 A - Brown
5 A - Beige

Starting with fuse position 23, fuses in the fuse holder are identified with 223 in the wiring diagram.

13 - Fold Relay Panel

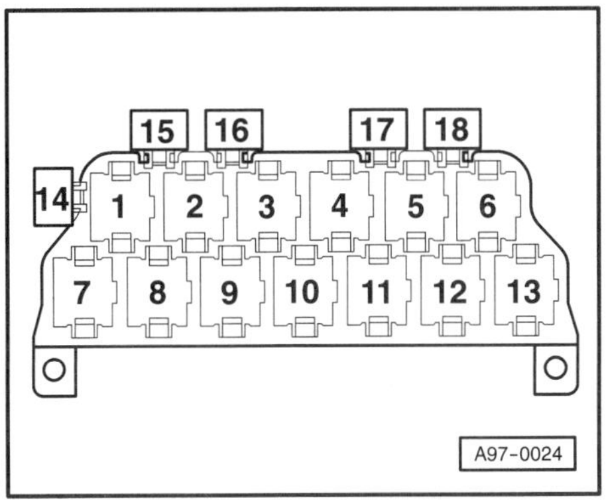

Relay Location:

4 -Starting Interlock Relay, J207

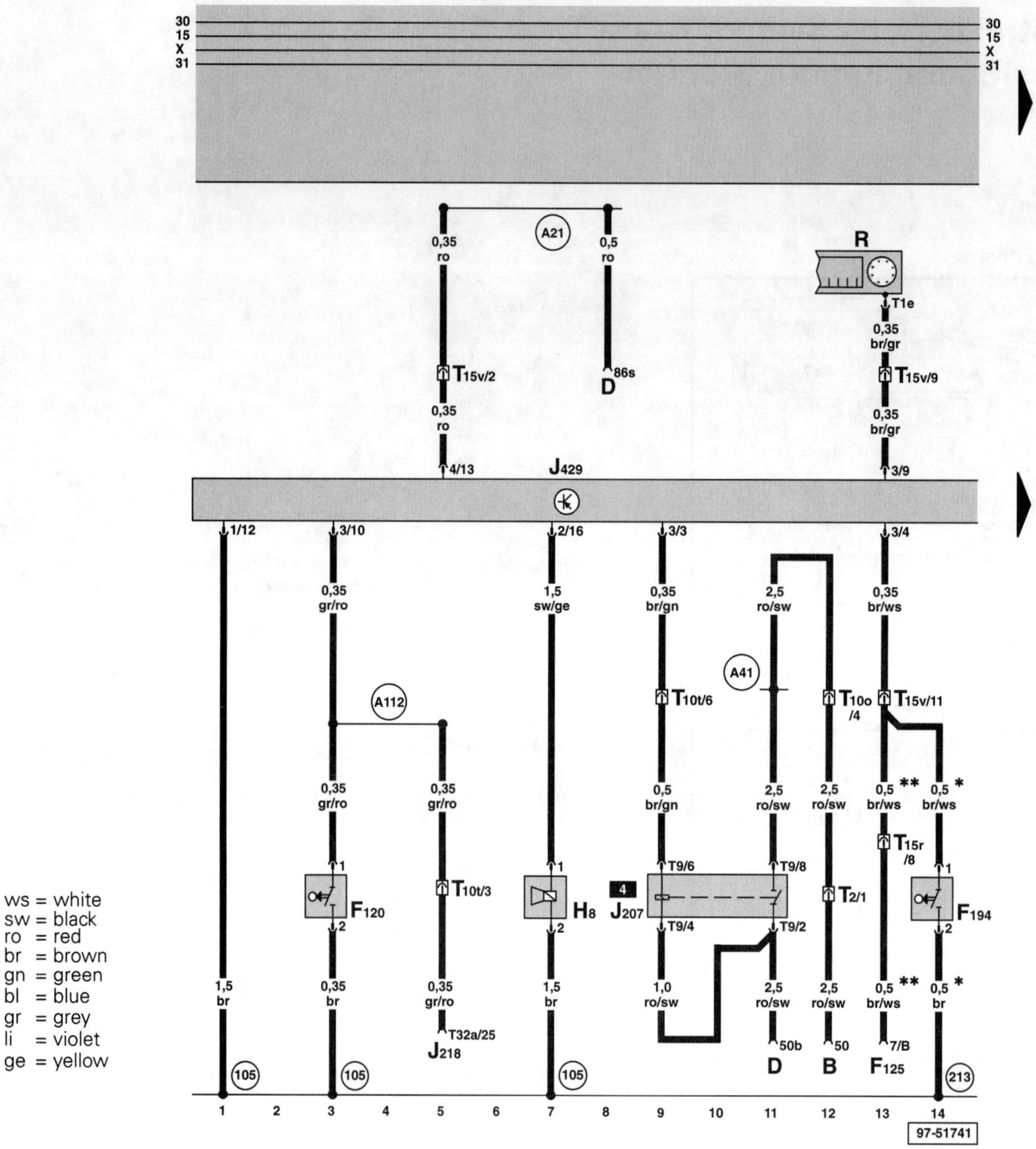

B - Starter
D - Ignition/Starter Switch
F120 - Hood Alarm Switch
F125 - Multi-Function Transmission Range (TR) Switch
F194 - Clutch Pedal Position (CPP) Switch
H8 - Alarm Horn
J207 - Starting Interlock Relay
J218 - Instrument Cluster Combination Processor
J429 - Control module for central locking
R - Radio
T1e - Single connector, on Radio
T2 - Connector 2 pin, grey, in engine compartment, right
T9 - Connector 9 pin, brown, on starting interlock relay
T10o - Connector 10 pin, brown, connector station electronic box plenum chamber
T10t - Connector 10 pin, orange, connector station A pillar, right
T15r - Connector 15 pin, blue, connector station A pillar, right
T15v - Connector 15 pin, orange, connector station A pillar, right
T32a - Connector 32 pin, green, on instrument cluster

(105) - Ground connection -1-, in central locking system wiring harness
(213) - Ground connection, in instrument panel wiring harness
(A21) - Wire connection (86s), in instrument panel wiring harness
(A41) - Plus connection (50), in instrument panel wiring harness
(A112) - Connection (driver's cab lock), in instrument panel wiring harness

* - Vehicles with manually shifted transmission
** - Vehicles with automatic transmission

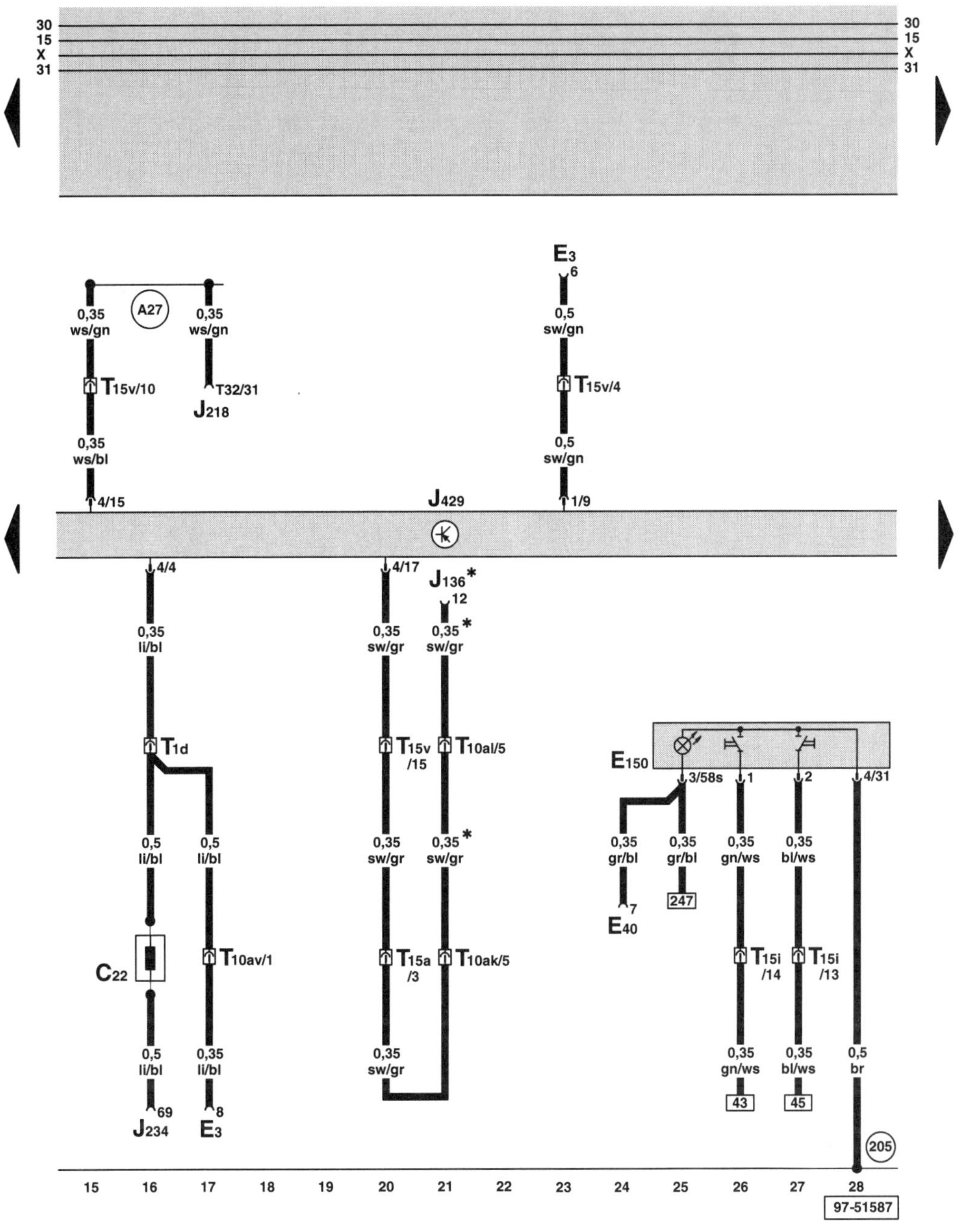

ws = white
sw = black
ro = red
br = brown
gn = green
bl = blue
gr = grey
li = violet
ge = yellow

C22 - Interference filter
E3 - Emergency Flasher Switch
E40 - Left Front Window Switch
E150 - Switch for interior lock, driver side
J136 - Memory Seat Control Module
J218 - Instrument Cluster Combination Processor
J234 - Airbag Control Module
J429 - Control module for central locking
T1d - Single connector, brown, connector station A pillar right
T10ak - Connector 10 pin, green, connector station A pillar, left
T10al - Connector 10 pin, brown, below driver's seat
T10av - Connector 10 pin, yellow, connector station A pillar, right
T15a - Connector 15 pin, black, connector station A pillar, left
T15i - Connector 15 pin, brown, connector station A pillar, left
T15v - Connector 15 pin, orange, connector station A pillar, right
T32 - Connector 32 pin, blue, on instrument cluster

(205) - Ground connection, in driver's door wiring harness

(A27) - Wire connection (vehicle speed signal), in instrument panel wiring harness

* - Vehicles with Memory Seat

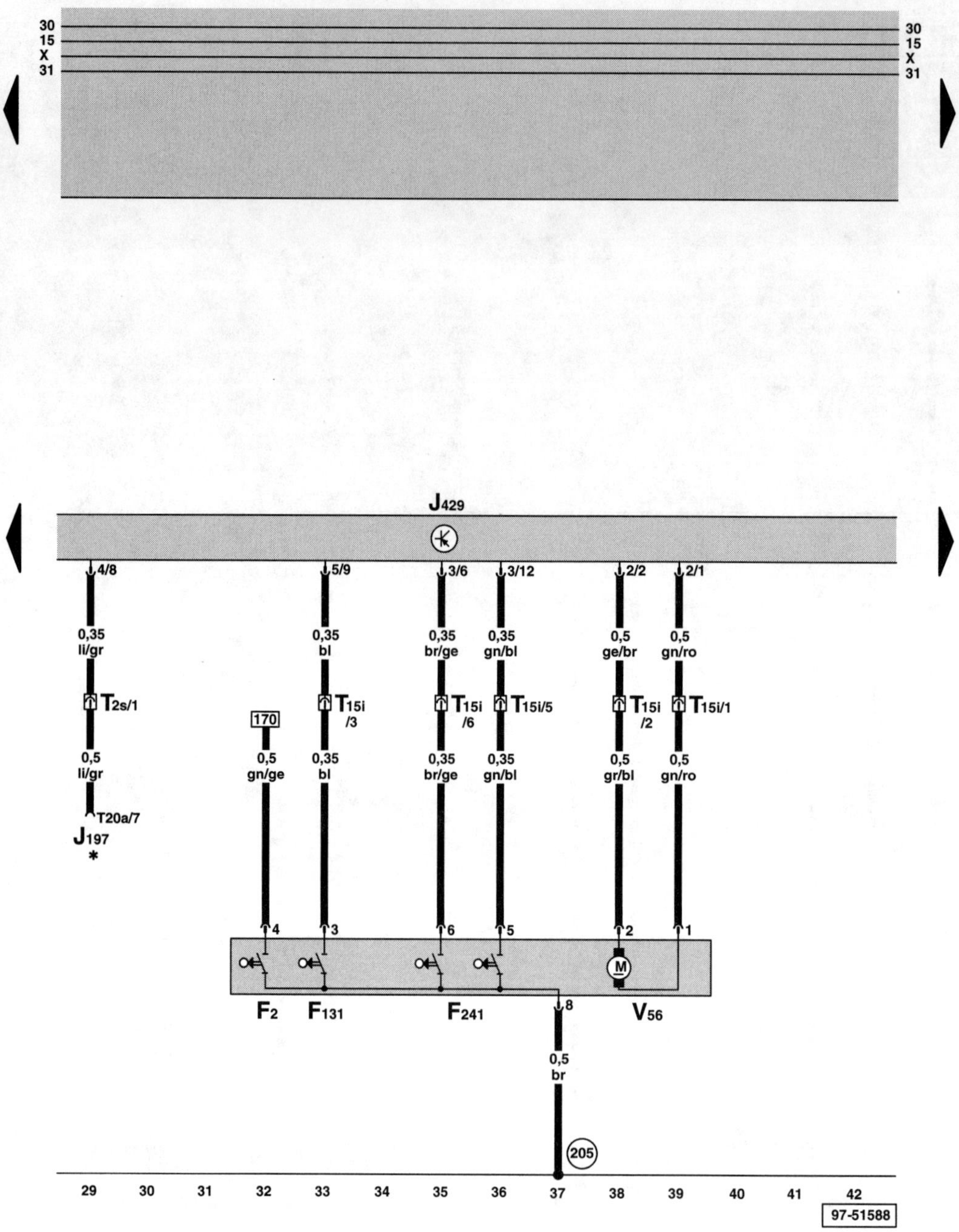

ws = white
sw = black
ro = red
br = brown
gn = green
bl = blue
gr = grey
li = violet
ge = yellow

F2 - Door Contact Switch - Driver's side
F131 - Left Front Central Locking System Activator/Switch
F241 - Contact switch for lock cylinder, driver side
J197 - Level Control System Control Module
J429 - Control module for central locking
T2s - Connector 2 pin, black, connector station A pillar, right
T15i - Connector 15 pin, brown, connector station A pillar, left
T20a - Connector 20 pin, red, on Control Module for Level Control
V56 - Driver's Door Central Locking System Motor

(205) - Ground connection, in driver's door wiring harness

* - Vehicles with Level Control System

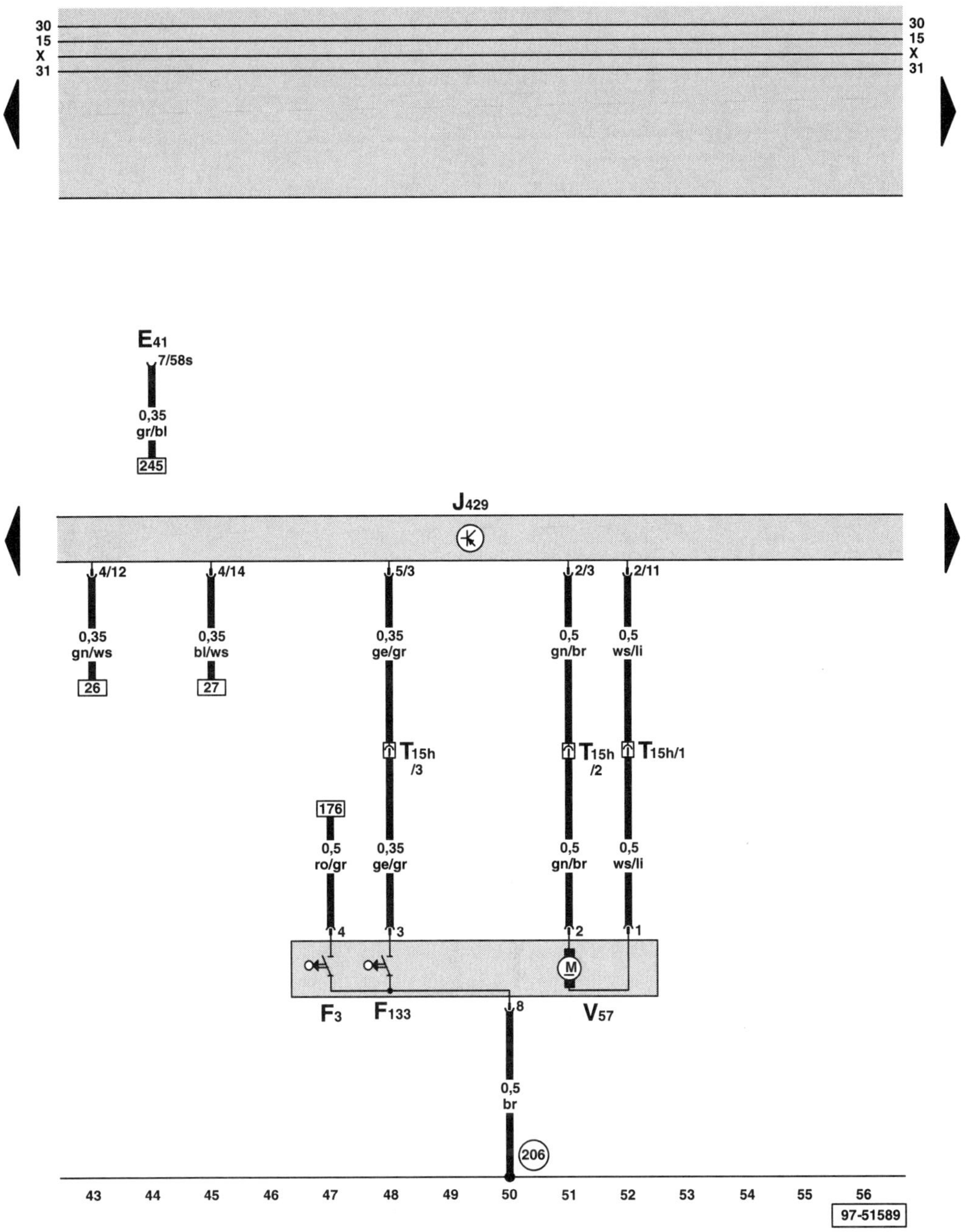

ws = white
sw = black
ro = red
br = brown
gn = green
bl = blue
gr = grey
li = violet
ge = yellow

E41 - Right Front Window Switch
F3 - Door Contact Switch - Passenger's side
F133 - Right Front Central Locking System Activator/Switch
J429 - Control module for central locking
T15h - Connector 15 pin, brown, connector station A pillar, right
V57 - Passenger's Door Central Locking System Motor

(206) - Ground connection, in passenger's door wiring harness

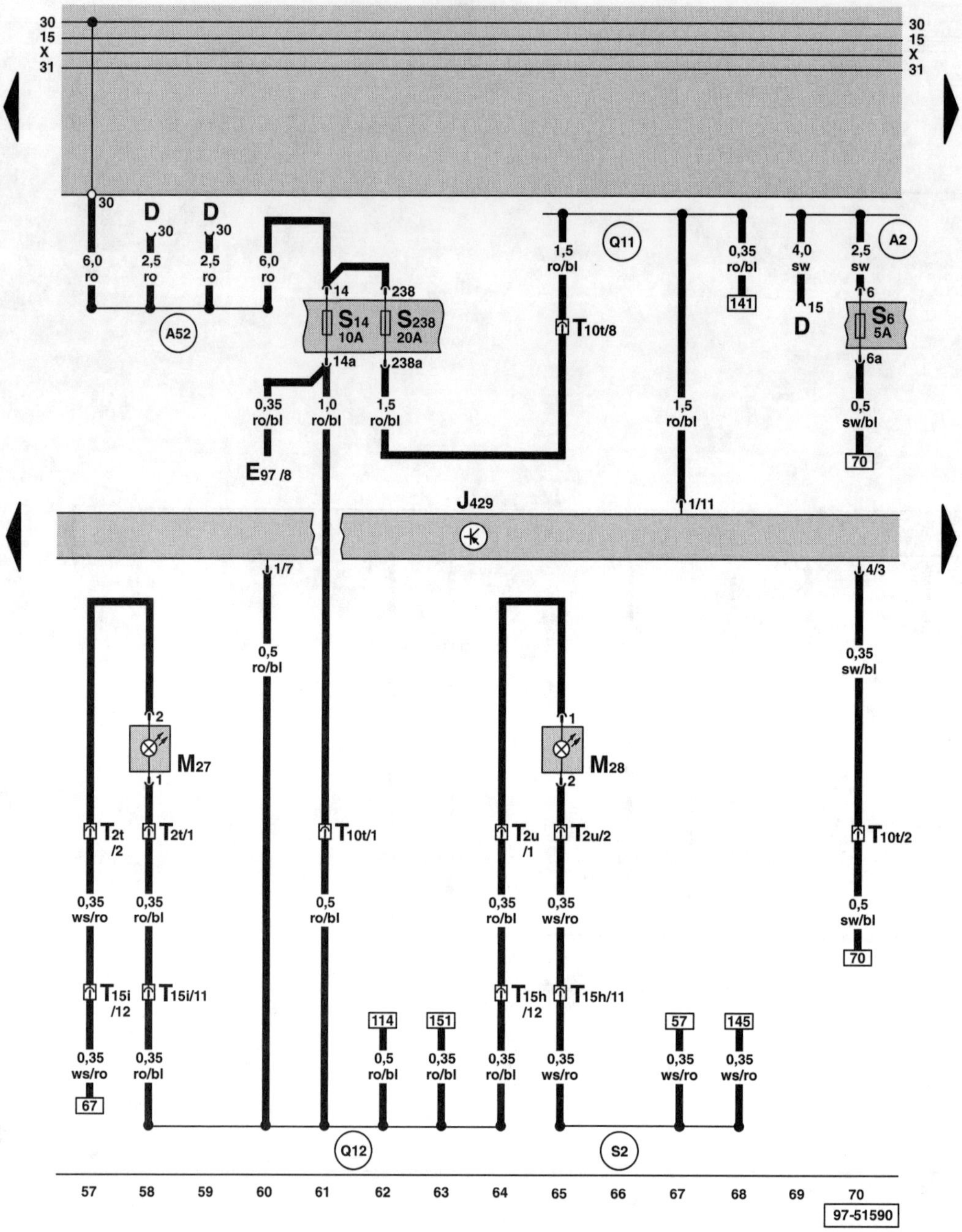

ws = white
sw = black
ro = red
br = brown
gn = green
bl = blue
gr = grey
li = violet
ge = yellow

D - Ignition/Starter Switch
E97 - Memory Program Switch
J429 - Control module for central locking
M27 - Left Door Warning Light
M28 - Right Door Warning Light
S6 - Fuse
S14 - Fuse
S238 - Fuse 38 in fuse holder
T2t - Connector 2 pin, black, in driver's door
T2u - Connector 2 pin, black, in front passenger's door
T10t - Connector 10 pin, orange, connector station A pillar, right
T15h - Connector 15 pin, brown, connector station A pillar, right
T15i - Connector 15 pin, brown, connector station A pillar, left

(A2) - Plus connection (15), in instrument panel wiring harness
(A52) - Plus connection -2- (30), in instrument panel wiring harness
(Q11) - Plus connection (30a), in power window/central locking system and door contact switch wiring harness
(Q12) - Plus connection (30a), in power window/central locking system and door contact switch wiring harness
(S2) - Wire connection, in central locking system wiring harness

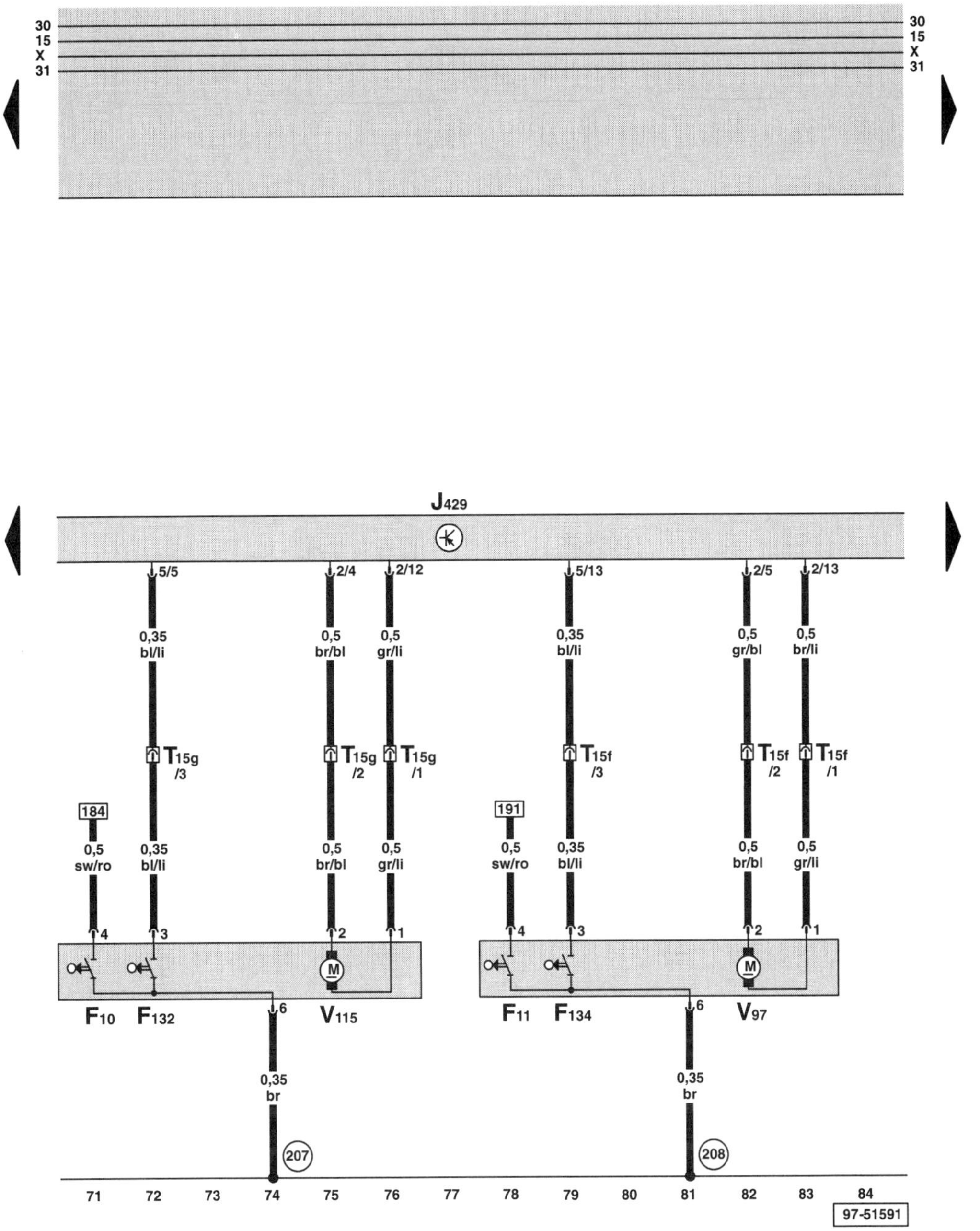

ws = white
sw = black
ro = red
br = brown
gn = green
bl = blue
gr = grey
li = violet
ge = yellow

F10 - Left Rear Door Contact Switch
F11 - Right Rear Door Contact Switch
F132 - Left Rear Central Locking System Activator/Switch
F134 - Right Rear Central Locking System Activator/Switch
J429 - Control module for central locking
T15f - Connector 15 pin, blue, in connector station B pillar, right
T15g - Connector 15 pin, blue, in connector station B pillar, left
V97 - Double Cab Right Rear Door Central Locking System Motor
V115 - Left Rear Central Locking System Motor

(207) - Ground connection, in left rear door wiring harness
(208) - Ground connection, in right rear door wiring harness

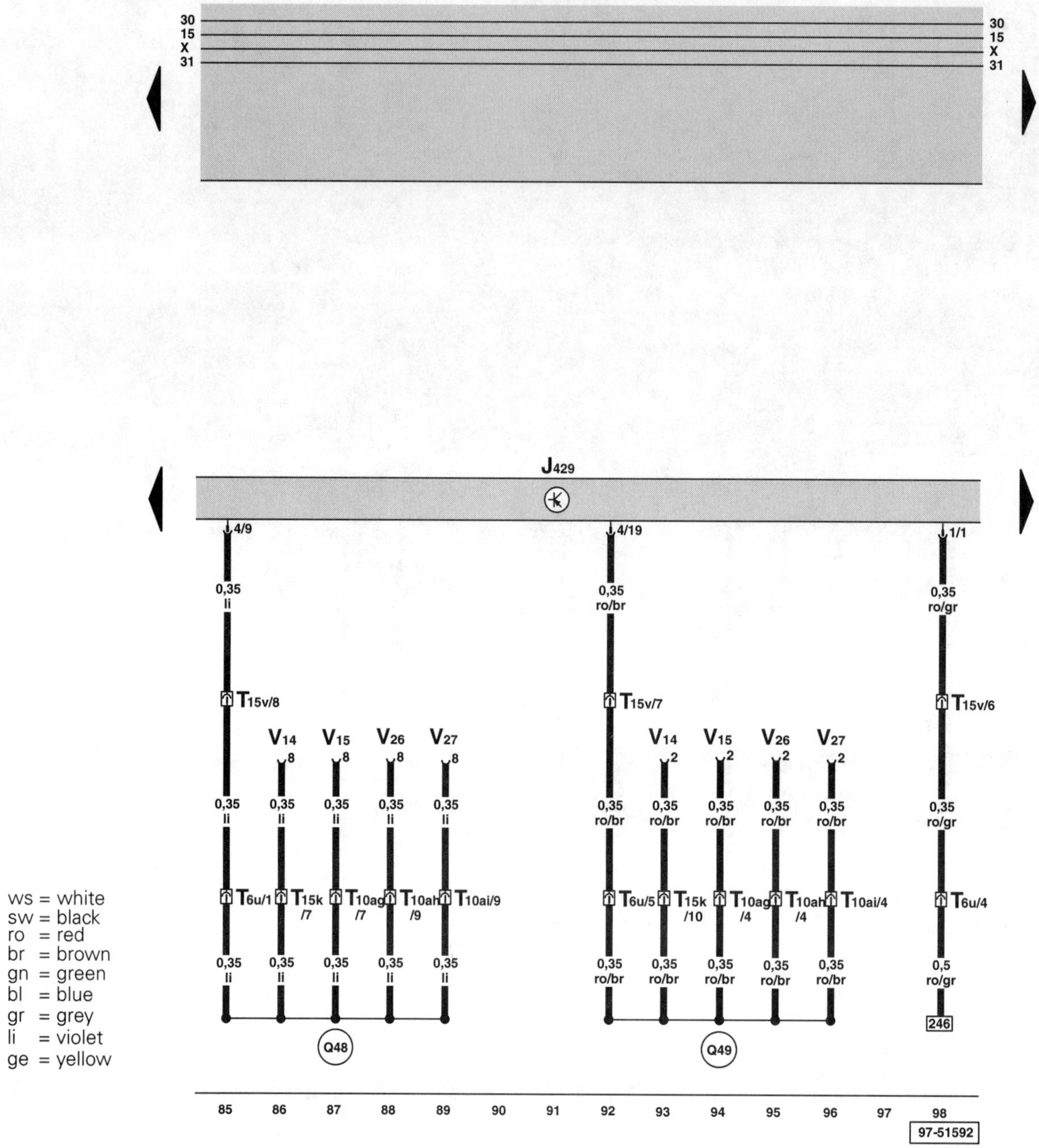

J429 - Control module for central locking
T6u - Connector 6 pin, grey, connector station A pillar, left
T10ag - Connector 10 pin, red, connector station A pillar, right
T10ah - Connector 10 pin, red, in B pillar, left
T10ai - Connector 10 pin, red, in B pillar, right
T15k - Connector 15 pin, red, connector station A pillar, left
T15v - Connector 15 pin, orange, connector station A pillar, right
V14 - Left Window Motor
V15 - Right Window Motor
V26 - Left Rear Door Window Motor
V27 - Right Rear Door Window Motor

(Q48) - Connection (open), in wiring harness interior

(Q49) - Connection (closed), in wiring harness interior

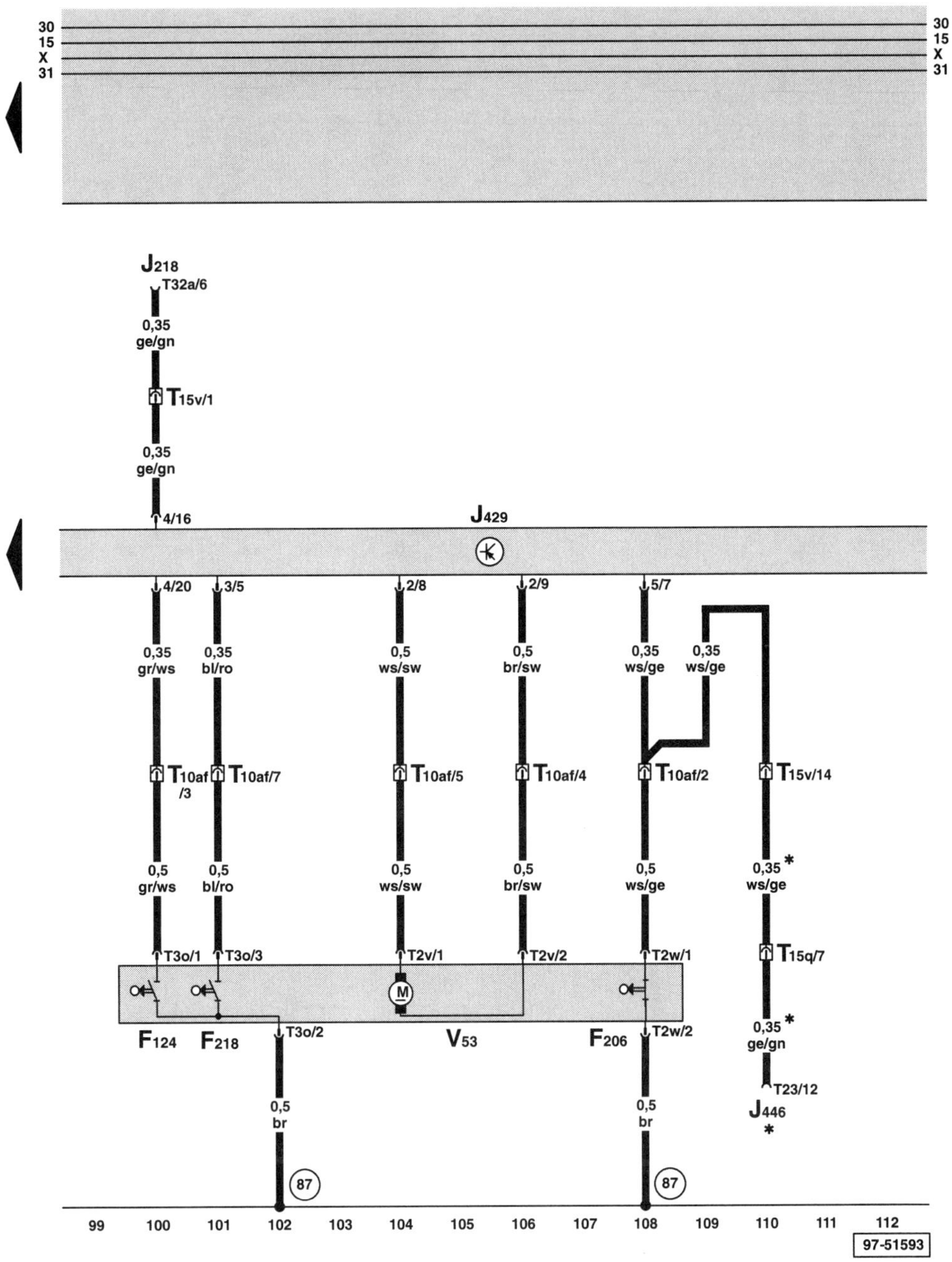

ws = white
sw = black
ro = red
br = brown
gn = green
bl = blue
gr = grey
li = violet
ge = yellow

F124 - Trunk Lock Alarm/Central Locking Switch
F206 - Trunk Lid Closed Position Switch
F218 - Switch for central locking, trunk lid
J218 - Instrument Cluster Combination Processor
J429 - Control module for central locking
J446 - Control module for parking aid
T2v - Connector 2 pin, black, luggage compartment
T2w - Connector 2 pin, grey, luggage compartment
T3o - Connector 3 pin, black, in rear lid
T10af - Connector 10 pin, violet, connector station A pillar, left
T15q - Connector 15 pin, white, connector station A pillar, right
T15v - Connector 15 pin, orange, connector station A pillar, right
T23 - Connector 23 pin, black, on control module for parking aid
T32a - Connector 32 pin, green, on instrument cluster
V53 - Decklid Central Locking System Motor

(87) - Ground connection -2-, in rear wiring harness

* - Vehicles with parking aid

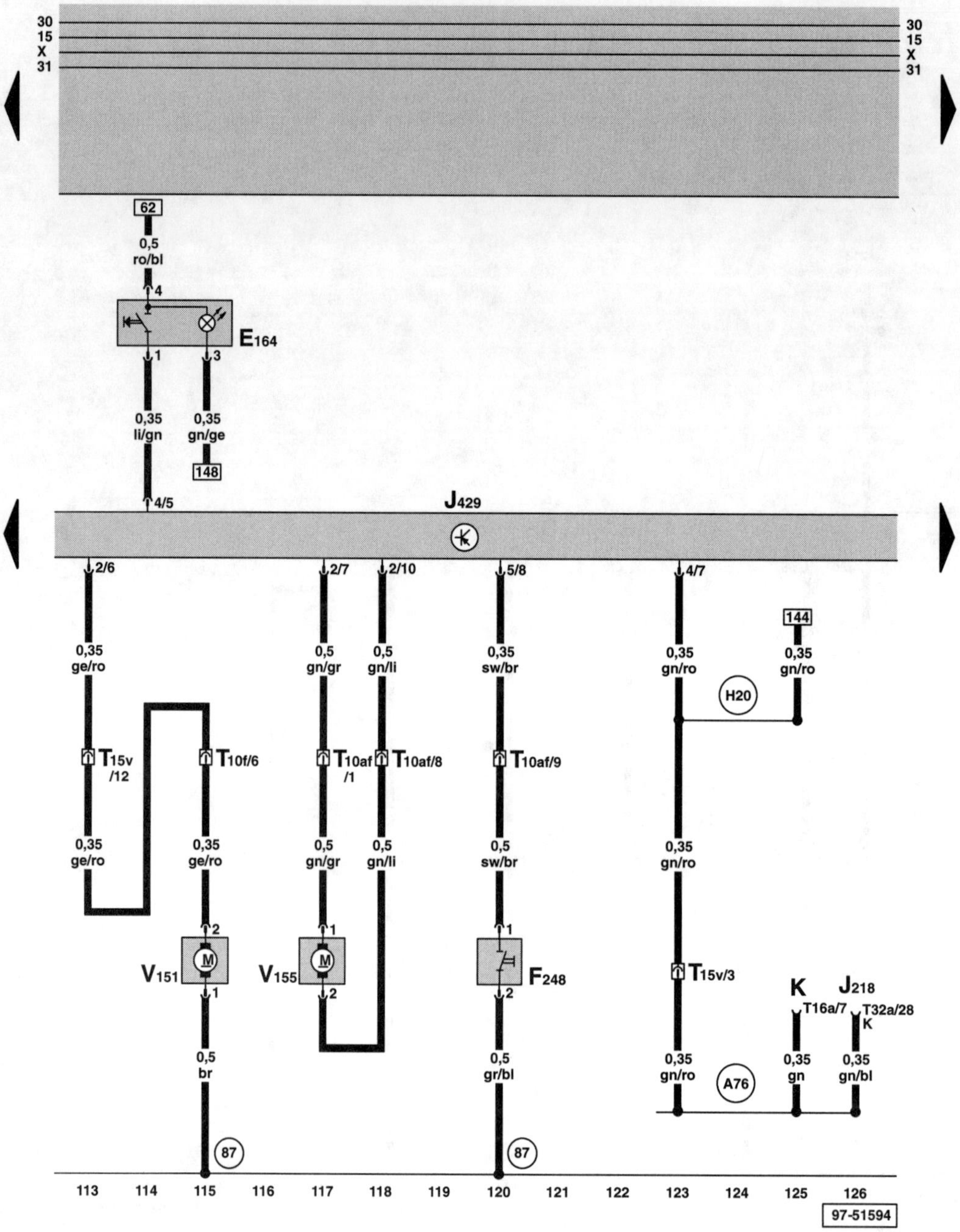

ws = white
sw = black
ro = red
br = brown
gn = green
bl = blue
gr = grey
li = violet
ge = yellow

E164 - Luggage Compartment Release Switch
F248 - Switch for unlocking, rear lid lock cylinder
J218 - Instrument Cluster Combination Processor
J429 - Control module for central locking
K - Data Link Connector (DLC) Rapid Data Transfer
T10f - Connector 10 pin, brown, connector station A pillar, left
T10af - Connector 10 pin, violet, connector station A pillar, left
T15v - Connector 15 pin, orange, connector station A pillar, right
T16a - Connector 16 pin, data link connector
T32a - Connector 32 pin, green, on instrument cluster
V151 - Motor for remote unlock rear lid
V155 - Motor for fuel tank lid unlock

(87) - Ground connection -2-, in rear wiring harness
(A76) - Connector (K-diagnosis wire), in instrument panel wiring harness
(H20) - Wire connection -1-, in On-Board Diagnostic (OBD) wiring harness

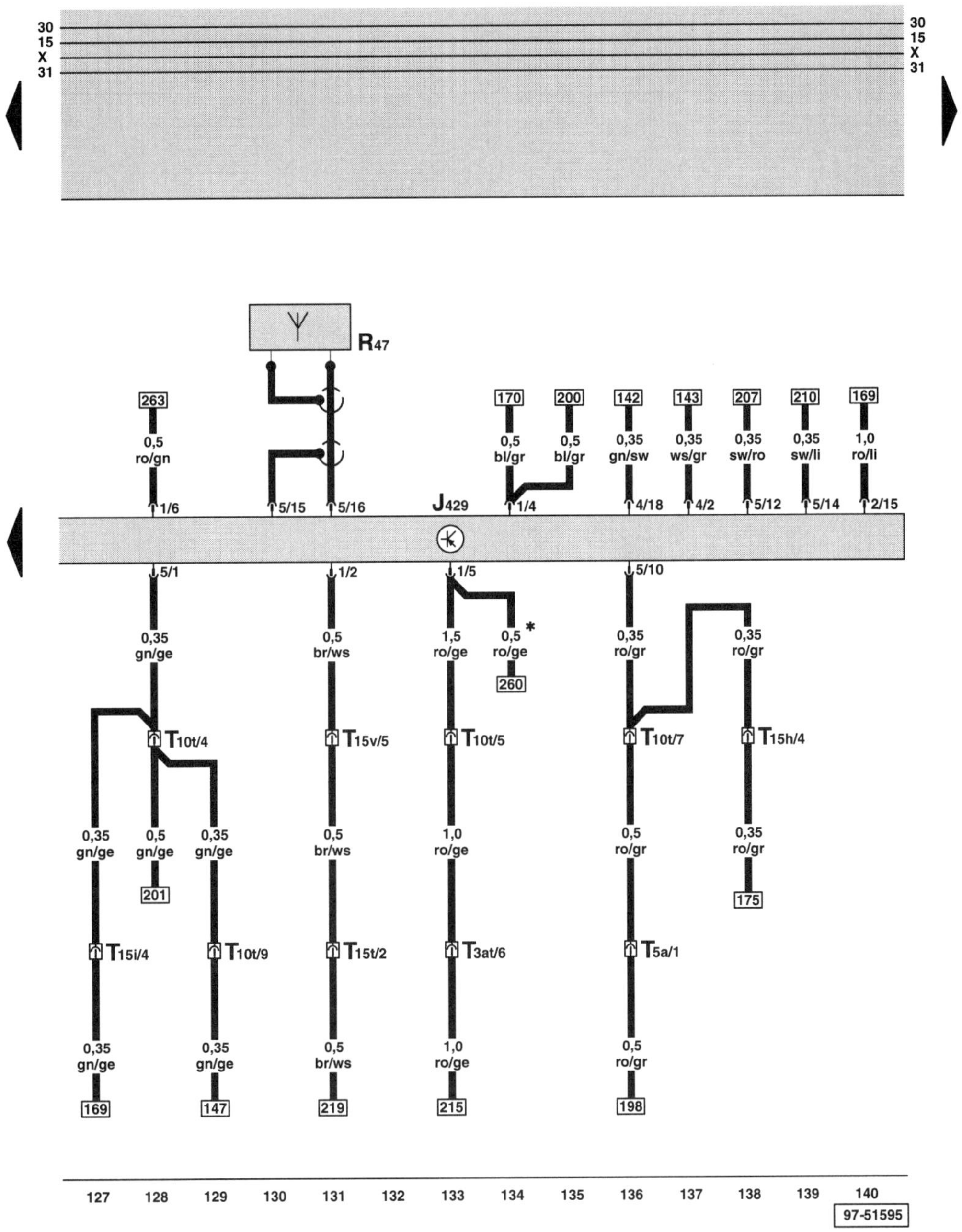

ws = white
sw = black
ro = red
br = brown
gn = green
bl = blue
gr = grey
li = violet
ge = yellow

J429 - Control module for central locking
R47 - Window antenna for central locking and anti-theft warning system
T3at - Connector 3 pin, grey, connector station A pillar, left
T5a - Connector 5 pin, black, behind instrument panel, right
T10t - Connector 10 pin, orange, connector station A pillar, right
T15h - Connector 15 pin, brown, connector station A pillar, right
T15i - Connector 15 pin, brown, connector station A pillar, left
T15t - Connector 15 pin, grey, connector station A pillar, left
T15v - Connector 15 pin, orange, connector station A pillar, right

* - Wagon

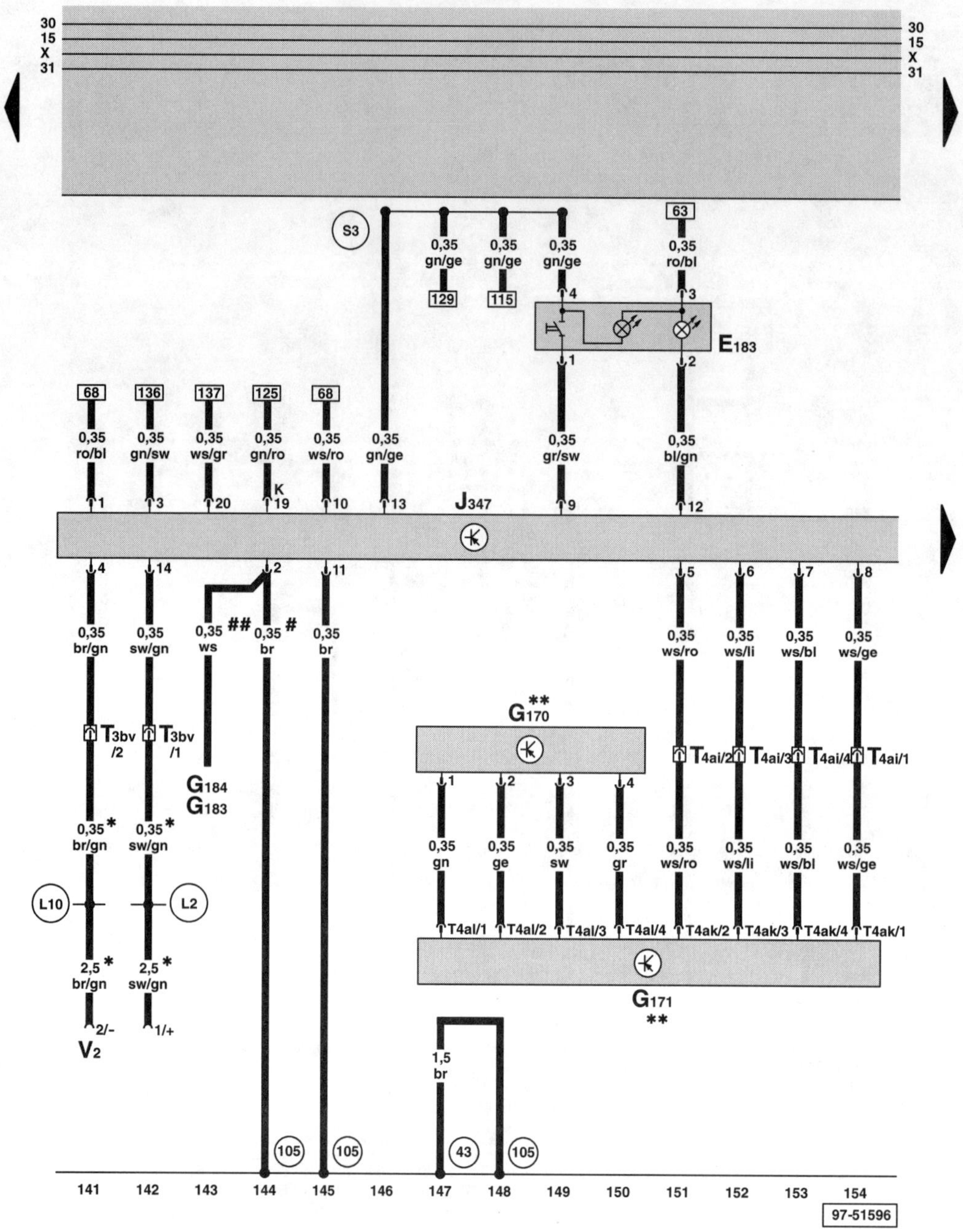

ws = white
sw = black
ro = red
br = brown
gn = green
bl = blue
gr = grey
li = violet
ge = yellow

E183 - Switch for passenger compartment monitoring
G170 - Interior Monitoring sensor, left for anti-theft warning system
G171 - Interior Monitoring sensor, right for anti-theft warning system
G183 - Sensor for broken window glass rear left
G184 - Sensor for broken window glass rear right
J347 - Control Module for Interior Monitoring sensors
T3bv - Connector 3 pin, orange, connector station A pillar right
T4ai - Connector 4 pin, on C pillar, right
T4ak - Connector 4 pin, on interior monitoring sensor, right
T4al - Connector 4 pin, on interior monitoring sensor, right
V2 - Fresh Air Blower

(43) - Ground connection, on right A-pillar, lower part
(105) - Ground connection -1-, in central locking system wiring harness
(L2) - Wire connection, in A/C wiring harness
(L10) - Wire connection -2-, in A/C wiring harness
(S3) - Wire connection -1- (switch/control module), central locking system wiring harness

* - Vehicles with Sunroof with Solar Cells
** - Vehicles with head airbag
# - Sedan
- Wagon

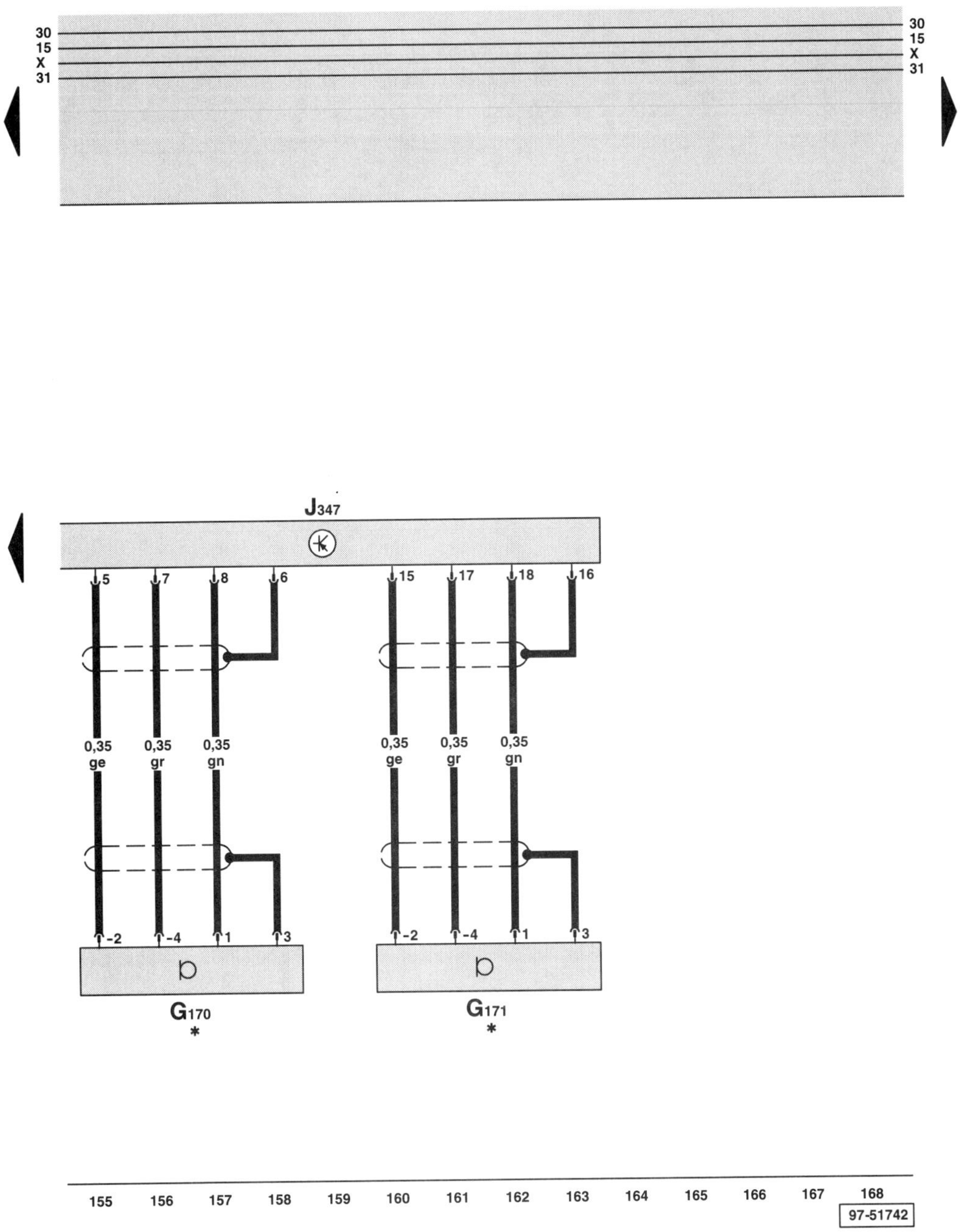

ws = white
sw = black
ro = red
br = brown
gn = green
bl = blue
gr = grey
li = violet
ge = yellow

G170 - Interior Monitoring sensor, left for anti-theft warning system
G171 - Interior Monitoring sensor, right for anti-theft warning system
J347 - Control Module for Interior Monitoring sensors

* - Vehicles without head airbag

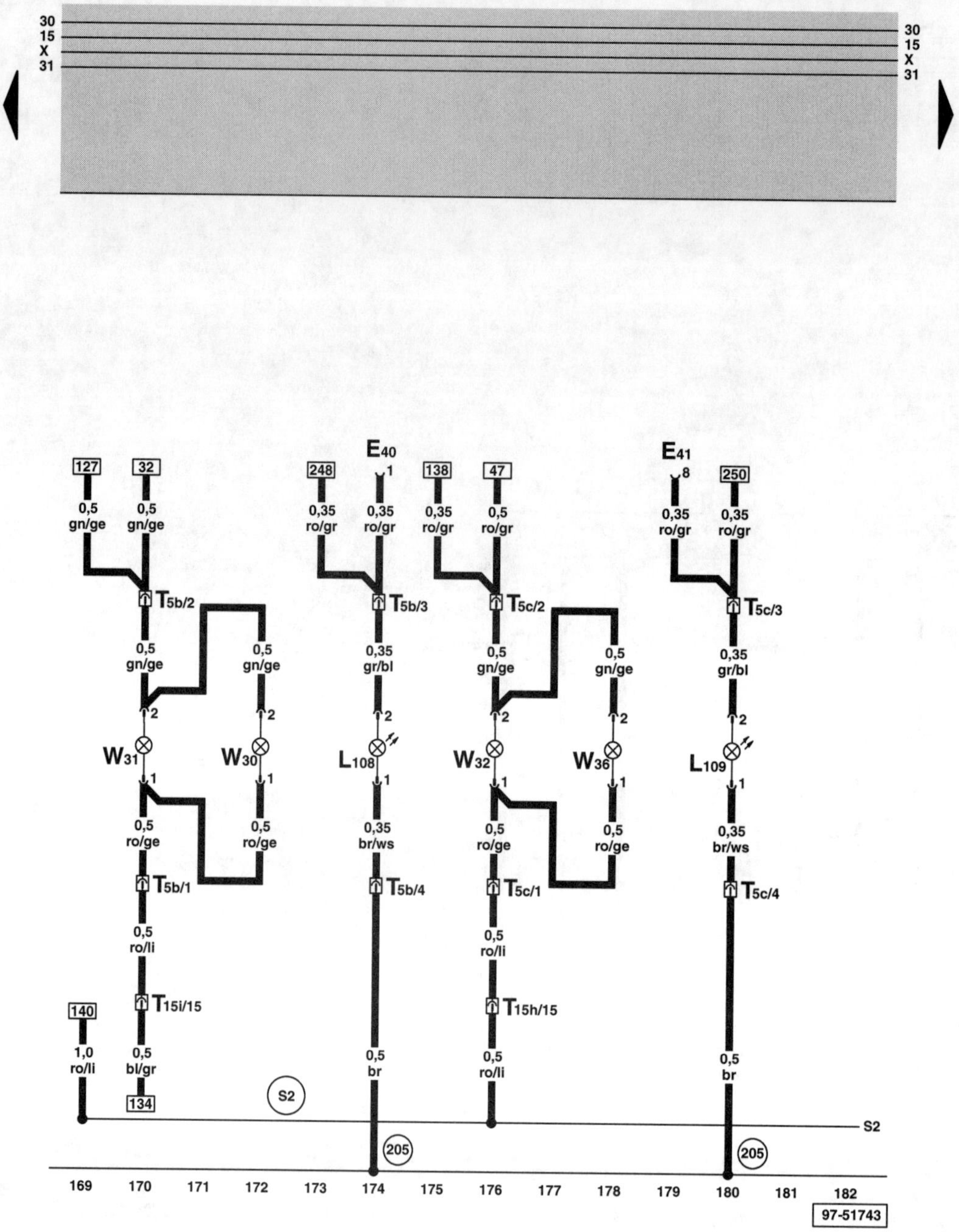

ws = white
sw = black
ro = red
br = brown
gn = green
bl = blue
gr = grey
li = violet
ge = yellow

E40 - Left Front Window Switch
E41 - Right Front Window Switch
L108 - Door opener lighting, driver side
L109 - Door opener lighting, passenger side
T5b - Connector 5 pin, black, in driver's door
T5c - Connector 5 pin, black, in passenger's door
T15h - Connector 15 pin, brown, connector station A pillar, right
T15i - Connector 15 pin, brown, connector station A pillar, left
W30 - Door Warning Light
W31 - Entry light front, left
W32 - Entry light front, right
W36 - Door warning light, passenger side

(205) - Ground connection, in driver's door wiring harness

(S2) - Wire connection, in central locking system wiring harness

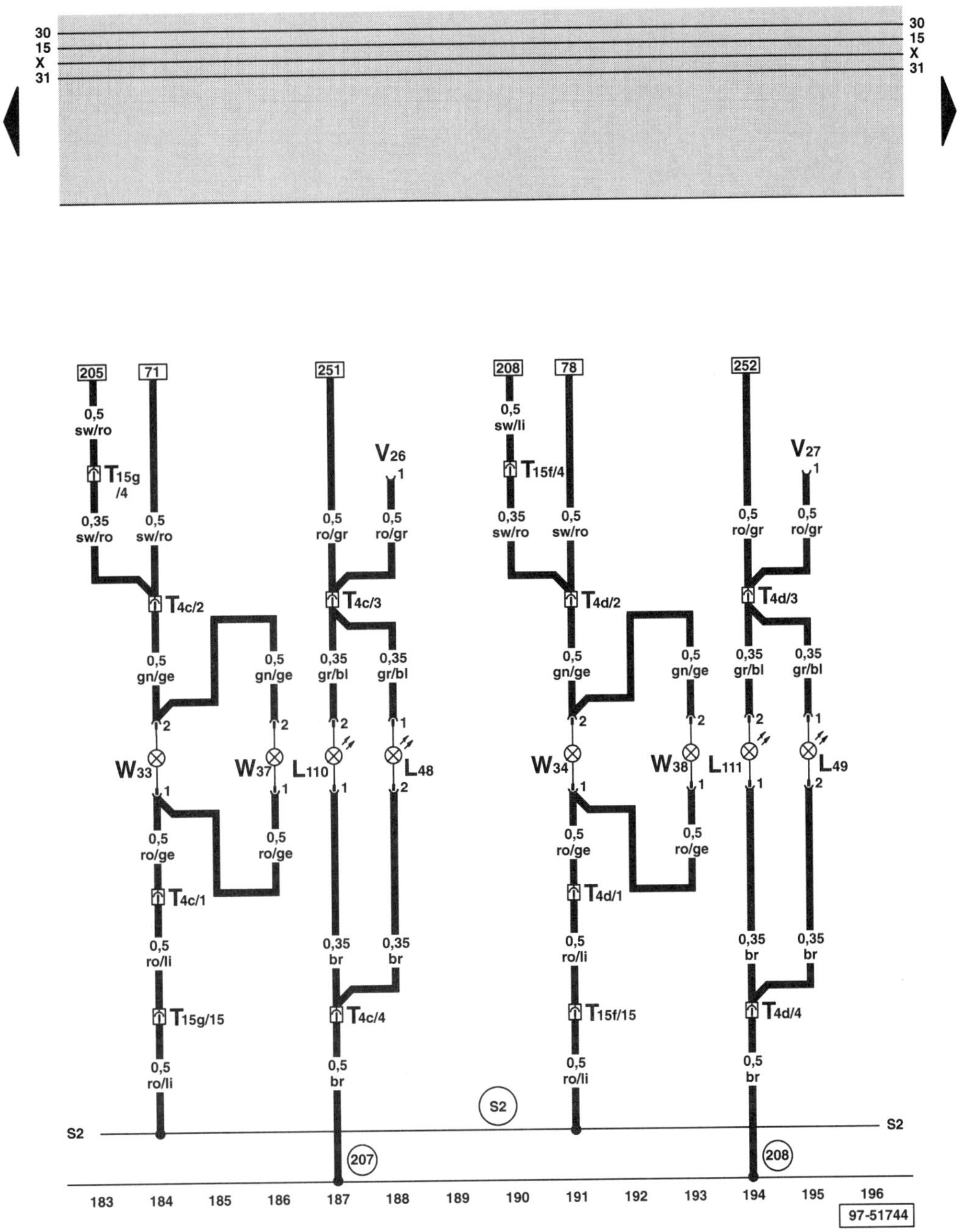

ws = white
sw = black
ro = red
br = brown
gn = green
bl = blue
gr = grey
li = violet
ge = yellow

L48 - Left Rear Ashtray Light
L49 - Right Rear Ashtray Light
L110 - Door opener lighting, left rear
L111 - Door opener lighting, right rear
T4c - Connector 4 pin, black, in left rear door
T4d - Connector 4 pin, black, in right rear door
T15f - Connector 15 pin, blue, in connector station B pillar, right
T15g - Connector 15 pin, blue, in connector station B pillar, left
V26 - Left Rear Door Window Motor
V27 - Right Rear Door Window Motor
W33 - Entry light rear, left
W34 - Entry light rear, right
W37 - Door warning light rear, left
W38 - Door warning light rear, right

(207) - Ground connection, in left rear door wiring harness

(208) - Ground connection, in right rear door wiring harness

(S2) - Wire connection, in central locking system wiring harness

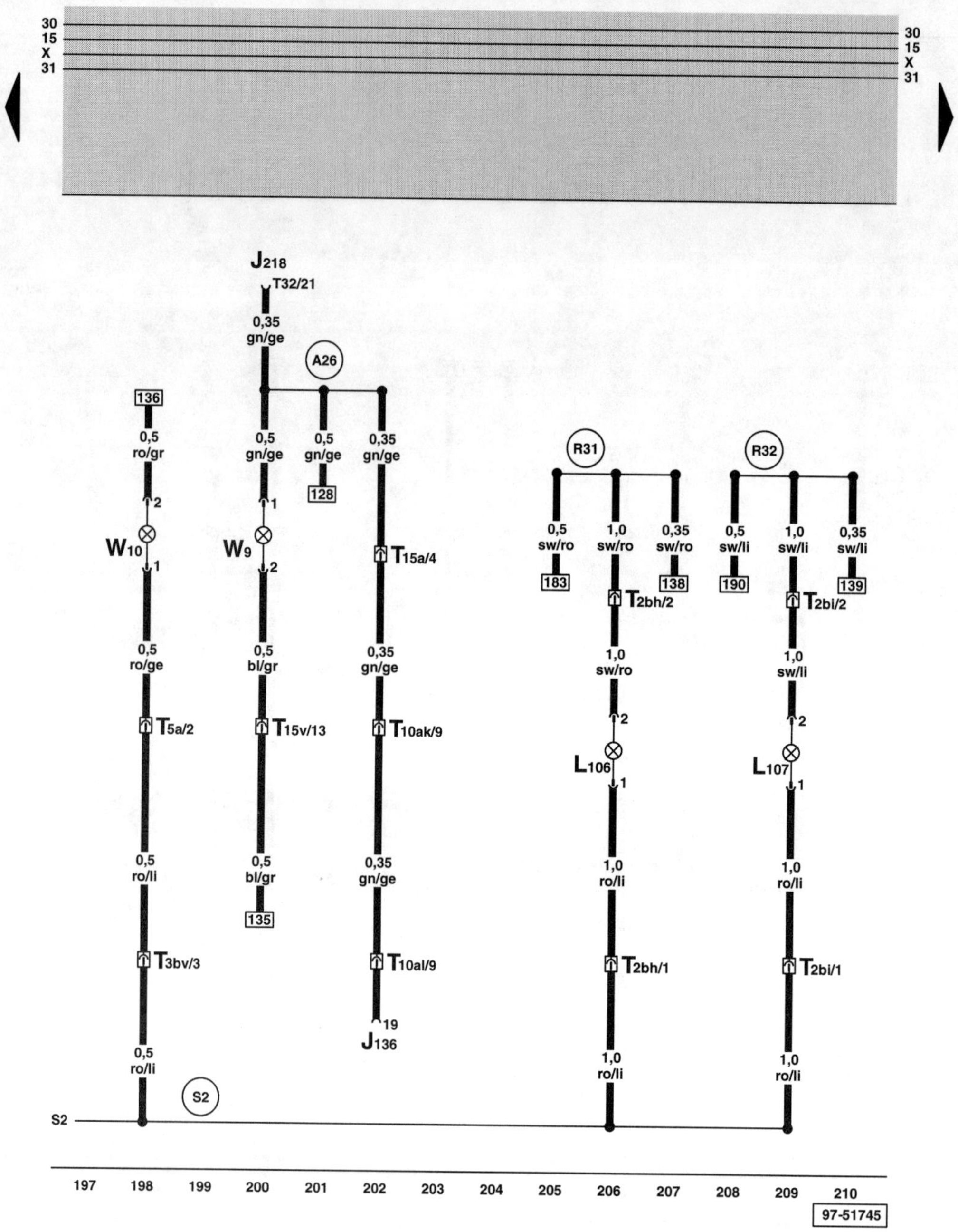

ws = white
sw = black
ro = red
br = brown
gn = green
bl = blue
gr = grey
li = violet
ge = yellow

- J136 - Memory Seat Control Module
- J218 - Instrument Cluster Combination Processor
- L106 - Footwell lighting, left rear
- L107 - Footwell lighting, right rear
- T2bh - Connector 2 pin, black, below driver's seat
- T2bi - Connector 2 pin, black, below passenger's seat
- T3bv - Connector 3 pin, orange, connector station A pillar, right
- T5a - Connector 5 pin, black, behind instrument panel, right
- T10ak - Connector 10 pin, green, connector station A pillar, left
- T10al - Connector 10 pin, brown, below driver's seat
- T15a - Connector 15 pin, black, connector station A pillar, left
- T15v - Connector 15 pin, orange, connector station A pillar, right
- T32 - Connector 32 pin, blue, on instrument cluster
- W9 - Left Footwell Light
- W10 - Right Footwell Light

- (A26) - Wire connection (driver's door contact switch), in instrument panel wiring harness
- (R31) - Wire connection -1-, in left rear door wiring harness
- (R32) - Wire connection -2-, in right rear door wiring harness
- (S2) - Wire connection, in central locking system wiring harness

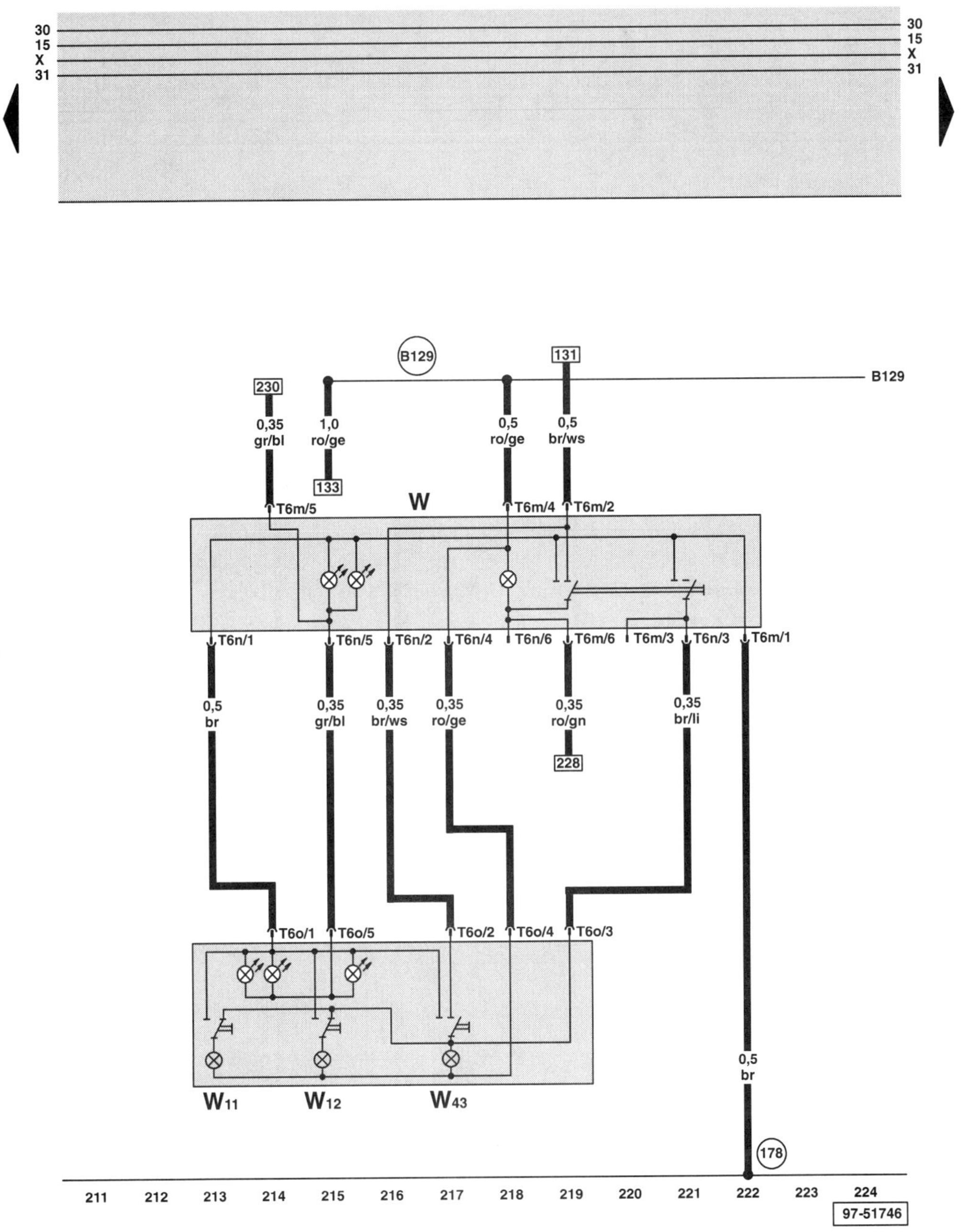

ws = white
sw = black
ro = red
br = brown
gn = green
bl = blue
gr = grey
li = violet
ge = yellow

T6m - Connector 6 pin, on interior light, front
T6n - Connector 6 pin, on interior light, front
T6o - Connector 6 pin, on interior light, rear
W - Front Interior Light
W11 - Left Rear Reading Light
W12 - Right Rear Reading Light
W43 - Rear courtesy lamp

(178) - Ground connection, in sunroof wiring harness

(B129) - Connector (interior light 31L), in wiring harness interior

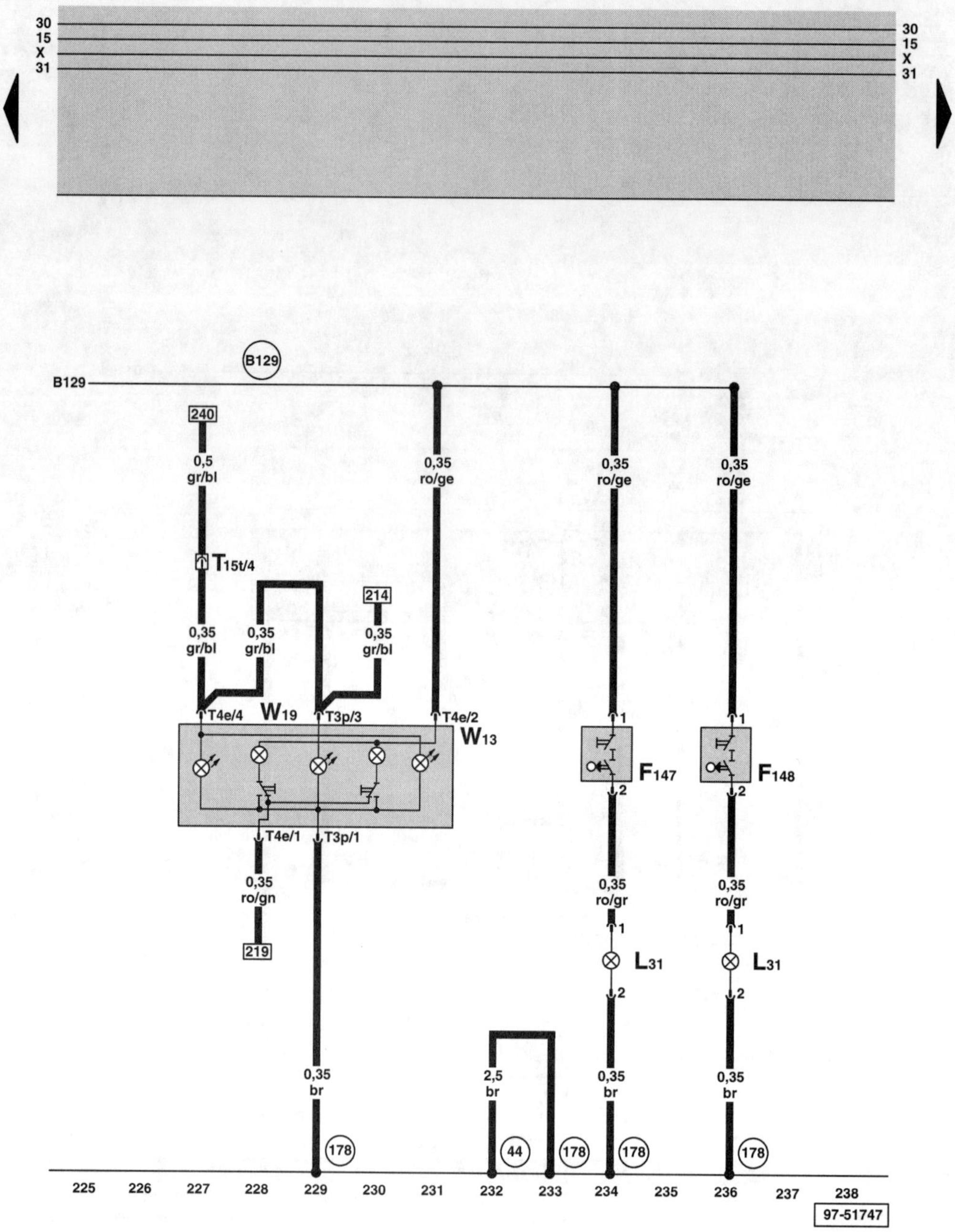

ws = white
sw = black
ro = red
br = brown
gn = green
bl = blue
gr = grey
li = violet
ge = yellow

F147 - Left Make-Up Mirror Light Switch
F148 - Right Make-Up Mirror Light Switch
L31 - Make-up Mirror Light
T3p - Connector 3 pin, on reading lamp front right
T4e - Connector 4 pin, on reading lamp front left
T15t - Connector 15 pin, grey, connector station A pillar, left
W13 - Right Front Map/Reading Light
W19 - Left Front Reading Light

(44) - Ground connection, on left A-pillar, lower part
(178) - Ground connection, in sunroof wiring harness
(B129) - Connector (interior light 31L), in wiring harness interior

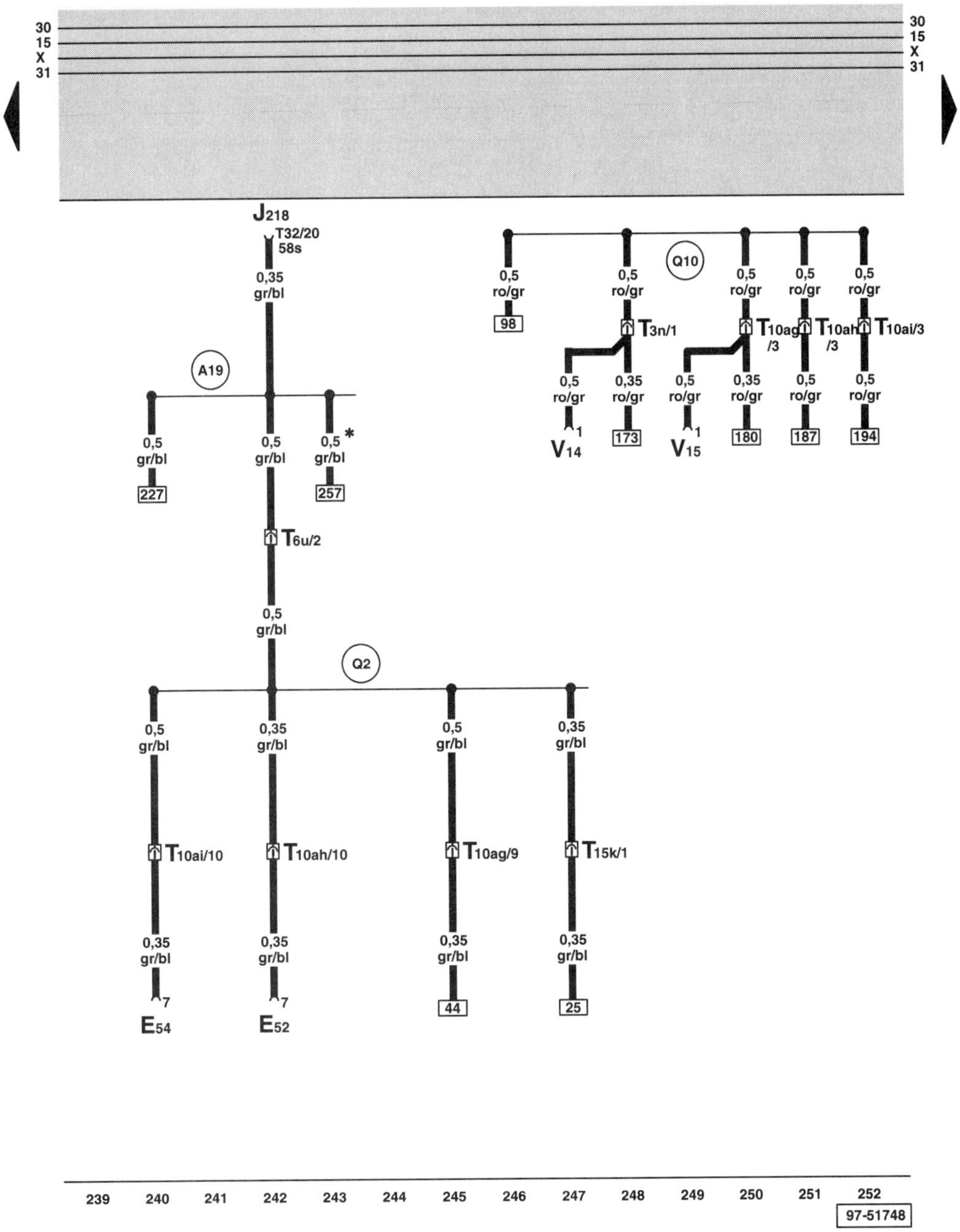

ws = white
sw = black
ro = red
br = brown
gn = green
bl = blue
gr = grey
li = violet
ge = yellow

E52 - Left Rear Window Switch, (In LR Door)
E54 - Right Rear Window Switch, (In RR Door)
J218 - Instrument Cluster Combination Processor
T3n - Connector 3 pin, red, connector station A pillar, left
T6u - Connector 6 pin, grey, connector station A pillar, left
T10ag - Connector 10 pin, red, connector station A pillar, right
T10ah - Connector 10 pin, red, in B pillar, left
T10ai - Connector 10 pin, red, in B pillar, right
T15k - Connector 15 pin, red, connector station A pillar, left
T32 - Connector 32 pin, blue, on instrument cluster
V14 - Left Window Motor
V15 - Right Window Motor

(A19) - Wire connection (58s), in instrument panel wiring harness

(Q2) - Plus connection (58s), in power window wiring harness

(Q10) - Plus connection (87), in power window/central locking system and door contact switch wiring harness

* - Wagon

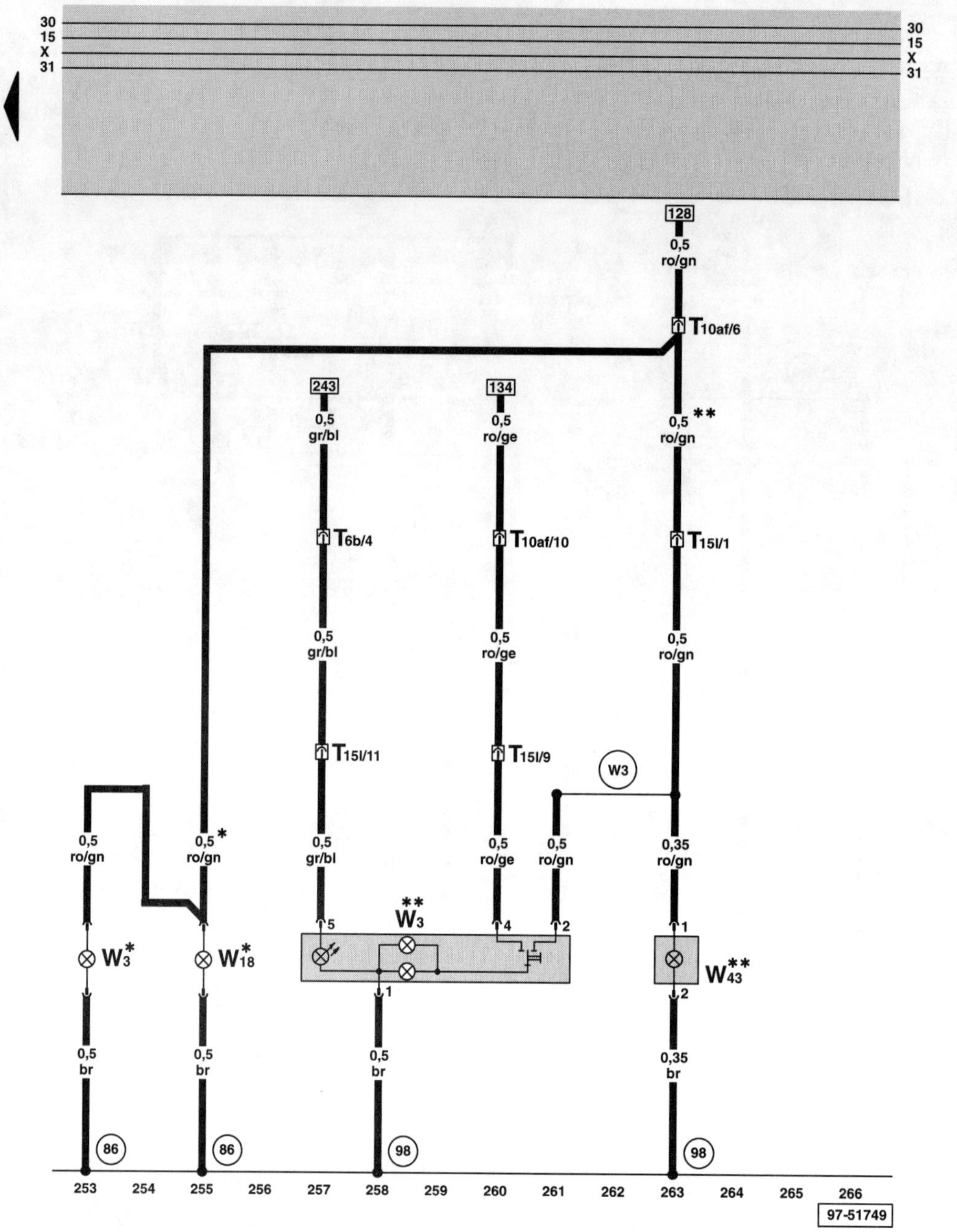

ws = white
sw = black
ro = red
br = brown
gn = green
bl = blue
gr = grey
li = violet
ge = yellow

T6b - Connector 6 pin, white, connector station A pillar, left
T10af - Connector 10 pin, violet, connector station A pillar, left
T15l - Connector 15 pin, black, in luggage compartment
W3 - Luggage Compartment Light
W18 - Left Luggage Compartment Light
W43 - Rear courtesy lamp

(86) - Ground connection -1-, in rear wiring harness
(98) - Ground connection, in rear lid wiring harness
(W3) - Wire connection, in rear wiring harness

* - only Sedan
** - only Wagon

Anti-Lock Brake with Electronic Differential Lock (EDS) and Anti-Slip Regulation (ASR) and Electronic Stability Program (ESP, Traction Control)

2000 m. y.

Fuse Panel

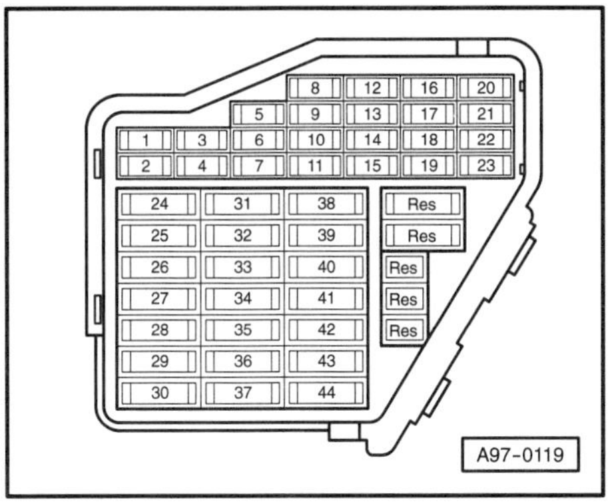

Fuse Colors:

30 A - Green
25 A - White
20 A - Yellow
15 A - Blue
10 A - Red
7,5 A - Brown
5 A - Beige

Starting with fuse position 23, fuses in the fuse holder are identified with 223 in the wiring diagram.

8 - Fold Relay Panel

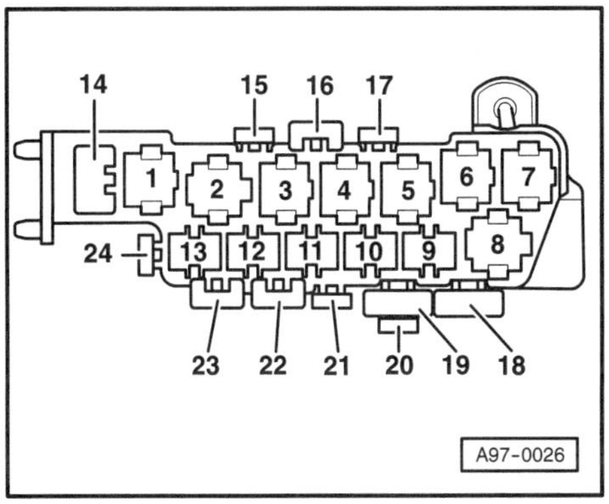

Fuse Location:

4 -ABS Solenoid Valve Relay, J106

7 -ABS Return Flow Pump Relay, J105

18 -Fuse 1 for control module - ABS, S123

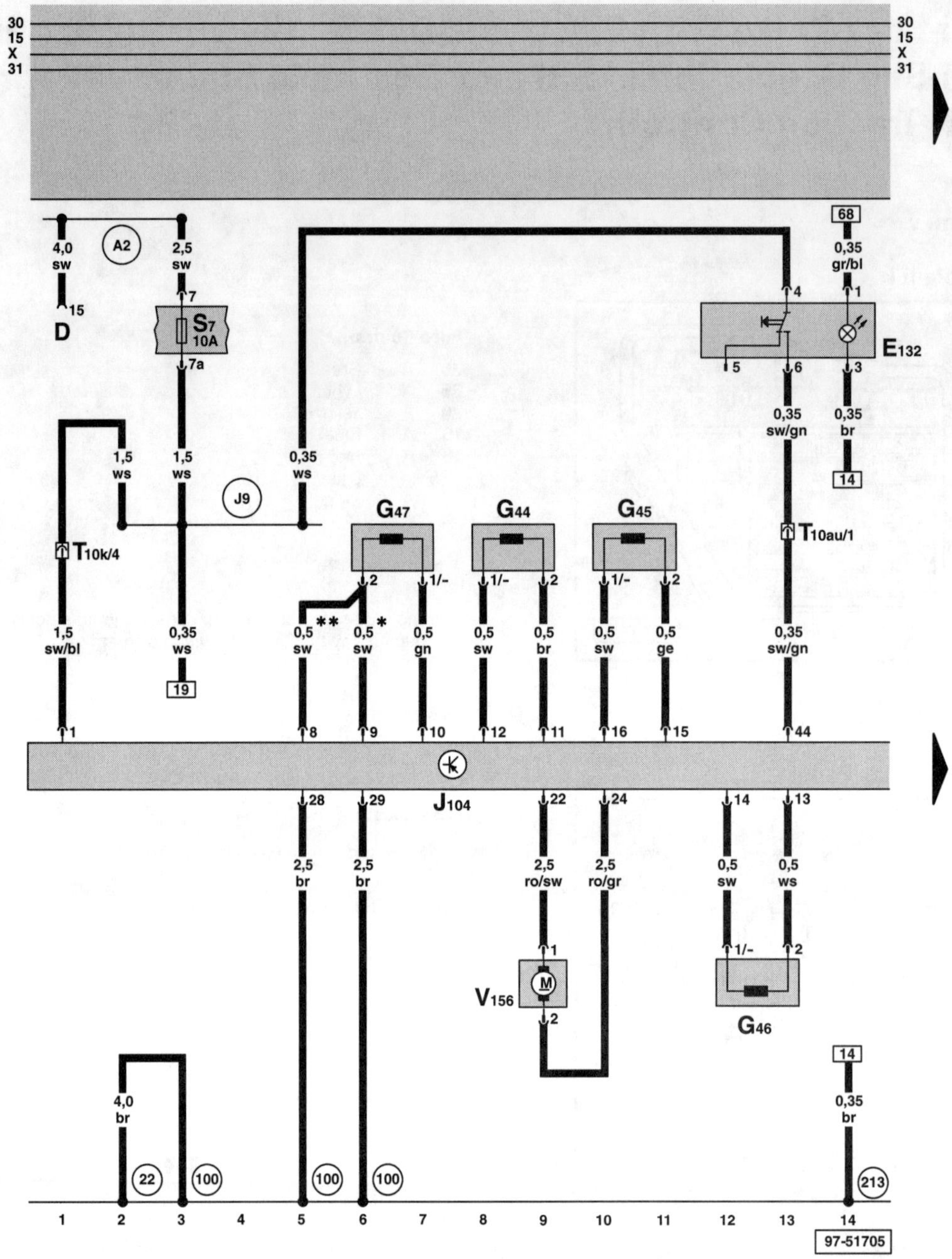

ws = white
sw = black
ro = red
br = brown
gn = green
bl = blue
gr = grey
li = violet
ge = yellow

D - Ignition/Starter Switch
E132 - Anti-Slip Control Switch
G44 - Right Rear ABS Wheel Speed Sensor
G45 - Right Front ABS Wheel Speed Sensor
G46 - Left Rear ABS Wheel Speed Sensor
G47 - Left Front ABS Wheel Speed Sensor
J104 - ABS Control Module (w/EDL)
S7 - Fuse
T10k - Connector 10 pin, orange, connector station A pillar, left
T10au - Connector 10 pin, grey, connector station A pillar, right
V156 - Hydraulic pump for traction control

(22) - Ground connection, on hydraulic unit
(100) - Ground connection -1-, in ABS wiring harness
(213) - Ground connection, in instrument panel wiring harness
(A2) - Plus connection (15), in instrument panel wiring harness
(J9) - Wire connection -1- (15a), in ABS wiring harness
* - Vehicles with front-wheel-drive
** - Vehicles with all-wheel-drive

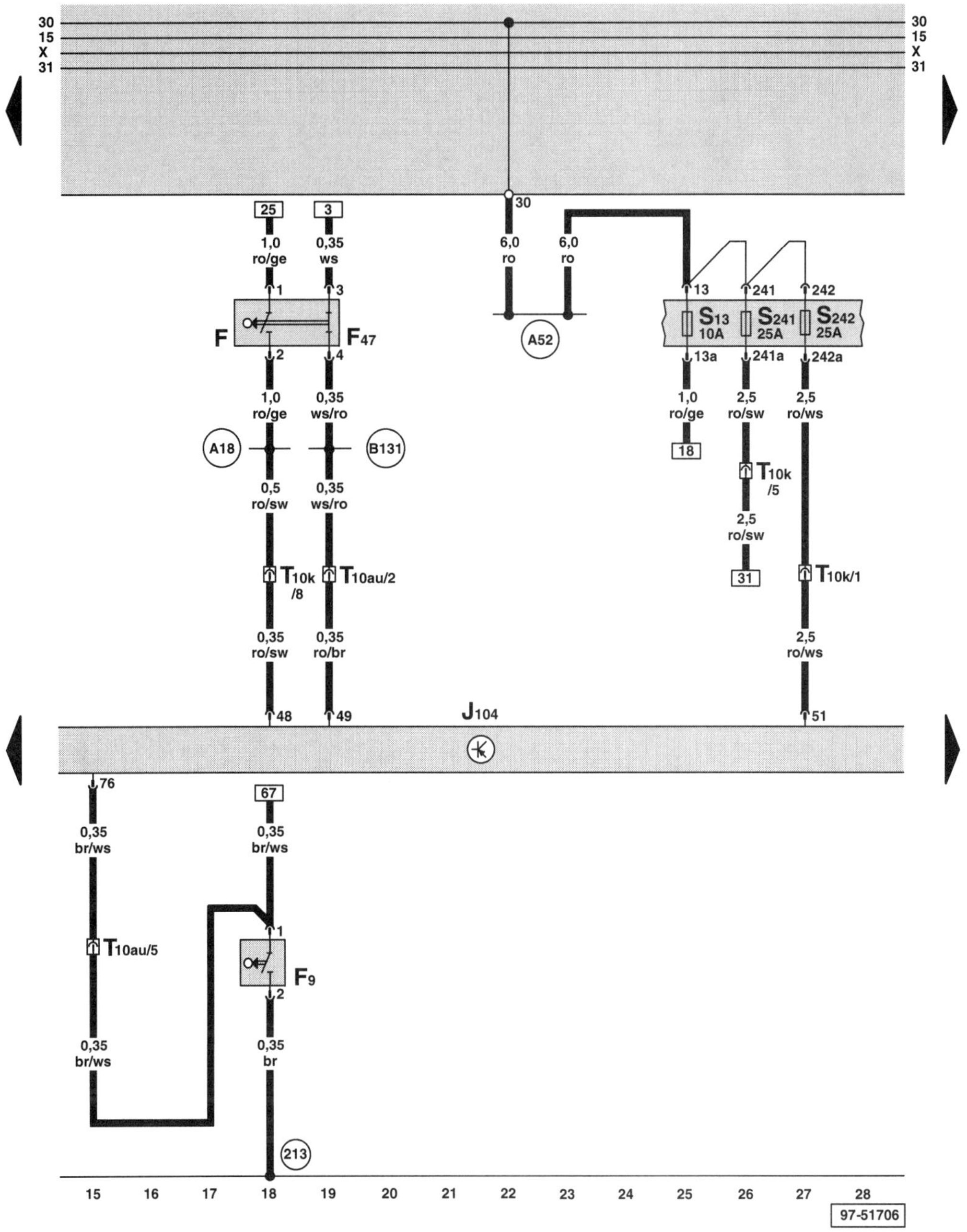

ws = white
sw = black
ro = red
br = brown
gn = green
bl = blue
gr = grey
li = violet
ge = yellow

F - Brake Light Switch
F9 - Parking Brake Warning Light Switch
F47 - Vacuum Vent Valve, Brake
J104 - ABS Control Module (w/EDL)
S13 - Fuse
S241 - Fuse 41 in fuse holder
S242 - Fuse 42 in fuse holder
T10k - Connector 10 pin, orange, connector station A pillar, left
T10au - Connector 10 pin, grey, connector station A pillar, right

(213) - Ground connection, in instrument panel wiring harness
(A18) - Wire connection (54), in instrument panel wiring harness
(A52) - Plus connection -2- (30), in instrument panel wiring harness
(B131) - Connector (54), in wiring harness interior

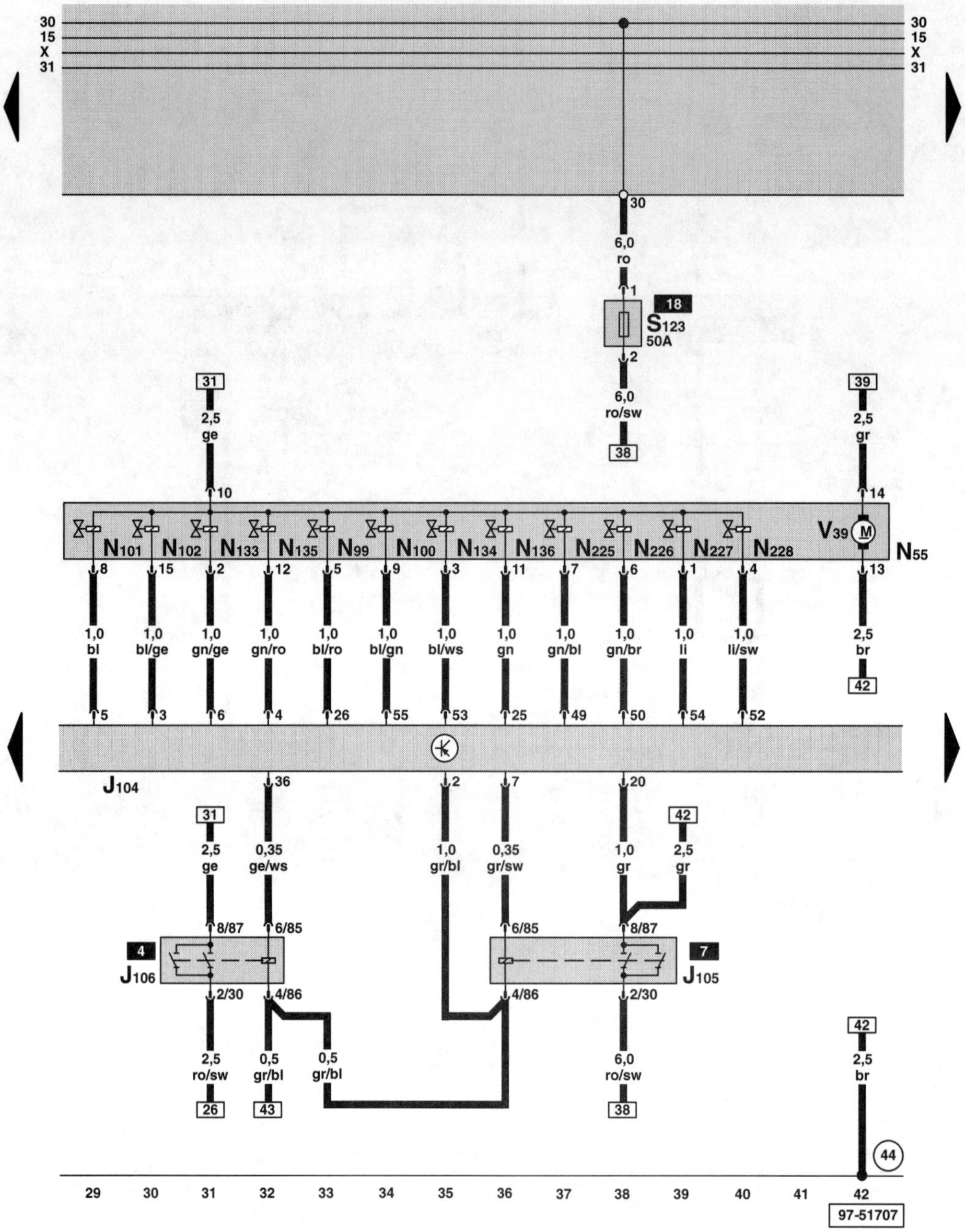

ws = white
sw = black
ro = red
br = brown
gn = green
bl = blue
gr = grey
li = violet
ge = yellow

J104 - ABS Control Module (w/EDL)
J105 - ABS Return Flow Pump Relay
J106 - ABS Solenoid Valve Relay
N55 - ABS Hydraulic Unit
N99 - Right Front ABS Inlet Valve
N100 - Right Front ABS Outlet Valve
N101 - Left Front ABS Inlet Valve
N102 - Left Front ABS Outlet Valve
N133 - Right Rear ABS Inlet Valve
N134 - Left Rear ABS Inlet Valve
N135 - Right Rear ABS Outlet Valve
N136 - Left Rear ABS Outlet Valve
N225 - Pilot valve -1- traction control (01418)
N226 - Pilot valve -2- traction control (01419)
N227 - High pressure switch valve -1- traction control (01420)
N228 - High pressure switch valve -2- traction control (01421)
S123 - Fuse for 1 for control module - ABS

V39 - ABS Return Flow Pump

(44) - Ground connection, on left A-pillar, lower part

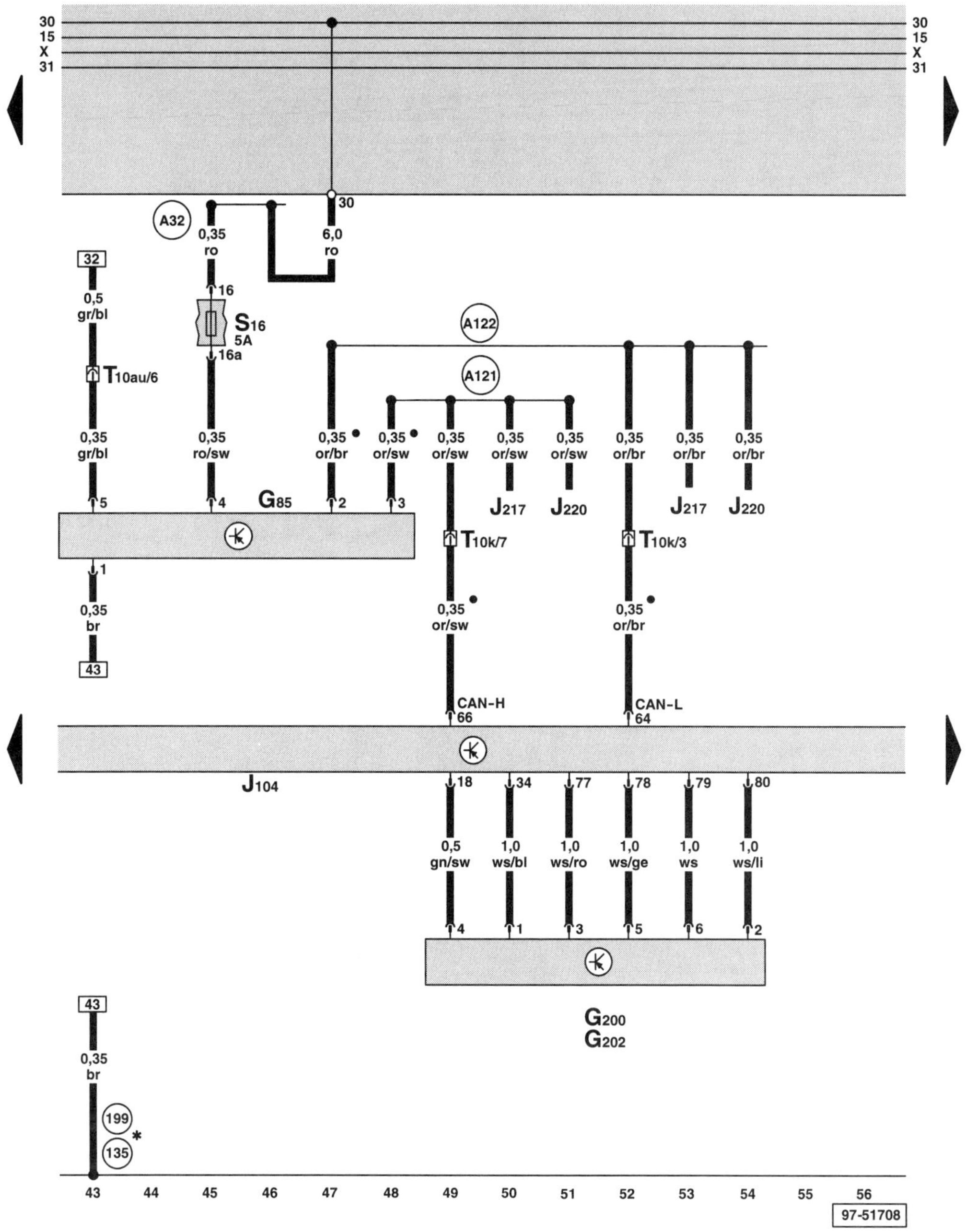

ws = white
sw = black
ro = red
br = brown
gn = green
bl = blue
gr = grey
li = violet
ge = yellow
or = orange

G85 - Steering Angle Sensor
G200 - Sensor for transverse acceleration
G202 - Yaw sensor
J104 - ABS Control Module (w/EDL)
J217 - Transmission Control Module (TCM)
J220 - Motronic Engine Control Module (ECM)
S16 - Fuse
T10k - Connector 10 pin, orange, connector station A pillar, left
T10au - Connector 10 pin, grey, connector station A pillar, right

(135) - Ground connection -2-, in instrument panel wiring harness
(199) - Ground connection -3-, in instrument panel wiring harness
(A32) - Plus connection (30), in instrument panel wiring harness
(A121) - Connection (high bus) in instrument panel wiring harness
(A122) - Connection (low bus) in instrument panel wiring harness
• - CAN-Bus (Data-Bus)
* - Both are possible

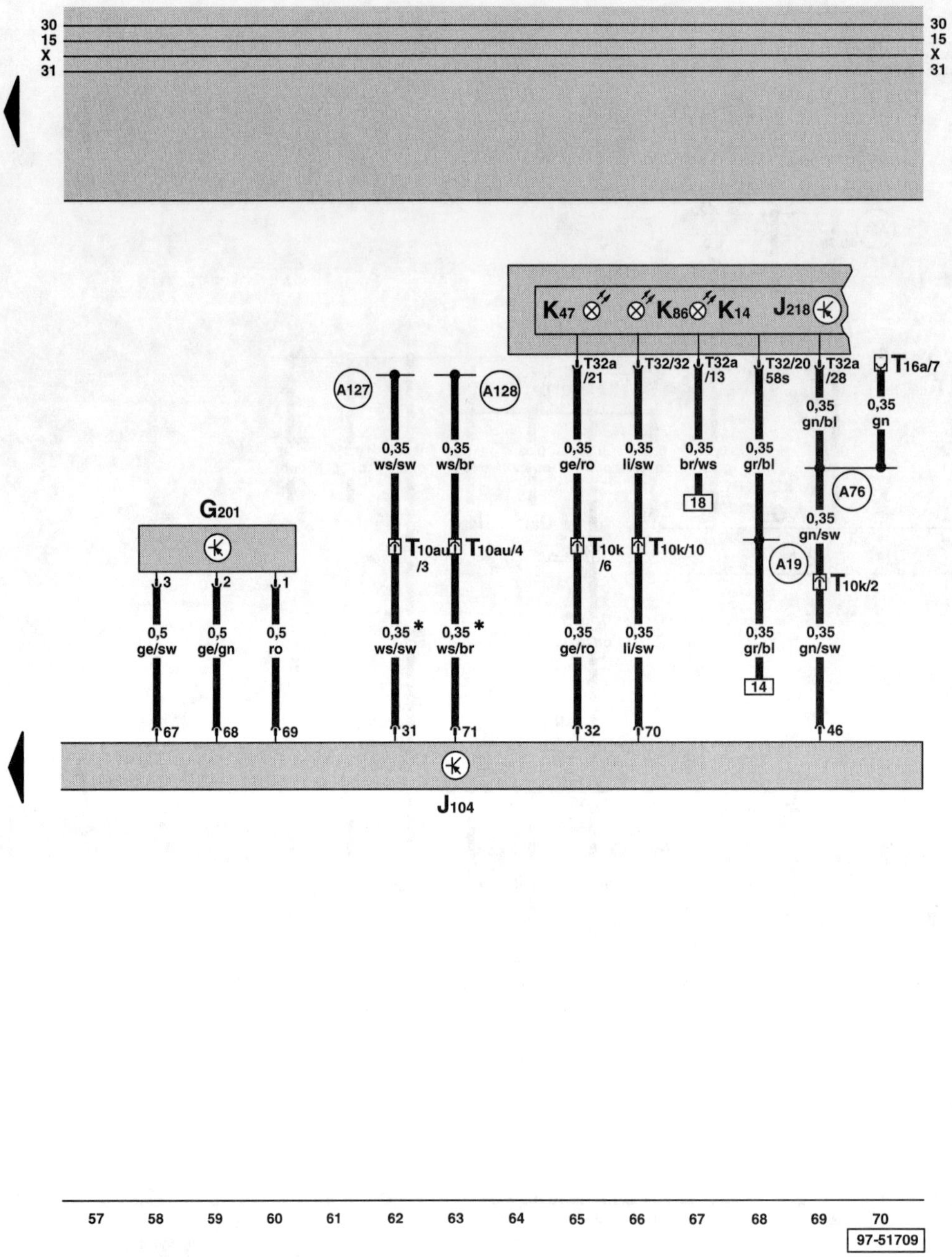

ws = white
sw = black
ro = red
br = brown
gn = green
bl = blue
gr = grey
li = violet
ge = yellow

G201 - Sensor for brake pressure
J104 - ABS Control Module (w/EDL)
J218 - Instrument Cluster Combination Processor
K14 - Parking Brake Indicator Light
K47 - ABS Warning Light
K86 - Traction Control Indicator Light
T10k - Connector 10 pin, orange, connector station A pillar, left
T10au - Connector 10 pin, grey, connector station A pillar, right
T16a - Connector 16 pin, data link connector
T32 - Connector 32 pin, blue, on instrument cluster
T32a - Connector 32 pin, green, on instrument cluster

(A19) - Wire connection (58s), in instrument panel wiring harness
(A76) - Connector (K-diagnosis wire), in instrument panel wiring harness
(A127) - Connection (road wheel speed sensor, left rear) in harness for instrument cluster
(A128) - Connection (road wheel speed sensor, right rear) in harness for instrument cluster

* - Vehicles with Navigation

Anti-Lock Brake with Electronic Differential Lock (EDS) and Anti-Slip Regulation (ASR)

2000 m. y.

Fuse Panel

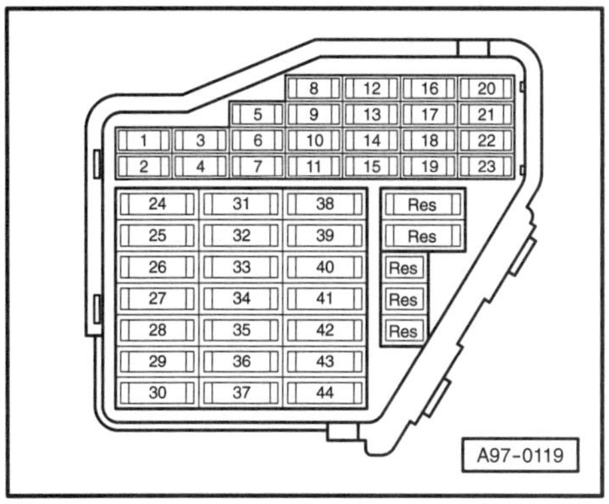

Fuse Colors:

30 A - Green
25 A - White
20 A - Yellow
15 A - Blue
10 A - Red
7,5 A - Brown
5 A - Beige

Starting with fuse position 23, fuses in the fuse holder are identified with 223 in the wiring diagram.

8 - Fold Relay Panel

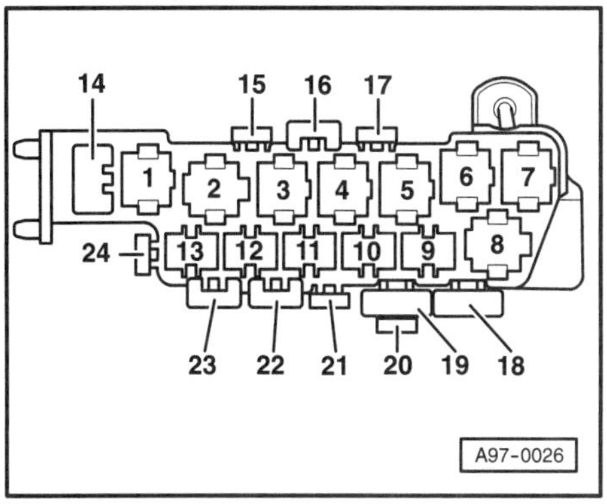

Fuse Location:

18 -Fuse 1 for control module - ABS, S123

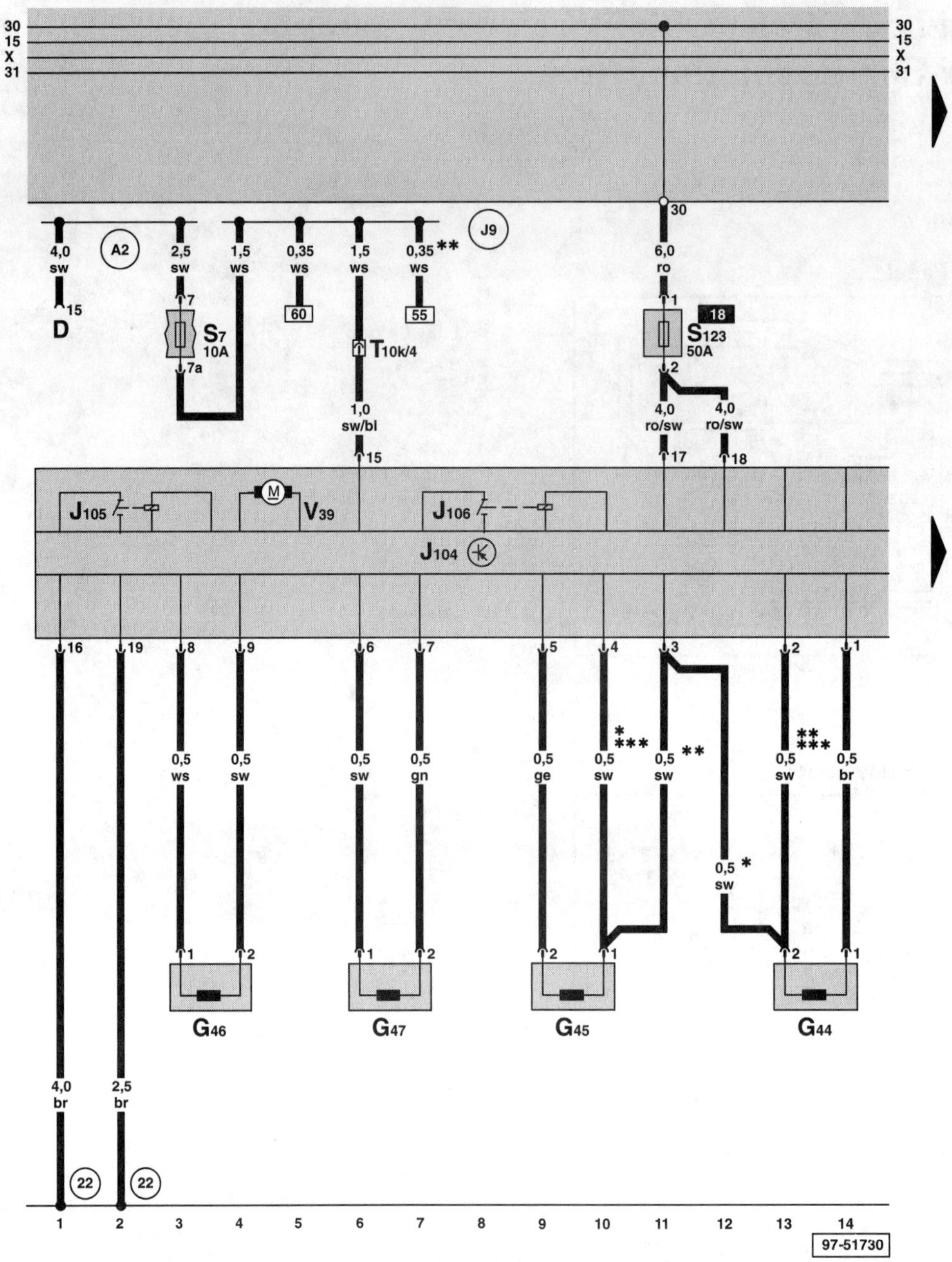

ws = white
sw = black
ro = red
br = brown
gn = green
bl = blue
gr = grey
li = violet
ge = yellow

D - Ignition/Starter Switch
G44 - Right Rear ABS Wheel Speed Sensor
G45 - Right Front ABS Wheel Speed Sensor
G46 - Left Rear ABS Wheel Speed Sensor
G47 - Left Front ABS Wheel Speed Sensor
J104 - ABS Control Module (w/EDL)
J105 - ABS Return Flow Pump Relay
J106 - ABS Solenoid Valve Relay
S7 - Fuse
S123 - Fuse for 1 for control module - ABS
T10k - Connector 10 pin, orange, connector station A pillar, left
V39 - ABS Return Flow Pump

(22) - Ground connection, on hydraulic unit
(A2) - Plus connection (15), in instrument panel wiring harness
(J9) - Wire connection -1- (15a), in ABS wiring harness

* - Vehicles without Anti-Slip Control
** - Vehicles with Anti-Slip Control
*** - Vehicles with all-wheel-drive

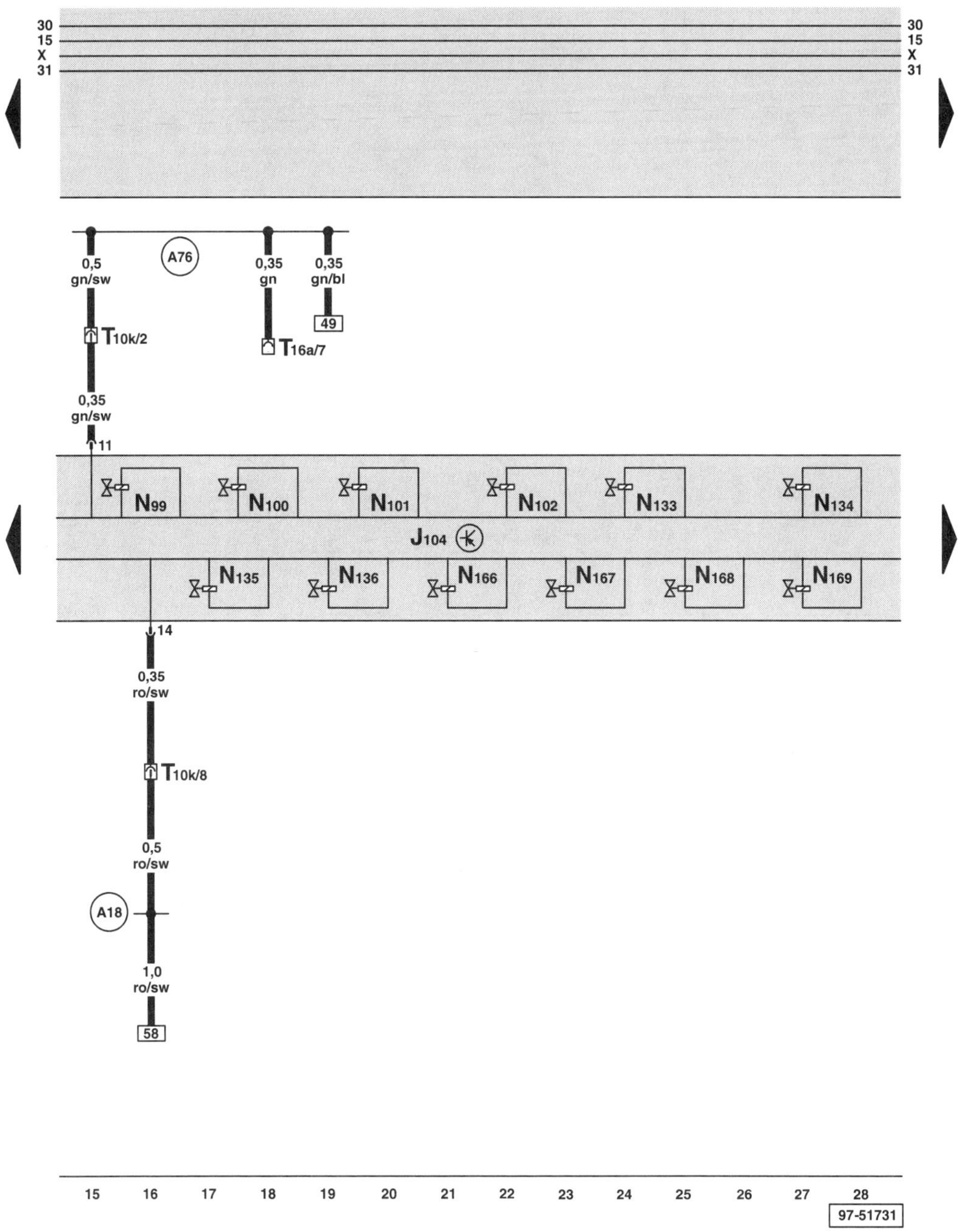

ws = white
sw = black
ro = red
br = brown
gn = green
bl = blue
gr = grey
li = violet
ge = yellow

J104 - ABS Control Module (w/EDL)
N99 - Right Front ABS Inlet Valve
N100 - Right Front ABS Outlet Valve
N101 - Left Front ABS Inlet Valve
N102 - Left Front ABS Outlet Valve
N103 - Rear ABS Inlet Valve
N133 - Right Rear ABS Inlet Valve
N134 - Left Rear ABS Inlet Valve
N135 - Right Rear ABS Outlet Valve
N136 - Left Rear ABS Outlet Valve
N166 - Right Front Traction Control Switch-Over Valve
N167 - Right Front Traction Control Outlet Valve
N168 - Left Front Traction Control Switch-Over Valve
N169 - Left Front Traction Control Outlet Valve
T10k - Connector 10 pin, orange, connector station A pillar, left
T16a - Connector 16 pin, Data Link Connector (DLC)

(A18) - Wire connection (54), in instrument panel wiring harness

(A76) - Connector (K-diagnosis wire), in instrument panel wiring harness

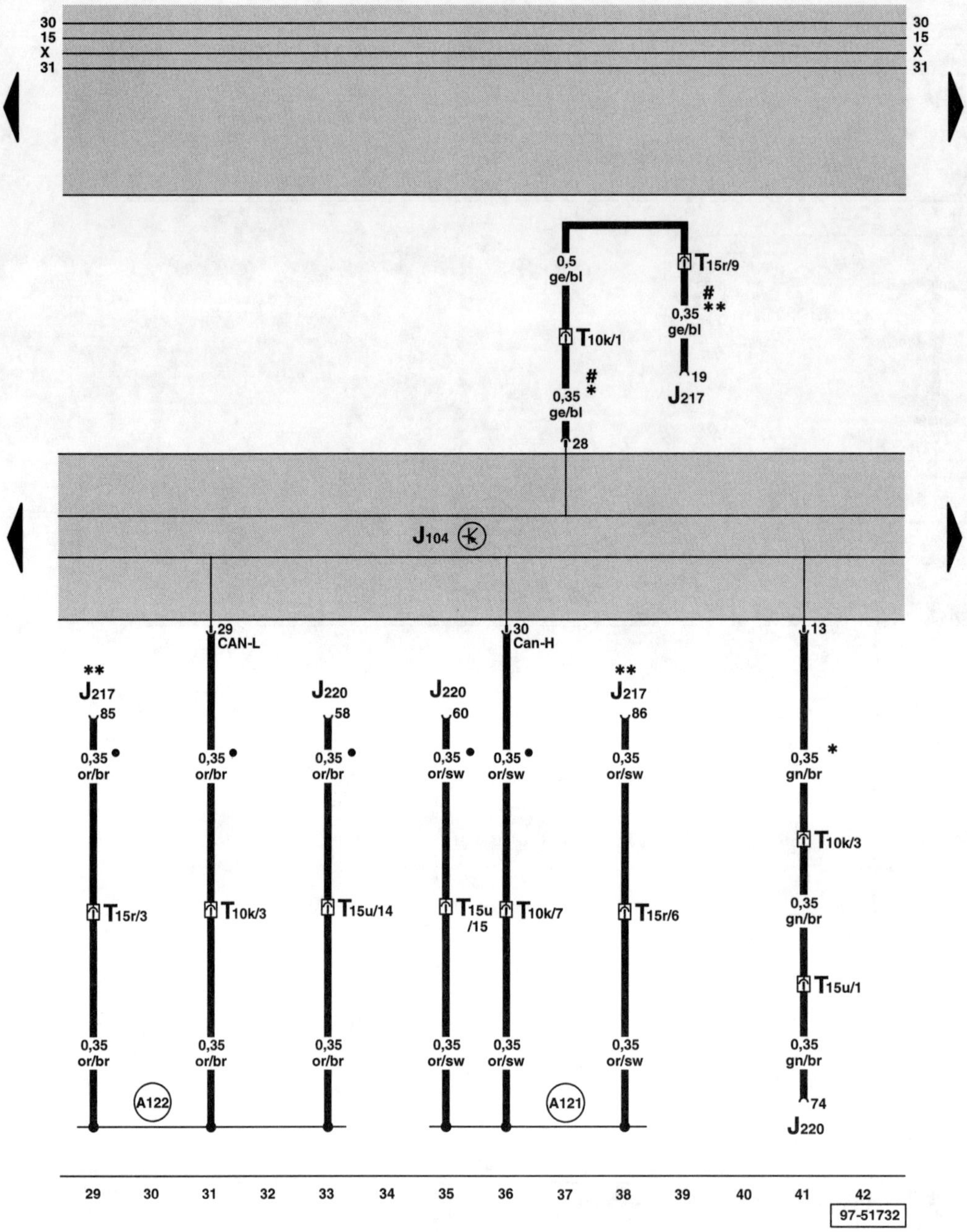

ws = white
sw = black
ro = red
br = brown
gn = green
bl = blue
gr = grey
li = violet
ge = yellow
or = orange

J104 - ABS Control Module (w/EDL)
J217 - Transmission Control Module (TCM)
J220 - Motronic Engine Control Module (ECM)
T10k - Connector 10 pin, orange, connector station A pillar, left
T15r - Connector 15 pin, blue, connector station A pillar, right
T15u - Connector 15 pin, red, connector station electronic box plenum chamber

(A121) - Connection (high bus) in instrument panel witring harness
(A122) - Connection (low bus) in instrument panel wiring harness
• - CAN-Bus (Data-Bus)
* - Vehicles with Anti-Slip-Control
** - Vehicles with automatic transmission
# - Vehicles without CAN-Bus (Data-Bus)

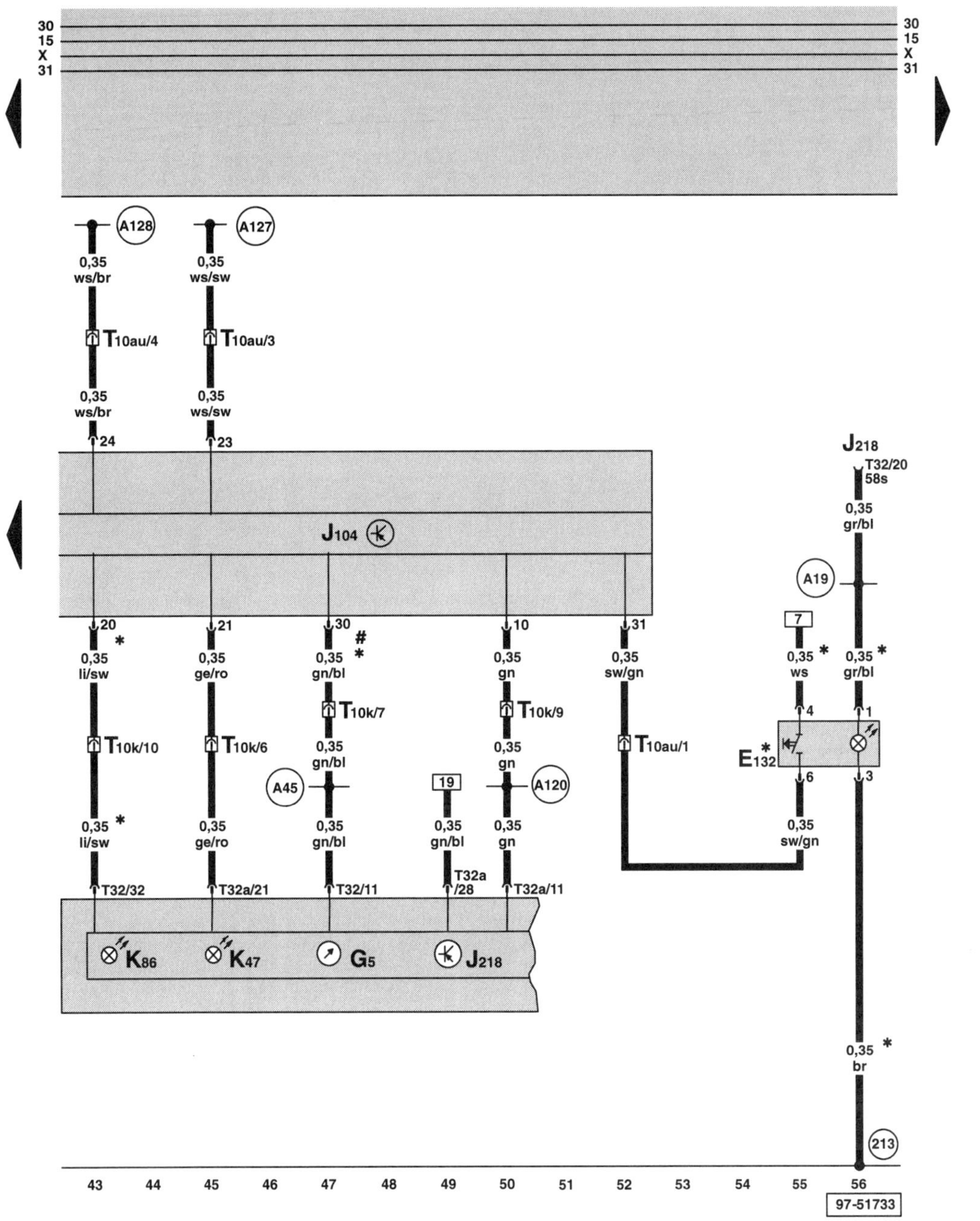

ws = white
sw = black
ro = red
br = brown
gn = green
bl = blue
gr = grey
li = violet
ge = yellow

E132 - Anti-Slip Control Switch
G5 - Tachometer
J104 - ABS Control Module (w/EDL)
J218 - Instrument Cluster Combination Processor
K47 - ABS Warning Light
K86 - Traction Control Indicator Light
T10k - Connector 10 pin, orange, connector station A pillar, left
T10au - Connector 10 pin, grey, connector station A pillar, right
T32 - Connector 32 pin, blue, on instrument cluster
T32a - Connector 32 pin, green, on instrument cluster

(213) - Ground connection, in instrument panel wiring harness
(A19) - Wire connection (58s), in instrument panel wiring harness
(A45) - Wire connection (RPM-signal), in instrument panel wiring harness
(A120) - Connection (standing time) in instrument panel wiring harness
(A127) - Connection (road wheel speed sensor, left rear) in harness for instrument cluster
(A128) - Connection (road wheel speed sensor, right rear) in harness for instrument cluster

* - Vehicles with Anti-Slip-Control
# - Vehicles without CAN-Bus (Data-Bus)

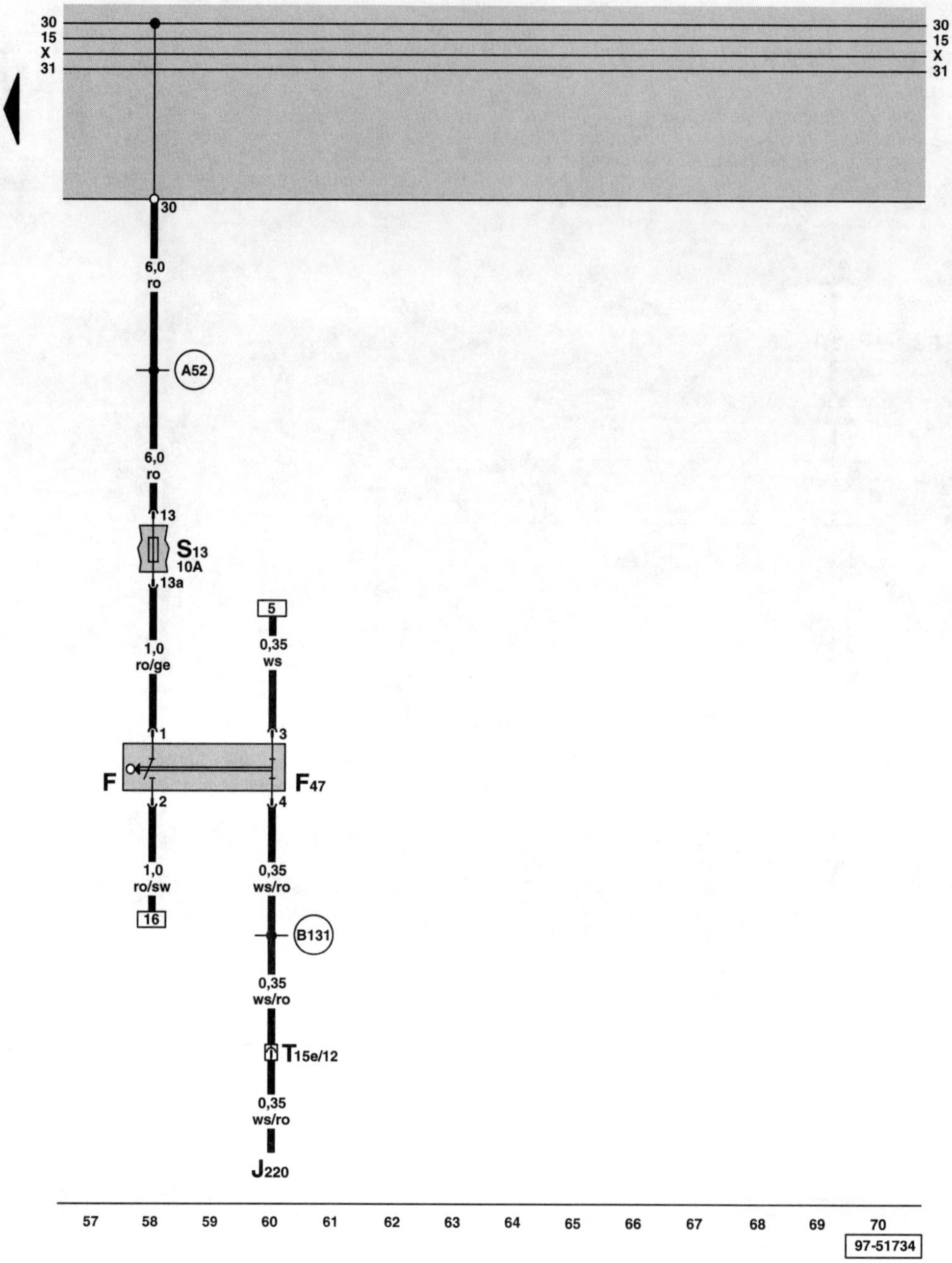

ws = white
sw = black
ro = red
br = brown
gn = green
bl = blue
gr = grey
li = violet
ge = yellow

F - Brake Light Switch
F47 - Vacuum Vent Valve, Brake
J220 - Motronic Engine Control Module (ECM)
S13 - Fuse
T15e - Connector 15 pin, white, connector station electronic box plenum chamber

(A52) - Plus connection -2- (30), in instrument panel wiring harness
(B131) - Connector (54), in wiring harness interior

Airbag

2000 m. y.

Fuse Panel

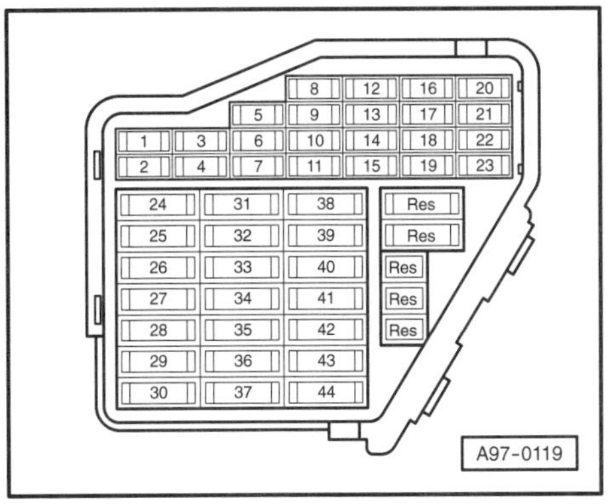

Fuse Colors:

30 A - Green
25 A - White
20 A - Yellow
15 A - Blue
10 A - Red
7,5 A - Brown
5 A - Beige

Starting with fuse position 23, fuses in the fuse holder are identified with 223 in the wiring diagram.

Micro Central Electric Panel

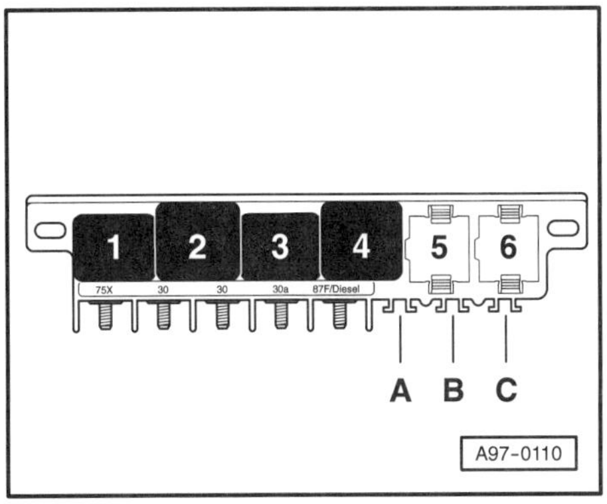

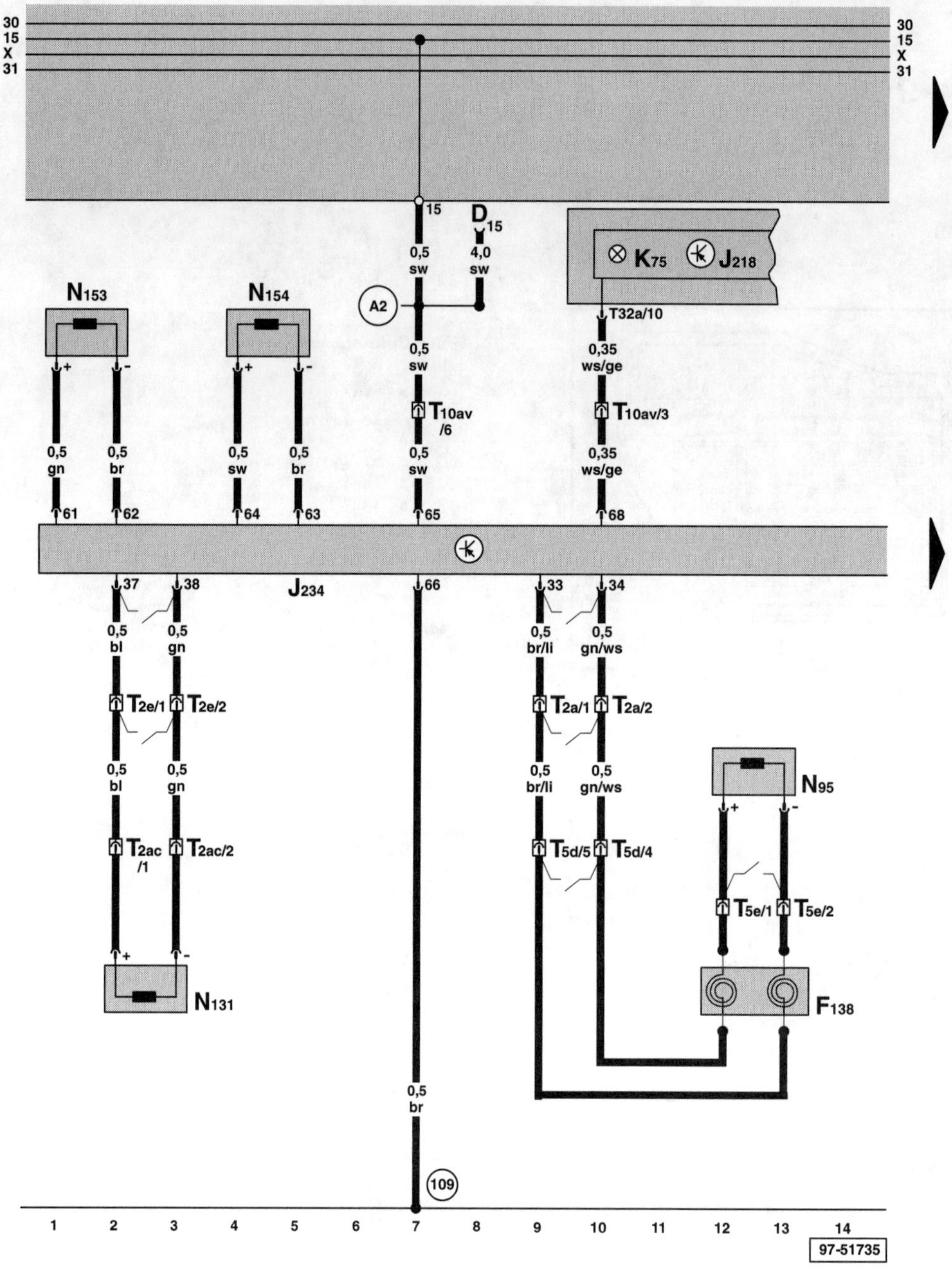

ws = white
sw = black
ro = red
br = brown
gn = green
bl = blue
gr = grey
li = violet
ge = yellow

D - Ignition/Starter Switch
F138 - Airbag Spiral Spring/Return Spring With Slip Ring
J218 - Instrument Cluster Combination Processor
J234 - Airbag Control Module
K75 - Airbag Malfunction Indicator Lamp (MIL)
N95 - Driver's Side Airbag Igniter
N131 - Passenger's Side Airbag Igniter 1
N153 - Left Seat Belt Tensioner Igniter
N154 - Right Seat Belt Tensioner Igniter
T2a - Connector 2 pin, yellow, connector station A pillar, left
T2e - Connector 2 pin, yellow, connector station A pillar, right
T2ac - Connector 2 pin, yellow, behind instrument panel, right
T5d - Connector 5 pin, behind steering column switch cover
T5e - Connector 5 pin, in steering wheel
T10av - Connector 10 pin, yellow, connector station A pillar, right
T32a - Connector 32 pin, green, on instrument cluster

(109) - Ground connection, in airbag wiring harness

(A2) - Plus connection (15), in instrument panel wiring harness

Bridge closes when disconnecting connector (safety feature)

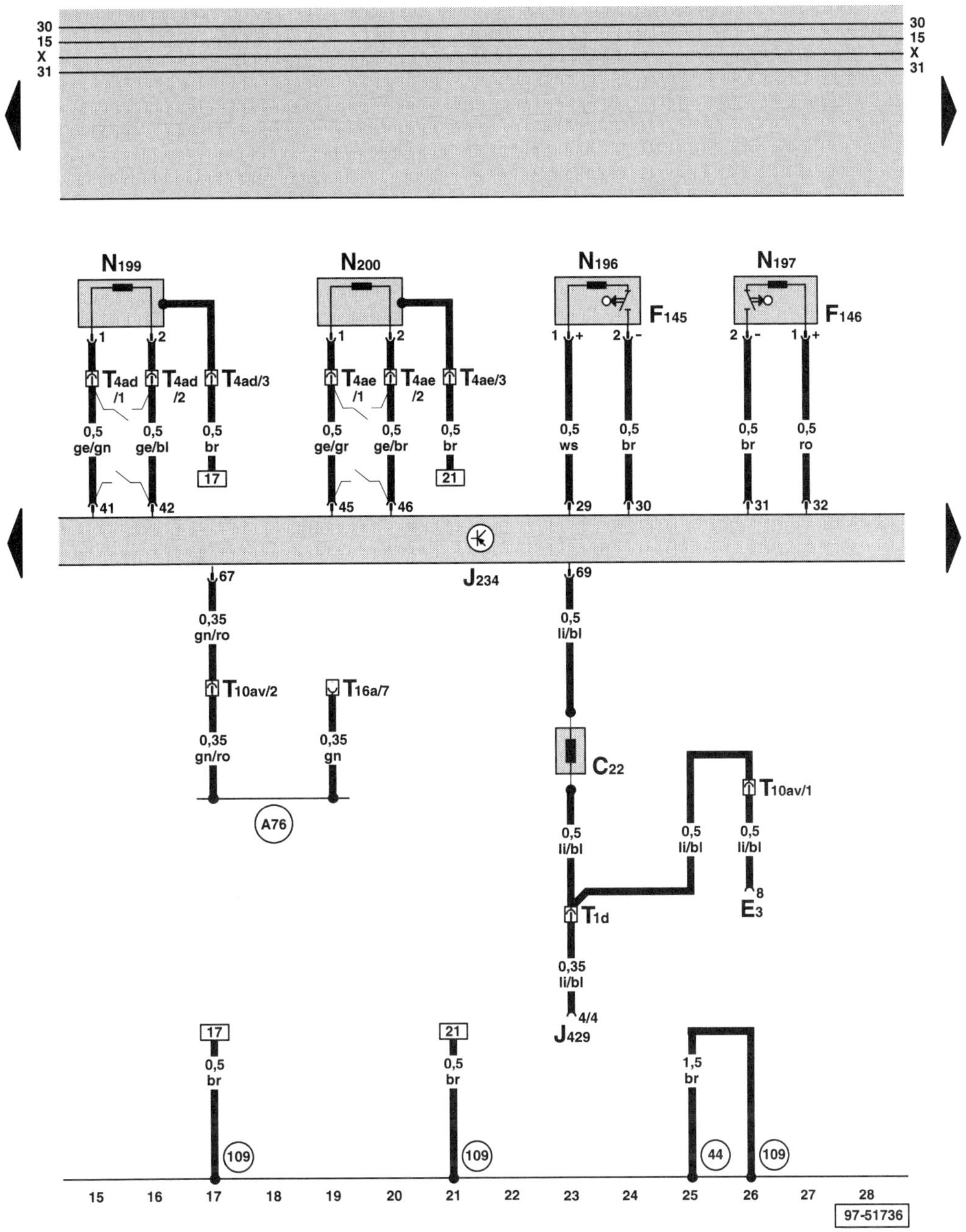

ws = white
sw = black
ro = red
br = brown
gn = green
bl = blue
gr = grey
li = violet
ge = yellow

C22 - Interference filter
E3 - Emergency Flasher Switch
F145 - Left Rear B-pillar Microswitch
F146 - Right Rear B-pillar Microswitch
J234 - Airbag Control Module
J429 - Control module for central locking
N196 - Igniter for belt tensioner rear, driver's side
N197 - Igniter for belt tensioner rear, passenger's side
N199 - Igniter for side airbag, driver's side
N200 - Igniter for side airbag, passenger's side
T1d - Single connector, brown, connector station A pillar right
T4ad - Connector 4 pin, yellow, below driver's seat
T4ae - Connector 4 pin, yellow, below passenger's seat
T10av - Connector 10 pin, yellow, connector station A pillar, right
T16a - Connector 16 pin, Data Link Connector (DLC)

(44) - Ground connection, on left A-pillar, lower part

(109) - Ground connection, in airbag wiring harness

(A76) - Connector (K-diagnosis wire), in instrument panel wiring harness

- Bridge closes when disconnecting connector (safety feature)

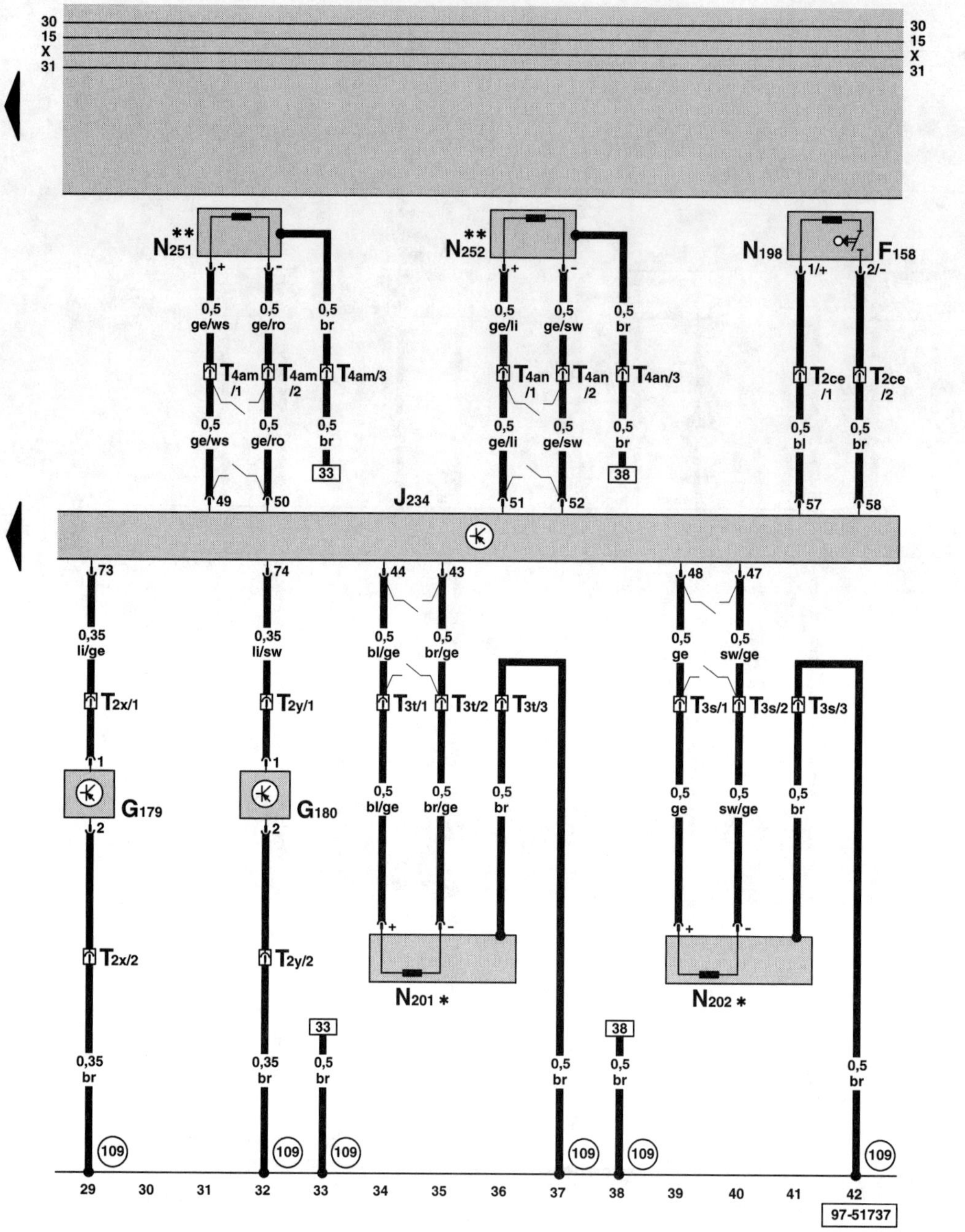

ws = white
sw = black
ro = red
br = brown
gn = green
bl = blue
gr = grey
li = violet
ge = yellow

F158 - Seat Belt Tensioner Switch (1)
G179 - Crash sensor for side airbag, driver's side
G180 - Crash sensor for side airbag, passenger's side
J234 - Airbag Control Module
N198 - Igniter for belt tensioner, rear, center
N201 - Igniter for side airbag rear, driver's side
N202 - Igniter for side airbag rear, passenger's side
N251 - Igniter for head airbag, driver side
N252 - Igniter for head airbag, passenger side
T2x - Connector 2 pin, yellow, below driver's seat
T2y - Connector 2 pin, yellow, below passenger's seat
T2ce - Connector 2 pin, yellow, in rear seat
T4am - Connector 4 pin, yellow, on C pillar, left
T4an - Connector 4 pin, yellow, on C pillar, right
T3s - Connector 3 pin, orange, in rear seat
T3t - Connector 3 pin, orange, in rear seat

(109) - Ground connection, in airbag wiring harness

- Bridge closes when disconnecting connector (safety feature)

* - Vehicles with sideairbag rear

** - Vehicles with headairbag

Radio-Stereo, Concert with and without CD-Changer

2000 m. y.

Fuse Panel

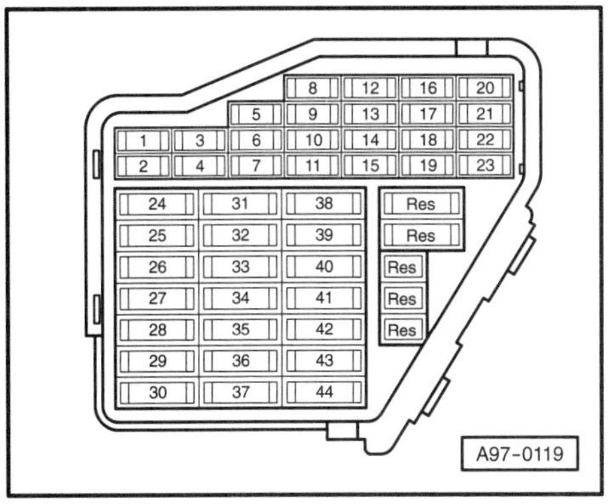

Fuse Colors:

30 A - Green
25 A - White
20 A - Yellow
15 A - Blue
10 A - Red
7,5 A - Brown
5 A - Beige

Starting with fuse position 23, fuses in the fuse holder are identified with 223 in the wiring diagram.

Micro Central Electric Panel

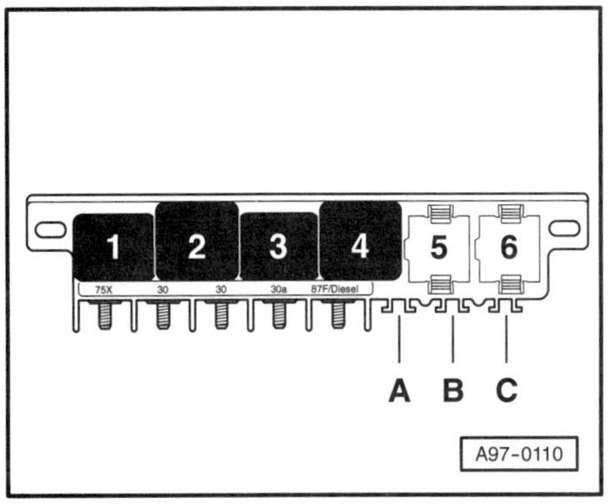

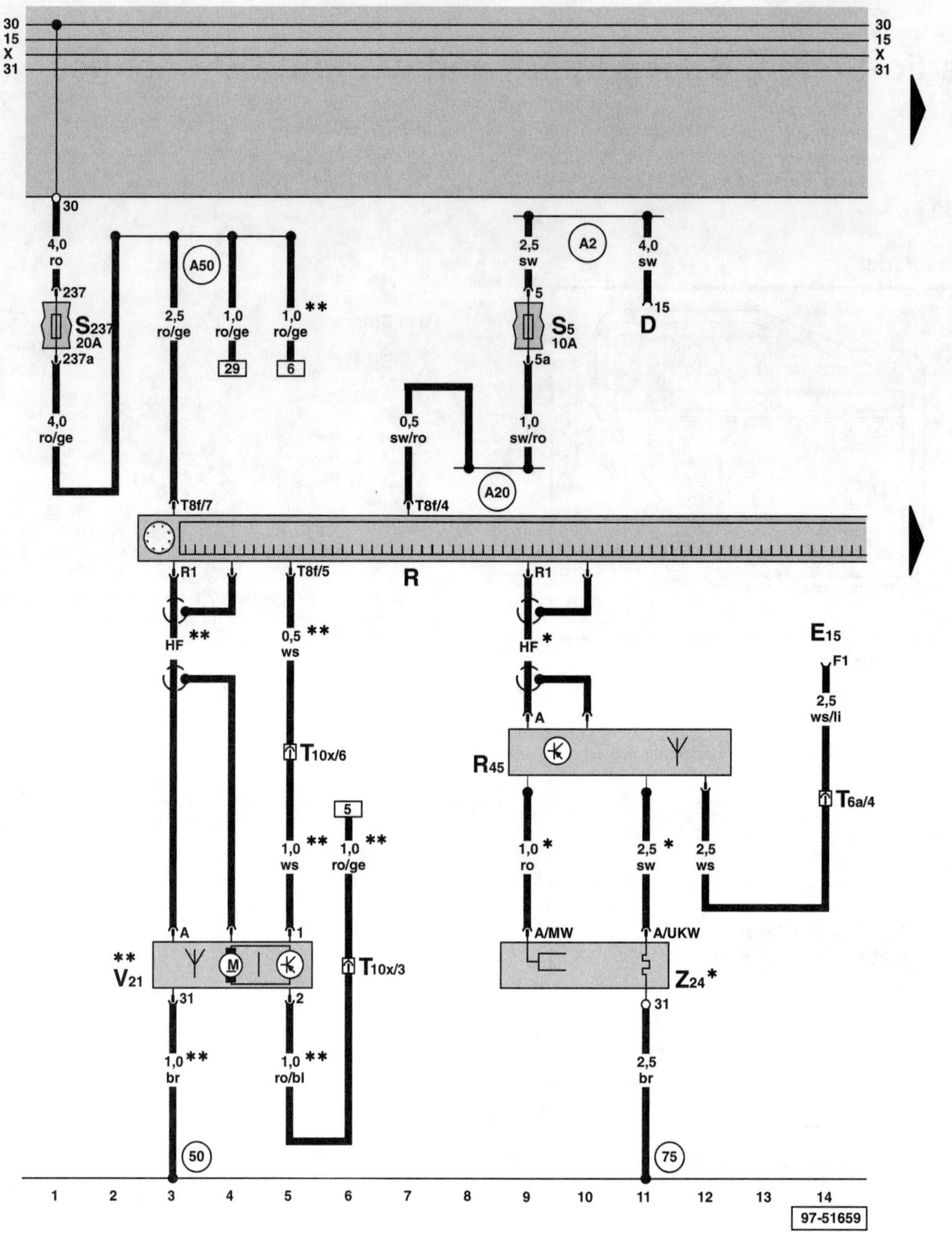

ws = white
sw = black
ro = red
br = brown
gn = green
bl = blue
gr = grey
li = violet
ge = yellow

D - Ignition/Starter Switch
E15 - Rear Window Defogger Switch
R - Radio
R1 - Antenna Connection
R45 - Telephone/Radio Antenna Amplifier
S5 - Fuse
S237 - Fuse 37 in fuse holder
T6a - Connector 6 pin, blue, connector station A pillar, left
T8f - Connector 8 pin, black, connector radio III
T10x - Connector 10 pin, green, connector station A pillar, right
V21 - Power Antenna
Z24 - Rear Window Defogger With Window Antenna

(50) - Ground connection, in luggage compartment, left
(75) - Ground connection, on right rear pillar
(A2) - Plus connection (15), in instrument panel wiring harness
(A20) - Wire connection (15a), in instrument panel wiring harness
(A50) - Plus connection (30as), in instrument panel wiring harness
* - Vehicles with rear Window Antenna
** - Vehicles with Automatic Antenna

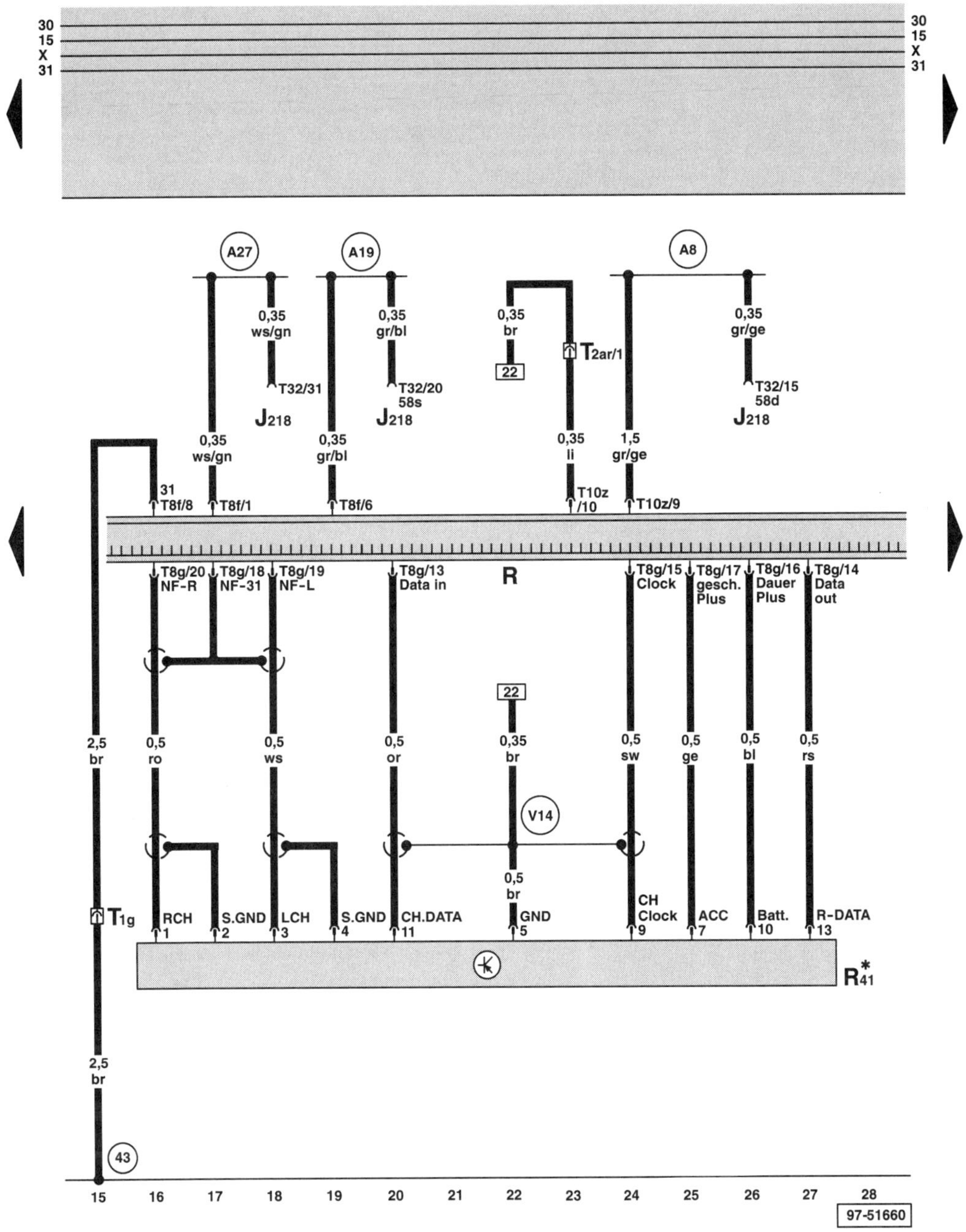

ws = white
sw = black
ro = red
br = brown
gn = green
bl = blue
gr = grey
li = violet
ge = yellow
or = orange
rs = pink

J218 - Instrument Cluster Combination Processor
R - Radio
R41 - CD Changer Unit
T1g - Single connector, black, connector station A pillar, right
T2ar - Connector 2 pin, black, behind instrument panel, center
T8f - Connector 8 pin, black, connector radio III
T8g - Connector 8 pin, blue, connector radio I - 3
T10z - Connector 10 pin, red, connector radio IV
T32 - Connector 32 pin, blue, on instrument cluster

(43) - Ground connection, on right A-pillar, lower part
(A8) - Plus connection (58d), in instrument panel wiring harness
(A19) - Wire connection (58s), in instrument panel wiring harness
(A27) - Wire connection (vehicle speed signal), in instrument panel wiring harness
(V14) - Wire connection (shielding), in CD-changer wiring harness
* - Vehicles with CD Changer

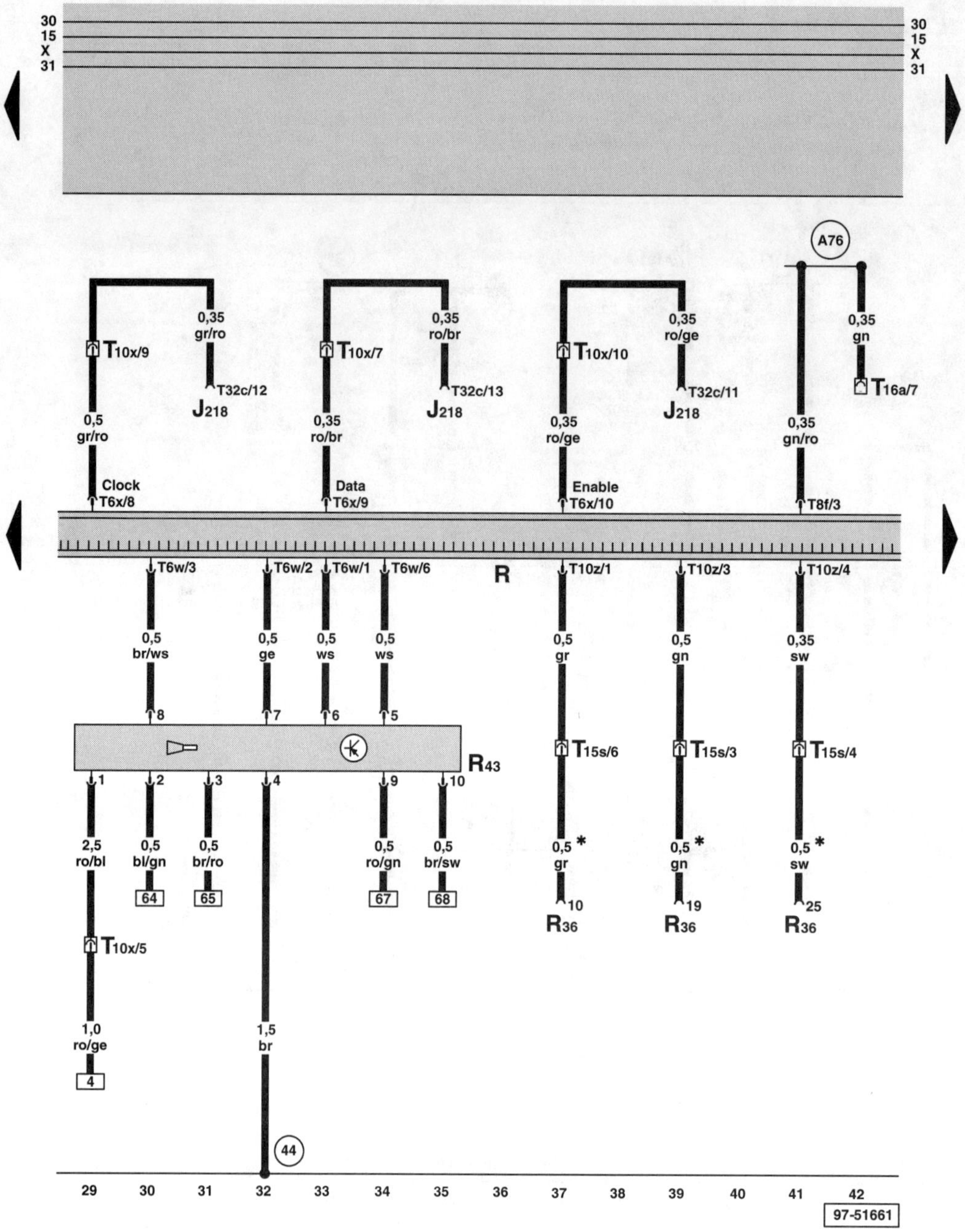

ws = white
sw = black
ro = red
br = brown
gn = green
bl = blue
gr = grey
li = violet
ge = yellow

J218 - Instrument Cluster Combination Processor
R - Radio
R36 - Telephone Transceiver
R43 - Right Rear Woofer/Amplifier
T6w - Connector 6 pin, yellow, connector radio I - 1
T6x - Connector 6 pin, green, connector radio I - 2
T8f - Connector 8 pin, black, connector radio III
T10x - Connector 10 pin, green, connector station A pillar, right
T10z - Connector 10 pin, red, connector radio IV
T15s - Connector 15 pin, blue, connector station A pillar, left
T16a - Connector 16 pin, Data Link Connector (DLC)
T32c - Connector 32 pin, grey, on instrument cluster

(44) - Ground connection, on left A-pillar, lower part

(A76) - Connector (K-diagnosis wire), in instrument panel wiring harness

* - Vehicles with Telephone

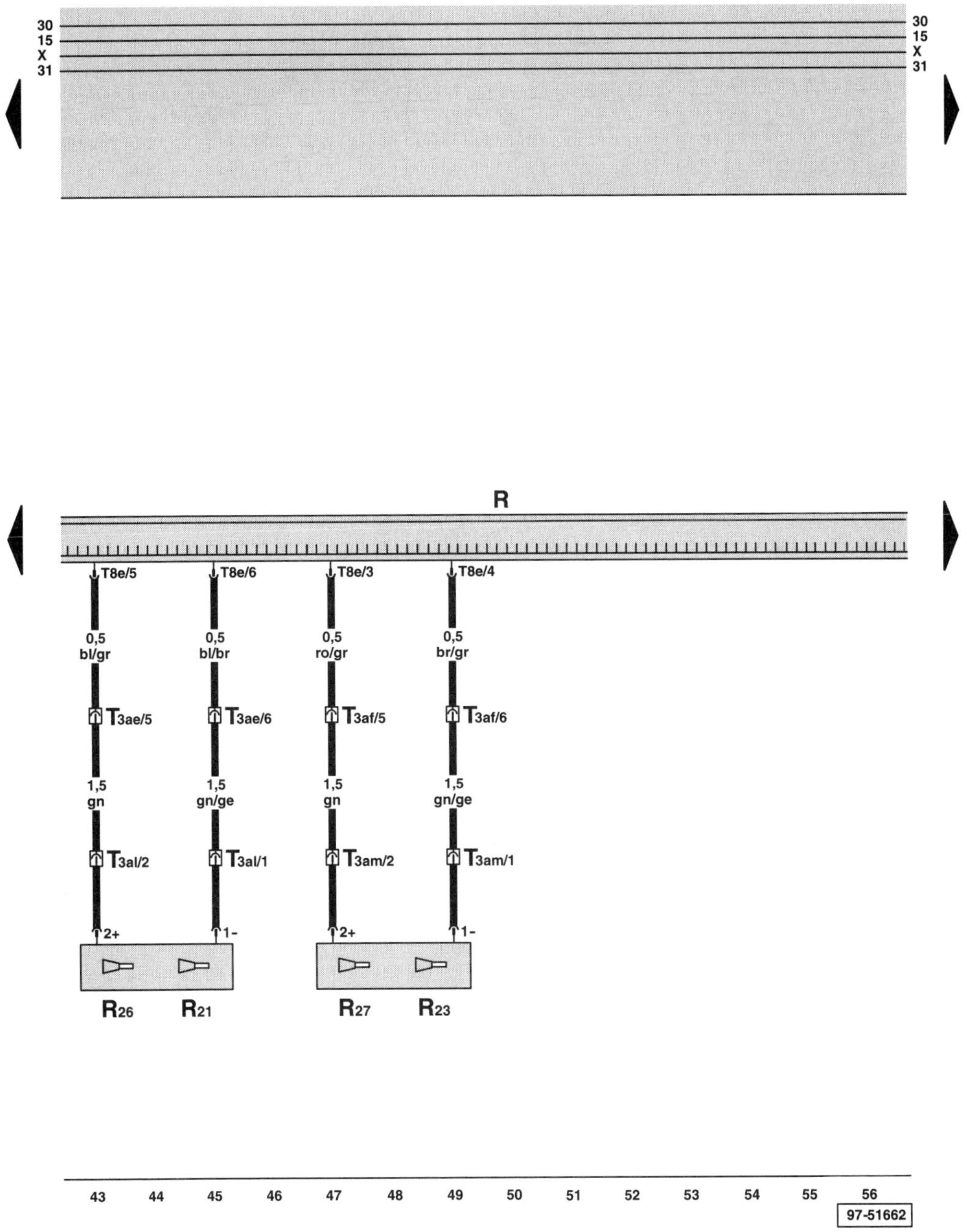

ws = white
sw = black
ro = red
br = brown
gn = green
bl = blue
gr = grey
li = violet
ge = yellow

R - Radio
R21 - Left Front Woofer
R23 - Right Front Woofer
R26 - Left Front Midrange Speaker
R27 - Right Front Midrange Speaker
T3ae - Connector 3 pin, brown, connector station A pillar, left
T3af - Connector 3 pin, brown, connector station A pillar, right
T3al - Connector 3 pin, black, in driver's door
T3am - Connector 3 pin, black, in passenger's door
T8e - Connector 8 pin, brown, connector radio II

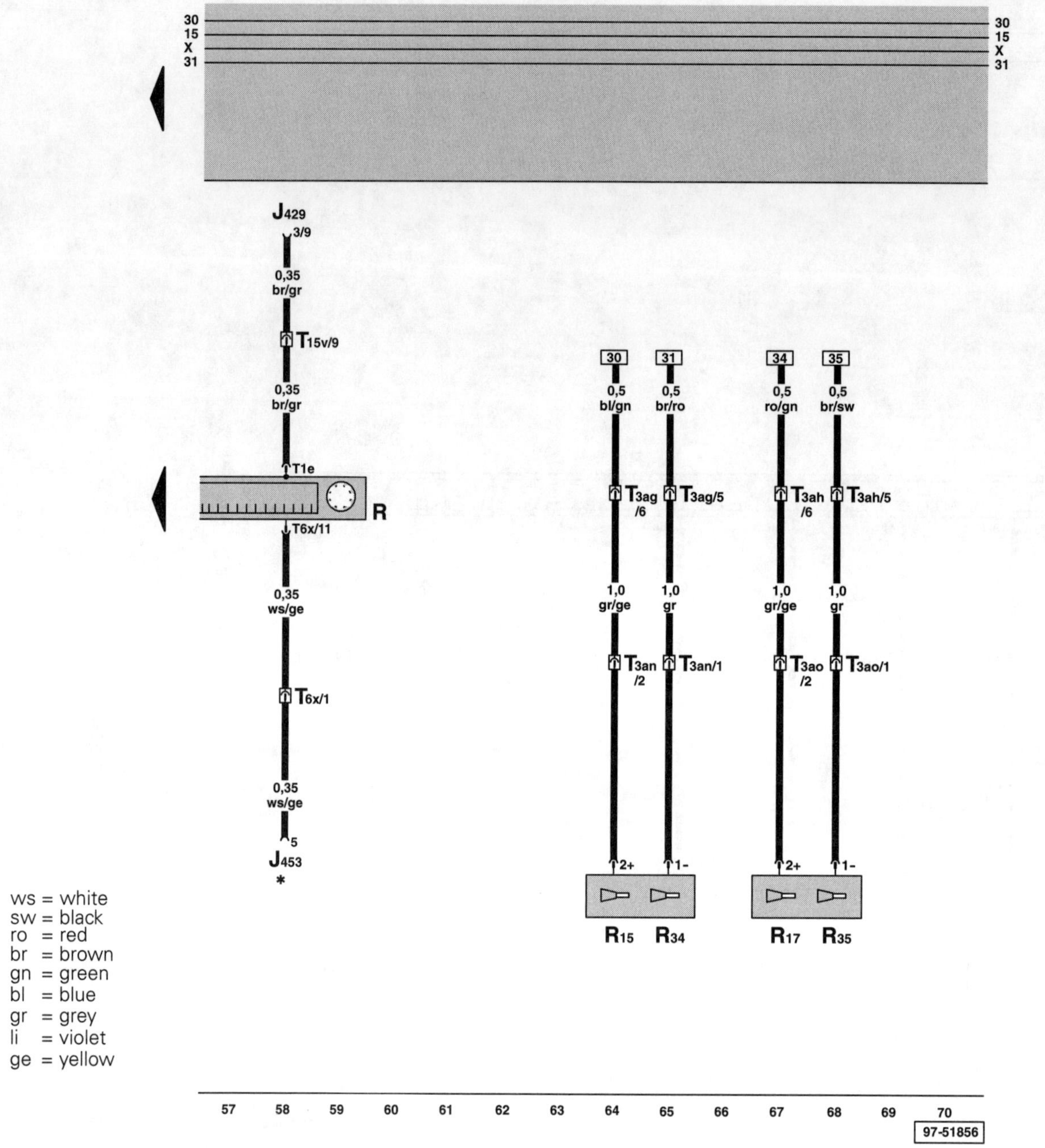

J429 - Control module for central locking
J453 - Control module for multi-function steering wheel
R - Radio
R15 - Left Rear Woofer
R17 - Right Rear Woofer
R34 - Left Rear Midrange Speaker
R35 - Right Rear Midrange Speaker
T1e - Single connector, on Radio
T3ag - Connector 3 pin, blue, in B pillar left
T3ah - Connector 3 pin, blue, in B pillar right
T3an - Connector 3 pin, black, in left rear door
T3ao - Connector 3 pin, black, in right rear door
T6x - Connector 6 pin, green, connector radio I - 2
T15v - Connector 15 pin, orange, connector station A pillar, right

* - Vehicles with multi-function steering wheel

Radio Concert, Symphony, Bose with and without CD Changer

2000 m. y.

Fuse Panel

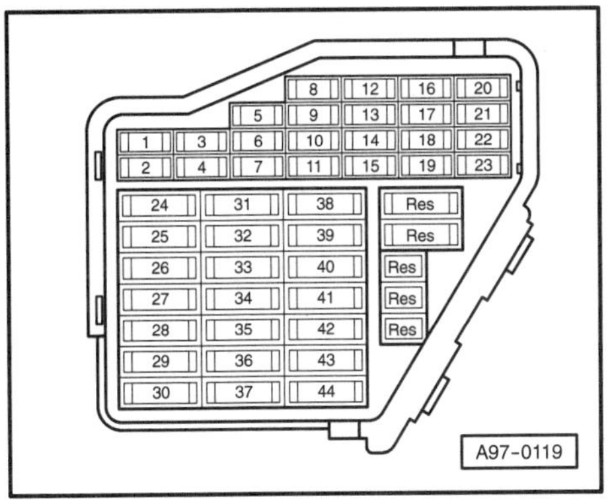

Fuse Colors:

30 A - Green
25 A - White
20 A - Yellow
15 A - Blue
10 A - Red
7,5 A - Brown
5 A - Beige

Starting with fuse position 23, fuses in the fuse holder are identified with 223 in the wiring diagram.

Micro Central Electric Panel

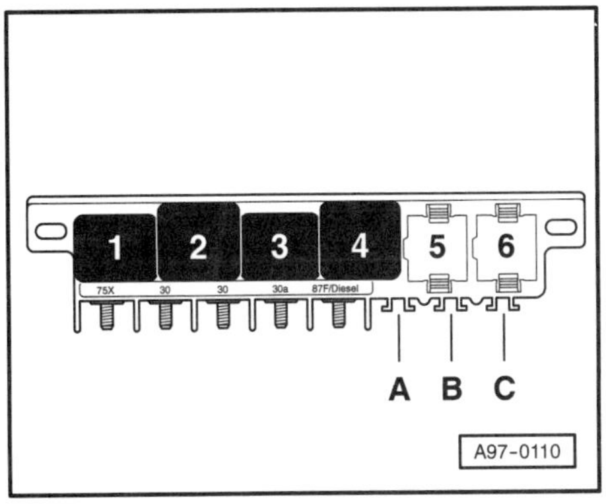

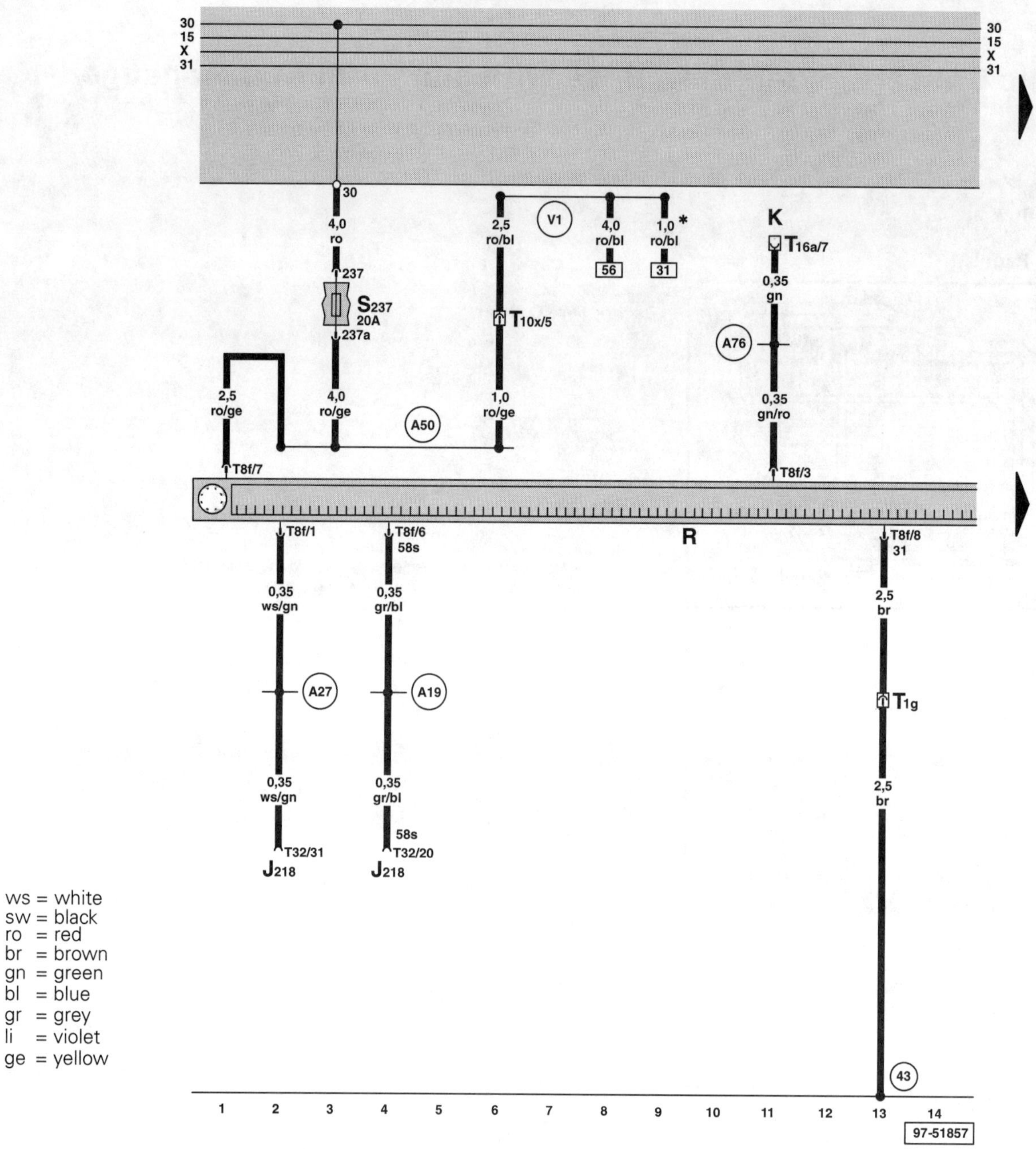

J218	-	Instrument Cluster Combination Processor
R	-	Radio
S237	-	Fuse 37 in fuse holder
T1g	-	Single connector, black, connector station A pillar, right
T8f	-	Connector 8 pin, black, connector radio III
T10x	-	Connector 10 pin, green, connector station A pillar, right
T16a	-	Connector 16 pin, Data Link Connector (DLC)
T32	-	Connector 32 pin, blue, on instrument cluster

(43)	-	Ground connection, on right A-pillar, lower part
(A19)	-	Wire connection (58s), in instrument panel wiring harness
(A27)	-	Wire connection (vehicle speed signal), in instrument panel wiring harness
(A50)	-	Plus connection (30as), in instrument panel wiring harness
(A76)	-	Connector (K-diagnosis wire), in instrument panel wiring harness
(V1)	-	Plus connection (30), in rear speaker wiring harness
*	-	Vehicles with Automatic Antenna

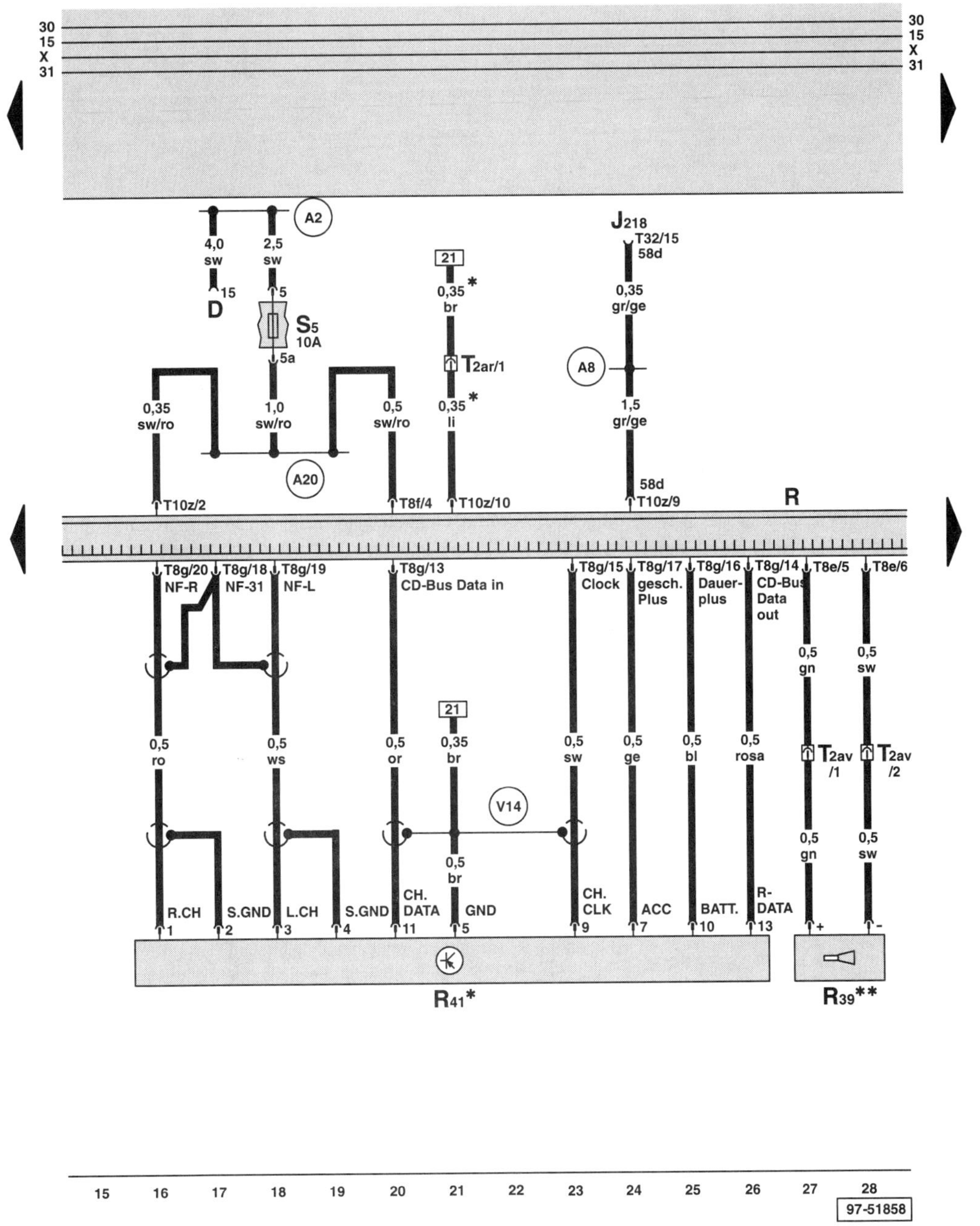

ws = white
sw = black
ro = red
br = brown
gn = green
bl = blue
gr = grey
li = violet
ge = yellow
or = orange

D - Ignition/Starter Switch
J218 - Instrument Cluster Combination Processor
R - Radio
R39 - Speaker (telephone, navigation)
R41 - CD Changer Unit
S5 - Fuse
T2ar - Connector 2 pin, black, behind instrument panel, center
T2av - Connector 2 pin, black, for loudspeaker, B pillar, left
T8e - Connector 8 pin, brown, connector radio II
T8f - Connector 8 pin, black, connector radio III
T8g - Connector 8 pin, blue, connector radio I - 3
T10z - Connector 10 pin, red, connector radio IV
T32 - Connector 32 pin, blue, on instrument cluster

(A2) - Plus connection (15), in instrument panel wiring harness

(A8) - Plus connection (58d), in instrument panel wiring harness

(A20) - Wire connection (15a), in instrument panel wiring harness

(V14) - Wire connection (shielding), in CD-changer wiring harness

* - Vehicles with CD Changer

** - Vehicles with Telephone

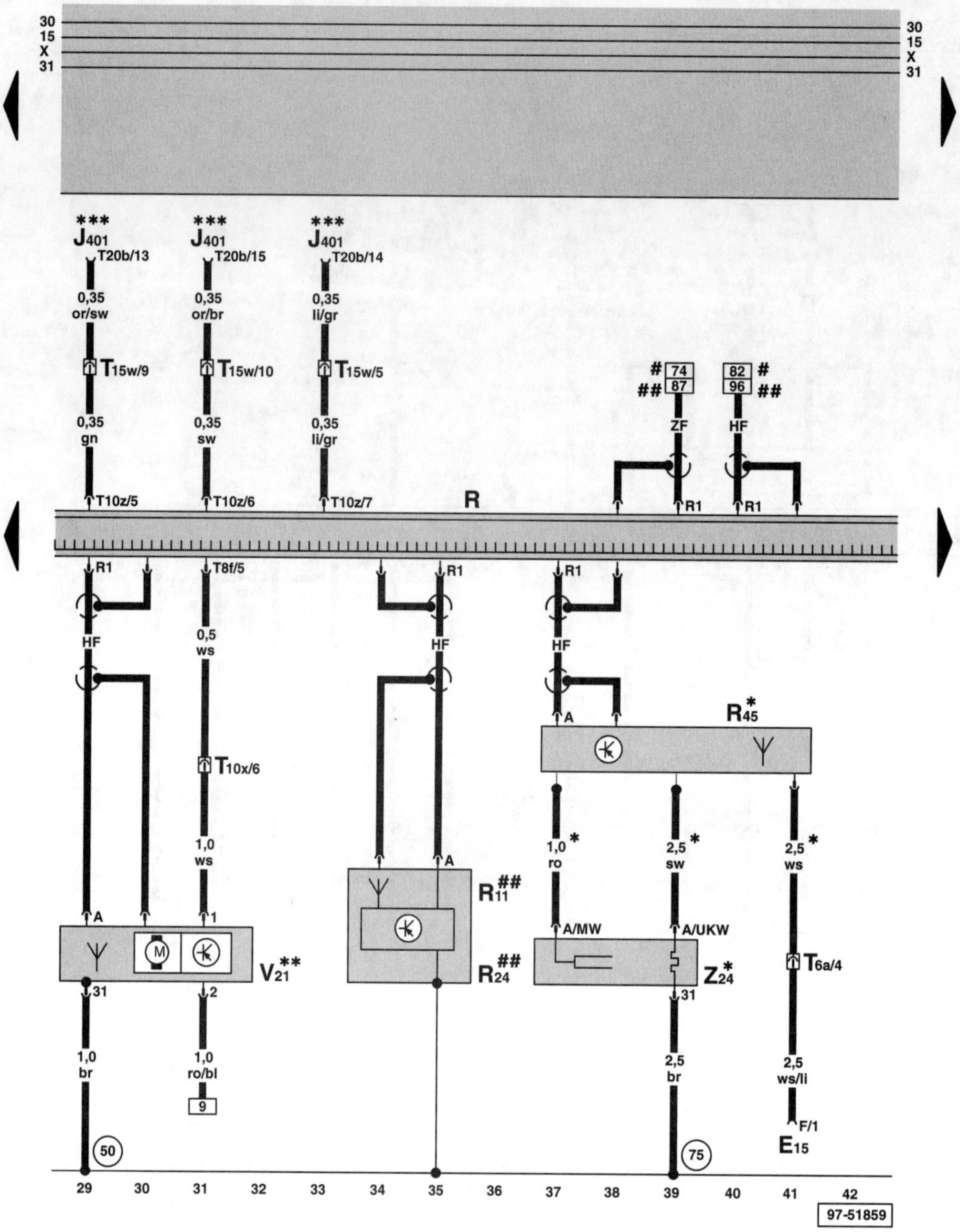

ws = white
sw = black
ro = red
br = brown
gn = green
bl = blue
gr = grey
li = violet
ge = yellow
or = orange

E15 - Rear Window Defogger Switch
J401 - Control unit for navigation with CD drive
R - Radio
R1 - Antenna Connection
R11 - Antenna
R24 - Antenna Amplifier
R45 - Telephone/Radio Antenna Amplifier
T6a - Connector 6 pin, blue, connector station A pillar, left
T8f - Connector 8 pin, black, connector radio III
T10x - Connector 10 pin, green, connector station A pillar, right
T10z - Connector 10 pin, red, connector radio IV
T15w - Connector 15 pin, green, connector station A pillar, right
T20b - Connector 20 pin, black, on control unit for navigation
V21 - Power Antenna
Z24 - Rear Window Defogger With Window Antenna

(50) - Ground connection, in luggage compartment, left

(75) - Ground connection, on right rear pillar

* - Vehicles with rear Window Antenna
** - Vehicles with Automatic Antenna
*** - Vehicles with Navigation
# - Sedan
- Wagon

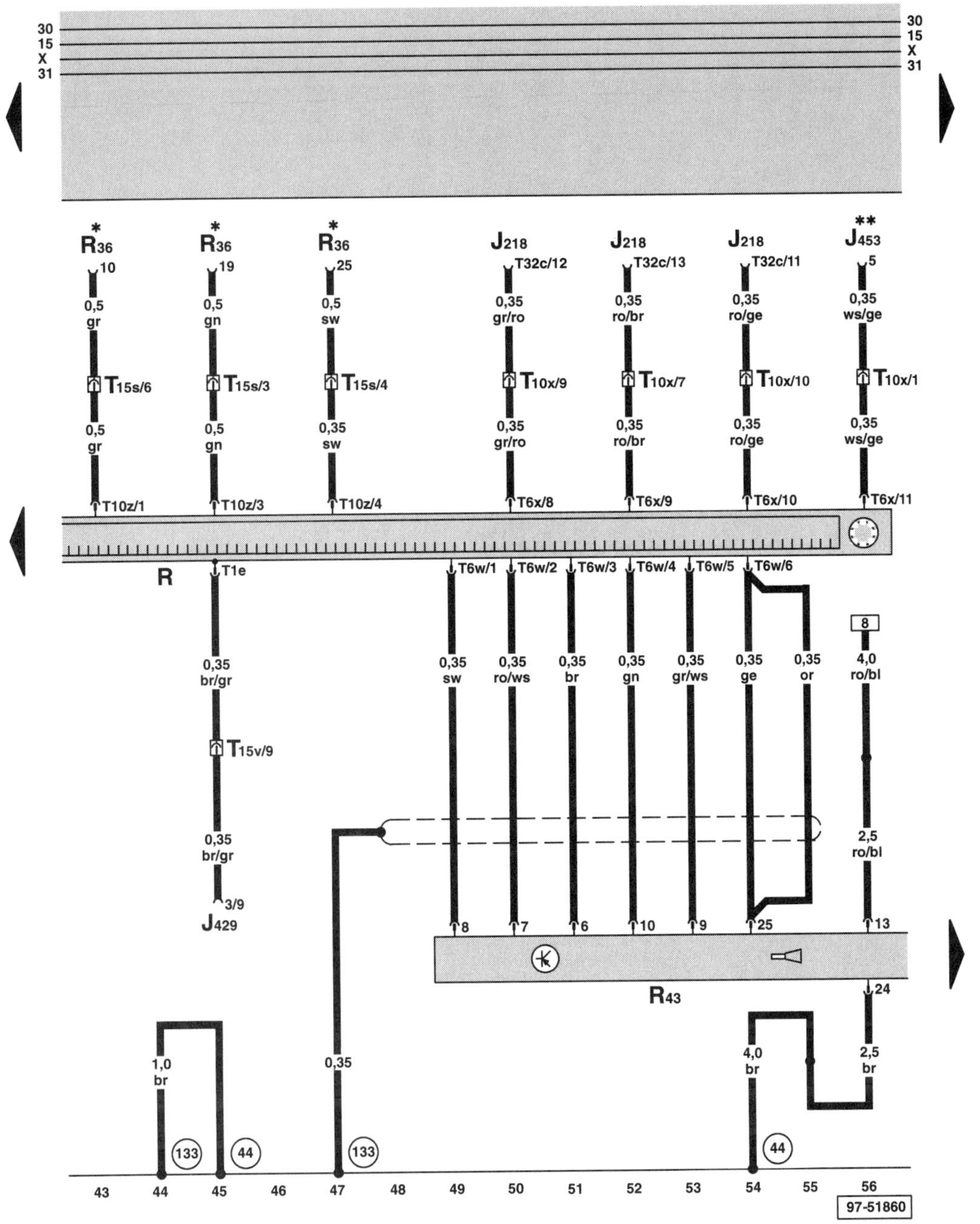

ws = white
sw = black
ro = red
br = brown
gn = green
bl = blue
gr = grey
li = violet
ge = yellow
or = orange

J218 - Instrument Cluster Combination Processor
J429 - Control module for central locking
J453 - Control module for multi-function steering wheel
R - Radio
R36 - Telephone Transceiver
R43 - Right Rear Woofer/Amplifier
T1e - Single connector, on Radio
T6w - Connector 6 pin, yellow, connector radio I - 1
T6x - Connector 6 pin, green, connector radio I - 2
T10x - Connector 10 pin, green, connector station A pillar, right
T10z - Connector 10 pin, red, connector radio IV
T15s - Connector 15 pin, blue, connector station A pillar, left
T15v - Connector 15 pin, orange, connector station A pillar, right
T32c - Connector 32 pin, grey, on instrument cluster

(44) - Ground connection, on left A-pillar, lower part
(133) - Ground connection, in speaker wiring harness

* - Vehicles with Telephone
** - Vehicles with multi-function steering wheel

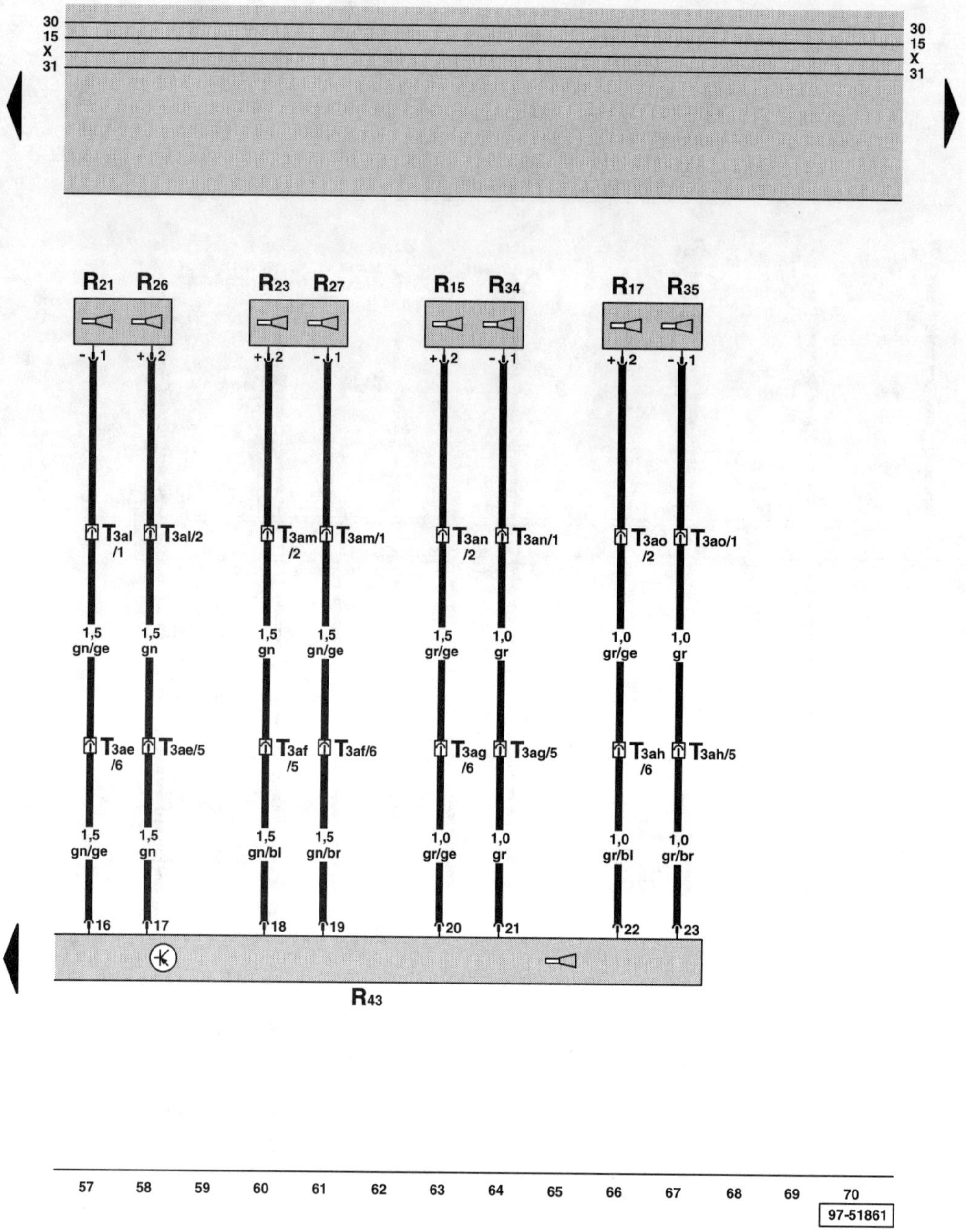

ws = white
sw = black
ro = red
br = brown
gn = green
bl = blue
gr = grey
li = violet
ge = yellow

R15 - Left Rear Woofer
R17 - Right Rear Woofer
R21 - Left Front Woofer
R23 - Right Front Woofer
R26 - Left Front Midrange Speaker
R27 - Right Front Midrange Speaker
R34 - Left Rear Midrange Speaker
R35 - Right Rear Midrange Speaker
R43 - Right Rear Woofer/Amplifier
T3ae - Connector 3 pin, brown, connector station A pillar, left
T3af - Connector 3 pin, brown, connector station A pillar, right
T3ag - Connector 3 pin, blue, in B pillar left
T3ah - Connector 3 pin, blue, in B pillar right
T3al - Connector 3 pin, black, in driver's door
T3am - Connector 3 pin, black, in passenger's door
T3an - Connector 3 pin, black, in left rear door
T3ao - Connector 3 pin, black, in right rear door

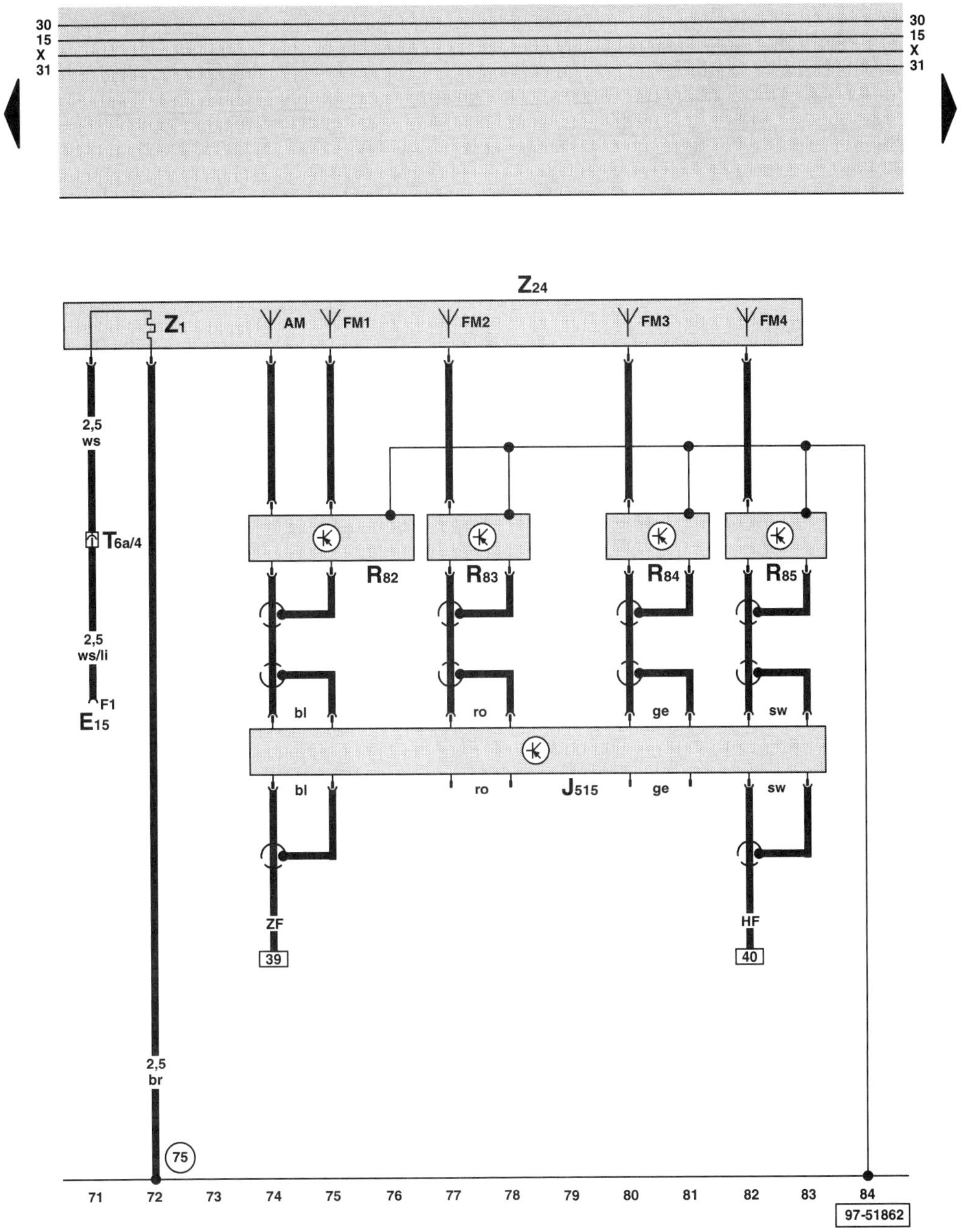

ws = white
sw = black
ro = red
br = brown
gn = green
bl = blue
gr = grey
li = violet
ge = yellow

E15 - Rear Window Defogger Switch
J515 - Control unit for aerial selection (aerial diversity function)
R82 - TV Antenna Amplifier 1
R83 - TV Antenna Amplifier 2
R84 - TV Antenna Amplifier 3
R85 - TV Antenna Amplifier 4
T6a - Connector 6 pin, blue, connector station A pillar, left
Z1 - Rear Window Defogger/Heat Element
Z24 - Rear Window Defogger With Window Antenna

(75) - Ground connection, on right rear pillar

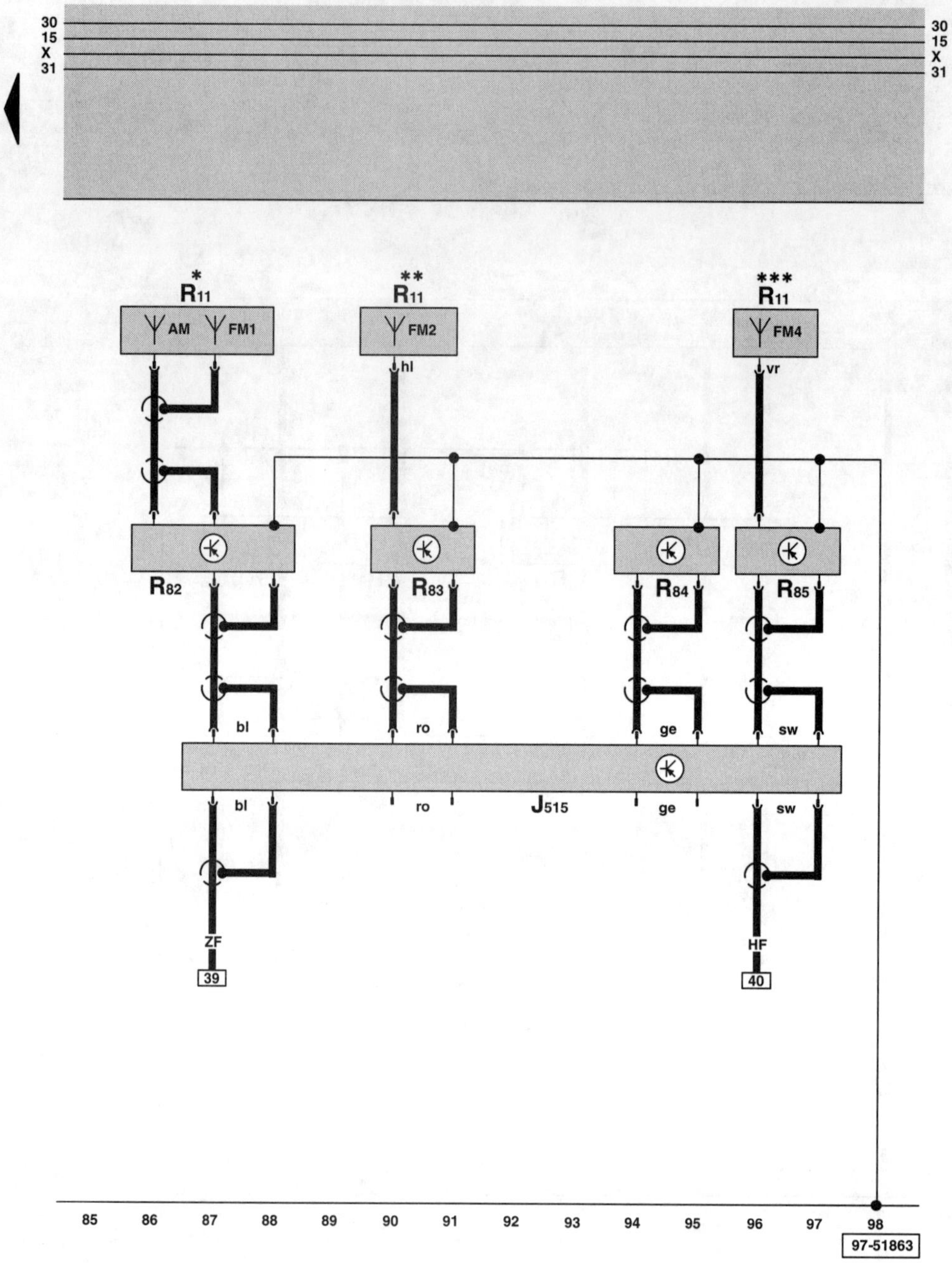

ws = white
sw = black
ro = red
br = brown
gn = green
bl = blue
gr = grey
li = violet
ge = yellow

J515 - Control unit for aerial selection (aerial diversity function)
R11 - Antenna
R82 - TV Antenna Amplifier 1
R83 - TV Antenna Amplifier 2
R84 - TV Antenna Amplifier 3
R85 - TV Antenna Amplifier 4

* - Roof Antenna
** - Window rear left
*** - Window rear right

Telephone

2000 m. y.

Fuse Panel

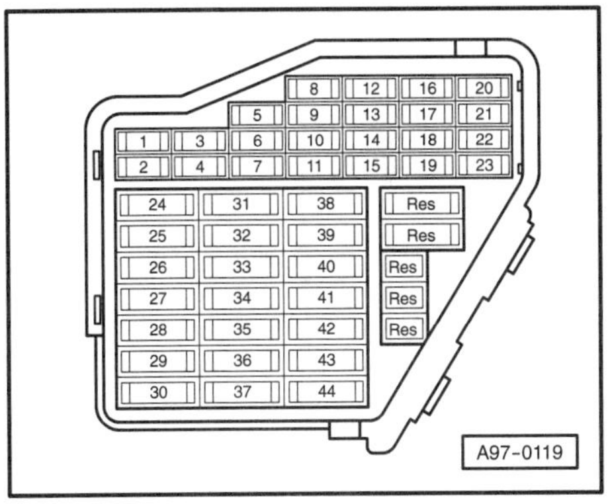

Fuse Colors:

30 A - Green
25 A - White
20 A - Yellow
15 A - Blue
10 A - Red
7,5 A - Brown
5 A - Beige

Starting with fuse position 23, fuses in the fuse holder are identified with 223 in the wiring diagram.

Micro Central Electric Panel

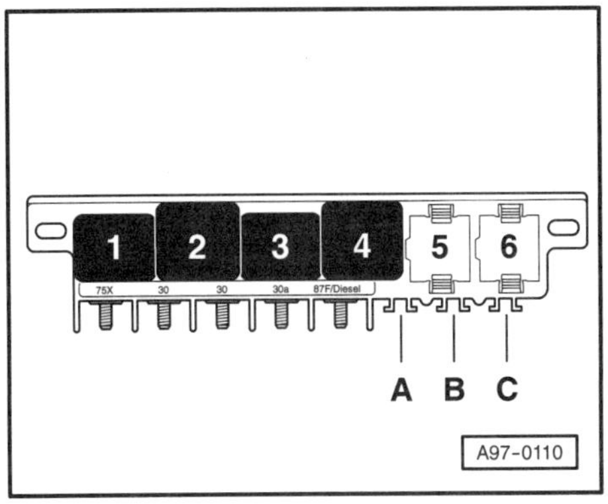

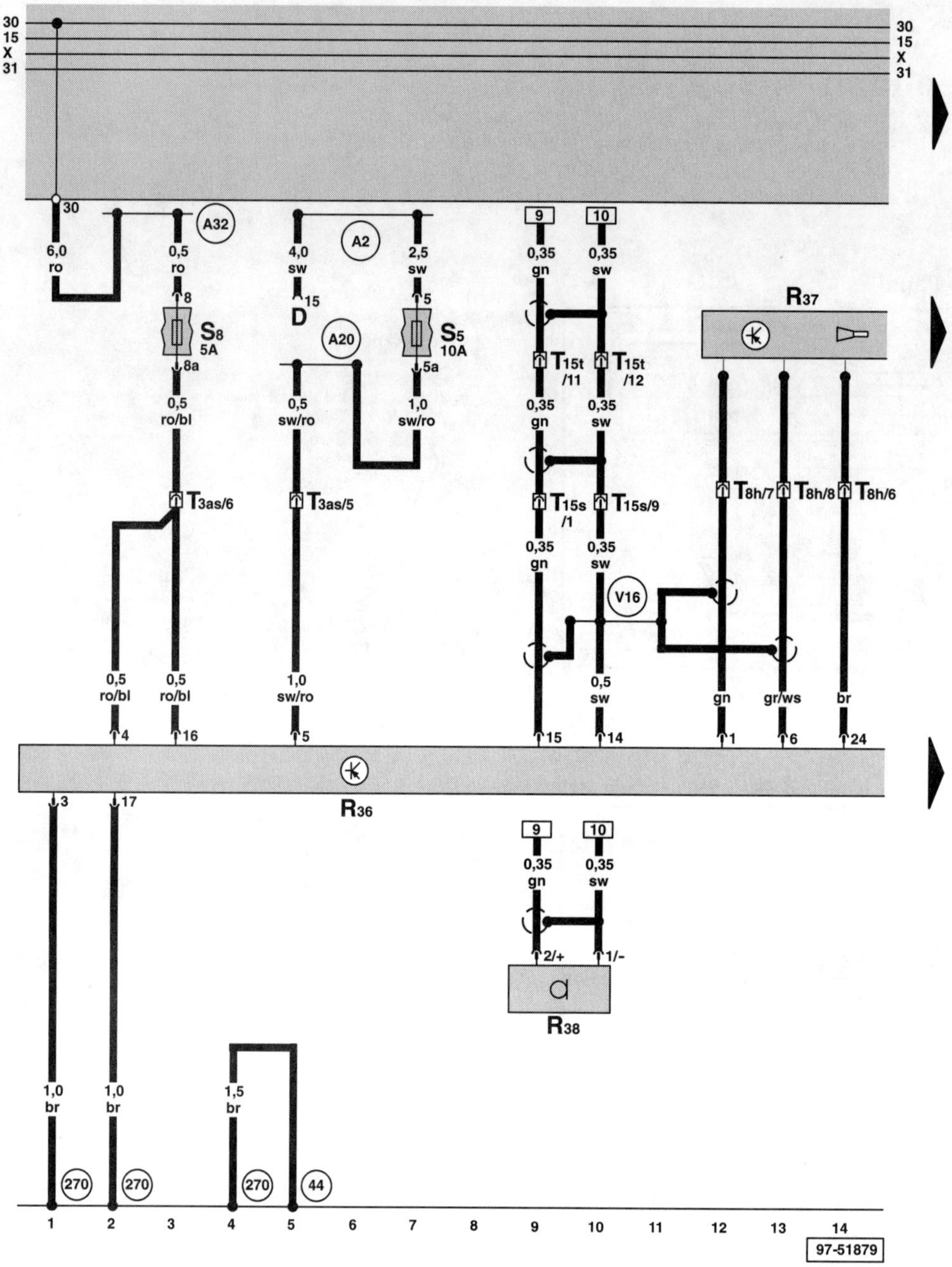

ws = white
sw = black
ro = red
br = brown
gn = green
bl = blue
gr = grey
li = violet
ge = yellow

D - Ignition/Starter Switch
R36 - Telephone Transceiver
R37 - Telephone Handset
R38 - Telephone Microphone
S5 - Fuse
S8 - Fuse
T3as - Connector 3 pin, blue, connector station A pillar, left
T8h - Connector 8 pin, Telephone Handset, in console
T15s - Connector 15 pin, blue, connector station A pillar, left
T15t - Connector 15 pin, grey, connector station A pillar, left

(44) - Ground connection, on left A-pillar, lower part
(270) - Ground connector, in wiring harness telephone
(A2) - Plus connection (15), in instrument panel wiring harness
(A20) - Wire connection (15a), in instrument panel wiring harness
(A32) - Plus connection (30), in instrument panel wiring harness
(V16) - Connector (shielding), in wiring harness telephone

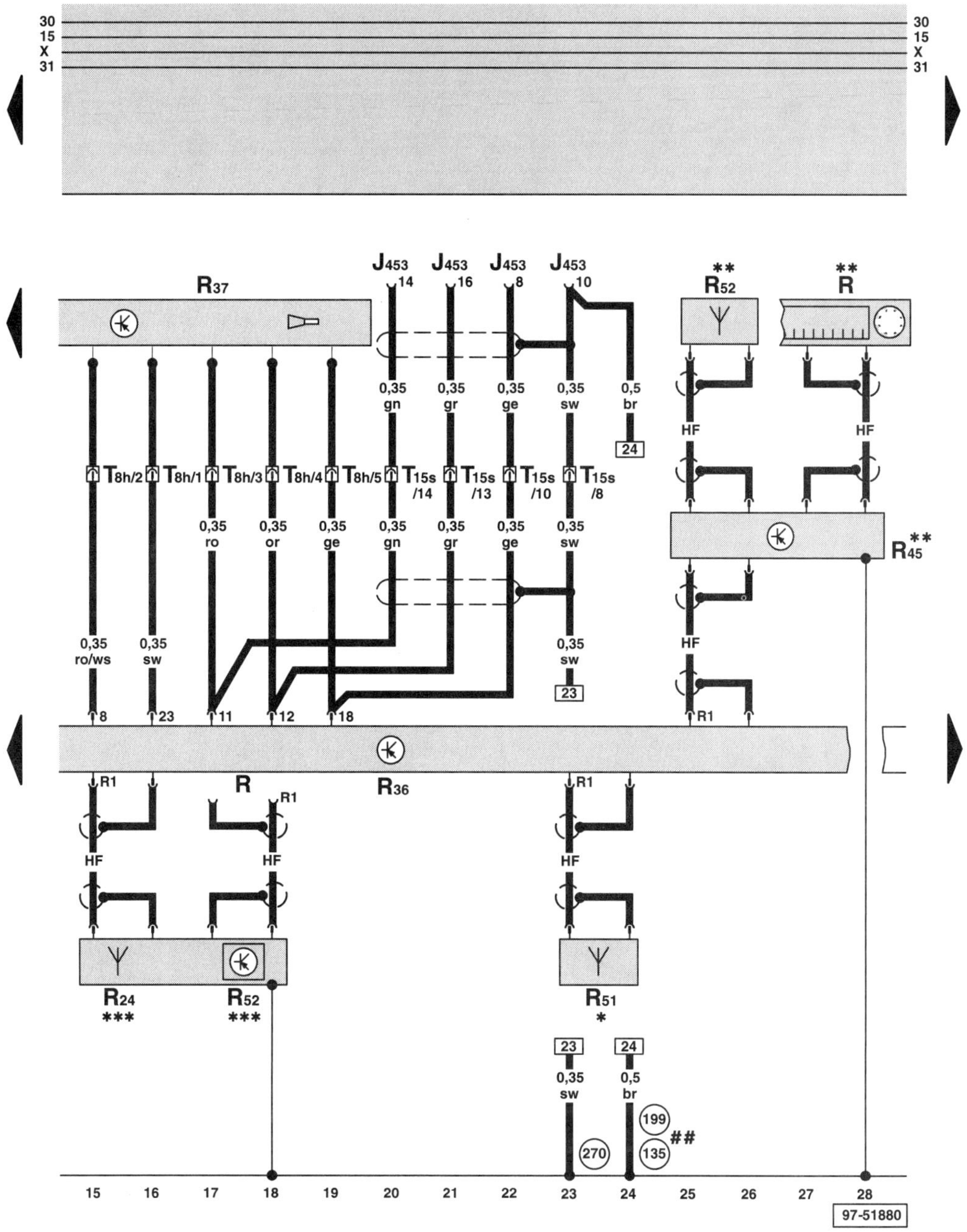

ws = white
sw = black
ro = red
br = brown
gn = green
bl = blue
gr = grey
li = violet
ge = yellow
or = orange

J453 - Control module for multi-function steering wheel
R - Radio
R1 - Antenna Connection
R24 - Antenna Amplifier
R36 - Telephone Transceiver
R37 - Telephone Handset
R45 - Telephone/Radio Antenna Amplifier
R51 - Radio/Telephone Antenna
R52 - Antenna for radio/telephone/navigation (GPS)
T8h - Connector 8 pin, Telephone Handset, in console
T15s - Connector 15 pin, blue, connector station A pillar, left

(135) - Ground connection -2-, in instrument panel wiring harness
(199) - Ground connection -3-, in instrument panel wiring harness
(270) - Ground connector, in wiring harness telephone

* - Vehicles with rod antenna
** - Vehicles with insulating glass
*** - Roof Antenna Wagon
# - Vehicles with multi-function steering wheel
- Both are possible

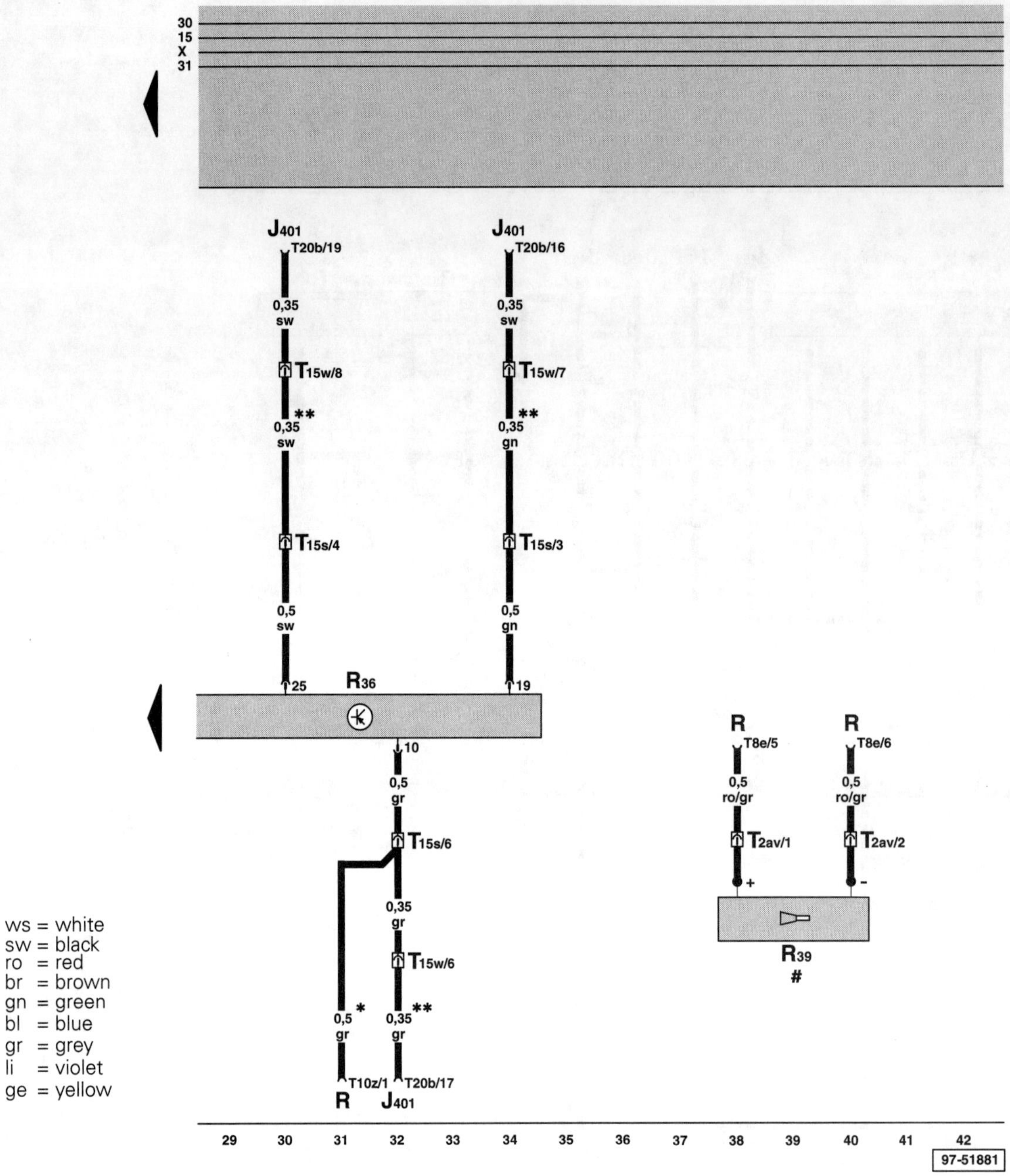

J401 - Control unit for navigation with CD drive
R - Radio
R36 - Telephone Transceiver
R39 - Speaker (telephone, navigation)
T2av - Connector 2 pin, black, for loudspeaker, B pillar, left
T8e - Connector 8 pin, brown, connector radio II
T8f - Connector 8 pin, black, connector radio III
T10z - Connector 10 pin, red, connector radio IV
T15s - Connector 15 pin, blue, connector station A pillar, left
T15w - Connector 15 pin, green, connector station A pillar, right
T20b - Connector 20 pin, black, on control unit for navigation

* - Vehicles with Radio
** - Vehicles with Navigation
# - Vehicles with Radio-Bose

Navigation

2000 m. y.

Fuse Panel

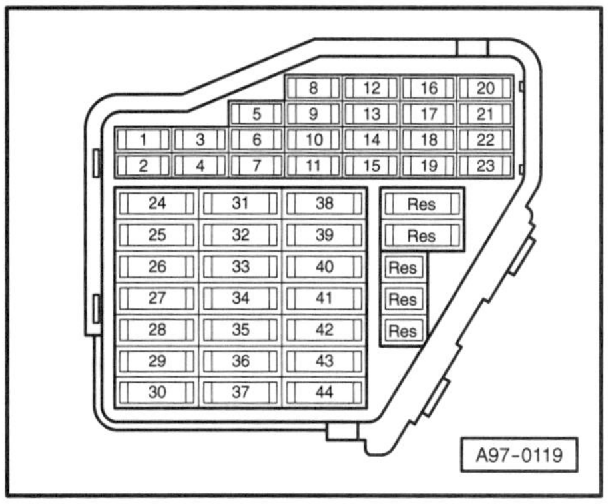

Fuse Colors:

30 A - Green
25 A - White
20 A - Yellow
15 A - Blue
10 A - Red
7,5 A - Brown
5 A - Beige

Starting with fuse position 23, fuses in the fuse holder are identified with 223 in the wiring diagram.

Micro Central Electric Panel

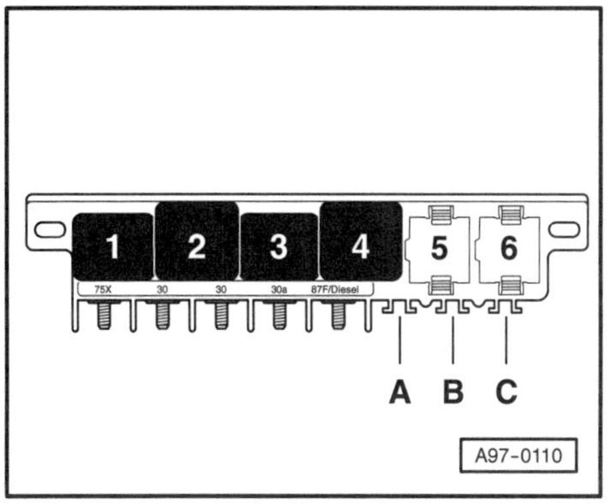

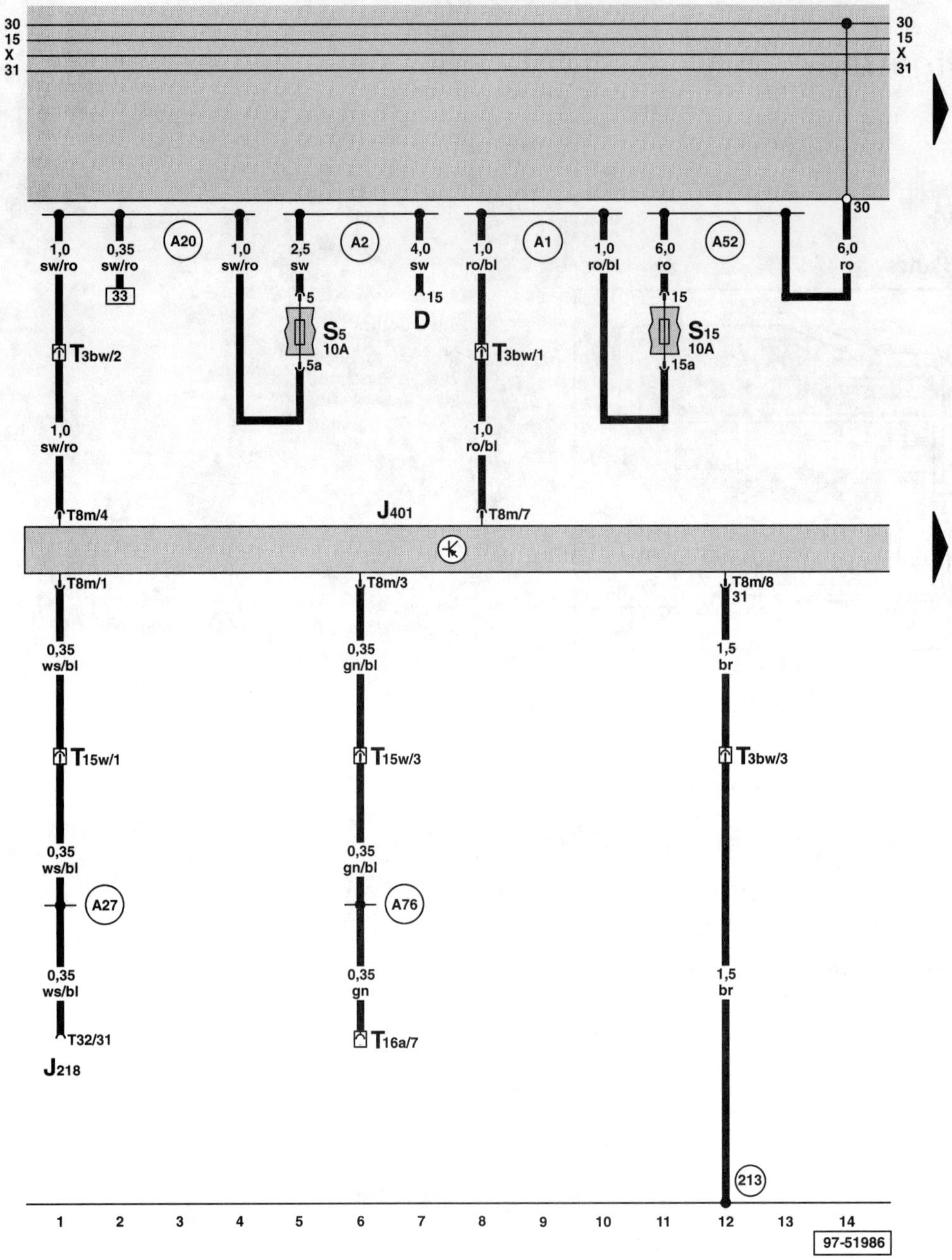

ws = white
sw = black
ro = red
br = brown
gn = green
bl = blue
gr = grey
li = violet
ge = yellow

- D - Ignition/Starter Switch
- J218 - Instrument Cluster Combination Processor
- J401 - Control unit for navigation with CD drive
- S5 - Fuse
- S15 - Fuse
- T3bw - Connector 3 pin, green, connector station A pillar right
- T8m - Connector 8 pin, black, on control unit for navigation
- T15w - Connector 15 pin, green, connector station A pillar, right
- T16a - Connector 16 pin, Data Link Connector (DLC)
- T32 - Connector 32 pin, blue, on instrument cluster

- (213) - Ground connection, in instrument panel wiring harness
- (A1) - Plus connection (30a), in instrument panel wiring harness
- (A2) - Plus connection (15), in instrument panel wiring harness
- (A20) - Wire connection (15a), in instrument panel wiring harness
- (A27) - Wire connection (vehicle speed signal), in instrument panel wiring harness
- (A52) - Plus connection -2- (30), in instrument panel wiring harness
- (A76) - Connector (K-diagnosis wire), in instrument panel wiring harness

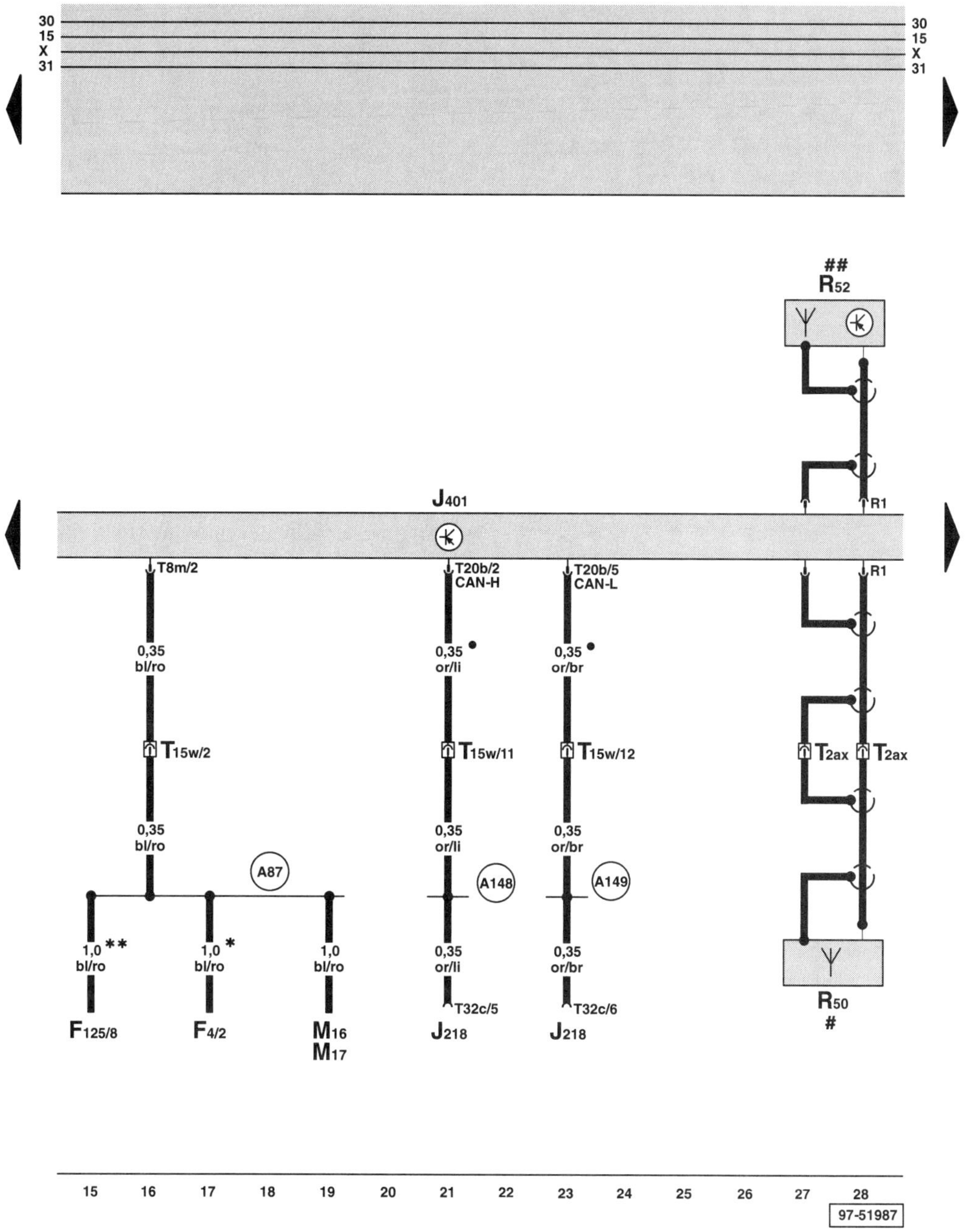

ws = white
sw = black
ro = red
br = brown
gn = green
bl = blue
gr = grey
li = violet
ge = yellow
or = orange

F4 - Back-Up Light Switch
F125 - Multi-Function Transmission Range (TR) Switch
J218 - Instrument Cluster Combination Processor
J401 - Control unit for navigation with CD drive
M16 - Left Back-Up Light
M17 - Right Back-Up Light
R1 - Antenna Connection
R50 - Navigation antenna
R52 - Antenna for radio/telephone/navigation (GPS)
T2ax - Connector 2 pin, in rear lid
T8m - Connector 8 pin, black, on control unit for navigation
T15w - Connector 15 pin, green, connector station A pillar, right
T20b - Connector 20 pin, black, on control unit for navigation
T32c - Connector 32 pin, grey, on instrument cluster

(A87) - Connector (RF), in instrument panel wiring harness
(A148) - Connection (high bus, navigation), in instrument panel wiring harness
(A149) - Connection (low bus, navigation), in instrument panel wiring harness

• - CAN-Bus (Data-Bus)
* - Vehicles with manually shifted transmission
** - Vehicles with automatic transmission
# - Sedan
- Wagon

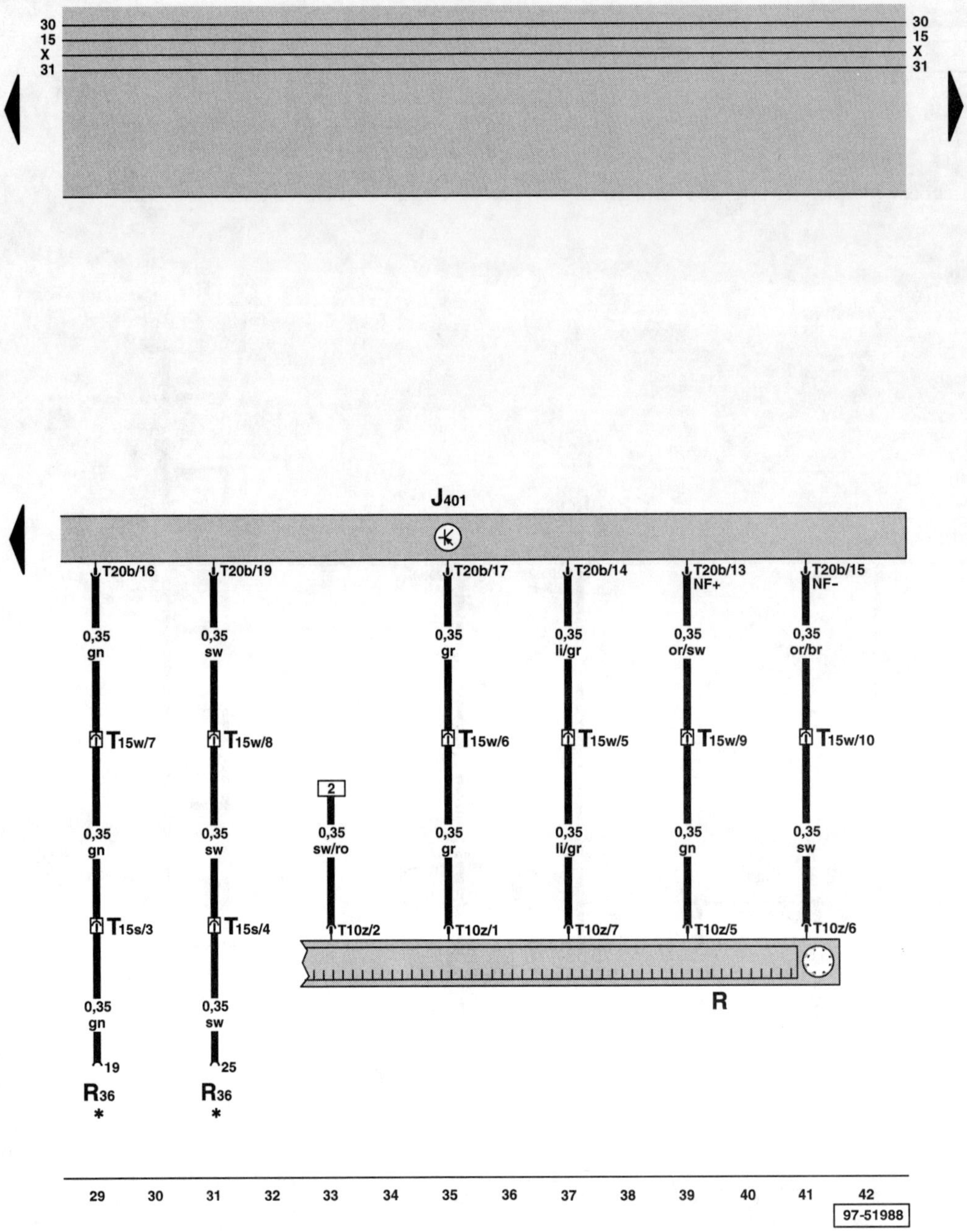

ws = white
sw = black
ro = red
br = brown
gn = green
bl = blue
gr = grey
li = violet
ge = yellow
or = orange

J401 - Control unit for navigation with CD drive
R - Radio
R36 - Telephone Transceiver
T10z - Connector 10 pin, red, connector radio IV
T15s - Connector 15 pin, blue, connector station A pillar, left
T15w - Connector 15 pin, green, connector station A pillar, right
T20b - Connector 20 pin, black, on control unit for navigation

* - Vehicles with Telephon

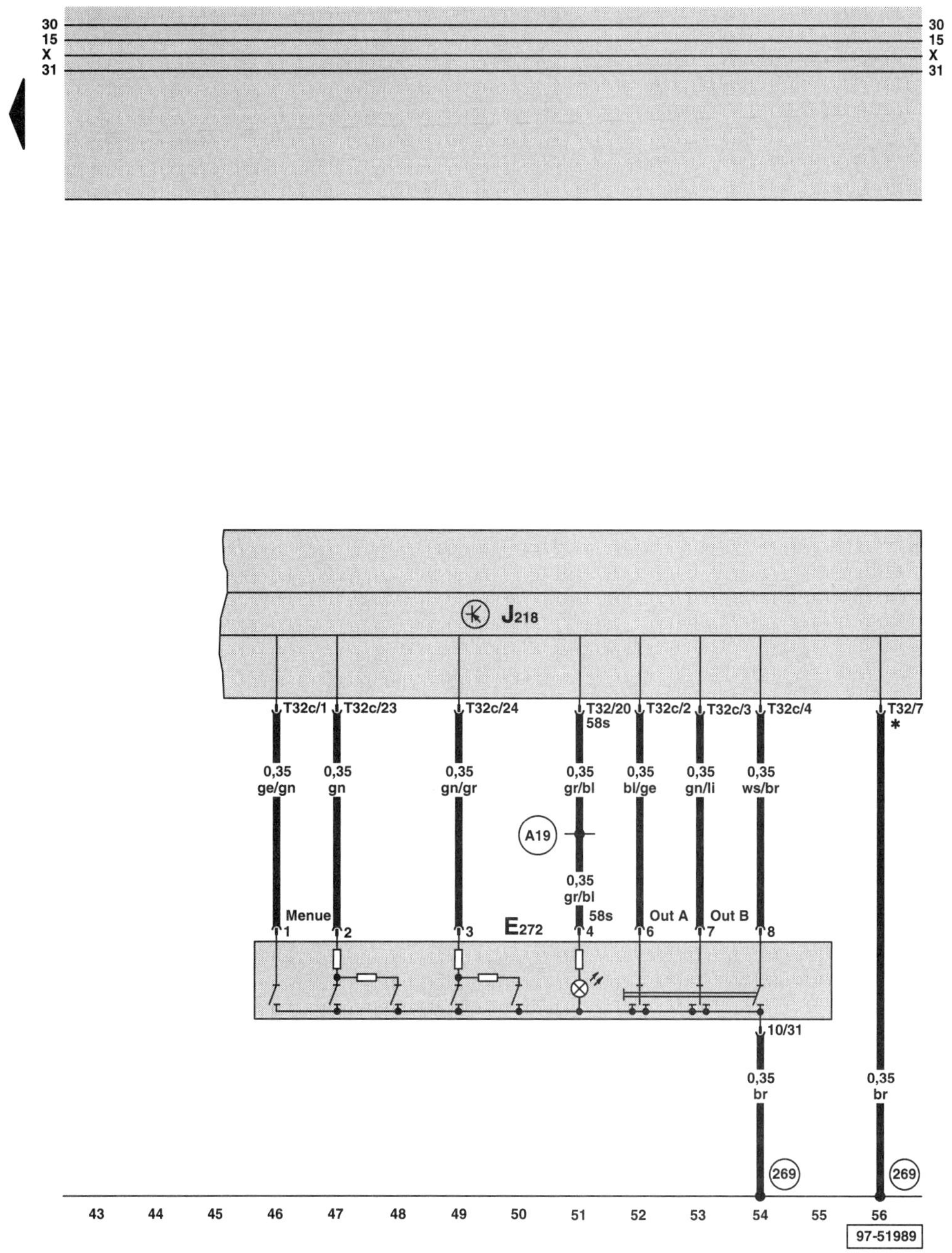

ws = white
sw = black
ro = red
br = brown
gn = green
bl = blue
gr = grey
li = violet
ge = yellow

E272 - Board Computer Function Selector Switch II, in console
J218 - Instrument Cluster Combination Processor
T32 - Connector 32 pin, blue, on instrument cluster
T32c - Connector 32 pin, grey, on instrument cluster

(269) - Ground connector (sensor ground) -1-, in instrument panel wiring harness
(A19) - Wire connection (58s), in instrument panel wiring harness

* - Exit Sensor ground

THIS PAGE INTENTIONALLY LEFT
BLANK

Automatic Climate Control

2000 m. y.

Fuse Panel

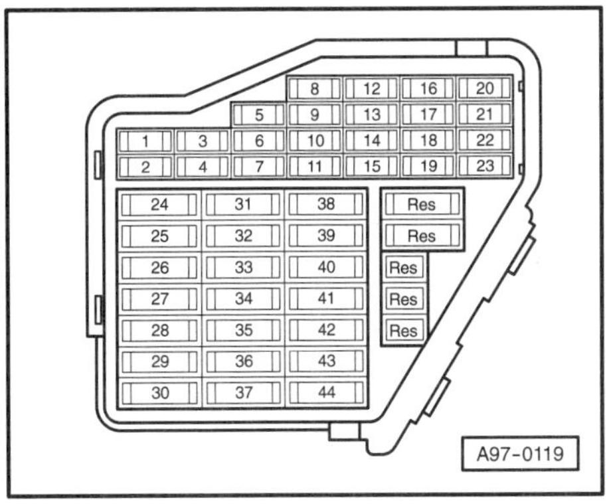

Fuse Colors:

30 A - Green
25 A - White
20 A - Yellow
15 A - Blue
10 A - Red
7,5 A - Brown
5 A - Beige

Starting with fuse position 23, fuses in the fuse holder are identified with 223 in the wiring diagram.

Micro Central Electric Panel

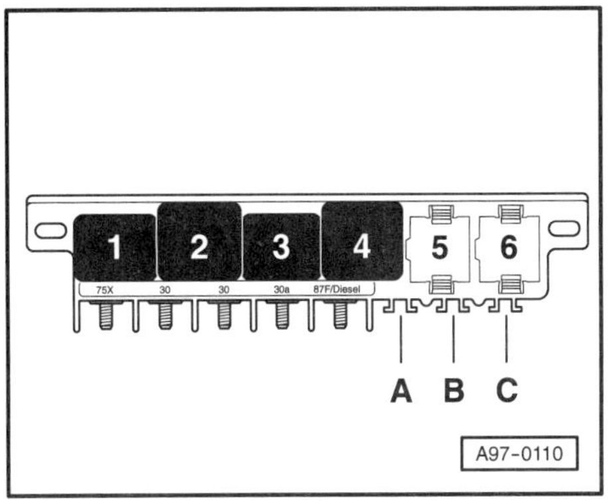

13 - Fold Relay Panel

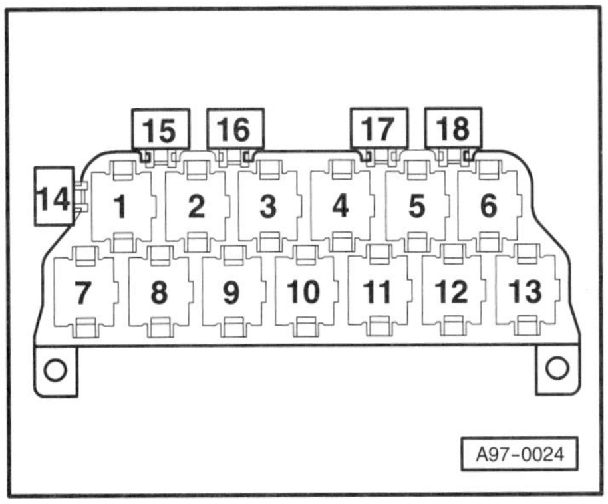

Relay Location:

3 -Solar Cell Separation Relay, J309

5 -A/C Clutch Relay, J44

8 - Fold Relay Panel

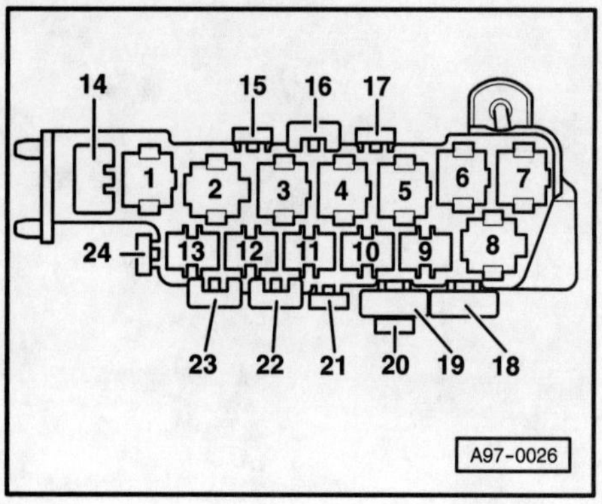

Fuse Location:

20 -Coolant Fan Fuse, S42

21 -Control module fuse for coolant fan, S142

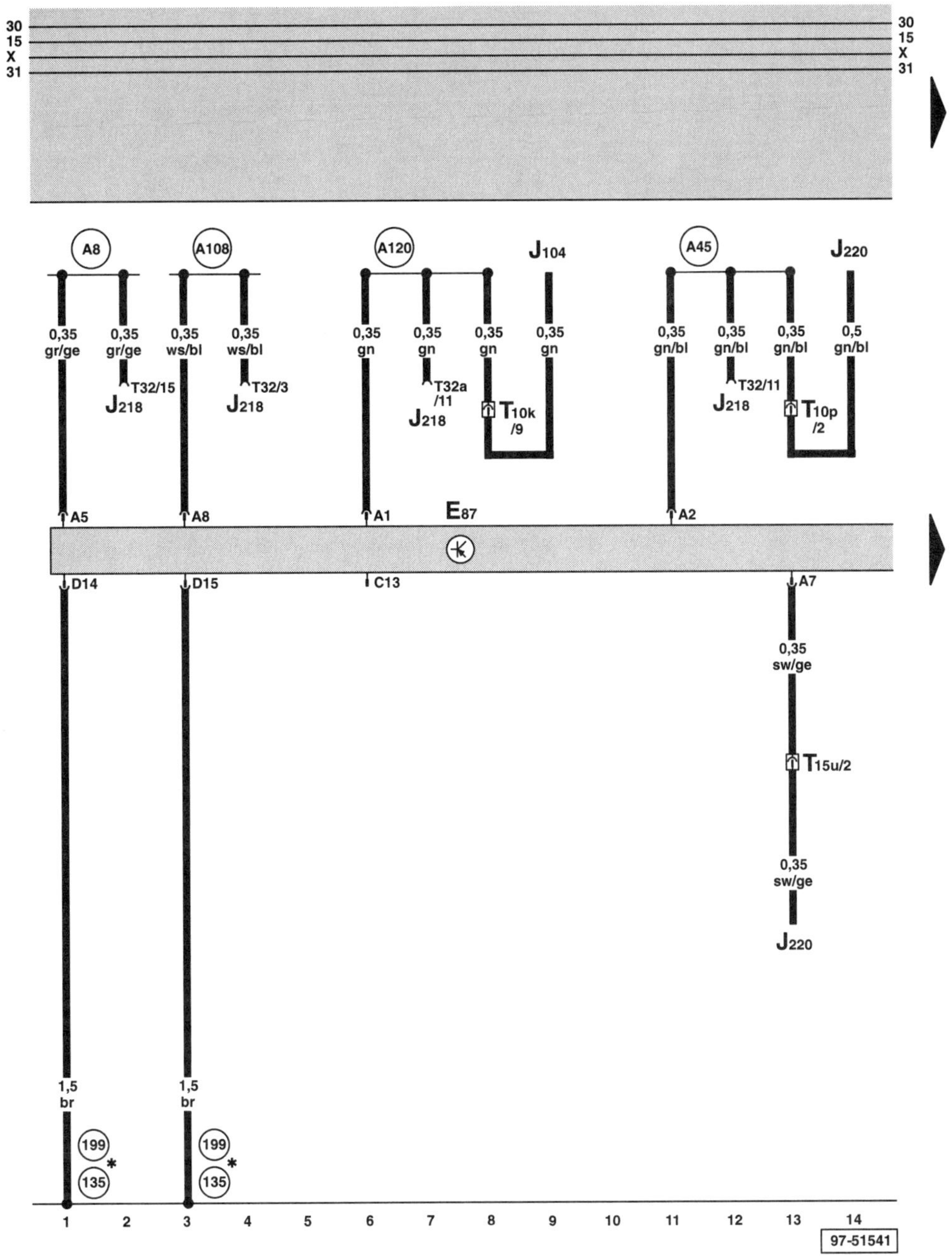

ws = white
sw = black
ro = red
br = brown
gn = green
bl = blue
gr = grey
li = violet
ge = yellow

E87 - A/C Control Head
J104 - ABS Control Module (w/EDL)
J218 - Instrument Cluster Combination Processor
J220 - Motronic Engine Control Module (ECM)
T10k - Connector 10 pin, orange, connector station A pillar, left
T10p - Connector 10 pin, black, connector station electronic box plenum chamber
T15u - Connector 15 pin, red, connector station electronic box plenum chamber
T32 - Connector 32 pin, blue, on instrument cluster
T32a - Connector 32 pin, green, on instrument cluster

(135) - Ground connection -2-, in instrument panel wiring harness
(199) - Ground connection -3-, in instrument panel wiring harness
(A8) - Plus connection (58d), in instrument panel wiring harness
(A45) - Wire connection (RPM-signal), in instrument panel wiring harness
(A108) - Connector (speed signal), in instrument panel wiring harness
(A120) - Connection (standing time) in instrument panel wiring harness

* - Both are possible

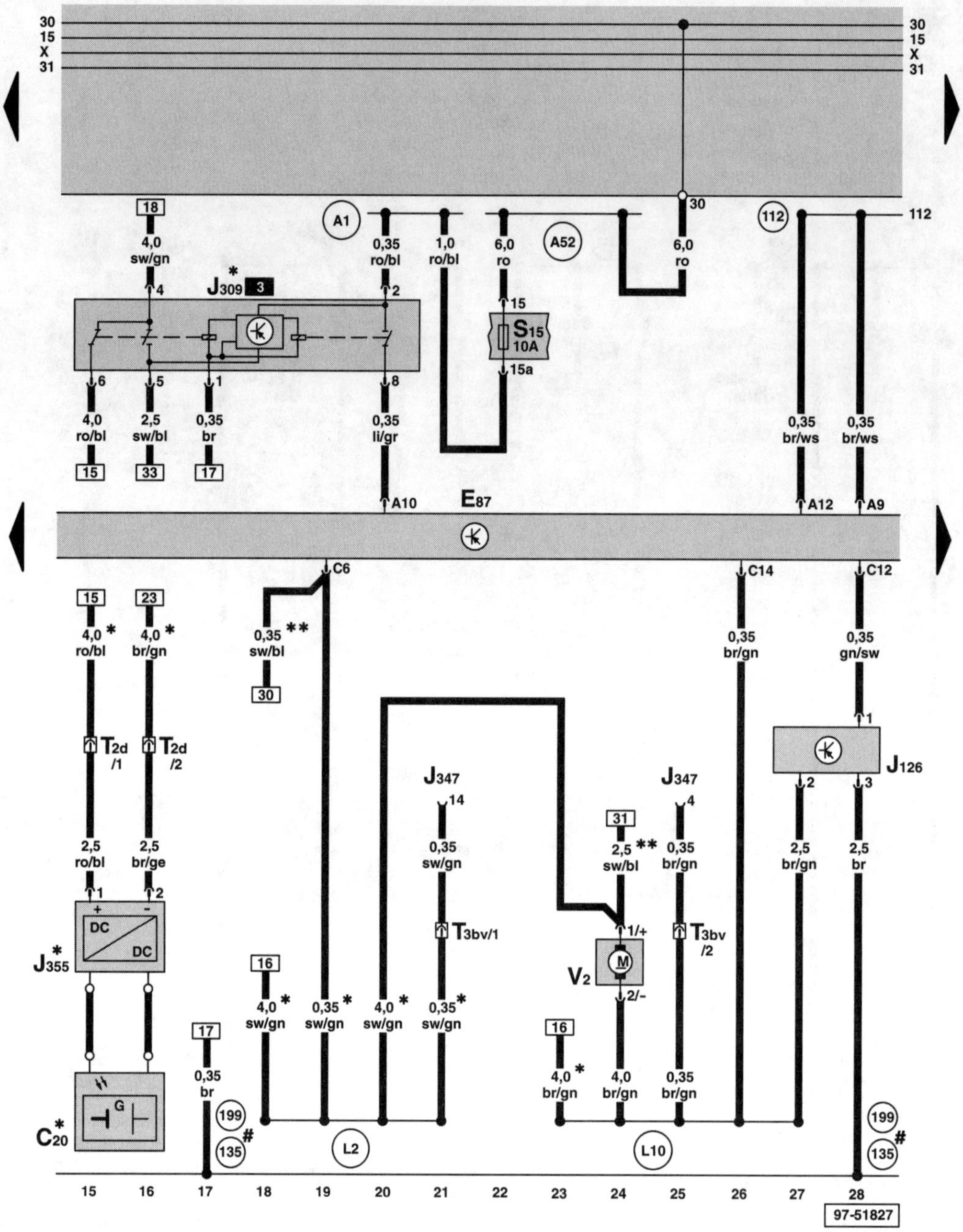

ws = white
sw = black
ro = red
br = brown
gn = green
bl = blue
gr = grey
li = violet
ge = yellow

C20 - Solar Cells
E87 - A/C Control Head
J126 - Control module for fresh air blower
J309 - Solar Cell Separation Relay
J347 - Control Module for Interior Monitoring sensors
J355 - Control Module for solar operation
S15 - Fuse
T2d - Connector 2 pin, blue, connector station A pillar, left
T3bv - Connector 3 pin, orange, connector station A pillar right
V2 - Fresh Air Blower

(112) - Ground connection -2-, in A/C wiring harness
(135) - Ground connection -2-, in instrument panel wiring harness
(199) - Ground connection -3-, in instrument panel wiring harness
(A1) - Plus connection (30a), in instrument panel wiring harness
(A52) - Plus connection -2- (30), in instrument panel wiring harness
(L2) - Wire connection, in A/C wiring harness
(L10) - Wire connection -2-, in A/C wiring harness

* - Vehicles with sunroof with Solar Cells
** - Vehicles without sunroof with Solar Cells
# - Both are possible

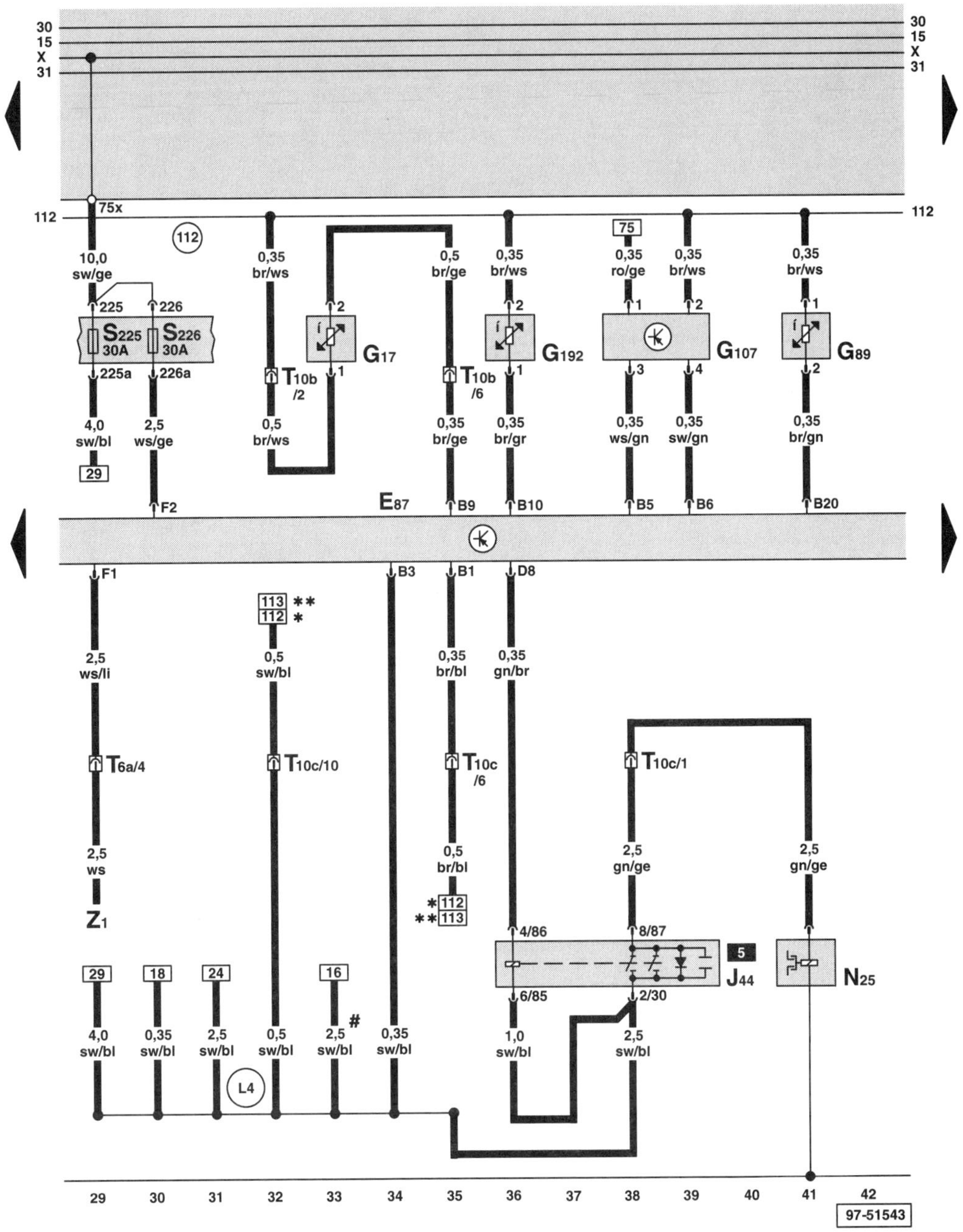

ws = white
sw = black
ro = red
br = brown
gn = green
bl = blue
gr = grey
li = violet
ge = yellow

E87 - A/C Control Head
G17 - Outside Air Temperature Sensor
G89 - Fresh Air Intake Duct Temperature Sensor
G107 - Sunlight Photo Sensor (3241)
G192 - Sensor for outlet temperature, center
J44 - A/C Clutch Relay
N25 - A/C Clutch
S225 - Fuse 25 in fuse holder
S226 - Fuse 26 in fuse holder
T6a - Connector 6 pin, blue, connector station A pillar, left
T10b - Connector 10 pin, red, connector station A pillar, left
T10c - Connector 10 pin, grey, connector station A pillar, left
Z1 - Rear Window Defogger/Heat Element

(112) - Ground connection -2-, in A/C wiring harness

(L4) - Wire connection (75al), in A/C wiring harness

* - Vehicles with Coolant FC (Fan Control) Control Module
** - Vehicles with Coolant Fan Control (FC) Relay
# - Vehicles with Sunroof with Solar Cells

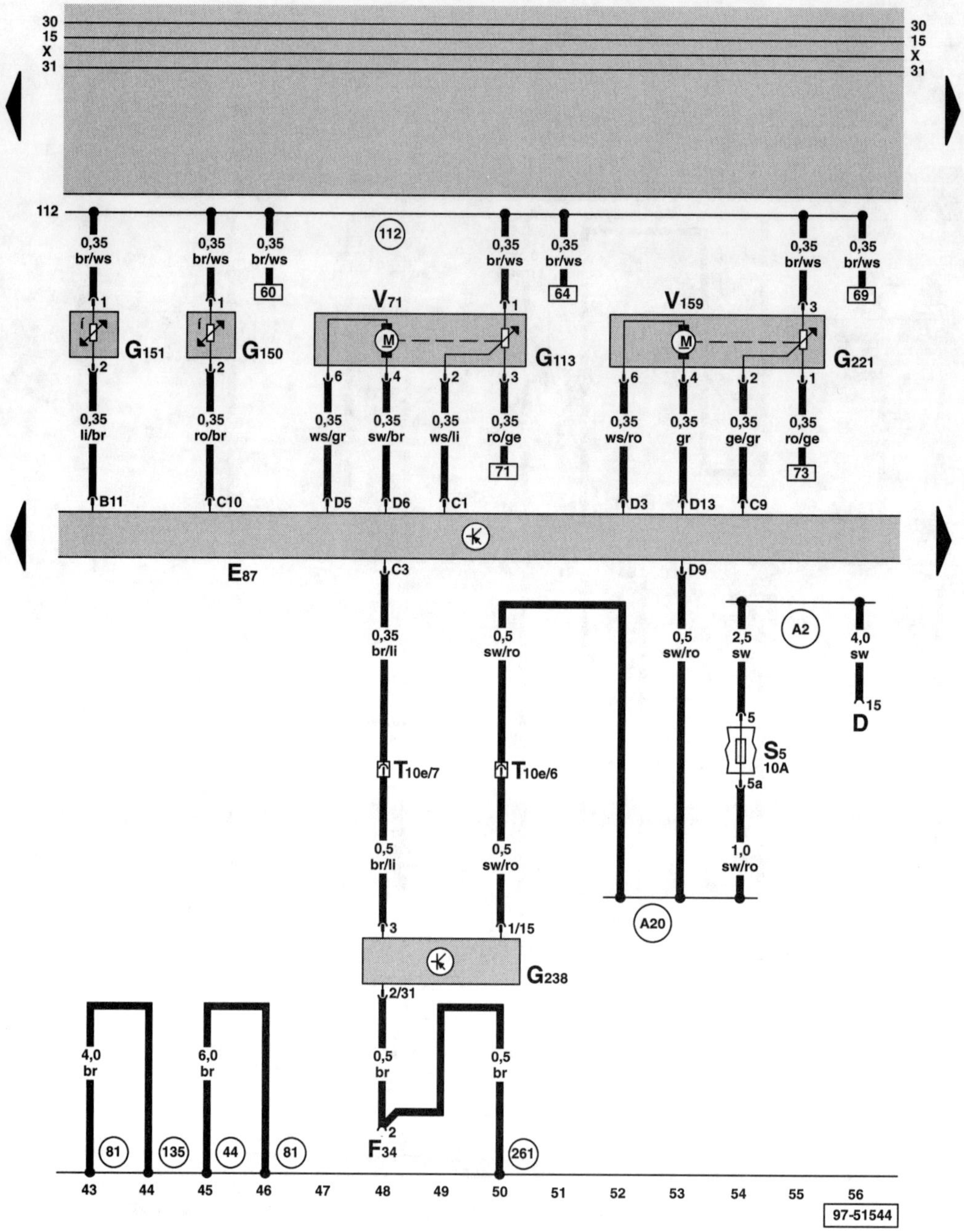

ws = white
sw = black
ro = red
br = brown
gn = green
bl = blue
gr = grey
li = violet
ge = yellow

- D - Ignition/Starter Switch
- E87 - A/C Control Head
- F34 - Brake Fluid Level Warning Switch
- G113 - Back Pressure Flap Motor Position Sensor
- G150 - Left Vent Temperature Sensor
- G151 - Right Vent Temperature Sensor
- G221 - Potentiometer-actuator for temperature flap right
- G238 - Sensor for air quality
- S5 - Fuse
- T10e - Connector 10 pin, violet, connector station A pillar, left
- V71 - Air Flow Flap Motor
- V159 - Actuator for temperature flap right

- (44) - Ground connection, on left A-pillar, lower part
- (81) - Ground connection -1-, in instrument panel wiring harness
- (112) - Ground connection -2-, in A/C wiring harness
- (135) - Ground connection -2-, in instrument panel wiring harness
- (261) - Ground connection, in wiring harness heated spray jet
- (A2) - Plus connection (15), in instrument panel wiring harness
- (A20) - Wire connection (15a), in instrument panel wiring harness

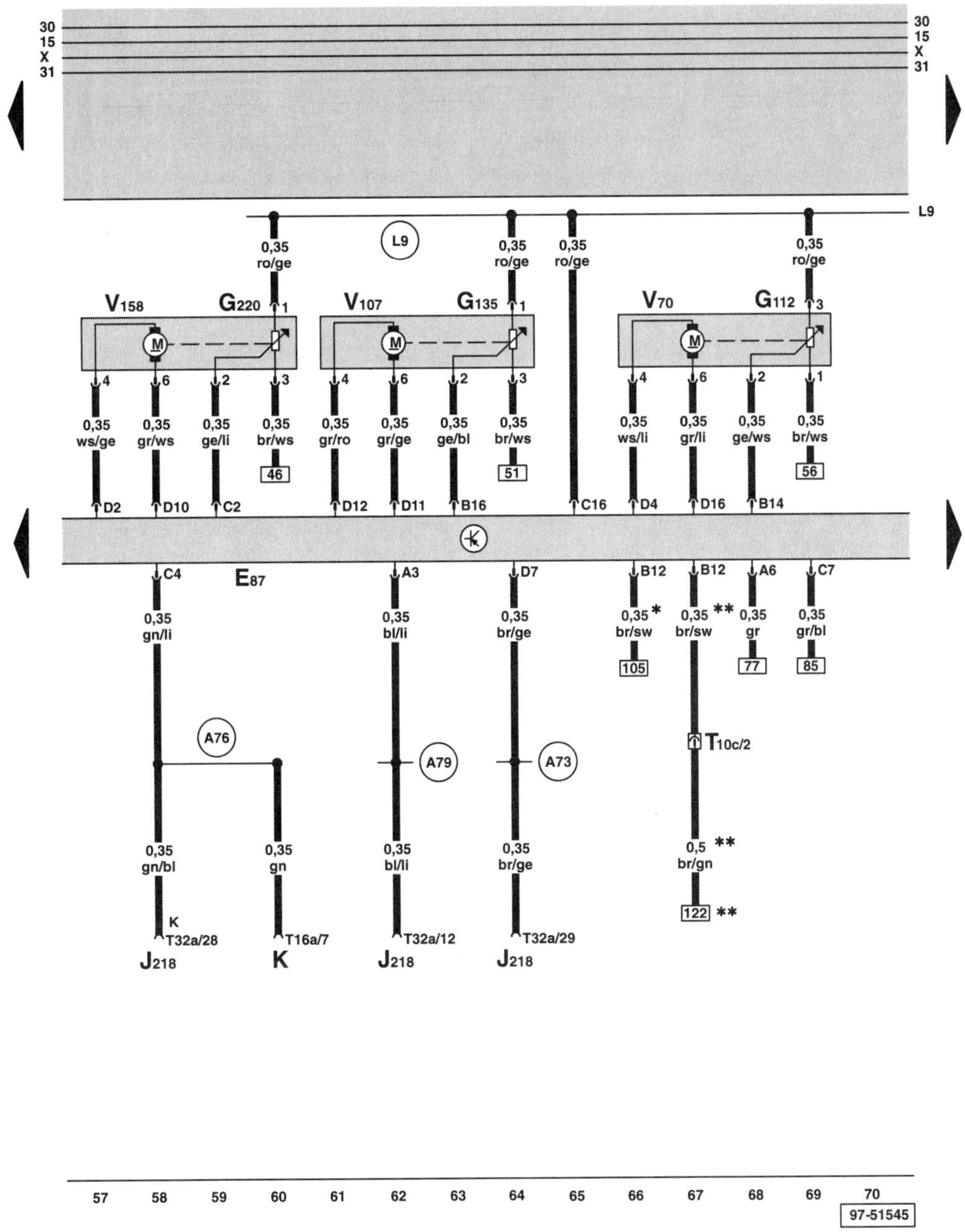

ws = white
sw = black
ro = red
br = brown
gn = green
bl = blue
gr = grey
li = violet
ge = yellow

E87 - A/C Control Head
G112 - Central Flap Motor Position Sensor
G135 - Defroster Flap Motor Position Sensor
G220 - Potentiometer-actuator for temperature flap left
J218 - Instrument Cluster Combination Processor
T10c - Connector 10 pin, grey, connector station A pillar, left
T16a - Connector 16 pin, Data Link Connector
T32a - Connector 32 pin, green, on instrument cluster
V70 - Central Air Flap Motor
V107 - Defroster Flap Motor
V158 - Actuator for temperature flap left

(A73) - Connector (outside temperature indicator), in instrument panel wiring harness
(A76) - Connector (K-diagnosis wire), in instrument panel wiring harness
(A79) - Connector (engine overheat warning light), in instrument panel wiring harness
(L9) - Wire connection -1-, in A/C wiring harness

* - Vehicles with Coolant FC (Fan Control) Control Module
** - Vehicles with Coolant Fan (FC) Relay

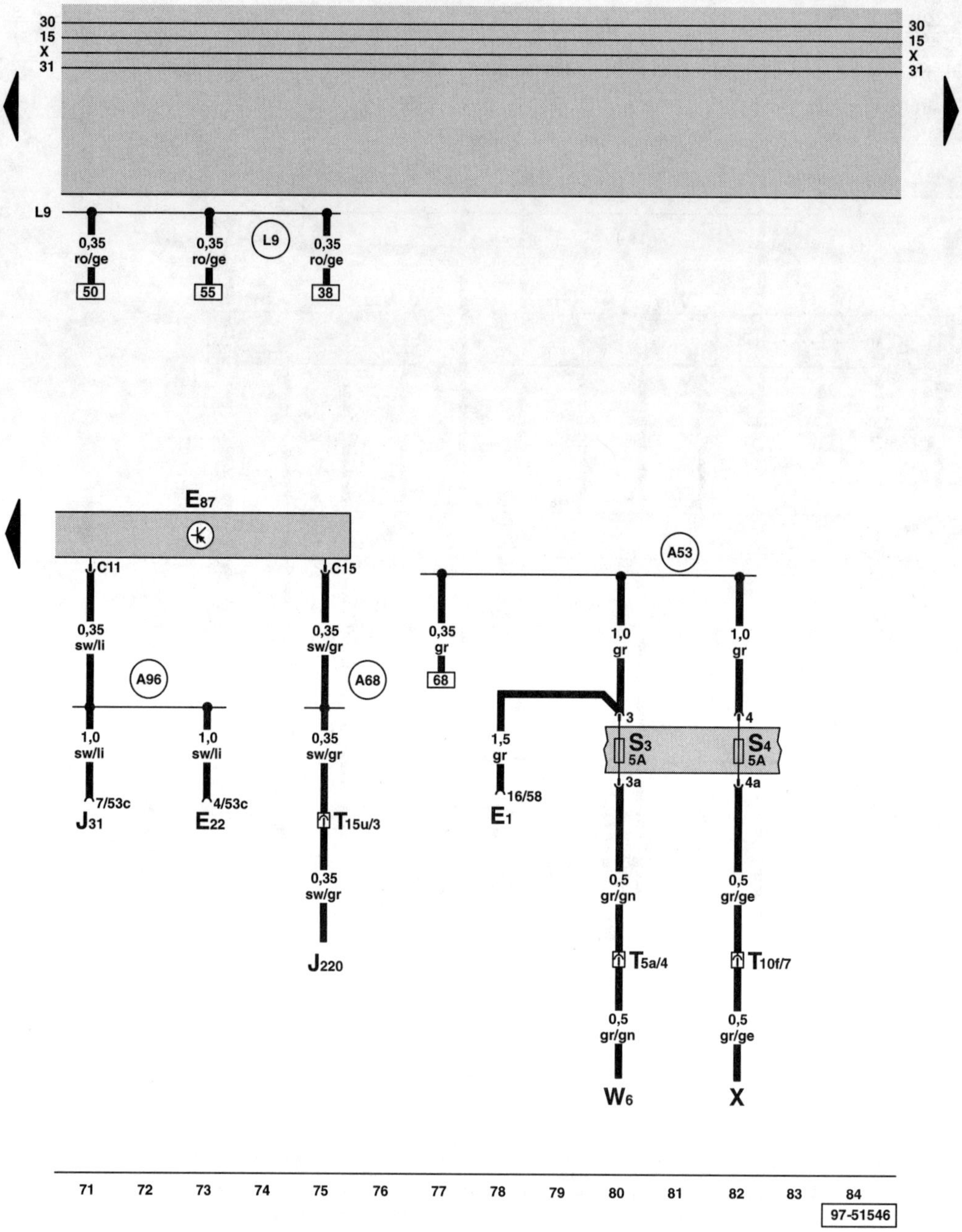

ws = white
sw = black
ro = red
br = brown
gn = green
bl = blue
gr = grey
li = violet
ge = yellow

E1 - Light switch
E22 - Windshield Wiper/Washer Switch
E87 - A/C Control Head
J31 - Wiper/Washer Intermittent Relay
J220 - Motronic Engine Control Module (ECM)
S3 - Fuse
S4 - Fuse
T5a - Connector 5 pin, black, behind instrument panel, right
T10f - Connector 10 pin, brown, connector station A pillar, left
T15u - Connector 15 pin, red, connector station electronic box plenum chamber
W6 - Glove Compartment Light
X - License Plate Light

(A53) - Plus connection (58b), in instrument panel wiring harness
(A68) - Connector (C15, A/C), in instrument panel wiring harness
(A96) - Connector (53a), in instrument panel wiring harness
(L9) - Wire connection -1-, in A/C wiring harness

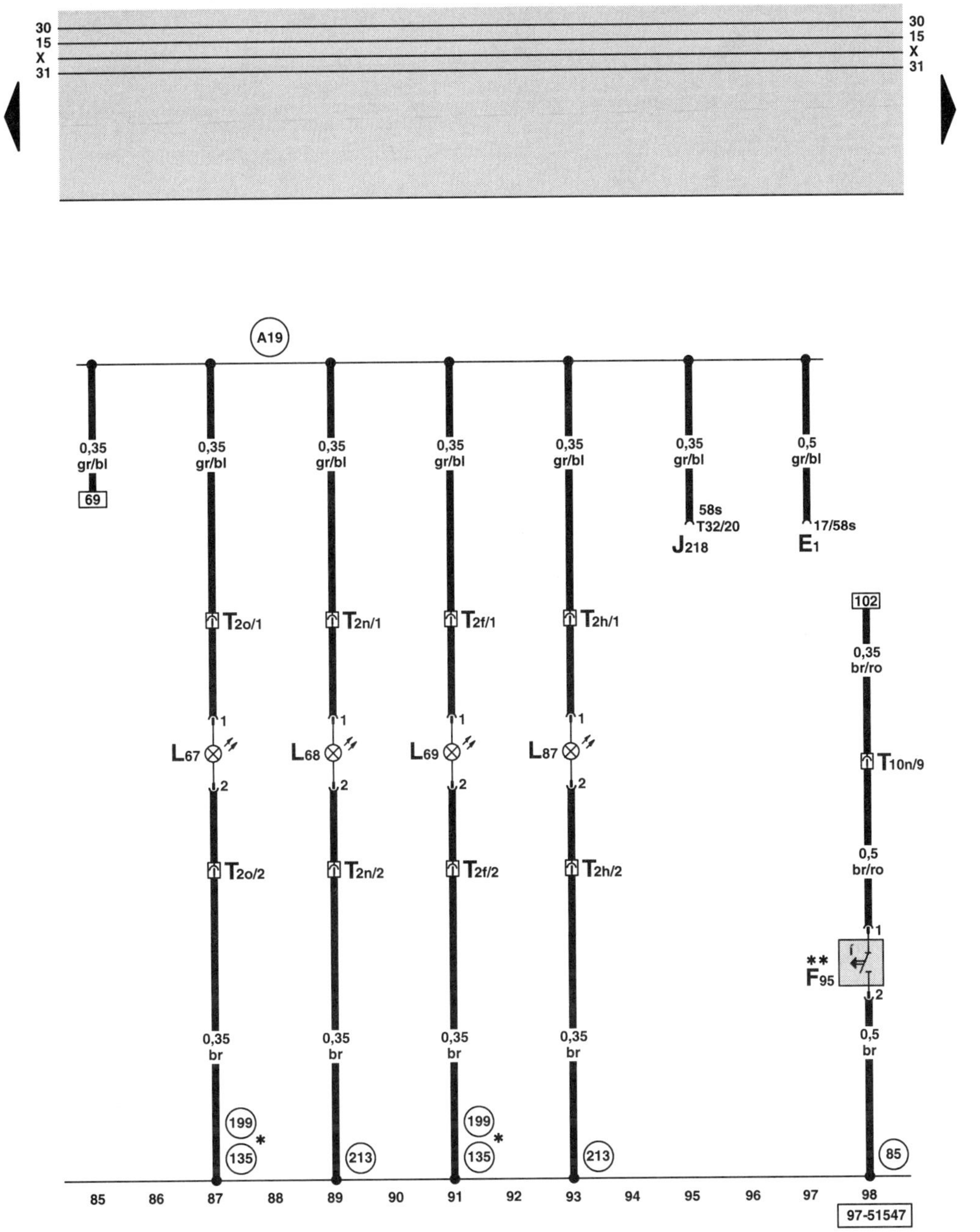

ws = white
sw = black
ro = red
br = brown
gn = green
bl = blue
gr = grey
li = violet
ge = yellow

E1 - Light switch
F95 - After-Run Coolant Thermal Switch
J218 - Instrument Cluster Combination Processor
L67 - Left Instrument Panel Vent Illumination
L68 - Center Instrument Panel Vent Illumination
L69 - Right Instrument Panel Vent Illumination
L87 - Lighting for rearseat outlet center
T2f - Connector 2 pin, brown, behind instrument panel , right
T2h - Connector 2 pin, black, in console
T2n - Connector 2 pin, brown, behind instrument panel, center
T2o - Connector 2 pin, behind instrument panel, left
T10n - Connector 10 pin, orange, connector station electronic box plenum chamber
T32 - Connector 32 pin, blue, on instrument cluster

(85) - Ground connection -1-, in engine compartment wiring harness
(135) - Ground connection -2-, in instrument panel wiring harness
(199) - Ground connection -3-, in instrument panel wiring harness
(213) - Ground connection, in instrument panel wiring harness
(A19) - Wire connection (58s), in instrument panel wiring harness

* - Both are possible
** - 2,7 L Engine, code APB only

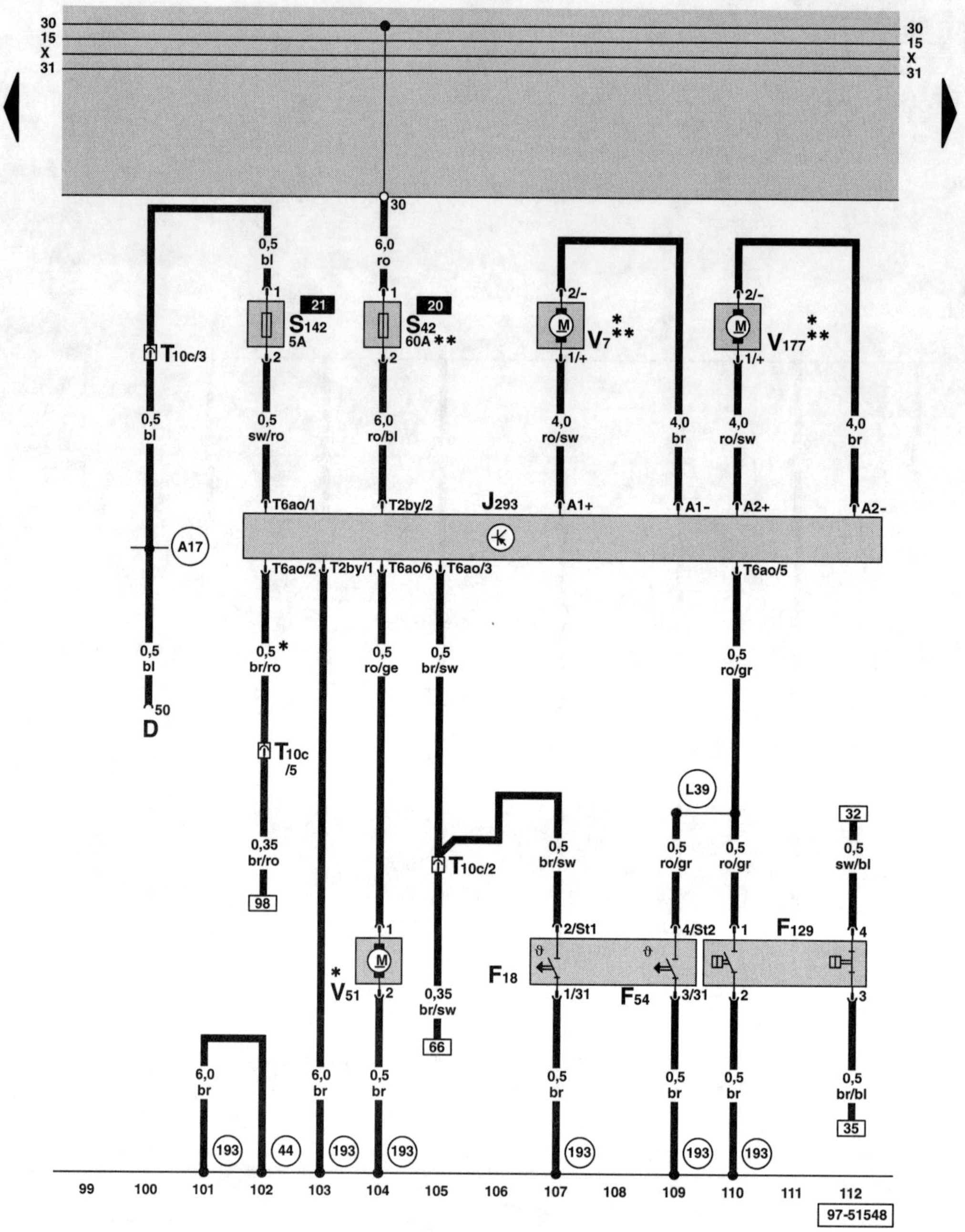

ws = white
sw = black
ro = red
br = brown
gn = green
bl = blue
gr = grey
li = violet
ge = yellow

D - Ignition/Starter Switch
F18 - Coolant Fan Control (FC) Thermal Switch
F54 - Coolant Fan Control (FC) Thermal Switch
F129 - A/C Pressure Switch
J293 - Coolant FC (Fan Control) Control Module
S42 - Coolant Fan Fuse
S142 - Control module fuse for coolant fan
T2by - Connector 2 pin, on Coolant FC (Fan Control) Control Module
T6ao - Connector 6 pin, on Coolant FC (Fan Control) Control Module
T10c - Connector 10 pin, grey, connector station A pillar, left
V7 - Coolant Fan
V51 - After-Run Coolant Pump
V177 - Coolant fan -2-

(44) - Ground connection, on left A-pillar, lower part

(193) - Ground connection -1-, in coolant fan wiring harness

(A17) - Wire connection (61), in instrument panel wiring harness

(L39) - Wire connection -1- (A/C Pressure Switch), in Climatronic wiring harness

* - Vehicles with 6 cylinder engine

** - Vehicles with 8 cylinder engine

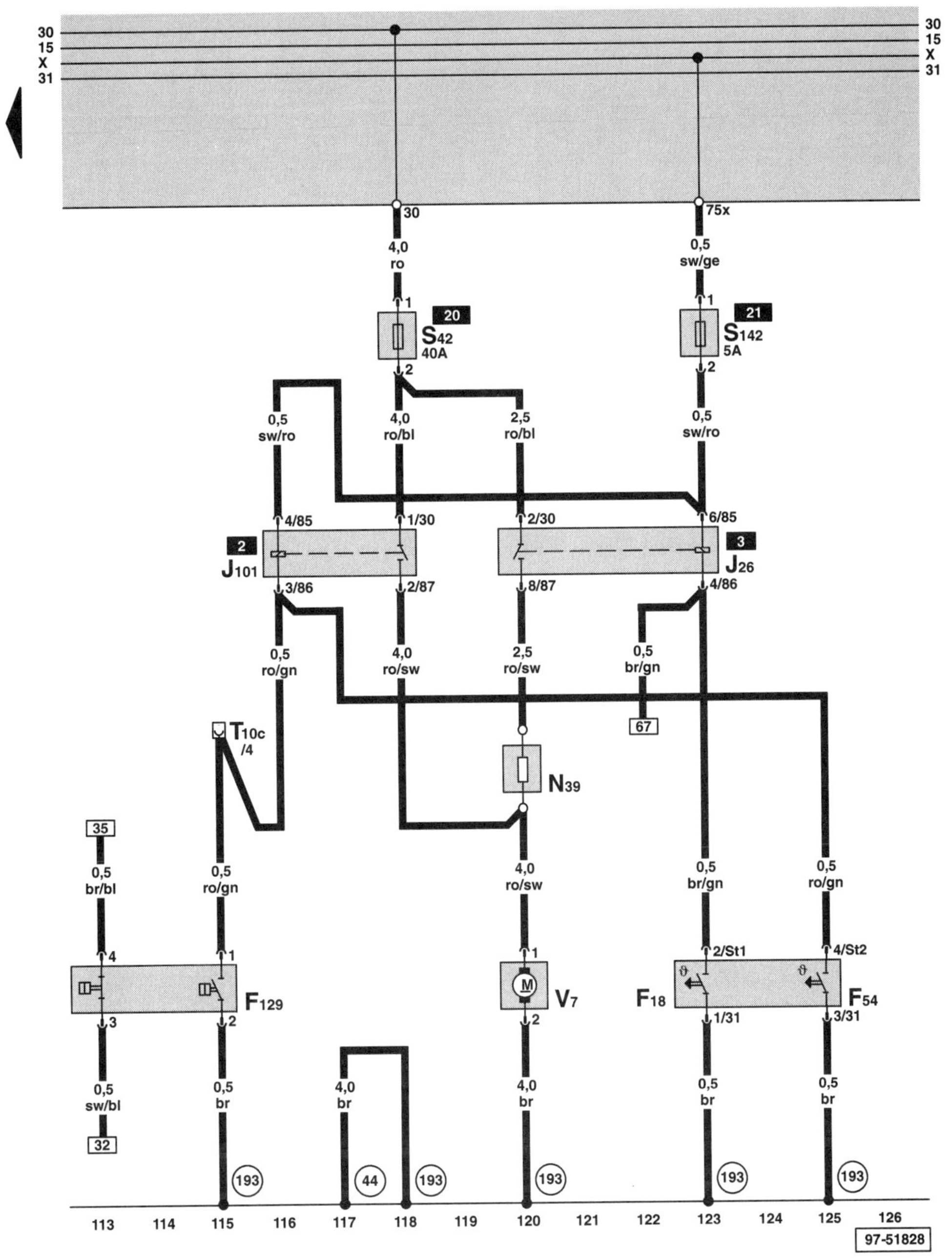

ws = white
sw = black
ro = red
br = brown
gn = green
bl = blue
gr = grey
li = violet
ge = yellow

F18 - Coolant Fan Control (FC) Thermal Switch
F54 - Coolant Fan Control (FC) Thermal Switch
F129 - A/C Pressure Switch
J26 - Coolant Fan control (FC) Relay
J101 - Second Speed Coolant Fan Control (FC) Relay
N39 - Coolant Fan Control (FC) Series Resistance
S42 - Coolant Fan Fuse
S142 - Control module fuse for coolant fan
T10c - Connector 10 pin, grey, connector station A pillar, left
V7 - Coolant Fan

(44) - Ground connection, on left A-pillar, lower part

(193) - Ground connection -1-, in coolant fan wiring harness

THIS PAGE INTENTIONALLY LEFT BLANK

Rear Wiring, Wagon

2000 m. y.

Fuse Panel

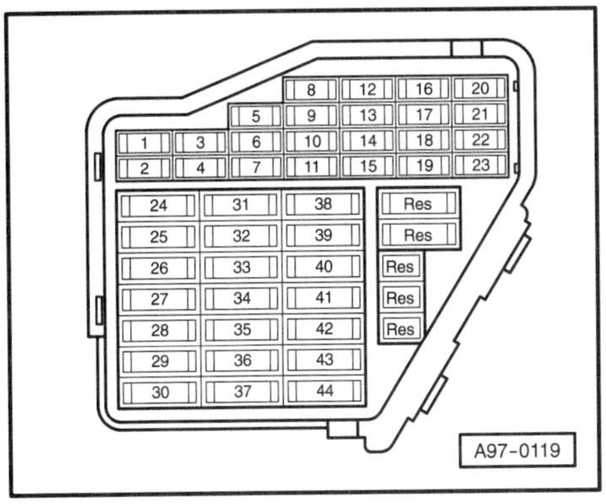

Fuse Colors:

30 A - Green
25 A - White
20 A - Yellow
15 A - Blue
10 A - Red
7,5 A - Brown
5 A - Beige

Starting with fuse position 23, fuses in the fuse holder are identified with 223 in the wiring diagram.

13 - Fold Relay Panel

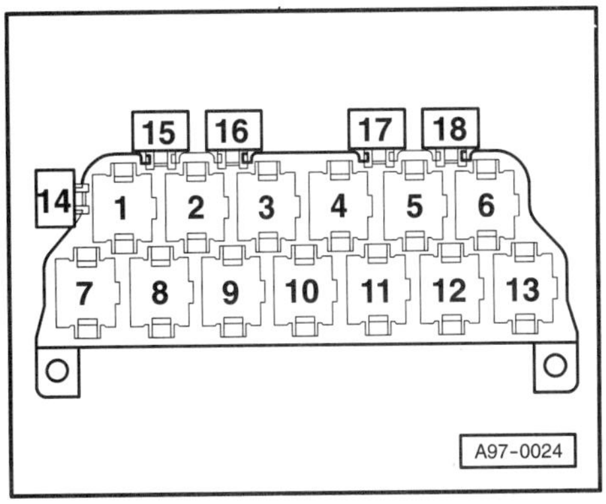

Relay Location:

9 -Lamp Control Module, J123

10 -Lamp Control Module, J123

8 - Fold Relay Panel

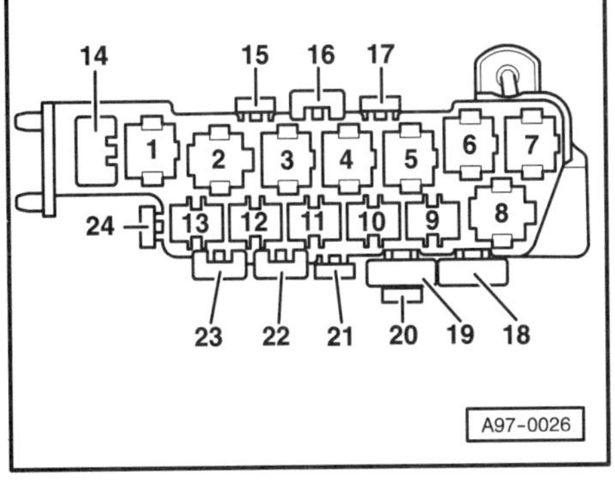

Relay Location:

15 -Fuse for Luggage Compartment Socket, S184

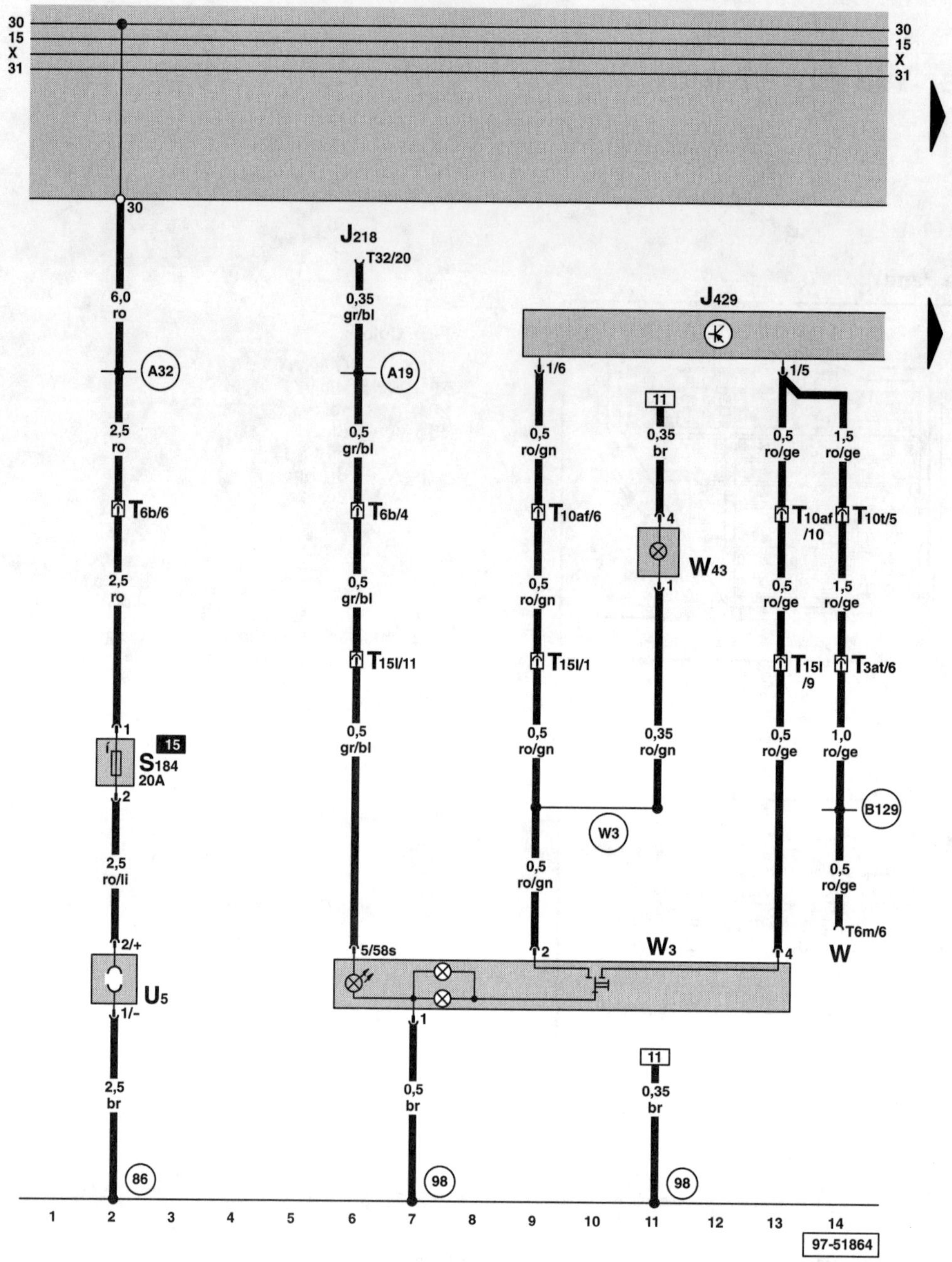

ws = white
sw = black
ro = red
br = brown
gn = green
bl = blue
gr = grey
li = violet
ge = yellow

J218 - Instrument Cluster Combination Processor
J429 - Control module for central locking
S184 - Fuse for Luggage Compartment Socket
T3at - Connector 3 pin, grey, connector station A pillar, left
T6b - Connector 6 pin, white, connector station A pillar, left
T6m - Connector 6 pin, on interior light, front
T10t - Connector 10 pin, orange, connector station A pillar, right
T10af - Connector 10 pin, violet, connector station A pillar, left
T15l - Connector 15 pin, black, in luggage compartment
T32 - Connector 32 pin, blue, on instrument cluster
U5 - 12 V Socket
W - Front Interior Light
W3 - Luggage Compartment Light
W43 - Rear courtesy lamp

(86) - Ground connection -1-, in rear wiring harness

(98) - Ground connection, in rear lid wiring harness

(A19) - Wire connection (58s), in instrument panel wiring harness

(A32) - Plus connection (30), in instrument panel wiring harness

(B129) - Connector (interior light 31L), in wiring harness interior

(W3) - Wire connection, in rear wiring harness

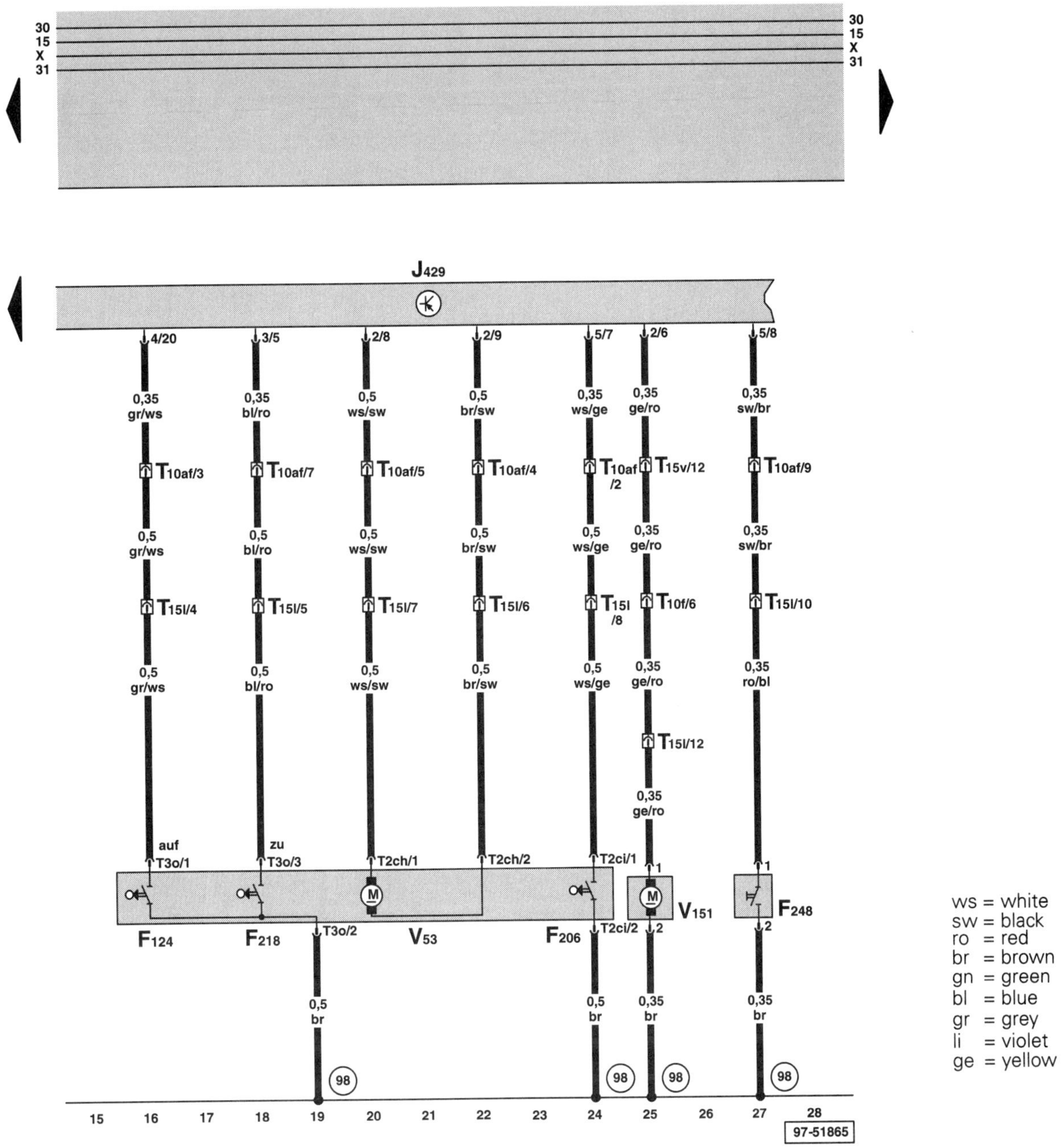

ws = white
sw = black
ro = red
br = brown
gn = green
bl = blue
gr = grey
li = violet
ge = yellow

F124 - Trunk Lock Alarm/Central Locking Switch
F206 - Trunk Lid Closed Position Switch
F218 - Switch for central locking, trunk lid
F248 - Switch for unlocking, rear lid lock cylinder
J429 - Control module for central locking
T2ch - Connector 2 pin, black, in rear lid
T2ci - Connector 2 pin, grey, in rear lid
T3o - Connector 3 pin, black, in rear lid
T10f - Connector 10 pin, brown, connector station A pillar, left
T10af - Connector 10 pin, violet, connector station A pillar, left
T15l - Connector 15 pin, black, in luggage compartment
T15v - Connector 15 pin, orange, connector station A pillar, right
V53 - Decklid Central Locking System Motor
V151 - Motor for remote unlock rear lid

(98) - Ground connection, in rear lid wiring harness

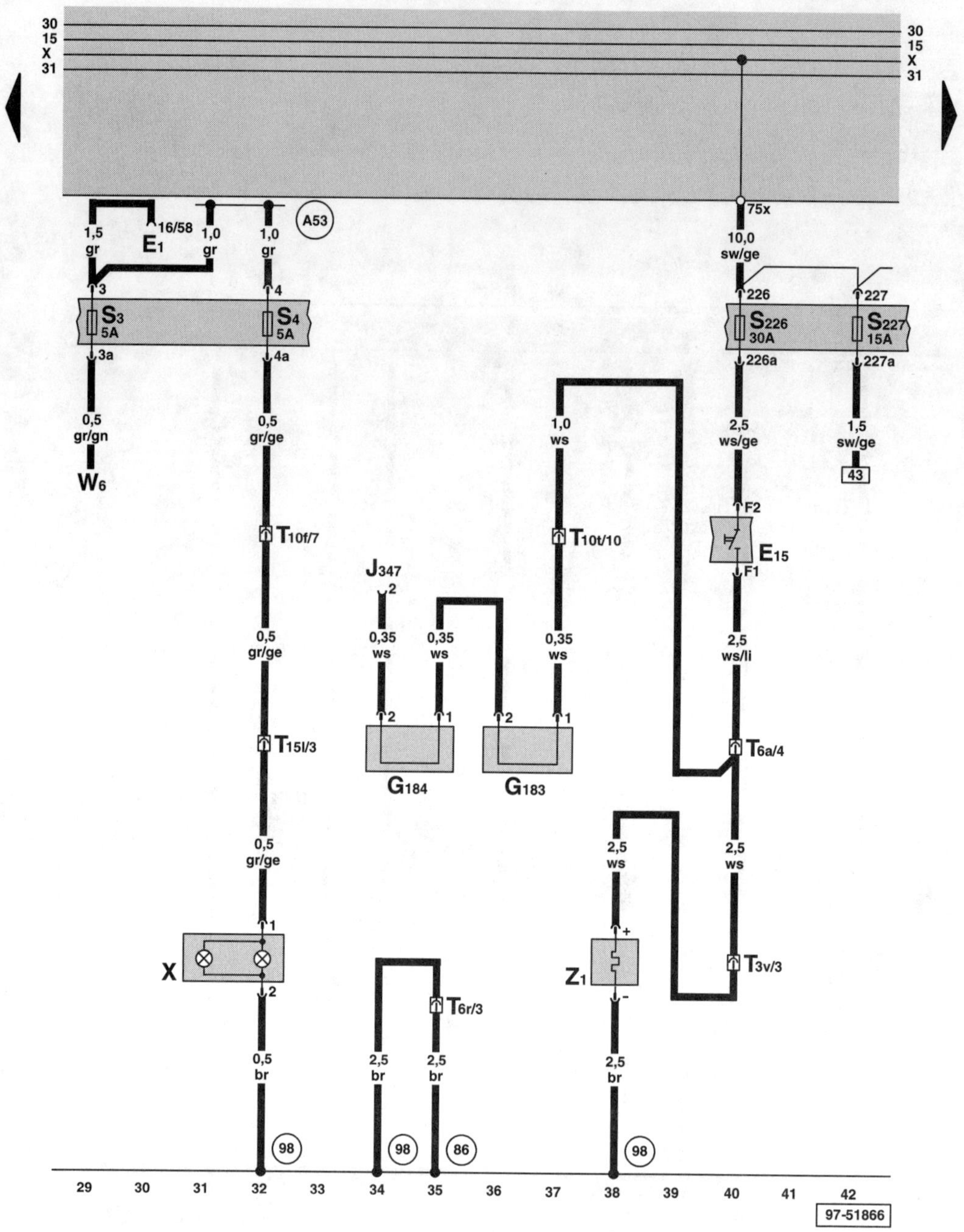

ws = white
sw = black
ro = red
br = brown
gn = green
bl = blue
gr = grey
li = violet
ge = yellow

E1 - Light switch
E15 - Rear Window Defogger Switch
G183 - Sensor for broken window glass rear left
G184 - Sensor for broken window glass rear right
J347 - Control Module for Interior Monitoring sensors
S3 - Fuse
S4 - Fuse
S226 - Fuse 26 in fuse holder
S227 - Fuse 27 in fuse holder
T3v - Connector 3 pin, black, in luggage compartment
T6a - Connector 6 pin, blue, connector station A pillar, left
T6r - Connector 6 pin, brown, in luggage compartment
T10f - Connector 10 pin, brown, connector station A pillar, left
T10t - Connector 10 pin, orange, connector station A pillar, right
T15l - Connector 15 pin, black, in luggage compartment
W6 - Glove Compartment Light
X - License Plate Light
Z1 - Rear Window Defogger/Heat Element

(86) - Ground connection -1-, in rear wiring harness
(98) - Ground connection, in rear lid wiring harness
(A53) - Plus connection (58), in instrument panel wiring harness

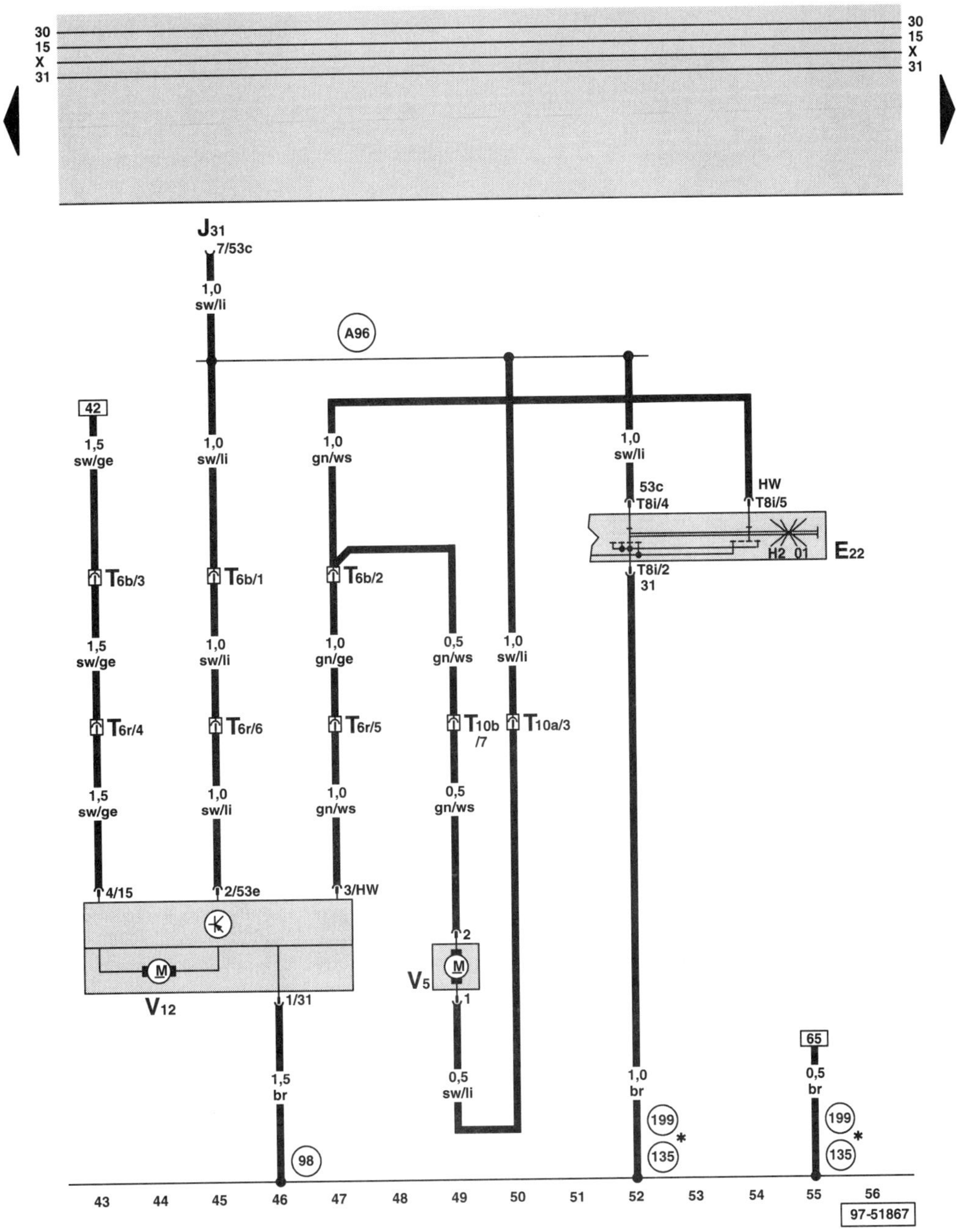

ws = white
sw = black
ro = red
br = brown
gn = green
bl = blue
gr = grey
li = violet
ge = yellow

E22 - Windshield Wiper/Washer Switch
J31 - Wiper/Washer Intermittent Relay
T6b - Connector 6 pin, white, connector station A pillar, left
T6r - Connector 6 pin, brown, in luggage compartment
T8i - Connector 8 pin, on windshield wiper/washer switch
T10a - Connector 10 pin, pink, connector station A pillar, left
T10b - Connector 10 pin, red, connector station A pillar, left
V5 - Windshield Washer Pump
V12 - Rear Window Wiper Motor

(98) - Ground connection, in rear lid wiring harness
(135) - Ground connection -2-, in instrument panel wiring harness
(199) - Ground connection -3-, in instrument panel wiring harness
(A96) - Connector (53a), in instrument panel wiring harness
* - Both are possible

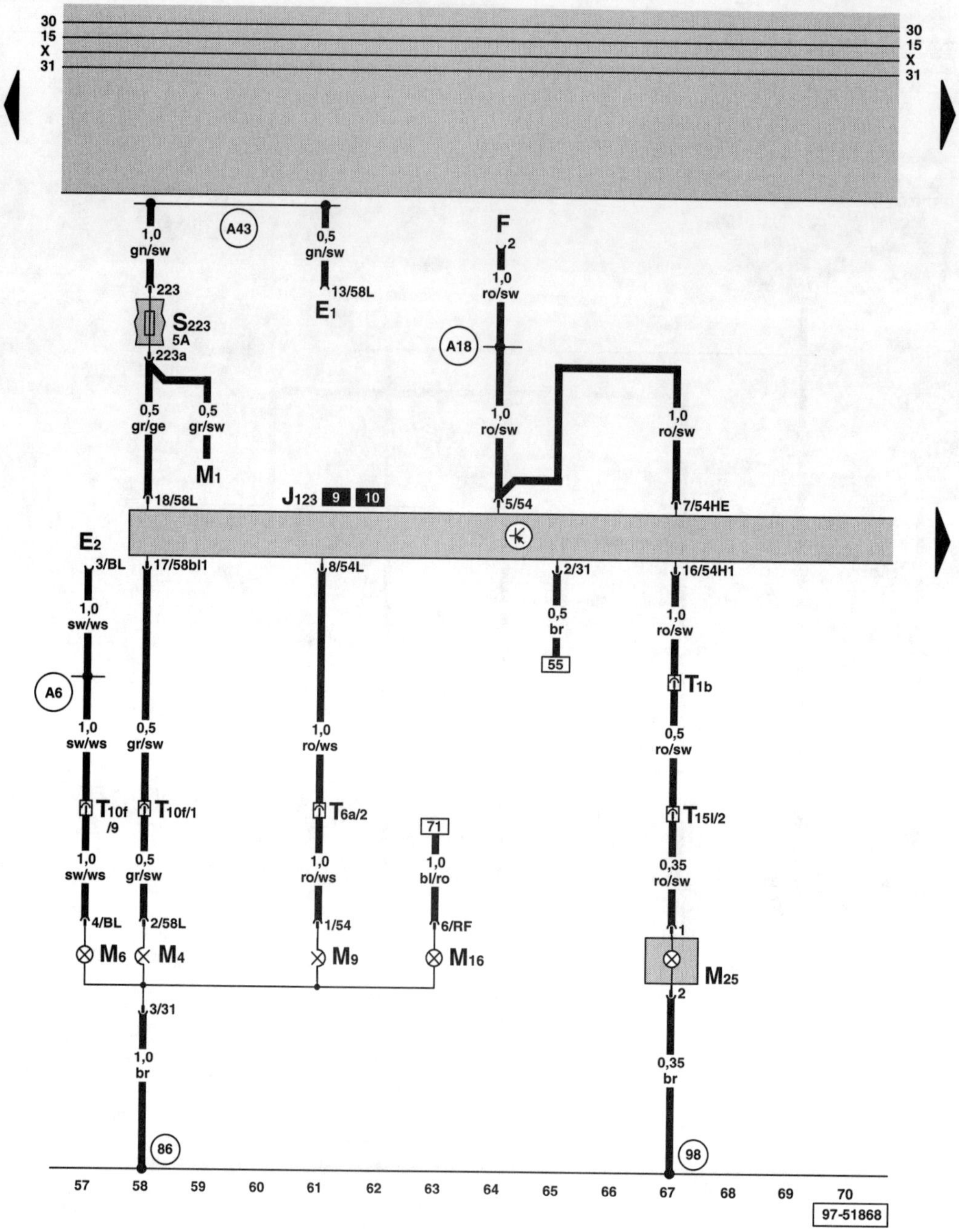

ws = white
sw = black
ro = red
br = brown
gn = green
bl = blue
gr = grey
li = violet
ge = yellow

E1 - Light switch
E2 - Turn Signal Switch
F - Brake Light Switch
J123 - Lamp Control Module
M1 - Left Parking Light
M4 - Left Tail Light
M6 - Left Rear Turn Signal Light
M9 - Left Brake Light
M16 - Left Back-Up Light
M25 - High-mount Brake Light
S223 - Fuse 23 in fuse holder
T1b - Single connector, black, connector station A pillar right
T6a - Connector 6 pin, blue, connector station A pillar, left
T10f - Connector 10 pin, brown, connector station A pillar, left
T15l - Connector 15 pin, black, in luggage compartment

(86) - Ground connection -1-, in rear wiring harness
(98) - Ground connection, in rear lid wiring harness
(A6) - Plus connection (left turn signal), in instrument panel wiring harness
(A18) - Wire connection (54), in instrument panel wiring harness
(A43) - Wire connection (57l), in instrument panel wiring harness

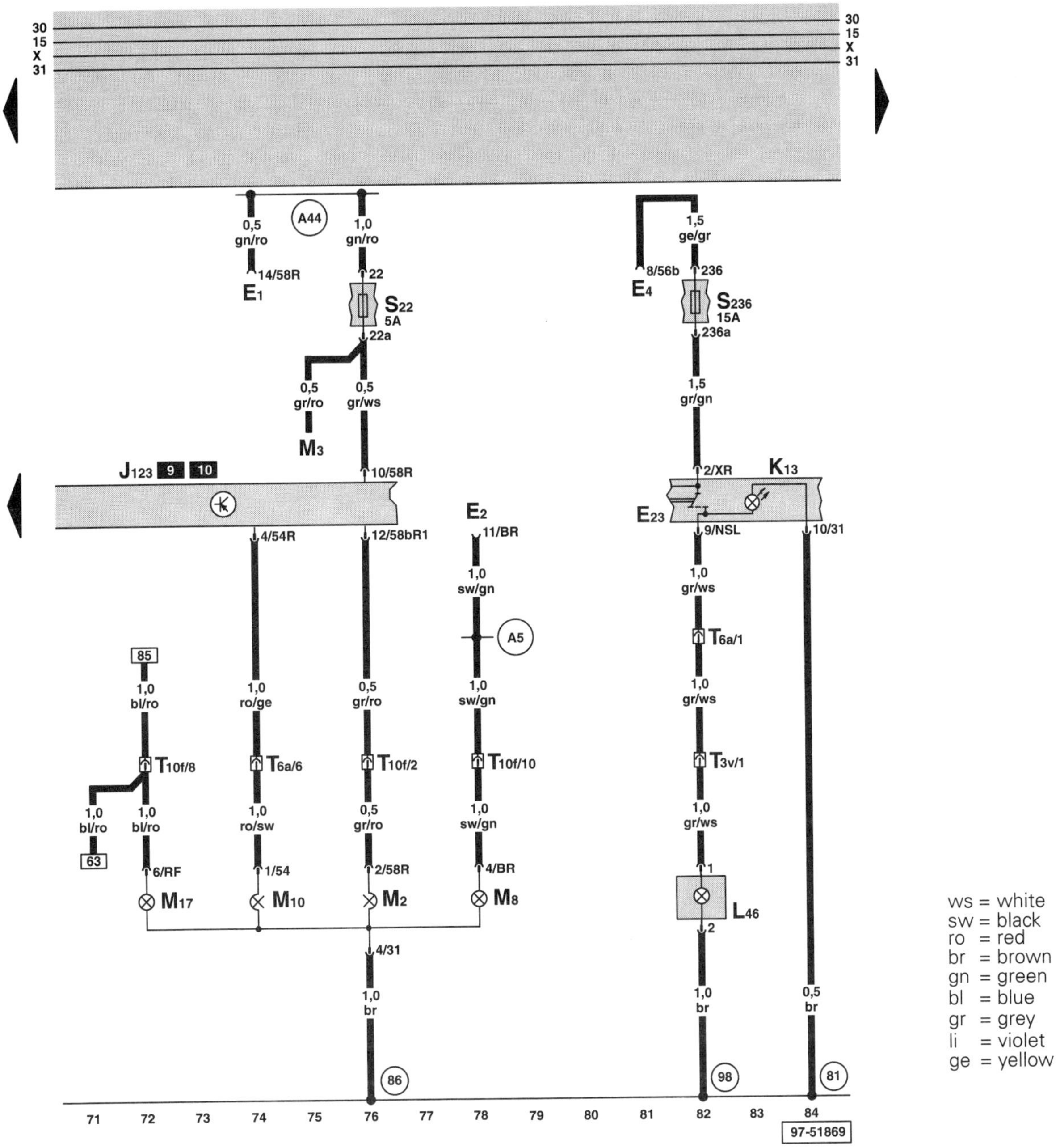

E1 - Light switch
E2 - Turn Signal Switch
E4 - Headlight Dimmer/Flasher Switch
E23 - Fog Light Switch
J123 - Lamp Control Module
K13 - Rear Fog Light Indicator Light
L46 - Left Rear Fog Light
M2 - Right Tail Light
M3 - Right Parking Light
M8 - Right Rear Turn Signal Light
M10 - Right Brake Light
M17 - Right Back-Up Light
S22 - Fuse
S236 - Fuse 36 in fuse holder
T3v - Connector 3 pin, black, in luggage compartment
T6a - Connector 6 pin, blue, connector station A pillar, left
T10f - Connector 10 pin, brown, connector station A pillar, left

(81) - Ground connection -1-, in instrument panel wiring harness
(86) - Ground connection -1-, in rear wiring harness
(98) - Ground connection, in rear lid wiring harness
(A5) - Plus connection (right turn signal), in instrument panel wiring harness
(A44) - Wire connection (57r), in instrument panel wiring harness

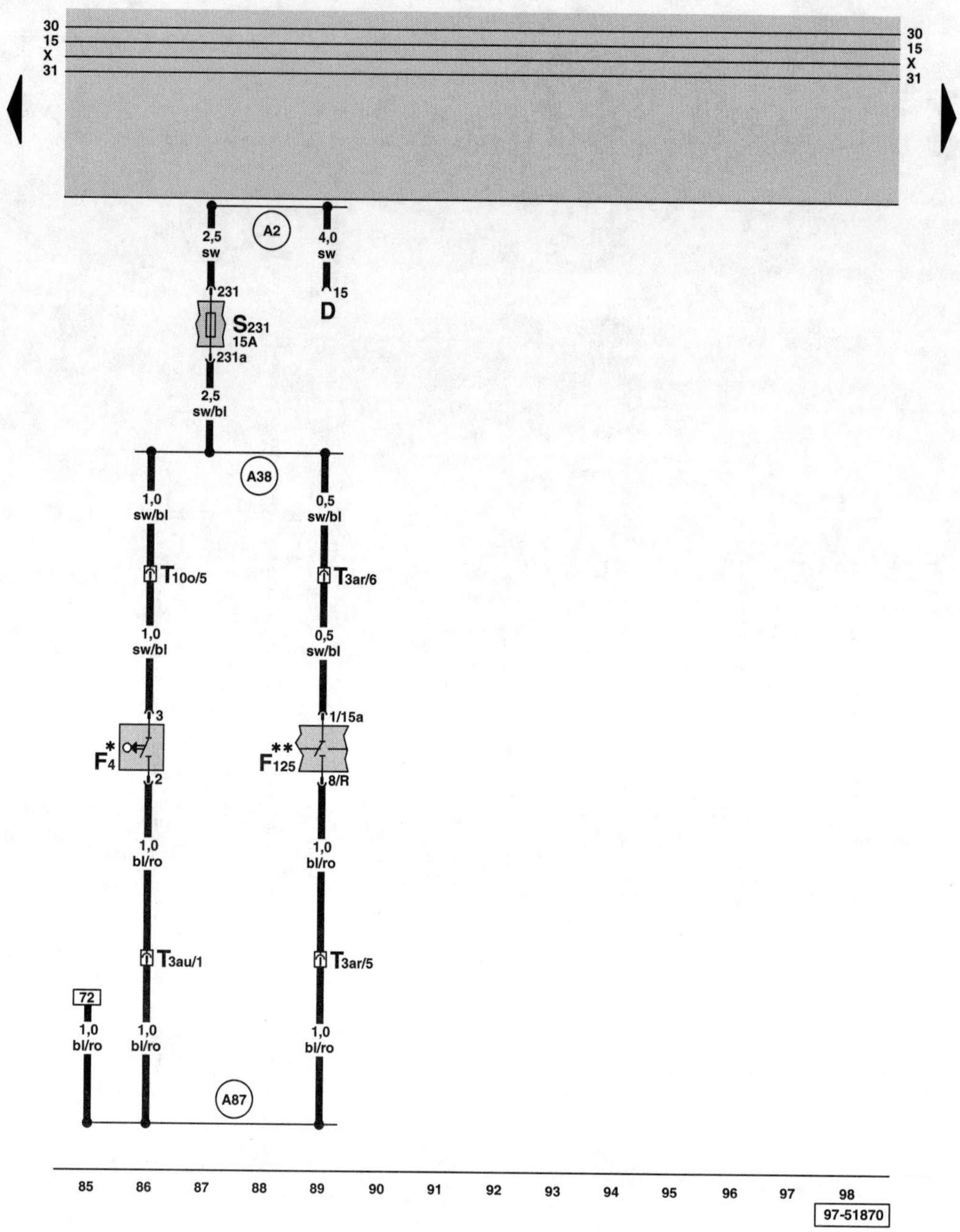

ws = white
sw = black
ro = red
br = brown
gn = green
bl = blue
gr = grey
li = violet
ge = yellow

D - Ignition/Starter Switch
F4 - Back-Up Light Switch
F125 - Multi-Function Transmission Range (TR) Switch
S231 - Fuse 31 in fuse holder
T3ar - Connector 3 pin, blue, connector station A pillar, right
T3au - Connector 3 pin, red, connector station electronic box plenum chamber
T10o - Connector 10 pin, brown, connector station electronic box plenum chamber

(A2) - Plus connection (15), in instrument panel wiring harness
(A38) - Plus connection -2- (15a), in instrument panel wiring harness
(A87) - Connector (RF), in instrument panel wiring harness
* - Vehicles with manually shifted transmission
** - Vehicles with automatic transmission 01V, 01L

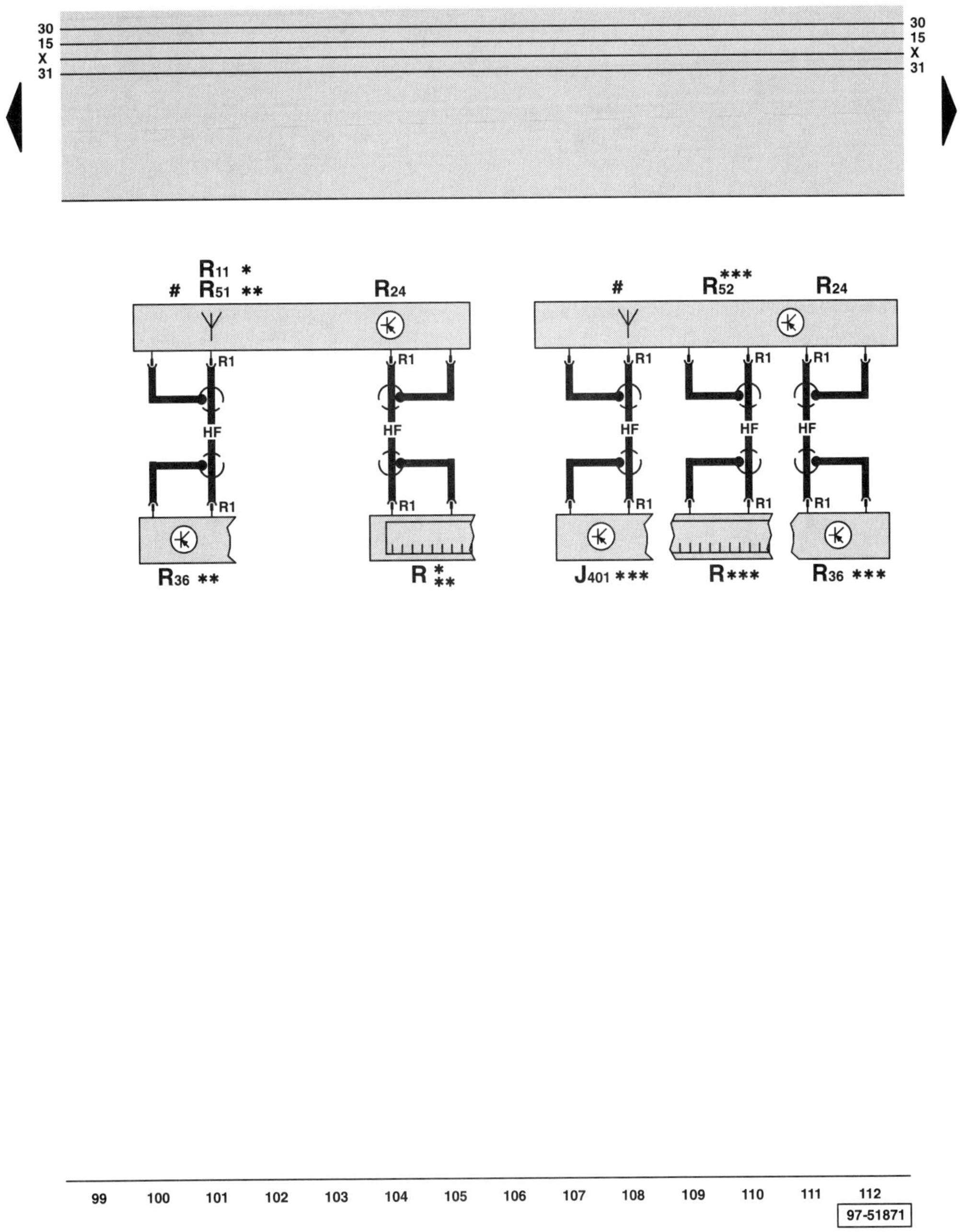

ws = white
sw = black
ro = red
br = brown
gn = green
bl = blue
gr = grey
li = violet
ge = yellow

J401 - Control unit for navigation with CD drive
R - Radio
R1 - Antenna Connection
R11 - Antenna
R24 - Antenna Amplifier
R36 - Telephone Transceiver
R51 - Radio/Telephone Antenna
R52 - Antenna for radio/telephone/navigation (GPS)

* - Vehicles with Radio
** - Vehicles with Radio and Telephone
*** - Vehicles with Radio, Telephone and Navigation
# - Roof Antenna

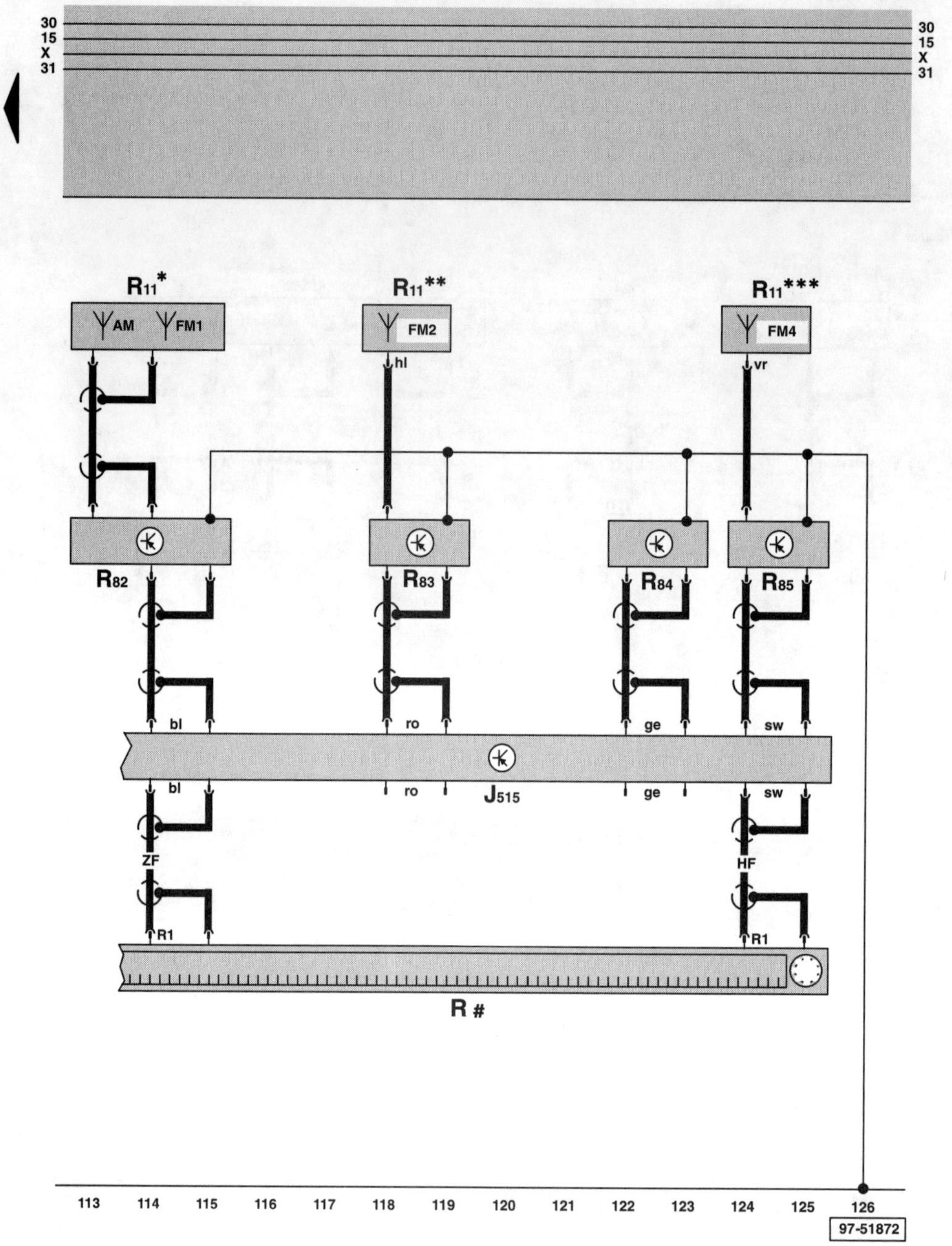

ws = white
sw = black
ro = red
br = brown
gn = green
bl = blue
gr = grey
li = violet
ge = yellow

J515 - Control unit for aerial selection (aerial diversity function)
R - Radio
R11 - Antenna
R82 - TV Antenna Amplifier 1
R83 - TV Antenna Amplifier 2
R84 - TV Antenna Amplifier 3
R85 - TV Antenna Amplifier 4

* - Roof Antenna
** - Window rear left
*** - Window rear right
# - Radio Symphony

Driver's Information System

2000 m. y.

Fuse Panel

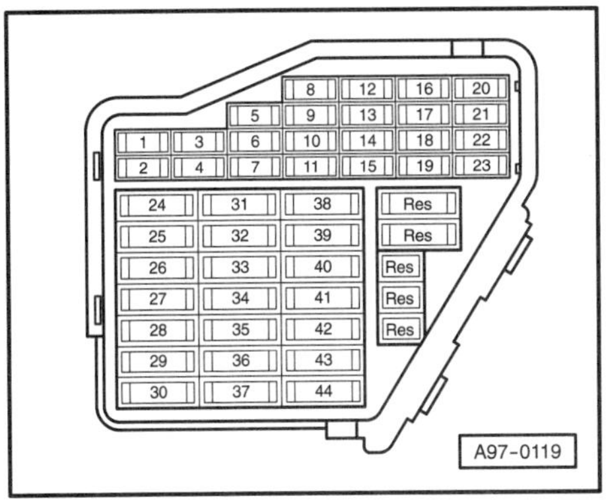

Fuse Colors:

30 A - Green
25 A - White
20 A - Yellow
15 A - Blue
10 A - Red
7,5 A - Brown
5 A - Beige

Starting with fuse position 23, fuses in the fuse holder are identified with 223 in the wiring diagram.

13 - Fold Relay Panel

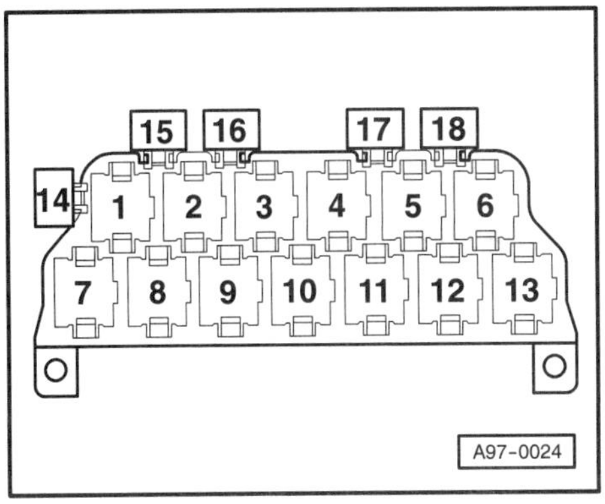

Relay Location:

9 -Lamp Control Module, J123

10 -Lamp Control Module, J123

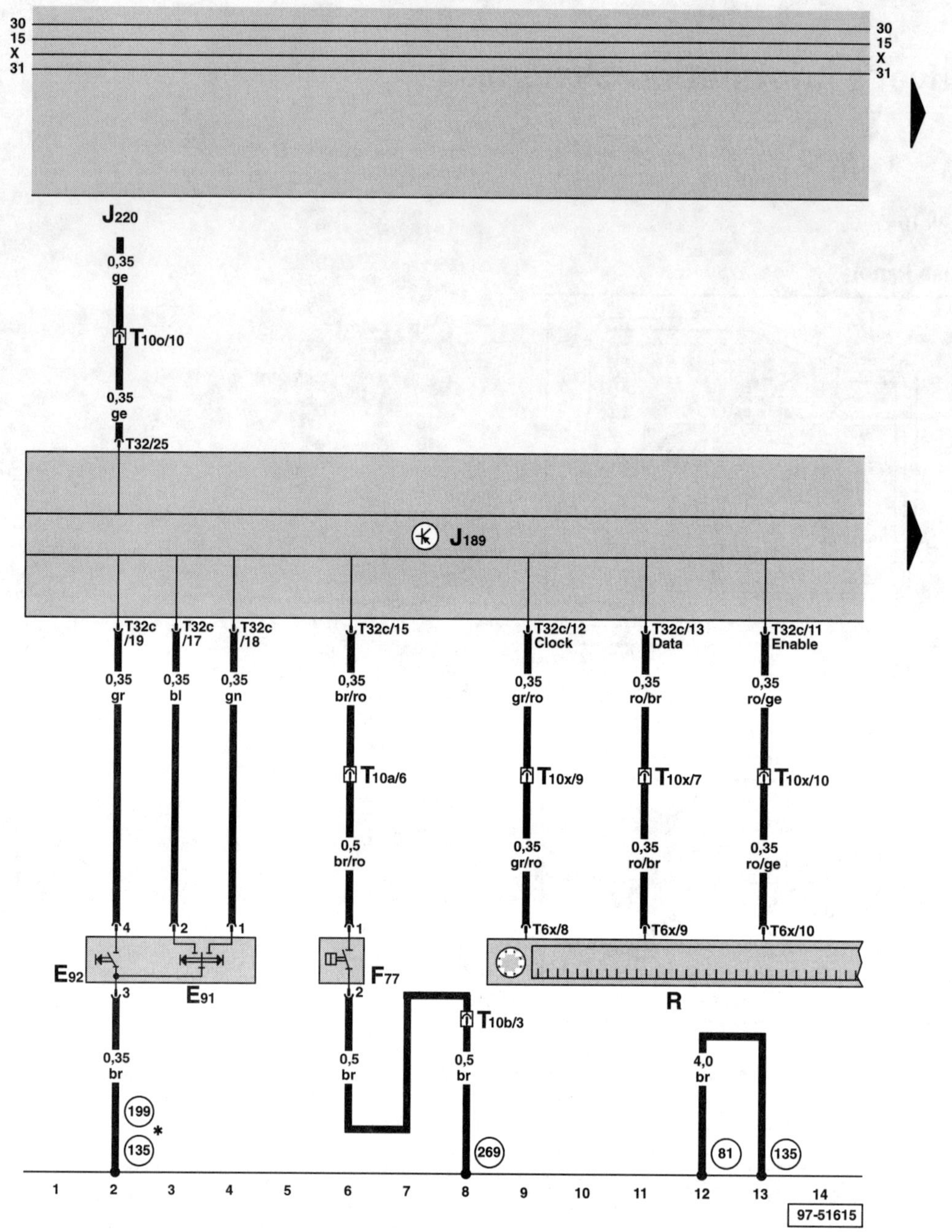

ws = white
sw = black
ro = red
br = brown
gn = green
bl = blue
gr = grey
li = violet
ge = yellow

E91 - Board Computer Function Selector Switch
E92 - Board Computer Reset Button
F77 - Windshield Washer Fluid Level Warning Switch
J189 - Auto Check System
J220 - Motronic Engine Control Module (ECM)
R - Radio
T6x - Connector 6 pin, green, connector radio I - 2
T10a - Connector 10 pin, pink, connector station A pillar, left
T10b - Connector 10 pin, red, connector station A pillar, left
T10o - Connector 10 pin, brown, connector station electronic box plenum chamber
T10x - Connector 10 pin, green, connector station A pillar, right
T32 - Connector 32 pin, blue, on instrument cluster
T32c - Connector 32 pin, grey, on instrument cluster

(81) - Ground connection -1-, in instrument panel wiring harness
(135) - Ground connection -2-, in instrument panel wiring harness
(199) - Ground connection -3-, in instrument panel wiring harness
(269) - Ground connector (sensor ground) -1-, in instrument panel wiring harness
* - Both are possible

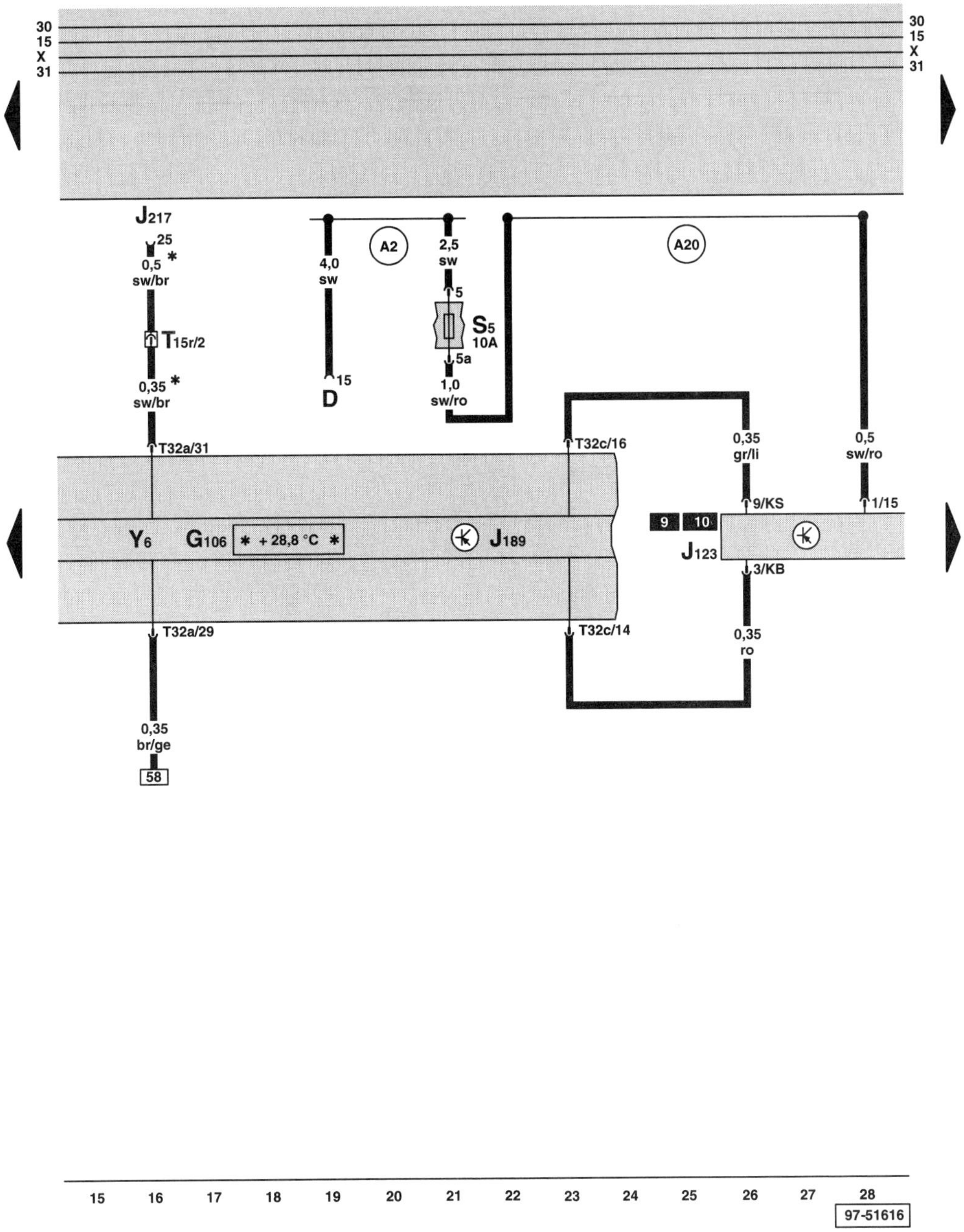

ws = white
sw = black
ro = red
br = brown
gn = green
bl = blue
gr = grey
li = violet
ge = yellow

D - Ignition/Starter Switch
G106 - Outside Air Temperature Display
J123 - Lamp Control Module
J189 - Auto Check System
J217 - Transmission Control Module (TCM)
S5 - Fuse
T15r - Connector 15 pin, blue, connector station A pillar, right
T32a - Connector 32 pin, green, on instrument cluster
T32c - Connector 32 pin, grey, on instrument cluster
Y6 - Transmission Range (TR) Selector Lever Display

(A2) - Plus connection (15), in instrument panel wiring harness

(A20) - Wire connection (15a), in instrument panel wiring harness

* - Vehicles with automatic transmission 01V

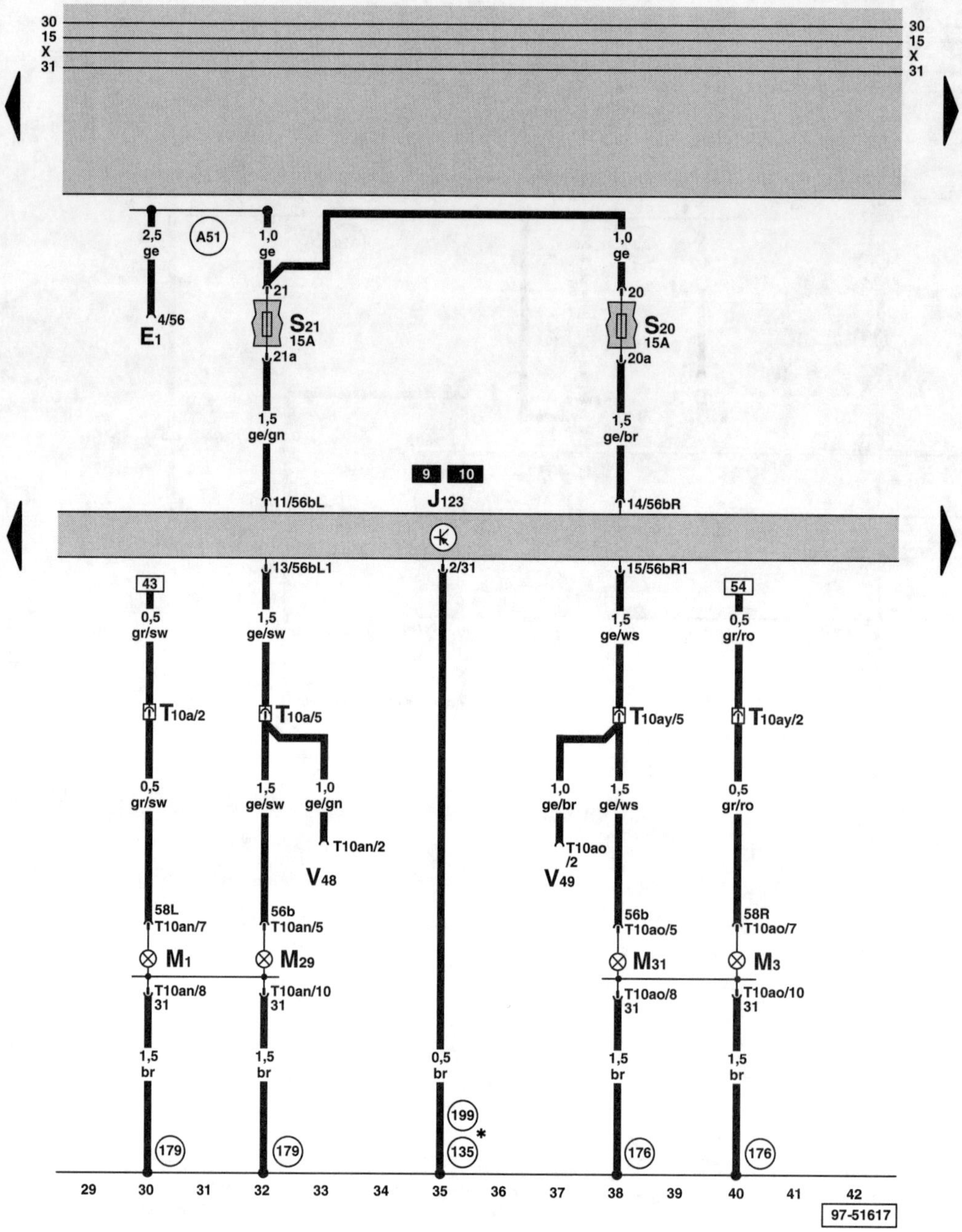

ws = white
sw = black
ro = red
br = brown
gn = green
bl = blue
gr = grey
li = violet
ge = yellow

E1 - Light switch
J123 - Lamp Control Module
M1 - Left Parking Light
M3 - Right Parking Light
M29 - Left Low Beam Headlight
M31 - Right Low Beam Headlight
S20 - Fuse
S21 - Fuse
T10a - Connector 10 pin, pink, connector station A pillar, left
T10an - Connector 10 pin, on headlight, left
T10ao - Connector 10 pin, on headlight, right
T10ay - Connector 10 pin, pink, connector station A pillar, right
V48 - Left Headlight Beam Adjusting Motor
V49 - Right Headlight Beam Adjusting Motor

(135) - Ground connection -2-, in instrument panel wiring harness
(176) - Ground connection, in right headlight wiring harness
(179) - Ground connection, in left headlight wiring harness
(199) - Ground connection -3-, in instrument panel wiring harness
(A51) - Wire connection (56), in instrument panel wiring harness

* - Both are possible

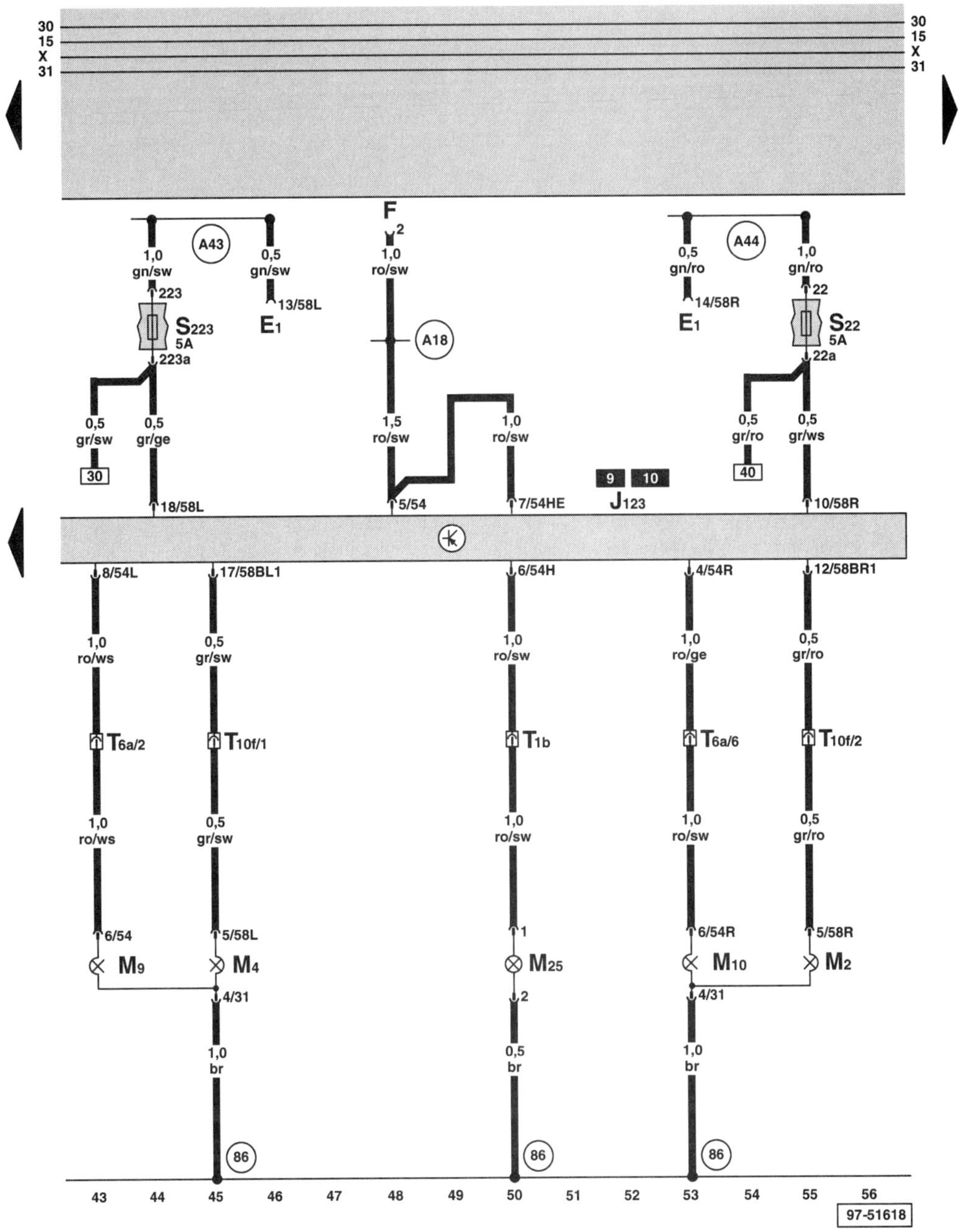

ws = white
sw = black
ro = red
br = brown
gn = green
bl = blue
gr = grey
li = violet
ge = yellow

E1 - Light switch
F - Brake Light Switch
J123 - Lamp Control Module
M2 - Right Tail Light
M4 - Left Tail Light
M9 - Left Brake Light
M10 - Right Brake Light
M25 - High-mount Brake Light
S22 - Fuse
S223 - Fuse 23 in fuse holder
T1b - Single connector, black, connector station A pillar right
T6a - Connector 6 pin, blue, connector station A pillar, left
T10f - Connector 10 pin, brown, connector station A pillar, left

(86) - Ground connection -1-, in rear wiring harness
(A18) - Wire connection (54), in instrument panel wiring harness
(A43) - Wire connection (57l), in instrument panel wiring harness
(A44) - Wire connection (57r), in instrument panel wiring harness

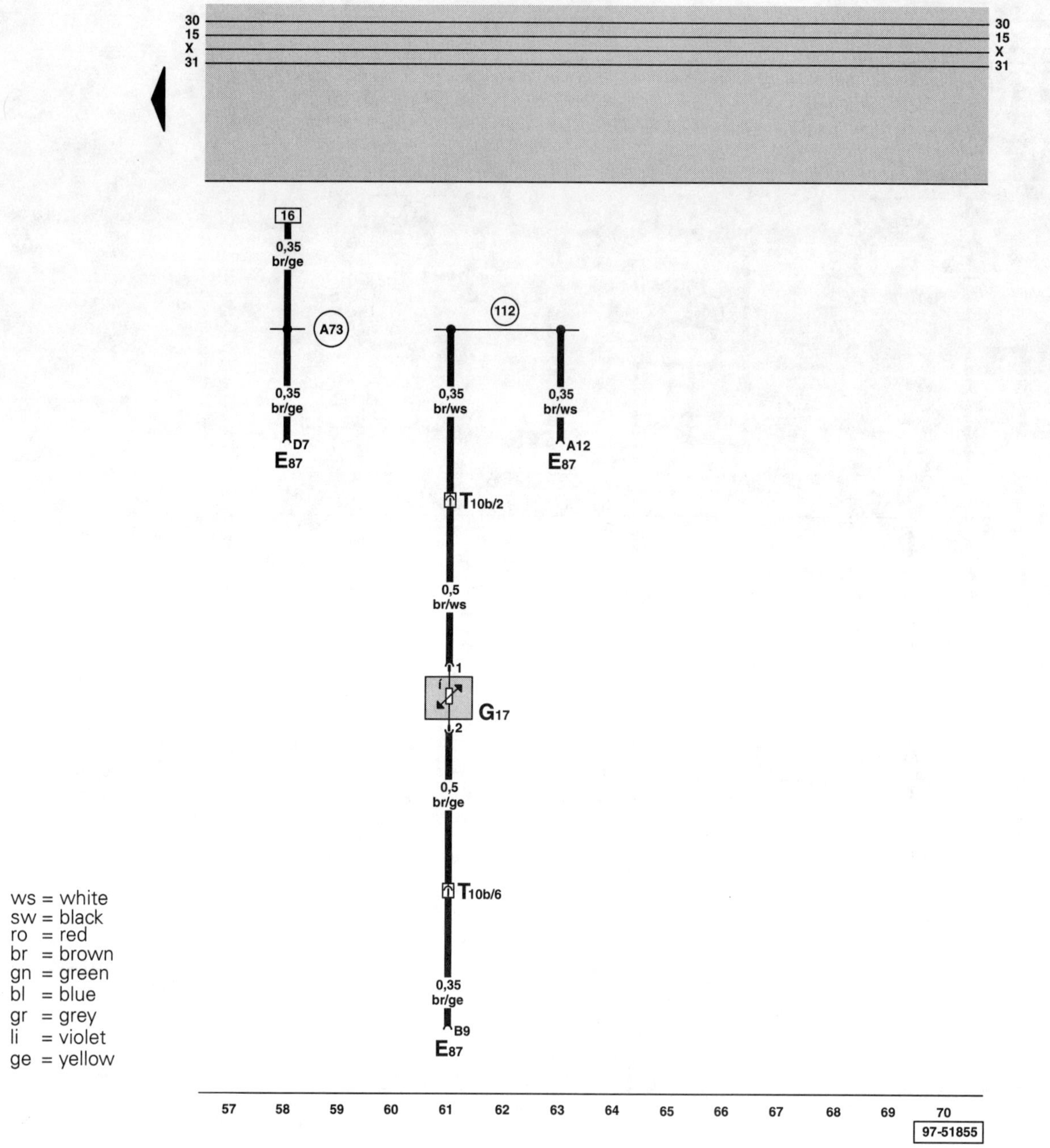

E87 - A/C Control Head
G17 - Outside Air Temperature Sensor
T10b - Connector 10 pin, red, connector station A pillar, left

(112) - Ground connection -2-, in A/C wiring harness
(A73) - Connector (outside temperature indicator), in instrument panel wiring harness

Daytime Running Lights with Fog Light and Rear Fog Light (Canada)

2000 m. y.

Fuse Panel

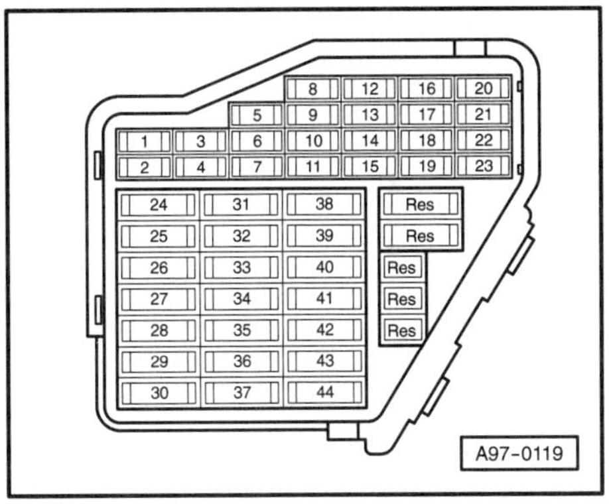

Fuse Colors:

30 A - Green
25 A - White
20 A - Yellow
15 A - Blue
10 A - Red
7,5 A - Brown
5 A - Beige

Starting with fuse position 23, fuses in the fuse holder are identified with 223 in the wiring diagram.

Micro Central Electric Panel

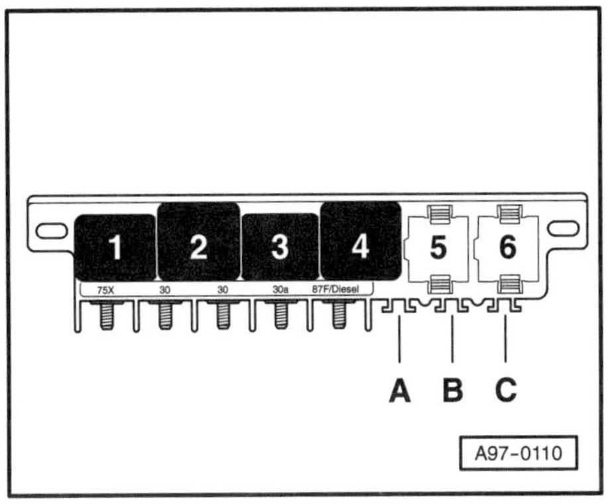

Relay Location:

2 -Load Reduction Relay, J59

13 - Fold Relay Panel

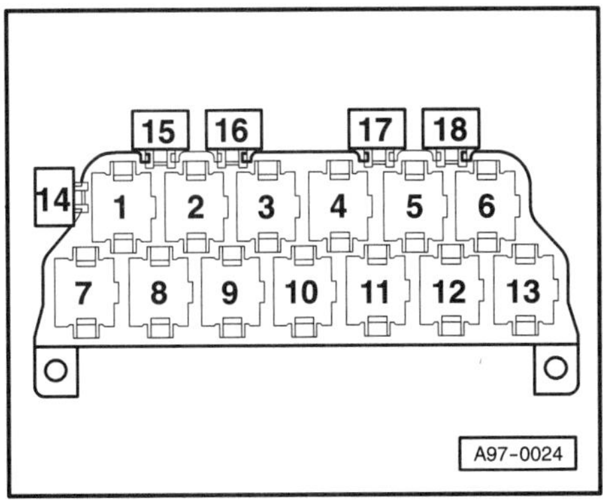

Relay Location:

6 -Fog Light Relay, J5

9 -Lamp Control Module, J123

10 -Lamp Control Module, J123

17 -Fog Light Circuit Breaker, S28

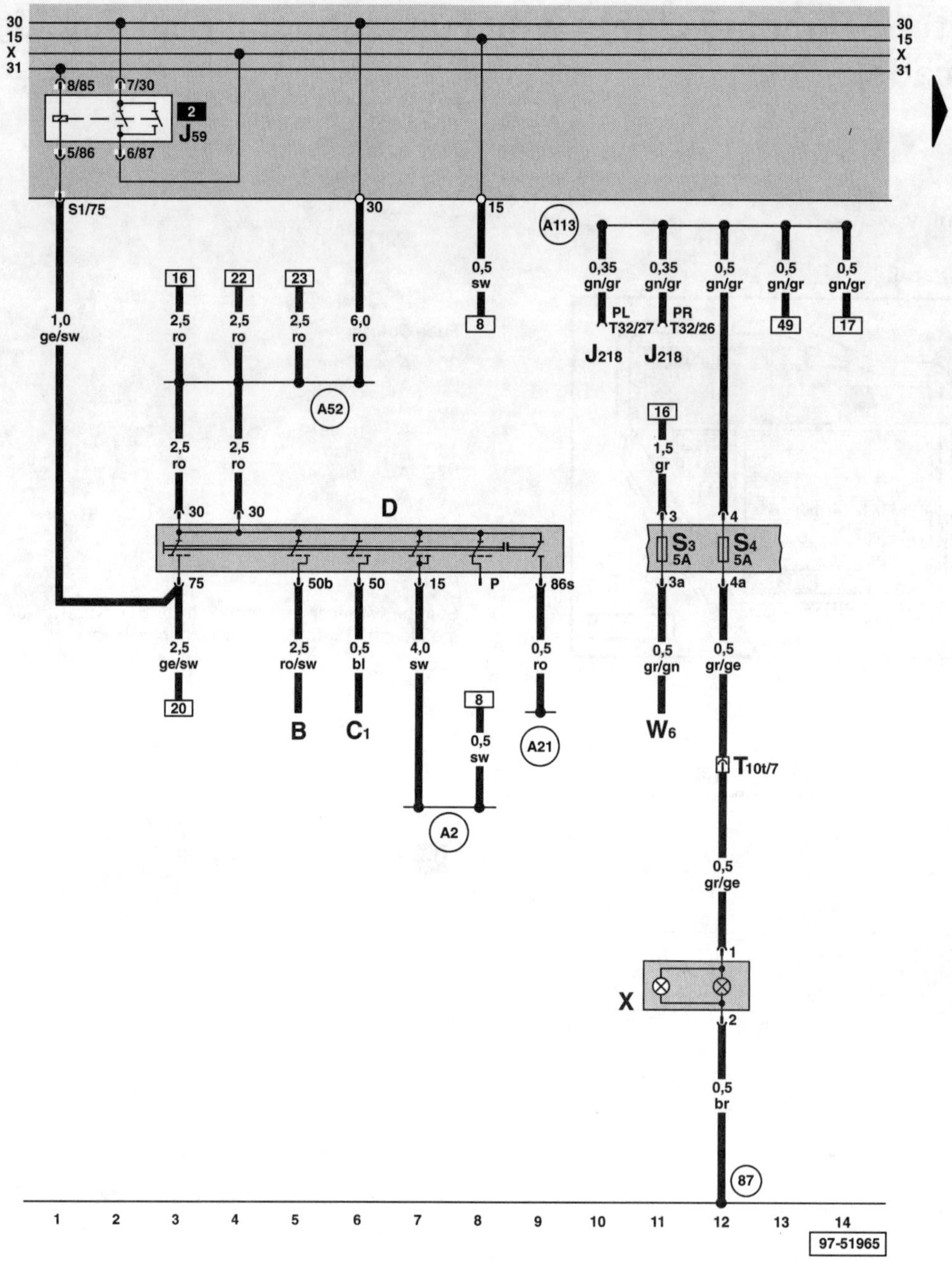

ws = white
sw = black
ro = red
br = brown
gn = green
bl = blue
gr = grey
li = violet
ge = yellow

B - Starter
C1 - Voltage Regulator (VR)
D - Ignition/Starter Switch
J59 - Load Reduction Relay
J218 - Instrument Cluster Combination Processor
S3 - Fuse
S4 - Fuse
T10t - Connector 10 pin, orange, connector station A pillar, right
T32 - Connector 32 pin, blue, on instrument cluster
W6 - Glove Compartment Light
X - License Plate Light

(87) - Ground connection -2-, in rear wiring harness
(A2) - Plus connection (15), in instrument panel wiring harness
(A21) - Wire connection (86s), in instrument panel wiring harness
(A52) - Plus connection -2- (30), in instrument panel wiring harness
(A113) - Connection 2 (daytime running lights), in instrument panel wiring harness

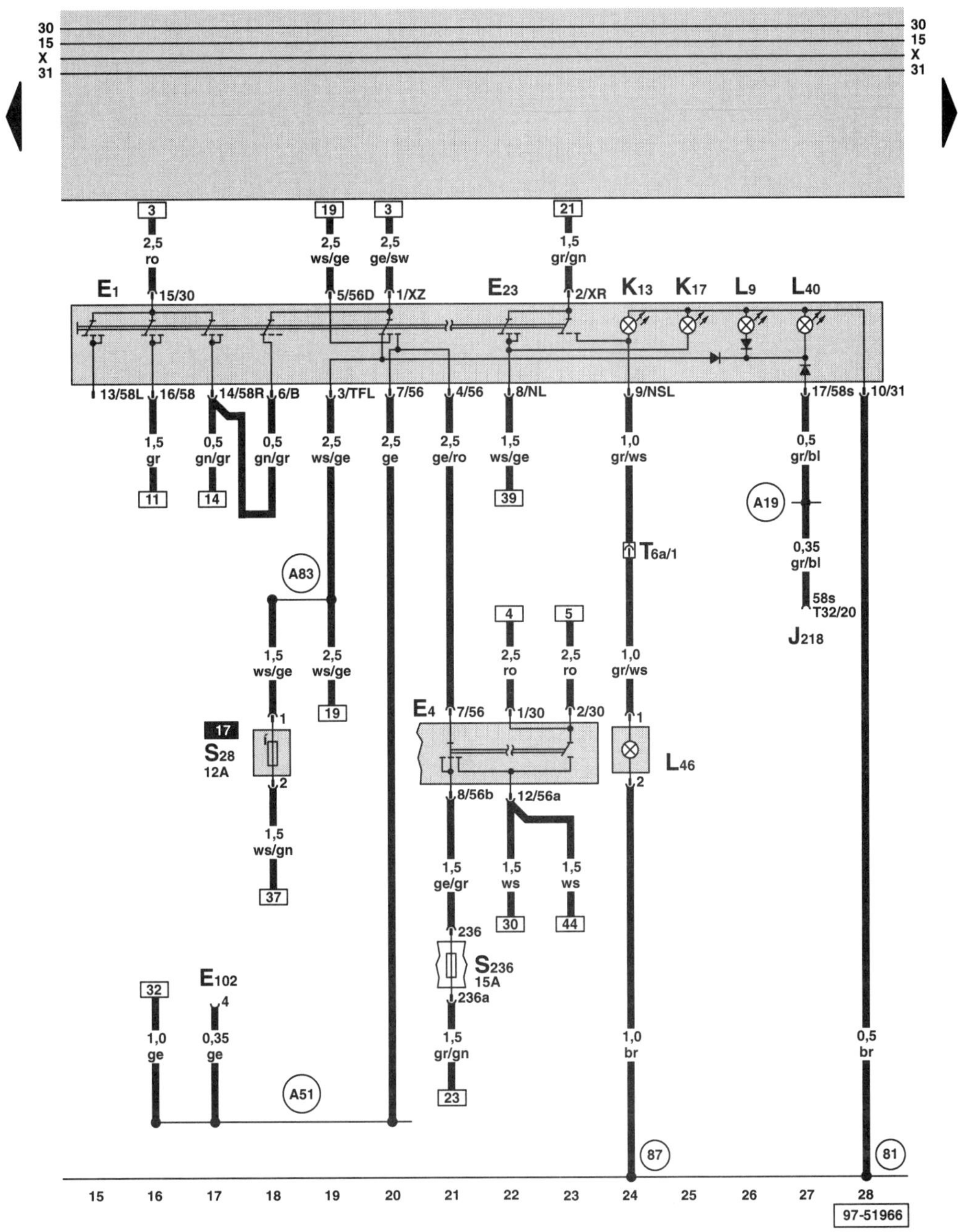

ws = white
sw = black
ro = red
br = brown
gn = green
bl = blue
gr = grey
li = violet
ge = yellow

E1 - Light switch
E4 - Headlight Dimmer/Flasher Switch
E23 - Fog Light Switch
E102 - Headlight Adjuster
J218 - Instrument Cluster Combination Processor
K13 - Rear Fog Light Indicator Light
K17 - Fog Light Indicator Light
L9 - Headlight Switch Light
L40 - Front And Rear Fog Light Switch Light
L46 - Left Rear Fog Light
S28 - Fog Light Fuse
S236 - Fuse 36 in fuse holder
T6a - Connector 6 pin, blue, connector station A pillar, left
T32 - Connector 32 pin, blue, on instrument cluster

(81) - Ground connection -1-, in instrument panel wiring harness

(87) - Ground connection -2-, in rear wiring harness

(A19) - Wire connection (58s), in instrument panel wiring harness

(A51) - Wire connection (56), in instrument panel wiring harness

(A83) - Connector (dayttime running lights), in instrument panel wiring harness

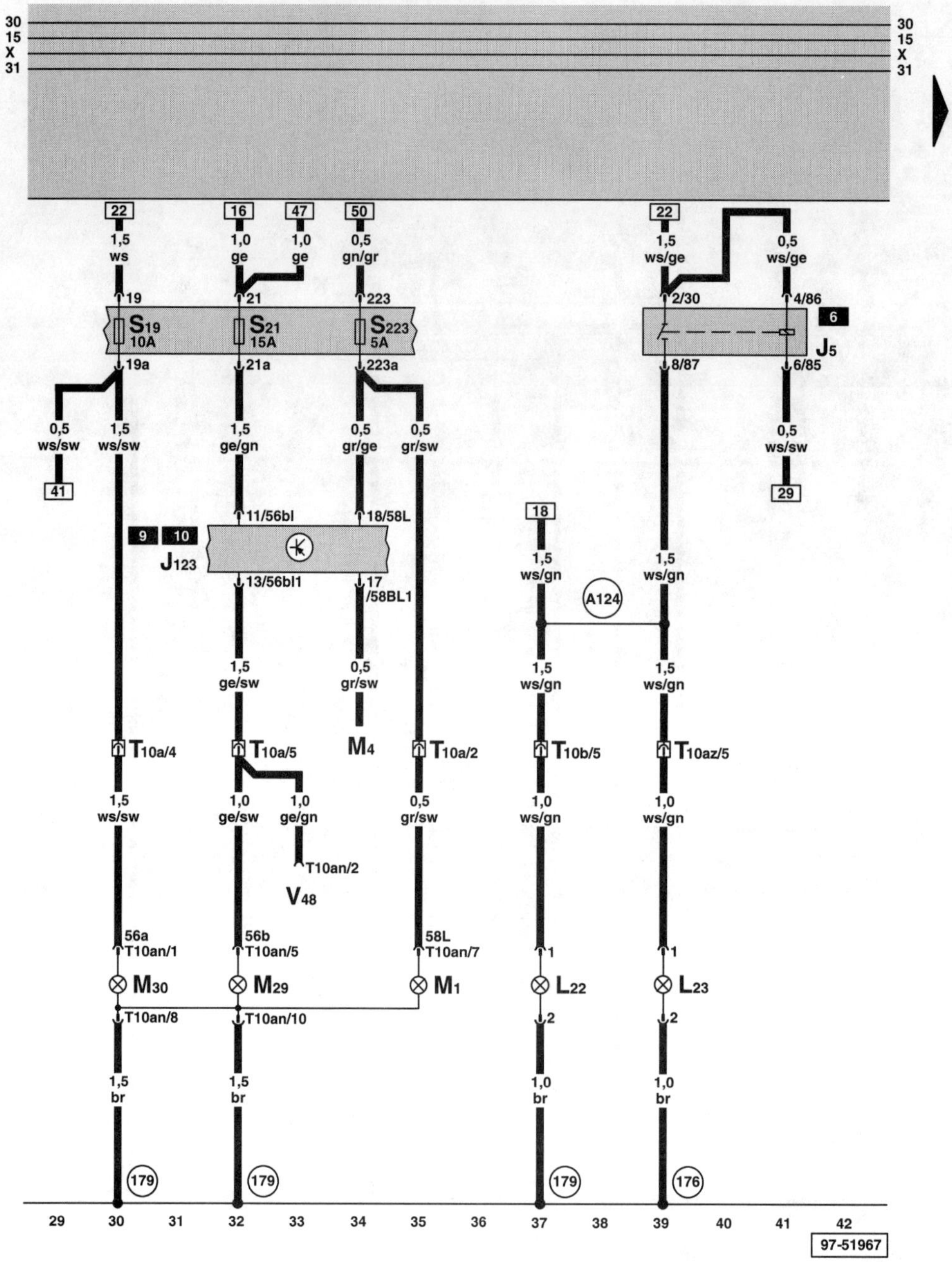

ws = white
sw = black
ro = red
br = brown
gn = green
bl = blue
gr = grey
li = violet
ge = yellow

J5 - Fog Light Relay
J123 - Lamp Control Module
L22 - Left Front Fog Light
L23 - Right Front Fog Light
M1 - Left Parking Light
M4 - Left Tail Light
M29 - Left Low Beam Headlight
M30 - Left High Beam Headlight
S19 - Fuse
S21 - Fuse
S223 - Fuse 23 in fuse holder
T10a - Connector 10 pin, pink, connector station A pillar, left
T10b - Connector 10 pin, red, connector station A pillar, left
T10an - Connector 10 pin, on headlight, left
T10az - Connector 10 pin, red, connector station A pillar, right
V48 - Left Headlight Beam Adjusting Motor

(176) - Ground connection, in right headlight wiring harness

(179) - Ground connection, in left headlight wiring harness

(A124) - Connection 3 (daytime running lights), in instrument panel wiring harness

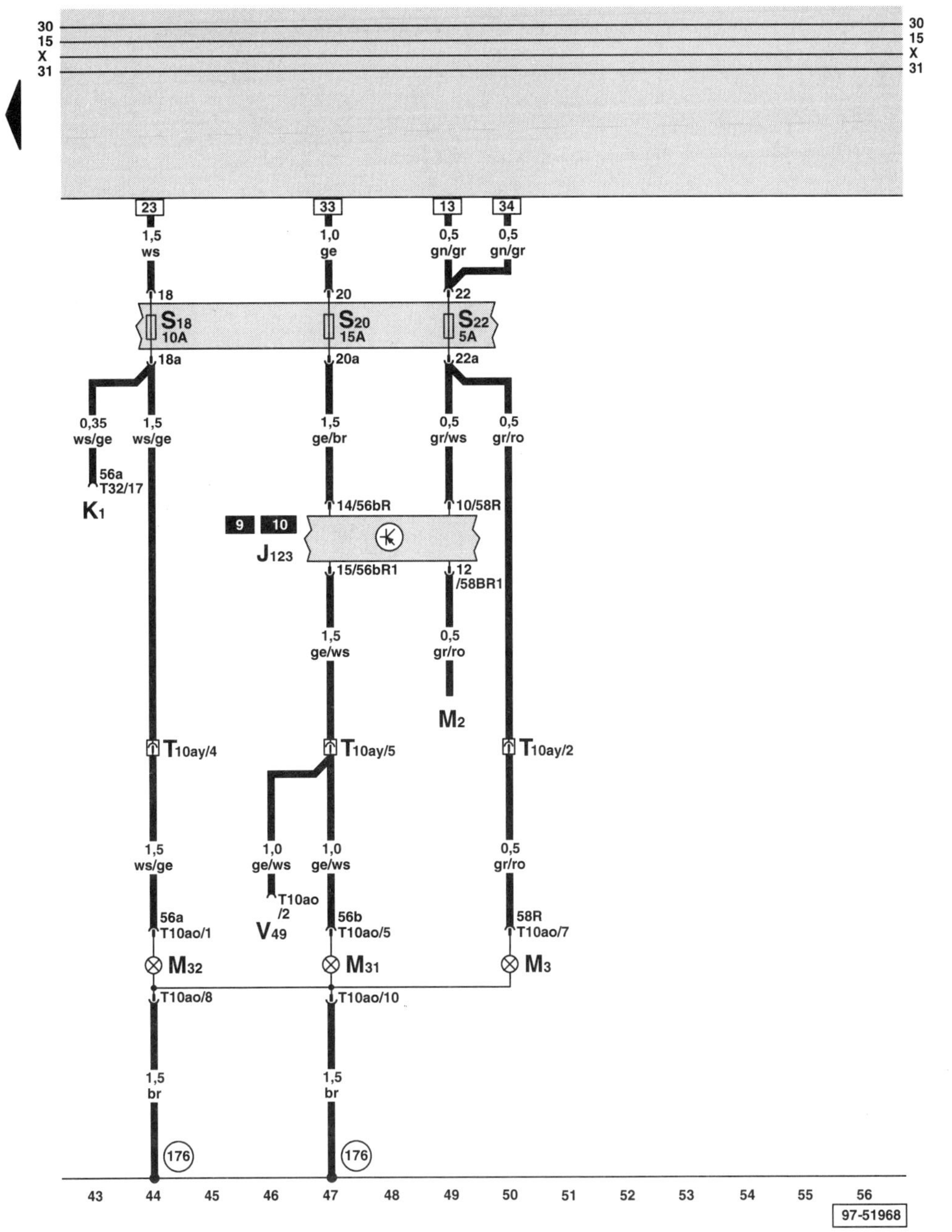

ws = white
sw = black
ro = red
br = brown
gn = green
bl = blue
gr = grey
li = violet
ge = yellow

J123 - Lamp Control Module
K1 - Headlight High Beam Indicator Light
M2 - Right Tail Light
M3 - Right Parking Light
M31 - Right Low Beam Headlight
M32 - Right High Beam Headlight
S18 - Fuse
S20 - Fuse
S22 - Fuse
T10ao - Connector 10 pin, on headlight, right
T10ay - Connector 10 pin, pink, connector station A pillar, right
T32 - Connector 32 pin, blue, on instrument cluster
V49 - Right Headlight Beam Adjusting Motor

(176) - Ground connection, in right headlight wiring harness

THIS PAGE INTENTIONALLY LEFT BLANK

Headlight with High Intensity Gas discharge Lamps and automatic Headlight Beam Adjusting

2000 m. y.

Fuse Panel

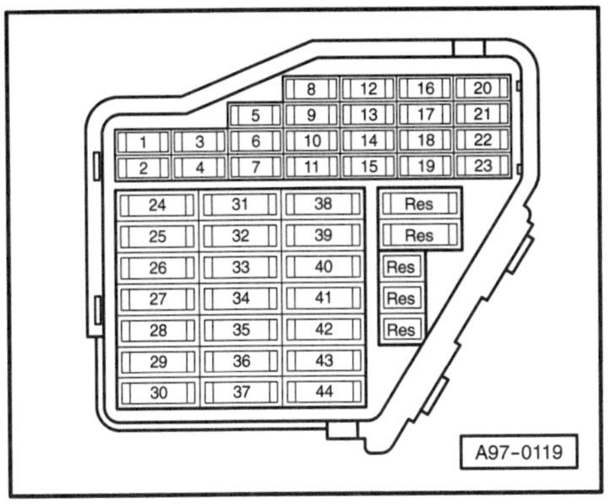

Fuse Colors:

30	A	-	Green
25	A	-	White
20	A	-	Yellow
15	A	-	Blue
10	A	-	Red
7,5	A	-	Brown
5	A	-	Beige

Starting with fuse position 23, fuses in the fuse holder are identified with 223 in the wiring diagram.

Micro Central Electric Panel

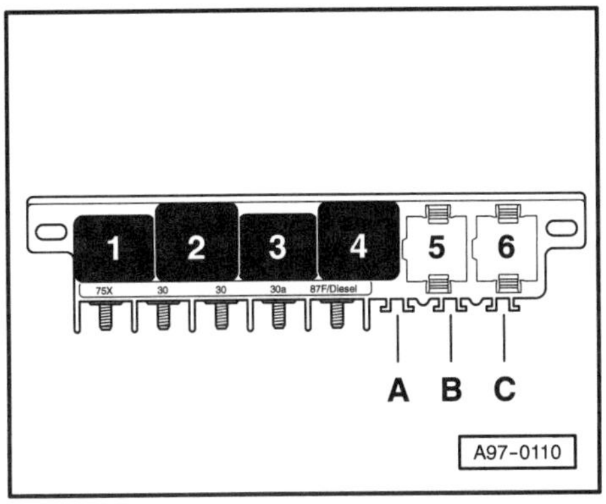

Relay Location:

2 -Load Reduction Relay, J59

13 - Fold Relay Panel

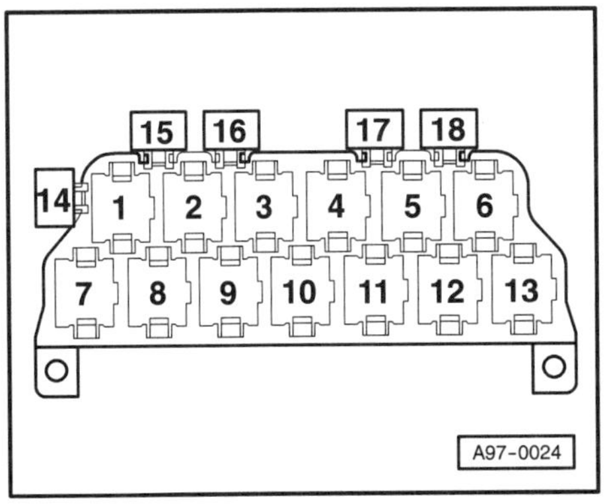

Relay Location:

9 -Lamp Control Module, J123

10 -Lamp Control Module, J123

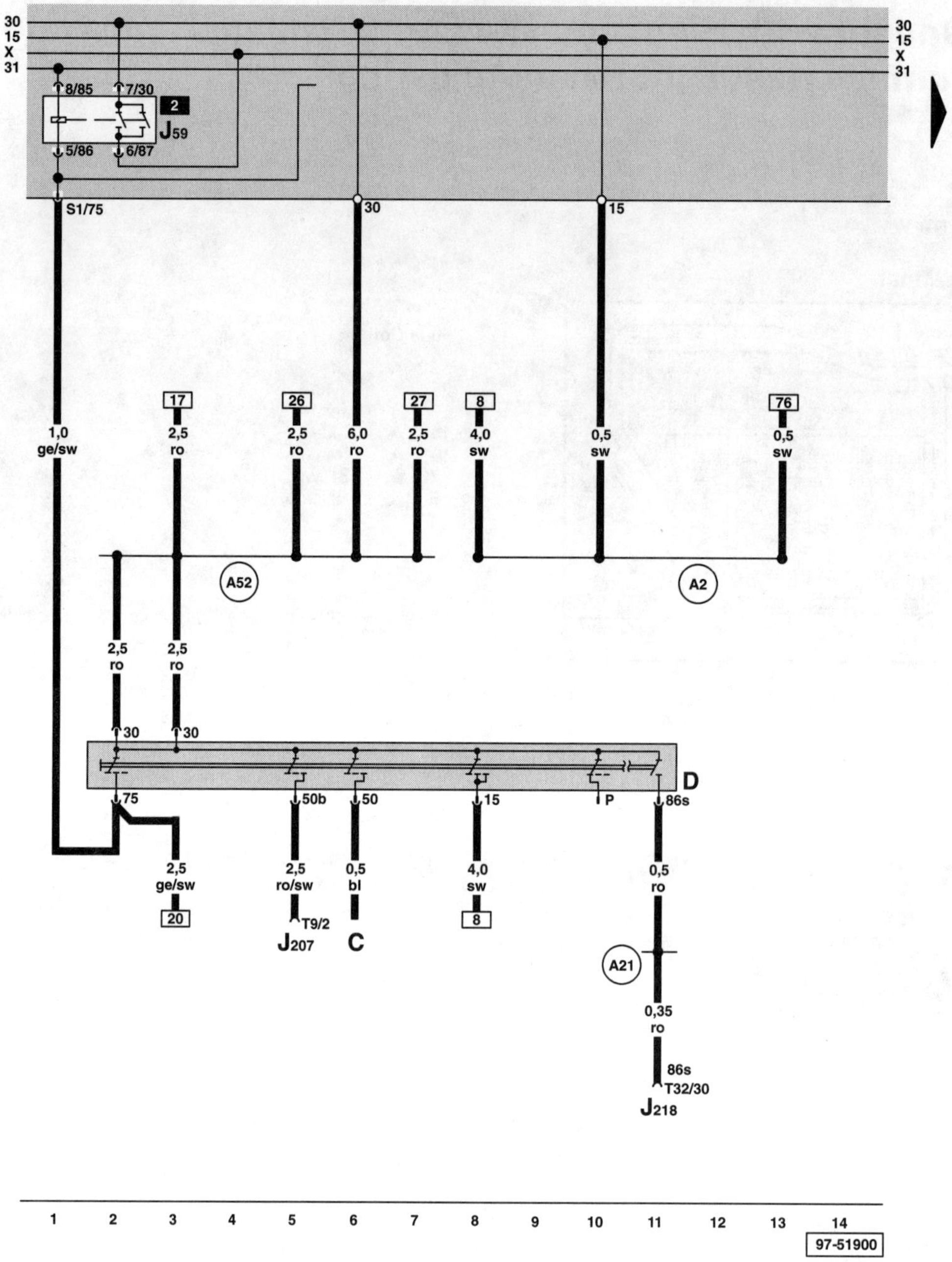

ws = white
sw = black
ro = red
br = brown
gn = green
bl = blue
gr = grey
li = violet
ge = yellow

C - Generator (GEN)
D - Ignition/Starter Switch
J59 - Load Reduction Relay
J207 - Starting Interlock Relay
J218 - Instrument Cluster Combination Processor
T9 - Connector 9 pin, brown, on starting interlock relay
T32 - Connector 32 pin, blue, on instrument cluster

(A2) - Plus connection (15), in instrument panel wiring harness

(A21) - Wire connection (86s), in instrument panel wiring harness

(A52) - Plus connection -2- (30), in instrument panel wiring harness

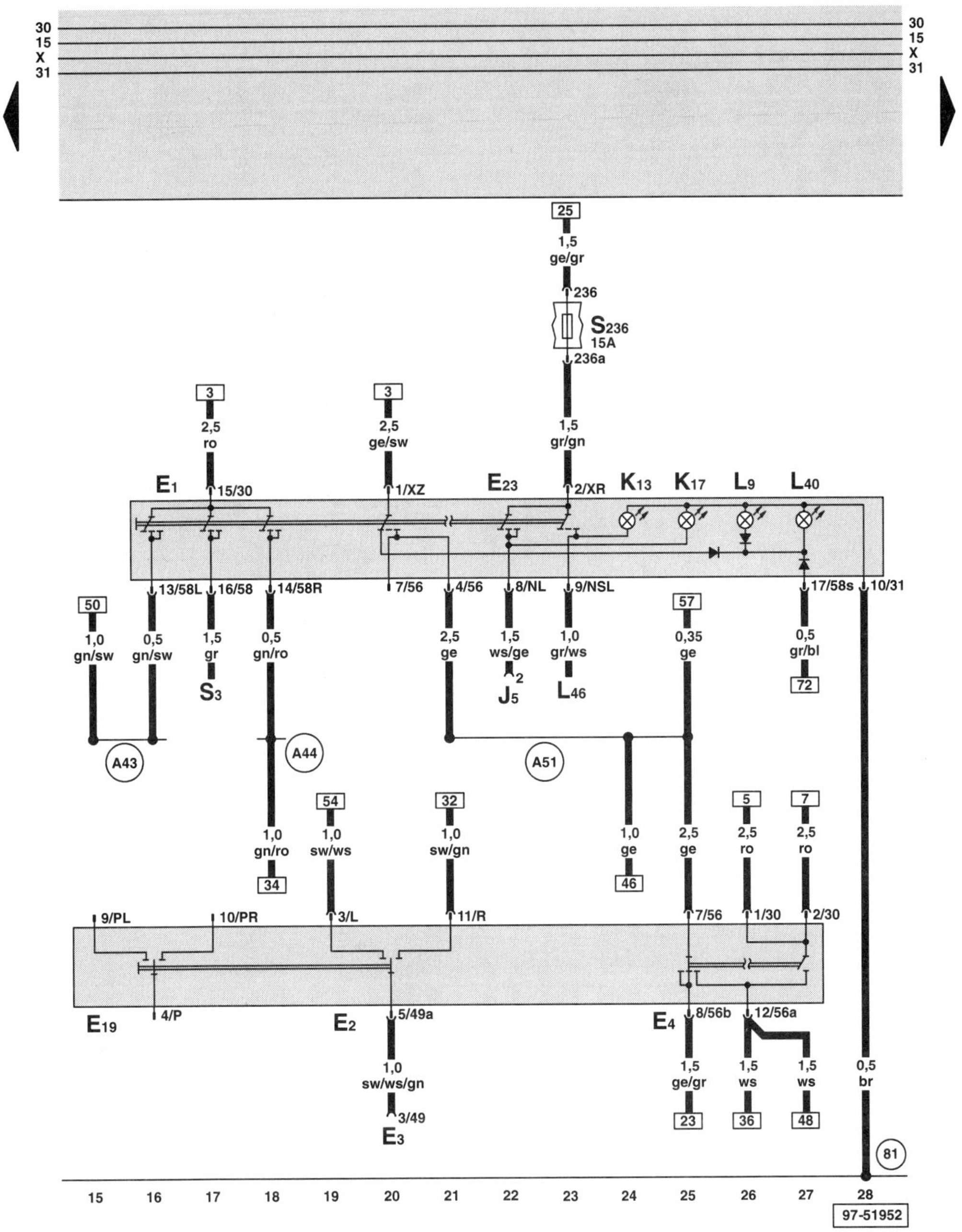

ws = white
sw = black
ro = red
br = brown
gn = green
bl = blue
gr = grey
li = violet
ge = yellow

E1 - Light switch
E2 - Turn Signal Switch
E3 - Emergency Flasher Switch
E4 - Headlight Dimmer/Flasher Switch
E19 - Park Light Switch
E23 - Fog Light Switch
J5 - Fog Light Relay
K13 - Rear Fog Light Indicator Light
K17 - Fog Light Indicator Light
L9 - Headlight Switch Light
L40 - Front And Rear Fog Light Switch Light
L46 - Left Rear Fog Light
S3 - Fuse
S236 - Fuse 36 in fuse holder

(81) - Ground connection -1-, in instrument panel wiring harness
(A43) - Wire connection (57l), in instrument panel wiring harness
(A44) - Wire connection (57r), in instrument panel wiring harness
(A51) - Wire connection (56), in instrument panel wiring harness

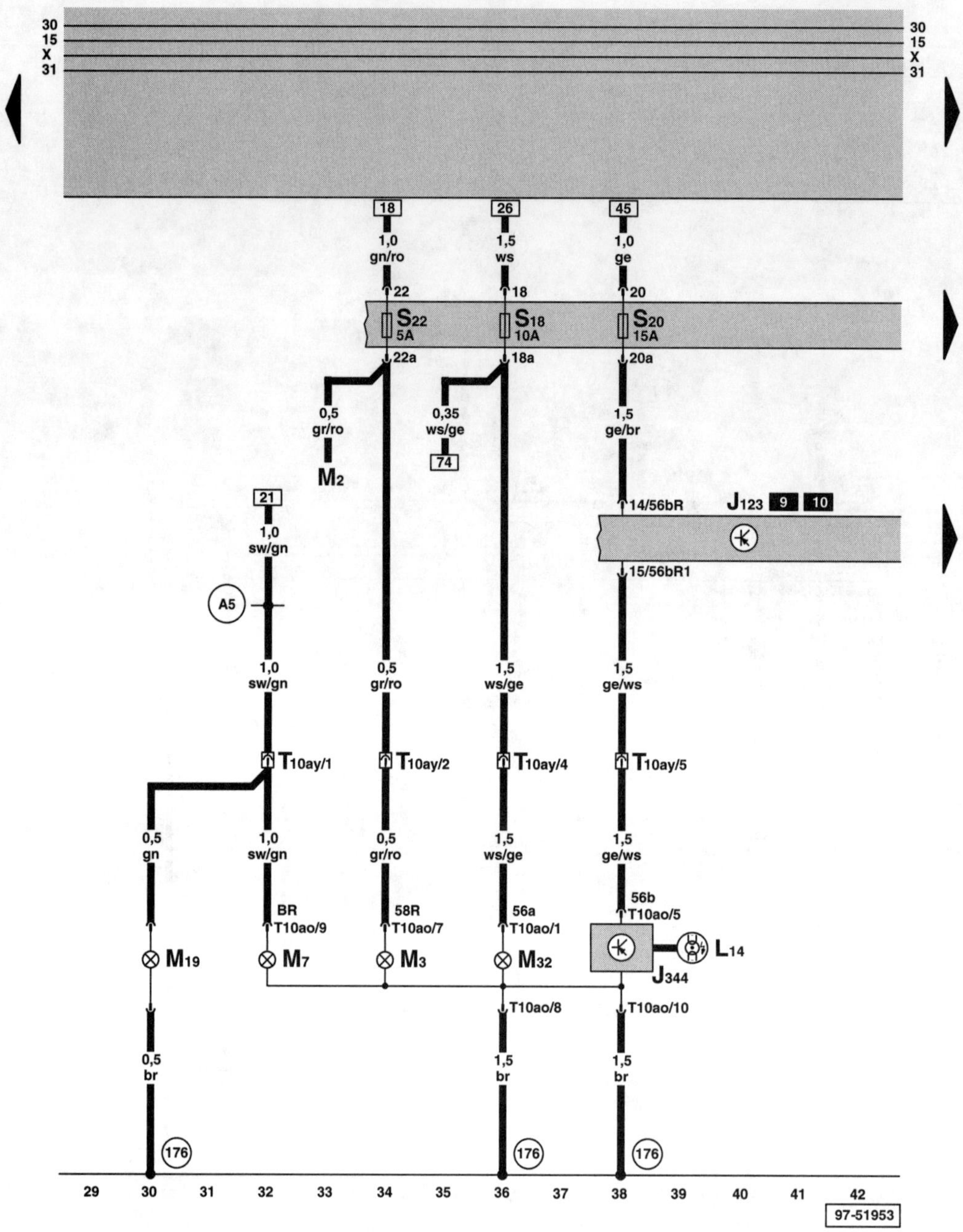

ws = white
sw = black
ro = red
br = brown
gn = green
bl = blue
gr = grey
li = violet
ge = yellow

- J123 - Lamp Control Module
- J344 - Control Module for High-intensity gas discharge lamp
- L14 - High intensity gas discharge lamp right
- M2 - Right Tail Light
- M3 - Right Parking Light
- M7 - Right Front Turn Signal Light
- M19 - Right, Side Turn Signal Light
- M32 - Right High Beam Headlight
- S18 - Fuse
- S20 - Fuse
- S22 - Fuse
- T10ao - Connector 10 pin, on headlight, right
- T10ay - Connector 10 pin, pink, connector station A pillar, right

- (176) - Ground connection, in right headlight wiring harness
- (A5) - Plus connection (right turn signal), in instrument panel wiring harness

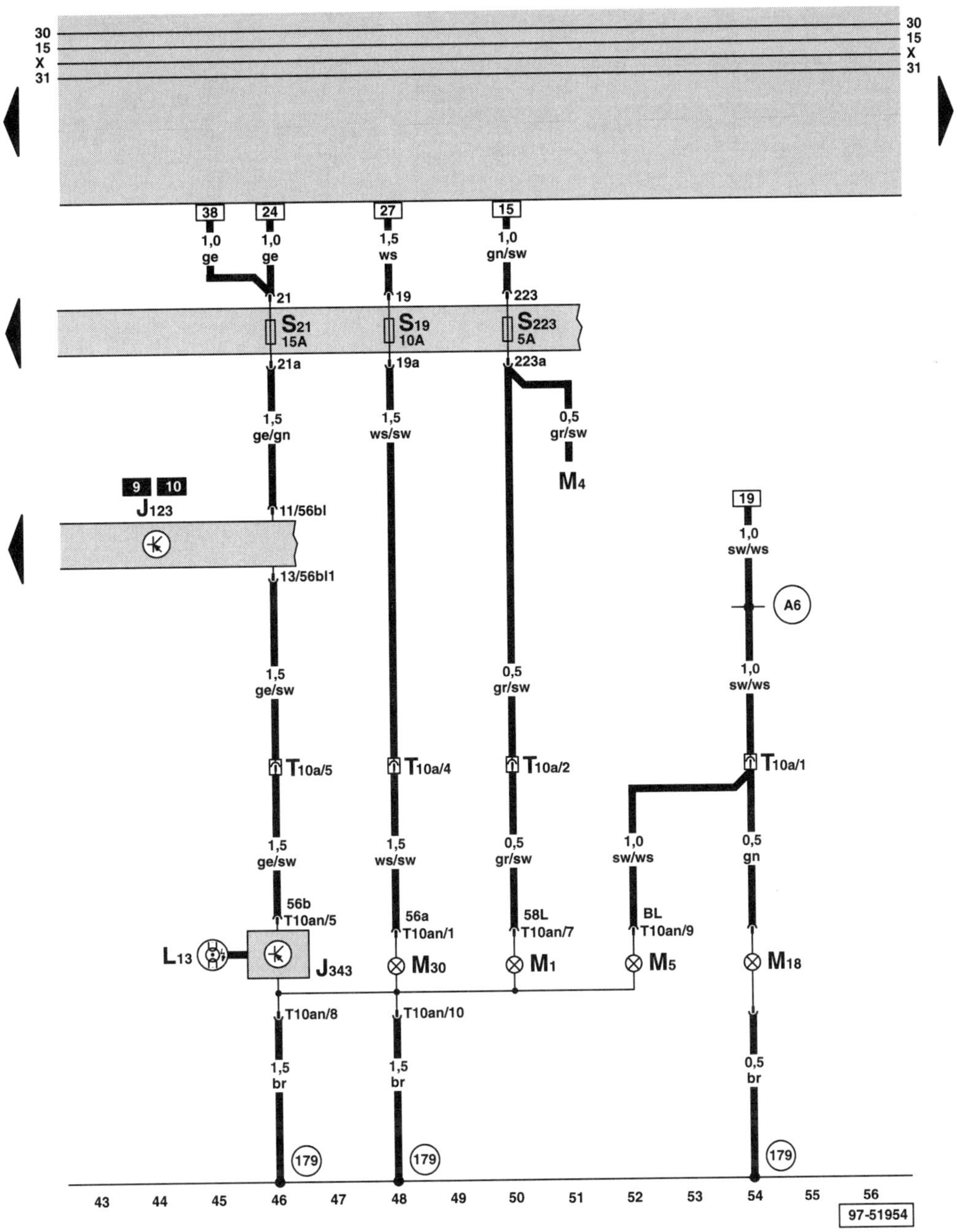

ws = white
sw = black
ro = red
br = brown
gn = green
bl = blue
gr = grey
li = violet
ge = yellow

J123 - Lamp Control Module
J343 - Control Module for High-intensity gas discharge lamp
L13 - High intensity gas discharge lamp left
M1 - Left Parking Light
M4 - Left Tail Light
M5 - Left Front Turn Signal Light
M18 - Left, Side Turn Signal Light
M30 - Left High Beam Headlight
S19 - Fuse
S21 - Fuse
S223 - Fuse 23 in fuse holder
T10a - Connector 10 pin, pink, connector station A pillar, left
T10an - Connector 10 pin, on headlight, left

(179) - Ground connection, in left headlight wiring harness

(A6) - Plus connection (left turn signal), in instrument panel wiring harness

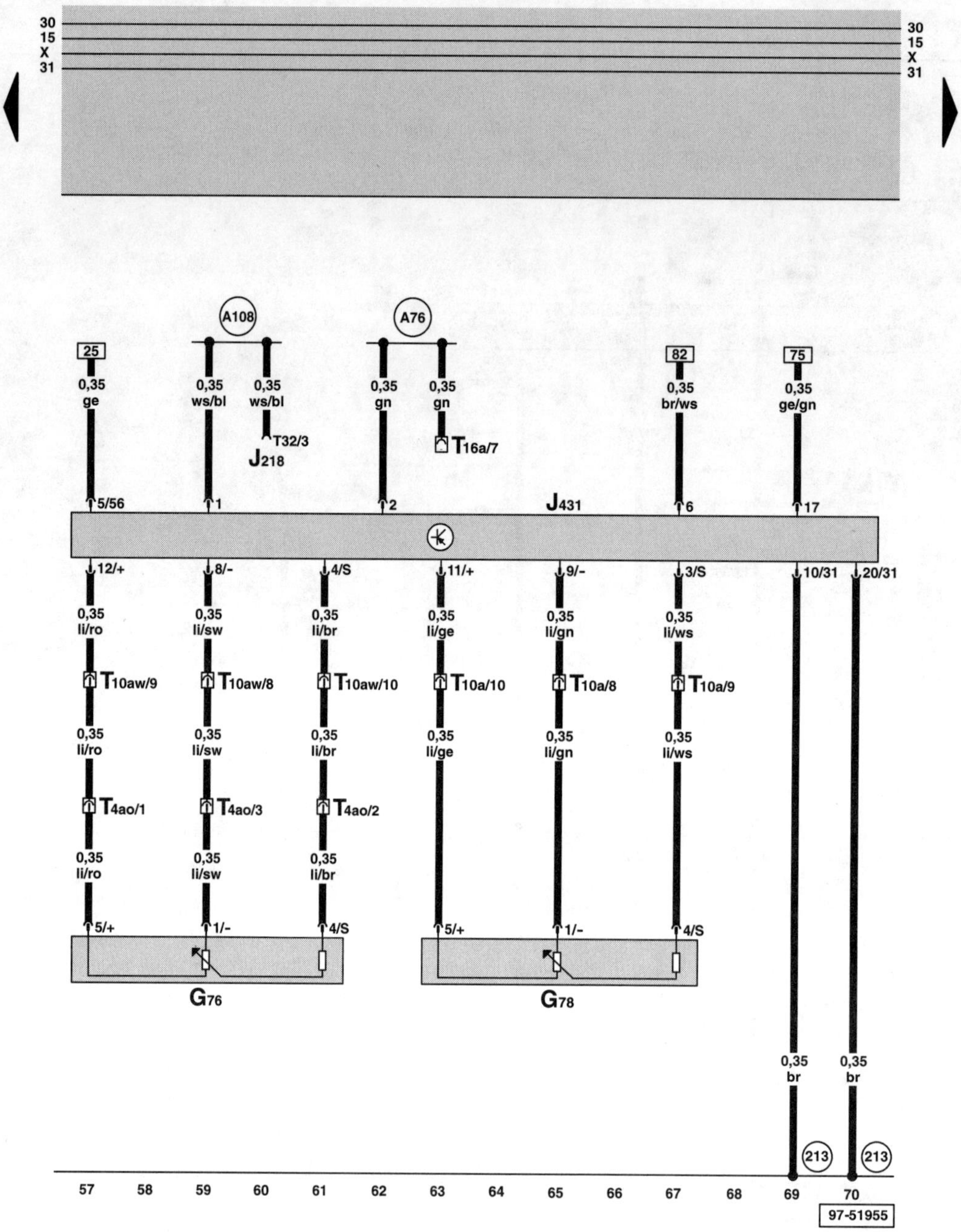

ws = white
sw = black
ro = red
br = brown
gn = green
bl = blue
gr = grey
li = violet
ge = yellow

G76 - Left Rear Level Control System Sensor
G78 - Left Front Level Control System Sensor
J218 - Instrument Cluster Combination Processor
J431 - Control module for Headlight Beam Adjusting
T4ao - Connector 4 pin, black, below rear seat, left
T10a - Connector 10 pin, pink, connector station A pillar, left
T10aw- Connector 10 pin, black, connector station A pillar, right
T16a - Connector 16 pin, Data Link Connector (DLC)
T32 - Connector 32 pin, blue, on instrument cluster

(213) - Ground connection, in instrument panel wiring harness
(A76) - Connector (K-diagnosis wire), in instrument panel wiring harness
(A108) - Connector (speed signal), in instrument panel wiring harness

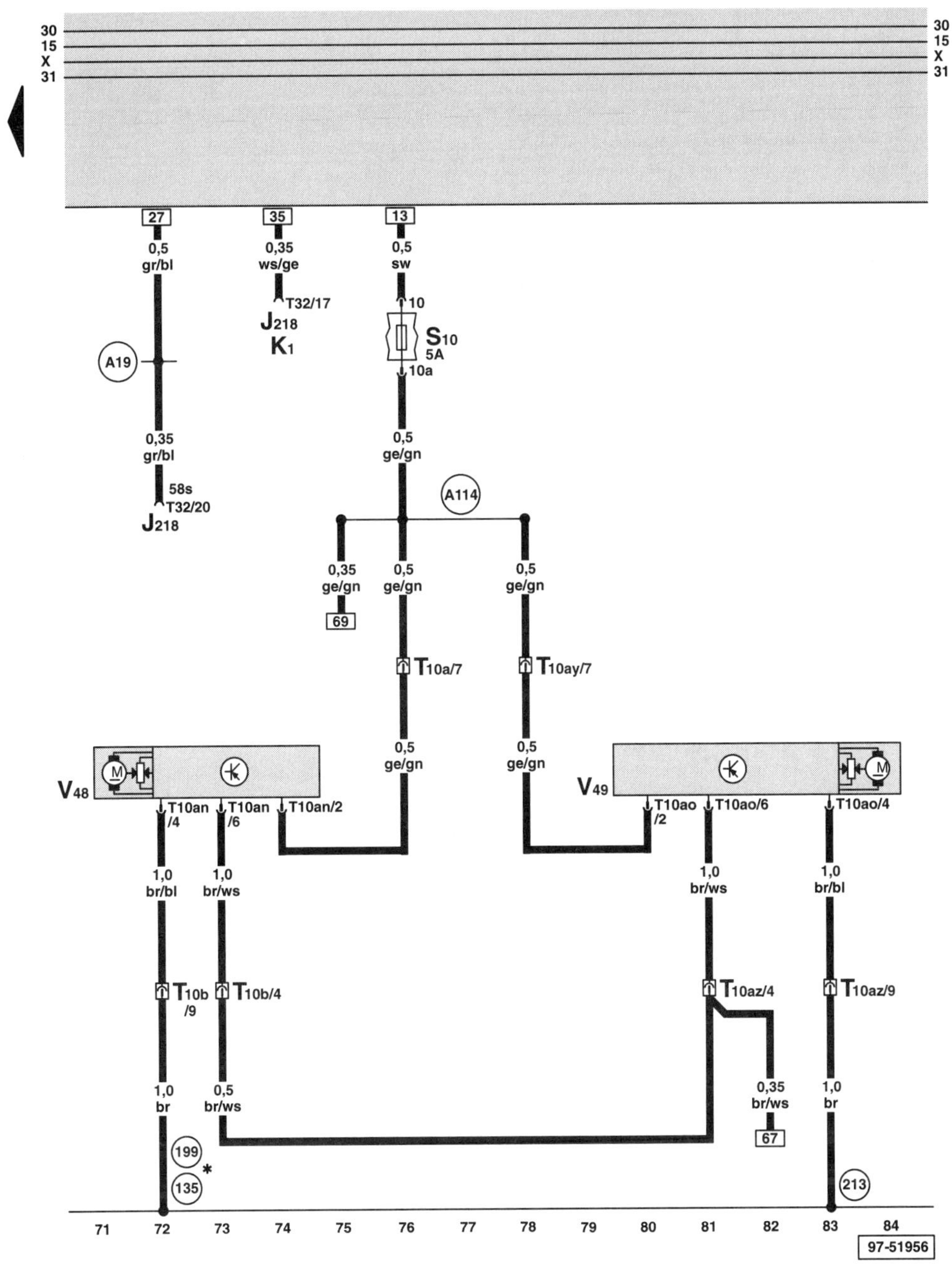

ws = white
sw = black
ro = red
br = brown
gn = green
bl = blue
gr = grey
li = violet
ge = yellow

J218 - Instrument Cluster Combination Processor
K1 - Headlight High Beam Indicator Light
S10 - Fuse
T10a - Connector 10 pin, pink, connector station A pillar, left
T10b - Connector 10 pin, red, connector station A pillar, left
T10an - Connector 10 pin, on headlight, left
T10ao - Connector 10 pin, on headlight, right
T10ay - Connector 10 pin, pink, connector station A pillar, right
T10az - Connector 10 pin, red, connector station A pillar, right
T32 - Connector 32 pin, blue, on instrument cluster
V48 - Left Headlight Beam Adjusting Motor
V49 - Right Headlight Beam Adjusting Motor

(135) - Ground connection -2-, in instrument panel wiring harness
(199) - Ground connection -3-, in instrument panel wiring harness
(213) - Ground connection, in instrument panel wiring harness
(A19) - Wire connection (58s), in instrument panel wiring harness
(A114) - Connection (headlight beam adjusting) in instrument panel wiring harness

* - Both are possible

THIS PAGE INTENTIONALLY LEFT BLANK

Headlight Washer

2000 m. y.

Fuse Panel

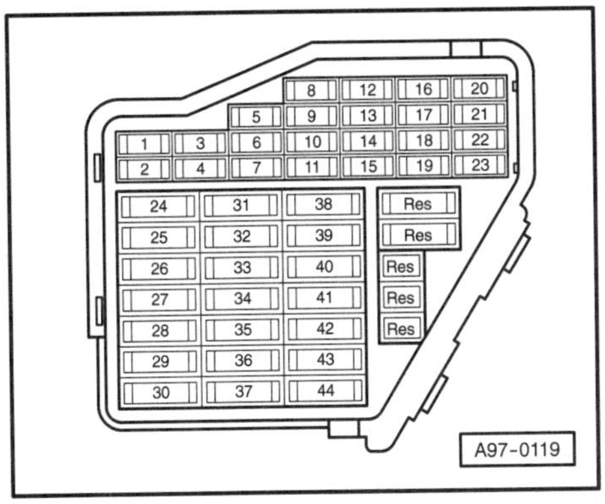

Fuse Colors:

30 A - Green
25 A - White
20 A - Yellow
15 A - Blue
10 A - Red
7,5 A - Brown
5 A - Beige

Starting with fuse position 23, fuses in the fuse holder are identified with 223 in the wiring diagram.

Micro Central Electric Panel

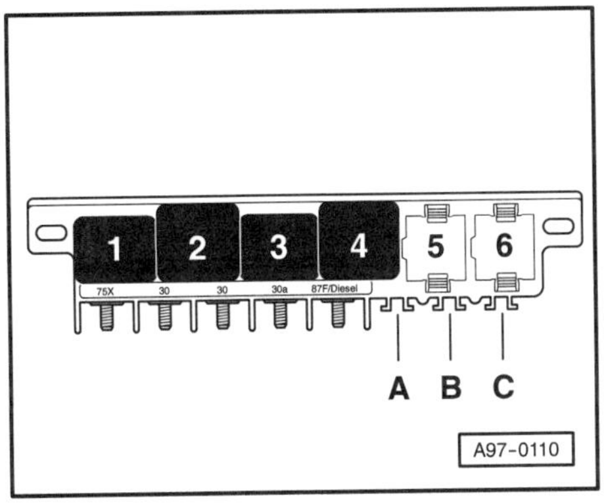

Relay Location:

5 -Wiper/Washer Intermittent Relay, J31

6 -Wiper/Washer Intermittent Relay, J31

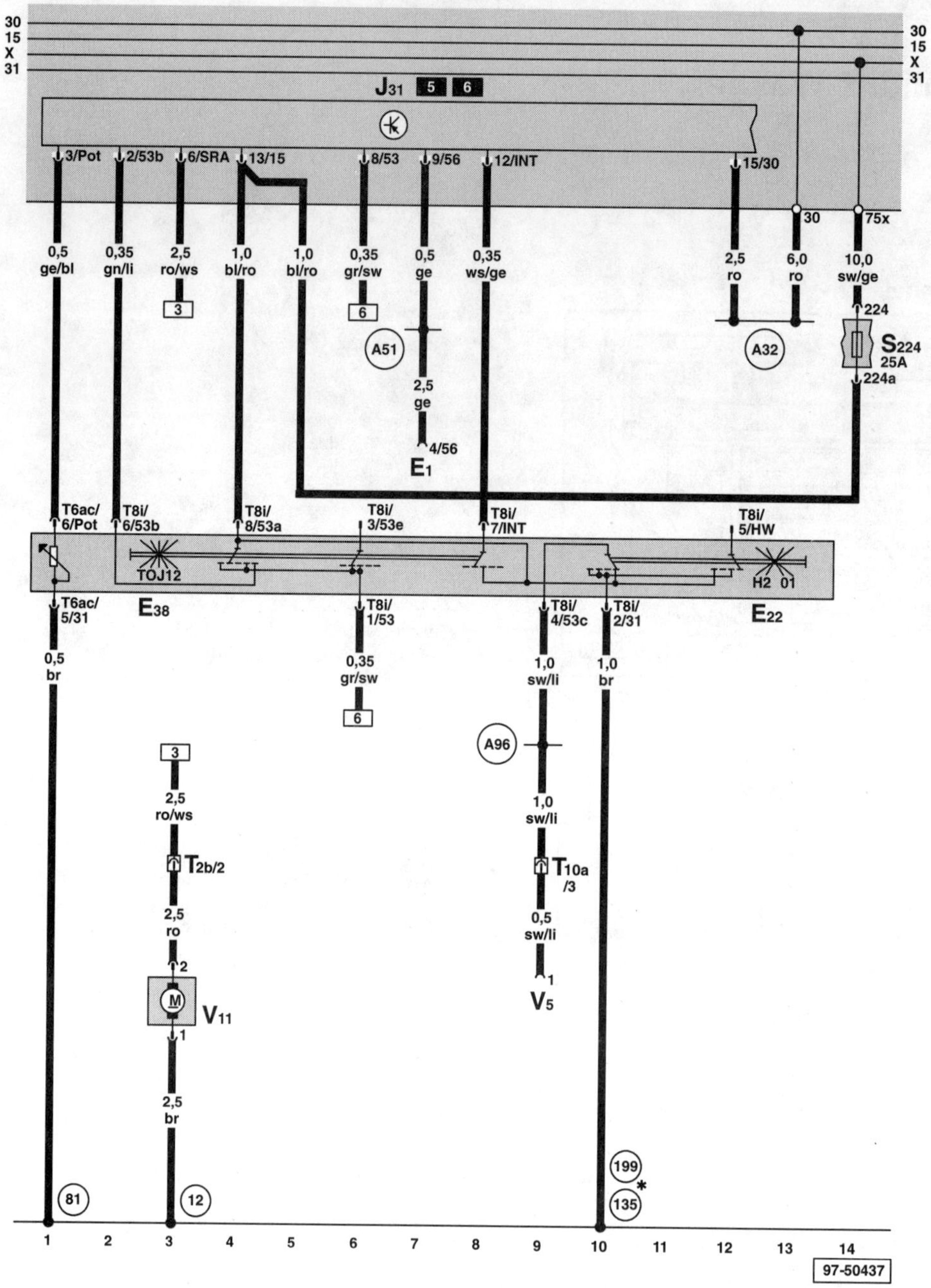

ws = white
sw = black
ro = red
br = brown
gn = green
bl = blue
gr = grey
li = violet
ge = yellow

E1 - Light switch
E22 - Windshield Wiper/Washer Switch
E38 - Windshield Wiper Intermittent Regulator
J31 - Wiper/Washer Intermittent Relay
S224 - Fuse 24 in fuse holder
T2b - Connector 2 pin, black, connector station A pillar, left
T6ac - Connector 6 pin, brown, on windshield wiper intermittent regulator
T8i - Connector 8 pin, on windshield wiper/washer switch
T10a - Connector 10 pin, pink, connector station A pillar, left
V5 - Windshield Washer Pump
V11 - Headlight Washer Pump

(12) - Ground connection, in engine compartment, left
(81) - Ground connection -1-, in instrument panel wiring harness
(135) - Ground connection -2-, in instrument panel wiring harness
(199) - Ground connection -3-, in instrument panel wiring harness
(A32) - Plus connection (30), in instrument panel wiring harness
(A51) - Wire connection (56), in instrument panel wiring harness
(A96) - Connector (53a), in instrument panel wiring harness

* - Both are possible

Power Seats and Mirror Adjustment with Memory, Mirror with and without fold-away Function

2000 m. y.

Fuse Panel

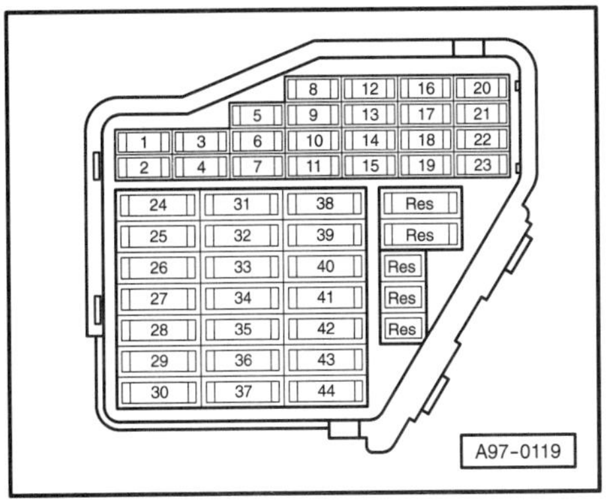

Fuse Colors:

30	A	-	Green
25	A	-	White
20	A	-	Yellow
15	A	-	Blue
10	A	-	Red
7,5	A	-	Brown
5	A	-	Beige

Starting with fuse position 23, fuses in the fuse holder are identified with 223 in the wiring diagram.

13 - Fold Relay Panel

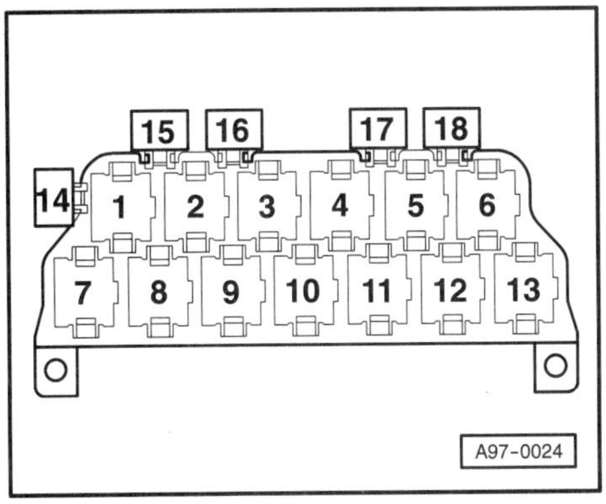

Relay Location:

11 -Control Module for Mirror fold-away Function, J351

12 -Control Module for Mirror fold-away Function, J351

8 - Fold Relay Panel

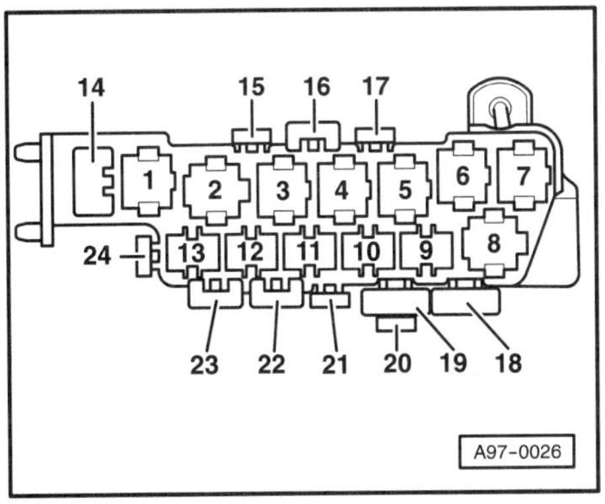

Fuse Location:

14 -Power Seat Circuit Breaker 1, S44

24 -Power Seat Circuit Breaker 2, S80

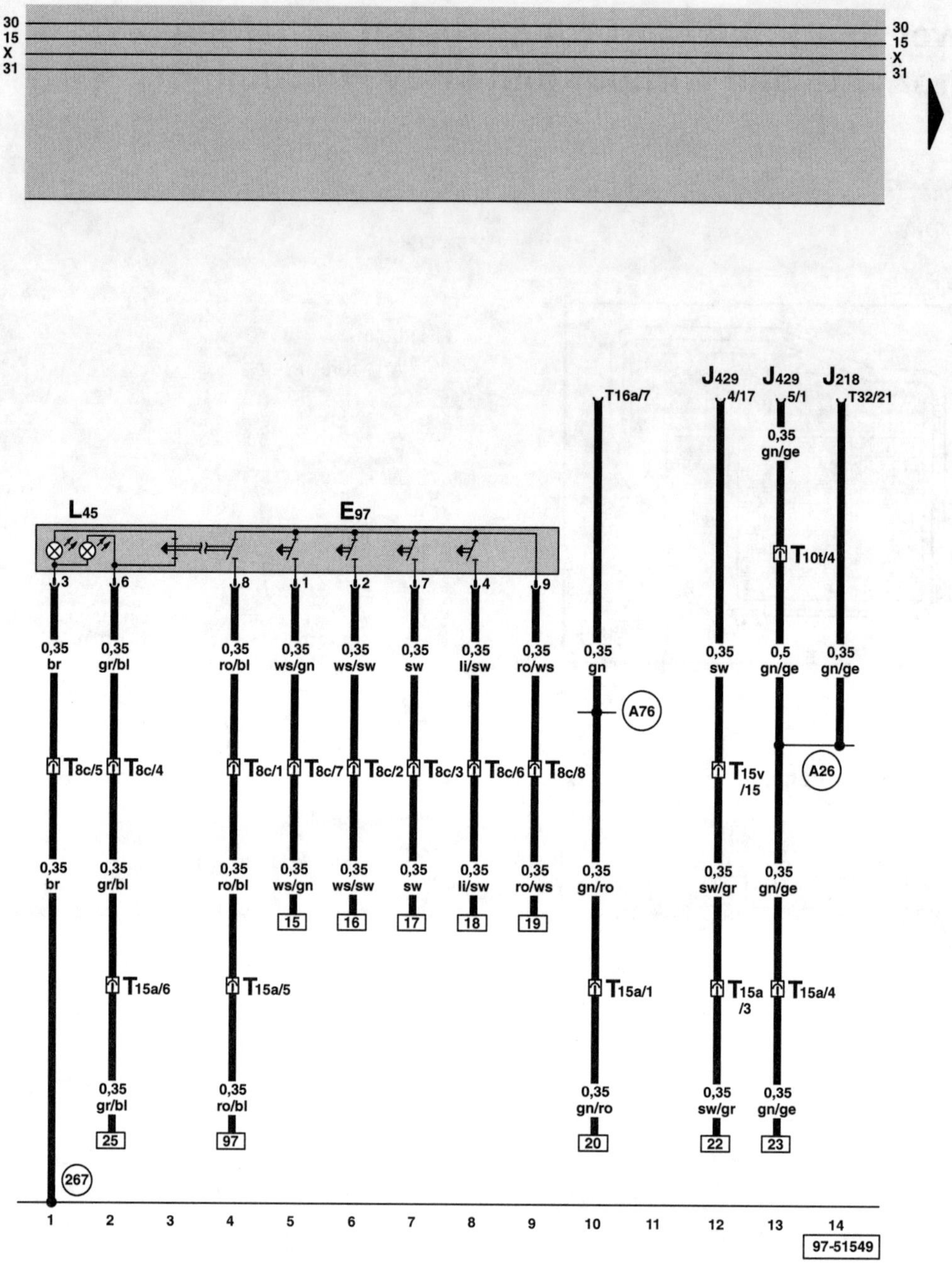

ws = white
sw = black
ro = red
br = brown
gn = green
bl = blue
gr = grey
li = violet
ge = yellow

E97 - Memory Program Switch
J218 - Instrument Cluster Combination Processor
J429 - Control module for central locking
L45 - Memory Operating Unit Light
T8c - Connector 8 pin, black, in driver's door
T10t - Connector 10 pin, orange, connector station A pillar, right
T15a - Connector 15 pin, black, connector station A pillar, left
T15v - Connector 15 pin, orange, connector station A pillar, right
T16a - Connector 16 pin, Data Link Connector (DLC)
T32 - Connector 32 pin, blue, on instrument cluster

(267) - Ground connector -2-, in wiring harness door cable - driver side
(A26) - Wire connection (driver's door contact switch), in instrument panel wiring harness
(A76) - Connector (K-diagnosis wire), in instrument panel wiring harness

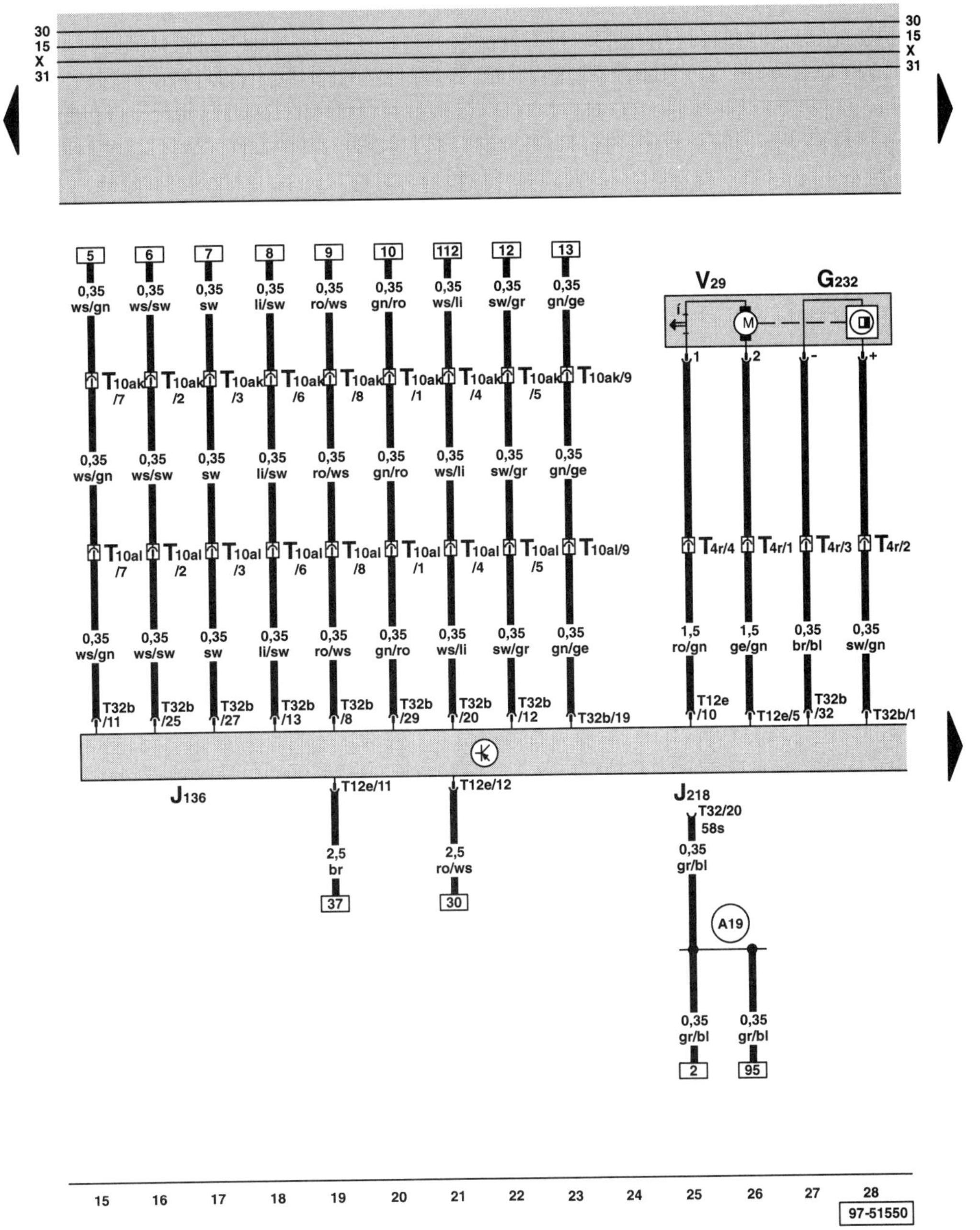

G232 - Sensor for slope adjustment, driver seat
J136 - Memory Seat Control Module
J218 - Instrument Cluster Combination Processor
T4r - Connector 4 pin, black, in driver's seat
T10ak - Connector 10 pin, green, connector station A pillar, left
T10al - Connector 10 pin, brown, below driver's seat
T12e - Connector 12 pin, black, in driver's seat
T32 - Connector 32 pin, blue, on instrument cluster
T32b - Connector 32 pin, blue, in driver's seat
V29 - Driver's Seat Front Height Adjusting Motor

(A19) - Wire connection (58s), in instrument panel wiring harness

ws = white
sw = black
ro = red
br = brown
gn = green
bl = blue
gr = grey
li = violet
ge = yellow

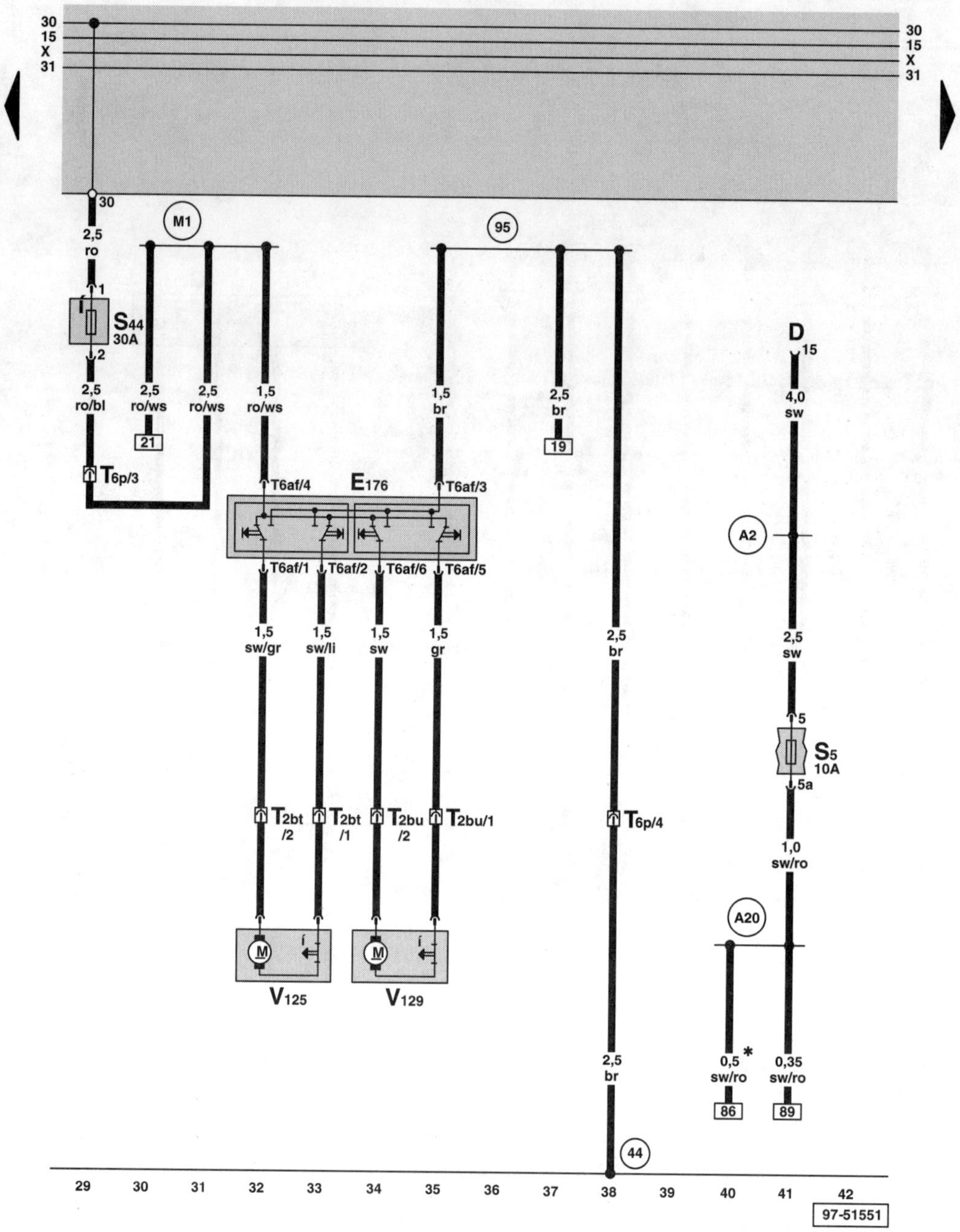

ws = white
sw = black
ro = red
br = brown
gn = green
bl = blue
gr = grey
li = violet
ge = yellow

D - Ignition/Starter Switch
E176 - Driver's Seat Lumbar Support Adjustment Switch
S5 - Fuse
S44 - Power Seat Circuit Breaker 1
T2bt - Connector 2 pin, black, in driver's seat
T2bu - Connector 2 pin, black, in driver's seat
T6p - Connector 6 pin, red, below driver's seat
T6af - Connector 6 pin, black, in driver's seat
V125 - Driver's Seat Lumbar Support Curvature Adjustment Motor
V129 - Driver's Seat Lumbar Support Height Adjustment Motor

(44) - Ground connection, on left A-pillar, lower part
(95) - Ground connection -1-, in power seat wiring harness
(A2) - Plus connection (15), in instrument panel wiring harness
(A20) - Wire connection (15a), in instrument panel wiring harness
(M1) - Plus connection -1- (30), in power seats wiring harness
* - Mirror with fold-away function

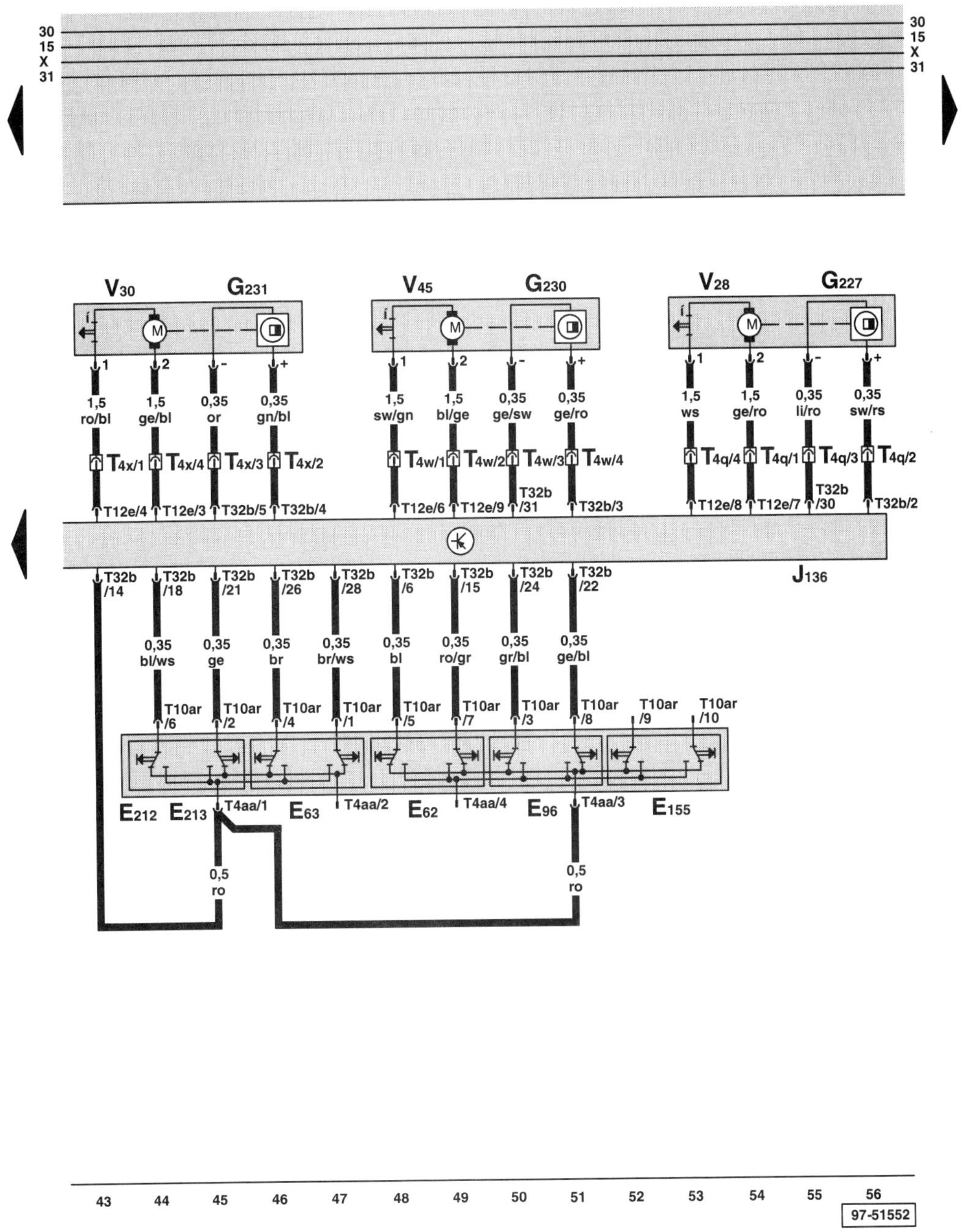

ws = white
sw = black
ro = red
br = brown
gn = green
bl = blue
gr = grey
li = violet
ge = yellow
or = orange
rs = pink

E62 - Driver's Seat Front Height Adjusting Switch
E63 - Driver's Seat Rear Height Adjusting Switch
E96 - Driver's Backrest Adjustment Switch
E155 - Driver's Headrest Adjustment Switch
E212 - Button for length adjustment, forward, driver seat
E213 - Button for length adjustment, backward, driver seat
G227 - Sensor for fore and aft adjustment, driver seat
G230 - Sensor for seat back adjustment, driver seat
G231 - Sensor for height adjustment, driver seat
J136 - Memory Seat Control Module
T4q - Connector 4 pin, black, in driver's seat
T4w - Connector 4 pin, black, in driver's seat
T4x - Connector 4 pin, black, in driver's seat
T4aa - Connector 4 pin, black, in driver's seat
T10ar - Connector 10 pin, black, in driver's seat
T12e - Connector 12 pin, black, in driver's seat
T32b - Connector 32 pin, blue, in driver's seat

V28 - Driver's Seat Fore/Aft Adjusting Motor
V30 - Driver's Seat Rear Height Adjusting Motor
V45 - Driver's Backrest Adjusting Motor

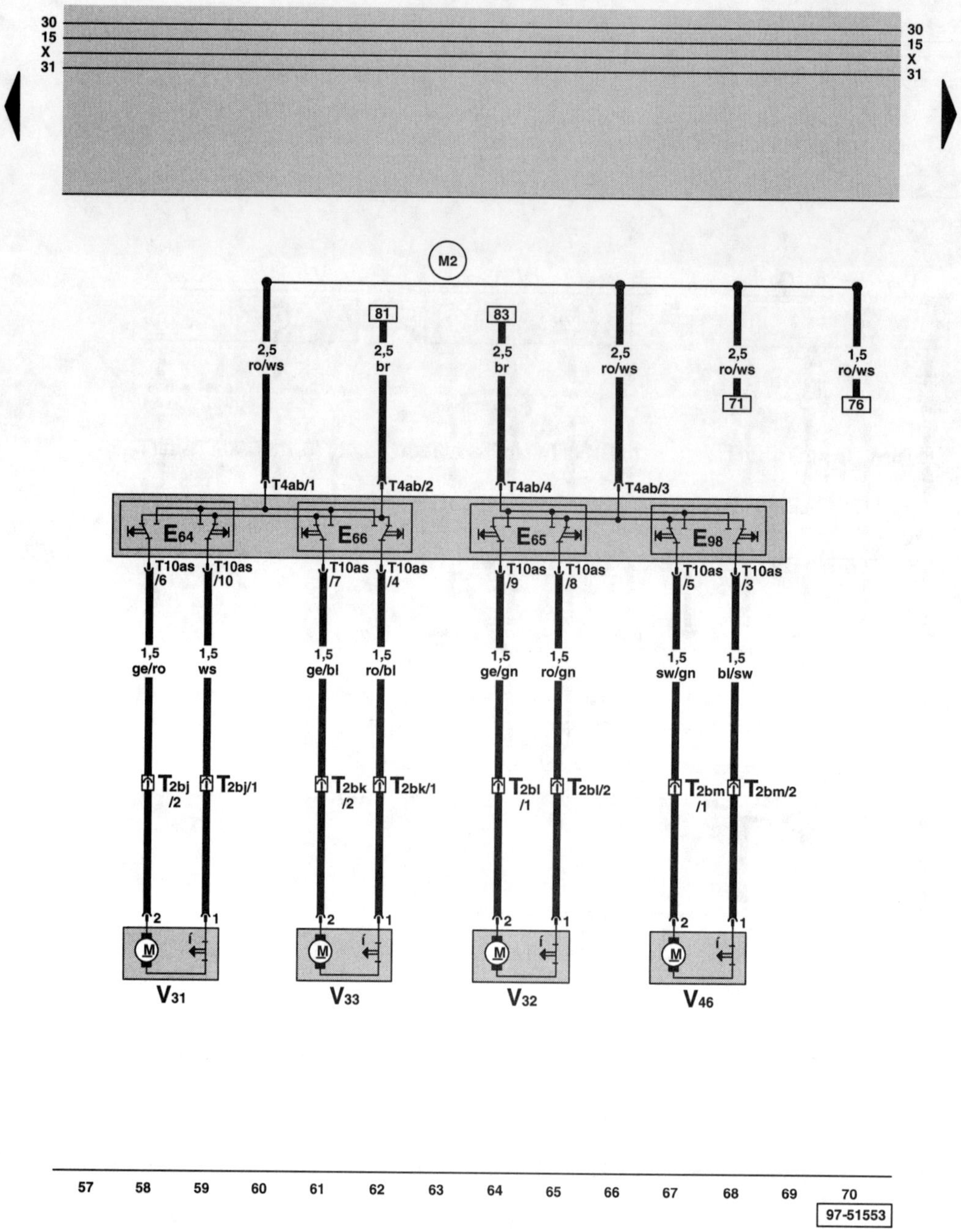

ws = white
sw = black
ro = red
br = brown
gn = green
bl = blue
gr = grey
li = violet
ge = yellow

E64 - Passenger's Seat Fore/Aft Adjusting Switch
E65 - Passenger's Seat Front Height Adjusting Switch
E66 - Passenger's Seat Rear Height Adjusting Switch
E98 - Passenger Backrest Adjustment Switch
T2bj - Connector 2 pin, black, in passenger's seat
T2bk - Connector 2 pin, black, in passenger's seat
T2bl - Connector 2 pin, black, in passenger's seat
T2bm - Connector 2 pin, black, in passenger's seat
T4ab - Connector 4 pin, black, in passenger's seat
T10as - Connector 10 pin, black, in passenger's seat
V31 - Passenger's Seat Fore/Aft Adjusting Motor
V32 - Passenger's Seat Front Height Adjusting Motor
V33 - Passenger's Seat Rear Height Adjusting Motor
V46 - Passenger's Backrest Adjusting Motor

(M2) - Plus connection -2- (30), in power seats wiring harness

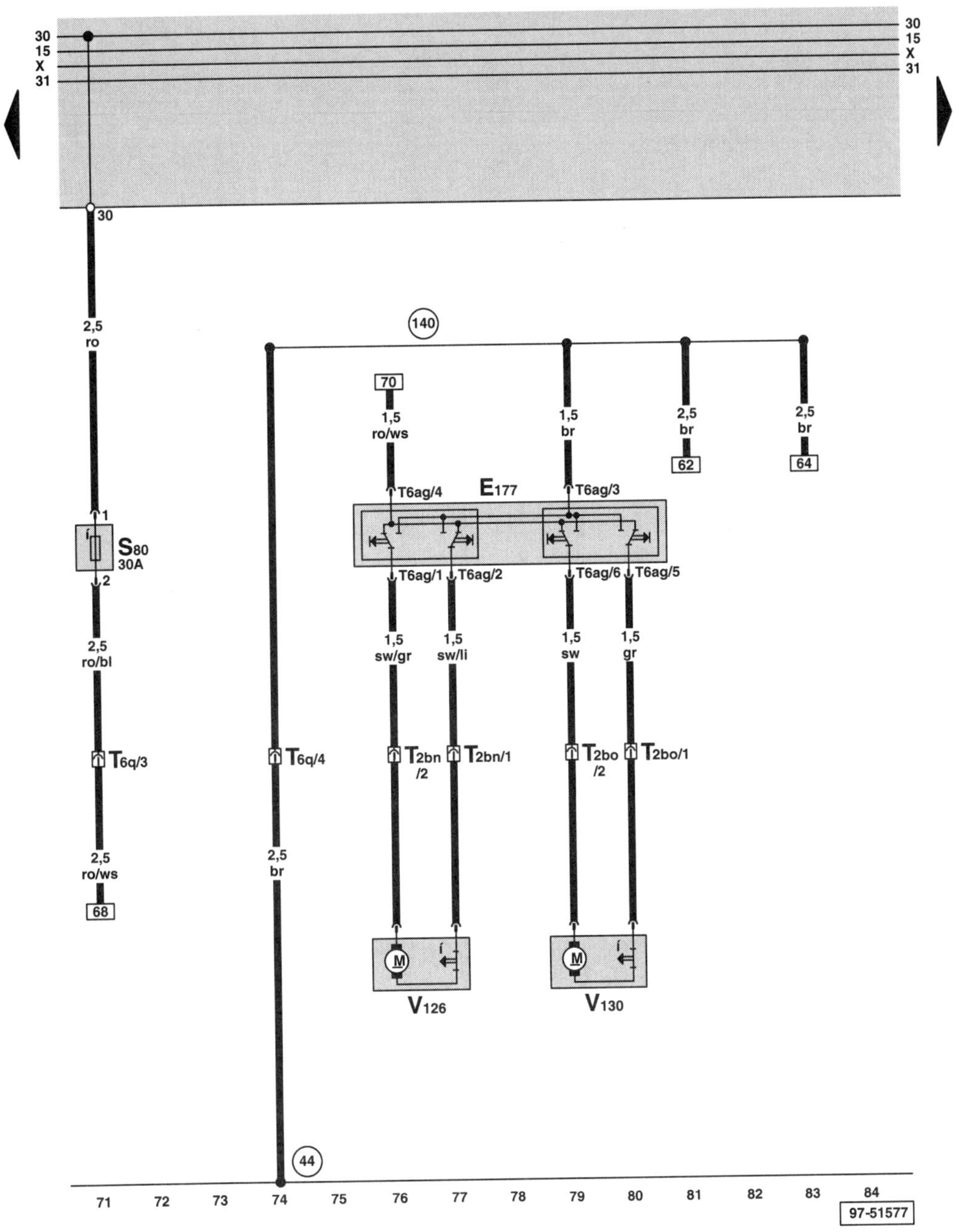

ws = white
sw = black
ro = red
br = brown
gn = green
bl = blue
gr = grey
li = violet
ge = yellow

E177 - Front Passenger Seat Lumbar Support Adjustment Switch
S80 - Power Seat Circuit Breaker 2
T2bn - Connector 2 pin, black, in passenger's seat
T2bo - Connector 2 pin, black, in passenger's seat
T6q - Connector 6 pin, red, below passenger's seat
T6ag - Connector 6 pin, black, in passenger's seat
V126 - Front Passenger Seat Lumbar Support Curvature Adjustment Motor
V130 - Front Passenger Seat Lumbar Support Height Adjustment Motor

(44) - Ground connection, on left A-pillar, lower part
(140) - Ground connection -2-, in power seat wiring harness

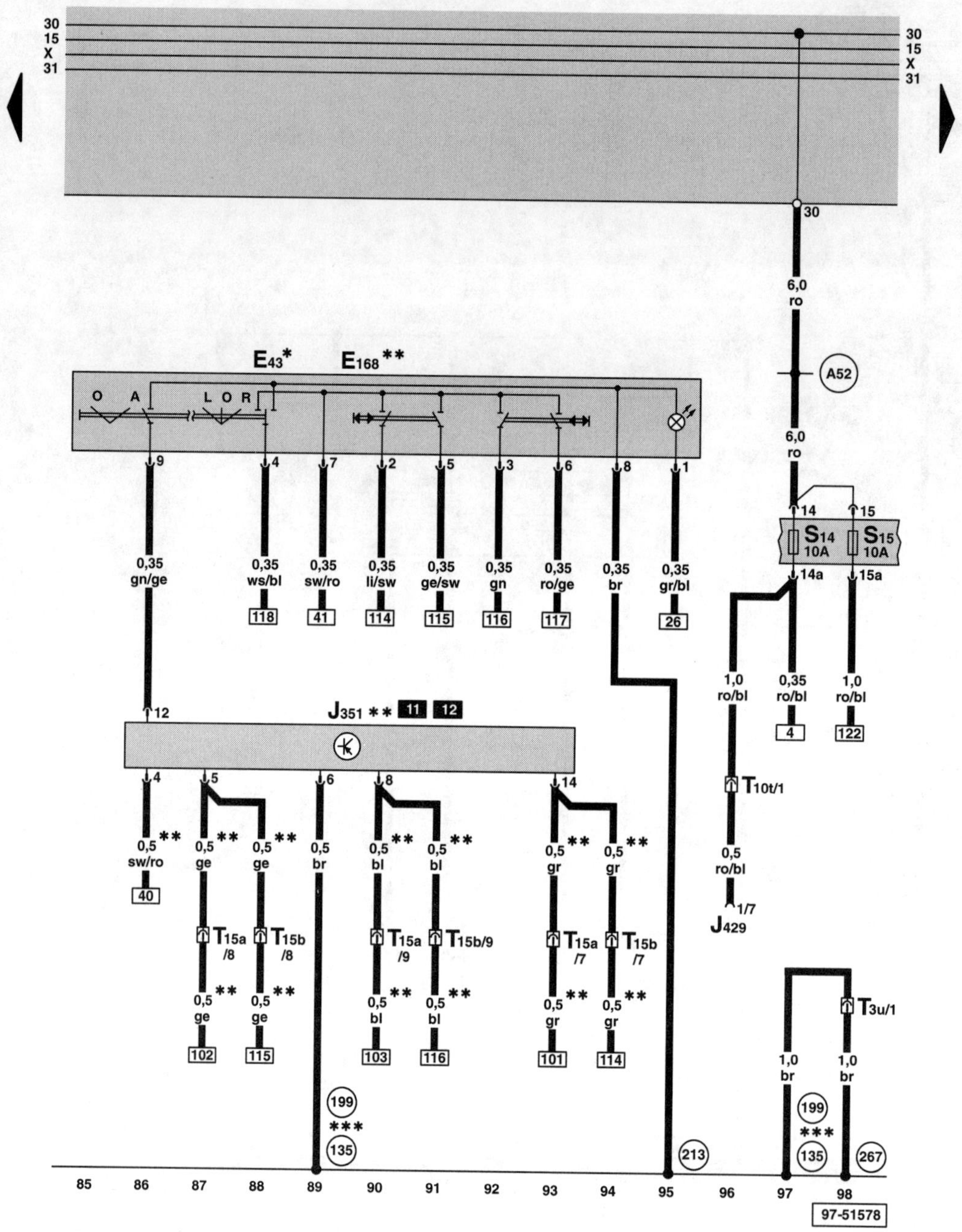

ws = white
sw = black
ro = red
br = brown
gn = green
bl = blue
gr = grey
li = violet
ge = yellow

E43 - Mirror Adjustment Switch
E168 - Switch for mirror adjustment with fold-away Function
J351 - Control Module for Mirror fold-away Function
J429 - Control module for central locking
S14 - Fuse
S15 - Fuse
T3u - Connector 3 pin, black, connector station A pillar, left
T10t - Connector 10 pin, orange, connector station A pillar, right
T15a - Connector 15 pin, black, connector station A pillar, left
T15b - Connector 15 pin, black, connector station A pillar, right

(135) - Ground connection -2-, in instrument panel wiring harness
(199) - Ground connection -3-, in instrument panel wiring harness
(213) - Ground connection, in instrument panel wiring harness
(267) - Ground connector -2-, in wiring harness door cable - driver side
(A52) - Plus connection -2- (30), in instrument panel wiring harness

* - Mirror without fold-away function
** - Mirror with fold-away function
*** - Both are possible

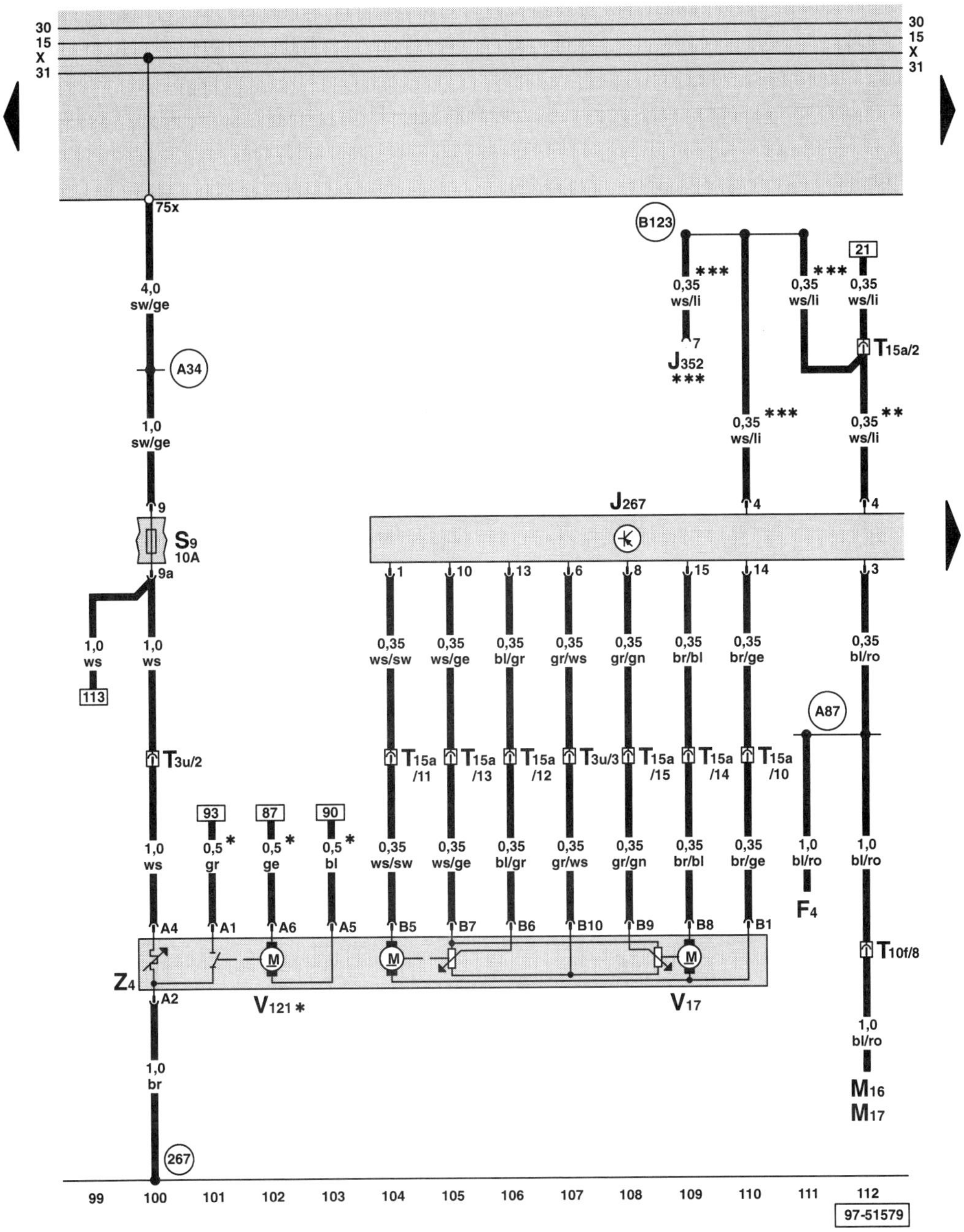

ws = white
sw = black
ro = red
br = brown
gn = green
bl = blue
gr = grey
li = violet
ge = yellow

F4 - Back-Up Light Switch
J267 - Mirror Memory Control Module
J352 - Control Module for Steering column-belt height adjustment
M16 - Left Back-Up Light
M17 - Right Back-Up Light
S9 - Fuse
T3u - Connector 3 pin, black, connector station A pillar, left
T10f - Connector 10 pin, brown, connector station A pillar, left
T15a - Connector 15 pin, black, connector station A pillar, left
V17 - Driver's Side Mirror Adjustment Motor
V121 - Motor for fold-in Mirror driver's side
Z4 - Driver's Side Heated Mirror

(267) - Ground connector -2-, in wiring harness door cable - driver side
(A34) - Wire connection (75x), in instrument panel wiring harness
(A87) - Connector (RF), in instrument panel wiring harness
(B123) - Connection (75a heated seat, mirror switch seat memory), i n wiring harness interior

* - Mirror with fold-away function
** - Vehicles without Power Steering Column Adjustment
*** - Vehicles with Power Steering Column Adjustment

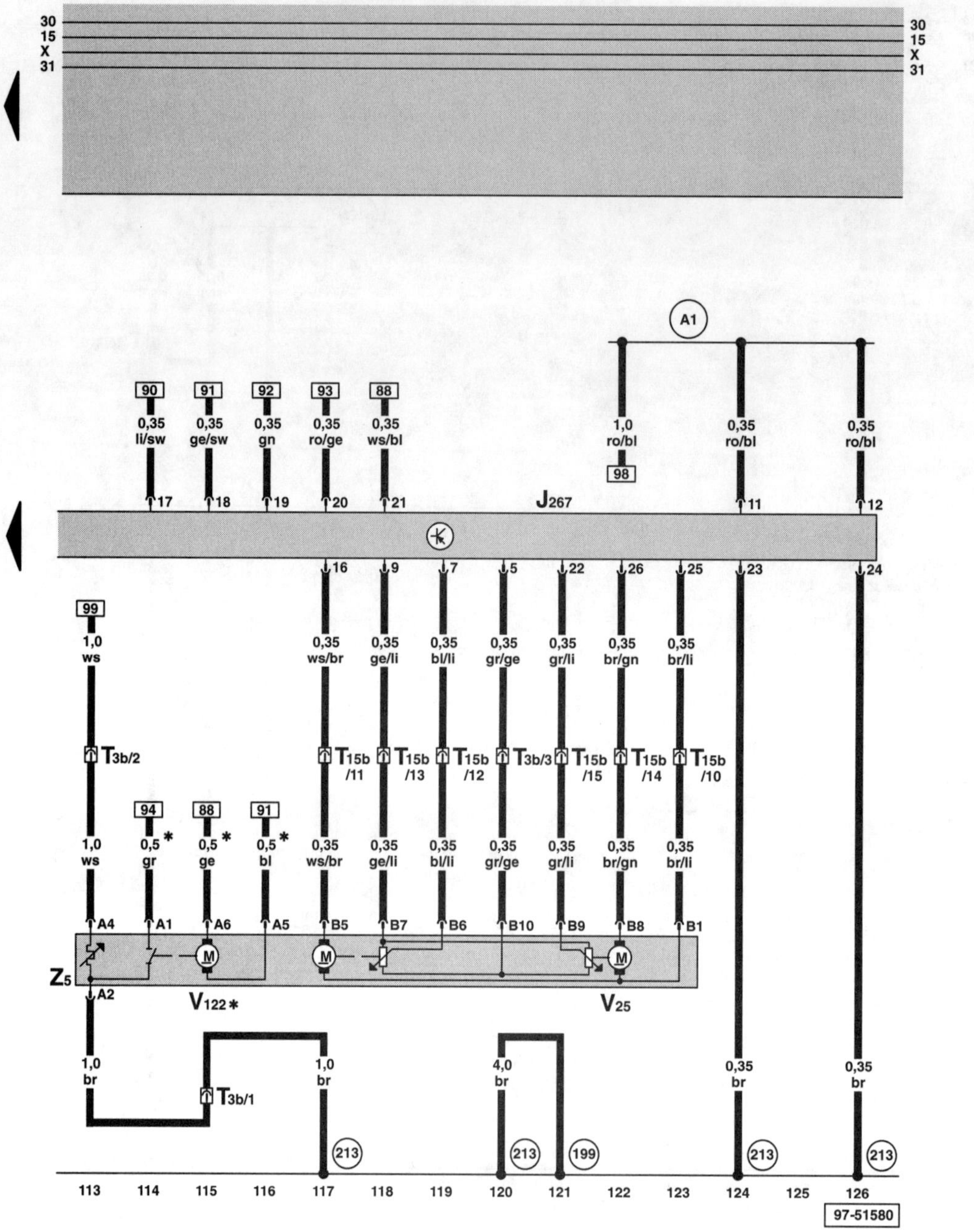

ws = white
sw = black
ro = red
br = brown
gn = green
bl = blue
gr = grey
li = violet
ge = yellow

J267 - Mirror Memory Control Module
T3b - Connector 3 pin, black, connector station A pillar, right
T15b - Connector 15 pin, black, connector station A pillar, right
V25 - Passenger's Side Mirror Adjustment Motor
V122 - Motor for fold-in Mirror passenger side
Z5 - Passenger's Side Heated Mirror

(199) - Ground connection -3-, in instrument panel wiring harness
(213) - Ground connection, in instrument panel wiring harness
(A1) - Plus connection (30a), in instrument panel wiring harness

* - Mirror with fold-away function

Power Seat and Mirror Adjustment, Mirror with and without fold-away Function, Heated Mirror

2000 m. y.

Fuse Panel

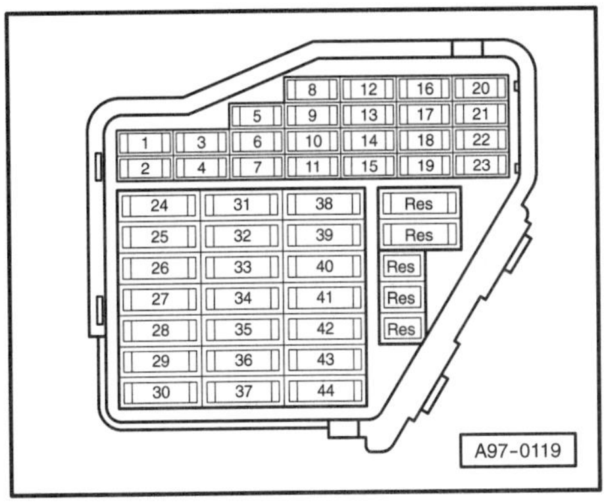

Fuse Colors:

30 A - Green
25 A - White
20 A - Yellow
15 A - Blue
10 A - Red
7,5 A - Brown
5 A - Beige

Starting with fuse position 23, fuses in the fuse holder are identified with 223 in the wiring diagram.

13 - Fold Relay Panel

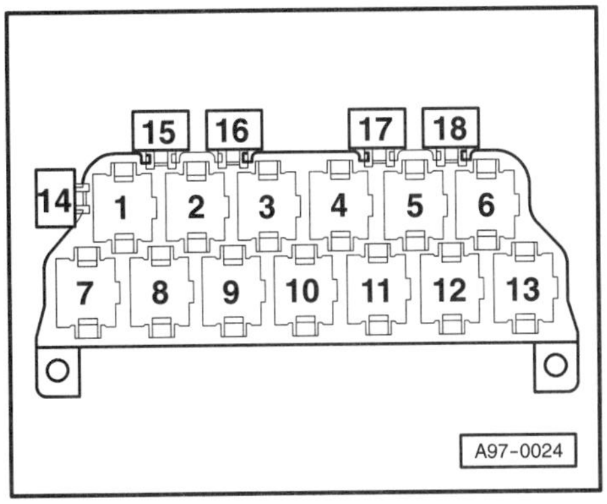

Relay Location:

11 -Control Module for Mirror fold-away Function, J351

12 -Control Module for Mirror fold-away Function, J351

8 - Fold Relay Panel

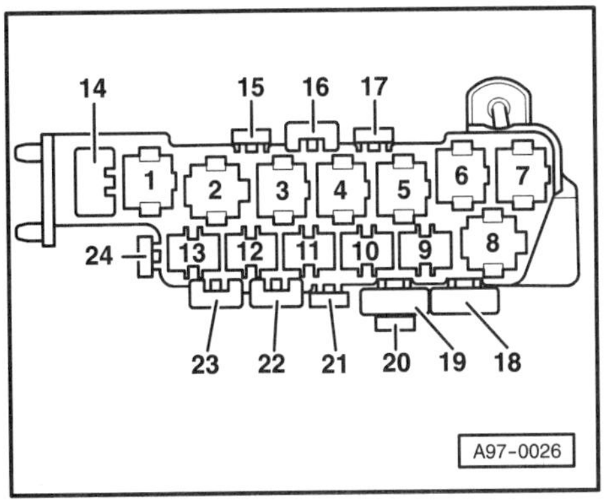

Fuse Location:

14 -Power Seat Circuit Breaker 1, S44

24 -Power Seat Circuit Breaker 2, S80

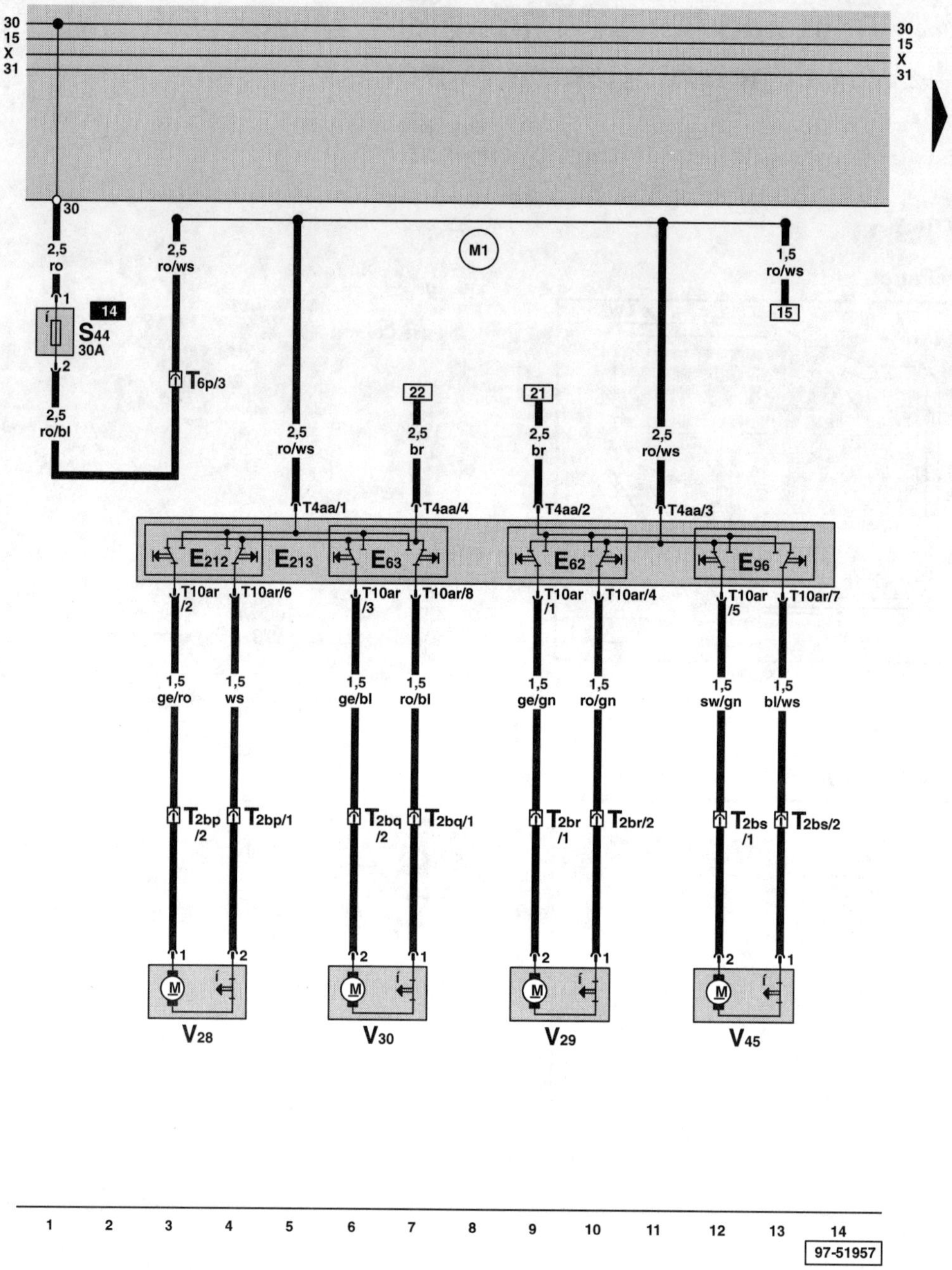

ws = white
sw = black
ro = red
br = brown
gn = green
bl = blue
gr = grey
li = violet
ge = yellow

E62 - Driver's Seat Front Height Adjusting Switch
E63 - Driver's Seat Rear Height Adjusting Switch
E96 - Driver's Backrest Adjustment Switch
E212 - Button for length adjustment, forward, driver seat
E213 - Button for length adjustment, backward, driver seat
S44 - Power Seat Circuit Breaker 1
T2bp - Connector 2 pin, black, in driver's seat
T2bq - Connector 2 pin, black, in driver's seat
T2br - Connector 2 pin, black, in driver's seat
T2bs - Connector 2 pin, black, in driver's seat
T4aa - Connector 4 pin, black, in driver's seat
T6p - Connector 6 pin, red, below driver's seat
T10ar - Connector 10 pin, black, in driver's seat
V28 - Driver's Seat Fore/Aft Adjusting Motor
V29 - Driver's Seat Front Height Adjusting Motor
V30 - Driver's Seat Rear Height Adjusting Motor
V45 - Driver's Backrest Adjusting Motor

(M1) - Plus connection -1- (30), in power seats wiring harness

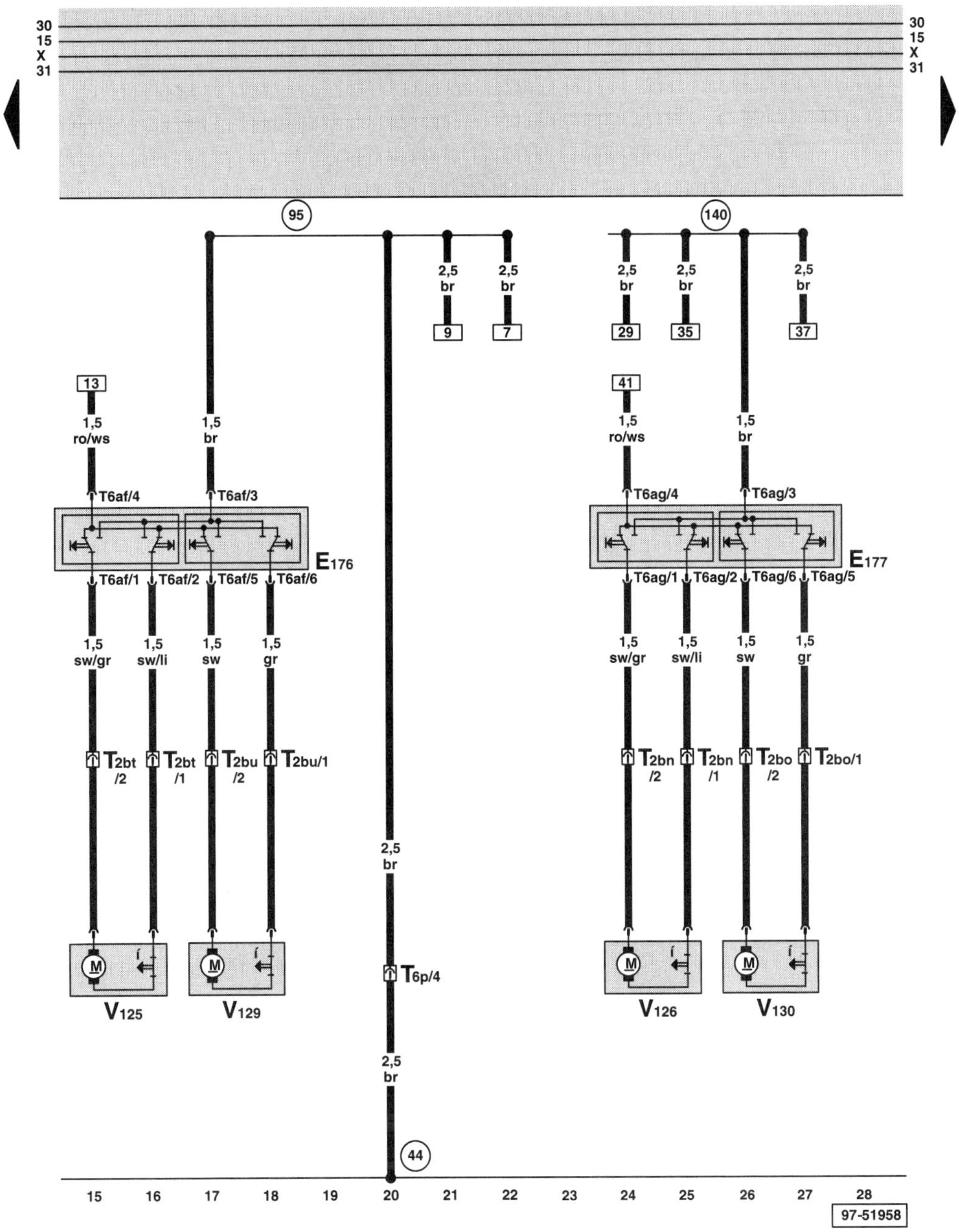

ws = white
sw = black
ro = red
br = brown
gn = green
bl = blue
gr = grey
li = violet
ge = yellow

E176 - Driver's Seat Lumbar Support Adjustment Switch
E177 - Front Passenger Seat Lumbar Support Adjustment Switch
T2bn - Connector 2 pin, black, in passenger's seat
T2bo - Connector 2 pin, black, in passenger's seat
T2bt - Connector 2 pin, black, in driver's seat
T2bu - Connector 2 pin, black, in driver's seat
T6p - Connector 6 pin, red, below driver's seat
T6af - Connector 6 pin, black, in driver's seat
T6ag - Connector 6 pin, black, in passenger's seat
V125 - Driver's Seat Lumbar Support Curvature Adjustment Motor
V126 - Front Passenger Seat Lumbar Support Curvature Adjustment Motor
V129 - Driver's Seat Lumbar Support Height Adjustment Motor
V130 - Front Passenger Seat Lumbar Support Height Adjustment Motor

(44) - Ground connection, on left A-pillar, lower part
(95) - Ground connection -1-, in power seat wiring harness
(140) - Ground connection -2-, in power seat wiring harness

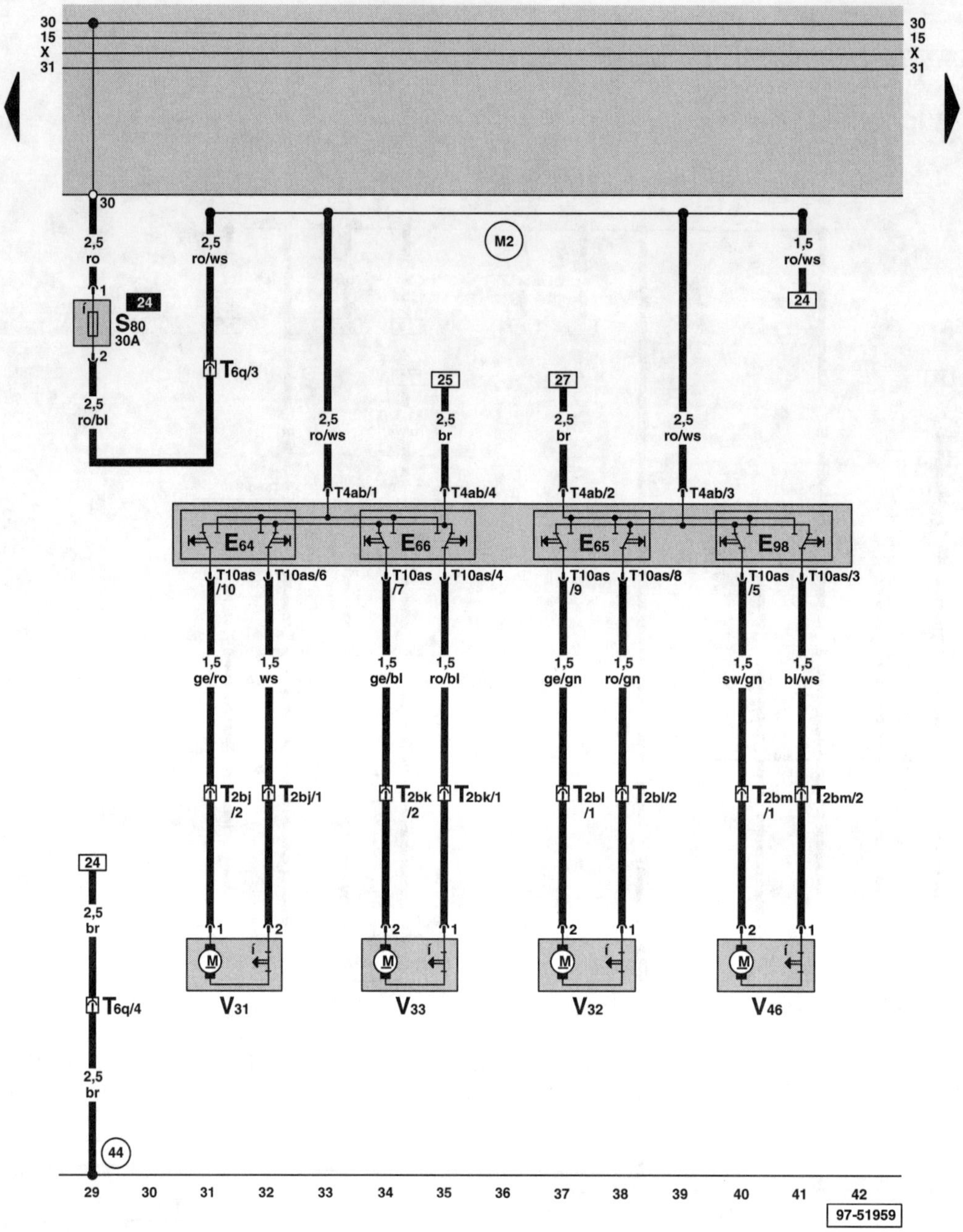

ws = white
sw = black
ro = red
br = brown
gn = green
bl = blue
gr = grey
li = violet
ge = yellow

E64 - Passenger's Seat Fore/Aft Adjusting Switch
E65 - Passenger's Seat Front Height Adjusting Switch
E66 - Passenger's Seat Rear Height Adjusting Switch
E98 - Passenger Backrest Adjustment Switch
S80 - Power Seat Circuit Breaker 2
T2bj - Connector 2 pin, black, in passenger's seat
T2bk - Connector 2 pin, black, in passenger's seat
T2bl - Connector 2 pin, black, in passenger's seat
T2bm - Connector 2 pin, black, in passenger's seat
T4ab - Connector 4 pin, black, in passenger's seat
T6q - Connector 6 pin, red, below passenger's seat
T10as - Connector 10 pin, black, in passenger's seat
V31 - Passenger's Seat Fore/Aft Adjusting Motor
V32 - Passenger's Seat Front Height Adjusting Motor
V33 - Passenger's Seat Rear Height Adjusting Motor
V46 - Passenger's Backrest Adjusting Motor

(44) - Ground connection, on left A-pillar, lower part

(M2) - Plus connection -2- (30), in power seats wiring harness

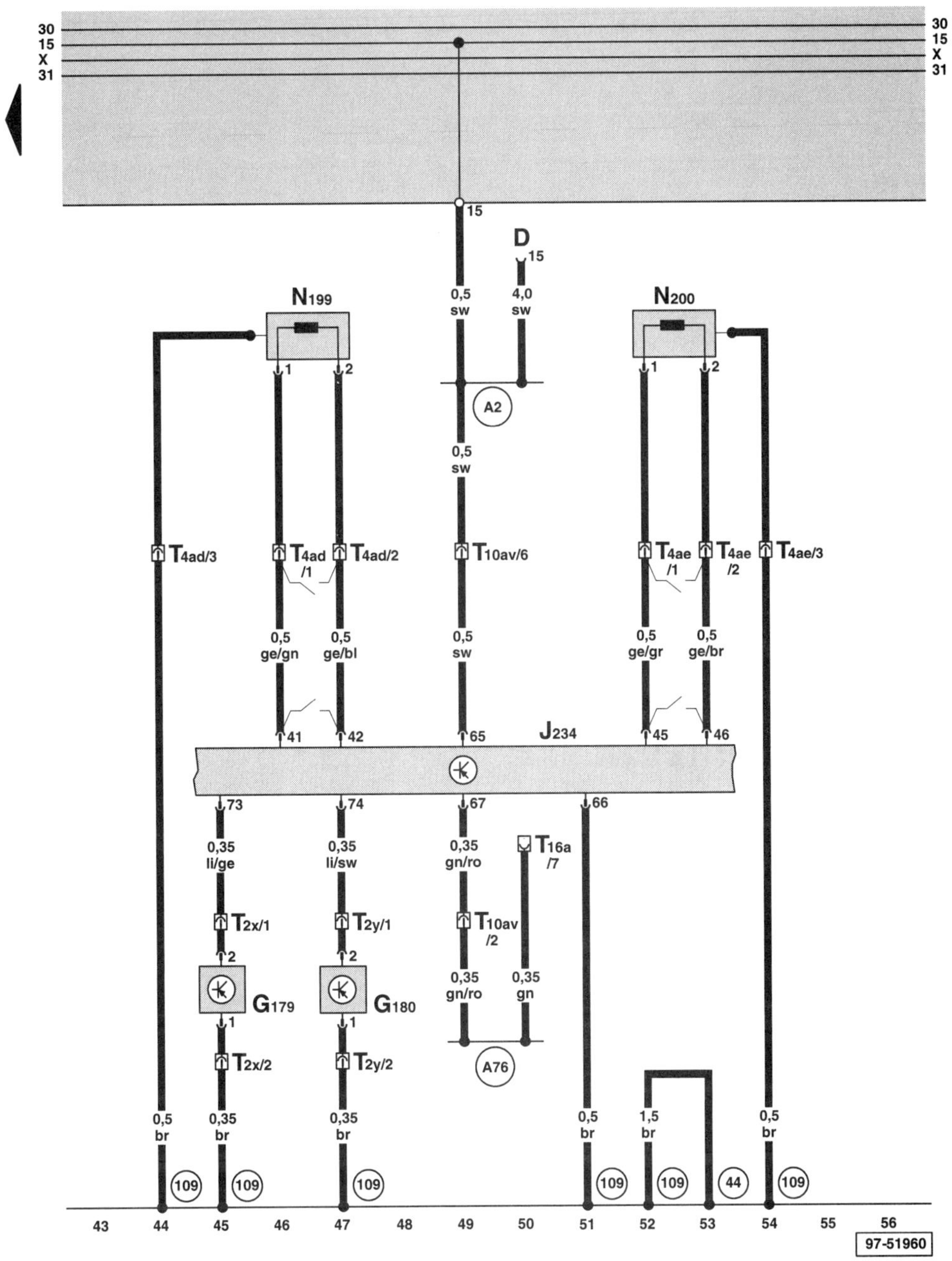

ws = white
sw = black
ro = red
br = brown
gn = green
bl = blue
gr = grey
li = violet
ge = yellow

D - Ignition/Starter Switch
G179 - Crash sensor for side airbag, driver's side
G180 - Crash sensor for side airbag, passenger's side
J234 - Airbag Control Module
N199 - Igniter for side airbag, driver's side
N200 - Igniter for side airbag, passenger's side
T2x - Connector 2 pin, yellow, below driver's seat
T2y - Connector 2 pin, yellow, below passenger's seat
T4ad - Connector 4 pin, yellow, below driver's seat
T4ae - Connector 4 pin, yellow, below passenger's seat
T10av - Connector 10 pin, yellow, connector station A pillar, right
T16a - Connector 16 pin, Data Link Connector (DLC)

(44) - Ground connection, on left A-pillar, lower part
(109) - Ground connection, in airbag wiring harness
(A2) - Plus connection (15), in instrument panel wiring harness
(A76) - Connector (K-diagnosis wire), in instrument panel wiring harness
- Bridge closes when disconnecting connector (safety feature)

THIS PAGE INTENTIONALLY LEFT BLANK

Power Sports Seats

2000 m. y.

Fuse Panel

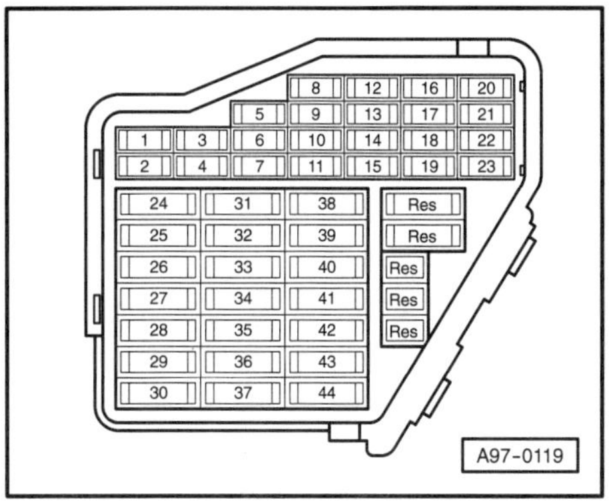

Fuse Colors:

30 A - Green
25 A - White
20 A - Yellow
15 A - Blue
10 A - Red
7,5 A - Brown
5 A - Beige

Starting with fuse position 23, fuses in the fuse holder are identified with 223 in the wiring diagram.

8 - Fold Relay Panel

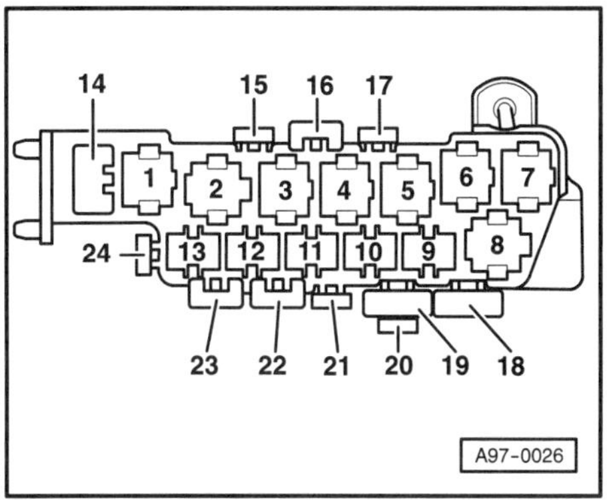

Fuse Location:

14 -Power Seat Circuit Breaker 1, S44

24 -Power Seat Circuit Breaker 2, S80

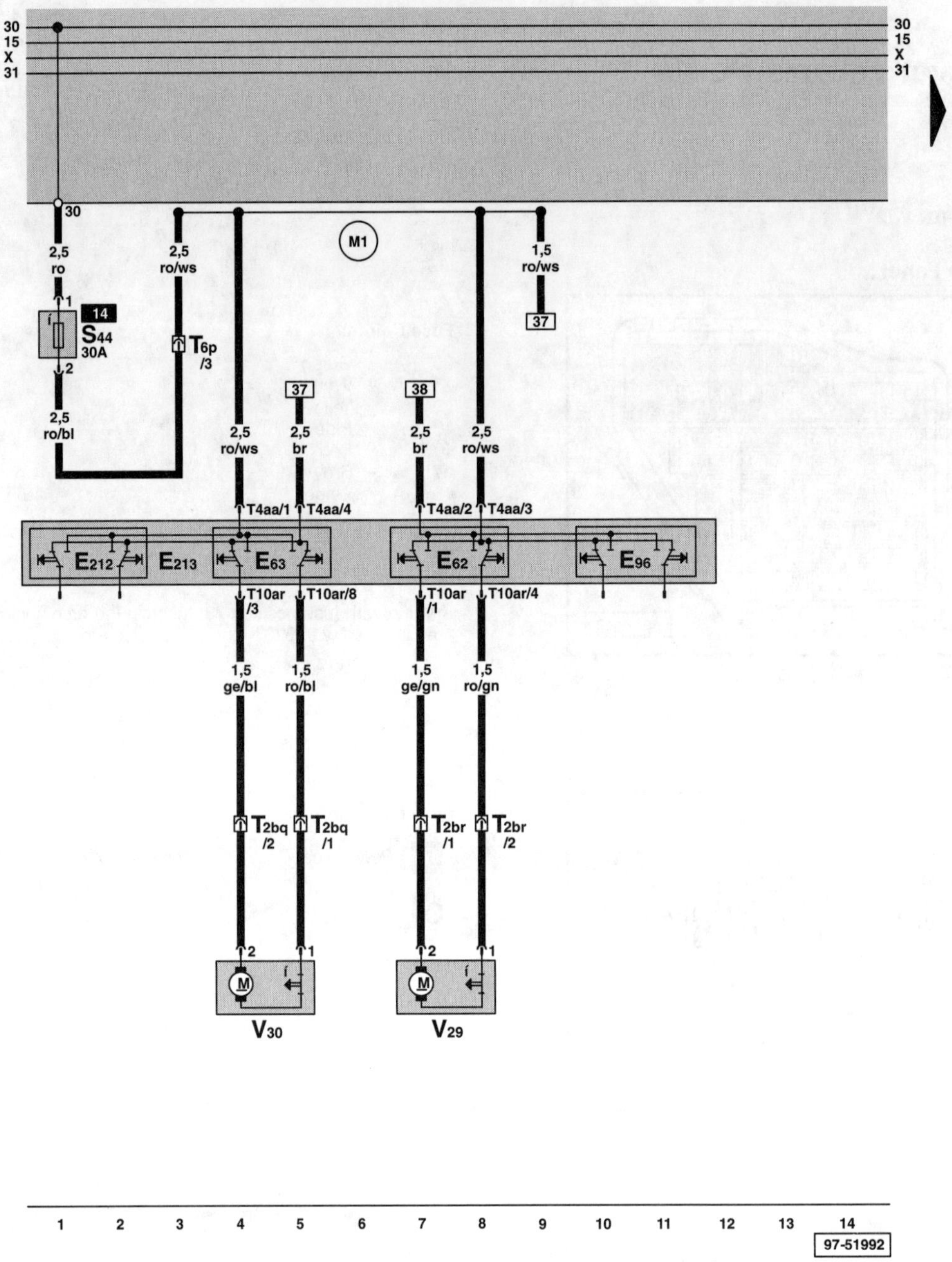

ws = white
sw = black
ro = red
br = brown
gn = green
bl = blue
gr = grey
li = violet
ge = yellow

E62 - Driver's Seat Front Height Adjusting Switch
E63 - Driver's Seat Rear Height Adjusting Switch
E96 - Driver's Backrest Adjustment Switch
E212 - Button for length adjustment, forward, driver seat
E213 - Button for length adjustment, backward, driver seat
S44 - Power Seat Circuit Breaker 1
T2bp - Connector 2 pin, black, in driver's seat
T2br - Connector 2 pin, black, in driver's seat
T4aa - Connector 4 pin, black, in driver's seat
T6p - Connector 6 pin, red, below driver's seat
T10ar - Connector 10 pin, black, in driver's seat
V29 - Driver's Seat Front Height Adjusting Motor
V30 - Driver's Seat Rear Height Adjusting Motor

(M1) - Plus connection -1- (30), in power seats wiring harness

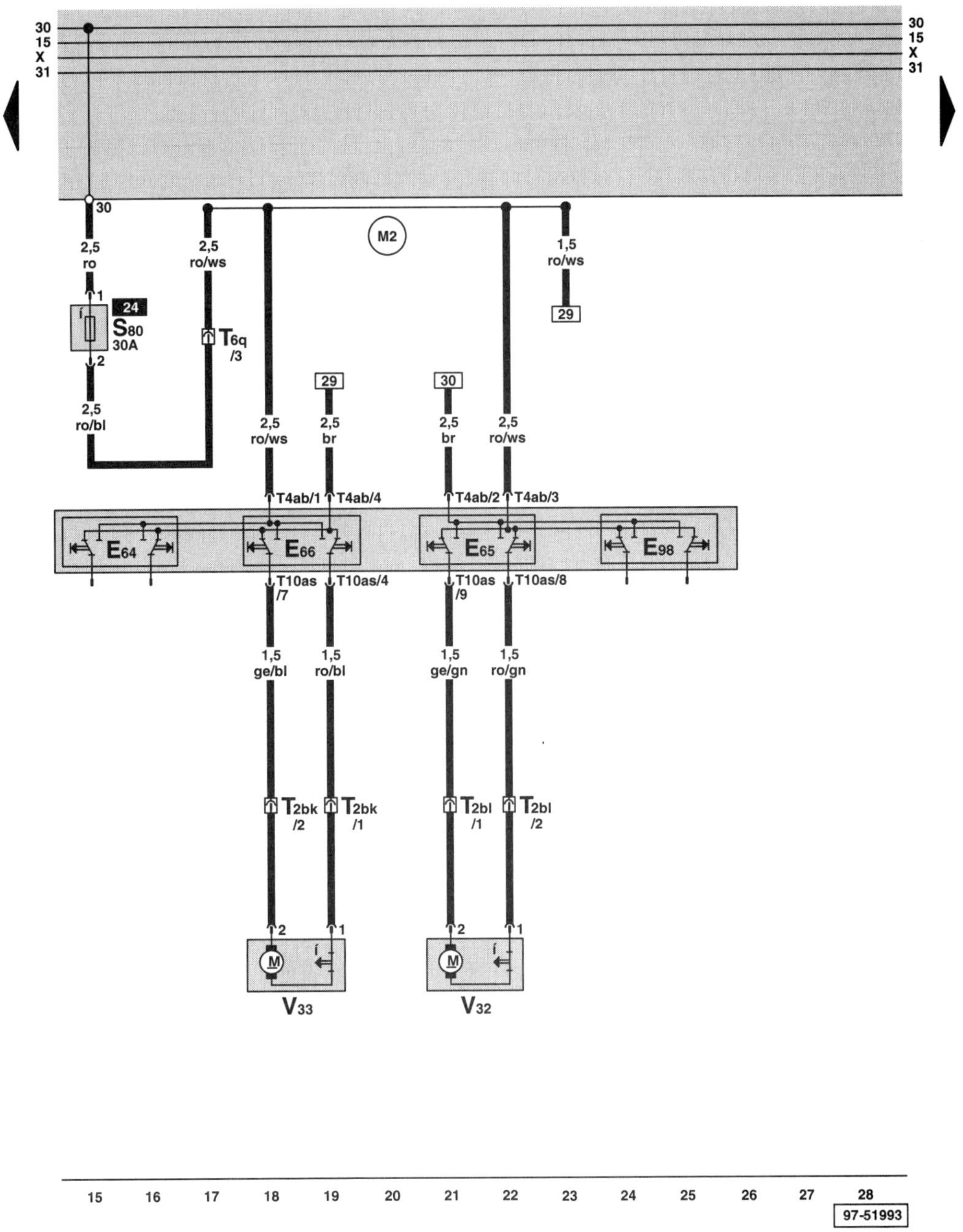

ws = white
sw = black
ro = red
br = brown
gn = green
bl = blue
gr = grey
li = violet
ge = yellow

E64 - Passenger's Seat Fore/Aft Adjusting Switch
E65 - Passenger's Seat Front Height Adjusting Switch
E66 - Passenger's Seat Rear Height Adjusting Switch
E98 - Passenger Backrest Adjustment Switch
S80 - Power Seat Circuit Breaker 2
T2bk - Connector 2 pin, black, in passenger's seat
T2bl - Connector 2 pin, black, in passenger's seat
T4ab - Connector 4 pin, black, in passenger's seat
T6q - Connector 6 pin, red, below passenger's seat
T10as - Connector 10 pin, black, in passenger's seat
V32 - Passenger's Seat Front Height Adjusting Motor
V33 - Passenger's Seat Rear Height Adjusting Motor

(M2) - Plus connection -2- (30), in power seats wiring harness

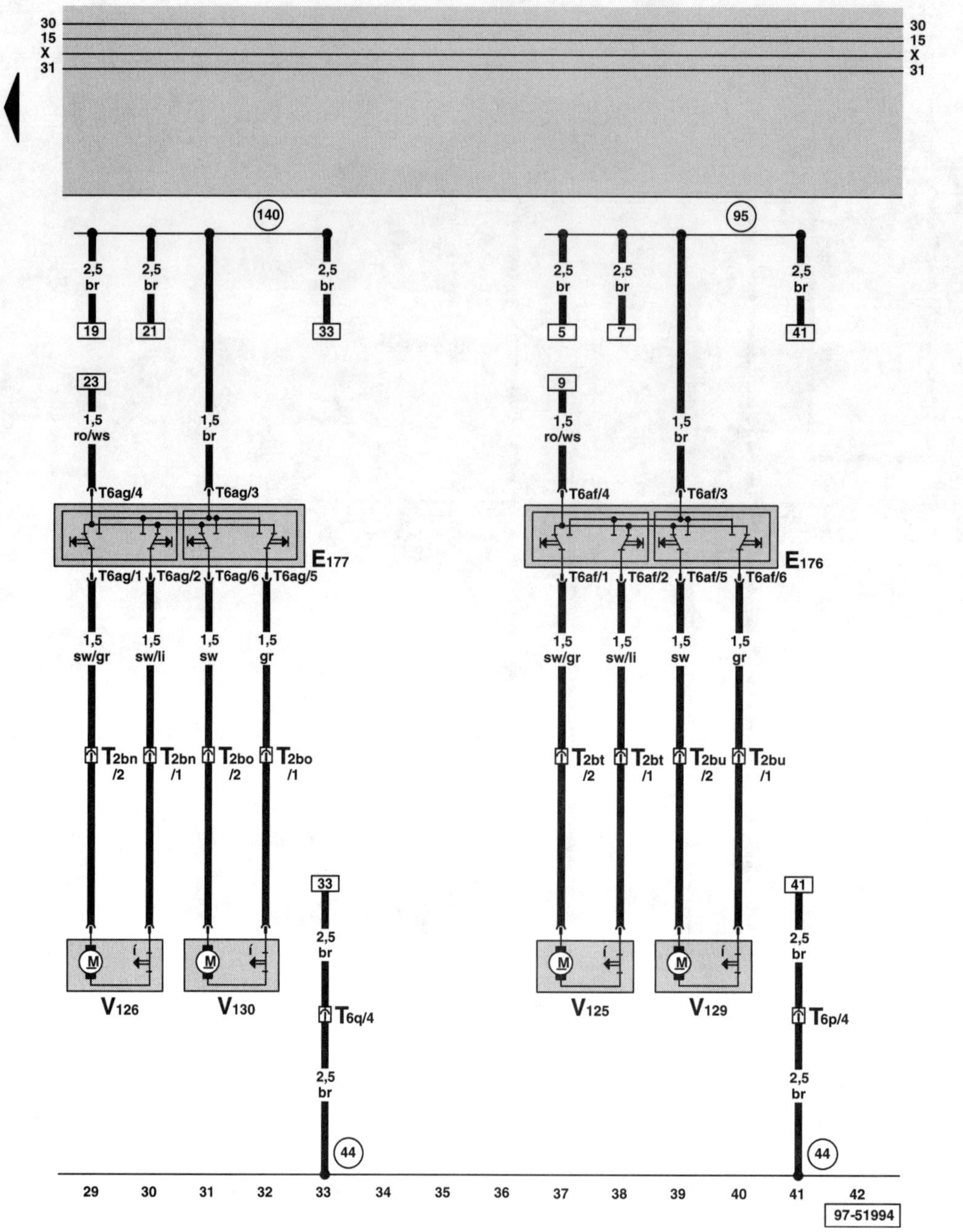

ws = white
sw = black
ro = red
br = brown
gn = green
bl = blue
gr = grey
li = violet
ge = yellow

E176 - Driver's Seat Lumbar Support Adjustment Switch
E177 - Front Passenger Seat Lumbar Support Adjustment Switch
T2bn - Connector 2 pin, black, in passenger's seat
T2bo - Connector 2 pin, black, in passenger's seat
T2bt - Connector 2 pin, black, in driver's seat
T2bu - Connector 2 pin, black, in driver's seat
T6p - Connector 6 pin, red, below driver's seat
T6q - Connector 6 pin, red, below passenger's seat
T6af - Connector 6 pin, black, in driver's seat
T6ag - Connector 6 pin, black, in passenger's seat
V125 - Driver's Seat Lumbar Support Curvature Adjustment Motor
V126 - Front Passenger Seat Lumbar Support Curvature Adjustment Motor
V129 - Driver's Seat Lumbar Support Height Adjustment Motor
V130 - Front Passenger Seat Lumbar Support Height Adjustment Motor

(44) - Ground connection, on left A-pillar, lower part
(95) - Ground connection -1-, in power seat wiring harness
(140) - Ground connection -2-, in power seat wiring harness

Heated Seats

2000 m. y.

Fuse Panel

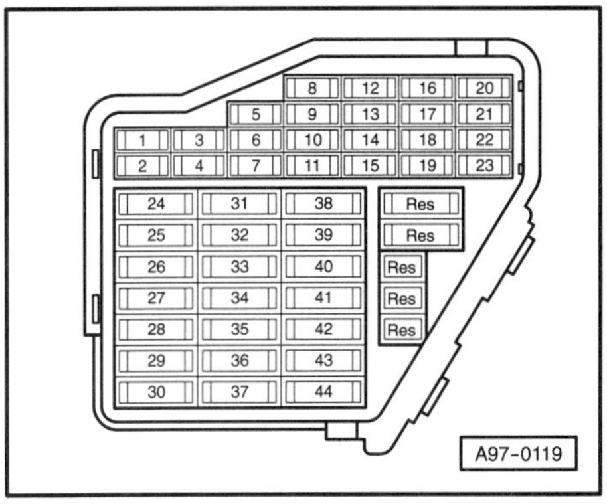

Fuse Colors:

30	A	-	Green
25	A	-	White
20	A	-	Yellow
15	A	-	Blue
10	A	-	Red
7,5	A	-	Brown
5	A	-	Beige

Starting with fuse position 23, fuses in the fuse holder are identified with 223 in the wiring diagram.

Micro Central Electric Panel

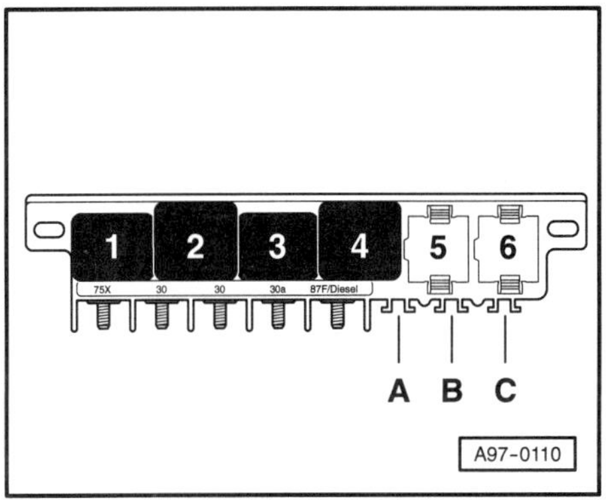

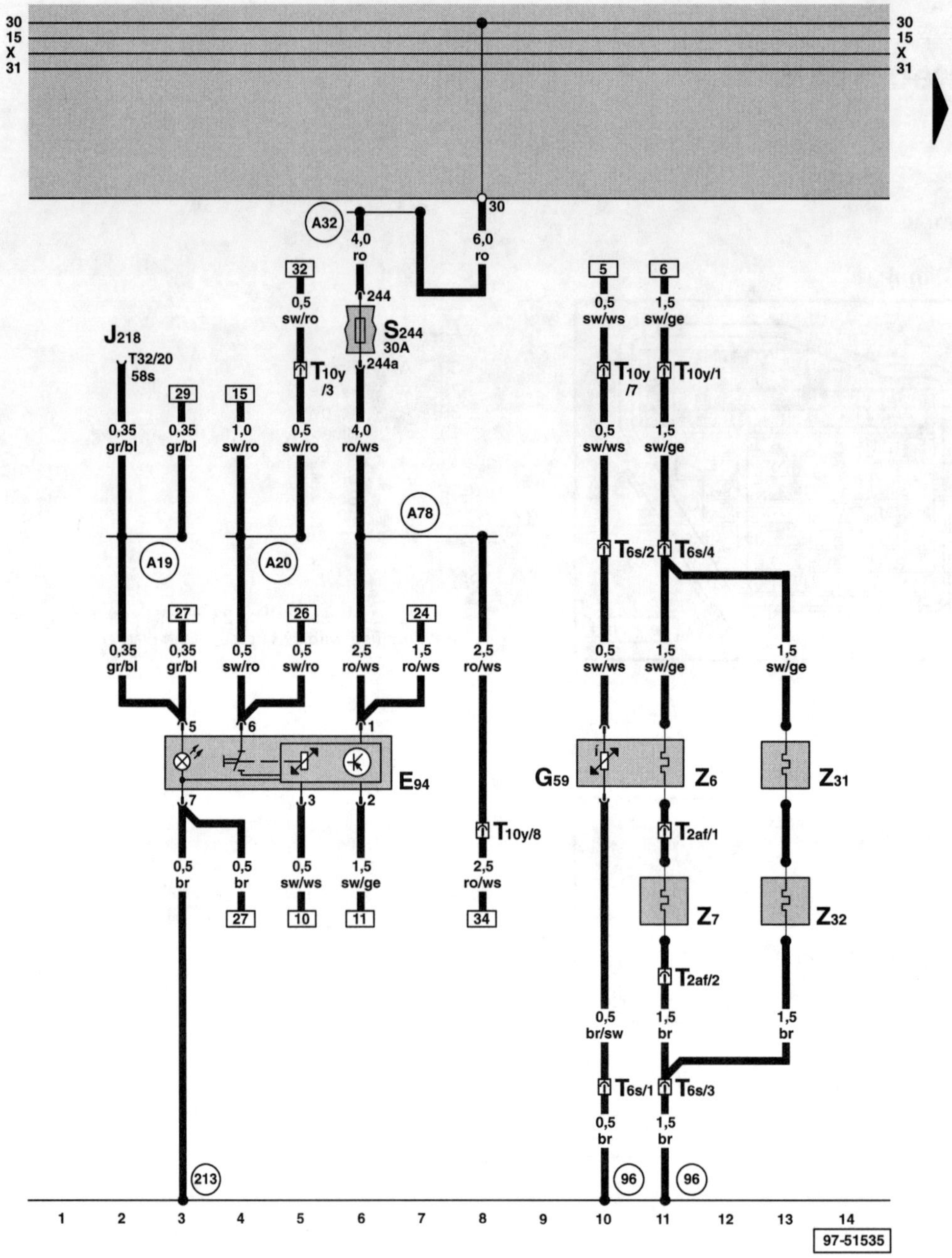

ws = white
sw = black
ro = red
br = brown
gn = green
bl = blue
gr = grey
li = violet
ge = yellow

E94 - Adjuster for heated driver's seat
G59 - Driver's Heated Seat Temperature Sensor
J218 - Instrument Cluster Combination Processor
S244 - Fuse 44 in fuse holder
T2af - Connector 2 pin, black, below driver's seat
T6s - Connector 6 pin, green, below driver's seat
T10y - Connector 10 pin, green, connector station A pillar, right
T32 - Connector 32 pin, blue, on instrument cluster
Z6 - Driver's Seat Heating Element
Z7 - Driver's Backrest Heating Element
Z31 - Driver's Seat Side Bolster Heater
Z32 - Driver's Seat Backrest Side Bolster Heater

(96) - Ground connection -1-, in heated seats wiring harness
(213) - Ground connection, in instrument panel wiring harness
(A19) - Wire connection (58s), in instrument panel wiring harness
(A20) - Wire connection (15a), in instrument panel wiring harness
(A32) - Plus connection (30), in instrument panel wiring harness
(A78) - Connector (30a,S244), in instrument panel wiring harness

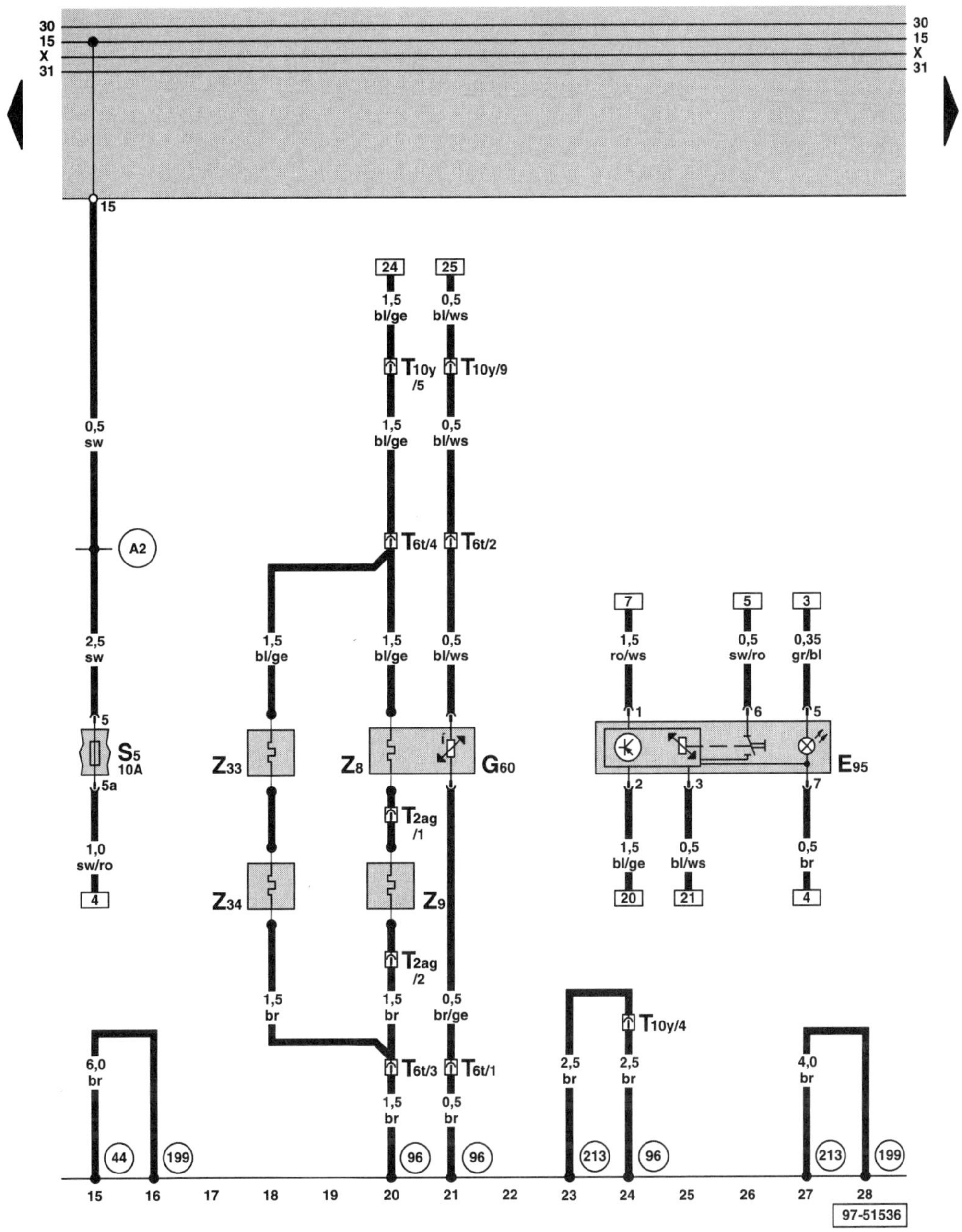

ws = white
sw = black
ro = red
br = brown
gn = green
bl = blue
gr = grey
li = violet
ge = yellow

E95 - Adjuster for heated passenger's seat
G60 - Passenger's Heated Seat Temperature Sensor
S5 - Fuse
T2ag - Connector 2 pin, black, below passenger's seat
T6t - Connector 6 pin, green, below passenger's seat
T10y - Connector 10 pin, green, connector station A pillar, right
Z8 - Passenger's Seat Heating Element
Z9 - Passenger's Backrest Heating Element
Z33 - Passenger's Seat Side Bolster Heater
Z34 - Passenger's Seat Backrest Side Bolster Heater

(44) - Ground connection, on left A-pillar, lower part
(96) - Ground connection -1-, in heated seats wiring harness
(213) - Ground connection, in instrument panel wiring harness
(199) - Ground connection -3-, in instrument panel wiring harness
(A2) - Plus connection (15), in instrument panel wiring harness

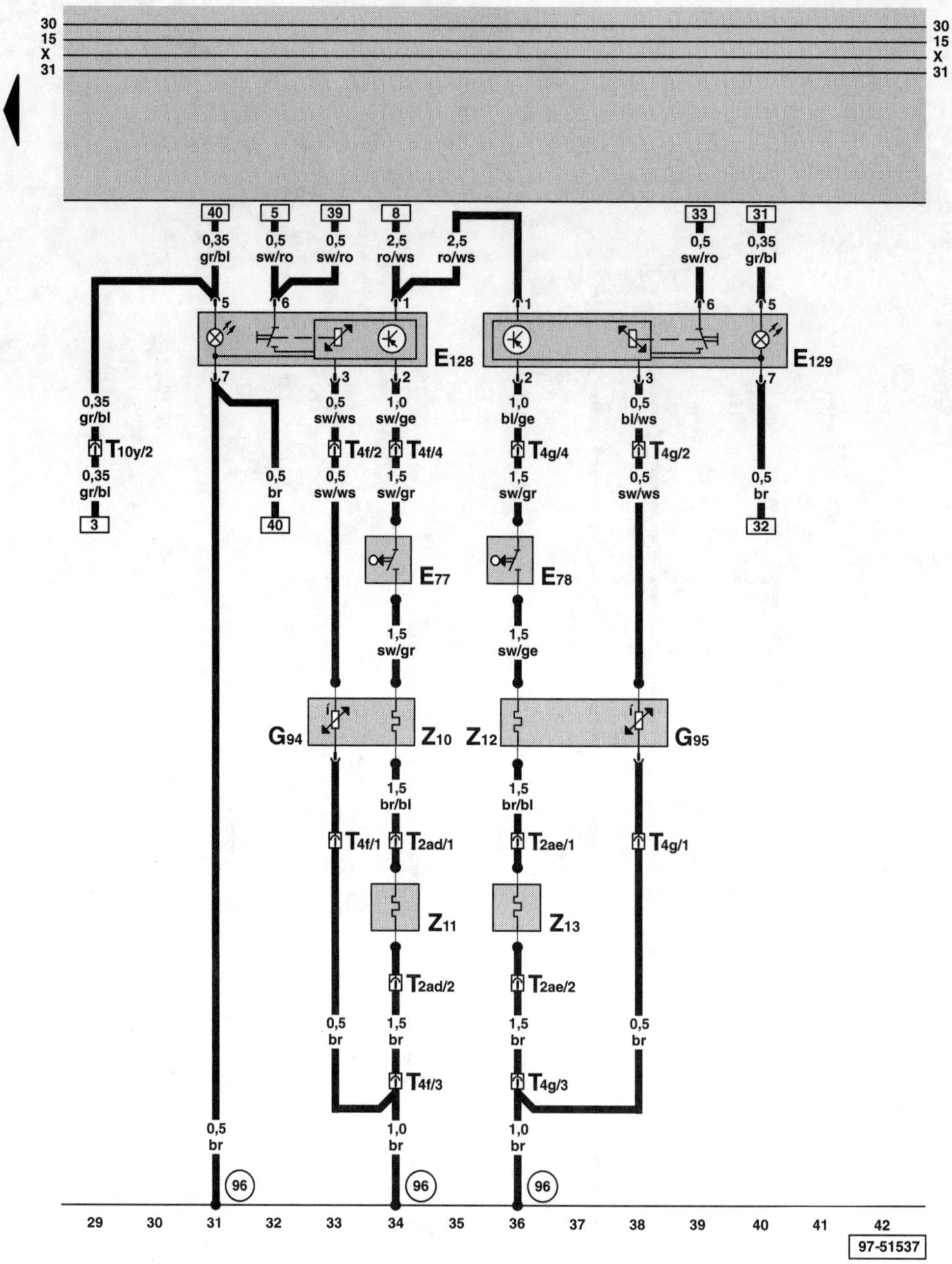

ws = white
sw = black
ro = red
br = brown
gn = green
bl = blue
gr = grey
li = violet
ge = yellow

- E77 - Left Rear Heated Seat Switch
- E78 - Right Rear Heated Seat Switch
- E128 - Left Rear Heated Seat Regulating Switch
- E129 - Right Rear Heated Seat Regulating Switch
- G94 - Left Rear Heated Seat Temperature Sensor
- G95 - Right Rear Heated Seat Temperature Sensor
- T2ad - Connector 2 pin, black, below rear seat, left
- T2ae - Connector 2 pin, black, below rear seat, right
- T4f - Connector 4 pin, black, below rear seat, left
- T4g - Connector 4 pin, black, below rear seat, right
- T10y - Connector 10 pin, green, connector station A pillar, right
- Z10 - Left Rear Seat Heating Element
- Z11 - Left Rear Backrest Heating Element
- Z12 - Right Rear Seat Heating Element
- Z13 - Right Rear Backrest Heating Element

(96) - Ground connection -1-, in heated seats wiring harness

Power Windows

2000 m. y.

Fuse Panel

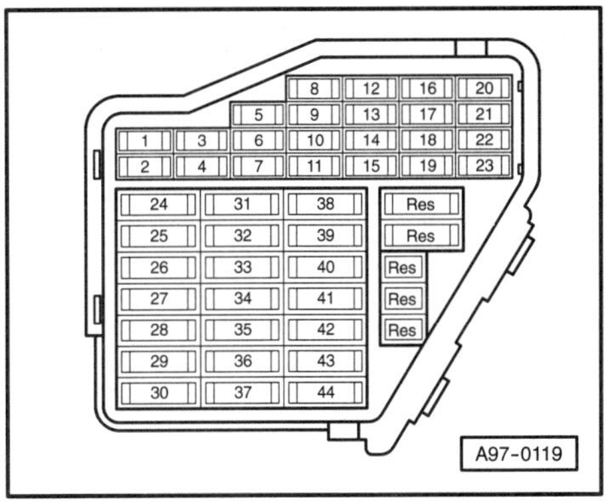

Fuse Colors:

30 A - Green
25 A - White
20 A - Yellow
15 A - Blue
10 A - Red
7,5 A - Brown
5 A - Beige

Starting with fuse position 23, fuses in the fuse holder are identified with 223 in the wiring diagram.

Micro Central Electric Panel

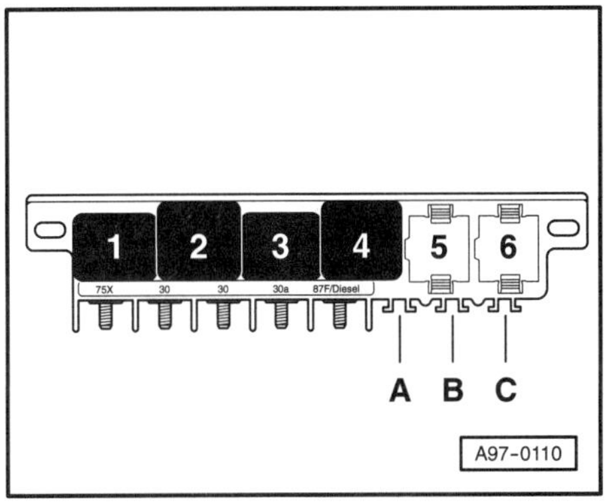

8 - Fold Relay Panel

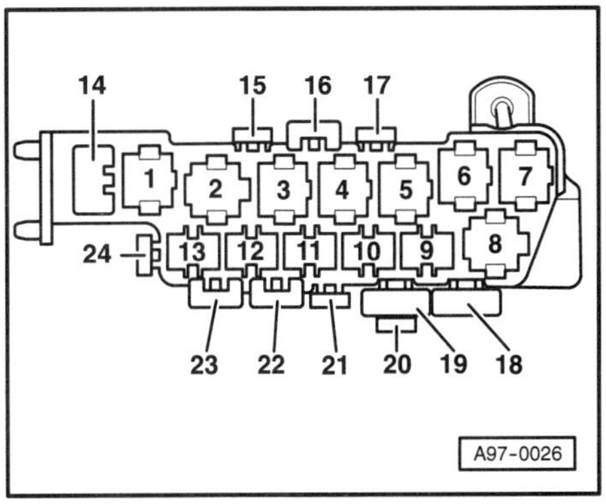

Fuse Location:

16 -Power Window Circuit Breaker Front, S37

17 -Power Window Circuit Breaker Rear, S43

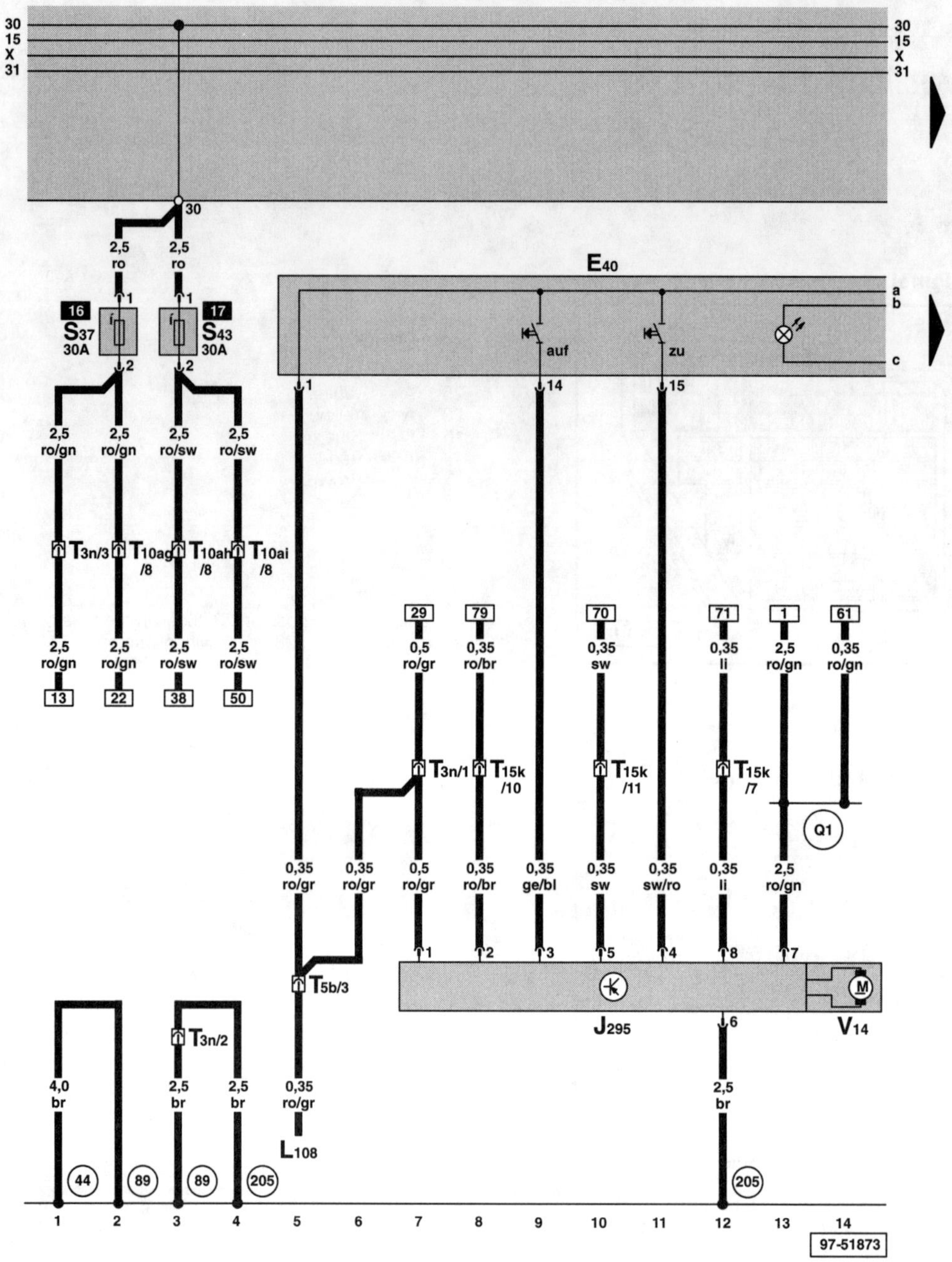

ws = white
sw = black
ro = red
br = brown
gn = green
bl = blue
gr = grey
li = violet
ge = yellow

- E40 - Left Front Window Switch
- J295 - Left Front Power Window Control Module
- L108 - Door opener lighting, driver side
- S37 - Power Window Circuit Breaker Front
- S43 - Power Window Circuit Breaker Rear
- T3n - Connector 3 pin, red, connector station A pillar, left
- T5b - Connector 5 pin, black, in driver's door
- T10ag - Connector 10 pin, red, connector station A pillar, right
- T10ah - Connector 10 pin, red, in B pillar, left
- T10ai - Connector 10 pin, red, in B pillar, right
- T15k - Connector 15 pin, red, connector station A pillar, left
- V14 - Left Window Motor

- (44) - Ground connection, on left A-pillar, lower part
- (89) - Ground connection -1-, in power window wiring harness
- (205) - Ground connection, in driver's door wiring harness
- (Q1) - Plus connection (30), in power window wiring harness

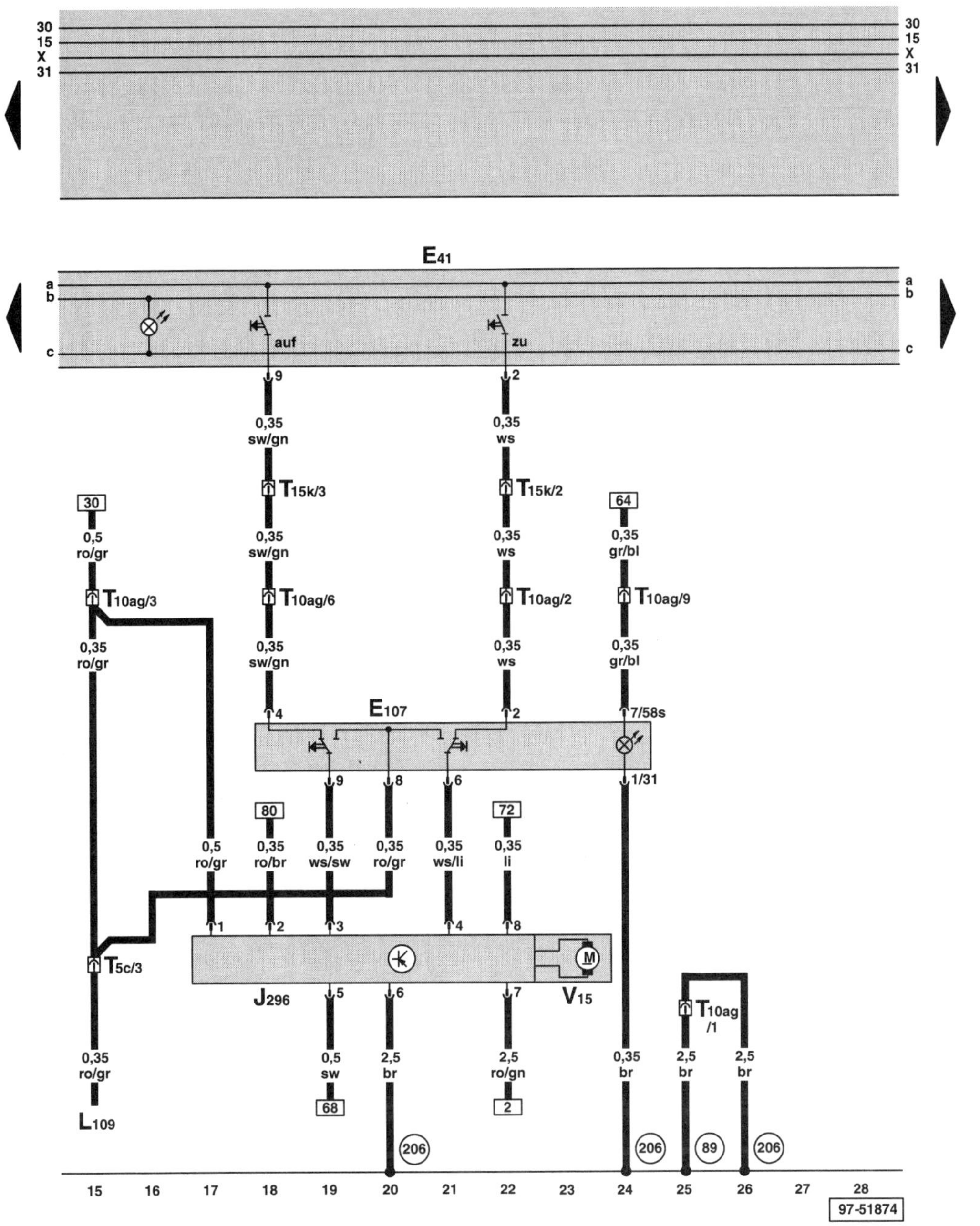

ws = white
sw = black
ro = red
br = brown
gn = green
bl = blue
gr = grey
li = violet
ge = yellow

E41 - Right Front Window Switch
E107 - Switch for window regulator, in passenger door
J296 - Right Front Power Window Control Module
L109 - Door opener lighting, passenger side
T5c - Connector 5 pin, black, in passenger's door
T10ag - Connector 10 pin, red, connector station A pillar, right
T15k - Connector 15 pin, red, connector station A pillar, left
V15 - Right Window Motor

(89) - Ground connection -1-, in power window wiring harness
(206) - Ground connection, in passenger's door wiring harness

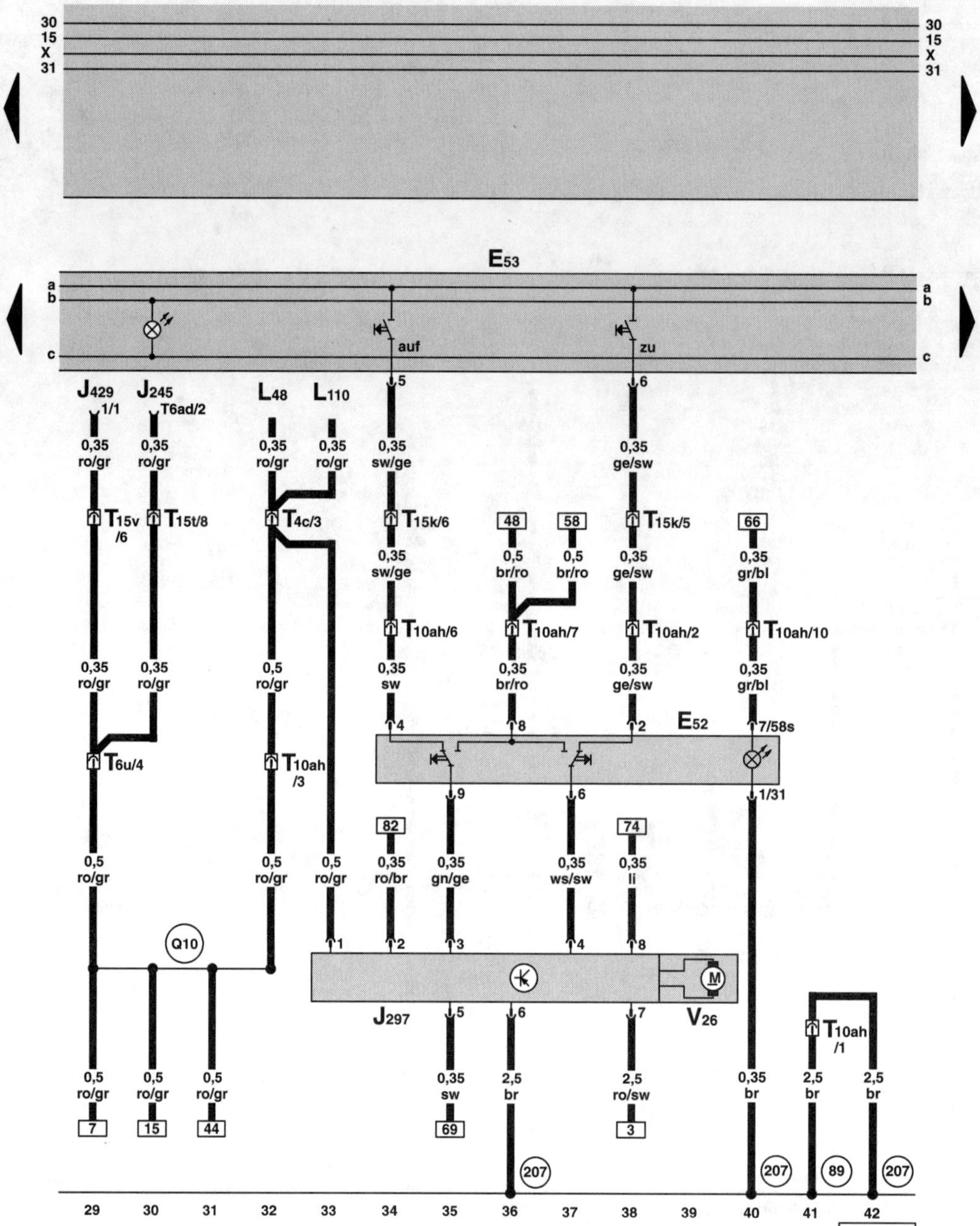

ws = white
sw = black
ro = red
br = brown
gn = green
bl = blue
gr = grey
li = violet
ge = yellow

97-51875

E52 - Left Rear Window Switch, (In LR Door)
E53 - Left Rear Window Switch, (In Console)
J245 - Power Sunroof Control Module
J297 - Left Rear Power Window Control Module
J429 - Control module for central locking
L48 - Left Rear Ashtray Light
L110 - Door opener lighting, left rear
T4c - Connector 4 pin, black, in left rear door
T6u - Connector 6 pin, grey, connector station A pillar, left
T6ad - Connector 6 pin, blue, on power sunroof control module
T10ah - Connector 10 pin, red, in B pillar, left
T15k - Connector 15 pin, red, connector station A pillar, left
T15t - Connector 15 pin, grey, connector station A pillar, left
T15v - Connector 15 pin, orange, connector station A pillar, right
V26 - Left Rear Door Window Motor

(89) - Ground connection -1-, in power window wiring harness

(207) - Ground connection, in left rear door wiring harness

(Q10) - Plus connection (87), in power window/central locking system and door contact switch wiring harness

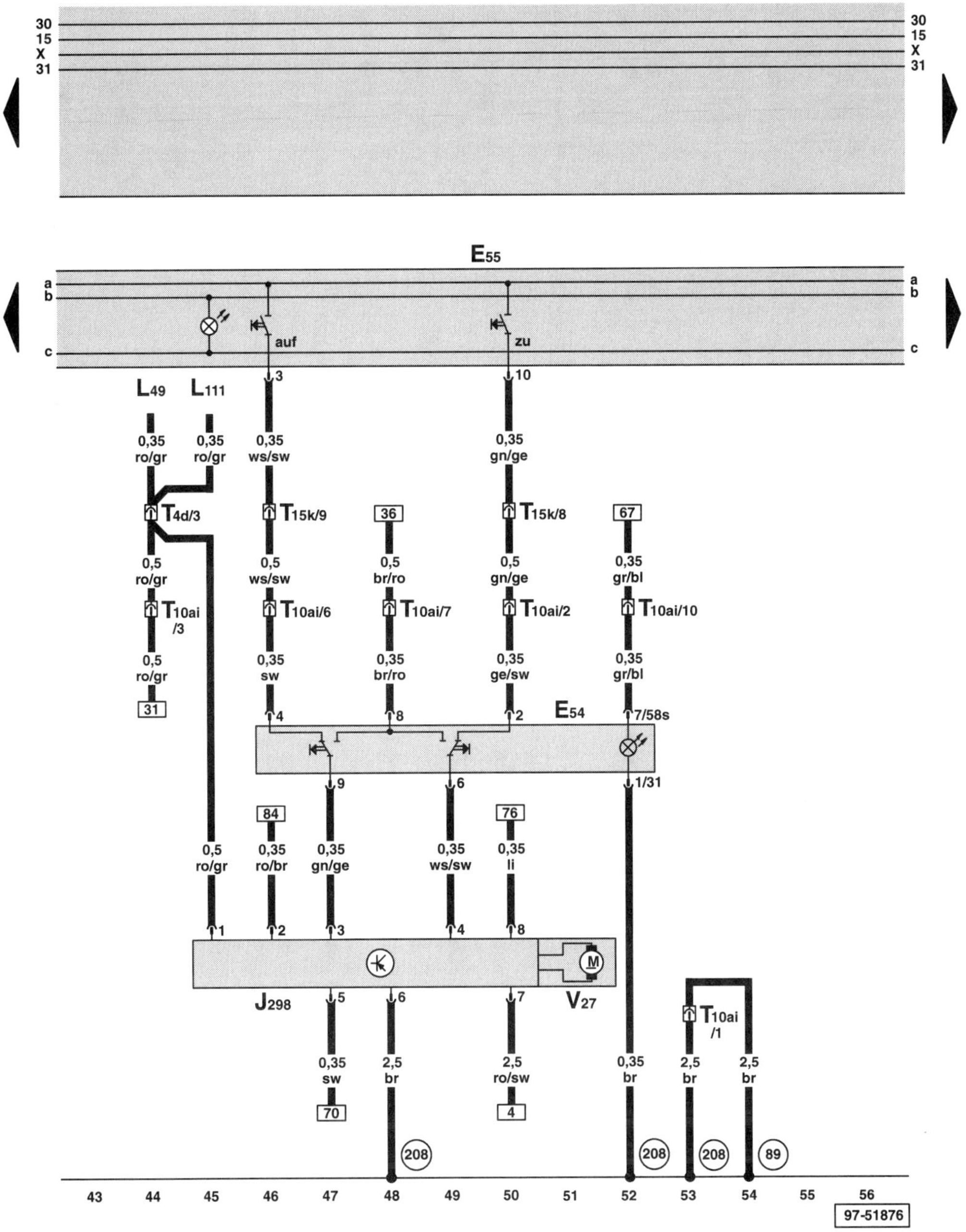

ws = white
sw = black
ro = red
br = brown
gn = green
bl = blue
gr = grey
li = violet
ge = yellow

E54 - Right Rear Window Switch, (In RR Door)
E55 - Right Rear Window Switch, (In Console)
J298 - Right Rear Power Window Control Module
L49 - Right Rear Ashtray Light
L111 - Door opener lighting, right rear
T4d - Connector 4 pin, black, in right rear door
T10ai - Connector 10 pin, red, in B pillar, right
T15k - Connector 15 pin, red, connector station A pillar, left
V27 - Right Rear Door Window Motor

(89) - Ground connection -1-, in power window wiring harness

(208) - Ground connection, in right rear door wiring harness

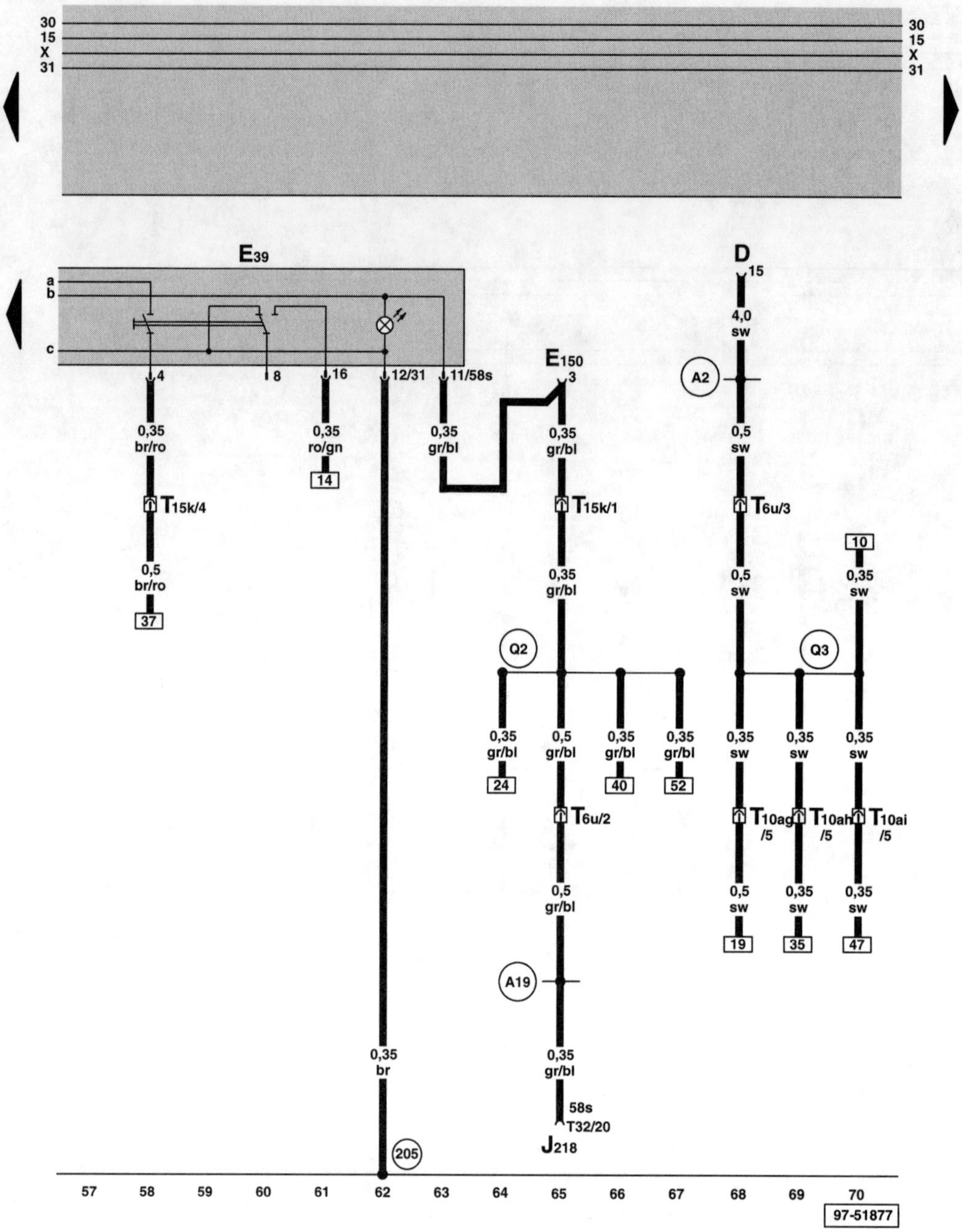

ws = white
sw = black
ro = red
br = brown
gn = green
bl = blue
gr = grey
li = violet
ge = yellow

D - Ignition/Starter Switch
E39 - Window Lockout Switch
E150 - Switch for interior lock, driver side
J218 - Instrument Cluster Combination Processor
T6u - Connector 6 pin, grey, connector station A pillar, left
T10ag - Connector 10 pin, red, connector station A pillar, right
T10ah - Connector 10 pin, red, in B pillar, left
T10ai - Connector 10 pin, red, in B pillar, right
T15k - Connector 15 pin, red, connector station A pillar, left
T32 - Connector 32 pin, blue, on instrument cluster

(205) - Ground connection, in driver's door wiring harness
(A2) - Plus connection (15), in instrument panel wiring harness
(A19) - Wire connection (58s), in instrument panel wiring harness
(Q2) - Plus connection (58), in power window wiring harness
(Q3) - Plus connection (75s), in power window wiring harness

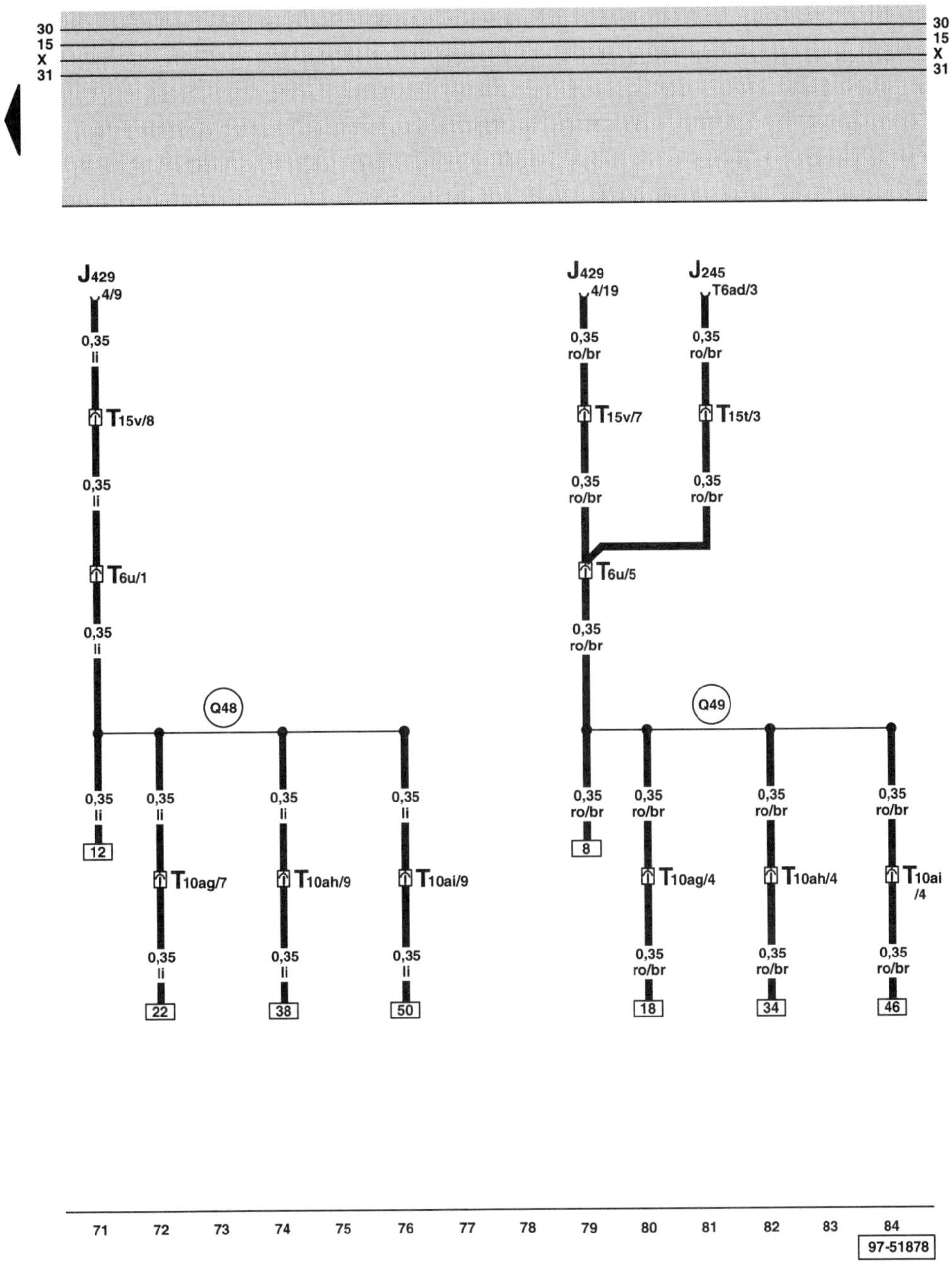

ws = white
sw = black
ro = red
br = brown
gn = green
bl = blue
gr = grey
li = violet
ge = yellow

J245 - Power Sunroof Control Module
J429 - Control module for central locking
T6u - Connector 6 pin, grey, connector station A pillar, left
T6ad - Connector 6 pin, blue, on power sunroof control module
T10ag - Connector 10 pin, red, connector station A pillar, right
T10ah - Connector 10 pin, red, in B pillar, left
T10ai - Connector 10 pin, red, in B pillar, right
T15t - Connector 15 pin, grey, connector station A pillar, left
T15v - Connector 15 pin, orange, connector station A pillar, right

Q48 - Connection (open), in wiring harness interior

Q49 - Connection (closed), in wiring harness interior

Coolant Fan

2000 m. y.

Fuse Panel

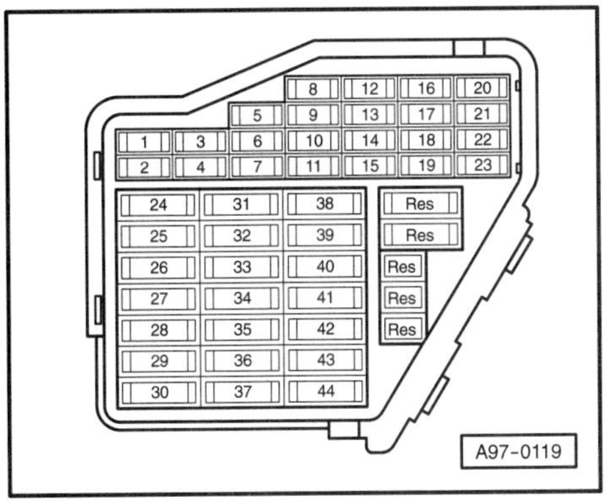

Fuse Colors:

30 A - Green
25 A - White
20 A - Yellow
15 A - Blue
10 A - Red
7,5 A - Brown
5 A - Beige

Starting with fuse position 23, fuses in the fuse holder are identified with 223 in the wiring diagram.

8 - Fold Relay Panel

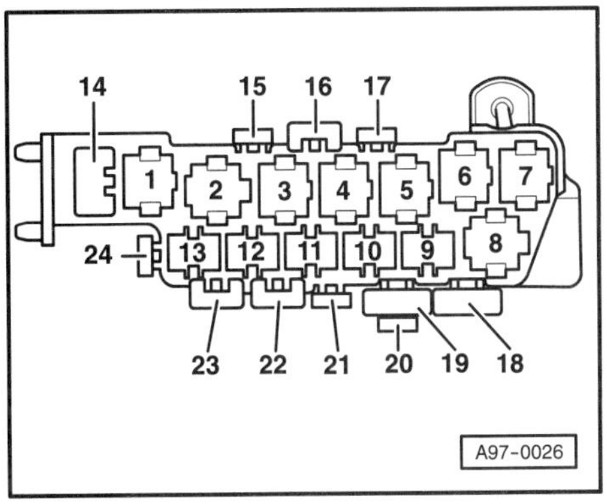

Fuse Location:

20 -Coolant Fan Fuse, S42

21 -Control module fuse for coolant fan, S142

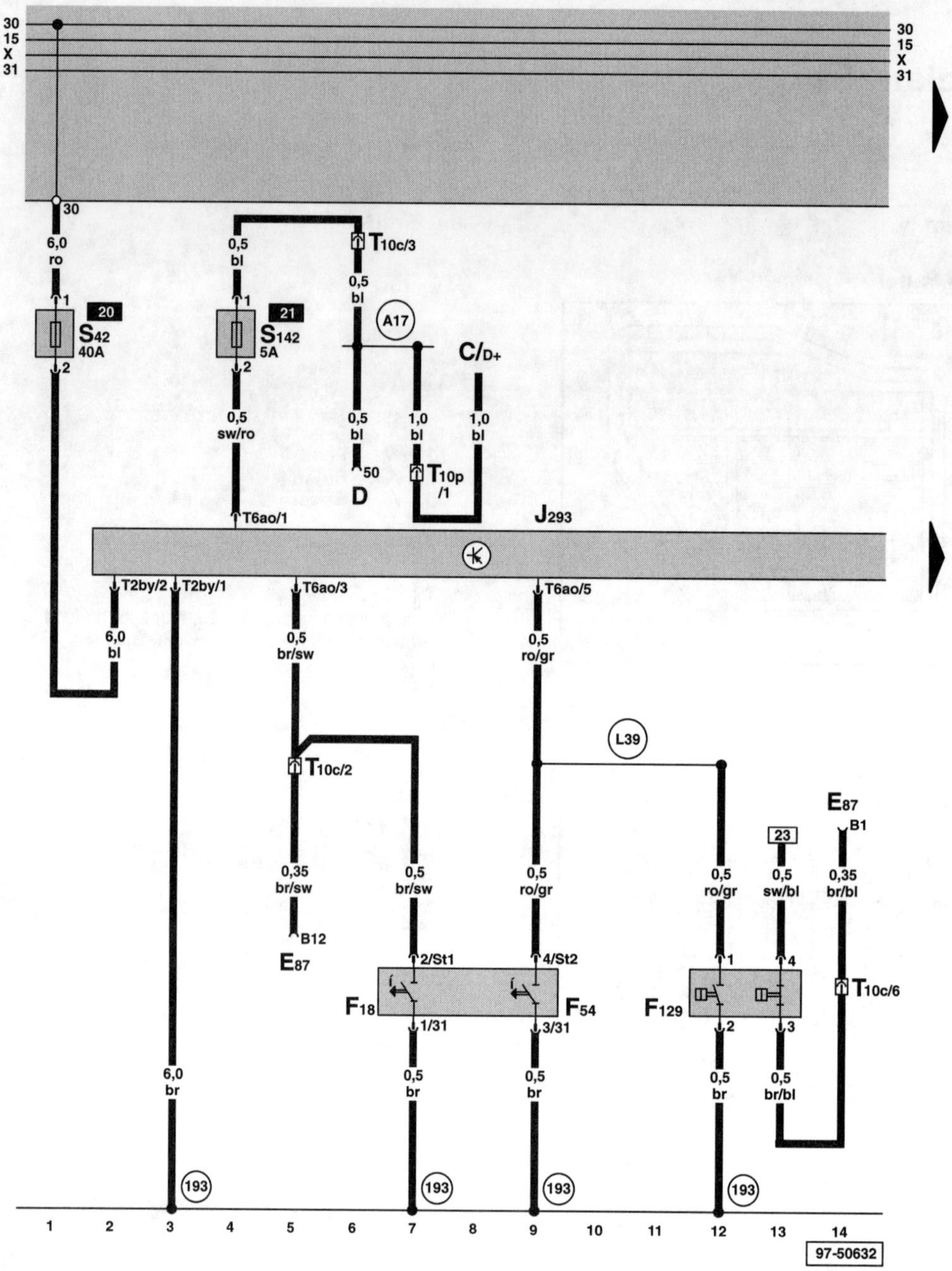

ws = white
sw = black
ro = red
br = brown
gn = green
bl = blue
gr = grey
li = violet
ge = yellow

C	-	Generator (GEN)
D	-	Ignition/Starter Switch
E87	-	A/C Control Head
F18	-	Coolant Fan Control (FC) Thermal Switch
F54	-	Coolant Fan Control (FC) Thermal Switch
F129	-	A/C Pressure Switch
J293	-	Coolant FC (Fan Control) Control Module
S42	-	Coolant Fan Fuse
S142	-	Control module fuse for coolant fan
T2by	-	Connector 2 pin, on Coolant FC (Fan Control) Control Module
T6ao	-	Connector 6 pin, on Coolant FC (Fan Control) Control Module
T10c	-	Connector 10 pin, grey, connector station A pillar, left
T10p	-	Connector 10 pin, black, connector station electronic box plenum chamber

(193)	-	Ground connection -1-, in coolant fan wiring harness
(A17)	-	Wire connection (61), in instrument panel wiring harness
(L39)	-	Wire connection -1- (A/C Pressure Switch), in Climatronic wiring harness

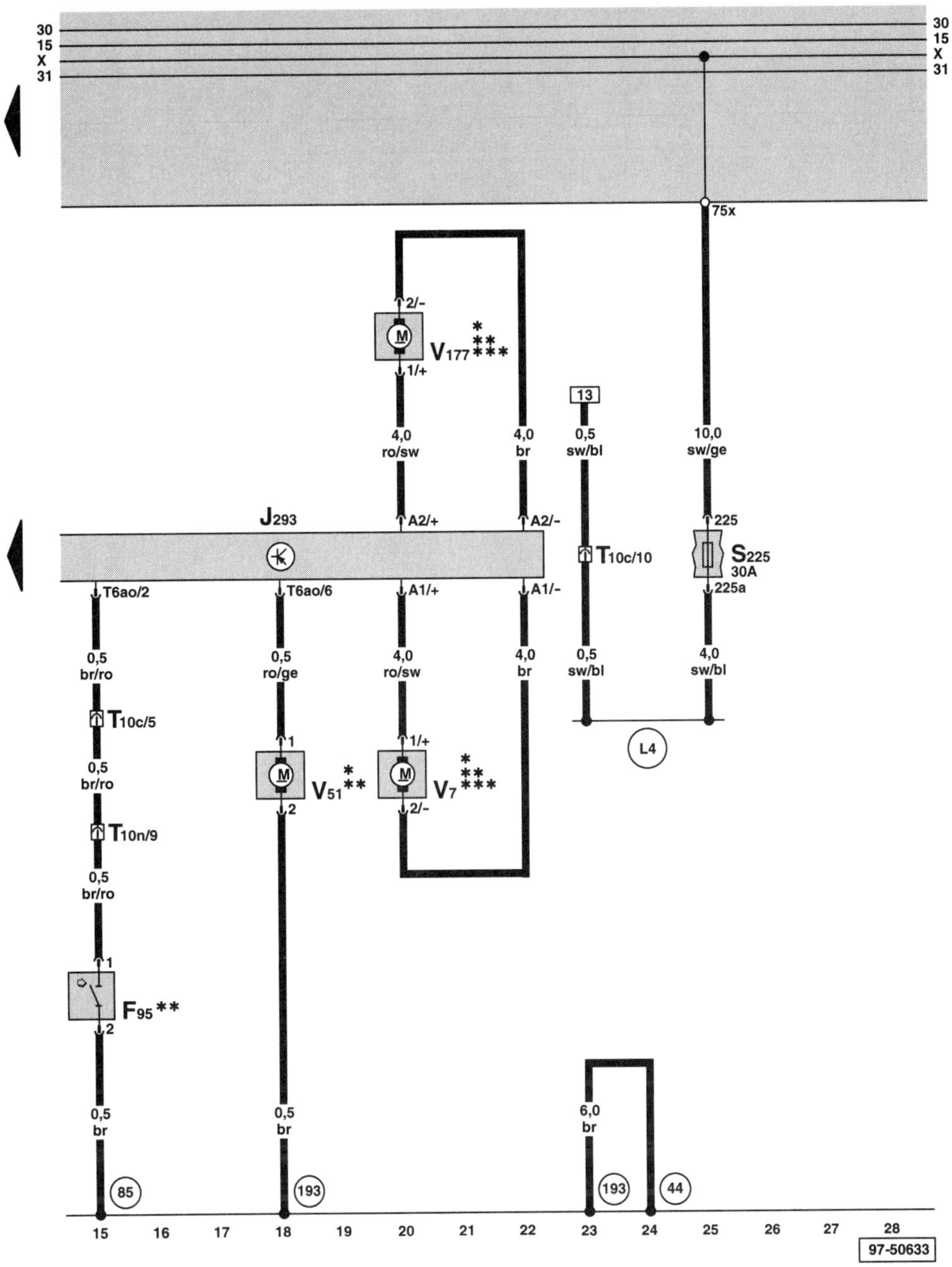

ws = white
sw = black
ro = red
br = brown
gn = green
bl = blue
gr = grey
li = violet
ge = yellow

F95 - After-Run Coolant Thermal Switch
J293 - Coolant FC (Fan Control) Control Module
S225 - Fuse 25 in fuse holder
T6ao - Connector 6 pin, on Coolant FC (Fan Control) Control Module
T10c - Connector 10 pin, grey, connector station A pillar, left
T10n - Connector 10 pin, orange, connector station electronic box plenum chamber
V7 - Coolant Fan
V51 - After-Run Coolant Pump
V177 - Coolant fan -2-

(44) - Ground connection, on left A-pillar, lower part
(85) - Ground connection -1-, in engine compartment wiring harness
(193) - Ground connection -1-, in coolant fan wiring harness
(L4) - Wire connection (75al), in A/C wiring harness
* - Vehicles with 6 cylinder engine ATQ
** - Vehicles with 6 cylinder engine APB
*** - Vehicles with 8 cylinder engine ART

THIS PAGE INTENTIONALLY LEFT BLANK

CAN-Bus Wiring

2000 m. y.

Fuse Panel

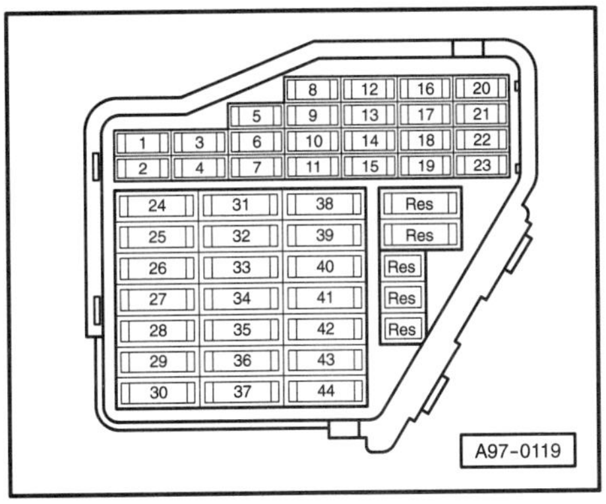

Fuse Colors:

30 A - Green
25 A - White
20 A - Yellow
15 A - Blue
10 A - Red
7,5 A - Brown
5 A - Beige

Starting with fuse position 23, fuses in the fuse holder are identified with 223 in the wiring diagram.

Micro Central Electric Panel

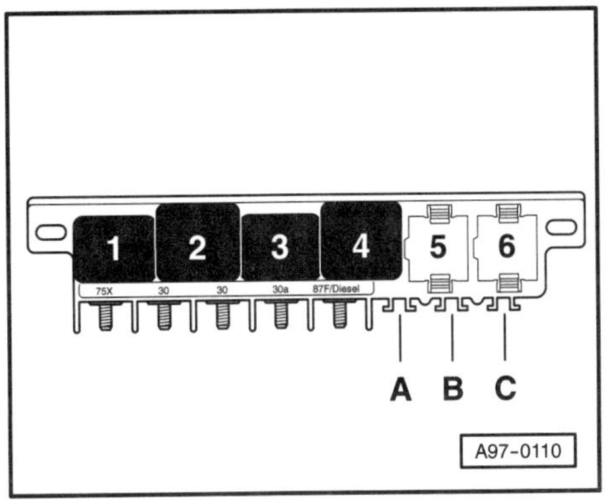

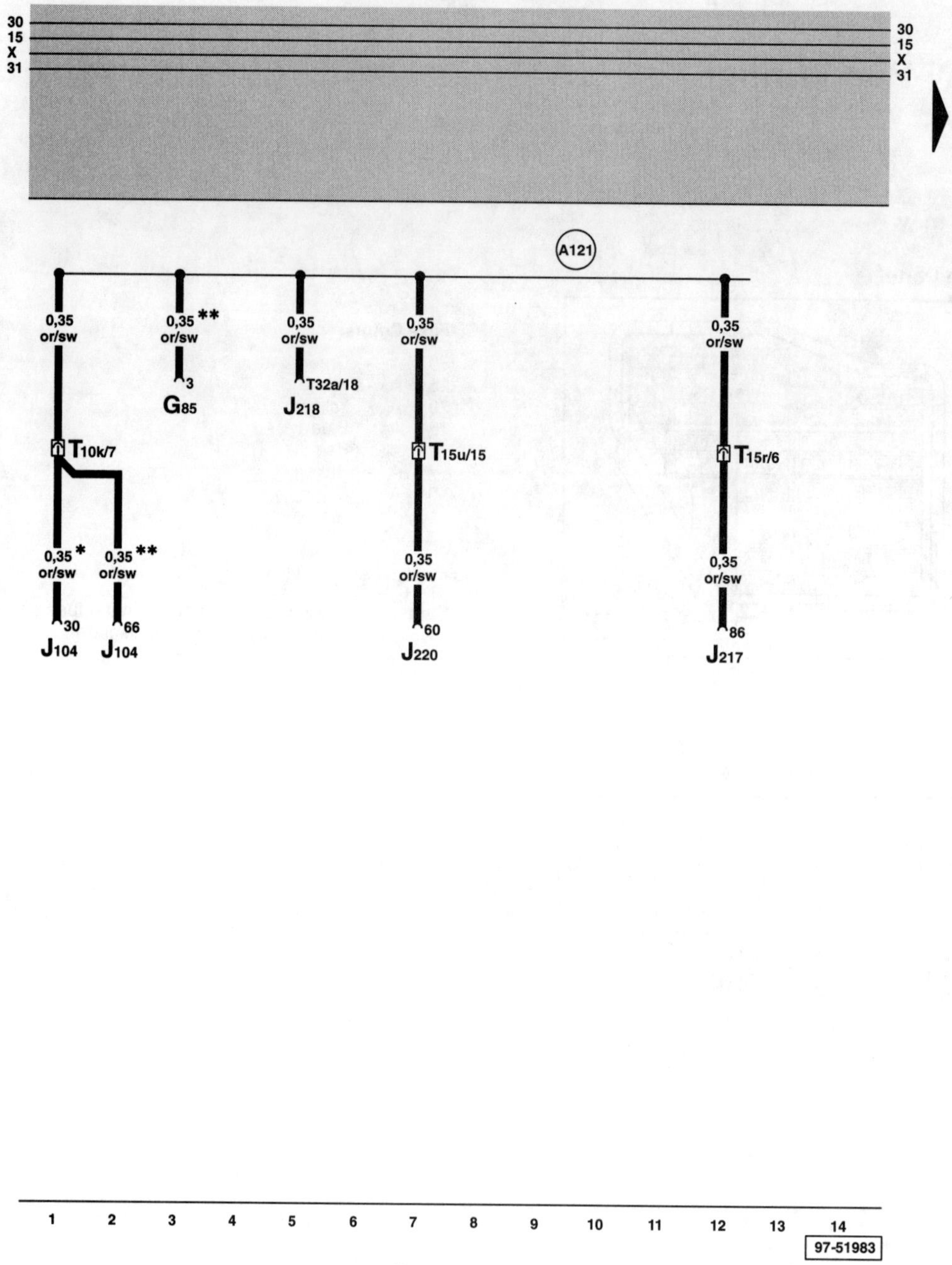

ws = white
sw = black
ro = red
br = brown
gn = green
bl = blue
gr = grey
li = violet
ge = yellow
or = orange

G85 - Steering Angle Sensor
J104 - ABS Control Module (w/EDL)
J217 - Transmission Control Module (TCM)
J218 - Instrument Cluster Combination Processor
J220 - Motronic Engine Control Module (ECM)
T10k - Connector 10 pin, orange, connector station A pillar, left
T15r - Connector 15 pin, blue, connector station A pillar, right
T15u - Connector 15 pin, red, connector station electronic box plenum chamber
T32a - Connector 32 pin, green, on instrument cluster

(A121) - Connection (high bus) in instrument panel wiring harness

* - Vehicles with ABS with EDS and ASR
** - Vehicles with Electronic Stability Program

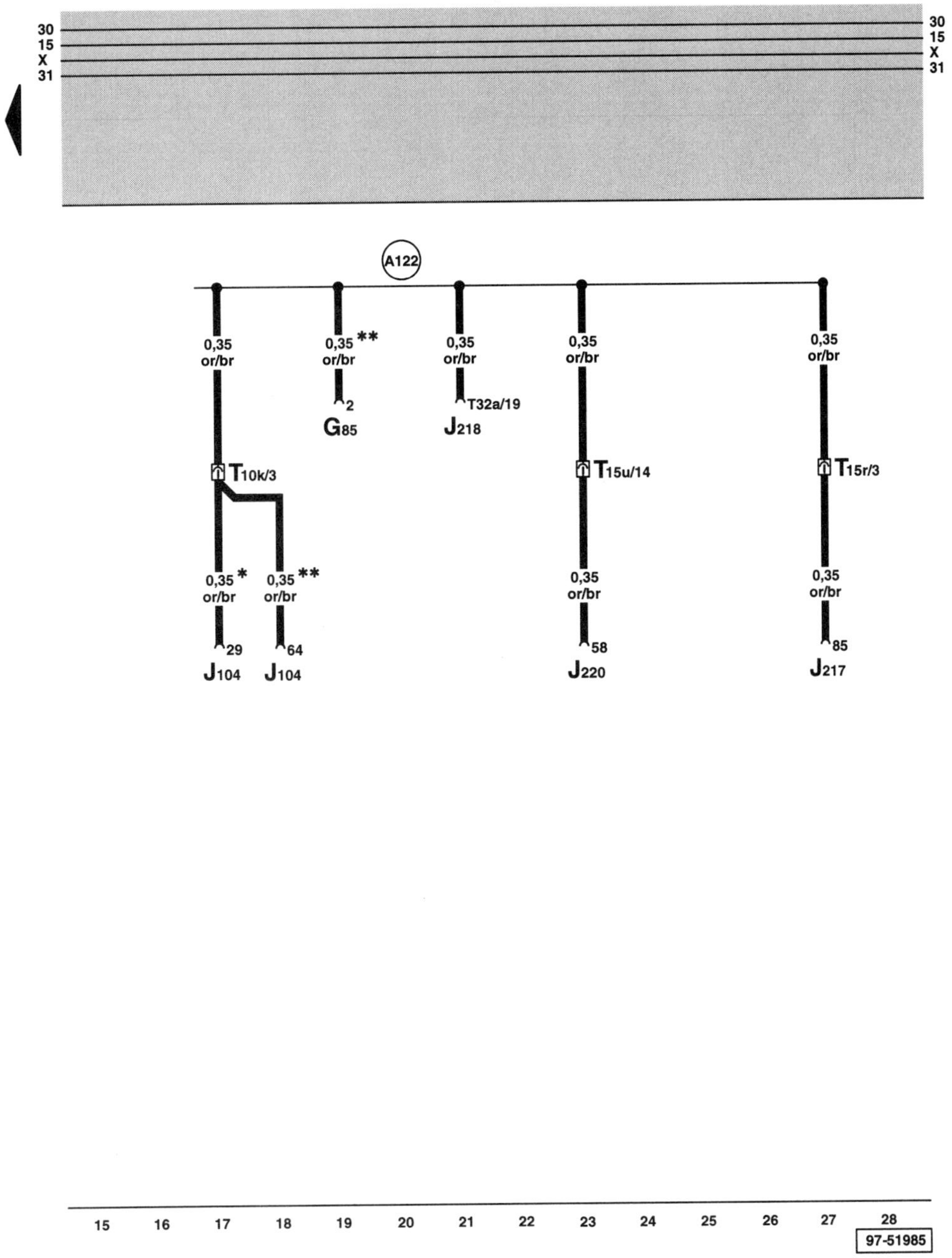

ws = white
sw = black
ro = red
br = brown
gn = green
bl = blue
gr = grey
li = violet
ge = yellow
or = orange

G85 - Steering Angle Sensor
J104 - ABS Control Module (w/EDL)
J217 - Transmission Control Module (TCM)
J218 - Instrument Cluster Combination Processor
J220 - Motronic Engine Control Module (ECM)
T10k - Connector 10 pin, orange, connector station A pillar, left
T15r - Connector 15 pin, blue, connector station A pillar, right
T15u - Connector 15 pin, red, connector station electronic box plenum chamber
T32a - Connector 32 pin, green, on instrument cluster

(A122) - Connection (low bus) in instrument panel wiring harness

* - Vehicles with ABS with EDS and ASR
** - Vehicles with Electronic Stability Program

Servotronic

2000 m. y.

Fuse Panel

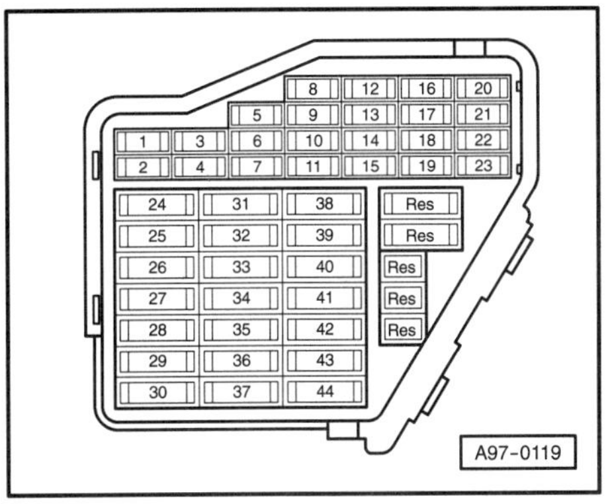

Fuse Colors:

30 A - Green
25 A - White
20 A - Yellow
15 A - Blue
10 A - Red
7,5 A - Brown
5 A - Beige

Starting with fuse position 23, fuses in the fuse holder are identified with 223 in the wiring diagram.

13 - Fold Relay Panel

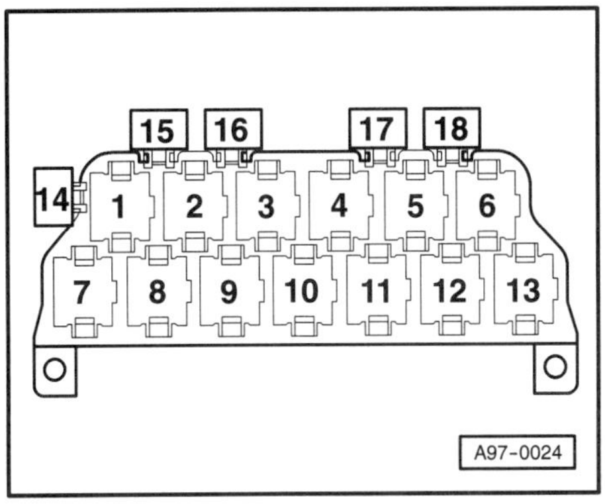

Fuse Location:

13 -Servotronic Control Module, J236

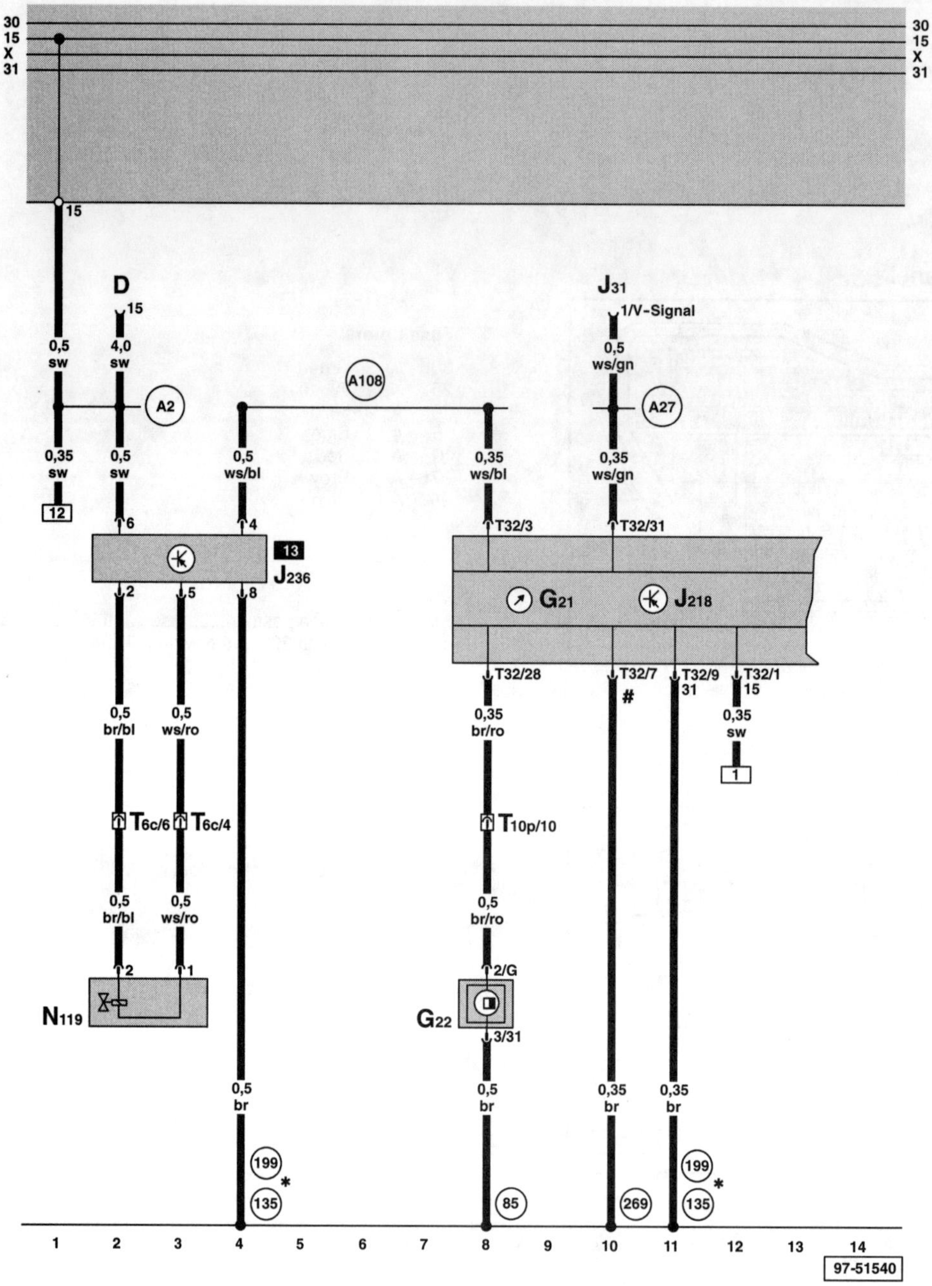

ws = white
sw = black
ro = red
br = brown
gn = green
bl = blue
gr = grey
li = violet
ge = yellow

D	-	Ignition/Starter Switch
G21	-	Speedometer
G22	-	Speedometer Vehicle Speed Sensor (VSS)
J31	-	Wiper/Washer Intermittent Relay
J218	-	Instrument Cluster Combination Processor
J236	-	Servotronic Control Module
N119	-	Servotronic Solenoid Valve
T6c	-	Connector 6 pin, black, connector station A pillar, left
T10p	-	Connector 10 pin, black, connector station electronic box plenum chamber
T32	-	Connector 32 pin, blue, on instrument cluster

(85)	-	Ground connection -1-, in engine compartment wiring harness
(135)	-	Ground connection -2-, in instrument panel wiring harness
(199)	-	Ground connection -3-, in instrument panel wiring harness
(269)	-	Ground connector (sensor ground) -1-, in instrument panel wiring harness
(A2)	-	Plus connection (15), in instrument panel wiring harness
(A27)	-	Wire connection (vehicle speed signal), in instrument panel wiring harness
(A108)	-	Connector (speed signal), in instrument panel wiring harness
*	-	Both are possible
#	-	Exit Sensor ground

Parking Aid

2000 m. y.

Fuse Panel

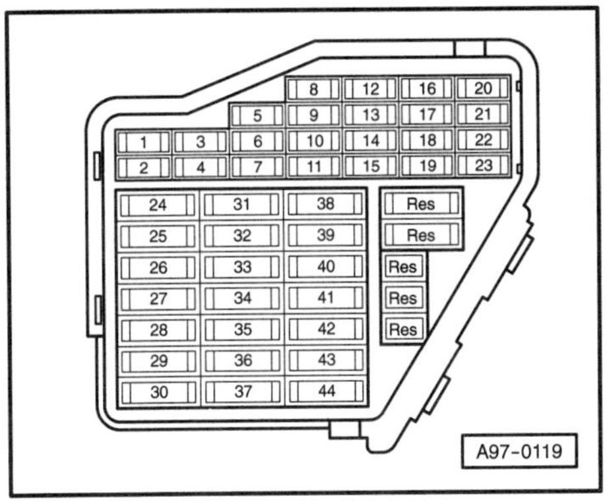

Fuse Colors:

30 A - Green
25 A - White
20 A - Yellow
15 A - Blue
10 A - Red
7,5 A - Brown
5 A - Beige

Starting with fuse position 23, fuses in the fuse holder are identified with 223 in the wiring diagram.

Micro Central Electric Panel

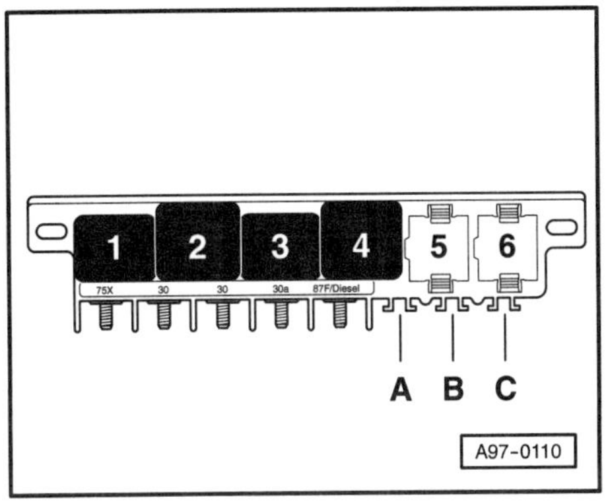

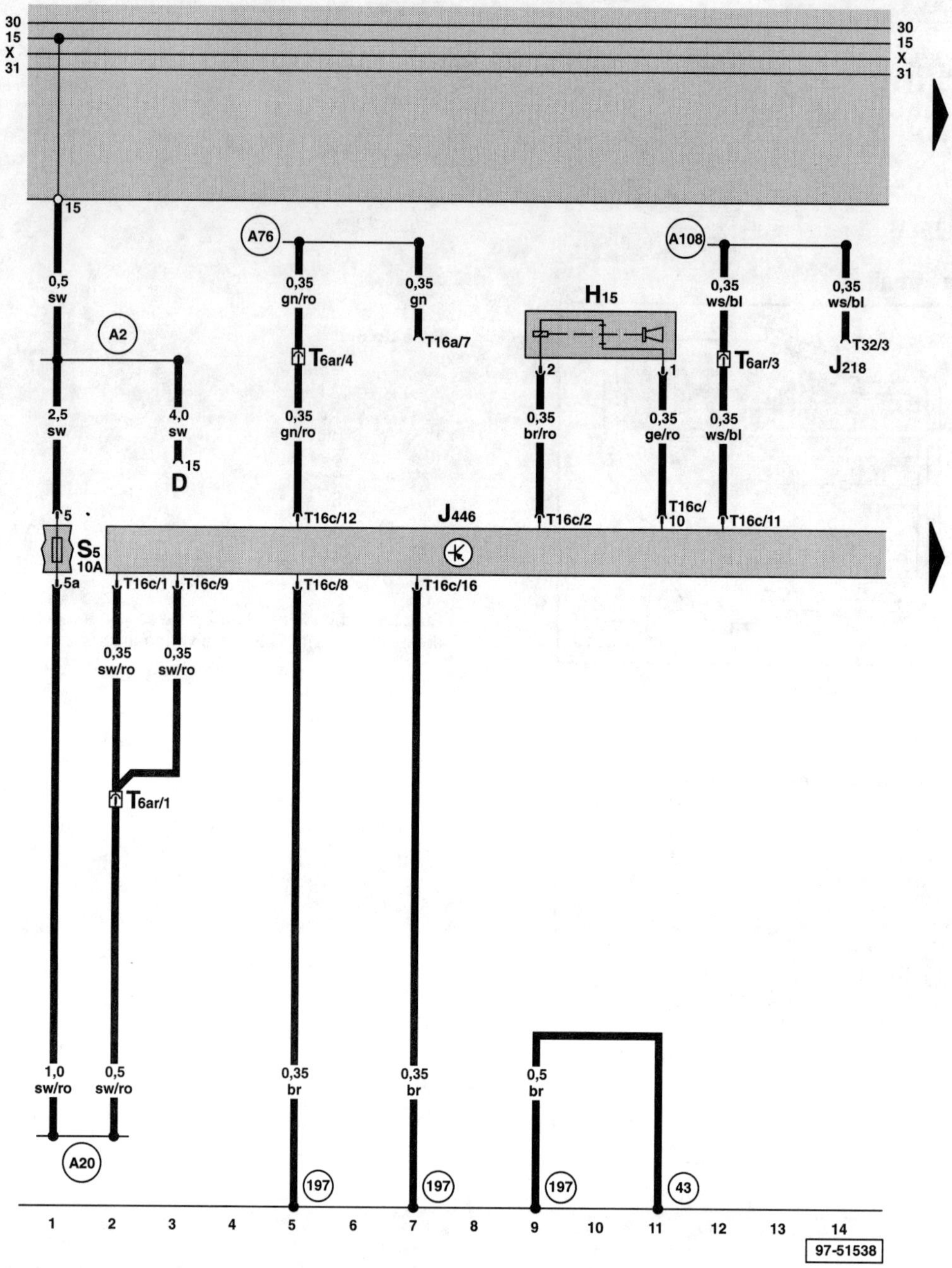

ws = white
sw = black
ro = red
br = brown
gn = green
bl = blue
gr = grey
li = violet
ge = yellow

D	-	Ignition/Starter Switch
H15	-	Warning buzzer for parking
J218	-	Instrument Cluster Combination Processor
J446	-	Control module for parking aid
S5	-	Fuse
T6ar	-	Connector 6 pin, white, connector station A pillar, right
T16a	-	Connector 16 pin, Data Link Connector (DLC)
T16c	-	Connector 16 pin, on Control Module for parking aid
T32	-	Connector 32 pin, blue, on instrument cluster

(43)	-	Ground connection, on right A-pillar, lower part
(197)	-	Ground connection -4-, in wiring harness rear
(A2)	-	Plus connection (15), in instrument panel wiring harness
(A20)	-	Wire connection (15a), in instrument panel wiring harness
(A76)	-	Connector (K-diagnosis wire), in instrument panel wiring harness
(A108)	-	Connector (speed signal), in instrument panel wiring harness

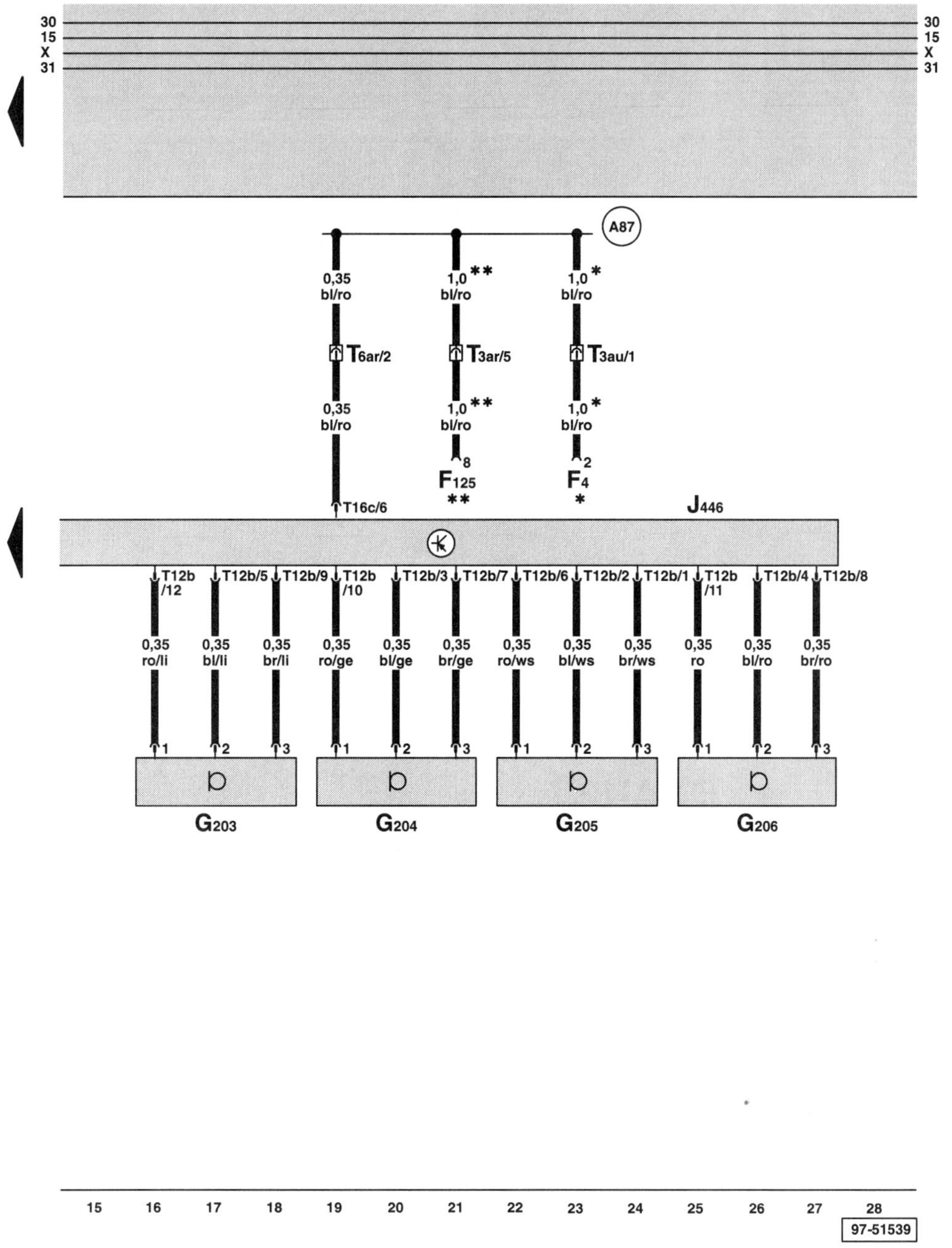

ws = white
sw = black
ro = red
br = brown
gn = green
bl = blue
gr = grey
li = violet
ge = yellow

F4 - Back-Up Light Switch
F125 - Multi-Function Transmission Range (TR) Switch
G203 - Parking sensor, rear left
G204 - Parking sensor, rear left, center
G205 - Parking sensor, rear right, center
G206 - Parking sensor, rear right
J446 - Control module for parking aid
T3ar - Connector 3 pin, blue, connector station A pillar, right
T3au - Connector 3 pin, red, connector station electronic box plenum chamber
T6ar - Connector 6 pin, white, connector station A pillar, right
T12b - Connector 12 pin, black, on Control Module for parkind aid
T16c - Connector 16 pin, on Control Module for parking aid

(A87) - Connector (RF), in instrument panel wiring harness

* - Vehicles with manually shifted transmission
** - Vehicles with automatic transmission

THIS PAGE INTENTIONALLY LEFT BLANK

Level Control System

2000 m. y.

Fuse Panel

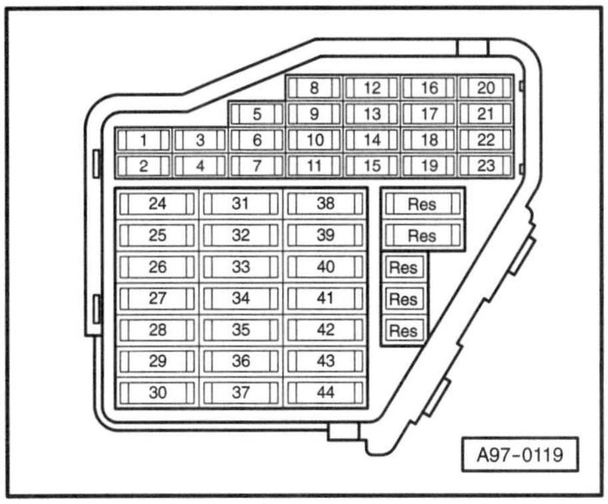

Fuse Colors:

30 A - Green
25 A - White
20 A - Yellow
15 A - Blue
10 A - Red
7,5 A - Brown
5 A - Beige

Starting with fuse position 23, fuses in the fuse holder are identified with 223 in the wiring diagram.

Micro Central Electric Panel

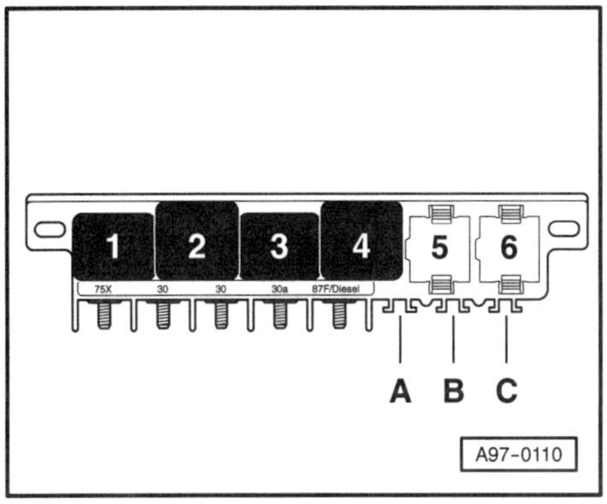

8 - Fold Relay Panel

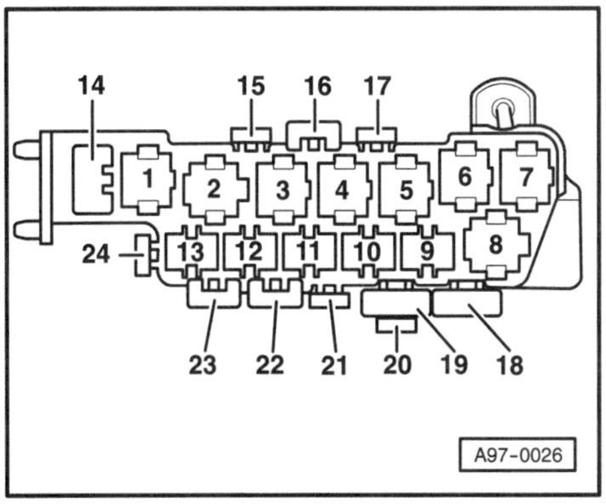

Fuse Location:

23 -Self-Levelling System Fuse, S110

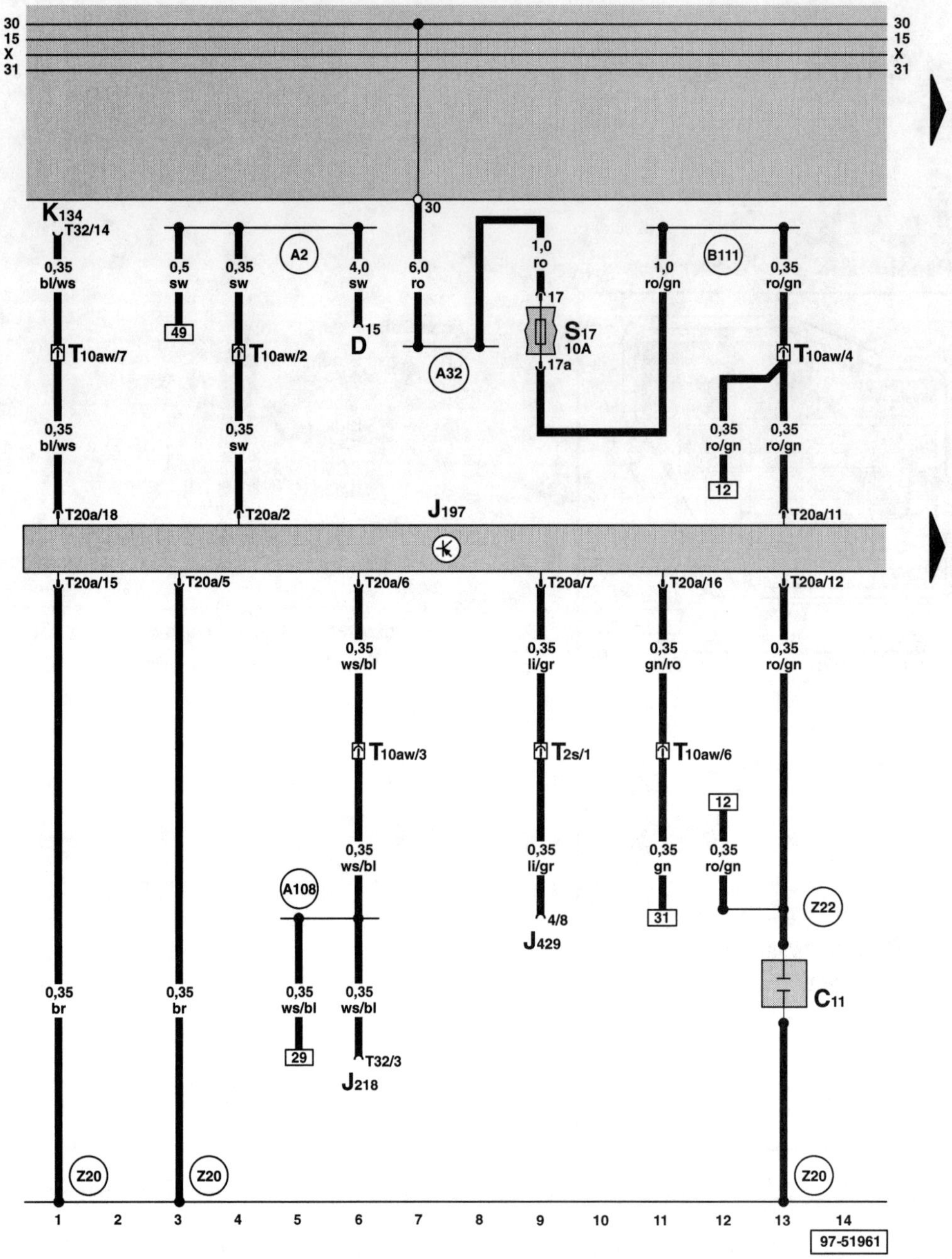

ws = white
sw = black
ro = red
br = brown
gn = green
bl = blue
gr = grey
li = violet
ge = yellow

C11 - Combustion Air Blower Capacitor
D - Ignition/Starter Switch
J197 - Level Control System Control Module
J218 - Instrument Cluster Combination Processor
J429 - Control module for central locking
K134 - Indicator lamp for level control system
S17 - Fuse
T2s - Connector 2 pin, black, connector station A pillar, right
T10aw- Connector 10 pin, black, connector station A pillar, right
T20a - Connector 20 pin, red, on Control Module for Level Control
T32 - Connector 32 pin, blue, on instrument cluster

(A2) - Plus connection (15), in instrument panel wiring harness
(A32) - Plus connection (30), in instrument panel wiring harness
(A108) - Connector (speed signal), in instrument panel wiring harness
(B111) - Connector (30a), in wiring harness interior
(Z20) - Wire connection, in self-leveling wiring harness
(Z22) - Plus connection (30), in self-levelling System wiring harness

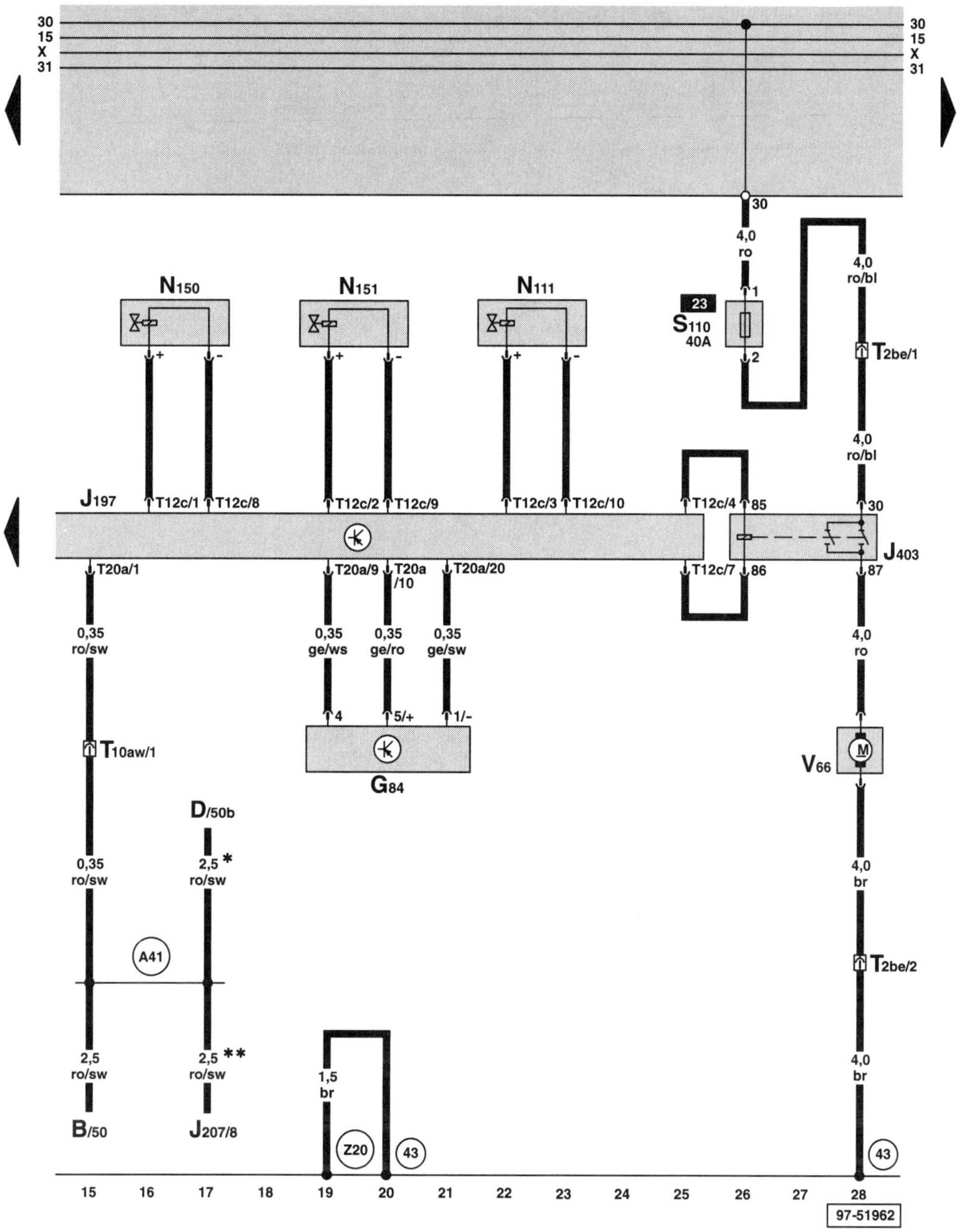

ws = white
sw = black
ro = red
br = brown
gn = green
bl = blue
gr = grey
li = violet
ge = yellow

B - Starter
D - Ignition/Starter Switch
G84 - Level Control System Sensor
J197 - Level Control System Control Module
J207 - Starting Interlock Relay
J403 - Relay for compressor level control system
N111 - Level Control System Solenoid Valve
N150 - Left Rear Suspension Strut Valve
N151 - Right Rear Suspension Strut Valve
S110 - Self-Levelling System Fuse
T2be - Connector 2 pin, near Control Module for Level Control
T10aw- Connector 10 pin, black, connector station A pillar, right
T12c - Connector 12 pin, red, on Control Module for Level Control
T20a - Connector 20 pin, red, on Control Module for Level Control
V66 - Level Control System Compressor Motor

(43) - Ground connection, on right A-pillar, lower part
(A41) - Plus connection (50), in instrument panel wiring harness
(Z20) - Wire connection, in self-leveling wiring harness
* - Vehicles with manually shifted transmission
** - Vehicles with automatic transmission

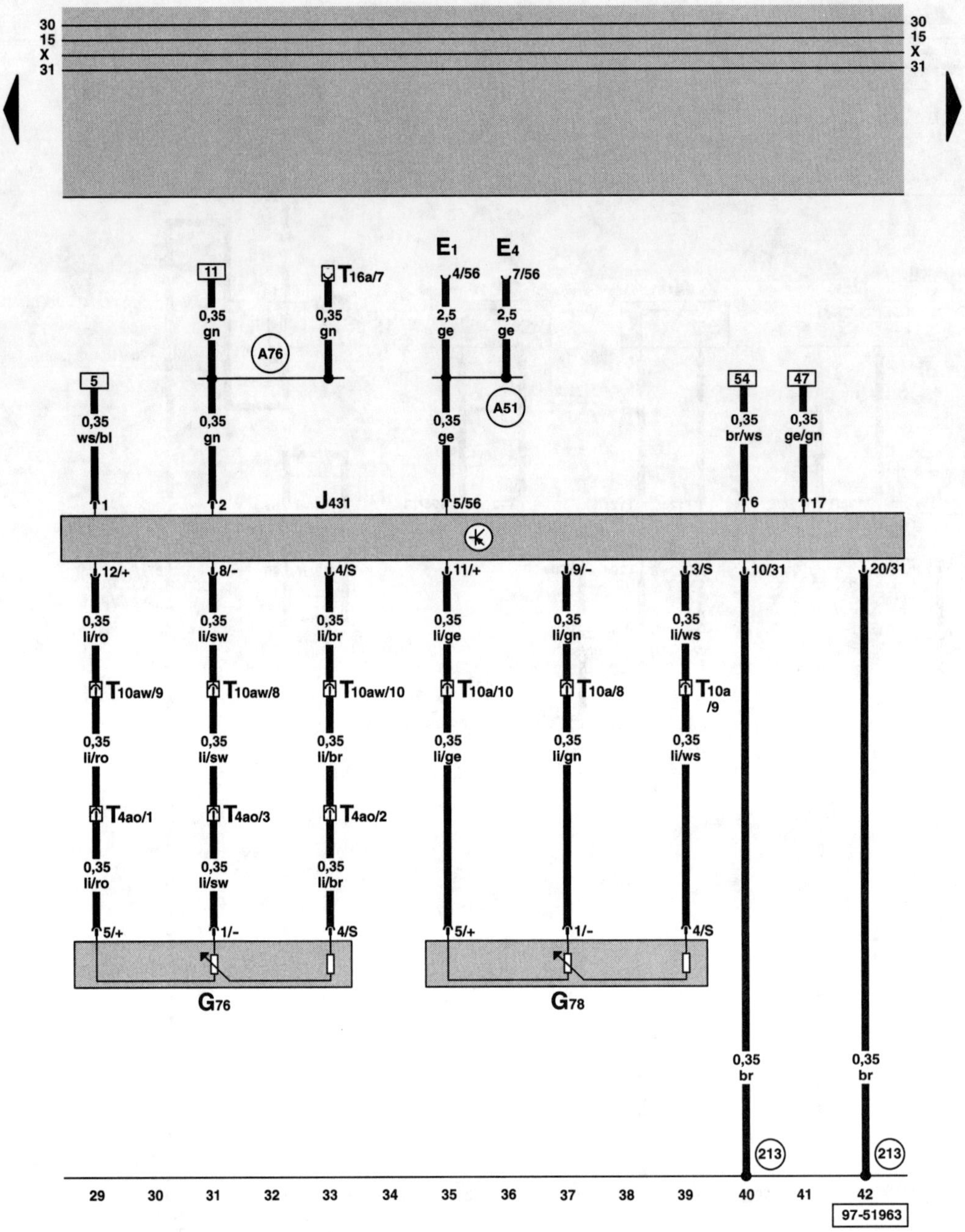

ws = white
sw = black
ro = red
br = brown
gn = green
bl = blue
gr = grey
li = violet
ge = yellow

E1 - Light switch
E4 - Headlight Dimmer/Flasher Switch
G76 - Left Rear Level Control System Sensor
G78 - Left Front Level Control System Sensor
J431 - Control module for Headlight Beam Adjusting
T4ao - Connector 4 pin, black, below rear seat, left
T10a - Connector 10 pin, pink, connector station A pillar, left
T10aw- Connector 10 pin, black, connector station A pillar, right
T16a - Connector 16 pin, Data Link Connector (DLC)

(213) - Ground connection, in instrument panel wiring harness

(A51) - Wire connection (56), in instrument panel wiring harness

(A76) - Connector (K-diagnosis wire), in instrument panel wiring harness

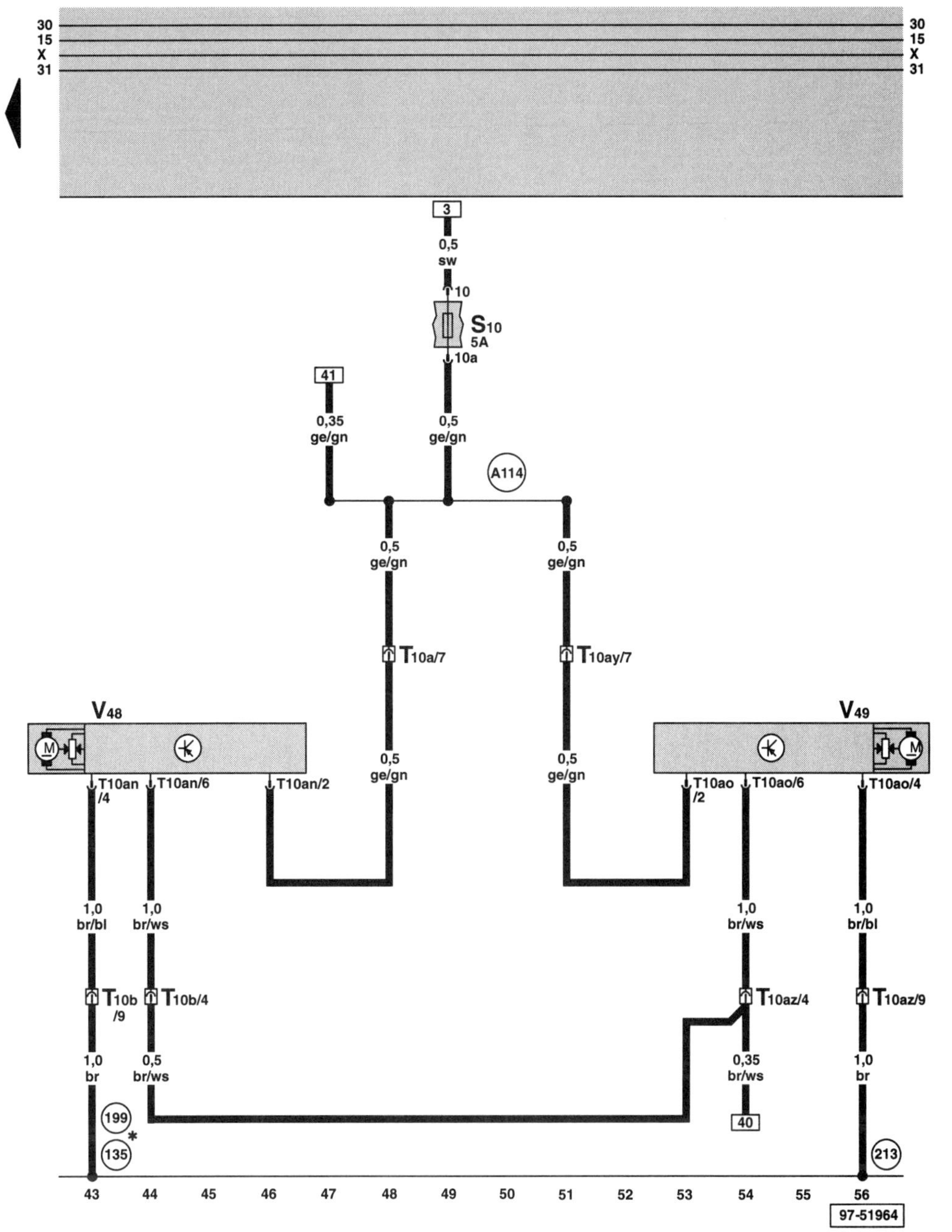

ws = white
sw = black
ro = red
br = brown
gn = green
bl = blue
gr = grey
li = violet
ge = yellow

S10 - Fuse
T10a - Connector 10 pin, pink, connector station A pillar, left
T10b - Connector 10 pin, red, connector station A pillar, left
T10an - Connector 10 pin, on headlight, left
T10ao - Connector 10 pin, on headlight, right
T10ay - Connector 10 pin, pink, connector station A pillar, right
T10az - Connector 10 pin, red, connector station A pillar, right
V48 - Left Headlight Beam Adjusting Motor
V49 - Right Headlight Beam Adjusting Motor

(135) - Ground connection -2-, in instrument panel wiring harness
(199) - Ground connection -3-, in instrument panel wiring harness
(213) - Ground connection, in instrument panel wiring harness
(A114) - Connection (headlight beam adjusting) in instrument panel wiring harness
* - Both are possible

Cruise Control with Mechanical Throttle Actuation

2000 m. y.

Fuse Panel

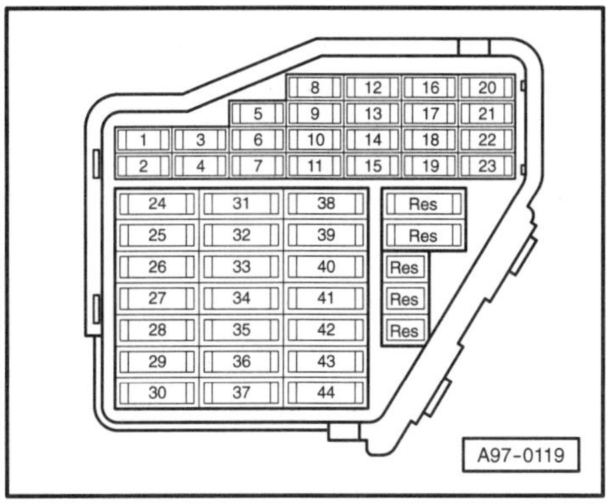

Fuse Colors:

30 A - Green
25 A - White
20 A - Yellow
15 A - Blue
10 A - Red
7,5 A - Brown
5 A - Beige

Starting with fuse position 23, fuses in the fuse holder are identified with 223 in the wiring diagram.

Micro Central Electric Panel

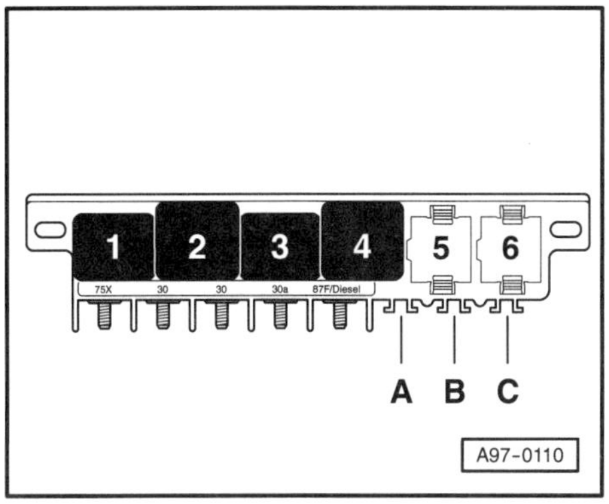

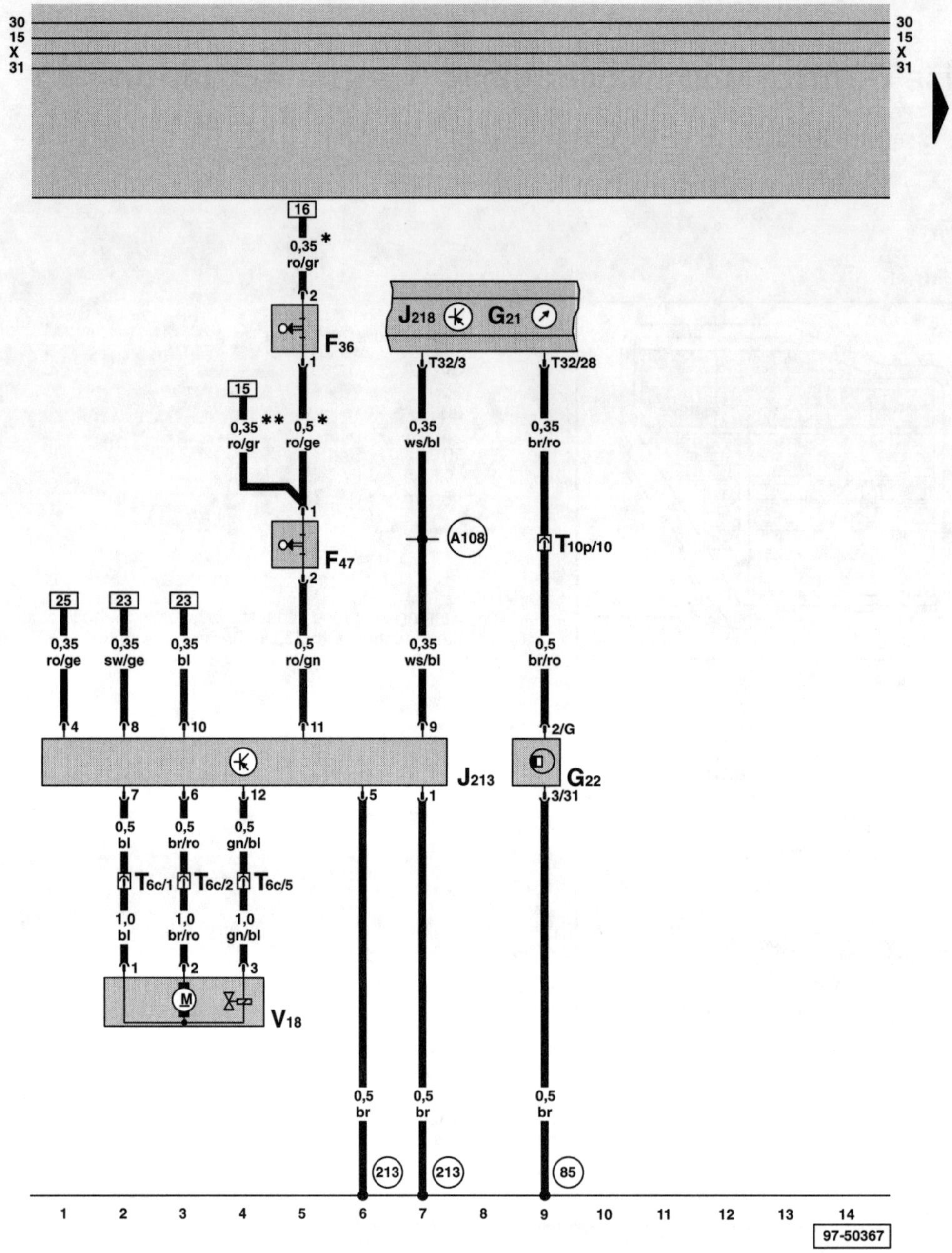

ws = white
sw = black
ro = red
br = brown
gn = green
bl = blue
gr = grey
li = violet
ge = yellow

F36 - Clutch Vacuum Vent Valve Switch
F47 - Vacuum Vent Valve, Brake
G21 - Speedometer
G22 - Speedometer Vehicle Speed Sensor (VSS)
J213 - Cruise Control, Control Module
J218 - Instrument Cluster Combination Processor
T6c - Connector 6 pin, black, connector station A pillar, left
T10p - Connector 10 pin, black, connector station electronic box plenum chamber
T32 - Connector 32 pin, blue, on instrument cluster
V18 - Cruise Control Vacuum Pump

(85) - Ground connection -1-, in engine compartment wiring harness
(213) - Ground connection, in instrument panel wiring harness
(A108) - Connector (speed signal), in instrument panel wiring harness
* - Vehicles with manually shifted transmission
** - Vehicles with automatic transmission

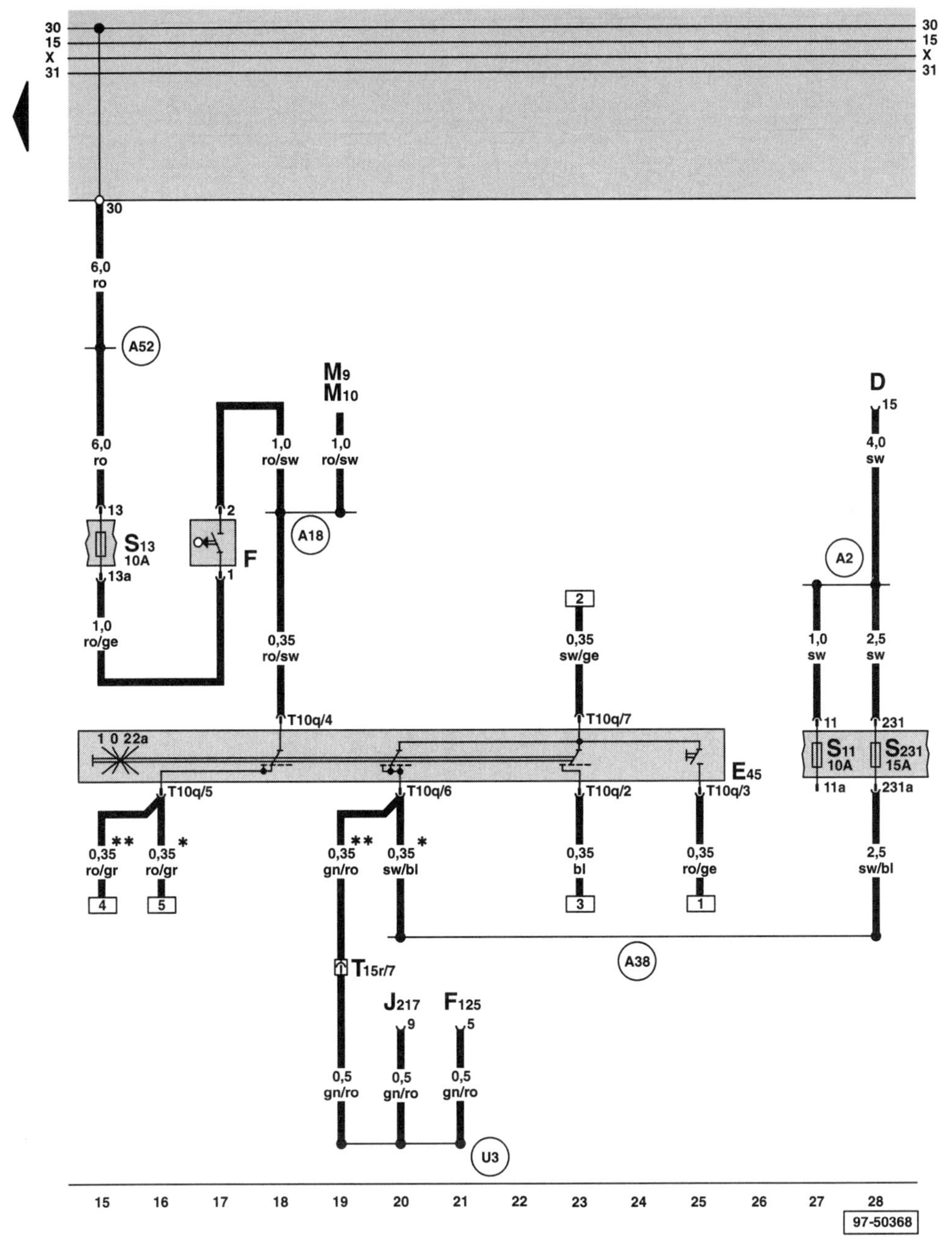

ws = white
sw = black
ro = red
br = brown
gn = green
bl = blue
gr = grey
li = violet
ge = yellow

D - Ignition/Starter Switch
E45 - Cruise Control Switch
F - Brake Light Switch
F125 - Multi-Function Transmission Range (TR) Switch
J217 - Transmission Control Module (TCM)
M9 - Left Brake Light
M10 - Right Brake Light
S11 - Fuse
S13 - Fuse
S231 - Fuse 31 in fuse holder
T10q - Connector 10 pin, black, on cruise control switch
T15r - Connector 15 pin, blue, connector station A pillar, right

(A2) - Plus connection (15), in instrument panel wiring harness
(A18) - Wire connection (54), in instrument panel wiring harness
(A38) - Plus connection (15a II), in instrument panel wiring harness
(A52) - Plus connection -2- (30), in instrument panel wiring harness
(U3) - Wire connection -1-, in automatic transmission wiring harness

* - Vehicles with manually shifted transmission
** - Vehicles with automatic transmission

THIS PAGE INTENTIONALLY LEFT BLANK

Heated Steering Wheel

2000 m. y.

Fuse Panel

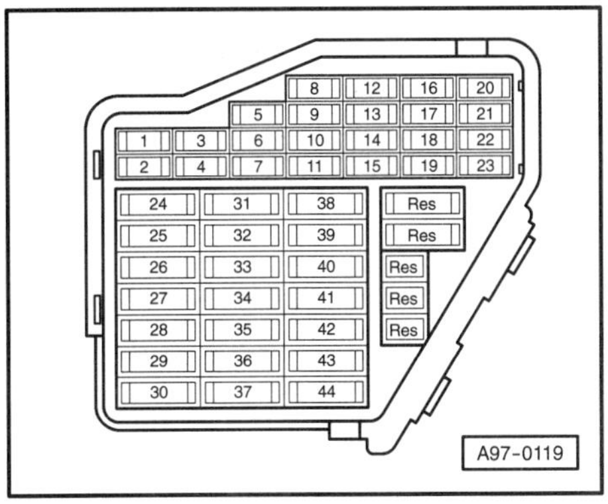

Fuse Colors:

30 A - Green
25 A - White
20 A - Yellow
15 A - Blue
10 A - Red
7,5 A - Brown
5 A - Beige

Starting with fuse position 23, fuses in the fuse holder are identified with 223 in the wiring diagram.

Micro Central Electric Panel

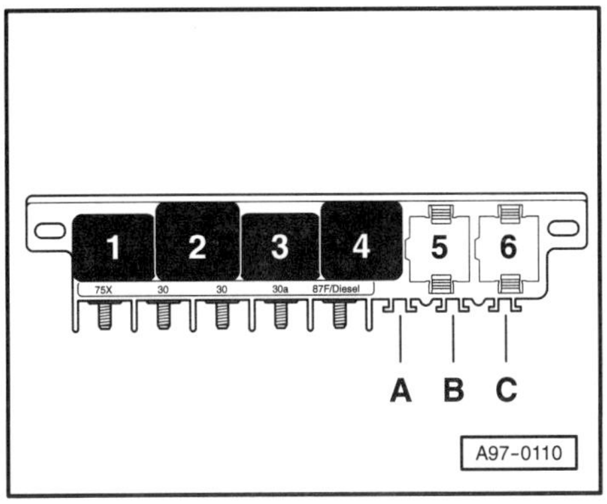

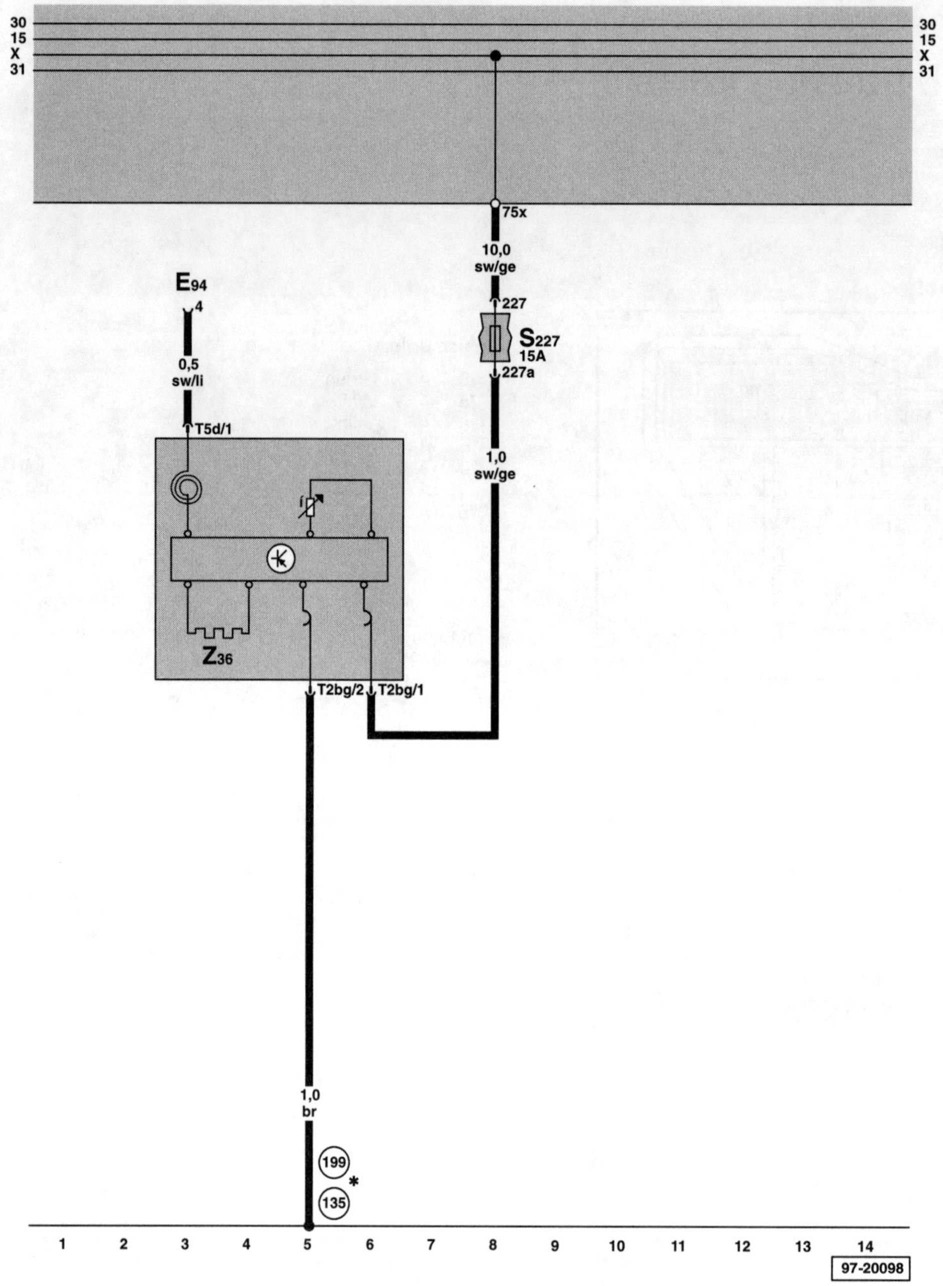

ws = white
sw = black
ro = red
br = brown
gn = green
bl = blue
gr = grey
li = violet
ge = yellow

E94 - Adjuster for heated driver's seat
S227 - Fuse 27 in fuse holder
T2bg - Connector 2 pin, behind steering column switch cover
T5d - Connector 5 pin, behind steering column switch cover
Z36 - Heated steering wheel

(135) - Ground connection -2-, in instrument panel wiring harness
(199) - Ground connection -3-, in instrument panel wiring harness

* - Both are possible

Steering Wheel with Tiptronic Switch

2000 m. y.

Fuse Panel

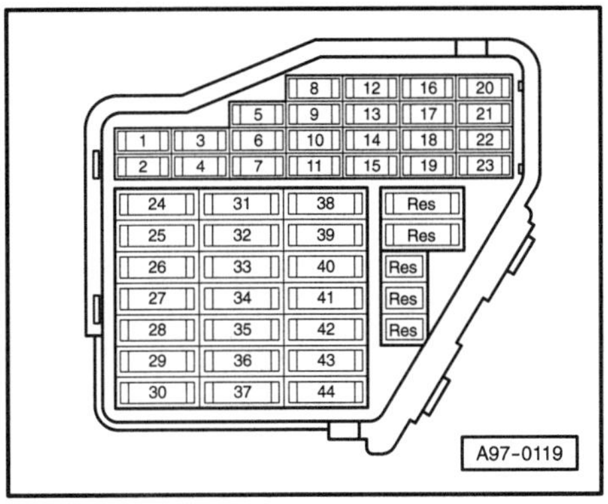

Fuse Colors:

30 A - Green
25 A - White
20 A - Yellow
15 A - Blue
10 A - Red
7,5 A - Brown
5 A - Beige

Starting with fuse position 23, fuses in the fuse holder are identified with 223 in the wiring diagram.

Micro Central Electric Panel

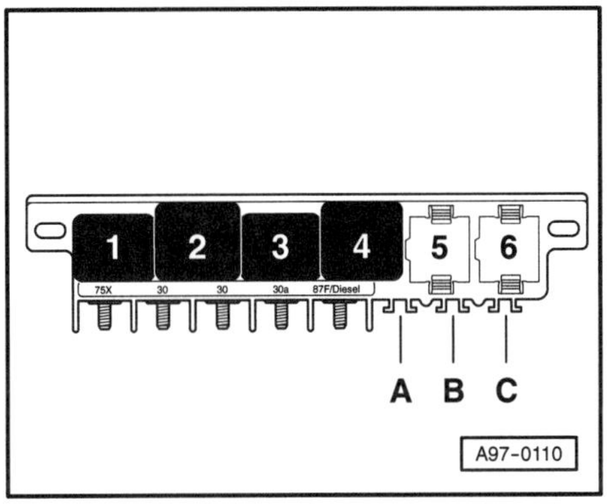

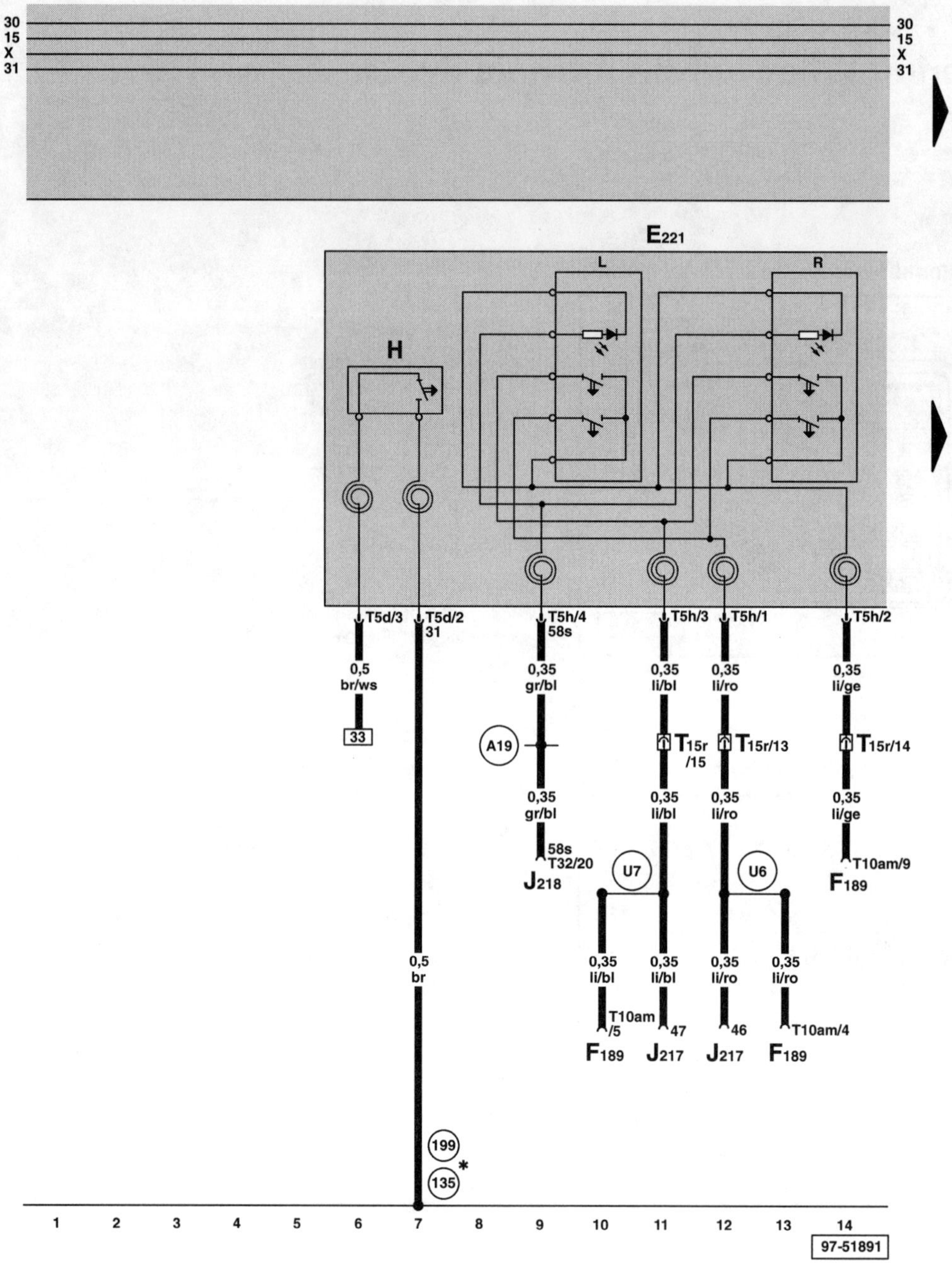

ws = white
sw = black
ro = red
br = brown
gn = green
bl = blue
gr = grey
li = violet
ge = yellow

E221 - Control unit in steering wheel
F189 - Tiptronic Switch
H - Horn Button
J217 - Transmission Control Module (TCM)
J218 - Instrument Cluster Combination Processor
T5d - Connector 5 pin, behind steering column switch cover
T5h - Connector 5 pin, black, in steering wheel
T10am- Connector 10 pin, on selector lever
T15r - Connector 15 pin, blue, connector station A pillar, right
T32 - Connector 32 pin, blue, on instrument cluster

(135) - Ground connection -2-, in instrument panel wiring harness
(199) - Ground connection -3-, in instrument panel wiring harness
(A19) - Wire connection (58s), in instrument panel wiring harness
(U6) - Wire connection -2-, in automatic transmission wiring harness
(U7) - Wire connection -3-, in automatic transmission wiring harness
* - Both are possible

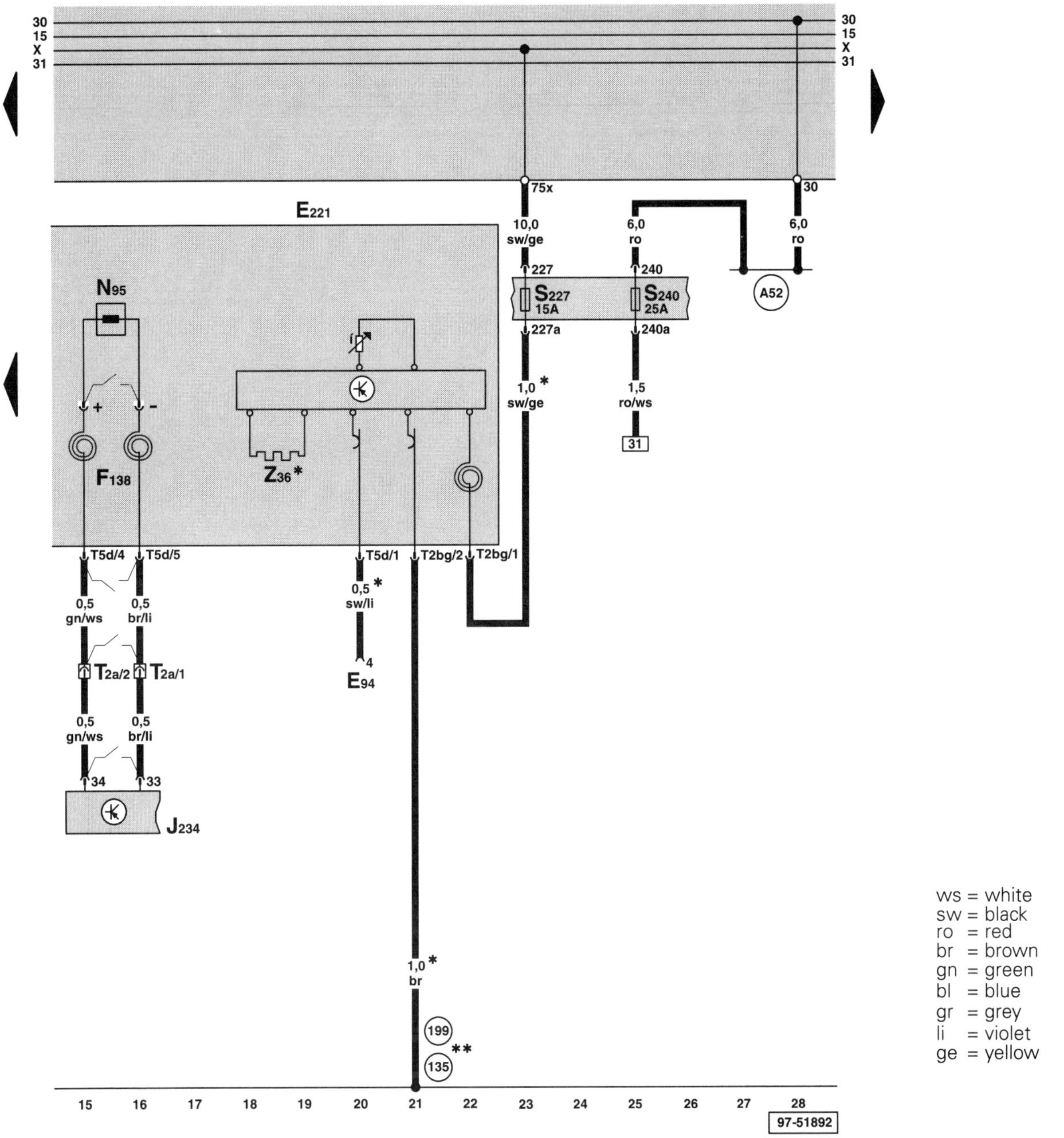

E94 - Adjuster for heated driver's seat
E221 - Control unit in steering wheel
F138 - Airbag Spiral Spring/Return Spring With Slip Ring
J234 - Airbag Control Module
N95 - Driver's Side Airbag Igniter
S227 - Fuse 27 in fuse holder
S240 - Fuse 40 in fuse holder
T2a - Connector 2 pin, yellow, connector station A pillar, left
T2bg - Connector 2 pin, behind steering column switch cover
T5d - Connector 5 pin, behind steering column switch cover
Z36 - Heated steering wheel

(135) - Ground connection -2-, in instrument panel wiring harness
(199) - Ground connection -3-, in instrument panel wiring harness
(A52) - Plus connection -2- (30), in instrument panel wiring harness

- Bridge closes when disconnecting connector (safety feature)
* - Vehicles with heated steering wheel
** - Both are possible

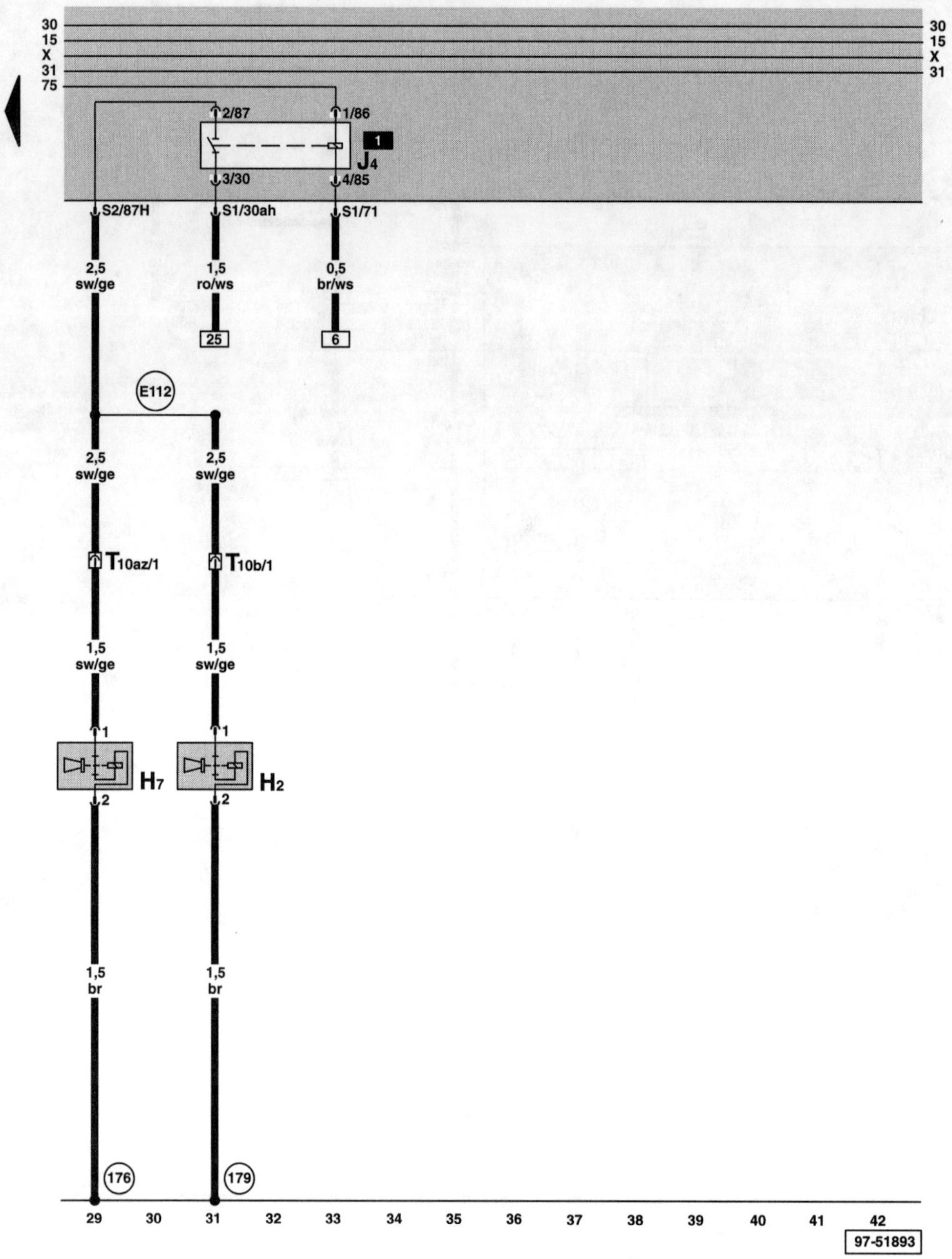

ws = white
sw = black
ro = red
br = brown
gn = green
bl = blue
gr = grey
li = violet
ge = yellow

H2 - High Tone Horn
H7 - Low Tone Horn
J4 - Dual Horn Relay
T10b - Connector 10 pin, red, connector station A pillar, left
T10az - Connector 10 pin, red, connector station A pillar, right

(176) - Ground connection, in right headlight wiring harness
(179) - Ground connection, in left headlight wiring harness
(E112) - Connector (Horn - 87h), in instrument panel wiring harness

Power Steering Column Adjustment with and without Memory

2000 m. y.

Fuse Panel

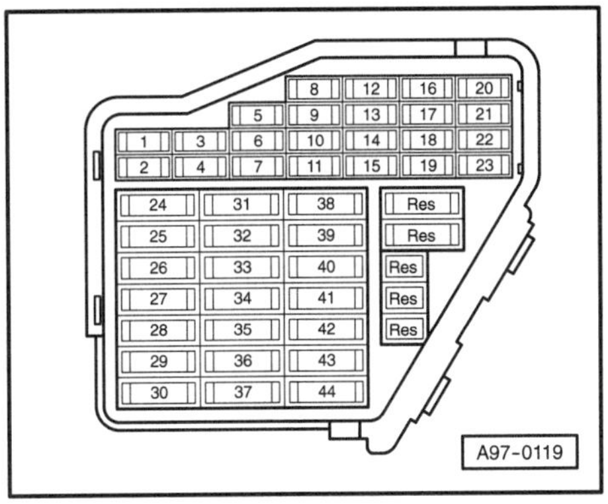

Fuse Colors:

30 A - Green
25 A - White
20 A - Yellow
15 A - Blue
10 A - Red
7,5 A - Brown
5 A - Beige

Starting with fuse position 23, fuses in the fuse holder are identified with 223 in the wiring diagram.

Micro Central Electric Panel

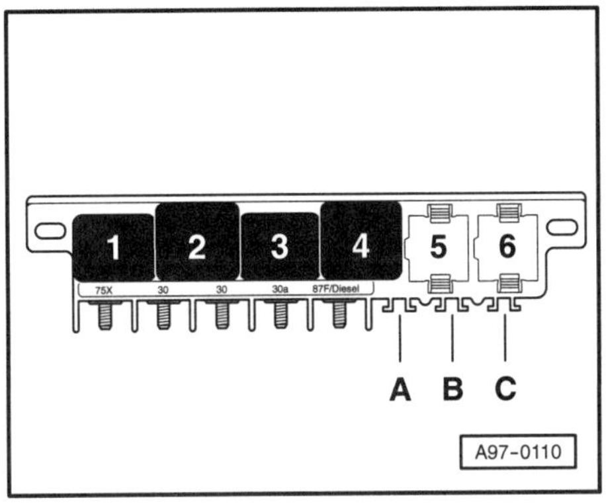

Fuse Location:

A -Fuse for steering column adjustment, S275

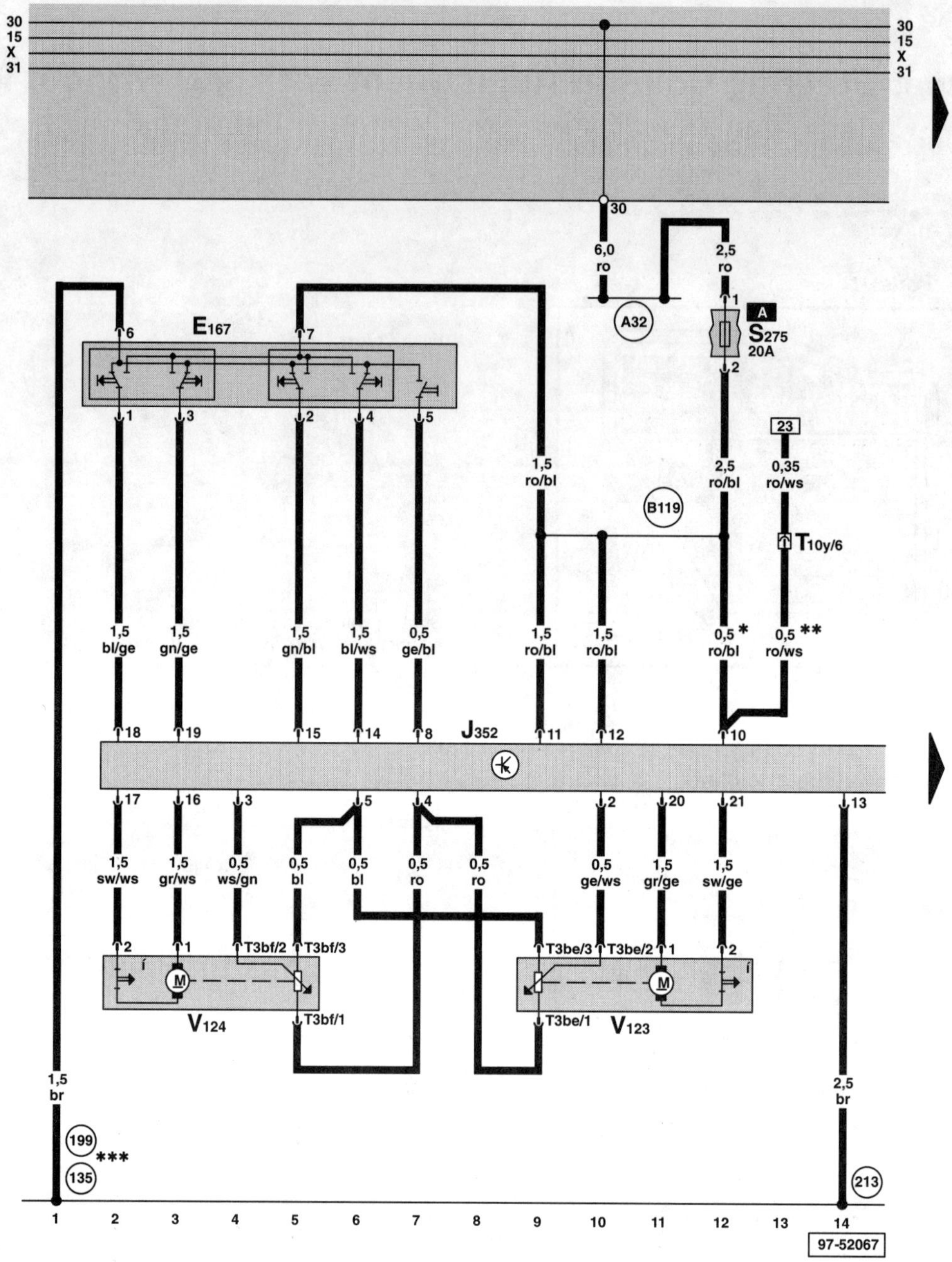

ws = white
sw = black
ro = red
br = brown
gn = green
bl = blue
gr = grey
li = violet
ge = yellow

E167 - Switch for steering column adjustment
J352 - Control Module for Steering column-belt height adjustment
S275 - Fuse for steering column adjustment
T3be - Connector 3 pin, blue, Motor for steerring column adjustment, vertical
T3bf - Connector 3 pin, blue, Motor for steerring column adjustment, axial
T10y - Connector 10 pin, green, connector station A pillar, right
V123 - Motor for steerring column adjustment, vertical
V124 - Motor for steerring column adjustment, axial

(135) - Ground connection -2-, in instrument panel wiring harness
(199) - Ground connection -3-, in instrument panel wiring harness
(213) - Ground connection, in instrument panel wiring harness
(A32) - Plus connection (30), in instrument panel wiring harness
(B119) - Connector (30a, steering column./belt height adjustment), in wiring harness interior

* - Vehicles without Memory
** - Vehicles with Memory
*** - Both are possible

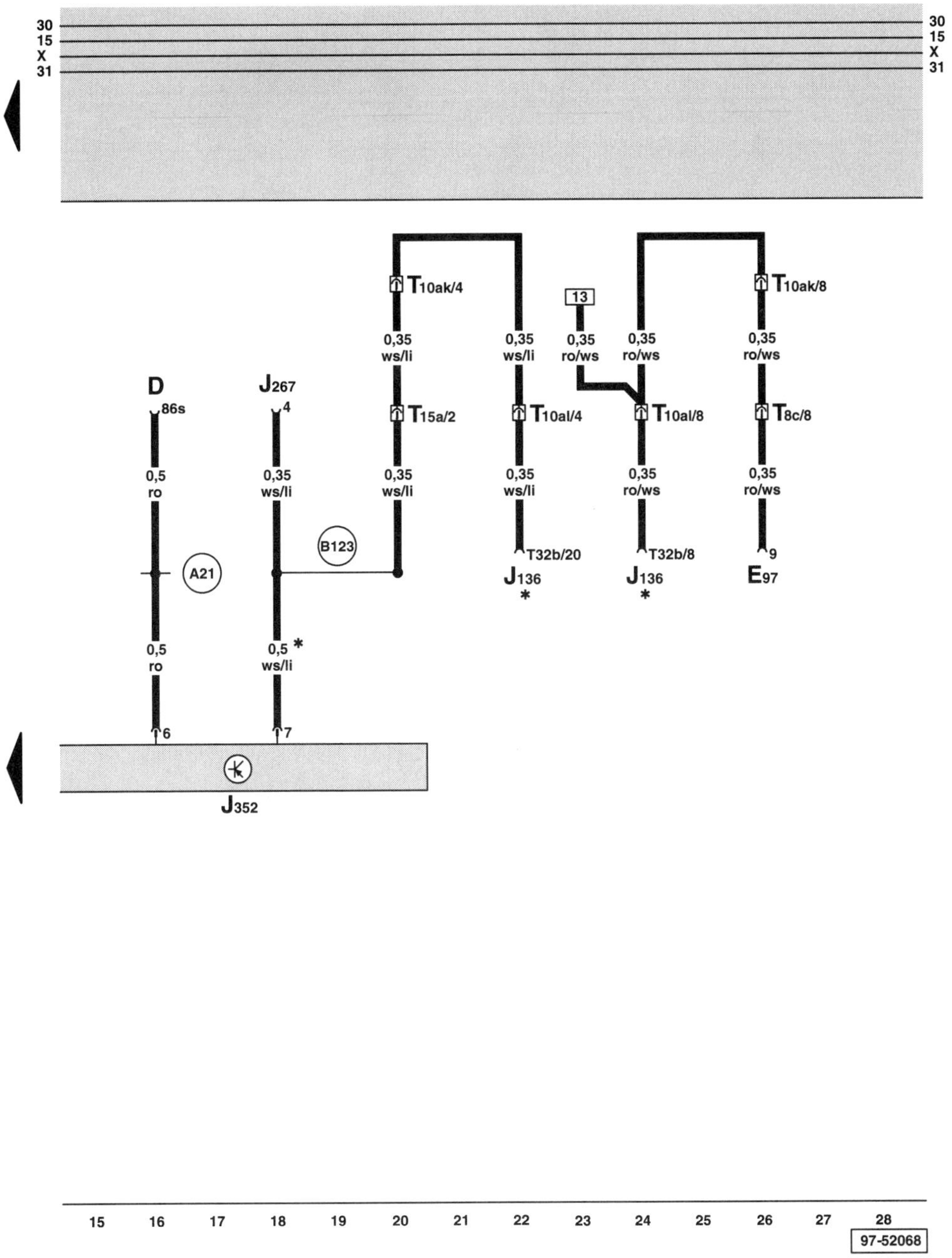

ws = white
sw = black
ro = red
br = brown
gn = green
bl = blue
gr = grey
li = violet
ge = yellow

D - Ignition/Starter Switch
E97 - Memory Program Switch
J136 - Memory Seat Control Module
J267 - Mirror Memory Control Module
J352 - Control Module for Steering column-belt height adjustment
T8c - Connector 8 pin, black, in driver's door
T10ak - Connector 10 pin, green, connector station A pillar, left
T10al - Connector 10 pin, brown, below driver's seat
T15a - Connector 15 pin, black, connector station A pillar, left
T32b - Connector 32 pin, blue, in driver's seat

(A21) - Wire connection (86s), in instrument panel wiring harness

(B123) - Connection (75a heated seat, mirror switch seat memory), in wiring harness interior

* - Vehicles with Memory

Automatic Dimmer Inside and Outside Mirror

2000 m. y.

Fuse Panel

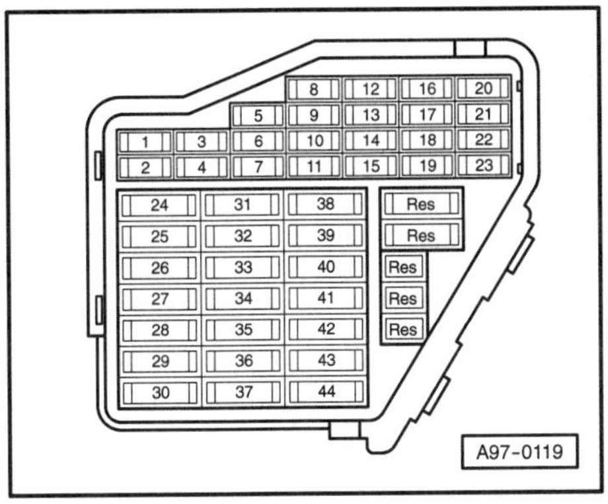

Fuse Colors:

30 A - Green
25 A - White
20 A - Yellow
15 A - Blue
10 A - Red
7,5 A - Brown
5 A - Beige

Starting with fuse position 23, fuses in the fuse holder are identified with 223 in the wiring diagram.

Micro Central Electric Panel

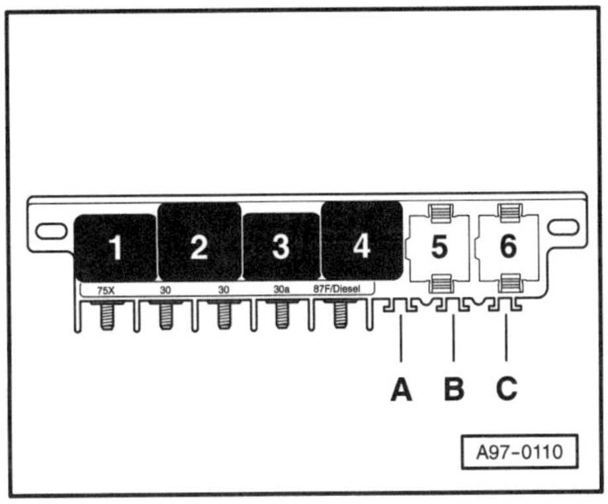

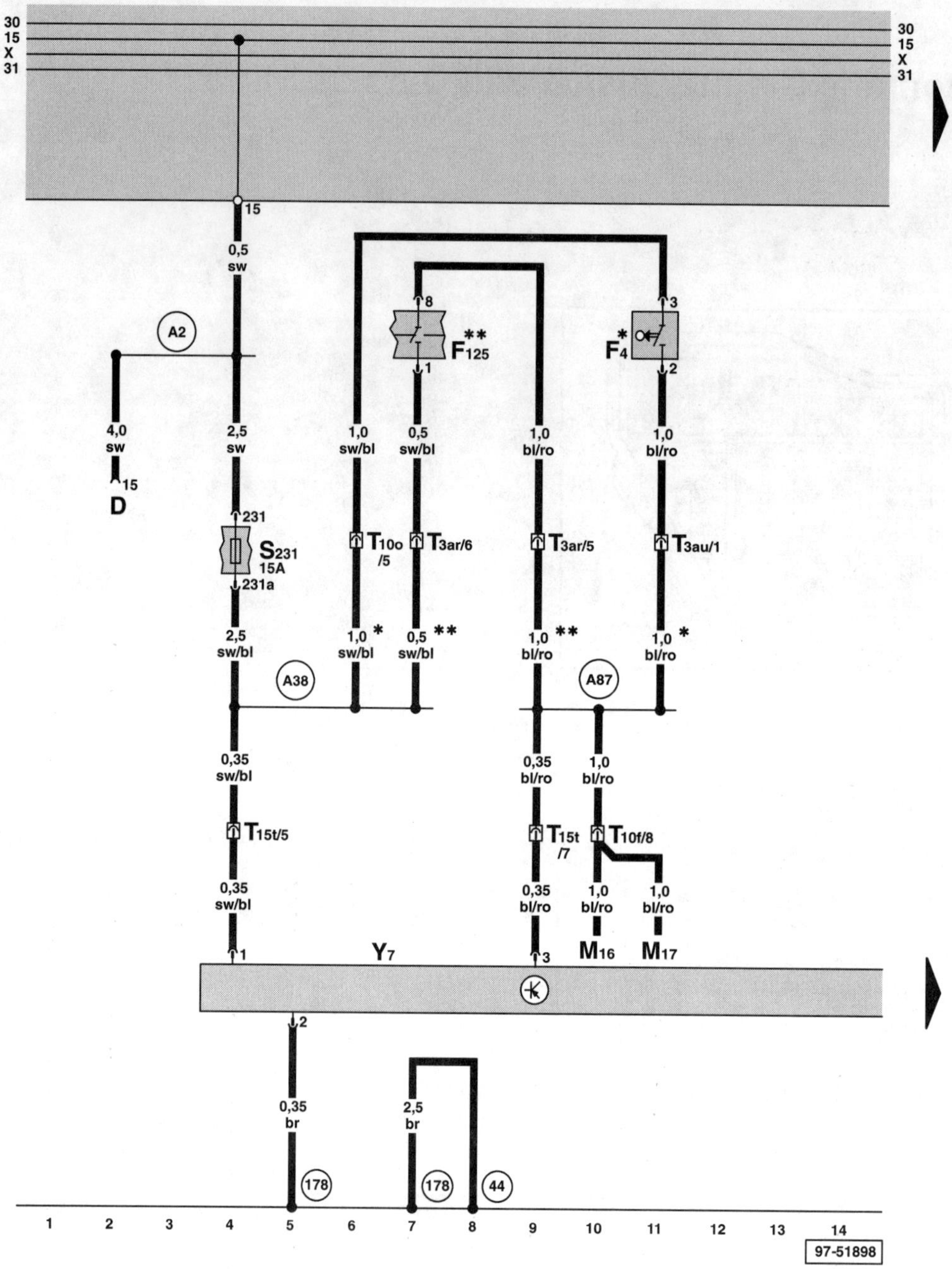

ws = white
sw = black
ro = red
br = brown
gn = green
bl = blue
gr = grey
li = violet
ge = yellow

D - Ignition/Starter Switch
F4 - Back-Up Light Switch
F125 - Multi-Function Transmission Range (TR) Switch
M16 - Left Back-Up Light
M17 - Right Back-Up Light
S231 - Fuse 31 in fuse holder
T3ar - Connector 3 pin, blue, connector station A pillar, right
T3au - Connector 3 pin, red, connector station electronic box plenum chamber
T10f - Connector 10 pin, brown, connector station A pillar, left
T10o - Connector 10 pin, brown, connector station electronic box plenum chamber
T15t - Connector 15 pin, grey, connector station A pillar, left
Y7 - Automatic day/night interior mirror

(44) - Ground connection, on left A-pillar, lower part
(178) - Ground connection, in sunroof wiring harness
(A2) - Plus connection (15), in instrument panel wiring harness
(A38) - Plus connection (15a II), in instrument panel wiring harness
(A87) - Connector (RF), in instrument panel wiring harness

* - Vehicles with manually shifted transmission
** - Vehicles with automatic transmission

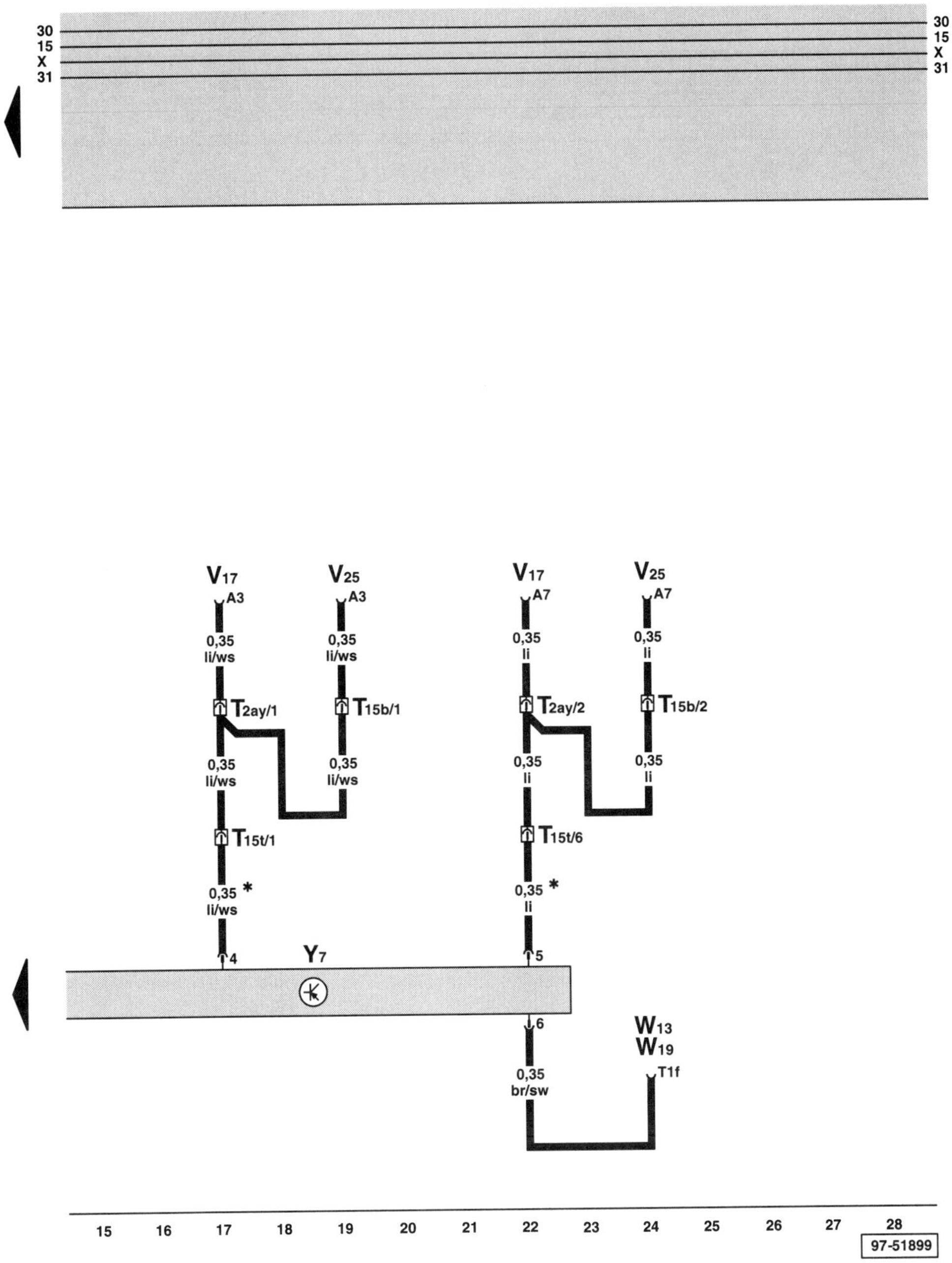

ws = white
sw = black
ro = red
br = brown
gn = green
bl = blue
gr = grey
li = violet
ge = yellow

T1f - Single connector, black, on front interior light
T2ay - Connector 2 pin, green, connector station A pillar, left
T15b - Connector 15 pin, black, connector station A pillar, right
T15t - Connector 15 pin, grey, connector station A pillar, left
V17 - Driver's Side Mirror Adjustment Motor
V25 - Passenger's Side Mirror Adjustment Motor
W13 - Right Front Map/Reading Light
W19 - Left Front Reading Light
Y7 - Automatic day/night interior mirror

* - Vehicles with automatic dimmer outside mirror

THIS PAGE INTENTIONALLY LEFT

BLANK

Garage Door Opener

2000 m. y.

Fuse Panel

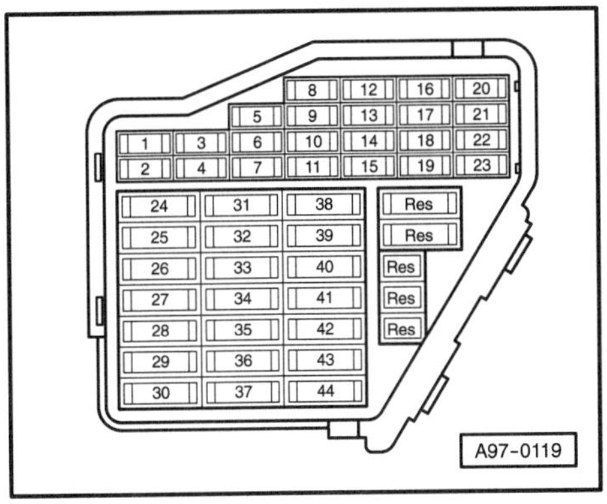

Fuse Colors:

30 A - Green
25 A - White
20 A - Yellow
15 A - Blue
10 A - Red
7,5 A - Brown
5 A - Beige

Starting with fuse position 23, fuses in the fuse holder are identified with 223 in the wiring diagram.

Micro Central Electric Panel

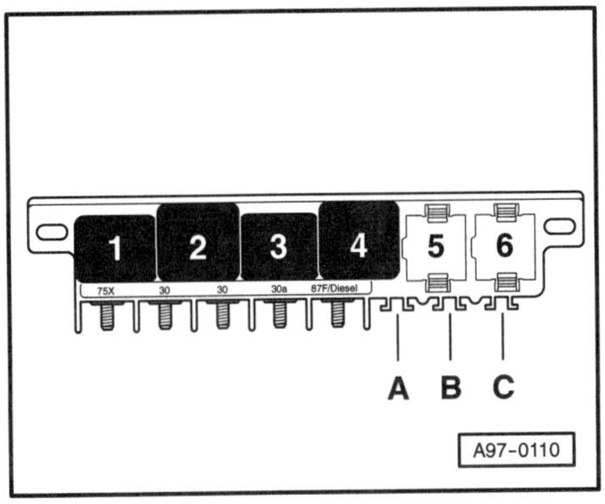

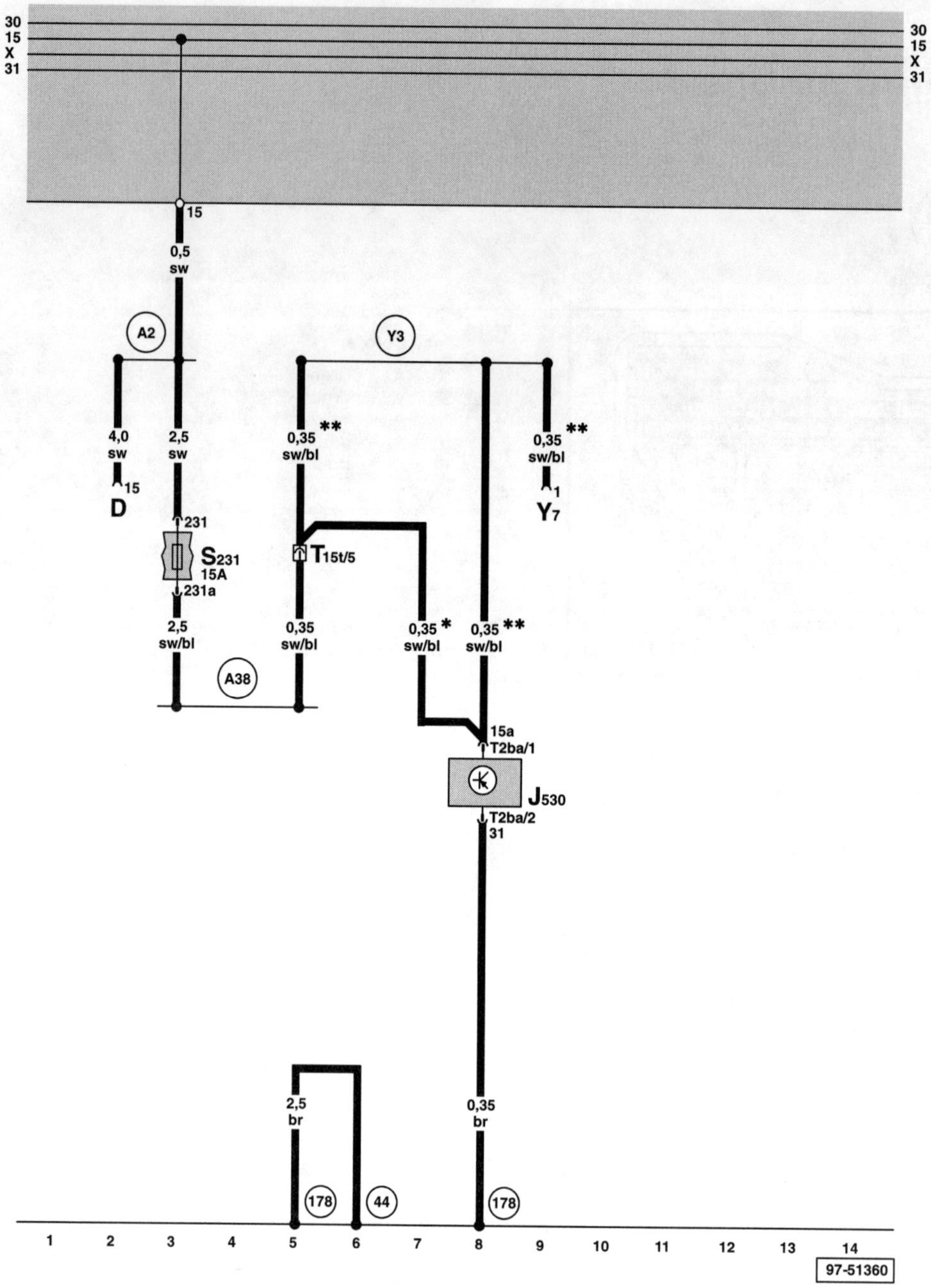

ws = white
sw = black
ro = red
br = brown
gn = green
bl = blue
gr = grey
li = violet
ge = yellow

D - Ignition/Starter Switch
J530 - Control module garage door opener
S231 - Fuse 31 in fuse holder
T2ba - Connector 2 pin, in passenger's seat
T15t - Connector 15 pin, grey, connector station A pillar, left
Y7 - Automatic day/night interior mirror

(44) - Ground connection, on left A-pillar, lower part
(178) - Ground connection, in sunroof wiring harness
(A2) - Plus connection (15), in instrument panel wiring harness
(A38) - Plus connection (15a II), in instrument panel wiring harness
(Y3) - Plus connection (15), in interior light/sunroof wiring harness

* - Vehicles without Automatic day/night interior mirror
** - Vehicles with Automatic day/night interior mirror

Multi-function Steering Wheel

2000 m. y.

Fuse Panel

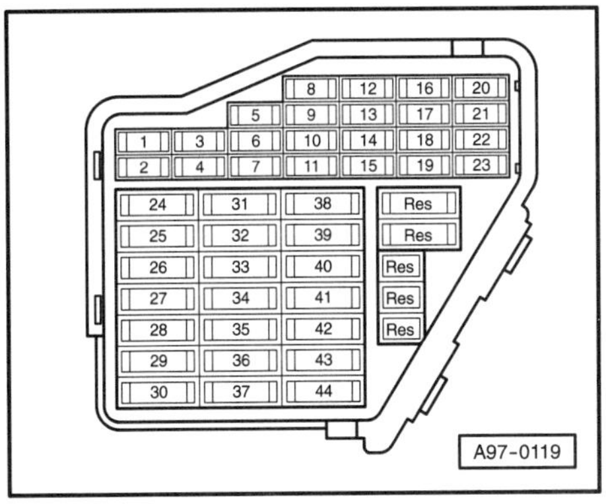

Fuse Colors:

30 A - Green
25 A - White
20 A - Yellow
15 A - Blue
10 A - Red
7,5 A - Brown
5 A - Beige

Starting with fuse position 23, fuses in the fuse holder are identified with 223 in the wiring diagram.

Micro Central Electric Panel

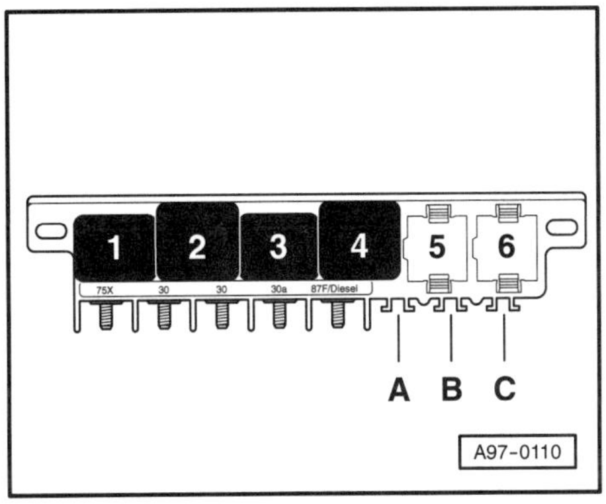

Relay Location:

1 -Dual Horn Relay, J4

13 - Fold Relay Panel

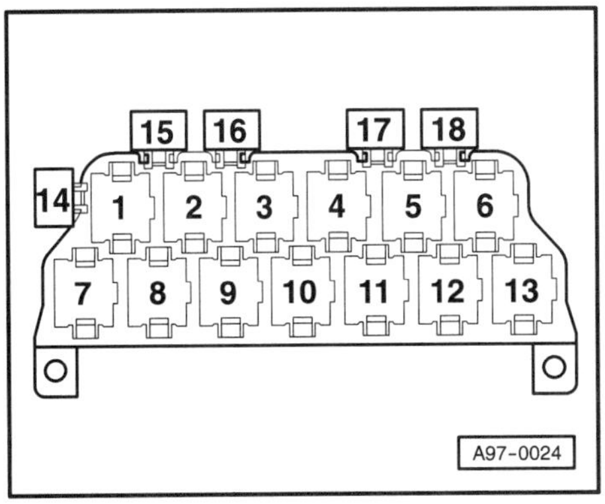

Relay Location:

7 -Control module for multi-function steering wheel, J453

8 -Control module for multi-function steering wheel, J453

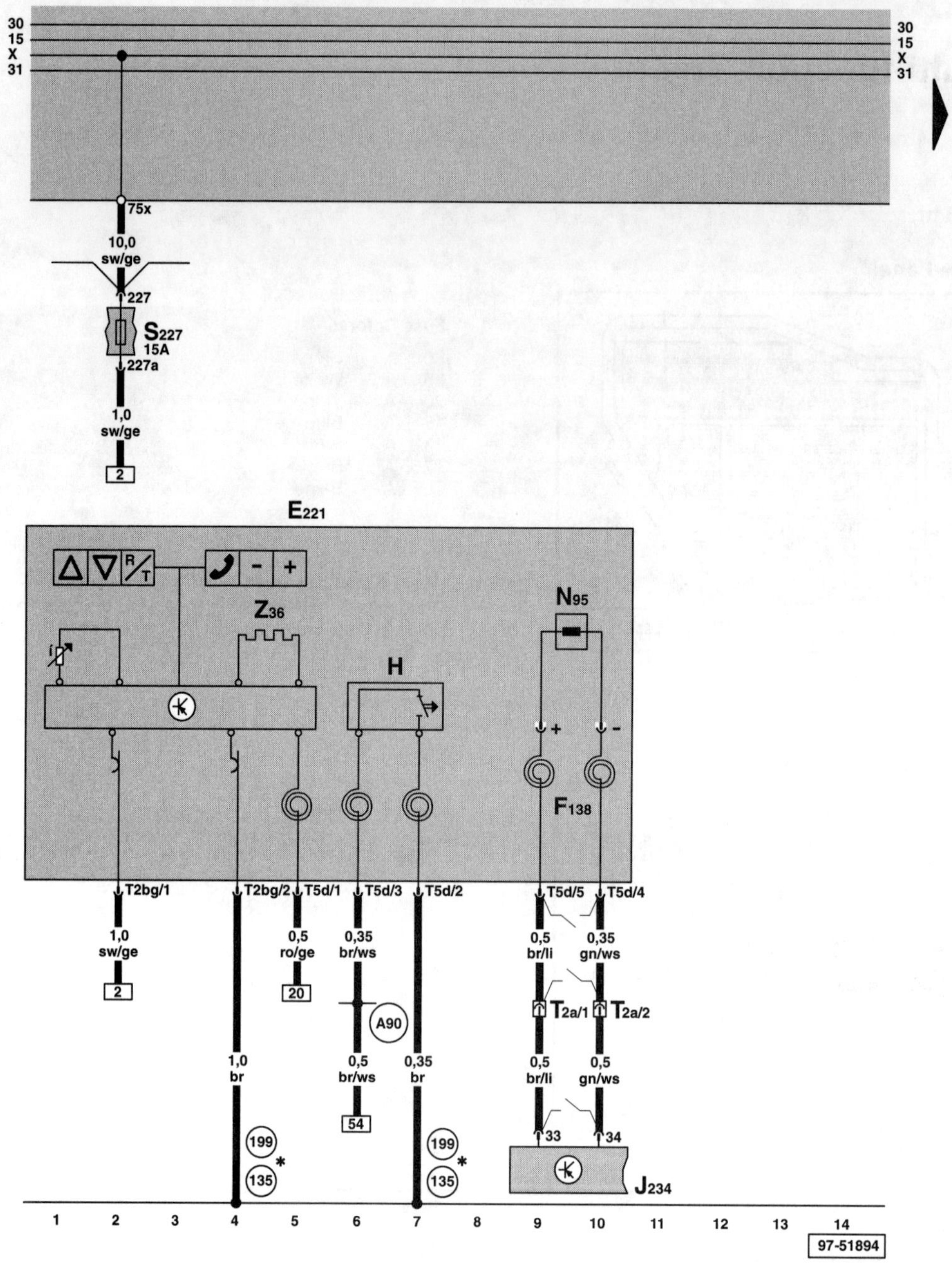

ws = white
sw = black
ro = red
br = brown
gn = green
bl = blue
gr = grey
li = violet
ge = yellow

E221 - Control unit in steering wheel
F138 - Airbag Spiral Spring/Return Spring With Slip Ring
H - Horn Button
J234 - Airbag Control Module
N95 - Driver's Side Airbag Igniter
S227 - Fuse 27 in fuse holder
T2a - Connector 2 pin, yellow, connector station A pillar, left
T2bg - Connector 2 pin, behind steering column switch cover
T5d - Connector 5 pin, behind steering column switch cover
Z36 - Heated steering wheel

(135) - Ground connection -2-, in instrument panel wiring harness
(199) - Ground connection -3-, in instrument panel wiring harness
(A90) - Connector (two-tone horn), in instrument panel wiring harness
Bridge symbol - Bridge closes when disconnecting connector (safety feature)
* - Both are possible

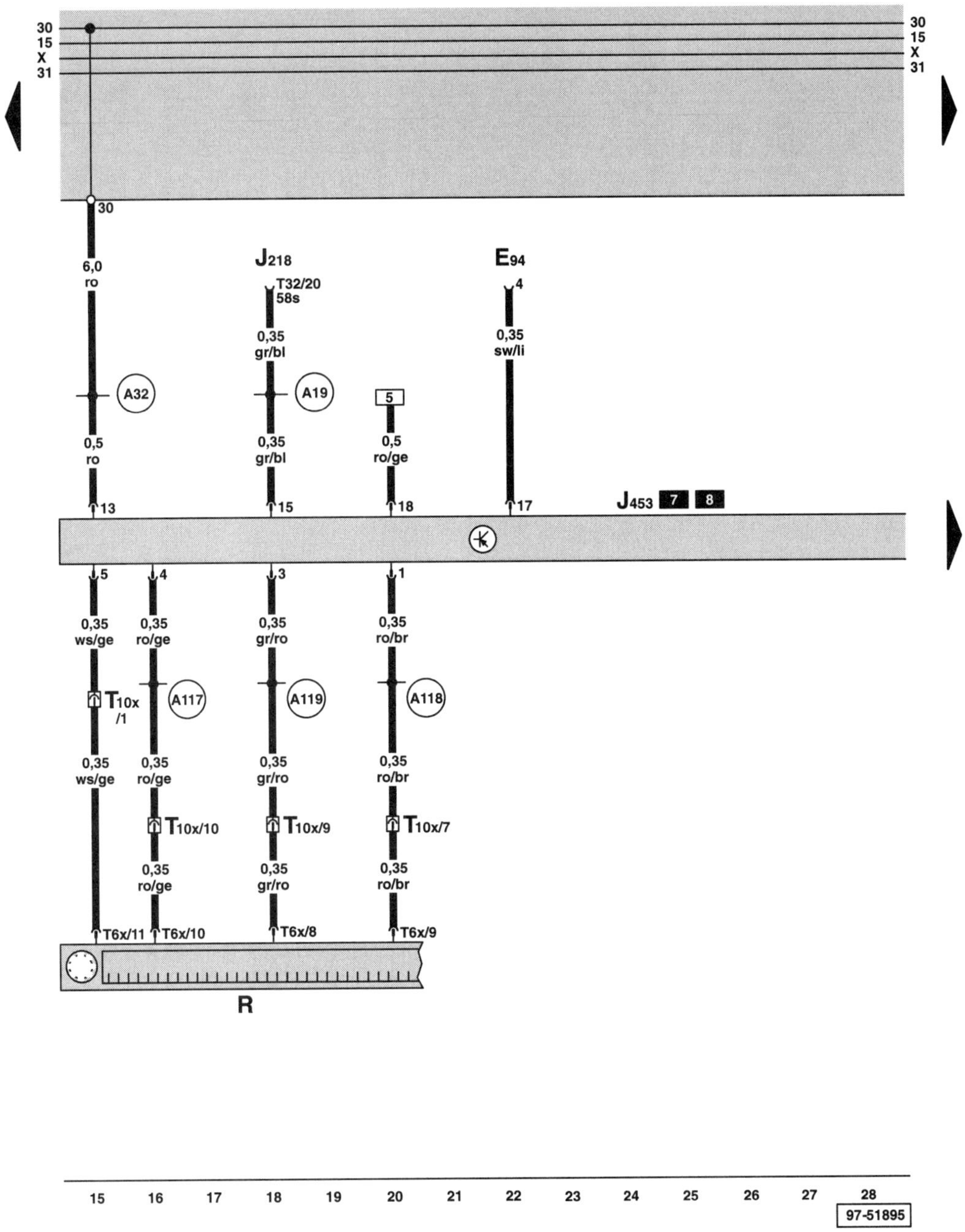

ws = white
sw = black
ro = red
br = brown
gn = green
bl = blue
gr = grey
li = violet
ge = yellow

E94 - Adjuster for heated driver's seat
J218 - Instrument Cluster Combination Processor
J453 - Control module for multi-function steering wheel
R - Radio
T6x - Connector 6 pin, green, connector radio I - 2
T10x - Connector 10 pin, green, connector station A pillar, right
T32 - Connector 32 pin, blue, on instrument cluster

A19 - Wire connection (58s), in instrument panel wiring harness
A32 - Plus connection (30), in instrument panel wiring harness
A117 - Connection (enable) in instrument panel wiring harness
A118 - Connection (data) in instrument panel wiring harness
A119 - Connection (clock) in instrument panel wiring harness

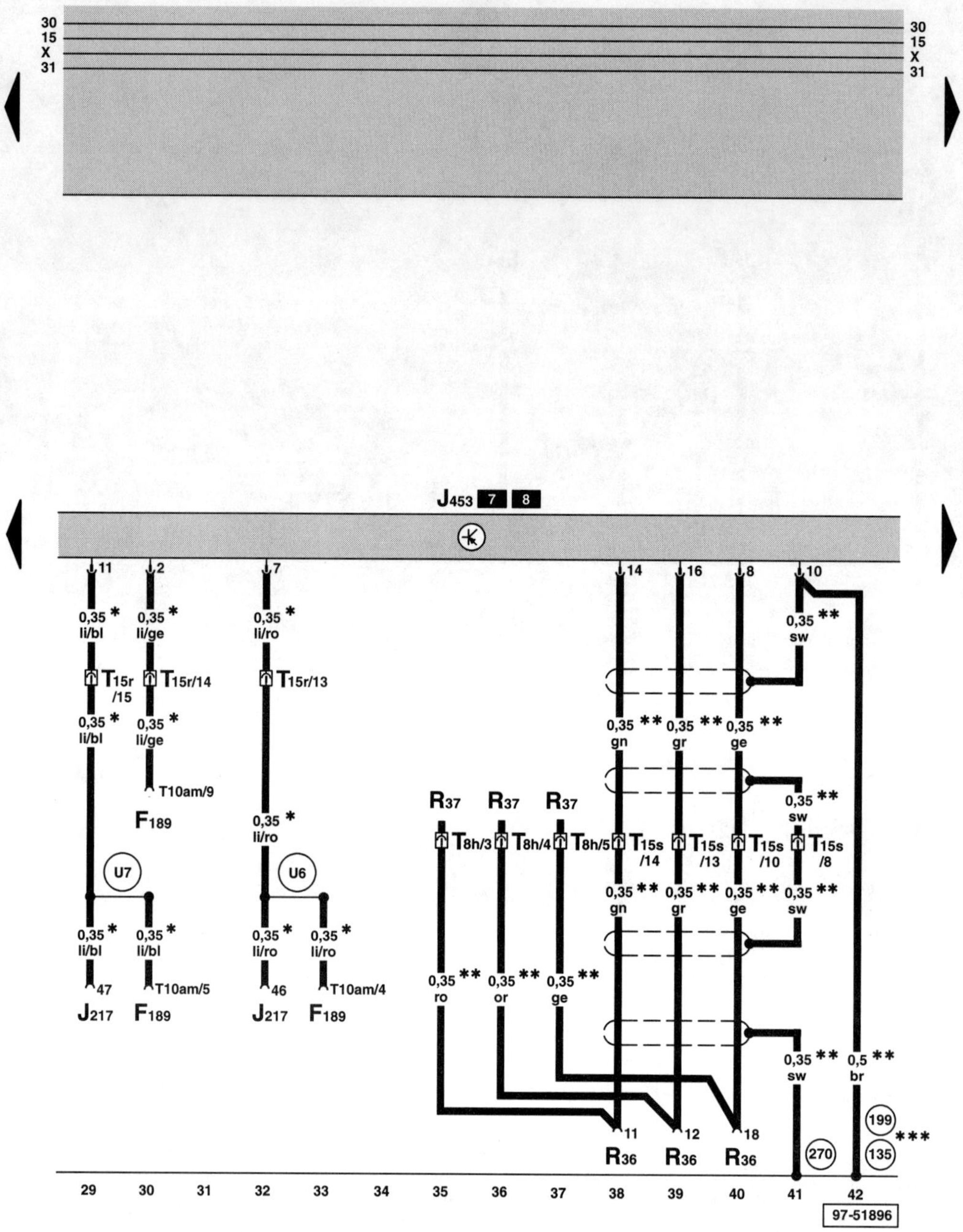

ws = white
sw = black
ro = red
br = brown
gn = green
bl = blue
gr = grey
li = violet
ge = yellow
or = orange

F189 - Tiptronic Switch
J217 - Transmission Control Module (TCM)
J453 - Control module for multi-function steering wheel
R36 - Telephone Transceiver
R37 - Telephone Handset
T8h - Connector 8 pin, Telephone Handset, in console
T10am- Connector 10 pin, on selector lever
T15r - Connector 15 pin, blue, connector station A pillar, right
T15s - Connector 15 pin, blue, connector station A pillar, left

(135) - Ground connection -2-, in instrument panel wiring harness
(199) - Ground connection -3-, in instrument panel wiring harness
(270) - Ground connector, in wiring harness telephone
(U6) - Wire connection -2-, in automatic transmission wiring harness
(U7) - Wire connection -3-, in automatic transmission wiring harness

* - Vehicles with Tiptronic
** - Vehicles with Telephone
*** - Both are possible

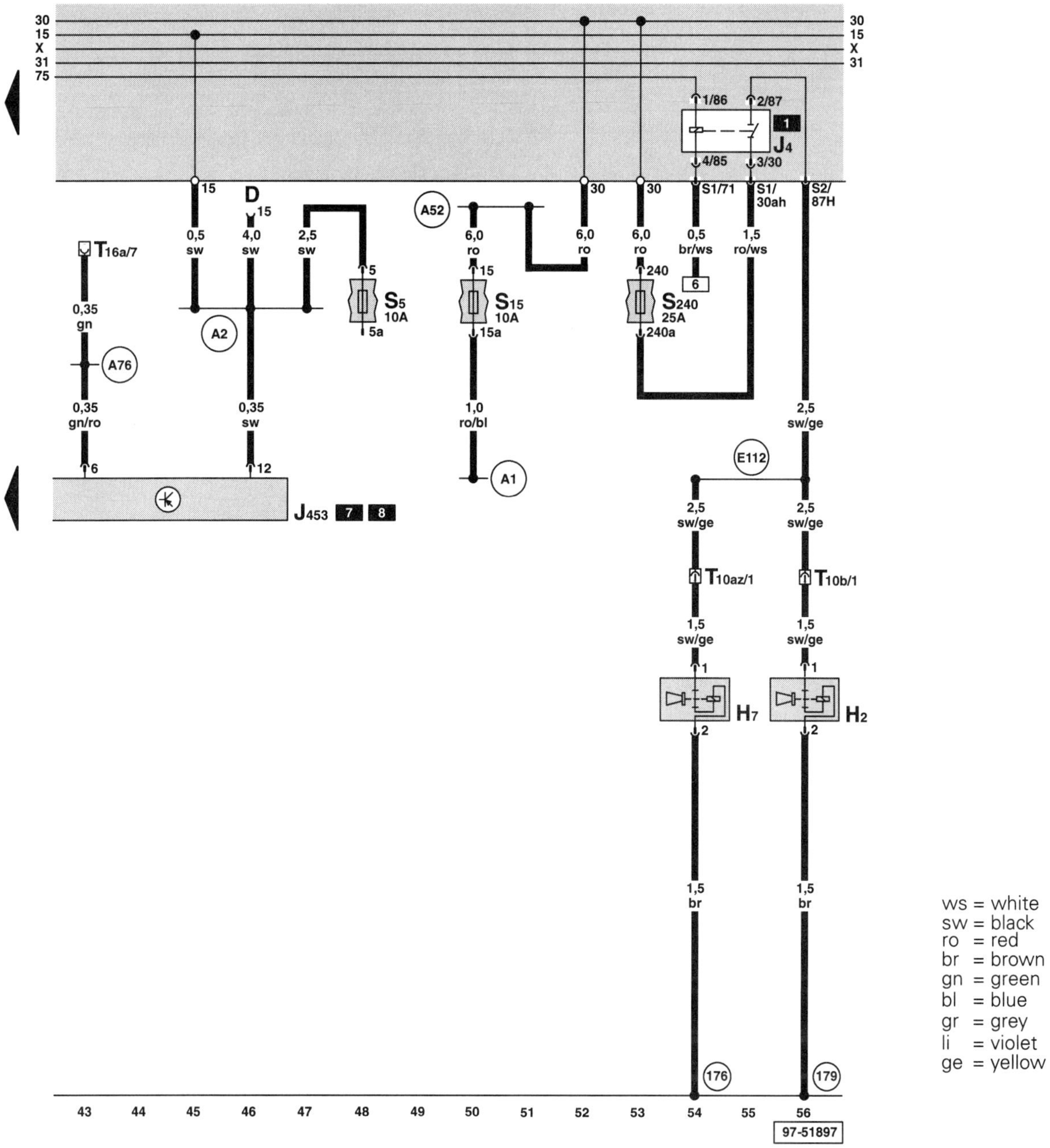

D - Ignition/Starter Switch
H2 - High Tone Horn
H7 - Low Tone Horn
J4 - Dual Horn Relay
J453 - Control module for multi-function steering wheel
S5 - Fuse
S15 - Fuse
S240 - Fuse 40 in fuse holder
T10b - Connector 10 pin, red, connector station A pillar, left
T10az - Connector 10 pin, red, connector station A pillar, right
T16a - Connector 16 pin, Data Link Connector (DLC)

(176) - Ground connection, in right headlight wiring harness
(179) - Ground connection, in left headlight wiring harness
(A1) - Plus connection (30a), in instrument panel wiring harness
(A2) - Plus connection (15), in instrument panel wiring harness
(A52) - Plus connection -2- (30), in instrument panel wiring harness
(A76) - Connector (K-diagnosis wire), in instrument panel wiring harness
(E112) - Connector (Horn - 87h), in instrument panel wiring harness

THIS PAGE INTENTIONALLY LEFT BLANK

Power Rear Window Shade

2000 m. y.

Fuse Panel

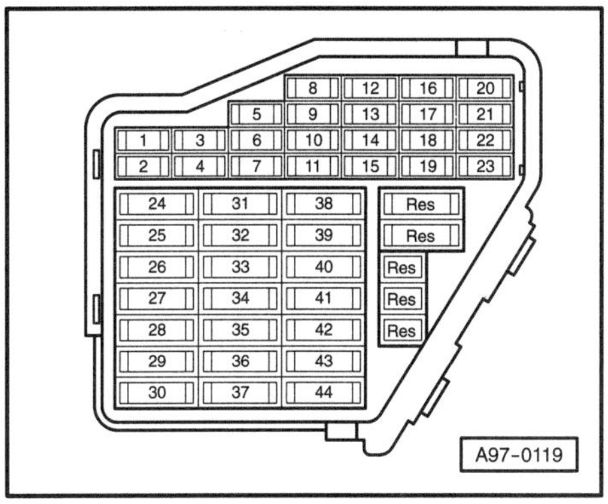

Fuse Colors:

30 A - Green
25 A - White
20 A - Yellow
15 A - Blue
10 A - Red
7,5 A - Brown
5 A - Beige

Starting with fuse position 23, fuses in the fuse holder are identified with 223 in the wiring diagram.

Micro Central Electric Panel

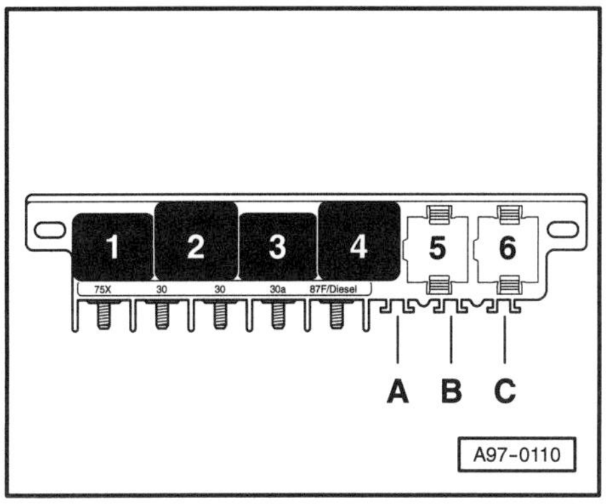

Relay Location:

B -Rear Shade Fuse, S100

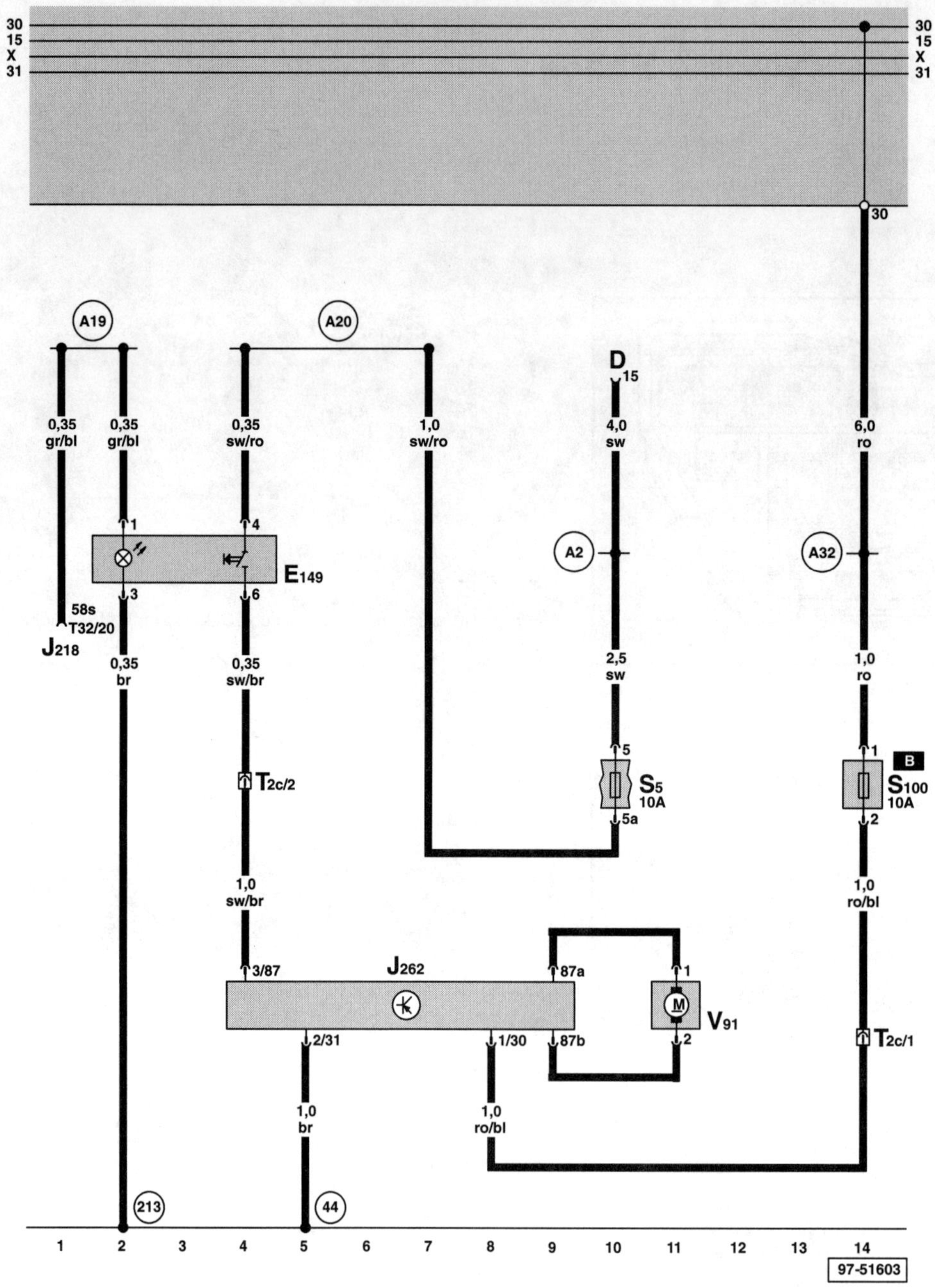

ws = white
sw = black
ro = red
br = brown
gn = green
bl = blue
gr = grey
li = violet
ge = yellow

D - Ignition/Starter Switch
E149 - Rear Window Shade Switch
J218 - Instrument Cluster Combination Processor
J262 - Rear Window Shade Control Module
S5 - Fuse
S100 - Rear Shade Circuit Breaker
T2c - Connector 2 pin, brown, connector station A pillar, left
T32 - Connector 32 pin, blue, on instrument cluster
V91 - Rear Window Shade Motor

(44) - Ground connection, on left A-pillar, lower part
(213) - Ground connection, in instrument panel wiring harness
(A2) - Plus connection (15), in instrument panel wiring harness
(A19) - Wire connection (58s), in instrument panel wiring harness
(A20) - Wire connection (15a), in instrument panel wiring harness
(A32) - Plus connection (30), in instrument panel wiring harness

Power Sunroof with Automatic Preselection

2000 m. y.

Fuse Panel

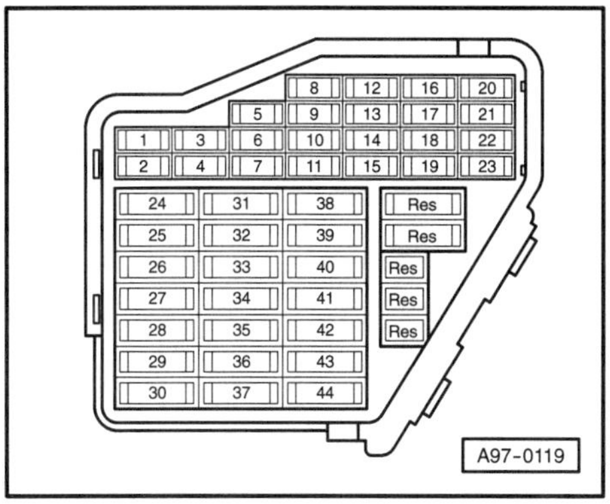

Fuse Colors:

30 A - Green
25 A - White
20 A - Yellow
15 A - Blue
10 A - Red
7,5 A - Brown
5 A - Beige

Starting with fuse position 23, fuses in the fuse holder are identified with 223 in the wiring diagram.

Micro Central Electric Panel

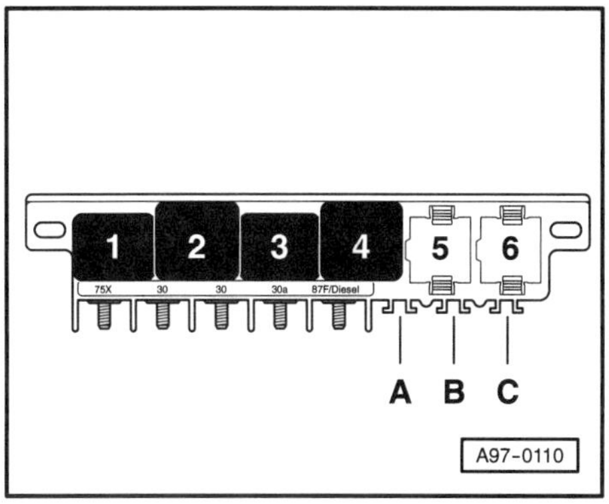

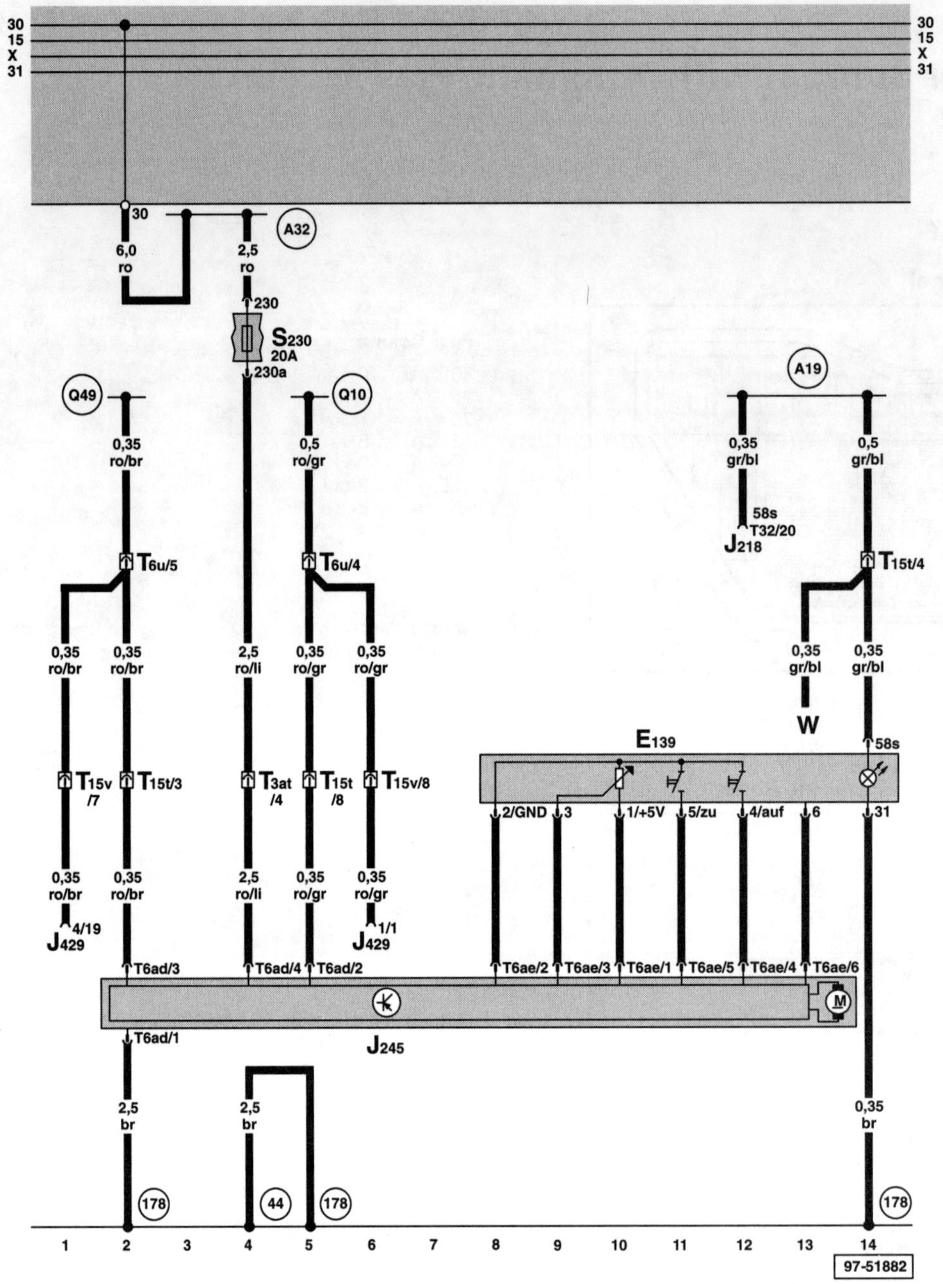

ws = white
sw = black
ro = red
br = brown
gn = green
bl = blue
gr = grey
li = violet
ge = yellow

E139 - Sunroof Regulator
J218 - Instrument Cluster Combination Processor
J245 - Power Sunroof Control Module
J429 - Control module for central locking
S230 - Fuse 30 in fuse holder
T3at - Connector 3 pin, grey, connector station A pillar, left
T6ad - Connector 6 pin, blue, on power sunroof control module
T6ae - Connector 6 pin, on power sunroof control module
T6u - Connector 6 pin, grey, connector station A pillar, left
T15t - Connector 15 pin, grey, connector station A pillar, left
T15v - Connector 15 pin, orange, connector station A pillar, right
T32 - Connector 32 pin, blue, on instrument cluster
W - Front Interior Light

(44) - Ground connection, on left A-pillar, lower part

(178) - Ground connection, in sunroof wiring harness

(A19) - Wire connection (58s), in instrument panel wiring harness

(A32) - Plus connection (30), in instrument panel wiring harness

(Q10) - Plus connection (87), in power window/central locking system and door contact switch wiring harness

(Q49) - Connection (closed), in wiring harness interior

Component Locations

Section
CL

WARNING!
Before proceeding with any work, read the Warnings and Cautions in Section WD.

General Information **Page**

- Overview . . . . . CL 1
- How to use Component Locations . . . . . CL 2

Component Location and Identification **Page**

- General . . . . . CL 2
- Component Location and Identification Table . . . . . CL 3
- Fuse and relay panels, control modules . . . . . CL 13
- Automatic headlight beam adjusting, with HID headlight system . . . . . CL 19
- Central locking system, Sedan and Avant (m.y. 2000) . . . . . CL 22
- Connector station, A-pillar, left side . . . . . CL 26
- Connector station, A-pillar, right side . . . . . CL 28
- Relay panel behind instrument panel on central carrier . . . . . CL 29
- E-Box in plenum chamber . . . . . CL 30
- Connector station in plenum chamber . . . . . CL 30
- Parking aid, Sedan and Avant (m.y. 2000) . . . . . CL 31
- Level control system, Sedan and Avant (m.y. 2000) . . . . . CL 35
- Ground connection locations . . . . . CL 37

General Information

Overview

The wiring (current flow) diagrams in this manual list references to hundreds of different electrical components. Since new systems are added every model year, space to locate all components becomes limited. This can result in different locations for old components, and new components situated where they can be accomodated. For this reason a **Component and Identification Table** has been included along with an illustrated section for selected components. The table is designed to utilize the standard alpha-numeric component designation (code) used on the wiring diagrams. For details about locating designations on wiring diagrams, see **Section WD**.

In general, the capital letter in the designation refers to the specific component classification according to the following guidelines:

- **A** Battery, Chargers, Converters.
- **B** Starter.
- **C** Generator (alternator), Voltage Regulator, Suppressors, Filters, Capacitors, Solar Cells.
- **D** Ignition Switch, Immobilizer and related.
- **E** Switches (generally manually operated).
- **F** Switches (generally mechanically controlled).
- **G** Sending Units, Sensors.
- **H** Horns, Buzzers, Signaling devices.
- **J** Relays, Electronic Control Units, Electronic Control Modules, Diodes, Stabilizers.

(continued on following page)

(continued from previous page)

- **K** Instrument Cluster, Warning and Indicator Lights and LEDs, Illumination Lights.
- **L** Headlights, Exterior Emergency Vehicle Lights, Interior Lights, Component Illumination Lights.
- **M** Parking, Tail, Brake, Running, Marker, Lights/Lamps.
- **N** Output Devices/Stages, Injectors, Solenoid Valves, Coils, Control Elements, Series Resistance, Ignitors, Detonators.
- **O** Ignition Distributor, Breaker Points.
- **P** Spark Plug Connectors.
- **Q** Spark Plugs, Glow Plugs, Flame Plugs.
- **R** Radio, Antenna, Speakers, Amplifiers, CD Players/Changers, Telephone, Controls, FAX, Television, Navigation, GPS, Interfaces.
- **S** Fuses, Circuit Breakers, Fusible Links/Strips.
- **SA** Fuse in holder.
- **SB** Fuse in holder.
- **SC** Fuse in holder.
- **T** Terminal Connectors, Plugs.
- **TV** Circuit Distribution Points.
- **U** Sockets, Outlets, Lighters.
- **V** Motors, Pumps, Fans
- **W** Interior Lights/Lamps.
- **X** Rear Fog, Back-up, and License Plate Lights.
- **Y** Clocks, Digital LED and LCD Displays.
- **Z** Heaters, Heating Elements.

The components selected for this section concentrate on major items from the F, G, J, N, and V classifications, although components from other classifications are included.

How to Use Component Locations

This section is divided into two main parts. The first part is a **Component and Identification Table** with components listed by the alpha-numeric designation in the column marked Code. Other columns provide component name and application along with location and any additional information. Some components have an OBD II designation which is included where applicable. The last column denotes the page number of an illustration where applicable.

The second part contains 2 illustrations for the selected component. The first illustration shows the general area of the vehicle where the component is located, while the second illustration shows the component in greater detail.

Component Location and Identification

General

Every effort has made to insure the accuracy of table entries and illustrations in this section. However, equipment levels vary and production variations occur, sometimes even within the same model year. If in doubt, particularly concerning electronic control units and relays, record the numbers on the component and verify identification with an Authorized Audi Parts Department or qualified Independent Parts Professional.

Component Location and Identification Table

Code	Component	OBD	Application	Location and additional information	Illust.
B	Starter		all	Mounted to lower section of all engines near transmission, includes starter solenoid.	
C	Generator (alternator)	Gen	all	Mounted to lower front of all engines; belt driven.	
C1	Voltage regulator	VR	all	Built into generator assembly.	
C6	Distributor Ignition Capacitor (suppressor)	DI	2.8L as applicable	In engine compartment, mounted to ignition coil assembly.	
C20	Solar cells		as applicable	Built into sliding sunroof panel.	
C22	Interference filter		as applicable	In wiring harness between Airbag control module (J234) and control module for central locking (J429).	
E15	Rear window defroster switch		all	Built into A/C control head (E87). Includes timer control unit for heated rear window.	
E87	A/C control head		all	In passenger compartment, below radio, integral with controls.	CL13 CL14
E139	Sunroof regulator		as applicable	In passenger compartment, in overhead panel.	
E150	Door lock switch		all	In front door trim panel, driver's side door	CL22 CL25
E183	Bypass switch for passenger compartment m monitoring		as applicable	In passenger compartment, on lower left B pillar trim	CL22 CL24
E221	Multi-Function steering wheel control unit		as applicable	In steering wheel, integral with Z36, H, N95, and F138.	
E266	Parking aid button		as applicable	In upper section of front console.	CL31 CL32
E266	Parking aid button		as applicable (allroad)	In lower section of center console.	CL31 CL32
F	Brake light switch		all	On pedal cluster, above brake vacuum vent valve switch F47 for cruise control as applicable.	
F1	Oil pressure switch		2.7L, 2.8L	On oil filter housing, front left side of engine.	
F2	Door contact switch-driver's side		all	Part of assembly with driver's door central locking system motor V56.	CL22 CL25
F3	Door contact switch-passenger's side		all	Part of assembly with passenger door central locking system motor V57.	
F4	Back-up light switch		manual transmission	On transmission case, right side.	CL31 CL34
F8	Kick down switch		with automatic transmission, without ME7	In engine compartment on bulkhead, integrated with accelerator pedal cable.	
F9	Parking brake warning light switch		all	On parking brake lever mechanism beneath center console.	
F10	Left rear door contact switch		all	Part of assembly with left rear central locking system motor V115.	
F11	Right rear door contact switch		all	Part of assembly with right rear central locking system motor V97.	
F18	Coolant fan control thermal switch	FC	all	On radiator, lower right, combined switches F18 and F54.	
F34	Brake fluid level warning switch		all	Built into brake master cylinder fluid reservoir in fresh air plenum.	
F47	Brake vacuum vent valve switch		all	On pedal cluster, mounted below brake light switch F.	
F54	Coolant fan control thermal switch	FC	all	On radiator, lower right, combined switches F18 and F54.	
F66	Engine coolant level warning switch	ECL	all	Bottom of coolant reservoir/expansion tank.	
F93	After-run thermal coolant switch		2.7L	In coolant pipe at rear of cylinder bank 2.	
F120	Hood alarm switch		all	Part of hood latch assembly.	CL22 CL23
F124	Switch for rear lid		all	Behind trim for rear lid	CL22 CL23
F125	Multi-function transmission range switch	TR	with automatic transmission	Left side of transmission housing, with harness connector.	CL31 CL34

Code	Component	OBD	Application	Location and additional information	Illust.
F129	A/C pressure switch		all	Right side of A/C condenser.	
F131	Switch, door lock activator, driver's side		all	In driver's side door lock activator assembly, in driver's side door	CL22 CL25
F146	Right-rear B-pillar micro switch		all	Passenger side, underneath B-pillar lower trim.	
F189	Tiptronic switch		with automatic transmission	Integrated in gear selector cover/trim assembly.	
F206	Switch for rear lid		all	Behind trim for rear lid	CL22 CL23
F216	Contact switch for fog light shut-off		as applicable (vehicles with trailer coupling)	In electrical system coupling for trailer, rear.	CL31 CL33
F218	Switch for rear lid		all	Behind trim for rear lid	CL22 CL23
F241	Switch, door lock activator, driver's side		all	In driver's side door lock activator assembly, in driver's side door	CL22 CL25
F243	Switch, door lock activator, driver's side		all	In driver's side door lock activator assembly, in driver's side door	CL22 CL25
G	Fuel level sensor		all	Part of main fuel pump module with fuel pump G6, access from under rear seat cushion.	
G2	Engine coolant temperature sensor (gauge)	ECT	all	In coolant pipe behind intake manifold, with ECT sensor G62, combined in single housing. Harness connector T3y.	
G6	Fuel pump	FP	all	Fuel pump module includes main fuel pump G6 and fuel level sensor G, access under rear seat cushion.	
G8	Engine oil temperature sensor (gauge)		as applicable	Front of cylinder block, left side.	
G17	Outside air temperature sensor		all	In front of A/C condenser.	
G22	Speedometer vehicle speed sensor (VSS)	VSS	with automatic transmission	On transmission, left-side, near drive flange, access with shield removed.	
G28	Engine speed sensor	RPM	all	On transmission housing, left side, near drive flange. Harness connector on bulkhead, connector T3w.	
G31	Charge air pressure sensor		2.7L	In air intake boot, on intake manifold, front, attached to throttle valve housing.	
G38	Transmission vehicle speed sensor	VSS	with automatic transmission	On transmission, left-side, near drive flange.	
G39	Heated oxygen sensor	HO2S (B1S1)	all	Main oxygen sensor for bank 1 (right side), mounted in exhaust pipe just behind manifold (pre-cat sensor).	
G40	Camshaft position sensor	CMP	2.7L, 2.8L	Back of left side cylinder head (bank 2), mounted to outside of cylinder head.	
G40	Camshaft position (CMP) sensor 1	CMP	4.2L	Back of left side cylinder head (bank 2), mounted to outside of cylinder head.	
G42	Intake air temperature sensor	IAT	all	Mounted in intake air ducting at back of intake manifold.	
G44	Right rear ABS wheel speed sensor		front wheel drive	Pressed into back of rear wheel bearing housing/carrier.	
G44	Right rear ABS wheel speed sensor		quattro	Mounted in back of wheel bearing housing/carrier.	
G45	Right front ABS wheel speed sensor		all	Mounted at front of wheel bearing housing.	
G46	Left rear ABS wheel speed sensor		front wheel drive	Pressed into back of rear wheel bearing housing/carrier.	
G46	Left rear ABS wheel speed sensor (quattro)		quattro	Mounted in back of wheel bearing housing/carrier.	
G47	Left front ABS wheel speed sensor			Mounted at front of wheel bearing housing.	
G61	Knock sensor	KS1	2.7L, 2.8L	Near right side cylinder head (bank 1), beneath intake manifold. Harness connector T3x.	
G62	Engine coolant temperature sensor	ECT		In coolant pipe behind intake manifold, with ECT sensor G62, combined in single housing. Beneath intake manifold. Harness connector T3y.	
G66	Knock sensor (KS) 2	KS2		Near left-side cylinder head (bank 2), beneath intake manifold. Harness connector T3y.	

Code	Component	OBD	Application	Location and additional information	Illust.
G70	Mass air flow (MAF) sensor	MAF		In intake air duct near top of air cleaner housing.	
G76	Left rear level control system sensor		allroad	In left rear wheel housing with link rod attached to lower suspension arm.	CL19 CL20
G77	Right rear level control system sensor		allroad	In right rear wheel housing with link rod attached to lower suspension arm.	
G78	Left-front level control system sensor		allroad	In left front wheel housing with link rod attached to lower suspension arm.	CL19 CL21
G79	Throttle Position sensor	TPS		Under dash, driver's side, on accelerator pedal integral with G185.	
G84	Level control system sensor		as applicable	In center of rear axle.	CL35 CL36
G85	Steering angle sensor		ESP	Mounted on the steering column, integral with airbag spiral spring F138	
G89	Fresh air intake duct temperature sensor		all	Behind instrument panel, above blower motor housing, near fresh air intake duct.	
G93	Transmission fluid temperature sensor		with automatic transmission	In transmission valve body assemble, integral with wiring harness.	
G107	Sunlight photo sensor		all	In upper center of padded dash panel, on outer skin.	
G108	Heated oxygen sensor (HO2S) 2	HO2S (B2S1)		Main oxygen sensor for bank 2 (left side), mounted in exhaust pipe just behind manifold (pre-cat sensor).	
G112	Central flap motor position sensor			Part of central air flap motor V70, lowest of three motor/sensor units on right side of evaporator housing.	
G113	Back pressure flap motor position sensor			Part of air flow (backpressure) flap motor V71, near blower motor fresh air intake	
G130	Oxygen sensor behind three way catalytic converter (TWC)	HO2S (B1S2)	all	Post-Cat Oxygen sensor for bank 1 (right side), mounted in exhaust system behind (downstream) Three Way Catalytic Converter.	
G131	Oxygen sensor 2 behind three way catalytic converter (TWC)	HO2S (B2S2)	all	Post-Cat Oxygen sensor for bank 2 (left side), mounted in exhaust system behind (downstream) Three Way Catalytic Converter.	
G135	Defroster flap motor position sensor		all	Part of defroster flap motor V107, upper-most of three motor/sensor units on right side of evaporator housing.	
G150	Left vent temperature sensor		all	In ducting behind instrument panel vents.	
G151	Right vent temperature sensor		all	In ducting behind instrument panel vents.	
G163	Camshaft position (CMP) sensor 2	CMP	4.2L	Front of right side cylinder head (bank 1), behind camshaft drive belt cover.	
G169	Fuel level sensor 2		quattro	In fuel tank, access from under rear seat cushion, left side.	
G170	Interior monitoring sensor		as applicable	In left B pillar behind trim panel.	CL22 CL23
G171	Interior monitoring sensor		as applicable	In right B pillar behind trim panel.	CL22 CL23
G179	Crash sensor for side airbag, driver's side		all	Under driver's seat/carpet, mounted on back side of floorpan crossmember.	
G180	Crash sensor for side airbag, passenger's side		all	Under front passenger seat/carpet, on back side of floorpan crossmember.	
G182	Sensor for transmission rpm		with automatic transmission	Version 1: Inside transmission, mounted on valve body. Version 2: Inside transmission, mounted on lower part of transmission housing above valve body.	
G183	Sensor for broken window glass, rear, left		Avant	Under left-rear window trim.	
G184	Sensor for broken window glass, rear, right		Avant	Under right-rear window trim.	
G185	Sensor 2 for accelerator pedal position		as applicable	Under dash, driver's side, on accelerator pedal, integral with G79.	
G186	Throttle drive (power accelerator actuation)		as applicable	Integral with throttle valve control module, J338.	

Code	Component	OBD	Application	Location and additional information	Illust.
G187	Angle sensor 1 for throttle drive (power accelerator actuation)		as applicable	Integral with throttle valve control module, J338.	
G188	Angle sensor 2 for throttle drive (power accelerator actuation)		as applicable	Integral with throttle valve control module, J338.	
G195	Sender for transmission output		with automatic transmission	On rear of transmission, externally mounted on right side.	
G200	Sensor for transverse acceleration		ESP	Under rear seat bench, center, integral with G202.	
G201	Sensor for brake booster		ESP	Threaded into hydraulic pump for brake system.	
G202	Yaw sensor (Rotation Rate Sensor)		ESP	Under rear seat bench, center, integral with G200.	
G203 G204, G205, G206	Rear parking aid sensors		as applicable	In lower part of rear bumper, behind cover, rear bumper fascia.	CL31 CL33
G220	Potentiometer/actuator for temperature flap, left		all	Part of V158, actuator for temperature flap on left side of evaporator housing	
G221	Potentiometer/actuator for temperature flap, right		all	Part of V159, actuator for temperature flap on right side of evaporator housing.	
G235	Sensor 1 for exhaust temperature	EGT	2.7L	In exhaust manifold, upstream of turbocharger, passenger's side, bank 1.	
G236	Sensor 2 for exhaust temperature	EGT	2.7L	In exhaust manifold, upstream of turbocharger, driver's side, bank 2.	
G237	Fuel level sensor 3		quattro	In fuel tank, access from inside luggage compartment (behind rear seat backrest).	
G238	Sensor for air quality		as applicable	Under plenum cover, attached to cabin air filter housing.	
G252, G253, G254, G255	Front parking aid sensors		as applicable	In upper part of front bumper, behind cover, front bumper fascia.	CL31 CL32
G266	Oil level thermal sensor		as applicable, for SRI	Under engine, on oil pan.	
G294	Brake booster pressure sensor		all	Under plenum cover, directly in front of brake booster.	
H2	High tone horn		all	Inside of bumper cover, left side.	
H7	Low tone horn		all	Inside of bumper cover, right side.	
H8	Alarm horn		Sedan	In luggage compartment, left rear, behind trim, CD player, navigation system, etc. (when so equipped.)	CL22 CL24
H8	Alarm horn		Avant	In cargo compartment, right rear, behind trim panel	CL22 CL24
H15	Warning buzzer for parking aid, rear		as applicable (Sedan)	Below rear shelf lining, right side	CL31 CL32
H15	Warning buzzer for parking aid, rear		as applicable (Avant)	In cargo compartment, behind left rear panel trim.	CL31 CL33
H22	Warning buzzer for parking aid, front		as applicable	In center console, left side.	CL31 CL32
J2	Turn signal/emergency flasher relay		all	Built into emergency flasher switch, E3.	
J4	Dual horn relay		all	In micro central electric panel, position 1.	
J5	Fog light relay		all	In 13-fold relay panel, position 6.	
J17	Fuel pump relay	FP relay	all	Micro central electric panel position4.	
J26	Coolant fan relay	FC relay	all	In 8-fold relay panel, position 3.	
J31	Wiper washer intermittent relay		all	In micro central electric panel, positions 5 and 6.	
J44	A/C clutch relay		all	In 13-fold relay panel, position 5.	
J59	Load reduction relay		all	In micro central electric panel, position 2.	
J101	Coolant fan relay, 2nd speed	FC relay	all	In 8-fold relay panel, position 2.	

Code	Component	OBD	Application	Location and additional information	Illust.
J104	ABS/EDL control module		as applicable	Left side of engine compartment.	
J104	ABS/EDL control module		as applicable	In passenger compartment, right side footwell, under carpet.	CL13 CL14
J105	ABS return flow relay		as applicable	Integral with J104.	
J106	ABS solenoid valve relay		as applicable	Integral with J104.	
J123	Lamp control module		as applicable	In 13-fold relay panel, positions 9 and 10.	
J126	Control module for fresh air blower		all	Mounted in heating and A/C duct assembly between evaporator housing and blower motor.	
J136	Memory seat control module		as applicable	Mounted to driver's seat cushion frame.	CL13 CL16
J189	Auto check system control module		all	In instrument cluster, driver's side dash, behind steering wheel.	
J197	Level control system control module		allroad	In sound absorbing protection box for air supply components, in cargo compartment, left rear.	CL13 CL16 CL35 CL36
J201	Protection diode		early, with automatic transmission	In wiring harness, between J307 and S3.	
J207	Starter interlock relay		as applicable	In 13-fold relay panel, position 4.	
J213	Cruise control, control module		w/mechanical throttle actuation	Behind glovebox/instrument panel, near blower motor housing.	
J217	Transmission Control Module	TCM	with automatic transmission	Inside protective box in floor cavity beneath front passenger seat and carpet, in front of floorpan crossmember.	CL13 CL14
J218	Instrument cluster combination processor		all	In instrument cluster, driver's side dash, behind steering wheel.	
J220	Engine Control Module, Motronic	ECM	all	In E-box in fresh air plenum, left side. Remove 5 screws and cover for access.	CL13 CL17 CL30
J234	Airbag control module		all	On center tunnel, ahead of console, near ventilation ducting and bulkhead.	CL13 CL17
J236	Servotronic control module		as applicable	In 13-fold relay panel, position 13.	
J245	Power Sunroof Control Module		as applicable	In passenger compartment, integral with sunroof drive motor and mechanism.	
J262	Rear window shade control module		as applicable	In luggage compartment below rear window shelf, left center.	CL13 CL15
J267	Mirror memory control module		as applicable	Behind glove compartment on central support, position 1.	CL13 CL18
J271	Motronic Engine Control Module (ECM) power supply relay	ECM relay	all	In E-box in fresh air plenum, left side, to the right of the ECM. Remove 5 screws and cover for access.	
J293	Coolant fan control control module	FC	all	In engine compartment, front lower left, mounted on underside of frame rail.	
J295	Left front power window control module		all	Integrated with window motor and window regulator assembly.	
J296	Right front power window control module		all	Integrated with window motor and window regulator assembly.	
J297	Left rear power window control module		all	Integrated with window motor and window regulator assembly.	
J298	Right rear power window control module		all	Integrated with window motor and window regulator assembly.	
J299	Secondary air injection (AIR) pump relay		as applicable	In E-box in fresh air plenum, left side. Remove 5 screws and cover for access.	
J307	Selector lever light relay		early, with automatic transmission	In 13-fold relay panel, position 1.	
J309	Solar cell separation relay		as applicable	In 13-fold relay panel, position 3.	
J338	Throttle valve control module		all	On throttle body at back of intake manifold.	

Code	Component	OBD	Application	Location and additional information	Illust.
J343	Control module for high intensity gas discharge (HID) lamp, left side		as applicable	Integral with left side HID lamp, L13	
J344	Control module for high intensity gas discharge (HID) lamp, right side		as applicable	Integral with right side HID lamp, L14	
J347	Control module for ultra-sound sensors		as applicable	Mounted to inside left rear door jamb below C-pillar, behind rear seat side trim.	CL13 CL16 CL22 CL24
J351	Control Module for mirror fold-away		as applicable	In 13-fold relay panel, positions 11 and 12.	
J352	Control module for steering column belt height adjustment		4.2L	Behind glove compartment on central support, position 3.	CL13 CL18
J355	Control module for solar operation		as applicable	Located between the sunroof lining and the solar sunroof panel.	
J401	Control unit for navigation, with CD drive		as applicable	In left side compartment of luggage compartment.	
J403	Relay for level control system compressor		allroad	In sound absorbing protection box for air supply components, in cargo compartment, left rear.	CL35 CL36
J412	Telephone and operating electronics control module		as applicable	In rear section of center console, near handbrake lever.	
J429	Control module for central locking		all	Inside protective box in floor cavity beneath driver's seat and carpet, in front of floorpan crossmember.	CL13 CL16 CL22 CL25
J431	Headlight range control module		as applicable	Behind glove box on central support, position 2.	CL13 CL14 CL19 CL20
J446	Control module for parking aid		as applicable (Sedan)	In luggage compartment, behind left rear side panel trim.	CL13 CL16 CL31 CL33
J446	Control module for parking aid		as applicable (Avant)	In cargo compartment, behind left rear side panel trim.	CL13 CL16 CL31 CL33
J453	Control module for multi-function steering wheel		as applicable	In 13-fold relay panel, positions 7 and 8.	
J502	Tire pressure monitoring control module		as applicable	Behind luggage compartment side panel	CL13 CL15
J515	Antenna selection control module (diversity antenna)		Avant	Above headliner. near antenna amplifier R24.	
J515	Antenna selection control module (diversity antenna)		Sedan	In passenger compartment, under rear parcel shelf trim.	
J530	Control module for garage door opener		as applicable	In driver's side sun vizor	
K134	Indicator lamp for level control system		as applicable	In instrument cluster.	CL35 CL36
N	Ignition coil 1		2.7L	Mounted above first cylinder.	
N	Ignition coil		2.8L	Engine center, in front of intake manifold integral with coils N128, and N158, and power output stage N122.	
N	Ignition coil 1		4.2L	Mounted above first cylinder, integral with amplifier.	
N30	Cylinder 1 fuel injector		2.7L, 2.8L, 4.2L	Mounted above first cylinder, on fuel rail in "V" of engine.	
N31	Cylinder 2 fuel injector		2.7L, 2.8L, 4.2L	Mounted above second cylinder, on fuel rail in "V" of engine.	
N32	Cylinder 3 fuel injector		2.7L, 2.8L, 4.2L	Mounted above third cylinder, on fuel rail in "V" of engine.	
N33	Cylinder 4 fuel injector		2.7L, 2.8L, 4.2L	Mounted above fourth cylinder, on fuel rail in "V" of engine.	
N83	Cylinder 5 fuel injector		2.7L, 2.8L, 4.2L	Mounted above fifth cylinder, on fuel rail in "V" of engine.	
N84	Cylinder 6 fuel injector		2.7L, 2.8L, 4.2L	Mounted above sixth cylinder, on fuel rail in "V" of engine.	
N85	Cylinder 7 fuel injector		4.2L	Mounted above seventh cylinder, on fuel rail in "V" of engine.	
N86	Cylinder 8 fuel injector		4.2L	Mounted above eighth cylinder, on fuel rail in "V" of engine.	

Code	Component	OBD	Application	Location and additional information	Illust.
N39	Coolant fan control series resistance	FC	all	Left front, on bottom side of longitudinal chassis member.	
N75	Waste gate bypass regulator valve		2.7L	On top of intake manifold, center, to right of air intake duct.	
N80	Evaporative emission canister purge regulator valve	EVAP	all	Engine compartment, right side, next to upper air cleaner housing	
N110	Shift lock solenoid		with automatic transmission	Attached to gear selector mechanism beneath center console.	
N111	Relay for level control system compressor		allroad	In sound absorbing protection box for air supply components, in cargo compartment, left rear.	CL35 CL36
N112	Secondary air injection solenoid valve	AIR	as applicable	At back of intake manifold, mounted above intake air duct, left side.	
N122	Power output stage 1		2.7L	On right side of engine compartment near intake air ducting, under trim cover.	
N122	Power output stage		2.8L	Engine center, in front of intake manifold integral with coils N, N128, and N158.	
N128	Ignition coil 2		4.2L	Mounted above second cylinder, integral with amplifier.	
N128	Ignition coil 2		2.7L	Mounted above second cylinder.	
N128	Ignition coil 2		2.8L	Engine center, in front of intake manifold integral with coils N, and N158, and power output stage N122.	
N131	Passenger's Side Airbag Igniter 1		all	Behind instrument panel, passenger side, above glove box, integral with passenger's side airbag unit.	
N144	Electro-hydraulic engine mount solenoid valve, left		4.2L	Left side lower engine mount, Integral with engine mount.	
N145	Electro-hydraulic engine mount solenoid valve, right		4.2L	Right side lower engine mount, Integral with engine mount.	
N148	Left-front suspension strut valve		allroad	Under vehicle, center rear, above sway bar, integral with N149, N150, N151, N311	
N149	Right-front suspension strut valve		allroad	Under vehicle, center rear, above sway bar, integral with N148, N150, N151, N311.	
N150	Left-rear suspension strut valve		allroad	Under vehicle, center rear, above sway bar, integral with N148, N149, N151, N311, or in sound absorbing protection box for air supply components, in cargo compartment, left rear.	CL35 CL36
N151	Right-rear suspension strut valve		allroad	Under vehicle, center rear, above sway bar, integral with N148, N149, N150, N311, in sound absorbing protection box for air supply components, in cargo compartment, left rear.	CL35 CL36
N153	Left seat belt tensioner igniter		all	mounted behind lower B-pillar trim	
N154	Right seat belt tensioner igniter		all	mounted behind lower B-pillar trim	
N156	Intake manifold tuning valve	IMT	2.8L	At back of intake manifold, mounted above intake air duct, left side (also identified as intake manifold change-over valve).	
N158	Ignition coil 3		4.2L	Mounted above third cylinder, integral with amplifier.	
N158	Ignition coil 3		2.7L	Mounted above third cylinder.	
N158	Ignition coil		2.8L	Engine center, in front of intake manifold integral with coils N, and N128, and power output stage N122.	
N163	Ignition coil 4		4.2L	Mounted above fourth cylinder, integral with amplifier.	
N163	Ignition coil 4		2.7L	Mounted above fourth cylinder.	
N164	Ignition coil 5		4.2L	Mounted above fifth cylinder, integral with amplifier.	
N164	Ignition coil 5		2.7L	Mounted above fifth cylinder.	
N189	Ignition coil 6		4.2L	Mounted above sixth cylinder, integral with amplifier.	
N189	Ignition coil 6		2.7L	Mounted above sixth cylinder.	
N190	Ignition coil 7		4.2L	Mounted above seventh cylinder, integral with amplifier.	
N191	Ignition coil 8		4.2L	Mounted above eighth cylinder, integral with amplifier.	
N192	Power output stage 2		2.7L	On right side of engine compartment near intake air ducting, under trim cover.	
N205	Valve 1 for camshaft adjustment		2.8L	Back of right side (bank 1) cylinder head.	
N208	Valve 2 for camshaft adjustment		2.8L	Front of left side (bank 2) cylinder head.	

Code	Component	OBD	Application	Location and additional information	Illust.
N261	Intake manifold tuning (IMT) valve 2	IMT	4.2L	On front of intake manifold.	
N311	Level control accumulator valve (allroad)		allroad	Under vehicle, center rear, above sway bar, integral with N148, N149, N150, N151	
R24	Antenna amplifier, radio		Avant	On rear roof section, integral with antenna base.	
R38	Telephone microphone		as applicable	In cover for interior light switch.	
R45	Antenna amplifier, radio with telephone		Sedan	Near rear window glass.	
R47	Antenna for central locking and anti-theft alarm system		all	Sedan, on rear bumper carrier, left side Avant, on rear bumper carrier, right side	CL22 CL24
R51	Antenna amplifier, radio with telephone		Avant as applicable	On rear roof section, integral with antenna base.	
R52	Antenna amplifier, radio, with telephone and navigation (GPS)		Avant as applicable	On rear roof section, integral with antenna base.	
T3w	Connector for RPM sensor G28		all	On engine compartment bulkhead, left side, beneath coolant expansion tank, color = gray.	
T3x	Connector for knock sensor 1 G61		all	On engine compartment bulkhead, right side, near intake air duct color = blue.	
T3y	Connector for knock sensor 2 G66		all	On engine compartment bulkhead, left side, beneath coolant expansion tank, color = blue.	
T4s	Connector for O2S G130 (B2S1)		all	On engine compartment bulkhead, left side, color = green	
T4t	Connector for O2S G131(B2S2)		all	On engine compartment bulkhead, left side, color = brown	
T4u	Connector for HO2S G39 (B1S1)		all	On engine compartment bulkhead, right side, color = black	
T4v	Connector for HO2S 2 G108 (B1S2)		all	On engine compartment bulkhead, right side, color = black	
T16a	Data Link Connector	DLC	all	In passenger compartment, below and under dash knee bar, to the left of the brake pedal.	
T16b	16-pin harness connector		with automatic transmission	For automatic transmission 01V, 01L valve body components.	
V2	Fresh air blower		all	Inside fan/evaporator housing behind glovebox/instrument panel, right side	
V12	Rear window wiper motor		Avant	In rear lid, includes electronic control module.	
V14	Left window motor		all	Integral with window motor and window regulator assembly	
V15	Right window motor		all	Integral with window motor and window regulator assembly	
V18	Cruise control vacuum pump		w/mechanical throttle actuation	Beneath ABS hydraulic modulator assembly.	
V21	Power antenna		as applicable	In rear fender, includes control module.	
V26	Left rear window motor		all	Integral with window motor and window regulator assembly	
V27	Right rear window motor		all	Integral with window motor and window regulator assembly	
V39	ABS return flow pump		all	Integral with ABS/EDL/ASR/ESP hydraulic modulator assembly, in engine compartment, left front.	
V48	Left head light beam adjusting motor		as applicable	In headlamp housing, driver's side, below headlamp.	CL19 CL21
V49	Right head light beam adjusting motor		as applicable	In headlamp housing, passenger side, below headlamp.	CL19 CL21
V51	After-run coolant pump		all	Under vehicle, behind left side of oil pan.	
V53	Deck lid central locking system motor		all	Under deck lid trim, center, next to V151.	CL22 CL23
V56	Driver's door central locking system motor		all	Integral with driver's door lock activator assembly in driver's door.	CL22 CL25
V57	Passenger's door central locking system motor		all	Integral with door contact switch F3.	
V66	Level control system compressor motor		allroad	Under vehicle, driver's side rear, between exhaust and spare tire well, or, in sound absorbing protection box for air supply components, in cargo compartment, left rear.	CL35 CL36

Code	Component	OBD	Application	Location and additional information	Illust.
V70	Central air flap motor		all	Lowest of three motor/position sensor units on right side of evaporator housing, includes air flap position sensor G112.	
V71	Air flow flap motor		all	Near blower motor fresh air intake, includes position sensor G113 (also identified as backpressure flap motor).	
V97	Right rear central locking system motor			Integral with door contact switch F11.	
V101	Secondary air injection pump motor	AIR	as applicable	In engine compartment, lower right front.	
V107	Defroster flap motor		all	Upper most of three motor/position sensor units on right side of evaporator housing, includes position sensor G135.	
V115	Left rear central locking system motor		all	Integral with door contact switch F10.	
V139	Actuator for rear lid		all	Behind trim for rear lid	CL22 CL23
V144	Leak detection pump (LDP)		all	Back of left rear wheelhousing, behind wheelhousing liner.	
V151	Motor for remote unlock rear lid		all	Under deck lid trim, center, next to V53.	CL20 CL21
V155	Motor for fuel tank lid unlock		all	Luggage compartment, right side, behind trim panel.	CL22 CL23
V158	Actuator for temperature flap, left		all	Left side of evaporator housing, includes potentiometer G220.	
V159	Actuator for temperature flap, right		all	Central of three motor/position sensor units on right side of evaporator housing, includes potentiometer G221.	
V161	Driver's door central locking system motor		all	Integral with driver's door lock activator assembly in driver's door.	CL22 CL25
V192	Brake vacuum pump		as applicable	engine compartment, left side, in front of ABS/EDL/ESP unit	
Y6	Transmission range selector lever display	TR	all with automatic transmission	In instrument cluster.	
Y7	Automatic day/night interior mirror control module		as applicable	Integral with interior mirror.	
Z1	Heated rear window elements		as applicable	Embedded in rear window glass	
Z19	Oxygen sensor heater	HO2S (B1S1)	all	Heater for main oxygen sensor for bank 1 (right side), mounted in exhaust pipe just behind manifold (pre-cat sensor). Integral with sensor.	
Z24	Heated rear window elements with antenna		as applicable	Embedded in rear window glass	
Z28	Oxygen sensor heater 2	HO2S (B2S1)	all	Heater for main oxygen sensor for bank 2 (left side), mounted in exhaust pipe just behind manifold (pre-cat sensor). Integral with sensor.	
Z29	Oxygen sensor heater	HO2S (B1S2)	all	Heater for post-cat oxygen sensor for bank 1 (right side), mounted in exhaust system behind (downstream) Three Way Catalytic Converter. Integral with sensor.	
Z30	Oxygen sensor heater 2	HO2S (B2S2)	all	Heater for post-cat oxygen sensor for bank 2 (left side), mounted in exhaust system behind (downstream) Three Way Catalytic Converter. Integral with sensor.	
Z36	Steering wheel heating element		as applicable	Integral with steering wheel. Includes temperature sensor and control module,	
-	3-fold relay and fuse panel		all	Behind storage compartment on central support, driver's side	CL13 CL14
-	3-fold relay and fuse panel		all	In E-Box, in plenum chamber	CL11 CL15 CL30
-	8-fold relay panel		all	Behind storage compartment on central support, driver's side	CL13 CL15
-	13-fold relay and fuse panel			Behind storage compartment on central support, driver's side	CL13 CL17

Code	Component	OBD	Application	Location and additional information	Illust.
-	Connector station in E-Box		all	In E-Box in plenum chamber	CL13 CL17 CL30
-	Connector station, A pillar, left side		all	Lower section of driver's side A pillar, behind trim panels.	CL26
-	Connector station, A pillar, right side		all	Lower section of passenger side A pillar, behind trim panels.	CL28
-	E-Box in plenum chamber		all	In plenum chamber (compartment under hood behind bulkhead), left side.	CL30
-	Fuse panel		all	On instrument panel, left side	CL13 CL15
-	Ground connections		all	Various locations	CL37
-	Micro central electric panel		all	On relay panel, behind storage compartment on central support, driver's side	CL13 CL15
-	Relay/fuse panel		as applicable	Behind instrument panel on central carrier.	CL29

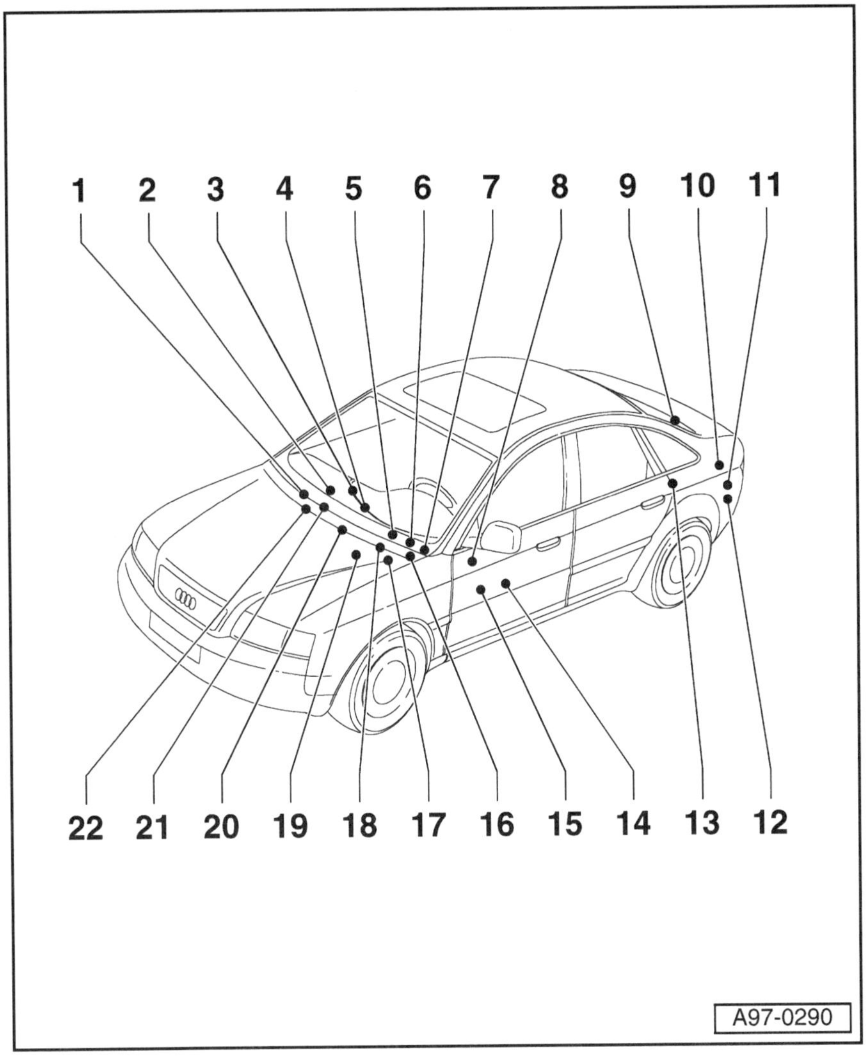

Fuse and relay panels, control modules

1. J431, Headlight range control module
2. J104, ABS/EDL control module
3. J217, Transmission control module
4. E87, A/C control head
5. 3-fold relay and fuse panel
6. 8-fold relay panel
7. Micro central electric panel
8. Fuse panel
9. J262, Rear window shade control module
10. J502, Tire pressure monitoring control module
11. J446, Control module for parking aid
12. J197, Level control system control module
13. J347, Control module for ultrasound sensors
14. J136, Memory seat control module
15. J429, Control module for central locking
16. 13-fold relay and fuse panel
17. J220, Motronic engine control module
18. Connector station E-Box
19. 3-fold relay panel
20. J234, Airbag control module
21. J352, Control module for steering column-belt height adjustment
22. J267, Mirror memory control module

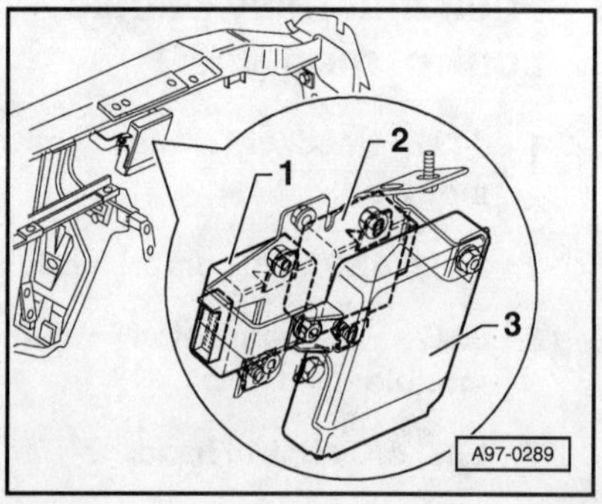

1. J431, Headlight range control module

- Behind glove compartment on central support, location **2**

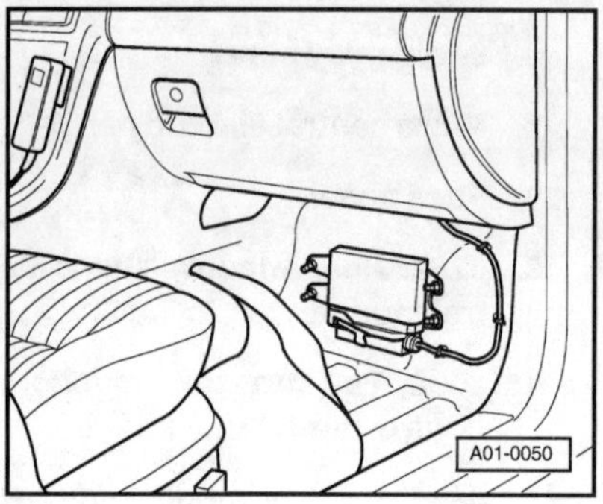

2. J104, ABS/EDL control module

- In passenger footwell, below carpet

3. J217, Transmission control module

- In protection box below carpet in front of passenger seat

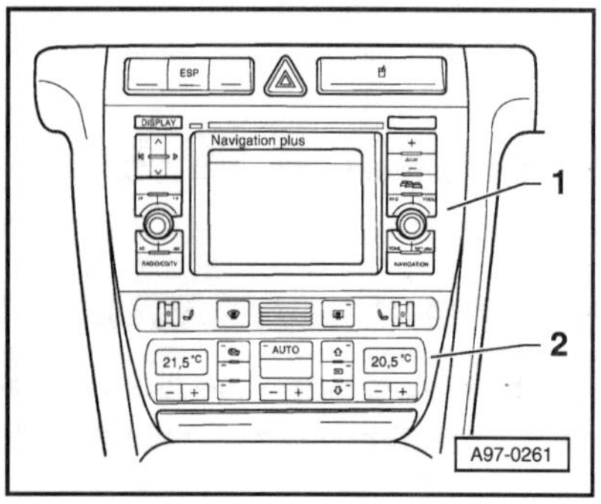

4. E87, A/C control head

- In center console, position **2**

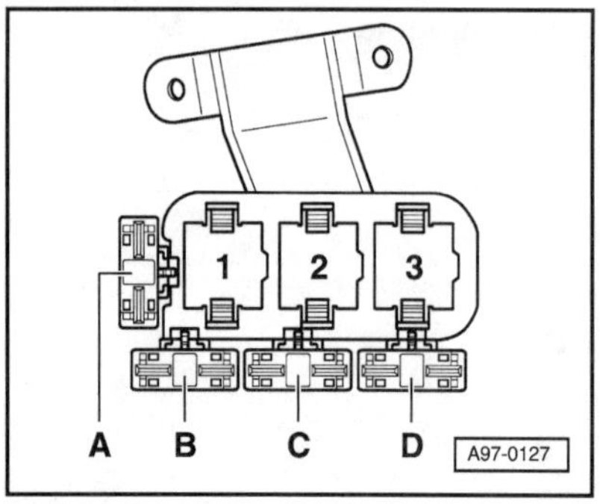

5. 3-fold relay and fuse panel

- Behind storage compartment on central support, driver's side

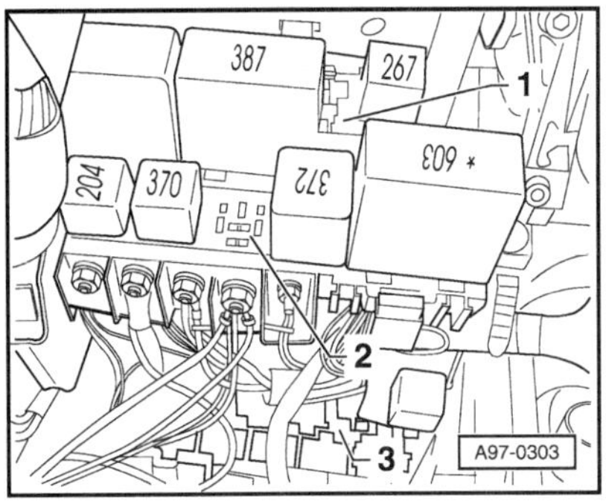

6. 8-fold relay panel

- Behind storage compartment, driver's side, position **3**

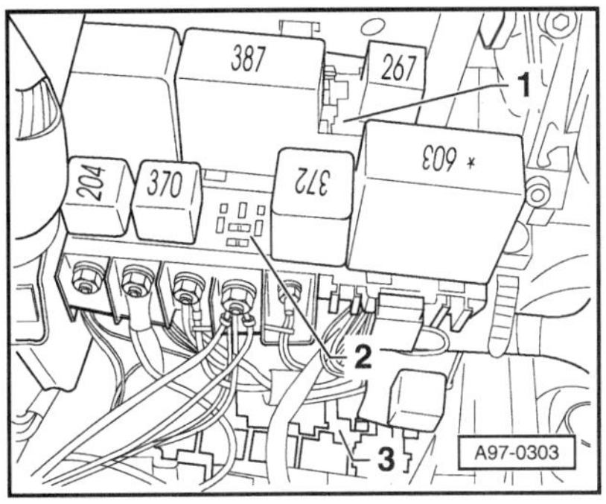

7. Micro central electric panel

- On relay panel behind storage compartment, driver's side, position **2**

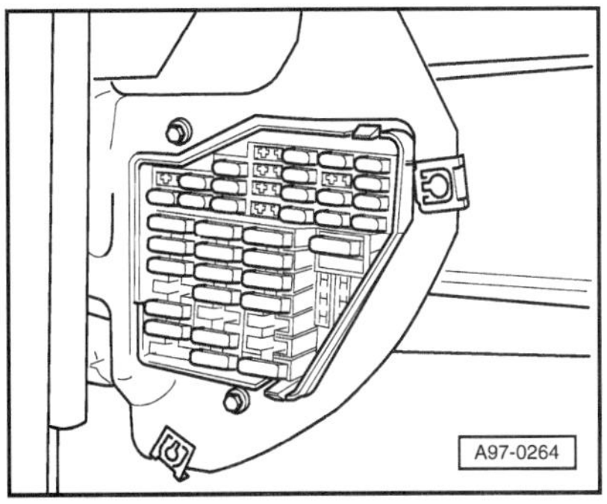

8. Fuse panel

- On instrument panel, left side

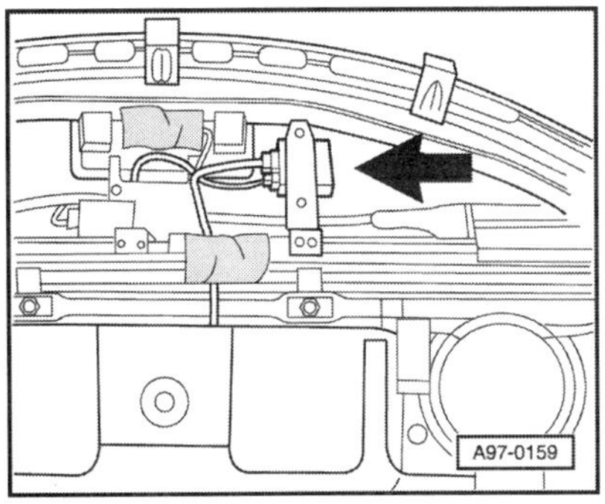

9. J262, Rear window shade control module

- Below rear window shelf, center left

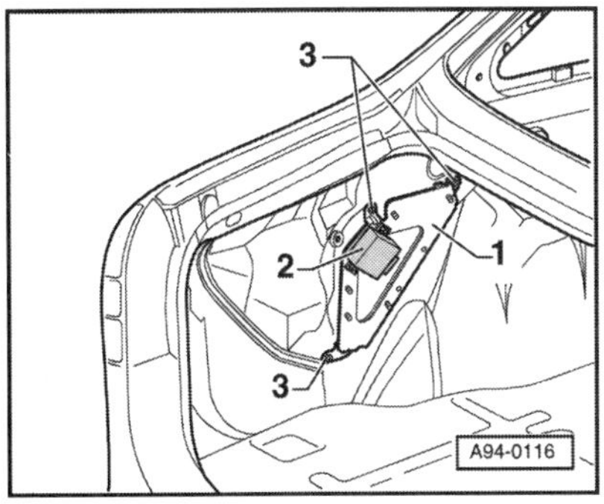

10. J502, Tire pressure monitoring control module

- Behind luggage compartment side panel trim, position **2**

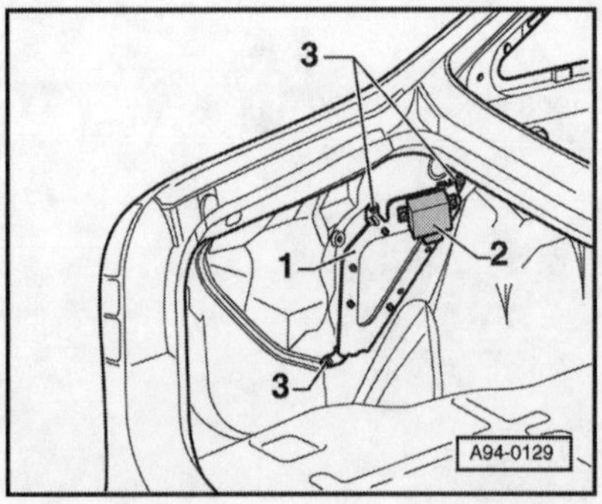

11. J446, Control module for parking aid

- Behind luggage compartment side trim, position **2**

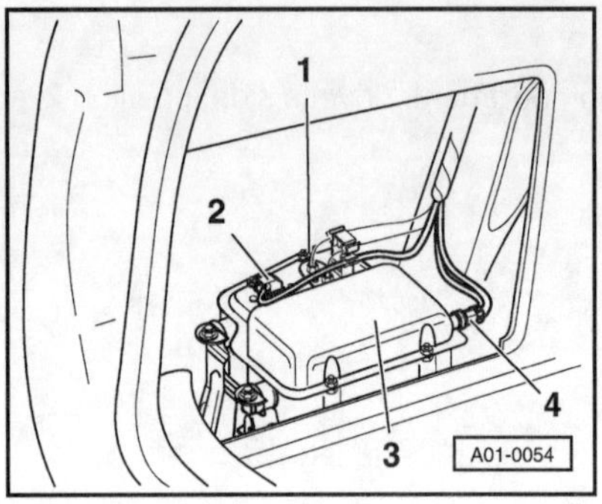

12. J197, Level control system control module

- Behind left luggage compartment side panel trim, in sound absorbing protection box, position **3**

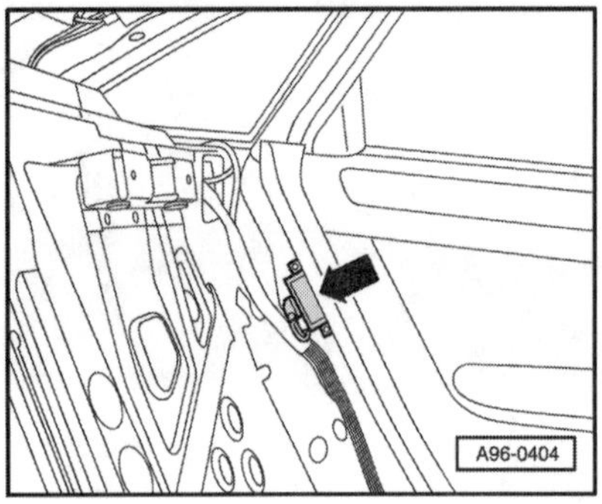

13. J347, Control module for ultra-sound sensors

- Behind left C-pillar trim

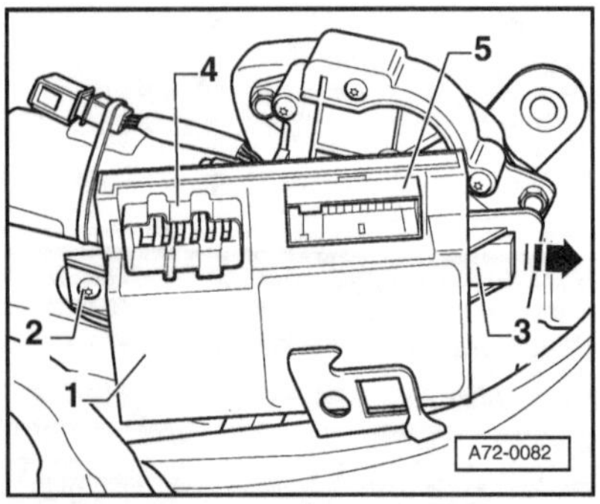

14. J136, Memory seat control module

- Below driver's seat, on seat frame, position **1**

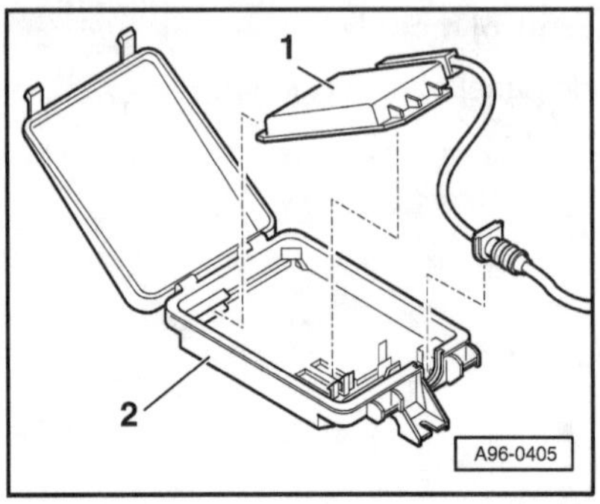

15. J429, Control module for central locking

- In protective box, below driver's seat

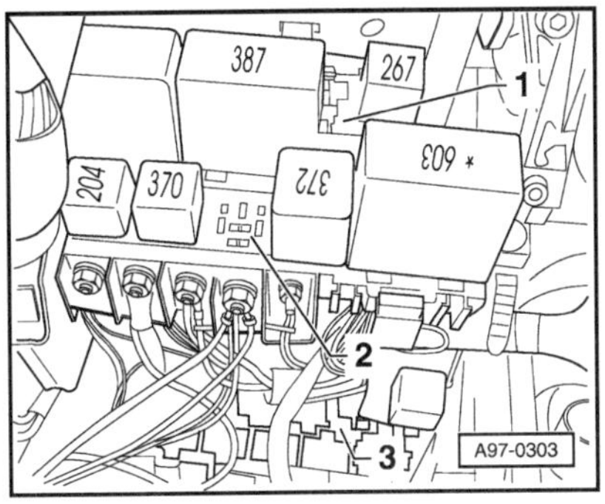

16. 13-fold relay and fuse panel

- Behind storage compartment, driver's side position **1**

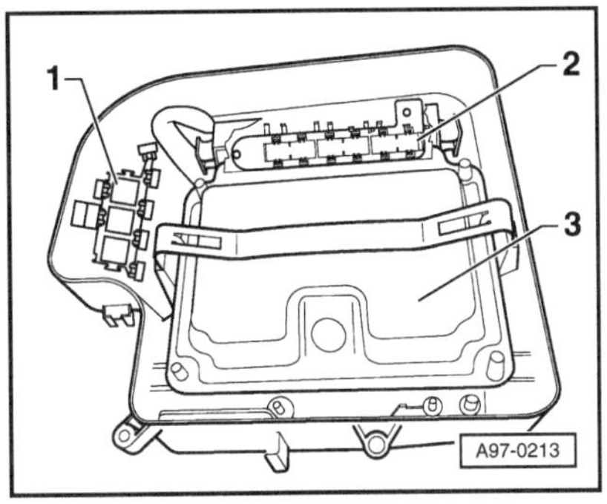

17. J220, Motronic engine control module

- In E-Box in plenum chamber, position **3**

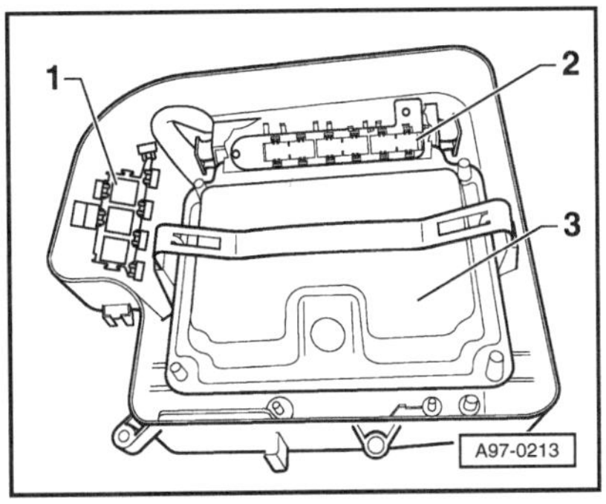

18. Connector station E-Box

- In E-Box in plenum chamber, position **2**

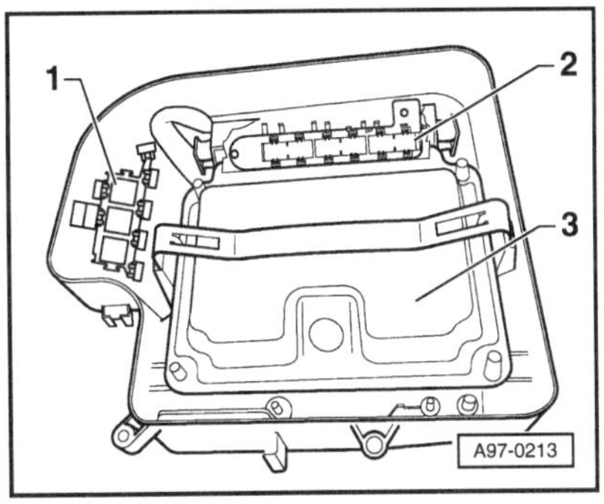

19. 3-fold relay panel

- In E-Box in plenum chamber, position **1**

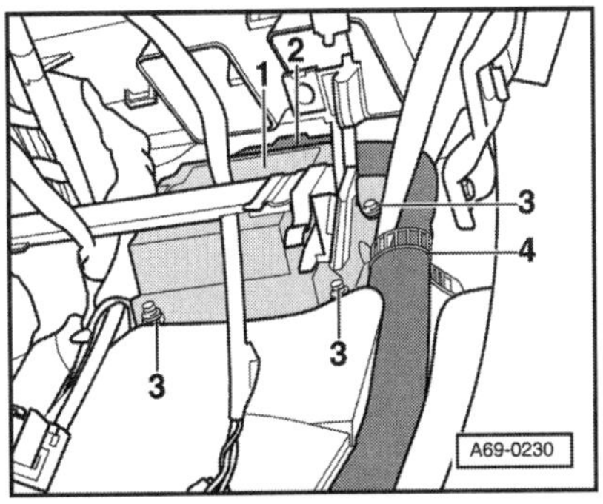

20. J234, Airbag control module

- Below center console on tunnel, position **1**

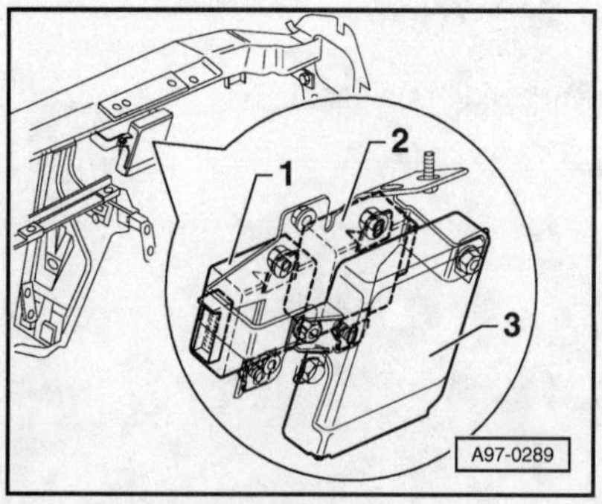

21. J352, Control module for steering column belt-height adjustment

- Behind glove compartment on central support, position **3**

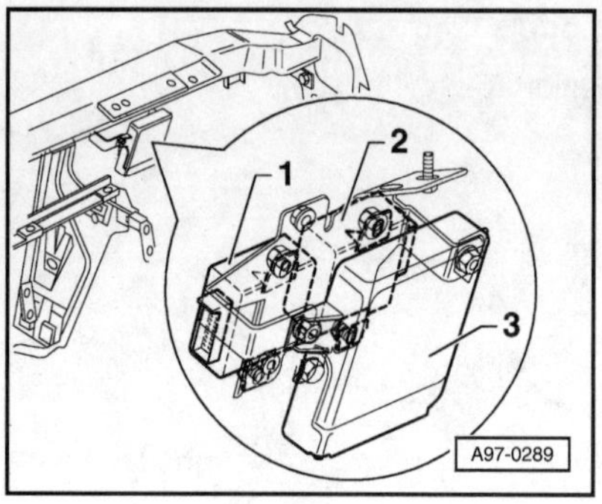

22. J267, Mirror memory control module

- Behind glove compartment on central support, position **1**

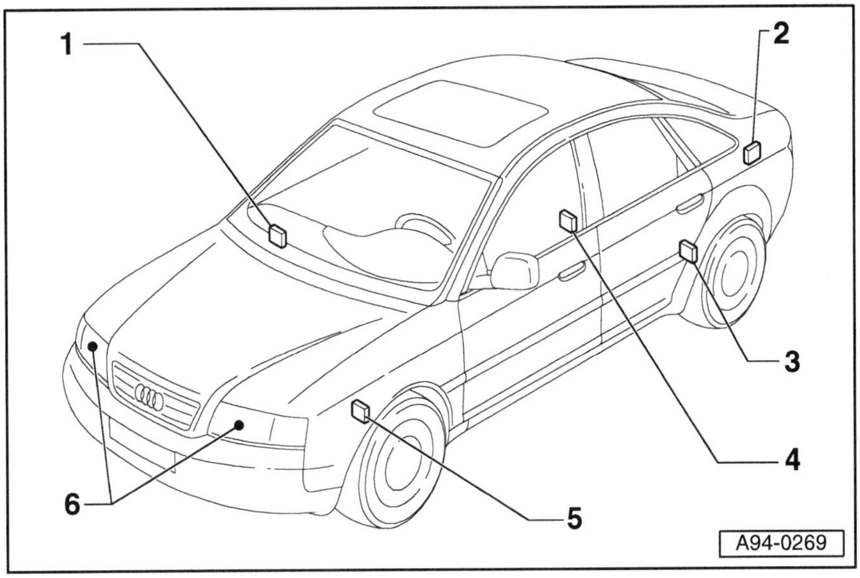

Automatic headlight beam adjusting, with High Intensity Gas Discharge (HID) headlight system)

1. **J431, Headlight range control module (Sedan and Avant from production 08/98)**
2. **J431, Headlight range control module**
   - **a. Sedan up to production 07/98**
   - **b. Avant up to production 07/98**
3. **G76, Left rear level control system sensor (quattro)**
4. **G76, Left rear level control system sensor (front wheel drive)**
5. **G78, Left front level control system sensor (quattro and front wheel drive)**
6. **Automatic headlight beam adjusting motors**
   - **a. V48 and V49, left and right static motor**
   - **b. V48 and V49, left and right dynamic motor**

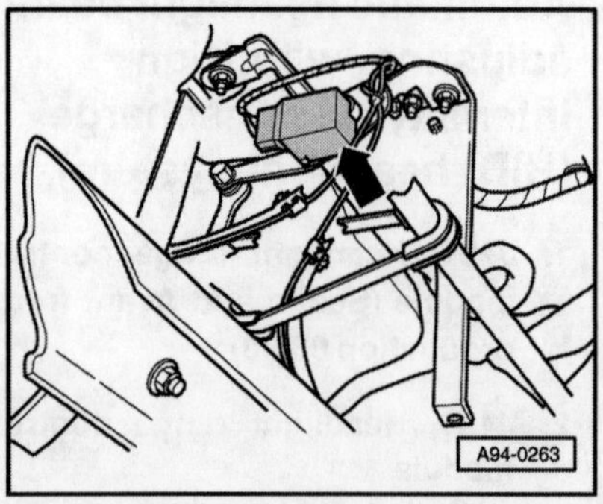

1. J431, Headlight range control module (Sedan and Avant from production 08/98)

- Behind glove box on support for instrument panel

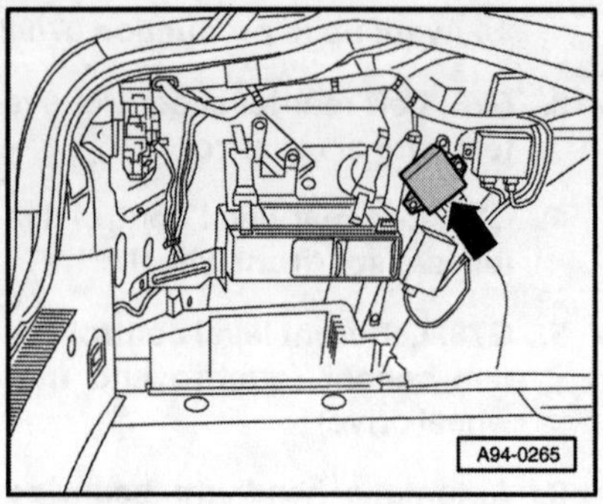

2a. J431, Headlight range control module (Sedan up to production 07/98)

- In luggage compartment, behind left rear side panel trim

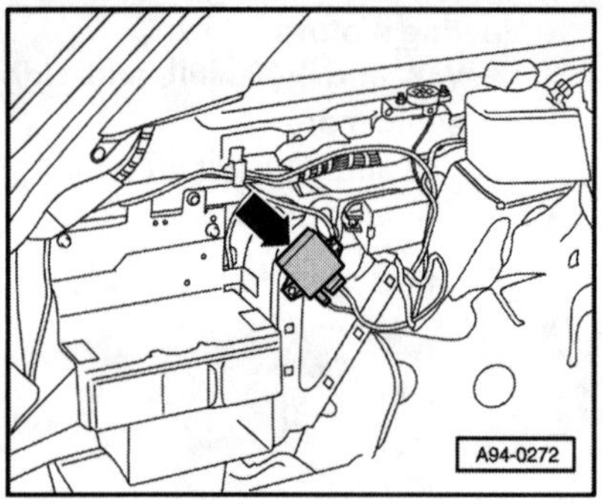

2b. J431, Headlight range control module (Avant up to production 07/98)

- In luggage compartment, behind left rear side panel trim

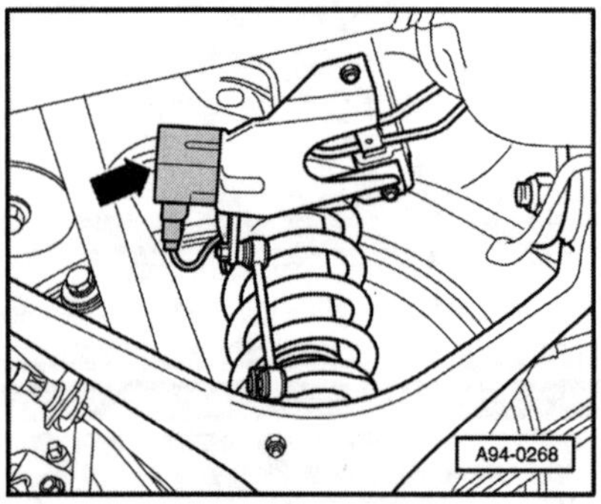

3. G76, Left rear level control system sensor (quattro)

- On axle, left rear

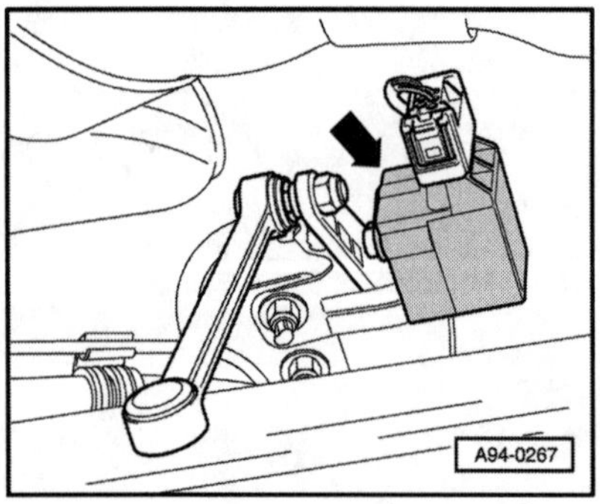

4. G76, Left rear level control system sensor (front wheel drive)

- In center of rear axle

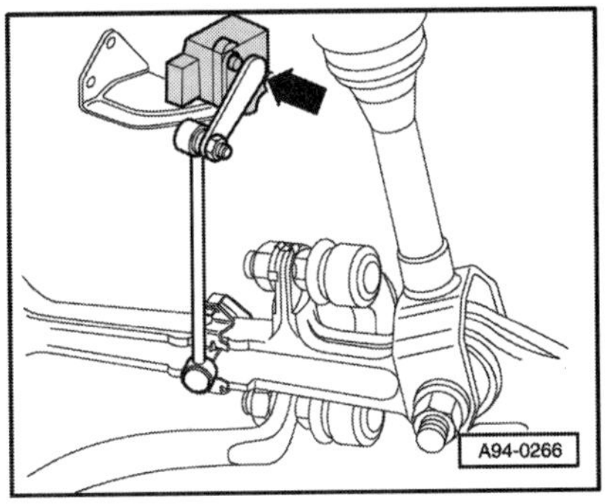

5. G78, Left front level control system sensor (quattro and front wheel drive)

- On supporting link, left front axle

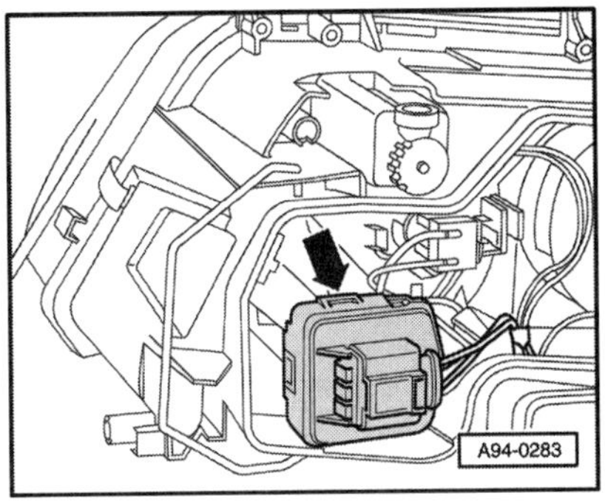

6a. V48, V49,Left/ Right automatic static headlight beam adjusting motor

- In headlight housing

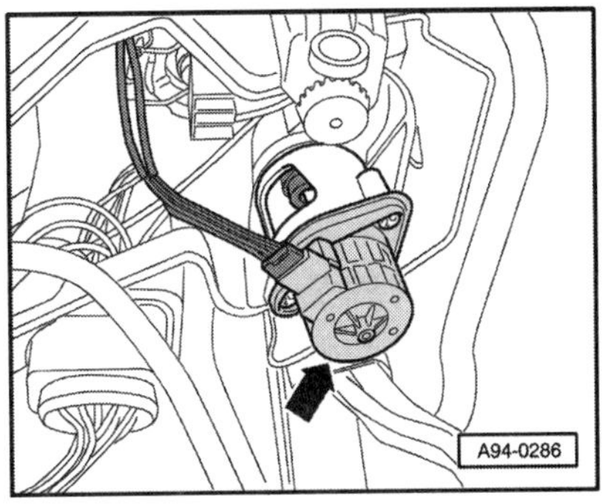

6b. V48, V49, Left/Right automatic dynamic headlight beam adjusting motor

- In headlight housing

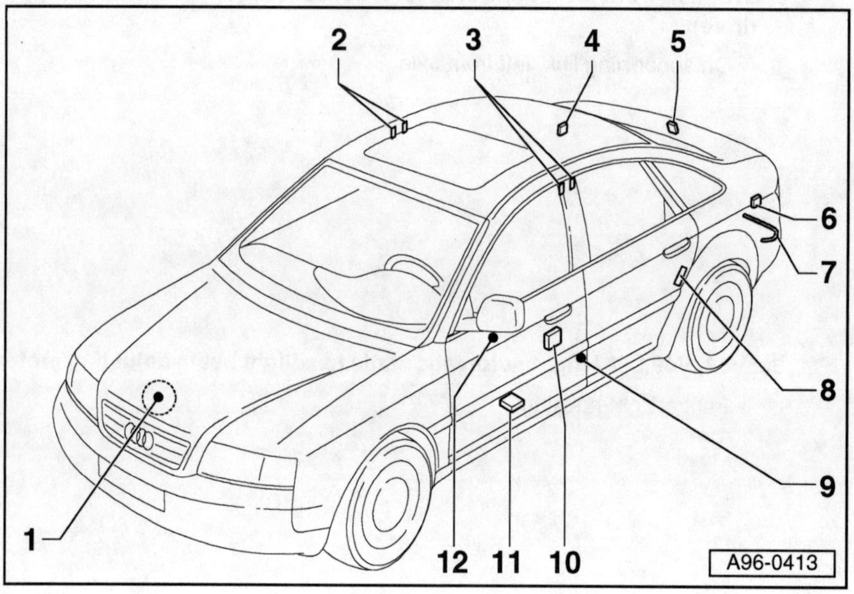

Central locking system, Sedan and Avant (m.y. 2000)

1. **F120, Hood alarm switch**
2. **G170, Ultra-sound sensor, left, for anti-theft warning system**
3. **G171, Ultra-sound sensor, right, for anti-theft warning system**
4. **V155, Motor for fuel tank lid unlock**
5. **F124, F206, F218, V53, V139, Actuator (assembly) for rear lid**
6. **H8, Alarm horn**
   a. Sedan
   b. Avant
7. **R47, Central locking and anti-theft alarm system antenna**
8. **J347, Control module for ultra-sound sensors**
9. **E183, Switch (bypass) for passenger compartment monitoring**
10. **F2, F131, F241, F243, V56, V161, Door lock activator (assembly), driver's side.**
11. **J429, Control module for central locking**
12. **E150, Switch for interior lock, driver's side**

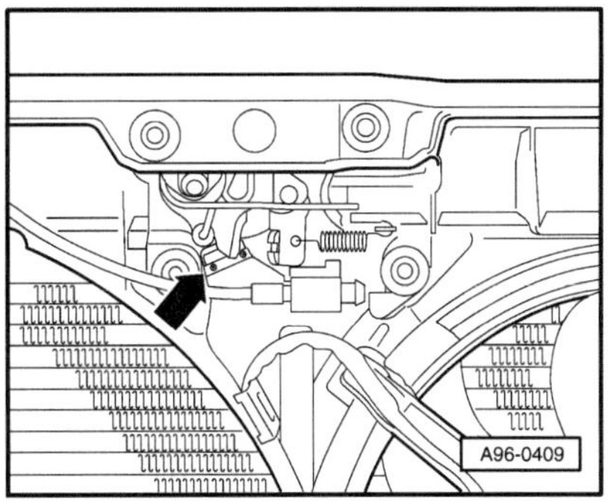

1. **F120, Hood alarm switch**
    - On front hood lock assembly

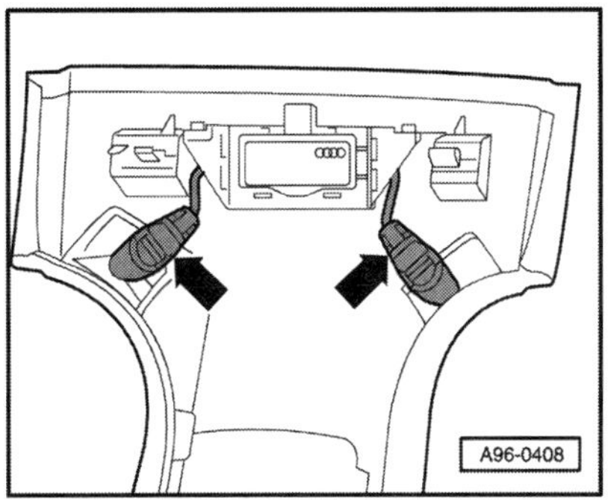

2. **G170, Ultra-sound sensor, left, for anti-theft warning system**
    - Behind upper B-pillar trim, left side

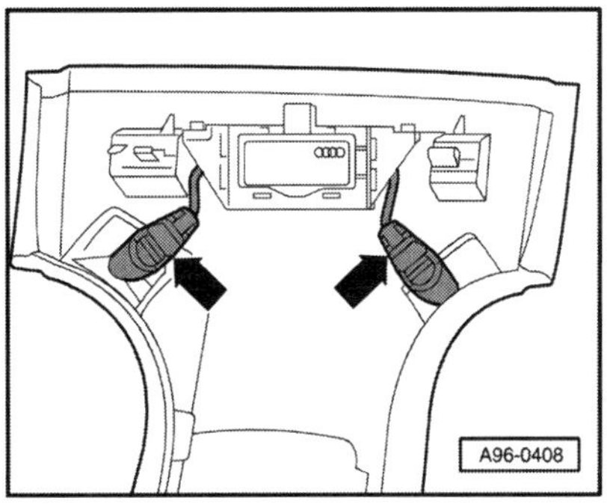

3. **G171, Ultra-sound sensor, right, for anti-theft warning system**
    - Behind upper B-pillar trim, right side

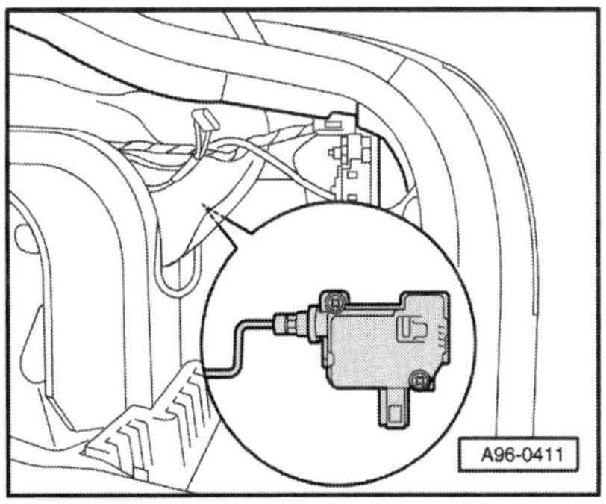

4. **V155, Motor for fuel tank lid unlock**
    - In luggage compartment, behind right rear side panel trim

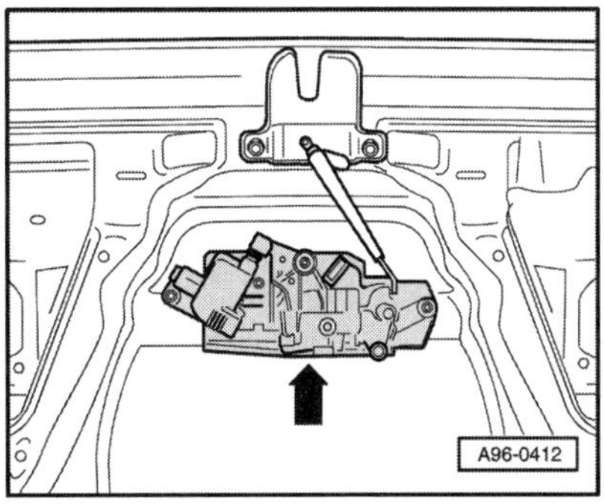

5. **F124, F206, F218, V53, V139, Actuator (assembly) for rear lid**
    - Behind trim for rear lid

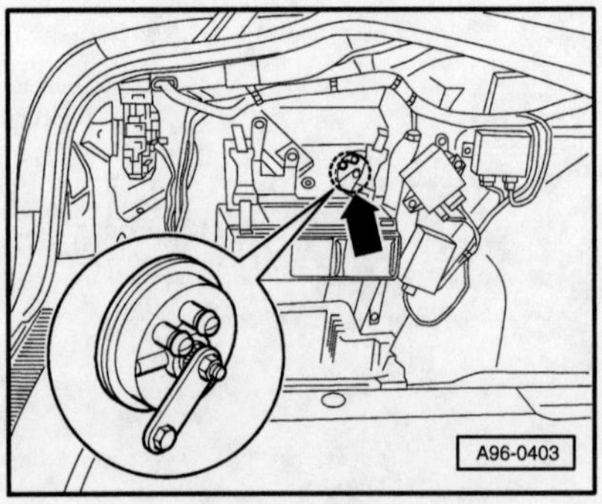

6a. H8, Alarm horn (Sedan)

- In luggage compartment, behind left rear panel trim

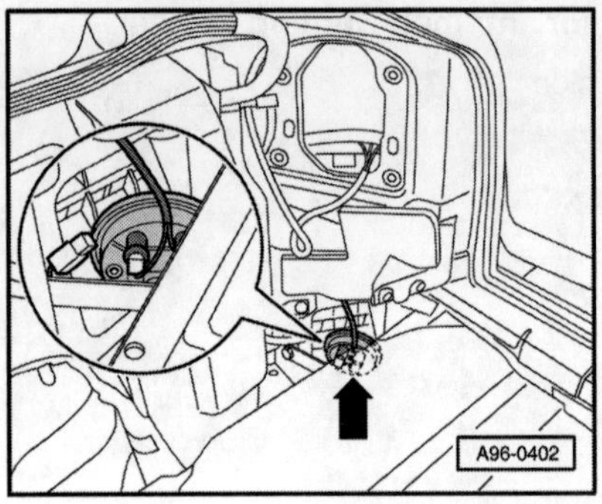

6b. H8, Alarm horn (Avant)

- In cargo compartment, behind right rear panel trim

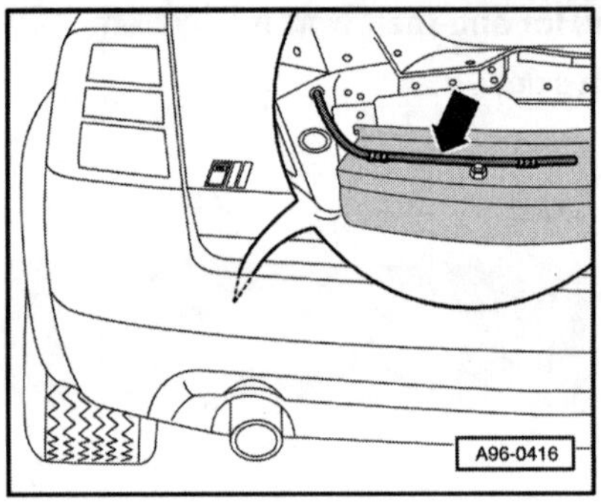

7. R47, Central locking and anti-theft alarm system antenna

- Sedan: on rear bumper carrier, left side
- Avant: on rear bumper carrier, right side

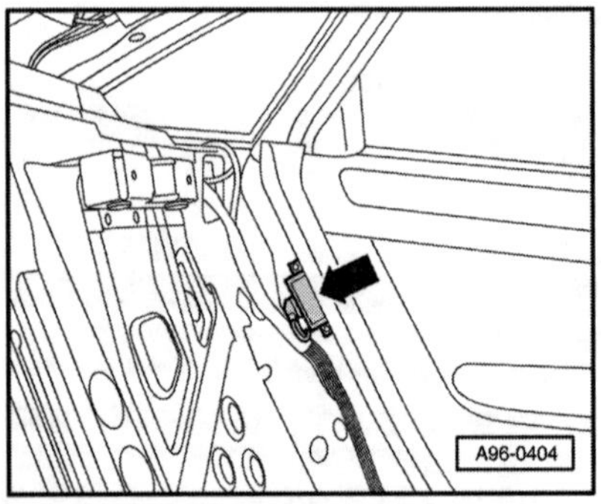

8. J347, Control module for ultra-sound sensors

- On left rear C-pillar trim

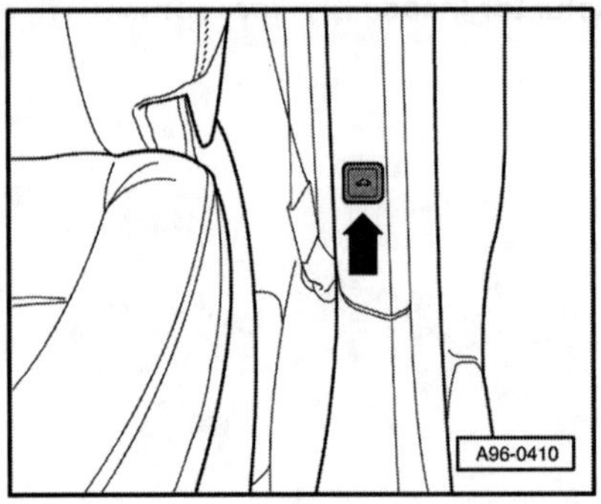

9. E183, Switch (bypass) for passenger compartment monitoring

- In passenger compartment, on lower left B-pillar trim

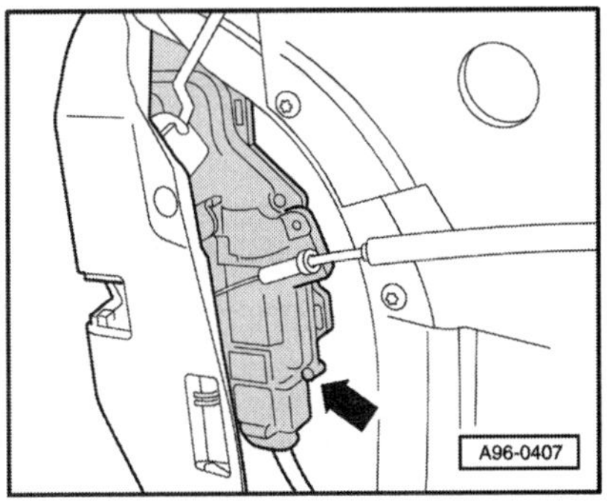

10. F2, F131, F241, F243, V56, V161, Door lock activator (assembly), driver's side.

- In driver's door

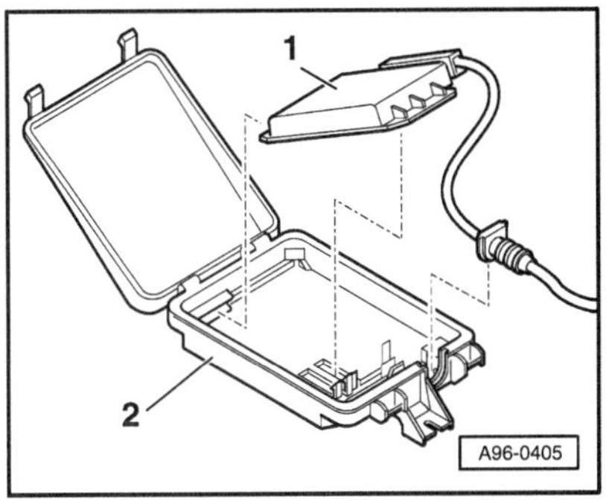

11. J429, Control module for central locking

- In passenger compartment below driver's seat
- In protective box (2)

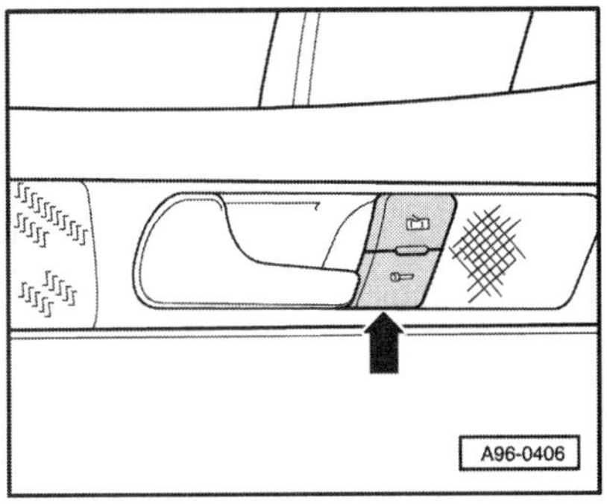

12. E150, Switch for interior lock, driver's side

- In front door trim panel, driver's side

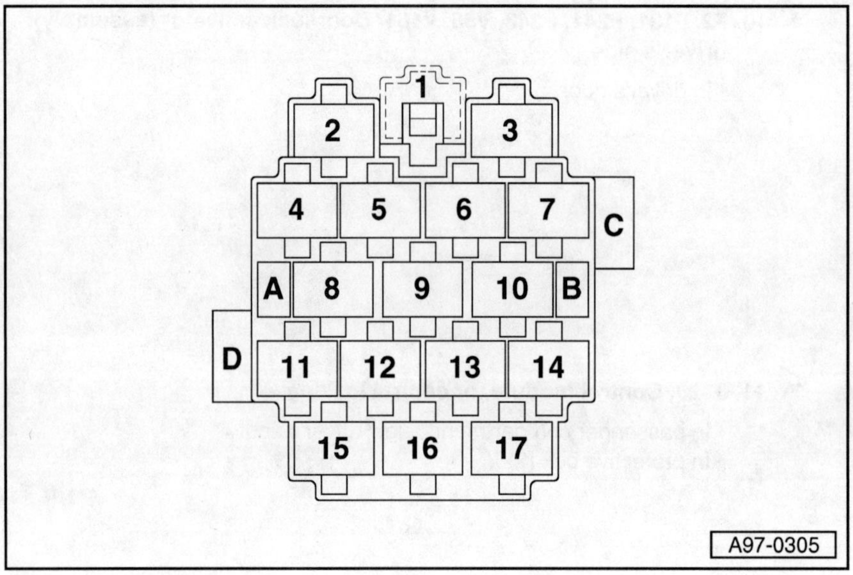

Connector station, A-pillar, left side

1. **Black**
   - From: Instrument panel harness
   - To: Mirror adjustment harness
2. **Light Brown**
   - From: Central locking system, anti-theft alarm system harness
   - To: Door lock, anti-theft alarm system harness
3. **Red**
   - From: Power window (opening) harness
   - To: Power window (opening) harness in door
4. **Black**
   - From: Cruise control harness
   - To: Instrument panel harness
5. **Blue**
   - From: Speaker harness
   - To: Speaker harness
6. **Grey**
   - From: Power window (opening) harness
   - To: Power window (opening) harness
7. **Blue**
   - From: Rear harness
   - To: Instrument panel harness
8. **Brown**
   - From: Rear harness
   - To: Instrument panel harness
9. **Red**
   - From: Gas discharge headlight (HID) left
   - To: Instrument panel harness
10. **Violet**
    - From: Gas discharge headlight (HID) left
    - To: Instrument panel harness

(continued on following page)

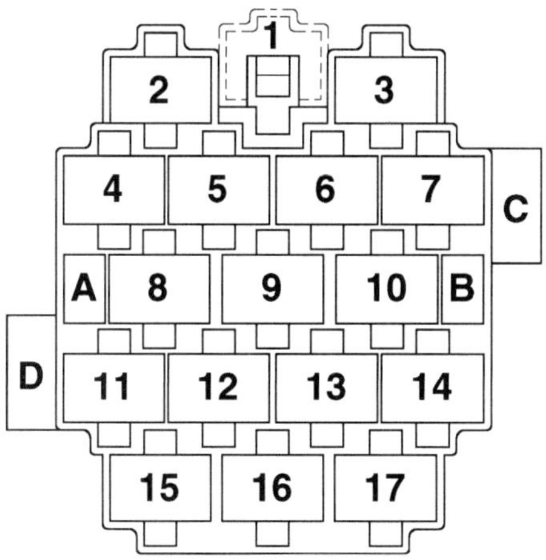

(continued from previous page)

11. Blue
- From: Rear harness
- To: Instrument panel harness

12. White
- From: Heated seat, power seat, seat belt warning harness
- To: Instrument panel harness

13. Brown
- From: Power window (opening), sunroof harness
- To: Instrument panel harness

14. Grey
- From: Radiator fan, A/C compressor harness
- To: Instrument panel harness

15. Orange
- From: ABS harness
- To: Instrument panel harness

16. Violet
- From: Interior light harness
- To: Instrument panel harness

17. Black
- From: Wiper motor harness
- To: Instrument panel harness

A. Black
- From: Rear harness
- To: Instrument panel harness

B. Black
- From: Headlight washer assembly harness
- To: Instrument panel harness

C. Blue
- From: Instrument panel harness
- To: Temperature gauge harness

D. Yellow
- From: Driver's side airbag harness
- To: Instrument panel harness

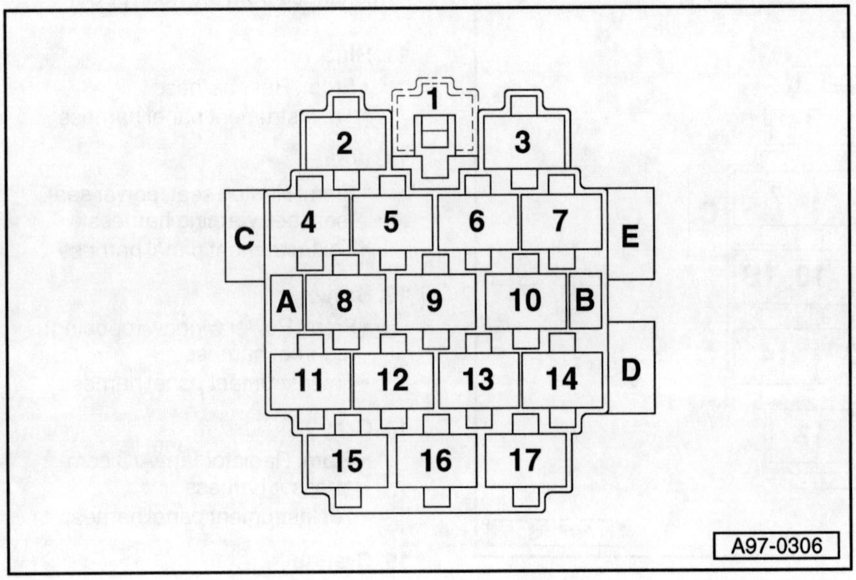

Connector station, A-pillar, right side

1. **Black**
   - From: Instrument panel harness
   - To: Mirror adjustment harness
2. **Red**
   - From: Power window (opening) harness
   - To: Power window (opening) harness in door
3. **Yellow**
   - From: Central locking system, anti-theft alarm system harness
   - To: Door lock, anti-theft alarm system harness
4. **White**
   - From: Parking aid harness
   - To: Instrument panel harness
5. **Blue**
   - From: Gas discharge headlight (HID) left
   - To: Instrument panel harness
6. **Green**
   - From: Speaker harness
   - To: Speaker harness
7. **Black**
   - From: Central locking system, anti-theft alarm system harness
   - To: Instrument panel harness
8. **Violet**
   - From: Gas discharge headlight (HID) right
   - To: Instrument panel harness
9. **Red**
   - From: Gas discharge headlight (HID) right
   - To: Instrument panel harness
10. **Blue**
    - From: Automatic transmission harness
    - To: Instrument panel harness
11. **Grey**
    - From: ABS harness
    - To: Instrument panel harness
12. **Yellow**
    - From: Driver's side airbag harness
    - To: Instrument panel harness
13. **Black**
    - From: Reading light harness
    - To: Instrument panel harness

(continued on following page)

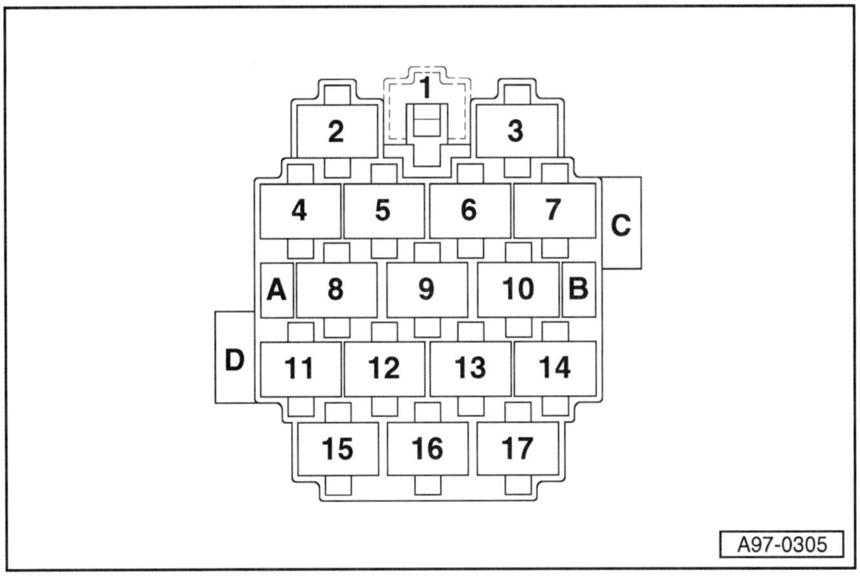

(continued from previous page)

14. Grey
- From: Automatic transmission harness
- To: Instrument panel harness

15. Yellow
- From: Daytime running light harness
- To: Instrument panel harness

16. Brown
- From: Leak detection pump, automatic day/night mirror, alarm system harness
- To: Instrument panel harness

A. N/A

B. N/A

C. N/A

D. White
- From: Seat belt tensioner, passenger side airbag harness
- To: Instrument panel harness

E. Blue (N/A)

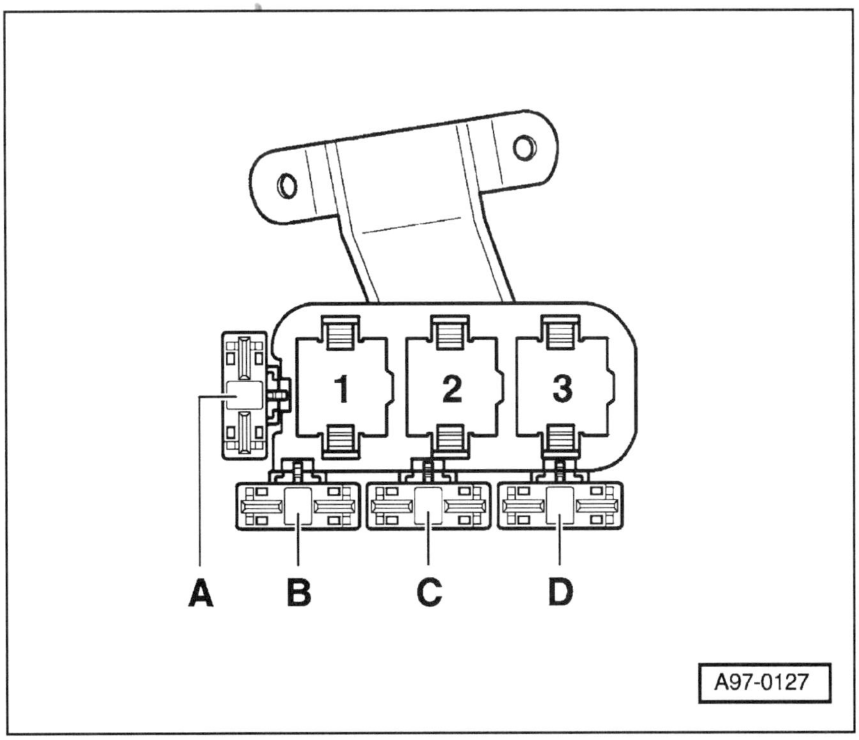

Relay panel behind instrument panel on central carrier

1. **J569, Brake booster relay**

C. **S279, Brake booster relay fuse**

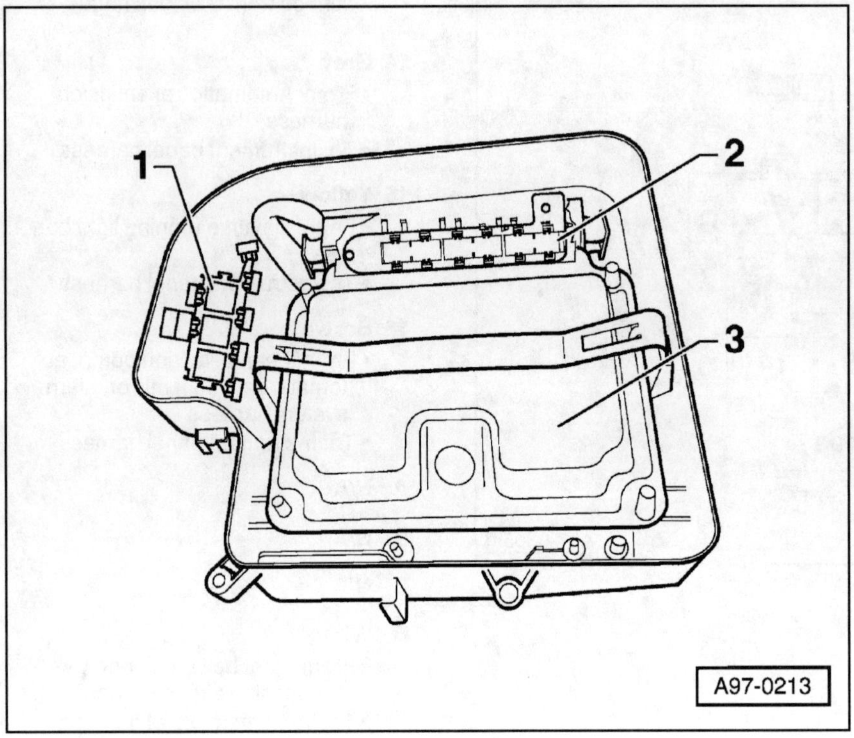

E-Box in plenum chamber

1. **3-Fold relay and fuse panel**
2. **Connector station in plenum chamber E-Box**
3. **J220, Motronic Engine Control Module, ECM**

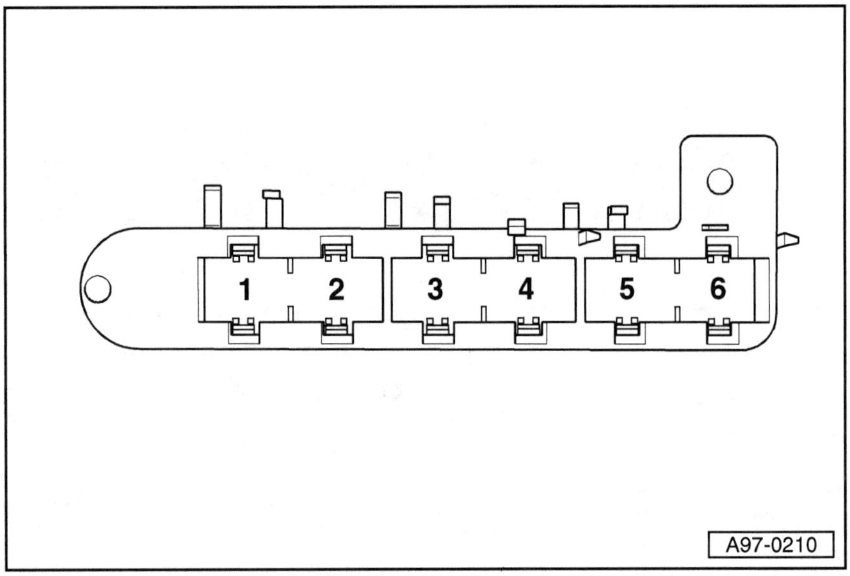

Connector station in plenum chamber E-Box

1. **White**
   - From: Instrument panel harness
   - To: Mirror adjustment harness
2. **Blue**
   - From: Power window (opening) harness
   - To: Power window (opening) harness in door
3. **Orange**
   - From: Central locking system, anti-theft alarm system harness
   - To: Door lock, anti-theft alarm system harness
4. **Red**
   - From: Parking aid harness
   - To: Instrument panel harness
5. **Brown**
   - From: Gas discharge headlight (HID) left
   - To: Instrument panel harness
6. **Black**
   - From: Speaker harness
   - To: Speaker harness

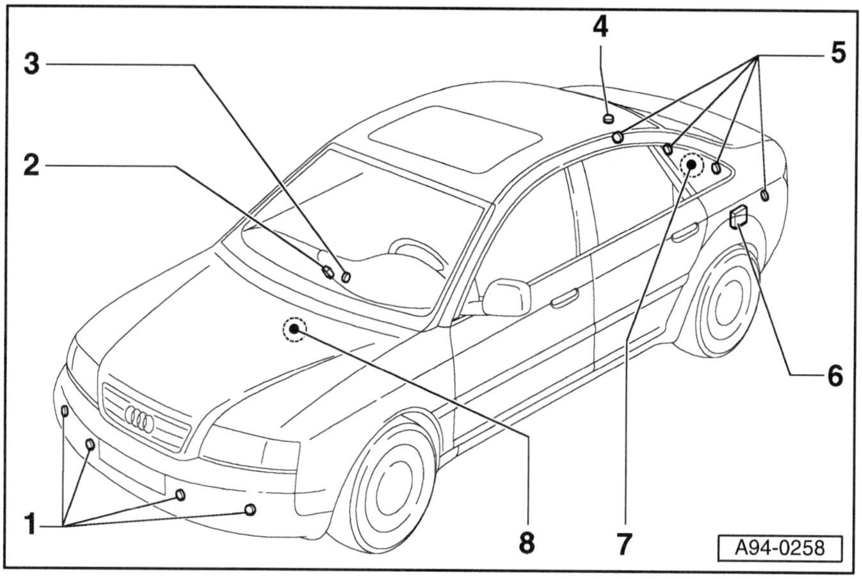

Parking aid, Sedan and Avant (m.y. 2000)

1. G252, G253, G254, G255, Front parking aid sensors
2. E266, Parking aid button
   a. Sedan and avant
   b. Allroad quattro
3. H22, Front parking aid warning buzzer
4. H15, Warning buzzer for parking, rear
   a. Sedan
   b. Avant
5. G203, G204, G205, G206, Rear parking aid sensors
6. J446, Control module for parking aid
   a. Sedan
   b. Avant
7. F216, Contact switch for fog light shut-off (vehicles with trailer coupling)
8. Transmission switch
   a. F4, Back-up light switch (manual transmission only)
   b. F125, Multi-function transmission range switch (automatic transmission only)

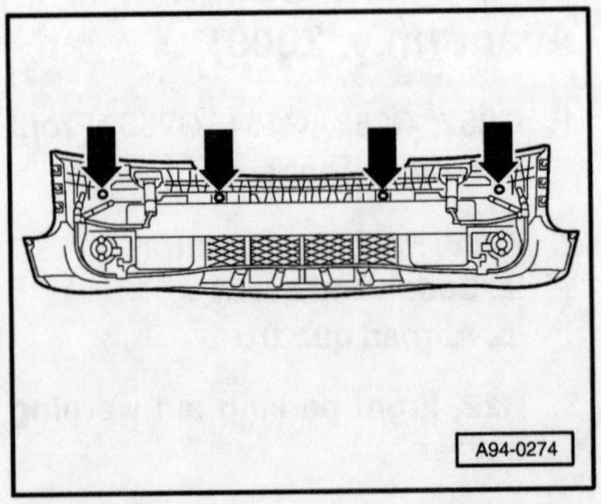

1. G252, G253, G254, G255, Front parking aid sensors

- In upper part of front bumper

2a. E266, Parking aid button

- In upper section of center console

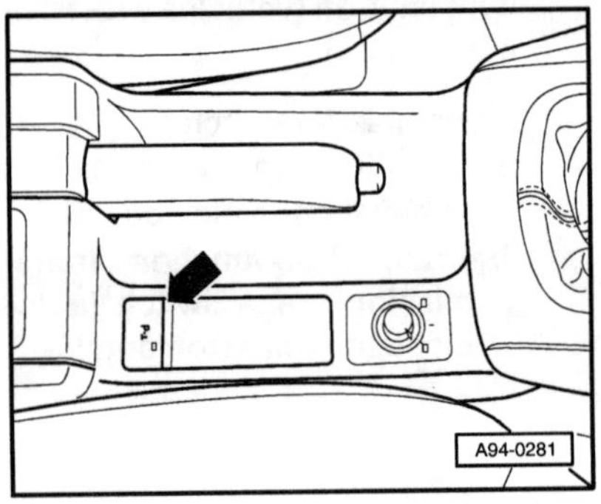

2b. E266, Parking aid button (allroad quattro)

- In lower section of center console

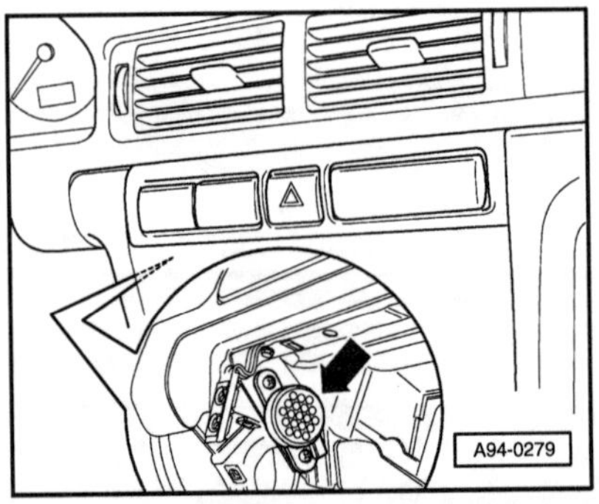

3. H22, Front parking aid warning buzzer

- In center console, left side

4a. H15, Warning buzzer for parking, rear (Sedan)

- Below rear window shelf lining, right side

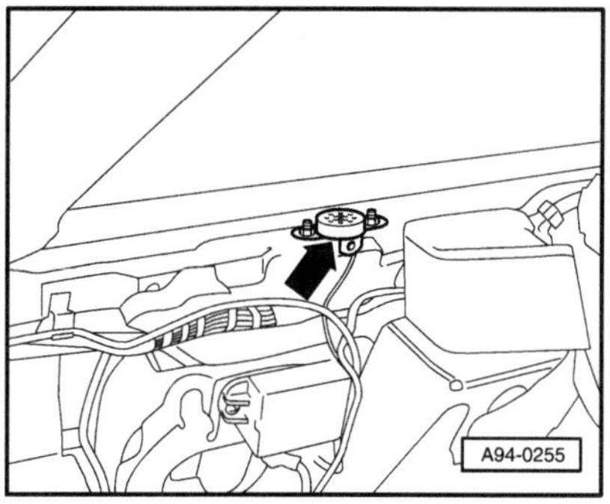

4b. H15, Warning buzzer for parking, rear (Avant)

- In cargo compartment, behind left rear panel trim

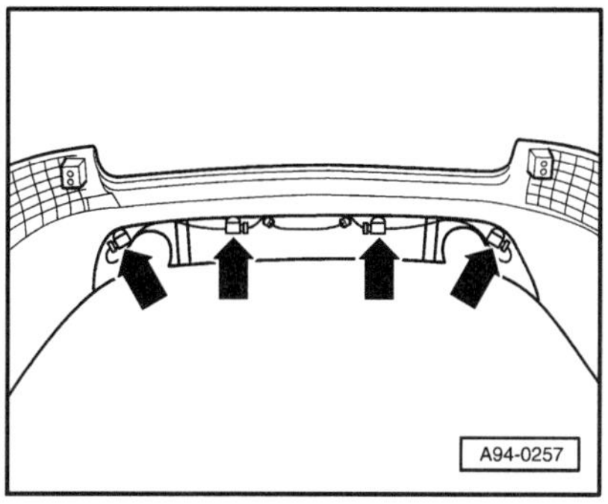

5. G203, G204, G205, G206, Rear parking aid sensors

- In lower part of rear bumper

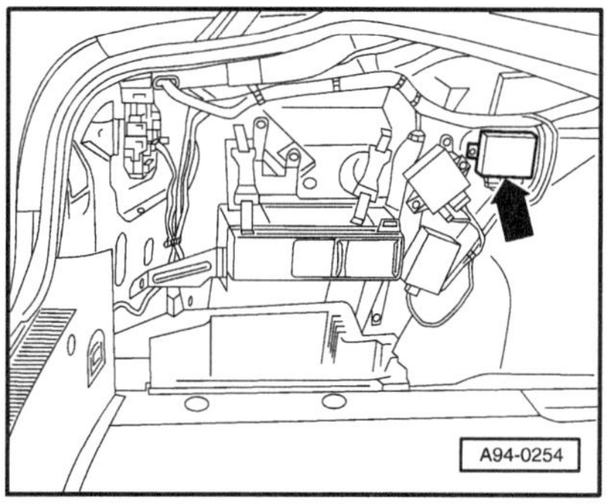

6a. J446, Control module for parking aid (Sedan)

- In luggage compartment, behind left rear side panel trim

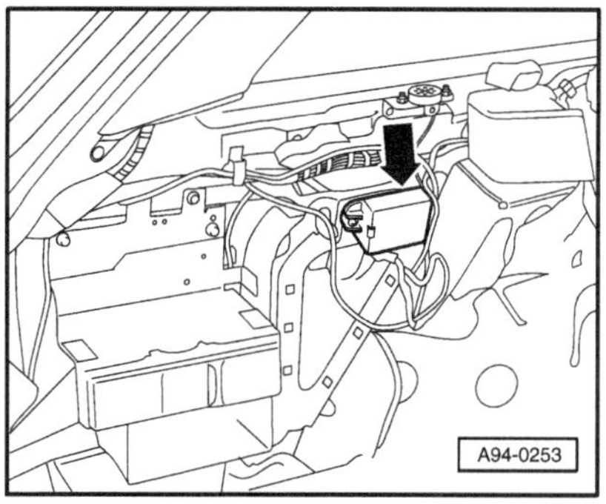

6b. J446, Control module for parking aid (Avant)

- In cargo compartment, behind left rear side panel trim

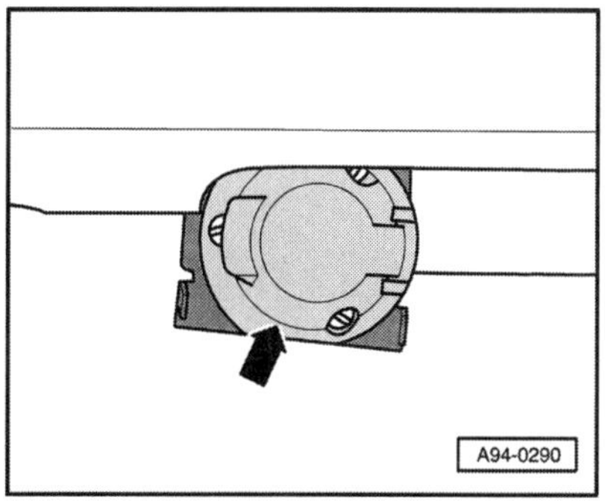

7. F216, Contact switch for fog light shut-off (vehicles with trailer coupling)

- In electrical system coupling for trailer, rear

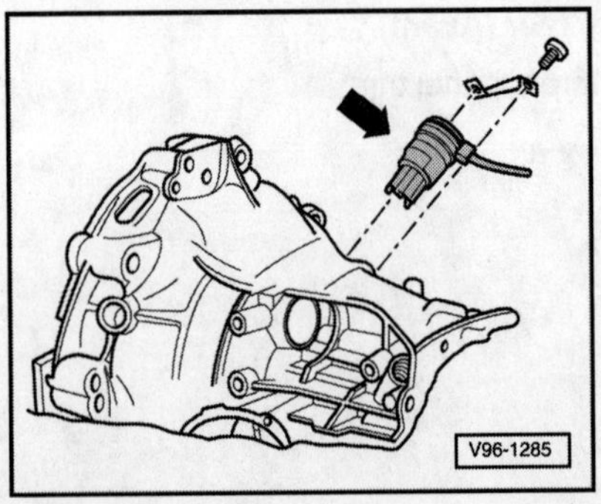

◄ **8a. F4, Back-up light switch (manual transmission only)**

- In transmission case, on right side

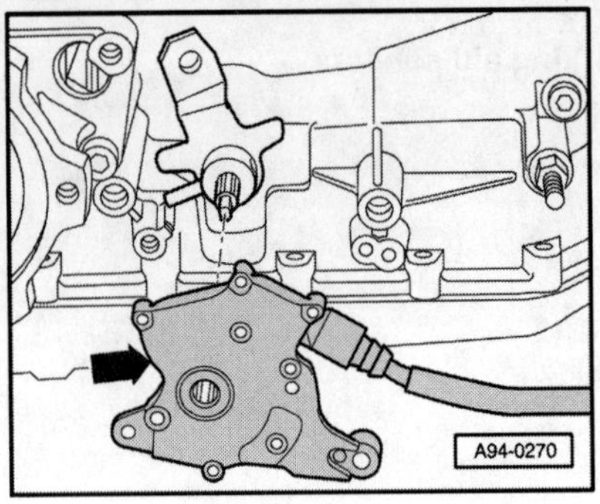

◄ **8b. F125, Multi-function transmission range switch (automatic transmission only)**

- In transmission case, on shift shaft, left side

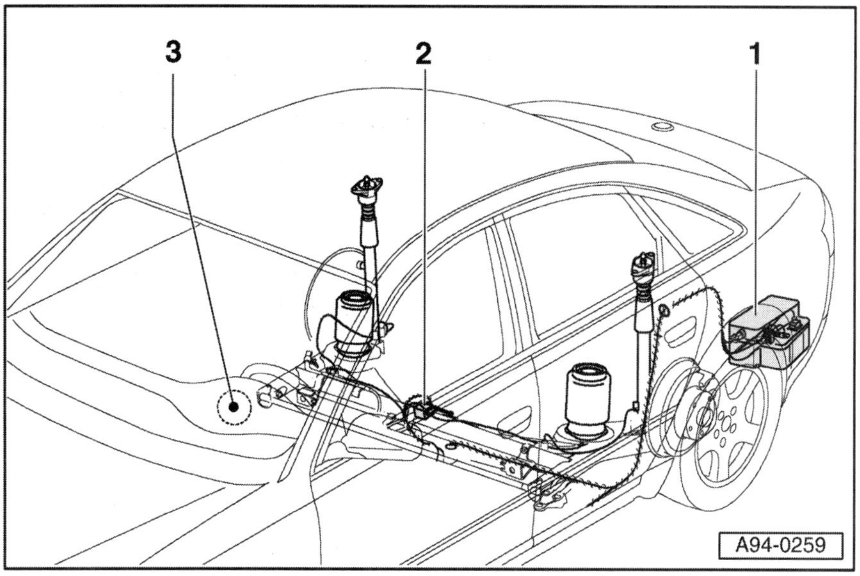

Level control system, Sedan and Avant (m.y. 2000)

1. **Sound-absorbing protection box for air supply component assembly**
   a. **Box contains individual components:**
      - **J197**, Level control system control module
      - **J403**, Relay for level control system compressor
      - **V66**, Level control system compressor motor
      - **N111**, Level control system solenoid valve.
      - **N150**, Left rear suspension strut valve
      - **N151**, Right rear suspension strut valve
2. **G84, Level control system sensor, ESP**
   a. **Front wheel drive**
   b. **Quattro**
3. **K134, Indicator lamp for level control system**

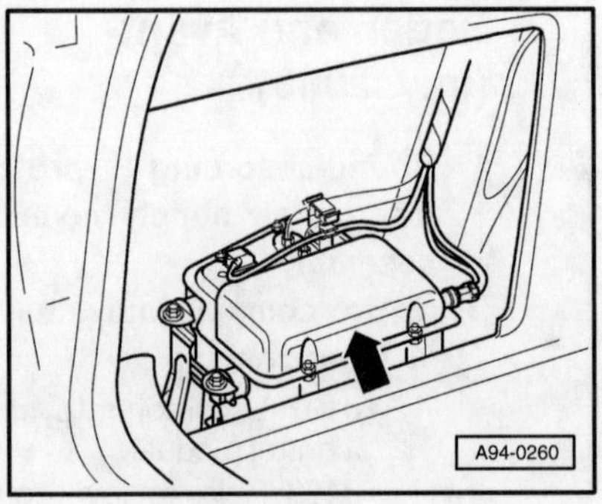

1. Sound-absorbing protection box for air supply component assembly,

- In shelf area in left side of luggage/cargo compartment (see below for individual components)

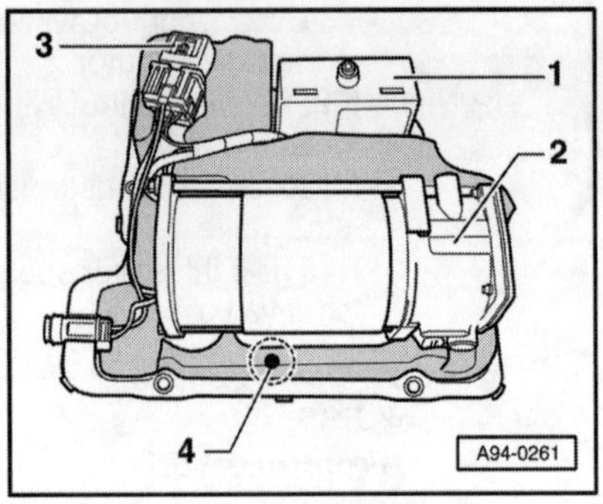

1a. Sound-absorbing protection box for air supply; individual components:

1. **J197**, Level control system control module
2. **V66**, Level control system compressor motor
3. **J403**, Relay for level control system compressor
4. Under foam sound absorber:
   N111, Level control system solenoid valve.
   N150, Left rear suspension strut valve
   N151, Right rear suspension strut valve

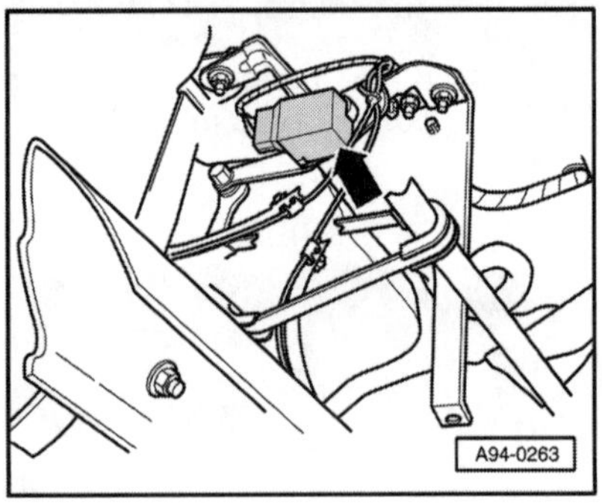

2a. G84, Level control system sensor, ESP (front wheel drive)

- In center of rear axle

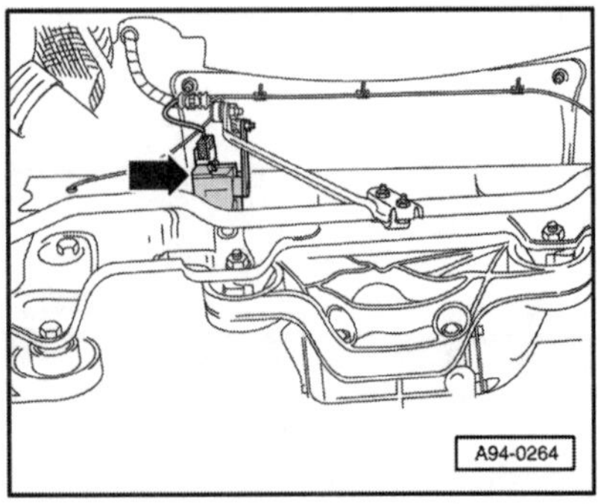

2b. G84, Level control system sensor, ESP (quattro)

- In center of rear axle

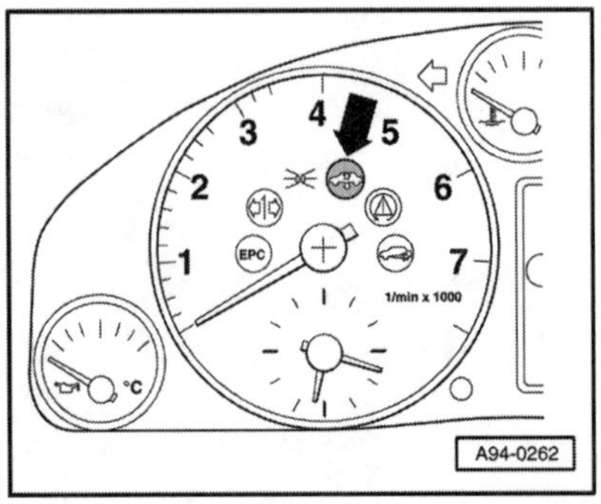

3. K134, Indicator lamp for level control system

- In instrument cluster

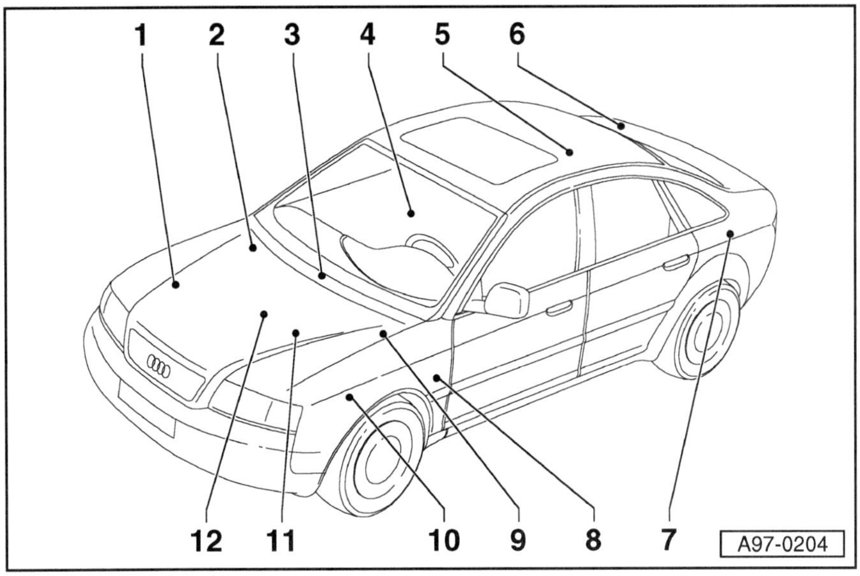

Ground connection locations

1. Ground connection, in engine compartment, right side
2. Ground connection, on right A-pillar, lower section
3. Ground strap, battery to body
4. Ground connection, on right B-pillar
5. Ground connection, on right D-pillar
6. Ground connection, near right side taillight
7. Ground connection, luggage/cargo compartment left rear
8. Ground connection, on left A-pillar, lower section
9. Ground connection, on ABS hydraulic unit
10. Ground connection, in engine compartment, left side
11. Ground strap, engine to body
12. Ground connection, near ignition coil

Notes

Scan Tool

Section
ST

WARNING!
Before proceeding with any work, read the Warnings and Cautions in Section WD.

General Information **Page**
- Overview . . . . . ST 2
- Data Link Connector, Scan Tool Connection . . . . . ST 3
- V.A.G. 1551/1552 Operating Overview Flow Chart . . . . . ST 54

Diagnostic Trouble Codes **Page**
- Diagnostic Scan Tool Suppliers . . . . . ST 28
- Diagnostic Trouble Code Table . . . . . ST 4

Control Module Coding **Page**
- Overview . . . . . ST 29
- Acoustic Parking System Coding Table . . . . . ST 46
- Airbag System Coding Table . . . . . ST 39
- Anti-Lock Braking System (ABS/EDL/ASR) Coding Table . . . . . ST 35
- Automatic Climate Control Coding Table . . . . . ST 42
- Automatic Vertical Headlight Aim Control coding Table . . . . . ST 47
- Central Locking Coding Table . . . . . ST 37
- Electronic Stability Program (ESP) Coding Tables . . . . . ST 36
- Engine Control Module (ECM) Coding Tables . . . . . ST 30
- Instrument Cluster Coding Tables . . . . . ST 45
- Interior Monitoring Coding Tables . . . . . ST 38
- Memory Seat (Driver and Passenger) Coding Table . . . . . ST 37
- Multi-Function Steering Wheel Coding Table . . . . . ST 43
- Radio and Sound System Coding Table . . . . . ST 48
- Suspension Level Control System Coding Table . . . . . ST 37
- Transmission Control Module (TCM) Coding Tables . . . . . ST 34

Readiness Codes **Page**
- Overview . . . . . ST 49
- Readiness Code Quick Reference Tables . . . . . ST 49
- Engine code: APB . . . . . ST 50
- Engine code: AHA . . . . . ST 51
- Engine code: ATQ (through m.y. 2000) . . . . . ST 52
- Engine code: ART . . . . . ST 53

General Information

Overview

Vehicles covered by this manual are equipped with the Government required diagnostic system known as On-Board Diagnostics II (OBD II). This system monitors operation and function of engine management system activity and automatic transmission operation to insure compliance with specified emission levels.

Vehicle emission levels are continuosly monitored by the OBD II system and malfunctions are recognized and recorded. A Malfunction Indicator Light (MIL) in the instrument cluster alerts the driver to the fault and the need to have the system checked for fault codes. These codes follow a standard format and are known as Diagnostic Trouble Codes (DTCs).

Each powertrain malfunction is assigned two codes. The first code is a numerical code assigned by the factory (VAG code). The second code is referred to as a P-code and follows a structure required by law and defined by the Society of Automotive Engineers (SAE). This standard uses a letter to designate the system and four numbers to further identify and detail the malfunction as listed below.

First digit structure is as follows:

- Pxxxx for powertrain
- Bxxxx for body
- Cxxxx for chassis
- Uxxxx for future systems

Second digit structure is:

- P0xxx Government required codes
- P1xxx Manufacturer codes for additional emission system function; not required but reported to the government
- P3xxx Addition supplementary codes for expanded engine management supplied by the manufacturer and reported to the government

Third digit structure is:

- Px1xx measurement of air and fuel
- Px2xx measurement of air and fuel
- Px3xx ignition system
- Px4xx additional emission control
- Px5xx speed and idle regulation
- Px6xx computer and output signals
- Px7xx transmission
- Px8xx transmission
- Px9xx control modules, input and output signals

The fourth and fifth digits designate the individual component and system.

DTCs and data can be retrieved with Volkswagen/Audi Factory Scan Tools such as the VAG 1551, VAG 1552, or the newer diagnostic computers VAS 5051 and VAS 5052, through a Data Link Connector (DLC). Several aftermarket scan tools and computer programs are also capable of retrieving this information in this factory mode. The factory mode also allows the scan tool to be used for other system diagnostic functions and information retrieval.

◄ Audi factory diagnostic testers and scan tools. At the top is the newest portable diagnostic tester VAS 5052, with the VAS 5051 below it. Both have an array of diagnostic cables and test leads and the VAS 5051 is cart-mounted with a laser printer. The VAS 5051 and 5052 replace the older scan tools VAG 1551 and 1552 shown in the lower two illustrations. Testers are not shown to scale.

Some DTC information can also be retrieved in a generic mode. The generic mode is not as complete as the factory mode, but allows commercially available scan tools to be used to read DTCs. Generic scan tool mode does not have the capability to retrieve the detailed information of a manufacturer-specific scan tool.

Data Link Connector, Scan Tool Connection

The Data Link Connector (DLC) is of the standard OBD II type. For all versions of the Audi A6, it is located to the left of the steering column, at the lower edge of the knee pad. On vehicles equipped with a manual transmission, this means that the DLC is above the clutch pedal. If diagnosis involves driving the vehicle with a scan tool connected, be sure to rout the scan tool cable in such a way so as to avoid disconnecting it while driving.

> ***WARNING —***
> *If it becomes necessary to road test the vehicle with the scan tool connected, test and measuring instruments must be secured to the rear seat and operated by a second person from this location for safety reasons. If test and measuring instruments are operated from the front passenger seat, there is a danger that the person sitting there could be injured in the event of an accident and/or airbag deployment.*

> ***WARNING —***
> - *DO NOT attempt to connect a VAG 1551/1552 scan tool, VAS 5051/5052 diagnostic tool, aftermarket scan tool or scan tool program to any vehicle if an incorrect radio has been installed.*
> - *Installation of aftermarket radios or Audi radios not produced for use in model year 1998 and newer vehicles will:*
>   - *Affect various vehicle systems connected to the CAN-BUS.*
>   - *Cause severe damage to VAG 1551 and VAG 1552 scan tools.*
>   - *Cause severe damage to VAS 5051 and VAS 5052 diagnostic tools.*
>   - *Cause severe damage to aftermarket scan tools.*
>   - *Cause severe damage to computers running scan tool programs.*
> - *Inspect radio; if radio is not correct for the vehicle or if there is any doubt, disconnect radio wiring harness before connecting scan tools.*

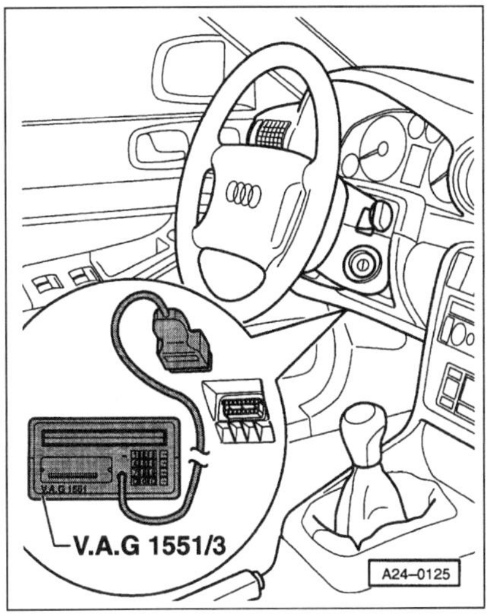

Data link connector is recessed into the lower edge of the knee pad on the left side of the dash. Scan Tool VAG 1551 is shown with high speed data cable VAG 1551/3.

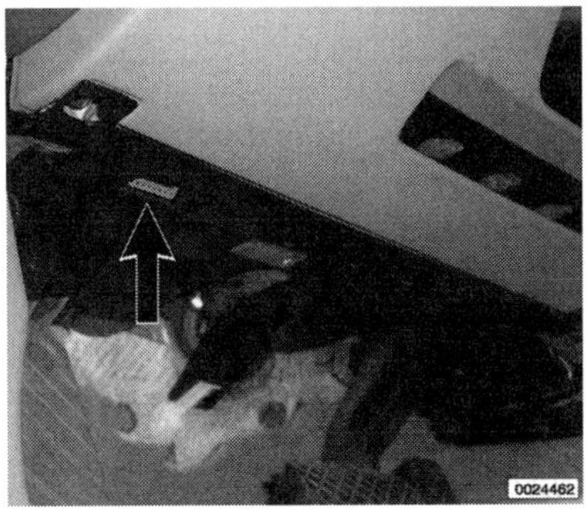

Access to the DLC (arrow) is most easily accomplished from below.

Diagnostic Trouble Code Table

NOTE —

The following table contains a list of available scan tool codes for all Volkswagen and Audi vehicles available at the time of publication. Not all of the codes apply to the vehicles covered by this manual.

Diagnostic Trouble Codes (Powertrain)

VAG code	SAE P-code	Description
16394	P0010	-A- Camshaft Pos. Actuator Circ. Bank 1 Malfunction
16397	P0013	-B- Camshaft Pos. Actuator Circ. Bank 1 Malfunction
16404	P0020	-A- Camshaft Pos. Actuator Circ. Bank 2 Malfunction
16407	P0023	-B- Camshaft Pos. Actuator Circ. Bank 2 Malfunction
16449	P0065	Air Assisted Injector Control Range/Performance
16450	P0066	Air Assisted Injector Control Low Input/Short to ground
16451	P0067	Air Assisted Injector Control Input/Short to B+
16484	P0100	Mass or Volume Air Flow Circuit
16485	P0101	Mass or Volume Air Flow Circ Range/Performance
16486	P0102	Mass or Volume Air Flow Circ Low Input
16487	P0103	Mass or Volume Air Flow Circ High Input
16489	P0105	Manifold Abs.Pressure or Bar.Pressure Voltage supply
16490	P0106	Manifold Abs.Pressure or Bar.Pressure Range/Performance
16491	P0107	Manifold Abs.Pressure or Bar.Pressure Low Input
16492	P0108	Manifold Abs.Pressure or Bar.Pressure High Input
16496	P0112	Intake Air Temp.Circ Low Input
16497	P0113	Intake Air Temp.Circ High Input
16500	P0116	Engine Coolant Temp.Circ Range/Performance
16501	P0117	Engine Coolant Temp.Circ Low Input
16502	P0118	Engine Coolant Temp.Circ High Input
16504	P0120	Throttle/Pedal Pos.Sensor A Circ Malfunction
16505	P0121	Throttle/Pedal Pos.Sensor A Circ Range/Performance
16506	P0122	Throttle/Pedal Pos.Sensor A Circ Low Input
16507	P0123	Throttle/Pedal Pos.Sensor A Circ High Input
16509	P0125	Insufficient Coolant Temp.for Closed Loop Fuel Control
16512	P0128	Coolant Thermostat/Valve Temperature below control range
16514	P0130	O2 Sensor Circ.,Bank1-Sensor1 Malfunction
16515	P0131	O2 Sensor Circ.,Bank1-Sensor1 Low Voltage
16516	P0132	O2 Sensor Circ.,Bank1-Sensor1 High Voltage
16517	P0133	O2 Sensor Circ.,Bank1-Sensor1 Slow Response
16518	P0134	O2 Sensor Circ.,Bank1-Sensor1 No Activity Detected

Diagnostic Trouble Codes (Powertrain)

VAG code	SAE P-code	Description
16519	P0135	O2 Sensor Heater Circ.,Bank1-Sensor1 Malfunction
16520	P0136	O2 Sensor Circ.,Bank1-Sensor2 Malfunction
16521	P0137	O2 Sensor Circ.,Bank1-Sensor2 Low Voltage
16522	P0138	O2 Sensor Circ.,Bank1-Sensor2 High Voltage
16523	P0139	O2 Sensor Circ.,Bank1-Sensor2 Slow Response
16524	P0140	O2 Sensor Circ.,Bank1-Sensor2 No Activity Detected
16525	P0141	O2 Sensor Heater Circ.,Bank1-Sensor2 Malfunction
16534	P0150	O2 Sensor Circ.,Bank2-Sensor1 Malfunction
16535	P0151	O2 Sensor Circ.,Bank2-Sensor1 Low Voltage
16536	P0152	O2 Sensor Circ.,Bank2-Sensor1 High Voltage
16537	P0153	O2 Sensor Circ.,Bank2-Sensor1 Slow Response
16538	P0154	O2 Sensor Circ.,Bank2-Sensor1 No Activity Detected
16539	P0155	O2 Sensor Heater Circ.,Bank2-Sensor1 Malfunction
16540	P0156	O2 Sensor Circ.,Bank2-Sensor2 Malfunction
16541	P0157	O2 Sensor Circ.,Bank2-Sensor2 Low Voltage
16542	P0158	O2 Sensor Circ.,Bank2-Sensor2 High Voltage
16543	P0159	O2 Sensor Circ.,Bank2-Sensor2 Slow Response
16544	P0160	O2 Sensor Circ.,Bank2-Sensor2 No Activity Detected
16545	P0161	O2 Sensor Heater Circ.,Bank2-Sensor2 Malfunction
16554	P0170	Fuel Trim,Bank1 Malfunction
16555	P0171	Fuel Trim,Bank1 System too Lean
16556	P0172	Fuel Trim,Bank1 System too Rich
16557	P0173	Fuel Trim,Bank2 Malfunction
16558	P0174	Fuel Trim,Bank2 System too Lean
16559	P0175	Fuel Trim,Bank2 System too Rich
16566	P0182	Fuel temperature sender-G81 Short to ground
16567	P0183	Fuel temperature sender-G81 Interruption/Short to B+
16581	P0197	Engine Oil Temperature Circuit Low Input
16582	P0198	Engine Oil Temperature Circuit High Input
16585	P0201	Cyl.1, Injector Circuit Fault in electrical circuit
16586	P0202	Cyl.2, Injector Circuit Fault in electrical circuit
16587	P0203	Cyl.3, Injector Circuit Fault in electrical circuit
16588	P0204	Cyl.4, Injector Circuit Fault in electrical circuit
16589	P0205	Cyl.5 Injector Circuit Fault in electrical circuit
16590	P0206	Cyl.6 Injector Circuit Fault in electrical circuit
16591	P0207	Cyl.7 Injector Circuit Fault in electrical circuit

Diagnostic Trouble Codes (Powertrain)

VAG code	SAE P-code	Description
16592	P0208	Cyl.8 Injector Circuit Fault in electrical circuit
16599	P0215	Engine Shut-Off Solenoid Malfunction
16600	P0216	Injector/Injection Timing Control Malfunction
16603	P0219	Engine Overspeed Condition
16605	P0221	Throttle Pos. Sensor -B- Circuit Range/Performance
16606	P0222	Throttle Pos. Sensor -B- Circuit Low Input
16607	P0223	Throttle Pos. Sensor -B- Circuit High Input
16609	P0225	Throttle Pos. Sensor -C- Circuit Voltage supply
16610	P0226	Throttle Pos. Sensor -C- Circuit Range/Performance
16611	P0227	Throttle Pos. Sensor -C- Circuit Low Input
16612	P0228	Throttle Pos. Sensor -C- Circuit High Input
16614	P0230	Fuel Pump Primary Circuit Fault in electrical circuit
16618	P0234	Turbocharger Overboost Condition Control limit exceeded
16619	P0235	Turbocharger Boost Sensor (A) Circ Control limit not reached
16620	P0236	Turbocharger Boost Sensor (A) Circ Range/Performance
16621	P0237	Turbocharger Boost Sensor (A) Circ Low Input
16622	P0238	Turbocharger Boost Sensor (A) Circ High Input
16627	P0243	Turbocharger Wastegate Solenoid (A) Open/Short Circuit to Ground
16629	P0245	Turbocharger Wastegate Solenoid (A) Low Input/Short to ground
16630	P0246	Turbocharger Wastegate Solenoid (A) High Input/Short to B+
16636	P0252	Injection Pump Metering Control (A) Range/Performance
16645	P0261	Cyl.1 Injector Circuit Low Input/Short to ground
16646	P0262	Cyl.1 Injector Circuit High Input/Short to B+
16648	P0264	Cyl.2 Injector Circuit Low Input/Short to ground
16649	P0265	Cyl.2 Injector Circuit High Input/Short to B+
16651	P0267	Cyl.3 Injector Circuit Low Input/Short to ground
16652	P0268	Cyl.3 Injector Circuit High Input/Short to B+
16654	P0270	Cyl.4 Injector Circuit Low Input/Short to ground
16655	P0271	Cyl.4 Injector Circuit High Input/Short to B+
16657	P0273	Cyl.5 Injector Circuit Low Input/Short to ground
16658	P0274	Cyl.5 Injector Circuit High Input/Short to B+
16660	P0276	Cyl.6 Injector Circuit Low Input/Short to ground
16661	P0277	Cyl.6 Injector Circuit High Input/Short to B+
16663	P0279	Cyl.7 Injector Circuit Low Input/Short to ground
16664	P0280	Cyl.7 Injector Circuit High Input/Short to B+
16666	P0282	Cyl.8 Injector Circuit Low Input/Short to ground

Diagnostic Trouble Codes (Powertrain)

VAG code	SAE P-code	Description
16667	P0283	Cyl.8 Injector Circuit High Input/Short to B+
16684	P0300	Random/Multiple Cylinder Misfire Detected
16685	P0301	Cyl.1 Misfire Detected
16686	P0302	Cyl.2 Misfire Detected
16687	P0303	Cyl.3 Misfire Detected
16688	P0304	Cyl.4 Misfire Detected
16689	P0305	Cyl.5 Misfire Detected
16690	P0306	Cyl.6 Misfire Detected
16691	P0307	Cyl.7 Misfire Detected
16692	P0308	Cyl.8 Misfire Detected
16697	P0313	Misfire Detected Low Fuel Level
16698	P0314	Single Cylinder Misfire
16705	P0321	Ign./Distributor Eng.Speed Inp.Circ Range/Performance
16706	P0322	Ign./Distributor Eng.Speed Inp.Circ No Signal
16708	P0324	Knock Control system error
16709	P0325	Knock Sensor 1 Circuit Electrical Fault in Circuit
16710	P0326	Knock Sensor 1 Circuit Range/Performance
16711	P0327	Knock Sensor 1 Circ Low Input
16712	P0328	Knock Sensor 1 Circ High Input
16716	P0332	Knock Sensor 2 Circ Low Input
16717	P0333	Knock Sensor 2 Circ High Input
16719	P0335	Crankshaft Pos. Sensor (A) Circ Malfunction
16720	P0336	Crankshaft Pos. Sensor (A) Circ Range/Performance/Missing tooth
16721	P0337	Crankshaft Pos.Sensor (A) Circ Low Input
16724	P0340	Camshaft Pos. Sensor (A) Circ Incorrect allocation
16725	P0341	Camshaft Pos.Sensor Circ Range/Performance
16726	P0342	Camshaft Pos.Sensor Circ Low Input
16727	P0343	Camshaft Pos.Sensor Circ High Input
16729	P0345	Camshaft Pos.Sensor Circuit Bank 2
16730	P0346	Camshaft Pos.Sensor Circuit Bank 2 Range/performance
16735	P0351	Ignition Coil (A) Cyl.1 Prim./Sec. Circ Malfunction
16736	P0352	Ignition Coil (B) Cyl.2 Prim./Sec. Circ Malfunction
16737	P0353	Ignition Coil (C) Cyl.3 Prim./Sec. Circ Malfunction
16738	P0354	Ignition Coil (D) Cyl.4 Prim./Sec. Circ Malfunction
16739	P0355	Ignition Coil (E) Cyl.5 Prim./Sec. Circ Malfunction
16740	P0356	Ignition Coil (F) Cyl.6 Prim./Sec. Circ Malfunction

Diagnostic Trouble Codes (Powertrain)

VAG code	SAE P-code	Description
16741	P0357	Ignition Coil (G) Cyl.7 Prim./Sec. Circ Malfunction
16742	P0358	Ignition Coil (H) Cyl.8 Prim./Sec. Circ Malfunction
16751	P0367	Camshaft Pos.Sensor (B) Circ (Bank 1) Low Input
16752	P0368	Camshaft Pos.Sensor (B) Circ (Bank 1) High Input
16764	P0380	Glow Plug/Heater Circuit (A) Electrical Fault in Circuit
16776	P0392	Camshaft Pos.Sensor (B) Circ (Bank 2) Low Input
16777	P0393	Camshaft Pos.Sensor (B) Circ (Bank 2) High Input
16784	P0400	Exhaust Gas Recirc.Flow Malfunction
16785	P0401	Exhaust Gas Recirc.Flow Insufficient Detected
16786	P0402	Exhaust Gas Recirc.Flow Excessive Detected
16787	P0403	Exhaust Gas Recirc. Contr. Circ Malfunction
16788	P0404	Exhaust Gas Recirc. Contr. Circ Range/Performance
16789	P0405	Exhaust Gas Recirc. Sensor (A) Circ Low Input
16790	P0406	Exhaust Gas Recirc. Sensor (A) Circ High Input
16791	P0407	Exhaust Gas Recirc. Sensor (B) Circ Low Input
16792	P0408	Exhaust Gas Recirc. Sensor (B) Circ High Input
16794	P0410	Sec.Air Inj.Sys Malfunction
16795	P0411	Sec.Air Inj.Sys. Incorrect Flow Detected
16796	P0412	Sec.Air Inj.Sys.Switching Valve A Circ Malfunction
16799	P0415	Sec.Air Inj.Sys.Switching Valve B Circ Malfunction
16802	P0418	Sec. Air Inj. Sys. Relay (A) Contr. Circ Malfunction
16803	P0419	Sec.Air Inj. Sys.Relay (B) Contr. Circ Malfunction
16804	P0420	Catalyst System,Bank1 Efficiency Below Threshold
16805	P0421	Warm-up Catalyst,Bank1 Below Threshold
16806	P0422	Main Catalyst,Bank1 Below Threshold
16811	P0427	Catalyst Temperature Sensor, Bank 1 Low Input/Short to ground
16812	P0428	Catalyst Temperature Sensor, Bank 1 High Input/Open/Short Circuit to B+
16815	P0431	Warm-up Catalyst,Bank2 Efficiency Below Threshold
16816	P0432	Main Catalyst,Bank2 Efficiency Below Threshold
16820	P0436	Catalyst Temperature Sensor, Bank 2 Range/Performance
16821	P0437	Catalyst Temperature Sensor, Bank 2 Low Input/Short to ground
16822	P0438	Catalyst Temperature Sensor, Bank 2 High Input/Open/Short Circuit to B+
16824	P0440	EVAP Emission Contr.Sys. Malfunction
16825	P0441	EVAP Emission Contr.Sys.Incorrect Purge Flow
16826	P0442	EVAP Emission Contr.Sys.(Small Leak) Leak Detected
16827	P0443	EVAP Emiss. Contr. Sys. Purge Valve Circ Electrical Fault in Circuit

Diagnostic Trouble Codes (Powertrain)

VAG code	SAE P-code	Description
16836	P0452	EVAP Emission Contr.Sys.Press.Sensor Low Input
16837	P0453	EVAP Emission Contr.Sys.Press.Sensor High Input
16839	P0455	EVAP Emission Contr.Sys.(Gross Leak) Leak Detected
16840	P0456	EVAP Emission Contr.Sys.(Very Small Leak) Leak Detected
16845	P0461	Fuel Level Sensor Circ Range/Performance
16846	P0462	Fuel Level Sensor Circuit Low Input
16847	P0463	Fuel Level Sensor Circuit High Input
16885	P0501	Vehicle Speed Sensor Range/Performance
16887	P0503	Vehicle Speed Sensor Intermittent/Erratic/High Input
16889	P0505	Idle Control System Malfunction
16890	P0506	Idle Control System RPM Lower than Expected
16891	P0507	Idle Control System Higher than Expected
16894	P0510	Closed Throttle Pos.Switch Malfunction
16897	P0513	Incorrect Immobilizer key
16915	P0531	A/C Refrigerant Pressure Sensor Circuit Range/Performance
16916	P0532	A/C Refrigerant Pressure Sensor Circuit Low Input
16917	P0533	A/C Refrigerant Pressure Sensor Circuit High Input
16929	P0545	Exhaust gas temperature sensor circuit - Bank 1 Low
16930	P0546	Exhaust gas temperature sensor circuit - Bank 1 High
16932	P0548	Exhaust gas temperature sensor circuit - Bank 2 Low
16933	P0549	Exhaust gas temperature sensor circuit - Bank 2 High
16934	P0550	Power Steering Pressure Sensor Circuit
16935	P0551	Power Steering Pressure Sensor Circuit Range/Performance
16944	P0560	System Voltage Malfunction
16946	P0562	System Voltage Low Voltage
16947	P0563	System Voltage High Voltage
16952	P0568	Cruise Control Set Signal Incorrect Signal
16955	P0571	Cruise/Brake Switch (A) Circ Malfunction
16984	P0600	Serial Comm. Link (Data Bus) Message Missing
16985	P0601	Internal Contr.Module Memory Check Sum Error
16986	P0602	Control Module Programming Error/Malfunction
16987	P0603	Internal Contr.Module (KAM) Error
16988	P0604	Internal Contr.Module Random Access Memory (RAM) Error
16989	P0605	Internal Contr.Module ROM Test Error
16990	P0606	ECM/PCM Processor
17022	P0638	Throttle Actuator Control (Bank 1) Range/Performance

Diagnostic Trouble Codes (Powertrain)

VAG code	SAE P-code	Description
17026	P0642	Knock Control Module Malfunction
17029	P0645	A/C Clutch Relay Control Circuit
17034	P0650	MIL Control Circuit Electrical Fault in Circuit
17038	P0654	Engine RPM Output Circuit Electrical Fault in Circuit
17040	P0656	Fuel Level Output Circuit Electrical Fault in Circuit
17044	P0660	Intake Manifold Tuning Circuit - Bank 1
17045	P0661	Intake Manifold Tuning Circuit - Bank 1 Low
17046	P0662	Intake Manifold Tuning Circuit - Bank 1 High
17069	P0685	ECM/PCM Power Relay Control Circuit Open
17070	P0686	ECM/PCM Power Relay Control Circuit Low
17071	P0687	ECM/PCM Power Relay Control Circuit High
17084	P0700	Transm.Contr.System Malfunction
17086	P0702	Transm.Contr.System Electrical
17087	P0703	Torque Converter/Brake Switch B Circ Malfunction
17089	P0705	Transm.Range Sensor Circ.(PRNDL Inop.) Malfunction
17090	P0706	Transm.Range Sensor Circ Range/Performance
17091	P0707	Transm.Range Sensor Circ Low Input
17092	P0708	Transm.Range Sensor Circ High Input
17094	P0710	Transm.Fluid Temp.Sensor Circ. Malfunction
17095	P0711	Transm.Fluid Temp.Sensor Circ. Range/Performance
17096	P0712	Transm.Fluid Temp.Sensor Circ. Low Input
17097	P0713	Transm.Fluid Temp.Sensor Circ. High Input
17099	P0715	Input Turbine/Speed Sensor Circ. Malfunction
17100	P0716	Input Turbine/Speed Sensor Circ. Range/Performance
17101	P0717	Input Turbine/Speed Sensor Circ. No Signal
17105	P0721	Output Speed Sensor Circ Range/Performance
17106	P0722	Output Speed Sensor Circ No Signal
17109	P0725	Engine Speed Inp.Circ. Malfunction
17110	P0726	Engine Speed Inp.Circ. Range/Performance
17111	P0727	Engine Speed Inp.Circ. No Signal
17114	P0730	Gear Incorrect Ratio
17115	P0731	Gear 1 Incorrect Ratio
17116	P0732	Gear 2 Incorrect Ratio
17117	P0733	Gear 3 Incorrect Ratio
17118	P0734	Gear 4 Incorrect Ratio
17119	P0735	Gear 5 Incorrect Ratio

Diagnostic Trouble Codes (Powertrain)

VAG code	SAE P-code	Description
17124	P0740	Torque Converter Clutch Circ Malfunction
17125	P0741	Torque Converter Clutch Circ Performance or Stuck Off
17132	P0748	Pressure Contr.Solenoid Electrical
17134	P0750	Shift Solenoid A malfunction
17135	P0751	Shift Solenoid A Performance or Stuck Off
17136	P0752	Shift Solenoid A Stuck On
17137	P0753	Shift Solenoid A Electrical
17140	P0756	Shift Solenoid B Performance or Stuck Off
17141	P0757	Shift Solenoid B Stuck On
17142	P0758	Shift Solenoid B Electrical
17145	P0761	Shift Solenoid C Performance or Stuck Off
17146	P0762	Shift Solenoid C Stuck On
17147	P0763	Shift Solenoid C Electrical
17152	P0768	Shift Solenoid D Electrical
17157	P0773	Shift Solenoid E Electrical
17162	P0778	Pressure Control Solenoid B Electrical
17174	P0790	Normal/Performance Switch Circ Malfunction
17182	P0798	Pressure Control Solenoid C Electrical
17195	P0811	Excessive clutch slippage
17225	P0841	Transmission Fluid Pressure Sensor/Switch A Range/Performance
17230	P0846	Transmission Fluid Pressure Sensor/Switch B Range/Performance
17508	P1100	O2 Sensor Heating Circuit, Bank1-Sensor2 Output too Low
17509	P1101	O2 Sensor Circ.,Bank1-Sensor1Voltage too Low/Air Leak
17510	P1102	O2 Sensor Heating Circ.,Bank1-Sensor1 Short to B+
17511	P1103	O2 Sensor Heating Circ.,Bank1-Sensor1 Output too Low
17512	P1104	Bank1-Sensor2 Voltage too Low/Air Leak
17513	P1105	O2 Sensor Heating Circ.,Bank1-Sensor2 Short to B+
17514	P1106	O2 Sensor Circ.,Bank2-Sensor1 Voltage too Low/Air Leak
17515	P1107	O2 Sensor Heating Circ.,Bank2-Sensor1 Short to B+
17516	P1108	O2 Sensor Heating Circ.,Bank2-Sensor1 Output too Low
17517	P1109	O2 Sensor Circ.,Bank2-Sensor2 Voltage too Low/Air Leak
17518	P1110	O2 Sensor Heating Circ.,Bank2-Sensor2 Short to B+
17519	P1111	O2 Control (Bank 1) System too lean
17520	P1112	O2 Control (Bank 1) System too rich
17521	P1113	Bank1-Sensor1 Internal Resistance too High
17522	P1114	Bank1-Sensor2 Internal Resistant too High

Diagnostic Trouble Codes (Powertrain)

VAG code	SAE P-code	Description
17523	P1115	O2 Sensor Heater Circ.,Bank1-Sensor1 Short to Ground
17524	P1116	O2 Sensor Heater Circ.,Bank1-Sensor1 Open
17525	P1117	O2 Sensor Heater Circ.,Bank1-Sensor2 Short to Ground
17526	P1118	O2 Sensor Heater Circ.,Bank1-Sensor2 Open
17527	P1119	O2 Sensor Heater Circ.,Bank2-Sensor1 Short to Ground
17528	P1120	O2 Sensor Heater Circ.,Bank2-Sensor1 Open
17529	P1121	O2 Sensor Heater Circ.,Bank2-Sensor2 Short to Ground
17530	P1122	O2 Sensor Heater Circ.,Bank2-Sensor2 Open
17531	P1123	Long Term Fuel Trim Add.Air.,Bank1 System too Rich
17532	P1124	Long Term Fuel Trim Add.Air.,Bank1 System too Lean
17533	P1125	Long Term Fuel Trim Add.Air.,Bank2 System too Rich
17534	P1126	Long Term Fuel Trim Add.Air.,Bank2 System too Lean
17535	P1127	Long Term Fuel Trim mult.,Bank1 System too Rich
17536	P1128	Long Term Fuel Trim mult.,Bank1 System too Lean
17537	P1129	Long Term Fuel Trim mult.,Bank2 System too Rich
17538	P1130	Long Term Fuel Trim mult.,Bank2 System too Lean
17539	P1131	Bank2-Sensor1 Internal Resistance too High
17540	P1132	O2 Sensor Heating Circ.,Bank1+2-Sensor1 Short to B+
17541	P1133	O2 Sensor Heating Circ.,Bank1+2-Sensor1 Electrical Malfunction
17542	P1134	O2 Sensor Heating Circ.,Bank1+2-Sensor2 Short to B+
17543	P1135	O2 Sensor Heating Circ.,Bank1+2-Sensor2 Electrical Malfunction
17544	P1136	Long Term Fuel Trim Add.Fuel,Bank1 System too Lean
17545	P1137	Long Term Fuel Trim Add.Fuel,Bank1 System too Rich
17546	P1138	Long Term Fuel Trim Add.Fuel,Bank2 System too Lean
17547	P1139	Long Term Fuel Trim Add.Fuel,Bank2 System too Rich
17548	P1140	Bank2-Sensor2 Internal Resistance too High
17549	P1141	Load Calculation Cross Check Range/Performance
17550	P1142	Load Calculation Cross Check Lower Limit Exceeded
17551	P1143	Load Calculation Cross Check Upper Limit Exceeded
17552	P1144	Mass or Volume Air Flow Circ Open/Short to Ground
17553	P1145	Mass or Volume Air Flow Circ Short to B+
17554	P1146	Mass or Volume Air Flow Circ Supply Malfunction
17555	P1147	O2 Control (Bank 2) System too lean
17556	P1148	O2 Control (Bank 2) System too rich
17557	P1149	O2 Control (Bank 1) Out of range
17558	P1150	O2 Control (Bank 2) Out of range

Diagnostic Trouble Codes (Powertrain)

VAG code	SAE P-code	Description
17559	P1151	Bank1, Long Term Fuel Trim, Range 1 Leanness Lower Limit Exceeded
17560	P1152	Bank1, Long Term Fuel Trim, Range 2 Leanness Lower Limit Exceeded
17561	P1153	Lambda probe behind catalyst, exchanged
17562	P1154	Manifold Switch Over Malfunction
17563	P1155	Manifold Abs.Pressure Sensor Circ. Short to B+
17564	P1156	Manifold Abs.Pressure Sensor Circ. Open/Short to Ground
17565	P1157	Manifold Abs.Pressure Sensor Circ. Power Supply Malfunction
17566	P1158	Manifold Abs.Pressure Sensor Circ. Range/Performance
17568	P1160	Manifold Temp.Sensor Circ. Short to Ground
17569	P1161	Manifold Temp.Sensor Circ. Open/Short to B+
17570	P1162	Fuel Temp.Sensor Circ. Short to Ground
17571	P1163	Fuel Temp.Sensor Circ. Open/Short to B+
17572	P1164	Fuel Temperature Sensor Range/Performance/Incorrect Signal
17573	P1165	Bank1, Long Term Fuel Trim, Range 1 Rich Limit Exceeded
17574	P1166	Bank1, Long Term Fuel Trim, Range 2 Rich Limit Exceeded
17579	P1171	Throttle Actuation Potentiometer Sign.2 Range/Performance
17580	P1172	Throttle Actuation Potentiometer Sign.2 Signal too Low
17581	P1173	Throttle Actuation Potentiometer Sign.2 Signal too High
17582	P1174	Fuel Trim, Bank 1 Different injection times
17584	P1176	O2 Correction Behind Catalyst,B1 Limit Attained
17585	P1177	O2 Correction Behind Catalyst,B2 Limit Attained
17586	P1178	Linear 02 Sensor / Pump Current Open Circuit
17587	P1179	Linear 02 Sensor / Pump Current Short to ground
17588	P1180	Linear 02 Sensor / Pump Current Short to B+
17589	P1181	Linear 02 Sensor / Reference Voltage Open Circuit
17590	P1182	Linear 02 Sensor / Reference Voltage Short to ground
17591	P1183	Linear 02 Sensor / Reference Voltage Short to B+
17592	P1184	Linear 02 Sensor / Common Ground Wire Open Circuit
17593	P1185	Linear 02 Sensor / Common Ground Wire Short to ground
17594	P1186	Linear 02 Sensor / Common Ground Wire Short to B+
17595	P1187	Linear 02 Sensor / Compens. Resistor Open Circuit
17596	P1188	Linear 02 Sensor / Compens. Resistor Short to ground
17597	P1189	Linear 02 Sensor / Compens. Resistor Short to B+
17598	P1190	Linear 02 Sensor / Reference Voltage Incorrect Signal
17604	P1196	O2 Sensor Heater Circ.,Bank1-Sensor1 Electrical Malfunction
17605	P1197	O2 Sensor Heater Circ.,Bank2-Sensor1 Electrical Malfunction

Diagnostic Trouble Codes (Powertrain)

VAG code	SAE P-code	Description
17606	P1198	O2 Sensor Heater Circ.,Bank1-Sensor2 Electrical Malfunction
17607	P1199	O2 Sensor Heater Circ.,Bank2-Sensor2 Electrical Malfunction
17609	P1201	Cyl.1-Fuel Inj.Circ. Electrical Malfunction
17610	P1202	Cyl.2-Fuel Inj.Circ. Electrical Malfunction
17611	P1203	Cyl.3-Fuel Inj.Circ. Electrical Malfunction
17612	P1204	Cyl.4-Fuel Inj.Circ. Electrical Malfunction
17613	P1205	Cyl.5-Fuel Inj.Circ. Electrical Malfunction
17614	P1206	Cyl.6-Fuel Inj.Circ. Electrical Malfunction
17615	P1207	Cyl.7-Fuel Inj.Circ. Electrical Malfunction
17616	P1208	Cyl.8-Fuel Inj.Circ. Electrical Malfunction
17617	P1209	Intake valves for cylinder shut-off Short circuit to ground
17618	P1210	Intake valves for cylinder shut-off Short to B+
17619	P1211	Intake valves for cylinder shut-off Open circuit
17621	P1213	Cyl.1-Fuel Inj.Circ. Short to B+
17622	P1214	Cyl.2-Fuel Inj.Circ. Short to B+
17623	P1215	Cyl.3-Fuel Inj.Circ. Short to B+
17624	P1216	Cyl.4-Fuel Inj.Circ. Short to B+
17625	P1217	Cyl.5-Fuel Inj.Circ. Short to B+
17626	P1218	Cyl.6-Fuel Inj.Circ. Short to B+
17627	P1219	Cyl.7-Fuel Inj.Circ. Short to B+
17628	P1220	Cyl.8-Fuel Inj.Circ. Short to B+
17629	P1221	Cylinder shut-off exhaust valves Short circuit to ground
17630	P1222	Cylinder shut-off exhaust valves Short to B+
17631	P1223	Cylinder shut-off exhaust valves Open circuit
17633	P1225	Cyl.1-Fuel Inj.Circ. Short to Ground
17634	P1226	Cyl.2-Fuel Inj.Circ. Short to Ground
17635	P1227	Cyl.3-Fuel Inj.Circ. Short to Ground
17636	P1228	Cyl.4-Fuel Inj.Circ. Short to Ground
17637	P1229	Cyl.5-Fuel Inj.Circ. Short to Ground
17638	P1230	Cyl.6-Fuel Inj.Circ. Short to Ground
17639	P1231	Cyl.7-Fuel Inj.Circ. Short to Ground
17640	P1232	Cyl.8-Fuel Inj.Circ. Short to Ground
17645	P1237	Cyl.1-Fuel Inj.Circ. Open Circ.
17646	P1238	Cyl.2-Fuel Inj.Circ. Open Circ.
17647	P1239	Cyl.3-Fuel Inj.Circ. Open Circ.
17648	P1240	Cyl.4-Fuel Inj.Circ. Open Circ.

Diagnostic Trouble Codes (Powertrain)

VAG code	SAE P-code	Description
17649	P1241	Cyl.5-Fuel Inj.Circ. Open Circ.
17650	P1242	Cyl.6-Fuel Inj.Circ. Open Circ.
17651	P1243	Cyl.7-Fuel Inj.Circ. Open Circ.
17652	P1244	Cyl.8-Fuel Inj.Circ. Open Circ.
17653	P1245	Needle Lift Sensor Circ. Short to Ground
17654	P1246	Needle Lift Sensor Circ. Range/Performance
17655	P1247	Needle Lift Sensor Circ. Open/Short to B+
17656	P1248	Injection Start Control Deviation
17657	P1249	Fuel consumption signal Electrical Fault in Circuit
17658	P1250	Fuel Level Too Low
17659	P1251	Start of Injection Solenoid Circ Short to B+
17660	P1252	Start of Injection Solenoid Circ Open/Short to Ground
17661	P1253	Fuel consumption signal Short to ground
17662	P1254	Fuel consumption signal Short to B+
17663	P1255	Engine Coolant Temp.Circ Short to Ground
17664	P1256	Engine Coolant Temp.Circ Open/Short to B+
17665	P1257	Engine Coolant System Valve Open
17666	P1258	Engine Coolant System Valve Short to B+
17667	P1259	Engine Coolant System Valve Short to Ground
17688	P1280	Fuel Inj.Air Contr.Valve Circ. Flow too Low
17691	P1283	Fuel Inj.Air Contr.Valve Circ. Electrical Malfunction
17692	P1284	Fuel Inj.Air Contr.Valve Circ. Open
17693	P1285	Fuel Inj.Air Contr.Valve Circ. Short to Ground
17694	P1286	Fuel Inj.Air Contr.Valve Circ. Short to B+
17695	P1287	Turbocharger bypass valve open
17696	P1288	Turbocharger bypass valve short to B+
17697	P1289	Turbocharger bypass valve short to ground
17704	P1296	Cooling system malfunction
17705	P1297	Connection turbocharger - throttle valve pressure hose
17708	P1300	Misfire detected Reason: Fuel level too low
17721	P1319	Knock Sensor 1 Circ. Short to Ground
17728	P1320	Knock Sensor 2 Circ. Short to Ground
17729	P1321	Knock Sensor 3 Circ. Low Input
17730	P1322	Knock Sensor 3 Circ. High Input
17731	P1323	Knock Sensor 4 Circ. Low Input
17732	P1324	Knock Sensor 4 Circ. High Input

Diagnostic Trouble Codes (Powertrain)

VAG code	SAE P-code	Description
17733	P1325	Cyl.1-Knock Contr. Limit Attained
17734	P1326	Cyl.2-Knock Contr. Limit Attained
17735	P1327	Cyl.3-Knock Contr. Limit Attained
17736	P1328	Cyl.4-Knock Contr. Limit Attained
17737	P1329	Cyl.5-Knock Contr. Limit Attained
17738	P1330	Cyl.6-Knock Contr. Limit Attained
17739	P1331	Cyl.7-Knock Contr. Limit Attained
17740	P1332	Cyl.8-Knock Contr. Limit Attained
17743	P1335	Engine Torque Monitoring 2 Control Limit Exceeded
17744	P1336	Engine Torque Monitoring Adaptation at limit
17745	P1337	Camshaft Pos.Sensor,Bank1 Short to Ground
17746	P1338	Camshaft Pos.Sensor,Bank1 Open Circ./Short to B+
17747	P1339	Crankshaft Pos./Engine Speed Sensor Cross Connected
17748	P1340	Crankshaft-/Camshaft Pos.Sens.Signals Out of Sequence
17749	P1341	Ignition Coil Power Output Stage 1 Short to Ground
17750	P1342	Ignition Coil Power Output Stage 1 Short to B+
17751	P1343	Ignition Coil Power Output Stage 2 Short to Ground
17752	P1344	Ignition Coil Power Output Stage 2 Short to B+
17753	P1345	Ignition Coil Power Output Stage 3 Short to Ground
17754	P1346	Ignition Coil Power Output Stage 3 Short to B+
17755	P1347	Bank2,Crankshaft-/Camshaft os.Sens.Sign. Out of Sequence
17756	P1348	Ignition Coil Power Output Stage 1 Open Circuit
17757	P1349	Ignition Coil Power Output Stage 2 Open Circuit
17758	P1350	Ignition Coil Power Output Stage 3 Open Circuit
17759	P1351	Camshaft Position Sensor, Bank 1 Range/Performance at Start
17762	P1354	Modulation Piston Displ.Sensor Circ. Malfunction
17763	P1355	Cyl. 1, ignition circuit Open Circuit
17764	P1356	Cyl. 1, ignition circuit Short to B+
17765	P1357	Cyl. 1, ignition circuit Short to ground
17766	P1358	Cyl. 2, ignition circuit Open Circuit
17767	P1359	Cyl. 2, ignition circuit Short Circuit to B+
17768	P1360	Cyl. 2, ignition circuit Short Circuit to Ground
17769	P1361	Cyl. 3, ignition circuit Open Circuit
17770	P1362	Cyl. 3, ignition circuit Short Circuit to B+
17771	P1363	Cyl. 3, ignition circuit Short Circuit to ground
17772	P1364	Cyl. 4 ignition circuit Open Circuit

Diagnostic Trouble Codes (Powertrain)

VAG code	SAE P-code	Description
17773	P1365	Cyl. 4 ignition circuit Short circuit to B+
17774	P1366	Cyl. 4 ignition circuit Short circuit to ground
17775	P1367	Cyl. 5, ignition circuit Open Circuit
17776	P1368	Cyl. 5, ignition circuit Short Circuit to B+
17777	P1369	Cyl. 5, ignition circuit short to ground
17778	P1370	Cyl. 6, ignition circuit Open Circuit
17779	P1371	Cyl. 6, ignition circuit Short Circuit to B+
17780	P1372	Cyl. 6, ignition circuit short to ground
17781	P1373	Cyl. 7, ignition circuit Open Circuit
17782	P1374	Cyl. 7, ignition circuit Short Circuit to B+
17783	P1375	Cyl. 7, ignition circuit short to ground
17784	P1376	Cyl. 8, ignition circuit Open Circuit
17785	P1377	Cyl. 8, ignition circuit Short Circuit to B+
17786	P1378	Cyl. 8, ignition circuit short to ground
17794	P1386	Internal Control Module Knock Control Circ.Error
17795	P1387	Internal Contr. Module altitude sensor error
17796	P1388	Internal Contr. Module drive by wire error
17797	P1389	Internal Control Module 2, Defect
17799	P1391	Camshaft Pos.Sensor,Bank2 Short to Ground
17800	P1392	Camshaft Pos.Sensor,Bank2 Open Circ./Short to B+
17801	P1393	Ignition Coil Power Output Stage 1 Electrical Malfunction
17802	P1394	Ignition Coil Power Output Stage 2 Electrical Malfunction
17803	P1395	Ignition Coil Power Output Stage 3 Electrical Malfunction
17804	P1396	Engine Speed Sensor Missing Tooth
17805	P1397	Engine speed wheel Adaptation limit reached
17806	P1398	Engine RPM signal, TD Short to ground
17807	P1399	Engine RPM signal, TD Short Circuit to B+
17808	P1400	EGR Valve Circ Electrical Malfunction
17809	P1401	EGR Valve Circ Short to Ground
17810	P1402	EGR Valve Circ Short to B+
17811	P1403	EGR Flow Deviation
17812	P1404	EGR Flow Basic Setting not carried out
17814	P1406	EGR Temp.Sensor Range/Performance
17815	P1407	EGR Temp.Sensor Signal too Low
17816	P1408	EGR Temp.Sensor Signal too High
17817	P1409	Tank Ventilation Valve Circ. Electrical Malfunction

Diagnostic Trouble Codes (Powertrain)

VAG code	SAE P-code	Description
17818	P1410	Tank Ventilation Valve Circ. Short to B+
17819	P1411	Sec.Air Inj.Sys.,Bank2 Flow too Flow
17820	P1412	EGR Different.Pressure Sensor Signal too Low
17821	P1413	EGR Different.Pressure Sensor Signal too High
17822	P1414	Sec.Air Inj.Sys.,Bank2 Leak Detected
17825	P1417	Fuel Level Sensor Circ Signal too Low
17826	P1418	Fuel Level Sensor Circ Signal too High
17828	P1420	Sec.Air Inj.Valve Circ Electrical Malfunction
17829	P1421	Sec.Air Inj.Valve Circ Short to Ground
17830	P1422	Sec.Air Inj.Sys.Contr.Valve Circ Short to B+
17831	P1423	Sec.Air Inj.Sys.,Bank1 Flow too Low
17832	P1424	Sec.Air Inj.Sys.,Bank1 Leak Detected
17833	P1425	Tank Vent.Valve Short to Ground
17834	P1426	Tank Vent.Valve Open
17840	P1432	Sec.Air Inj.Valve Open
17841	P1433	Sec.Air Inj.Sys.Pump Relay Circ. open
17842	P1434	Sec.Air Inj.Sys.Pump Relay Circ. Short to B+
17843	P1435	Sec.Air Inj.Sys.Pump Relay Circ. Short to ground
17844	P1436	Sec.Air Inj.Sys.Pump Relay Circ. Electrical Malfunction
17847	P1439	EGR Potentiometer Error in Basic Setting
17848	P1440	EGR Valve Power Stage Open
17849	P1441	EGR Valve Circ Open/Short to Ground
17850	P1442	EGR Valve Position Sensor Signal too high
17851	P1443	EGR Valve Position Sensor Signal too low
17852	P1444	EGR Valve Position Sensor range/performance
17853	P1445	Catalyst Temp.Sensor 2 Circ. Range/Performance
17854	P1446	Catalyst Temp.Circ Short to Ground
17855	P1447	Catalyst Temp.Circ Open/Short to B+
17856	P1448	Catalyst Temp.Sensor 2 Circ. Short to Ground
17857	P1449	Catalyst Temp.Sensor 2 Circ. Open/Short to B+
17858	P1450	Sec.Air Inj.Sys.Circ Short to B+
17859	P1451	Sec.Air Inj.Sys.Circ Short to Ground
17860	P1452	Sec.Air Inj.Sys. Open Circ.
17861	P1453	Exhaust gas temperature sensor 1 open/short to B+
17862	P1454	Exhaust gas temperature sensor short 1 to ground
17863	P1455	Exhaust gas temperature sensor 1 range/performance

Diagnostic Trouble Codes (Powertrain)

VAG code	SAE P-code	Description
17864	P1456	Exhaust gas temperature control bank 1 limit attained
17865	P1457	Exhaust gas temperature sensor 2 open/short to B+
17866	P1458	Exhaust gas temperature sensor 2 short to ground
17867	P1459	Exhaust gas temperature sensor 2 range/performance
17868	P1460	Exhaust gas temperature control bank 2 limit attained
17869	P1461	Exhaust gas temperature control bank 1 Range/Performance
17870	P1462	Exhaust gas temperature control bank 2 Range/Performance
17873	P1465	Additive Pump Short Circuit to B+
17874	P1466	Additive Pump Open/Short to Ground
17875	P1467	EVAP Canister Purge Solenoid Valve Short Circuit to B+
17876	P1468	EVAP Canister Purge Solenoid Valve Short Circuit to Ground
17877	P1469	EVAP Canister Purge Solenoid Valve Open Circuit
17878	P1470	EVAP Emission Contr.LDP Circ Electrical Malfunction
17879	P1471	EVAP Emission Contr.LDP Circ Short to B+
17880	P1472	EVAP Emission Contr.LDP Circ Short to Ground
17881	P1473	EVAP Emission Contr.LDP Circ Open Circ.
17882	P1474	EVAP Canister Purge Solenoid Valve electrical malfunction
17883	P1475	EVAP Emission Contr.LDP Circ Malfunction/Signal Circ.Open
17884	P1476	EVAP Emission Contr.LDP Circ Malfunction/Insufficient Vacuum
17885	P1477	EVAP Emission Contr.LDP Circ Malfunction
17886	P1478	EVAP Emission Contr.LDP Circ Clamped Tube Detected
17908	P1500	Fuel Pump Relay Circ. Electrical Malfunction
17909	P1501	Fuel Pump Relay Circ. Short to Ground
17910	P1502	Fuel Pump Relay Circ. Short to B+
17911	P1503	Load signal from Alternator Term. DF Range/performance/Incorrect Signal
17912	P1504	Intake Air Sys.Bypass Leak Detected
17913	P1505	Closed Throttle Pos. Does Not Close/Open Circ
17914	P1506	Closed Throttle Pos.Switch Does Not Open/Short to Ground
17915	P1507	Idle Sys.Learned Value Lower Limit Attained
17916	P1508	Idle Sys.Learned Value Upper Limit Attained
17917	P1509	Idle Air Control Circ. Electrical Malfunction
17918	P1510	Idle Air Control Circ. Short to B+
17919	P1511	Intake Manifold Changeover Valve circuit electrical malfunction
17920	P1512	Intake Manifold Changeover Vaive circuit Short to B+
17921	P1513	Intake Manifold Changeover Valve2 circuit Short to B+
17922	P1514	Intake Manifold Changeover Valve2 circuit Short to ground

Diagnostic Trouble Codes (Powertrain)

VAG code	SAE P-code	Description
17923	P1515	Intake Manifold Changeover Valve circuit Short to Ground
17924	P1516	Intake Manifold Changeover Valve circuit Open
17925	P1517	Main Relay Circ. Electrical Malfunction
17926	P1518	Main Relay Circ. Short to B+
17927	P1519	Intake Camshaft Contr.,Bank1 Malfunction
17928	P1520	Intake Manifold Changeover Valve2 circuit Open
17929	P1521	Intake Manifold Changeover Valve2 circuit electrical malfunction
17930	P1522	Intake Camshaft Contr.,Bank2 Malfunction
17931	P1523	Crash Signal from Airbag Control Unit range/performance
17932	P1524	Fuel Pump Relay Circuit, Open/Short to ground
17933	P1525	Intake Camshaft Contr.Circ.,Bank1 Electrical Malfunction
17934	P1526	Intake Camshaft Contr.Circ.,Bank1 Short to B+
17935	P1527	Intake Camshaft Contr.Circ.,Bank1 Short to Ground
17936	P1528	Intake Camshaft Contr.Circ.,Bank1 Open
17937	P1529	Camshaft Control Circuit Short to B+
17938	P1530	Camshaft Control Circuit Short to ground
17939	P1531	Camshaft Control Circuit open
17940	P1532	Idle Control System, Leanmodus, RPM lower than expected
17941	P1533	Intake Camshaft Contr.Circ.,Bank2 Electrical Malfunction
17942	P1534	Intake Camshaft Contr.Circ.,Bank2 Short to B+
17943	P1535	Intake Camshaft Contr.Circ.,Bank2 Short to Ground
17944	P1536	Intake Camshaft Contr.Circ.,Bank2 Open
17945	P1537	Engine Shutoff Solenoid Malfunction
17946	P1538	Engine Shutoff Solenoid Open/Short to Ground
17947	P1539	Clutch Vacuum Vent Valve Switch Incorrect signal
17948	P1540	Vehicle Speed Sensor High Input
17949	P1541	Fuel Pump Relay Circ Open
17950	P1542	Throttle Actuation Potentiometer Range/Performance
17951	P1543	Throttle Actuation Potentiometer Signal too Low
17952	P1544	Throttle Actuation Potentiometer Signal too High
17953	P1545	Throttle Pos.Contr Malfunction
17954	P1546	Boost Pressure Contr.Valve Short to B+
17955	P1547	Boost Pressure Contr.Valve Short to Ground
17956	P1548	Boost Pressure Contr.Valve Open
17957	P1549	Boost Pressure Contr.Valve Short to Ground
17958	P1550	Charge Pressure Deviation

Diagnostic Trouble Codes (Powertrain)

VAG code	SAE P-code	Description
17959	P1551	Barometric Pressure Sensor Circ. Short to B+
17960	P1552	Barometric Pressure Sensor Circ. Open/Short to Ground
17961	P1553	Barometric/manifold pressure signal ratio out of range
17962	P1554	Idle Speed Contr.Throttle Pos. Basic Setting Conditions not met
17963	P1555	Charge Pressure Upper Limit exceeded
17964	P1556	Charge Pressure Contr. Negative Deviation
17965	P1557	Charge Pressure Contr. Positive Deviation
17966	P1558	Throttle Actuator Electrical Malfunction
17967	P1559	Idle Speed Contr.Throttle Pos. Adaptation Malfunction
17968	P1560	Maximum Engine Speed Exceeded
17969	P1561	Quantity Adjuster Deviation
17970	P1562	Quantity Adjuster Upper Limit Attained
17971	P1563	Quantity Adjuster Lower Limit Attained
17972	P1564	Idle Speed Contr.Throttle Pos. Low Voltage During Adaptation
17973	P1565	Idle Speed Control Throttle Position lower limit not attained
17974	P1566	Load signal from A/C compressor range/performance
17975	P1567	Load signal from A/C compressor no signal
17976	P1568	Idle Speed Contr.Throttle Pos. mechanical Malfunction
17977	P1569	Cruise control switch Incorrect signal
17978	P1570	Contr.Module Locked
17979	P1571	Left Eng. Mount Solenoid Valve Short to B+
17980	P1572	Left Eng. Mount Solenoid Valve Short to ground
17981	P1573	Left Eng. Mount Solenoid Valve Open circuit
17982	P1574	Left Eng. Mount Solenoid Valve Electrical fault in circuit
17983	P1575	Right Eng. Mount Solenoid Valve Short to B+
17984	P1576	Right Eng. Mount Solenoid Valve Short to ground
17985	P1577	Right Eng. Mount Solenoid Valve Open circuit
17986	P1578	Right Eng. Mount Solenoid Valve Electrical fault in circuit
17987	P1579	Idle Speed Contr.Throttle Pos. Adaptation not started
17988	P1580	Throttle Actuator B1 Malfunction
17989	P1581	Idle Speed Contr.Throttle Pos. Basic Setting Not Carried Out
17990	P1582	Idle Adaptation at Limit
17991	P1583	Transmission mount valves Short to B+
17992	P1584	Transmission mount valves Short to ground
17993	P1585	Transmission mount valves Open circuit
17994	P1586	Engine mount solenoid valves Short to B+

Diagnostic Trouble Codes (Powertrain)

VAG code	SAE P-code	Description
17995	P1587	Engine mount solenoid valves Short to ground
17996	P1588	Engine mount solenoid valves Open circuit
18007	P1599	Idle Control System, Leanmodus, RPM higher than expected
18008	P1600	Power Supply (B+) Terminal 15 Low Voltage
18009	P1601	Relay for Power Supply (B+) Terminal 30, Range/Performance
18010	P1602	Power Supply (B+) Terminal 30 Low Voltage
18011	P1603	Internal Control Module Malfunction
18012	P1604	Internal Control Module Driver Error
18013	P1605	Rough Road/Acceleration Sensor Electrical Malfunction
18014	P1606	Rough Road Spec Engine Torque ABS-ECU Electrical Malfunction
18015	P1607	Vehicle speed signal Error message from instrument cluster
18016	P1608	Steering angle signal Error message from steering angle sensor
18017	P1609	Crash shut-down activated
18018	P1610	Control Module Defective
18019	P1611	MIL Call-up Circ./Transm.Contr.Module Short to Ground
18020	P1612	Electronic Control Module Incorrect Coding
18021	P1613	MIL Call-up Circ Open/Short to B+
18022	P1614	MIL Call-up Circ./Transm.Contr.Module Range/Performance
18023	P1615	Engine Oil Temperature Sensor Circuit range/performance
18024	P1616	Glow Plug/Heater Indicator Circ. Short to B+
18025	P1617	Glow Plug/Heater Indicator Circ. Open/Short to Ground
18026	P1618	Glow Plug/Heater Relay Circ. Short to B+
18027	P1619	Glow Plug/Heater Relay Circ. Open/Short to Ground
18028	P1620	Engine coolant temperature signal open/short to B+
18029	P1621	Engine coolant temperature signal short to ground
18030	P1622	Engine coolant temperature signal range/performance
18031	P1623	Data Bus Powertrain No Communication
18032	P1624	MIL Request Sign.active
18033	P1625	Data-Bus Powertrain Implausible Message from Transm.Contr.
18034	P1626	Data-Bus Powertrain Missing Message from Transm.Contr.
18035	P1627	Data-Bus Powertrain missing message from fuel injection pump
18036	P1628	Data-Bus Powertrain missing message from steering sensor
18037	P1629	Data-Bus Powertrain missing message from distance control
18038	P1630	Accelera.Pedal Pos.Sensor 1 Signal too Low
18039	P1631	Accelera.Pedal Pos.Sensor 1 Signal too High
18040	P1632	Accelera.Pedal Pos.Sensor 1 Power Supply Malfunction

Diagnostic Trouble Codes (Powertrain)

VAG code	SAE P-code	Description
18041	P1633	Accelera.Pedal Pos.Sensor 2 Signal too Low
18042	P1634	Accelera.Pedal Pos.Sensor 2 Signal too High
18043	P1635	Data Bus Powertrain missing message f.air condition control
18044	P1636	Data Bus Powertrain missing message from Airbag control
18045	P1637	Data Bus Powertrain missing message f.central electr.control
18046	P1638	Data Bus Powertrain missing message from clutch control
18047	P1639	Accelera.Pedal Pos.Sensor 1+2 Range/Performance
18048	P1640	Internal Contr.Module (EEPROM) Error
18049	P1641	Please check DTC Memory of Air Condition ECU
18050	P1642	Please check DTC Memory of Airbag ECU
18051	P1643	Please check DTC Memory of central electric ECU
18052	P1644	Please check DTC Memory of clutch ECU
18053	P1645	Data Bus Powertrain missing message f.all wheel drive contr.
18054	P1646	Please Check DTC Memory of all wheel drive ECU
18055	P1647	Please check coding of ECUs in Data Bus Powertrain
18056	P1648	Data Bus Powertrain Malfunction
18057	P1649	Data Bus Powertrain Missing message from ABS Control Module
18058	P1650	Data Bus Powertrain Missing message fr.instrument panel ECU
18059	P1651	Data Bus Powertrain missing messages
18060	P1652	Please check DTC Memory of transmission ECU
18061	P1653	Please check DTC Memory of ABS Control Module
18062	P1654	Please check DTC Memory of control panel ECU
18063	P1655	Please check DTC Memory of ADR Control Module
18064	P1656	A/C clutch relay circuit short to ground
18065	P1657	A/C clutch relay circuit short to B+
18066	P1658	Data Bus Powertrain Incorrect signal from ADR Control Module
18067	P1659	Cooling Fan Control Circuit 1, short to B+
18068	P1660	Cooling Fan Control Circuit 1, short to ground
18080	P1672	Cooling Fan Control Circuit 1, open/short to ground
18081	P1673	Speed signal, error message from ABS ECU
18082	P1674	Data Bus Powertrain, Implausible signal from Instrument Panel ECU
18084	P1676	Drive by Wire-MIL Circ. Electrical Malfunction
18085	P1677	Drive by Wire-MIL Circ. Short to B+
18086	P1678	Drive by Wire-MIL Circ. Short to Ground
18087	P1679	Drive by Wire-MIL Circ. Open
18089	P1681	Contr.Unit Programming, Programming not Finished

Diagnostic Trouble Codes (Powertrain)

VAG code	SAE P-code	Description
18090	P1682	Data Bus Powertrain, Implausible signal from Brake Control Module
18091	P1683	Data Bus Powertrain, Implausible signal from Airbag Control Module
18092	P1684	Contr.Unit Programming Communication Error
18093	P1685	Data Bus Powertrain, Implausible signal from Steering Angle Sensor
18094	P1686	Contr.Unit Error Programming Error
18095	P1687	Data Bus Powertrain, Implausible signal from A/C Control Module
18096	P1688	Data Bus Powertrain, Implausible signal from All Wheel Drive Control Module
18097	P1689	Data Bus Powertrain, Implausible signal from Central Electric Control Module
18098	P1690	Malfunction Indication Light Malfunction
18099	P1691	Malfunction Indication Light Open
18100	P1692	Malfunction Indication Light Short to Ground
18101	P1693	Malfunction Indication Light Short to B+
18102	P1694	Malfunction Indication Light Open/Short to Ground
18104	P1696	Data Bus Powertrain, Implausible signal from Steering Column ECU
18105	P1697	Please Check DTC Memory Steering Angle Sensor
18106	P1698	Please Check DTC Memory Steering Column ECU
18107	P1699	Data Bus Powertrain, Missing signal from Steering Column ECU
18112	P1704	Kick Down Switch Malfunction
18113	P1705	Gear/Ratio Monitoring Adaptation limit reached
18119	P1711	Wheel Speed Signal 1 Range/Performance
18124	P1716	Wheel Speed Signal 2 Range/Performance
18129	P1721	Wheel Speed Signal 3 Range/Performance
18131	P1723	Starter Interlock Circ. Open
18132	P1724	Starter Interlock Circ. Short to Ground
18134	P1726	Wheel Speed Signal 4 Range/Performance
18136	P1728	Different Wheel Speed Signals Range/Performance
18137	P1729	Starter Interlock Circ. Short to B+
18141	P1733	Tiptronic Switch Down Circ. Short to Ground
18147	P1739	Tiptronic Switch up Circ. Short to Ground
18148	P1740	Clutch temperature control
18149	P1741	Clutch pressure adaptation at limit
18150	P1742	Clutch torque adaptation at limit
18151	P1743	Clutch slip control signal too high
18152	P1744	Tiptronic Switch Recognition Circ. Short to Ground
18153	P1745	Transm.Contr.Unit Relay Short to B+
18154	P1746	Transm.Contr.Unit Relay Malfunction

Diagnostic Trouble Codes (Powertrain)

VAG code	SAE P-code	Description
18155	P1747	Transm.Contr.Unit Relay Open/Short to Ground
18156	P1748	Transm.Contr.Unit Self-Check
18157	P1749	Transm.Contr.Unit Incorrect Coded
18158	P1750	Power Supply Voltage Low Voltage
18159	P1751	Power Supply Voltage High Voltage
18160	P1752	Power Supply Malfunction
18161	P1753	Tiptronic Switch, Malfunction
18162	P1754	Tiptronic Switch Up Circuit, Open/Short to B+
18163	P1755	Tiptronic Switch Down Circuit, Open/Short to B+
18164	P1756	Tiptronic Switch Recognition Circuit, Open/Short to B+
18165	P1757	Power Supply Voltage, Open
18168	P1760	Shift Lock Malfunction
18169	P1761	Shift Lock Short to Ground
18170	P1762	Shift Lock Short to B+
18171	P1763	Shift Lock Open
18172	P1764	Transmission temperature control
18173	P1765	Hydraulic Pressure Sensor 2 adaptation at limit
18174	P1766	Throttle Angle Signal Stuck Off
18175	P1767	Throttle Angle Signal Stuck On
18176	P1768	Hydraulic Pressure Sensor 2 Too High
18177	P1769	Hydraulic Pressure Sensor 2 Too Low
18178	P1770	Load Signal Range/Performance
18179	P1771	Load Signal Stuck Off
18180	P1772	Load Signal Stuck On
18181	P1773	Hydraulic Pressure Sensor 1 Too High
18182	P1774	Hydraulic Pressure Sensor 1 Too Low
18183	P1775	Hydraulic Pressure Sensor 1 adaptation at limit
18184	P1776	Hydraulic Pressure Sensor 1 range/performance
18185	P1777	Hydraulic Pressure Sensor 2 range/performance
18186	P1778	Solenoid EV7 Electrical Malfunction
18189	P1781	Engine Torque Reduction Open/Short to Ground
18190	P1782	Engine Torque Reduction Short to B+
18192	P1784	Shift up/down Wire Open/Short to Ground
18193	P1785	Shift up/down Wire Short to B+
18194	P1786	Reversing Light Circ. Open
18195	P1787	Reversing Light Circ. Short to Ground

Diagnostic Trouble Codes (Powertrain)

VAG code	SAE P-code	Description
18196	P1788	Reversing Light Circ. Short to B+
18197	P1789	Idle Speed Intervention Circ. Error Message from Engine Contr.
18198	P1790	Transmission Range Display Circ. Open
18199	P1791	Transmission Range Display Circ. Short to Ground
18200	P1792	Transmission Range Display Circ. Short to B+
18201	P1793	Output Speed Sensor 2 Circ. No Signal
18203	P1795	Vehicle Speed Signal Circ. Open
18204	P1796	Vehicle Speed Signal Circ. Short to Ground
18205	P1797	Vehicle Speed Signal Circ. Short to B+
18206	P1798	Output Speed Sensor 2 Circ. Range/Performance
18207	P1799	Output Speed Sensor 2 Circ. RPM too High
18221	P1813	Pressure Contr.Solenoid 1 Electrical
18222	P1814	Pressure Contr.Solenoid 1 Open/Short to Ground
18223	P1815	Pressure Contr.Solenoid 1 Short to B+
18226	P1818	Pressure Contr.Solenoid 2 Electrical
18227	P1819	Pressure Contr.Solenoid 2 Open/Short to Ground
18228	P1820	Pressure Contr.Solenoid 2 Short to B+
18231	P1823	Pressure Contr.Solenoid 3 Electrical
18232	P1824	Pressure Contr.Solenoid 3 Open/Short to Ground
18233	P1825	Pressure Contr.Solenoid 3 Short to B+
18236	P1828	Pressure Contr.Solenoid 4 Electrical
18237	P1829	Pressure Contr.Solenoid 4 Open/Short to Ground
18238	P1830	Pressure Contr.Solenoid 4 Short to B+
18242	P1834	Pressure Contr.Solenoid 5 Open/Short to Ground
18243	P1835	Pressure Contr.Solenoid 5 Short to B+
18249	P1841	Engine/Transmission Control Modules Versions do not match
18250	P1842	Please check DTC Memory of instrument panel ECU
18251	P1843	Please check DTC Memory of ADR Control Module
18252	P1844	Please check DTC Memory of central electric control ECU
18255	P1847	Please check DTC Memory of brake system ECU
18256	P1848	Please check DTC Memory of engine ECU
18257	P1849	Please check DTC Memory of transmission ECU
18258	P1850	Data-Bus Powertrain Missing Message from Engine Contr.
18259	P1851	Data-Bus Powertrain Missing Message from Brake Contr.
18260	P1852	Data-Bus Powertrain Implausible Message from Engine Contr.
18261	P1853	Data-Bus Powertrain Implausible Message from Brake Contr.

Diagnostic Trouble Codes (Powertrain)

VAG code	SAE P-code	Description
18262	P1854	Data-Bus Powertrain Hardware Defective
18263	P1855	Data-Bus Powertrain Software version Contr.
18264	P1856	Throttle/Pedal Pos.Sensor A Circ. Error Message from Engine Contr.
18265	P1857	Load Signal Error Message from Engine Contr.
18266	P1858	Engine Speed Input Circ. Error Message from Engine Contr.
18267	P1859	Brake Switch Circ. Error Message from Engine Contr.
18268	P1860	Kick Down Switch Error Message from Engine Contr.
18269	P1861	Throttle Position (TP) sensor Error Message from ECM
18270	P1862	Data Bus Powertrain Missing message from instr. panel ECU
18271	P1863	Data Bus Powertrain Missing Message from St. Angle Sensor
18272	P1864	Data Bus Powertrain Missing message from ADR control module
18273	P1865	Data Bus Powertrain Missing message from central electronics
18274	P1866	Data Bus Powertrain Missing messages
18275	P1867	Data Bus Powertrain Missing message from Steering Column Electronic Systems Control Module J527

Diagnostic scan tool suppliers

> ***NOTE —***
> *The following suppliers offer diagnostic scan tools and/or scan tool programs that can be used to retrieve diagnostic trouble codes and system data.*

Assenmacher Specialty Tools, Inc.
6440 Odell Place
Boulder, CO 80301
(800)525-2943
www.asttool.com

Baum Tools Unlimited, Inc.
PO Box 5867
Sarasota, FL 34277
(800)848-6657
www.baumtools.com

Equipment Solutions (Factory tool supplier)
PO Box 1450
Kenosha, WI 53141-1450
800-892-9650

Mac Tools
4635 Hilton Corporate Drive
Columbus, OH 43232
(800)622-8655
www.mactools.com

Ross-Tech
888 Sumneytown Pike
Lansdale, PA 19446
215-361-8942
www.ross-tech.com

Snap-On Technologies, Inc.
2801 80th St.
Kenosha, WI 53141-1410
(262)656-5200
www.snapon.com

Zelenda Automotive Inc.
66-02 Austin Street
Forest Hills, NY 11374
888-892-8348
www.zelenda.com

Control Module Coding

Control modules are produced in a few major variations and electronically matched to the exact system needs through a process known as coding. Coding is automatically displayed when the control module responds to the scan tool inquiry and can also be called up through Function 01 of the Function Menu of the VAG 1551/1552. The coding process requires a properly set-up scan tool or scan tool computer program and is accessed through Function 07.

Most codes contain 5 digits and many require "assembly" from the tables in this section. Careful attention must be paid to assemble accurate codes. Failure to code, or insertion of an incorrect code can cause any or all of the following conditions:

- Driveability problems
- Increased fuel consumption
- Increased emissions
- Storage of non-existent or erroneous DTCs resulting in Malfunction Indicator Light (MIL) illumination.
- Necessary functions not being performed (oxygen sensor regulation, activation of the EVAP canister system, non-functioning of anti-slip regulation (ASR) etc.).
- Necessary functions being performed incorrectly (mileage recorded instead of kilometers, warning light failure, etc.).
- Reduction in component service life.

NOTE —

- *New or replacement control units will usually require coding of the control module before they will function properly.*
- *In some instances, a control module will require input of a 7 digit code, but the table lists only a 5 digit code. In these instances, enter 00 before the correct code to satisfy the 7 digit requirement.*
- *After a new code is entered it will usually be necessary to switch the ignition off and back on for the new code to be recorded.*
- *In some cases more information may be needed to clarify information in the* **Control Module Coding** *tables. Refer to Official Factory Repair Manual for more information.*

WARNING —

- *DO NOT attempt to connect a VAG 1551/1552 scan tool, VAS 5051/5052 diagnostic tool, aftermarket scan tool or scan tool program to any vehicle if an incorrect radio has been installed.*
- *Installation of aftermarket radios or Audi radios not produced for use in model year 1998 and newer vehicles will*
  - *Affect various vehicle systems connected to the CAN-BUS.*
  - *Cause severe damage to VAG 1551 and VAG 1552 scan tools.*
  - *Cause severe damage to VAS 5051 and VAS 5052 diagnostic tools.*
  - *Cause severe damage to aftermarket scan tools.*
  - *Cause severe damage to computers running scan tool programs.*
- *Inspect radio; if radio is not correct for the vehicle or if there is any doubt, disconnect radio wiring harness before connecting scan tools.*

Engine Control Module (ECM) Coding Tables

Engine Control Module Coding Variations, 2.7L BiTurbo (engine code: APB)			
Country or Emission Level	**Drive or Additional Functions**	**Transmission**	**Vehicle Type**
00 =	**0 =**	**0 =**	**0 =**
01 =	**1 =**	**1 =** 6-Speed Manual	**1 =**
02 =	**2 =**	**2 =**	**2 =** C Class (C5 Audi A6 and S6)
03 =	**3 =**	**3 =**	**3 =**
04 =	**4 =**	**4 =**	**4 =**
05 =	**5 =** Front Wheel Drive **with ASR** Front Wheel Drive **with ESP**	**5 =** 5-Speed Automatic 01V (Tiptronic)	**5 =**
06 = Emission Standard TLEV	**6 =** All Wheel Drive (quattro) without ESP	**6 =**	**6 =**
07 =	**7 =** All Wheel Drive (quattro) with ESP	**7 =**	**7 =**
Note: Disregard empty table cells. ECM will only recognize the values listed above as valid. Access through Address Word 01.			

Example, engine code: APB				
S6				2
6-Speed Manual Transmission			1	
Quattro without ESP		6		
TLEV Emission Standard	06			
The correct code number for the above vehicle is:	**06612**			

Engine Control Module Coding Variations, 2.8L 5V V6 (engine code: AHA)			
Country or Emission Level	**Drive or Additional Functions**	**Transmission**	**Vehicle Type**
00 =	**0** = Front Wheel Drive without Anti-Slip Regulation (ASR)	**0** = 5-Speed Manual	**0** =
01 =	**1** = Front Wheel Drive with Anti-Slip Regulation (ASR)	**1** =	**1** =
02 =	**2** = All Wheel Drive (quattro) without Anti-Slip Regulation (ASR)	**2** =	**2** = C Class (C5 Audi A6)
03 =	**3** = All Wheel Drive (quattro) with Anti-Slip Regulation (ASR)	**3** =	**3** =
04 =	**4** =	**4** =	**4** =
05 =	**5** =	**5** = 5-Speed Automatic 01V (Tiptronic)	**5** =
06 = USA OBD II, with Evaporative Emissions System (EVAP) and Leak Detection Pump (LDP)	**6** =	**6** =	**6** =
Notes: Disregard empty table cells. ECM will only recognize the values listed above as valid. Access through Address Word 01.			

Example, engine code: AHA				
A6 Sedan				2
5-Speed Automatic Transmission			5	
All Wheel Drive (quattro) without ASR		2		
With EVAP and LDP systems	06			
The correct code number for the above vehicle is:	**06252**			

Engine Control Module Coding Variations, 2.8L 5V V6 (engine code: ATQ)			
Country or Emission Level	**Drive or Additional Functions**	**Transmission**	**Vehicle Type**
00 =	**0** =	**0** = 5-Speed Manual	**0** =
01 =	**1** =	**1** =	**1** =
02 =	**2** = All Wheel Drive (quattro) without Electronic Stability Program (ESP)	**2** =	**2** = C Class (C5 Audi A6)
03 =	**3** =	**3** =	**3** =
04 =	**4** =	**4** =	**4** =
05 =	**5** = Front Wheel Drive with Anti-Slip Regulation (ASR)	**5** = 5-Speed Automatic 01V (Tiptronic)	**5** =
06 =	**6** =	**6** =	**6** =
07 = Emission Standard LEV	**7** = All Wheel Drive (quattro) with Electronic Stability Program (ESP)	**7** =	**7** =

Note: Some model years display a 7 digit code instead of a 5 digit code. If this is the case with your vehicle, enter 00 and then the correct 5 digit code. Disregard empty table cells. ECM will only recognize the values listed above as valid. Access through Address Word 01.

Example, engine code: ATQ				
A6 Sedan				2
5-Speed Automatic Transmission			5	
All Wheel Drive (quattro) without Electronic Stability Program (ESP)		2		
LEV Emission Standard	07			
The correct code number for the above vehicle is:	**07252**			

Engine Control Module Coding Variations, 2.8L 5V V6 (engine code: ART)			
Country or Emission Level	**Drive or Additional Functions**	**Transmission**	**Vehicle Type**
00 =	**0 =**	**0 =**	**0 =**
01 =	**1 =**	**1 =**	**1 =**
02 =	**2 =**	**2 =**	**2 =** C Class (C5 Audi A6)
03 =	**3 =**	**3 =**	**3 =**
04 =	**4 =**	**4 =**	**4 =**
05 =	**5 =**	**5 =** 5-Speed Automatic 01V (Tiptronic)	**5 =**
06 = Emission Standard TLEV	**6 =** All Wheel Drive (quattro) without Electronic Stability Program (ESP)	**6 =**	**6 =**

Note: Some model years display a 7 digit code instead of a 5 digit code. If this is the case with your vehicle, enter 00 and then the correct 5 digit code. Disregard empty table cells. ECM will only recognize the values listed above as valid. Access through Address Word 01.

Example, engine code: ART				
A6 Sedan				2
5-Speed Automatic Transmission			5	
All Wheel Drive (quattro) without Electronic Stability Program (ESP)		6		
TLEV Emission Standard	06			
The correct code number for the above vehicle is:	**06652**			

Transmission Control Module (TCM) Coding Tables

Transmission Control Module Coding Variations, Automatic Transmission Type 01V[2]

Engine Power Control (cable throttle or E-gas drive-by-wire)	**Vehicle Equipment/Version**	**Market[1]**	**Dynamic Shift Program (DSP)**	**TCM Code[3, 4]**
With cable throttle	All engine codes with cable throttle All transmission codes	Front Wheel Drive RdW & USA/Canada All Wheel Drive (quattro) RdW & USA/Canada	Active	00000
With E-gas	Not a valid code for E-gas vehicles	n/a	n/a	00000
With E-gas	All engine codes with E-gas all transmission codes	All Wheel Drive (quattro) (RdW)	Active	00001
With E-gas	All engine codes with E-gas all transmission codes	All Wheel Drive (quattro) (USA/Canada)	Active	00002
With E-gas	All engine codes with E-gas all transmission codes	Front Wheel Drive (RdW)	Active	00003
With E-gas	All engine codes with E-gas all transmission codes	Front Wheel Drive (USA/Canada)	Active	00004

Notes:
1. USA and Canada represent a specific market; RdW indicates not USA/Canada i.e. the rest of the world.
2. Access through Address Word 02.
3. TCM will only recognize the listed values as valid.
4. Some model years display a 7 digit code instead of a 5 digit code. If this is the case with your vehicle, enter 00 and then the correct 5 digit code.

Transmission Control Module Coding Variations, Automatic Transmission Type 01L[2]

Engine Code	**Vehicle Equipment/Version**	**Market[1]**	**Dynamic Shift Program (DSP)**	**TCM Code[3, 4]**
ART (V8)	Not a valid code	n/a	n/a	00000
ART (V8)	All transmission codes	RdW	Active	00001
ART (V8)	All transmission codes	USA/Canada	Active	00002

Notes:
1. USA and Canada represent a specific market; RdW indicates not USA/Canada i.e. the rest of the world.
2. Access through Address Word 02.
3. TCM will only recognize the listed values as valid.
4. Some model years display a 7 digit code instead of a 5 digit code. If this is the case with your vehicle, enter 00 and then the correct 5 digit code.

Anti-Lock Braking System, Electronic Differential Lock, Anti-Slip Regulation, (ABS/EDL/ASR) Coding Table

ABS Control Module Coding Variations[1], ABS/EDL/ASR					
Engine	**Transmission[1]**	**Data Transmission**	**Control Module Part Number**	**Code[2]**	**Note**
2.8L	Manual transmission	Separate wires	8E0 614 111 R 8E0 614 111 S 8E0 614 111 C 8E0 614 111 G	00000	Pin 28 on ABS control module (J104) is not used
2.8L	Automatic transmission	Separate wire	8E0 614 111 R 8E0 614 111 S 8E0 614 111 C 8E0 614 111 G	00003	12 volts at pin 28 on ABS control module (J104)
2.7L	Manual transmission	CAN-BUS	8E0 614 111 AH 8E0 614 111 AJ	00032	
2.7L	Automatic transmission	CAN-BUS	8E0 614 111 AH 8E0 614 111 AJ	00031	
2.8L	Manual transmission	CAN-BUS	8E0 614 111 AH 8E0 614 111 AJ	00032	
2.8L	Automatic transmission	CAN-BUS	8E0 614 111 AH 8E0 614 111 AJ	00031	

Notes:
1. Access through Address Word 03.
2. Only the codes listed above are recognized as valid.

Electronic Stability Program (ESP) Coding Tables

ESP Control modules that can be coded require a log-in code before accessing Function 03 for the actual procedure. The log-in procedure is accessed through function 11 using the following table.

ESP Control Module Log-In Table[1]

Engine/Transmission	Brake Caliper Type[2]	Log-In Number[3]
V6 with Manual transmission	FN3 (Teves/ATE)	07595
V6 with Automatic transmission	FN3 (Teves/ATE)	07597
V6/V8 & Manual transmission	HP2 (dual piston caliper)	07495
V6/V8 & Automatic transmission	HP2 (dual piston caliper)	07497
2.7L allroad quattro w/manual transmission	HP2 (dual piston caliper)	07495
2.7L allroad quattro w/automatic transmission	HP2 (dual piston caliper)	07497

Notes:
1. Access through Address Word 03, function 11.
2. Requires visual inspection for identification.
3. Only the codes listed above are recognized as valid.

ESP Control Module Coding Variations[1]

Engine	Transmission	Brake Caliper Type	Control Module Part Number[4]	Code[2,3]	Note
			4B0 907 389		Control module can not be coded
V6	Manual transmission	FN3 (Teves/ATE)	8D0 407 389 A 8D0 407 389 D 8D0 407 389 E	06295	
V6	Automatic transmission	FN3 (Teves/ATE)	8D0 407 389 A 8D0 407 389 D 8D0 407 389 E	06297	
V6/V8	Manual transmission	HP2 (dual piston caliper)	8D0 407 389 A 8D0 407 389 D 8D0 407 389 E	06395	
V6/V8	Automatic transmission	HP2 (dual piston caliper)	8D0 407 389 A 8D0 407 389 D 8D0 407 389 E	06397	
2.7L allroad quattro	Manual transmission	HP2 (dual piston caliper)	8D0 407 389 A 8D0 407 389 D 8D0 407 389 E	06395	
2.7L allroad quattro	Automatic transmission	HP2 (dual piston caliper)	8D0 407 389 A 8D0 407 389 D 8D0 407 389 E	06397	

Notes:
1. Access through Address Word 03.
2. Only the codes listed above are recognized as valid.
3. If the control module is coded with the coding 00000 or a code number that is not correct for the vehicle, the ABS warning light K47 and the ESP control lamp K155 will come on continuously and the DTC01044 (control module incorrectly coded) will be stored in the DTC memory of the control module.
4. Part numbers are listed for reference only and are subject to change.

Suspension Level Control System Coding Table

Suspension Level Control System Electronic Control Unit Coding		
Vehicle Type	**Equipment Level**	**Code**
allroad quattro	all versions	25500
Note: Only the code listed above is recognized as valid. Access through Address Word 34.		

Central Locking Coding Table

Central Locking Control Module Coding Variations			
Code Level	**Locking Mode**	**Code Number**	**Note**
Standard Code	Standard operation with central locking by way of manual switch.	15885	Control modules are coded to this standard as delivered
Optional Code	Vehicle automatically locks upon reaching a speed of approximately 10 MPH (15 KPH), and unlocks when key is removed from ignition lock.	15917	
Note: Only the code listed above is recognized as valid. Access through Address Word 35.			

Memory Seat (Driver and Passenger) Coding Table

Suspension Level Control System Electronic Control Unit Coding			
System	**Address Word**	**Function**	**Code**
Driver's seat, (ECM 4B0 959 760 B)	36	Match remote control key memory by way of allocation to personalized buttons	00001
Driver's seat, (ECM 4B0 959 760 B)	36	Remote control key memory off	00003
Front passenger seat	06	Remote control key memory off	00003
Note: Only the codes listed above are recognized as valid. Control modules are shipped uncoded as delivered. Note different Address Words.			

Interior Monitoring Coding Tables

Interior Monitoring Control Module Coding - Vehicles Without Side Curtain Airbag Protection			
No Current Significance	**Vehicle Type**	**Mode**	**Body Type**
00 = (always 00)	**0** = C Class (C5 Audi A6)	**0** = Static priming mode (always 0)	**0** =
01 =	**1** =	**1** =	**1** = Sedan
02 =	**2** =	**2** =	**2** = Avant
Note: Disregard empty table cells. ECM will only recognize the values listed above as valid. Control modules are shipped uncoded as delivered.ccess through Address Word 45.			

Example, Vehicles Without Side Curtain Airbag Protection				
Sedan				1
Static priming mode			0	
A6		0		
(Always 00)	00			
The correct code number for the above vehicle is:	**00001**			

Interior Monitoring Control Module Coding - Vehicles With Side Curtain Airbag Protection			
No Current Significance	**Vehicle Type**	**No Current Significance**	**Body Type**
00 = (always 00)	**0** =	**0** = (always 0)	**0** =
01 =	**1** = C Class (C5 Audi A6)	**1** =	**1 = Sedan**
02 =	**2** =	**2** =	**2 = Avant**
Note: Disregard empty table cells. ECM will only recognize the values listed above as valid. Control modules are shipped uncoded as delivered. Access through Address Word 45.			

Example, Vehicles With Side Curtain Airbag Protection				
Avant				2
(Always 0)			0	
A6		1		
(Always 00)	00			
The correct code number for the above vehicle is:	**00102**			

Airbag System Coding Table

Airbag System Electronic Control Unit Coding Variations[6]		
System and Vehicle Equipment Level	**ECU Part Number[2]**	**ECU Code[3, 4, 5]**
Vehicles produced through calender week 35/1998, Airbag 7[1]		
Driver and front passenger front airbags, driver and front passenger side airbags, front seat belt tensioners, 2 rear seat belt tensioners, rear side airbags disabled	4B0 959 655 C 4B0 959 655 D 4B0 959 655 J 4B0 959 655 K	00104
Driver and front passenger front airbags, driver and front passenger side airbags, front seat belt tensioners, 3 rear seat belt tensioners, rear side airbags disabled	4B0 959 655 C 4B0 959 655 D 4B0 959 655 J 4B0 959 655 K	00106
Driver and front passenger front airbags, driver and front passenger side airbags, front seat belt tensioners, 2 rear seat belt tensioners rear side airbags enabled	4B0 959 655 C 4B0 959 655 D 4B0 959 655 J 4B0 959 655 K	00204
Driver and front passenger front airbags, driver and front passenger side airbags, front seat belt tensioners, 3 rear seat belt tensioners rear side airbags enabled	4B0 959 655 C 4B0 959 655 D 4B0 959 655 J 4B0 959 655 K	00206
Driver and front passenger front airbags, driver and front passenger side airbags, front seat belt tensioners, 2 rear seat belt tensioners belt fastened sensor	4B0 959 655 C 4B0 959 655 D 4B0 959 655 J 4B0 959 655 K	00205
Driver and front passenger front airbags, driver and front passenger side airbags, front seat belt tensioners, 3 rear seat belt tensioners belt fastened sensor	4B0 959 655 C 4B0 959 655 D 4B0 959 655 J 4B0 959 655 K	00207
Driver and front passenger front airbags, driver and front passenger side airbags, front seat belt tensioners, 2 rear seat belt tensioners belt fastened sensor, front seat occupied sensor	4B0 959 655 C 4B0 959 655 D 4B0 959 655 J 4B0 959 655 K	00115
Driver and front passenger front airbags, driver and front passenger side airbags, front seat belt tensioners, 3 rear seat belt tensioners belt fastened sensor, front seat occupied sensor	4B0 959 655 C 4B0 959 655 D 4B0 959 655 J 4B0 959 655 K	00117
Vehicles produced from calender week 36/1998 through 22/2000, Airbag 8[1]		
Driver and front passenger front airbags, front seat belt tensioners, 2 rear seat belt tensioners		00004
Driver and front passenger front airbags, front seat belt tensioners, 3 rear seat belt tensioners		00006
Driver and front passenger front airbags, driver and front passenger side airbags, front seat belt tensioners, 2 rear seat belt tensioners rear side airbags disabled		00104
Driver and front passenger front airbags, driver and front passenger side airbags, front seat belt tensioners, 3 rear seat belt tensioners rear side airbags disabled		00106
Driver and front passenger front airbags, driver and front passenger side airbags, rear side airbags front seat belt tensioners, 2 rear seat belt tensioners rear side airbags enabled		00204
Driver and front passenger front airbags, driver and front passenger side airbags, rear side airbags front seat belt tensioners, 3 rear seat belt tensioners rear side airbags enabled		00206

Airbag System Electronic Control Unit Coding Variations[6]		
System and Vehicle Equipment Level	**ECU Part Number[2]**	**ECU Code[3, 4, 5]**
Driver and front passenger front airbags, front seat belt tensioners, 2 rear seat belt tensioners, key switch for deactivating front passenger airbag		01004
Driver and front passenger front airbags, front seat belt tensioners, 3 rear seat belt tensioners, key switch for deactivating front passenger airbag		01006
Driver and front passenger front airbags, driver and front passenger side airbags, front seat belt tensioners, 2 rear seat belt tensioners, key switch for deactivating front passenger airbag		01104
Driver and front passenger front airbags, driver and front passenger side airbags, front seat belt tensioners, 3 rear seat belt tensioners, key switch for deactivating front passenger airbag		01106
Driver and front passenger front airbags, driver and front passenger side airbags, rear side airbags, front seat belt tensioners, 2 rear seat belt tensioners, key switch for deactivating front passenger airbag		01204
Driver and front passenger front airbags, driver and front passenger side airbags, rear side airbags, front seat belt tensioners, 3 rear seat belt tensioners, key switch for deactivating front passenger airbag		01206
Driver and front passenger front airbags, driver and front passenger side airbags, front seat belt tensioners, 2 rear seat belt tensioners, side curtain protection, rear side airbags deactivated		00304
Driver and front passenger front airbags, driver and front passenger side airbags, front seat belt tensioners, 3 rear seat belt tensioners, side curtain protection, rear side airbags deactivated		00306
Driver and front passenger front airbags, driver and front passenger side airbags, front seat belt tensioners, 2 rear seat belt tensioners, side curtain protection, rear side airbags enabled		00404
Driver and front passenger front airbags, driver and front passenger side airbags, front seat belt tensioners, 3 rear seat belt tensioners, side curtain protection, rear side airbags enabled		00406
Driver and front passenger front airbags, driver and front passenger side airbags, head airbags, front seat belt tensioners, 2 rear seat belt tensioners, key switch for deactivating front passenger airbag		01304
Driver and front passenger front airbags, driver and front passenger side airbags, head airbags, front seat belt tensioners, 3 rear seat belt tensioners, key switch for deactivating front passenger airbag		01306
Driver and front passenger front airbags, driver and front passenger side airbags, rear side airbags, front seat belt tensioners, 2 rear seat belt tensioners, key switch for deactivating front passenger airbag		01404
Driver and front passenger front airbags, driver and front passenger side airbags, rear side airbags, front seat belt tensioners, 3 rear seat belt tensioners, key switch for deactivating front passenger airbag		01406

Airbag System Electronic Control Unit Coding Variations[6]		
System and Vehicle Equipment Level	**ECU Part Number[2]**	**ECU Code[3, 4, 5]**
Vehicles produced from calender week 23/2000, Airbag 8[1]		
Driver and front passenger front airbags, front seat belt tensioners, 2 rear seat belt tensioners front belt buckle check		00005
Driver and front passenger front airbags, front seat belt tensioners, 3 rear seat belt tensioners front belt buckle check		00007
Driver and front passenger front airbags, driver and front passenger side airbags, front seat belt tensioners, 2 rear seat belt tensioners front belt buckle check		00105
Driver and front passenger front airbags, driver and front passenger side airbags, front seat belt tensioners, 3 rear seat belt tensioners front belt buckle check		00107
Driver and front passenger front airbags, driver and front passenger side airbags, front seat belt tensioners, 2 rear seat belt tensioners front belt buckle check		00205
Driver and front passenger front airbags, driver and front passenger side airbags, front seat belt tensioners, 3 rear seat belt tensioners front belt buckle check		00207
Driver and front passenger front airbags, driver and front passenger side airbags, side curtain protection front seat belt tensioners, 2 rear seat belt tensioners front belt buckle check		00305
Driver and front passenger front airbags, driver and front passenger side airbags, side curtain protection front seat belt tensioners, 3 rear seat belt tensioners front belt buckle check		00307
Driver and front passenger front airbags, driver and front passenger side airbags, rear side airbags, side curtain protection front seat belt tensioners, 2 rear seat belt tensioners front belt buckle check		00405
Driver and front passenger front airbags, driver and front passenger side airbags, rear side airbags, side curtain protection front seat belt tensioners, 3 rear seat belt tensioners front belt buckle check		00407

Notes:

1. For technical reasons, the chassis number range 4BWN 024500 to 4BWN 033300 may feature non-operational seat occupied and belt fastened sensors in certain cases. If a belt buckle in the above chassis number range has to be replaced, fit a buckle with no belt fastened sensor. In the event of a seat occupied sensor malfunction, disconnect the sensor and code the Airbag Control Module J234 in the same way as for a vehicle with no seat occupied sensor.
2. Part numbers, where given, are listed for reference only and are subject to change.
3. Only select from the listed codes.
4. Control modules are shipped uncoded as delivered.
5. Airbag warning light will remain on if control module has not been coded.
6. Access through Address Word 15.

Automatic Climate Control Coding Table

The climate control system electronic control module is integral with the A/C control head E87. Several different versions are installed depending vehicle equipment level. When replacing a climate control system control module. consider the following points:

- Control modules are shipped uncoded as delivered.
- If a new control head is installed and its display panel flashes (for about 2 minutes) when the ignition is switched on, the A/C control head (control module) is uncoded. It must be correctly coded and basic settings must be carried out in order to function properly.
- Code A/C Control Head E87 according to vehicle type, model year, engine, vehicle equipment and country variant.
- Two different A/C versions have been installed in A6 vehicles. version 1 and version 2. Differences will influence control module codes. Always consult an authorized Audi parts department or knowledgeable independent parts specialist for correct current information.
- Between week 36 and week 44 of 1998 the production of air conditioning unit version "1" has been converted to air conditioning version "2" as a running change. Both versions can be differentiated through specific characteristics.

Automatic Climate Control, Control Module Coding Variations[1]				
Designations	**Code Assembly Position**[3, 4, 5]			
	1 + 2	**3**	**4**	**5**
Unallocated				
Currently unassigned, always 00	00			
Country Version				
RdW[2] (except USA and Japan)		0		
USA		1		
Japan (not permissible for USA)		2		
Engine Cylinders				
4-cylinder			4	
6-cylinder			6	
8-cylinder			8	
Vehicle Type and Engine Version				
Left hand steering, gasoline engine				0
Right hand steering, gasoline engine (not USA)				1
Left hand steering, Diesel engine (not USA)				3
Right hand steering, Diesel engine (not USA)				4

Notes:
1. Access through Address Word 08.
2. USA and Canada represent a specific market; RdW indicates not USA/Canada, i.e. the rest of the world.
3. Only select from the listed codes.
4. Code will always be 5 digits.
5. Always consult an authorized Audi parts department or knowledgeable independent parts specialist for correct, current information.

Multi-Function Steering Wheel Coding Table

Multi-Function Steering Wheel, Control Module Coding Variations[1]				
Designations	**Code Assembly Position**[3, 4]			
	1 + 2	**3**	**4**	**5**
Control unit index code C[2]				
Standard				
Always 00	00			
Language Version				
English		1		
Tiptronic				
Without			0	
Steering Wheel Version				
Radio and Telephone operation				1
Radio operation (6-button radio)				2
Control unit index code D[2]				
Standard				
Always 00	00			
Language Version				
English		1		
Tiptronic				
Without			0	
Steering Wheel Version				
Radio and Telephone operation				1
Radio operation (4-button radio)				2
Control unit index code J[2]				
Standard				
Always 00	00			
Language Version				
English		1		
Tiptronic				
With			1	
Steering Wheel Version				
Radio and Telephone operation				1
Radio operation (6-button radio)				2
Control unit index code L[2]				

Multi-Function Steering Wheel, Control Module Coding Variations[1]				
Designations	**Code Assembly Position[3, 4]**			
	1 + 2	**3**	**4**	**5**
Standard				
Always 00	00			
Language Version				
English		1		
Tiptronic				
With			1	
Steering Wheel Version				
Radio and Telephone operation				1
Radio operation (4-button radio)				2

Notes:

1. Access through Address Word 16.
2. Index letter is suffix letter of part number as in 4B0 907 487 D; the letter D is the index.
3. Only select from the listed codes.
4. Code will always be 5 digits.

Instrument Cluster Coding Tables

Instrument Cluster Control Module Coding Variations[6]				
Designations	**Code Assembly Position[5]**			
	1 + 2[1]	**3[2]**	**4[3]**	**5[4]**
Available Equipment				
None	00			
Brake pad wear warning	01			
Seat belt warning	02			
Washer fluid level warning	04			
Market Version				
Germany (D)		0		
Europe (EU)		1		
USA (USA)		2		
Canada (CDN)		3		
Great Britain (GB)		4		
Japan (JP)		5		
Saudi Arabia (SA)		6		
Australia (AUS)		7		
Number of Cylinders				
6			6	
8			8	
Engine Version				
Gasoline V6				2
Gasoline V8				3

Notes:

1. Access through Address Word 17.
2. Values in this column are added together where applicable to determine value. For example, if vehicle in question has seat belt warning (02) and washer fluid warning (04), values are added together (02 + 04 = 06) to get a value of 06 for positions 1 and 2 of the instrument cluster code. If vehicle has all of the equipment listed, add all of the values together.
3. Only select from the listed codes.
4. Only select from the listed codes.
5. Code will be 5 digits. For example, an A6 with brake pad wear warning, seat belt warning, and washer fluid level warning, for the USA market, with a 2.8L V6 gasoline engine will have a code of 07262.

Acoustic Parking System Coding Tables

Acoustic Parking System Control Module Coding Variations[1]					
Designations	**Code Assembly Position[2, 3]**				
	1	**2**	**3**	**4**	**5**
Standard					
Always 0	0				
Transmission Type					
Standard		0			
Standard		1			
APS Signal					
Without active acoustic signal			0		
With active acoustic signal (from factory)			1		
Vehicle equipment					
Sedan (RdW)				0	
Avant (station wagon) (RdW)				1	
Sedan (USA)				2	
Avant (USA)				3	
S-version Sedan (RdW)				4	
S-version Avant (station wagon) (RdW)				5	
S-version Sedan (USA)				6	
S-version Avant (USA)				7	
Vehicle Type					
A3					3
A4					4
A6					6
A8					8

Notes:
1. Access through Address Word 76.
2. Code will be 5 digits.
3. Only select from the listed codes.

Automatic Vertical Headlight Aim Coding

Automatic Vertical Headlight Aim Control, Control Module Coding Variations[1]		
Model Year	**Vehicle Type**	**Code[2]**
1998, 1999	A6 Sedan/Avant, front wheel drive	00005
1998, 1999	A6/S6 Sedan/Avant, quattro	00006
2000	A6 Sedan/Avant, front wheel drive	00010
2000	A6/S6 Sedan/Avant, quattro	00020
Notes: 1. Access through Address Word 55. 2. Only the codes listed above are recognized as valid.		

Radio and Sound System Coding Table

Radio Head Unit Coding Variations[1]					
Designations	Code Assembly Position[2, 3]				
	1	2	3	4	5
Country Identification					
Standard	0				
Sound Matching					
Standard		0			
Number of Passive Loudspeakers					
No passive loudspeakers (BOSE System)			0		
1 (single) passive loudspeaker, left front (BOSE System with telephone)			1		
2 passive loudspeakers (front) and 2 active loudspeakers (rear)			2		
2 passive loudspeakers (front) and no active loudspeakers (rear)			5		
Sound System Adjustment (system type)					
Standard system (non-BOSE)				0	
BOSE Sound System				1	
Sound System Configuration					
Antenna with remote power supply					1
Antenna with remote power supply, CD changer unit					3
Antenna with remote power supply, secondary radio display in instrument cluster					5
Antenna with remote power supply, CD changer unit, secondary radio display in instrument cluster					7

Notes:

1. Access through Address Word 56.
2. Code will be 5 digits.
3. Only select from the listed codes.

Readiness code values[2]								
Digit position[1]								Diagnostic Function
1	2	3	4	5	6	7	8	
							0	Three Way Catalyst Efficiency
						0		Catalyst Heating (currently always 0)
					0			Evaporative Emissions System (Fuel tank venting system)
				0				Secondary Air Injection System
			0					Air Conditioning (R12, no current diagnostic function, always 0)
		0						Oxygen Sensors
	0							Oxygen Sensor Heaters
0								Exhaust Gas Recirculation (EGR)

Notes:
1. Value can only be a 1 or a 0.
   A value of 1 indicates that test has not yet run or has run and failed.
   A value of 0 indicates that test has run and passed or function is not applicable.
2. Readiness code is properly set when all 8 positions are zero, i.e. "00000000".

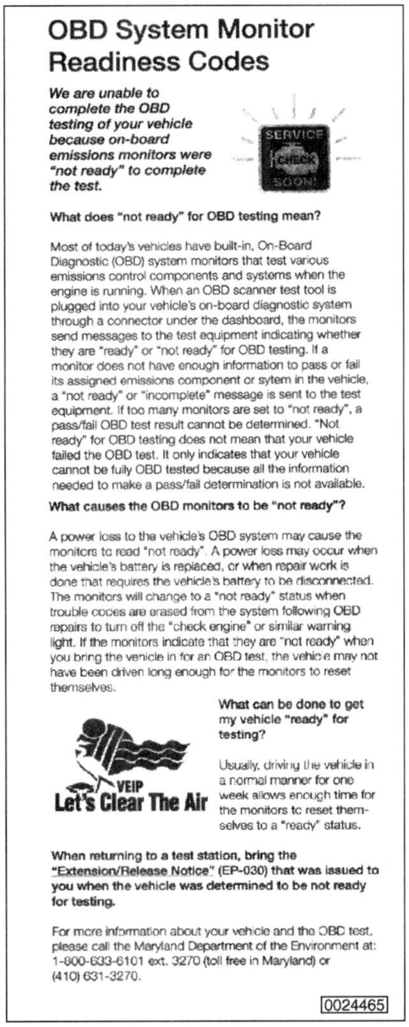
OBD System Monitor Readiness Codes

We are unable to complete the OBD testing of your vehicle because on-board emissions monitors were "not ready" to complete the test.

SERVICE CHECK SOON!

What does "not ready" for OBD testing mean?

Most of today's vehicles have built-in, On-Board Diagnostic (OBD) system monitors that test various emissions control components and systems when the engine is running. When an OBD scanner test tool is plugged into your vehicle's on-board diagnostic system through a connector under the dashboard, the monitors send messages to the test equipment indicating whether they are "ready" or "not ready" for OBD testing. If a monitor does not have enough information to pass or fail its assigned emissions component or sytem in the vehicle, a "not ready" or "incomplete" message is sent to the test equipment. If too many monitors are set to "not ready", a pass/fail OBD test result cannot be determined. "Not ready" for OBD testing does not mean that your vehicle failed the OBD test. It only indicates that your vehicle cannot be fully OBD tested because all the information needed to make a pass/fail determination is not available.

What causes the OBD monitors to be "not ready"?

A power loss to the vehicle's OBD system may cause the monitors to read "not ready". A power loss may occur when the vehicle's battery is replaced, or when repair work is done that requires the vehicle's battery to be disconnected. The monitors will change to a "not ready" status when trouble codes are erased from the system following OBD repairs to turn off the "check engine" or similar warning light. If the monitors indicate that they are "not ready" when you bring the vehicle in for an OBD test, the vehicle may not have been driven long enough for the monitors to reset themselves.

VEIP Let's Clear The Air

What can be done to get my vehicle "ready" for testing?

Usually, driving the vehicle in a normal manner for one week allows enough time for the monitors to reset themselves to a "ready" status.

When returning to a test station, bring the "Extension/Release Notice" (EP-030) that was issued to you when the vehicle was determined to be not ready for testing.

For more information about your vehicle and the OBD test, please call the Maryland Department of the Environment at: 1-800-633-6101 ext. 3270 (toll free in Maryland) or (410) 631-3270.

0024465

Readiness Codes

Overview

The inspection and maintenance (I/M) legislation requiring On Board Diagnostics (OBD II) includes provisions to prevent tampering or manipulating an Inspection/Maintenance (I/M) test. The readiness code will be stored after complete diagnostic monitoring of all components and systems has been carried out. This function was put into place to prevent manipulating an I/M emission test procedure by clearing faults codes or disconnecting the ECM or battery.

An additional benefit of this legislation is the ability of the engine management system to check for proper operation of up to 8 functions or systems. For the Motronic M5.9.2 and ME7.X engine management systems used on A6s, the exact number will depend on the equipment level of each particular vehicle.

Most of the time, the readiness code will set itself after the appropriate conditions have been met. These may include a cold or a hot start, operation at a certain load or speed for a certain period of time, or operation at different temperatures. After repairs, it is usually advantageous to verify the proper operation of a component or system. Sometimes this can be accomplished by driving the vehicle and observing the Malfunction Indicator Light (MIL), but a more efficient way to confirm a repair is by setting the readiness codes.

◄ Significance of the 8 digit readiness code

Setting the readiness code requires Audi scan tools VAG 1551, VAG 1552, or the new diagnostic computers VAS 5051 and VAS 5052, or an equivalent aftermarket scan tool or scan tool computer program. A thorough understanding of your equipment operation and it's capabilities is also required. Readiness codes are then generated according to instructions provided by the factory and the scan tool manufacturer.

◄ Recently, some states, such as Maryland, check Readiness Codes in lieu of a tailpipe emissions test on OBD II vehicles. This informational card is supplied to drivers if a Readiness Code is not set to 00000000.

The readiness code can be displayed in any of several areas within the ECM's diagnostic memory depending on the individual scan tool. It will always have a format consisting of 8 digits. If the 8 digit code contains characters other than the number 0, such as 10100110, than the readiness code is NOT set. When the code displays all zeros, i.e. 00000000, the readiness code IS set and no further action is required. While the format is required to be the same for all engines, certain equipment may not be used on each engine. For example, most A6 engines do not use EGR and as such, digit position number 8 will always show 0. Other Audi engines that use EGR will show either a 1 or a 0 as appropriate.

Readiness Code Quick Reference Tables

The following tables outline the procedures for setting the readiness code for each A6 engine type. The code can be read from function 08 in the appropriate display group. Most ECM versions will also display the readiness code separately through function 15.

Quick Reference:
Readiness code setting, engine code APB

Step	Procedure VAS 5051/ 5052 Diagnostic Status	Engine Condition	Function/ Display-Group	Channel 1	Channel 2	Channel 3	Channel 4
1 2	Check DTC Memory Clear DTC Memory	Engine OFF Ignition ON	02 05	-	-	-	-
3	Adaptation Throttle Valve Control Module	Engine OFF Ignition ON	04/060	3% - 20%	80 - 97%	8	ADP.OK
4	Adaptation Kick Down (Only with Automatic Transmission)	Engine OFF Ignition ON Press gas pedal to the floor	04/063	-	78 - 94%	Kick Down	ADP.OK
5	Fuel System Leakage Check	Engine ON @ Idle	04/071	-	-	Check END	Syst. OK
6	EVAP Canister Purge Regulator Valve Check	Engine ON @ Idle	04/070	0% - 100 %	-	-	EVAP OK
7	Lambda Control	Engine ON @ Idle	04/107	640 – 920 RPM	-18 - +18%	-18 - +18%	Syst. OK
8 9	Heated Oxygen Sensor	Engine ON 1800 - 2200/min*	04/041 04/042	0 - 0.5 kOhm	Htg.bC ON	0 - 0.5 kOhm	Htg.aC ON
10	Operational Readiness of Oxygen Sen. behind Cat. Conv.	Engine ON @ Idle	04/036	0–1V	B1-S2 OK	0–1V	B2-S2 OK
11 12	Response Time Front Heated Oxygen Sensor	Engine ON 1520 - 2280/min**	04/034 04/035	1520 - 2280 RPM**	more than 260°C	0,4 - 1,0 s	B1-S1 OK B2-S1 OK
13 14	Oxygen Sensor Control System	Engine ON @ Idle	04/037 04/038	0 - 30 %	0 - 1 V	-150 - +150 ms	Syst. OK
15 16	Oxygen Sensor aging behind Cat. Conv.	Engine ON @ Idle	04/043 04/044	700 – 3000 U/min	more than 350°C	0 – 1 V	B1-S2 OK B2-S2 OK
17 18	Three Way Catalytic Converter	Engine ON 1880 - 2280/min**	04/046 04/047	1880 - 2280 RPM**	more than 315°C	0 – 0.80	CAT B1 OK CAT B2 OK
19 20	Secondary Air System Only AT vehicles	Engine ON @ Idle	04/077 04/078	640 – 920 RPM	0 – 5 g/s	-40 - +10 %	Syst. OK
21	Check Readiness Code Setting	Engine ON @ Idle	08/086	00000000	00000000	00000000	00000000

Requirement: Coolant Temperature more than 80°C
*** to warm up engine if necessary, test is complete when values are present in channel 1 and 3** **** only during Diagnostic Performance**

Confirm Readiness Code generating with a VAS 5051 Printout!

Quick Reference:
Readiness code setting, engine code AHA

Step	Procedure VAS 5051/ 5052 Diagnostic Status	Engine Condition	Function/ Display-Group	Channel 1	Channel 2	Channel 3	Channel 4
0	Check DTC Memory Clear DTC Memory	Engine OFF Ignition ON	02 05	-	-	-	-
1	Check Readiness Code Setting	Engine OFF Ignition ON	08/086	01101101	11111111	00111111	00011111
2	Adaptation Throttle Valve Control Module	Engine OFF Ignition ON	04/098	3.255- 4.745V	0.902- 4.863V	Idle	ADP.OK
3	Pre and Post Heated Oxygen Sensor, Readiness	Engine ON 2000 - 2500/min	08/030	111	111	111	111
4	Response Time Front Heated Oxygen Sensor (B1-S1)	Engine ON 1800 - 2200/min	04/034	1800 - 2200 RPM	0.8 - 2.5 ms*	0.00 - 2.1 s	B1-S1 OK
5	Response Time Front Heated Oxygen Sensor (B2-S1)	Engine ON 1800 - 2200/min	04/035	1800 - 2200 RPM	0.8 - 2.5 ms*	0.00 - 2.1 s	B2-S1 OK
6	Three Way Catalytic Converter (CATB1)	Engine ON 1800 - 2200/min	04/046	0.00 - 0.28	0 - 2	58.6s	CAT B1 OK
7	Three Way Catalytic Converter (CATB2)	Engine ON 1800 - 2200/min	04/047	0.00 - 0.28	0 - 2	58.6s	CAT B2 OK
8	EVAP Canister Purge Regulator Valve Check	Engine ON @ Idle	04/070	- %	-5.5-+6.3 %	max. 0.2 g/s	EVAP OK
9	Fuel System, Leakage Check**	Engine ON @ Idle	04/071	Reed cl.	-	Check. END	Syst. OK
10	Secondary Air System	Engine ON @ Idle	04/77	- %	- %	-°C	Syst. OK
11	Check Readiness Code Setting	Engine ON @ Idle	08/086	00000000	00000000	00000000 / 00000100	00000000 / 00000100

* only during Diagnostic Performance
** could take up to 1 min. to "Test ON"

Confirm Readiness Code generating with a VAS 5051 Printout!

Quick Reference:
Readiness code setting, engine code ATQ (through m.y. 2000)

Step	Procedure VAS 5051/ 5052 Diagnostic Status	Engine Condition	Function/ Display-Group	Channel 1	Channel 2	Channel 3	Channel 4
1 2	Check DTC Memory Clear DTC Memory	Engine OFF Ignition ON	02 05	-	-	-	-
3	Adaptation Throttle Valve Control Module	Engine OFF Ignition ON	04/060	8% - 18%	80 - 90%	0 - 8	ADP.OK
4	Adaptation Kick Down (Only with Automatic Transmission)	Engine OFF Ignition ON Press gas pedal to the floor	04/063	12% - 97%	4 - 94%	Kick Down	ADP.OK
5	Fuel System Leakage Check	Engine ON @ Idle	04/071	-	-	Check END	Syst. OK
6	EVAP Canister Purge Regulator Valve Check	Engine ON @ Idle	04/070	0% - 100 %	-	-	EVAP OK
7	Lambda Control	Engine ON @ Idle	04/107	630 – 820 RPM	-10-+10%	-10-+10%	Syst. OK
8 9	Heated Oxygen Sensor	Engine ON 1800 - 2200/min*	04/041 04/042	0–0.9 kOhm	Htg.bC ON	0–0.9 kOhm	Htg.aC ON
10	Operational Readiness of Oxygen Sen. behind Cat. Conv.	Engine ON @ Idle	04/036	0–1V	B1-S2 OK	0–1V	B2-S2 OK
11 12	Response Time Front Heated Oxygen Sensor	Engine ON 1000 - 2400/min**	04/034 04/035	1000 - 2400 RPM**	more than 380°C	0.1 – 1.8s	B1-S1 OK B2-S1 OK
13 14	Oxygen Sensor Control System	Engine ON @ Idle	04/037 04/038	12 % - 26 %	0 V – 1 V	-800ms – +800ms	Syst. OK
15 16	Oxygen Sensor aging behind Cat. Conv.	Engine ON @ Idle	04/043 04/044	630 – 820 RPM	more than 380°C	0 – 1 V	B1-S2 OK B2-S2 OK
17 18	Three Way Catalytic Converter	Engine ON 1800 - 2400/min**	04/046 04/047	1800 - 2400 RPM**	more than 420°C	0 - 0.58*** 0 - 0.62****	CAT B1 OK CAT B2 OK
19 20	Secondary Air System	Engine ON @ Idle	04/077 04/078	630 – 820 RPM	0 – 5 g/s	-80 - +20 %	Syst. OK
21	Check Readiness Code Setting	Engine ON @ Idle	08/086	00000000	00000000	00000000	00000000

Quick Reference:
Readiness code setting, engine code ART

Step	Procedure VAS 5051/ 5052 Diagnostic Status	Engine Condition	Function/ Display-Group	Channel 1	Channel 2	Channel 3	Channel 4
1 2	Check DTC Memory Clear DTC Memory	Engine OFF Ignition ON	02 05	-	-	-	-
3	Adaptation Throttle Valve Control Module	Engine OFF Ignition ON	04/060	10% - 20%	80% - 90%	0 - 8	ADP.OK
4	Adaptation Kick Down (Only with Automatic Transmission)	Engine OFF Ignition ON Press gas pedal to the floor	04/063	12% - 92%	2% - 47%	Kick Down	ADP.OK
5	Fuel System Leakage Check	Engine ON @ Idle	04/071	-	-	Check END	Syst. OK
6	EVAP Canister Purge Regulator Valve Check	Engine ON @ Idle	04/070	0% - 100%	-	-	EVAP OK
7	Lambda Control	Engine ON @ Idle	04/107	650 – 730 RPM	-10% - +10%	-10% - +10%	Syst. OK
8 9	Heated Oxygen Sensor ***	Engine ON 1800 - 2200/min*	04/041 04/042	0 - 0.9 kOhm	Htg.bC ON	0 - 0.9 kOhm	Htg.aC ON
10	Operational Readiness of Oxygen Sen. behind Cat. Conv.	Engine ON @ Idle	04/036	0V - 1V	B1-S2 OK	0V - 1V	B2-S2 OK
11 12	Response Time Front Heated Oxygen Sensor	Engine ON @ Idle	04/034 04/035	650 - 730 RPM	more than 200°C	0.1 - 1.0s	B1-S1 OK B2-S1 OK
13 14	Oxygen Sensor Control System	Engine ON @ Idle	04/037 04/038	13.5% - 22.5%	0.1V – 1.0V	-800ms – +800ms	Syst. OK
15 16	Oxygen Sensor aging behind Cat. Conv.	Engine ON @ Idle	04/043 04/044	650 – 730 RPM	more than 230°C	0 – 1V	B1-S2 OK B2-S2 OK
17 18	Three Way Catalytic Converter	Engine ON 1900 - 2400/min**	04/046 04/047	1900 - 2400 RPM**	more than 350°C	0 – 0.25	CAT B1 OK CAT B2 OK
19 20	Secondary Air System	Engine ON 1800 - 2400/min**	04/077 04/078	1800 - 2400 RPM**	more than 13 g/s	- 70 - + 30%	Syst. OK
21	Check Readiness Code Setting	Engine ON @ Idle	08/086	00000000	00000000	00000000	00000000

Requirement: Coolant Temperature more than 80°C

* to warm up engine if necessary, test is complete when values are present in channel 1 and 3 ** only during Diagnostic Performance

Confirm Readiness Code generating with a VAS 5051 Printout!

V.A.G. 1551/1552 OPERATING OVERVIEW FLOW CHART

DE MENU

apid Data =>
link Code
elf Test
ealership Code
T - Diagnosis

ADDRESS WORDS MENU (V9)

01 - Engine Electronics =>
11 - Engine Electronics II
41 - Diesel Pump Electronics
51 - Electric Drive
61 - Battery Control
71 - Battery Charger
02 - Transmission Electronics
12 - Clutch Electronics
22 - 4 Wheel Drive Electronics
03 - Brake Electronics
13 - Distance Control
33 - OBD II
14 - Suspension Electronics
24 - Anti - Slip Control
34 - Level Control
44 - Steering Assist
54 - Rear Spoiler
15 - Airbag
25 - Anti-Theft Immobilizer Sensor
35 - Central Locking
45 - Interior Monitor
55 - Headlamp Vert. Aim Control
65 - Tire Pressure Monitoring
75 - Emergency Call Module
16 - Steering Wheel Electronics
26 - Automatic Roof Control
36 - Seat Adjustment Driver's Side
46 - Comfort System (AC CM)
56 - Radio
66 - Seat & Mirror Adjustment
76 - Parking Aid
17 - Instrument Cluster
37 - Navigation System
47 - Sound System
57 - Airbag (Cabriolet only)
67 - Language Control
08 - A/C & Heating Electronics
18 - Auxiliary Heater
09 - Electronic Control Module
19 - Diagnostic Interface for Data Bus
29 - Left Light Control
39 - Right Light Control
49 - Auto Light Switch
00 - Automatic Test Sequence

FUNCTIONS MENU

01 - Check Control Module Versions
02 - Check DTC Memory
03 - Output Diagnostic Test Mode
04 - Basic Setting
Display Group xxx =>
05 - Erase DTC Memory
06 - End Output
07 - Code Control Module
08 - Read Measuring Value Block
Display Group xxx =>
09 - Read Individual Measuring Values
Channel xx
10 - Adaptation
Channel xx
11 - Log-In Procedure
15 - Readiness Code

DISPLAY GROUP FORMAT

Display Group 000 (only)
xx xx xx xx xx xx xx xx xx xx

Display Group 001 (and higher)
xxxxx xxxxx xxxxx xxxxx

Electrical Repairs

Section
ER

General Information **Page**

- Overview . . . . . ER 2
- Introduction . . . . . ER 3
- Wiring harness design . . . . . ER 4
- Crimped terminals, visual checks . . . . . ER 21
- Seals, visual checks. . . . . ER 23
- Wiring repairs. . . . . ER 27
- Stabilant 22A . . . . . ER 32
- Wiring harness repair kit VAS 1978. . . . . ER 34
- Summary . . . . . ER 35
- Acknowledgements . . . . . ER 36

NOTE —

This information is provided for reference only. Instructions for testing, adjustment, and repair of various electrical and electronic systems can be found in the appropriate Service and/or Repair Manual. Specific tools and procedures for proper wiring harness repair using the VAS 1978 Wiring Harness Repair Kit can be found in the Instruction Manual provided and supplied with that kit.

Overview

If repairs must be made to an A6 wiring harness, Audi specifies only the use of the VAS 1978 Wiring Harness Repair Kit.

The VAS 1978 kit is designed to meet the requirements for high quality wiring harness repairs. The kit contains the tools, equipment, and spare components needed to maintain the Audi factory standard. Also included in the kit are detailed instructions for each tool and piece of equipment along with specific information concerning disconnecting, unlocking, disassembling, and repairing most of the different types of multipoint connectors currently in use.

Most wire terminals are not available separately as individual pieces. This is due, in part, to the difficulty in crimping wires to the very small terminals. Currently, if a new wire terminal is needed, the correct replacement is selected from the kit complete with a pre-crimped length of yellow wire. The old terminal is cut off and the two lengths of wire are connected, crimped and sealed, thus eliminating the necessity of handling the terminal end altogether.

Since wiring harness repairs are so critical, always use extra care. Repairs using non-approved components and procedures such as soldering, can compromise the integrity of the wiring harness resulting in an unsatisfactory repair. If the wiring harness cannot be restored to its original configuration, it must be replaced. Additionally, in some cases, repairs to certain safety-related wiring harnesses may not be possible and will require harness replacement. If in doubt, consult your local authorized Audi Dealer for the most current information.

The complete VAS 1978 (**shown below**) includes tools, equipment and wiring harness repair parts, and is housed in a large plastic tool box with handles that is easily moved to the vehicle when needed.

Information on obtaining consistent high quality wiring repairs is detailed in this section.

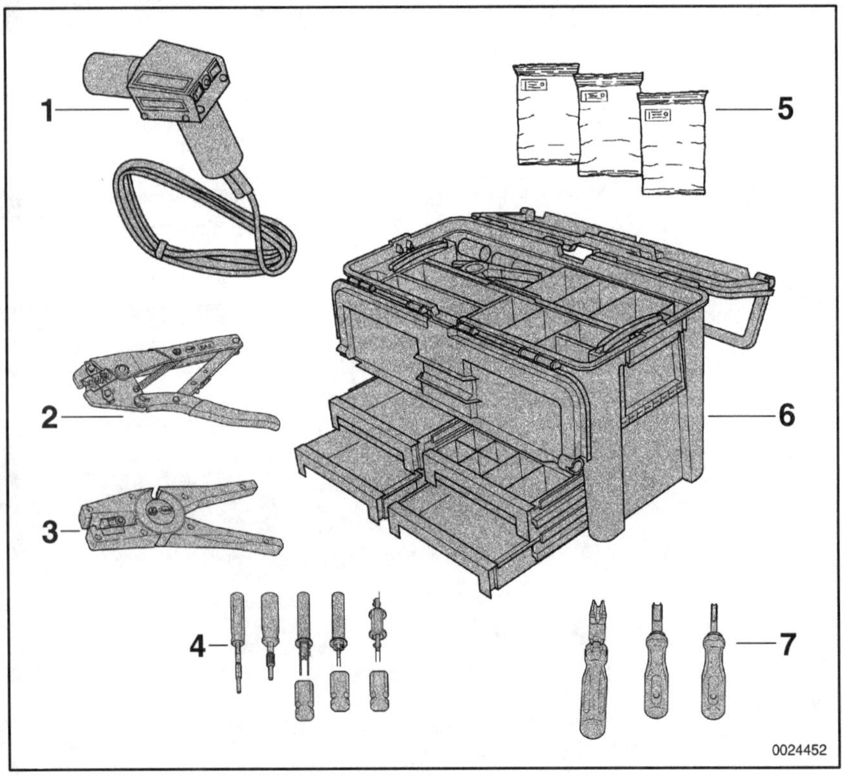

Audi special tool VAS 1978, major components

1. **Multiple temperature heat gun (120 volt)**
2. **Adjustable setting, multiple die crimping pliers**
3. **Insulation removal pliers with wire cutters**
4. **Terminal extraction tools**
5. **Bagged pre-crimped terminal ends with wire pigtails**
6. **VAS 1978 repair kit tool box with additional repair parts and storage space**
7. **Terminal and seal insertion tools**

- **Instruction Manual (not shown)**

Wiring harnesses in Audi vehicles have undergone a tremendous evolution over the past 30 years. Changing customer expectations, regulatory requirements, and advances in technology have resulted in broad proliferation of electrical and electronic components. Wiring harnesses have become much more complex. The number of terminals alone has increased from just over 200 in a typical 1966 model, to over 1,400 terminals in the latest models.

Cost has increased just as dramatically, from about 8% to more than 26% of the total manufacturing cost of the vehicle, making the wiring harness the single most expensive assembly in our cars. If that fact is difficult to accept, consider that it takes about 1-1/4 hours of direct labor to assemble an engine, but well over 5 hours are required to construct the wiring harness!

The difficulty associated with repairing the wiring harness has, unfortunately, kept pace. Smaller wires and terminals, carrying extremely low-level signals, with increasingly limited access, and more complicated, interlinked electronic control modules and sensors have all contributed to the "confusion factor."

While we have continuously improved wiring harness design, layout and assembly processes, about half of all warranty repairs address electrical malfunctions, and roughly half of *those*, and one third of all tow-ins, are caused by problems associated with the wiring harness.

With that in mind, it is easy to see that the wiring harness can no longer be considered a simple, bound collection of wires and connectors. Every length of wire, every connection, every attachment, every twist and turn, is carefully engineered to suit the application and its environment. When repairs are necessary, they will be successful *only* when the quality of the repair and the components *completely* restores the wiring harness to its original integrity.

To assure quality repairs, Audi has developed a simplified harness repair concept, and developed the **VAS 1978 Wiring Harness Repair Kit**. This kit provides everything that may be needed for proper, high-quality harness repairs. Within the framework of the new repair concept, any authorized repair can be carried out using the tools and hardware supplied in the kit. The harness must *never* be repaired by any other means.

The purpose of this Self-Study Program is to provide essential background technical information about all aspects of the wiring harnesses used in Audi vehicles. It describes why they are designed and built the way they are. It goes on to describe the repair fundamentals that, if respected, will maintain those high standards.

This information is intended to assist you in completing critical wiring harness repairs properly, in evaluating the quality of those repairs, and in preventing recurrent problems by getting it ***fixed right the first time!***

Wiring Harness Design

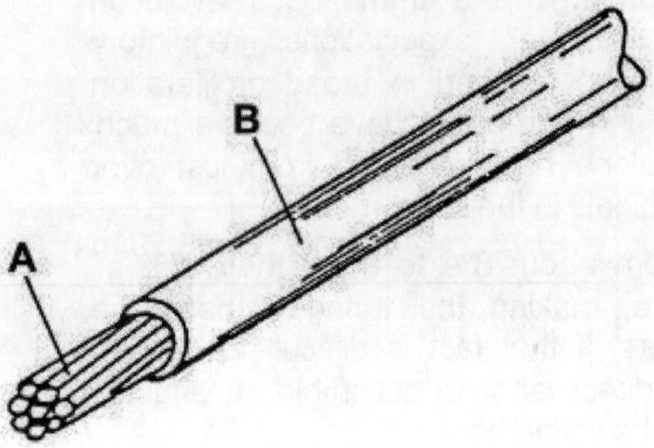

Wire and Cable

An automotive wiring harness employs wires and cables. Each is made up of at least seven conductor strands (A), and covered with an insulating material (B).

Within Audi, wire and cable are classified in three general categories, and then specified according to size, type of insulation, and color.

Wires

Wires are stranded conductors ranging in size from 0.35 mm^2 to 6.0 mm^2, or approximately 22 to 9 gage (AWG), that make up most of a vehicle's wiring harness. Smaller sizes save cost and weight. The larger sizes accommodate higher current loads with less resistance. Wire larger than 6 mm^2 or 9 gage (AWG) is generally considered "cable."

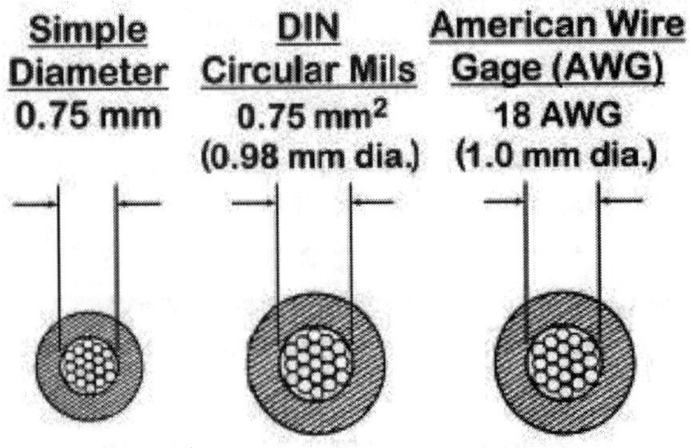

Wire size

Wire size (cross-sectional area) is the primary factor that will determine the current-carrying capacity of a conductor. Standard sizes differ slightly, depending on which measuring standard is used.

Audi uses wire and cable that has been manufactured according to the German industrial or "DIN" standards. Wire size is defined by cross-sectional area, and is measured in *circular mils* (mm^2). As in the example shown, "0.75 mm" refers to the area (0.75 mm^2), not the diameter.

Sizes of wire and cable originating in North America are described according to the American Wire Gage (AWG) standard. The standard wire sizes in both systems are roughly equivalent.

In practical terms, this discussion is significant only because there are slight differences between AWG sizes and the "equivalent" DIN sizes. When carrying out wiring harness repairs, it may be necessary to check and confirm that two wires sizes are indeed equivalent.

The table at the right lists standard DIN wire sizes, their approximate AWG equivalents, and the actual size (cross-sectional area) of AWG-sized wires.

Wire sizes

DIN (mm^2)	AWG	AWG actual size
0.35	22	(0.32 mm^2)
0.50	20	(0.51 mm^2)
0.75	18	(0.79 mm^2)
1.0	17	(1.0 mm^2)
1.5	16	(1.3 mm^2)
2.0	14	(2.1 mm^2)
2.5	13	(2.6 mm^2)
4.0	11	(4.2 mm^2)
6.0	9	(6.6 mm^2)

Cables

Cables are large-gage stranded wiring with high current-carrying capacity. Such cables are used to connect the battery to the starter, the generator and the fuse/relay panel, and for other similar high-current-load applications.

Cable sizes

DIN (mm^2)	AWG	AWG actual size
10.0	7	(10.3 mm^2)
12.0	6	(13.3 mm^2)
16.0	5	(16.9 mm^2)

Special-case wire and cable

Some wire characteristics are modified for improved performance, or to meet specific environmental requirements such as resistance to extreme heat or cold, improved abrasion resistance, greater flexibility, etc.

Ignition secondary or "high-tension" cables — such as spark plug wires — fall into just such a special category. Their design and construction is determined by the unique requirements of the ignition system — carrying very high voltage, operating reliably in harsh under-hood conditions, and suppressing radio interference, to name just a few.

Wiring Harness Design

Insulation

Insulation is non-conductive material that covers the conductors of a wire or cable to guard against unwanted electrical contact — short circuits. The type of insulation used in a particular application is part of the wire or cable specification.

"Standard Duty" or "Thick Wall"
Wiring with this type of insulation was commonly used until the mid-1980s. It has a relatively thick layer of insulation, usually made from polyvinyl chloride (PVC). It may still be found on cables or wiring within some components.

"Thin Wall"
Thin wall refers to reduced insulation thickness. Wiring of this type was introduced in Audi vehicles to save weight, and space, and to reduce the amount of PVC material used.

"Special Duty"
Some wire and cable is designed and selected for a specific use, and equipped with insulation meeting the demands of that "special duty." Double-insulated wire, high-heat resistant wire and cable, and shielded wire are some examples.

Color codes

Wire color codes are used on wiring diagrams to help identify different circuits. The code identifies the wire's base color, and often a tracer color.

Base color
The "base" color is the dominant wire color that covers at least 60% of the circumference of the insulation.

Tracer color
The second or tracer color is the color of the contrasting identification stripe used on some wires to help identify the circuit.

Identifying wire colors
Wire colors may be identified on wiring diagrams by color codes using German abbreviations for the color names. For example, "ro/ws" indicates a red wire with a white tracer. Look for the color code chart on the diagram you are using.

The table below lists the German color abbreviations, the German color names, and their English equivalents*.

Color codes

Abbr.	Color (German)	Color (English)
ws	weiss	white
ge	gelb	yellow
ws/gn	weiss/grün	white/green
ro	rot	red
li	lila	purple -or- violet
bl	blau	blue
gn	grün	green
gr	grau	gray
br	braun	brown
sw	schwarz	black

* Other colors such as Orange, Pink and the natural insulation color may also be used, but they are not officially released into the system.

Connectors

Connectors are used to make separable (non-permanent) connections between the wiring harness and an electrical or electronic device, or between two wiring harnesses. Each connector half consists of one or more terminals and a housing.

Terminals

Terminals are the electrical contacts in a connector, or at the mating connection to an electrical or electronic device. The only function of a terminal is to conduct electricity as efficiently as possible.

The terminal design, material, surface protection and method of attachment or "termination" (see page 10) all depend on the cost, reliability and durability requirements of the application.

The main features of a typical terminal in an Audi wiring harness are:

A – Contact

B – Exit radius

C – Core crimp

D – Entry radius

E – Insulation crimp

F – Runner
The stub that remains after machine crimping of production terminals supplied on a reel.

G – Primary lock
Ensures that the terminal locks into the housing; usually a lock tab on one or more sides; its purpose is to simplify harness fabrication.

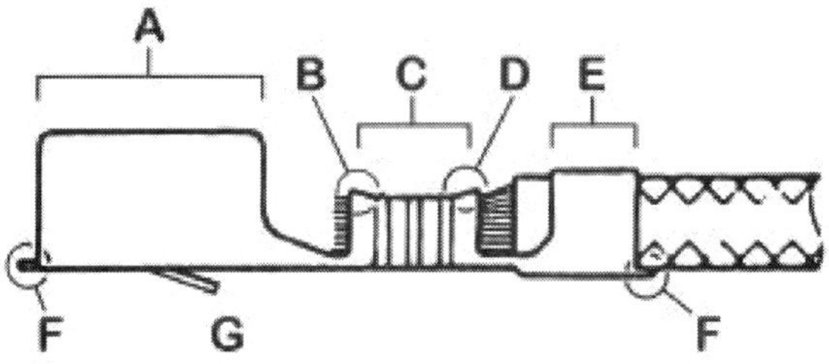

Wiring Harness Design

Types of terminals

The terminals used in wiring harnesses for Audi vehicles are classified according to four general types.

Blade and receptacle terminals
Terminal widths from 0.63 mm to 9.5 mm are in general use. Two distinct types are predominantly used in harnesses for Audi vehicles:

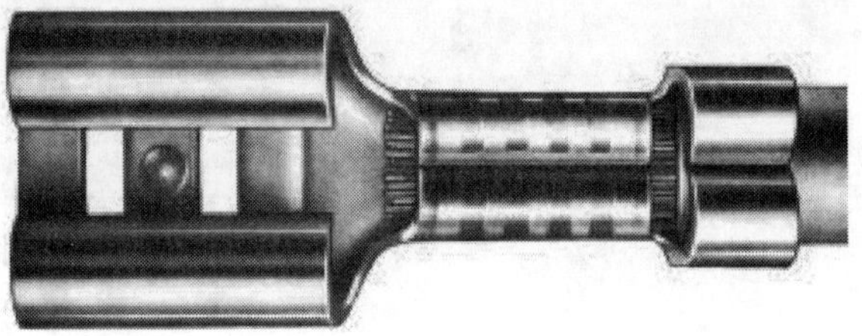

Older generation "Faston" terminals get their name from the original name used by the supplier (AMP) for this product line. The name became common usage, and is still used internally at Audi to refer to this type of terminal, even though other manufacturers have introduced some slightly different designs. "Faston" terminals in 2.8 mm and 6.3 mm widths are still frequently encountered.

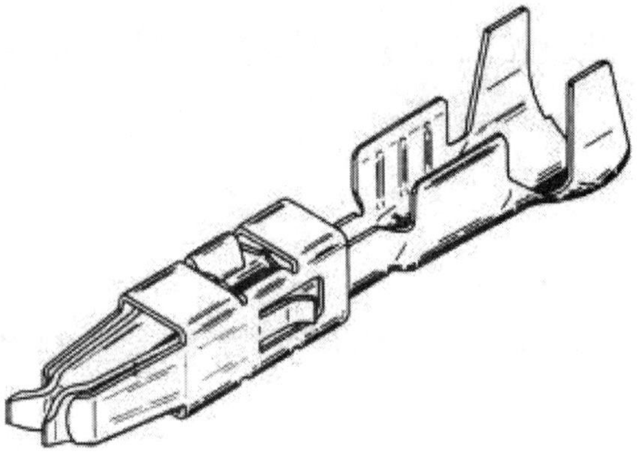

The newer generation "Power-Timer" terminal uses a two-part receptacle. A contact support spring helps to maintain pressure on the contacts, so that the metal terminal does not "take a set," relax its tension, and loosen over time. Here again, AMP's product name is commonly used to refer to this type of terminal, even though there are other manufacturers with similar designs.

Pin and socket terminals
Round pin terminals of two standard diameters (1.6 mm and 3.5 mm) are used in Audi vehicles.

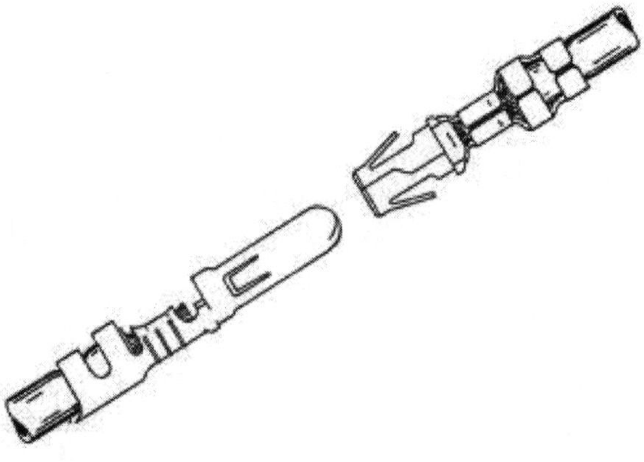

Ring terminals
Ring terminals include versions with or without internal teeth at the through-hole. Some have special shapes to keep the terminal in a certain position — at the battery positive (B+) terminal on the starter, for example.

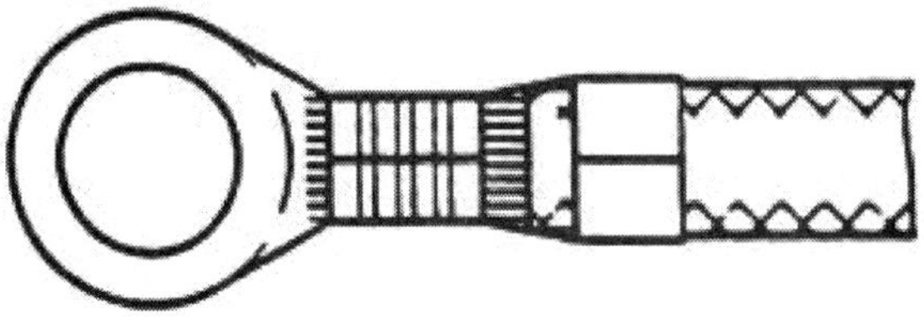

Special purpose terminals
These include splice connectors and other types of terminals used at crimped splice junctions in the wiring harness.

The barrel-type crimp connector found in the VAS 1978 Wiring Harness Repair Kit is an example. Others include the bridge terminals in voltage distribution ("TV") connectors, and the bus bar in the fuse block on the battery.

Housings

A connector housing insulates the terminal(s) and serves the mechanical function of safeguarding and maintaining the connection. The housing shape, the type of engagement, and the materials used depend on functional requirements, and on the operating environment.

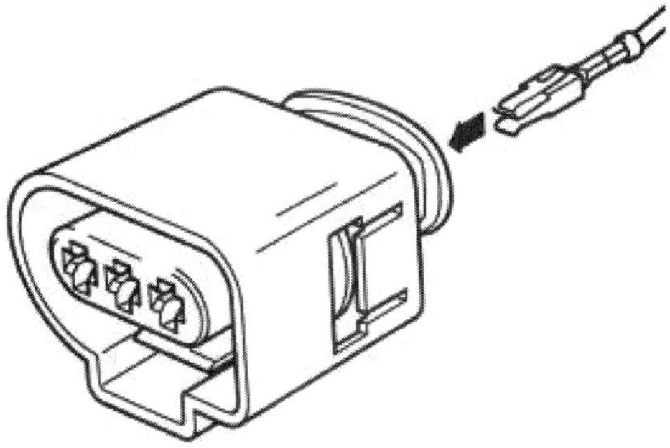

Seals
Where terminals and connectors may be subjected to a harsh environment, the connector housings are equipped with seals to protect the internal connections from contamination and corrosion. They may be in the form of a gasket that fits between two mating connector housings, or single-wire seals (arrow) that seal the openings at the back of the housing, or both. For more information on single-wire seals, refer to pages 13 and 26.

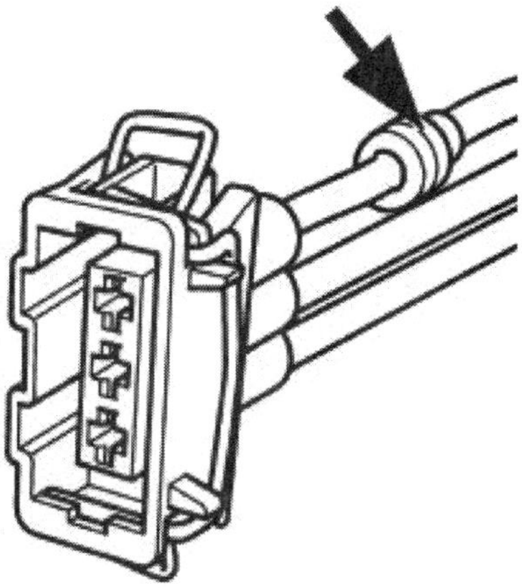

Wiring Harness Design

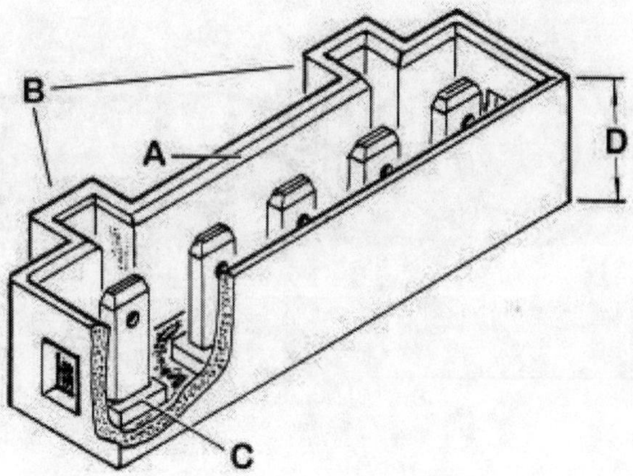

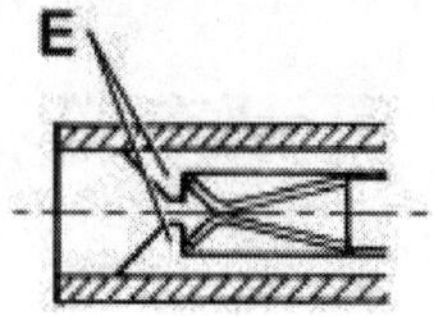

The main features of a typical connector housing are:

A – Beveled edge
Helps in aligning and mating the two halves of the housing; the edge also acts as a sealing surface when a gasket is used

B – Polarization features
Design of the housing ensures that the connection is only possible when the two halves, and therefore the mating terminals, are in correct alignment with respect to each other; in other words, it can only be connected one way

C – Terminal socket
For pin- or blade-type terminals, the socket supports and guides the terminal; it also serves to increase the insulated distance between terminals

D – Housing length
The housing must be long enough to shield the terminals when disconnected, and prevent them from making contact with anything other than the intended terminals in the mating connector half

E – Insertion guides
An aid to proper alignment in the "female" side of a pin and socket terminal connector; they guide the blade or pin into proper position while making the connection

F – Secondary lock
A mechanical interlock that keeps terminal(s) in place; normally a separate piece or a molded hinge that fits into position after the terminals have been inserted (inset); the main functions of the secondary lock are to simplify assembly, and to ensure that a terminal is not pushed out while joining two connector housings.

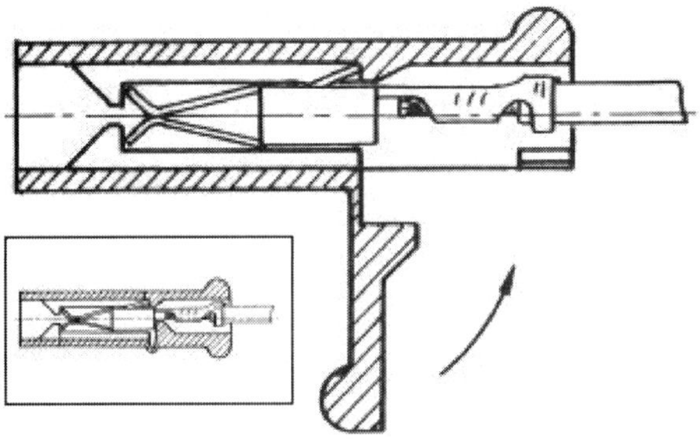

G – Housing detent
A mechanical retaining feature with mating parts built into the each half of the connector housing, such as a ball and dimple (as shown); the mating housings can be separated, without damage, simply by pulling; more force is required to connect or disconnect this type than is required for a connector with a separate locking feature

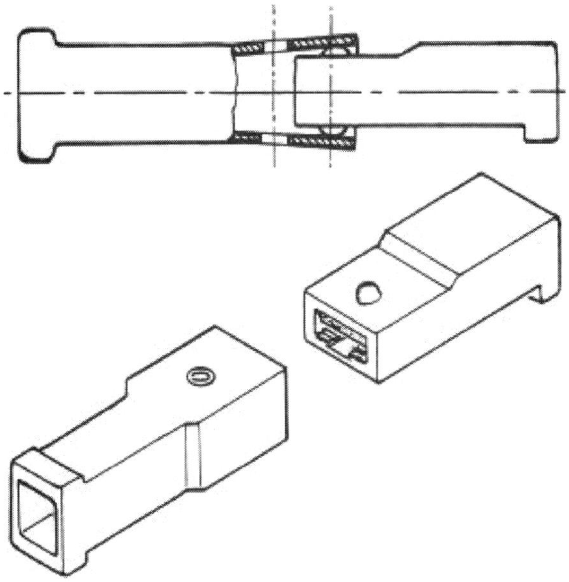

Housing lock

This is a mechanical interlock that keeps the two halves of the housing securely connected. Unless it has been released, the two halves cannot be disconnected without mechanical damage. By design, the lock can be opened without special tools, although a screwdriver or pliers may be required.

The most common types of housing lock currently used in Audi wiring harnesses are versions of the so-called "K-Jetronic" connector that feature a latching spring (arrow). Depressing the spring with a thumb opens the latching segments, so that the mating connector housings can be freely disconnected.

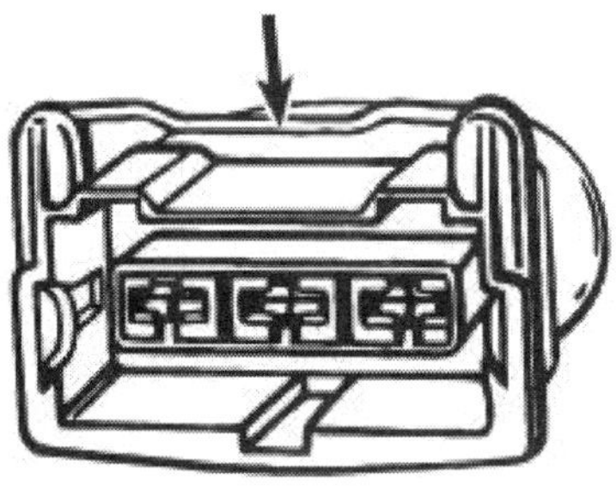

Wiring Harness Design

Termination Methods

Termination describes the attachment of some kind of terminal to the end-point of a wire or cable, on the wiring harness or in an electrical device. Termination also refers to the formation of splices or other junctions within a wiring harness.

Terminations are either permanent or non-permanent. Permanent terminations are **crimped**, **welded** or **soldered**.

Insulation displacement, **discrete force** (clamping), or **wire wrapping** are examples of non-permanent methods.

As we discuss termination types, we will see that some are acceptable, and some are not. Proper termination method is critical. To ensure quality production and long-term reliability, Audi adheres to rigorous production standards that are reviewed and enforced with regular performance audits.

Crimping

Crimping means using force to compress and mechanically deform or "crimp" the terminal around the uninsulated strands of the wire or cable. Compressing the terminal and wire accomplishes two important results:

1. Forces contaminants out of the crimp area, achieving more direct contact and better conductivity

2. Causes terminal and wire strands to "flow," filling the space between strands, significantly increasing contact area, and conductivity

The ideal amount of compression or crimping force — and therefore the degree of deformation — is a balance of mechanical strength and electrical conductivity. *Too little* crimping force means greater resistance (reduced conductivity) and less strength — the wire may pull out. *Too much* crimping force means there is greater potential for wire fatigue, and breakage.

As shown by the graph below, the amount of crimping force that produces the **ideal deformation "C"** lies between the **peak mechanical strength "M"** and the **peak electrical conductivity "E."** This ideal result is achieved when the crimped terminal and the wire inside are compressed by approximately 20%, for both open- and closed-barrel terminals.

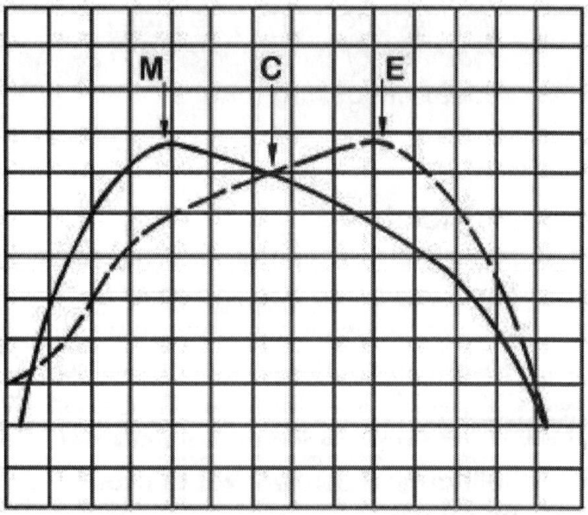

Crimping compression force →

We can see that, for crimping to work effectively, the wire and terminal must be matched in size. This ensures that both the terminal and wire will be compressed during crimping, resulting in a good, reliable crimped connection.

Open-barrel terminals

An open-barrel terminal has crimping "wings" that are folded over to compress the wire during crimping. When folded in sequence, they form an overlapping crimp. When folded simultaneously, they form a "B" crimp, as shown.

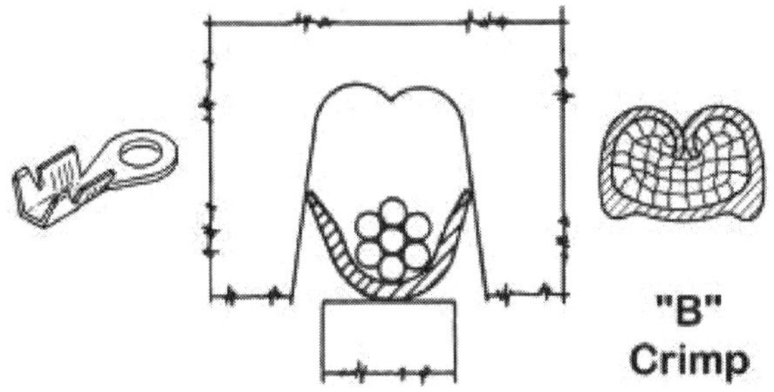

Closed-barrel terminals

Closed-barrel terminals have a round of oval shaped crimp area. Crimping deforms the barrel using, for example, a nest and indent die, as shown.

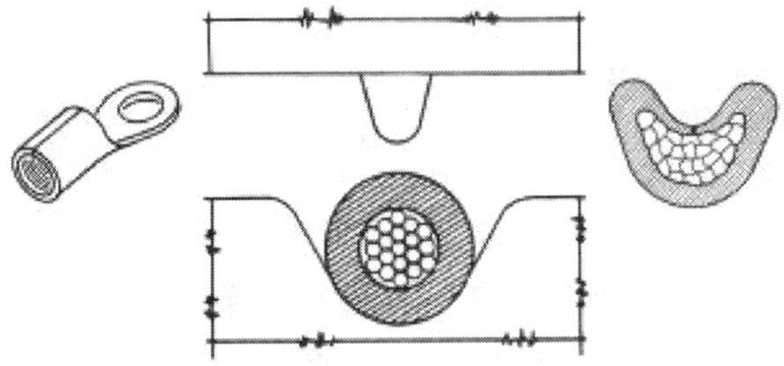

Repair practices

We can see that the quality of crimped terminations in wiring harnesses is critical! It is important to always use the proper tools and components.

Repair wires and connectors must both be of the correct, matched size.

Inspection of several features can be used to assess the quality of a crimped termination. Refer to *Crimped Terminals — Visual Checks*, page 24.

Welding

Welding fuses the uninsulated strands of wire or cable. The flow of material and expulsion or burning off of contaminants improves contact and conductivity much like crimping. Like proper crimping, a welded termination reduces the size of the conductor strands by about 20%

Improper welding causes faults ranging from welds that pull apart, to those that break due to heat embrittlement.

Wiring Harness Design

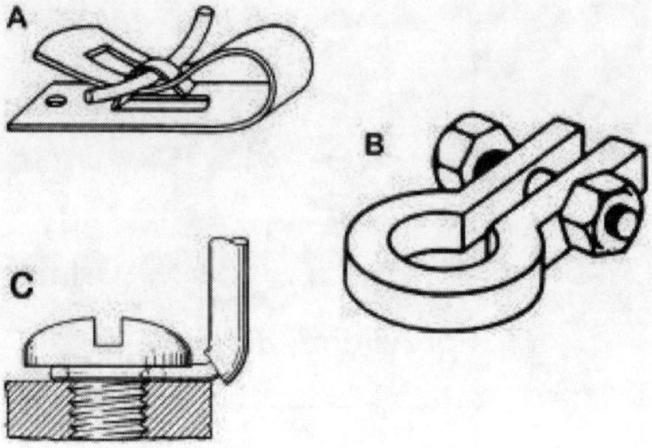

Discrete force (clamping)

Clamping involves mechanically joining an uninsulated wire or cable to a terminal, or to another uninsulated wire or cable, using a clamp or clamping feature. Examples are spring clip terminals (A), battery terminal clamps (B), and threaded fasteners (C).

Splices

A splice is a junction of common circuits within the wiring harness, used to reduce weight and cost by eliminating redundant wire or cable leads.

Example

The five Ground (–) circuits for the headlight, parking light, side marker light, turn signal and fog light on the right side are all routed to a splice near the right front corner of the car. From that splice, a single, larger diameter wire runs the 8 feet to a chassis ground connection, thus saving 32 feet of wire.

Splices may be welded or crimped – never soldered!

Light bulb sockets

Bulb sockets are "connector" housings that make electrical contact with, and physically hold, a light bulb. They are considered part of the wiring harness.

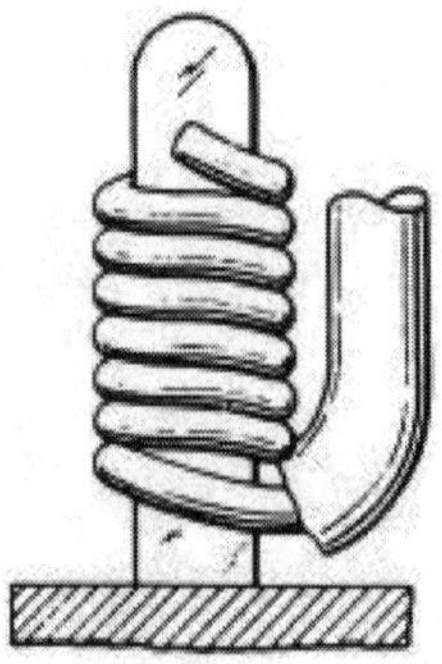

Wire wrapping

Lacquered or coated wire is wrapped tightly around a binding post terminal. The edges of the post cut into the wire to make the connection. This method may be used inside some devices, but not in Audi wiring harnesses.

Seals

Multi-wire seal

This type is essentially a flat piece of rubber or other elastomer with holes for the size and number of wires. A ribbed outer edge fits a recess in the connector.

Single-wire seal

These are generally made of silicone, allowing them to remain flexible at both temperature extremes. The seal slips onto the wire before it is terminated. In production, the terminal's insulation crimp wings secure the seal collar.

For wiring harness repair, new single-wire seals are installed differently. Refer to the VAS 1978 Instruction Manual for details.

When making harness repairs using the VAS 1978 kit, new single-wire seals are installed, but *not* crimped to the terminal.

Visual inspection can help determine the condition of a seal, and whether or not it has been damaged. Refer to *Seals — Visual Checks*, page 26.

Gasket

A gasket is generally a flat rubber or elastomer washer, fit between the two parts of a connector, seated or retained by features of the mating part.

When connector halves are disconnected, their seals or gaskets may get dislodged and be lost. Watch for any seals or gaskets that get separated from connectors.

A clean, undamaged edge is required to maintain an effective seal. It is important during repairs to examine the connector carefully for any sign of damage to the sealing edge that contacts the gasket.

Radial seal

A seal is secured inside one connector half by a retainer. It is compressed by the mating connector, which makes the seal. Always make sure that the seal and retainer are in place for reassembly.

Wiring Harness Design

Soldering *must not* be used to repair wiring harnesses in Audi vehicles.

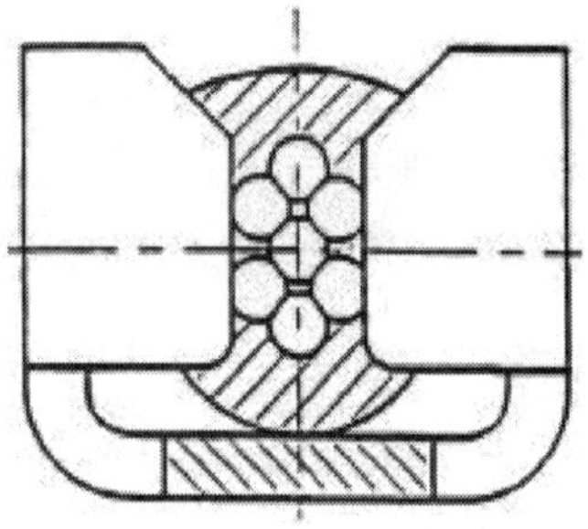

Most insulation displacement terminals *must not* be used for wiring harness service and repair in Audi vehicles.

Only tools, materials and repair procedures that have been specifically approved under stringent Audi guidelines may be used.

Termination "Don'ts"

Soldering

Soldering adds conductive material that flows into position between conductors to make electrical contact and hold them together. It is most often used for attaching terminals to conductors inside an electrical or electronic device.

While soldering is not inherently faulty, soldered terminations tend to be reliable and economical *only* when prepared in a controlled manufacturing environment, and *only* when the attached components are not subject to flexing or vibration. Thus, soldering is *not* recommended for wiring harness repair.

Insulation displacement terminals

Using this type of termination, a terminal with lancing or shearing features (like 3M's Scotchlok® for example) pierces the insulation to contact the core conductor strands, *in the process* of terminating the wire or assembling the terminal in a housing. Stripping the insulation from the wire is unnecessary.

Several shortcomings of this method can affect the quality of the repair. The most common is corrosion, either because the termination is not adequately sealed, or because of the inherent potential for corrosion between dissimilar materials.

Insulation displacement terminals are released under stringent guidelines for special wiring harness applications. Only the tools, materials and procedures specified for service or repair of these insulation displacement terminals may be used in Audi vehicles.

Wiring Harness Fabrication

Terminated lengths of wire and cable are bundled together into a wiring harness to facilitate efficient routing in a vehicle, and to provide protection. The materials and components used to fabricate the wiring harness are specifically selected to help prevent mechanical, thermal or chemical damage. Many of these key components are described below.

Tape wrap

Several types of wrapping tape are used to secure wires and cables in the desired form. The type of tape (PVC, nylon reinforced, cloth, coated cloth, metallic, etc.) and the type of wrap are specified in an Engineering Part Drawing.

As part of a repair, the harness should be re-wrapped from the connector end, to help ensure even wire lengths. Try to restore the original wrap configuration, using the same type of tape material, if possible. For repairs using yellow tape from the VAS 1978 kit, it should be used to wrap the harness in the same style as the original. See the descriptions below.

Open spiral wrapping
Open spiral wrapping leaves space between the wraps. This allows for greater flexibility of the wiring harness.

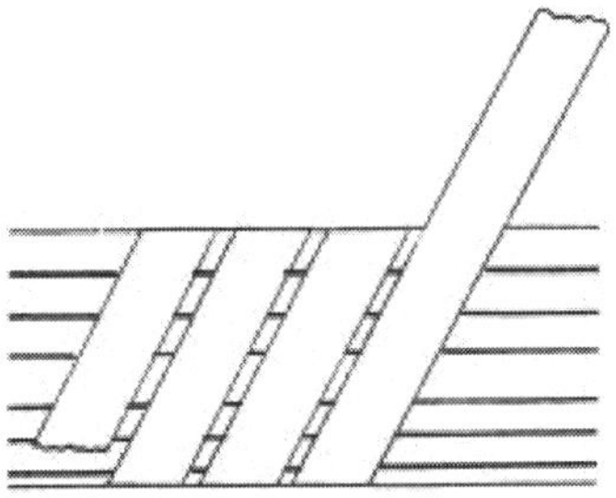

Overlapped wrapping
In this case, the wraps overlap by about 1/3rd. This results in a very rigid wiring harness, with less flexibility. This method is used to ensure that the harness is routed in a specific manner.

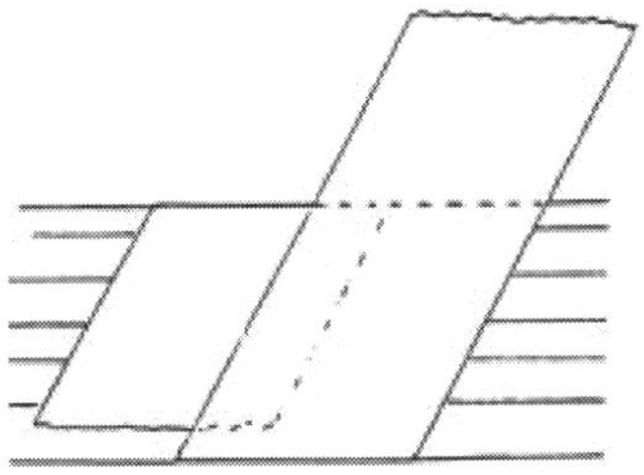

Spot taping
This technique is used during fabrication of the wiring harness to hold dimensions, specifically for branches and breakouts. Spot tapes may also be also applied to keep wiring from tangling.

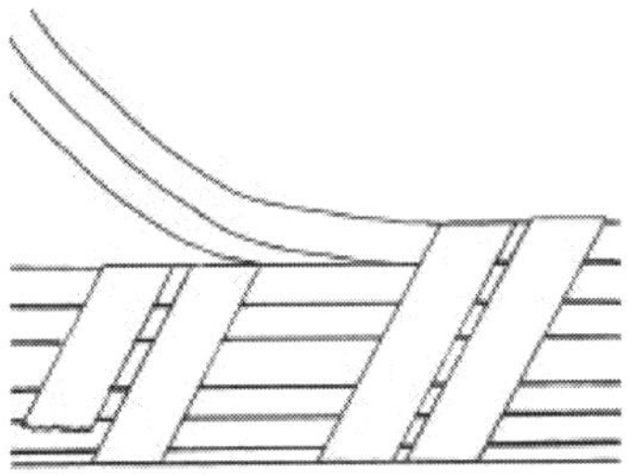

Wiring Harness Design

Tubing

Tubing (also called sheathing or a wire "loom") protects wires branching off from the main harness to other devices, and protects the harness from abrasion and other hazards. Like other parts of the harness, an Engineering Part Drawing specifies the materials, dimensions and locations. Typical versions are:

Convoluted tube
Also called convoluted conduit, this is a semi-rigid ribbed construction used for mechanical protection and for its appearance. Both continuous-tube and slit-tube versions are used.

Thermal sheath
A thermal sheath is reinforced tubing made of reflective, high-temperature resistant materials. It may be flexible to semi-rigid in design.

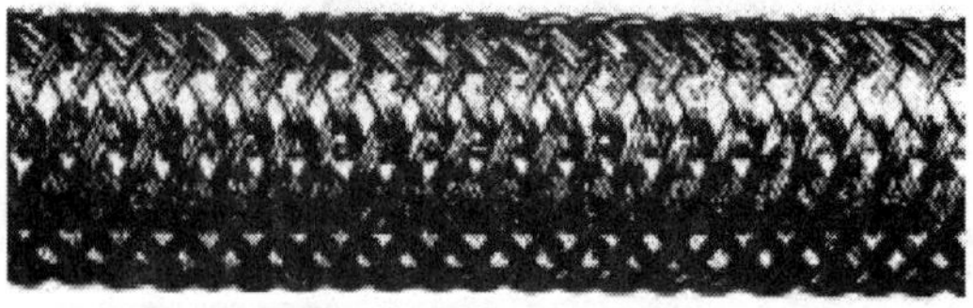

Woven armor sheath
This is sheathing woven of synthetics or metal for mechanical protection. It may be flexible to rigid, depending on design.

Carefully remove tubing for repairs, so that it might be re-used. If this is not possible, always re-install the same type of tubing, in the same manner as the original.

Shrink tube
Shrink tube or "heat-shrink" tubing is sheathing that will shrink to form a tight seal around the component when heat is applied. Shrink tubing is not reusable.

Fasteners

A wide range of anchors, clips, cable ties and other special-purpose hold-downs are used to route and retain the wiring harness in a vehicle, and to fasten it to the bodywork or to major components.

During repairs, always replace or re-install all fasteners and retainers to match the original harness configuration.

Socket anchor

The socket base fits onto a welded stud on the body, or some other retainer that is cast or molded into a component.

Splay anchor / Tree anchor

These anchors flex and expand to retain themselves after being pushed through a hole in the bodywork or a component.

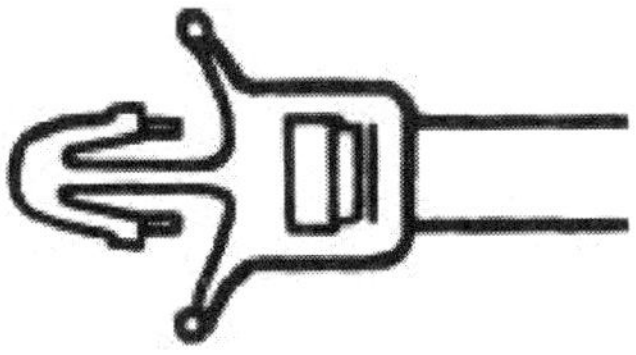

Clips

Clips hold the wiring harness to some feature of the bodywork, or a component. Most clips must be opened deliberately to release the harness.

Cable ties / Tie-wraps

Cable ties are used in production, as shown here, for example. They must always be replaced, but must *never* be over-tightened, as this can easily damage the harness. *Never* use cable ties to secure the harness to flexible parts like coolant hoses.

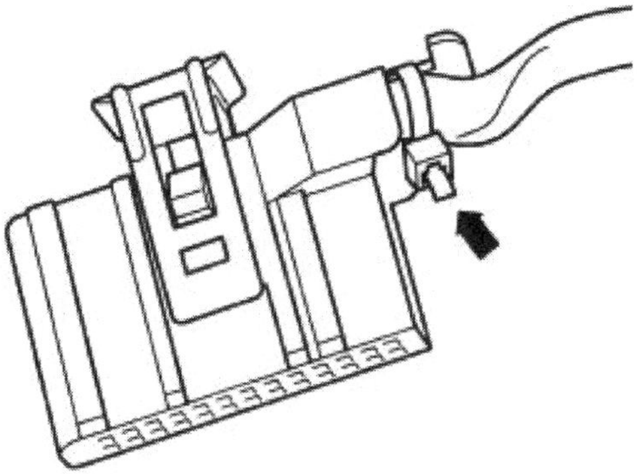

Fasteners are for mounting and routing only. Do not add any extra ones. Avoid pinching, stretching or other distortion of the harness, and check for any previous damage.

Wiring Harness Design

Fixed-point fasteners or "fix-points"

As for any other fasteners used to attach the harness, always make sure that the fixed-point fasteners have been re-attached in the same locations following repairs.

These fasteners are attached to the wiring harness. The location of such fasteners is specified in an Engineering Part Drawing. Fix-points are an aid to assembly, helping the worker on the assembly line to route the harness correctly and avoid strain or tension, hanging loops, or routing too close to potential hazards. It may be necessary to detach them for increased access during harness repairs.

Conduit

Also called a wire or cable channel, a conduit is a formed enclosure specifically designed to hold, route or protect the wiring harness in a defined area of the motor vehicle.

Grommets

Grommets may be easily damaged. Care must be taken when removing and installing them, both on the wiring harness and in the vehicle, to ensure that they are not damaged, and that they are re-installed correctly.

Grommets are made of rubber or synthetics and are intended to seal the passage of a harness through a hole in the bodywork, and to prevent the harness from chafing at the edges of the hole. Some grommets are molded in place and are, therefore, not reusable.

Mounting brackets

Brackets and other wiring harness mountings may be either separate components or features of the bodywork, or features of other components. They retain the wiring harness in a specific position, route the harness in a specific direction, or hold the harness away from a specific component or area. The harness may be attached to a bracket either temporarily or permanently.

Crimped Terminals – Visual Checks

Crimped Terminals – Visual Checks

Terminals must not be bent, cracked, or otherwise deformed. Check the overall shape. Inspect the locking tabs and, if necessary, re-bend them. If a terminal cannot be restored to original condition, it must be replaced. The key features and basic checks described below will help ensure quality repair.

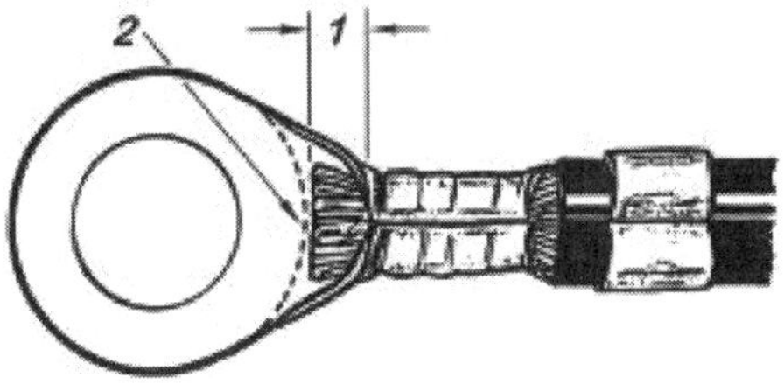

Wire length in the crimped terminal

1 – The ends of the core strands must be visible beyond the core crimp.

2 – The ends of the core strands must not extend into the area where they will interfere with a fastener.

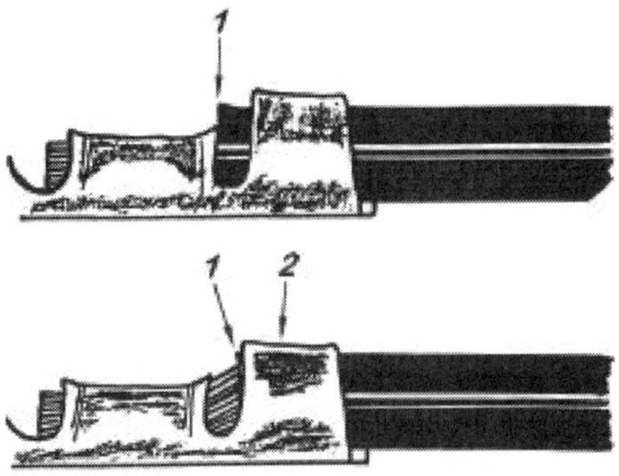

Insulation in the crimped terminal

1 – Insulation must not be drawn into the core crimp, or it might insulate the conductor strands. Stripped insulation must be at least even with the insulation crimp wing.

2 – The insulation crimp wing must grip the insulation securely to provide strain relief. If not, wire fatigue may cause the termination to fail.

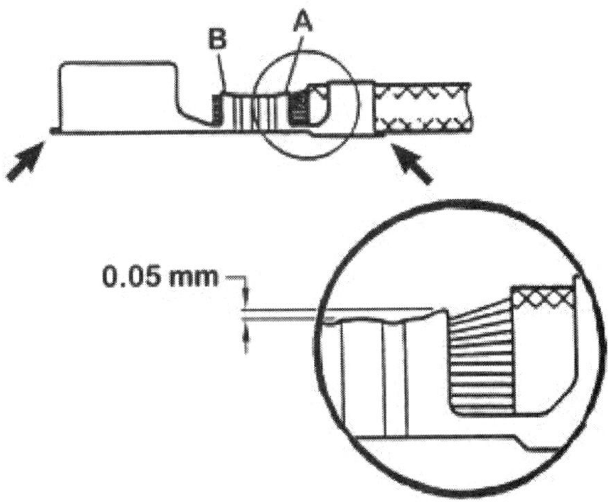

Crimp accuracy

If crimping tools become worn, they may be deformed such that crimping cuts the wire strands, or causes other damage.

Check to see that the runners (arrows) are visible, and have not been eliminated by the crimping operation (also see p. 5).

Check for a slight radius (flaring) at the entry (A) and exit (B) of the core crimp.

Inset

The 0.05 mm shown is the minimum dimension for the radius flaring, as specified by the Audi Terminal Audit.

Crimped Terminals – Visual Checks

Fully closed crimp
The terminal crimp wings support each other in a correct "B" crimp, so evenness and symmetry in the crimp is important.

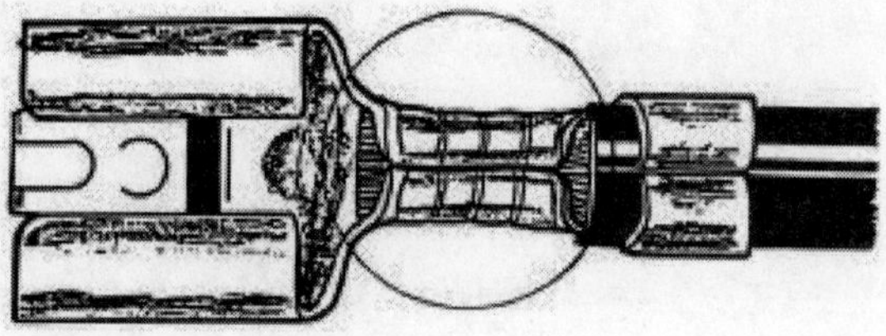

Left: the terminal crimp wings should be folded evenly and uniformly, indicating a good cross-section match, achieved with properly adjusted crimp-tools

Right: there must be no gap between the crimp wings

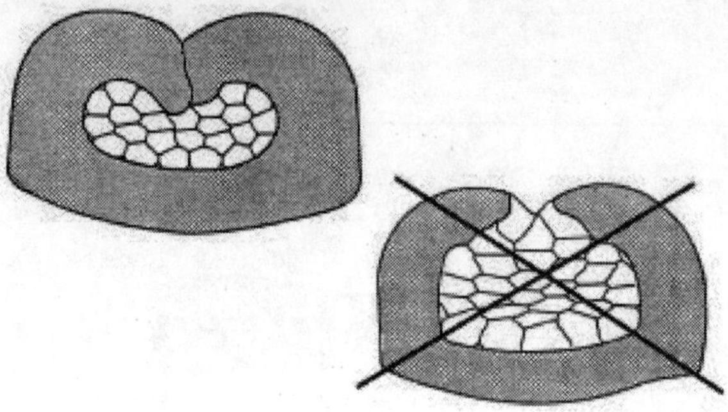

Over-crimping
Burrs at the bottom of the core crimp area are created by excessive crimping pressure, which in turn generates too much compression force acting to deform the metal. This "over-crimping" creates sharp edges and weak spots in the terminal walls.

1 – Cracks

2 – Burr length

3 – Burr width

4 – Terminal material thickness

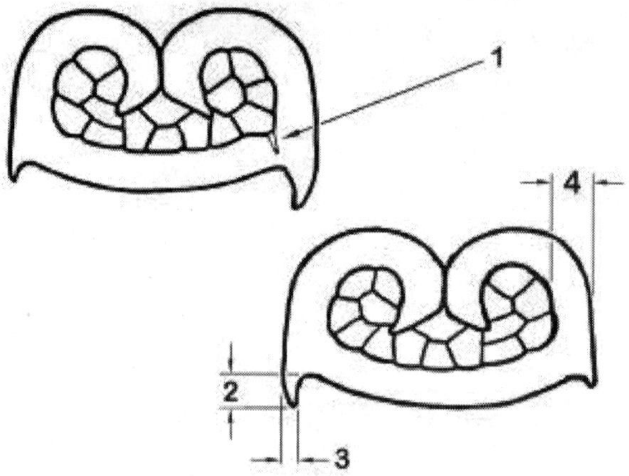

General guidelines
When burrs have been created by the crimping process, neither the length of the burr (2) nor the width of the burr (3) may be greater than one-half of the terminal's material thickness (4), as specified by the Audi Terminal Audit.

Terminals that have been subjected to excessive crimping compression force invariably show cracks (1) that lead to reduced strength and possible failure.

Seals – Visual Checks

Always inspect wire and connector seals to be sure they are in place, clean and undamaged, and that the mating connector housing is intact, undamaged, and able to seal properly.

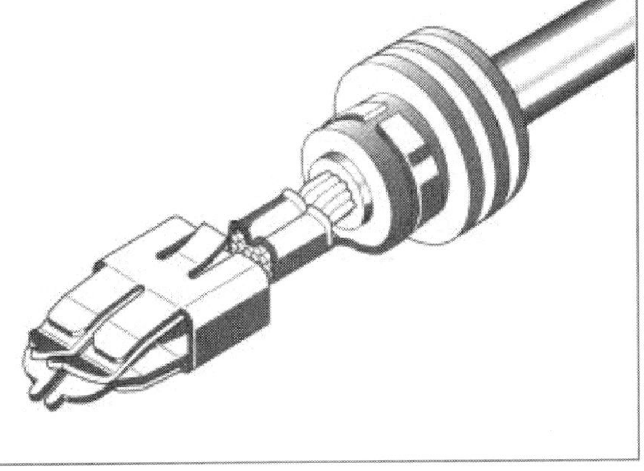

Production

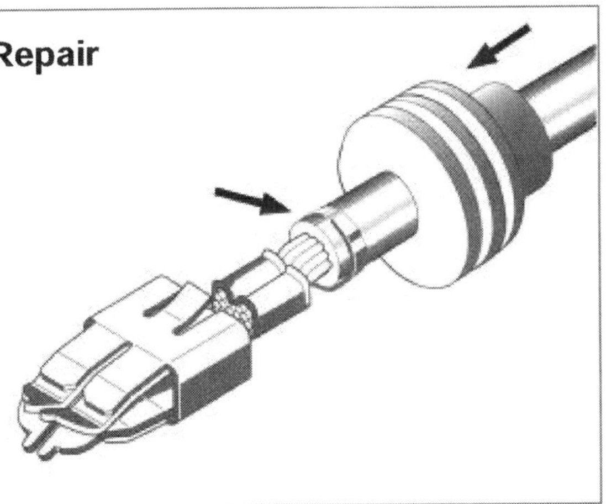

Repair

For wiring harness repair, new single-wire seals are installed differently. Refer to the VAS 1978 Instruction Manual for details.

Seals — Visual Checks

Most of the information in this section pertains to single-wire seals, but a few general guidelines apply to any kind of wire or connector seal.

As for any seal, the sealing surfaces and the seal itself must be clean and free of damage. For flat gaskets or radial seals, this also means inspecting the sealing parts of the connector housing to ensure that they are intact and undamaged.

Single-wire Seals

Single-wire seals are used with certain types of connectors. In production, the seal is installed with the wire, and held in place by a larger set of crimp wings that are part of the terminal, as shown. The pages that follow show various ways of inspecting these factory-installed seals for damage.

When making repairs to a harness using the VAS 1978 kit, the pre-crimped repair wires have smaller crimp wings used to secure the wire's insulation (left arrow). There is no provision on the terminal for attaching the single-wire seal.

Instead, the new seal is installed over the wire (right arrow) — in the *opposite* orientation, with the *larger* part facing the terminal — *before* attaching the repair wire to the harness.

Once the terminal is re-installed in the connector housing, the new seal is pressed into place using special tools supplied in the kit. More complete installation instructions, with details on pressing in a new seal, are included in the VAS 1978 Instruction Manual.

Wire and insulation condition

The end of the wire, which is crimped to a terminal and carries the seal, must be free of damage and distortion. Check for unevenness in the outer insulation that may prevent a proper seal.

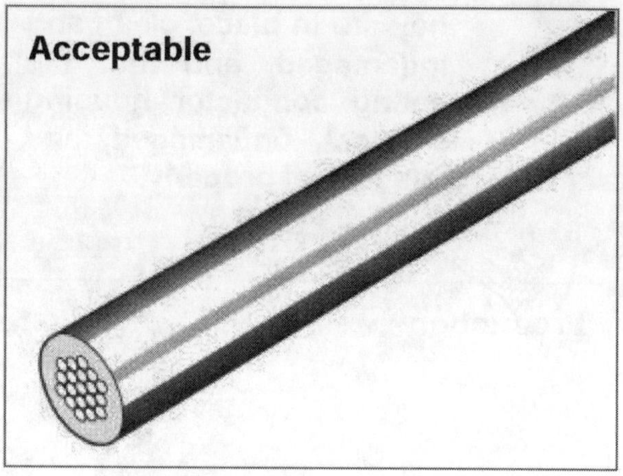

Inspect especially where a cable tie or other fastener has been used to secure the harness. If it has been installed too tightly, and the insulation has been damaged, it may be necessary to make the repair in a different place, where there is undamaged wire.

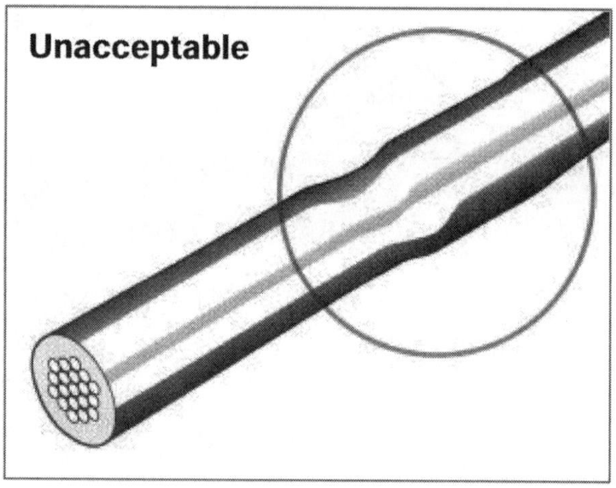

Seals – Visual Checks

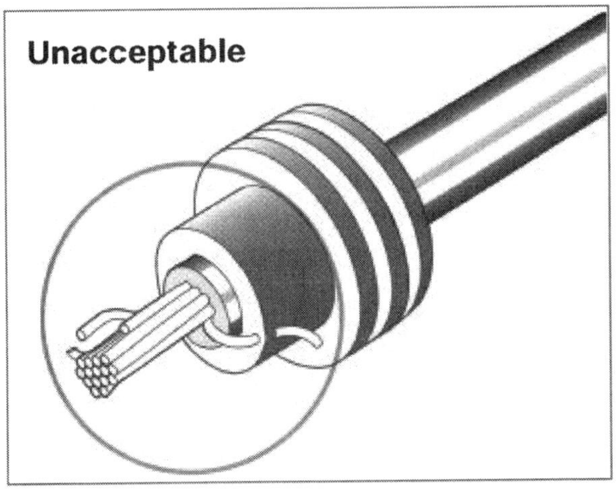

Seal installation (production)

When inspecting the condition of a seal that was installed in production:

a) Check to be sure that the wire strands have not been folded back and trapped under the seal.

b) Make sure that no wire strands are poking through the seal.

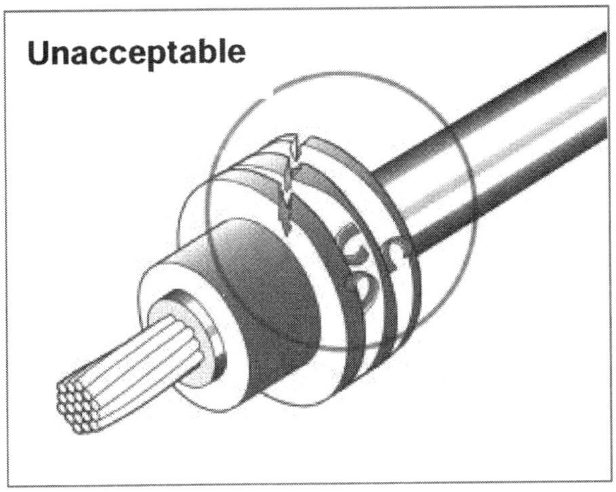

c) Check the outer edges or sealing surfaces of the seal to make sure that the seal itself is not damaged.

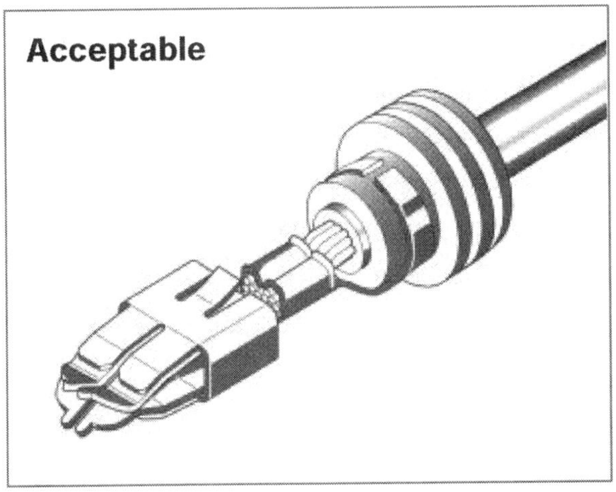

Seal attachment to terminal (production)

When inspecting the condition of a seal that was installed in production and attached to the terminal by crimp wings:

a) Make sure that the seal is properly secured.

b) Make sure that the seal itself is not damaged.

Seals – Visual Checks

c) Check to see that the seal is not too far from the end of the wire, and thus not adequately secured by the terminal crimp wings.

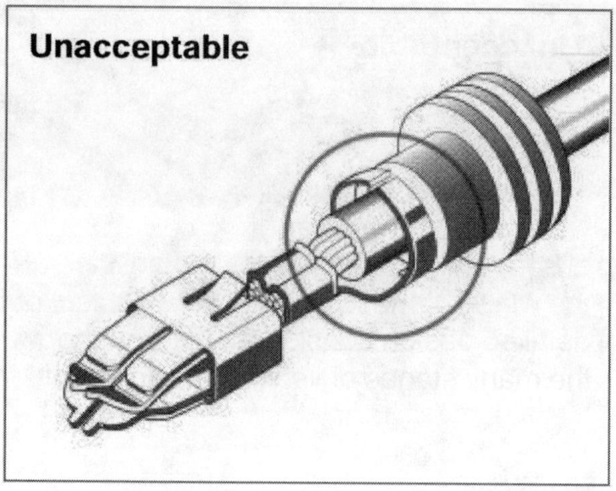

d) Check the crimp wings to be sure that the seal is well attached and secure, but not damaged in the sealing area itself.

e) Also, check to be sure the crimp wings have not been over-crimped to the point of damaging the seal.

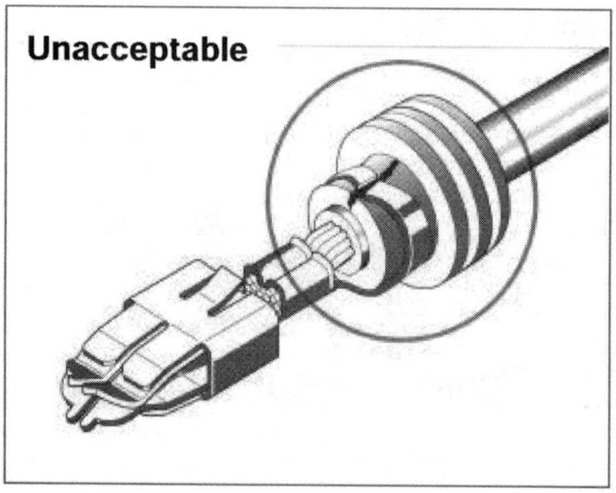

Wiring Repairs

Fundamentals of wiring harness repair

Looking back, we know that the modern wiring harness is a very complex assembly of specialized components. Every inch of the wiring harness has been the subject of detailed design consideration throughout the many stages of vehicle development. Considering this, it is clear that quality repairs require a lot more than finding the damage and splicing in some new wire.

All repairs must be carried out using *only* the VAS 1978 Wiring Harness Repair Kit!

Each and every component of the original wiring harness plays an important role. Some parts enhance electrical reliability; some are there for physical protection.

Either way, when the harness has been damaged, those requirements still exist after a repair. **Rule #1**: *all* parts of the harness must be completely "rebuilt" and restored to *original* condition.

1. *All* parts of the harness must be completely "rebuilt" and restored to *original* condition.

Harness routing and attachment is equally important. Repairs must take into account and duplicate every detail of the way the harness was mounted and protected. Do not improvise. Do not leave anything out.

2. The repaired harness must *always* be mounted and routed exactly as it was before repair — no changes.

This rule applies in reverse as well. Do not add fasteners where none were used before. If there is a part of the harness that seems unsecured, it is for a reason. Perhaps some degree of flexibility is required to prevent damage.

Rule #2: the repaired harness must *always* be mounted and routed exactly as it was before repair — without changes.

3. *Every* repair adds some resistance. When carrying out repairs, make as few repairs as possible, using only the proper tools and components.

Any addition of terminals, connectors, splices, etc. — in short, any repair — runs the risk of degrading the harness. Every new feature adds *some* resistance. No matter how skillfully done, it can never match the conductivity of an unbroken conductor. **Rule #3**: make as few repairs as possible, using only the proper tools and components.

4. Repairs must be carried out using *only* the tools, wires, terminals and other components supplied in the VAS 1978 Wiring Harness Repair Kit.

Harnesses are produced to exacting standards, so harness repair is no place to cut corners on repair methods, or use old parts scavenged from the dark recesses of the tool box! The focus of the new repair concept is restoring the harness to its original condition, using exactly the right tools and exactly the right components.

Introduced in 1997, the VAS 1978 Wiring Harness Repair Kit features an organized and easy-to-find assortment of tools and color-coded components that support all *authorized* Audi wiring harness repairs. **Rule #4**: repairs must be carried out using only the tools, wires, terminals and other components supplied in the VAS 1978 Wiring Harness Repair Kit.

5. When a top-quality repair cannot be achieved, for whatever reason, the wiring harness must be replaced.

The well-known "Murphy's Law" states that anything that possibly *can* go wrong, *will* go wrong. Sometimes a quality repair is just not possible – the harness is too badly damaged, or it cannot be fully restored to original condition, or good connections cannot be made, even though all repair instructions have been followed.

Rule #5: when a top-quality repair cannot be achieved, for whatever reason, the wiring harness must be replaced.

CAUTION!

BEFORE working on the electrical system:

- **Determine the correct code for the anti-theft radio.**
- **Stop the engine, and be sure the ignition is switched OFF. Failure to do so may damage the Engine Control Module (ECM).**
- **Disconnect the battery Ground (GND) strap.**

When connecting and disconnecting electrical test equipment (e.g. LED voltage tester, multimeter, etc.):

- **Switch the ignition OFF.**
- **Use the correct adapters from the VW 1594 connector test kit.**

When wiring harness repairs are complete:

- **Always determine what problem has caused the harness damage (i.e. sharp-edged body features, faulty electrical components, corrosion, etc.) and eliminate the cause.**
- **Carry out a functional test after each repair.**
- **Check Diagnostic Trouble Code (DTC) memory, correct indicated malfunctions as necessary, and erase DTC memory.**
- **Check and re-establish control module "Basic Settings" as necessary.**

Basic repair methodology

To repair a wiring harness, the first step (after locating the damaged area) should be to assess the extent of the damage:

- How extensive is the damage? Does it extend through the bulkhead or special grommets that cannot be repaired?
- What other components may have been affected by the damage to the circuit? Have the plastic connector housings been damaged? Are the electronic components damaged?
- Can the harness be repaired to a level that matches factory quality? Will the repair last? Will it be weatherproof?

Only after answering these questions, is it possible to determine whether the harness can be repaired and "rebuilt" to the original quality, or whether it must be replaced.

In either case, determine the cause of the damage, and take action as necessary to ensure that the repair or the replacement harness will not be damaged again.

The following pages list several key points to keep in mind while using the VAS 1978 kit to make harness repairs. For more detailed questions about special cases, or the latest information about critical repairs involving safety systems, refer to the Technician Helpline.

Wiring Repairs

All repairs must be carried out using *only* the VAS 1978 Wiring Harness Repair Kit!

Repair wires and connectors must both be of the correct, matched size.

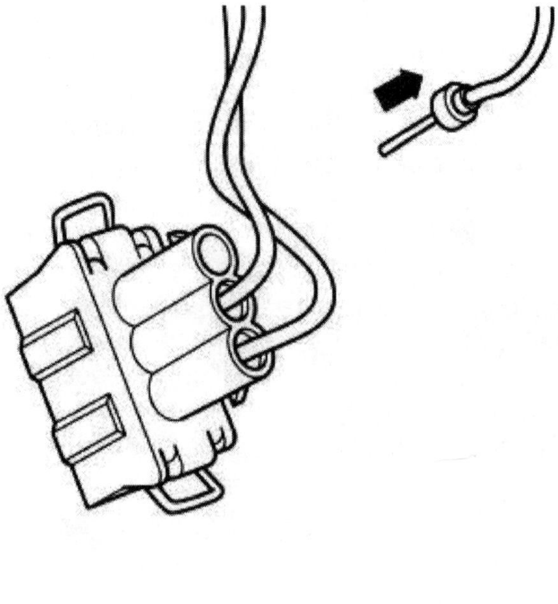

***Points to remember*:**

- *When making wiring harness repairs, use only the tools and components from the VAS 1978 Wiring Harness Repair Kit.*
- *Replace any harness that cannot be repaired and restored to its original configuration using the VAS 1978 kit.*
- *When making repairs by splicing a new length of wire, choose a replacement wire that is the same size as the original, or greater.*
- *Choose from among the color-coded connector terminals and other kit components to match them to the proper wire size.*
- *If it is necessary to join an original wire and a replacement wire using a butt-joint crimp connector, the wires and the crimp connector must be matched in size to ensure a reliable crimp.*
- *Check for special surface treatment on connector terminals. If, for example, the original terminal was gold-plated, a gold-plated replacement must be used.*
- *If heat-resistant wire was used in the original harness (typically with a softer, matte-finish insulation), the equivalent wire must also be used for the repair.*
- *Always replace single-wire seals, where applicable, using color-coded seals that match the wire size. Refer to the VAS 1978 Instruction Manual.*

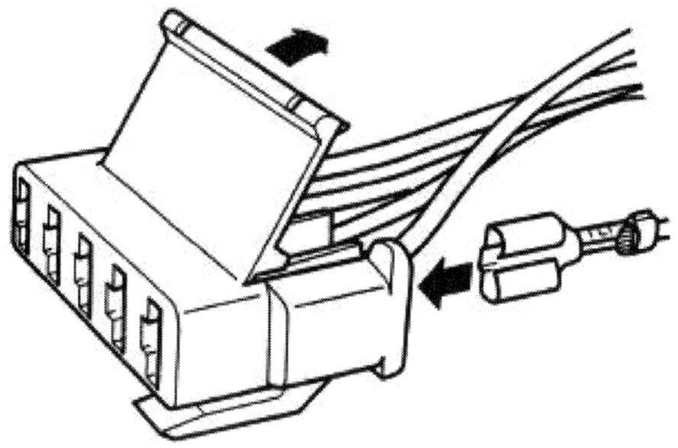

- *Secondary locking features are critical components that help maintain positive terminal contact. Always fully insert the terminal in the housing before engaging the lock.*
- *Always inspect all features of the connectors and terminal housings for damage before reassembly.*
- *Always wrap the repaired area with the yellow tape provided in the kit — both to replace the original harness wrap, and to identify the repaired area.*
- *Do not locate a wire repair where the harness will flex, such as at a doorjamb or trunk hinge. Replace damaged sections with new wire, and locate the new connections where they won't flex.*
- *When repairing more than one wire, stagger the crimped connections. Do not position them directly side by side.*

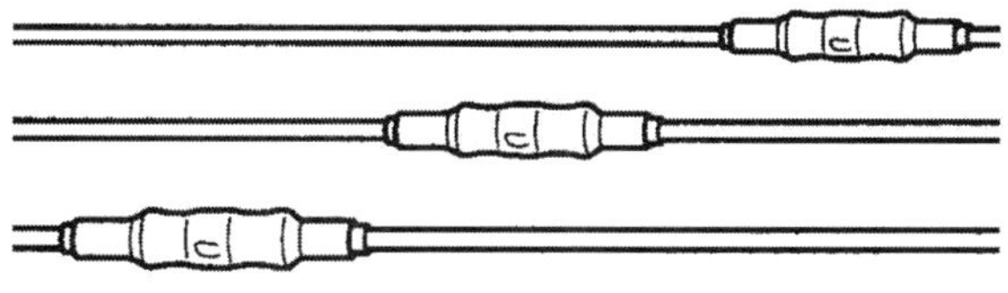

- *When cutting a replacement wire to fit, be certain there is sufficient length to ensure that the repaired section is not in tension after being reinstalled.*
- *Shielded cable should not be repaired.*
- *Never use soldered connections to attach terminals, make any kind of termination, or otherwise repair the wiring harness.*

Soldering *must not* be used to repair wiring harnesses in Audi vehicles.

Additional information
The instruction manual that accompanies the VAS 1978 kit is an excellent guide for how to use the kit and its contents. It also has detailed instructions on disassembly of many types of connectors commonly used on Audi vehicles.

Stabilant 22A

Do NOT **apply Stabilant 22A to oxygen sensor terminals, or other very small terminals (0.64 to 1.5 mm).**

For questions about specific applications, refer to AESIS repair procedures, Technical Bulletins, or consultants on the Audi Technician Helpline.

Stabilant 22A

The problem

Poor or intermittent electrical contact between terminals of a component and a harness connector can cause problems that result in customer complaints, and may lead to unnecessary repairs.

The main cause of these problems is what is known as "fretting" corrosion. Metals expand and contract with changes in temperature, and metal terminals do the same. On a very small scale, temperature changes can cause the mating terminals to "fret" back and forth against each other.

The fretting is actually a good thing, up to a point. It helps scrape away the oxide molecules created by natural atmospheric corrosion. Over time, however, a terminal can "ride up" on the accumulated oxide material, and actually break the circuit. What makes troubleshooting so difficult is that the very act of disconnecting and connecting terminals "cleans" them, and temporarily resolves the problem!

A solution

Stabilant 22A is a compound that is used to enhance contact, and to help avoid the problems described above. It does not, by its chemical make-up alone, increase conductivity or improve electrical contact. The chemical compound itself is, in fact, non-conductive in most circumstances!

In a thin layer, under pressure between mating electrical terminals, the compound becomes conductive. While it is not an *electrical* insulator, it seals and protects the connection, and helps delay oxidation and the onset of fretting corrosion.

Regardless of the scientific details, the real "bottom line" is that Stabilant 22A, when used properly, enhances contact.

Stabilant 22A should be used whenever a harness connector has been disconnected or when component has been replaced.

Do NOT apply Stabilant 22A to oxygen sensor terminals, or other very small terminals (0.64 to 1.5 mm).

For questions about specific applications, refer to AESIS repair procedures, Technical Bulletins, or consultants on the Audi Technician Helpline.

Be sure to clean the terminals thoroughly, then apply Stabilant 22A to contacts in both the male and female connectors. Use a toothpick or equivalent to overcome surface tension and work small amounts past the openings of a multi-pin connector. Avoid "gobbing." Allow Stabilant 22A to dry before reconnecting.

Stabilant 22A is good for _one use only_. It must be re-applied each time terminals are disconnected.

Use only Stabilant 22A. Other materials may not be compatible.

Use Stabilant 22A sparingly. A very thin coat is useful. Too much is harmful.

Points to Remember ...

When diagnosing the cause of a customer complaint, always inspect and clean the electrical connectors before replacing a component.

Wiring Harness Repair Kit VAS 1978

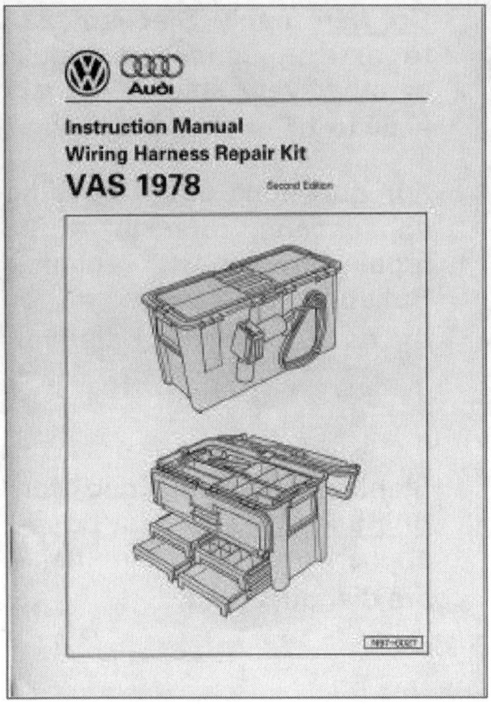

Wiring Harness Repair Kit VAS 1978

The VAS 1978 kit has been designed to meet the requirements of the latest Audi wiring diagram repair concept. It is equipped with the tools, equipment and spare components needed to make the necessary repairs in a manner endorsed by Audi. Replacement tools, terminals and other harness repair components are available separately.

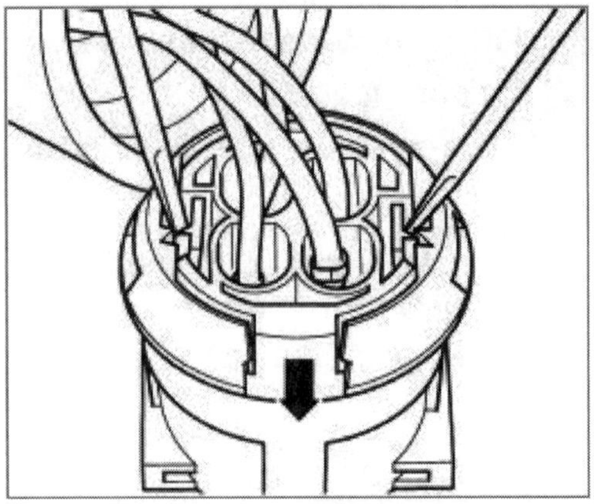

The kit includes detailed instructions for disconnecting, unlocking, disassembling and repairing most of the different types of multi-point connector currently in use in Audi vehicles. The illustrations shown here are examples.

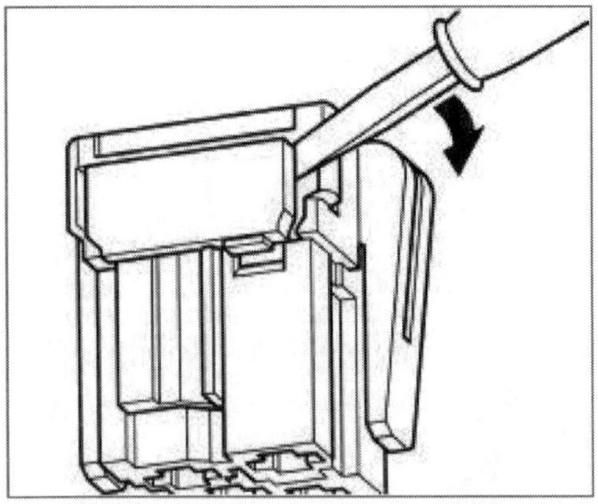

Points to Remember ...
When diagnosing the cause of a customer complaint, always inspect, clean and test the electrical connectors before replacing a component.

Summary

We can summarize the key things we have learned in this course as follows:

1. The wiring harness is a complicated and expensive part of the vehicle. About half of all electrical system warranty claims involve repair or replacement of the wiring harness.

   The wiring harness is complicated, and proper repairs are critical ...

2. Each part of the harness is designed, engineered and built to meet specific needs. This includes its routing and attachments. A proper repair ensures that all elements of the harness are fully restored to original configuration.

   All parts of the harness must be restored to original condition ...

3. Because proper repair of the wiring harness is critical, all repairs should be made using only the tools and components found in the VAS 1978 Wiring Harness Repair Kit, and in accordance with the *Instruction Manual* supplied with the kit.

   Use only the VAS 1978 Wiring Harness Repair Kit ...

4. If the wiring harness cannot be fully restored to its original configuration and function using the contents of the VAS 1978 kit, then an adequate and acceptable repair is not possible and the harness must be replaced.

   If it can't be done with VAS 1978, it cannot be done right — and the harness must be replaced!

5. Unauthorized means of repair such as soldering, and non-OEM wires and components, must *never* be used to repair an Audi wiring harness.

   No welding, no soldering, no hardware store connectors ...

6. There are *no* exceptions!

   No exceptions!

Acknowledgements

We especially want to recognize the contribution of:

J. J. Kavanagh
QA Resident Engineer for NAR
VWAG

His extensive study of wiring harness design and the challenges associated with reliable harness repair forms the basis for this publication.

Product illustrations and component information from several Audi suppliers have been used as aids in preparing this course material. Those suppliers are:

AMP	AMP
GH	Grote & Hartmann
PE	Delphi/Packard Electric Systems
Schlemmer	Josef Schlemmer GmbH
Schunk	Schunk Untraschalltechnik GmbH
Siemens	Siemens Bordnetze GmbH & Co. KG
Stapla	Stapla Ultraschall Technik GmbH

Many DIN, ISO and Audi standards, recommended practices and test specifications have also been used (specifically, VW 603 30 and Audi Wire Harness Audit Procedures).

External publications and sources considered in the preparation of this information are:

Committee of Sheet Steel Producers, *Cracking Down on Corrosion – Cooperative Efforts toward Vehicle Durability* (AISI SO 731-1078-20M-SR), Washington, D.C.: American Iron and Steel Institute

Denny, Hugh W., *Grounding for the Control of EMI*, 1st edition, Don White Consultants, Inc., 1983

Electronic Connector Study Group, Inc., *Connectors and Interconnection Handbook*, Volumes 1-5, 4th printing, Fort Washington, PA: Electronic Connector Study Group, Inc., 1985

Society of Automotive Engineers, Inc., SAE Handbooks, Warrendale, PA: SAE, Inc.